LEARNSMART®

The market leading **adaptive study tool** proven to strengthen memory recall, increase class retention and boost grades.

> Moves students beyond memorizing

> Allows instructors to align content with their goals

> Allows instructors to spend more time teaching higher-level concepts

SMARTBOOK®

The first—and only—adaptive reading experience designed to transform the way students read.

> Engages students with a personalized reading experience

> Ensures students retain knowledge

LEARNSMART PREP®

An adaptive course preparation tool that quickly and efficiently helps students prepare for college-level work.

> Levels out student knowledge

> Keeps students on track

LEARNSMART ACHIEVE®

A learning system that continually adapts and provides learning tools to teach students the concepts they don't know.

> Adaptively provides learning resources

> A time management feature ensures students master course material to complete their assignments by the due date

WWW.LEARNSMARTADVANTAGE.COM

WARDLAW'S CONTEMPORARY NUTRITION, TENTH EDITION

Published by McGraw-Hill Education, 2 Penn Plaza, New York, NY 10121.
Copyright © 2016 by McGraw-Hill Education. All rights reserved. Printed in the United States of America. Previous editions © 2013, 2011, and 2009. No part of this publication may be reproduced or distributed in any form or by any means, or stored in a database or retrieval system, without the prior written consent of McGraw-Hill Education, including, but not limited to, in any network or other electronic storage or transmission, or broadcast for distance learning.

Some ancillaries, including electronic and print components, may not be available to customers outside the United States.

This book is printed on acid-free paper.

1 2 3 4 5 6 7 8 9 0 RMN/RMN 1 0 9 8 7 6 5

ISBN 978-0-07-802137-4
MHID 0-07-802137-5

Senior Vice President, Products & Markets: *Kurt L. Strand*
Vice President, General Manager, Products & Markets: *Marty Lange*
Vice President, Content Design & Delivery: *Kimberly Meriwether David*
Managing Director: *Michael S. Hackett*
Brand Manager: *Amy L. Reed/Marija Magner*
Director, Product Development: *Rose Koos*
Director of Digital Content: *Michael G. Koot, PhD*
Product Developer: *Angela R. FitzPatrick*
Marketing Manager: *Kristine Rellihan*
Digital Product Analyst: *Christine Carlson*
Director, Content Design & Delivery: *Linda Avenarius*
Program Manager: *Angela R. FitzPatrick*
Content Project Managers: *April R. Southwood/Christina Nelson*
Buyer: *Jennifer Pickel*
Design: *Trevor Goodman/Matt Backhaus*
Content Licensing Specialists: *John Leland/Leonard Behnke*
Cover Image: *© Lauren Burke/Exactostock*
Compositor: *Laserwords Private Limited*
Printer: *R. R. Donnelley*

All credits appearing on page or at the end of the book are considered to be an extension of the copyright page.

Library of Congress Cataloging-in-Publication Data
Wardlaw, Gordon M.
  Contemporary nutrition / Gordon M. Wardlaw, PH.D., Formerly of Department of Human Nutrition, College of Education and Human Ecology, The Ohio State University, Anne M. Smith, PH.D., R.D., L.D., Department of Human Sciences, College of Education and Human Ecology, The Ohio State University, Angela L. Collene, M.S., R.D., L.D., Department of Biological and Allied Health Sciences, Getty College of Arts and Sciences, Ohio Northern University.—Tenth edition.
     pages cm
  Includes index.
  ISBN 978-0-07-802137-4 (alk. paper)
  1. Nutrition–Textbooks. I. Smith, Anne M., 1955- II. Collene, Angela. III. Title.
  QP141.W378 2016
  612.3–dc23

                                                                                    2014026222

The Internet addresses listed in the text were accurate at the time of publication. The inclusion of a website does not indicate an endorsement by the authors or McGraw-Hill Education, and McGraw-Hill Education does not guarantee the accuracy of the information presented at these sites.

www.mhhe.com

Brief Contents

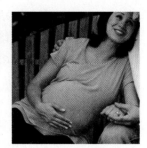

Dear Students,

Welcome to the fascinating world of nutrition! We are all nutrition experts, in a sense, because we all eat—several times a day. At the same time, though, nutrition can seem a bit confusing. One reason for all the confusion is that it seems like "good nutrition" is a moving target: different authorities have different ideas of how we should eat, and nutrition recommendations are subject to change! Are eggs good for us or not? Should we eat foods that contain gluten or not? Second, there are so many choices. Did you know that the average supermarket carries about 40,000 food and beverage products? Food manufacturers and grocery chains have one purpose—to make a profit. Typically, the most aggressively marketed items are not the healthiest. This has made shopping very complicated. In addition, as a nation, we eat out a lot. When we eat foods that someone else has prepared for us, we surrender control over what is in our food, where the food came from, and how much of it goes on our plates. There is a lot yet to learn, and you are undoubtedly interested in what you should be eating and how the food you eat affects you.

Wardlaw's Contemporary Nutrition is designed to accurately convey changing and seemingly conflicting messages to all kinds of students. Our students commonly have misconceptions about nutrition, and many have a limited background in biology or chemistry. We teach complex scientific concepts at a level that will enable you to apply the material to your own life.

This marks the tenth edition of Wardlaw's Contemporary Nutrition. As in previous editions, we have written it to help you make informed choices about the food you eat. We will take you through explanations of the nutrients in food and their relationship to health but will also make you aware of the multitude of other factors that drive food choices. To guide you, we refer to many reputable research studies, books, policies, and websites throughout the book. With this information at your fingertips, you will be well equipped to make your own informed choices about what and how much to eat. There is much to learn, so let's get started!

Anne Smith
Angela Collene

WARDLAW'S CONTEMPORARY

Nutrition

TENTH EDITION

Anne M. Smith Ph.D., R.D., L.D.
Department of Human Sciences,
College of Education and Human Ecology
The Ohio State University

Angela L. Collene M.S., R.D., L.D.
Department of Biological and Allied Health Sciences,
Getty College of Arts and Sciences
Ohio Northern University

Mc
Graw
Hill
Education

About the Authors

ANNE M. SMITH, Ph.D., R.D., L.D., is an associate professor at The Ohio State University. She was the recipient of the 1995 Outstanding Teacher Award from the College of Human Ecology, the 2008 Outstanding Dietetic Educator Award from the Ohio Dietetic Association, the 2006 Outstanding Faculty Member Award from the Department of Human Nutrition, and the 2011 Distinguished Service Award from the College of Education and Human Ecology for her commitment to under-graduate education in nutrition. Dr. Smith's research in the area of vitamin and mineral metabolism has appeared in prominent nutrition journals, and she was awarded the 1996 Research Award from the Ohio Agricultural Research and Development Center. She is a member of the American Society for Nutrition and the Academy of Nutrition and Dietetics.

ANGELA L. COLLENE, M.S., R.D., L.D., began her career at her alma mater, The Ohio State University, as a research dietitian for studies related to diabetes and aging. Other professional experiences include community nutrition lecturing and counseling, owner of a personal chef business, and many diverse and rewarding science writing and editing projects. Her interests include novel approaches to gly-cemic control, weight management, and—quite predictably for the mother of three little girls—maternal and child nutrition. Mrs. Collene currently teaches nutrition to nursing students at Ohio Northern University in Ada, Ohio, and is a member of the Academy of Nutrition and Dietetics.

With the tenth edition of *Wardlaw's Contemporary Nutrition*, we remember its founding author, Gordon M. Wardlaw. We are saddened that he lost his battle with cancer on January 19, 2014. He was a brilliant example of a man who acts deci-sively, follows his dreams, and creates a legacy.

Dr. Wardlaw had a passion for the science of nutrition and the research that supports it. With his clear teaching style and no-nonsense nutrition advice, he demonstrated an exceptional ability to translate scientific principles into practical knowledge both inside and outside the classroom. This skill is what made his book truly "contemporary." He was tire-less when it came to staying current and relevant. Each edition benefited from his genuine interest in responding to feed-back from students, colleagues, and instructors. He retired from teaching and writing in 2005 to pursue another passion—building his dream home in California. There, in a home he built with his own two hands, he spent the last years of his life with his family and friends.

It has been a privilege for both of us to join Dr. Wardlaw as co-authors of this textbook. For Anne Smith, he was an extraordinary colleague, mentor, and friend, who patiently taught her every step of the textbook writing process, begin-ning with the sixth edition. Angela Collene was blessed to have been one of his graduate students at The Ohio State University, and she first began to assist with revisions to the fifth edition. Like so many other students, colleagues, and friends, we remember Dr. Wardlaw as a source of vast knowledge, good humor, and inspiration. The best way we know to honor our dear friend is to carry on his legacy of outstanding textbooks in introductory nutrition. *Wardlaw's Contem-porary Nutrition* will continue to evolve and reflect current trends and breakthroughs in nutrition science, but Dr. Wardlaw's fingerprints will remain on every page.

Acknowledgements

Special Acknowledgements

It takes a cooperative and dedicated team to produce a successful textbook. We are privileged to work with many talented people that have made contributions to the tenth edition of *Wardlaw's Contemporary Nutrition*. We are indebted to the energetic staff at McGraw-Hill Education for their commitment to this textbook. We are grateful to Amy Reed, our Brand Manager, for her leadership and support of the *Wardlaw's Contemporary Nutrition* team. Our biggest *"thank you"* goes to our Product Developer, Angela FitzPatrick. Angie enthusiastically coordinated the editorial team and guided us through every phase of this revision of the textbook. We are grateful for her analysis of reviews, meticulous editing skills, and day-to-day encouragement and patience while we wrote the manuscript. We are grateful to our Content Project Manager, April Southwood, and her staff for the careful coordination of the numerous production efforts needed to create the very appealing and accurate tenth edition. We appreciate the thorough work and attention to detail of our copy editor, Marilynn Taylor; proofreaders, Debbie Budde-Bandy and Gina Delaney; and photo editors, John Leland and Mary Reeg. We thank our designer, Trevor Goodman, who not only guided the visual appeal of this edition but also worked to ensure each page was where it needed to be. Finally, we thank our colleagues, friends, and families for their continued support, encouragement, and genuine interest in nutrition.

Thank you to reviewers

Our goal is to provide students and educators with the most accurate, up-to-date, and useful textbook possible. As with earlier editions, the quality of the tenth edition of *Wardlaw's Contemporary Nutrition* is largely dependent on the thorough, professional assistance of nutrition educators from academic institutions across the nation. We are indebted to these colleagues who reviewed the ninth edition, evaluated new material for the tenth edition, participated in instructional symposia, and responded to surveys. The advice and suggestions from these colleagues have been used in every chapter and have resulted in a textbook that is current and inviting.

Kwaku Addo
University of Kentucky

Judy Kaufman, Ph.D.
Monroe Community College

Lee-Ellen C. Kirkhorn, Ph.D., RN
University of Wisconsin-Eau Claire

Mara Manis
Hillsborough Community College

Suresh Mathews
Auburn University

Paul Moore
Appalachian State University

Lee Murphy, MS-MPH, RD, LDN
University of Tennessee-Knoxville

Marsha Spence
University of Tennessee

Tammy Stephenson, Ph.D.
University of Kentucky

Priya Venkatesan
Pasadena City College

Sue Ellen Warren
El Camino College

Dana Wassmer
Cosumnes River College

Personalized, Adaptive Learning

Adaptive learning resources offer a personalized learning experience.

McGraw-Hill LearnSmart® is one of the most effective and successful adaptive learning resources available on the market today. More than 2 million students have answered more than 1.3 billion questions in LearnSmart since 2009, making it the most widely used and intelligent adaptive study tool that is proven to strengthen memory recall, keep students in class, and boost grades. Students using LearnSmart are 13% more likely to pass their classes and 35% less likely to drop out.

LearnSmart continuously adapts to each student's needs by building an individual learning path so students study smarter and retain more knowledge. Turnkey reports provide valuable insight to instructors, so precious class time can be spent on higher-level concepts and discussion.

Fueled by LearnSmart—the most widely used and intelligent adaptive learning resource—**McGraw-Hill SmartBook®** is the first and only adaptive reading experience available today.

Distinguishing what a student knows from what he or she does not and honing in on concepts he or she is most likely to forget, SmartBook personalizes content for each student in a continuously adapting reading experience. Reading is no longer a passive and linear experience but an engaging and dynamic one in which students are more likely to master and retain important concepts, coming to class better prepared. Valuable reports provide instructors insight as to how students are progressing through textbook content and are useful for shaping in-class time or assessment.

As a result of the adaptive reading experience found in SmartBook, students are more likely to retain knowledge, stay in class, and get better grades.

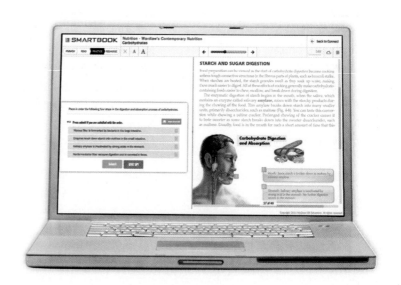

Digital efficacy study shows results!

Digital efficacy study final analysis shows students experience higher success rates when required to use McGraw-Hill LearnSmart®.

- Passing rates increased by an average of **11.5%** across the schools and by a weighted average of **7%** across all students.
- Retention rates increased by an average of **10%** across the schools and by a weighted average of **8%** across all students.

Study details:

- Included two state universities and four community colleges
- Control sections assigned chapter assignments consisting of testbank questions and the experimental sections assigned LearnSmart, both through McGraw-Hill Connect®.
- Both types of assignments were counted as a portion of the grade, and all other course materials and assessments were consistent.
- 358 students opted into the LearnSmart sections and 332 into the sections where testbank questions were assigned.

"After collecting data for five semesters, including two eight-week intensive courses, the trend was very clear: students who used LearnSmart scored higher on exams and tended to achieve a letter grade higher than those who did not."

Gabriel Guzman
Triton College

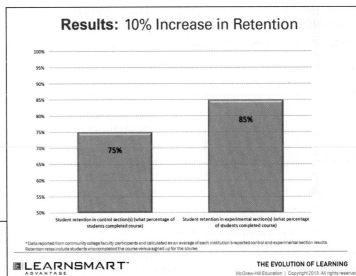

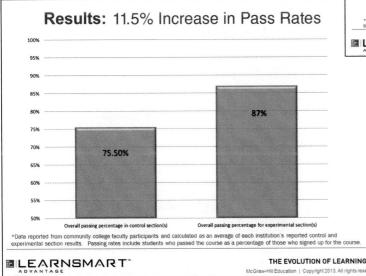

"LearnSmart has helped me to understand exactly what concepts I do not yet understand. I feel like after I complete a module, I have a deeper understanding of the material and a stronger base to then build on to apply the material to more challenging concepts."

Student

Connect Improves Performance

Easily assign key features from the book to increase engagement and learning

|NUTRITION

McGraw-Hill Connect® Nutrition is a digital teaching and learning environment that saves students and instructors time while improving performance over a variety of critical outcomes. From in-site tutorials, to tips and best practices, to live help from colleagues and specialists—you are never left alone to maximize Connect's potential.

- **Auto-graded assessments and tutorials**

You can easily create customized assessments that will be automatically graded. All Connect content is created by the authors, so it is pedagogical, instructional, and at the appropriate level. Interactive questions using high-quality art from the textbook and animations and videos from a variety of sources take you way beyond multiple choice. Assignable features from the book, such as *Newsworthy Nutrition, What Would You Choose?*, and *Rate Your Plate,* increase student engagement and understanding.

- **Gather assessment information**

All Connect questions are tagged to a learning outcome, specific section and topic, and Bloom's level so you can easily track assessment data!

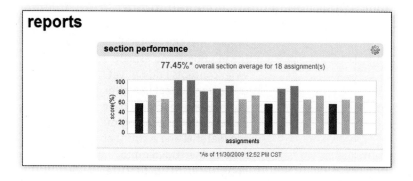

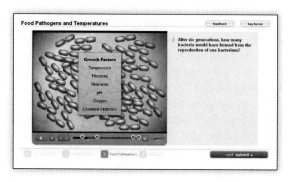

▲ Connect Assessment: Animation Tutorial

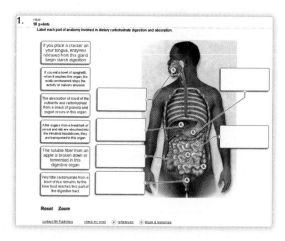

▲ Connect Assessment: Labeling Interactive

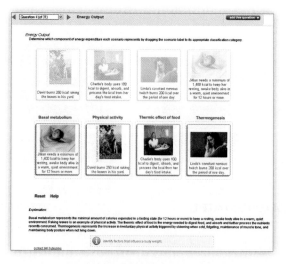

▲ Connect Assessment: Classification Interactive

McGraw-Hill Connect® Insight

McGraw-Hill Connect® Insight is a powerful data analytics tool that allows instructors to leverage aggregated information about their courses and students to provide a more personalized teaching and learning experience.

NutritionCalc Plus is a suite of dietary self-assessment tools available within any Connect Nutrition course or as a stand-alone online. NutritionCalc Plus is the most extensive, most accurate database available—over 29,000 foods—from ESHA Research, the leading provider of nutrient data. Create and save recipes allowing students to accurately reflect their personal food preferences. Students can add up to three profiles. The flexibility of multiple profiles allows students to analyze their own nutritional status as well as the unique nutritional needs of others. NutritionCalc Plus includes a MyPlate report.

McGraw-Hill Create™

McGraw-Hill Create™ is a self-service website that allows you to create customized course materials using McGraw-Hill's comprehensive, cross-disciplinary content and digital products. You can even access third party content such as readings, articles, cases, videos, and more. Arrange the content you've selected to match the scope and sequence of your course. Personalize your book with a cover design and choose the best format for your students–eBook, color print, or black-and-white print. And, when you are done, you'll receive a PDF review copy in just minutes!

Campus

McGraw-Hill Campus® is a groundbreaking service that puts world-class digital learning resources just a click away for all your faculty and students. All your faculty—whether or not they use a McGraw-Hill title—can instantly browse, search and access the entire library of McGraw-Hill instructional resources and services including eBooks, test banks, PowerPoint slides, animations and learning objects—from any Learning Management System (LMS), at no additional cost to your institution. Users also have single sign-on access to McGraw-Hill digital platforms, including Connect, ALEKS, Create, and Tegrity.

McGraw-Hill Tegrity® Campus is a fully automated lecture capture solution used in traditional, hybrid, "flipped classes" and online courses to record lesson, lectures, and skills.

Connecting Students to Today's Nutrition

Understanding Our Audience

We have written *Wardlaw's Contemporary Nutrition* assuming that our students have a limited background in college-level biology, chemistry, or physiology. We have been careful to include the essential science foundation needed to adequately comprehend certain topics in nutrition, such as protein synthesis in Chapter 6. The science in this text has been presented in a simple, straightforward manner so that undergraduate students can master the material and apply it to their own lives. The Concept Maps that provide a visual depiction of macronutrient functions and characteristics are an additional aid to help students grasp nutrition science.

Featuring the Latest Guidelines and Research

Nutrition is a dynamic field. A vast quantity of research constantly reshapes our knowledge of nutritional science. The tenth edition has been carefully updated to reflect current scientific understanding, as well as the latest health and nutrition guidelines. For everyday diet planning, students will learn about the Dietary Guidelines for Americans 2010, MyPlate, and Healthy People 2020. In discussions about specific nutrition concerns, the most recent data and recommendations from the American Heart Association, American Diabetes Association, Institute of Medicine, and American Psychological Association have been included in this edition.

Newsworthy Nutrition, a feature in each chapter, highlights the use of the scientific method in recently published research studies that relate to the chapter topics. In addition, assignable questions in Connect take learning a step farther by asking students to read primary literature and apply what they have learned.

Newsworthy Nutrition

Calcium supplements decrease risk of hip fracture in women

The Women's Health Initiative (WHI) clinical trial randomly assigned 36,282 postmenopausal women in the United States to a placebo or 1000 milligram supplement of calcium carbonate plus 400 IU of vitamin D daily for 7.0 years. The hypothesis of this study was that calcium plus vitamin D supplementation would reduce hip fracture. This study also examined the health benefits and risks of calcium and vitamin D supplementation on total fractures, cardiovascular disease, cancer, and total mortality. Women in both groups were also allowed to take personal calcium and vitamin D supplements during the study, which was accounted for during the analysis of the data. Regarding bone health, the women who took the assigned calcium and vitamin D supplement but took no personal calcium supplements had a 75 percent lower risk of hip fracture than women who got the placebo and took no personal calcium supplements. The results also showed that women assigned to take calcium and vitamin D had no higher risk of heart disease, heart attacks, stroke, colorectal cancer, or total mortality. The supplement group did have a 17 percent increased risk of kidney stones. The authors concluded that long-term use of calcium and vitamin D appears to confer a substantial reduction in the risk of hip fracture among postmenopausal women. Because the risk reduction was significant only in women not taking any additional calcium supplements, it is recommended that women strive to consume the RDA for calcium (1200 milligrams per day) and vitamin D (600 IU/per day up to age 70 and 800 IU per day over 70) from food and supplements combined.

Source: RL Prentice, "Health risks and benefits from calcium and vitamin D supplementation: Women's Health Initiative clinical trial and cohort study," Osteoporosis International, 24, 2, 2013, 567. Copyright © 2013 Springer-Verlag London LTD. All rights reserved. Used with permission.

connect NUTRITION · Check out Connect at www.mcgrawhillconnect.com to further explore calcium and vitamin D supplements.

In the new margin feature, *Medicine Cabinet*, we present information on common medications used to treat diseases that have a nutrition connection. These features highlight the ways medications can affect nutritional status, as well as ways food and nutrients can affect how medications work.

Medicine Cabinet

Some people take **diuretics** to lower their blood pressure. Diuretics cause the kidneys to excrete more urine but at the same time may increase urinary excretion of minerals. This is nutritionally relevant for regulation of blood levels of potassium, magnesium (see Chapter 11), and zinc (see Chapter 13). People who take potassium-wasting diuretics need to carefully monitor their dietary intake of this mineral. Increased intake of fruits and vegetables or potassium chloride supplements are prescribed by physicians.

Examples:

Hydrochlorothiazide (Microzide)

Furosemide (Lasix)

Connecting with a Personal Focus

Applying Nutrition on a Personal Level

Throughout the tenth edition, we reinforce the fact that each person responds differently to nutrients. To further convey the importance of applying nutrition to their personal lives, we include many examples of people and situations that resonate with college students. We also stress the importance of learning to intelligently sort through the seemingly endless range of nutrition messages to recognize reliable information and to sensibly apply it to their own lives. Our goal is to provide students the tools they need to eat healthy and make informed nutrition decisions after they leave this course. Many of these features can be assigned and graded through Connect Nutrition to help students learn and apply the information and engage with the text.

Rate Your Plate

Working for Denser Bones

Osteoporosis and related low bone mass affect many adults in North America, especially older women. One-third of all women experience fractures because of this disease, amounting to about 2 million bone fractures per year.

Osteoporosis is a disease you can do something about. Some risk factors cannot be changed, but others, such as poor calcium intake, can. Is this true for you? To find out, complete this tool for estimating your current calcium intake. For all the following foods, write the number of servings you eat in a day. Total the number of servings in each category and then multiply the total number of servings by the amount of calcium for each category. Finally, add the total amount for each category to estimate your calcium intake for that day.

Does your intake meet your RDA set for calcium?

Food	Serving Size	Number of Servings	Calcium (mg)	Total Calcium (mg)
Plain low-fat yogurt	1 cup	______		
Fat-free dry milk powder	½ cup	______		
	Total servings	______	X 400	= ______ mg
Canned sardines (with bones)	3 ounces	______		
Fruit-flavored yogurt	1 cup	______		
Milk: fat-free, reduced-fat, whole, chocolate, buttermilk	1 cup	______		
Calcium-fortified soy rice, or almond milk (e.g., Silk)	1 cup	______		
Parmesan cheese (grated)	¼ cup	______		
Swiss cheese	1 ounce	______		
	Total servings	______	X 300	= ______ mg
Cheese (all other hard cheese)	1 ounce	______		
Pancakes	3	______		
	Total servings	______	X 200	= ______ mg
Canned pink salmon	3 ounces	______		
Tofu (processed with calcium)	4 ounces	______		
	Total servings	______	X 150	= ______ mg
Collards or turnip greens, cooked	½ cup	______		
Ice cream or ice milk	½ cup	______		
Almonds	1 ounce	______	X 75	= ______ mg
	Total servings			
Chard, cooked	½ cup	______		
Cottage cheese	½ cup	______		
Corn tortilla	1 medium	______		
Orange	1 medium	______		
	Total servings	______	X 50	= ______ mg
Kidney, lima, or navy beans, cooked	½ cup	______		
Broccoli	½ cup	______		
Carrot, raw	1 medium	______		
Dates or raisins	¼ cup	______		
Egg	1 large	______		
Whole-wheat bread	1 slice	______		
Peanut butter	2 tablespoons	______		
	Total servings	______	X 25	= ______ mg
Calcium-fortified orange juice	6 ounces	______		
Calcium-fortified snack bars	1 each	______		
Calcium-fortified breakfast bars	½ bar	______		
	Total servings	______	X 200	= ______ mg
Calcium-fortified chocolate candies	1 each	______		
Calcium supplements*	1 each	______	X 500	
	Total servings	______	Total calcium	= ______ mg
			intake	= ______ mg

Other calcium sources to consider include many breakfast cereals (100–250 mg per cup) and some vitamin/mineral supplements (200–500 mg or more per tablet).

*Amount varies, so check the label for the amount in a specific product and then adjust the calculation as needed.

Adapted from *Topics in Clinical Nutrition*, "Putting Calcium into Perspective for Your Clients," G. Wardlaw and N. Weese, 11:1, © 1995 Aspen Publishers, Inc.

395

CASE STUDY Getting the Most Nutrition from Your Food

In the dietary supplements aisle of the grocery store, the choices are endless—and expensive. Julie, a college sophomore, just read the Academy of Nutrition and Dietetics' position paper on nutrient supplementation for her class. She learned that dietary supplements, such as a balanced multivitamin and mineral supplement, can be a good back-up plan to ensure adequate nutrition, but the jury is still out when it comes to demonstrating a benefit of dietary supplements for long-term health. About one-third of Americans regularly take nutrient supplements, but it is usually the people who already consume a healthy diet who take them. Getting more than the recommended amount of a nutrient does not confer additional health benefits. In fact, too much of some vitamins and minerals can lead to toxicity.

Julie decides she would rather focus on getting her nutrients from foods. How can she get the most vitamins and minerals out of the foods she eats? Answer the following questions and check your responses at the end of the chapter.

1. What factors can damage or reduce vitamins in food?
2. To maximize vitamin content, what should Julie keep in mind as she selects fresh produce for purchase?
3. How does food processing affect vitamin and mineral content? Does it make a difference if Julie chooses products with whole grains or refined grains?
4. When storing fruits and vegetables in her apartment, what steps can Julie take to minimize nutrient losses?
5. Which cooking methods are best for preserving vitamin content?

What Would You Choose?

Since being accepted into the business program, your courses are getting tougher and you are staying up late doing homework. You also need work experience, so you are putting in 20 hours a week at the bookstore. Between classes, homework, work, and hanging out with friends, you are only getting about 4 to 5 hours of sleep a night and barely have time to eat. You are feeling exhausted and looking for a good, quick energy source. Which of the following would you choose to get energized? (Supplement Facts for these products are shown at the end of the chapter.)

- **a** 2-ounce 5-Hour Energy® shot
- **b** Nature Made® B-Complex with Vitamin C Dietary Supplement
- **c** Centrum® Adults Multivitamin and Mineral Supplement
- **d** Several small meals throughout the day and 7 to 9 hours of sleep each night

connect NUTRITION

Think about your choice as you read Chapter 8, then see What the Dietitian Chose at the end of the chapter. To learn more about nutrients and energy metabolism, check out the Connect site: www.mcgrawhillconnect.com.

Challenging Students to Think Critically

The pages of *Wardlaw's Contemporary Nutrition* contain numerous opportunities for students to learn more about themselves and their diet and to use their new knowledge of nutrition to improve their health. These pedagogical elements include Critical Thinking, Case Studies, Nutrition and Your Health, *What Would You Choose?*, and *Newsworthy Nutrition.* Many of the thought-provoking topics highlighted in these features are expanded upon in the online resources found in Connect Nutrition.

Nutrition and Your Health
Lipids and Cardiovascular Disease

The typical forms of cardiovascular disease—coronary heart disease and strokes—are associated with inadequate blood circulation in the heart and brain related to buildup of this plaque. Blood supplies the heart muscle, brain, and other body organs with oxygen and nutrients. When blood flow via the coronary arteries surrounding the heart is interrupted, the heart muscle can be damaged. A heart attack, or **myocardial infarction,** may result (review Fig. 5-17). This may cause the heart to beat irregularly or to stop. About 25% of people do not survive their first heart attack. If blood flow to parts of the brain is interrupted long enough, part of the brain dies, causing a **cerebrovascular accident,** or stroke.

A heart attack can strike with the sudden force of a sledgehammer, with pain radiating up the neck or down the arm. It can sneak up at night, masquerading as indigestion, with slight pain or pressure in the chest. Crushing chest pain is a more common symptom in men (see Further Reading 12). Many times, the symptoms are so subtle in women that death occurs before she or the health professional realizes that a heart attack is taking place. If there is any suspicion at all that a heart attack is taking place, the person should first call 911

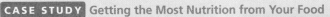

Connecting to Engaging Visuals

Attractive, Accurate Artwork

Illustrations, photographs, and tables in the text were created to help students more easily master complex scientific concepts.

- Many illustrations were updated or replaced to inspire student inquiry and comprehension and to promote interest and retention of information. Many were also redesigned to use brighter colors and a more attractive, contemporary style.

- In many figures, color-coding and directional arrows make it easier to follow events and reinforce interrelationships. Process descriptions appear in the body of the figures. This pairing of the action and an explanation walks students step-by-step through the process and increases teaching effectiveness.

The final result is a striking visual program that holds readers' attention and supports the goals of clarity, ease of comprehension, and critical thinking. The attractive layout and design of this edition are clean, bright, and inviting. This creative presentation of the material is geared toward engaging today's visually oriented students.

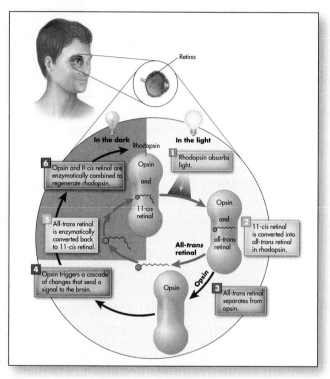

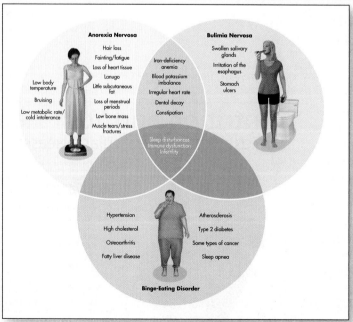

Connecting with the Latest Updates

Chapter-by-Chapter Revisions

Chapter 1: *Nutrition, Food Choices, and Health*

- Chapter 1 has a new title to reflect the discussion of food choices and their influence on overall health.

- Figure 1-2 is a new illustration depicting the marketing expenditures used to promote food and beverages to children and teens in the United States. The pledge to limit food advertising to children as part of the Children's Food and Beverage Advertising Initiative (CFBAI) is also discussed.

- An update on restaurant menu label regulations is included.

- A grilled chicken sandwich is now used as the example food for calculating calories.

- A review of the scientific method behind nutrition recommendations has been moved to Chapter 1 (Section 1.5: How do we know what we know about nutrition?), along with an introduction to our *Newsworthy Nutrition* feature that is used throughout the book. The new Figure 1-5 illustrates the steps used for testing all types of hypotheses.

- Figure 1-7 showing the percentage of adults who were obese in 2011, as well as other obesity statistics, has been updated.

- A new *Newsworthy Nutrition feature,"* Health Status of Baby Boomers Appears Lower Than Previous Generation," has been included.

- New evidence that most college students do not gain the "freshmen fifteen" is included in the Nutrition and Your Health on Eating Well in College.

- Further Readings have been updated with six new articles.

Chapter 2: *Guidelines for Designing a Healthy Diet*

- Chapter 2 has been reorganized to begin with discussion of the 2010 Dietary Guidelines and 2008 Physical Activity Guidelines for Americans and end with a description of the Dietary Reference Intakes.

- "Food Labels and Diet Planning" has been moved to be the new Nutrition and Your Health followed by a new Case Study, "Using the Nutrition Facts Label to Make Food Choices."

- A new *What Would You Choose?* feature illustrates nutrient density and energy density of various beverage choices.

- The term *proportionality* has been introduced in Chapter 2 to describe the concept of eating more nutrient-dense foods.

- There are six new Further Readings.

Chapter 3: *The Human Body: A Nutrition Perspective*

- Of all the chapters in the book, Chapter 3 has the greatest number of illustrations. We realize that many of our students have very little background in human anatomy and physiology, and we know that developing a keen understanding of digestion and absorption early in the semester will help students understand nutrition in future chapters. Therefore, we have gone to great lengths to help students visualize the body systems, especially the digestive system, and the processes that enable the body to use the nutrients in food. For example, Figure 3-2 now illustrates the organ systems as we highlight the links between nutrition and each body system. Figure 3-5 now more clearly illustrates the exchange of nutrients and wastes between the body's internal and external environments. Figure captions, such as Figure 3-8, now more clearly explain the structures and step-by-step processes of digestion, absorption, and metabolism.

- Several new terms have been introduced into Chapter 3 to broaden students' understanding of the relationship between nutrition and human physiology.

- The presentation of body systems has been reorganized slightly to allow for greatest emphasis on the digestive system in Section 3.9

- We know that students look for quick summaries of important information, so we have made special efforts to make tables more informative and readable. For example, Table 3-2 has been expanded to clarify the roles of secretions of the GI tract.

- Discoveries in the field of nutritional genomics highlight the important relationship between nutrition and genetics. Section 3.11 has been expanded to show students the importance of tailoring nutrition recommendations to each individual.

- In the Nutrition and Your Health section, we have included many updates to help students sort fact from fiction when it comes to dietary strategies to prevent or treat common ailments of the digestive system. The section on irritable bowel syndrome has been revised and expanded. In addition, we have added information to answer many of students' questions about celiac disease and nonceliac gluten sensitivity.

- A new *Medicine Cabinet* feature highlights the nutrition implications of proton pump inhibitors and H_2 blockers.

Chapter 4: *Carbohydrates*

- A new *What Would You Choose?* feature focuses on using the MyPlate food guide to select a meal that provides several sources of carbohydrates.

- The Carbohydrate Concept Map has been enhanced to include definitions of each carbohydrate category.

- The Carbohydrates in Foods section has been rewritten to include subsections (from MyPlate) on whole grains, vegetables, fruits, and dairy.

- Information on alternative sweeteners has been updated and now includes luo han guo, the extract of the monk fruit.

- The section on blood glucose response to food has been updated to emphasize the concept of glycemic index rather than glycemic load of foods.
- Information on sugar intake has been rewritten to highlight the impact of excessive sugar intake on diet quality.
- A new *Newsworthy Nutrition* feature on the decreased consumption of added sugars in the United States is included.
- The Nutrition and Your Health on diabetes has been updated to include new recommendations for diagnosis from the American Diabetes Association.
- The *Medicine Cabinet* margin feature includes information about drugs used to manage diabetes.
- The Further Readings include 12 new references.

Chapter 5: *Lipids*

- Discussion of the essential fatty acids has been moved to the beginning of the chapter.
- Figure 5-1 has been updated to include numbered carbons on the fatty acids to emphasize the omega-3, -6, and -9 fatty acids.
- The triglyceride structure in Figure 5-5a now includes three specific fatty acids.
- The potential for mercury in fish is discussed relative to the recommendation to increase our intake of fatty fish.
- The hydrogenation illustration in Figure 5-10 is now more realistic, showing the metal tanks in which the process takes place.
- The discussion of *trans* fats in foods has been updated and includes information on proposed legislation for foodservice establishments.
- More details are provided in the new illustration of fat digestion and absorption in Figure 5-12.
- The composition of each lipoprotein is now shown in Figure 5-14.
- More details are shown in the phospholipid membrane in Figure 5-16.
- Recommendations for fat intake have been expanded to include information on rich plant sources of the omega-3 alpha-linolenic acid.
- The *Newsworthy Nutrition* feature focuses on new research on the primary prevention of cardiovascular disease with a Mediterranean diet.
- The Nutrition and Your Health section on the development of cardiovascular disease now includes a description of the action of macrophages and foam cells in the production of plaque in the arterial walls.
- The beneficial effects of dark chocolate on LDL and HDL cholesterol levels are highlighted.
- The *Medicine Cabinet* feature summarizes medications commonly used to treat and prevent atherosclerosis.
- The Further Readings include six new references.

Chapter 6: *Proteins*

- The synthesis of a peptide bond is illustrated in Figure 6-2.
- A cookbook, with recipes and ingredients, is used as an analogy to describe protein synthesis.
- New trends in meat and poultry consumption in the United States are compared to those around the world.
- The section on soy and nut allergy has been updated and includes a new illustration (Figure 6-10).
- Gluten sensitivity is highlighted in a margin note.
- A *Newsworthy Nutrition* feature summarizes new research on the link between red meat consumption and increased premature mortality.
- Recent trends in vegetarianism are discussed, including the Meatless Monday initiative.
- The Further Readings include nine new references.

Chapter 7: *Energy Balance and Weight Control*

- The prevalence and recent trends in obesity are discussed and illustrated in the updated Figure 7-1.
- The discussion of the effect of lean body mass on basal metabolism has been expanded and illustrated in the new Figure 7-5.
- Information on expected weight loss from energy restriction has been revised to reflect the latest evidence.
- Body Mass Index categories are listed in Table 7-1.
- The metabolic effects of abdominal obesity are discussed, as well as the effect of excess body fat on the appetite regulating hormone, leptin.
- An expanded discussion of portion control and the basics of the Volumetrics Diet are included.
- A new section, Controlling Hunger, including Conquering the Weight-Loss Plateau, has been added.
- Information on calorie estimations on exercise machines is featured in a margin box.
- Mindful eating is included in the Behavior Modification section.
- Updated information on the use of online or mobile/smartphone applications is included.
- Research on the new weight loss drug, Lorcaserin, is summarized in the *Newsworthy Nutrition* feature.
- Recent advances in bariatric surgery are presented.
- A new section, Gaining Weight as Muscle, Not Fat, is included.
- The risks of the hCG diet are emphasized.
- The Further Readings include 12 new references.

Chapter 8: *Vitamins*

- *The What Would You Choose?* feature explores the use of dietary supplements as a source of energy.
- Newly redesigned figures throughout the chapter illustrate food sources of the vitamins with reference to the food groups pictured on MyPlate.

- A new Figure 8-1 summarizes the functional roles of the vitamins.
- The discussions of each micronutrient have been reorganized and expanded to touch upon functions, deficiency diseases, food sources, daily requirements, and toxicity in a consistent, easy-to-outline format.
- Vegetables rich in lutein and zeaxanthin are listed in Table 8-2.
- The sources and production of all of the various forms of vitamin D have been added.
- A new Figure 8-8 provides a summary and overview of the role of vitamin D in blood calcium regulation.
- Mushrooms are highlighted as the newest addition to foods naturally high in vitamin D.
- The names of the vitamin coenzymes are introduced and illustrated in Figure 8-17.
- The advantages of whole grains are emphasized and illustrated in Figure 8-18.
- A new *Newsworthy Nutrition* feature explores the relationship between B vitamins and cognitive function.
- A new Figure 8-29 illustrates the steps involved in digestion and absorption of vitamin B-12.
- Several new *Medicine Cabinet* features discuss the nutritional implications of anticoagulant medications, methotrexate, various medications that affect vitamin B-12 absorption, and use of vitamin A derivatives for acne treatment.
- Tables 8-4 and 8-5 have been updated to enhance readability and quickly summarize pertinent information on the functions, requirements, food sources, deficiency, and toxicity of the vitamins.
- The discussion of choline has been revised and expanded to reflect the latest research.
- The new Nutrition and Your Health section on Nutrition and Cancer reflects the latest recommendations from the American Cancer Society and American Institute for Cancer Research.

Chapter 9: *Water and Minerals*

- Newly redesigned figures throughout the chapter illustrate food sources of water and each mineral from MyPlate food groups.
- The presentation of water's many roles in the body has been expanded and reorganized.
- A new Figure 9-9 illustrates the steps involved in hormonal regulation of water balance to help students grasp this multi-step process.
- Information on dairy alternatives is presented within the section about avoiding too much calcium.
- Recent findings on the effectiveness of calcium supplements are discussed, including the *Newsworthy Nutrition* feature on the ability of calcium supplements to decrease the risk of hip fractures in women.
- The T-score scale used for diagnosing osteoporosis is illustrated in Figure 9-21.

- Within the section on copper, two genetic defects of copper metabolism, Wilson's disease and Menkes disease, are introduced.
- A *Newsworthy Nutrition* feature presents the results of the SELECT trial on use of antioxidant supplements to prevent cancer.
- A description of controversy surrounding fluoridation of the water supply is now included in the section on fluoride.
- Two new *Medicine Cabinet* features explain the nutritional relevance of medications used to control blood pressure and bone loss.

Chapter 10: *Nutrition: Fitness and Sports*

- A new section on achieving and maintaining fitness shows how to use the FITT principle to design an effective fitness program at any level.
- A discussion of fat adaptation is now included in the section on energy sources for exercising muscles.
- The dangers of cutting weight and the female athlete triad are explored, pointing students to information on eating disorders presented in Chapter 11.
- The discussion of the controversial topic of protein needs of athletes has been updated with the latest research.
- New content on the relevance of B vitamins and antioxidants for athletic performance has been included.
- Tables 10-6 and 10-10 have been updated to reflect new sports nutrition products on the market.
- We realized that much of our discussion of sports nutrition in previous editions was strictly applicable to endurance activities, such as long-distance running and cycling. Advances in sports nutrition research allow for the application of dietary strategies for a wider variety of sports. Therefore, we have revised and reorganized the chapter to present dietary strategies to enhance physical performance in muscular strength and power sports in addition to aerobic, endurance activities.

Chapter 11: *Eating Disorders*

- Research now shows the importance of genetic factors in addition to environmental triggers in the development of eating disorders. Discussions of origins of eating disorders throughout the chapter have been updated accordingly.
- With the 2013 release of the *Diagnostic and Statistical Manual of Mental Disorders,* 5th edition, binge-eating disorder is now classified as a distinct eating disorder. Now, a new Section 11.4 is devoted to presenting the most current research on binge-eating disorder. Figure 11-1 now illustrates physical effects of binge-eating disorder in addition to those of anorexia nervosa and bulimia nervosa.
- Tables 11-1 and 11-2 have been updated to include *DSM-5* revised diagnostic criteria for anorexia nervosa and bulimia nervosa. References to *purging* have been revised to reflect the preferred term, *compensatory behaviors*. The new Figure 11-3 presents a striking example of one of the physical effects of self-induced vomiting.

- Although eating disorders are still most prevalent among young women, additional information is presented on eating disorders that occur among men and older women.
- References to famous figures who have dealt with eating disorder have been updated to be more relevant to a younger generation of students.
- In the discussions of treatment of anorexia nervosa, bulimia nervosa, and binge-eating disorder, the most up-to-date information has been broken down into components of nutrition, psychological, and pharmacological therapy.
- A new *Medicine Cabinet* feature discusses the use of antidepressant medications for treatment of eating disorders.
- The section on other, lesser-known eating disorders, such as pica, purging disorder, night eating syndrome, and subthreshold eating disorders, has been updated to reflect new information in *DSM-5*. In addition, we have described several emerging disordered eating patterns, such as diabulimia and orthorexia.

Chapter 12: *Undernutrition Throughout the World*

- The concept of *nutrition security* is introduced in the discussion of world hunger and food security.
- Global undernutrition is discussed and illustrated region-by-region in Figure 12-1.
- The United Nations Millennium Development Goals (MDGs) and targets are presented in Table 12-1 and are used as benchmarks throughout the chapter.
- A *Newsworthy Nutrition* feature summarizes research on the relationship between maternal iodine deficiency and lower educational outcomes in their offspring.
- Information on the impact of assistance programs in the United States has been updated.
- Progress on the eradication of "food deserts," including use of the Food Access Research Atlas, is discussed.
- Statistics on the impact of war and political unrest on poverty and hunger have been updated.
- Discussion of the positive impact of global efforts to combat HIV/AIDS has been added.
- The reductions in undernutrition that are occurring in the developing world are discussed relative to progress in achieving the MDGs.
- Recent debates over the safety of genetically modified foods are considered.
- Examples of the effects of undernutrition during pregnancy and childhood are given for specific countries.
- The Further Readings include 10 new references.

Chapter 13: *Safety of Our Food Supply*

- Recent examples of foodborne illness outbreaks are discussed and are summarized in Table 13-1.
- The dramatic increase in the use and effects of antibiotics in livestock is discussed.

- The concerns of the American Heart Association regarding the GRAS listing for sodium and the concerns of other groups about FDA procedures for regulating and monitoring the safety of food additives are considered.
- Positive health effects of coffee consumption are discussed.
- Table 13-10 lists updated information on what you can do to reduce exposure to pesticides.
- The Dirty Dozen™ and Clean Fifteen lists are now included.
- New sections, Environmental Contaminants in Fish and Sustainable Seafood, have been added.
- The Hazard Analysis Critical Control Point (HACCP) tool in the battle against foodborne illness is now discussed, as well as food safety tips for what to do when the power goes out.
- The Further Readings include eight new references.

Chapter 14: *Nutrition During Pregnancy and Breastfeeding*

- Chapter 14 now opens with a new Section 14.1 on Nutrition and Fertility. In this section, we discuss the importance of weight management and other nutritional strategies for both men and women to optimize chances of conception. Polycystic ovary syndrome is defined, and management options are discussed.
- Updated information on the fetal origins hypothesis of health and disease has been introduced. The concept is first defined in Section 14.1, and current research that relates the fetal origins hypothesis to various aspects of nutritional status is woven throughout the chapter.
- In Section 14.2, the discussion of harmful exposures during pregnancy has been expanded to include more information on the effects of maternal use of both legal and illegal drugs.
- Our discussion of success in pregnancy places special emphasis on starting pregnancy at a healthy BMI and gaining weight within the recommendations of the Institute of Medicine. To highlight this point, a new *Newsworthy Nutrition* feature explains a link between gestational weight gain and future cognitive abilities of the offspring.
- In Section 14.4, we now include information on the importance of adequate vitamin D intake during pregnancy.
- The terminology regarding *hypertensive disorders of pregnancy* (formerly called *pregnancy-induced hypertension*) has been updated to reflect current medical literature.
- Section 14.7, including Table 14-5, has been revised to show even more advantages of breastfeeding for both the mother and the infant.
- A thorough discussion of folate and neural tube defects has been moved from Chapter 8 to the Nutrition and Your Health section of Chapter 14.
- We have updated terminology used to describe *congenital hypothyroidism* (formerly called *cretinism*) to reflect current medical literature. In addition, a new Figure 14-11 illustrates an infant with this disorder.

Chapter 15: *Nutrition from Infancy Through Adolescence*

- In Section 15.1, we have updated Figure 15-1 to illustrate the application of revised growth charts from WHO and the CDC. The definitions of *underweight, healthy weight, overweight,* and *obesity,* which are based on BMI-for-age, have been updated to reflect current medical literature.

- The section on failure to thrive has been extended to include an updated definition and a more thorough discussion of possible causes.

- Information on fiber and water needs of infants has been expanded in Section 15.2.

- Although covered in Chapter 14, we reemphasize the superiority of breastfeeding over formula feeding for infant nutrition in Section 15.3.

- The importance of food safety for infants and young children is highlighted throughout the chapter.

- In Sections 15.4 and 15.5, we have reorganized the presentation of strategies to improve nutritional status of preschool and school-age children. We included a new Table 15-5 to show the energy needs of young children. Throughout these sections, we have incorporated more hands-on examples of how to apply scientific recommendations at home. Table 15-7 has been updated to clearly identify foods that are potential choking hazards for young children.

- New information on avoidant/restrictive food intake disorder reflects the updated *DSM-5.*

- The latest research on the links between nutrition and autism spectrum disorders is presented in Section 15.4.

- We have updated statistics and recommendations for childhood obesity and comorbid conditions within Section 15.4. We point to specific strategies, such as eating breakfast and avoiding sugar-sweetened beverages, to prevent or correct this growing public health problem.

- An updated discussion of nutrition education and the school breakfast and lunch programs has been added.

- Special emphases on reducing intake of fast foods and caffeinated beverages now appear in Section 15.6 on nutrition for teenagers. In addition, we address the extremely harmful practice of underage alcohol consumption.

- A *Newsworthy Nutrition* feature on possible links between diet and acne is found in Section 15.6.

- The Nutrition and Your Health section on food allergies and intolerances has been updated with new statistics and terminology to reflect current medical literature.

Chapter 16: *Nutrition During Adulthood*

- A new Figure 16-1 illustrates demographic changes that will impact health care in coming years.

- We present current research on nutrition and longevity that underscores the importance of maintaining a healthy weight, consuming a plant-based diet, and continuing to be physically active.

- In Section 16.2, the latest research on protein and sodium recommendations for older adults is presented.

- Our discussion of physiological factors that affect nutritional status of older adults in Section 16.3 and Table 16-2 has been reorganized and expanded to enhance students' understanding of the nutritional needs of the world's fastest-growing population group. In particular, we have updated our discussions of oral health and physical activity. To address the unique concerns of older adults, such as sarcopenic obesity and declining bone health, the exercise recommendations for older adults now include endurance, strength-training, flexibility, and balance exercises.

- Table 16-3 now includes a larger variety of complementary and alternative therapies commonly used by older adults.

- The Nutrition and Your Health section now includes information on nutrition implications of alcohol consumption. The term *alcohol use disorders* now replaces older references to *alcohol abuse* and *alcohol dependence,* reflecting the updates of *DSM-5.* The diagnostic criteria for alcohol use disorders have been included in the feature.

Contents

Preface iv

Part One Nutrition: A Key to Health

Part Two Energy Nutrients and Energy Balance

Part Three Vitamins, Minerals, and Water

Part Four Nutrition: Beyond the Nutrients

Part Five Nutrition: A Focus on Life Stages

Student Learning Outcomes

Chapter 1 is designed to allow you to:

1.1 Describe how our food habits are affected by the flavor, texture, and appearance of food; routines and habits; early experiences and customs; advertising; nutrition and health concerns; restaurants; social changes; economics; and physiological processes affected by meal size and composition.

1.2 Identify diet and lifestyle factors that contribute to the 15 leading causes of death in North America.

1.3 Define the terms *nutrition, carbohydrate, protein, lipid (fat), alcohol, vitamin, mineral, water, phytochemical, kilocalorie (kcal),* and *fiber*.

1.4 Determine the total calories (kcal) of a food or diet using the weight and calorie content of the energy-yielding nutrients, convert English to metric units, and calculate percentages, such as percent of calories from fat in a diet.

Chapter 1
Nutrition, Food Choices, and Health

What Would You Choose?

We begin each chapter with this activity to get you thinking. We ask you to make a choice that is right for you using the concepts discussed in the chapter. At the end of each chapter, we provide the logic behind what a dietitian would recommend.

You were awake last night until 2:30 A.M. finishing a class project. Unfortunately, your Psychology 101 class meets at 9:00 this morning. When your alarm goes off at 7:30 A.M., you decide to sleep those extra 20 minutes it would take to sit down and enjoy breakfast at the dining hall. What's your best time-saving breakfast option? What factors may hold you back from making the correct choice?

a Skip breakfast but plan to consume a few extra calories at lunch and dinner.

b Eat a low-fat granola bar and iced coffee from the vending machines in your dorm.

c Fix yourself a quick bowl of Wheaties with a banana and low-fat milk along with a yogurt, all from your dorm room "pantry."

d Pick up a ham, egg, and cheese bagel.

connect |NUTRITION Think about your choice as you read Chapter 1, then see **What the Dietitian Chose** at the end of the chapter. To learn more about breakfast choices, check out the Connect site: www.mcgrawhillconnect.com.

Research has clearly shown that a lifestyle that includes a diet rich in fruits, vegetables, and whole grains, coupled with regular exercise, can enhance our quality of life in the short term and keep us healthy for many years to come. Unfortunately, this healthy lifestyle is not always easy to follow. When it comes to "nutrition," it is clear that some of our diets are out of balance with our metabolism, physiology, and physical activity level.

We begin this chapter with some questions. What influences your daily food choices? How important are factors such as taste, appearance, convenience, cost, or value? Is nutrition one of the factors you consider? Are your food choices influencing your quality of life and long-term health? By making optimal dietary choices, we can bring the goal of a long, healthy life within reach. This is the primary theme of this chapter and throughout this book.

The ultimate goal of this book is to help you find the best path to good nutrition. The information presented is based on emerging science that is translated into everyday actions that improve health. After completion of your nutrition course, you should understand the knowledge behind the food choices you make and recommend to others. We call this achievement of making food choices that are right for you "nutrition literacy."

1.5 Understand the scientific method as it is used in developing hypotheses and theories in the field of nutrition, including the determination of nutrient needs.

1.6 List the major characteristics of the North American diet, the food habits that often need improvement, and the key "Nutrition and Weight Status" objectives of the *Healthy People 2020* report.

1.7 Describe a basic plan for health promotion and disease prevention, and what to expect from good nutrition and a healthy lifestyle.

1.8 Identify food and nutrition issues relevant to college students.

1.1 Why Do You Choose the Food You Eat?

In your lifetime, you will eat about 70,000 meals and 60 tons of food. Many factors—some internal, some external—influence our food choices. This chapter begins with a discussion of these factors and ends with a conversation specifically about eating well as a college student. In between, we examine the powerful effect of dietary habits in determining overall health and take a close look at the general classes of nutrients—as well as the calories—supplied by the food we eat. We also discuss the major characteristics of the North American diet, the food habits that often need improvement, and the key "Nutrition and Weight Status" objectives in the *Healthy People 2020* report. A review of the scientific process behind nutrition recommendations is also included, along with an introduction to our "Newsworthy Nutrition" feature that you will see throughout the book.

Understanding what drives us to eat and what affects food choice will help you understand the complexity of factors that influence eating, especially the effects of our routines and food advertising (Fig. 1-1). You can then appreciate why foods may have different meanings to different people and thus why food habits and preferences of others may differ from yours.

WHAT INFLUENCES YOUR FOOD CHOICES?

Food means so much more to us than nourishment—it reflects much of what we think about ourselves. In the course of our lives, we spend the equivalent of 4 years eating. The Bureau of Labor Statistics estimated that in 2011, Americans spent the equivalent of 19 days eating and drinking. If we live to be 80 years old, that will add up to 4.1 years of eating and drinking. Overall, our daily food choices stem from a complicated mix of biological and social influences (see Fig. 1-1). Let's examine some of the key reasons we choose what we eat.

Flavor, texture, and appearance are the most important factors determining our food choices. Creating more flavorful foods that are both healthy and profitable is a major focus of the food industry. These foods are often referred to as "healthy" choices or "better for you" products. The challenge to the food industry is to match the "taste" of the foods we prefer with the nutrition and health characteristics of these products.

Early influences that expose us to various people, places, and events have a continuing impact on our food choices. Many ethnic diet patterns begin as we are introduced to foods during childhood. Parents can lay a strong foundation knowing that early exposure to food choices during infancy, toddler, and preschool years is important in influencing later health behaviors. Developing healthy patterns during childhood will go a long way to ensure healthy preferences and choices when we are teenagers and adults.

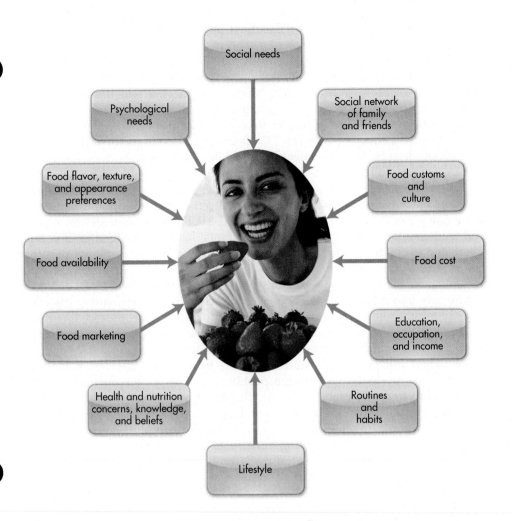

FIGURE 1-1 ◄ Food choices are affected by many factors. Which have the greatest impact on your food choices?

Routines and habits are tied to some food choices. Food habits, food availability, and convenience strongly influence choices. Most of us eat from a core group of foods with about 100 basic items accounting for 75% of our total food intake. Recent surveys indicate that the most commonly purchased foods in America are milk (about 30 gallons yearly), ready-to-eat cereal, bottled water (about 25 gallons per year), soft drinks (nearly 50 gallons per year), and bread. It is no surprise that milk and cereal are both on the top-five list because many Americans eat them together as their daily breakfast. Bread is also on the list because it is typically consumed at every meal in America, making it one of the most common forms of grain eaten. Bottled water has become the drink of choice for business meetings and other large-group gatherings, including outdoor activities. Despite the popularity of water and milk, Americans still drink nearly twice as many carbonated soft drinks per year as either water or milk. The large amount of sugar in many soft drinks is of particular concern because studies have found an association between consumption of sugar-sweetened drinks and obesity in children (see Chapter 15).

Advertising is a major media tool for capturing the food interest of the consumer. Consumers have more food choices than ever and these choices are well advertised in newspapers, magazines, billboards, radio, television, and now the Internet. The food industry in the United States spends billions on advertising. Some of this advertising is helpful, as it promotes the importance of food components such as calcium and fiber in our diets. However, the food industry also advertises highly sweetened cereals, cookies, cakes, and soft drinks because they bring in the greatest profits. Researchers at Yale University found that in 2009, the 20 largest

▲ Cereal and milk are two of the most commonly purchased foods in America largely because they are eaten together for breakfast every day by many.

Can healthy choices be legislated? Public health authorities would like to influence health behaviors by banning or taxing unhealthy foods or ingredients. However, industries and many concerned citizens oppose the expanding reach of the government. This struggle gained national attention in the summer of 2014. Mounting evidence for the link between consumption of sugar-sweetened beverages and obesity had prompted the proposal of a ban on the sale of sugary drinks in containers larger than 16 ounces in some establishments. With the backing of then-Mayor Michael Bloomberg, this ban was approved by the New York City Board of Health and would have taken effect in 2013. However, the beverage industry staunchly opposed the ban and the New York Supreme Court invalidated the law. In June 2014, the New York Court of Appeals ruled that this ban over-reached the health department's regulatory authority. Whose responsibility is it to determine which foods are good for you?

fast-food restaurant chains spent $3.8 billion on advertising, primarily television ads. Food advertising and marketing have been shown to have a definite effect on weight gain in children and adolescents. A 2012 Federal Trade Commission (FTC) report found that although food marketing to youth dropped from $2.1 billion in 2006 to $1.79 billion in 2009 (Fig. 1-2), much of the decline was the result of less spending on expensive TV advertising and more spending on cheaper online and mobile media (see Further Reading 9). Recent studies in several Western countries indicate that the association between TV advertising of foods and drinks, and childhood obesity is especially prevalent in the United States (see Further Readings 3 and 10). Concern for the negative effect of advertising and marketing on the diets and health of children has led to several strategies, including the Children's Food and Beverage Advertising Initiative (CFBAI), a self-regulatory program launched by the Council of Better Business Bureaus in 2006. CFBAI participants are 16 packaged-food companies and quick-serve restaurants that have pledged to limit their advertising to children to foods meeting science-based nutrition criteria or to not engage in child-directed advertising (see Further Reading 13). Research also indicates that mass media influences the onset of eating disorders through its depiction of extremely thin models as stereotypes of attractive bodies. The eating disorders that may result from this type of marketing lead to body distortion and dissatisfaction. Eating disorders will be introduced in the Eating Well in College section at the end of this chapter and discussed at length in Chapter 11.

Restaurant dining plays a significant role in our food choices. Restaurant food is often calorie-dense, in large portions, and of poorer nutritional quality compared to foods made at home. Fast-food and pizza restaurant menus typically emphasize meat, cheese, fried foods, and carbonated beverages. In response to recent consumer demands, restaurants have placed healthier items on their menus and many are listing nutritional content on their menus. Mandatory posting of the calorie content of restaurant items will go into effect soon as a result of the health care reform bill that President Barack Obama signed into law in March 2010. The law requires chain restaurants with 20 or more locations to post the calorie content of their offerings on menus or menu boards with other nutritional information available upon request. Food and Drug Administration (FDA) released the proposed regulations in November 2012. While many restaurants have placed calorie information on their menus, most have delayed their calorie disclosures until menu labeling regulations are approved by the FDA.

Time and convenience have become significant influences affecting food choices. These stem from a lifestyle that limits the amount of time spent in food preparation (see Further Reading 4). In the 2011 Academy of Nutrition and Dietetics Trends Survey, 62% of American adults indicated that it took too much time to keep track of their diet (see Further Reading 2). Restaurants and supermarkets have responded to our demanding work schedules and long hours away from home by supplying prepared meals, microwavable entrees, and various quick-prep frozen products.

Economics play a role in our food choices. The 2012 Food and Health Survey indicates that after taste, cost is now the number two reason why people choose the food they do. While the average American now spends less on food than in the past, young adults and those with higher incomes spend the most on food. As income increases, so do meals eaten away from home and preferences for foods such as cookies, chocolate, cheese, and meat. Also keep in mind that as calorie intake increases, so does the food bill. Tips for eating well on a college student's budget are discussed in the Nutrition and Your Health feature at the end of the chapter.

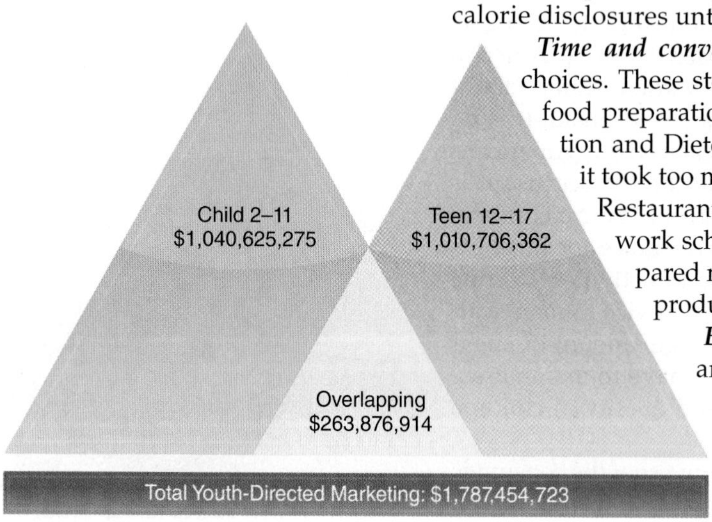

Child 2–11 $1,040,625,275
Teen 12–17 $1,010,706,362
Overlapping $263,876,914
Total Youth-Directed Marketing: $1,787,454,723

FIGURE 1-2 ▲ This chart shows the marketing expenditures used to promote food and beverages to children and teens in the United States in 2009. From Further Reading 9.

Source: Reading 9: 9. Federal Trade Commission: A Review of Food Marketing to Children and Adolescents, Follow-Up Report, December 2012. www.ftc.gov/os/2012/12/121221foodmarketingreport.pdf

Last but not least, *nutrition*—or what we think of as "healthy foods"—also directs our food purchases. North Americans who tend to make health-related food choices are often well-educated, middle-class professionals. These same people are generally health-oriented, have active lifestyles, and focus on weight control. The recent National Health Interview Survey showed that 74% of women in the study habitually or always read the nutrition labels, whereas only 58% of the men read labels on food products. Label reading was associated with a lower body mass index (BMI, body weight relative to height), especially in women. Women who read labels had a BMI of 1.48 points lower than women who did not read nutrition labels. This translates to a difference of 8.6 pounds for a woman of average height (see Further Reading 16).

WHY ARE YOU SO HUNGRY?

Two drives, **hunger** and **appetite,** influence our desire to eat. These drives differ dramatically. Hunger is primarily our physical, biological drive to eat and is controlled by internal body mechanisms. For example, as foods are digested and absorbed by the stomach and small intestine, these organs send signals to the liver and brain to reduce further food intake.

Appetite, our primarily psychological drive to eat, is affected by many of the external food choice mechanisms we discussed in the last section, such as environmental and psychological factors and social customs (see Fig. 1-1). Appetite can be triggered simply by seeing a tempting dessert or smelling popcorn popping at the movie theater. Fulfilling either or both drives by eating sufficient food normally brings a state of **satiety,** a feeling of satisfaction that temporarily halts our desire to continue eating.

A region of the brain helps regulate satiety. Imagine a tug-of-war in the brain. The *feeding center* and the *satiety center* work in opposite ways to promote adequate availability of nutrients at all times. When stimulated, cells in the feeding center signal us to eat. As we eat, cells in the satiety center are stimulated and we stop eating. For example, when we haven't eaten for a while, stimulation of the feeding center signals us to eat. When the nutrient content in the blood rises after a meal, the satiety center is stimulated, and we no longer have a strong desire to seek food. Admittedly, this concept of a tug-of-war between the feeding and satiety centers is an oversimplification of a complex process. The various feeding and satiety messages from body cells to the brain do not single-handedly determine what we eat. We often eat because food comforts us (see Further Reading 20). Almost everyone has encountered a mouthwatering dessert and devoured it, even on a full stomach. It smells, tastes, and looks good. We might eat because it is the right time of day, we are celebrating, or we are seeking emotional comfort to overcome the blues. After a meal, memories of pleasant tastes and feelings reinforce appetite. If stress or depression sends you to the refrigerator, you are mostly seeking comfort, not food calories. Appetite may not be a physical process, but it does influence food intake. We will discuss more about this mechanism, including the effect of meal size and composition on satiety, in Chapter 7 on energy balance and weight control.

PUTTING OUR FOOD CHOICES INTO PERSPECTIVE

The next time you pick up a candy bar or reach for a second helping, remember the internal and external influences on eating behavior. You should now understand that daily food intake is a complicated mix of biological and social influences. Body cells, nutrients in the blood, hormones, brain chemicals, and our social and family customs all influence food choices. When food is abundant, appetite—not hunger—most likely triggers eating. Satiety associated with consuming a meal may reside primarily in our psychological frame of mind. Also, because satiety regulation is

hunger The primarily physiological (internal) drive to find and eat food, mostly regulated by internal cues to eating.

appetite The primarily psychological (external) influences that encourage us to find and eat food, often in the absence of obvious hunger.

satiety State in which there is no longer a desire to eat; a feeling of satisfaction.

CRITICAL THINKING

Sarah is majoring in nutrition and is well aware of the importance of a healthy diet. She has recently been analyzing her diet and is confused. She notices that she eats a great deal of high-fat foods, such as peanut butter, cheese, chips, ice cream, and chocolate, and few fruits, vegetables, and whole grains. She also has become hooked on her daily cappuccino with lots of whipped cream. What three factors may be influencing Sarah's food choices? What advice would you give her on how to have her diet match her needs?

not perfect, body weight can fluctuate. We become accustomed to a certain amount of food at a meal. Providing less than that amount leaves us wanting more. One way to use this observation for weight-loss purposes is to train your eye to expect less food by slowly decreasing serving sizes to more appropriate amounts. Your appetite then readjusts as you expect less food. Keep track of what triggers your eating for a few days. Is it primarily hunger or appetite? The Rate Your Plate activity in this chapter also asks you to keep track of what influences your food intake on a daily basis.

✔ CONCEPT CHECK 1.1

1. What are the factors that influence our food choices?
2. How do hunger and appetite differ in the way they influence our desire to eat?
3. What factors influence satiety?

1.2 How Is Nutrition Connected to Good Health?

Fortunately, the foods we eat can support good health in many ways depending on their components. You just learned, however, that lifestyle habits and other factors may have a bigger impact on our food choices than the food components themselves. Unfortunately, many North Americans suffer from diseases that could have been prevented if they had known more about the foods and, more importantly, had applied this knowledge to plan meals and design their diet. We will now look at the effect these choices are having on our health both today and in the future.

WHAT IS NUTRITION?

Nutrition is the science that links foods to health and disease. It includes the processes by which the human organism ingests, digests, absorbs, transports, and excretes food substances.

NUTRIENTS COME FROM FOOD

What is the difference between food and **nutrients?** Food provides the energy (in the form of calories) as well as the materials needed to build and maintain all body cells. Nutrients are the substances obtained from food that are vital for growth and maintenance of a healthy body throughout life. For a substance to be considered an **essential nutrient,** three characteristics are needed:

- First, at least one specific biological function of the nutrient must be identified in the body.
- Second, omission of the nutrient from the diet must lead to a decline in certain biological functions, such as production of blood cells.
- Third, replacing the omitted nutrient in the diet before permanent damage occurs will restore those normal biological functions.

WHY STUDY NUTRITION?

As we mentioned in the preface, we are all nutrition experts because we all eat several times a day. Nutrition knowledge can be confusing, however, and seem like a moving target. Recommendations may seem to differ depending on their source,

nutrients Chemical substances in food that contribute to health, many of which are essential parts of a diet. Nutrients nourish us by providing calories to fulfill energy needs, materials for building body parts, and factors to regulate necessary chemical processes in the body.

essential nutrient In nutritional terms, a substance that, when left out of a diet, leads to signs of poor health. The body either cannot produce this nutrient or cannot produce enough of it to meet its needs. If added back to a diet before permanent damage occurs, the affected aspects of health are restored.

▲ Many foods are rich sources of nutrients.

and there are so many choices when shopping for food or eating out. We just learned that nutrition is only one of many factors that influence our eating habits. There is a lot to learn, and we know that you are interested in what you should be eating and how the food you eat affects you. Studying nutrition will help you erase any misconceptions you have about food and nutrition and make informed choices about the foods you eat and their relationship to health.

Nutrition is a lifestyle factor that is a key to developing and maintaining an optimal state of health for you. A poor diet and a sedentary lifestyle are known to be **risk factors** for life-threatening **chronic** diseases such as **cardiovascular (heart) disease, hypertension, diabetes,** and some forms of **cancer** (Table 1-1). Together, these and related disorders account for two-thirds of all deaths in North America (Fig. 1-3) (see Further Reading 11). Not meeting nutrient needs in younger years makes us more likely to suffer health consequences, such as bone fractures from the disease **osteoporosis,** in later years. At the same time, taking too much of a nutrient—such as a vitamin A supplement—can be harmful. Another dietary problem, drinking too much alcohol, is associated with many health problems.

U.S. government scientists have calculated that a poor diet combined with a lack of sufficient physical activity contributes to hundreds of thousands of fatal cases of cardiovascular disease, cancer, and diabetes each year among adults in the United States. Thus, the combination of poor diet and too little physical activity may be the second leading cause of death in the United States. In addition, **obesity,** which the American Medical Association recently declared as a disease, is considered the second leading cause of preventable death in North America (smoking is the first). When they occur together, obesity and smoking cause even more health problems.

▲ Major health problems can be avoided by a healthy diet, moderate calories, and adequate physical activity.

glucose A six-carbon sugar that exists in a ring form; found as such in blood and in table sugar bound to fructose; also known as *dextrose*, it is one of the simple sugars.

TABLE 1-1 ▶ Glossary Terms to Aid Your Introduction to Nutrition*

Cancer	A condition characterized by uncontrolled growth of abnormal cells.
Cardiovascular (heart) disease	A general term that refers to any disease of the heart and circulatory system. This disease is generally characterized by the deposition of fatty material in the blood vessels (hardening of the arteries), which in turn can lead to organ damage and death. Also termed coronary heart disease (CHD), as the vessels of the heart are the primary sites of the disease.
Cholesterol	A waxy lipid found in all body cells; it has a structure containing multiple chemical rings. Cholesterol is found only in foods of animal origin.
Chronic	Long-standing, developing over time. When referring to disease, this term indicates that the disease process, once developed, is slow and lasting. A good example is cardiovascular disease.
Diabetes	A group of diseases characterized by high blood **glucose.** Type 1 diabetes involves insufficient or no release of the hormone insulin by the pancreas and therefore requires daily insulin therapy. Type 2 diabetes results from either insufficient release of insulin or general inability of insulin to act on certain body cells, such as muscle cells. Persons with type 2 diabetes may or may not require insulin therapy.
Hypertension	A condition in which blood pressure remains persistently elevated. Obesity, inactivity, alcohol intake, excess salt intake, and genetics may each contribute to the problem.
Kilocalorie (kcal)	Unit that describes the energy content of food. Specifically, a kilocalorie (kcal) is the heat energy needed to raise the temperature of 1000 grams (1 liter) of water 1 degree Celsius. Although kcal refers to a 1000-calorie unit of measurement, it is commonly referred to as calories. *Calories* is a familiar term for the energy content of a food, so we will use it in this book.
Obesity	A condition characterized by excess body fat.
Osteoporosis	Decreased bone mass related to the effects of aging (including estrogen loss during menopause in women), genetic background, and poor diet.
Risk factor	A term used frequently when discussing the factors contributing to the development of a disease. A risk factor is an aspect of our lives, such as heredity, lifestyle choices (e.g., smoking), or nutritional habits.

*Many bold terms are also defined in the page margins within each chapter and in the glossary at the end of this book.

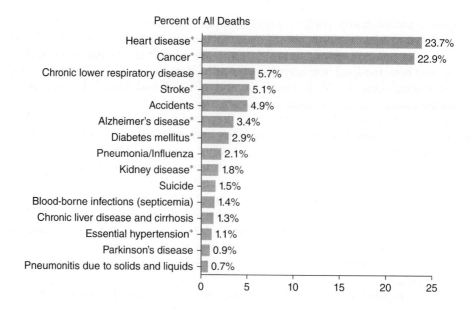

FIGURE 1-3 ▶ Fifteen leading causes of death in the United States

From Centers for Disease Control and Prevention, National Vital Statistics Report, Preliminary Data for 2011, October 10, 2012 (see Further Reading 11). Canadian statistics are quite similar.

*Causes of death in which diet plays a part.

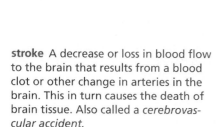

stroke A decrease or loss in blood flow to the brain that results from a blood clot or other change in arteries in the brain. This in turn causes the death of brain tissue. Also called a *cerebrovascular accident.*

carbohydrate A compound containing carbon, hydrogen, and oxygen atoms. *Sugars, starches,* and *fibers* are forms of carbohydrates.

lipid A compound containing much carbon and hydrogen, little oxygen, and sometimes other atoms. Lipids do not dissolve in water and include fats, oils, and cholesterol.

protein Food and body compounds made of amino acids; proteins contain carbon, hydrogen, oxygen, nitrogen, and sometimes other atoms, in a specific configuration.

vitamin Compound needed in very small amounts in the diet to help regulate and support chemical reactions in the body.

mineral Element used to promote chemical reactions and form body structures.

water The universal solvent; chemically, H_2O. The body is composed of about 60% water.

kilocalorie (kcal) Heat energy needed to raise the temperature of 1000 grams (1 liter) of water 1 degree Celsius; also written as *Calories.*

Obesity and chronic diseases are often preventable. An important key to good health and more health care savings is to realize that the cost of prevention, usually when we are children and young adults, is a small fraction of the cost of treating these diseases when we are older.

The good news is that the increased interest in health, fitness, and nutrition shown by Americans has been associated with long-term decreasing trends for heart disease, cancer, and **stroke** (the three leading causes of death) that continued in 2011. Mortality from heart disease, the leading cause of death, has been declining steadily since 1980. As you gain understanding about your nutritional habits and increase your knowledge about optimal nutrition, you will have the opportunity to dramatically reduce your risk for many common health problems. Recent research has shown that a healthy diet along with other healthy lifestyle factors can protect women from sudden cardiac death. A healthy diet was defined as one with a high proportion of vegetables, fruits, nuts, omega-3 fats, and fish (see Further Reading 6). For additional help, the U.S. federal government provides two websites that contain links to many sources of health and nutrition information (**www.healthfinder.gov** and **www.nutrition.gov**). Other useful sites are **www.webmd.com** and **www.eatright.org**.

✔ CONCEPT CHECK 1.2

1. How do we define nutrition?
2. What are the three leading causes of death in which diet plays a part?

1.3 What Are the Classes and Sources of Nutrients?

To begin the study of nutrition, let's start with an overview of the six classes of nutrients. You are probably already familiar with the terms **carbohydrates, lipids** (fats and oils), **proteins, vitamins,** and **minerals.** These nutrients, plus **water,** make up the six classes of nutrients found in food.

Nutrients can then be assigned to three functional categories: (1) those that primarily provide us with calories to meet energy needs (expressed in **kilocalories [kcal]**); (2) those important for growth, development, and maintenance; and (3) those that act to keep body functions running smoothly. Some function overlap exists among

TABLE 1-2 ▶ **Major Functions of the Various Classes of Nutrients**

Nutrient Classes That Provide Energy	Nutrient Classes That Promote Growth, Development, and Maintenance	Nutrient Classes That Regulate Body Processes
Most carbohydrates	Proteins	Proteins
Proteins	Lipids	Some lipids
Most lipids	Some vitamins	Some vitamins
	Some minerals	Some minerals
	Water	Water

these categories (Table 1-2). The energy-yielding nutrients (carbohydrates, lipids, and protein) along with water are needed in relatively large amounts, so they are called **macronutrients.** Vitamins and minerals are needed in such small amounts in the diet that they are called **micronutrients.**

CARBOHYDRATES

Chemically, carbohydrates can exist in foods as simple sugars and complex carbohydrates. **Simple sugars,** frequently referred to as *sugars*, are relatively small molecules. These sugars are found naturally in fruits, vegetables, and dairy products. Table sugar, known as sucrose, is an example of a simple sugar that is added to many foods we eat. Glucose, also known as blood sugar or dextrose, is an example of a simple sugar in your blood. **Complex carbohydrates** are formed when many simple sugars are joined together. For example, plants store carbohydrates in the form of **starch,** a complex carbohydrate made up of hundreds of glucose units. Breads, cereals, grains, and starchy vegetables are the main sources of complex carbohydrates.

During digestion, complex carbohydrates are broken down into single sugar molecules (such as glucose) and absorbed via **cells** lining the small intestine into the bloodstream (see Chapter 3 for more on digestion and absorption). However, the **bonds** between the sugar molecules in certain complex carbohydrates, called **fiber,** cannot be broken down by human digestive processes. Fiber passes through the small intestine undigested to provide bulk for the stool (feces) formed in the large intestine (colon).

Aside from enjoying their taste, we need sugars and other carbohydrates in our diets primarily to help satisfy the calorie needs of our body cells. Carbohydrates provide a major source of calories for the body, on average 4 kcal per gram. Glucose, a simple sugar that the body can derive from most carbohydrates, is a major source of calories for most cells. When insufficient carbohydrate is consumed, the body is forced to make glucose from proteins—not a healthy change. Chapter 4 focuses on carbohydrates.

LIPIDS

Lipids (mostly fats and oils) in the foods we eat also provide energy. Lipids yield more calories per gram than do carbohydrates—on the average, 9 kcal per gram—because of differences in their chemical composition. They are also the main form for energy storage in the body.

Lipids dissolve in certain chemical solvents (e.g., ether and benzene) but not in water. In this book, the more familiar terms *fats* and *oils* will generally be used, rather than lipids. Generally, fats are lipids that are solid at room temperature, and oils are lipids that are liquid at room temperature. We obtain fats and oils from animal and plant sources. Animal fats, such as butter or lard, are solid at room temperature. Plant oils, such as corn or olive oil, tend to be liquid at room temperature.

macronutrient A nutrient needed in gram quantities in a diet.

micronutrient A nutrient needed in milligram or microgram quantities in a diet.

simple sugar Carbohydrate composed of only one (monosaccharide) or two (disaccharide) sugar units.

complex carbohydrate Carbohydrate composed of many sugar units (polysaccharide). Examples include glycogen, starch, and fiber.

starch A complex carbohydrate made of multiple units of glucose attached together in a form the body can digest.

cell The structural basis of plant and animal organization. Cells have the ability to take up compounds from and excrete compounds into their surroundings.

bond A linkage between two atoms, such as that formed by the sharing of electrons.

fiber Substances in plant foods not digested by the processes that take place in the human stomach or small intestine. These add bulk to feces. Fiber naturally found in foods is also called *dietary fiber.*

▲ Salmon is a fatty fish that is a healthy source of essential fatty acids.

enzyme A compound that speeds the rate of a chemical reaction but is not altered by the reaction. Almost all enzymes are proteins (some are made of genetic material).

amino acid The building block for proteins containing a central carbon atom with nitrogen and other atoms attached.

chemical reaction An interaction between two chemicals that changes both chemicals.

inorganic Any substance lacking carbon atoms bonded to hydrogen atoms in the chemical structure.

To promote heart health, most people would benefit from using more plant oils in place of solid fats (see Chapter 5).

Certain fats are essential nutrients that must come from our diet. These key fats that the body cannot produce, called essential fatty acids, perform several important functions in the body: they help regulate blood pressure and play a role in the synthesis and repair of vital cell parts. However, we need only about 4 tablespoons of a common plant oil (such as canola or soybean oil) each day to supply these essential fatty acids. A serving of fatty fish, such as salmon or tuna, at least twice a week is another healthy source of fats. The unique fatty acids in these fish complement the healthy aspects of common vegetable oils. This will be explained in greater detail in Chapter 5, which focuses on lipids.

PROTEINS

Proteins are the main structural material in the body. For example, proteins constitute a major part of bone and muscle; they are also important components in blood, body cells, **enzymes,** and immune factors. Proteins can also provide calories for the body—on average, 4 kcal per gram. Typically, however, the body uses little protein for the purpose of meeting daily calorie needs. Proteins are formed when **amino acids** are bonded together. Some of these are essential nutrients.

Protein in our diet comes from animal and plant sources. The animal products meat, poultry, fish, dairy products, and eggs are significant sources of protein in most diets. Beans, grains, and some vegetables are good plant protein sources and are important to include in vegetarian diets.

Most North Americans eat up to two times as much protein as the body needs to maintain health. This amount of extra protein in the diet reflects the standard of living and the dietary habits of most North Americans. It is generally not harmful for healthy persons with no evidence of heart or kidney disease, diabetes, or family history of colon cancer or kidney stones. The excess is used for calorie needs and carbohydrate production but ultimately can be converted to and stored as fat. Chapter 6 focuses on proteins.

VITAMINS

The main function of vitamins is to enable many **chemical reactions** to occur in the body. Some of these reactions help release the energy trapped in carbohydrates, lipids, and proteins. Remember, however, that vitamins themselves contain no usable calories for the body.

The 13 vitamins are divided into two groups: four are **fat-soluble** because they dissolve in fat (vitamins A, D, E, and K); nine are **water-soluble** because they dissolve in water (the B vitamins and vitamin C). The two groups of vitamins have different sources, functions, and characteristics. Water-soluble vitamins are found mainly in fruits and vegetables, whereas dairy products, nuts, seeds, oils, and breakfast cereals are good sources of fat-soluble vitamins. Cooking destroys water-soluble vitamins much more readily than it does fat-soluble vitamins. Water-soluble vitamins are also excreted from the body much more readily than are fat-soluble vitamins. Thus, the fat-soluble vitamins, especially vitamin A, have the ability to accumulate in excessive amounts in the body, which then can lead to toxicity. Vitamins are discussed in Chapter 8.

MINERALS

Minerals are structurally simple, **inorganic** substances that do not contain carbon atoms. Minerals such as sodium and potassium typically function independently in the body, whereas minerals such as calcium and phosphorus combine to function in tissue, such as bone mineral. Because of their simple structure, minerals are not destroyed during cooking, but they can still be lost if they dissolve in the water used for cooking and that water is then discarded. Minerals are critical players in nervous system functioning, water balance, structural (e.g., skeletal) systems, and many other cellular processes but produce no calories as such for the body.

The 16 or more essential minerals required in the diet for good health are divided into two groups: **major minerals** and **trace minerals** because dietary needs and concentrations in the body vary enormously. If daily needs are less than 100 milligrams, the mineral is classified as a trace mineral; otherwise, it is a major mineral. Minerals that function based on their electrical charge when dissolved in water are also called **electrolytes;** these include sodium, potassium, and chloride. Many major minerals are found naturally in dairy products and fruits, whereas many trace minerals are found in meats, poultry, fish, and nuts. Minerals are covered in Chapter 9.

electrolytes Substances that separate into ions in water and, in turn, are able to conduct an electrical current. These include sodium, chloride, and potassium.

WATER

Water makes up the sixth class of nutrients. Although sometimes overlooked as a nutrient, water (chemically, H_2O) has numerous vital functions in the body. It acts as a **solvent** and lubricant, as a vehicle for transporting nutrients and waste, and as a medium for temperature regulation and chemical processes. For these reasons, and because the human body is approximately 60% water, the average man should consume about 3 liters—equivalent to 3000 grams or about 13 cups—of water and/or other fluids every day. Women need closer to 2200 grams or about 9 cups per day. Fluid needs vary widely, however, based on differences in body mass and environmental conditions. Because thirst is a late sign of dehydration, urine color can be used as another measure of hydration status. Urine color should be no more yellow than lemonade.

solvent A liquid substance in which other substances dissolve.

Water is not only available from the obvious sources, but it is also the major component in some foods, such as many fruits and vegetables (e.g., lettuce, grapes, and melons). The body even makes some water as a by-product of **metabolism.** Water is examined in detail in Chapter 9.

metabolism Chemical processes in the body by which energy is provided in useful forms and vital activities are sustained.

OTHER IMPORTANT COMPONENTS IN FOOD

Another group of compounds in foods from plant sources, especially within the fruit and vegetable groups, is what scientists call **phytochemicals.** Although these plant components are not considered essential nutrients in the diet, many of these substances provide significant health benefits. Considerable research attention is focused on various phytochemicals in reducing the risk for certain diseases. For example, evidence from animal and laboratory studies indicates that compounds in blueberries and strawberries prevent the growth of certain cancer cells. Although certain phytochemicals are now available as dietary supplements, research suggests that their health benefits are best obtained through the consumption of whole foods. Foods with high phytochemical content are sometimes called "superfoods" because of the health benefits they are thought to confer. There is no legal definition of the term *superfood,* however, and there is concern that it is being overused in marketing certain foods. Table 1-3 lists some noteworthy phytochemicals with their common food sources. Tips for boosting the phytochemical content of your diet will be discussed in Chapter 2.

phytochemical A chemical found in plants. Some phytochemicals may contribute to a reduced risk of cancer or cardiovascular disease in people who consume them regularly.

Some related compounds under study are also found in animal products, such as sphingolipids (meat and dairy products) and conjugated linoleic acid (meat and cheese). These are not phytochemicals per se because they are not from plant sources, but they have been shown to have health benefits.

SOURCES OF NUTRIENTS

Now that we know the six classes of nutrients, it is important to understand that the quantities of the various nutrients that people consume in different foods vary widely. On a daily basis, we consume about 500 grams, or about 1 pound, of

▲ Blueberries are sometimes considered "superfoods" because they are rich in health-promoting phytochemicals.

genes A specific segment on a chromosome. Genes provide the blueprints for the production of all body proteins.

TABLE 1-3 ▶ Food Sources of Some Phytochemical Compounds Under Study

Food Sources	Phytochemical
Garlic, onions, leeks	Allyl sulfides/organosulfurs
Garlic, onions, licorice, legumes	Saponins
Orange, red, and yellow fruits and vegetables (egg yolks are a source as well)	Carotenoids (e.g., lycopene)
Oranges, lemons, grapefruit	Monoterpenes
Chili peppers	Capsaicin
Flaxseed, berries, whole grains	Lignans
Cruciferous vegetables (broccoli, cabbage, kale)	Indoles
Cruciferous vegetables, especially broccoli	Isothiocyanates
Soybeans, other legumes, cucumbers, other fruits and vegetables	Phytosterols
Citrus fruit, onions, apples, grapes, red wine, tea, chocolate, tomatoes	Flavonoids
Soybeans, other legumes	Isoflavones
Tea	Catechins
Blueberries, strawberries, raspberries, grapes, apples, bananas, nuts	Polyphenols
Red, blue, and purple plants (blueberries, eggplant)	Anthocyanosides
Onions, bananas, oranges (small amounts)	Fructooligosaccharides
Grapes, peanuts, red wine	Resveratrol

protein, fat, and carbohydrate. In contrast, the typical daily mineral intake totals about 20 grams (about 4 teaspoons), and the daily vitamin intake totals less than 300 milligrams (1/15 of a teaspoon). Although we require a gram or so of some minerals, such as calcium and phosphorus, we need only a few milligrams or less of other minerals each day. For example, we need about 10 milligrams of zinc per day, which is just a few specks of the mineral.

The nutrient content of the foods we eat also differs from the nutrient composition of the human body. This is because growth, development, and later maintenance of the human body are directed by the genetic material (DNA) inside body cells. This genetic blueprint determines how each cell uses the essential nutrients to perform body functions. These nutrients can come from a variety of sources. Cells are not concerned about whether available amino acids come from animal or plant sources. The carbohydrate glucose can come from sugars or starches. The food that you eat provides cells with basic materials to function according to the directions supplied by the genetic material (**genes**) housed in body cells. Genetics and nutrition will be discussed in Chapter 3.

✔ CONCEPT CHECK 1.3

1. What are the six classes of nutrients?
2. What are the three general functions of nutrients in the body?

1.4 What Math Concepts Will Aid Your Study of Nutrition?

CALORIES

We obtain the energy we need for involuntary body functions and voluntary physical activity from various calorie sources: carbohydrates (4 kcal per gram), fats (9 kcal per gram), and proteins (4 kcal per gram). Foods generally provide more than one calorie source. Plant oils, such as soybean or canola oil, are one exception; these are 100% fat at 9 kcal per gram.

Alcohol is also a potential source of calories, supplying about 7 kcal per gram. It is not considered an essential nutrient, however, because it is not required for human function. Still, alcoholic beverages, such as beer—also rich in carbohydrate—are a contributor of calories to the diets of many adults.

The body releases the energy from the chemical bonds in carbohydrate, protein, and fat (and alcohol) in order to:

- Build new compounds.
- Perform muscular movements.
- Promote nerve transmission.
- Maintain electrolyte balance within cells.

Chapter 7 describes how that energy is released from the chemical bonds in energy-yielding nutrients and then used by body cells to support the processes just described.

The energy in food is often expressed in terms of calories on food labels. As defined earlier, a calorie is the amount of heat energy it takes to raise the temperature of 1 gram of water 1 degree Celsius (1°C, centigrade scale). (Chapter 7 has a diagram of the bomb calorimeter that can be used to measure calories in foods.) A calorie is a tiny measure of heat, so food energy is more conveniently expressed in terms of the kilocalorie (kcal), which equals 1000 calories. (If the "c" in calories is capitalized, this also signifies kilocalories.) A kcal is the amount of heat energy it takes to raise the temperature of 1000 grams (1 liter) of water 1°C. The abbreviation *kcal* is used throughout this book. On food labels, the word *calorie* (without a capital "C") is also used loosely to mean *kilocalorie*. Any values given on food labels in calories are actually in kilocalories (Fig. 1-4). A suggested intake of 2000 calories per day on a food label is technically 2000 kcal.

CALCULATING CALORIES

Use the 4-9-4 estimates for the calorie content of carbohydrate, fat, and protein introduced over the last few pages to determine calorie content of a food. Consider these foods:

Carbohydrate
4 kcal per gram

Fat
9 kcal per gram

Protein
4 kcal per gram

Alcohol
7 kcal per gram

▲ Calorie content of energy nutrients and alcohol. The weights illustrate their relative energy potential per gram.

alcohol Ethyl alcohol or ethanol (CH_3CH_2OH) is the compound in alcoholic beverages.

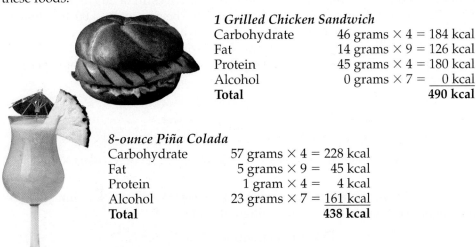

1 Grilled Chicken Sandwich

Carbohydrate	46 grams × 4 =	184 kcal
Fat	14 grams × 9 =	126 kcal
Protein	45 grams × 4 =	180 kcal
Alcohol	0 grams × 7 =	0 kcal
Total		**490 kcal**

8-ounce Piña Colada

Carbohydrate	57 grams × 4 =	228 kcal
Fat	5 grams × 9 =	45 kcal
Protein	1 gram × 4 =	4 kcal
Alcohol	23 grams × 7 =	161 kcal
Total		**438 kcal**

WHOLE WHEAT BREAD

Nutrition Facts

Serving Size 1 slice (36g) Servings Per Container 19

Amount Per Serving

Calories 80 Calories from Fat 10

% Daily Value*		% Daily Value*	
Total Fat 1g	**2%**	**Total Carbohydrate** 15g	**5%**
Saturated Fat 0g	**0%**	Dietary Fiber 2g	**8%**
Trans Fat less than 1g **			
Cholesterol 0mg	**0%**	Sugars less than 1g	
Sodium 200mg	**8%**	**Protein** 3g	
Vitamin A 0%	Vitamin C 0%	Calcium 0%	Iron 4%

*Percent Daily Values (DV) are based on a 2,000 calorie diet. Your daily values may be higher or lower depending on your calorie needs:

	Calories:	2,000	2,500
Total Fat	Less than	65g	80g
Sat Fat	Less than	20g	25g
Cholesterol	Less than	300mg	300mg
Sodium	Less than	2,400mg	2,400mg
Total Carbohydrate		300g	375g
Dietary Fiber		25g	30g

** Intake of *trans* fat should be as low as possible.

INGREDIENTS: WHOLE WHEAT, WATER, ENRICHED WHEAT FLOUR [FLOUR, MALTED BARLEY, NIACIN, REDUCED IRON, THIAMINE MONONITRATE (VITAMIN B1) AND RIBOFLAVIN (VITAMIN B2)], CORN SYRUP, PARTIALLY HYDROGENATED COTTONSEED OIL, SALT, YEAST.

FIGURE 1-4 ▲ Use the nutrient values on the Nutrition Facts panel to calculate calorie content of a food. Based on carbohydrate, fat, and protein content, a serving of this food (Whole Wheat Bread) contains 81 kcal ([15 × 4] + [1 × 9] + [3 × 4] = 81). The label lists 80, suggesting that the calorie value was rounded down.

You can also use the 4-9-4 estimates to determine what portion of total calorie intake is contributed by the various calorie-yielding nutrients. Assume that one day you consume 290 grams of carbohydrates, 60 grams of fat, and 70 grams of protein. This consumption yields a total of 1980 kcal ([290 × 4] + [60 × 9] + [70 × 4]5 = 1980). The percentage of your total calorie intake derived from each nutrient can then be determined:

% of kcal as carbohydrate = (290 × 4) ÷ 1980 = 0.59 (× 100 = 59%)
% of kcal as fat = (60 × 9) ÷ 1980 = 0.27 (× 100 = 27%)
% of kcal as protein = (70 × 4) ÷ 1980 = 0.14 (× 100 = 14%)

Check your calculations by adding the percentages together. Do they total 100%?

PERCENTAGES

You will use a few mathematical concepts in studying nutrition. Besides performing addition, subtraction, multiplication, and division, you need to know how to calculate percentages and convert English units of measurement to metric units.

The term *percent* (%) refers to a part of the total when the total represents 100 parts. For example, if you earn 80% on your first nutrition examination, you will have answered the equivalent of 80 out of 100 questions correctly. This equivalent also could be 8 correct answers out of 10; 80% also describes 16 of 20 (16/20 = 0.80 or 80%). The decimal form of percents is based on 100% being equal to 1.00. It is difficult to succeed in a nutrition course unless you know what a percentage means and how to calculate one. Percentages are used frequently when referring to menus and nutrient composition. The best way to master this concept is to calculate some percentages. Some examples follow:

Question	Answer
What is 6% of 45?	6% = 0.06, so 0.06 × 45 = 2.7
What percent of 99 is 3?	3/99 = 0.03 or 3% (0.03 × 100)

Joe ate 15% of the adult Recommended Dietary Allowance for iron (RDA = 8 milligrams) at lunch. How many milligrams did he eat?

0.15 × 8 milligrams = 1.2 milligrams

THE METRIC SYSTEM

The basic units of the metric system are the meter, which indicates length; the gram, which indicates weight; and the liter, which indicates volume. Appendix F in this textbook lists conversions from the metric system to the English system (pounds, feet, and cups) and vice versa. Here is a brief summary:

A gram (g) is about 1/30 of an ounce (28 grams to the ounce).
5 grams of sugar or salt is about 1 teaspoon.
A pound (lb) weighs 454 grams.
A kilogram (kg) is 1000 grams, equivalent to 2.2 pounds.
To convert your weight to kilograms, divide it by 2.2.
   A 154-pound man weighs 70 kilograms (154/2.2 = 70).
A gram can be divided into 1000 milligrams (mg) or 1,000,000 micrograms (µg or mcg).
   10 milligrams of zinc (approximate adult need) would be a few grains of zinc.
Liters are divided into 1000 units called milliliters (ml).
One teaspoon equals about 5 milliliters (ml), 1 cup is about 240 milliliters, and
   1 quart (4 cups) equals almost 1 liter (L) (0.946 liter to be exact).
A centimeter is 1/100 of a meter. 2.54 centimeters equals 1 inch.

If you plan to work in any scientific field, you will need to learn the metric system. *For now, remember that a kilogram equals 2.2 pounds, an ounce weighs 28 grams, 2.54 centimeters equals 1 inch, and a liter is almost the same as a quart.* In addition, know the fractions that the following prefixes represent: *micro* (1/1,000,000), *milli* (1/1000), *centi* (1/100), and *kilo* (1000).

✔ CONCEPT CHECK 1.4

1. What are the energy (calorie) values for each of the "energy nutrients"?

1.5 How Do We Know What We Know About Nutrition?

How do we know what we know about nutrient needs? In a word, research. Like other sciences, the research that sets the foundation for nutrition knowledge has developed using the *scientific method,* a testing procedure designed to detect and eliminate error.

THE SCIENTIFIC METHOD

The first step of the scientific method is the observation of a natural phenomenon (Fig. 1-5). Scientists then suggest possible explanations, called **hypotheses,** about its cause. At times, historical events have provided clues to important relationships in nutrition science, such as the link between the need for vitamin C and the development of the disease **scurvy** (see Chapter 8). Another approach is for scientists to study diet and disease patterns among various populations, a research method called **epidemiology.**

Thus, hypotheses about the role of diet in various health problems can be suggested by historical and epidemiological findings. *Proving* the role of particular dietary components, however, requires controlled experiments. The data gathered from experiments may either support or refute each hypothesis. If the results of many experiments support a hypothesis, scientists accept the hypothesis as a **theory.** Often, the results from one experiment suggest a new set of questions.

The most rigorous type of controlled experiment follows a randomized, **double-blind,** placebo-controlled study design. In this type of study, a group of participants—the experimental group—follows a specific protocol (e.g., consuming a certain food

hypotheses Tentative explanations by a scientist to explain a phenomenon.

scurvy The deficiency disease that results after a few weeks to months of consuming a diet that lacks vitamin C; pinpoint sites of bleeding on the skin are an early sign.

epidemiology The study of how disease rates vary among different population groups.

theory An explanation for a phenomenon that has numerous lines of evidence to support it.

double-blind study An experimental design in which neither the participants nor the researchers are aware of each participant's assignment (test or placebo) or the outcome of the study until it is completed. An independent third party holds the code and the data until the study has been completed.

FIGURE 1-5 ▶ The scientific method. Scientists consistently follow these steps when testing all types of hypotheses. Scientists do not accept a nutrition or other scientific hypothesis until it has been thoroughly tested using the scientific method.

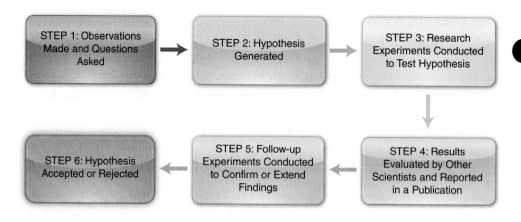

control group Participants in an experiment who are not given the treatment being tested.

placebo Generally a fake medicine or treatment used to disguise the treatments given to the participants in an experiment.

case-control study A study in which individuals who have a disease or condition, such as lung cancer, are compared with individuals who do not have the condition.

or nutrient), and participants in a corresponding **control group** follow their normal habits or consume a **placebo.** People are randomly assigned to each group. Scientists then observe the experimental group over time to see if there is any effect not found in the control group. A **case-control study** compares individuals who have a disease or condition, such as lung cancer, to individuals who do not have the condition.

TESTING A HYPOTHESIS: CAN A HIGH-FAT DIET CAUSE WEIGHT LOSS?

The following example shows how the scientific method (Fig. 1-5) was used to test a hypothesis about the effects of low-calorie, high-fat diets on weight loss.

Step 1. Observations Made and Questions Asked. In the mid-1950s, physicians note that in short-term experiments, people eating a low-calorie, high-fat diet lost weight more quickly than people eating a low-calorie, high-carbohydrate diet.

Step 2. Hypothesis Generated. Low-calorie, high-fat diets (e.g., Atkins diet) lead to more weight loss over time than low-calorie, high-carbohydrate diets.

Step 3. Research Experiments Conducted. For 1 year, researchers followed 63 people assigned to either a low-calorie, high-fat diet or a low-calorie, high-carbohydrate diet. At the end of the study, weight loss did not differ significantly between the two groups.

Step 4. Results Evaluated by Other Scientists and Published. A peer review indicated that the study was conducted in an unbiased, scientific manner and the results appeared valid. The study was published in *The New England Journal of Medicine* (348:2082, 2003).

Step 5. Follow-up Experiments Conducted to Confirm or Extend the Findings. A study published in 2005 described what happened when 160 people were assigned to a specific diet for a year. One diet was a low-calorie, high-fat diet, and another was a low-calorie, high-carbohydrate diet. Again, at the end of 1 year, weight loss in these two groups did not differ significantly. Peer reviewers indicated that the study was conducted scientifically. It was published in the *Journal of the American Medical Association* (293:43, 2005). A more recent 2-year study compared people who ate one of four reduced-calorie diets that was either low or high in fat, average or high in protein, or low or high in carbohydrates. After 2 years, the amount of weight loss was similar among participants. This study was published in *The New England Journal of Medicine* (360:859, 2009).

Step 6. Accept or Reject Hypothesis? Based on the currently available research studies, the hypothesis is not accepted. Reduced-calorie diets result in weight loss regardless of whether they emphasize fat, protein, or carbohydrate.

TYPES OF EXPERIMENTS

Human experiments provide the most convincing evidence about relationships between nutrients and health, but they are often not practical or ethical. Thus, much of what we know about human nutritional needs and functions has been

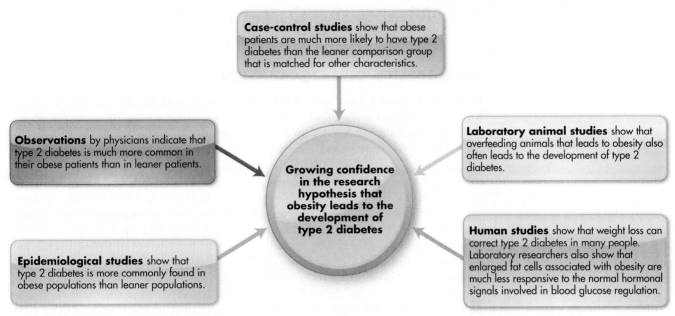

Case-control studies show that obese patients are much more likely to have type 2 diabetes than the leaner comparison group that is matched for other characteristics.

Observations by physicians indicate that type 2 diabetes is much more common in their obese patients than in leaner patients.

Laboratory animal studies show that overfeeding animals that leads to obesity also often leads to the development of type 2 diabetes.

Growing confidence in the research hypothesis that obesity leads to the development of type 2 diabetes

Epidemiological studies show that type 2 diabetes is more commonly found in obese populations than leaner populations.

Human studies show that weight loss can correct type 2 diabetes in many people. Laboratory researchers also show that enlarged fat cells associated with obesity are much less responsive to the normal hormonal signals involved in blood glucose regulation.

FIGURE 1-6 ▲ Data from a variety of sources can come together to support a research hypothesis. This diagram shows how various types of research data support the hypothesis that obesity leads to the development of type 2 diabetes.

gleaned from animal experiments. The use of animal experiments to study the role of nutrition in certain human diseases depends on the availability of an **animal model** in which a disease in laboratory animals closely mimics a particular human disease. Often, if no animal model is available and human experiments are ruled out, scientific knowledge cannot advance beyond what can be learned from epidemiological studies.

animal model Use of animals to study disease to understand more about human disease.

Once an experiment is complete, scientists summarize the findings and seek to publish the results in scientific journals. Generally, before articles are published in scientific journals, they are critically reviewed by other scientists familiar with the subject, which helps to ensure that only high-quality, objective research findings are published.

Keep in mind that one experiment is never enough to prove a particular hypothesis or provide a basis for nutritional recommendations. Rather, through follow-up studies, the results obtained in one laboratory must be confirmed by similar experiments conducted in other laboratories and, possibly, under varying circumstances. Only then can we really trust and use the results. As shown in Figure 1-6, the more lines of evidence available to support an idea, the more likely it is to be true.

Epidemiological studies may suggest hypotheses, but controlled experiments are needed to rigorously test hypotheses before nutrition recommendations can be made. For example, epidemiologists found that smokers who regularly consumed fruits and vegetables had a lower risk for lung cancer than smokers who ate few fruits and vegetables. Some scientists proposed that beta-carotene, a pigment present in many fruits and vegetables, may be responsible for reducing the damage that tobacco smoke creates in the lungs. However, in double-blind studies involving heavy smokers, the risk of lung cancer was found to be *higher* for those who took beta-carotene supplements than for those who did not (this is not true for the small amount of beta-carotene found naturally in foods). Soon after these results were reported, two other large studies using beta-carotene supplements were stopped by the federal funding agency on the basis that these supplements are ineffective in preventing both lung cancer and cardiovascular disease.

Newsworthy Nutrition

Throughout the textbook, we have highlighted the use of the scientific method in research studies in a feature we call "Newsworthy Nutrition." These are recently published studies that relate to chapter topics and that have made a significant impact on our nutrition knowledge. You will find the first "Newsworthy Nutrition" study in section 1.6 on the current state of the North American diet and health.

connect
NUTRITION

Visit **www.mcgrawhillconnect.com** to further explore the research presented in these Newsworthy Nutrition features using interactive learning tools assigned by your instructor.

✓ CONCEPT CHECK 1.5

1. What are the six steps used in the scientific method?

1.6 What Is the Current State of the North American Diet and Health?

DOES OBESITY THREATEN OUR FUTURE?

There is no doubt that the obesity epidemic threatens the future health of Americans. It is estimated that more than 35 percent of adults are currently obese, with *obesity* defined as having an excessive amount of body fat relative to lean tissue. Considered more broadly, two-thirds of adults and one-third of children are overweight or obese. According to the Centers for Disease Control (CDC), the average American adult male weighs 194.7 pounds, and the average American adult woman weighs 164.7 pounds. Where you live is also a factor, with obesity rates varying by state. The most recent report from the Trust for America's Health and the Robert Wood Johnson Foundation indicates that in 2011, no state had a prevalence of obesity less than 20%, with 11 states and the District of Columbia having a prevalence between 20% to not quite 25%. Their report, *F as in Fat: How Obesity Threatens America's Future 2012* (see Further Reading 19), also revealed that 12 states now have obesity rates above 30%, compared to 2006, when only one state was above 30%. The 2012 report is based on self-reported, state-by-state obesity data from the CDC (Fig. 1-7) (see Further Reading 5), and predicts that if the prevalence of obesity continues to grow at current rates, the effects on the health and wealth of our nation will be catastrophic by the year 2030. The report calls for a national commitment to the prevention of obesity and outlines a scenario by which a 9% to 14% (depending on the state) reduction of the states' obesity rates by 2030 could greatly decrease obesity-related diseases and health care costs.

It is well documented that this extra weight of more than 4.5 billion extra pounds has and will continue to have dangerous consequences. In Section 1.2, we already pointed out that obesity plays a role in chronic illness, including heart disease, stroke, high blood pressure, high cholesterol, diabetes, arthritis, and certain cancers. It is estimated that obesity kills more than 200,000 Americans a year. Because of its role in so many chronic disorders, obesity is an expensive condition with more than $190 billion spent annually on health care related to obesity. Obesity also results in the loss of tremendous work productivity. Because of numerous medical conditions, obese individuals are absent from work more often than

FIGURE 1-7 ▶ Percentage of adults who are obese,* by state, 2012.

*Body mass index (BMI) ≥ 30, or about 30 pounds overweight for a 5′ 4″ person, based on self-reported weight and height.

Source: CDC, Behavioral Risk Factor Surveillance System.

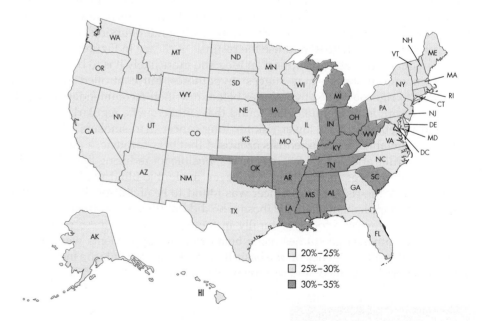

Newsworthy Nutrition

Health Status of Baby Boomers Appears Lower Than Previous Generation's

'Baby boomers' are the 78 million children born in the United States between 1946 and 1964. They make up more than a quarter of the U.S. population. Because medicine has advanced during the baby boomers' lifetimes and led to longer life expectancy, scientists at the West Virginia University School of Medicine developed a hypothesis that the baby boomer generation is healthier than prior generations. They studied the overall health status of aging baby boomers (average age 54 years) compared with the previous generation (at the same age) by analyzing data from the National Health and Nutrition Examination Surveys. Positive results for the baby boomers included longer life expectancy, less likelihood of smoking cigarettes, and lower rates of emphysema and heart attacks than the previous generation. The "self-rated" health of the baby boomers, however, was lower, with only 13.2% reporting "excellent" health compared with 32% of those in the previous generation. More specifically, the research team found that the baby boomers were more likely to suffer from several chronic diseases, including diabetes, hypertension, and hypercholesterolemia, than those born earlier. Obesity was also more common among baby boomers (38.7% versus 29.4%), and more than half the baby boomers reported no regular physical activity (52.2% versus 17.4%). The results of this study, therefore, do not support the hypothesis that baby boomers are healthier than prior generations. The authors of this published report conclude that the increased rates of chronic disease they found in baby boomers demonstrate a clear need for expanded efforts at disease prevention and health promotion in this generation.

Source: King DE, and others: The status of baby boomers' health in the United States: The healthiest generation? *Journal of the American Medical Association Internal Medicine,* 173:385, 2013 (see Further Reading 12).

connect | NUTRITION Check out the Connect site www.mcgrawhillconnect.com to further explore the health status of baby boomers.

those of healthy weight. Health economists estimate that obesity-related absenteeism costs employers as much as $6.4 billion a year, whereas loss of on-the-job productivity due to pain, shortness of breath, or other obstacles costs another $30 billion. It has become obvious that the answers to the obesity crisis are not simple. From a nutrition perspective, however, the problem can be clearly stated. Most of us continue to eat too much, especially foods with a high number of calories and a low number of nutrients, and we do not engage in enough physical activity. Read about recent research on the health status, including obesity, of the baby boomer generation in the Newsworthy Nutrition above.

ASSESSING THE CURRENT NORTH AMERICAN DIET

With the aim of finding out what North Americans eat, federal agencies conduct surveys to collect data about food and nutrient consumption, as well as connections between diet and health. In the United States, the U.S. Department of Health and Human Services monitors food consumption with the National Health and Nutrition Examination Survey (NHANES). In Canada, this information is gathered by Health Canada in conjunction with Agriculture and Agrifood Canada. Survey data from 2009 to 2010 indicate that North American adults consume about 15% of their calorie intake as proteins, 52% as carbohydrates, and 33% as fats. These percentages, which do not consider alcohol, fall within the ranges recommended by the Food and Nutrition Board (FNB) of the National Academy of Sciences. The FNB advocates that

▲ African-Americans and Hispanics have a greater chance of developing hypertension than do other ethnic groups and therefore should replace salty snack foods with healthier low-sodium options.

salt Compound of sodium and chloride in a 40:60 ratio.

10% to 35% of calories come from protein, 45% to 65% from carbohydrate, and 20% to 35% from fat. These standards apply to people in both the United States and Canada.

Food-consumption data also indicate that animal sources supply about two-thirds of protein intake for most North Americans, whereas plant sources supply only about one-third. In many other parts of the world, it is just the opposite: plant proteins—from rice, beans, corn, and other grains and vegetables—dominate protein intake. About half the carbohydrate in North American diets comes from simple sugars; the other half comes from starches (such as in pastas, breads, and potatoes). About 60% of dietary fat comes from animal sources and 40% from plant sources.

Results from national nutrition surveys and other studies show that North Americans consume more calories than ever before and from a wide variety of foods. Individuals, however, often do not choose the foods that will meet all their nutrient needs. Food availability data from 1909 to 2007 confirm that the major contributors to increased energy intake over the last century are oils, shortening, meat, cheese, and frozen desserts. Since 1970, there has been an increase in added sweeteners, and carbonated beverage consumption has increased at the expense of milk.

In the next section, we discuss recommendations to consume a variety of nutrient-dense foods within and across the food groups, especially whole grains, fruits, vegetables, low-fat or fat-free milk or milk products, and lean meats and other protein sources. These foods will provide nutrients that are often overlooked, including various B vitamins, vitamin C (especially for smokers), vitamin D, vitamin E, calcium, potassium, magnesium, iron, fiber, and many phytochemicals. Daily intake of a balanced multivitamin and mineral supplement is another strategy to help meet nutrient needs but does not make up for a poor diet, particularly for calcium, potassium, and fiber intake. Also keep in mind that use of nutrient supplements should be discussed with your health professional to avoid potentially harmful side effects (dietary supplements are discussed in Chapter 8).

Routinely, experts also recommend that we pay more attention to balancing calorie intake with needs. An excess intake of calories is usually tied to overindulgence in sugar, fat, and alcoholic beverages. African-Americans and Hispanics have a greater chance of developing hypertension than do other ethnic groups in North America and therefore may need to decrease the amount of **salt** (sodium chloride) and alcohol in their diets. These substances are two of the many factors linked to hypertension. Moderation of salt and alcohol intake—along with certain fats, cholesterol, and total calorie intake—is a recommended practice for all adults.

Many North Americans would benefit from a healthier balance of food in their diets. Moderation is the key for some foods, such as sugared soft drinks and fried foods. For other foods, such as fruits and vegetables, increased quantity and variety are warranted. Few adults currently meet the new recommendation to "fill half your plate with fruits and vegetables" promoted in the U.S. Department of Agriculture (USDA) MyPlate guidelines for total servings of vegetables and fruits.

HEALTH OBJECTIVES FOR THE UNITED STATES FOR THE YEAR 2020

Health promotion and disease prevention have been public health strategies in North America for the past three decades. One part of this strategy is *Healthy People 2020,* a report issued in December 2010 by the U.S. Department of Health and Human Services' (DHHS) Public Health Service. Every 10 years, DHHS issues a collection of health objectives for the nation. These objectives are developed by experts in federal agencies and target major public health concerns, setting goals for the coming decade. *Healthy People 2020* sets forth more than 600 health objectives across 42 topic areas and outlines national standards to eliminate health disparities, improve access to health education and quality health care, and strengthen public health services. The vision for 2020 is a society in which all people live long, healthy lives. Important features of *Healthy People 2020* include a focus on health equity and social determinants of health, and an interactive, personalized website at **www.HealthyPeople.gov.**

The overarching goals of *Healthy People 2020* are to:

- Attain high-quality, longer lives free of preventable disease, disability, injury, and premature death.
- Achieve health equity, eliminate disparities, and improve health of all groups.
- Create social and physical environments that promote good health for all.
- Promote quality of life, healthy development, and healthy behaviors across all life stages.

▲ Many nutrition-related objectives are part of the *Healthy People 2020* report. The report outlines health promotion and disease prevention objectives for the United States for the year 2020.

Healthy People 2020, like earlier versions, includes a topic area specific to nutrition. This topic is called Nutrition and Weight Status, and its objectives target individual behaviors, as well as the policies and environments that support these behaviors. Nutrition and weight status are important because a healthful diet helps us reduce our risks for numerous health conditions that burden the public health system, including heart disease, high blood pressure, diabetes, osteoporosis, and some cancers. Good nutrition for children is also emphasized in this report because of its importance for growth and development. The goal of this topic is to promote health and reduce chronic disease risk through the consumption of healthful diets and achievement and maintenance of healthy body weights. This goal also includes increasing household food security and eliminating hunger.

The Nutrition and Weight Status objectives are based on strong science that supports the health benefits of eating a healthful diet and maintaining a healthy body weight. A healthful diet is described as one that includes:

- Consuming a variety of nutrient-dense foods within and across the food groups, especially whole grains, fruits, vegetables, low-fat or fat-free milk or milk products, and lean meats and other protein sources.
- Limiting intake of solid fats, cholesterol, added sugars, sodium (salt), and alcohol.
- Limiting intake of calories to meet needs for calories.

The objectives also emphasize that individual behaviors should be addressed in any efforts to change diet and weight, as well as the policies and environments that support these behaviors in settings such as schools, worksites, health care organizations, and communities.

Table 1-4 is a list of the six categories of objectives for the Nutrition and Weight Status topic, along with the 22 specific objectives. Table 1-5 provides a more detailed sample of nine of the specific Nutrition and Weight Status objectives along with the current status of these objectives and their targets for 2020.

TABLE 1-4 ▶ *Healthy People 2020:* Nutrition and Weight Status Categories and Objectives

Category 1: Healthier Food Access
1. Increase the number of states with nutrition standards for child care.
2. Increase the proportion of schools that offer nutritious foods and beverages outside of school meals.
3. Increase the number of states that have incentive policies for food retail outlets to provide foods that are encouraged by the 2010 Dietary Guidelines for Americans.
4. Increase the proportion of Americans who have retail access to foods recommended by the 2010 Dietary Guidelines for Americans.
Category 2: Health Care and Worksite Settings
5. Increase the proportion of primary care physicians who measure patients' body mass index (BMI).
6. Increase the proportion of physician office visits that include nutrition or weight counseling or education.
7. Increase the proportion of worksites that offer nutrition and weight-management classes and counseling.

(continued)

TABLE 1-4 ▶ *Healthy People 2020:* **Nutrition and Weight Status Categories and Objectives** *(continued)*

Category 3: Weight Status
8. Increase the proportion of adults who are at a healthy weight.
9. Reduce the proportion of adults who are obese.
10. Reduce the proportion of children and adolescents who are considered obese.
11. Prevent inappropriate weight gain in youth and adults.

Category 4: Food Insecurity
12. Eliminate very low food security among children.
13. Reduce household food insecurity and, in so doing, reduce hunger.

Category 5: Food and Nutrient Consumption
14. Increase the contribution of fruits to the diets of the population ages 2 years and older.
15. Increase the variety and contribution of vegetables to the diets of the population ages 2 years and older.
16. Increase the contribution of whole grains to the diets of the population ages 2 years and older.
17. Reduce consumption of calories from solid fats and added sugars in the population ages 2 years and older.
18. Reduce consumption of saturated fat in the population ages 2 years and older.
19. Reduce consumption of sodium in the population ages 2 years and older.
20. Increase consumption of calcium in the population ages 2 years and older.

Category 6: Iron Deficiency
21. Reduce iron deficiency among young children and females of childbearing age.
22. Reduce iron deficiency among pregnant females.

▲ An increase in the consumption of some foods, such as fruits and vegetables, can lead to a healthier balance of food in the North American diet.

TABLE 1-5 ▶ **A Sample of Nutrition and Weight Status Objectives from** *Healthy People 2020* **along with Details about the Current Status and Targets for 2020**

	Target	Current Estimate
Increase the proportion of adults at a healthy weight.	33.9%	30.8%
Reduce the proportion of overweight or obese children and adolescents.	14.6%	16.2%
Increase the contribution of the following to the diets of the population ages 2 years and older (per 1000 calories).		
• Fruits	0.9 cup	0.5 cup
• Total vegetables	1.1 cups	0.8 cup
• Whole grains	0.6 ounce	0.3 ounce
Reduce consumption of calories from solid fats (% total calorie intake).	16.7%	18.9%
Reduce consumption of calories from added sugars (% total calorie intake).	10.8%	15.7%
Increase consumption of calcium in the population ages 2 years and older.	1300 mg	1118 mg
Reduce iron deficiency among females of childbearing age.	9.4%	10.4%

Note: In later chapters, we will explore additional nutrition-related objectives, such as those addressing osteoporosis, various forms of cancer, diabetes prevention and treatment, food allergies, cardiovascular disease, low birth weight, nutrition during pregnancy, breastfeeding, eating disorders, physical activity, and alcohol use.

Other new topic areas highlight changes in the health needs of specific segments of the population: Early and Middle Childhood, Adolescence, and Older Adults. Because young people develop habits, including eating and physical activity behaviors, that are likely to persist throughout life, new objectives promote strengthened health education in schools and communities, and fostering an environment in which young people can develop healthy habits. Older adults are the fastest-growing segment of the American population and are at high risk for experiencing the chronic health problems that so severely impact our health care system. The objectives for Older Adults include improving access to health care, helping older adults to manage their own health conditions, and ensuring proper training and support of professionals and nonprofessionals who care for this population.

A scientifically exciting new topic area in the *Healthy People 2020* report is **genomics.** Nine of the 10 leading causes of death have a strong genetic component. Genetic testing is becoming a valuable tool for improving diagnosis and treatment of chronic diseases, especially for cancers of the breast and colon. In combination with family history, genetic testing can help health care professionals guide patients in treatment options, including lifestyle changes. The relationship between genetics and nutrition will be discussed in Chapter 3.

✔ CONCEPT CHECK 1.6

1. Surveys indicate that we could improve our diets by focusing on which type of food sources?

2. The consumption of which types of foods should be reduced to attain and maintain good health?

1.7 What Can You Expect from Good Nutrition and a Healthy Lifestyle?

The obesity epidemic and the prevalence of chronic diseases in the United States illustrate that something is not right with many of our diets and/or lifestyles. The strong association between obesity and poor health is clear. The reverse is also well documented: when an obese or overweight person loses just 5% to 10% of body weight, that person's risks of many chronic diseases are greatly reduced.

HEALTHY WEIGHT

Because weight gain is one of the greatest lifelong nutrition challenges we face, we encourage you to seek a lifestyle that will make gaining weight more difficult and maintaining a healthy weight easier. Believe it or not, preventing obesity in the first place is the easiest approach. Unfortunately, many aspects of our society make it hard for us not to gain weight. The earlier (preferably in childhood) we develop lifestyle habits of good nutrition, regular physical activity, and the avoidance of addictions to salt, fat, sweets, high-calorie foods, and sedentary lifestyles, the better. As you enter the workforce, seek out employers who offer wellness programs that encourage weight management and weight loss among their employees. Aim to live in a city or town that has opportunities for physical activity such as bike paths, walking trails, and parks, as well as access to fresh fruits and vegetables through farmers' markets and community gardens. Seek out and join running or walking clubs. Make a habit of shopping at grocery stores that offer a good selection of fruits, vegetables, and other healthy foods. When dining out, choose restaurants that have tasty but healthy options on their menu.

While we still are choosing foods with too many calories, many other dietary habits have improved during the past decade. Today, we can choose from a tremendous variety of food products as a result of continual innovation by food manufacturers. We are eating more breakfast cereals, pizza, pasta entrees, stir-fried meats and vegetables served on rice, salads, tacos, burritos, and fajitas than ever before. Sales of whole milk are down, and sales of fat-free and 1% low-fat milk have increased. Consumption of frozen vegetables rather than canned vegetables is also on the rise. Despite the alarming problem of overweight and obesity, our cultural diversity, varied cuisines, and general lack of nutrient deficiencies should be points of pride for North Americans.

LONGER, HEALTHIER LIVES

Today, North Americans live longer than ever and enjoy better general health. Many also have more money and more diverse food and lifestyle choices to consider. The nutritional consequences of these trends are varied. Deaths from cardiovascular disease, for example, have dropped dramatically since the late 1960s, partly because of better medical care and diets. Affluence, however, has also led to sedentary lifestyles and high intakes of animal fat, cholesterol, salt, and alcohol. This lifestyle pattern has led to problems such as cardiovascular disease, hypertension, diabetes, and, of course, obesity. Greater efforts are needed by the general public to lower intake of animal fat and cholesterol and to improve variety in our diets, especially from fruits, vegetables, and whole grains. With better technology and greater choices, we can have a much better diet today than ever before—if we know what choices to make.

THE TOTAL DIET

Nutrition experts generally agree that there are no "good" or "bad" foods, but some foods provide relatively few nutrients in comparison to calorie content. In Chapter 2, you will learn that an individual's total diet is the proper focus in a nutritional evaluation. It continues to be the position of the Academy of Nutrition and Dietetics that "the total diet or overall pattern of food eaten is the most important focus of healthy eating" (see Further Reading 1). Health experts have prepared many reports and outlined numerous objectives to get us closer to being a *Healthy People* as soon as 2020. In Chapter 2, we will discuss the "Dietary Guidelines" that were published in 2010 and the interactive programs available on "**ChooseMyPlate.gov**" that were unveiled in 2011. As you reexamine your nutritional habits, remember your health is largely your responsibility. Your body has a natural ability to heal itself. Offer it what it needs, and it will serve you well. Confusing and conflicting health messages hinder diet change.

Prevention of disease is an important investment of one's time, including during the college years. Nutrition science does not have all the answers, but as you will see, enough is known to help you set a path to good health and put diet-related recommendations you hear in the future into perspective. Table 1-6 summarizes several diet, physical activity, and general lifestyle recommendations to promote your health and prevent chronic diseases. In addition to the diet and physical activity recommendations in Table 1-6, you can optimize your health by getting adequate sleep (7 to 9 hours per night), consuming sufficient water (9 to 13 cups per day from foods and beverages), reducing stress, using medications prudently, and, of course, abstaining from use of illicit drugs. Having close relationships with others and maintaining a positive outlook on life are also linked to lower risk for disease. Finally, consultation with health care professionals on a regular basis is important. This is because early diagnosis is especially useful for controlling the damaging effects of many diseases. In total, these contribute to maximal health and prevention of the diseases listed. The final section in this chapter, Nutrition and Your Health: Eating Well in College, elaborates on several nutrition issues very relevant to most college students, including the "freshman 15," vegetarianism, fuel for athletes, eating disorders, and alcohol and binge drinking. This section gives you a "sneak peek" at issues that will be covered more fully later in the book.

▲ Access to fresh fruits and vegetables through farmers' markets and community gardens is important to a healthy lifestyle.

TABLE 1-6 ▶ **Recommendations for Health Promotion and Disease Prevention**

Diet

Consuming enough essential nutrients, including fiber, while moderating energy, solid fat, cholesterol, added sugar, and alcohol intake can result in:

- Increased bone mass during childhood and adolescence

- Prevention of some adult bone loss and osteoporosis, especially in older adults

- Fewer dental caries

- Prevention of digestive problems, such as constipation

- Decreased susceptibility to some cancers

- Decreased degradation of the retina (especially through intake of green and orange vegetables)

- Lower risk of obesity and related diseases, such as type 2 diabetes and cardiovascular disease

- Reduced risk for deficiency disease (e.g., anemia) through adequate intake of iron, folate, and other nutrients

Physical Activity

Adequate, regular physical activity (at least 30 minutes on most or all days) helps reduce the risk of:

- Obesity

- Type 2 diabetes

- Cardiovascular disease

- Some adult bone loss and loss of muscle tone

- Premature aging

- Certain cancers

Lifestyle

Minimizing alcohol intake (no more than two drinks per day for men and one drink for women and all adults age 65 years and older) helps prevent:

- Liver disease

- Accidents

Not smoking cigarettes or cigars helps prevent:

- Lung cancer and other lung disease

- Kidney disease

- Cardiovascular disease

- Degenerative eye diseases

▲ Regular physical activity complements a healthy diet. Whether it is all at once or in segments throughout the day, incorporate 30 to 60 minutes or more of such activity into your daily routine.

✓ CONCEPT CHECK 1.7

1. What are some diet, physical activity, and lifestyle recommendations for health promotion and disease prevention?

Nutrition and Your Health
Eating Well in College

The college years are a time for freedom and a chance to make personal lifestyle decisions. Studies show that the diets of college students are not optimal. Typically, students fall short of diet recommendations for whole grains, vegetables, fruits, milk, and meat, opting instead to max out on fats, sweets, and alcohol. This information is disturbing because young adulthood is the time when many health behaviors are formed that will persist throughout life.

What is it about the college lifestyle that makes it so difficult to build healthy habits? In this section, we will discuss several topics and provide possible solutions.

Food Choices

College students face changes in academic requirements, interpersonal relationships, and living environment. These stressful situations contribute to poor health behaviors. For example, when you are writing papers and cramming for exams, balanced meals are all-too-easily replaced by high-fat and high-calorie fast-foods; convenience items; and sugary, caffeinated beverages. Physical activity is sacrificed in favor of study time. In a recent study of college students living on and off campus, two-thirds of the students reported skipping meals, with "no time to prepare" the major reason (see Further Reading 7).

Also consider that on campus, you are faced with a wide variety of dining choices. Dining halls, fast-food establishments, bars, and vending machines combine to offer food 24 hours per day. While it is certainly possible to make wise food choices at each of these outlets, the temptations of convenience, taste, and value (i.e., inexpensive, oversized portions) may persuade the college student to select unhealthy options.

Meals and snacks are also times to socialize. You may unintentionally eat a big lunch at noon without regard to hunger if your peers are meeting in the dining hall to catch up. While chatting, it is easy to lose track of portions and to overeat. In addition, food may be a source of familiarity and comfort in a new and stressful place.

Weight Control and the "Freshman 15"

Studies show that most college students gain weight during their first year (see Further Reading 8). The "freshman 15" is a term used to describe the weight gained by students during their first year of college. Although it is becoming evident that most freshman do not gain the 15 pounds, a recent study of over 7000 U.S. college students found that students pack on 2.4 to 3.5 pounds on average during their first year away from home (see Further Reading 21). Very few (just under 10%) of freshman gained 15 pounds or more, and 25% of freshman actually lost weight. The research also determined cumulative weight gain over the entire years at college and found that women gained approximately 9 pounds and men, 13 pounds, on average. The two lifestyle factors that made a difference in weight gain among the students were heavy drinking and working during college.

There are several reasons to maintain a healthy weight. Over the long term, risk of chronic diseases goes up as weight increases. In the short term, losing excess weight can improve how you feel and perform. Detecting "flab" around your midsection or feeling that your clothes are getting tighter are two good indicators that you are carrying excess weight. If weight loss is necessary, with some knowledge and perseverance, you can safely lose excess pounds.

Behavioral research clearly demonstrates that setting several small, achievable goals will spur motivation. As you will learn in Chapter 7, body weight is a balancing act between calories in and calories burned. Try keeping track of your calorie consumption for several days and comparing that to your energy needs, based on your age, gender, and activity level. You can use one of the equations presented in Chapter 7 or take advantage of the SuperTracker interactive tools on **www.ChooseMyPlate.gov** to estimate your energy needs.

A healthy rate of weight loss is 1 to 2 pounds per week. Greater rates of weight loss will not likely be sustained over time. Remember that the numbers on the scale are not as important as your body composition—the amount of fat in relation to lean mass. In order to lose weight, you must create an energy deficit, either by restricting energy intake below what you need to maintain your current weight or

CASE STUDY Typical College Student

Andy is like many other college students. He grew up on a quick bowl of cereal and milk for breakfast and a hamburger, French fries, and cola for lunch, either in the school cafeteria or at a local fast-food restaurant. At dinner, he generally avoided eating any of his salad or vegetables, and by 9 o'clock, he was deep into bags of chips and cookies. Andy has taken most of these habits to college. He prefers coffee for breakfast and possibly a chocolate bar. Lunch is still mainly a hamburger, French fries, and cola, but pizza and tacos now alternate more frequently than when he was in high school. One thing Andy really likes about the restaurants surrounding campus is that, for a few cents more, he can make his hamburger a double or get extra cheese and pepperoni on his pizza. This helps him stretch his food dollar; searching out large-portion value meals for lunch and dinner now has become part of a typical day.

Provide some dietary advice for Andy. Start with his positive habits and then provide some constructive criticism, based on what you now know.

Answer the following questions, and check your responses at the end of this chapter. As you make suggestions for Andy, think about your favorite food choices, why they are your favorites, and whether these are positive choices.

1. **Start with Andy's positive habits:** What healthy choices are being made when Andy eats at local restaurants?
2. **Now provide some constructive criticisms:**
   a. What are some of the negative aspects of items available at fast-food restaurants?
   b. Why is ordering the "value meals" a dangerous habit?
   c. What healthier substitutions could he make at each meal?
   d. List some healthier choices he could make at fast-food restaurants on campus.

by increasing your physical activity. For an adult with excess weight, an energy deficit of 500 kcal per day will result in weight loss of about 25 pounds over a year's time. As weight is lost, energy needs gradually decrease, such that further deficits will be required to lose additional weight.

Although many students skip breakfast, breakfast is the *most* important meal of the day. Starting the day off with a serving of lean protein (e.g., an egg, Canadian bacon, or protein shake), a fortified, whole-grain breakfast cereal, skim milk, and a serving of fruit puts you on the right path for meeting recommendations for fiber, calcium, and fruit intake. Even though it may seem that coffee gets your brain going in the morning, your brain is fueled best by carbohydrates, not caffeine. Studies also show that eating breakfast prevents overeating later in the day. Read more about breakfast choices in the *What the Dietitian Chose* recommendations at the end of this chapter.

One of the biggest contributors to weight gain for college students is consuming several hundred calories per day in the form of sugary or alcoholic beverages. One 12-ounce can of regular cola contains about 140 kcal. A 12-ounce can of regular

▲ Research (see Further Reading 13) has shown that gourmet coffee beverages, such as lattes and cappuccinos, can increase calorie consumption by about 200 kcal per day.

▶ **Five Simple Tips to Avert Weight Gain**

- *Eat breakfast.* Rev up your metabolism with a protein source such as an egg or low-fat yogurt, at least one serving of whole grains such as a breakfast cereal, and a fruit such as a banana.

- *Plan ahead.* Eat a balanced meal or snack every 3 to 4 hours.

- *Limit liquid calories.* Drink water instead of high-calorie soft drinks, fruit juice, alcohol, or coffee; if you drink alcohol, limit it to 1 or 2 drinks per day.

- *Stock the fridge.* Keep a stash of low-calorie, nutritious snacks, such as pretzels, light microwave popcorn, and fruit (fresh, canned, or dried).

- *Exercise regularly.* Find a friend to work out with you. Experts recommend 30 minutes of moderate exercise at least 5 days a week.

beer has 150 kcal. Consuming gourmet coffee beverages, such as lattes and cappuccinos, can increase average calorie consumption by about 200 kcal per day (see Further Reading 18). Even fruit juices have at least 100 kcal per 8-ounce glass. Furthermore, a 24-ounce mug of a soft drink makes you feel no fuller than an equal volume of water, yet the soft drink adds 300 kcal more. A convenient stash of water is the best way to quench your thirst.

Exercise is very important to any weight loss and weight maintenance plan, but sticking with it is hard to do. When you find yourself short on time, exercise is often the first thing that goes. To ensure your success at boosting daily activity, choose activities you enjoy such as working out with friends at the campus recreation center, participating in intramural sports, or taking an activity class such as dancing. Don't forget the brisk walking to and from classes. For more information on planning an exercise program, see Chapter 10.

Alcohol and Binge Drinking

Excessive alcohol consumption is a big problem on college campuses. Many college students consider drinking alcohol, legal or not, to be a "rite of passage" into adulthood. On campuses, binge drinking—consuming five or more drinks in a row for men or four drinks or more for women—has become an epidemic. A new level of "extreme drinking" goes far beyond binge drinking. For example, recent studies indicate that college students celebrate twenty-first birthdays with an average of about eight drinks more during the week of that birthday compared to a typical week (see Further Reading 14).

The statistics on the impact of binge drinking on college campuses are sobering. An estimated two of every five students on college campuses participate in binge drinking. Each year, 1400 college students between the ages of 18 and 24 die from alcohol-related unintentional injuries, including motor vehicle crashes. In addition to deaths and injuries, other problems stemming from binge drinking include unsafe sex and its consequences, long-term health problems, suicides, academic problems, legal troubles, and alcohol abuse or dependence. Thirty-one percent of college students meet the criteria for alcohol use disorders.

In addition, alcohol consumption definitely contributes to weight gain—by virtue of its own calories and the increased food consumption at events where drinking occurs. If you choose to drink alcohol, do so in moderation—no more than two drinks per day for men and one drink per day for women. Be aware of the warning signs and dangers of alcohol poisoning shown below. Special issues related to alcohol use by adolescents are covered in Chapter 15.

▶ **The warning signs and symptoms of alcohol poisoning**
- Semiconsciousness or unconsciousness
- Slow respiration of eight or fewer breaths per minute or lapses between breaths of more than 8 seconds
- Cold, clammy, pale, or bluish skin
- Strong odor of alcohol, which usually accompanies these symptoms

▲ Late-night pizza can add extra calories to the college student's daily intake.

Eating Disorders

As many as 30% of college students are at risk of developing an eating disorder. As you will learn in Chapter 11, disordered eating is a mild and short-term change in eating patterns that typically occurs in response to life stress, a desire to change appearance, or a bad habit. Sometimes, disordered eating habits may lead to an eating disorder, such as anorexia nervosa, bulimia nervosa, or binge-eating disorder. Chapter 11 includes advice on what to do if you suspect that your roommate or friend is suffering from an eating disorder.

Starving the body also starves the brain, which limits performance in academics and beyond. The negative consequences of disordered eating may last a lifetime. Ultimately, eating disorders do not arise from problems with food but rather from problems with self-esteem, control, and abusive relationships. Frequently, what begins as a diet spirals into a much larger problem. Eating disorders are not just diets gone bad: they require professional intervention. Left unchecked, eating disorders lead to serious adverse effects, such as loss of menstrual periods, thinning of bones, gastrointestinal problems, kidney problems, heart abnormalities, and eventually death.

Choosing a Vegetarian Lifestyle

Many college students experiment with or adopt a vegetarian eating pattern. Plant-based diets can meet nutrition needs and decrease risk of many chronic diseases, but they require appropriate planning at all life stages.

Protein is not typically deficient, even with a vegan diet, which contains no animal products. However, vegetarians, and especially vegans, may be at risk for deficiencies of several vitamins and minerals. Consuming a ready-to-eat breakfast cereal is an easy and inexpensive way to obtain these nutrients. See Chapter 6 for more information on vegetarian food planning.

Restaurants and campus dining services have responded to the growing interest in vegetarian meals by offering a variety of vegetarian options. For optimal health benefits, choose foods that are baked, steamed, or stir-fried rather than deep-fried; select whole grains rather than refined carbohydrates; and consume food fortified with vitamins and minerals. Even if you do not follow a plant-based diet all the time, choosing several plant-based meals each week can help with weight control and boost intake of fiber and healthy phytochemicals. You will learn in Chapter 2 that the ChooseMyPlate program recommends that the largest portion of your plate be filled with plant foods, including whole grains, fruits, and vegetables.

Fuel for Competition: Student Athletes

Students who compete in sports such as intramural and intercollegiate athletics need to consume more calories and nutrients. Despite an emphasis on a lean physique, athletes at all levels must take care not to severely restrict calories, as this could impact performance and health. Muscles require adequate carbohydrates for fuel and protein for growth and repair. Fat, as well, is an important source of stored energy for use during exercise. Low energy availability in women may lead to a loss of menstruation (amenorrhea), a condition costly to long-term bone health.

In addition to the calories needed to fuel the body, fluids are essential for health and performance. While water is adequate for events lasting less than 60 minutes, sports drinks are ideal for longer events because they supply carbohydrates to fuel fatigued muscles as well as electrolytes to replenish those lost in perspiration. Intentional fluid losses to "cut weight" for a competition are detrimental to health and performance.

Athletes also should take care not to be wooed by the supplement industry. Increasing food intake to meet the energy demands of athletic training is also sufficient to meet most vitamin and mineral needs. As an exception, athletes may be at risk for iron-deficiency anemia. Consuming a balanced multivitamin and mineral supplement is adequate for most people. Individual vitamin, mineral, amino acid, or herbal supplements are not advised, in spite of the hype of supplement makers.

Tips for Eating Well on a College Student's Budget

Because higher education can be hard on the wallet, it is good to know that it is possible to eat well on campus on a budget. If you live on campus, participate in a prepaid campus meal plan. These plans are generally designed to offer great food value with a variety of healthy foods. If you live off campus or have your own kitchen, plan ahead. Packing a lunch from home will save a lot of cash compared to grabbing lunch on the run and puts you more in control of healthy choices. For example, preparing a sandwich at home costs less than half as much as purchasing one at a fast-food restaurant or deli.

Never go grocery shopping on an empty stomach: everything will look good and you'll buy more. Also, go to the store with a list in hand and stick to it, because impulse buys tend to drain your wallet. Buy store-brand foods rather than name-brand items. Make use of canned and frozen fruits and

vegetables: they are just as nutritious as fresh fruits and vegetables, particularly if you choose low-sodium and low-sugar varieties. Rather than buying cartons of fruit juice, select cans of fruit juice concentrate and mix with water at home. Likewise, preparing other drinks, such as iced tea, from store-brand powder (look for sugar-free) will save over gallon jugs or vending machine containers of drinks. Canned (fruits, tuna) and dry (oatmeal) foods can be nutritious and last a long time, so you can avoid throwing out spoiled items. Finally, eggs and peanut butter are inexpensive and simple sources of protein.

In conclusion, *The College Student's Guide to Eating Well on Campus* (see Further Reading 15) and *The Dorm Room Diet* (see Further Reading 17) are excellent books with more details on ways college students can plan a healthy diet and stay fit.

Some calorie traps for college students:

	Number of Calories
Two handfuls of almonds	500
Two handfuls of granola	330
Personal-size pizza	500 to 600
1 cup of ice cream	300
Two handfuls of frosted cereal	250
Six-pack of regular beer	900

Source: Ann Litt, *The College Student's Guide to Eating Well on Campus*

▲ Many students adopt a vegetarian diet during college. Guidelines for planning a nutritious vegetarian diet, with items such as this veggie lasagna and spinach salad, are presented in Chapter 6.

Summary

1.1 The flavor, texture, and appearance of foods primarily influence our food choices. Several other factors also help determine food habits and choices: food availability and convenience; early childhood experiences and ethnic customs; nutrition and health concerns; advertising; restaurants; social changes; and economics. A variety of external (appetite-related) forces affect satiety (feeling of satisfaction that halts our desire to continue eating). Hunger cues combine with appetite cues, such as easy availability of food, to promote food intake.

1.2 Nutrition is a lifestyle factor that is a key to developing and maintaining an optimal state of health. A poor diet and a sedentary lifestyle are known to be risk factors for life-threatening chronic diseases such as heart disease, hypertension, diabetes, and cancer. Not meeting nutrient needs in younger years makes us more likely to suffer health consequences in later years. Too much of a nutrient also can be harmful. Drinking too much alcohol is another dietary problem associated with many health problems.

1.3 Nutrition is the study of how the body uses food substances to promote and support growth, maintenance, and reproduction of cells. Nutrients in foods fall into six classes: (1) carbohydrates, (2) lipids (mostly fats and oils), (3) proteins, (4) vitamins, (5) minerals, and (6) water. The first three, along with alcohol, provide calories for the body to use. Phytochemicals are plant chemicals that may contribute to a reduced risk of disease in people who consume them.

1.4 The body transforms the energy contained in carbohydrate, protein, and fat into other forms of energy that in turn allow the body to function. Fat provides, on average, 9 kcal per gram, whereas both protein and carbohydrate provide, on average, 4 kcal per gram. Calculating percentages and converting English units to metric units are important skills needed for the study of nutrition.

1.5 The scientific method is the procedure for testing the validity of possible explanations of a phenomenon, called hypotheses. Experiments are conducted to either support or refute a specific hypothesis. Once we have enough experimental information to support a specific hypothesis, it then can be called a theory. All of us need to be skeptical of new ideas in the nutrition field, waiting until many lines of experimental evidence support a concept before adopting any suggested dietary practice.

1.6 The obesity problem has worsened with 35% of people in the United States reported to be overweight or obese in 2013. This increase is a result of eating too much, especially foods with a high number of calories and a low number of nutrients, and not engaging in enough physical activity. Results from large nutrition surveys in the United States and Canada suggest that some of us need to concentrate on consuming foods that supply more of certain vitamins, minerals, and fiber. *Healthy People 2020* is a national initiative that includes Nutrition and Weight Status objectives related to eating a healthful diet and maintaining a healthy body weight. A healthful diet includes consuming a variety of nutrient-dense foods within and across the food groups, especially whole grains, fruits, vegetables, low-fat or fat-free milk or milk products, and lean meats and other protein sources; limiting intake of solid fats, cholesterol, added sugars, sodium (salt), and alcohol; and limiting intake of calories to meet energy needs.

1.7 A basic plan for health promotion and disease prevention includes eating a varied diet, performing regular physical activity, not smoking, not abusing nutrient supplements (if used), consuming adequate water and other fluids, getting enough sleep, limiting alcohol intake (if consumed), and limiting or appropriately coping with stress. The primary focus of nutrition planning should be on food, not on dietary supplements. The focus on foods to supply nutrient needs avoids the possibility of severe nutrient imbalances.

NAYH Studies show that the diets and other health habits of college students are not optimal. Students fall short of recommendations for servings of grains, vegetables, fruits, milk, and meat, opting instead for fats, sweets, and alcohol. This information is disturbing from a public health standpoint, because young adulthood is the time when many health behaviors are formed and will likely persist throughout life. Issues of particular importance on college campuses are weight control, making healthy meal choices, alcohol and binge drinking, and eating disorders.

Check Your Knowledge

1. Our primary psychological drive to eat that is affected by many external food-choice mechanisms is called
   a. hunger.   b. appetite.   c. satiety.   d. feeding.

2. Energy-yielding nutrients include
   a. vitamins, minerals, and water.
   b. carbohydrates, proteins, and fats.
   c. trace minerals and fat-soluble vitamins.
   d. iron, vitamin C, and potassium.

3. The *essential* nutrients
   a. must be consumed at every meal.
   b. are required for infants but not adults.
   c. can be made in the body when they are needed.
   d. cannot be made by the body and therefore must be consumed to maintain health.

4. Sugars, starches, and dietary fibers are examples of
   a. proteins.   c. carbohydrates.
   b. vitamins.   d. minerals.

5. Which nutrient classes are most important in the regulation of body processes?
   a. vitamins   c. minerals
   b. carbohydrates   d. both a and c

6. A kcal is a
   a. measure of heat energy.
   b. measure of fat in food.
   c. heating device.
   d. term used to describe the amount of sugar and fat in foods.

7. A food that contains 10 grams of fat would yield _____ kcal.
   a. 40   b. 70   c. 90   d. 120

8. If you consume 300 grams of carbohydrate in a day that you consume 2400 kcal, the carbohydrates will provide _____% of your total energy intake.
   a. 12.5   b. 30   c. 50   d. 60

9. Which of the following is true about the North American diet?
   a. Most of our protein comes from plant sources.
   b. About half of the carbohydrates come from simple sugars.
   c. Most of our fats come from plant sources.
   d. Most of our carbohydrates come from starches.

10. A behavior that will decrease the risk of weight gain in college is to
    a. skip breakfast.
    b. drink more liquid calories.
    c. stock your fridge with nutritious snacks.
    d. exercise infrequently.

Answer Key: 1. b (LO 1.1), 2. b (LO 1.3), 3. d (LO 1.3), 4. c (LO 1.3), 5. d (LO 1.3), 6. a (LO 1.4), 7. c (LO 1.4), 8. c (LO 1.4), 9. b (LO 1.5), 10. c (LO 1.7)

Study Questions (Numbers refer to Learning Outcomes)

1. Describe the process that controls hunger and satiety in the body. List other factors that influence our food choices. **(LO 1.1)**

2. Describe how your food preferences have been shaped by the following factors:
   a. Exposure to foods at an early age
   b. Advertising (what is the newest food you have tried?)
   c. Eating out
   d. Peer pressure
   e. Economic factors **(LO 1.1)**

3. What products in your supermarket reflect the consumer demand for healthier foods? For convenience? **(LO 1.1)**

4. Name one chronic disease associated with poor nutrition habits. Now list a few corresponding risk factors. **(LO 1.2)**

5. Describe two sources of fat and explain why the differences are important in terms of overall health. **(LO 1.3)**

6. Identify three ways that water is used in the body. **(LO 1.3)**

7. Explain the concept of calories as it relates to foods. What are the values used to calculate kcal from grams of carbohydrate, fat, protein, and alcohol? **(LO 1.4)**

8. A bowl of Panera's broccoli cheddar soup contains 21 grams carbohydrate, 13 grams fat, and 12 grams protein. Calculate the percentage of calories derived from fat. **(LO 1.4)**

9. According to national nutrition surveys, which nutrients tend to be underconsumed by many North Americans? Why do you think this is the case? **(LO 1.5)**

10. List four *Healthy People 2020* objectives for the United States. How would you rate yourself in each area? Why? **(LO 1.5 & 1.6)**

11. List five strategies to avoid weight gain during college. **(LO 1.7)**

What the Dietitian Chose

Even bleary-eyed procrastinators can fuel their bodies for a new day of higher education! *Skipping breakfast is not a smart plan.* After a period of fasting (i.e., overnight), the body and the brain need fuel to operate at peak efficiency. In addition, many research studies demonstrate that eating a sensible breakfast is a good way to control weight. Compared to those who eat breakfast, people who skip breakfast tend to crave higher-calorie foods, snack more throughout the day, and eat more at subsequent meals. Consider eating breakfast each day as part of your plan to fend off the freshman 15.

Grab-and-go food options, such as those available in vending machines or from fast-food establishments, are often high in calories but low in nutrients. For example, a low-fat granola bar has only 100 kcal and 3 grams of fat but offers little else in terms of nutrition. Also, a granola bar is not likely to stave off hunger for very long.

On the other hand, a fast-food breakfast sandwich will probably promote satiety, but its calorie, fat, and sodium contents are too high. This type of sandwich provides 550 kcal, 23 grams of fat (38% of a day's total kilocalories in the sandwich and 35% of the whole day's limit for fat), and 1490 mg sodium (just under the 1500 mg Adequate Intake for this nutrient). A fast-food breakfast sandwich can fit into an otherwise healthy diet on occasion but should not be part of your normal routine.

Keeping nutritious but convenient breakfast options accessible is a good strategy for any time-pressed college student. Whole-grain, fortified, ready-to-eat breakfast cereal with fat-free milk is a great choice: it is quick, provides a wide variety of vitamins and minerals, and boosts fiber intake. Adding a source of protein will help to support body processes and make you feel full for a longer time. Hard-boiled eggs or a handful of dry-roasted nuts provide protein in ready-to-eat form. A cup of low-fat yogurt offers

▲ With a little bit of planning, breakfast can be both quick and healthy.

protein with the added benefit of calcium. Fresh and dried fruit are portable and nutritious options for breakfast. Many fruit choices are loaded with potassium and vitamin C, plus they provide fiber. A 1.5-ounce box of raisins, which can be stored for months without a refrigerator, provides about 130 kcal, no fat or cholesterol, very little sodium, 2 grams of fiber, and 320 milligrams of potassium.

So, the bowl of Wheaties with a banana and low-fat milk along with a yogurt, all from your dorm room "pantry," would be the healthiest choice for most.

CASE STUDY SOLUTION   Typical College Student

1. The most positive aspect of Andy's diet is that it contains good sources of protein from animal products also rich in zinc and iron. Alternating his hamburger and fries for lunch with pizza or tacos is a move in the right direction.

2.a. On the downside, Andy's diet is low in dairy products, fruits, and vegetables. This will result in a low intake of calcium, several of the vitamins, and the phytochemical (plant-based) substances discussed in this chapter. His diet is also low in fiber because fast-food restaurants primarily use refined grain products rather than whole-grain products and have few fruit and vegetable choices on the menu. Most of the beverages are soft drinks loaded with sugar. Many menu items, especially fries and chicken nuggets, are high in fat.

2.b. Most value menu options apply to foods rich in fat (French fries) and sugar (soft drinks), resulting in excessive amounts of these two components.

2.c. He could choose a low-fat fruit and granola bar instead of the candy bar for breakfast, or he could take the time to eat a bowl of whole-grain breakfast cereal with fruit and low-fat or fat-free milk to increase fiber and calcium (in the latter case) intake. Juices such as orange juice for breakfast are beverage options that will add fruits or vegetables to his diet. Overall, Andy could improve his intake of fruits, vegetables, and dairy products by focusing on variety in food choice and balance among the food groups.

2.d. Andy could alternate between tacos and bean burritos to gain the benefits of plant proteins in a diet. To incorporate more veggies and fruits in the diet, Andy could add lettuce, tomato, and onion to his tacos and hamburgers. When choosing pizza, he can replace high-fat toppings (e.g., pepperoni), and extra cheese with veggies (e.g., green peppers and onions). Many sub shops and delis offer several low-fat sandwiches, emphasizing lean meats such as turkey and vegetable toppings. For his beverages, Andy could order milk at least half of the time at his restaurant visits and substitute diet soft drinks for the regular variety. This would help moderate his sugar intake.

Further Readings

1. Academy of Nutrition and Dietetics: Position of the Academy of Nutrition and Dietetics: Total Diet Approach to Healthy Eating. *Journal of the Academy of Nutrition and Dietetics* 113:307, 2013.

2. Academy of Nutrition and Dietetics: Nutrition and you: Trends 2011. **www.eatright.org/nutritiontrends/.** Accessed March 4, 2013.

3. Andreyeva T and others: Exposure to food advertising on television: Associations with children's fast food and soft drink consumption and obesity. *Economics and Human Biology* 9:221, 2011.

4. Blake CE and others: Behavioral contexts, food-choice coping strategies, and dietary quality of a multiethnic sample of employed parents. *Journal of American Dietetic Association* 111:401, 2011.

5. Centers for Disease Control and Prevention: *Overweight and Obesity.* May 24, 2012. **www.cdc.gov/obesity/.** Accessed March 6, 2013.

6. Chiuve SE and others: Adherence to a low-risk, healthy lifestyle and risk of sudden cardiac death among women. *Journal of the American Medical Association* 306:62, 2011.

7. Choi S and Lee Y: Relationship of college students' residence to frequency of meal skipping and snacking pattern. *Journal of the Academy of Nutrition and Dietetics* 112:A24, 2012.

8. Edmonds MJ and others: Body weight and percent body fat increase during the transition from high school to university in females. *Journal of the American Dietetic Association* 108:1033, 2008.

9. Federal Trade Commission: *A Review of Food Marketing to Children and Adolescents, Follow-Up Report,* December 2012. **www.ftc.gov/os/2012/1 2/121221foodmarketingreport.pdf** Accessed 2/23/2013.

10. Goris JM and others: Television food advertising and the prevalence of childhood overweight and obesity: A multicountry comparison. *Public Health Nutrition* 13:1003, 2010.

11. Hoyert DL and Xu J: *National Vital Statistics Reports,* deaths: Preliminary data for 2011: 61, 2011.

12. King DE and others: The status of baby boomers' health in the United States: The healthiest generation? *Journal of the American Medical Association Internal Medicine,* 173:385, 2013.

13. Kolish ED and Hernandez M: The Children's Food and Beverage Advertising Initiative, A Report on Compliance and Progress During 2011, December 2012.

14. Lewis MA and others: Examining the relationship between typical drinking behavior and 21st birthday drinking behavior among college students: Implications for event-specific prevention. *Addiction* 104:760, 2009.

15. Litt AS: *The College Student's Guide to Eating Well on Campus.* Glen Echo, MD: Tulip Hill Press, 2005.

16. Loureiro ML and others: The effects of nutritional labels on obesity. *Agricultural Economics* 43:333, 2012.

17. Oz D: *The Dorm Room Diet: The 10-Step Program for Creating a Healthy Lifestyle Plan That Really Works.* Newmarket Press, 2010.

18. Shields DH and others: Gourmet coffee beverage consumption among college women. *Journal of the American Dietetic Association* 104:650, 2004.

19. Trust for America's Health: *F as in Fat: How Obesity Threatens America's Future 2012.* Robert Wood Johnson Foundation, September 2012. **healthyamericans.org/report/100/** Accessed 2/27/2013.

20. Yanover T and Sacco WP: Eating beyond satiety and body mass index. *Eating and Weight Disorders* 13:119, 2008.

21. Zagorsky JL and Smith PK: The Freshman 15: A critical time for obesity intervention or media myth? *Social Science Quarterly* 92: 1389, 2011.

 **connect** NUTRITION   To get the most out of your study of nutrition, visit McGraw-Hill Connect at www.mcgrawhillconnect.com where you will find NutritionCalc Plus, LearnSmart, and many other dynamic tools.

Rate Your Plate

I. Examine Your Eating Habits More Closely

Choose 1 day of the week that is typical of your eating pattern. Using the first table found in Appendix C, list all foods and drinks you consumed for 24 hours. In addition, write down the approximate amounts of food you ate in units, such as cups, ounces, teaspoons, and tablespoons. Place the corresponding abbreviation in the *Reason for Choice* column to indicate why you picked that food or drink. There can be more than one reason for choosing a particular food or drink.

FLVR	Flavor/texture	ADV	Advertisement	PEER	Peers
CONV	Convenience	WTCL	Weight control	NUTR	Nutritive value
EMO	Emotions	HUNG	Hunger	$	Cost
AVA	Availability	FAM	Family/cultural	HLTH	Health

Application

Ask yourself what your most frequent reason is for eating or drinking. To what degree is health or nutritive value a reason for your food choices? Should you make these higher priorities?

II. Observe the Supermarket Explosion

Today's supermarkets carry up to 60,000 items. Think about your last grocery shopping trip and the items you purchased to eat. Following is a list of 20 newer food products added to supermarket shelves. For those items you have tried, use the key from Part I to identify why you chose these products.

_____ Prepackaged salad greens (variety packs other than iceberg lettuce)

_____ Gourmet or sprayable salad oils (e.g., walnut, almond, olive, or sesame oil)

_____ Precooked frozen turkey patties, precooked bacon

_____ Microwavable sandwiches (e.g., Hotpockets or frozen sandwiches)

_____ Microwavable meals in a bowl (e.g., mac and cheese, or soup)

_____ Refrigerated, precooked pasta (e.g., tortellini or fettucini) and accompanying sauces (e.g., pesto or tomato basil)

_____ Imported grain products (e.g., risotto, farfalline, gnocchi, or fusilli)

_____ Whole-grain pasta or rice

_____ Frozen dinners (list your favorite of any of the wide variety)

_____ Bottled waters (flavored or unflavored)

_____ Trendy juices (e.g., draft apple cider, acai, or pomegranate)

_____ Roasted and/or flavored coffees (e.g., beans, ground, instant, or k-cups)

_____ Instant hot cereal in a bowl (add water and go!)

_____ "Fast-shake" pancake mix (add water, shake, and ready to cook)

_____ Breakfast bars or cookies (e.g., granola or fruit-flavored bars)

_____ Meal replacement/fitness products (e.g., "energy" bars, high-protein bars, or sports drinks)

_____ Low-calorie muffin tops or bagel thins

_____ Packaged yogurt smoothies

_____ Milk substitutes (e.g., rice milk or soy milk)

Finally, identify three new food products not on this list that you have seen in the past year. Discuss the appeal of these products to the North American consumer.

Student Learning Outcomes

Chapter 2 is designed to allow you to:

2.1 Use variety, proportionality, and moderation, as well as nutrient and energy density, to develop a healthy eating plan.

2.2 List the purpose and key recommendations of the Dietary Guidelines and the 2008 Physical Activity Guidelines for Americans.

2.3 Design a meal that conforms to the MyPlate recommendations as well as the Mediterranean diet and/or other diet planning guides.

2.4 Describe the three states of nutritional health.

2.5 Outline the measurements used (ABCDEs) in nutritional assessment: *A*nthropometric, *B*iochemical, *C*linical, *D*ietary, and *E*nvironmental status.

Chapter 2
Guidelines for Designing a Healthy Diet

What Would You Choose?

In between classes, you stop at a nearby convenience store to pick up a cold drink that will quench your thirst and fill your growling stomach. Immediately, you reach for your favorite cola, but the Jones Soda Co.® display with its personalized labels catches your eye. You notice that these sodas are made with no high-fructose corn syrup but instead contain pure cane sugar. Is there a nutritional difference between the added sugars in these two types of sodas? Which of the following beverages provides the best of calories, nutrients, and energy for an afternoon snack?

a 20-ounce bottle of regular cola (e.g., Coca-Cola®) containing high-fructose corn syrup

b 20-ounce bottle of diet cola (e.g., Coke Zero®) containing an artificial sweetener

c 12-ounce bottle of Jones Soda Co.® Pure Cane Cola Soda

d ½ pint (8 fluid ounces) of low-fat (1%) chocolate milk

e 16-ounce bottle of water

■ connect |NUTRITION **Think about your choice as you read Chapter 2, then see What the Dietitian Chose** at the end of the chapter. To learn more about nutrient density and energy density, check out the Connect site: www.mcgrawhillconnect.com.

How many times have you heard wild claims about how healthful certain foods are for you? As consumers focus more on diet and disease, food manufacturers are asserting that their products have all sorts of health benefits. "Eat more olive oil and oat bran to lower blood cholesterol." "Drink pomegranate juice to guard your body against free radicals." Hearing these claims, you would think that food manufacturers have all the answers.

Advertising aside, nutrient intakes that are out of balance with our needs—such as excess calories, saturated fat, cholesterol, *trans* fat, salt, alcohol, and sugar intakes—are linked to many leading causes of death in North America, including obesity, hypertension, cardiovascular disease, cancer, liver disease, and type 2 diabetes. Physical inactivity is also too common. In this chapter, you will explore the components of a healthy diet and lifestyle—an approach that will minimize your risks of developing nutrition-related diseases. The goal is to provide you with a firm understanding of these concepts before you study the nutrients in detail.

2.6 Describe the specific nutrient recommendations categories within the Dietary Reference Intakes.

2.7 Identify reliable sources of nutrition information.

2.8 Describe the components of the Nutrition Facts panel and the various health claims and label descriptors that are allowed.

2.1 A Food Philosophy That Works

You may be surprised that what you should eat to minimize the risk of developing the nutrition-related diseases seen in North America is exactly what you have heard many times before: *Consume a variety of foods balanced by a moderate intake of each food.* Health professionals have recommended the same basic diet and health plan for many years:

- Control *how much* you eat.
- Pay attention to *what* you eat: choose whole grains, fruits, and vegetables.
- Stay physically active.

A healthful diet does not have to mean deprivation and misery; it simply requires some basic nutrition know-how and planning. Besides, eliminating favorite foods typically does not work for "dieters" in the long run. The best plan consists of learning the basics of a healthful diet: variety, moderation, and proportionality. Monitoring total calorie intake is also important for many of us, especially if unwanted weight gain is taking place.

As noted in Chapter 1, many nutrition experts agree that there are no exclusively "good" or "bad" foods. Even so, many North Americans have diets that miss the mark when it comes to the foundations of healthy eating. Diets overloaded with fatty meats, fried foods, sugared soft drinks, and refined starches can result in substantial risk for nutrition-related chronic diseases.

We are more likely to maintain a healthy diet and lifestyle when we emphasize a total diet pattern that is balanced and moderate. Let's now define *variety, moderation,* and *proportionality.* We will also introduce two very important concepts that will help us to make healthy food choices: nutrient density and energy density.

VARIETY MEANS EATING MANY DIFFERENT FOODS

Variety in your diet means choosing foods from all the food groups and subgroups, rather than eating the "same old thing" day after day. Variety makes meals more interesting and helps ensure that a diet contains sufficient nutrients. A variety of foods is best because no one food meets all your nutrient needs. For example, meat provides protein and iron but little calcium and no vitamin C. Eggs are a source of protein, but they provide little calcium because the calcium is mostly in the shell. Cow's milk contains calcium but very little iron. None of these foods contains fiber.

One way to balance your diet as you consume a variety of foods is to select foods from each of these five major food groups every day:

- Grains
- Vegetables
- Fruits
- Dairy
- Protein

MyPlate, a food guide plan discussed in Section 2.3, offers a visual reminder and advice to help you make smart choices from each of these food groups. A dinner consisting of a bean burrito, lettuce and tomato salad with oil-and-vinegar dressing, a glass of milk, and an apple covers all groups.

Carrots—a source of fiber and a pigment that form vitamin A—may be your favorite vegetable. However, if you choose carrots every day as your only vegetable source, you may miss out on the vitamin folate. Other vegetables, such as broccoli and asparagus, are rich sources of this nutrient. Hopefully, you're beginning to get a sense of how different foods and food groups vary in the nutrients they contain. For now, just recognize that you need a variety of foods in your diet because the required nutrients are scattered among many foods.

An added bonus of variety in the diet, especially within the fruit and vegetable groups, is the inclusion of a rich supply of phytochemicals. Many of these substances provide significant health benefits. Considerable research attention is focused on various phytochemicals in reducing the risk for certain diseases

▲ A menu full of fruits, vegetables, and whole-grain breads and cereals will help ward off disease and control body weight.

▲ Tomatoes are considered functional foods because they contain phytochemicals such as lycopene that are important for health.

(e.g., cancer). You can't just buy a bottle of phytochemicals: they are generally available only within whole foods. Current multivitamin and mineral supplements contain few or none of these beneficial plant chemicals.

Numerous population studies show reduced cancer risk among people who regularly consume fruits and vegetables (see Further Reading 8). Researchers suspect that some phytochemicals present in the fruits and vegetables block the cancer process. Links between cancer and nutrition are described more thoroughly in Chapter 8. Some phytochemicals have also been linked to a reduced risk of cardiovascular disease (see Further Readings 2 and 8). Because humans evolved eating a wide variety of plant-based foods, the body may have developed a need for these phytochemicals, along with the various nutrients present, to maintain optimal health.

Foods rich in phytochemicals are now part of a family of foods referred to as **functional foods.** A functional food provides health benefits beyond those supplied by the traditional nutrients it contains. For example, a tomato contains the phytochemical lycopene, so it can be called a functional food.

It will likely take many years for scientists to unravel all of the important effects of the myriad of phytochemicals in foods, and it is unlikely that all will ever be available or effective in supplement form. For this reason, leading nutrition and medical experts suggest that a diet rich in fruits, vegetables, and whole-grain breads and cereals is the most reliable way to obtain the potential benefits of phytochemicals.

Table 2-1 provides a number of suggestions for including more phytochemicals from fruits, vegetables, and whole grains in your diet, as do the websites

▲ Vegetables such as sliced cucumbers can be added to salads, sandwiches, pizza, tacos, soups, and much more to increase your phytochemical intake.

functional foods Foods that provide health benefits beyond those supplied by the traditional nutrients they contain.

TABLE 2-1 ▶ Tips for Boosting the Phytochemical Content of a Diet

- Include vegetables in main and side dishes. Add these to rice, omelets, potato salad, and pastas. Try broccoli or cauliflower florets, mushrooms, peas, carrots, corn, or peppers.
- Look for quick-to-fix grain side dishes in the supermarket. Pilafs, couscous, rice mixes, and tabbouleh are just a few that you'll find.
- Choose fruit-filled cookies, such as fig bars, instead of sugar-rich cookies. Use fresh or canned fruit as a topping for pudding, hot or cold cereal, pancakes, and frozen desserts.
- Put raisins, grapes, apple chunks, pineapples, grated carrots, zucchini, or cucumber into coleslaw, chicken salad, or tuna salad.
- Be creative at the salad bar: Try fresh spinach, leaf lettuce, red cabbage, zucchini, yellow squash, cauliflower, peas, mushrooms, or red or yellow peppers.
- Pack fresh or dried fruit for snacks away from home instead of grabbing a candy bar or going hungry.
- Add slices of cucumber or zucchini, spinach, or carrot slivers to the lettuce and tomato on your sandwiches.
- Each week try one or two vegetarian meals, such as beans and rice or pasta; vegetable stir fry; or spaghetti with tomato sauce.
- If your daily protein intake exceeds the recommended amounts, reduce the meat, fish, or poultry in casseroles, stews, and soups by one-third to one-half and add more vegetables and legumes.
- Keep a container of fresh vegetables in the refrigerator for snacks.
- Choose fruit or vegetable juices (preferably 100% juice varieties) instead of soft drinks.
- Substitute tea for coffee or soft drinks on a regular basis.
- Have a bowl of fresh fruit on hand.
- Switch from crisp head lettuce to leaf lettuce, such as romaine.
- Use salsa as a dip for chips in place of creamy dips.
- Choose whole-grain breakfast cereals, breads, and crackers.
- Add flavor to your plate with ginger, rosemary, basil, thyme, garlic, onions, parsley, and chives in place of salt.
- Incorporate soy products, such as tofu, soy milk, soy protein isolate, and roasted soybeans, into your meals (see Chapter 6).

CRITICAL THINKING

Andy would benefit from more variety in his diet. What are some practical tips he can use to increase his fruit and veg-etable intake?

nutrient density The ratio derived by dividing a food's nutrient content by its calorie content. When the food's contribution to our nutrient need for that nutrient exceeds its contribution to our calorie need, the food is con-sidered to have a favorable nutrient density.

www.fruitsandveggiesmorematters.org and www.fruitsandveggiesmatter.gov. A note of caution: Some research suggests that increasing dietary variety can lead to overeating. Thus, as one incorporates a wide variety of foods in a diet, attention to total calorie intake is also important to consider.

PROPORTIONALITY MEANS EATING MORE NUTRIENT-DENSE FOODS

Proportionality, also referred to as balance, is eating more of nutrient-dense foods and beverages such as fruits, vegetables, whole grains, and fat-free or low-fat milk products, and less of foods high in certain types of fat, sugars, cholesterol, salt, and alcohol. Balance also refers to matching your energy intake (how many total calo-ries you consume) with energy expenditure (calories burned by metabolism and physical activity) over time. A prolonged imbalance between energy intake and energy expenditure leads to fluctuations in body weight.

The **nutrient density** of a food is a characteristic used to determine its nutritional quality. Nutrient density of a food is determined by comparing its protein, vitamin, or mineral content with the amount of calories it provides. A food is deemed nutri-ent dense if it provides a large amount of a nutrient for a relatively small amount of calories when compared with other food sources. The higher a food's nutrient density is, the better it is as a nutrient source. Comparing the nutrient density of dif-ferent foods is an easy way to estimate their relative nutritional quality.

Generally, nutrient density is determined with respect to individual nutrients. For example, many fruits and vegetables have a high content of vitamin C compared with their modest calorie content; that is, they are nutrient-dense foods for vitamin C. Figure 2-1 shows that fat-free milk is much more nutrient dense than sugared soft drinks for many nutrients, especially protein, vitamin A, riboflavin, and calcium.

As noted previously, menu planning should focus mainly on the total diet—not on the selection of one critical food as the key to an adequate diet. Many low-cost, nutrient-dense foods—such as fat-free and low-fat milk, lean meats, legumes (beans), oranges, carrots, broccoli, whole-wheat bread, and whole-grain breakfast

FIGURE 2-1 ▶ Comparison of the nutrient contributions of a sugared soft drink with that of fat-free (i.e., nonfat or skim) milk. Choosing a glass of fat-free milk makes a significantly greater contribution to nutrient intake than does a sugared soft drink. An easy way to determine nutrient den-sity from this chart is to compare the lengths of the bars indicating vitamin or mineral contribution with the bar that represents calorie contribution. For the soft drink, no nutrient sur-passes calorie content. Fat-free milk, in contrast, has longer nutrient bars for protein, vitamin A, the vitamins thiamin and riboflavin, and the min-eral calcium than it does for calories. Including many nutrient-dense foods in your diet is a good way to meet nutrient needs without exceeding calo-rie needs.

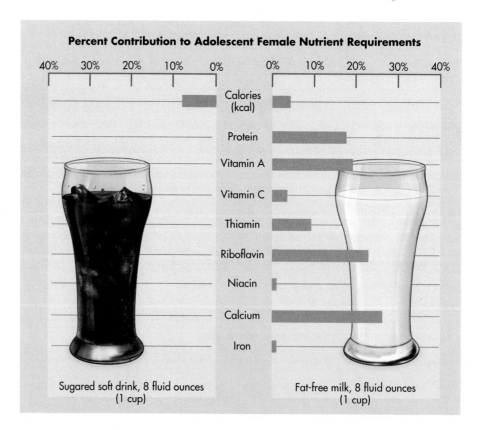

Percent Contribution to Adolescent Female Nutrient Requirements

| 40% 30% 20% 10% 0% | | 0% 10% 20% 30% 40% |

Calories (kcal)
Protein
Vitamin A
Vitamin C
Thiamin
Riboflavin
Niacin
Calcium
Iron

Sugared soft drink, 8 fluid ounces (1 cup)

Fat-free milk, 8 fluid ounces (1 cup)

cereals—do help balance less nutrient-dense foods—such as cookies and potato chips, which many people like to eat. The latter are often called empty-calorie foods because they tend to be high in sugar and/or fat but provide few other nutrients.

Eating nutrient-dense foods is especially important for people who consume diets relatively low in calories. This includes some older people and those following weight-loss diets. This is because nutrient needs remain high even though calorie needs may be diminished.

MODERATION REFERS MOSTLY TO PORTION SIZE

Eating in moderation requires paying attention to portion sizes and planning your day's diet so that you do not overconsume any nutrients. It is especially important to choose foods that help you limit intake of animal fat, added sugars, cholesterol, salt, and alcohol. Americans typically consume too much of these food components—and too many calories overall. For example, if you plan to eat a bacon cheeseburger (relatively high in fat, salt, and calories) at lunch, you should eat foods such as fruits and salad greens (less concentrated sources of these nutrients) at other meals that same day. If you prefer whole milk to low-fat or fat-free milk, reduce the fat elsewhere in your meals. Try low-fat salad dressings or use jam rather than butter or margarine on toast. Overall, it is more feasible to consume moderate portions of foods that supply lots of fat, salt, and sugar than to try to eliminate these foods altogether.

Let's be clear that moderation is important for all food components. For example, many North Americans do not consume enough vitamin E, which is found in plant oils, nuts, and some fruits and vegetables. However, taking large doses of vitamin E (e.g., from supplements) can lead to excessive bleeding because of its effects on blood clotting. You *can* get too much of a good thing!

Energy density is a measurement that best describes the calorie content of a food. Energy density of a food is determined by comparing the calorie (kcal) content with the weight of food. A food that is rich in calories but weighs relatively little is considered energy dense. Examples include nuts; cookies; fried foods in general; and even fat-free snacks, such as fat-free pretzels. Foods with low energy density include fruits, vegetables, and any food that incorporates lots of water during cooking, such as oatmeal (Table 2-2).

Researchers, including Dr. Barbara Rolls, the creator of the Volumetrics Diet, have shown that eating a meal with many foods of low energy density promotes satiety without contributing many calories (see Further Readings 5 and 6). This is probably because we typically consume a constant weight of food at a meal rather than a constant number of calories. How this constant weight of food is regulated is not known, but careful laboratory studies show that people consume fewer calories in a meal if most of the food choices are low in energy density, compared with foods high in energy density. Eating a diet low in energy density can aid in losing (or maintaining) weight.

Overall, foods with lots of water and fiber (i.e., low-energy-density foods) contribute few calories even though they help one feel full. Alternatively, foods with high energy density must be eaten in greater amounts to promote fullness. This is one more reason to eat a diet rich in fruits, vegetables, and whole-grain breads and cereals, a pattern that is typical of many ethnic diets throughout rural areas of the world. Unfortunately, low-energy-density foods are often more expensive than those that have high energy density. Read more about this situation in Further Reading 3, which is summarized in the Newsworthy Nutrition on the following page.

Many foods such as peanut butter are both energy and nutrient dense. Even energy-dense foods can have a place in your dietary pattern, but you will have to plan for them. For example, chocolate is a very energy-dense food, but a small portion at the end of a meal can supply a satisfying finale. In addition, foods with high energy density can help people with poor appetites, such as some older people, to maintain or gain weight.

▲ Focus on nutrient-rich foods as you strive to meet your nutrient needs. The more colorful the food on your plate, the greater the content of nutrients and phytochemicals.

energy density A comparison of the calorie (kcal) content of a food with the weight of the food. An energy-dense food is high in calories but weighs very little (e.g., potato chips), whereas a food low in energy density has few calories but weighs a lot, such as an orange.

▲ Salsa is full of phytochemicals and a great, very low-energy-density alternative to higher calorie chip dips. It helps to balance out the high-energy-density tortilla chips.

TABLE 2-2 ▶ **Energy Density of Common Foods (Listed in Relative Order)**

Very Low Energy Density (less than 0.6 kcal per gram)	Low Energy Density (0.6–1.5 kcal per gram)	Medium Energy Density (1.5–4 kcal per gram)	High Energy Density (greater than 4 kcal per gram)
Lettuce	Whole milk	Eggs	Graham crackers
Tomatoes	Oatmeal	Ham	Fat-free sandwich cookies
Strawberries	Cottage cheese	Pumpkin pie	Chocolate
Broccoli	Beans	Whole-wheat bread	Chocolate chip cookies
Salsa	Bananas	Bagels	Tortilla chips
Grapefruit	Broiled fish	White bread	Bacon
Fat-free milk	Fat-free yogurt	Raisins	Potato chips
Carrots	Ready-to-eat breakfast cereals	Cream cheese Cake with frosting	Peanuts Peanut butter
Vegetable soup	With 1% low-fat milk	Pretzels	Mayonnaise
Celery	Plain baked potato	Rice cakes	Butter or margarine
Cabbage	Cooked rice		Vegetable oils
Melon	Spaghetti noodles		

Data adapted from Rolls B, *The Ultimate Volumetrics Diet.* New York: HarperCollins, 2012.

Newsworthy Nutrition

Diet quality impacts waistlines and wallets

Diets with low energy density are associated not only with higher intakes of nutrients but also with higher food costs. Researchers used food-frequency questionnaires to assess dietary intake and food costs among 164 men and women in Seattle, Washington. Dietary data were compared with components of socioeconomic status. Household income and especially education were strong predictors of dietary energy density. The association between food spending and energy density was particularly strong for women. Nutrition advice to consume foods with lower energy density may be difficult to follow for people with limited financial resources. The increasing cost of consuming foods with low energy density and high nutrient density may play a role in the correlations between socioeconomic status (e.g., income and education) and rates of obesity and diet-related chronic diseases.

Source: Monsivais P, Drewnowski A. Lower-energy-density diets are associated with higher monetary costs per kilocalorie and are consumed by women of higher socioeconomic status. *Journal of the American Dietetic Association* 109:804, 2009. Further Reading 3.

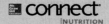 Check out the Connect site **www.mcgrawhillconnect.com** to further explore energy density and food costs.

✔ CONCEPT CHECK 2.1

1. What do variety, proportionality, and moderation mean, and how do they work together to result in a healthy diet?
2. How do the concepts of nutrient density and energy density differ?

2.2 Dietary and Physical Activity Guidelines

Since the early twentieth century, researchers have worked to translate the science of nutrition into practical terms so that people with no special training could estimate whether their nutritional needs were being met. Early food guidance systems aimed to reduce risk for nutrient deficiencies, but severe deficiency diseases are no longer common. Marginal deficiencies of calcium, iron, folate and other B vitamins, vitamin C, vitamin D, vitamin E, potassium, magnesium, and fiber are still a problem; but for many North Americans, major health problems stem from overconsumption of one or more of the following: calories, saturated fat, cholesterol, *trans* fat, alcohol, and sodium.

The following sections of this chapter describe guidelines and tools for planning healthy lifestyles. You will notice how those core concepts of variety, proportionality, and moderation keep showing up throughout our discussions of the Dietary Guidelines, MyPlate, and the Physical Activity Guidelines.

DIETARY GUIDELINES—THE BASIS FOR MENU PLANNING

The 2010 **Dietary Guidelines for Americans** provide nutrition and physical activity advice based on the latest and strongest scientific information to improve the health of all Americans age 2 and older (see Further Readings 10 and 12). The USDA and U.S. Department of Health and Human Services (DHHS) have published Dietary Guidelines since 1980 to aid diet planning. In light of the current epidemic of overweight and obesity—pressing health issues that now affect two-thirds of adults and one-third of children and adolescents—the message of calorie balance is woven throughout this eighth edition of the Dietary Guidelines (see Further Reading 7).

The 2010 Dietary Guidelines include powerful emphases on reduction of total calories, sugar-sweetened beverages, saturated fat, and sodium. In addition, the report calls for a much-needed increase in physical activity among all population groups. These latest Guidelines also include recommendations for those at risk of developing chronic diseases. The health of children is highlighted. These recommendations are also more culturally sensitive to reflect the growing diversity and varied health concerns of the American population. Finally, the 2010 Dietary Guidelines recognize the prevalence of food insecurity (see Chapter 12), aiming to help populations with limited access to food optimize the nutritional content of meals within their resource constraints.

The Dietary Guidelines give direction for the development of educational materials, aid policy makers, and serve as the basis for consumer nutrition messages (see Further Reading 12). Overall, Americans should use this information along with related tools, such as MyPlate and the Physical Activity Guidelines for Americans, to form lifestyle patterns that optimize health.

The report identifies 20 key recommendations (outlined in Figure 2-2) to support three major goals:

- Balance calories with physical activity to manage weight.
- Consume more of certain foods and nutrients, such as fruits, vegetables, whole grains, fat-free and low-fat dairy products, and seafood.
- Consume fewer foods with sodium (salt), saturated fats, *trans* fats, cholesterol, added sugars, and refined grains.

The full report contains background on the development of the Dietary Guidelines, many informative tables and charts to support the recommendations, and a comprehensive list of consumer behaviors and key strategies for achieving each recommendation. This information is available at **www.health.gov/dietaryguidelines**.

Balancing Calories to Manage Weight. The balance between calories consumed (from foods and beverages) and calories expended (through physical activity and

Dietary Guidelines for Americans
General goals for nutrient intakes and diet composition set by the USDA and the U.S. Department of Health and Human Services.

BALANCING CALORIES TO MANAGE WEIGHT

- Prevent and/or reduce overweight and obesity through improved eating and physical activity behaviors.
- Control total calorie intake to manage body weight. For people who are overweight or obese, this will mean consuming fewer calories from foods and beverages.
- Increase physical activity and reduce time spent in sedentary behaviors.
- Maintain appropriate calorie balance during each stage of life—childhood, adolescence, adulthood, pregnancy and breastfeeding, and older age.

FOODS AND FOOD COMPONENTS TO REDUCE

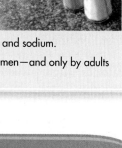

- Reduce daily sodium intake to less than 2300 milligrams and further reduce intake to 1500 milligrams among persons who are 51 and older and those of any age who are African-American or have hypertension, diabetes, or chronic kidney disease. The 1500-mg recommendation applies to about half of the U.S. population, including children, and the majority of adults.
- Consume less than 10% of calories from saturated fatty acids by replacing them with monounsaturated and polyunsaturated fatty acids.
- Consume less than 300 milligrams per day of dietary cholesterol.
- Keep *trans* fatty acid consumption as low as possible by limiting foods that contain synthetic sources of *trans* fats, such as partially hydrogenated oils, and by limiting other solid fats.
- Reduce the intake of calories from solid fats and added sugars.
- Limit the consumption of foods that contain refined grains, especially refined grain foods that contain solid fats, added sugars, and sodium.
- If alcohol is consumed, it should be consumed in moderation—up to one drink per day for women and two drinks per day for men—and only by adults of legal drinking age.

FOODS AND NUTRIENTS TO INCREASE

- Increase vegetable and fruit intake.
- Eat a variety of vegetables, especially dark-green and red and orange vegetables and beans and peas.
- Consume at least half of all grains as whole gains. Increase whole-grain intake by replacing refined grains with whole grains.
- Increase intake of fat-free or low-fat milk and milk products, such as milk, yogurt, cheese, or fortified soy beverages.
- Choose a variety of protein foods, which include seafood, lean meat and poultry, eggs, beans and peas, soy products, and unsalted nuts and seeds.
- Increase the amount and variety of seafood consumed by choosing seafood in place of some meat and poultry.
- Replace protein foods that are higher in solid fats with choices that are lower in solid fat and calories and/or sources of oils.
- Use oils to replace solid fats where possible.
- Choose foods that provide more potassium, dietary fiber, calcium, and vitamin D, which are nutrients of concern in American diets. These foods include vegetables, fruits, whole grains, and milk and milk products.

FIGURE 2-2 ▲ Key Recommendations from the 2010 Dietary Guidelines for Americans.

metabolic processes) determines body weight. Consuming too many calories without increasing physical activity will inevitably lead to weight gain, which exacts an enormous toll on individuals and communities. Many chronic diseases, especially cardiovascular disease, type 2 diabetes, and osteoporosis, could be alleviated by meeting nutrient needs within calorie limits.

The Dietary Guidelines encourage all Americans to achieve and maintain a healthy body weight. Knowing how many calories you need each day is a good place to start (Fig. 2-3). You can calculate your calorie needs on your own (see Chapter 7) or use an online calculator such as the one at **www.ChooseMyPlate.gov.** Once calorie needs are known, the next step is to become familiar with the calorie content of foods and beverages. Finally, monitoring weight over time will allow you to see how your food and physical activity choices are balancing out.

Foods and Food Components to Reduce. Typical American diets contain too much sodium (salt), **solid fats, added sugars,** and refined grains (Fig. 2-4). Solid fats are fats such as butter, beef fat, and shortening that are solid at room temperature. They can be found naturally in foods and are also added during food processing and cooking. Added sugars are sugars and other sweeteners that are added during food processing or cooking.

Solid fats and added sugars, now referred to as *SoFAS*, are energy dense (high in calories) and typically are a source of many **empty calories.** SoFAS also do not contain many important nutrients like vitamins, minerals, or dietary fiber. Diets predominated by these food components especially increase risk for obesity, type 2 diabetes, hypertension, cardiovascular disease, and cancer.

Moderate alcohol consumption is associated with reduced risk of cardiovascular disease, deaths, and cognitive decline. However, those who do not drink should not begin drinking to attain these health benefits because there are risks associated with even moderate alcohol consumption, including increased risk of breast cancer, violence, drowning, and injuries from falls and motor vehicle crashes. Heavy drinking is inherently risky and should be avoided altogether.

Foods and Nutrients to Increase. The Dietary Guidelines urge Americans to replace the problem foods we have discussed with nutrient-dense foods. Emphasize vegetables, fruits, whole grains, fat-free or low-fat milk and milk products, seafood, lean meats and poultry, eggs, beans and peas, and nuts and seeds. Instead of solid fats, choose foods made with plant oils.

These recommendations reflect the nutrient inadequacies of greatest public health concern: potassium, dietary fiber, calcium, and vitamin D (Fig. 2-4). Individuals should strive to meet these goals without exceeding their calorie needs. Focusing on vegetables, fruits, whole grains, lean sources of protein, and low-fat or fat-free dairy products will not only contribute to nutrient adequacy but also lower intake of problem nutrients, improve gastrointestinal function, aid in weight management, and decrease risk for a variety of chronic diseases.

A basic premise of the Dietary Guidelines is that nutrient needs should be met primarily through consuming foods. Foods provide an array of nutrients and other compounds that may benefit health. In certain cases, fortified foods and dietary supplements may be useful sources of one or more nutrients that otherwise might be consumed in less than recommended amounts. These are especially important for people whose typical food choices lead to a diet that cannot meet one or more nutrient recommendations, such as for vitamin D, vitamin E, or calcium. However, dietary supplements cannot and should not replace a healthy diet.

Building Healthy Eating Patterns. The Dietary Guidelines steer clear of a rigid prescription and, instead, promote an array of healthful options that can accommodate cultural, ethnic, traditional, and personal preferences, as well as food cost

	Calorie Range (kcal)		
Children	Sedentary	⟶	Active
2–3 years	1000	⟶	1400
Females			
4–8 years	1200	⟶	1800
9–13	1400	⟶	2200
14–18	1800	⟶	2400
19–30	1800	⟶	2400
31–50	1800	⟶	2200
51+	1600	⟶	2200
Males			
4–8 years	1200	⟶	2000
9–13	1600	⟶	2600
14–18	2000	⟶	3200
19–30	2400	⟶	3000
31–50	2200	⟶	3000
51+	2000	⟶	2800

FIGURE 2-3 ▲ Estimates of calorie needs (kcals).

solid fats Fats that are solid at room temperature, such as butter and margarine. Foods containing solid fats tend to be high in saturated fatty acids or *trans* fatty acids.

added sugars Sugars or syrups that are added to foods during processing or preparation.

empty calories Calories from solid fats and/or added sugars. Foods with empty calories supply energy but few or no other nutrients.

▲ Active means a lifestyle that includes physical activity equivalent to walking more than 3 miles per day at 3 to 4 miles per hour, in addition to the light physical activity associated with typical day-to-day life.

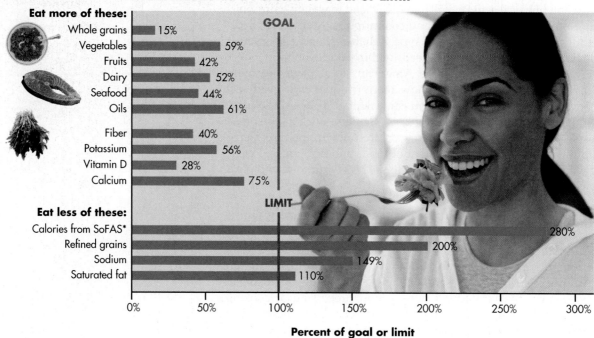

Usual Intake as a Percent of Goal or Limit

Eat more of these:

Whole grains	15%
Vegetables	59%
Fruits	42%
Dairy	52%
Seafood	44%
Oils	61%
Fiber	40%
Potassium	56%
Vitamin D	28%
Calcium	75%

Eat less of these:

Calories from SoFAS*	280%
Refined grains	200%
Sodium	149%
Saturated fat	110%

Percent of goal or limit

*SoFAS = solid fats and added sugars.
Note: Bars show average intakes for all individuals (ages 1 or 2 years or older, depending on the data source) as a percent of the recommended intake level or limit. Recommended intakes for food groups and limits for refined grains and solid fats and added sugars are based on amounts in the USDA 2000-calorie food pattern. Recommended intakes for fiber, potassium, vitamin D, and calcium are based on the highest AI or RDA for age 14 to 70 years. Limits for sodium are based on the UL and saturated fat on 10% of calories. The protein foods group is not shown here because, on average, intake is close to recommended levels.

FIGURE 2-4 ▲ Comparing American dietary habits to the Dietary Guidelines.

Source: Based on data from U.S. Department of Agriculture, Agricultural Research Service and U.S. Department of Health and Human Services, Centers for Disease Control and Prevention. *What We Eat in America*, NHANES 2001–2004 or 2005–2006.

eating pattern A combination of foods and beverages that constitutes an individual's complete dietary intake over time.

and availability factors. Well-studied examples of **eating patterns** consistent with the Dietary Guidelines include Dietary Approaches to Stop Hypertension (DASH), the USDA Food Patterns that accompany MyPlate, the Harvard Healthy Eating Plate, vegetarian eating patterns, and Mediterranean-style eating patterns. Common among these patterns are an abundance of vegetables and fruits, emphasis on whole grains, moderate amounts and varied sources of protein-rich foods, limited added sugars and solid fats, a high proportion of unsaturated fats compared to saturated fats, high potassium, and lower sodium.

The Dietary Guidelines and You. When applying the Dietary Guidelines, you need to consider your own state of health. Make specific changes and see whether they are effective for you. Note that results do not occur overnight and may not meet your expectations. Even when carefully following a diet low in saturated fat, some people continue to have high blood cholesterol. Other people can eat greater amounts of saturated fats and keep their blood cholesterol under control. Differences in genetic background are a key reason for these different responses. Each of us must take into consideration our individual nutritional needs and our risks of developing certain diseases. Plan your diet with your specific needs in mind, taking into account your current health status and family history.

While the Dietary Guidelines are not able to tailor a unique nutrition program for every North American citizen, they do provide adults with simple nutritional advice, which can be implemented by anyone willing to take a step toward good health. Table 2-3 provides examples of recommended diet changes based on the Dietary Guidelines. Although the cost of healthy eating is on the rise, you can make

Click on the Library tab in Connect for links to alternative menu planning tools, such as the Healthy Eating Plate from Harvard School of Public Health. **connect** | NUTRITION

TABLE 2-3 ▶ Recommended Diet Changes Based on the Dietary Guidelines

If You Usually Eat This,	Try This Instead	Benefit
White bread	Whole-wheat bread	• Higher nutrient density, due to less processing • More fiber
Sugary breakfast cereal	Low-sugar, high-fiber cereal with fresh fruit	• Higher nutrient density • More fiber • More phytochemicals
Cheeseburger with French fries	Hamburger and baked beans	• Less saturated fat and *trans* fat • Less cholesterol • More fiber • More phytochemicals
Potato salad	Three-bean salad	• More fiber • More phytochemicals
Doughnuts	Bran muffin or bagel with light cream cheese	• More fiber • Less fat
Regular soft drinks	Diet soft drinks	• Fewer calories
Boiled vegetables	Steamed vegetables	• Higher nutrient density, due to reduced loss of water-soluble vitamins
Canned vegetables	Fresh or frozen vegetables	• Higher nutrient density, due to reduced loss of heat-sensitive vitamins • Lower in sodium
Fried meats	Broiled meats	• Less saturated fat
Fatty meats, such as ribs or bacon	Lean meats, such as ground round, chicken, or fish	• Less saturated fat
Whole milk	Low-fat or fat-free milk	• Less saturated fat • Fewer calories • More calcium
Ice cream	Sherbet or frozen yogurt	• Less saturated fat • Fewer calories
Mayonnaise or sour cream salad dressing	Oil-and-vinegar dressings or light creamy dressings	• Less saturated fat • Less cholesterol • Fewer calories
Cookies	Popcorn (air popped with minimal margarine or butter)	• Fewer calories and *trans* fat
Heavily salted foods	Foods flavored primarily with herbs, spices, lemon juice	• Lower in sodium
Chips	Pretzels	• Less fat

▲ Choose low-sugar, high-fiber cereal with fresh fruit instead of sugary breakfast cereal.

good choices and stay within your budget—canned or frozen fruits and vegetables and non-fat dry milk are a few of the available lower-cost foods.

Diet recommendations for adults have been issued by other scientific groups, such as the American Heart Association, U.S. Surgeon General, National Academy of Sciences, American Cancer Society, Canadian Ministries of Health, and World Health Organization. All are consistent with the spirit of the Dietary Guidelines. These groups encourage people to modify their eating behaviors in ways that are both healthful and pleasurable.

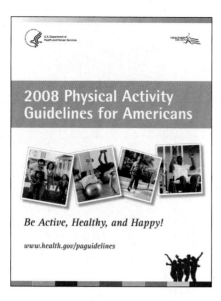

Be Active, Healthy, and Happy!

www.health.gov/paguidelines

▲ http://www.health.gov/PAGuidelines/

PHYSICAL ACTIVITY GUIDELINES FOR AMERICANS

In line with its goal for all Americans to live healthier, more prosperous, and more productive lives, the U.S. Department of Health and Human Services issued its first Physical Activity Guidelines for Americans in 2008 as a complement to the Dietary Guidelines (see Further Reading 9). The overarching idea is that regular physical activity—for people of all ages, races, ethnicities, and physical abilities—produces long-term health benefits. The guidelines are truly meant to inform the work of health professionals and policy makers, but consumer materials are available. *Be Active Your Way: A Guide for Adults*, available at **www.health.gov**, translates the guidelines into consumer-friendly, practical advice.

The key guidelines, listed in Table 2-4, provide measurable physical activity standards for Americans age 6 and older. Specific recommendations (not listed in Table 2-4) also apply to special population groups, including pregnant women, adults with disabilities, and people with chronic medical conditions. For adults, the guidelines emphasize that health benefits occur with at least 150 minutes per week of moderate-intensity physical activity. Adults may accumulate activity throughout the week in a variety of ways: extended sessions (e.g., 50 minutes on 3 days a week) or in short bursts throughout the week that amount to at least 150 minutes. Children and adolescents should strive to include 60 minutes of physical activity per day. For optimum benefits, include both aerobic and muscle-strengthening activities. Overall, physical activity should be enjoyable and safe for each individual. In 2012, a follow-up report, *Physical Activity Guidelines for Americans Midcourse Report: Strategies to Increase Physical Activity Among Youth,* was published and identified interventions that can help increase physical activity in youth across a variety of settings.

TABLE 2-4 ▶ Selected Recommendations of the 2008 Physical Activity Guidelines for Americans*

Key Guidelines for Children and Adolescents

• Children and adolescents should do 60 minutes or more of physical activity daily.

   • Aerobic: Most of the 60 or more minutes a day should be either moderate- or vigorous-intensity aerobic physical activity and should include vigorous-intensity activity at least 3 days a week.

   • Muscle-strengthening: As part of their 60 or more minutes of daily physical activity, children and adolescents should include muscle-strengthening physical activity on at least 3 days of the week.

   • Bone-strengthening: As part of their 60 or more minutes of daily physical activity, children and adolescents should include bone-strengthening physical activity on at least 3 days of the week.

• It is important to encourage young people to participate in physical activities that are appropriate for their age, are enjoyable, and offer variety.

Key Guidelines for Adults

• All adults should avoid inactivity. Some physical activity is better than none, and adults who participate in any amount of physical activity gain some health benefits.

• For substantial health benefits, adults should do at least 150 minutes a week of moderate-intensity, or 75 minutes a week of vigorous-intensity aerobic physical activity, or an equivalent combination of moderate- and vigorous-intensity aerobic activity. Aerobic activity should be performed in episodes of at least 10 minutes, and preferably, it should be spread throughout the week.

• For additional and more extensive health benefits, adults should increase their aerobic physical activity to 300 minutes a week of moderate-intensity, or 150 minutes a week of vigorous-intensity aerobic physical activity, or an equivalent combination of moderate- and vigorous-intensity activity. Additional health benefits are gained by engaging in physical activity beyond this amount.

• Adults should also do muscle-strengthening activities that are moderate or high intensity and involve all major muscle groups on 2 or more days a week, as these activities provide additional health benefits.

(continued)

TABLE 2-4 ▶ **Selected Recommendations of the 2008 Physical Activity Guidelines for Americans*** *(continued)*

Key Guidelines for Older Adults

- When older adults cannot do 150 minutes of moderate-intensity aerobic activity a week because of chronic conditions, they should be as physically active as their abilities and conditions allow.

- Older adults should do exercises that maintain or improve balance if they are at risk of falling.

- Older adults should determine their level of effort for physical activity relative to their level of fitness.

- Older adults with chronic conditions should understand whether and how their conditions affect their ability to do regular physical activity safely.

Key Guidelines for Safe Physical Activity

To do physical activity safely and reduce the risk of injuries and other adverse events, people should:

- Understand the risks and yet be confident that physical activity is safe for almost everyone.

- Choose to do types of physical activity that are appropriate for their current fitness level and health goals because some activities are safer than others.

- Increase physical activity gradually over time whenever more activity is necessary to meet guidelines or health goals. Inactive people should "start low and go slow" by gradually increasing how often and how long activities are done.

- Protect themselves by using appropriate gear and sports equipment; looking for safe environments; following rules and policies; and making sensible choices about when, where, and how to be active.

- Be under the care of a health care provider if they have chronic conditions or symptoms. People with chronic conditions and symptoms should consult their health care provider about the types and amounts of activity appropriate for them.

*The 2008 Physical Activity Guidelines for Americans also include recommendations for pregnant women, adults with disabilities, and people with chronic medical conditions. These are available at **www.health.gov**.

✓ CONCEPT CHECK 2.2

1. What are three of the major goals of the 2010 Dietary Guidelines for Americans?

2. How many minutes of moderate-intensity physical activity are advised per week in the 2008 Physical Activity Guidelines for Americans?

2.3 MyPlate—A Menu-Planning Tool

The titles, food groupings, and shapes of food guides have evolved since the first edition published by the U.S. Department of Agriculture (USDA) a century ago. The most recent food-guidance systems have incorporated physical activity and provided a means for individualization of dietary advice via interactive technology available on the Internet.

To keep pace with updated nutrition advice presented by the 2010 Dietary Guidelines for Americans and *Healthy People 2020* (see Chapter 1), MyPlate was released in 2011 as the leading depiction of healthy eating for Americans (Fig. 2-5). MyPlate, which replaced the familiar MyPyramid, shapes the key recommendations from the Dietary Guidelines into an easily recognizable and extremely applicable visual: a place setting.

FIGURE 2-5 ▲ MyPlate is a visual representation of the advice contained in the 2010 Dietary Guidelines for Americans.

Dishing Up MyPlate. Although it is not intended to stand alone as a source of dietary advice, MyPlate serves as a reminder of how to build a healthy plate at mealtimes (see Further Reading 11). It emphasizes important areas of the American diet that are in need of improvement. Recall from the discussion of the Dietary Guidelines that Americans need to increase the relative proportions of fruits, vegetables, whole grains, and fat-free or low-fat dairy products while simultaneously decreasing consumption of refined grains and high-fat meats.

The new MyPlate icon includes five food groups:

- **Fruits** and **vegetables** cover half of the plate. These foods are dense sources of nutrients and health-promoting phytochemicals despite their low calorie contents.
- **Grains** occupy slightly more than one-fourth of the plate. The message to make half your grains whole is stressed throughout accompanying consumer-education materials.
- The remaining space on the plate is reserved for sources of **protein.** Specifically, the Dietary Guidelines recommend lean meats and poultry, plant sources of protein, and inclusion of fish twice a week.
- A cup of **dairy** appears next to the plate. Depending on personalized calorie recommendations, users should have 2 to 3 cups per day of low-fat or fat-free dairy products or other rich sources of calcium.

MyPlate does not display a separate group for fats and oils, as they are mostly incorporated into other foods. MyPlate food guide recommends limiting solid fats and focusing instead on plant oils, which are sources of essential fatty acids and vitamin E.

Consumer Health Messages. Consumer research points to the need for simple, actionable health messages to capture the attention of the public and achieve successful behavior change. Accordingly, **ChooseMyPlate.gov** provides a series of succinct recommendations to help Americans make healthier food choices. Consumer messages include:

Balancing Calories

- Enjoy your food, but eat less.
- Avoid oversized portions.

Foods to Increase

- Make half your plate fruits and vegetables.
- Make at least half your grains whole.
- Switch to skim or 1% milk.

Foods to Reduce

- Compare sodium in foods such as soup, bread, and frozen meals—and choose the foods with lower numbers.
- Drink water instead of sugary drinks.

▲ In 1942, Canada released its first set of Official Food Rules. Since then, food guidance has evolved based on nutrition research and the changing needs of the population. Now, Health Canada publishes *Canada's Food Guide to Healthy Eating,* available at **http://www.hc-sc.gc.ca/fn-an/ food-guide-aliment/index-eng.php**

Daily Food Plans. On **www.ChooseMyPlate.gov,** you will find an interactive tool that estimates your calorie needs and suggests a food pattern based on your age, gender, height, and weight (Table 2-5). These Daily Food Plans provide useful information for each food group, including recommended daily amounts in common household measures. Modified daily food plans are also available for preschoolers, pregnant or breastfeeding mothers, and those interested in losing weight. Be sure to visit the site to generate your own Daily Food Plan.

The recommended numbers of servings are given in cups for vegetables, fruits, and dairy foods. Grains and protein foods are listed in ounces. See Figure 2-6 for a description of what counts as a MyPlate serving. Pay close attention to the stated serving size for each choice when following your Daily Food Plan to help control calorie intake.

TABLE 2-5 ▶ MyPlate Food-Intake Patterns Based on Calorie Needs

Daily Amount of Food from Each Group												
Calorie Level	1000	1200	1400	1600	1800	2000	2200	2400	2600	2800	3000	3200
Fruits	1 cup	1 cup	1.5 cups	1.5 cups	1.5 cups	2 cups	2 cups	2 cups	2 cups	2.5 cups	2.5 cups	2.5 cups
Vegetables[1,2]	1 cup	1.5 cups	1.5 cups	2 cups	2.5 cups	2.5 cups	3 cups	3 cups	3.5 cups	3.5 cups	4 cups	4 cups
Grains[3]	3 oz-eq	4 oz-eq	5 oz-eq	5 oz-eq	6 oz-eq	6 oz-eq	7 oz-eq	8 oz-eq	9 oz-eq	10 oz-eq	10 oz-eq	10 oz-eq
Protein Foods	2 oz-eq	3 oz-eq	4 oz-eq	5 oz-eq	5 oz-eq	5.5 oz-eq	6 oz-eq	6.5 oz-eq	6.5 oz-eq	7 oz-eq	7 oz-eq	7 oz-eq
Dairy[4]	2 cups	2 cups	2 cups	3 cups	3 cups	3 cups	3 cups	3 cups	3 cups	3 cups	3 cups	3 cups

oz-eq stands for ounce equivalent; tsp stands for teaspoon.

[1]Vegetables are divided into five subgroups (dark green, orange, legumes, starchy, and other). Over a week's time, a variety of vegetables should be eaten, especially green and orange vegetables.

[2]Dry beans and peas can be counted *either* as vegetables (dry beans and peas subgroup) *or* in the protein foods group. Generally, individuals who regularly eat meat, poultry, and fish would count dry beans and peas in the vegetable group. Individuals who seldom eat meat, poultry, or fish (vegetarians) would consume more dry beans and peas, and count some of them in the protein foods group until enough servings from that group are chosen for the day.

[3]At least half of the grain servings should be whole-grain varieties.

[4]Most of the dairy servings should be fat-free or low-fat.

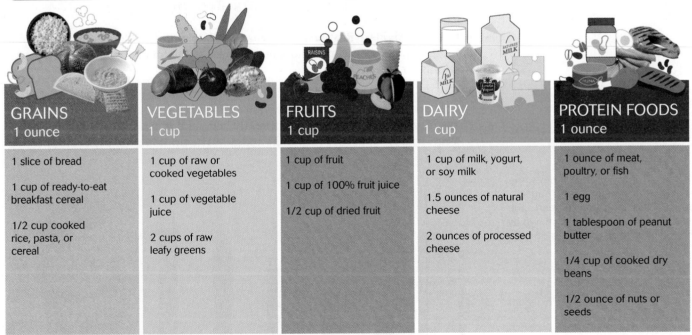

FIGURE 2-6 ▲ MyPlate: What counts as a serving?

How familiar are you with serving size measurements? Figure 2-7 shows a convenient guide to estimate common serving size measurements. Common household units are listed in Appendix F with their metric equivalents. Ounces and fluid ounces differ: ounces are a measure of weight, whereas fluid ounces are a measure of volume.

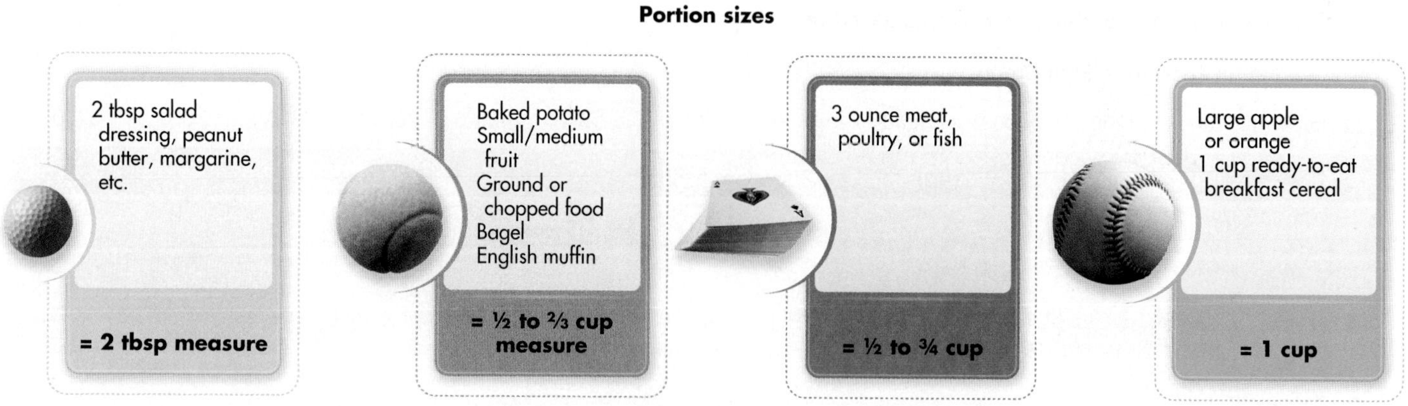

Portion sizes

2 tbsp salad dressing, peanut butter, margarine, etc.	Baked potato Small/medium fruit Ground or chopped food Bagel English muffin	3 ounce meat, poultry, or fish	Large apple or orange 1 cup ready-to-eat breakfast cereal
= 2 tbsp measure	**= ½ to ⅔ cup measure**	**= ½ to ¾ cup**	**= 1 cup**

FIGURE 2-7 ▲ A golf ball, tennis ball, deck of cards, and baseball are standard-size objects that make convenient guides for judging serving sizes.

MyPlate sets limits for empty calories, which come from solid fats and/or added sugars (SoFAS) (Fig. 2-8). SoFAS add calories to the diet but contribute few nutrients. Solid fats are solid at room temperature and include butter, beef fat, and shortening. Some solid fats, such as the marbling in a ribeye steak, are naturally present in foods. Others, such as the shortening used to make a flaky croissant, are added during food processing or preparation. Added sugars include sugars and syrups that are added to foods during processing or preparation. Examples of foods that are major contributors of empty calories in the American diet are cakes, cookies, pastries, soft drinks, energy drinks, cheese, pizza, ice cream, and processed meats. The Daily Food Plans available on **www.ChooseMyPlate.gov** make some allowance for empty calories throughout the day; depending on total energy needs, allowances range from 120 to 600 kcal per day.

Additional MyPlate Resources. ChooseMyPlate.gov also offers in-depth information regarding the Dietary Guidelines, as well as the following additional interactive tools for consumers.

- USDA's 10 Tips Nutrition Education series provides access to one-page printable documents for consumers and health educators. The materials cover over two dozen topics, such as "Kid-friendly veggies and fruits," "Healthy eating for vegetarians," and "Got your dairy today?" The printouts are perfect for posting

FIGURE 2-8 ► Empty calories come from solid fats and/or added sugars (SoFAS). This bar graph compares the amounts of empty calories, shown by the red bars, in various types of milk.

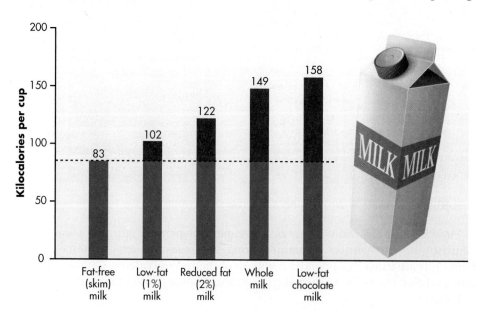

Kilocalories per cup

83	102	122	149	158
Fat-free (skim) milk	Low-fat (1%) milk	Reduced fat (2%) milk	Whole milk	Low-fat chocolate milk

on a refrigerator, and the suggestions can help you get started toward a healthy diet.

- Sample menus and recipes are available online for consumers who are ready to make a change and need a place to start.
- Food-A-Pedia allows users to locate calorie and food group information for specific food entries.
- The SuperTracker enables users to self-monitor food and activity.

Menu Planning with MyPlate. Overall, MyPlate exemplifies the foundations of a healthy diet you have already learned: variety, proportionality, and moderation. To achieve optimal nutrition, remember the following points when using MyPlate to plan your daily menus:

- The guide does not apply to infants or children under 2 years of age. Daily Food Plans for children from ages 2 to 8 are based on average height and weight for age and gender.
- Variety is a key to successful implementation of MyPlate. There is no single, perfect food that is absolutely essential to good nutrition. Each food is rich in some nutrients but deficient in at least one essential nutrient. Likewise, no food group is more important than another; each food group makes an important, distinctive contribution to nutritional intake (Table 2-6). Choose foods from each food group and also choose different foods within each food group. For a sample meal plan, see Table 2-7.
- The foods within a group may vary widely with respect to nutrients and calories. For example, the calorie content of 3 ounces of baked potato is 98 kcal, whereas that of 3 ounces of potato chips is 470 kcal. With respect to vitamin C, an orange has 70 mg and an apple has 10 mg.
- Choose primarily low-fat and fat-free items from the dairy group. By reducing calorie intake in this way, you can select more items from other food groups. If milk causes intestinal gas and bloating, emphasize yogurt and cheese (see Chapter 4 for details on the problem of lactose intolerance).

Solid fats contribute almost 20% of total calories in typical American diets, but they have little to offer in terms of essential nutrients and dietary fiber. Instead of solid fats, choose foods containing plant oils.

TABLE 2-6 ▶ Nutrient Contributions of MyPlate Food Groups

Food Category	Major Nutrient Contributions
Grains	Carbohydrate Vitamins such as thiamin Minerals such as iron Fiber*
Vegetables	Carbohydrate Vitamins such as plant pigments that form vitamin A Minerals such as magnesium Fiber
Fruits	Carbohydrate Vitamins such as folate and vitamin C Minerals such as potassium Fiber
Dairy	Carbohydrate Protein Vitamins such as vitamin D Minerals such as calcium and phosphorus
Protein Foods	Protein Vitamins such as vitamin B-6 Minerals such as iron and zinc

*Whole-grain varieties.

TABLE 2-7 ▶ **Putting MyPlate into Practice**

Meal	Food Group
Breakfast	
1 small orange	Fruits
¾ cup Healthy Choice Low-fat Granola	Grains
with ½ cup fat-free milk	Dairy
½ toasted, small raisin bagel	Grains
with 1 tsp soft margarine	Oils
Optional: coffee or tea	
Lunch	
Turkey sandwich	
2 slices whole-wheat bread	Grains
2 oz turkey	Protein Foods
2 tsp mustard	
1 small apple	Fruits
2 oatmeal-raisin cookies (small)	Empty Calories
Optional: diet soft drink	
3 P.M. Study Break	
6 whole-wheat crackers	Grains
1 tbsp peanut butter	Protein Foods
½ cup fat-free milk	Dairy
Dinner	
Tossed salad	
1 cup romaine lettuce	Vegetables
½ cup sliced tomatoes	Vegetables
1½ tbsp Italian dressing	Oils
½ carrot, grated	Vegetables
3 oz broiled salmon	Protein Foods
½ cup rice	Grains
½ cup green beans	Vegetables
with 1 tsp soft margarine	Oils
Optional: coffee or tea	
Late-Night Snack	
1 cup "light" fruit yogurt	Dairy
Nutrient Breakdown	
1800 kcal	
Carbohydrate	56% of kcal
Protein	18% of kcal
Fat	26% of kcal

This menu meets nutrient needs for all vitamins and minerals for an average adult who needs 1800 kcal. For adolescents, teenagers, and older adults, add one additional serving of milk or other calcium-rich sources.

- Include plant foods that are good sources of proteins, such as beans and nuts, at least several times a week because many are rich in vitamins (such as vitamin E), minerals (such as magnesium), and fiber.

- For vegetables and fruits, try to include a dark-green or orange vegetable for vitamin A, and a vitamin-C-rich fruit, such as an orange, every day. Do not focus primarily on potatoes (e.g., French fries) for your vegetable choices. Surveys show that fewer than 5% of adults eat a full serving of a dark-green vegetable on any given day. Increased consumption of these foods is important because they contribute vitamins, minerals, fiber, and phytochemicals.
- Choose whole-grain varieties of breads, cereals, rice, and pasta because they contribute vitamin E and fiber. A daily serving of a whole-grain, ready-to-eat breakfast cereal is an excellent choice because the vitamins (such as vitamin B-6) and minerals (such as zinc) typically added to it, along with fiber, help fill in common nutritional gaps.
- Include some plant oils on a daily basis, such as those in salad dressing, and eat fish at least twice a week. This supplies you with health-promoting essential fatty acids.

▲ Typical restaurant meals contain oversized portions that do not align with MyPlate.

Limitations of MyPlate. Although MyPlate will promote important changes in American diets, it does have some limitations. Some critics say that the icon is too simple. For example, it does not immediately provide information about overall calories, serving sizes, or number of servings to choose from each food group. However, many of these details will vary by person. Users will need to access the accompanying materials available on **www.ChooseMyPlate.gov** to obtain a personally tailored Daily Food Plan.

The MyPlate icon does not address the types of foods to choose within each food group. Making appropriate food choices for weight management and prevention of diet-related chronic diseases requires consumers to have some nutrition knowledge. Fortunately, public health messages and online content related to MyPlate are available to educate Americans.

MyPlate shows how to build a healthy plate at mealtimes, but it does not adequately address the total diet, which, in reality, includes many snacks between meals. Consumer messages about healthy snacking will be a part of the consumer communications initiative over the next few years.

As with any public health campaign, it is possible that the people who need it most will overlook the MyPlate message. Educated consumers with access to interactive MyPlate tools likely already comply with many of the Dietary Guidelines. Populations with poor diets may be unlikely or unable to click through to find a personalized Daily Food Plan.

Overall, the MyPlate icon is an attractive and relevant tool that immediately shows us how to build a healthy plate at meals. The strength of MyPlate lies in its simplicity. It conveys the major messages that are needed when shopping, cooking, and eating and can be enhanced with the details provided on **www.ChooseMyPlate.gov.**

How Does Your Plate Rate? Regularly comparing your daily food intake with your personalized Daily Food Plan recommendations is a relatively simple way to evaluate the quality of your overall diet. Identify the nutrients that are low in your diet based on the nutrients found in each food group. For example, if you do not consume enough servings from the milk group, your calcium intake is most likely too low. Look for foods that you enjoy that supply calcium, such as calcium-fortified orange juice.

For a more detailed analysis of your current diet, use the SuperTracker tool on **www.ChooseMyPlate.gov.** Your NutritionCalc software also helps you compare your food choices to MyPlate. With a detailed dietary analysis, you can compare your intakes of individual nutrients to standards set by the Food and Nutrition Board and clearly see the areas that need improvement. Even small diet and exercise changes can have positive results.

Choose Your Foods booklets from the American Diabetes Association offer alternative menu-planning tools. These food lists organize foods based on calorie, protein, carbohydrate, and fat content. The result is a manageable framework for designing diets, especially for treatment of diabetes. For more information on the food lists, see Appendix B.

Food *quality* is just as important as food *quantity* when it comes to good nutrition. Consider the difference in calorie contents of the following two meals, both of which fit the proportions suggested by MyPlate:

Fried chicken fillet sandwich with mayonnaise on a white sandwich roll, 1 each
French fries, 1 medium order
Apple-filled pastry, 1 each
Whole milk, 1 cup
1293 kcal

Skinless grilled chicken breast, 3 ounces
Brown rice, prepared with reduced-fat tub margarine, 1 cup
Steamed green beans, 1 cup
Cubed watermelon, 1 cup
Skim milk, 1 cup
513 kcal

FIGURE 2-9 ▶ The Mediterranean Diet Pyramid is based on dietary patterns from the Mediterranean region, which has low rates of chronic diseases and high life expectancy.

Source: © 2009 Oldways Preservation & Exchange Trust, **www.oldwayspt.org**

Characteristics include:

- Foods from plant sources form the foundation of every meal.
- A variety of minimally processed and, wherever possible, seasonally fresh and locally grown foods is emphasized.
- Olive oil is the principal fat.
- Total fat ranges from less than 25% to over 35% of energy, with saturated fat no more than 7% to 8% of calories.
- Fish and seafood are consumed at least twice weekly.
- Lean or low-fat sources of protein, such as cheese, yogurt, poultry, and eggs, should be consumed in moderation.
- Red meats and sweet desserts are consumed less often.
- Regular physical activity is performed at a level that promotes a healthy weight, fitness, and well-being.
- Moderate wine drinking has health benefits.
- Water is the beverage of choice.

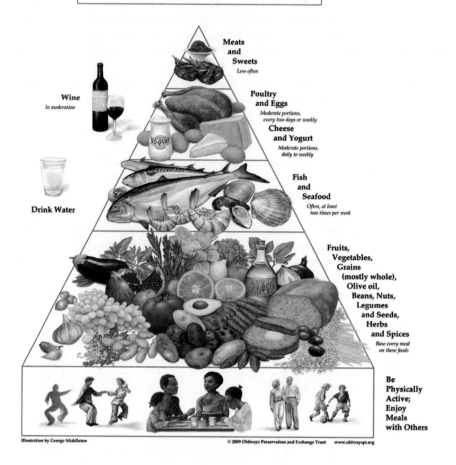

THE MEDITERRANEAN DIET PYRAMID

Updated in 2009, the Mediterranean Diet Pyramid (Fig. 2-9) is a useful alternative to MyPlate. It is based on the dietary patterns of the southern Mediterranean region, which has enjoyed the lowest recorded rates of chronic diseases and the highest adult life expectancy. An abundance of research supports the health benefits of following the Mediterranean Diet (see Further Reading 8).

Oldways—a respected, international, nonprofit culinary think tank—also publishes the Latin-American Diet Pyramid and was the force behind development of the Whole Grain stamp seen on food packages.

✔ CONCEPT CHECK 2.3

1. What is the website where you can find all of the tools associated with MyPlate?
2. What are the five major food groups represented on MyPlate?

2.4 States of Nutritional Health

The ultimate intent of the sound nutrition advice found in the Dietary Guidelines and the MyPlate food guide is to promote optimal nutritional status for individuals. The amount of each nutrient needed to maintain a state of desirable nutrition is the basis for published dietary intake recommendations. We have already discussed general dietary guidelines and cover more specific nutrient recommendations in Section 2.6. The body's nutritional health is determined by considering the **nutritional state** of each needed nutrient. Three general categories of nutritional status are recognized: desirable nutrition, undernutrition, and overnutrition. The common term **malnutrition** can refer to either **overnutrition** or **undernutrition.** Neither state is conducive to good health. Furthermore, it is possible to be both overnourished (e.g., consume excess calories) and undernourished (e.g., consume too few essential vitamins and minerals) at the same time.

The amount of each nutrient needed to maintain a state of desirable nutrition is the basis for published dietary intake recommendations.

DESIRABLE NUTRITION

The nutritional state for a particular nutrient is desirable when body tissues have enough of the nutrient to support normal metabolic functions and surplus stores that can be used in times of increased need. A desirable nutritional state can be achieved by obtaining essential nutrients from a variety of foods.

UNDERNUTRITION

Undernutrition occurs when nutrient intake does not meet nutrient needs. At first, any surpluses are put to use; then, as stores are exhausted, health begins to decline. Many nutrients are in high demand due to constant cell loss and regeneration in the body, such as in the gastrointestinal tract. For this reason, the stores of certain nutrients, including many of the B vitamins, are exhausted rapidly and therefore require a regular intake. In addition, some women in North America do not consume sufficient iron to compensate for monthly losses and eventually deplete their iron stores (Fig. 2-10).

Once availability of a nutrient falls sufficiently low, biochemical evidence indicates that the body's metabolic processes have slowed or stopped. At this state of deficiency, there are no outward **symptoms;** thus, it is termed a **subclinical** deficiency. A subclinical deficiency can go on for some time before clinicians are able to detect its effects.

Eventually, clinical symptoms will develop. Clinical evidence of a nutritional deficiency—perhaps in the skin, hair, nails, tongue, or eyes—can occur within months but may take years to develop. Often, clinicians do not detect a problem until a deficiency produces outward symptoms, such as small areas of bruising on the skin from a vitamin C deficiency.

nutritional state The nutritional health of a person as determined by anthropometric measurements (height, weight, circumferences, and so on), biochemical measurements of nutrients or their by-products in blood and urine, a clinical (physical) examination, a dietary analysis, and economic evaluation; also called nutritional status.

malnutrition Failing health that results from long-standing dietary practices that do not coincide with nutritional needs.

overnutrition A state in which nutritional intake greatly exceeds the body's needs.

undernutrition Failing health that results from a long-standing dietary intake that is not enough to meet nutritional needs.

symptom A change in health status noted by the person with the problem, such as stomach pain.

subclinical Stage of a disease or disorder not severe enough to produce symptoms that can be detected or diagnosed.

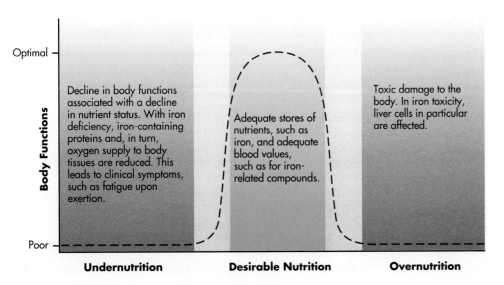

FIGURE 2-10 ◄ The general scheme of nutritional status. Green reflects good status, yellow marginal status, and red poor status (undernutrition or overnutrition). This general concept can be applied to all nutrients. Iron was chosen as an example because iron deficiency is the most common nutrient deficiency worldwide.

▲ The most common type of overnutrition in North America is the excess intake of calories, which often leads to obesity.

OVERNUTRITION

Prolonged consumption of more nutrients than the body needs can lead to overnutrition. In the short run (e.g., 1 to 2 weeks), overnutrition may cause only a few symptoms, such as stomach distress from excess iron intake. If an excess intake continues, however, some nutrients may accumulate to toxic amounts, which can lead to serious disease. For example, too much vitamin A during pregnancy can cause birth defects.

The most common form of overnutrition in developed nations is an excess intake of calories that leads to obesity. In the long run, outcomes of obesity include other serious diseases, such as type 2 diabetes and certain forms of cancer. Use the website **www.shapeup.org** to learn more about the importance of lifelong weight control.

For most vitamins and minerals, the gap between desirable intake and overnutrition is wide. Therefore, even if people take a typical balanced multivitamin and mineral supplement daily, they probably will not receive a harmful dose of any nutrient. The gap between desirable intake and overnutrition is the smallest for vitamin A and the minerals calcium, iron, and copper. Thus, if you take nutrient supplements, keep a close eye on your total vitamin and mineral intake from both food and supplements to avoid toxicity (see Chapter 8 for further advice on use of nutrient supplements).

✓ CONCEPT CHECK 2.4

1. What are the main differences between the undernutrition, desirable nutrition, and overnutrition states of nutritional health?

2.5 Measuring Your Nutritional State

To find out how nutritionally fit *you* are, a nutritional assessment—either whole or in part—needs to be performed (Table 2-8). Generally, this is performed by a physician, often with the aid of a registered dietitian.

ANALYZING BACKGROUND FACTORS

Because family health history plays an important role in determining nutritional and health status, it must be carefully recorded and critically analyzed as part of a nutritional assessment. Other related background information includes (1) a medical

TABLE 2-8 ▶ Conducting an Evaluation of Nutritional Health

Parameters	Example
Background	Medical history (e.g., current diseases, past surgeries, current weight, weight history, and current medications)
	Social history (e.g., marital status and living conditions)
	Family health history
	Education level
	Economic status
Nutritional	Anthropometric assessment: height, weight, skinfold thickness, arm muscle circumference, and other parameters
	Biochemical (laboratory) assessment of blood and urine: enzyme activities, concentrations of nutrients or their by-products
	Clinical assessment (physical examination): general appearance of skin, eyes, and tongue; rapid hair loss; sense of touch; and ability to walk
	Dietary assessment: usual intake or record of previous days' meals

history, especially for any disease states or treatments that could decrease nutrient absorption or ultimate use; (2) a list of medications taken; (3) a social history (e.g., marital status and living conditions); (4) level of education to determine the degree of complexity that can be used in written materials and oral discussions; and (5) economic status to determine the ability to purchase, transport, and cook food.

ASSESSING NUTRITIONAL STATUS USING THE ABCDEs

In addition to background factors, four nutritional-assessment categories complete the picture of nutritional status. **Anthropometric assessment** of height, weight (and weight changes), skinfold thicknesses, and body circumferences provide information about the current state of nutrition. Most measures of body composition are easy to obtain and are generally reliable. However, an in-depth examination of nutritional health is impossible without the more expensive process of **biochemical assessment.** This involves the measurement of the concentrations of nutrients and nutrient by-products in the blood, urine, and feces and the activities of specific blood enzymes.

A **clinical assessment** would follow, during which a health professional would search for any physical evidence (e.g., high blood pressure) of diet-related diseases or deficiencies. Then, a close look at the person's diet (**dietary assessment**), including a record of at least the previous few days' food intake, would help to determine any possible problem areas.

Finally, adding the **environmental assessment** (from the background analysis) provides further details about the living conditions, education level, and ability to purchase and prepare foods needed to maintain health. Now the true nutritional state of a person emerges. Taken together, these five assessments form the ABCDEs of nutritional assessment: anthropometric, biochemical, clinical, dietary, and environmental (Fig. 2-11).

anthropometric assessment Measurement of body weight and the lengths, circumferences, and thicknesses of parts of the body.

biochemical assessment Measurement of biochemical functions (e.g., concentrations of nutrient by-products or enzyme activities in the blood or urine) related to a nutrient's function.

clinical assessment Examination of general appearance of skin, eyes, and tongue; evidence of rapid hair loss; sense of touch; and ability to cough and walk.

dietary assessment Estimation of typical food choices relying mostly on the recounting of one's usual intake or a record of one's previous days' intake.

environmental assessment Includes details about living conditions, education level, and the ability of the person to purchase, transport, and cook food. The person's weekly budget for food purchases is also a key factor to consider.

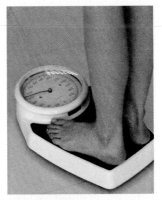

Anthropometric

Biochemical

Clinical

Dietary

Environmental

FIGURE 2-11 ◀ A complete nutritional assessment includes anthropometric, biochemical, clinical, and dietary information. Environmental status adds further information, rounding out the ABCDEs of nutritional assessment.

A practical example using the ABCDEs for evaluating nutritional status can be illustrated in a person who chronically abuses alcohol. Upon evaluation, the physician notes that this person needs medical attention, including nutrient repletion based on the following nutritional assessment results:

(A) Low weight for height, recent 10-pound weight loss, muscle wasting in the upper body
(B) Low amounts of the vitamins thiamin and folate in the blood
(C) Psychological confusion, facial sores, and uncoordinated movement
(D) Dietary intake of little more than wine and hamburgers for the last week
(E) Currently residing in a homeless shelter; $35.00 in wallet; unemployed

RECOGNIZING THE LIMITATIONS OF NUTRITIONAL ASSESSMENT

A long time may elapse between the initial development of poor nutritional health and the first clinical evidence of a problem. A diet high in animal and other solid fat often increases blood cholesterol but without producing any clinical evidence for years. However, when the blood vessels become sufficiently blocked by cholesterol and other materials, chest pain during physical activity or a **heart attack** may occur. An active area of nutrition research is the development of better methods for early detection of nutrition-related problems such as heart attack risk.

Another example of a serious health condition with delayed symptoms is low bone density resulting from a calcium deficiency—a particularly relevant issue for adolescent and young adult females. Many young women do not consume the needed amount of calcium but suffer no obvious effects in their younger years. However, the bone structures of these women with low calcium intakes do not reach full potential during the years of growth, making osteoporosis more likely later in life.

Furthermore, clinical symptoms of some nutritional deficiencies (e.g., diarrhea, inability to walk normally, and facial sores) are not very specific. These may have causes other than poor nutrition. The long time it takes for symptoms to develop and their potential to be vague often make it difficult to establish a link between an individual's current diet and nutritional state.

CONCERN ABOUT THE STATE OF YOUR NUTRITIONAL HEALTH IS IMPORTANT

Table 1-6 in Chapter 1 portrayed the close relationship between nutrition and health. The good news is that people who focus on maintaining nutritional health are apt to enjoy a long, vigorous life. For example, a recent study found that women with a healthy lifestyle had a decreased risk for heart attacks (80% reduction) compared to women without such healthy practices (see the Newsworthy Nutrition on the next page). The healthy habits included:

- Consumed a healthy diet
  - Varied
  - Rich in fiber
  - Included some fish
  - Low in animal fat and *trans* fat
- Maintained a healthy weight
- Occasionally consumed alcohol in small amounts
- Exercised for at least 30 minutes daily
- Avoided use of tobacco

Should all adults follow this example (with optional use of alcohol)?

heart attack Rapid fall in heart function caused by reduced blood flow through the heart's blood vessels. Often part of the heart dies in the process. Technically called a myocardial infarction.

Newsworthy Nutrition

Healthy diet lowers women's risk of sudden cardiac death

Sudden cardiac death (death occurring within 1 hour after symptom onset) is the cause of more than half of all heart-related deaths and usually occurs as the first sign of heart disease, especially in women. Lifestyle information from the Nurses' Health Study (81,722 women) was used to determine if adherence to a healthy lifestyle lowers the risk of sudden cardiac death among women. Low-risk lifestyle was considered as not smoking, not overweight, exercising 30 minutes per day or longer, and following the Mediterranean Diet. Risk of sudden cardiac death dropped by 92% with a combination of the four healthy lifestyles; and women who ate a diet most similar to the Mediterranean Diet with a high proportion of vegetables, fruits, nuts, omega-3 fats, and fish, along with moderate amounts of alcohol and small amounts of red meat, had a 40% less risk than women whose diets least resembled this diet. The conclusion is that a healthy diet along with other healthy lifestyle factors can protect women from sudden cardiac death.

Source: Chiuve SE and others: Adherence to a low risk, healthy lifestyle and risk of sudden cardiac death among women. *Journal of the American Medical Association,* 306:62, 2011 (see Further Reading 2).

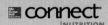 **connect** |NUTRITION   Check out Connect **www.mcgrawhillconnect.com** to further explore the effect of diet on disease risk.

Although nutrition recommendations are often made for the entire healthy population, each of us has individual needs based on our particular health status and genetic background. It would be more appropriate, but also more expensive, if recommendations were made on an individual basis once a person's health status is known.

✔ CONCEPT CHECK 2.5

1. What are the ABCDE steps used in assessing nutritional status?

2.6 Specific Nutrient Standards and Recommendations

The overarching goal of any healthy diet plan is to meet nutrient needs. To begin, we must determine what amount of each essential nutrient is necessary to maintain health. Most of the terms that describe nutrient needs fall under one umbrella term: **Dietary Reference Intakes (DRIs).** The development of DRIs is an ongoing, collaborative effort between the Food and Nutrition Board of the Institute of Medicine in the United States and Health Canada (see Further Reading 12). Included under the DRI umbrella are **Recommended Dietary Allowances (RDAs), Adequate Intakes (AIs), Estimated Energy Requirements (EERs),** and **Tolerable Upper Intake Levels (Upper Levels or ULs).**

As you begin your study of nutrition, all these acronyms can seem like an alphabet soup of abbreviations! To help you more easily sift through these nutrient standards, some basic knowledge about their use is summarized in Table 2-9.

RECOMMENDED DIETARY ALLOWANCE

An RDA is the daily amount of a nutrient that will meet the needs of nearly all individuals (about 97%) in a particular age and gender group. A person can compare his or her daily intake of specific nutrients to the RDA. Although an intake slightly above or below the RDA for a particular nutrient on any given day is no reason for

Dietary Reference Intakes (DRIs) Term used to encompass nutrient recommendations made by the Food and Nutrition Board of the Institute of Medicine. These include RDAs, AIs, EERs, and ULs.

Recommended Dietary Allowance (RDA) Nutrient intake amount sufficient to meet the needs of 97% to 98% of the individuals in a specific life stage.

Adequate Intake (AI) Nutrient intake amount set for any nutrient for which insufficient research is available to establish an RDA. AIs are based on estimates of intakes that appear to maintain a defined nutritional state in a specific life stage.

Estimated Energy Requirement (EER) Estimate of the energy (kcal) intake needed to match the energy use of an average person in a specific life stage.

Tolerable Upper Intake Level (UL) Maximum chronic daily intake level of a nutrient that is unlikely to cause adverse health effects in almost all people in a specific life stage.

TABLE 2-9 ▶ **Nutrient Standards Used in the United States and Canada**

RDA	Recommended Dietary Allowance. Use to evaluate your current intake for a specific nutrient. The further you stray above or below this value, the greater your chances of developing nutritional problems.
AI	Adequate Intake. Use to evaluate your current intake of nutrients but realize that an AI designation implies that further research is required before scientists can establish a more definitive recommendation.
EER	Estimated Energy Requirement. Use to estimate calorie needs of the average person within a specific height, weight, gender, age, and physical activity pattern.
UL	Upper Level. Use to evaluate the highest amount of daily nutrient intake unlikely to cause adverse health effects in the long run in almost all people (97% to 98%) in a population. This number applies to chronic use and is set to protect even very susceptible people in the healthy general population. As intake increases above the Upper Level, the potential for adverse effects generally increases.
DV	Daily Value. Use as a rough guide for comparing the nutrient content of a food to approximate human needs. Typically, the Daily Value used on food labels refers to ages 4 years through adulthood. It is based on a 2000-kcal diet. Some Daily Values also increase slightly with higher calorie intakes (see Fig. 2-13 in the section on food labeling).

concern, a significant deviation below (about 70%) or above (about three times or more for some nutrients) the RDA for an extended time can eventually result in a deficiency or toxicity of that nutrient, respectively.

ADEQUATE INTAKE

An RDA can be set for a nutrient only if there is sufficient information on the human needs for that particular nutrient. Today, there is not enough information on some nutrients, such as chromium, to set such a precise standard as an RDA. For this and other nutrients, the DRIs include a category called an AI. This standard is based on the dietary intakes of people that appear to be maintaining nutritional health. That amount of intake is assumed to be adequate, as no evidence of a nutritional deficiency is apparent.

ESTIMATED ENERGY REQUIREMENT

For calorie needs, we use the EER instead of an RDA or AI. In contrast to the RDAs, which are set somewhat higher than the average needs for nutrients, the EER is set for the average person. While a slight excess of vitamins and minerals is not harmful, a long-term excess of even a small amount of calories will lead to weight gain. Therefore, the calculation of EER needs to be more specific, taking into account age, gender, height, weight, and physical activity (e.g., sedentary or moderately active). In some cases, the additional calorie needs for growth and lactation are also included (see Chapters 7, 14, and 15 for the specific formulas used). Note that the EER is based on the "average" person. Thus, it can only serve as a starting point for estimating calorie needs.

TOLERABLE UPPER INTAKE LEVEL

A Tolerable Upper Intake Level (Upper Level or UL) has been set for some vitamins and minerals (see the inside cover). The UL is the highest amount of a nutrient unlikely to cause adverse health effects in the long run. As intake exceeds the UL, the risk of ill effects increases. These amounts generally should not be exceeded day after day, as toxicity could develop. For people eating a varied diet and/or using a balanced multivitamin and mineral supplement, exceeding the UL is unusual. Problems are more likely to arise with diets that promote excessive intakes of a limited variety of foods, with the use of many fortified foods, or with excessive doses of individual vitamins or minerals.

DAILY VALUE

A nutrition standard more relevant to everyday life is the Daily Value (DV). This is a generic standard used on food labels. It is applicable to both genders from 4 years of age through adulthood and is based on consuming a 2000-kcal diet. DVs are mostly set at or close to the highest RDA value or related nutrient standard seen in the various age and gender categories for a specific nutrient (see Appendix A). DVs have been set for vitamins, minerals, protein, and other dietary components. For fat and cholesterol, the DVs represent a maximum level, not a goal one should strive to reach. DVs allow consumers to compare their intake from a specific food to desirable (or maximum) intakes.

HOW SHOULD THESE NUTRIENT STANDARDS BE USED?

As nutrient intake increases, the Recommended Dietary Allowance (RDA) for the nutrient, if set, is eventually met and a deficient state is no longer present (Fig. 2-12). An individual's needs most likely will be met since RDAs are set high to include almost all people. Related to the RDA concept of meeting an individual's needs are the standards of AI and the EER. These can be used to estimate an individual's needs for some nutrients and calories, respectively. Still, keep in mind that these

Nutrition Facts		
Serving Size 1 donut (about 52g)		
Servings Per Container 12		
Amount Per Serving		
Calories 200	**Calories From Fat** 100	
		%Daily Value*
Total Fat 12g		18**%**
Saturated Fat 3g		15**%**
Trans Fat 4g		
Cholesterol 5mg		1**%**
Sodium 95mg		4**%**
Total Carbohydrate 22g		7**%**
Dietary Fiber <1g		1**%**
Sugars 10g		
Protein 2g		
Vitamin A 0%	●	Vitamin C 2%
Calcium 6%	●	Iron 4%
*Percent of Daily Values (DV) are based on a 2,000 calorie diet.		

▲ The Daily Value is the nutrient standard used on the Nutrition Facts portion of the food label. The percent Daily Value for each nutrient is based on consuming a 2000 kcal diet.

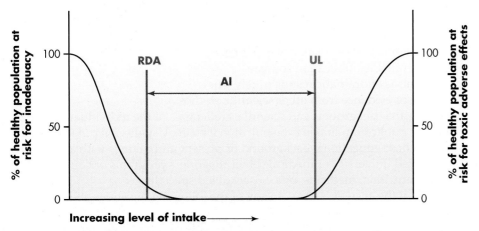

Recommended Dietary Allowance (RDA): The dietary intake level that is sufficient to meet the nutrient requirement of nearly all (97% to 98%) healthy individuals in a particular life stage and gender group. When set for a nutrient, aim for this intake.

Adequate Intake (AI): A recommended intake value based on observed or experimentally determined approximations or estimates of nutrient intake by a group (or groups) of healthy people that is assumed to be adequate; used when an RDA cannot be determined. When set for a nutrient, aim for this intake.

Tolerable Upper Intake Level (Upper Level or UL): The highest level of nutrient intake that is likely to pose no risk of adverse health effects for almost all individuals in the general population. As intake increases above the Upper Level, the risk of adverse effects increases.

FIGURE 2-12 ◄ This figure shows the relationship of the Dietary Reference Intakes (DRIs) to each other and the percentage of the population covered by each. At intakes between the RDA and the UL, the risk of either an inadequate diet or adverse effects from the nutrient in question is close to 0. The UL is then the highest level of nutrient intake likely to pose no risks of adverse health effects to almost all individuals in the general population. At intakes above the UL, the margin of safety to protect against adverse effects is reduced. The AI is set for some nutrients instead of an RDA. The Food and Nutrition Board states that there is no established benefit for healthy individuals if they consume nutrient intakes above the RDA or AI.

standards do not share the same degree of accuracy as the RDA. For example, EER may have to be adjusted upward if the individual is very physically active. Finally, as nutrient intake increases above the UL, poor nutritional health is again likely. However, this poor health is due now to the toxic effects of a nutrient, rather than those of a deficiency.

The type of standard set for nutrients depends on the quality of available evidence. A nutrient recommendation backed by lots of experimental research will be expressed as an RDA. For a nutrient that still requires more research, only an AI is presented. We use the EER as a starting point for determining calorie needs. Some nutrients also have a UL if information on toxicity or adverse health effects is available. Periodically, new DRIs become available as expert committees review and interpret the available research.

RDAs and related standards are intended mainly for diet planning. Specifically, a diet plan should aim to meet the RDA or AI as appropriate and not to exceed the UL over the long term (Fig. 2-12). Specific RDA, AI, EER, and UL standards are printed on the inside cover of this book. To learn more about these nutrient standards, visit the link for Food and Nutrition on the Institute of Medicine's website (**www.iom.edu**).

✔ CONCEPT CHECK 2.6

1. How do the definitions of RDA and AI differ?
2. Which DRI category includes the highest amounts of a nutrient unlikely to cause adverse health effects?

2.7 Evaluating Nutrition Information

The following suggestions should help you make healthful and logical nutrition decisions:

1. Apply the basic principles of nutrition along with the 2010 Dietary Guidelines for Americans and related resources to any nutrition claim, including those on websites. Do you note any inconsistencies? Do reliable references support the claims? Beware of the following:
   - Testimonials about personal experience
   - Disreputable publication sources
   - Promises of dramatic results (rarely true)
   - Lack of evidence from other scientific studies
2. Examine the background and scientific credentials of the individual, organizations, or publication making the nutritional claim. Usually, a reputable author is one whose educational background or present affiliation is with a nationally recognized university or medical center that offers programs or courses in the field of nutrition, medicine, or a closely allied specialty.
3. Be wary if the answer is "Yes" to any of the following questions about a health-related nutrition claim:
   - Are only advantages discussed and possible disadvantages ignored?
   - Are claims made about "curing" disease? Do they sound too good to be true?
   - Is extreme bias against the medical community or traditional medical treatments evident? Health professionals as a group strive to cure diseases in their patients, using what proven techniques are available. They do not ignore reliable cures.
   - Is the claim touted as a new or secret scientific breakthrough?

4. Note the size and duration of any study cited in support of a nutrition claim. The larger it is and the longer it went on, the more dependable its findings. Also consider the type of study: epidemiology versus case-control versus double-blind. Check out the group studied; a study of men or women in Sweden may be less relevant than one of men or women of Southern European, African, or Hispanic descent, for example. Keep in mind that "contributes to," "is linked to," or "is associated with" does not mean "causes."

5. Beware of news conferences and other hype regarding the latest findings. Much of this will not survive more detailed scientific evaluation.

6. When you meet with a nutrition professional, you should expect that he or she will do the following:
   • Ask questions about your medical history, lifestyle, and current eating habits.
   • Formulate a diet plan tailored to your needs, as opposed to simply tearing a form from a tablet that could apply to almost anyone.
   • Schedule follow-up visits to track your progress, answer any questions, and help keep you motivated.
   • Involve family members in the diet plan, when appropriate.
   • Consult directly with your physician and readily refer you back to your physician for those health problems a nutrition professional is not trained to treat.

7. Avoid practitioners who prescribe **megadoses** of vitamin and mineral supplements for everyone.

8. Examine product labels carefully. Be skeptical of any promotional information about a product that is not clearly stated on the label. A product is not likely to do something not specifically claimed on its label or package insert (legally part of the label).

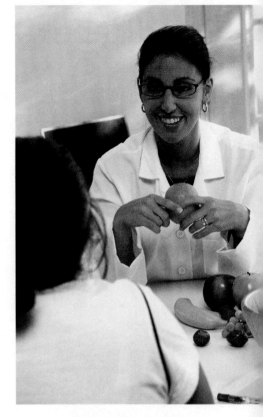

▲ Registered dietitians are a reliable source of nutrition advice.

megadose Large intake of a nutrient beyond estimates of needs or what would be found in a balanced diet; 2 to 10 times human needs is a starting point.

registered dietitian (R.D.) or registered dietitian nutritionist (R.D.N.) A person who has completed a baccalaureate degree program approved by the Accreditation Council for Education in Nutrition and Dietetics (ACEND), performed at least 1200 hours of supervised professional practice, passed a registration examination, and complied with continuing education requirements.

The best approach to finding answers about your nutritional state is to consult a physician or **registered dietitian** first (see Further Reading 1). You can find a registered dietitian in North America by consulting a local online or printed telephone directory, contacting the local dietetic association, calling the dietary department of a local hospital, or visiting **www.eatright.org** or **www.dietitians.ca**. Make sure that the person has the credentials "R.D." or "R.D.N." after his or her name. This indicates that the person has completed rigorous classroom and clinical training in nutrition and participates in continuing education. Appendix E also lists many reputable sources of nutrition advice for your use. Overall, nutrition is a rapidly advancing field, and there are always new findings. Finally, the following websites can help you evaluate ongoing nutrition and health claims:

acsh.org
American Council on Science and Health

www.quackwatch.org
Quackwatch: Your Guide to Quackery, Health Fraud, and Intelligent Decisions

www.ncahf.org
National Council Against Health Fraud

http://ods.od.nih.gov/
National Institutes of Health, Office of Dietary Supplements

www.fda.gov
U.S. Food and Drug Administration

Nutrition and Your Health
Food Labels and Diet Planning

Today, nearly all foods sold in stores must be in a package that has a label containing the following information: the product name, name and address of the manufacturer, amount of product in the package, and ingredients listed in descending order by weight. This food and beverage labeling is monitored in North America by government agencies such as the Food and Drug Administration (FDA) in the United States. The listing of certain food constituents is also required—specifically, on a Nutrition Facts panel (Fig. 2-13). Consumers can use the information in the Nutrition Facts panel to learn more about what they eat. The following components must be listed:

- Total calories (kcal)
- Calories from fat
- Total fat
- Saturated fat
- *Trans* fat
- Cholesterol
- Sodium
- Total carbohydrate

- Fiber
- Sugars
- Protein
- Vitamin A
- Vitamin C
- Calcium
- Iron

In addition to these required components, manufacturers can choose to list polyunsaturated and monounsaturated fat, potassium, and others. Listing these components becomes

required if the food is fortified with that nutrient or if a claim is made about the health benefits of the specific nutrient.

Remember that the Daily Value is a generic standard used on the food label. The percentage of the Daily Value (% Daily Value or % DV) is usually given for each nutrient per serving. These percentages are based on a 2000-kcal diet and must be adjusted for people who require considerably more or less than 2000 kcal per day with respect to fat and carbohydrate intake. DVs are mostly set at or close to the highest RDA value or related nutrient standard seen in the various age and gender categories for a specific nutrient.

Serving sizes on the Nutrition Facts panel must be consistent among similar foods. This means that all brands of ice cream, for example, must use the same serving size on their label. These serving sizes may differ from those of MyPlate because those on food labels are based on more typical portion sizes. In addition, food claims made on packages must follow legal definitions. A long list of definitions for nutrient claims allowed on food labels is given in Table 2-10. For example, if a product claims to be "low sodium," it must have 140 milligrams of sodium or less per serving.

Many manufacturers list the Daily Values set for dietary components such as fat, cholesterol, and carbohydrate on the Nutrition Facts panel. This can be useful as a reference point. As noted, they are based on 2000 kcal; if the label is large enough, amounts based on 2500 kcal are listed as well. As mentioned, DVs allow consumers to compare their intake from a specific food to desirable (or maximum) daily intakes.

PROPOSED CHANGES TO NUTRITION LABELS

In February 2014, the FDA proposed new rules for the Nutrition Facts panel. The changes are designed to promote healthier eating and combat obesity. The label makeover would be the first upgrade in 20 years and would include updating the Daily Values for various nutrients. Other changes, such as more realistic serving information, are aimed at making it easier for Americans to know how many calories they are consuming. The new label would have the calorie count printed in a larger size than the rest, and serving size information would more accurately reflect how much is consumed in one sitting. For example, the serving sizes for a soda will go from 8 ounces to 12 ounces, ice cream will increase from ½ cup to 1 cup, and the yogurt serving size will be reduced from 8 ounces to 6 ounces. To eliminate confusion about the number of servings in a container and calories in a serving, larger packages, such as a pint of ice cream, would have two columns on the labels, "per serving" and "per package." "Added sugars," as well as potassium and vitamin D amounts, would also be additions to the label, whereas calories from fat would be eliminated. More information as well as an illustration of the proposed label can be found at **http://www.fda.gov/ForConsumers/ConsumerUpdates/ucm387114.htm.** Once approved, the proposals would take at least 2 years to implement.

The Nutrition Facts label uses the term *calorie* to express energy content in some cases, but kilocalorie (kcal) values are actually listed.

Serving size

Serving size is listed in household units (and grams). Pay careful attention to serving size to know how many servings you are eating: e.g., if you eat double the serving size, you must double the % Daily Values and calories.

Servings per container

The number of servings of the size given in the serving size above that are in one package of the food.

% Daily Value

This shows how a single serving compares to the DV. Recall that the DVs for fat, saturated fat, cholesterol, protein, and fiber are based on a 2000-calorie diet.

Nutrient claims, such as "Good source," and health claims, such as "Reduce the risk of osteoporosis," must follow legal definitions.

Sugars DV

There is no % Daily Value for sugar. Limiting intake is the best advice.

Protein DV

% Daily Value for protein is generally not included due to expensive testing required to determine protein quality.

Nutrients

These nutrients must appear on most labels. Labels of foods that contain few nutrients, such as candy and soft drinks, may omit some nutrients. Some manufacturers list more nutrients. Other nutrients must be listed if manufacturers make a claim about them or if the food is fortified with them.

Name and address of the food manufacturer.

Daily Value Footnote

This footnote appears on many labels. It is omitted when there is too little space on the label to print it. The footnote reports the DVs used to compute the % Daily Value for a 2000- and 2500-calorie diet.

A Quick Guide to Nutrient Sources

% Daily Value
20% or more = Rich source
10%–19% = Good source

Ingredients are listed in descending order by weight.

Nutrition Facts

Serving Size 1 Pouch (61g)
Serving Per Container 6

Amount Per Serving

| Calories 250 | Calories from Fat 70 |

	% Daily value*
Total Fat 7g	**11**%
Saturated Fat 2.5g	**13**%
Trans Fat 1g	**
Cholesterol 5mg	**2**%
Sodium 400mg	**16**%
Total Carbohydrate 38g	**13**%
Dietary Fiber <1g	**3**%
Sugars 6g	
Protein 7g	

Vitamin A 0% • Vitamin C 0%
Calcium 12% • Iron 8%

*Percent Daily Values are based on a 2,000 calorie diet. Your daily values may be higher or lower depending on your calorie needs:

		Calories:	2,000	2,500
Total Fat	Less than		65g	80g
Sat Fat	Less than		20g	25g
Cholest	Less than		300mg	300mg
Sodium	Less than		2,400mg	2,400mg
Total Carb			300g	375g
Fiber			25g	30g

Calories per gram:

Fat 9   •   Carbohydrate 4   •   Protein 4

**Intake should be as low as possible.

INGREDIENTS: ENRICHED MACARONI PRODUCT (DURUM WHEAT FLOUR, GLYCERYL MONO-STEARATE, SALT, NIACIN, FERROUS SULFATE, THIAMIN MONONITRATE (VITAMIN B1), RIBOFLAVIN (VITAMIN B2), FOLIC ACID), CHEESE SAUCE MIX (WHEY, PARTIALLY, HYDROGENATED SOYBEAN OIL, MALTODEXTRIN, WHEY PROTEIN CONCENTRATE, CORN SYRUP SOLIDS, SALT, MILKFAT, SUGAR, SODIUM, NATURAL FLAVOR, CITRIC ACID, MONOSODIUM GLUTAMATE, MODIFIED FOOD STARCH, LACTIC ACID, YELLOW 5.

FIGURE 2-13 ▲ Food packages must list product name, name and address of the manufacturer, amount of product in the package, and ingredients. The Nutrition Facts panel is required on virtually all packaged food products. The % Daily Value listed on the label is the percent of the amount of a nutrient needed daily that is provided by a single serving of the product. Canadian food labels use a slightly different group of health claims and label descriptors.

TABLE 2-10 ▶ Definitions for Nutrient Claims Allowed on Food Labels

Sugar

- **Sugar free:** less than 0.5 grams (g) per serving.

- **No added sugar; without added sugar; no sugar added:**
  - No sugars were added during processing or packing, including ingredients that contain sugars (for example, fruit juices, applesauce, or jam).
  - Processing does not increase the sugar content above the amount naturally present in the ingredients. (A functionally insignificant increase in sugars is acceptable for processes used for purposes other than increasing sugar content.)
  - The food that it resembles and for which it substitutes normally contains added sugars.
  - If the food doesn't meet the requirements for a low- or reduced-calorie food, the product bears a statement that the food is not low calorie or calorie reduced and directs consumers' attention to the Nutrition Facts panel for further information on sugars and calorie content.

- **Reduced sugar:** at least 25% less sugar per serving than reference food

Calories

- **Calorie free:** fewer than 5 kcal per serving

- **Low calorie:** 40 kcal or less per serving and, if the serving is 30 grams or less or 2 tablespoons or less, per 50 grams of the food

- **Reduced or fewer calories:** at least 25% fewer kcal per serving than reference food

Fiber

- **High fiber:** 5 grams or more per serving. (Foods making high-fiber claims must meet the definition for low fat, or the level of total fat must appear next to the high-fiber claim.)

- **Good source of fiber:** 2.5 to 4.9 grams per serving

- **More or added fiber:** at least 2.5 grams more per serving than reference food

Fat

- **Fat free:** less than 0.5 gram of fat per serving

- **Saturated fat free:** less than 0.5 gram per serving, and the level of *trans* fatty acids does not exceed 0.5 gram per serving

- **Low fat:** 3 grams or less per serving and, if the serving is 30 grams or less or 2 tablespoons or less, per 50 grams of the food. 2% milk can no longer be labeled low fat, as it exceeds 3 grams per serving. *Reduced fat* will be the term used instead.

- **Low saturated fat:** 1 gram or less per serving and not more than 15% of kcal from saturated fatty acids

- **Reduced or less fat:** at least 25% less per serving than reference food

- **Reduced or less saturated fat:** at least 25% less per serving than reference food

Cholesterol

- **Cholesterol free:** less than 2 milligrams (mg) of cholesterol and 2 grams or less of saturated fat per serving

- **Low cholesterol:** 20 milligrams or less of cholesterol and 2 grams or less of saturated fat per serving or, if the serving is 30 grams or less or 2 tablespoons or less, per 50 grams of the food

- **Reduced or less cholesterol:** at least 25% less cholesterol than reference food and 2 grams or less of saturated fat per serving

Sodium

- **Sodium free:** less than 5 milligrams per serving

- **Very low sodium:** 35 milligrams or less per serving and, if the serving is 30 grams or less or 2 tablespoons or less, per 50 grams of the food

- **Low sodium:** 140 milligrams or less per serving or, if the serving is 30 grams or less or 2 tablespoons or less, per 50 grams of the food

- **Light in sodium:** at least 50% less per serving than reference food

- **Reduced or less sodium:** at least 25% less per serving than reference food

Other Terms

- **Fortified or enriched:** Vitamins and/or minerals have been added to the product in amounts in excess of at least 10% of that normally present in the usual product. *Enriched* generally refers to replacing nutrients lost in processing, whereas *fortified* refers to adding nutrients not originally present in the specific food.

- **Healthy:** An individual food that is low fat and low saturated fat and has no more than 360 to 480 milligrams of sodium or 60 milligrams of cholesterol per serving can be labeled "healthy" if it provides at least 10% of the Daily Value for vitamin A, vitamin C, protein, calcium, iron, or fiber.

- **Light or lite:** The descriptor *light* or *lite* can mean two things: first, that a nutritionally altered product contains one-third fewer kcal or half the fat of reference food (if the food derives 50% or more of its kcal from fat, the reduction must be 50% of the fat) and, second, that the sodium content of a low-calorie, low-fat food has been reduced by 50%. In addition, "light in sodium" may be used for foods in which the sodium content has been reduced by at least 50%. The term *light* may still be used to describe such properties as texture and color, as long as the label explains the intent; for example, "light brown sugar" and "light and fluffy."

- **Diet:** A food may be labeled with terms such as *diet, dietetic, artificially sweetened,* or *sweetened with nonnutritive sweetener* only if the claim is not false or misleading. The food can also be labeled *low calorie* or *reduced calorie.*

- **Good source:** *Good source* means that a serving of the food contains 10% to 19% of the Daily Value for a particular nutrient. If 5% or less, it is a *low source.*

(continued)

TABLE 2-10 ▶ Definitions for Nutrient Claims Allowed on Food Labels *(continued)*

- **High:** *High* means that a serving of the food contains 20% or more of the Daily Value for a particular nutrient.

- **Organic:** Federal standards for organic foods allow claims when much of the ingredients do not use chemical fertilizers or pesticides, genetic engineering, sewage sludge, antibiotics, or irradiation in their production. At least 95% of ingredients (by weight) must meet these guidelines to be labeled "organic" on the front of the package. If the front label instead says "made with organic ingredients," only 70% of the ingredients must be organic. For animal products, the animals must graze outdoors, be fed organic feed, and cannot be exposed to large amounts of antibiotics or growth hormones.

- **Natural:** The food must be free of food colors, synthetic flavors, or any other synthetic substance.

The following terms apply only to meat and poultry products regulated by USDA.

- **Extra lean:** less than 5 grams of fat, 2 grams of saturated fat, and 95 milligrams of cholesterol per serving (or 100 grams of an individual food)

- **Lean:** less than 10 grams of fat, 4.5 grams of saturated fat, and 95 milligrams of cholesterol per serving (or 100 grams of an individual food)

Many definitions are from FDA's *Dictionary of Terms,* as established in conjunction with the 1990 Nutrition Labeling and Education Act (NLEA).

Diet Planning with Labels

All of the tools discussed in this chapter greatly aid in menu planning. Menu planning can start with MyPlate. The totality of choices made within the groups can then be evaluated using the Dietary Guidelines. Individual foods that make up a diet can be examined more closely using the Daily Values listed on the Nutrition Facts panel of the product. For the most part, these Daily Values are in line with the Recommended Dietary Allowances and related nutrient standards. The Nutrition Facts panel is especially useful in identifying nutrient-dense foods (foods high in a specific nutrient, such as the vitamin folate, but low in the relative amount of calories provided) and the energy-dense foods (foods that fill you up without providing a lot of calories). Generally speaking, the more you learn about and use these tools together, the more they will benefit your diet. Research has shown that individuals who read the Nutrition Facts when shopping for food report healthier nutrient consumption compared to nonusers (see Further Reading 4).

Exceptions to Food Labeling

Foods such as fresh fruits, vegetables, and fish currently are not required to have Nutrition Facts labels. However, many grocers have voluntarily chosen to provide their customers with information about these products on posters or pamphlets that may contain recipes that can assist you in your endeavor to improve your diet.

The % Daily Value for protein is not mandatory on foods because protein deficiency is not a public health concern in the United States. If the % Daily Value for protein is given on a label, FDA requires that the product be analyzed for protein quality. This procedure is expensive and time-consuming, so many companies opt not to list a % Daily Value for protein. However, labels on food for infants and children under 4 years of age must include the % Daily Value for protein, as must the labels on any food carrying a claim about protein content.

Health Claims on Food Labels

As a marketing tool directed toward the health-conscious consumer, food manufacturers like to claim that their products have all sorts of health benefits. FDA has legal oversight over most food products and permits some health claims with certain restrictions. Overall, claims on foods fall into one of four categories:

- Health claims—closely regulated by FDA
- Preliminary health claims—regulated by FDA but evidence may be scant for the claim
- Nutrient claims—closely regulated by FDA (review Table 2-10)
- Structure/function claims—these are not FDA-approved or necessarily valid

Table 2-10 lists the definitions for nutrient claims on food labels. Currently, FDA limits the use of health messages to specific instances in which there is significant scientific agreement that a relationship exists between a nutrient, food, or food constituent and the disease. The claims allowed at this time

▲ Use the Nutrition Facts label to learn more about the nutrient content of the foods you eat. Nutrient content is expressed as a percent of Daily Value. Canadian food laws and related food labels have a slightly different format.

▲ Specific health claims can be made on food labels for whole-grain cereals. Cheerios® are made from oats, which can be singled out in reducing the risk of cardiovascular disease as long as the statement also says that the diet should also be low in saturated fat and cholesterol.

may show a link (a "may" or "might" qualifier must be used in the statement) between the following:

- A diet with enough calcium and vitamin D and a reduced risk of osteoporosis
- A diet low in total fat and a reduced risk of some cancers
- A diet low in saturated fat and cholesterol and a reduced risk of cardiovascular disease (typically referred to as heart disease on the label)
- A diet rich in fiber—containing grain products, fruits, and vegetables—and a reduced risk of some cancers
- A diet low in sodium and high in potassium and a reduced risk of hypertension and stroke
- A diet rich in fruits and vegetables and a reduced risk of some cancers
- A diet adequate in the synthetic form of the vitamin folate (called folic acid) and a reduced risk of neural tube defects (a type of birth defect)
- Use of sugarless gum and a reduced risk of tooth decay, especially when compared with foods high in sugars and starches
- A diet rich in fruits, vegetables, and grain products that contain fiber and a reduced risk of cardiovascular disease. Oats (oatmeal, oat bran, and oat flour) and psyllium are two fiber-rich ingredients that can be singled out in reducing the risk of cardiovascular disease, as long as the statement also says that the diet should also be low in saturated fat and cholesterol
- A diet rich in whole-grain foods and other plant foods, as well as low in total fat, saturated fat, and cholesterol, and a reduced risk of cardiovascular disease and certain cancers

- A diet low in saturated fat and cholesterol that also includes 25 grams of soy protein and a reduced risk of cardiovascular disease. The statement "one serving of the (name of food) provides _____ grams of soy protein" must also appear as part of the health claim
- Fatty acids from oils present in fish and a reduced risk of cardiovascular disease
- Margarines containing plant stanols and sterols and a reduced risk of cardiovascular disease

In addition, before a health claim can be made for a food product, it must meet two general requirements. First, the food must be a "good source" (before any fortification) of fiber, protein, vitamin A, vitamin C, calcium, or iron. (The legal definition of "good source" appears in Table 2-10.) Second, a single serving of the food product cannot contain more than 13 grams of fat, 4 grams of saturated fat, 60 milligrams of cholesterol, or 480 milligrams of sodium. If a food exceeds any one of these requirements, no health claim can be made for it, despite its other nutritional qualities. For example, even though whole milk is high in calcium, its label can't make the health claim about calcium and osteoporosis because whole milk contains 5 grams of saturated fat per serving. In another example, a health claim regarding fat and cancer can be made only if the product contains 3 grams or less of fat per serving, the standard for low-fat foods.

The FDA's 2003 Consumer Health Information for Better Nutrition Initiative also allows the use of qualified health claims when there is emerging evidence for a relationship between a food, food component, or dietary supplement and reduced risk of a disease or health-related condition. In this case, the evidence is not well enough established to meet the significant scientific agreement standard required for an FDA authorized health claim. A few examples of QUALIFIED HEALTH CLAIMS SUBJECT TO ENFORCEMENT DISCRETION are listed below. More information can be found at **http://www.fda.gov/Food/IngredientsPackagingLabeling/LabelingNutrition/ucm073992.htm.**

Qualified claims about cancer risk
- Tomatoes and/or tomato sauce and prostate, ovarian, gastric, and pancreatic cancers
- Calcium and colon/rectal cancer and calcium and recurrent colon/rectal polyps

Qualified claims about cardiovascular disease risk
- Nuts and heart disease
- Omega-3 fatty acids and coronary heart disease
- Monounsaturated fatty acids from olive oil and coronary heart disease

CASE STUDY Using the Nutrition Facts Label to Make Food Choices

On her way home from her afternoon nutrition class, Laura stops at the supermarket to pick up a few items she needs for the rest of the week. After picking up bananas, yogurt, bagels, and milk, she heads to the frozen food section to find a quick meal before a busy evening. She is in the mood for pasta and finds two brands of frozen cheese manicotti. Laura remembers to check out the Nutrition Facts panel for the two products to help her choose the healthiest option. Help Laura make the best choice by studying the Nutrition Facts labels for the two brands (Luigi's and Mario's) and answering the following questions.

1. Given that a serving of each product of manicotti weighs the same, which product has the highest energy density?
2. Compare the calories from fat for each product of manicotti. Divide the calories from fat by the total calories for each product. What is the percent calories from fat for Luigi's and Mario's?
3. Compare the saturated fat content to the total kilocalorie content of each manicotti product. What is the percent of the total kilocalories that is saturated fat for each?
4. Laura sees that the total calories for Mario's manicotti are much lower than the calories for Luigi's. If she ate two packages of Mario's manicotti, how many calories would she consume?
5. How many servings of Luigi's manicotti would Laura need to consume to get 100% of her Daily Value of sodium?
6. The Daily Value for vitamin C is 60 milligrams. What is the percent Daily Value for vitamin C in each manicotti product?
7. Which product has the higher nutrient density for vitamin C?
8. The Daily Value for calcium is 1000 milligrams. What is the percent Daily Value in each manicotti product?
9. Which product has the higher nutrient density for calcium?
10. Which of the two brands would you choose? What information on the Nutrition Facts label in the figure contributed to this decision?

Nutrition Facts (a) Luigi's Manicotti
Serving Size 1 Package (260g) — Servings Per Container 1

Calories 390, Calories from Fat 160. Total Fat 18g 27%, Saturated Fat 9g 45%, Trans Fat 2g. Cholesterol 45mg 14%, Sodium 880mg 36%, Total Carbohydrate 38g 13%, Dietary Fiber 4g 15%, Sugars 12g, Protein 17g. Vitamin A 10% • Vitamin C 4%. Calcium 40% • Iron 8%.

Nutrition Facts (b) Mario's Manicotti
Serving Size 1 Package (260g) — Servings Per Container 1

Calories 230, Calories from Fat 35. Total Fat 4g 6%, Saturated Fat 2g 10%, Trans Fat 1g. Cholesterol 15mg 4%, Sodium 590mg 24%, Total Carbohydrate 28g 9%, Dietary Fiber 3g 12%, Sugars 10g, Protein 19g. Vitamin A 10% • Vitamin C 10%. Calcium 35% • Iron 4%.

Summary (Numbers refer to numbered sections in the chapter.)

2.1 A healthy eating plan is based on consuming a *variety* of foods balanced by a *moderate* intake of each food and will minimize the risk of developing nutrition-related diseases.

Proportionality is eating foods with greater nutrient density and reflects the nutrient content of a food in relation to its calorie content. Nutrient-dense foods are relatively rich in nutrients, in comparison with calorie content.

Energy density of a food is determined by comparing calorie content with the weight of food. A food rich in calories but weighing relatively very little, such as nuts, cookies, fried foods in general, and most snack foods (including fat-free brands), is considered energy dense. Foods with low energy density include fruits, vegetables, and any food that incorporates lots of water during cooking, such as oatmeal.

2.2 Dietary Guidelines for Americans have been issued to help improve the health of all Americans ages 2 and older. The guidelines emphasize balancing calories to manage weight; performing regular physical activity; moderating consumption of fat, *trans* fat, cholesterol, sugar, salt, and alcohol; and eating plenty of whole-grain products, fruits, and vegetables.

2.3 MyPlate and accompanying online tools are designed to translate nutrient recommendations into a food plan that exhibits variety, proportionality, and moderation. The best results are obtained by using low-fat or fat-free dairy products; incorporating some vegetable proteins in the diet in addition to animal-protein foods; including citrus fruits and dark-green vegetables; and emphasizing whole-grain breads and cereals.

2.4 A person's nutritional state can be categorized as *desirable nutrition,* in which the body has adequate stores for times of increased needs; *undernutrition,* which may be present with or without clinical symptoms; and *overnutrition,* which can lead to vitamin and mineral toxicities and various chronic diseases.

2.5 Evaluation of nutritional state involves analyzing background factors, as well as anthropometric, biochemical, clinical, dietary, and environmental assessments. It is not always possible to detect nutritional inadequacies via nutritional assessment because symptoms of deficiencies are often nonspecific and may not appear for many years.

2.6 Recommended Dietary Allowances (RDAs) are set for many nutrients. These amounts yield enough of each nutrient to meet the needs of healthy individuals within specific gender and age categories. Adequate Intake (AI) is the standard used when not enough information is available to set a more specific RDA. Estimated Energy Requirements (EERs) set calorie needs for both genders at various ages and physical activity patterns. Tolerable Upper Intake Levels (Upper Levels or ULs) for nutrient intake have been set for some vitamins and minerals. All of these dietary standards fall under the term *Dietary Reference Intakes (DRIs).*

Daily Values are used as a basis for expressing the nutrient content of foods on the Nutrition Facts panel and are based for the most part on the RDAs.

2.7 Apply the basic principles of nutrition to evaluate any nutrition claim. Several indicators of nutrition misinformation include insufficient scientific evidence to support a product claim, lack of credible sources, promises of unbelievable results, or distrust of the medical community. To sort nutrition fact from fiction, seek the advice of a registered dietitian.

NAYH Food labels, especially the Nutrition Facts panels, are a useful tool to track your nutrient intake and learn more about the nutritional characteristics of the foods you eat. Any health claims listed must follow criteria set by FDA.

Check Your Knowledge (Answers to the following questions are below.)

1. Anthropometric measurements include
   a. height, weight, skinfolds, and body circumferences.
   b. blood concentrations of nutrients.
   c. a diet history of the previous days' intake.
   d. blood levels of enzyme activities.

2. Foods with *high* nutrient density offer the _____ nutrients for the _____ calories.
   a. least, lowest   c. most, lowest
   b. least, most   d. most, most

3. A meal of a bean burrito, tossed salad, and glass of milk represents foods from all MyPlate food groups except
   a. dairy.   c. vegetables.
   b. protein.   d. fruits.

4. The Dietary Guidelines for Americans were recently revised in
   a. 2000.   c. 2008.
   b. 2005.   d. 2010.

5. The 2010 Dietary Guidelines recommend that we increase which of the following foods?
   a. refined grains   c. seafood
   b. whole milk products d. added sugars

6. How many minutes of moderate-intensity physical activity are recommended for adults in the 2008 Physical Activity Guidelines for Americans?
   a. 150 minutes per week   c. 50 minutes every day
   b. 60 minutes every day   d. 30 minutes days a week

7. The term Daily Value is used on
   a. restaurant menus.
   b. food labels.
   c. medical charts.
   d. None of the above.

8. The Tolerable Upper Intake Level, or UL, is used to
   a. estimate calorie needs of the average person.
   b. evaluate the highest amount of daily nutrient intake unlikely to cause adverse health effects.
   c. evaluate your current intake for a specific nutrient.
   d. compare the nutrient content of a food to approximate human needs.

9. The current food label must list
   a. a picture of the product.
   b. a uniform and realistic serving size.
   c. the RDA for each age group.
   d. ingredients alphabetically.

10. The most common type of undernutrition in industrialized nations, such as the United States, is
    a. anorexia.   c. obesity.
    b. protein deficiency. d. iron deficiency.

Answer Key: 1. a (LO 2.5), 2. c (LO 2.1), 3. d (LO 2.3), 4. d (LO 2.2), 5. c (LO 2.2), 6. a (LO 2.2), 7. b (LO 2.8), 8. b (LO 2.6), 9. b (LO 2.8), 10. d (LO 2.4)

Study Questions (Numbers refer to Learning Outcomes)

1. How would you explain the concepts of nutrient density and energy density to a fourth-grade class **(LO 2.1)?**

2. Describe the intent of the Dietary Guidelines for Americans. Based on the discussion of the Dietary Guidelines for Americans, suggest two key dietary changes the typical North American adult should consider making. **(LO 2.2).**

3. What dietary changes would you need to make to comply with the healthy eating guidelines exemplified by MyPlate on a regular basis **(LO 2.3)?**

4. Describe what would happen to the status of a nutrient in the body for a person who transitions from an overnourished to an undernourished state **(LO 2.4).**

5. What steps would you follow to evaluate the nutritional state of an undernourished person **(LO 2.5)?**

6. How do RDAs and AIs differ from Daily Values in intention and application **(LO 2.6)?**

7. What would you list as the top five sources of reliable nutrition information? What makes these sources reliable **(LO 2.7)?**

8. Dietitians encourage all people to read labels on food packages to learn more about what they eat. What four nutrients could easily be tracked in your diet if you read the Nutrition Facts panels regularly on food products **(LO 2.8)?**

9. Define the USDA definition for the term "organic" **(LO 2.8).**

10. List some specific health claims can be made on food labels **(LO 2.8).**

What the Dietitian Chose

It is smart to reach for a carbohydrate-containing beverage when you need quick energy. Some of us also rely on caffeine in drinks to give us a boost. Many beverages on the market contain high concentrations of carbohydrate as "added sugar." This added sugar has recently been the subject of many debates because the consumption of added sugars has been on the rise since 1970, and excessive sugar intake has been linked to several adverse health conditions. The major sources of added sugars are sodas, energy drinks, and sports drinks.

The main issue here is the abundance of empty calories supplied by soft drinks. Studies show that liquid calories do not promote satiety in the same way that food calories do. In other words, consuming 250 kcal of cola does not mean that you will feel full enough to eat one less slice of pizza to compensate for those extra kcal. Excess calories from any source will lead to weight gain over time.

For these empty calories, portion size is a concern. A 20-ounce bottle of regular cola contains about 68 grams (about 15 teaspoons) of sweeteners. Both types of soft drink provide about 250 kcal. In terms of a soft drink, a 12-ounce soda, such as the Jones Root Beer Soda®, will have proportionally less sugar (48 grams) due to its smaller volume. If you consume one soft drink per day or less, the type of sweetener is not likely to make a big difference. If you consume several soft drinks per day, you would be better off choosing a beverage that does not contain calories, such as diet cola or water. Beverages with artificial sweeteners, such as Coke Zero®, typically are calorie-free and therefore provide you with no energy when you need a boost. You may feel a temporary lift from the caffeine in some of these products. Water

is the perfect drink to restore fluid losses and prevent dehydration, but it contains no calories for energy.

Your best choice when needing a quick energy boost in this case is low-fat chocolate milk. The nutrient-dense chocolate milk not only provides less sugar than the sodas on our list, but it is also the best source of total nutrition. Most convenience stores sell chocolate milk in pint bottles or half-pint (8-ounce) cartons (similar to those served in school cafeterias). The 8-ounce carton of chocolate milk provides a good amount of calories (150 kcal) as

▲ Chocolate milk is a very nutritious beverage that provides protein, vitamins, and minerals, along with calories.

well as 8 grams of protein, 2.5 grams of fat, and 25 grams of sugar. It also provides calcium (290 milligrams), vitamin A (490 IU), and vitamin D (2.8 micrograms), as well as other vitamins and minerals. A pint bottle (16 ounces) would provide twice the amount of calories and other nutrients. In addition, the chocolate provides a source of caffeine that you may be looking for.

CASE STUDY SOLUTION

1. Luigi's has the highest energy density because it has the highest calories per weight **(390 per 260 grams).**

2. The percent calories from fat for Luigi's is **41%**, and the percent calories from fat for Mario's is **15%**.

3. The percent of the total fat that is saturated fat is **50%** for Luigi's and **50%** for Mario's. They both have the same ratio of saturated fat to total fat, even though Luigi's has more grams of saturated fat per serving.

4. If she ate two packages of Mario's manicotti, she would consume 460 calories because **230 × 2 = 460.**

5. Laura would need to consume a little less than three servings of Luigi's manicotti to get 100% of her Daily Value of sodium because **100%/36% = 2.78.**

6. The percent Daily Value for vitamin C is **4%** for Luigi's manicotti and 10% for Mario's manicotti.

7. Mario's manicotti has a higher nutrient density for vitamin C.

8. The percent Daily Value for calcium is 40% for Luigi's manicotti and 35% for Mario's manicotti.

9. Mario's manicotti has a higher nutrient density for calcium (35% DV per 230 calories) than that of Luigi's (40% DV per 390 calories) because Mario's has significantly fewer calories, even though the DV is lower.

10. Mario's manicotti would be the best choice because it has greater nutrient density and lower energy density. This means that Laura will get more nutrients and fewer calories. If she wants to increase her calorie intake, it would be healthier to eat two servings of Mario's manicotti rather than one serving of Luigi's. This will increase her protein, vitamin, mineral, and fiber intake but will keep her total fat and saturated fat intake below that in just one serving of Luigi's.

Further Readings

1. Academy of Nutrition and Dietetics: Practice Paper of the Academy of Nutrition and Dietetics: Communicating accurate food and nutrition information. *Journal of the Academy of Nutrition and Dietetics* 112:759, 2012.

2. Chiuve SE and others: Adherence to a low risk, healthy lifestyle and risk of sudden cardiac death among women. *Journal of the American Medical Association* 306: 62, 2011.

3. Monsivais P and Drewnowski A: Lower-energy-density diets are associated with higher monetary costs per kilocalorie and are consumed by women of higher socioeconomic status. *Journal of the American Dietetic Association* 109:804, 2009.

4. Ollberding NJ and others: Food label use and its relation to dietary intake among U.S. adults. *Journal of the American Dietetic Association* 110:1233, 2010.

5. Perez-Escamilla R and others: Dietary energy density and body weight in adults and children: A systematic review. *Journal of the Academy of Nutrition and Dietetics* 112: 671, 2012.

6. Rolls B and Hermann M: *The ultimate volumetrics diet: Smart, simple, science-based strategies for losing weight and keeping it off.* New York: HarperCollins, 2012.

7. Rowe S and others: Translating the Dietary Guidelines for Americans 2010 to bring about real behavior change. *Journal of the American Dietetic Association* 111:28, 2011.

8. Sofi F and others: Adherence to Mediterranean Diet and health status: Meta-analysis. *British Medical Journal* 337:a1344, 2008.

9. U.S. Department of Health and Human Services: *2008 Physical Activity Guidelines for Americans.* 2008. **www.health.gov/PAGuidelines/** Accessed 3/13/2013.

10. U.S. Department of Health and Human Services: *Dietary Guidelines for Americans, 2010.* 2011. **www.health.gov/dietaryguidelines/2010** Accessed 3/4/2013.

11. United States Department of Agriculture: *USDA's MyPlate.* 2011. **www.choosemyplate.gov** Accessed 3/4/2013.

12. Watts ML and others: The art of translating nutritional science into dietary guidance: History and evolution of the Dietary Guidelines for Americans. *Nutrition Reviews* 69:404, 2011.

 **To get the most out of your study of nutrition, visit McGraw-Hill Connect at www.mcgrawhillconnect.com where you will find NutritionCalc Plus, LearnSmart, and many other dynamic tools.**

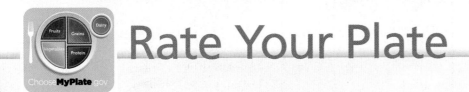

Rate Your Plate

Does your Diet Compare to MyPlate?

Using your food-intake record from Chapter 1, place each food item in the appropriate group of the accompanying MyPlate chart. That is, for each food item, indicate how many servings it contributes to each group based on the amount you ate (see Food Composition Table Supplement for serving sizes). Many of your food choices may contribute to more than one group. For example, spaghetti with meat sauce contributes to three categories; grains, vegetables, and proteins. After entering all the values, add the number of servings consumed in each group. Finally, compare your total in each food group with the recommended number of servings shown in Table 2-5 or obtained from the **www.ChooseMyPlate.gov** website. Enter a minus sign (−) if your total falls below the recommendation or a plus sign (+) if it equals or exceeds the recommendation.

Indicate the number of servings from MyPlate that each food yields:

Food or Beverage	Amount Eaten	Grains	Vegetables	Fruits	Dairy	Protein
Group totals						
Recommended servings						
Shortages/overages in numbers of servings						

 **Find more Rate Your Plate activities for this chapter in Connect at www.mcgrawhillconnect.com.**

Student Learning Outcomes

Chapter 3 is designed to allow you to:

3.1 Understand some basic roles of nutrients in human physiology.

3.2 Identify the functions of the common cellular components.

3.3 Define tissue, organ, and organ system.

3.4 Identify the role of the cardiovascular and lymphatic systems in nutrition.

3.5 List basic characteristics of the nervous system and its role in nutrition.

3.6 List basic characteristics of the endocrine system, especially the pancreas, and its role in nutrition.

3.7 List basic characteristics of the immune system and its role in nutrition.

3.8 Outline the overall processes of digestion and absorption in the mouth, stomach, small intestine, and large intestine, as well as the roles played by the liver, gallbladder, and pancreas.

3.9 List basic characteristics of the urinary system and its role in nutrition.

Chapter 3

The Human Body: A Nutrition Perspective

What Would You Choose?

For spring break, you are volunteering to help build a house with Habitat for Humanity. You are carpooling with some friends and staying in a retreat house. Unfortunately, the travel, budget, living accommodations, and your building schedule won't allow for home-cooked meals this week. Being out of your normal routine and relying on fast-food sandwiches and pizza have left you feeling constipated. This reminds you that you will need to make smarter choices at fast-food establishments. To decrease your constipation, which menu combination would you choose from a pizza buffet?

a 2 slices of pepperoni pizza and 2 cups of tossed salad

b 2 slices of veggie pizza and 1 cup of bean and pasta soup

c 2 slices of ham and pineapple pizza and 1 cup of pasta with Alfredo sauce

d 2 slices of cheese pizza and 1 garlic breadstick with marinara sauce

connect |NUTRITION **Think about your choices as you read Chapter 3, then see What the Dietitian Chose at the end of the chapter. To learn more about the connection between our diets and our bodies, check out the Connect site: www.mcgrawhillconnect.com.**

3.10 Understand the importance of the body storage areas for nutrients.

3.11 Understand the emerging field of nutritional genomics.

3.12 Identify the major nutrition-related gastrointestinal health problems and approaches to treatment.

Merely eating food won't nourish you. You must first digest the food by breaking it down into usable forms of the essential nutrients that can be absorbed into the bloodstream. Once nutrients are taken up by the bloodstream, they can be distributed to and used by body cells.

We rarely think about digesting and absorbing foods. Except for a few voluntary responses—such as deciding what and when to eat, how well to chew food, and when to eliminate the remains—most digestion and absorption processes control themselves. As suggested in the comic in this chapter, we don't consciously decide when the pancreas will secrete digestive substances into the small intestine or how quickly foodstuffs will be propelled down the intestinal tract. Hormones and nerve impulses control these functions. Your only awareness of these involuntary responses may be a hunger pang right before lunch or a "full" feeling after eating that last slice of pizza.

You've learned about cells, tissues, and organs before, but now let's look at the human body from a nutrition perspective. Refresh your memory of the basic anatomy (structure) and physiology (function) of the circulatory system, nervous system, endocrine system, immune system, and urinary system. In particular, as you focus on the digestive system, you will gain an in-depth understanding of how the food you eat nourishes your body.

3.1 Nutrition's Role in Human Physiology

tissues Collections of cells adapted to perform a specific function.

organ A group of tissues designed to perform a specific function; for example, the heart, which contains muscle tissue, nerve tissue, and so on.

organ system A collection of organs that work together to perform an overall function.

The everyday function of the human body relies on the coordination of many highly structured organ systems. Together, these system are composed of trillions of cells. Each cell is a self-contained, living entity. Cells of the same type normally join together, using intercellular substances to form **tissues,** such as muscle tissue. One, two, or more tissues then combine in a particular way to form more complex structures called **organs.** All organs contribute to nutritional health, and a person's overall nutritional state determines how well each organ functions. At a still higher level of coordination, several organs can cooperate for a common purpose to form an **organ system,** such as the digestive system.

Chemical processes (reactions) occur constantly in every living cell: The production of new substances is balanced by the breaking down of older ones. An example is the constant formation and degradation of bone. For this turnover of substances to occur, cells require a continuous supply of energy derived from dietary carbohydrate, protein, and/or fat. Cells also need water; building supplies, especially protein and minerals; and chemical regulators, such as the vitamins. Almost all cells also need a steady supply of oxygen. These substances enable the tissues, made of individual cells, to function properly.

Getting an adequate supply of all nutrients to the body's cells begins with a healthy diet. To ensure optimal use of nutrients, the body's cells, tissues, organs, and organ systems also must work efficiently.

This chapter covers the anatomy and physiology of the cell and major organ systems, especially as they relate to human nutrition. The information you are about to study is limited to the components of the various organ systems specifically influenced by the more than 45 essential nutrients discussed in this text.

☑ CONCEPT CHECK 3.1

1. Discuss the levels of organization of the human body.
2. Describe three ways essential nutrients support cell functions.

FRANK & ERNEST® by Bob Thaves

▲ Some popular (fad) diets suggest not combining meat and potatoes to improve digestion and that fruit should only be eaten before noon. These diets might also claim that foods get stuck in the body and in turn putrefy and create toxins. Are there any scientific reasons to suggest that the timing of our food intake should optimize digestion? Do certain food practices improve digestion and subsequent absorption? This chapter provides some answers.

Source: Frank and Ernest © Thaves. Used with the permission of the Thaves and the Cartoonist Group. All rights reserved.

3.2 The Cell: Structure, Function, and Metabolism

The cell is the basic structural and functional component of life. Living organisms are made of many different kinds of cells specialized to perform particular functions, and all cells are derived from preexisting cells. In the human body, all cells have certain common features. These cells have membranes and **organelles** that perform specialized functions (Fig. 3-1). There are at least 15 different organelles, but the nutritional relevance of just six of the organelles will be examined here. Metabolism, the chemical processes that take place in body cells, is also discussed.

organelles Compartments, particles, or filaments that perform specialized functions within a cell.

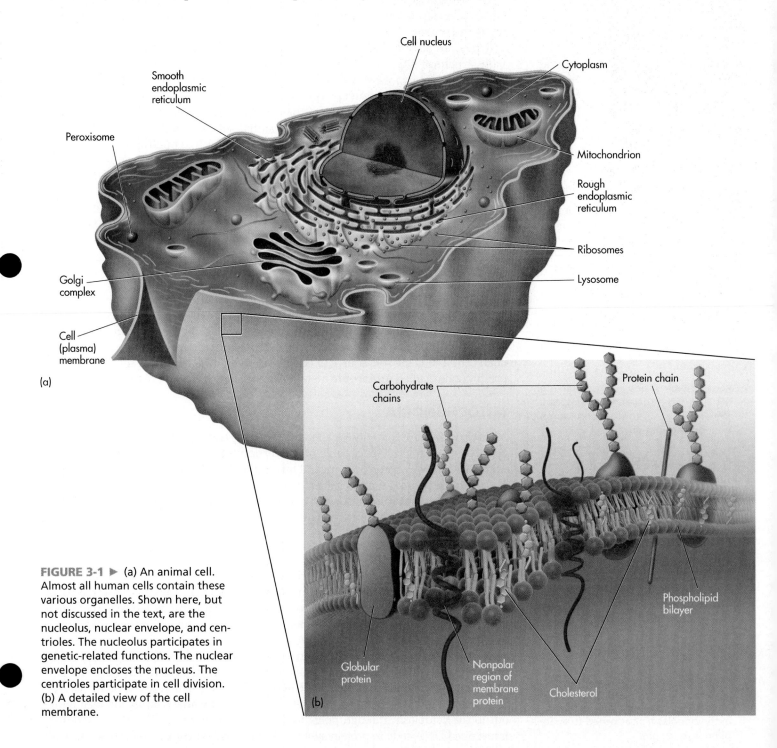

(a)

(b)

FIGURE 3-1 ▶ (a) An animal cell. Almost all human cells contain these various organelles. Shown here, but not discussed in the text, are the nucleolus, nuclear envelope, and centrioles. The nucleolus participates in genetic-related functions. The nuclear envelope encloses the nucleus. The centrioles participate in cell division. (b) A detailed view of the cell membrane.

CELL (PLASMA) MEMBRANE

There is an outside and inside to every cell, separated by the cell (plasma) membrane. (Please note that animal cells do not have cell walls.) The cell membrane itself is not an organelle, but it holds the cellular contents (cytoplasm and organelles) together and regulates the flow of substances into and out of the cell. Cell-to-cell communication also occurs by way of this membrane.

The cell membrane, illustrated in Figure 3-1(b), is a lipid bilayer (or double membrane) of **phospholipids** with their water-soluble heads facing both the interior of the cell and the exterior of the cell. Their water-insoluble tails are tucked into the interior of the cell membrane.

Cholesterol is another component of the cell membrane. It is fat soluble, so it is embedded within the bilayer. This cholesterol provides rigidity and thus stability to the membrane.

There are also various proteins embedded in the cell membrane. Proteins provide structural support, act as transport vehicles, and function as **enzymes** that affect chemical processes within the membrane (see Section 3.9 on digestion for more about enzymes). Some proteins form open channels that allow water-soluble substances to pass into and out of the cell. Proteins on the outside surface of the membrane act as receptors, snagging essential substances that the cell needs and drawing them into the cell. Other proteins act as gates that open and close to control the flow of various particles into and out of the cell.

In addition to lipids and proteins, the membrane also contains carbohydrates that mark the exterior of the cell. These carbohydrates are combined with either protein or fat, and they help send messages to the cell's organelles and act as identification markers for the cell. In addition, they detect invaders and initiate defensive actions. In sum, these carbohydrates provide tags that are important to cellular identity and interaction.

CYTOPLASM

The **cytoplasm** is the combination of fluid material and organelles within the cell, not including the nucleus. (A handful of these organelles will be described within the next two pages.) A small amount of energy for use by the cell can be produced by chemical processes that occur in the cytoplasm. This contributes to the survival of all cells and is the sole source of energy production in red blood cells. This energy production is called **anaerobic** metabolism because it doesn't require oxygen.

MITOCHONDRIA

Mitochondria are sometimes called the "power plants," or the "powerhouse" of the cell. These organelles are capable of converting the food energy in energy-yielding nutrients (carbohydrate, protein, and fat) to a form of energy that cells can use. This is an **aerobic** process that uses the oxygen we inhale, as well as water, enzymes, and some vitamins and minerals (see Chapter 8 for details). Except for red blood cells, all cells contain mitochondria; only the size, shape, and quantity vary.

CELL NUCLEUS

With the exception of the red blood cell, all cells have one or more nuclei. The **cell nucleus** is bounded by its own double membrane. The nucleus contains the genetic material responsible for controlling actions that occur in the cell. The genetic material includes many **chromosomes,** which contain several **genes** made up of **deoxyribonucleic acid (DNA).** DNA is the "code book" that contains directions for making substances, specifically proteins, the cell needs. This code book remains in the nucleus of the cell but sends its information to other cell organelles by way of a "messenger" molecule called **ribonucleic acid (RNA).** The information stored on the DNA in the nucleus is copied onto RNA through the process of **transcription** and then moves out to the cytoplasm through pores in the nuclear membrane. The RNA carries the transcribed DNA code to protein-synthesizing sites called **ribosomes.** There, the RNA

phospholipid Any of a class of fat-related substances that contain phosphorus, fatty acids, and a nitrogen-containing component. Phospholipids are an essential part of every cell.

enzyme A compound that speeds the rate of a chemical process but is not altered by that process. Almost all enzymes are proteins.

cytoplasm The fluid and organelles (except the nucleus) in a cell.

anaerobic Not requiring oxygen.

mitochondria Organelles that are the main sites of energy production in a cell. They contain the pathway for oxidizing fat for fuel, among other metabolic pathways.

aerobic Requiring oxygen.

cell nucleus An organelle bound by its own double membrane and containing chromosomes, the genetic information for cell protein synthesis and cell replication.

chromosome A single, large DNA molecule and its associated proteins; contains many genes to store and transmit genetic information.

gene A specific segment on a chromosome. Genes provide the blueprint for the production of cell proteins.

deoxyribonucleic acid (DNA) The site of hereditary information in cells; DNA directs the synthesis of cell proteins.

ribonucleic acid (RNA) The single-stranded nucleic acid involved in the transcription of genetic information and translation of that information into protein structure.

transcription Process by which genetic information stored as DNA within the nucleus is copied to RNA during protein synthesis.

ribosomes Cytoplasmic particles that mediate the linking together of amino acids to form proteins; may exist freely in the cytoplasm or attached to endoplasmic reticulum.

template is used in the process of **translation** to make a specific protein (see Chapter 6 for details on protein synthesis). Altogether, this process is known as **gene expression.**

All of the DNA in a cell is copied during cell replication. DNA is a double-stranded molecule, and when the cell begins to divide, each strand is separated and an identical copy of each is made. Thus, each new DNA contains one new strand of DNA and one strand from the original DNA. In this way, the genetic code is preserved from one cell generation to the next. (The mitochondria contain their own DNA, so they reproduce themselves within a cell independent of action in the cell's nucleus.)

ENDOPLASMIC RETICULUM (ER)

The outer membrane of the cell nucleus is continuous with a network of tubes called the **endoplasmic reticulum (ER).** Part of the endoplasmic reticulum (termed the rough [as opposed to smooth] endoplasmic reticulum) contains the ribosomes, where the RNA code is translated into proteins during protein synthesis. Many of these proteins play a central role in human nutrition. Parts of the endoplasmic reticulum are also involved in lipid synthesis, detoxification of toxic substances, and calcium storage and release in the cell.

GOLGI COMPLEX

The **Golgi complex** is a packaging site for proteins used in the cytoplasm or exported from the cell. It consists of sacs within the cytoplasm in which proteins are "packaged" as **secretory vesicles** for secretion by the cell.

LYSOSOMES

Lysosomes are the cell's digestive system. They are sacs that contain enzymes for the digestion of foreign material. Sometimes known as "suicide bags," they are responsible for digesting worn-out or damaged cell components. Certain cells associated with immune functions contain many lysosomes.

PEROXISOMES

Peroxisomes contain enzymes that detoxify harmful chemicals. Peroxisomes get their name from the fact that hydrogen peroxide (H_2O_2) is formed as a result of such enzyme action. Peroxisomes also contain a protective enzyme called *catalase*, which prevents excessive accumulation of hydrogen peroxide in the cell, which would be very damaging. Peroxisomes also play a minor role in metabolizing one possible source of energy for cells—alcohol.

CELL METABOLISM

Metabolism refers to the entire collection of chemical processes involved in maintaining life. It encompasses all the sequences of chemical reactions that occur in the body's cells. These biochemical reactions take place in the cell cytoplasm and organelles that we have just discussed. They enable us to release and use energy from foods, synthesize one substance from another, and prepare waste products for excretion.

The reactions of metabolism that take place within your body can be categorized into one of two types. One type of reaction, anabolic, puts different molecules together and, therefore, requires energy. The other type of reaction, catabolic, takes molecules apart and, therefore, releases energy. The metabolism of the nutrients, carbohydrates, proteins, and fats is interrelated and yields energy. The other nutrients, vitamins and minerals, contribute to the enzyme activity that supports metabolic reactions in the cell.

The metabolism of energy production begins in the cytoplasm with the initial anaerobic breakdown of glucose. The remaining aerobic steps of energy production take place in the mitochondria. Ultimately, the cells of the body use these interconnected processes to convert the energy found in food to energy stored in the high-energy compound **adenosine triphosphate (ATP).** You will learn more about the metabolism of energy sources in Chapter 8.

translation Process by which genetic information copied onto RNA dictates the amino acid sequence to form a protein.

gene expression Use of DNA information on a gene to produce a protein. Thought to be a major determination of cell development.

endoplasmic reticulum (ER) An organelle composed of a network of canals running through the cytoplasm. Part of the endoplasmic reticulum contains ribosomes.

Golgi complex The cell organelle near the nucleus that processes newly synthesized protein for secretion or distribution to other organelles.

secretory vesicles Membrane-bound vesicles produced by the Golgi complex; contain protein and other compounds to be secreted by the cell.

lysosome A cellular organelle that contains digestive enzymes for use inside the cell for turnover of cell parts.

peroxisome A cell organelle that destroys toxic products within the cell.

adenosine triphosphate (ATP) The main energy currency for cells. ATP energy is used to promote ion pumping, enzyme activity, and muscular contraction.

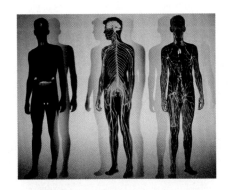

▲ The body is made up of numerous organ systems including the endocrine, nervous, and circulatory systems shown here.

1. What is an organelle?
2. Choose three organelles and explain their relevance to human nutrition.
3. What is the difference between anabolic and catabolic reactions? How does ATP fit into cellular metabolism?

3.3 Body Systems

epithelial tissue The surface cells that line the outside of the body and all external passages within it.

connective tissue Protein tissue that holds different structures in the body together. Some body structures are made up of connective tissue—notably, tendons and cartilage. Connective tissue also forms part of bone and the nonmuscular structures of arteries and veins.

muscle tissue A type of tissue adapted to contract to cause movement.

nervous tissue Tissue composed of highly branched, elongated cells that transport nerve impulses from one part of the body to another.

As noted earlier, when groups of similar cells work together to accomplish a specialized task, the arrangement is referred to as a tissue. Humans are composed of four primary types of tissue: **epithelial, connective, muscle,** and **nervous.** Epithelial tissue is composed of cells that cover surfaces both inside and outside the body. For example, the lining of the respiratory tract is made up of epithelial cells. Epithelial cells secrete important substances, absorb nutrients, and excrete waste. Connective tissue supports and protects the body, stores fat, and produces blood cells. Muscle tissue is designed for movement. Nervous tissue found in the brain and spinal cord is designed for communication. These four types of tissues then go on to form various organs and, ultimately, organ systems (Fig. 3-2).

We focus primarily on the digestive system in this chapter. The nutrients we consume in food are unavailable until they have been processed by the digestive system. Chemical and mechanical digestive processes alter food so that the nutrients can be released and absorbed into the body for distribution to body tissues.

Sometimes organs within a system can serve another system. For example, the basic function of the digestive system is to convert the food we eat into absorbable nutrients. At the same time, the digestive system serves the immune system by preventing dangerous pathogens from invading and causing illness in the body. As you study nutrition, you will note the multiple roles played by many organs (Fig. 3-2).

The overriding objective of this chapter is to understand the actions of nutrients as they affect different cells, tissues, organs, and organ systems. As we explore several key organ systems—cardiovascular, lymphatic, urinary, nervous, endocrine, immune, and digestive systems—look for the ways each system both *affects* and *is affected by* nutrition.

Also in this chapter, we introduce the study of interactions between genetics and nutrition. Throughout this book, discussions will point out how you can personalize nutrition advice based on your genetic background. In this way, you can identify and avoid the "controllable" risk factors that would contribute to development of genetically linked diseases present in your family.

1. List the four types of tissues and give an example of where you could find each in the body.
2. Examine Figure 3-2. Provide three examples of ways the organs of one system support the functions of another system.

Cardiovascular System

Major components
heart, blood vessels, and blood

Functions

• Carries blood and regulates blood supply

• Transports nutrients, waste products, hormones, and gases (oxygen and carbon dioxide) throughout the body

• Regulates blood pressure

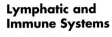

Lymphatic and Immune Systems

Major lymphatic components
lymph, lymphocytes, lymphatic vessels, and lymph nodes

Major immune components
white blood cells, lymph vessels and nodes, spleen, thymus gland, and other lymph tissues

Lymphatic functions

• Removes foreign substances from blood and lymph

• Maintains tissue fluid balance

• Aids fat absorption

Immune functions

• Provides defense against pathogens

• Formation of white blood cells

Urinary System

Major components
kidneys, urinary bladder, and the ducts that carry urine

Functions

• Removes waste products from the blood and forms urine

• Regulates blood acid–base (pH) balance, overall chemical balance, and water balance

Nervous System

Major components
brain, spinal cord, nerves, and sensory receptors

Functions

• Detects and interprets sensation

• Controls movements, physiological, and intellectual functions

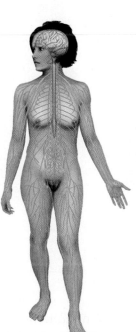

Endocrine System

Major components
endocrine glands, such as the pituitary, thyroid, and adrenal glands; hypothalamus; and pancreas

Functions

• Regulates metabolism, growth, reproduction, and many other functions by producing and releasing hormones

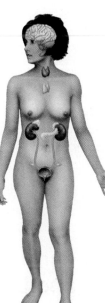

Digestive System

Major components
mouth, esophagus, stomach, intestines, and accessory organs (liver, gallbladder, and pancreas)

Functions

• Performs the mechanical and chemical processes of digestion of food, absorption of nutrients, and elimination of wastes

• Assists the immune system by destroying some pathogens and forming a barrier against foreign materials

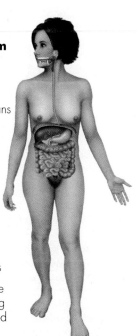

FIGURE 3-2 ▲ Organ Systems of the Body. (continued on next page)

Integumentary System

Major components
skin, hair, nails, and sweat glands

Functions

• Protects the body

• Regulates body temperature

• Prevents water loss

• Produces vitamin D

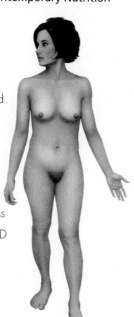

Skeletal System

Major components
bones, cartilage, ligaments, and joints

Functions

• Protects organs

• Supports body weight

• Allows body movement

• Produces blood cells

• Stores minerals

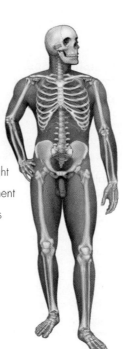

Muscular System

Major components
smooth, cardiac, and skeletal muscle

Functions

• Produces body movement, heartbeat, and body heat

• Propels food in the digestive tract

• Maintains posture

Respiratory System

Major components
lungs and respiratory passages

Functions

• Exchanges gases (oxygen and carbon dioxide) between the blood and the air

• Regulates blood acid–base (pH) balance

Reproductive System

Major components
gonads (ovaries and testes), genitals, and breasts

Functions

• Performs the processes of sexual maturation and reproduction

• Influences sexual functions and behaviors

• Produces human milk to nourish an infant

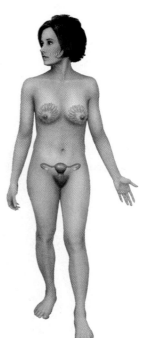

FIGURE 3-2 ▲ *(Continued)*

cardiovascular system The body system consisting of the heart, blood vessels, and blood. This system transports nutrients, waste products, gases, and hormones throughout the body and plays an important role in immune responses and regulation of body temperature.

lymphatic system A system of vessels and lymph that accepts fluid surrounding cells and large particles, such as products of fat absorption. Lymph eventually passes into the bloodstream from the lymphatic system.

lymph A clear fluid that flows through lymph vessels; carries most forms of fat after their absorption by the small intestine.

3.4 Cardiovascular System and Lymphatic System

The body has two separate organ systems that circulate fluids in the body: the **cardiovascular system** and the **lymphatic system.** Some texts group these two systems together as the *circulatory system*, but each system has distinct components and functions. The cardiovascular system consists of the heart and blood vessels. The lymphatic system consists of lymphatic vessels and a number of lymph tissues. Blood flows through the cardiovascular system, while **lymph** flows through the lymphatic system.

CARDIOVASCULAR SYSTEM

The heart is a muscular pump that normally contracts and relaxes 50 to 90 times per minute when the body is at rest. This continual pumping, measured by taking your pulse, keeps blood moving through the blood vessels. The blood that flows through the cardiovascular system is composed of **plasma,** red blood cells, white blood cells, platelets, and many other substances. It travels two basic routes. In the first route, blood circulates from the right side of the heart, through the lungs, and then back to the heart. In the lungs, blood picks up oxygen and releases carbon dioxide. After this exchange of gases has taken place, blood is said to be *oxygenated* and returns to the left side of your heart. In the second route, the oxygenated blood circulates from the left side of the heart to all other body cells, eventually returning back to the right side of the heart (Fig. 3-3). After blood has circulated throughout the body, it is *deoxygenated.* (As you review the cardiovascular system, recall from

plasma The fluid, extracellular portion of the circulating blood. This includes the blood serum plus all blood-clotting factors. In contrast, serum is the fluid that remains after clotting factors have been removed from plasma.

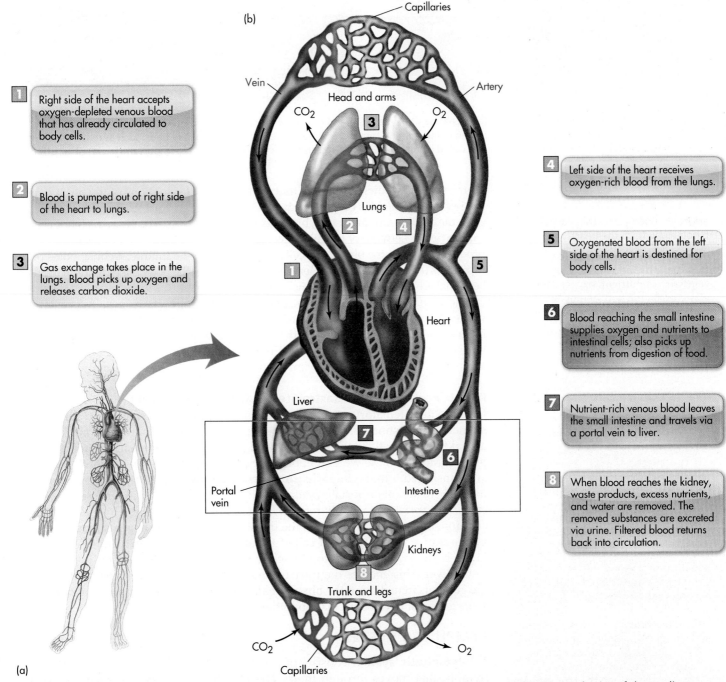

(b)

1 Right side of the heart accepts oxygen-depleted venous blood that has already circulated to body cells.

2 Blood is pumped out of right side of the heart to lungs.

3 Gas exchange takes place in the lungs. Blood picks up oxygen and releases carbon dioxide.

4 Left side of the heart receives oxygen-rich blood from the lungs.

5 Oxygenated blood from the left side of the heart is destined for body cells.

6 Blood reaching the small intestine supplies oxygen and nutrients to intestinal cells; also picks up nutrients from digestion of food.

7 Nutrient-rich venous blood leaves the small intestine and travels via a portal vein to liver.

8 When blood reaches the kidney, waste products, excess nutrients, and water are removed. The removed substances are excreted via urine. Filtered blood returns back into circulation.

Capillaries — Vein — Head and arms — CO_2 — 3 — O_2 — Artery — Lungs — 2 — 4 — 1 — 5 — Heart — Liver — 7 — 6 — Portal vein — Intestine — 8 — Kidneys — Trunk and legs — CO_2 — O_2 — Capillaries

(a)

FIGURE 3-3 ▲ Blood circulation through the body. (a) The heart and some examples of the major arteries and veins of the cardiovascular system. (b) The paths that blood takes from the heart to the lungs (1–3), back to the heart (4), and through the rest of the body (5–8). The red color indicates blood richer in oxygen; blue is for blood carrying more carbon dioxide. Keep in mind that arteries and veins go to all parts of the body.

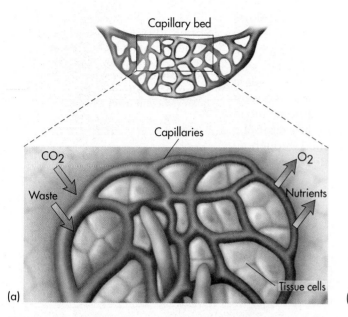

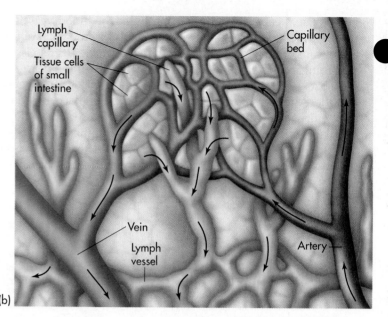

FIGURE 3-4 ▲ Capillary and lymph vessels. (a) Exchange of oxygen (O_2) and nutrients for carbon dioxide (CO_2) and other waste products occurs between the capillaries and the surrounding tissue cells. (b) Lymph vessels are also present in capillary beds, such as in the small intestine. Lymph vessels in the small intestine are also called lacteals. The lymph vessels have closed ends and are important for fat absorption.

artery A blood vessel that carries blood away from the heart.

capillary A microscopic blood vessel that connects the smallest arteries and veins; site of nutrient, oxygen, and waste exchange between body cells and the blood.

vein A blood vessel that carries blood to the heart.

hepatic portal circulation The portion of the circulatory system that uses a large vein (portal vein) to carry nutrient-rich blood from capillaries in the intestines and portions of the stomach to the liver.

hepatic portal vein Large vein leaving the intestine and stomach and connecting to the liver.

lymphatic system A system of vessels and lymph that accepts fluid surrounding cells and large particles, such as products of fat absorption. Lymph eventually passes into the bloodstream from the lymphatic system.

your previous studies of biology that *left* and *right* designations of the heart refer to the left and right sides of your body, not of the page in your textbook.)

In the cardiovascular system, blood leaves the heart via **arteries,** which branch into **capillaries,** a network of tiny blood vessels. Exchange of nutrients, oxygen, and waste products between the blood and cells occurs through the minute, weblike pores of the capillaries (Fig. 3-4). Capillaries service every region of the body via individual capillary beds that are only one cell layer thick. The blood then returns to the heart via the **veins.**

The cardiovascular system facilitates the exchange of oxygen, nutrients, and wastes between the body's internal and external environments. Other functions include delivery of hormones to their target cells, maintenance of a constant body temperature, and distribution of white blood cells throughout the body to protect against pathogens as part of the immune system (see Section 3.8, "Immune System").

Portal Circulation in the Gastrointestinal Tract. Water-soluble nutrients are transferred to the circulatory system through capillary beds. Once absorbed through the stomach or intestinal wall, nutrients reach one of two destinations. Some nutrients are taken up by cells in the intestines and portions of the stomach to nourish those organs. Most of these water-soluble nutrients from recently eaten foods, however, are transferred into the **hepatic portal circulation.** (The term *hepatic* refers to the liver. There are other portal systems in physiology, but the simpler terms *portal circulation* or *portal vein* usually refer to hepatic portal circulation.) To enter portal circulation, the nutrients pass from the intestinal capillaries into veins that eventually merge into a very large vein called the **hepatic portal vein.** Unlike most veins in the body—which carry blood back to the heart—this portal vein leads directly to the liver. This enables the liver to process absorbed nutrients before they enter the general circulation of the bloodstream. Overall, hepatic portal circulation represents a special form of circulation in the cardiovascular system.

LYMPHATIC SYSTEM

The **lymphatic system** consists of a network of lymphatic vessels and the fluid (lymph) that moves through them. The lymph vessels take up excess fluid that collects between cells and return it to the bloodstream. Lymph is similar to blood, consisting largely of blood plasma that has found its way out of capillaries and into the spaces between cells. It contains a full array of the various white blood cells that play an important role in the immune system. However, neither red blood cells nor platelets are present. Lymph is collected in tiny lymph vessels all over the body and moves through even larger vessels until it eventually enters the cardiovascular system through major veins near the heart. The lymphatic system does not have a pump (like the heart); its flow is driven by muscle contractions arising from normal body movements.

Lymphatic Circulation in the Gastrointestinal Tract. Besides contributing to the defense of the body against invading pathogens, lymphatic vessels that serve the small intestine play an important role in nutrition. These vessels pick up and transport the majority of products of fat digestion and fat absorption. These fat-related products are too large to enter the bloodstream directly and therefore are generally emptied into the bloodstream only after passing through the lymphatic system.

✔ CONCEPT CHECK 3.4

1. What is hepatic portal circulation?
2. Which nutrients are absorbed into the lymph? Why?
3. Describe how nutrients, oxygen, and wastes are exchanged between the body's internal and external environments.

3.5 Urinary System

The **urinary system** is composed of two kidneys, one on each side of the spinal column. Each kidney is connected to the bladder by a **ureter.** The bladder is emptied by way of the **urethra** (Fig. 3-5). The main function of the kidneys is to remove waste from the body. The kidneys are constantly filtering blood to control its composition. This results in the formation of urine, which is composed of water, dissolved waste products of metabolism such as **urea,** and excess and unneeded water-soluble vitamins and various minerals.

urinary system The body system consisting of the kidneys, urinary bladder, and the ducts that carry urine. This system removes waste products from the circulatory system and regulates blood acid–base balance, overall chemical balance, and water balance in the body.

ureter Tube that transports urine from the kidney to the urinary bladder.

urethra Tube that transports urine from the urinary bladder to the outside of the body.

urea Nitrogenous waste product of protein metabolism; major source of nitrogen in the urine.

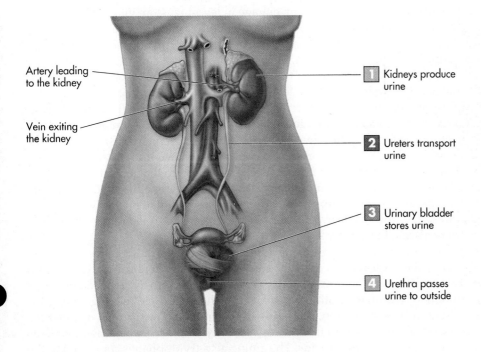

Artery leading to the kidney

Vein exiting the kidney

1 Kidneys produce urine

2 Ureters transport urine

3 Urinary bladder stores urine

4 Urethra passes urine to outside

FIGURE 3-5 ◀ Organs of the urinary system. The kidneys (1), bean-shaped organs located on either side of the spinal column, filter waste from the blood and form urine, which is transported to the bladder by the ureters (2) and stored in the bladder (3) as urine. The urethra (4) transports the urine to outside the body. The urinary system of the female is shown. The male's urinary system is the same, except that the urethra extends through the penis.

pH A measure of relative acidity or alkalinity of a solution. The pH scale is 0 to 14. A pH of 7 is neutral; a pH below 7 is acidic; and a pH above 7 is alkaline.

erythropoietin A hormone secreted mostly by the kidneys that enhances red blood cell synthesis and stimulates red blood cell release from bone marrow.

Together with the lungs, the kidneys also maintain the acid–base balance **(pH)** of the blood. The kidneys also convert a form of vitamin D into its active hormone form and produce a hormone that stimulates red blood cell synthesis (**erythropoietin;** see Chapter 10 for information on misuse of this hormone by some athletes). During times of fasting, the kidneys even produce glucose from certain amino acids. Thus, the kidneys perform many important functions related to nutrition and are a vital component of the body.

The proper function of the kidneys is closely tied to the strength of the cardio-vascular system, particularly its ability to maintain adequate blood pressure, and the consumption of sufficient fluid. Uncontrolled diabetes, hypertension, and drug abuse are harmful to the kidneys.

✔ CONCEPT CHECK 3.5

1. The kidneys filter wastes from the blood. Trace the path of waste products out of the body.
2. How are the kidneys involved in bone health?

3.6 Nervous System

nervous system The body system consisting of the brain, spinal cord, nerves, and sensory receptors. This system detects sensations, directs movements, and controls physiological and intellectual functions.

The **nervous system** is a regulatory system that centrally controls most body functions. The nervous system can detect changes occurring in various organs and the external environment and initiate corrective action when needed to maintain a constant internal body environment. The nervous system also regulates activities that change almost instantly, such as voluntary muscle contractions and the body's response to stress or danger. The body has many receptors that receive information about what is happening within the body and in the outside environment. For the most part, these receptors are found in our eyes, ears, skin, nose, and stomach. We act on information from these receptors via the nervous system.

neuron The structural and functional unit of the nervous system. Consists of a cell body, dendrites, and an axon.

The basic structural and functional unit of the nervous system is the **neuron.** Neurons are elongated, highly branched cells. The body contains about 100 billion neurons. Neurons respond to electrical and chemical signals, conduct electrical impulses, and release chemical regulators. Overall, neurons allow us to perceive what is occurring in our environment, engage in learning, store vital information in memory, and control the body's voluntary (and involuntary) actions.

The brain stores information, reacts to incoming information, solves problems, and generates thoughts. In addition, the brain plans a course of action based on the other sensory inputs. Responses to the stimuli are carried out mostly through the rest of the nervous system.

Simply put, the nervous system receives information through stimulation of various receptors, processes this information, and sends out signals through its various branches for an action that needs to be taken. Actual transmission of the signal occurs through a change in the concentration of two nutrients, sodium and potassium, in the neuron. There is an influx of sodium into the neuron and a loss of potassium as the message is sent. Concentrations of these minerals are then restored to normal amounts in the neuron after the signal passes, making it ready to conduct another message. You will learn more about the role of these electrolyte minerals in nerve transmission in Chapter 9.

synapse The space between one neuron and another neuron (or cell).

neurotransmitter A compound made by a nerve cell that allows for communication between it and other cells.

When the signal must bridge a gap **(synapse)** between the branches of different neurons, the message is generally converted into a chemical signal called a **neurotransmitter.** The neurotransmitter is then released into the gap, and its target may be another neuron or another type of cell, such as a muscle cell (Fig. 3-6). If the signal is sent to another neuron, this allows it to continue on to its final destination. The neurotransmitters used in this process are often made from common nutrients found in foods, such as amino acids. The amino acid tryptophan is converted to the

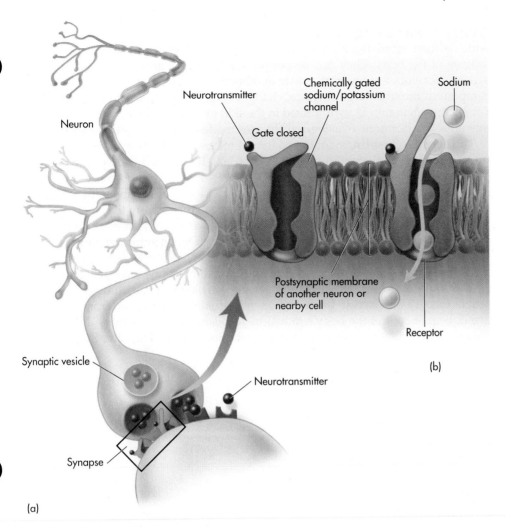

FIGURE 3-6 ◄ Transmission of a message from one neuron to another neuron or to another type of cell that relies on neurotransmitters. Figure (a) shows how vesicles containing neurotransmitters, formed within the neuron, fuse with the membrane of the neuron and the neurotransmitter is released into the synapse. Enlarged in (b), the neurotransmitter then binds to the receptors on the nearby neuron (or cell). In this way, the message is transmitted from one neuron to another or to the cell that ultimately performs the action directed by the message.

neurotransmitter serotonin, and the amino acid tyrosine is converted to the neurotransmitters **norepinephrine** and **epinephrine** (also called adrenaline).

Other nutrients also play a role in the nervous system. Calcium is needed for the release of neurotransmitters from neurons. Vitamin B-12 plays a role in the formation of the **myelin** sheath, which provides insulation around specific parts of most neurons. Finally, a regular supply of carbohydrate in the form of glucose is important for supplying fuel for the brain. The brain can use other calorie sources but generally relies on glucose.

✔ CONCEPT CHECK 3.6

1. How are sodium and potassium involved in nerve impulse transmission?
2. How are signals transmitted between one neuron and the next? Why are amino acids important in this process?
3. Which nutrient is the brain's preferred source of fuel?

3.7 Endocrine System

The **endocrine system** plays a major role in the regulation of metabolism, reproduction, water balance, and many other functions by producing hormones in the **endocrine glands** of the body and subsequently releasing them into the blood (Table 3-1). The term *hormone* comes from the Greek word for "to stir or excite."

norepinephrine A neurotransmitter from nerve endings and a hormone from the adrenal gland. It is released in times of stress and is involved in hunger regulation, blood glucose regulation, and other body processes.

epinephrine A hormone also known as *adrenaline;* it is released by the adrenal glands (located on each kidney) at times of stress. It acts to increase glycogen breakdown in the liver, among other functions.

myelin A lipid and protein combination (lipoprotein) that covers nerve fibers.

endocrine system The body system consisting of the various glands and the hormones these glands secrete. This system has major regulatory functions in the body, such as reproduction and cell metabolism.

endocrine gland A hormone-producing gland.

A true hormone is a regulatory compound that has a specific site of synthesis from which it then enters the bloodstream to reach target cells. Hormones are the messengers of the body. They can be permissive (turn on), antagonistic (turn off), or synergistic (work in cooperation with another hormone) in performing a task. Some compounds must undergo chemical changes before they can function as hormones. For example, vitamin D, synthesized in the skin or obtained from food, is converted into an active hormone by chemical changes made in the liver and kidneys.

The hormone **insulin,** synthesized in and released from the pancreas, helps control the amount of glucose in the blood (Fig. 3-7). Insulin is mostly produced when glucose in the blood rises to a certain level, usually after a meal. At this point, insulin is released and travels to the muscle, adipose, and liver cells of the body. Among its many functions, insulin allows for the movement of glucose from the blood into muscle and adipose cells. In the liver cells, insulin causes an increase in stored glycogen by stimulating the synthesis of glycogen from glucose. Once a sufficient amount of glucose has been cleared from the blood, the production of insulin lessens. The hormones epinephrine, norepinephrine, glucagon, and growth hormone have just the opposite effect on blood glucose. They all cause an increase in blood glucose through a variety of actions (Table 3-1). **Thyroid hormones,** synthesized in and released from the thyroid gland, help to control the body's rate of metabolism. Other hormones are especially important in regulating digestive processes (see Section 3.9, "Digestive System").

insulin A hormone produced by the pancreas. Among other processes, insulin increases the synthesis of glycogen in the liver and the movement of glucose from the bloodstream into body cells.

thyroid hormones Hormones produced by the thyroid gland that regulate growth and metabolic rate.

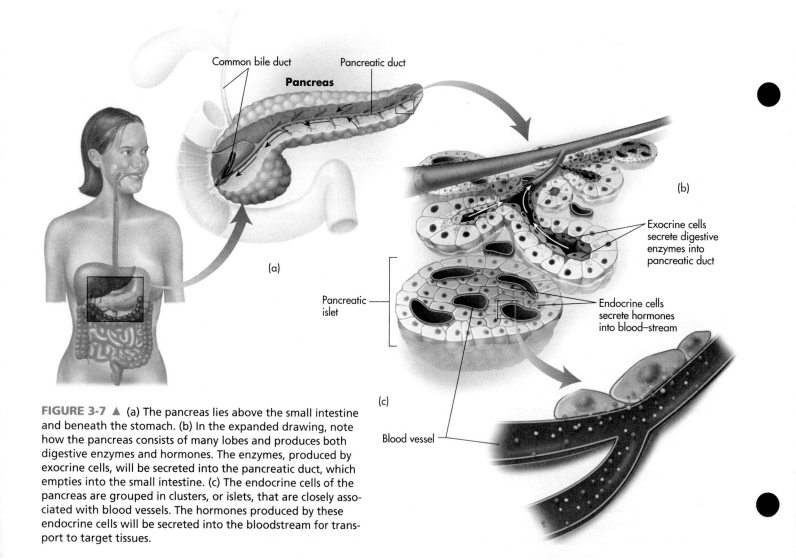

FIGURE 3-7 ▲ (a) The pancreas lies above the small intestine and beneath the stomach. (b) In the expanded drawing, note how the pancreas consists of many lobes and produces both digestive enzymes and hormones. The enzymes, produced by exocrine cells, will be secreted into the pancreatic duct, which empties into the small intestine. (c) The endocrine cells of the pancreas are grouped in clusters, or islets, that are closely associated with blood vessels. The hormones produced by these endocrine cells will be secreted into the bloodstream for transport to target tissues.

TABLE 3-1 ▶ Some Hormones of the Endocrine System with Nutritional Significance

Hormone	Gland/Organ	Target	Effect	Role in Nutrition
Insulin	Pancreas	Adipose, muscle, and liver cells	Decreased blood glucose	Uptake and storage of glucose, fat, and amino acids by cells
Glucagon	Pancreas	Liver	Increased blood glucose	Release of glucose from liver stores, release of fat from adipose tissue
Epinephrine, Norepinephrine	Adrenal glands	Heart, blood vessels, brain, lungs	Increased body metabolism and blood glucose	Release of glucose and fat into the blood
Growth hormone	Pituitary gland	Most cells	Promotion of amino acid uptake by cells, increased blood glucose	Promotion of protein synthesis and growth, increased fat use for energy
Thyroid hormones	Thyroid gland	Most organs	Increased oxygen consumption, overall growth, brain development of the nervous system	Protein synthesis, increased body metabolism

Hormones are not taken up by all cells in the body but only by those with the correct **receptor** protein. These binding sites, which generally are found on the cell membrane, are highly specific for a certain hormone. The hormone attaches to its receptor on the cell membrane. This binding activates additional compounds called second messengers within the cell to carry out the assigned task. This is true of insulin. A few hormones can penetrate the cell membrane and eventually bind to receptors on the DNA in the nucleus (e.g., thyroid hormone and estrogen).

receptor A site in a cell at which compounds (such as hormones) bind. Cells that contain receptors for a specific compound are partially controlled by that compound.

☑ CONCEPT CHECK 3.7

1. Examine Figure 3-7. How does the pancreas have both endocrine and exocrine roles?
2. What effect does insulin have on the storage of nutrients?
3. List at least three hormones that tend to increase blood sugar.
4. If a person has hypothyroidism, the thyroid gland produces low levels of thyroid hormone. Will a person with hypothyroidism tend to lose weight or gain weight? Explain your answer.

3.8 Immune System

The immune system provides a very clear example of the interrelationship between nutrition status and organ system function. In developing nations, where food shortages are common, malnutrition increases susceptibility to infectious diseases, such as diarrheal disease. Cells throughout the body—skin and intestinal cells—work in concert with the cells and tissues of the immune system to defend the body against infection.

We are born with some aspects of immune function, such as physical and chemical barriers against infection, the inflammatory response, and the ability of some **white blood cells** to engulf microorganisms by **phagocytosis.** These are termed **nonspecific** (or innate) **immunity** because they protect the body against invasion by any microorganism. The skin and the intestinal cells support the immune system by forming an important barrier against invading microorganisms. If the integrity of either one of these barriers is compromised, microorganisms can invade the body and cause illness. Substances secreted by the skin and intestinal cells can also destroy pathogens.

If the body's nonspecific immune defenses are unable to block a microorganism's entry into the bloodstream, cells and chemicals involved in **specific** (or adaptive) **immunity** will identify and destroy the invading pathogen. Specific immunity

white blood cells One of the formed elements of the circulating blood system; also called *leukocytes*. White blood cells are able to squeeze through intracellular spaces and migrate. They phagocytize bacteria, fungi, and viruses, as well as detoxify proteins that may result from allergic reactions, cellular injury, and other immune system cells.

phagocytosis Process in which a cell forms an indentation, and particles or fluids enter the indentation and are engulfed by the cell.

nonspecific immunity Defenses that stop the invasion of pathogens; requires no previous encounter with a pathogen; also called *innate immunity.*

specific immunity Function of white blood cells directed at specific antigens; also called *adaptive immunity.*

antibody Blood protein that binds foreign proteins found in the body; also called *immunoglobulin*. This helps to prevent and control infections.

antigen Any substance that induces a state of sensitivity and/or resistance to microorganisms or toxic substances after a lag period; foreign substance that stimulates a specific aspect of the immune system.

involves the process by which certain white blood cells produce **antibodies** (also called **immunoglobulins**) that target specific microorganisms or foreign proteins (known as **antigens**). After initial exposure to an antigen, a "memory" is created such that a second exposure to the substance will produce a more vigorous and rapid attack.

The turnover of many cells of the immune system is quite rapid—only a few hours or days. The constant resynthesis of cells requires steady nutrient intake. Nutrients that are important for the health of the immune system include protein, essential fatty acids, iron, copper, zinc, vitamins A, C, and D, and some B vitamins.

✅ CONCEPT CHECK 3.8

1. Contrast nonspecific (innate) and specific (adaptive) immunity.
2. What are the roles of antigens and antibodies in the immune response?
3. List three nutrients that support the immune system.

3.9 Digestive System

digestive system System consisting of the GI tract and accessory structures (liver, gallbladder, and pancreas). This system performs the mechanical and chemical processes of digestion, absorption of nutrients, and elimination of wastes.

digestion Process by which large ingested molecules are mechanically and chemically broken down to produce basic nutrients that can be absorbed across the wall of the GI tract.

absorption The process by which substances are taken up from the GI tract and enter the bloodstream or the lymph.

gastrointestinal (GI) tract The main sites in the body used for digestion and absorption of nutrients. It consists of the mouth, esophagus, stomach, small intestine, large intestine, rectum, and anus. Also called the *digestive tract.*

lumen The hollow opening inside a tube, such as the GI tract.

motility Generally, the ability to move spontaneously. It also refers to movement of food through the GI tract.

The foods and beverages we consume, for the most part, must undergo extensive alteration by the **digestive system** to provide us with usable nutrients. The processes of **digestion** and **absorption** take place in a long tube that is open at both ends and extends from the mouth to the anus. This tube is called the **gastrointestinal (GI) tract** (Fig. 3-8). The open space inside the GI tract is called the **lumen.** Nutrients from the food we eat must pass through the walls of the GI tract—from the lumen through the cells lining the GI tract—to be absorbed into the bloodstream. The organs that make up the GI tract, as well as some additional accessory organs located nearby, are collectively known as the digestive system.

In the digestive system, food is broken down mechanically and chemically. Mechanical digestion takes place as soon as you begin chewing your food and continues as muscular contractions simultaneously mix and move food through the length of the GI tract (as part of a process known as **motility**).

Chemical digestion refers to the chemical breakdown of foods by acid and enzymes secreted into the GI tract. Enzymes are a key part of digestion. Each enzyme is specific to one type of chemical process. For example, the enzyme that recognizes and digests table sugar (sucrose) ignores milk sugar (lactose). Besides working on only specific types of chemicals, enzymes are sensitive to acidic and alkaline conditions, temperature, and the types of vitamins and minerals they require to function. Digestive enzymes that work in the acidic environment of the stomach do not work well in the alkaline environment of the small intestine. The pancreas and small intestine produce most of the digestive enzymes; however, the mouth and the stomach also contribute their own enzymes to the process of digestion. The organs of the digestive system are able to fine-tune the production of each type of digestive enzyme in response to the nutritional makeup and amount of food consumed. Overall, the enzymes of the digestive system work together to hasten the breakdown of ingested food into absorbable nutrients (Fig. 3-9).

As food moves along the GI tract, nutrients are absorbed and the waste products reach the large intestine. In addition to nutrients from the food we eat, the bacteria that live in the large intestine produce some vitamins (vitamin K and biotin) that can be absorbed. The final role of the digestive system is elimination of wastes.

Most of the processes of digestion and absorption are under autonomic control; that is, they are involuntary. Almost all of the functions involved in digestion and absorption are controlled by signals from the nervous system, hormones from the endocrine system, and hormonelike compounds. Many common ailments arise from problems with the digestive system. Several of these digestive problems are discussed in the Nutrition and Your Health section at the end of this chapter.

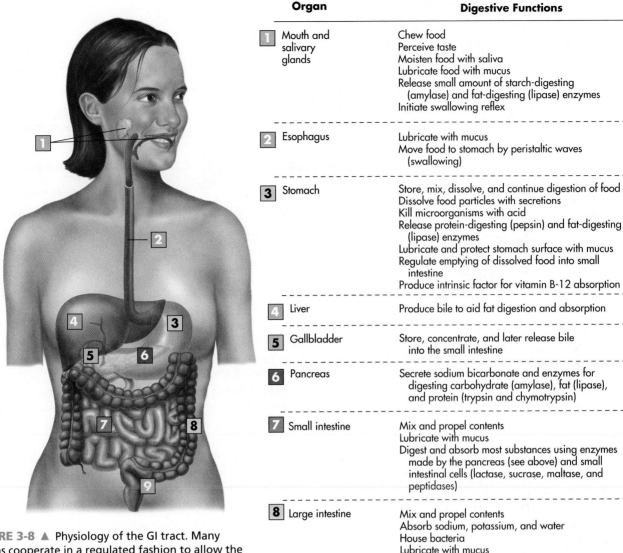

Organ	Digestive Functions
1 Mouth and salivary glands	Chew food Perceive taste Moisten food with saliva Lubricate food with mucus Release small amount of starch-digesting (amylase) and fat-digesting (lipase) enzymes Initiate swallowing reflex
2 Esophagus	Lubricate with mucus Move food to stomach by peristaltic waves (swallowing)
3 Stomach	Store, mix, dissolve, and continue digestion of food Dissolve food particles with secretions Kill microorganisms with acid Release protein-digesting (pepsin) and fat-digesting (lipase) enzymes Lubricate and protect stomach surface with mucus Regulate emptying of dissolved food into small intestine Produce intrinsic factor for vitamin B-12 absorption
4 Liver	Produce bile to aid fat digestion and absorption
5 Gallbladder	Store, concentrate, and later release bile into the small intestine
6 Pancreas	Secrete sodium bicarbonate and enzymes for digesting carbohydrate (amylase), fat (lipase), and protein (trypsin and chymotrypsin)
7 Small intestine	Mix and propel contents Lubricate with mucus Digest and absorb most substances using enzymes made by the pancreas (see above) and small intestinal cells (lactase, sucrase, maltase, and peptidases)
8 Large intestine	Mix and propel contents Absorb sodium, potassium, and water House bacteria Lubricate with mucus Synthesize some vitamins and short-chain fatty acids Form feces
9 Rectum	Hold feces and expel via the anus, which is the opening to the outside of the body

FIGURE 3-8 ▲ Physiology of the GI tract. Many organs cooperate in a regulated fashion to allow the digestion and absorption of nutrients in foods. Partially digested food spends about 2 to 3 hours in the stomach (longer for large meals). Passage through the small intestine takes 3 to 10 hours, followed by up to 72 hours in the large intestine. On average, digestion and absorption of a meal takes about 2 days. Food matter tends to pass more quickly through the GI tract of men than women.

The digestive system is composed of six separate organs; each organ performs one (or more) specific job(s). Let's look briefly at the role of each organ. These organs are shown in Figure 3-8. More detailed descriptions of digestive processes will be provided in later chapters as each nutrient is introduced.

MOUTH

The mouth performs many functions in the digestion of food. Besides chewing food to reduce it to smaller particles, the mouth also senses the taste of the foods we consume. The tongue, through the use of its taste buds, identifies foods on the basis of their specific flavor(s). Sweet, sour, salty, bitter, and **umami** comprise the primary taste sensations we experience. Surprisingly, the nose and our sense of smell greatly contribute to our ability to sense the taste of food. When we chew a food, chemicals

umami A brothy, meaty, savory flavor in some foods. Monosodium glutamate enhances this flavor when added to foods.

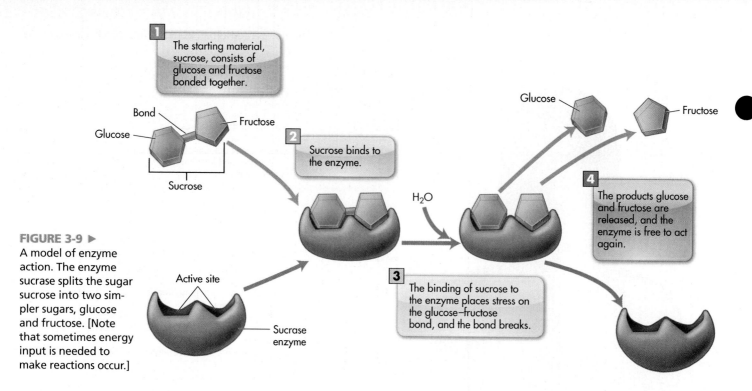

FIGURE 3-9 ▶ A model of enzyme action. The enzyme sucrase splits the sugar sucrose into two simpler sugars, glucose and fructose. [Note that sometimes energy input is needed to make reactions occur.]

1 The starting material, sucrose, consists of glucose and fructose bonded together.

Bond
Glucose
Fructose
Sucrose

2 Sucrose binds to the enzyme.

H_2O

3 The binding of sucrose to the enzyme places stress on the glucose–fructose bond, and the bond breaks.

4 The products glucose and fructose are released, and the enzyme is free to act again.

Glucose
Fructose

Active site
Sucrase enzyme

saliva Watery fluid, produced by the salivary glands in the mouth, that contains lubricants, enzymes, and other substances.

amylase A starch-digesting enzyme produced by salivary glands and the pancreas.

lipase Fat-digesting enzyme produced by the salivary glands, stomach, and pancreas.

mucus A thick fluid secreted by many cells throughout the body. It contains a compound that has both carbohydrate and protein parts. It acts as a lubricant and means of protection for cells.

are released that stimulate the nasal passages. Thus, it makes perfect sense that when we have a cold and our noses are stuffed up and congested, even our favorite foods will not taste as good as they normally do.

The taste of food, or the anticipation of it, signals the rest of the GI tract to prepare for the digestion of food. Once in the mouth, mechanical and chemical digestion begins. Salivary glands produce **saliva,** which functions as a solvent so that food particles can be further separated and tasted. In addition, saliva contains a starch-digesting enzyme, salivary **amylase** (see Chapter 4 for more on starch-digesting enzymes) and a fat-digesting enzyme, **lipase** (see Chapter 5). **Mucus,** another component of saliva, makes it easy to swallow a mouthful of food. The food then travels to the esophagus. The important secretions and products of digestion are listed in Table 3-2.

TABLE 3-2 ▶ **Important Secretions of the Digestive Tract**

Secretion	Site of Production	Purpose
Saliva	Mouth	• Contains enzymes that make a minor contribution to starch and fat digestion • Lubrication of food for swallowing
Mucus	Mouth, esophagus, stomach, small intestine, large intestine	• Protects GI tract cells • Lubricates food as it travels through the GI tract
Enzymes	Mouth, stomach, small intestine, pancreas	• Promote digestion of carbohydrates, fats, and proteins into forms small enough for absorption (examples: amylases, lipases, proteases)
Acid	Stomach	• Promotes digestion of protein • Destroys pathogens • Solubilizes some minerals • Activates some enzymes
Bile	Liver (stored in gallbladder)	• Aids fat digestion in the small intestine by suspending fat in water using **bile acids, cholesterol,** and **lecithin**
Bicarbonate	Pancreas, small intestine	• Neutralizes stomach acid when it reaches the small intestine
Hormones	Stomach, small intestine, pancreas	• Stimulate production and/or release of acid, enzymes, bile, and bicarbonate • Help regulate peristalsis and overall GI tract flow (examples: gastrin, secretin, insulin, cholecystokinin, glucagon)
Intrinsic factor	Stomach	• Facilitates absorption of vitamin B-12 in the small intestine

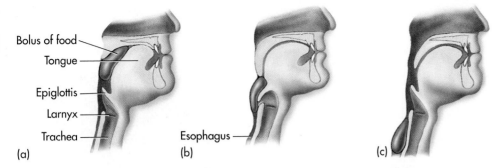

Bolus of food
Tongue
Epiglottis
Larnyx
Trachea
(a)

Esophagus
(b)

(c)

FIGURE 3-10 ▲ The process of swallowing. (a) During swallowing, food does not normally enter the trachea because the epiglottis closes over the larynx. (b) The closed epiglottis allows food to proceed down the esophagus. When a person chokes, food becomes lodged in the trachea, blocking airflow to the lungs. (c) The food moves down the esophagus and the airway reopens.

ESOPHAGUS

The **esophagus** is a long tube that connects the **pharynx** with the stomach. Near the pharynx is a flap of tissue (called the **epiglottis**) that prevents the **bolus** of swallowed food from entering the trachea (wind pipe) (Fig. 3-10). During swallowing, food lands on the epiglottis, folding it down to cover the opening of the trachea. Breathing also stops automatically. These responses ensure that swallowed food will only travel down the esophagus. If food instead travels down the trachea, choking may occur (the victim will not be able to speak, cough, or breathe). A group of techniques to treat such a person is called the Heimlich maneuver (see **www.heimlichinstitute.org** for details).

At the top of the esophagus, nerve fibers release signals to tell the GI tract that food has been consumed. This results in an increase in GI muscle action, called **peristalsis.** These continual waves of muscle contractions, followed by muscle relaxation, force the food along the digestive tract from the esophagus onward (Fig. 3-11).

At the end of the esophagus is the **lower esophageal sphincter,** a muscle that constricts (closes) after food enters the stomach. The main function of sphincters is to prevent the backflow of GI tract contents. Sphincters respond to various stimuli, such as signals from the nervous system, hormones, acidic versus alkaline conditions, and pressure that builds up around the sphincter. The primary function of the lower esophageal sphincter is to prevent the acidic contents of the stomach from flowing back up into the esophagus. Dysfunction of this sphincter can cause some of the health problems we will discuss in the Nutrition and Your Health section at the end of this chapter.

No digestion or absorption occurs in the esophagus; it serves merely to transport food from the mouth to the stomach. The cells of the esophagus secrete mucus to lubricate the passage of food, but no digestive enzymes are produced.

STOMACH

The stomach is a large sac that can hold up to 4 cups (or 1 quart) of food for several hours until all of the food is able to enter the small intestine. Stomach size varies individually and can be reduced surgically as a radical treatment for obesity (more on this in Chapter 7). While in the stomach, the food is mixed with gastric juice, which contains water, hydrochloric acid, and enzymes. (*Gastric* is a term pertaining to the stomach.) The acid in the gastric juice destroys the biological activity of proteins, converts inactive digestive enzymes to their active form, partially digests food protein, and makes dietary minerals soluble so that they can be absorbed. The mixing that takes place in the stomach produces a watery food mixture, called **chyme,** which slowly leaves the stomach a teaspoon (5 milliliters) at a time and enters the small intestine. Following a meal, the stomach contents are emptied into the small intestine over the course of 1 to 4 hours. The **pyloric sphincter,** located at the base of the stomach, controls the rate

protease Protein-digesting enzyme produced by the stomach, small intestine, and pancreas.

esophagus A tube in the GI tract that connects the pharynx with the stomach.

pharynx The organ of the digestive tract and respiratory tract located at the back of the oral and nasal cavities, commonly known as the throat.

epiglottis The flap that folds down over the trachea during swallowing.

bolus A moistened mass of food swallowed from the oral cavity into the pharynx.

peristalsis A coordinated muscular contraction used to propel food down the GI tract.

lower esophageal sphincter A circular muscle that constricts the opening of the esophagus to the stomach. Also called the *gastroesophageal sphincter* or the *cardiac sphincter.*

chyme A mixture of stomach secretions and partially digested food.

pyloric sphincter Ring of smooth muscle between stomach and small intestine.

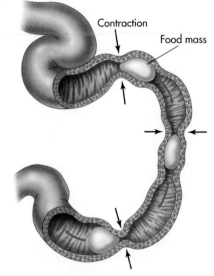

Contraction
Food mass

FIGURE 3-11 ▲ Peristalsis. Peristalsis is a progressive type of movement, propelling material from point to point along the GI tract. To begin this, a ring of contraction occurs where the GI wall is stretched, passing the food mass forward. The moving food mass triggers a ring of contraction in the next region, which pushes the food mass even farther along. The result is a ring of contraction that moves like a wave along the GI tract, pushing the food mass down the tract.

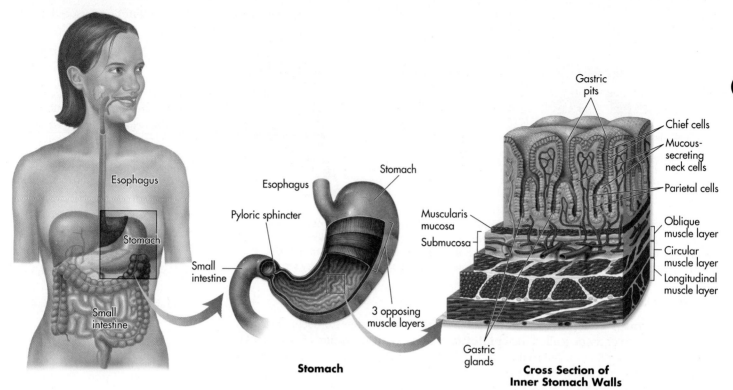

FIGURE 3-12 ▲ Physiology of the stomach. The interior surface mucous cells produce mucus for protection from stomach acid and enzymes. Parietal cells produce the hydrochloric acid (HCl), and chief cells produce the enzymes. Mucous neck cells, scattered among the cells in the gastric pits, also produce mucus. The exterior surface is made up of three layers of muscle that facilitate mechanical digestion.

intrinsic factor A proteinlike compound produced by the stomach that enhances vitamin B-12 absorption.

at which the chyme is released into the small intestine (Fig. 3-12). There is very little absorption of nutrients from the stomach, except for some water and alcohol.

You might wonder how the stomach prevents itself from being digested by the acid and enzymes it produces. First, the stomach has a thick layer of mucus that lines and protects it. The production of acid and enzymes also requires the release of a specific hormone (gastrin). This release happens primarily when we are eating or thinking about eating. Last, as the concentration of acid in the stomach increases, hormonal control causes acid production to taper off.

One other important function of the stomach is the production of a substance called **intrinsic factor.** This vital proteinlike compound is essential for the absorption of vitamin B-12.

SMALL INTESTINE

duodenum First segment of the small intestine that receives chyme from the stomach and digestive juices from the pancreas and gallbladder. This is the site of most chemical digestion of nutrients; approximately 10 inches in length.

jejunum Middle segment of the small intestine; approximately 4 feet in length.

ileum Last segment of the small intestine; approximately 5 feet in length.

The small intestine is considered "small" because of its narrow (1 inch [2.5 centimeters]) diameter. It is actually quite long—about 10 feet (3 meters), beginning at the stomach and extending to the large intestine (Fig. 3-13). The three parts of the small intestine are the **duodenum** (first 10 inches), the **jejunum** (second 4 feet), and the **ileum** (last 5 feet). Most of the digestion and absorption of food occurs in the small intestine. As chyme moves from the stomach into the first part of the small intestine, it is still very acidic. You just learned that the stomach secretes a thick layer of mucus to protect itself from the strong acid. However, if the small intestine were coated with mucus, digestion and absorption would be very limited. Therefore, the pancreas and intestinal cells secrete bicarbonate to neutralize the acid. The neutral pH also optimizes the activity of the digestive enzymes that work in the small intestine. The chyme is moved through the small intestine by peristaltic contractions so that it can be well mixed with the digestive juices of the small intestine (review Fig. 3-11). These juices contain many enzymes that function in the breakdown of carbohydrates, protein, and fat, as well as in the preparation of vitamins and minerals for absorption.

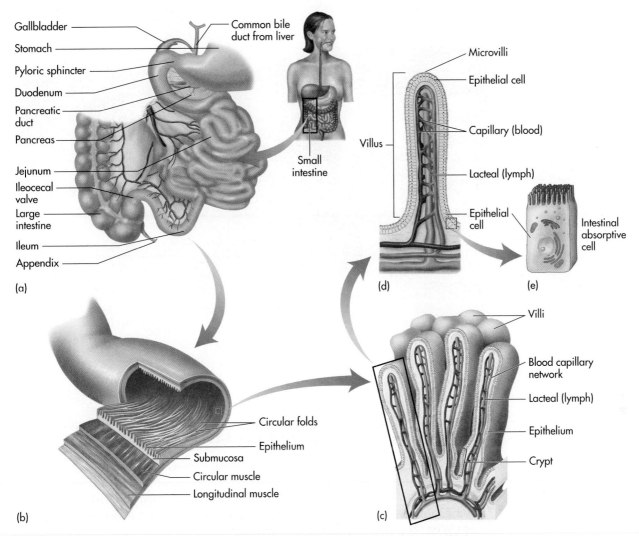

Gallbladder
Stomach
Pyloric sphincter
Duodenum
Pancreatic duct
Pancreas
Jejunum
Ileocecal valve
Large intestine
Ileum
Appendix

Common bile duct from liver

Small intestine

(a)

Circular folds
Epithelium
Submucosa
Circular muscle
Longitudinal muscle

(b)

Microvilli
Epithelial cell
Villus
Capillary (blood)
Lacteal (lymph)
Epithelial cell

Intestinal absorptive cell

(d)

(e)

Villi
Blood capillary network
Lacteal (lymph)
Epithelium
Crypt

(c)

FIGURE 3-13 ▲ Organization of the small intestine. (a) The three parts of the small intestine are the duodenum (10 in), jejunum (4 ft), and ileum (5 ft). (b) Several layers of muscle work together to mix and propel chyme through the small intestine. (c) The highly folded lining of the small intestine increases the surface area for absorption up to 600 times that of a simple tube. (d) The villi are covered with absorptive cells. Capillaries and lacteals inside each villus transport nutrients from the absorptive cells of the small intestine through the blood and lymph. (e) The mucosal side of the absorptive cells is covered with microvilli, which maximize nutrient absorption.

The physical structure of the small intestine is very important to the body's ability to digest and absorb the nutrients it needs. The lining of the small intestine is called the mucosa and is folded many times; within these folds are fingerlike projections called **villi.** These "fingers" are constantly moving, which helps them trap food to enhance absorption. Each individual villus (singular) is made up of many **absorptive cells** (also called enterocytes), and the mucosal surface of each of these cells is folded even further into **microvilli.** The combined folds, villi, and microvilli in the small intestine increase its surface area 600 times beyond that of a simple tube (Fig. 3-13).

The absorptive cells have a short life. New intestinal absorptive cells are constantly produced in the crypts of the small intestinal lining (Fig. 3-13) and appear daily along the surface of each villus "finger." This is probably because absorptive cells are subjected to a harsh environment, so renewal of the intestinal cell lining is necessary. This rapid cell turnover leads to high nutrient needs for the small intestine. Fortunately, many of the old cells can be broken down and have their component parts reused. The health of the cells is further enhanced by various hormones and other substances that participate in or are produced as part of the digestive process.

villi (singular, villus) The fingerlike protrusions into the small intestine that participate in digestion and absorption of food.

absorptive cells Also known as *enterocytes;* intestinal cells that line the villi and participate in nutrient absorption.

microvilli Extensive folds on the mucosal surface of the absorptive cells.

FIGURE 3-14 ▶ Nutrient absorption relies on four major absorptive processes. (1) Passive diffusion (in blue) is diffusion of nutrients across the absorptive cell membranes. (2) Facilitated diffusion (in green) uses a carrier protein to move nutrients down a concentration gradient. (3) Active absorption (in purple) involves a carrier protein as well as energy to move nutrients (against a concentration gradient) into absorptive cells. (4) Phagocytosis and pinocytosis (in green and brown) are forms of active transport in which the absorptive cell membrane forms an indentation that engulfs a nutrient to bring it into the cell.

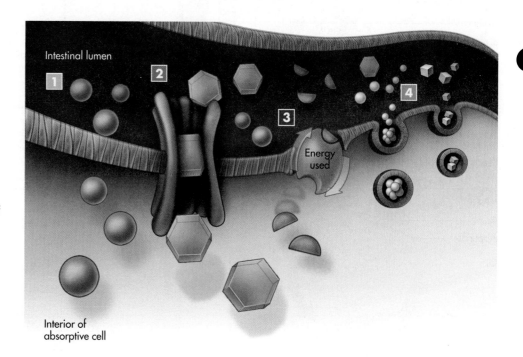

The small intestine absorbs nutrients through the intestinal wall through various means and processes, as illustrated in Figure 3-14:

- **Passive diffusion:** When the nutrient concentration is higher in the lumen of the small intestine than in the absorptive cells, the difference in nutrient concentration drives the nutrient into the absorptive cells by diffusion. Fats, water, and some minerals are examples of nutrients that move down a concentration gradient to be absorbed by passive diffusion.
- **Facilitated diffusion:** Some compounds require a carrier protein to follow a concentration gradient into absorptive cells. This type of absorption is called faciliated diffusion. Fructose is one example of a compound that makes use of such a carrier to allow for facilitated diffusion.
- **Active absorption:** In addition to the need for a carrier protein, some nutrients also require energy input to move from the lumen of the small intestine into the absorptive cells. This mechanism makes it possible for cells to take up nutrients even when they are consumed in low concentrations (i.e., against a concentration gradient). Some sugars, such as glucose, are actively absorbed, as are amino acids.
- **Phagocytosis** and **pinocytosis:** In a further means of active absorption, absorptive cells literally engulf compounds (phagocytosis) or liquids (pinocytosis). A cell membrane can form an indentation of itself so that when particles or fluids move into the indentation, the cell membrane surrounds and engulfs them. This process is used when an infant absorbs immune substances from human milk (see Chapter 14).

Once absorbed, water-soluble compounds such as glucose and amino acids are transported by the capillaries to the portal vein. Recall that the liver is the end of this process. Most fats are absorbed into the lymph vessels, which eventually empty into the bloodstream (review Figs. 3-3 and 3-4).

Undigested food cannot be absorbed into cells of the small intestine. Any undigested food that reaches the end of the small intestine must pass through the **ileocecal sphincter** on the way to the large intestine (Fig. 3-15). This sphincter prevents the contents of the large intestine from reentering the small intestine.

ileocecal sphincter The ring of smooth muscle between the end of the small intestine and the beginning of the large intestine.

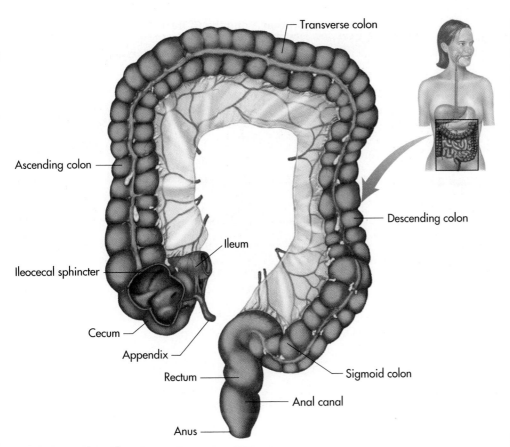

FIGURE 3-15 ◄ The parts of the large intestine include the cecum, ascending colon, transverse colon, descending colon, and sigmoid colon. Overall, the large intestine is about 3½ feet (1.1 meters) long.

LARGE INTESTINE

When the contents of the small intestine enter the large intestine, the material left bears little resemblance to the food originally eaten. Under normal circumstances, only a minor amount (5%) of carbohydrate, protein, and fat escapes absorption to reach the large intestine (Table 3-3).

The large intestine (sometimes called the *colon)* can be subdivided into five main segments: the **cecum, ascending colon, transverse colon, descending colon,** and **sigmoid colon** (Fig. 3-15). Physiologically, the large intestine differs from the small intestine in that there are no villi or digestive enzymes. The absence of villi means that little absorption takes place in the large intestine in comparison to the small intestine. Nutrients absorbed from the large intestine include water, some vitamins, some fatty acids, and the minerals sodium and potassium. Unlike the small intestine, the large intestine has a number of mucus-producing cells. The mucus secreted by these cells functions to hold the feces together and protect the large intestine from the bacterial activity within it.

The large intestine is home to a large population of bacteria (over 500 different species). While the stomach and small intestine have some bacterial activity, the large intestine is the organ most heavily colonized with bacteria. Starting at infancy, the diet plays a major part in determining the type of bacteria in our digestive tracts. The number and type of bacteria in the human colon recently has become of great interest. Research has shown that intestinal bacteria play a significant role in the maintenance of health, especially health of the colon. It is speculated that higher levels of beneficial organisms can reduce the activity of disease-causing bacteria. This is another illustration of the intestinal tract working as an important immune organ. The strains *bifidobacteria* and *lactobacilli* are typically associated with health, whereas *clostridia* are considered problematic. Bacteria in the large intestine are able to break down some of the remaining food products that enter

CRITICAL THINKING

The authors of some popular (fad) diet books contend that eating certain combinations of foods, such as meats and fruits together, hinders the digestive processes. Based on what you have learned so far about the enzymes and digestion, does this sound accurate?

Newsworthy Nutrition

Link between gut bacteria and health and disease

The benefits of probiotics and prebiotics have been confirmed in over 700 research studies. Probiotics, especially, are now recommended for the prevention and treatment of gastrointestinal tract disorders, including inflammation, infections, and allergy. Evidence is available to support the growing interest in these microorganisms and to aid the development of intervention strategies and practical guidelines for their use.

Source: Wallace TC and others: Human gut microbiota and its relationship to health and disease. *Nutrition Reviews* 69(7):392, 2011 (see Further Reading 16).

connect
NUTRITION

Check out the Connect site **www.mcgrawhillconnect.com** to further explore probiotics.

probiotic Product that contains specific types of bacteria. Use is intended to colonize the large intestine with the specific bacteria in the product. An example is yogurt.

prebiotic Substance that stimulates bacterial growth in the large intestines.

feces Mass of water, fiber, tough connective tissues, bacterial cells, and sloughed intestinal cells that passes through the large intestine and is excreted through the anus; also called stool.

rectum Terminal portion of the large intestine.

anus Last portion of the GI tract; serves as an outlet for the digestive system.

anal sphincters A group of two sphincters (inner and outer) that help control expulsion of feces from the body.

TABLE 3-3 ▶ Major Sites of Absorption Along the GI Tract

Organ	Primary Nutrients Absorbed
Stomach	Alcohol (20% of total)
	Water (minor amount)
Small intestine	Calcium, magnesium, iron, and other minerals
	Glucose
	Amino acids
	Fats
	Vitamins
	Water (70% to 90% of total)
	Alcohol (80% of total)
	Bile acids
Large intestine	Sodium
	Potassium
	Some fatty acids
	Gases
	Water (10% to 30% of total)

the large intestine, such as the milk sugar lactose (in lactose intolerant people), and some components of fiber. Some of the products of bacterial metabolism in the large intestine, which include various fatty acids and gases, can then be absorbed.

Foods containing certain live microorganisms such as *lactobacilli* have been linked to some health benefits, such as improving intestinal tract health. These microorganisms are called **probiotics** because once consumed, they take up residence in the large intestine and confer health benefits. You can find these probiotic microorganisms in certain forms of fluid milk, fermented milk, and yogurt, and in pill form (see Further Reading 16). A related term is **prebiotic.** These are substances that increase growth of probiotic microorganisms. One example is fructooligosaccharides (see Table 1-3 in Chapter 1 for dietary sources). The beneficial organisms of the large intestine and their use as probiotics are highlighted in "Newsworthy Nutrition" (see margin).

Some water remains in the material that enters the large intestine because the small intestine absorbs only 70% to 90% of the fluid it receives, which includes large amounts of GI-tract secretions produced during digestion. The remnants of a meal also contain some minerals and some fiber. Because water is removed from the large intestine, its contents become semisolid by the time they have passed through the first two-thirds of it. What remains in the **feces,** besides water and undigested fiber, is tough connective tissues (from animal foods); bacteria from the large intestine; and some body wastes, such as parts of dead intestinal cells.

RECTUM

The feces or stool remains in the last portion of the large intestine, the **rectum,** until muscular movements push it into the **anus** to be eliminated. The presence of feces in the rectum stimulates elimination. The anus contains two **anal sphincters** (internal and external), one of which is under voluntary control (external sphincter). Relaxation of this sphincter allows for elimination.

ACCESSORY ORGANS

The liver, **gallbladder,** and pancreas work with the GI tract and are considered accessory organs to the process of digestion (review Fig. 3-8). These accessory organs are not part of the GI tract through which food passes, but they play necessary roles in the process of digestion. These organs secrete digestive fluids into the GI tract and enable the process of converting food into absorbable nutrients.

The liver produces a substance called **bile.** The bile is stored and concentrated in the gallbladder until the gallbladder receives a hormonal signal to release the bile. This signal is induced by the presence of fat in the small intestine. Bile is released and delivered to the duodenum via a tube called the bile duct (Fig. 3-16).

In action, bile is like soap. Components of the bile enable large portions of fat to break into smaller bits so that they can be suspended in water (Chapter 5 will cover this process in detail). Interestingly, some of the bile constituents can be "recycled" in a process known as **enterohepatic circulation.** These components of bile are reabsorbed from the small intestine, returned to the liver via the portal vein, and reused.

In addition to bile, the liver releases a number of other unwanted substances that travel with the bile to the gallbladder and end up in the small intestine and eventually in the large intestine for excretion. The liver functions in this manner to remove unwanted substances from the blood. (Other by-products are excreted via the urine; see Section 3.5, "Urinary System.")

The pancreas has both endocrine and digestive functions. As a gland of the endocrine system, the pancreas manufactures hormones—insulin and glucagon—that are secreted into the blood to regulate blood glucose levels (review Fig. 3-7). As an organ of the digestive system, it produces "pancreatic juice," a mixture of water, bicarbonate, and a variety of digestive enzymes capable of breaking apart carbohydrates, proteins, and fats into small fragments. Bicarbonate is a base that neutralizes the acidic chyme as it moves from the stomach into the duodenum. As noted earlier, the small intestine does not have a protective layer of mucus because mucus would impede nutrient absorption. Instead, the neutralizing capacity of bicarbonate from the pancreas protects the walls of the small intestine from erosion by acid, which would otherwise lead to the formation of an ulcer (see the Nutrition and Your Health section at the end of this chapter).

gallbladder An organ attached to the underside of the liver; site of bile storage, concentration, and eventual secretion.

bile A liver secretion stored in the gallbladder and released through the common bile duct into the first segment of the small intestine. It is essential for the digestion and absorption of fat.

enterohepatic circulation A continual recycling of compounds such as bile acids between the small intestine and the liver.

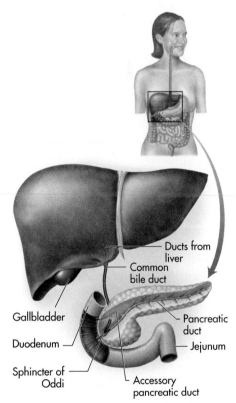

FIGURE 3-16 ▲ Bile is produced by the liver and stored in the gallbladder. The pancreas produces pancreatic juice, which contains water, bicarbonate, and digestive enzymes. When chyme reaches the small intestine, bile is released from the gallbladder through the common bile duct. In addition, pancreatic juices flow through the pancreatic duct, which joins the common bile duct at the sphincter of Oddi. Relaxation of this sphincter releases both bile and pancreatic juice into the duodenum to facilitate digestion.

✔ CONCEPT CHECK 3.9

1. Choose three secretions of the digestive system. Where is each secreted? What is the role of each in the process of digestion?
2. What are enzymes? Is bile an enzyme?
3. How do mucus and surface area affect absorption?
4. Which absorptive processes use energy? How does *concentration gradient* factor into this?
5. What did you have for lunch today? Trace the path of your meal through the digestive system. Where is each of the nutrients broken down as it passes through the GI tract? Where is each absorbed?

3.10 Nutrient Storage Capabilities

The human body must maintain reserves of nutrients; otherwise, we would need to eat continuously. Storage capacity varies for each different nutrient. Most fat is stored in adipose tissue, made up of cells designed specifically for this. Short-term storage of carbohydrate occurs in muscle and liver in the form of glycogen. The

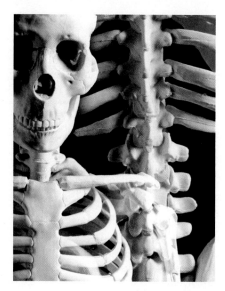

▲ The skeletal system provides a reserve of calcium for day-to-day needs when dietary intake is inadequate. Long-term use of this reserve, however, reduces bone strength.

epigenome The way that the genome is marked and packaged inside the cell nucleus.

epigenetics Changes in gene expression caused by mechanisms other than changes in the DNA sequence.

A genetic variation can directly affect the proteins encoded by our genes and result in different:

- nutrient requirements among individuals.
- susceptibilities to diseases.
- effects of environmental factors (such as our diet) on our genes and the proteins they make.

blood maintains a small reserve of glucose and amino acids. Many vitamins and minerals are stored in the liver, whereas other nutrient stores are found in other sites in the body.

When people do not meet certain nutrient needs, these nutrients are obtained by breaking down a tissue that contains high concentrations of the nutrient. For example, calcium is taken from bone and protein is taken from muscle. In cases of long-term deficiency, these nutrient losses weaken and harm these tissues.

Many people believe that if too much of a nutrient is obtained—for example, from a vitamin or mineral supplement—only what is needed is stored and the rest is excreted by the body. Though true for some nutrients, such as vitamin C, the large dosages of other nutrients frequently found in supplements, such as vitamin A and iron, can cause harmful side effects because they are not readily excreted. This is one reason why obtaining your nutrients primarily (or exclusively) from a balanced diet is the safest means to acquire the building blocks you need to maintain the good health of all organ systems.

✔ CONCEPT CHECK 3.10

1. What is the body's most efficient form of energy storage?
2. Why is it important to consume nutrients daily?
3. When it comes to vitamins and minerals, is consuming more than the RDA or AI a good way to ensure optimal nutrition status? Why or why not?

3.11 Nutrition and Genetics

Once nutrients and other dietary components are taken up by cells, they may interact with our genes and have an effect on gene expression. The growth, development, and maintenance of cells, and ultimately of the entire organism, are directed by genes present in the cells. Each gene essentially represents a recipe, noting the ingredients (amino acids) and how those ingredients should be put together (to make proteins). The products (proteins) of all the recipes in the cookbook (the human genome) would then make up the human organism. The genome and the **epigenome,** the way the genome is marked and packaged inside a cell's nucleus, control the expression of individual traits, such as height, eye color, and susceptibility to many diseases. **Epigenetics** refers to changes in gene expression caused by mechanisms other than changes in the underlying DNA sequence. While our genome contains the code for the proteins that can be made by our bodies, our epigenome is an extra layer of instructions that influences gene activity. In many cases, it is the epigenome that can be repaired by treatments or affected by diet, rather than the genes.

The causes of chronic diseases are complex and include a significant genetic component. Fortunately, scientific breakthroughs in our understanding of the links between nutrition and genetics are beginning to touch our lives. Genetic discoveries are leading to new drugs that disrupt disease processes at the molecular level and to tests that predict our risk for disease.

THE EMERGING FIELD OF NUTRITIONAL GENOMICS

In the near future, the accessibility of genetic information will enhance the ability of health professionals to personalize nutrition recommendations that can optimize nutritional status and improve the outcomes of nutrition-related diseases. Collectively, the interactions between genetics and nutrition are known as

nutritional genomics. It is evident that nutritional status can both *affect* and be *affected by* an individual's genetic makeup. **Nutrigenetics** is the branch of nutritional genomics that examines how variations in genes can affect nutritional health. For example, the efficiency of absorption, metabolism, and excretion of a particular nutrient is controlled by genes. On the other hand, **nutrigenomics** refers to the many ways dietary components affect gene expression—particularly as it relates to development and treatment of nutrition-related diseases, such as cardiovascular disease. Let us examine each of these branches of nutritional genomics more closely.

In Chapter 2, you learned how Dietary Reference Intakes are set. Recall that nutrient recommendations, such as RDAs, are not absolute but are actually estimates of a level of intake that is likely to meet the needs of most (97% to 98%) of the population. For example, the RDA for folic acid (a B vitamin) is 400 micrograms per day. For most of the population, consuming 400 micrograms per day of folic acid from foods or supplemental sources will supply enough of the vitamin to optimize its function in body processes, such as the formation of red blood cells and metabolism of amino acids. There are certain subgroups of the population, however, for whom 400 micrograms of folic acid is not sufficient. These people have a genetic variation that alters the production of an enzyme for amino acid metabolism, so their dietary requirements for folic acid may be 10 times higher than the RDA. Nutrigenetics researchers are actively examining how genetic variations like this can affect individual nutrient requirements, how we can identify these people, and how we can personalize nutrition advice based on this knowledge (see Further Readings 1 and 6).

With nutrigenomics, researchers are interested in finding out how nutrients or other dietary components can influence gene expression, particularly as it relates to development of chronic diseases. Traditionally, nutrition and lifestyle recommendations have been based on the results of observational or experimental research that links diet to disease within the population as a whole. Nutrigenomics research now highlights the fallacy of a "one-size-fits-all" approach to nutrition interventions for disease prevention and management. It is becoming clear that generalized nutrition recommendations may not apply to all individuals within a population group. Nutrients or other compounds in the diet can turn certain genes on or off, thus manipulating the production of proteins that can affect—positively or negatively—the development or progression of diseases. Current areas of research include obesity, cardiovascular disease, celiac disease, osteoporosis, and Alzheimer's disease. With a better understanding of the interactions between genes and our diet, it will not be long before dietary recommendations can be tailored to help those with various genetically linked diseases.

NUTRITIONAL DISEASES WITH A GENETIC LINK

Studies of families, including those with twins and adopted children, provide strong support for the effects of genetics in various disorders. In fact, family history is considered to be an important risk factor in the development of many nutrition-related diseases.

Cardiovascular Disease. There is strong evidence that cardiovascular disease is the result of gene–environment interactions. As discussed in Chapter 1, elevated blood cholesterol is one major risk factor for development of cardiovascular disease. The gene–diet interactions being discovered for cardiovascular disease, particularly the cases of high blood lipid levels, will likely be the first to lead to nutrition plans personalized to decrease cardiovascular disease risk. About one of every 500 people in North America has a defective gene that greatly delays cholesterol removal from the bloodstream. Another genetic variation can cause abnormally high levels of an amino acid called homocysteine, which increases cardiovascular disease risk. Diet

▲ Studies of twins have provided strong evidence for the interaction between genes and diet and their combined effects on disease risk.

nutritional genomics Study of interactions between nutrition and genetics; includes nutrigenetics and nutrigenomics.

nutrigenetics Study of the effects of genes on nutritional health, such as variations in nutrient requirements and responsiveness to dietary modifications.

nutrigenomics Study of how food impacts health through its interaction with our genes and its subsequent effect on gene expression.

CRITICAL THINKING

Wesley notices that at family gatherings, his parents, uncles, aunts, and older siblings typically drink excessive amounts of alcohol. His father has been arrested for driving while intoxicated, as has one of his aunts. Two of his uncles died before the age of 60 from alcohol use disorders. As Wesley approaches the age of legal drinking, he wonders if he is destined to fall into the pattern of heavy drinking. What advice would you give to Wesley concerning his future use of alcohol?

▲ Genetic testing for disease susceptibility will be more common in the future as the genes that increase the risk of developing various diseases are isolated and decoded.

The following weblinks will help you gather more information about genetic conditions and testing:

http://nutrigenomics.ucdavis.edu/
Center for Excellence for Nutritional Genomics. Website dedicated to promoting the new science of nutritional genomics

www.geneticalliance.org
Alliance of Genetic Support Groups

www.kumc.edu/gec/support
Information on genetic and rare conditions

www.cancer.gov/cancertopics/pdq/genetics
Genetics information from the National Cancer Institute

http://www.genome.gov/
National Human Genome Research Institute (at the National Institutes of Health) website. Describes the latest research findings, discusses some ethical issues, and provides a talking glossary.

http://history.nih.gov/exhibits/genetics/
Revolution in Progress: Human Genetics and Medical Research

changes can help these people, but medications and even surgery are needed to address these problems.

Obesity. Most obese North Americans have at least one obese parent. This strongly suggests a genetic link. Findings from many human studies suggest that a variety of genes (likely 60 or more) are involved in the regulation of body weight. For example, specific gene variations have been linked to the propensity to overeat and the rate of energy use.

Still, although some individuals may be genetically predisposed to store body fat, whether they do so depends on how many calories they consume relative to their needs. A common concept in nutrition is that *nurture*—how people live and the environmental factors that influence them—allows *nature*—each person's genetic potential—to be expressed. Although not every person with a genetic tendency toward obesity becomes obese, those genetically predisposed to weight gain have a higher lifetime risk than individuals without a genetic predisposition to obesity.

Diabetes. Both of the two common types of diabetes—type 1 and type 2—have genetic links. Evidence for these genetic links comes from studies of families, including twins, and from the high incidence of diabetes among certain population groups (e.g., South Asians or Pima Indians). Diabetes, in fact, is a complex disease with more than 200 genes identified as possible causes. Only sensitive and expensive testing can determine who is at risk. Type 2 diabetes is the most common form of diabetes (90% of all cases) and also has a strong link to obesity. Typically, a genetic tendency for type 2 diabetes is expressed once a person becomes obese but often not before, again illustrating that nurture affects nature.

Cancer. A few types of cancer (e.g., some forms of colon and breast cancer) have a strong genetic link, and genetics may play a role in others, such as **prostate** cancer. Because obesity increases the risk of several forms of cancer, a long-standing excess calorie intake is also a risk factor. Although genes are an important determinant in the development of cancer, environmental and lifestyle factors, such as excessive sun exposure and a poor diet, also contribute significantly to the risk profile.

YOUR GENETIC PROFILE

From this discussion, you can see that your genes can greatly influence your risk of developing certain diseases. By recognizing your potential for developing a particular disease, you can avoid behavior that further raises your risk. How can you figure out your genetic profile? Genetic testing can be valuable if it confirms that you carry a gene for a disease that you can do something about in terms of protecting against it. Testing is also of interest when you do not know your family medical history or there are gaps in your family tree. It typically costs about $1000 to have your DNA read to reveal the diseases to which you are most susceptible. Many tests are covered by health insurance plans. The Genetic Information Nondiscrimination Act (GINA), which became a law in May 2008, prohibits health insurers from raising premiums or denying coverage based on genetic information and applies to people who have genes that carry the risk of disease. DNA testing includes providing a saliva sample from which your DNA will be separated. Certain areas of the genome are then read and measured in the process known as genotyping. After this 1- to 2-month process, you receive your DNA profile, which will supplement what you already know about your family history. You can also compare your DNA profile to future findings from DNA research.

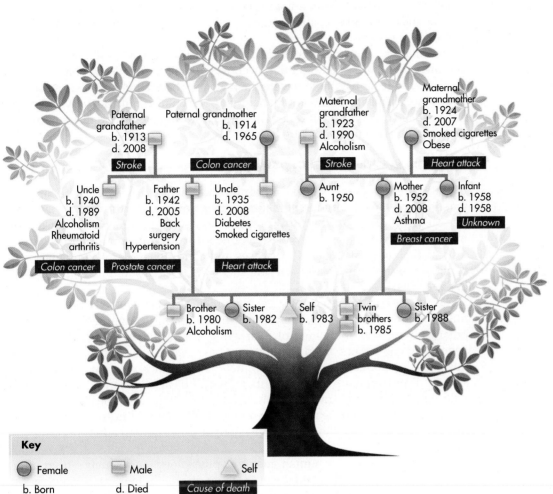

FIGURE 3-17 ◀ Example of a family tree for Justin, designated as "Self" at the trunk of the tree. The gender of each family member is identified by color (blue squares for males and orange circles for females). Dates of birth (b) and death (d) are listed below each family member. If deceased, the cause of death is highlighted using white text against a red background. Other medical conditions the family members experienced are noted beneath each name. Create your own family tree of frequent diseases using the diagram in "Rate Your Plate" and this figure as a guide. Then show your family tree to your physician to get a more complete picture of what the information means for your health.

Paternal grandfather
b. 1913
d. 2008
Stroke

Paternal grandmother
b. 1914
d. 1965
Colon cancer

Maternal grandfather
b. 1923
d. 1990
Alcoholism
Stroke

Maternal grandmother
b. 1924
d. 2007
Smoked cigarettes
Obese
Heart attack

Uncle
b. 1940
d. 1989
Alcoholism
Rheumatoid arthritis
Colon cancer

Father
b. 1942
d. 2005
Back surgery
Hypertension
Prostate cancer

Uncle
b. 1935
d. 2008
Diabetes
Smoked cigarettes
Heart attack

Aunt
b. 1950

Mother
b. 1952
d. 2008
Asthma
Breast cancer

Infant
b. 1958
d. 1958
Unknown

Brother
b. 1980
Alcoholism

Sister
b. 1982

Self
b. 1983

Twin brothers
b. 1985

Sister
b. 1988

Key

⬤ Female   ◻ Male   △ Self
b. Born   d. Died   Cause of death

Many of the genes involved in common diseases, such as diabetes, are still unknown. Although there are genetic tests available for some diseases, your family history of certain diseases is still a much better indicator of your genetic profile and risk of disease. Put together a family tree of illnesses and deaths by compiling a few key facts on your primary relatives: siblings, parents, aunts and uncles, and grandparents, as suggested in the "Rate Your Plate" section. In general, the greater number of your relatives who had a genetically transmitted disease and the closer they are related to you, the greater your risk. If there is a significant family history of a certain disease, lifestyle changes may be appropriate. For example, women with a family history of breast cancer should avoid becoming obese, minimize alcohol use, and obtain mammograms regularly.

Figure 3-17 shows an example of a family tree (also called a *genogram*). High-risk conditions include two or more first-degree relatives in a family with a specific disease (first-degree relatives include one's parents, siblings, and offspring). Another sign of risk of inherited disease is development of the disease in a first-degree relative before the age of 50 to 60 years. In the family depicted in Figure 3-17, prostate cancer killed the man's father. Knowing this, the man should be tested regularly for prostate cancer. His sisters should have frequent mammograms and other preventive practices because their mother died of breast cancer. Because heart attack and stroke are also common in the family, all the children should adopt a lifestyle that minimizes the risk of developing these conditions, such as avoiding excessive animal fat and salt intake. Colon cancer is also evident, so careful screening throughout life is important.

Information about our genetic makeup will increasingly influence our dietary and lifestyle choices. Throughout this book, we will discuss "controllable" risk factors that could contribute to development of genetically linked diseases present in your family. This information will help you personalize nutrition advice based on your genetic background and identify and avoid the risk factors that could lead to the diseases present in your family.

PERSONALIZING NUTRITION ADVICE

Nutrition professionals already recognize that dietary advice must be tailored to personal and cultural preferences. Research is now paving the way for even more personalized nutrition that incorporates the results of genetic testing to determine which diet will work most effectively for each person. There are genetic tests available for at least 1500 diseases and conditions. Many companies—most of them online—are already offering dietary advice and supplements based on genetic tests.

As we discussed in Chapter 2, caution is always needed when evaluating nutrition information and claims. Some DNA-testing companies are responsible organizations, and some are not. Marketing schemes may belittle the science of genetics to consumers. Even though there have been some great advancements in nutritional genomics in recent years, there is still much to learn. Not only is the science of nutritional genomics in its infancy, but application of this technology will require advanced training for health practitioners. Most health professionals agree that genetic testing complements carefully planned nutrient recommendations and dietary guidelines, but on its own, it is not quite ready for "prime time" just yet.

Chapter 3's review of human anatomy, physiology, and genetics from a nutrition perspective sets the stage for a more in-depth look at the nutrients in Chapters 4–9. As you learn about recommendations for specific nutrients or prevention of diseases, keep in mind that individual variations in genetic profile will prove to be important considerations when tailoring nutrition advice.

✓ CONCEPT CHECK 3.11

1. What is the difference between nutrigenetics and nutrigenomics?
2. List two nutrition-related diseases that are strongly affected by genetics.
3. Predict how nutritional genomics will affect nutrition recommendations in the future.

Nutrition and Your Health
Common Problems with Digestion

When suffering from persistent heartburn or GERD, see a doctor if you have:

- Difficulty swallowing or pain when swallowing
- Heartburn that has persisted for more than 10 years
- Initial onset of heartburn after age 50
- Heartburn that resists treatment with medications
- Sudden, unexplained weight loss
- Chest pain
- Blood loss or anemia
- Blood in stool or vomit

nausea, gagging, cough, or hoarseness. GERD is characterized by the occurrence of such symptoms of acid reflux two or more times per week. People who have GERD experience occasional relaxation of the gastroesophageal sphincter. Typically, it should be relaxed only during swallowing, but in individuals with GERD, it is relaxed at other times as well. For some people, slow movement of gastric contents from the stomach to the small intestine complicates the problem (see Further Reading 14).

The majority of heartburn sufferers say that it significantly affects their quality of life, particularly their enjoyment of many favorite foods. On a more serious note, however, if left untreated, heartburn can damage the lining of the esophagus, leading to chronic esophageal inflammation and an increased risk of esophageal cancer. Heartburn sufferers should follow the general recommendations given in Table 3-4. For occasional

FIGURE 3-18 ▶ Heartburn results from stomach acid refluxing into the esophagus.

Esophagus

Stomach

Relaxed lower esophageal sphincter

Inflamed esophagus

Acid reflux

Without fanfare, the digestive system does the important work of extracting nutrients from the food you eat to supply the needs of your body's trillions of cells. It is not until something goes awry that you notice digestion at all. In this section, you learn about nutritional strategies to cope with heartburn, ulcers, constipation, hemorrhoids, irritable bowel syndrome (IBS), diarrhea, gallstones, and celiac disease. Three other digestive disorders—lactose maldigestion, diverticulosis, and cystic fibrosis—will be discussed in Chapters 4 and 5.

Heartburn

About half of North American adults experience occasional heartburn, also known as acid reflux (Fig. 3-18). This gnawing pain in the upper chest is caused by the movement of acid from the stomach into the esophagus. The recurrent and therefore more serious form of the problem is called **gastroesophageal reflux disease (GERD).** Unlike the stomach, the esophagus has very little mucus to protect it, so acid quickly erodes the lining of the esophagus, causing pain. Symptoms may also include

gastroesophageal reflux disease (GERD) Disease that results from stomach acid backing up into the esophagus. The acid irritates the lining of the esophagus, causing pain.

CASE STUDY　Gastroesophageal Reflux Disease

Caitlin is a 20-year-old college sophomore. Over the last few months, she has been experiencing regular bouts of heartburn. This usually happens after a large lunch or dinner. Occasionally, she has even bent down after dinner to pick up something and had some stomach contents travel back up her esophagus and into her mouth. This especially frightened Caitlin, so she visited the University Health Center.

The nurse practitioner at the center told Caitlin it was good that she came in for a checkup because she suspects Caitlin has a disease called gastroesophageal reflux disease (GERD). She tells Caitlin that this can lead to serious problems, such as a rare form of cancer, if not controlled. She provides Caitlin with a pamphlet describing GERD and schedules an appointment with a physician for further evaluation.

Answer the following questions, and check your responses at the end of this chapter.

1. What dietary and lifestyle habits may have contributed to Caitlin's symptoms of GERD?
2. What is the dietary and lifestyle management advice that will help Caitlin cope with this health problem?
3. What types of medications have been especially useful for treating this problem?
4. Overall, how will Caitlin cope with this health problem, and will it ever go away?
5. Why is management of GERD so important?

▲ Caitlin was wise to see a health professional about her persistent heartburn.

TABLE 3-4 ▶ Nutrition and Lifestyle Recommendations for Care of Heartburn and Ulcers

	Heartburn	Peptic Ulcers
Avoid smoking.	√	√
Avoid large doses of aspirin, ibuprofen, and other NSAID compounds unless a physician advises otherwise.[a]	√	√
Achieve or maintain a healthy body weight.	√	√
Eat small, low-fat meals.	√	√
Limit alcohol consumption.	√	√
Limit consumption of caffeine (e.g., coffee, some soft drinks).	√	√
Consume a nutritionally complete diet with adequate fiber (see Chapter 4 for sources of fiber).	√	√
Avoid foods that worsen symptoms.[b]		
-Acidic foods (e.g., orange juice, tomato products)	√	√
-Highly spiced foods (e.g., chili, cayenne, and black pepper)	√	√
-Carbonated beverages	√	√
-Chocolate	√	√
-Onions and garlic	√	
-Peppermint and spearmint	√	
Avoid tight-fitting clothing.	√	
Elevate the head of the bed 6 to 8 inches.	√	
Avoid eating at least 3 to 4 hours before lying down.	√	
Wash hands often and follow food safety guidelines (see Chapter 13).		√

[a] For people who must use these medications, FDA has approved an NSAID combined with a medication to reduce gastric damage. The medication reduces gastric acid production and enhances mucus secretion.

[b] These foods do not cause heartburn or ulcers, but they may irritate sites of existing damage in the esophagus or stomach.

heartburn, quick relief can be found with over-the-counter (OTC) antacids. Taking antacids will reduce the acid in the stomach but will not stop the acid reflux. For more persistent (few days a week or everyday) heartburn or GERD, the H₂ blockers or **proton pump inhibitors (PPIs),** discussed in the Medicine Cabinet in the next section, may be needed. PPIs provide long-lasting relief by reducing stomach acid production and should be taken before the first meal of the day because they take longer to work. Medications that improve GI motility may also be useful. If the proper medications are not effective at controlling GERD, surgery may be needed to strengthen the weakened esophageal sphincter (see Further Reading 4).

Both pregnancy and obesity can lead to heartburn. These conditions result in increased pressure on the lower esophageal sphincter. In addition, estrogen and progesterone relax the lower esophageal sphincter, making heartburn more likely.

Ulcers

A peptic **ulcer** occurs when the lining of the esophagus, stomach, or small intestine is eroded by the acid secreted by the stomach cells (Fig. 3-19). A disruption of the layer of mucus that usually protects the stomach allows acid and protein-digesting enzymes to damage the stomach lining. This can cause pain, blood loss, and even perforation. Acid can also erode the lining of the esophagus and the first part of the small intestine, the duodenum. At any given time, about 4.5 million people in the United States are affected by peptic ulcers. In young people, most ulcers occur in the small intestine, whereas in older people, they occur primarily in the stomach.

How do you know if you have a peptic ulcer? Some people experience no symptoms at all, but most notice stomach pain about 2 hours after eating. Stomach acid acting on a meal irritates the ulcer after most of the meal has moved from the site of the ulcer. Other symptoms may include weight loss, lack of appetite, nausea and vomiting, or bloating. Vomiting blood or what looks like coffee grounds or the appearance of black, tarry stools are signs of bleeding in the GI tract. Any evidence of GI bleeding warrants immediate medical attention.

Not long ago, the major cause of ulcer disease was thought to be excess acid. Therefore, neutralizing and curtailing the secretion of stomach acid were the logical treatment choices. Although acid is still a significant player in ulcer formation, researchers now recognize that the chief culprits are infection of the stomach by the acid-resistant bacteria, *Helicobacter pylori (H. pylori),* and heavy use of medications that impair mucus production by the stomach. Stress is regarded as a predisposing factor for ulcers, especially if the person is infected with *H. pylori* or has certain anxiety disorders. Cigarette smoking is also known to cause ulcers, increase ulcer complications such as bleeding, and lead to ulcer treatment failure.

The *H. pylori* bacteria is found in more than 80% of patients with stomach and duodenal ulcers. The bacteria is common but results in ulcer disease in only 10% to 15% of those infected. Although the mechanism by which *H. pylori* causes ulcers is not well understood, treatment of the infection with antibiotics heals the ulcers and prevents their recurrence. Two Australian physicians were awarded the Nobel Prize in 2005 "for their discovery of the bacterium *H. pylori* and its role in gastritis and peptic ulcer disease."

ulcer Erosion of the tissue lining, usually in the stomach or the upper small intestine. As a group, these are generally referred to as *peptic ulcers.*

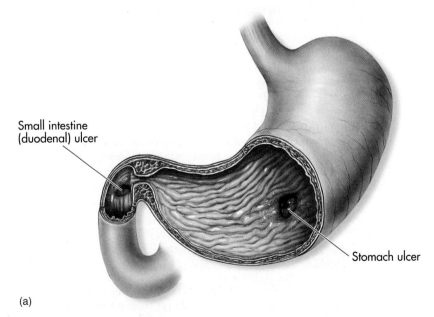
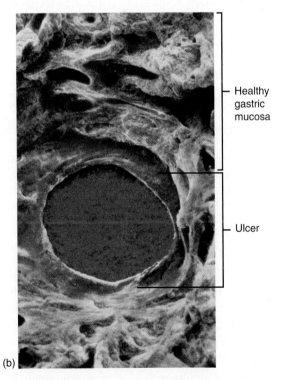

FIGURE 3-19 ▲ (a) A peptic ulcer in the stomach or small intestine. *H. pylori* bacteria and NSAIDs (e.g., aspirin) cause ulcers by impairing mucosal defense, especially in the stomach. In the same way, smoking, genetics, and stress can impair mucosal defense, as well as cause an increase in the release of pepsin and stomach acid. All of these factors can contribute to ulcers. (b) Close-up of a stomach ulcer. This needs to be treated or eventual perforation of the stomach is possible.

Non-steroidal anti-inflammatory drugs (NSAIDs) are medications for painful inflammatory conditions such as arthritis. Aspirin, ibuprofen, and naproxen are the most commonly used NSAIDs. NSAIDs reduce the mucus secreted by the stomach. Newer medications, called "Cox-2 inhibitors" (e.g., celecoxib [Celebrex]), have been used as a replacement for NSAIDs because they are less likely to cause stomach ulcers. They do offer some advantages over NSAIDs, but they may not be totally safe for some people, especially those with a history of cardiovascular disease or strokes.

The primary risk associated with an ulcer is the possibility that it will erode entirely through the stomach or intestinal wall. The GI contents could then spill into the body cavities, causing a massive infection. In addition, an ulcer may damage a blood vessel, leading to substantial blood loss. For these reasons, it is important to never ignore the early warning signs of ulcer development, including a persistent gnawing or burning near the stomach that may occur immediately following a meal or awaken you at night.

Today, a combination approach is used for ulcer therapy. People infected with *H. pylori* are given antibiotics and stomach acid-blocking medications called PPI. There is a 90% cure rate for *H. pylori* infections in the first week of this treatment. Recurrence is unlikely if the infection is cured, but an incomplete cure almost certainly leads to repeated ulcer formation (see Further Reading 13).

Antacid medications may also be part of ulcer care, as is a class of medicines called **H₂ blockers,** which prevent **histamine**-related acid secretion in the stomach. Medications that coat the ulcer are also commonly used.

Are dietary changes effective for prevention or treatment of peptic ulcers? Many people think that eating spicy or acidic foods can cause ulcers. Contrary to popular belief, these foods do not cause ulcers. However, once an ulcer has developed, these foods may irritate damaged tissues. Thus, for some people, avoidance of spicy or acidic foods may help to relieve symptoms.

NSAIDs Nonsteroidal anti-inflammatory drugs; includes aspirin, ibuprofen (Advil®), and naproxen (Aleve®).

histamine A breakdown product of the amino acid histidine that stimulates acid secretion by the stomach and has other effects on the body, such as contraction of smooth muscles, increased nasal secretions, relaxation of blood vessels, and changes in relaxation of airways.

In the past, milk and cream were thought to help cure ulcers. Clinicians now know that milk and cream are two of the worst foods for a person with ulcers because the calcium in these foods stimulates acid secretion and actually inhibits ulcer healing.

Overall, medical treatment of *H. pylori* infection has so revolutionized ulcer therapy that dietary changes are of minor importance. People with ulcers should refrain from smoking and minimize the use of NSAIDs. Current dietary therapy approaches simply recommend avoidance of foods that tend to worsen ulcer symptoms (Table 3-4).

Constipation

Constipation, difficult or infrequent evacuation of the bowels, is commonly reported by adults. Slow movement of fecal material through the large intestine causes constipation. As more fluid is absorbed during the extended time the feces stay in the large intestine, they become dry and hard.

Constipation can result when people regularly ignore their normal bowel reflexes for long periods. People may ignore normal urges when it is inconvenient to interrupt occupational or social activities. Muscle spasms of an irritated large intestine can also slow the movement of feces and contribute to constipation. Calcium, iron supplements, and medications such as antacids can also cause constipation.

Eating foods with plenty of fiber, such as whole-grain breads, cereals, and beans, along with drinking adequate fluid to avoid dehydration, is the best method for treating mild cases of constipation (see Further Reading 10). Fiber stimulates peristalsis by drawing water into the large intestine and helping form a bulky, soft fecal output. Dried fruits are a good source of fiber and therefore can also help stimulate the bowel. Additional fluid should be consumed to facilitate fiber's action in the large intestine. Also, people with constipation may need to develop more regular bowel habits; allowing the same time each day for a bowel movement can help train the large intestine to respond routinely. Finally, relaxation facilitates regular bowel movements, as does regular physical activity.

Medicine Cabinet

Proton pump inhibitors (PPIs) are medications that inhibit the ability of gastric cells to secrete hydrogen ions (i.e., protons). Low doses of this class of medications may be available without a prescription. Because stomach acid is important for the absorption of vitamin B-12, prolonged use of PPIs could impair vitamin B-12 status.

Examples:

- Omeprazole (Prilosec OTC®)
- Rabeprazole (Aciphex®)
- Lansoprazole (Prevacid®)
- Esomeprazole (Nexium®)

H₂ blockers impede the stimulating effect of histamine on acid-producing cells in the stomach.

Examples:

- Cimetidine (Tagamet®)
- Nizatidine (Axid®)
- Ranitidine (Zantac®)
- Famotidine (Pepcid®)

CRITICAL THINKING

Amelia has heard that taking laxatives after overeating prevents deposition of body fat from excess calorie intake. Laxatives hasten emptying of the large intestine and increase fluid losses. Based on what you have learned about digestion and absorption of nutrients, do you think Amelia can count on laxatives to prevent fat gain from excess calorie intake?

▲ Dried fruits are a natural source of fiber and can help prevent constipation when consumed with an adequate amount of fluid.

Laxatives can lessen constipation. Some laxatives work by irritating the intestinal nerve junctions to stimulate the peristaltic muscles, while others that contain fiber draw water into the intestine to enlarge fecal output. The larger output stretches the peristaltic muscles, making them rebound and then constrict. Regular use of laxatives, however, should be supervised by a physician. Overall, the bulk-forming fiber laxatives are the safest to use.

Hemorrhoids

Hemorrhoids, also called *piles,* are swollen veins of the rectum and anus. The blood vessels in this area are subject to intense pressure, especially during bowel movements. Added stress to the vessels from pregnancy, obesity, prolonged sitting, violent coughing or sneezing, or straining during bowel movements, particularly with constipation, can lead to a hemorrhoid. Hemorrhoids can develop unnoticed until a strained bowel movement precipitates symptoms, which may include pain, itching, and bleeding.

Itching, caused by moisture in the anal canal, swelling, or other irritation, is perhaps the most common symptom. Pain, if present, is usually aching and steady. Bleeding may result from a hemorrhoid and appear in the toilet as a bright red streak in the feces. The sensation of a mass in the anal canal after a bowel movement is symptomatic of an internal hemorrhoid that protrudes through the anus.

Anyone can develop a hemorrhoid, and about half of adults over age 50 do. Diet, lifestyle, and possibly heredity play a role. For example, a low-fiber diet can lead to hemorrhoids as a result of straining during bowel movements. If you think you

laxative A medication or other substance that stimulates evacuation of the intestinal tract.

hemorrhoid A pronounced swelling of a large vein, particularly veins found in the anal region.

have a hemorrhoid, you should consult your physician. Rectal bleeding, although usually caused by hemorrhoids, may also indicate other problems, such as cancer.

A physician may suggest a variety of self-care measures for hemorrhoids. Pain can be lessened by applying warm, soft compresses or sitting in a tub of warm water for 15 to 20 minutes. Dietary recommendations are the same as those for treating constipation, emphasizing the need to consume adequate fiber and fluid. OTC remedies, such as Preparation H®, can also offer relief from symptoms.

Irritable Bowel Syndrome

Many adults (25 million or more in the United States alone) have IBS, noted as a combination of cramps, gassiness, bloating, and irregular bowel function (diarrhea, constipation, or alternating episodes of both). It is more common in younger women than in younger men. In older adults, the ratio is closer to 50:50. The disease leads to about 3.5 million visits to physicians in the United States each year.

Symptoms associated with IBS include visible abdominal distention, pain relief after a bowel movement, increased stool frequency, loose stools with pain onset, mucus in stool, and a feeling of incomplete elimination even after a bowel movement. Although IBS can be uncomfortable and upsetting, it is harmless as it carries no risk for cancer or other serious digestive problems.

It is difficult to pinpoint an exact cause for IBS. Recent studies indicate that alterations in some of the hormones that regulate the movement of food matter through the GI tract may be to blame. Also, inflammatory responses in the GI tract could be involved for some people with IBS. The majority of people who suffer from IBS perceive that their symptoms are food-related, but there is little evidence of actual food allergies or intolerances (see Further Reading 5). When it comes to specific foods, poorly digested carbohydrates are a prime suspect (see Further Reading 9). Fructose, sugar alcohols, and other carbohydrates may lead to diarrhea or excessive gas if they reach the large intestine undigested. Depression and stress are also associated with IBS; up to 50% of sufferers report a history of verbal or sexual abuse.

Given the diversity of symptoms and possible causes, therapy must be individualized. Historically, a high-fiber diet has been recommended, but recent research shows little evidence of benefit. Increasing soluble fiber may be helpful for some people with IBS, but insoluble fiber seems to worsen symptoms (see Chapter 4). Elimination diets that focus on avoiding dairy products and gas-forming foods, such as legumes, certain vegetables (cabbage, beans, and broccoli), and some fruits (grapes, raisins, cherries, and cantaloupe), can alleviate symptoms. Herbal formulations, certain probiotics, and cognitive behavioral therapy have been shown to decrease symptoms of IBS and improve overall quality of life (see Further Readings 3 and 8). The patient should limit or eliminate caffeine-containing foods and beverages. Low-fat and more frequent, small meals may help because large meals can trigger contractions of the large intestine. Other strategies include a reduction in stress, psychological counseling, and certain antidepressant and other medications. Hypnosis has been shown to relieve symptoms in severe cases.

Following a diet that eliminates certain foods or entire food groups can limit nutritional adequacy. Indeed, research indicates that intakes of some nutrients, including calcium and vitamin A, are inadequate among people with IBS. For help with dietary monitoring, identifying problem foods, and planning a nutritionally adequate diet, a registered dietitian can be a valuable resource for a person with IBS.

Diarrhea

Diarrhea, a GI tract disease that generally lasts only a few days, is defined as increased fluidity, frequency, or amount of bowel movements compared to a person's usual pattern. Most cases of diarrhea result from infections in the intestines, with bacteria and viruses the usual offending agents. These microorganisms produce substances that cause the intestinal cells to secrete fluid rather than absorb fluid. Another form of diarrhea can be caused by consumption of substances that are not readily absorbed, such as the sugar alcohol sorbitol found in sugarless gum (see Chapter 4) or large amounts of a high-fiber source such as bran. When consumed in large amounts, the unabsorbed substance draws much water into the intestines, in turn leading to diarrhea.

The goal of diet therapy for any form of diarrhea is to prevent dehydration. Increasing intake of water and electrolytes is the first line of defense against dehydration. Prompt treatment of dehydration—within 24 to 48 hours—is critical, especially for infants and older adults (see Chapters 15 and 16). Diarrhea that lasts more than 7 days in adults should be investigated by a physician as it can be a sign of a more serious intestinal disease, especially if there is also blood in the stool.

For diarrhea caused by infection, other changes to the regular diet are usually not necessary. Some sources recommend temporarily decreasing intake of caffeine, fat, fiber, and poorly absorbed carbohydrates (see Further Reading 11), but other sources show that maintaining a regular diet speeds recovery. Foods containing probiotics may assist recovery. For diarrhea caused by a poorly absorbed substance, such as excess sugar alcohols or lactose, avoidance of the offending substance is the key to relief.

Gallstones

Gallstones are a major cause of illness and surgery, affecting 10% to 20% of U.S. adults. Gallstones are pieces of solid material that develop in the gallbladder when substances in the bile—primarily cholesterol (80% of gallstones)—form crystal-like particles. They may be as small as a grain of sand or as large as a golf ball (Fig. 3-20). These stones are caused by a combination of factors, with excess weight being the primary modifiable factor, especially in women 20 to 60 years old. Other factors include genetic background (e.g., Native Americans), advanced age (>60 years for both women and men), pregnancy, reduced activity of the gallbladder (contracts less than normal), altered bile composition (e.g., too much cholesterol or not enough bile salts), diabetes, and diet (e.g., low-fiber diets). In addition, gallstones may develop during rapid weight loss or prolonged fasting (as the liver metabolizes more fat, it secretes more cholesterol into the bile).

Attacks due to gallstones include intermittent pain in the upper right abdomen, gas and bloating, nausea or vomiting, or

FIGURE 3-20 ▲ Gallbladder and gallstones seen after surgical removal from the body. Size and composition of the stones vary from one case to another.

other health problems. Medications are available to dissolve gallstones, but these take a long time to work, and the recurrence of gallstones after therapy is common. Therefore, surgical removal of the gallbladder is the most common method for treating gallstones (500,000 surgeries per year in the United States).

Prevention of gallstones revolves around avoiding becoming overweight, especially for women. Avoiding rapid weight loss (>3 pounds per week), limiting animal protein and focusing more on plant protein intake (especially some nut intake), and following a high-fiber diet can help as well. Regular physical activity is also recommended, as is moderate to no caffeine and alcohol intake (see Further Reading 12).

Celiac Disease and Gluten Sensitivity

Celiac disease (sometimes called celiac sprue) affects about 1% of the U.S. population. Development of celiac disease depends on two factors: a genetic predisposition and dietary exposure to a protein called gluten. Gluten is a type of protein found in certain grains: wheat, rye, and barley. Protein-digesting enzymes in the GI tract break down some of the peptide bonds in gluten, but digestion is incomplete, leaving some small peptides in addition to individual amino acids. The small peptides that arise from digestion can be absorbed into the cells lining the small intestine. When people with a genetic predisposition for celiac disease are exposed to these small peptides from gluten, they experience an inflammatory reaction. Although many people think celiac disease is a food allergy, it is actually an autoimmune response: the immune system attacks and destroys its own cells. (You will learn more about food allergies in Chapters 6 and 15.)

The immune response that occurs after exposure to gluten targets the cells of the small intestine, causing a flattening of the villi, which thereby reduces the absorptive surface (Fig. 3-21). The production of some digestive enzymes is decreased, and the ability of the small intestine to absorb nutrients is impaired. Malabsorption leads to a variety of GI complaints: diarrhea, bloating, cramps, and flatulence. In fact, it is common for celiac disease to be misdiagnosed as IBS. However, the pathology underlying celiac disease has far worse consequences than

celiac disease Chronic, immune-mediated disease precipitated by exposure to dietary gluten in genetically predisposed people.

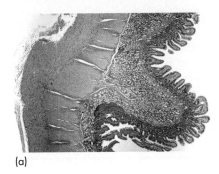
(a)

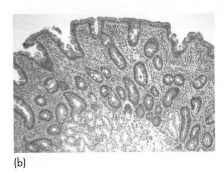
(b)

FIGURE 3-21 ◄ Biopsy findings in patients with and without celiac disease. (a) Normal small intestinal cells. Note the intact villi, which increase the surface area of the small intestine to maximize nutrient absorption. (b) The small intestinal cells of a person with celiac disease. The villi appear flattened.

IBS. Over time, malabsorption of nutrients can lead to fatigue, weight loss (or poor growth in children), anemia, infertility, and even bone loss (see Further Readings 7 and 15).

If celiac disease is suspected, the first step in making a formal diagnosis is a blood test for the presence of antibodies to gluten. This may be followed by one or more biopsies of the small intestine to confirm the pathological defects. There is also a genetic test for celiac disease, but having the gene does not always predict development of the disease.

Strict dietary avoidance of food products containing wheat, rye, and barley is the only proven way to manage the disease. Compliance with a gluten-free diet requires careful reading of food labels. In 2004, Congress passed the Food Allergen Labeling and Consumer Protection Act, which requires food manufacturers to identify the presence of eight major food allergens in products. Wheat is among these food allergens, but rye and barley are not. Therefore, people following a gluten-free diet must learn to carefully interpret the list of ingredients to identify sources of gluten. Within the grains group, rice, potato flour, cornmeal, buckwheat, arrowroot, and soy are gluten-free, but ingredients such as wheat, rye, barley, bran, graham flour, semolina, spelt, and malt are sources of gluten and must be avoided. Oats do not traditionally contain gluten, but contamination in the field or during food processing could introduce gluten into this grain as well.

People with celiac disease quickly learn that wheat, barley, and rye can be hidden ingredients in any food group: wheat and its derivatives are used to thicken sauces and condiments, as flavoring agents in dairy products and many other processed foods, and in breading for deep-fried vegetables and meats. It is helpful that many food manufacturers now voluntarily disclose the presence or absence of gluten in their products. However, not all products clearly identify gluten. Dining out is yet another challenge: even a dusting of wheat flour can have adverse effects for a person with celiac disease.

After several weeks on a gluten-free diet, the small intestinal lining regenerates. Gastrointestinal symptoms subside and nutrient absorption improves. So far, the gluten-free diet is the only proven way to manage celiac disease, but research on other treatments is underway. Food scientists are working toward developing strains of wheat, barley, and rye that do not contain gluten. From a gastroenterological perspective, others approaches are to supply digestive enzymes that will break down the gluten proteins before they stimulate an autoimmune response or to use polymers that will bind to gluten in the GI tract and prevent it from being absorbed. From an immunological perspective, researchers are looking at medications that could block the immune responses that damage the small intestine.

A related issue is **nonceliac gluten sensitivity** (sometimes called gluten intolerance, or NCGS). Some people experience symptoms of celiac disease after ingestion of gluten, but they do not have the small intestinal pathology of celiac disease, nor do they express the antibodies typical of CD. Some reports indicate that for each person who is diagnosed with CD, as many as six others have NCGS. Aside from GI symptoms, patients with NCGS may also report fatigue, headache, muscle and joint pain, and/or sleep disorders. Symptoms subside with a gluten-free diet but reappear when gluten is reintroduced. The medical community recognizes NCGS as a verifiable condition, but the immunological mechanism that causes it is not well understood. There is no diagnostic test for the condition—only the effectiveness of the gluten-free diet in alleviating symptoms (see Further Reading 2). Many questions remain: Is NCGS a permanent condition? Is there a level of gluten intake that would not trigger symptoms? It seems that there are multiple immunological reactions to gluten that are predicted by different genetic traits (see Further Reading 7).

Overall, the prevalence and awareness of celiac disease and NCGS seem to be on the rise. A cause for the increased prevalence has not been pinpointed, but some scientists speculate that changes in wheat production or widespread use of wheat in the food supply may be to blame. Others suspect that an infection or exposure to some environmental toxin could increase gut permeability and lead to gluten sensitization.

Summary

The conditions discussed here can be very serious, possibly leading to malnutrition, internal bleeding, and life-threatening infections. It is important to seek competent medical advice if you or someone you know suspects a GI disorder. However, you should feel empowered to know that you can control some risks and complement medical treatment with nutrition and other lifestyle changes. Overall, keeping body weight within a healthy range, meeting recommendations for fiber and fluid intake, and avoiding smoking and overuse of NSAID medications are useful strategies that can help you cope with several common disorders of the GI tract.

nonceliac gluten sensitivity One or more of a variety of immune-related conditions with symptoms similar to celiac disease that are precipitated by the ingestion of gluten in people who do not have celiac disease.

Summary (Numbers refer to numbered sections in the chapter.)

3.1 Cells join together to make up tissues, tissues unite to form organs, and organs work together as an organ system.

3.2 The basic structural unit of the human body is the cell. Almost all cells contain the same organelles, but cell structure varies according to the type of job cells must perform.

3.3 Epithelial, connective, muscle, and nervous tissues are the four primary types of tissues in the human body. Each type of organ system is affected by nutrient intake.

3.4 From the cells of the GI tract, water-soluble nutrients are absorbed into capillaries and fat-soluble nutrients are absorbed into lymph vessels, which eventually connect to the bloodstream. Blood delivers nutrients and oxygen to cells and picks up wastes as it circulates around the body.

3.5 The urinary system, including the kidneys, is responsible for filtering the blood, removing body wastes, and maintaining the chemical composition of the blood.

3.6 The nervous system allows for communication and regulation. Vitamin B-12 is part of the insulation that surrounds neurons. Transmission of nerve impulses relies on sodium and potassium. Neurotransmitters are made from amino acids.

3.7 The endocrine system produces hormones—protein-based chemical messengers—to regulate metabolic reactions and the levels of nutrients in the blood.

3.8 With assistance from the skin and the gastrointestinal tract, the immune system protects the body from pathogens. Optimal immune system function relies on protein; essential fatty acids; vitamins A, C, and D; some B vitamins; and the minerals iron, zinc, and copper.

3.9 The GI tract consists of the mouth, esophagus, stomach, small intestine, large intestine (colon), rectum, and anus. Spaced along the GI tract are sphincters that regulate the flow of foodstuffs. Peristalsis moves food matter along the GI

tract. Nerves, hormones, and other substances control the activity of sphincters and peristaltic muscles.

Digestive enzymes are secreted by the mouth, stomach, small intestine, and pancreas. Little digestion and absorption occur in the stomach or large intestine, but some protein is digested in the stomach.

Most absorption occurs through the cells of the villi, which line the small intestine. Absorptive processes include passive diffusion, facilitated diffusion, active transport, phagocytosis, and pinocytosis.

Some water and mineral absorption takes place in the large intestine. Some constituents of undigested carbohydrates are broken down by bacteria in the large intestine. Some of these products are absorbed; remaining undigested materials are eliminated in the feces.

The liver, gallbladder, and pancreas participate in digestion and absorption. Products from these organs, such as enzymes and bile, enter the small intestine and help in digesting protein, fat, and carbohydrate.

3.10 Limited stores of nutrients are present in the blood for immediate use. Some nutrients, such as minerals and fat-soluble vitamins, can be stored extensively in bone, adipose, and liver tissues. Excessive storage of nutrients can be toxic. Conversely, breakdown of vital tissues can supply nutrients in times of need but eventually leads to ill health.

3.11 Nutritional genomics includes the study of how genes influence nutritional status (nutrigenetics) and how nutrients and other dietary components influence genetic expression (nutrigenomics). Genograms and genetic testing can be useful tools when planning a diet for optimal health.

NAYH Common GI tract diseases, such as heartburn, constipation, and irritable bowel syndrome, can be treated with a combination of diet changes and medications.

Check Your Knowledge (Answers to the following questions are on the next page.)

1. The stomach is protected from digesting itself by producing
   a. bicarbonate.
   b. a thick layer of mucus.
   c. hydroxyl ions to neutralize acid.
   d. antipepsin that destroys enzymes.

2. The lower esophageal sphincter is located between the
   a. stomach and esophagus.
   b. stomach and duodenum.
   c. ileum and the cecum.
   d. colon and the anus.

3. A muscular contraction that propels food down the GI tract is called
   a. a sphincter.
   b. enterohepatic circulation.
   c. gravitational pull.
   d. peristalsis.

4. Bicarbonate ions (HCO_3^-) from the pancreas
   a. neutralize acid in the stomach.
   b. are synthesized in the pyloric sphincter.
   c. neutralize bile in the duodenum.
   d. neutralize acid in the duodenum.

5. Most digestive processes occur in the
   a. mouth.   c. small intestine.   e. liver.
   b. stomach.   d. large intestine.

6. Bile is formed in the _____ and stored in the _____.
   a. stomach, pancreas   c. liver, gallbladder
   b. duodenum, kidney   d. gallbladder, liver

7. Much of the digestion that occurs in the large intestine is caused by
   a. lipase.   c. saliva.
   b. pepsin.   d. bacteria.

8. Treatment of ulcers may include
   a. H₂ blockers.   c. antibiotics.
   b. proton pump inhibitors.   d. All of these.

9. The study of how food impacts health through interaction with genes is
   a. nutrigenomics.   c. immunology.
   b. epidemiology.   d. nutrigenetics.

10. Energy production that takes place in the cytoplasm is anaerobic metabolism because it does not require
   a. water.   c. anabolic steroids.
   b. oxygen.   d. anaerobic bacteria.

Answer Keys: 1. b (LO 3.8), 2. a (LO 3.8), 3. d (LO 3.8), 4. d (LO 3.8), 5. c (LO 3.8), 6. c (LO 3.8), 7. d (LO 3.8), 8. d (LO 3.8), 9. a (LO 3.11), 10. b (LO 3.2)

Study Questions (Numbers refer to Learning Outcomes)

1. Identify at least one function of the 12 organ systems related to nutrition. (LO 3.3)

2. Draw and label parts of the cell and explain the function of each organelle as it relates to human nutrition. (LO 3.2)

3. Trace the flow of blood from the right side of the heart and back to the same site. How is blood routed through the small intestine? Which class of nutrients enters the body via the blood? Via the lymph? (LO 3.4)

4. Explain why the small intestine is better suited than the other GI tract organs to carry out the absorptive process. (LO 3.8)

5. Identify the five basic tastes. Give an example of one food that exemplifies each of these basic taste sensations. (LO 3.8)

6. What is one role of acid in the process of digestion? Where is it secreted? (LO 3.8)

7. Contrast the processes of active absorption and passive diffusion of nutrients. (LO 3.8)

8. Identify two accessory organs that empty their contents into the small intestine. How do the digestive substances secreted by these organs contribute to the digestion of food? (LO 3.8)

9. In which organ systems would the following substances be found?
   chyme (LO 3.8), plasma (LO 3.4), lymph (LO 3.4), urine (LO 3.9)

10. Describe the nutrition-related diseases for which genetics or family history is considered to be an important risk factor. (LO 3.11)

CASE STUDY SOLUTIONS

1. Overeating at mealtimes appears to bring on Caitlin's regular bouts of heartburn.
2. Typical dietary advice includes consuming smaller, more frequent meals low in fat, not overeating at mealtimes, waiting about 2 hours after meals before lying down, and elevating the head of the bed about 6 inches (review Table 3-4). These recommendations reduce the risk of stomach contents forcing their way back up the esophagus. In addition, losing excess weight (if overweight) can alleviate GERD.
3. If dietary advice does not control symptoms, the primary medication used to control GERD inhibits acid production in the stomach (see the discussion in the Nutrition and Your Health section in this chapter on the proton pump inhibitors such as omeprazole [Prilosec®] also used to treat peptic ulcers). If this and other medical therapy fail to control the problem, surgery to strengthen the lower esophageal sphincter is possible.
4. Caitlin's GERD can be treated but most likely will be a lifelong condition. Even surgery will generally not cure the problem. Lifetime diet and lifestyle management, and most likely medications, will still be needed to manage the problem.
5. Lifelong management of GERD is important because long-standing GERD increases the risk of esophageal cancer.

What the Dietitian Chose

Constipation results from slow movement of fecal material through the large intestine. Increasing fiber and water in your diet can often relieve constipation without the aid of an over-the-counter laxative. To boost your fiber intake, choose whole grains, fruits, vegetables, and legumes (beans). This can be a tough task when most of the menu options you will find in fast-food restaurants and convenience stores are low in fiber.

As you select your pizza, look for slices with extra vegetables or fruit. This will increase the fiber content by about 1 gram per slice. Pizza with whole-grain crust would add another 1 gram of fiber per slice. These are small improvements, so you will probably still need to look for additional fiber sources.

What about your side dish? Unless they are made with whole grains, the pasta and breadsticks will not add much fiber to your meal. A breadstick and a cup of pasta with Alfredo sauce provide about 1 and 2 grams of fiber, respectively. These carbohydrate-rich add-ons are supplying extra calories and would be better off skipped. The salad provides about 1 to 3 grams of fiber, depending on its size and ingredients. The core ingredients of tossed salad—lettuce, cucumbers, tomato—are mostly water and not particularly high in fiber. Toppings such as cheese, egg, or diced meat will not help relieve constipation. Instead, dried fruit (e.g., dried cranberries or raisins) and nuts supply some additional fiber.

▲ Whole-grain pasta and beans used in salads and soups are good sources of fiber that can prevent intestinal issues such as constipation.

An even better choice, however, is the soup! A cup of bean and pasta soup is just what the doctor ordered to help relieve constipation. It provides about 6 grams of fiber and some extra water, as well.

Lastly, be sure to drink plenty of water. Dehydration is an often overlooked cause of constipation. Water helps lubricate the digestive tract and adds bulk to the feces when absorbed by fiber in the large intestine.

Further Readings

1. Baumler, MD: Nutrigenetics—building a platform for dietitians to offer personalized nutrition. *Today's Dietitian* 14:48, 2012.

2. Bizzaro N and others: Cutting-edge issues in celiac disease and in gluten intolerance. *Clinical Reviews in Allergy and Immunology* 42:279, 2012.

3. Camilleri M: Probiotics and irritable bowel syndrome: Rationale, putative mechanisms, and evidence of clinical efficacy. *Journal of Clinical Gastroenterology* 40:264, 2006.

4. DeVault KR and Castell DO: Updated guidelines for the diagnosis and treatment of gastroesophageal reflux disease. *American Journal of Gastroenterology* 100:190, 2005.

5. El-Salhy M and others: The role of diet in the pathogenesis and management of irritable bowel syndrome (review). *International Journal of Molecular Medicine* 29:723, 2012.

6. Fenech M and others: Nutrigenetics and nutrigenomics: Viewpoints on the current status and applications in nutrition research and practice. *Journal of Nutrigenetics and Nutrigenomics* 4:69, 2011.

7. Ferretti GF and others: Celiac disease, inflammation, and oxidative damage: A nutrigenetic approach. *Nutrients* 4:243, 2012.

8. Lynch A and Webb C: What are the most effective nonpharmacologic therapies for irritable bowel syndrome? *Journal of Family Practice* 57:57, 2008.

9. Marcason W: What is the FODMAP diet? *Journal of the Academy of Nutrition and Dietetics* 112:12, 2012.

10. Muller-Lisser SA and others: Myths and misconceptions about constipation. *American Journal of Gastroenterology* 100:232, 2005.

11. National Digestive Diseases Information Clearinghouse: Diarrhea (2011). Available at **http://digestive.niddk.nih.gov/ddiseases/pubs/diarrhea/index.aspx.** Accessed July 21, 2014.

12. National Digestive Diseases Information Clearinghouse: Gallstones (2007). Available at **http://digestive.niddk.nih.gov/ddiseases/pubs/gallstones/index.aspx.** Accessed July 21, 2014.

13. National Digestive Diseases Information Clearinghouse: *H. pylori* and peptic ulcers (2010). Available at **http://digestive.niddk.nih.gov/ddiseases/pubs/hpylori/index.aspx.** Accessed July 21, 2014.

14. National Digestive Diseases Information Clearinghouse: Heartburn, Gastroesophageal Reflux (GER), and Gastroesophageal Reflux Disease (GERD) (2007). Available at **http://digestive.niddk.nih.gov/ddiseases/pubs/gerd/index.aspx.** Accessed July 21, 2014.

15. Presutti RJ and others: Celiac disease. *American Family Physician* 76:1795, 2007.

16. Wallace TC and others: Human gut microbiota and its relationship to health and disease. *Nutrition Reviews* 69:392, 2011.

connect INUTRITION   To get the most out of your study of nutrition, visit McGraw-Hill Connect at www.mcgrawhillconnect.com where you will find NutritionCalc Plus, LearnSmart, and many other dynamic tools.

Rate Your Plate

I. Are You Taking Care of Your Digestive Tract?

People need to think about the health of their digestive tracts. There are symptoms we need to notice, as well as habits we need to practice to protect it. The following assessment is designed to help you examine your habits and symptoms associated with the health of your digestive tract. Put a *Y* in the blank to the left of the question to indicate yes and an *N* to indicate no.

_____ 1. Are you currently experiencing greater than normal stress and tension?

_____ 2. Do you have a family history of digestive tract problems (e.g., ulcers, hemorrhoids, recurrent heartburn, and constipation)?

_____ 3. Do you experience pain in your stomach region about 2 hours after you eat?

_____ 4. Do you smoke cigarettes?

_____ 5. Do you take aspirin frequently?

_____ 6. Do you have heartburn at least once per week?

_____ 7. Do you commonly lie down after eating a large meal?

_____ 8. Do you drink alcoholic beverages more than two or three times per day?

_____ 9. Do you experience abdominal pain, bloating, or gas 1½ to 2 hours after consuming milk products?

_____ 10. Do you often have to strain while having a bowel movement?

_____ 11. Do you consume less than 9 (women) or 13 (men) cups of a combination of water and other fluids per day?

_____ 12. Do you perform physical activity for less than 60 minutes on most or all days of the week (e.g., jog, swim, walk briskly, row, and stair climb)?

_____ 13. Do you eat a diet relatively low in fiber (recall that significant fiber is found in whole fruits, vegetables, legumes, nuts and seeds, whole-grain breads, and whole-grain cereals)?

_____ 14. Do you frequently have diarrhea?

_____ 15. Do you frequently use laxatives or antacids?

Add up the number of yes answers and record the total. If your score is from 8 to 15, your habits and symptoms put you at risk for experiencing future digestive tract problems. Take particular note of the habits to which you answered yes. Consider trying to cooperate more with your digestive tract.

II. Create Your Family Tree for Health-Related Concerns

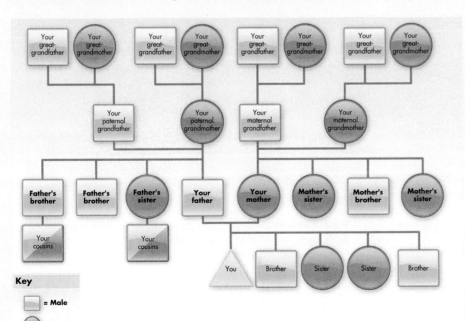

Adapt this diagram to your family tree. Under each heading, list year born, year died (if applicable), major diseases that developed during the person's lifetime, and cause of death (if applicable). Figure 3-17 shows one such example.

You are likely to be at risk for any diseases listed. Creating a plan for preventing such diseases when possible, especially those that developed in your family members before age 50 to 60 years, is advised. Speak with your physician about any concerns arising from this exercise.

Student Learning Outcomes

Chapter 4 is designed to allow you to:

4.1 Explain the most important nutritional role of carbohydrates and how they are created.

4.2 Identify the basic structures of the major carbohydrates: monosaccharides, disaccharides, polysaccharides (e.g., starches), and fiber.

4.3 Describe food sources of carbohydrates and list some alternative sweeteners.

4.4 Explain how carbohydrates are taken in and used by the body, including the processes of digestion, absorption, metabolism, and glucose regulation.

4.5 List the functions of carbohydrates in the body, the problems that result from not eating enough carbohydrates and the beneficial effects of fiber on the body.

Chapter 4
Carbohydrates

Think about your choice as you read Chapter 4, then see **What the Dietitian Chose** at the end of the chapter.

What Would You Choose?

It is lunchtime and you have a 30-minute break between classes. You stop at Doug's Deli for a quick sandwich. In making a sandwich choice, you remember that meals should resemble the MyPlate food pattern and that several of the food groups provide sources of carbohydrates. All of the sandwiches provide about 350 kcal. The fiber content ranges from about 1 gram to about 7.5 grams. Which sandwich would you choose to maximize the number of servings from a variety of healthy carbohydrate sources?

a Turkey and swiss on rye, with tomato slices, sliced cucumbers, romaine lettuce, and mustard

b Ham and swiss on sourdough, made with extra-lean ham and mayonnaise

c Tuna salad on whole wheat, made with tuna, grated carrots, onions, and mayonnaise and served with alfalfa sprouts, romaine lettuce, and cucumber slices

d Hot dog on a white bun, with relish, mustard, and ketchup

e Soy burger on whole-wheat English muffin, with tomato, pickle slices, romaine lettuce, and mayonnaise

f Peanut butter and jelly on soft white bread, made with strawberry jelly and smooth peanut butter

connect NUTRITION Think about your choice as you read Chapter 4, then see **What the Dietitian Chose** at the end of the chapter. To learn more about the connection between our diets and our bodies, check out the Connect site: www.mcgrawhillconnect.com.

What did you eat to obtain the energy you are using right now? Chapters 4–6 will examine this question by focusing on the main nutrients the human body uses for fuel. These nutrients are carbohydrates (on average, 4 kcal per gram) and fats and oils (on average, 9 kcal per gram). Although protein (on average, 4 kcal per gram) *can* be used for energy needs, the body typically reserves this nutrient for other processes.

It is likely that you have recently consumed fruits, vegetables, dairy products, cereal, breads, and pasta. These foods supply carbohydrates. Although some carbohydrate sources are more beneficial than others, carbohydrates should be a major part of our diets. Many people think carbohydrate-rich foods cause weight gain, but they do not any more so than fat or protein. In fact, pound for pound, carbohydrates are much less fattening than fats and oils. Furthermore, high-carbohydrate foods, especially fiber-rich foods such as fruits, vegetables, whole-grain breads and cereals, and legumes, provide many important health benefits in addition to the calories they contain. Almost all carbohydrate-rich foods, except pure sugars, provide several essential nutrients and should generally constitute 45% to 65% of our daily calorie intake. Let's take a closer look at carbohydrates.

4.6 State the RDA for carbohydrate and various guidelines for carbohydrate intake.

4.7 Identify the consequences of diabetes and explain appropriate dietary measures that will reduce the adverse effects of this health problem.

4.1 Carbohydrates—Our Most Important Energy Source

glycogen A carbohydrate made of multiple units of glucose with a highly branched structure. It is the storage form of glucose in humans and is synthesized (and stored) in the liver and muscles.

Carbohydrates are a main fuel source for some cells, especially those in the brain, nervous system, and red blood cells. Muscles also rely on a dependable supply of carbohydrates to fuel intense physical activity. Carbohydrates provide on average 4 kcal per gram and are a readily available fuel for all cells, both in the form of blood glucose and in the form of **glycogen** stored in the liver and muscles. The glycogen stored in the liver can be used to maintain blood glucose concentrations in times when you have not eaten for several hours or the diet does not supply enough carbohydrates. Regular intake of carbohydrates is important because liver glycogen stores are depleted in about 18 hours if no carbohydrates are consumed. After that point, the body is forced to produce carbohydrates, largely from breakdown of proteins in the body. This eventually leads to health problems, including the loss of muscle tissue. To obtain adequate energy, the Food and Nutrition Board of the Institute of Medicine recommends that 45% to 65% of the calories we consume each day be from carbohydrates. (See the Acceptable Macronutrient Distribution Range table in the DRI charts on the last five pages and the inside back cover of this book.)

Despite their important role as a calorie source, some forms of carbohydrate promote health more than others. As you will see in this chapter, whole-grain breads and cereals have greater health benefits than refined and processed forms of carbohydrate. Choosing the healthiest carbohydrate sources most often, while moderating intake of less healthful sources, contributes to a healthy diet. It is difficult to eat so little carbohydrate that body fuel needs are not met, but it is easy to overconsume the simple carbohydrates that can contribute to health problems. Let's explore this concept further as we look at carbohydrates in detail.

Green plants synthesize the carbohydrates in our foods. Leaves capture the sun's solar energy in their cells and transform it into chemical energy. This energy is then stored in the chemical bonds of the carbohydrate glucose as it is produced from carbon dioxide in the air and water in the soil. This complex process is called **photosynthesis** (Fig. 4-1).

▲ Fruits such as oranges and pears are an excellent source of carbohydrates, especially simple sugars and fiber.

photosynthesis Process by which plants use energy from the sun to synthesize energy-yielding compounds, such as glucose.

$$\text{6 carbon dioxide} + \text{6 water} + \text{solar energy} \rightarrow \text{glucose} + \text{6 oxyger}$$
$$(CO_2) \qquad (H_2O) \qquad\qquad (C_6H_{12}O_6) \qquad (O_2)$$

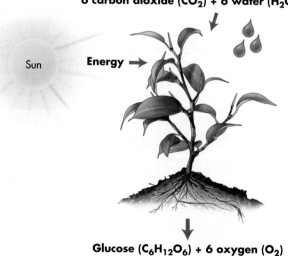

6 carbon dioxide (CO_2) + 6 water (H_2O)

Sun

Energy →

Glucose ($C_6H_{12}O_6$) + 6 oxygen (O_2)

FIGURE 4-1 ▲ A summary of photosynthesis. Plants use carbon dioxide, water, and energy to produce glucose. Glucose is then stored in the leaf and can also undergo further metabolism to form starch and fiber in the plant. With the addition of nitrogen from soil or air, glucose can also be transformed into protein.

Translated into English, this reads: 6 molecules of carbon dioxide combine with 6 molecules of water to form 1 molecule of **glucose.** Converting solar energy into chemical bonds in the sugar is a key part of the process. Six molecules of oxygen are then released into the air.

✔ CONCEPT CHECK 4.1

1. Why are carbohydrates considered our most valuable energy source?

4.2 Forms of Carbohydrates

As the name suggests, most carbohydrate molecules are composed of carbon, hydrogen, and oxygen atoms. Simple forms of carbohydrates are called **sugars.** Larger, more complex forms are primarily called either **starches** or **fibers,** depending on their digestibility by human GI tract enzymes. Starches are digestible, whereas fibers are not. The Concept Map on page 125 summarizes the forms and characteristics of carbohydrates.

The simple carbohydrates contain only one or two sugar units and are called monosaccharides and disaccharides, respectively. Food labels lump all of these sugars under one category, listing them as "sugars."

MONOSACCHARIDES—GLUCOSE, FRUCTOSE, AND GALACTOSE

Monosaccharides are the **simple sugar** units (*mono* means one) that serve as the basic unit of all carbohydrate structures. The most common monosaccharides in foods are glucose, fructose, and galactose (Fig. 4-2).

Glucose is the major monosaccharide found in the body. Glucose is also known as *dextrose,* and glucose in the bloodstream may be called blood sugar. Glucose is an important source of energy for human cells, although foods contain very little carbohydrate as this single sugar. Most glucose comes from the digestion of starches and **sucrose** (common table sugar) from our food. The latter is made up of the monosaccharides glucose and fructose. For the most part, sugars and other carbohydrates in foods are eventually converted into glucose in the liver. This glucose then goes on to serve as a source of fuel for cells.

Fructose, also called *fruit sugar,* is another common monosaccharide. Fructose is found naturally in fruits and forms half of each sucrose molecule. After it is consumed, fructose is absorbed by the small intestine and then transported to the liver, where it is quickly metabolized. Much is converted to glucose, but the rest goes on to form other compounds, such as fat, if fructose is consumed in very high amounts. Most of the free fructose in our diets comes from the use of **high-fructose corn syrup (HFCS)** in soft drinks, candies, jams, jellies, and many other fruit products and desserts (see the discussion on nutritive sweeteners in Section 4.3). Fructose is also found naturally in fruits and forms half of each sucrose molecule.

The sugar **galactose** has nearly the same structure as glucose. Large quantities of pure galactose do not exist in nature. Instead, galactose is usually found bonded to glucose in **lactose,** a sugar found in milk and other milk products. After lactose is digested and absorbed, galactose arrives in the liver. There it is either transformed

glucose A six-carbon monosaccharide that usually exists in a ring form; found as such in blood and in table sugar bonded to fructose; also known as *dextrose.*

sugar A simple carbohydrate with the chemical composition $(CH_2O)_n$. The basic unit of all sugars is glucose, a six-carbon ring structure. The primary sugar in the diet is sucrose, which is made up of glucose and fructose.

starch A carbohydrate made of multiple units of glucose attached together in a form the body can digest; also known as *complex carbohydrate.*

fiber Substances in plant foods not digested by the processes that take place in the stomach or small intestine. These add bulk to feces. Fibers naturally found in foods are also called dietary fiber.

monosaccharide Simple sugar, such as glucose, that is not broken down further during digestion.

simple sugar Monosaccharide or disaccharide in the diet.

sucrose Fructose bonded to glucose; table sugar.

fructose A six-carbon monosaccharide that usually exists in a ring form; found in fruits and honey; also known as *fruit sugar.*

high-fructose corn syrup Corn syrup that has been manufactured to contain between 42% and 90% fructose.

galactose A six-carbon monosaccharide that usually exists in a ring form; closely related to glucose.

lactose Glucose bonded to galactose; also known as *milk sugar.*

Monosaccharides

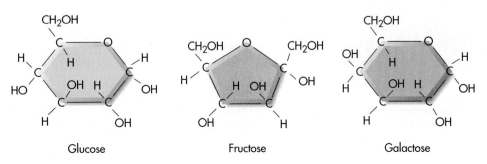

Glucose   Fructose   Galactose

FIGURE 4-2 ▲ Chemical forms of the important monosaccharides.

Disaccharides

Sucrose: glucose + fructose
Lactose: glucose + galactose
Maltose: glucose + glucose

Sucrose

FIGURE 4-3 ▲ Chemical form of the disaccharide sucrose.

disaccharide Class of sugars formed by the chemical bonding of two monosaccharides.

maltose Glucose bonded to glucose.

fermentation The conversion of carbohydrates in to alcohols, acids, and carbon dioxide without the use of oxygen.

polysaccharides Carbohydrates containing many glucose units, from 10 to 1000 or more.

amylose A digestible straight-chain type of starch composed of glucose units.

amylopectin A digestible branched-chain type of starch composed of glucose units.

▲ Root vegetables such as potato, yams, and tapioca are high in amylopectin starch.

into glucose or further metabolized into glycogen. This is a good example of the body not using all nutrients in their original states. Some of these substances are broken down and later reassembled into the same or a different substance when and where they are needed. Much of the galactose in the diet is metabolized to glucose. When later required for the production of milk in the mammary gland of a lactating female, galactose is resynthesized from glucose to help form the milk sugar lactose. Therefore, it is not necessary for a lactating women to drink milk to make milk!

DISACCHARIDES: SUCROSE, LACTOSE, AND MALTOSE

Disaccharides are formed when two monosaccharides combine (*di* means two). The disaccharides in food are sucrose, lactose, and **maltose.** All contain glucose.

Sucrose forms when the two sugars glucose and fructose bond together (Fig. 4-3). Sucrose is found naturally in sugarcane, sugar beets, honey, and maple sugar. These products are processed to varying degrees to make brown, white, and powdered sugars. Animals do not produce sucrose or much of any carbohydrate except glycogen.

Lactose forms when glucose bonds with galactose during the synthesis of milk. Therefore, our major food source for lactose is milk products. The discussion in Section 4.4 on lactose maldigestion and lactose intolerance looks at the problems that result when a person can't readily digest lactose.

Maltose results when starch is broken down to just two glucose molecules bonded together. Maltose plays an important role in the beer and liquor industry. In the production of alcoholic beverages, starches in various cereal grains are first converted to simpler carbohydrates by enzymes present in the grains. The products of this step—maltose, glucose, and other sugars—are then mixed with yeast cells in the absence of oxygen. The yeast cells convert most of the sugars in to alcohol (ethanol) and carbon dioxide through a process called **fermentation.** Little maltose remains in the final product. Few other food products or beverages contain maltose. In fact, most maltose that we ultimately digest in the small intestine is produced during our own digestion of starch.

COMPLEX CARBOHYDRATES

In many foods, many single-sugar units are bonded together to form a chain, known as a polysaccharide (*poly* means many). **Polysaccharides,** also called *complex carbohydrates* or *starch,* may contain 1000 or more glucose units and are found chiefly in grains, vegetables, and fruits. When the Nutrition Facts panel on food labels lists "Other Carbohydrates," this primarily refers to starch content.

Plants store carbohydrates in two forms of starch digestible by humans: **amylose** and **amylopectin.** Amylose, a long, straight chain of glucose units, comprises about 20% of the digestible starch found in vegetables, beans, breads, pasta, and rice. Amylopectin is a highly branched chain and makes up the remaining 80% of digestible starches in the diet (Fig. 4-4). Cellulose (a fiber) is another complex carbohydrate in plants. Although similar to amylose, it cannot be digested by humans, as discussed in the next section.

The enzymes that break down starches to glucose and other related sugars act only at the end of a glucose chain. Amylopectin, because it is branched, provides many more sites (ends) for enzyme action. Therefore, amylopectin is digested more rapidly and raises blood glucose much more readily than amylose (see the discussion of glycemic index in Section 4.5).

FIGURE 4-4 ▼ Some common starches, amylose and amylopectin, and glycogen. We consume essentially no glycogen. All glycogen found in the body is made by our cells, primarily in the liver and muscles.

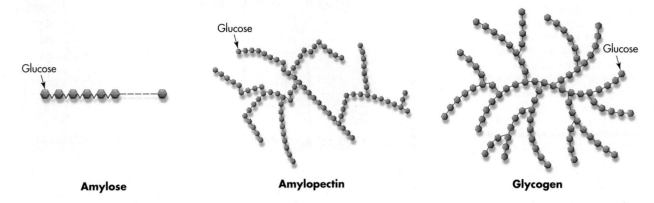

Amylose   **Amylopectin**   **Glycogen**

As noted earlier, animals—including humans—store glucose in the form of glycogen. Glycogen consists of a chain of glucose units with many branches, providing even more sites for enzyme action than amylopectin (review Fig. 4-4). Because of its branched structure that can be broken down quickly, glycogen is an ideal storage form of carbohydrate in the body.

The liver and muscles are the major storage sites for glycogen. Because the amount of glucose immediately available in body fluids can provide only about 120 kcal, the carbohydrate energy stored as glycogen—amounting to about 1800 kcal—is extremely important. Of this 1800 kcal, liver glycogen (about 400 kcal) can readily contribute to blood glucose. Muscle glycogen stores (about 1400 kcal) cannot raise blood glucose but instead supply glucose for muscle use, especially during high-intensity and endurance exercise. Although animals store glycogen in their muscles, animal products such as meats, fish, and poultry are not good sources of carbohydrates because glycogen stores quickly degrade after the animal dies.

FIBER

Fiber as a class is mostly made up of polysaccharides, but fibers differ from starches insofar as the chemical bonds that join the individual sugar units cannot be digested by human enzymes in the GI tract. This prevents the small intestine from absorbing the sugars because they cannot be released from the various fibers. Fiber is not a single substance but a group of substances with similar characteristics. The group is composed of the carbohydrates **cellulose, hemicelluloses, pectins,** gums, and **mucilages,** as well as the noncarbohydrate **lignin.** In total, these constitute all the nonstarch polysaccharides in foods. Nutrition Facts labels generally do not list these individual forms of fiber but instead lump them together under the term **dietary fiber.**

Cellulose, hemicelluloses, and lignin form the structural parts of plants. Bran layers form the outer covering of all grains, so **whole grains** (i.e., unrefined) are good sources of bran fiber (Fig. 4-5). Bran fiber is rich in hemicelluloses and lignin. (The woody fibers in broccoli are partly lignin.) Because the majority of these fibers neither readily dissolve in water nor are easily metabolized by intestinal bacteria, they are called **nonfermentable** or insoluble fibers. Insoluble fiber is found in wheat bran, nuts, fruit skins, and some vegetables. Insoluble fiber acts as a natural laxative because it speeds up the transit time of food through the GI tract.

cellulose An undigestible nonfermentable straight-chain polysaccharide made of glucose molecules.

hemicellulose A nonfermentable fiber containing xylose, galactose, glucose, and other monosaccharides bonded together.

pectin A viscous fiber containing chains of galacturonic acid and other monosaccharides; characteristically found between plant cell walls.

mucilages A viscous fiber consisting of chains of galactose, mannose, and other monosaccharides; characteristically found in seaweed.

lignins A nonfermentable fiber made up of a multiringed alcohol (noncarbohydrate) structure.

dietary fiber Fiber found in food.

whole grains Grains containing the entire seed of the plant, including the bran, germ, and endosperm (starchy interior). Examples are whole wheat and brown rice.

nonfermentable fiber A fiber that is not easily metabolized by intestinal bacteria; also called *insoluble fiber.*

FIGURE 4-5 ▶ Viscous (soluble) and nonfermentable (insoluble) fiber. (a) The skin of an apple consists of the nonfermentable fiber cellulose, which provides structure for the fruit. The viscous fiber pectin "glues" the fruit cells together. (b) The outside layer of a wheat kernel is made of layers of bran—primarily hemicellulose, a non-fermentable fiber—making this whole grain a good source of fiber. Overall, fruits, vegetables, whole-grain breads and cereals, and beans are rich in fiber.

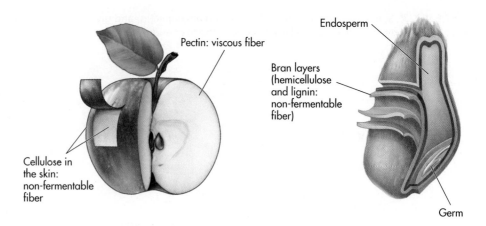

Pectin: viscous fiber

Cellulose in the skin: non-fermentable fiber

Endosperm

Bran layers (hemicellulose and lignin: non-fermentable fiber)

Germ

viscous fiber A fiber that is readily fermented by bacteria in the large intestine; also called *soluble fiber*.

functional fiber Fiber added to foods that has been shown to provide health benefits.

Pectins, gums, and mucilages are contained around and inside plant cells. These fibers either dissolve or swell when put into water and are therefore called **viscous** or soluble fibers. They also are readily fermented by bacteria in the large intestine. These fibers are found in beans, oats, oat bran, and some fruits and vegetables, as well as salad dressings, some frozen desserts, jams, and jellies as gum arabic, guar gum, locust bean gum, and various pectin forms. Soluble fiber slows the rate of absorption by attracting water into the GI tract, reduces blood cholesterol, and controls blood glucose.

Most foods contain mixtures of soluble and insoluble fibers. Food labels do not generally distinguish between the two types, but manufacturers have the option to do so. Often, if food is listed as a good source of one type of fiber, it usually contains some of the other type of fiber as well. The definition of fiber has recently been expanded to include both the dietary fiber, which describes the nondigestible carbohydrates and lignin that are naturally occurring and intact in plants, and **functional fiber,** which consists of the isolated nondigestible carbohydrates that are added to food because they have beneficial physiological effects in human beings. The commercially produced functional or isolated fibers include resistant starch, polydextrose, indigestible dextrins, and inulin, which is one of the most popular varieties currently added to products. The health benefits of many of these fibers are still unclear and are therefore a hot discussion topic (see Further Readings 7 and 10).

Many of these fibers fall into a category of functional fiber called the prebiotics. Prebiotics include a group of short-chain carbohydrates or oligosaccharides, resistant to digestion but fermented by bacteria in the colon. They are thought to stimulate the growth or activity of beneficial bacteria in the large intestine and therefore promote the host's health.

☑ CONCEPT CHECK 4.2

1. What are the names and definitions of the monosaccharides and disaccharides, and what happens to them when they are digested and absorbed?

2. What is a polysaccharide and what are the differences between the plant polysaccharides?

3. What are the name and locations of the storage form of glucose in the body?

4. What makes fiber a very different kind of carbohydrate, and what is the difference between nonfermentable, viscous, and functional fiber?

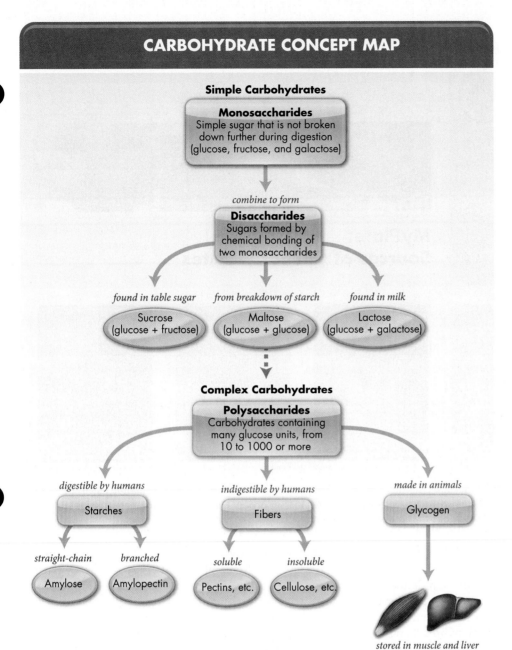

CARBOHYDRATE CONCEPT MAP

Simple Carbohydrates

Monosaccharides
Simple sugar that is not broken down further during digestion (glucose, fructose, and galactose)

combine to form

Disaccharides
Sugars formed by chemical bonding of two monosaccharides

found in table sugar
Sucrose (glucose + fructose)

from breakdown of starch
Maltose (glucose + glucose)

found in milk
Lactose (glucose + galactose)

Complex Carbohydrates

Polysaccharides
Carbohydrates containing many glucose units, from 10 to 1000 or more

digestible by humans
Starches

straight-chain
Amylose

branched
Amylopectin

indigestible by humans
Fibers

soluble
Pectins, etc.

insoluble
Cellulose, etc.

made in animals
Glycogen

stored in muscle and liver

4.3 Carbohydrates in Foods

Four of the groups on MyPlate—grains, vegetables, fruits, and dairy—contain the most nutrient-dense sources of carbohydrates (Fig. 4-6). In planning a healthy, carbohydrate-rich diet, you need to emphasize foods from these groups that do not contain too much sugar or fat. Most of the carbohydrate in our diets comes from starches. Because plants store glucose in the form of starches, plant-based foods, such as beans, potatoes, and the grains used to make breads, cereals, and pasta, are the best sources of starch. A diet rich in these starches provides plenty of carbohydrates, as well as many micronutrients and phytochemicals. Fiber is also found in many of the same foods as starch, so a diet rich in whole grains, beans, and potatoes also can provide significant amounts of dietary fiber. Soluble fibers (pectin, gums, and mucilages) are found in the skins and flesh of many fruits and berries; as thickeners and stabilizers in jams, yogurts, sauces, and fillings; and in products that contain psyllium and seaweed. Fiber is also available as a supplement or as an additive to certain foods (functional fiber) such that individuals with relatively low intakes of natural dietary fiber can still obtain the health benefits of fiber.

Unfortunately, the top five carbohydrate sources for U.S. adults in recent years have been white bread; soft drinks; cookies and cakes (including doughnuts);

MyPlate:
Sources of Carbohydrates

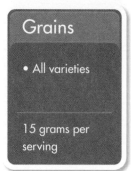

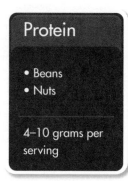

Grains	Vegetables	Fruits	Dairy	Protein
• All varieties	• All varieties	• All varieties	• Milk • Yogurt	• Beans • Nuts
15 grams per serving	5 grams per serving	18 grams per serving	12 grams per serving	4–10 grams per serving

FIGURE 4-6 ▲ Sources of carbohydrates from MyPlate. The fill of the background color (none, 1/3, 2/3, or completely covered) within each group in the plate indicates the average nutrient density for carbohydrates in that group. Overall, the grain group, vegetable group, fruit group, and dairy group contain many foods that are nutrient-dense sources of carbohydrate.

▲ How does this lunch of a turkey, cheese, lettuce, and tomato sandwich on whole-grain bread with an apple and a glass of low-fat milk compare to MyPlate?

sugars, syrups, and jams; and potatoes. Remember that these foods are on the Dietary Guidelines list of Foods and Food Components to Reduce. Many North Americans are starting to take a closer look at their carbohydrate sources and improve them from a nutritional standpoint by including more whole-grain versions of breads, pasta, rice, and cereals, as well as fruits and vegetables. The decline in soft drink consumption in the United States over the past 8 years is particularly encouraging, with shoppers turning to water, energy drinks, and coffee instead. Data from *Beverage Digest* indicates that Americans still drink an average of 44 gallons of soda a year, but this is an 18.5% drop from the peak in 1998.

The top 10 foods with the highest percentage of carbohydrate by weight are shown on page 127. While these total amounts of carbohydrates in foods are important to know, it is more significant to understand the percentage of calories from carbohydrates when planning a healthy diet. The food components that yield the highest percentage of calories from carbohydrates are table sugar, honey, jam, jelly, fruit, and plain baked potatoes (Fig. 4-7). Corn flakes, rice, bread, and noodles are next, all containing at least 75% of calories as carbohydrates. Foods with moderate amounts of carbohydrate calories are peas, broccoli, oatmeal, dry beans and other legumes, cream pies, French fries, and fat-free milk. In these foods, the carbohydrate content is diluted either by protein, as in the case of fat-free milk, or by fat, as in the case of cream pies. Foods with essentially no carbohydrates include beef, eggs, chicken, fish, vegetable oils, butter, and margarine.

WHOLE GRAINS

The 2010 Dietary Guidelines recommend that we consume at least half of all grains as whole grains; increase whole-grain intake by replacing refined grains with whole grains; and limit the consumption of foods that contain refined grains, especially refined

Food Sources of Carbohydrates

Food Item	Carbohydrate (grams)	% RDA
RDA	130	100%
Baked potato, 1 each	51	39%
Cola drink, 12 fluid oz	39	30%
Plain M&Ms, ½ oz	30	23%
Banana, 1 each	28	22%
Cooked rice, ½ cup	22	17%
Cooked corn, ½ cup	21	16%
Light yogurt, 1 cup	19	15%
Kidney beans, ½ cup	19	15%
Spaghetti noodles, ½ cup	19	15%
Orange, 1 each	16	12%
Seven-grain bread, 1 slice	12	9%
Fat-free milk, 1 cup	12	9%
Pineapple chunks, ½ cup	10	8%
Cooked carrots, ½ cup	8	6%
Peanuts, 1 oz	6	5%

Key:
- Grains
- Vegetables
- Fruits
- Dairy
- Protein
- Empty calories

ChooseMyPlate.gov

FIGURE 4-7 ◄ Food sources of carbohydrate compared to the RDA of 130 grams for carbohydrate.

Top 10 foods with the highest percentage of carbohydrate

Table sugar	99.9%
Hard candies & gummies	98–99%
Sugary cereals	90–93%
Dried fruits	75–90%
Cakes & cookies	84%
Low-fat crackers, rice cakes & potato chips	82%
Jams & preserves	64–68%
Hash browns	35%
French fries	27%
Baked potato	21%
Low-fat salad dressings	32%
Pizza	22–30%

grain foods that contain solid fats, added sugars, and sodium. The Guidelines define whole grain as the entire grain seed or kernel made of three components: bran, germ, and endosperm. When the term *whole grain* is used on a food package, it also means that the product contains a minimum of 51% whole-grain ingredients by weight per serving. In contrast, refined grains typically contain only the starchy endosperm portion of the seed or kernel. These highly processed grains are low in fiber because much of the fiber in whole grains is found in the outer layers, which are removed in processing.

Although more fiber is one of the primary advantages of whole grains, many benefits of whole grains are thought to be due to the combined effects of several compounds. These compounds include fiber, minerals, trace minerals, vitamins, carotenoids, and other phytochemicals and are mainly contained in the bran and germ parts of the grains. The 2010 Dietary Guidelines recommend consuming two to three servings of whole grains per day. Several studies have shown that this is enough to impart numerous health benefits, including reducing risks of cardiovascular disease, diabetes, metabolic syndrome, some cancers, and obesity.

It is estimated that a staggering 9 in 10 people don't meet the whole-grain recommendation of three servings per day. Although Americans like grains and consume an average of 6.4 ounces of them daily, only 0.6 ounces are in the form of whole grains. This average consumption of less than one serving of whole grains per day falls far short of the recommendations. There are several explanations for why Americans appear unexcited about whole grains. The reasons include differences in the taste, texture, cost, and availability of whole grains, compared to products made with refined flour.

In addition to our preference for refined grains, many consumers who are trying to choose a whole-grain product are very confused by the deceptive marketing messages on the labels of grain products. For example, a label that says a cereal is "made with whole grains" does not guarantee that the cereal is 100% whole grain. Terms such as *cracked wheat bread, stoneground wheat, enriched wheat flour, 12-grain bread,* and *multigrain* are confusing when, in fact, these products may contain little to

Is juicing healthier than eating whole fruits or vegetables?

Juicing has become a popular way to consume fruits and vegetables and can be a fun way to drink the fruits and vegetables you do not enjoy eating. The good news is that most of the vitamins, minerals, and phytochemicals remain in the juice that is extracted from the fruits and vegetables during the juicing process. Also, you control the amount of added sugars and preservatives in the end product. The downside of the juicing process is that the natural fiber found in whole fruits and vegetables is typically lost. This means that the juice is not healthier than whole fruit. In fact, there is no scientific evidence to support claims that juice extracts provide any health benefits or that nutrients are better absorbed from juice. If you enjoy juicing, remember that juices from both fruits and vegetables contain a significant amount of natural sugar and, if consumed in excess, those calories can quickly add up! To make your juice healthier, add some of the pulp back to obtain valuable fiber that can help you feel full. Try using the pulp as a creative way to add fiber to recipes, such as tuna salad, quick breads, pancakes, or pasta dishes.

Is White Whole Wheat Really Whole Grain?

White whole wheat is a whole grain. It has the nutritional benefits of whole wheat but a milder taste, softer texture, and the lighter color of white bread. Traditional whole wheat is made from red wheat, which has a darker color and strongly flavored phenolic compounds. Switching to white whole wheat may be an acceptable option for those who prefer the taste and texture of white bread.

▲ The Whole Grains Council developed the Whole Grain Stamp for use on grain products to help us identify whole-grain foods. There are two versions of the stamp. The 100% stamp is used if all grain ingredients are whole grains and there is a minimum of 16 grams (a full serving) of whole grain per serving. The Basic Stamp is used if a product contains at least 8 grams (a half serving) of whole grain, but it may also contain some refined grain.

Source: Whole Grains Council, An Oldways Program.

no whole grain. Multigrain cereals may contain multiple grains, but many of them may be refined with just a small amount of whole grains added to a substantial amount of refined grains. Some "whole-grain breads" are white bread in disguise because brown coloring is added to enriched white flour (see Further Reading 6).

With all of the confusing buzzwords, it is crucial to look beyond the front-of-the-package claims and examine the list of ingredients. To confirm that products contain 100% whole grain, look for *whole* as the first word on the ingredient list. Sugary breakfast cereals that claim to be whole grain may list the first ingredient as a whole grain such as corn, rice, oat, or wheat, but then the next several ingredients may be various forms of sugar that together weigh more than the whole grain (see Further Readings 13 and 15).

Although whole grains make up only about 10% to 15% of grains on the grocery shelves, it is getting easier to find whole-wheat products and other whole-grain pastas. See Table 4-1 for information on several whole grains that are available, including their potential health benefits. To help simplify the process of finding whole-grain foods, the Whole Grains Council developed the Whole Grain Stamp for food manufacturers to use on grain products. There are two versions of the stamp. The 100% Stamp is used on a product if all of its grain ingredients are whole grains. It must also have a minimum of 16 grams (a full serving) of whole grain per serving. The Basic Stamp may be used if a product contains at least 8 grams (a half serving) of whole grain, but it may also contain some refined grain. The Basic Stamp is also used even if a product contains large amounts of whole grain (23 grams, 37 grams, 41 grams, etc.) but also contains extra bran, germ, or refined flour. This reminds us that even if a product has whole grains, it can also contain a lot of sugar or solid fats. As of November 2012, the Whole Grain Stamp is on over 8000 different products and in over 41 countries. Choosing products with the 100% Whole Grain Stamp will help us reach the minimum goal to get three servings of whole grains per day (see Further Reading 6). Schools are doing their part to increase whole-grain consumption by children by meeting requirements of the Healthy, Hunger-Free Kids Act of 2010. Foods must contain at least 50% whole grains to meet the whole-grain–rich criteria for the federal school meal and child nutrition.

VEGETABLES

Vegetables are a valuable source of carbohydrates in the form of starch and fiber. They are naturally low in fat and calories and come packed with many other nutrients vital for health, including potassium, folate, vitamin A, and vitamin C. Eating the recommended amount of vegetables has been shown to reduce the risk of several chronic diseases. As we discussed with whole grains, the fiber in vegetables may reduce the risk of heart disease, obesity, and type 2 diabetes. Within the MyPlate guidelines, any vegetable or 100% vegetable juice counts as a member of the Vegetable Group. Vegetables are organized into five subgroups (dark green vegetables, starchy vegetables, red and orange vegetables, beans and peas, and other vegetables) based on their nutrient content. Vegetable choices may be raw or cooked; fresh, frozen, canned, or dried/dehydrated; and whole, cut-up, or mashed but should be selected from among the vegetable subgroups. Although it is not necessary to eat vegetables from each subgroup daily, the amounts listed from each subgroup should be eaten over a week as a way to reach your daily intake recommendation.

The amount of vegetables you need depends on your age, sex, and level of physical activity. For example, recommended total daily amounts for women 19 to 50 years old are 2.5 cups. In general, 1 cup of raw or cooked vegetables or vegetable juice or 2 cups of raw leafy greens can be considered as "1 cup" from the Vegetable Group. Recommended weekly amounts from each vegetable subgroup are given as amounts to eat weekly. For example, the recommendations for women 19 to 50 years old are 1.5 cups of dark green vegetables, 5.5 cups of red and orange vegetables, 1.5 cups of beans and peas, 5 cups of starchy vegetables, and 4 cups of other vegetables such as cauliflower or mushrooms.

TABLE 4-1 ▶ Know Your Whole Grains

Grain	Characteristics	How to Spot Whole Grain	Health Benefits
Barley	Highest in fiber with a tough hull, which is difficult to remove without losing some of the bran; very slow-cooking	Look for whole barley, hulled barley, or hull-less barley. Pearled barley is not technically a whole grain because small amounts of the bran are missing.	Barley fiber may lower cholesterol even more effectively than oat fiber.
Buckwheat	High levels of the antioxidant rutin and a high level of protein; cousin of rhubarb, not technically a grain at all—and certainly not a kind of wheat	If buckwheat is on ingredient list, it is almost invariably whole buckwheat.	Rutin improves circulation and prevents LDL cholesterol from blocking blood vessels.
Corn	Known for its sweet flavor	Avoid labels that say "degerminated" and look for the words *whole corn.*	Corn has highest level of antioxidants of any grain or vegetable.
Oats	High in protein and almost never have their bran and germ removed in processing. Sweet flavor make them popular for breakfast cereals.	*Oats, oatmeal,* or *oat groats* on an ingredient list are almost invariably whole oats. Steel-cut oats are the entire oat kernel sliced into smaller pieces to help water penetrate and cook the grain. In the United States, most oats are steamed and flattened to produce "old-fashioned" or regular oats, quick oats, and instant oats, which cook more quickly.	Oat fiber is especially effective in lowering cholesterol.
Quinoa	High in high-quality protein, quinoa is a small, light-colored round grain, similar in appearance to sesame seeds.	*Quinoa* on ingredient list almost invariably means whole quinoa.	Complete protein that contains all the essential amino acids.
Rice	Many whole-grain varieties, including brown, black, purple, or red. Brown rice is lower in fiber than most other whole grains but rich in many nutrients. White rice is refined, with the germ and bran removed.	Brown rice is always whole grain, as are most other colored rices, such as black rice or red rice.	One of the most easily digested grains; ideal for those on a restricted diet or who are gluten-intolerant.
Rye	Rye is unusual among grains for the high level of fiber in its endosperm—not just in its bran.	Look for whole rye or rye berries in the ingredient list.	Rye fiber promotes a rapid feeling of fullness, making rye a good choice for people trying to lose weight. Rye products generally have a lower glycemic index than products made from wheat and most other grains, making them a good choice for diabetics.
Wheat	Dominates the grains we eat because it contains large amounts of gluten, a stretchy protein that enables bakers to create satisfying risen breads.	Look for the term *whole wheat* (in Canada, for the term *whole-grain whole wheat*).	Since wheat is the most common grain eaten in the United States, most U.S. studies of "whole grains" can attest to the benefits of whole wheat. These benefits include reduced risk of stroke, type 2 diabetes, heart disease, inflammatory disease, and asthma; and better weight maintenance and healthier blood pressure levels.

Source: Adapted from the Whole Grains Council, Whole Grains A to Z, at **http://wholegrainscouncil.org/whole-grains-101/whole-grains-a-to-z.**

FRUITS

Fruits provide carbohydrates primarily in the form of natural sugar and fiber. Eating fruits provides health benefits similar to those discussed for vegetables; people who eat more fruits as part of an overall healthy diet are more likely to have a reduced risk of several chronic diseases. Dietary fiber from fruits helps reduce blood cholesterol levels, may lower risk of heart disease, and is important for proper bowel function. The fiber in fruits helps with weight maintenance by providing a feeling of fullness with fewer calories. Remember that whole or cut-up fruits are great sources of dietary fiber, compared to fruit juices that contain little or no fiber.

DAIRY

Foods in the Dairy Group provide carbohydrates in the form of lactose. Dairy products also provide other key nutrients, including calcium, potassium, vitamin D, and protein, which are responsible for many health benefits, especially improved bone health. While calcium-fortified soymilk (soy beverage) is part of the Dairy Group, foods made from milk that are primarily fat, such as cream cheese, cream, and butter, are not part of this group. The lactose content of dairy products varies somewhat. There are approximately 5 to 8 grams of lactose in the following high-lactose foods: ½ cup of fluid milk (whole, reduced-fat, fat-free, and buttermilk), half-and-half, plain yogurt, sour cream, and heavy cream; and ¾ cup of cottage cheese, ricotta cheese, ice cream, and ice milk. The following low-lactose foods contain about 0 to 2 grams of lactose: ½ cup milk treated with lactase enzyme and sherbet; and 1 ounce aged cheese (blue, brick, cheddar, Colby, Swiss, and Parmesan) and processed cheese. Some individuals experience the condition of lactose intolerance and maldigestion when they consume high-lactose foods. This situation is discussed in Section 4.4, "Making Carbohydrates Available for Body Use." The amount of food from the Dairy Group you need to eat depends on your age, with the recommended daily amounts being 3 cups for all males and females 9 years and older. Typically, 1 cup of milk, yogurt, or soymilk (soy beverage), 1½ ounces of natural cheese, or 2 ounces of processed cheese are considered as 1 cup from the Dairy Group.

NUTRITIVE SWEETENERS

The various substances that impart sweetness to foods fall into two broad classes: nutritive sweeteners, which can provide calories for the body, and alternative sweeteners, which for the most part provide no calories. As given in Table 4-2, the alternative sweeteners are much sweeter on a per-gram basis than the nutritive sweeteners. The taste and sweetness of sucrose make it the benchmark against which all other sweeteners are measured. Sucrose is obtained from sugarcane and sugar beet plants. Both sugars and sugar alcohols provide calories along with sweetness. Sugars are found in many different food products, whereas sugar alcohols have rather limited uses.

Sugars. All of the monosaccharides (glucose, fructose, and galactose) and disaccharides (sucrose, lactose, and maltose) discussed earlier are designated *nutritive sweeteners* because they provide calories. Many forms of sugar, including those naturally present in food and those added during processing or preparation, add up to an average intake of about 30 teaspoons of total sugar per day. The 2010 Dietary Guidelines recommend that we reduce the intake of calories from added sugars. Added sugars are defined as caloric sweeteners added to foods during processing or preparation or before consumption. It is estimated that Americans consume an average of 20 teaspoons of added sugar per day, far exceeding the recommendation of no more than 8 teaspoons daily for adults. Data from a recent National Health and Nutrition Examination Survey (NHANES) indicate that 14.6% of our total energy intake is from added sugar and that 60% of it comes from sugar-sweetened beverages, grain-based desserts, and fruit drinks.

HFCS is a sweetener now used in all kinds of foods, from soft drinks to barbecue sauce. HFCS is made by an enzymatic process that converts some of the glucose in cornstarch into fructose, which tastes sweeter than glucose. It is called "high-fructose" corn syrup because it contains 55% fructose, compared to sucrose, which contains only 50% fructose. In the United States, corn is abundant and inexpensive compared to sugarcane or sugar beets, much of which is imported. Food manufacturers prefer HFCS because of its low cost and broad range of food processing applications, and because it is easy to transport, has better shelf-stability, and improves food properties. An average American consumes about 60 pounds of HFCS each year. There has been much confusion and controversy surrounding the use and possible health effects of HFCS. Much of the controversy stemmed from a scientific commentary that was published in 2004 and suggested a link between HFCS consumption and obesity.

▲ There are many forms of sugar on the market. Together they contribute to our daily intake of approximately 100 grams (20 teaspoons) of sugars in our diets.

TABLE 4-2 ▶ **The Sweetness of Sugars (Nutritive) and Alternative Sweeteners**

Type of Sweetener	Relative Sweetness* (Sucrose = 1)	Typical Sources
Sugars		
Lactose	0.2	Dairy products
Maltose	0.4	Sprouted seeds
Glucose	0.7	Corn syrup
Sucrose	1.0	Table sugar, most sweets
Invert sugar[†]	1.3	Some candies, honey
Fructose	1.2–1.8	Fruit, honey, some soft drinks
Sugar Alcohols		
Sorbitol	0.6	Dietetic candies, sugarless gum
Mannitol	0.7	Dietetic candies
Xylitol	0.9	Sugarless gum
Maltitol	0.9	Baked goods, chocolate, candies
Alternative Sweeteners		
Stevia (Truvia®)	100 to 300	Tabletop sweetener, food ingredient
Luo han guo (Nectresse®)	150–300	Intended for use as a tabletop sweetener, a food ingredient, and a component of other sweetener blends
Aspartame (Equal®)	180	Diet soft drinks, diet fruit drinks, sugarless gum, tabletop sweetener
Acesulfame-K (Sunette®)	200	Sugarless gum, diet drink mixes, tabletop sweetener, puddings, gelatin desserts
Saccharin (Sweet'N Low®)	300	Diet soft drinks, tabletop sweetener
Sucralose (Splenda®)	600	Diet soft drinks, tabletop sweetener, sugarless gums, jams, frozen desserts
Neotame	7000 to 13,000	Tabletop sweetener, baked goods, frozen desserts, jams
Advantame	20,000	Tabletop sweetener, soft drinks, baked goods, frozen desserts, syrups, jams

*On a per gram basis.

[†]Sucrose broken down into glucose and fructose.

Source: From the Academy of Nutrition and Dietetics, 2012, and other sources (see Further Reading 11).

▲ Soft drinks are typical sources of either sugars or alternative sweeteners, depending on the type of soft drink chosen.

Since then, the scientific community has examined this issue and concluded that there are no metabolic or endocrine response differences between HFCS and sucrose related to obesity or any other adverse health outcome (see Further Reading 12).

In addition to sucrose and HFCS, brown sugar, turbinado sugar (sold as raw sugar), honey, maple syrup, agave nectar, and other sugars are also added to foods. Brown sugar is essentially sucrose containing some molasses that is not totally removed from the sucrose during processing or is added to the sucrose crystals. Turbinado sugar is a partially refined version of raw sucrose. Maple syrup is made by boiling down and concentrating the sap that runs during the late winter in sugar maple trees. Because pure maple syrup is expensive, most pancake syrup is primarily corn syrup and HFCS with maple flavor added.

Honey is a product of plant nectar that has been altered by bee enzymes. The enzymes break down much of the nectar's sucrose into fructose and glucose. Honey offers essentially the same nutritional value as other simple sugars—a source of energy and little else. However, honey is not safe to feed to infants because it can contain spores of the bacterium *Clostridium botulinum* that causes fatal food-borne illness. Unlike the acidic environment of an adult's stomach, which inhibits the growth of the bacteria, an infant's stomach does not produce much acid, making infants susceptible to the threat that this bacterium poses.

Sugar Alcohols. Food manufacturers and consumers have numerous options for obtaining sweetness while consuming less sugar and calories. A primary purpose of sugar alcohols and alternative sweeteners is to enable people with diabetes to enjoy the flavor of sweetness while controlling sugars in their diets; they also provide noncaloric or very-low-calorie sugar substitutes for persons trying to lose (or control) body weight.

Sugar alcohols, or polyols, such as **sorbitol** and **xylitol** are used as nutritive sweeteners but contribute fewer calories (about 2.6 kcal per gram) than sugars. They are also absorbed and metabolized to glucose more slowly than are simple sugars. Because of this, they remain in the intestinal tract for a longer time and in large quantities can cause diarrhea. In fact, any products that may be consumed in amounts that result in a daily ingestion of 50 grams of sugar alcohols must bear this labeling statement: "Excess consumption may have a laxative effect."

Sugar alcohols must be listed on labels. If only one sugar alcohol is used in a product, its name must be listed. However, if two or more are used in one product, they are grouped together under the heading "sugar alcohols." The caloric value of each sugar alcohol used in a food product is calculated so that when one reads the total amount of calories a product provides, it includes the sugar alcohols in the overall amount.

Sugar alcohols are used in sugarless gum, breath mints, and candy. Unlike sucrose, sugar alcohols are not readily metabolized by bacteria to acids in the mouth and thus do not promote tooth decay.

ALTERNATIVE SWEETENERS

Unlike sugar alcohols, alternative sweeteners yield little or no calories when consumed in amounts typically used in food products. In addition, they are not metabolized by bacteria in the mouth, so they do not promote dental caries. Eight alternative sweeteners (Table 4-2) are currently available in the United States.

For each sweetener, FDA determines an **Acceptable Daily Intake (ADI)** guideline. ADIs are set at a level 100 times less than the level at which no harmful effects were noted in animal studies. Alternative sweeteners can be used safely by adults and children, and are considered safe during pregnancy.

Saccharin. The oldest alternative sweetener, saccharin represents about half of the alternative sweetener market in North America (typically packaged in pink packets, including Sweet 'N Low®). Based on laboratory animal studies, saccharin was once thought to increase the risk of bladder cancer, but it is no longer listed as a potential cause of cancer. The FDA lists the ADI for saccharin at 5 milligrams per kilogram, or the equivalent of 9 to 12 packets of the sweetener.

Aspartame. Aspartame is in widespread use throughout the world (typically packaged in blue packets, including Equal®). It has been approved for use by more than 90 countries, and its use has been endorsed by the World Health Organization, the American Medical Association, the American Diabetes Association, and other groups.

The components of aspartame are the amino acids phenylalanine and aspartic acid, along with methanol. Recall that amino acids are the building blocks of proteins, so aspartame is more like a protein than a carbohydrate. Like protein, aspartame yields about 4 kcal per gram, but because it is about 200 times sweeter than sucrose, only a small amount is needed to obtain the desired sweetness. Like other proteins, however, aspartame is damaged when heated for a long time and thus would lose its sweetness if used in products requiring cooking. Aspartame is used in beverages, gelatin desserts, chewing gum, and toppings.

The ADI of aspartame is 50 milligrams per kilogram of body weight. This is equivalent to about 18 cans of aspartame-containing diet soft drink or 8 packets of

sorbitol Alcohol derivative of glucose that yields about 3 kcal per gram but is slowly absorbed from the small intestine; used in some sugarless gums and dietetic foods.

xylitol Alcohol derivative of the five-carbon monosaccharide xylose.

Acceptable Daily Intake (ADI) Estimate of the amount of a sweetener that an individual can safely consume daily over a lifetime. ADIs are given as milligrams per kilograms of body weight per day.

saccharin Alternative sweetener that yields no energy to the body; 300 times sweeter than sucrose.

aspartame Alternative sweetener made of two amino acids and methanol; about 200 times sweeter than sucrose.

INGREDIENTS: SORBITOL, GUM BASE, MANNITOL, GLYCEROL, HYDROGENATED GLUCOSE SYRUP, XYLITOL, ARTIFICIAL AND NATURAL FLAVORS, ASPARTAME, RED 40, YELLOW 6 AND BHT (TO MAINTAIN FRESHNESS). PHENYLKETONURICS: CONTAINS PHENYLALANINE.

Sugarless Gum

▲ Sugar alcohols and the alternative sweetener aspartame are used to sweeten this product. Note the warning for people with phenylketonuria (PKU) that this product is made with aspartame and, thus, contains phenylalanine.

Equal®. Aspartame appears to be safe for pregnant women and children, but some scientists suggest cautious use by these groups, especially young children, who need ample calories to grow. Although questions have been raised about a connection between aspartame and brain tumors, the National Cancer Institute conducted an extensive review of its data and declared, in 2009, that there is no link between aspartame and brain tumor development.

Persons with an uncommon disease called **phenylketonuria (PKU),** which interferes with the metabolism of phenylalanine, should avoid aspartame because of its high phenylalanine content. People with PKU will find a mandatory warning label on products containing aspartame (see margin on page 132).

Sucralose. Sucralose (Splenda®) is made by adding three chlorines to sucrose. It cannot be broken down or absorbed, so it yields no calories, and it is about 600 times sweeter than sucrose. Because it does not break down under high heat conditions, it can be used in cooking and baking. Sucralose is approved for use as an additive to foods such as soft drinks, gum, baked goods, syrups, gelatins, frozen dairy desserts such as ice cream, jams, processed fruits, and fruit juices, and for tabletop use. The ADI for sucralose is 5 milligrams per kilogram (about 6 cans of diet soda for an adult).

Neotame. Neotame was recently approved by FDA for use as a general-purpose sweetener but is used in very few foods. Depending on its food application, Neotame is approximately 7000 to 13,000 times sweeter than table sugar. Neotame is heat stable and can be used as a tabletop sweetener as well as in cooking applications. Neotame is safe for use by the general population, including children, pregnant and lactating women, and people with diabetes. Although similar to aspartame, neotame does not require labeling for people with PKU because it is not broken down in the body to its amino acid components.

Acesulfame-K. Acesulfame-K is an organic acid linked to potassium (K) and is 200 times sweeter than sucrose. It was originally approved by FDA in 1988 and is sold as Sunette®. Acesulfame-K can be used in baking because it does not lose its sweetness when heated. In the United States, it is currently approved for use as a general-purpose sweetener. The ADI for acesulfame-K is 15 milligrams per kilogram, or the equivalent of 6 cans of diet soda.

Stevia. Stevia, sold as Truvia® and Sweet Leaf®, is an alternative sweetener derived from a South American shrub. Stevia extracts are 100 to 300 times sweeter than sucrose but provide no energy. It has been used in teas and as a sweetener in Japan since the 1970s, and FDA stated in December 2008 that stevia is considered generally recognized as safe (GRAS) for use in foods.

Luo han guo. Luo han guo is the extract of the monk fruit. It was approved by FDA in 2009 and is sold as the sweetener Nectresse™ and Monk Fruit in the Raw™. It is between 150 and 300 times sweeter than sucrose.

Advantame. In May 2014, the FDA announced that a new food additive, advantame, is safe for use as a general-purpose sweetener and flavor enhancer in food. Advantame is stable at higher temperatures and can be used as a tabletop sweetener as well as in cooking applications. Chemically, advantame is similar to aspartame but is much sweeter. Because only a small amount is needed to achieve the same level of sweetness (it is 20,000 times sweeter than sucrose), foods that contain advantame do not need to include alerts for people with PKU.

phenylketonuria (PKU) Disease caused by a defect in the liver's ability to metabolize the amino acid phenylalanine into the amino acid tyrosine; untreated, toxic by-products of phenylalanine build up in the body and lead to mental retardation.

sucralose Alternative sweetener that has chlorines in place of three hydroxyl (—OH) groups on sucrose; 600 times sweeter than sucrose.

neotame General-purpose, nonnutritive sweetener that is approximately 7000 to 13,000 times sweeter than table sugar. It has a chemical structure similar to aspartame's.

acesulfame K Alternative sweetener that yields no energy to the body; 200 times sweeter than sucrose.

stevia Alternative sweetener derived from South American shrub; 100 to 300 times sweeter than sucrose.

▲ A variety of alternative sweeteners are available.

✔ CONCEPT CHECK 4.3

1. Which food groups are the primary sources of carbohydrates in our diets?
2. What specific foods contain the highest percentage of calories from carbohydrates?
3. What are the common nutritive sweeteners?
4. Which alternative sweeteners are approved for use in food?

4.4 Making Carbohydrates Available for Body Use

As discussed in Chapter 3, simply eating a food does not supply nutrients to body cells. Digestion and absorption must occur first.

STARCH AND SUGAR DIGESTION

Food preparation can be viewed as the start of carbohydrate digestion because cooking softens tough connective structures in the fibrous parts of plants, such as broccoli stalks. When starches are heated, the starch granules swell as they soak up water, making them much easier to digest. All of these effects of cooking generally make carbohydrate-containing foods easier to chew, swallow, and break down during digestion.

The enzymatic digestion of starch begins in the mouth, when the saliva, which contains an enzyme called salivary **amylase,** mixes with the starchy products during the chewing of the food. This amylase breaks down starch into many smaller units, primarily disaccharides, such as maltose (Fig. 4-8). You can taste this conversion while chewing a saltine cracker. Prolonged chewing of the cracker causes it to taste sweeter as some starch breaks down into the sweeter disaccharides, such as maltose. Usually, food is in the mouth for such a short amount of time that this

amylase Starch-digesting enzyme from the salivary glands or pancreas.

FIGURE 4-8 ▶ Carbohydrate digestion and absorption. Enzymes made by the mouth, pancreas, and small intestine participate in the process of digestion. Most carbohydrate digestion and absorption take place in the small intestine. Chapter 3 covered the physiology of digestion and absorption in detail.

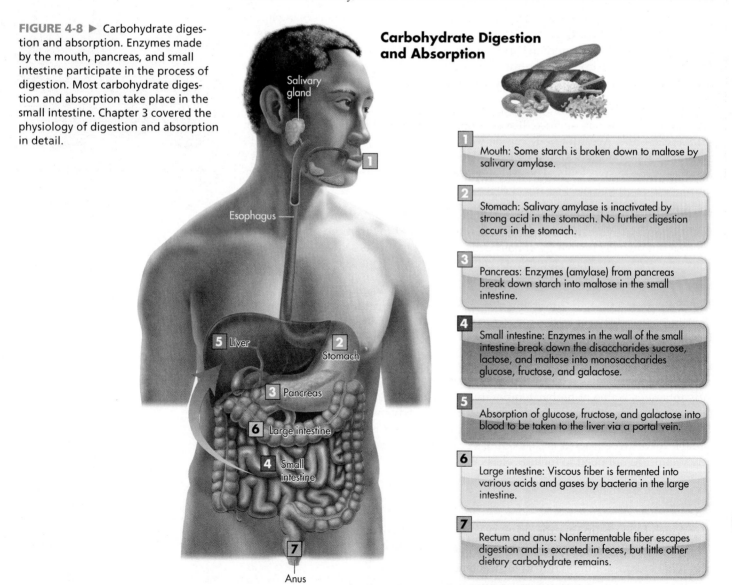

Carbohydrate Digestion and Absorption

1. Mouth: Some starch is broken down to maltose by salivary amylase.

2. Stomach: Salivary amylase is inactivated by strong acid in the stomach. No further digestion occurs in the stomach.

3. Pancreas: Enzymes (amylase) from pancreas break down starch into maltose in the small intestine.

4. Small intestine: Enzymes in the wall of the small intestine break down the disaccharides sucrose, lactose, and maltose into monosaccharides glucose, fructose, and galactose.

5. Absorption of glucose, fructose, and galactose into blood to be taken to the liver via a portal vein.

6. Large intestine: Viscous fiber is fermented into various acids and gases by bacteria in the large intestine.

7. Rectum and anus: Nonfermentable fiber escapes digestion and is excreted in feces, but little other dietary carbohydrate remains.

phase of digestion is negligible. In addition, once the food moves down the esophagus and reaches the stomach, the acidic environment inactivates salivary amylase.

When the carbohydrates reach the small intestine, the more alkaline environment of the intestine is better suited for further carbohydrate digestion. The pancreas releases enzymes, such as pancreatic amylase, to aid the last stage of starch digestion. After amylase action, the original carbohydrates in a food are now present in the small intestine as the monosaccharides glucose and fructose, originally present as such in food, and disaccharides (maltose from starch breakdown, lactose mainly from dairy products, and sucrose from food and that added at the table).

The disaccharides are digested to their monosaccharide units once they reach the wall of the small intestine, where the specialized enzymes on the absorptive cells digest each disaccharide into monosaccharides. The enzyme **maltase** acts on maltose to produce two glucose molecules. **Sucrase** acts on sucrose to produce glucose and fructose. **Lactase** acts on lactose to produce glucose and galactose.

LACTOSE MALDIGESTION AND LACTOSE INTOLERANCE

Deficient production of the enzyme lactase will impair the digestion of lactose. The most common form of this condition is **primary lactose maldigestion,** a normal pattern of physiology that often begins to develop after early childhood, at about ages 3 to 5 years. This primary form of lactose maldigestion is estimated to be present in about 75% of the world's population, although not all of these individuals experience symptoms. **Secondary lactose maldigestion** is a temporary condition in which lactase production is decreased in response to another condition, such as intestinal diarrhea. Rarely, lactase production is absent from birth, a condition known as **congenital lactase deficiency.** Any of these types of lactose maldigestion can lead to symptoms of gas, abdominal bloating, cramps, and diarrhea when lactose is consumed. The bloating and gas are caused by bacterial fermentation of lactose in the large intestine. The diarrhea is caused by undigested lactose in the large intestine as it draws water from the circulatory system into the large intestine. When significant symptoms develop after lactose intake, it is then called **lactose intolerance.** It is important to note that lactose maldigestion and resultant lactose intolerance are not equivalent to a milk allergy.

It is hypothesized that approximately 3000 to 5000 years ago, a genetic mutation occurred in regions that relied on milk and dairy foods as a main food source, allowing those individuals (mostly in northern Europe, pastoral tribes in Africa, and the Middle East) to retain the ability to maintain high lactase output for their entire lifetime. In North America, only about 25% of adults show signs of decreased lactose digestion in the small intestine. Asian Americans, African Americans, and Latino/Hispanic Americans are the population groups most likely to suffer lactose maldigestion, and the occurrence increases as people age. Many of these individuals can consume moderate amounts of lactose with minimal or no gastrointestinal discomfort because of eventual lactose breakdown by bacteria in the large intestine. Studies have shown that nearly all individuals with decreased lactase production can tolerate ½ to 1 cup of milk with meals and that most individuals adapt to intestinal gas production resulting from the fermentation of lactose by bacteria in the large intestine. Thus, it is unnecessary for these people to greatly restrict their intake of lactose-containing foods, such as milk and milk products, which are important for maintaining bone health. Obtaining enough calcium and vitamin D from the diet is much easier if milk and milk products are included in a diet.

Combining lactose-containing foods with other foods also helps because certain properties of foods can have positive effects on rates of digestion. For example, fat in a meal slows digestion, leaving more time for lactase action. Hard cheese and yogurt are also more easily tolerated than milk. Much of the lactose is lost in the production of cheese, and the active bacteria cultures in yogurt digest the lactose with their lactase. In addition, products such as lactose-free or lactose-reduced milk (Lactaid® and Dairy Ease®) are made by treating regular milk with the lactase enzyme. Lactase supplements are also available to assist lactose maldigesters when they decide to consume products

maltase An enzyme made by absorptive cells of the small intestine; this enzyme digests maltose to two glucoses.

sucrase An enzyme made by absorptive cells of the small intestine; this enzyme digests sucrose to glucose and fructose.

lactase An enzyme made by absorptive cells of the small intestine; this enzyme digests lactose to glucose and galactose.

lactose maldigestion (primary and secondary) Primary lactose maldigestion occurs when production of the enzyme lactase declines for no apparent reason. Secondary lactose maldigestion occurs when a specific cause, such as long-standing diarrhea, results in a decline in lactase production. When significant symptoms develop after lactose intake, it is then called lactose intolerance.

congenital lactase deficiency Birth defect resulting in the inability to produce lactase, such that a lactose-free diet is required from birth.

lactose intolerance A condition in which symptoms such as abdominal gas, bloating, and diarrhea appear as a result of severe lactose maldigestion.

▲ Use of yogurt helps lactose maldigesters meet calcium needs.

CASE STUDY Problems with Milk Intake

Myeshia is a 19-year-old African-American female who recently read about the health benefits of calcium and decided to increase her intake of dairy products. To start, she drank a cup of 1% milk at lunch. Not long afterward, she experienced bloating, cramping, and increased gas production. She suspected that the culprit of this pain was the milk she consumed, especially because her parents and her sister complain of the same problem. She wanted to determine if other milk products were, in fact, the cause of her discomfort, so the next day she substituted a cup of yogurt for the glass of milk at lunch. Subsequently, she did not have any pain.

Answer the following questions and check your response at the end of this chapter.

1. Why did Myeshia believe that she was sensitive to milk?
2. What component of milk is likely causing the problems that Myeshia experiences after drinking milk?
3. Why does this component cause intestinal discomfort in some individuals?
4. What is the name of this condition?
5. What groups of people are most likely to experience this condition?
6. Why did consuming yogurt not cause the same effects for Myeshia?
7. Are there any other products on the market that can replace regular milk or otherwise alleviate symptoms for individuals with this problem?
8. Can people with this condition ever drink regular milk?
9. What nutrients may be inadequate in the diet if dairy products are not consumed?
10. Why do some individuals have trouble tolerating milk products during or immediately after an intestinal viral infection?

containing lactose. Many plant-based "milks," including soy milk, almond milk, and rice milk, are naturally lactose-free and can be used as an alternative to regular milk.

CARBOHYDRATE ABSORPTION

Monosaccharides found naturally in foods and those formed as by-products of starch and disaccharide digestion in the mouth and small intestine generally follow an active absorption process. Recall from Chapter 3 that this is a process that requires a specific carrier and energy input for the substance to be taken up by the absorptive cells in the small intestine. Glucose and its close relative, galactose, undergo active absorption. They are pumped into the absorptive cells along with sodium.

Fructose is taken up by the absorptive cells via facilitated diffusion. In this case, a carrier is used, but no energy input is needed. This absorptive process is thus slower than that seen with glucose or galactose. So, large doses of fructose are not readily absorbed and can contribute to diarrhea as the monosaccharide remains in the small intestine and attracts water.

Once glucose, galactose, and fructose enter the absorptive cells, some fructose is metabolized into glucose. The single sugars in the absorptive cells are then transferred to the portal vein that goes directly to the liver. The liver then metabolizes those sugars by transforming the monosaccharides galactose and fructose into glucose and:

- Releasing it directly into the bloodstream for transport to organs such as the brain, muscles, kidneys, and adipose tissues
- Producing glycogen for storage of carbohydrate
- Producing fat (minor amount, if any)

Of these three options, producing fat occurs when carbohydrates are consumed in high amounts and overall calorie needs are exceeded.

Unless an individual has a disease that causes malabsorption or an intolerance to a carbohydrate such as lactose (or fructose), only a minor amount of some sugars (about 10%) escapes digestion. Any undigested carbohydrate travels to the large intestine and is fermented there by bacteria. The acids and gases produced by bacterial metabolism of the undigested carbohydrate are absorbed into the bloodstream.

Scientists suspect that some of these products of bacterial metabolism promote the health of the large intestine by providing it with a source of calories.

FIBER AND INTESTINAL HEALTH

Bacteria in the large intestine ferment soluble fibers into such products as acids and gases. The acids, once absorbed, also provide calories for the body. In this way, soluble fibers provide about 1.5 to 2.5 kcal per gram. Although the intestinal gas (flatulence) produced by this bacterial fermentation is not harmful, it can be painful and sometimes embarrassing. Over time, however, the body tends to adapt to a high-fiber intake, eventually producing less gas. You may notice that some potentially gas-forming foods are good sources of soluble fiber.

Because insoluble fiber is an indigestible carbohydrate, it remains in the intestinal tract and supplies mass to the feces, making elimination much easier. When enough fiber is consumed, the stool is large and soft because many types of plant fibers attract water. The larger size stimulates the intestinal muscles to contract, which aids elimination. Consequently, less pressure is necessary to expel the stool. When too little fiber is eaten, the opposite can occur: very little water is present in the feces, making it small and hard. Constipation may result, which forces one to exert excessive pressure in the large intestine during defecation. This high pressure can force parts of the large intestine (colon) wall out from between the surrounding bands of muscle, forming many small pouches called **diverticula** (Fig. 4-9). **Hemorrhoids,** discussed in the Nutrition and Your Health section of Chapter 3, may also result from excessive straining during defecation. Diverticula are asymptomatic in about 80% of affected people; that is, they are not noticeable. The asymptomatic form of this disease is called **diverticulosis.** If feces (and bacteria) become trapped within diverticuli, they may become inflamed or infected, a painful condition known as **diverticulitis.** In the short term, intake of fiber then should be reduced to limit further bacterial activity. Once the inflammation subsides, a high-fiber diet is resumed to ease stool elimination and reduce the risk of a future attack.

On the other hand, very high intakes of fiber—for example, 60 grams per day—can pose some health risks and therefore should be followed only under the guidance of a physician. Increased fluid intake is extremely important with a high-fiber diet. Inadequate fluid intake can leave the stool very hard and painful to eliminate. In more severe cases, the combination of excess fiber and insufficient fluid may contribute to blockages in the intestine, which may require surgery. Aside from problems with the passage of materials through the GI tract, a high-fiber diet may also decrease the availability of nutrients. Certain components of fiber may bind to essential minerals, keeping them from being absorbed. For example, when fiber is consumed in large amounts, zinc and iron absorption may be hindered.

Over the past 30 years, many population studies have shown a link between increased fiber intake and a decrease in colon cancer development. Most of the research on diet and colon cancer is focusing on the potential preventive effects of fruits, vegetables, whole-grain breads and cereals, and beans (rather than just fiber). Overall, the health benefits to the colon that stem from a high-fiber diet are partially due to the nutrients that are commonly present in most high-fiber foods, such as vitamins, minerals, phytochemicals, and, in some cases, essential fatty acids. Thus, it is more advisable to increase fiber intake using fiber-rich foods rather than mostly relying on fiber supplements.

✔ CONCEPT CHECK 4.4

1. In what form are carbohydrates absorbed, and what happens to these compounds after absorption?
2. What are the names and locations of the enzymes that digest carbohydrates?
3. Why do some individuals feel discomfort after they consume large amounts of lactose? How can they avoid these symptoms?
4. What are the beneficial effects of fiber in the intestinal tract?

▲ Beano® is a dietary supplement that contains natural digestive enzymes. Such products can be used to reduce intestinal gas produced by bacterial metabolism of undigested sugars in beans and some vegetables in the large intestine.

diverticula Pouches that protrude through the exterior wall of the large intestine.

hemorrhoid A pronounced swelling of a large vein, particularly veins found in the anal region.

diverticulosis The condition of having many diverticula in the large intestine.

diverticulitis An inflammation of the diverticula caused by acids produced by bacterial metabolism inside the diverticula.

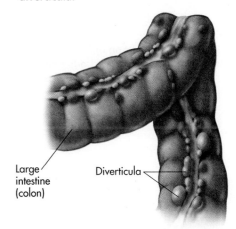

Large intestine (colon)   Diverticula

FIGURE 4-9 ▲ Diverticula in the large intestine. A low-fiber diet increases the risk of developing diverticula. About one-third of people over age 45 have diverticulosis, while two-thirds of people over 85 do.

4.5 Putting Carbohydrates to Work in the Body

As just discussed, all of the digestible carbohydrate that we eat is eventually converted into glucose. Glucose is then the form of carbohydrate that goes on to function in body metabolism. The other sugars can generally be converted into glucose, and the starches are broken down to yield glucose, so the functions described here apply to most carbohydrates. The functions of glucose in the body start with supplying calories to fuel the body.

PROVIDING ENERGY

The main function of glucose is to supply calories for use by the body. Certain tissues in the body, such as red blood cells, can use only glucose and other simple carbohydrate forms for fuel. Most parts of the brain and central nervous system also derive energy only from glucose, unless the diet contains almost none. In that case, much of the brain can use partial breakdown products of fat—called **ketone bodies**—for energy needs. Other body cells, including muscle cells, can use simple carbohydrates as fuel, but many of these cells can also use fat or protein for energy needs.

A diet that supplies enough digestible carbohydrates to prevent breakdown of proteins for energy needs is considered *protein sparing*. Under normal circumstances, digestible carbohydrates in the diet mostly end up as blood glucose, and protein is reserved for functions such as building and maintaining muscles and vital organs. However, if you don't eat enough carbohydrates, your body is forced to make glucose from body proteins, draining the pool of amino acids available in cells for other critical functions. During long-term starvation, the continuous withdrawal of proteins from the muscles, heart, liver, kidneys, and other vital organs can result in weakness, poor function, and even failure of body systems.

The wasting of protein that occurs during long-term fasting can be life threatening. This has prompted companies that make formulas for rapid weight loss to include sufficient carbohydrates in the products to decrease protein breakdown and thereby protect vital tissues and organs, including the heart. Most of these very low-calorie products are powders that can be mixed with different types of fluids and are consumed five or six times per day. When considering any weight-loss products, be sure that your total diet provides at least the RDA for carbohydrate.

In addition to the loss of protein, when you don't eat enough carbohydrates, the metabolism of fats is inefficient. In the absence of adequate carbohydrates, fats are not broken down completely in metabolism and instead form ketone bodies. This condition, known as **ketosis,** should be avoided because it disturbs the body's normal acid–base balance and leads to other health problems. This is a good reason to question the long-term safety of the low-carbohydrate diets that have been popular.

REGULATING BLOOD GLUCOSE

Under normal circumstances, a person's blood glucose concentration is regulated within a narrow range. When carbohydrates are digested and taken up by the absorptive cells of the small intestine, the resulting monosaccharides are transported directly to the liver. One of the liver's roles, then, is to guard against excess glucose entering the bloodstream after a meal. The liver works together with the pancreas to regulate blood glucose.

When the concentration of glucose in the blood is high, such as during and immediately after a meal, the pancreas releases the hormone **insulin** into the bloodstream. Insulin delivers two different messages to various body cells to cause the level of glucose in the blood to fall. First, insulin directs the liver to store glucose as glycogen. Second, insulin directs muscle, adipose, and other cells to remove glucose from the

ketone bodies Partial breakdown products of fat that contain three or four carbons.

ketosis The condition of having a high concentration of ketone bodies and related breakdown products in the bloodstream and tissues.

insulin A hormone produced by the pancreas. Among other processes, insulin increases the synthesis of glycogen in the liver and the movement of glucose from the bloodstream into body cells.

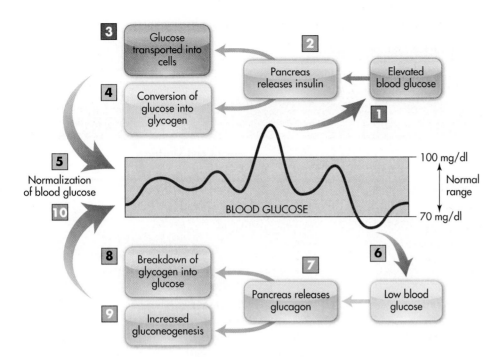

FIGURE 4-10 ◄ Regulation of blood glucose. Insulin and glucagon are key factors in controlling blood glucose. When blood glucose rises above the normal range of 70 to 100 milligrams per deciliter (mg/dl) (1), insulin is released (2) to lower it (3), and (4) blood glucose then falls back into the normal range (5). When blood glucose falls below the normal range (6), glucagon is released (7), which has the opposite effect of insulin (8), and (9) this then restores blood glucose to the normal range (10). Other hormones, such as epinephrine, norepinephrine, cortisol, and growth hormone, also contribute to blood glucose regulation.

bloodstream by taking it into those cells. By triggering both glycogen synthesis in the liver and glucose movement out of the bloodstream into certain cells, insulin keeps the concentration of glucose from rising too high in the blood (Fig. 4-10).

On the other hand, when a person has not eaten for a few hours and blood glucose begins to fall, the pancreas releases the hormone **glucagon.** This hormone has the opposite effect of insulin. It prompts the breakdown of liver glycogen into glucose, which is then released into the bloodstream. In this way, glucagon keeps blood glucose from falling too low.

A different mechanism increases blood glucose during times of stress. **Epinephrine** (adrenaline) is the hormone responsible for the "flight or fight" reaction. Epinephrine and a related compound are released in large amounts from the adrenal glands (located on each kidney) and various nerve endings in response to a perceived threat, such as a car approaching head-on. These hormones cause glycogen in the liver to be quickly broken down into glucose. The resulting rapid flood of glucose from the liver into the bloodstream helps promote quick mental and physical reactions.

To maintain blood glucose within an acceptable range, the body relies on a complex regulatory system. This provides a safeguard against extremely high blood glucose (**hyperglycemia**) or low blood glucose (**hypoglycemia**). In essence, the actions of insulin on blood glucose are balanced by the actions of glucagon, epinephrine, and other hormones. If hormonal balance is not maintained, such as during over- or underproduction of insulin or glucagon, major changes in blood glucose concentrations occur. The disease type 1 diabetes is an example of the underproduction of insulin. The failure of blood glucose regulation will be discussed in the Nutrition and Your Health section at the end of this chapter.

The Glycemic Index and Blood Glucose. Our bodies react uniquely to different sources of carbohydrates, with a serving of a high-fiber food, such as baked beans, resulting in lower blood glucose levels compared to the same size serving of mashed potatoes. We are interested in the effects of various foods on blood glucose because foods that result in a high blood glucose cause a large release of insulin from the pancreas. When this type of high insulin output occurs frequently, it leads to many deleterious effects on the body. Some of these undesirable effects are high blood triglycerides, increased fat deposition in the adipose tissue, increased tendency for

glucagon A hormone made by the pancreas that stimulates the breakdown of glycogen in the liver into glucose; this ends up increasing blood glucose. Glucagon also performs other functions.

epinephrine A hormone also known as *adrenaline;* it is released by the adrenal glands (located on each kidney) and various nerve endings in the body. It acts to increase glycogen breakdown in the liver, among other functions.

hyperglycemia High blood glucose, above 125 milligrams per 100 milliliters of blood.

hypoglycemia Low blood glucose, below 40 to 50 milligrams per 100 milliliters of blood for nondiabetics.

▲ The glycemic index of a food is not always a clear indicator of its nutritional value. For example, potato chips have a low GI, but they are not very nutrient dense.

You might wonder why the glycemic index of white bread and whole-wheat bread are similar. This is because whole-wheat flour is typically so finely ground that it is quickly digested. Thus, the effect of fiber in slowing digestion and related absorption of glucose is no longer present. Some experts suggest we focus more on min-imally processed (e.g., coarsely ground, steel-cut, or rolled) grains, such as with whole-wheat flour and oatmeal, to get the full benefits of these fiber sources in reducing blood glucose levels.

glycemic index (GI) The blood glucose response of a given food, compared to a standard (typically, glucose or white bread). Glycemic index is influenced by starch structure; fiber content; food processing; physical structure; and mac-ronutrients in the meal, such as fat.

TABLE 4-3 ▶ Glycemic Index (GI) of Common Foods*

Reference food glucose = 100
Low GI foods—below 55
Intermediate GI foods—between 55 and 70
High GI foods—more than 70

	Glycemic Index (GI)		Glycemic Index (GI)
Pastas/Grains		**Breads and Muffins**	
White rice, long grain	56	Bagel	72
White rice, short grain	72	Whole-wheat bread	69
Spaghetti	41	White bread	70
Vegetables		**Fruits**	
Carrots, boiled	49	Apple	38
Sweet corn	55	Banana	55
Potato, baked	85	Orange	44
Dairy Foods		**Beverages**	
Milk, skim	32	Orange juice	46
Yogurt, low-fat	33	Gatorade	78
Ice cream	61	Coca-Cola	63
Legumes		**Snack Foods**	
Baked beans	48	Potato chips	54
Kidney beans	27	Vanilla wafers	77
Navy beans	38	Jelly beans	80
Sugars			
Honey	73		
Sucrose	65		
Fructose	23		

*Based on a serving of food that supplies 50 grams of CHO

Source: Foster-Powell K and others: International table of glycemic index and glycemic load. *American Journal of Clinical Nutrition* 76:5, 2002.

blood to clot, increased fat synthesis in the liver, and a more rapid return of hunger after a meal (insulin rapidly lowers the macronutrients in the blood as it stimulates their storage, signaling hunger). Over time, this increase in insulin output may also cause the muscles to become resistant to the action of insulin and eventually lead to type 2 diabetes in some people.

The **glycemic index (GI)** is a measurement of how a carbohydrate-containing food raises blood glucose and, therefore, is helpful for planning a diet to avoid hyperglycemia (high blood glucose) and high insulin output. Glycemic index is a ratio of the blood glucose response to a given food compared to a reference food of either glucose or white bread (Table 4-3). Foods are ranked based on this compari-son with a high GI food raising blood glucose more than a medium or low GI food.

The GI of a food is influenced by starch structure, fiber content, food processing, physical structure, and other macronutrients in the meal, such as fat. Thus, the GI of a food is different when the food is eaten alone compared to when it is combined with other foods. In general, the more highly processed foods have a higher GI. For example, juice has a higher GI than whole fruit. Foods with particularly high glycemic index values are potatoes, especially baking potatoes (due to higher amylopectin con-tent compared to red potatoes), mashed potatoes (due to greater surface area exposed), short-grain white rice, honey, and jelly beans. Combining a low GI food, such as an apple, kidney beans, milk, or salad with dressing, with a high GI food balances the effect on blood glucose. Low GI foods include beans, all nonstarchy vegetables and beans, most fruits, and many whole-grain breads and cereals. Typically, consuming foods with a large amount of viscous (soluble) fiber, such as oat fiber, slows glucose absorption from the small intestine and so contributes to better glucose regulation.

The most important way to address the problem of high GI foods is to avoid eating too much of them at any one meal. This greatly minimizes their effects on blood glucose and the related increased insulin release. At least once per meal, consider substituting a low GI food for one with a higher value, such as long-grain rice or spaghetti for a baked potato. Keep in mind that the GI value only describes the type rather than the amount of carbohydrate in a food. Portion sizes are still important to manage to control blood glucose and maintain weight. Maintaining a healthy body weight and performing regular physical activity further reduces the effects of a high GI diet. Substituting low GI carbohydrates for high GI foods is very helpful in the treatment of diabetes. Be aware, however, that many foods with high nutrient density, such as oatmeal, may have a higher GI than foods with little nutritional value such as chocolate.

FIBER: REDUCING CHOLESTEROL ABSORPTION AND OBESITY RISK

Aside from its role in maintaining bowel regularity, the consumption of fiber has many additional health benefits. Recall that good sources of viscous fiber are apples, bananas, oranges, carrots, barley, oats, and kidney beans. A high intake of viscous fiber also inhibits absorption of cholesterol and cholesterol-rich bile acids from the small intestine, thereby reducing blood cholesterol and possibly reducing the risk of cardiovascular disease and gallstones. The beneficial bacteria in the large intestine degrade soluble fiber and produce certain fatty acids that probably also reduce cholesterol synthesis in the liver. In addition, the slower glucose absorption that occurs with diets high in viscous fiber is linked to a decrease in insulin release. One of the effects of insulin is to stimulate cholesterol synthesis in the liver, so this reduction in insulin may contribute to the ability of viscous fiber to lower blood cholesterol. Overall, a fiber-rich diet containing fruits, vegetables, beans, and whole-grain breads and cereals (including whole-grain breakfast cereals) is advocated as part of a strategy to reduce cardiovascular disease (coronary heart disease and stroke) risk. And again, this is something that a low-carbohydrate diet can't promise.

A diet high in fiber likely controls weight and reduces the risk of developing obesity (see Further Reading 14). Due to their bulky nature, high-fiber foods require more time to chew and fill us up without yielding many calories. Increasing intake of foods rich in fiber is one strategy for feeling satisfied or full after a meal (review the discussion of energy density in Chapter 2).

▲ Oatmeal is a rich source of viscous fiber. FDA allows a health claim for the benefits of oatmeal to lower blood cholesterol that arise from the effects of this viscous fiber.

✔ CONCEPT CHECK 4.5

1. What is the primary role of carbohydrates in the body?
2. How does the body respond when too little carbohydrate is consumed?
3. What are the mechanisms by which blood glucose levels are maintained within a narrow range?
4. How can the glycemic index be applied in meal planning?
5. What are some of the important functions of fiber?

4.6 Carbohydrate Needs

The RDA for carbohydrates is 130 grams per day for adults. This is based on the amount needed to supply adequate glucose for the brain and nervous system, without having to rely on ketone bodies from incomplete fat breakdown as a calorie source. Somewhat exceeding this amount is fine; the Food and Nutrition Board

▲ The 2010 Dietary Guidelines define whole grain as the entire grain seed or kernel made of three components: bran, germ, and endosperm.

▲ When buying bread, most people think they are buying a whole-wheat product when they see the name "wheat bread" on the label. The flour is from the wheat plant, so manufacturers correctly list enriched white (refined) flour as wheat flour on food labels. However, if the label does not list "whole-wheat flour" first, then the product is not primarily a whole-wheat bread and thus does not contain as much fiber as it could. Careful reading of labels is important in the search for more fiber. Look for the term *whole-grain* or *whole-wheat* flour on the label for breads that are an excellent source of fiber.

recommends that carbohydrate intake should range from 45% to 65% of total calorie intake. The Nutrition Facts panel on food labels uses 60% of calorie intake as the standard for recommended carbohydrate intake. This would be 300 grams of carbohydrate when consuming a 2000-calorie diet.

North American adults consume about 180 to 330 grams of carbohydrates per day, which supply about 50% of calorie intake. Worldwide, however, carbohydrates account for about 70% of all calories consumed and, in some countries, up to 80% of the calories consumed. One recommendation on which almost all experts agree, including the 2010 Dietary Guidelines, is that one's carbohydrate intake should be based primarily on fruits, vegetables, whole-grain breads and cereals, and beans, rather than on refined grains, potatoes, and sugars.

The 2010 Dietary Guidelines for Americans recommend that we choose fiber-rich fruits, vegetables, and whole grains often. More specifically, three or more ounces of grains, roughly one-half of one's grains, should be whole. Remember that the 2010 Dietary Guidelines define whole grain as the entire grain seed or kernel made of three components: the bran, germ, and endosperm, which must be in nearly the same relative proportions as the original grain if cracked, crushed, or flaked (see Further Readings 13 and 15).

HOW MUCH FIBER DO WE NEED?

An Adequate Intake for fiber has been set based on the ability of fiber to reduce risk of cardiovascular disease (and likely many cases of diabetes). The Adequate Intake for fiber for adults is 25 grams per day for women and 38 grams per day for men. The goal is to provide at least 14 grams per 1000 kcal in a diet. After age 50, the Adequate Intake falls to 21 grams per day and 30 grams per day, respectively. The Daily Value used for fiber on food and supplement labels is 25 grams for a 2000-kcal diet. In North America, fiber intake averages 13 grams per day for women and 17 grams per day for men, and the average whole-grain intake is less than one serving per day. This low intake is attributed to the lack of knowledge on the benefits of whole grains and the inability to recognize whole-grain products at the time of purchase. Thus, most of us should increase our fiber intake. At least three servings of whole grains per day are recommended. Eating a high-fiber cereal (at least 3 grams of fiber per serving) for breakfast is one easy way to increase fiber intake (Fig. 4-11; see Further Readings 7 and 10).

The "Rate Your Plate" exercise shows a diet containing 25 or 38 grams of fiber within moderate calorie intakes. Diets to meet the fiber recommendations are possible and enjoyable if you incorporate plenty of whole-wheat bread, fruits, vegetables, and beans. Use the "Rate Your Plate" exercise to estimate the fiber content of your diet. What is *your* fiber score?

Remember that very high intakes of fiber can be unhealthy and that fluid intake must be increased with a high-fiber diet. A high-fiber diet may also decrease the absorption of essential minerals, especially zinc and iron.

In the final analysis, keep in mind that any nutrient can lead to health problems when consumed in excess. High carbohydrate, high fiber, and low fat do not mean zero calories. Carbohydrates help moderate calorie intake in comparison with fats, but high-carbohydrate foods also contribute to total calorie intake.

HOW MUCH SUGAR IS TOO MUCH?

The main problems with consuming an excess amount of sugar are that it provides empty calories and increases the risk for dental decay.

Diet Quality Declines When Sugar Intake Is Excessive. Overcrowding the diet with sweet treats can leave little room for important, nutrient-dense foods, such as fruits and vegetables. Children and teenagers are at the highest risk for overconsuming empty calories in place of nutrients essential for growth. Many children and teenagers are drinking an excess of sugared soft drinks and other sugar-containing

Nutrition Facts

Serving Size 1 cup (55g/2.0 oz.)
Servings Per Container 10

Amount Per Serving	Cereal	Cereal with ½ Cup Vitamins A & D Skim Milk
Calories	170	210
Calories from Fat	10	10
	% Daily Value**	
Total Fat 1.0g*	**2**%	**2**%
Sat. Fat 0g	**0**%	**0**%
Trans Fat 0g		*
Cholesterol 0mg	**0**%	**0**%
Sodium 300mg	**13**%	**15**%
Potassium 340mg	**10**%	**16**%
Total Carbohydrate 43g	**14**%	**16**%
Dietary Fiber 7g	**28**%	**28**%
Sugars 16g		
Other Carbohydrate 20g		
Protein 4g		
Vitamin A	15%	20%
Vitamin C	20%	22%
Calcium	2%	15%
Iron	65%	65%
Vitamin D	10%	25%
Thiamin	25%	30%
Riboflavin	25%	35%
Niacin	25%	25%
Vitamin B$_6$	25%	25%
Folic acid	30%	30%
Vitamin B$_{12}$	25%	35%
Phosphorus	20%	30%
Magnesium	20%	25%
Zinc	25%	25%
Copper	10%	10%

*Amount in cereal. One half cup skim milk contributes an additional 40 calories, 65mg sodium, 6g total carbohydrate (6g sugars), and 4g protein.
**Percent Daily Values are based on a 2,000 calorie diet. Your daily values may be higher or lower depending on your calorie needs:

	Calories:	2,000	2,500
Total Fat	Less than	65g	80g
Sat Fat	Less than	20g	25g
Cholesterol	Less than	300mg	300mg
Sodium	Less than	2,400mg	2,400mg
Potassium		3,500mg	3,500mg
Total Carbohydrate		300g	375g
Dietary Fiber		25g	30g

Calories per gram:
Fat 9 • Carbohydrate 4 • Protein 4

*Intake of *trans* fat should be as low as possible.

Ingredients: Wheat bran with other parts of wheat, raisins, sugar, corn syrup, salt, malt flavoring, glycerin, iron, niacinamide, zinc oxide, pyridoxine hydrochloride (vitamin B$_6$), riboflavin (vitamin B$_2$), vitamin A palmitate, thiamin hydrochloride (vitamin B$_1$), folic acid, vitamin B$_{12}$, and vitamin D.

Nutrition Facts

Serving Size: ¾ Cup (30g)
Servings Per Package: About 17

Amount Per Serving	Cereal	Cereal With ½ Cup Skim Milk
Calories	170	210
Calories from Fat	0	5
	%Daily Value**	
Total Fat 0g*	**0**%	**1**%
Saturated Fat 0g	**0**%	**1**%
Trans Fat 0g		*
Cholesterol 0mg	**0**%	**1**%
Sodium 60mg	**2**%	**4**%
Potassium 80mg	**2**%	**8**%
Total Carbohydrate 35g	**9**%	**11**%
Dietary Fiber 1g	**4**%	**4**%
Sugars 20g		
Other Carbohydrate 13g		
Protein 3g		
Vitamin A	25%	30%
Vitamin C	0%	2%
Calcium	0%	15%
Iron	10%	10%
Vitamin D	10%	20%
Thiamin	25%	25%
Riboflavin	25%	35%
Niacin	25%	25%
Vitamin B$_6$	25%	25%
Folic acid	25%	25%
Vitamin B$_{12}$	25%	30%
Phosphorus	4%	15%
Magnesium	4%	8%
Zinc	10%	10%
Copper	2%	2%

*Amount in Cereal. One-half cup skim milk contributes an additional 65mg sodium, 6g total carbohydrate (6g sugars), and 4g protein.
**Percent Daily Values are based on a 2,000 calorie diet. Your daily values may be higher or lower depending on your calorie needs:

	Calories:	2,000	2,500
Total Fat	Less than	65g	80g
Sat. Fat	Less than	20g	25g
Cholesterol	Less than	300mg	300mg
Sodium	Less than	2,400mg	2,400mg
Potassium		3,500mg	3,500mg
Total Carbohydrate		300g	375g
Dietary Fiber		25g	30g

Calories per gram:
Fat 9 • Carbohydrate 4 • Protein 4

*Intake of *trans* fat should be as low as possible.

Ingredients: Wheat, Sugar, Corn Syrup, Honey, Caramel Color, Partially Hydrogenated Soybean Oil, Salt, Ferric Phosphate, Niacinamide (Niacin), Zinc Oxide, Vitamin A (Palmitate), Pyridoxine Hydrochloride (Vitamin B6), Riboflavin, Thiamin Mononitrate, Folic Acid (Folate), Vitamin B12 and Vitamin D.

FIGURE 4-11 ◄ Reading the Nutrition Facts on food labels helps us choose more nutritious foods. Based on the information from these nutrition labels, which cereal is the better choice for breakfast? Consider the amount of fiber in each cereal. Did the ingredient lists give you any clues? (Note: Ingredients are always listed in descending order by weight on a label.) When choosing a breakfast cereal, it is generally wise to focus on those that are rich sources of fiber. Sugar content can also be used for evaluation. However, sometimes this number does not reflect added sugar but simply the addition of fruits, such as raisins, complicating the evaluation.

▲ Whole-grain foods, such as granola, are excellent sources of fiber.

beverages and much less milk than ever before. This exchange of soft drinks for milk can compromise bone health because milk contains calcium and vitamin D, both essential for bone health.

With regard to sugar intake, an upper limit of 25% of total calorie intake from "added sugars" has been set by the Food and Nutrition Board. Remember that "added sugars" are sugars added to foods during processing and preparation. Diets

The 2010 Dietary Guidelines for Americans provide the following recommendations regarding carbohydrate intake as part of a healthy eating pattern while staying within their calorie needs:

- Limit the consumption of foods that contain refined grains, especially refined grain foods that contain solid fats, added sugars, and sodium.
- Increase vegetable and fruit intake.
- Eat a variety of vegetables, especially dark-green and red and orange vegetables, and beans and peas.
- Consume at least half of all grains as whole grains. Increase whole-grain intake by replacing refined grains with whole grains.
- Choose foods that provide more potassium, dietary fiber, calcium, and vitamin D, which are nutrients of concern in American diets. These foods include vegetables, fruits, whole grains, and milk and milk products.

▲ Cookies and cakes are one of the top five carbohydrate sources for U.S. adults.

An excess intake of sugared soft drinks has recently been linked to a risk for both weight gain and type 2 diabetes in adults.

that go beyond this upper limit are likely to be deficient in vitamins and minerals. The World Health Organization (WHO) continues to reiterate a "strong recommendation" that added sugars provide no more than 10% of total daily calorie intake. A moderate intake of about 10% of calorie intake corresponds to a maximum of approximately 50 grams (or 12 teaspoons) of sugars per day, based on a 2000-kcal diet. Because of the association between excessive consumption of sugars and several metabolic abnormalities and adverse health conditions, the American Heart Association recommends reductions in the intake of added sugars such that the upper limit of intake for most American women is no more than 100 calories (25 grams) per day from added sugars and no more than 150 calories (37.5 grams) per day for most American men (see Further Reading 9). In March 2014, the WHO proposed a similar "conditional recommendation" that a sugar intake of less than 5% of daily calories (25 grams for an adult of normal BMI) should be the target in order to avoid health risks such as weight gain and tooth decay.

Most of the sugars we eat come from foods and beverages to which sugar has been added during processing and/or manufacture. On average, North Americans eat about 82 grams of added sugars daily, amounting to about 15% of calorie intake. Major sources of added sugars include soft drinks; cakes; cookies; fruit drinks; and dairy desserts, such as ice cream (Fig. 4-12). Following recommendations to limit intakes of added sugars is easier if sweet desserts such as cakes, cookies, and ice cream (full and reduced fat) are consumed sparingly (Table 4-4). Today's low-fat and fat-free snack products usually contain lots of added sugar to produce a product with an acceptable taste. The result is a high-calorie food with at least as much, if not more calories than the high-fat food product it was designed to replace.

Supersizing sugar-rich beverages has also become common and has led to more sugar consumption; for example, in the 1950s, a typical serving size of a soft drink was a 6½-ounce bottle, and now a 20-ounce plastic bottle is a typical serving. This one change in serving size contributes 170 extra kcal to the diet—all from added sugars. Most convenience stores now offer cups that will hold 64 ounces of soft drinks. The New York City Board of Health recently tried to ban the sale of sugary drinks in containers larger than 16 ounces in restaurants and other establishments

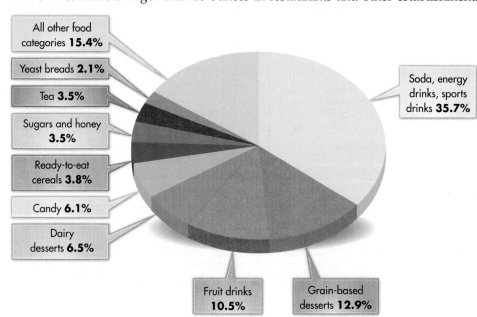

All other food categories **15.4%**

Yeast breads **2.1%**

Tea **3.5%**

Sugars and honey **3.5%**

Ready-to-eat cereals **3.8%**

Candy **6.1%**

Dairy desserts **6.5%**

Soda, energy drinks, sports drinks **35.7%**

Fruit drinks **10.5%**

Grain-based desserts **12.9%**

FIGURE 4-12 ▲ Sources of added sugars in the diets of the U.S. population ages 2 years and olders.

Sources: Dietary Guidelines for Americans Chapter 3.

National Cancer Institute. Sources of added sugars in the diets of the U.S. population ages 2 years and older, NHANES 2005–2006. Risk Factor Monitoring and Methods. Cancer Control and Population Sciences. **http://riskfactor.cancer.gov/diet/foodsources/added_sugars/table5a.html.** *Accessed August 11, 2010.*

TABLE 4-4 ▶ **Suggestions for Reducing Simple-Sugar Intake**

At the Supermarket

- Read ingredient labels. Identify all the added sugars in a product. Select items lower in total sugar when possible.
- Buy fresh fruits or fruits packed in water, juice, or light syrup rather than in heavy syrup.
- Buy fewer foods that are high in sugar, such as prepared baked goods, candies, sugared cereals, sweet desserts, soft drinks, and fruit-flavored punches. Substitute vanilla wafers, graham crackers, bagels, English muffins, diet soft drinks, and other low-sugar alternatives.
- Buy reduced-fat microwave popcorn to replace candy for snacks.

In the Kitchen

- Reduce the sugar in foods prepared at home. Try new low-sugar recipes or adjust your own. Start by reducing the sugar gradually until you've decreased it by one-third or more.
- Experiment with spices, such as cinnamon, cardamom, coriander, nutmeg, ginger, and mace, to enhance the flavor of foods.
- Use home-prepared items with less sugar instead of commercially prepared ones that are higher in sugar.

At the Table

- Reduce your use of white and brown sugars, honey, molasses, syrups, jams, and jellies.
- Choose fewer foods high in sugar, such as prepared baked goods, candies, and sweet desserts.
- Reach for fresh fruit instead of cookies or candy for dessert and between-meal snacks.
- Add less sugar to foods—coffee, tea, cereal, and fruit. Cut back gradually to a quarter or half the amount. Consider using sugar alternatives to substitute for some sugar.
- Reduce the number of sugared soft drinks, punches, and fruit juices you drink. Substitute water, diet soft drinks, and whole fruits.

▲ Many foods we enjoy are sweet. These should be eaten in moderation.

in an effort to lower the intake of excess calories from sugar. Controversy over the authority of the health board to limit or ban a legal item led to a New York appeals court ruling in July 2013 that the city Board of Health exceeded its legal authority and acted unconstitutionally by trying to put a size limit on soft drinks served in city restaurants. Health messages about the sugary beverages appear to be having some positive effects, resulting in a decrease in soft drink sales over the past 8 years. Market research indicates that youth are choosing more water, energy drinks, and coffee in place of soft drinks. Read more about trends in added sugar consumption and the decline in soda consumption in the Newsworthy Nutrition and Further Reading 16.

Sugar and Hyperactivity. There is a widespread notion that high sugar intake causes hyperactivity in children, typically part of the syndrome called *attention deficit hyperactivity disorder (ADHD)*. However, most researchers find that sucrose may have the opposite effect. A high-carbohydrate meal, if also low in protein and fat, has a calming effect and induces sleep; this effect may be linked to changes in the synthesis of certain neurotransmitters in the brain, such as serotonin. If there is a problem, it is probably the excitement or tension in situations in which sugar-rich foods are served, such as at birthday parties and on Halloween.

Sugar and Oral Health. Sugars in the diet (and starches readily fermented in the mouth, such as crackers and white bread) also increase the risk of developing **dental caries.** Recall that caries, also known as cavities, are formed when sugars and other carbohydrates are metabolized into acids by bacteria that live in the mouth. These acids dissolve the tooth enamel and underlying structure. Bacteria

dental caries Erosions in the surface of a tooth caused by acids made by bacteria as they metabolize sugars.

Newsworthy Nutrition

Consumption of added sugars is decreasing in the United States

Although the consumption of added sugars has been linked to chronic diseases including obesity, diabetes, and heart disease, recent consumption of added sugars in the United States had not been determined or compared with current guidelines. The objective of this study was to examine trends in intakes of added sugars in the United States over 8 years. The study design was a large cross-sectional study of 42,316 U.S. residents 2 years of age or older using dietary data from NHANES 1999–2008 and data for added-sugar contents from the MyPyramid Equivalents Database. The results of this study indicate that between 1999–2000 and 2007–2008, the absolute intake of added sugars decreased from a mean of 100.1 grams per day to 76.7 grams per day. The bulk of this decrease (from 37.4 grams per day to 22.8 grams per day) was attributed to a decrease in soda consumption. The only source of added sugars to increase over the study period were energy drinks, although consumption peaked at only 0.15 grams per day. The percentage of total energy from added sugars also decreased from 18.1% to 14.6%. The authors conclude that although consumption of added sugars declined during the 8 years between 1999 and 2008 because of reduced soda consumption, the mean intakes of added sugars continue to exceed the recommendation to limit the consumption of added sugars to 5% to 15% of total energy intake, depending on energy needs.
Source: Welsh JA and others: Consumption of added sugars is decreasing in the United States. *American Journal of Clinical Nutrition* 94:726, 2011.

connect NUTRITION Checkout the Connect site **www.mcgrawhillconnect.com** to further explore trends in the consumption of added sugars.

also use the sugars to make plaque, a sticky substance that both adheres acid-producing bacteria to teeth and diminishes the acid-neutralizing effect of saliva.

The worst offenders in terms of promoting dental caries are sticky and gummy foods high in sugars, such as caramel, because they stick to the teeth and supply the bacteria with a long-lived carbohydrate source. Frequent consumption of liquid sugar sources (e.g., fruit juices) can also cause dental caries. Snacking regularly on sugary foods is also likely to cause caries because it gives the bacteria on the teeth a steady source of carbohydrates from which to continually make acid. Sugared gum chewed between meals is a prime example of a poor dental habit. Still, sugar-containing foods are not the only foods that promote acid production by bacteria in the mouth. As mentioned, if starch-containing foods (e.g., crackers and bread) are held in the mouth for a long time, the starch will be broken down to sugars by enzymes in the mouth; bacteria can then produce acid from these sugars. Overall, the sugar and starch contents of a food and its ability to remain in the mouth largely determine its potential to cause caries.

Fluoridated water and toothpaste have contributed to fewer dental caries in North American children over the past 20 years due to fluoride's tooth-strengthening effect (see Chapter 9). Research has also indicated that certain foods—such as cheese, peanuts, and sugar-free chewing gum—can help reduce the amount of acid on teeth. In addition, rinsing the mouth after meals and snacks reduces the acidity in the mouth. Certainly, good nutrition, habits that do not present an overwhelming challenge to oral health (e.g., chewing sugar-free gum), and routine visits to the dentist all contribute to improved dental health.

CRITICAL THINKING

John and Mike are identical twins who like the same games, sports, and foods. However, John likes to chew sugar-free gum and Mike doesn't. At their last dental visit, John had no cavities but Mike had two. Mike wants to know why John, who chews gum after eating, doesn't have cavities and he does. How would you explain this to him?

✔ CONCEPT CHECK 4.6

1. What is the recommended intake of total carbohydrate per day, and how do typical dietary intakes compare?
2. How much fiber is recommended each day?
3. How can we reduce our consumption of added sugars?
4. What is the link between sugar and oral health?

Nutrition and Your Health
Diabetes—When Blood Glucose Regulation Fails

Improper regulation of blood glucose results in either hyperglycemia (high blood glucose) or hypoglycemia (low blood glucose). High blood glucose is most commonly associated with diabetes (technically, *diabetes mellitus*), a disease that affects almost 26 million adults, or 11.3% of the North American population 20 years or older (see Further Reading 4). It is estimated that over 37% of these people do not know that they have the disease. Diabetes contributed to 231,404 deaths in 2007 in North America. Diabetes is currently increasing in epidemic proportions in the United States, with 1.9 million new cases diagnosed in people 20 years or older in 2010. The American Diabetes Association (ADA) recommends testing fasting blood glucose in adults over age 45 every 3 years to screen for diabetes. Diabetes was diagnosed for decades using a fasting blood glucose of 126 milligrams per 100 milliliters of blood or greater. In 2010, the ADA along with several international diabetes organizations recommended the use of the hemoglobin A1c (HbA1c) test to diagnose diabetes with a threshold of greater than 6.5%. The HbA1c is a more sensitive, long-term indicator of poor blood glucose control than the fasting blood glucose level. When blood glucose is too high, the glucose builds up in the blood and combines with hemoglobin (protein in red blood cells), making it glycated. The amount of glycated

hemoglobin, or HbA1c, reflects the last several weeks of blood glucose levels (see Further Reading 3).

Diabetes

There are two major forms of diabetes: **type 1** (formerly called insulin-dependent or juvenile-onset diabetes) and **type 2 diabetes** (formerly called noninsulin-dependent or adult-onset diabetes) (Table 4-5). The change in names to type 1 and type 2 diabetes stems from the fact that many type 2 diabetics eventually must also rely on insulin injections as a part of their treatment. In addition, many children today have type 2 diabetes. A third form, called gestational diabetes, occurs in some pregnant women (see Chapter 14). It is usually treated with an insulin regimen and diet, and resolves after delivery of the baby. However, women who have gestational diabetes during pregnancy are at high risk for developing type 2 diabetes later in life.

Traditional symptoms of diabetes are excessive urination, excessive thirst, and excessive hunger. No one symptom is diagnostic of diabetes, and other symptoms—such as unexplained weight loss, exhaustion, blurred vision, tingling in hands and feet, frequent infections, poor wound healing, and impotence—often accompany traditional symptoms.

Type 1 Diabetes

Type 1 diabetes often begins in late childhood, around the ages of 10 to 14 years, but can occur at any age. The occurrence of the disease has been rising 3% to 4% per year in children and teens. Children usually are admitted to the hospital with abnormally high blood glucose after eating, as well as evidence of ketosis.

The onset of type 1 diabetes is generally associated with decreased release of insulin from the pancreas. As insulin in the blood declines, blood glucose increases, especially after eating. Figure 4-13 shows a typical glucose response observed in a patient with this form of diabetes after consuming about 75 grams of glucose. When blood glucose levels are high, the kidneys let excess glucose spill into the urine, resulting in frequent urination of urine high in sugar.

Most cases of type 1 diabetes begin with an immune system disorder, which causes destruction of the insulin-producing cells in the pancreas. The disease may stem from genetic, autoimmune, or environmental factors. Most likely, a virus or protein foreign to the body sets off the destruction. In response to their damage, the affected pancreatic cells release other proteins, which stimulate a more furious attack. Eventually, the pancreas loses its ability to synthesize insulin, and the clinical stage of the disease begins. Hyperglycemia and other symptoms develop slowly and only after 90% or more of the insulin-secreting cells have been destroyed. HbA1c, fasting blood glucose, or an

type 1 diabetes A form of diabetes characterized by total insulin deficiency due to destruction of insulin-producing cells of the pancreas. Insulin therapy is required.

type 2 diabetes A form of diabetes characterized by insulin resistance and often associated with obesity. Insulin therapy can be used but is often not required.

TABLE 4-5 ▶ Comparison of Type 1 and Type 2 Diabetes

	Type 1 Diabetes	Type 2 Diabetes
Occurrence	5% to 10% of cases of diabetes	90% of cases of diabetes
Cause	Autoimmune destruction of the pancreas	Insulin resistance
Risk factors	Moderate genetic predisposition	Strong genetic predisposition Obesity and physical inactivity Ethnicity Metabolic syndrome Prediabetes
Characteristics	Distinct symptoms (frequent thirst, hunger, and urination) Ketosis Weight loss	Mild symptoms, especially in early phases of the disease (fatigue and nighttime urination) Ketosis does not generally occur.
Treatment	Insulin Diet Exercise	Diet Exercise Oral medications to lower blood glucose Insulin (in advanced cases)
Complications	Cardiovascular disease Kidney disease Nerve disease Blindness Infections	Cardiovascular disease Kidney disease Nerve damage Blindness Infections
Monitoring	Blood glucose Urine ketones HbA1c*	Blood glucose HbA1c

*Hemoglobin A1c.

oral glucose tolerance test can be used to diagnose diabetes. Remember that HbA1c is the recommended measure because over time, blood glucose attaches to hemoglobin in red blood cells and more so when blood glucose remains elevated. A HbA1c value of over 7% indicates poor blood glucose control.

Type 1 diabetes is treated primarily by insulin therapy, either with injections or with an insulin infusion pump. The pump dispenses insulin at a steady rate into the body, with greater amounts delivered after each meal. Dietary therapy includes balancing carbohydrate intake with the insulin regimen and physical activity schedule to manage blood glucose levels. The amount and timing of carbohydrates eaten should be consistent from day to day to maintain blood glucose control. Insulin should be adjusted to match carbohydrate intake in persons who adjust their mealtime insulin doses or who are using an insulin pump. There are several methods available to estimate carbohydrate content of foods, including carbohydrate counting, exchange lists, and the glycemic index of foods (see Further Readings 2 and 5). For more information on nutrition treatment of diabetes, see Further Readings 1–3. If one does not eat often enough, the injected insulin can cause a severe drop in blood glucose or hypoglycemia because it acts on whatever glucose is available. The diet should be moderate in simple carbohydrates, include ample fiber

Symptoms of Diabetes

The symptoms of diabetes may occur suddenly and include one or more of the following:
- Extreme thirst
- Frequent urination
- Drowsiness, lethargy
- Sudden vision changes
- Increased appetite
- Sudden weight loss
- Sugar in urine
- Fruity, sweet, or winelike odor on breath
- Heavy, labored breathing
- Stupor, unconsciousness

and polyunsaturated fat, but be low in both animal and *trans* fats, and supply an amount of calories in balance with needs. Providing adequate calories and nutrients to promote growth and development in children is crucial for young diabetics.

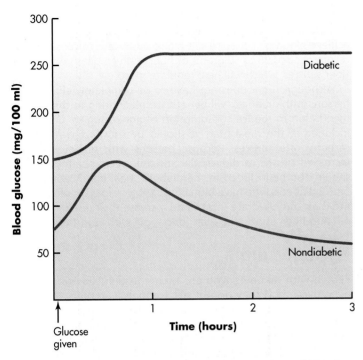

FIGURE 4-13 ▲ Glucose tolerance test. A comparison of blood glucose concentrations in untreated diabetic and healthy nondiabetic persons after consuming a 75-gram test load of glucose.

The hormone imbalances that occur in people with untreated type 1 diabetes—chiefly, not enough insulin—lead to mobilization of body fat, taken up by liver cells. Ketosis is the result because the fat is partially broken down to ketone bodies. Ketone bodies can rise excessively in the blood, and eventually spill into the urine. These pull sodium and potassium ions and water with them into the urine. This series of events also causes frequent urination and can contribute to a

▲ Insulin pumps alleviate the discomfort of injecting insulin under the skin multiple times per day.

chain reaction that eventually leads to dehydration; ion imbalance; coma; and even death, especially in patients with poorly controlled type 1 diabetes. Treatment includes provision of insulin, fluids, and minerals such as sodium and potassium.

Several degenerative complications result from poor blood glucose regulation, specifically long-term hyperglycemia, including cardiovascular disease, blindness, kidney disease, and nerve damage. The high blood sugar concentration physically deteriorates small blood vessels (capillaries) and nerves. When improper nerve stimulation occurs in the intestinal tract, intermittent diarrhea and constipation result. Because of nerve deterioration in the extremities, many people with diabetes lose the sensation of pain associated with injuries or infections. They do not have as much pain, so they often delay treatment of hand or foot problems. This delay, combined with a rich environment for bacterial growth (bacteria thrive on glucose), sets the stage for damage and death of tissues in the extremities, sometimes leading to the need for amputation of feet and legs. Elevated blood glucose also leads to glycation of other proteins and fats in the body, forming what are called advanced glycation endproducts (AGEs). These have been shown to be toxic to cells, especially those of the immune system and kidneys.

Current research has shown that the development of blood vessel and nerve complications of diabetes can be slowed with aggressive treatment directed at keeping blood glucose within the normal range. Maintaining near-normal HbA1c (6% or less) greatly reduces the risk of death and developing other diseases in people with diabetes. A person with diabetes generally must work closely with a physician and registered dietitian to make the correct alterations in diet and medications and to perform physical activity safely. Physical activity enhances glucose uptake by muscles independent of insulin action, which in turn can lower blood glucose. This outcome is beneficial, but people with type 1 diabetes need to be aware of their blood glucose response to physical activity and compensate appropriately to avoid hypoglycemia.

Type 2 Diabetes

Type 2 diabetes usually begins after age 30. This is the most common type of diabetes, accounting for about 90% to 95% of the cases diagnosed in North America. The disease is progressive and is present, in many cases, long before it is diagnosed. Hyperglycemia develops slowly such that the classic symptoms are not noticed in the early stages of the disease, even though the risk for complications is rising. Risk factors for type 2 diabetes are both genetic and environmental and include a family history of diabetes; older age; obesity, especially intraabdominal obesity; physical inactivity; prior history of gestational diabetes; prediabetes; and race or ethnicity. Latino/Hispanic Americans, African Americans, Asian Americans, Native Americans, and those from Pacific Islands are at particular risk. The overall number of people affected is also on the rise, primarily because of widespread inactivity and obesity in our population. There has been a substantial increase in type 2 diabetes in children, due mostly to an increase in overweight in this population (coupled with limited physical activity).

Type 2 diabetes arises when the insulin receptors on the cell surfaces of certain body tissues, especially muscle tissue, become insulin resistant. In this case, insulin secretion may be normal,

Are You Prediabetic?

Often people with type 2 diabetes did not develop the disease suddenly. It may develop for years before symptoms are noticed. Prediabetes is a condition in which the concentration of blood glucose drifts up higher than normal. By the time symptoms are noticeable, organs and tissues may be damaged. Simple tests of your fasting blood glucose level can determine if you are prediabetic. Early detection of diabetes risk can help prevent diabetes if you make lifestyle changes. If you have a family history of diabetes or if your habits (being physically inactive and overweight, and having a poor diet) put you at risk, it is fortunate to discover if your blood glucose is still in the prediabetic stage. Prediabetes, also called *impaired fasting glucose,* is diagnosed if the fasting blood glucose is 100 milligrams per deciliter to 125 milligrams per deciliter or the HbA1c is 5.7% to 6.4%.

decreased, or elevated, but blood glucose is not readily transferred into cells. The person develops high blood glucose as a result of the glucose remaining in the bloodstream. The pancreas attempts to increase insulin output to compensate, but there is a limit to its ability to do this. Thus, rather than insufficient insulin production, there is an abundance of insulin, particularly during the onset of the disease. As the disease develops, pancreatic function can fail, leading to reduced insulin output. Because of the genetic link for type 2 diabetes, those who have a family history should schedule regular blood glucose tests and be careful to avoid risk factors such as obesity and inactivity.

Many cases of type 2 diabetes (about 80%) are associated with obesity (especially with fat located in the abdominal region), but high blood glucose is not directly caused by the obesity. In fact, some lean people also develop this type of diabetes. Obesity associated with oversized adipose cells increases the risk for insulin resistance by the body as more fat is added to these cells during weight gain.

Because type 2 diabetes is linked to obesity, achieving a healthy weight should be a primary goal of treatment, with even limited weight loss leading to better blood glucose regulation. Although many cases of type 2 diabetes can be relieved by reducing excess adipose tissue stores, many people are not able to lose weight. They remain affected with diabetes and may experience the degenerative complications seen in the type 1 form of the disease. Ketosis, however, is not usually seen in type 2 diabetes. Glucose-lowering medications and insulin are used as needed in patients with type 2 diabetes. New classes of drugs that mimic gut hormones are helping diabetic patients overcome the chronic problems that conventional treatments alone have been unable to control. (See the Medicine Cabinet for information about drugs used to manage diabetes.)

Regular patterns of meals and physical activity are important elements of therapy for type 2 diabetes. Physical activity helps the muscles take up more glucose. Dietary therapy should emphasize overall calorie control, increased intakes of fiber-rich foods and fish, and reduced intakes of added sugars and solid fats. Distributing carbohydrates throughout the day helps minimize the high and low swings in blood glucose concentrations. As in type 1 diabetes, carbohydrate counting, exchange lists, and glycemic index of foods are excellent tools to manage carbohydrate

intake. Using the glycemic index of foods combined with carbohydrate counting is an excellent way for motivated individuals to achieve blood glucose goals. It is also important to consume the recommended 25 to 30 grams of fiber, with emphasis on soluble fiber sources, which will help regulate glucose.

Although sugar does not have to be completely eliminated, persons with diabetes will benefit from adhering to the recommendation to reduce consumption of added sugars. If sugar is included in the meal plan, it should be substituted for other carbohydrate sources. Because persons with diabetes are at increased risk of cardiovascular disease, heart healthy choices should also be included in the diabetic meal plan. People with type 2 diabetes who have high blood triglycerides should moderate their carbohydrate intake and increase their intake of plant oils and fiber, as noted earlier for people with type 1 diabetes.

Hypoglycemia

People with diabetes who are taking insulin sometimes have hypoglycemia if they do not eat frequently enough. The first signs of diabetic hypoglycemia include shakiness, sweating, palpitations, anxiety, and hunger. Later symptoms are the result of insufficient glucose reaching the brain and include mental confusion, extreme fatigue, seizures, and unconsciousness. Symptoms should be treated immediately with consumption of glucose or food containing carbohydrate.

Metabolic Syndrome

Metabolic syndrome is characterized by the presence of several risk factors for diabetes and cardiovascular disease. A person with metabolic syndrome must have at least three of

For more information on diabetes, consult the following websites: **www.diabetes.org** and **www.ndep.nih.gov.**

▲ Regular exercise is a key part of a plan to prevent (and control) type 2 diabetes (see Further Reading 8).

metabolic syndrome A condition in which a person has poor blood glucose regulation, hypertension, increased blood triglycerides, and other health problems. This condition is usually accompanied by obesity, lack of physical activity, and a diet high in refined carbohydrates. Also called Syndrome X.

the following metabolic risk factors (or be on medication to treat these risk factors) to be diagnosed with metabolic syndrome: a large waistline from abdominal obesity (accumulation of fat around and within the midsection), high blood triglycerides, low HDL or "good" cholesterol, hypertension, and high fasting blood glucose (Fig. 4-14). Each aspect of metabolic syndrome is a unique health problem with its own treatment. In metabolic syndrome, however, these risk factors are clustered together, making a person twice as likely to develop cardiovascular disease and five times more likely to develop diabetes.

It is generally accepted that one key element unifies all the aspects of metabolic syndrome: *insulin resistance.* As you learned in this chapter, insulin is a hormone that directs tissues to pull glucose out of the blood and into cells for storage or fuel. With insulin resistance, the pancreas produces plenty of insulin, but the cells of the body do not respond to it effectively. Instead, excess glucose stays in the bloodstream. For a while, the pancreas may be able to compensate for the resistance of cells to insulin by overproducing insulin. Over time, however, the pancreas is unable to keep up the accelerated insulin production, and blood glucose levels remain elevated. With metabolic syndrome, blood glucose is not high enough to be classified as diabetes (≥126 milligrams per deciliter), but without intervention, it is likely to get worse and eventually lead to diabetes.

Genetics and aging contribute to the development of insulin resistance and the other elements of metabolic syndrome, but environmental factors such as diet and activity play an important role. Obesity, particularly abdominal obesity, is highly related to insulin resistance. More than half of adults in the United States are overweight, 30% are obese, and these numbers continue to climb year after year. Increases in body weight among children and adolescents are of great concern because childhood obesity

▲ Decreasing body weight and increasing physical activity are interventions to help prevent metabolic syndrome.

Medicine Cabinet

Insulin. Synthetic "human" insulin or their analogs used to treat diabetes are classified by how fast they start to work and how long their effects last.

- Rapid-acting insulin starts working within a few minutes and lasts for a couple of hours. **Examples:** Humalog, Novolog, and Apridra.
- Regular- or short-acting insulin takes about 30 minutes to work and lasts for 3 to 6 hours. **Examples:** Humulin R and Novolin R.
- Intermediate-acting insulin takes 2 to 4 hours to work, and its effects can last for up to 18 hours. **Examples:** Humulin N and Novolin N.
- Long-acting insulin takes 6 to 10 hours to reach the bloodstream, but it can keep working for an entire day. **Examples:** Lantus and Levemir.

Injectable diabetes medications are used to stimulate the release of insulin. These drugs mimic the gut hormones amylin and incretin and are used with injectable insulin (Symlin) or with metformin and sulfonylurea. **Examples:** Byetta and Victoza.

Oral diabetes medications are usually prescribed to people with type 2 diabetes because they help control blood sugar levels when the body still produces some insulin. Each category of these drugs works differently.

- **Sulfonylureas.** Lower blood sugar by stimulating the pancreas to release more insulin. **Examples:** Glucotrol, DiaBeta, Micronase, Glynase PresTab, and Amaryl.
- **Biguanides.** Improve insulin's ability to move sugar into cells, especially muscle cells, and prevent the liver from releasing stored sugar. **Example:** metformin (Glucophage, Riomet, Fortamet, and Glumetza).
- **Thiazolidinediones.** Improve insulin's effectiveness (improving insulin resistance) in muscle and in fat tissue, and lower the amount of sugar released by the liver. These drugs have been linked to adverse cardiovascular events. **Examples:** Actos and Avandia.
- **Alpha-glucosidase inhibitors.** Block enzymes that help digest starches to slow the rise in blood sugar. These diabetes pills may cause diarrhea or gas. **Examples:** Precose and Glyset.
- **Meglitinides.** Lower blood sugar by stimulating the pancreas to release more insulin when blood sugar is high. **Examples:** Prandin and Starlix.
- **Dipeptidyl peptidase IV (DPP-IV) inhibitors.** Lower blood sugar in patients with type 2 diabetes by increasing insulin secretion from the pancreas when blood sugars are high and reducing sugar production by the liver. **Examples:** Januvia, Nesina, Onglyza, and Tradjenta.

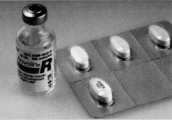

- **Combination therapy.** Combine two medications into one pill. **Example:** Glucovance combines glyburide, a sulfonylurea, and metformin.

Metabolic Syndrome Risk Indicators

For a patient to be diagnosed with metabolic syndrome, he or she must have **three** of the five risk factors listed below.

- **High blood pressure**
  130/85 mmHg or higher

- **Low HDL cholesterol**
  - Men with HDL level less than 40 mg/dl
  - Women with HDL level less than 50 mg/dl

- **Elevated glucose**
  Fasting level of 100 mg/dl or higher

- **Elevated triglycerides (blood fat)**
  150 mg/dl or higher

- **Abdominal obesity**
  - Men with waist circumference greater than 40 inches
  - Women with waist circumference greater than 35 inches

FIGURE 4-14 ▲ Metabolic syndrome is characterized by the presence of several risk factors for diabetes and cardiovascular disease.

places them at high risk for these health problems. This increase in body weight has precipitated a dramatic surge in cardiovascular disease and diabetes risk: an estimated 50 million Americans now have metabolic syndrome.

The insulin resistance that precedes and causes type 2 diabetes also leads to several other components of the metabolic syndrome. The chief culprits contributing to the high blood triglycerides of metabolic syndrome are excessively large meals full of foods rich in simple sugars and refined starches and low in fiber, coupled with little physical activity. Nutrition and lifestyle changes are key strategies in addressing all of the unhealthy conditions of metabolic syndrome as a whole. Suggested interventions include

- Decrease body weight. Even small improvements (e.g., 5% weight loss) for overweight and obese individuals can lessen disease risk. The most successful weight-loss and weight-maintenance programs include moderate dietary restriction combined with physical activity.
- Increase physical activity. To alleviate risks for chronic diseases, the 2010 Dietary Guidelines for Americans include a recommendation to do the equivalent of at least 150 minutes of moderate-intensity physical activity each week.
- Limit total fat consumption, especially animal and *trans* fat sources. Omega-3 fats, however, such as those found in fish and nuts, are considered healthy fats and should be included in your meal plan because they help combat several chronic diseases. We will explore the different types and sources of fats and their effects in Chapter 5.
- For those with particularly high risk for cardiovascular disease, medications to lower blood pressure, total cholesterol, and triglycerides may be warranted.

Summary (Numbers refer to numbered sections in the chapter.)

4.1 Carbohydrates are created in plants through photosynthesis. They are our main fuel source for body cells. Refined and processed products lack many of the health benefits provided by unprocessed sources of carbohydrates (e.g., whole grains, beans, fruits, and vegetables).

4.2 The common monosaccharides in food are glucose, fructose, and galactose. Once these are absorbed from the small intestine and delivered to the liver, much of the fructose and galactose is converted into glucose.

The major disaccharides are sucrose (glucose + fructose), maltose (glucose + glucose), and lactose (glucose + galactose). When digested, these yield their component monosaccharides.

One major group of polysaccharides consists of storage forms of glucose: starches in plants and glycogen in humans. These can be broken down by human digestive enzymes, releasing the glucose units. The main plant starches—straight-chain amylose and branched-chain amylopectin—are digested by enzymes in the mouth and small intestine. In humans, glycogen is synthesized in the liver and muscle tissue from glucose. Under the influence of hormones, liver glycogen is readily broken down to glucose, which can enter the bloodstream.

Fiber is composed primarily of the polysaccharides cellulose, hemicellulose, pectin, gum, and mucilage, as well as the noncarbohydrate lignins. These substances are not broken down by human digestive enzymes. However, soluble (also called viscous) fiber is fermented by bacteria in the large intestine.

4.3 Table sugar, honey, jelly, fruit, and plain baked potatoes are some of the most concentrated sources of carbohydrates. Other high-carbohydrate foods, such as pie and fat-free milk, are diluted by either fat or protein. Nutritive sweeteners in food include sucrose, high-fructose corn syrup, brown sugar, and maple syrup. Several alternative sweeteners are approved for use by FDA: saccharin, aspartame, sucralose, neotame, and acesulfame-K.

4.4 Some starch digestion occurs in the mouth. Carbohydrate digestion is completed in the small intestine. Some plant fibers are digested by the bacteria present in the large intestine; undigested plant fibers become part of the feces. Monosaccharides in the intestinal contents mostly follow an active absorption process. They are then transported via the portal vein that leads directly to the liver.

The ability to digest lactose often diminishes with age. Lactose maldigestion often develops early in childhood and is most common among people of African, Asian, or Hispanic descent. Undigested lactose travels to the large intestine, resulting in such symptoms as abdominal gas, pain, and diarrhea. The occurrence of severe symptoms after consuming lactose is called lactose intolerance. Most people with lactose maldigestion can tolerate cheese, yogurt, and moderate amounts of milk.

4.5 Carbohydrates provide calories (on average, 4 kcal per gram), protect against wasteful breakdown of food and body protein, and prevent ketosis. The RDA for carbohydrate is 130 grams per day. If carbohydrate intake is inadequate for the body's needs, protein is metabolized to provide glucose for energy needs. However, the price is loss of body protein, ketosis,

and eventually a general body weakening. For this reason, low-carbohydrate diets are not recommended for extended periods.

Blood glucose concentration is regulated within a narrow range of 70 to 99 milligrams per deciliter. Insulin and glucagon are hormones that control blood glucose concentration. When we eat a meal, insulin promotes glucose uptake by cells. When fasting, glucagon promotes glucose release from glycogen stores in the liver.

Insoluble (also called nonfermentable) fiber provides mass to the feces, thus easing elimination. In high doses, soluble fiber can help control blood glucose in diabetic people and lower blood cholesterol.

4.6 Diets high in complex carbohydrates are encouraged as a replacement for high-fat diets. A goal of about half of calories as complex carbohydrates is a good one, with about 45% to 65% of total calories coming from carbohydrates in general. Foods to consume are whole-grain cereal products, pasta, legumes, fruits, and vegetables. Many of these foods are rich in fiber.

Moderating sugar intake, especially between meals, reduces the risk of dental caries. Alternative sweeteners, such as aspartame, aid in reducing intake of sugars.

NAYH Diabetes is characterized by a persistent high blood glucose concentration. Regular physical activity and a balanced meal plan that emphasizes fiber and limits added sugars and solid fats are helpful in treating both type 1 and type 2 diabetes. Insulin is the main medication employed: it is required in type 1 diabetes and may be used in type 2 diabetes.

Check Your Knowledge (Answers to the following questions are below.)

1. Dietary fiber
   a. raises blood cholesterol levels.
   b. speeds up transit time for food through the digestive tract.
   c. causes diverticulosis.
   d. causes constipation.

2. When the pancreas detects excess glucose, it releases the
   a. enzyme amylase.
   b. monosaccharide glucose.
   c. hormone insulin.
   d. hormone glucagon.

3. Cellulose is a(n)
   a. indigestible fiber.
   b. simple carbohydrate.
   c. energy-yielding nutrient.
   d. animal polysaccharide.

4. Digested white sugar is broken into _____ and _____.
   a. glucose, lactose
   b. glucose, fructose
   c. sucrose, maltose
   d. fructose, sucrose

5. Starch is a
   a. complex carbohydrate.
   b. fiber.
   c. simple carbohydrate.
   d. gluten.

6. Fiber content of the diet can be increased by adding
   a. fresh fruits.
   b. fish and poultry.
   c. eggs.
   d. whole grains and cereals.
   e. Both a and d.

7. Which form of diabetes is most common?
   a. type 1
   b. type 2
   c. type 3
   d. gestational

8. The recommended daily intake for fiber is approximately _____ grams.
   a. 5   b. 30   c. 100   d. 450

9. Lactose intolerance is the result of
   a. drinking high-fat milk.
   b. eating a large amount of yogurt.
   c. low lactase activity.
   d. a high-fiber diet.

10. One of the components of metabolic syndrome is
    a. high body weight.
    b. high waist circumference.
    c. low blood sugar.
    d. low blood pressure.

Answer Key: 1. b (LO 4.5), 2. c (LO 4.5), 3. a (LO 4.5), 4. b (LO 4.2), 5. a (LO 4.2), 6. e (LO 4.3), 7. b (LO 4.7), 8. b (LO 4.6), 9. c (LO 4.4), 10. b (LO 4.7)

Study Questions (Numbers refer to Learning Outcomes)

1. Why do we need carbohydrates in the diet? **(LO 4.3)**

2. What are the three major monosaccharides and the three major disaccharides? Describe how each plays a part in the human diet. **(LO 4.2)**

3. Why are some foods that are high in carbohydrates, such as cookies and fat-free milk, not considered to be concentrated sources of carbohydrates? **(LO 4.3)**

4. Describe the digestion of the various types of carbohydrates in the body. **(LO 4.4)**

5. Describe the reason why some people are unable to tolerate high intakes of milk. **(LO 4.4)**

6. List three alternatives to simple sugars for adding sweetness to the diet. **(LO 4.3)**

7. Outline the basic steps in blood glucose regulation, including the roles of insulin and glucagon. **(LO 4.5)**

8. What are the important roles that fiber plays in the diet? **(LO 4.5)**

9. Summarize current carbohydrate intake recommendations. **(LO 4.6)**

10. What, if any, are the proven ill effects of sugar in the diet? **(LO 4.7)**

What the Dietitian Chose

Sandwiches can be a great option for a quick lunch that includes several, if not all, of the food groups on MyPlate. Most of us define a sandwich as meat and perhaps cheese between two slices of bread. This idea will guarantee plenty of grains, protein, and even dairy. All of the sandwiches on this menu provide 1½ ounces to 2½ ounces of protein. The hot dog and the peanut butter sandwiches are the lowest at 1½ ounces, and the soy burger is the highest with 2½ ounces of protein. All the others provide 2 ounces of protein. All of the sandwiches are equal in terms of their grain servings at 2 ounces.

The challenge arises in filling the other half of the plate, that is, getting some vegetables and perhaps fruit onto a sandwich. Additionally, we want to consider the nutritional quality of the grains, protein, and dairy items. If we evaluate the sandwiches at Doug's Deli based on their carbohydrate sources, we see that only three of the sandwiches offer servings of vegetables. The addition of two tomato slices, four slices of cucumber or pickles, and two lettuce leaves to the turkey sandwich and the soy burger provides ¾ cup of vegetables to each. The tuna salad sandwich provides the most vegetables, 1¼ cups, from the carrots and onions in the tuna salad and the ½ cup of sprouts, two leaves of lettuce, and four cucumber slices added to the sandwich.

We typically do not find fruits on a sandwich and that is true of the menu at Doug's Deli. Although we see strawberry jelly on the peanut butter sandwich, jelly does not count as a fruit serving and instead provides 32 kcal (~10% of kcal in the sandwich) from added sugars.

As we move to the Grains aspect of MyPlate, we want to examine the whole-grain content of the breads offered. Fortunately, two of the options we already highlighted for their vegetable content are served on whole-grain breads. The tuna salad on whole wheat and the soy burger on a whole-wheat English muffin both offer 2 ounces of whole grain for this quick lunch. The combination of vegetables and whole grain

provides a significant amount of dietary fiber. The tuna salad sandwich has 7 grams and the soy burger has 7.5 grams of dietary fiber which is close to 30% of the daily recommendation of 25 to 38 grams per day. The soy burger on whole wheat also has the benefit of having both soluble fiber from the soybean and insoluble from the whole wheat. The peanut butter sandwich also provides 3 grams of primarily soluble fiber from the peanut butter.

The turkey and Swiss sandwich is also a good choice with 4 grams of fiber, and it provides a ¾ cup serving of dairy. Unfortunately, the rye is not necessarily a whole-grain bread. The ham and Swiss provides 2 ounces of lean protein and ¾ cup of dairy but provides only 1.5 grams of fiber because it is on sourdough bread made from refined white flour. It is also low on the list of choices because of its high sodium content (about 1100 milligrams).

Although it is sometimes described as America's favorite, the hot dog on a white bun has the least to offer nutritionally. It provides the lowest amount of grain at 1½ ounces, and the grain is refined rather than whole. It also has the lowest protein content at 1½ ounces and only offers 1 gram of fiber. It is high in sodium at over 700 milligrams and contains 6 grams of animal fat.

In summary, the dietitian would choose either the meatless soy burger or the tuna salad sandwich at Doug's Deli to maximize the number of servings from a variety of healthy carbohydrate sources including vegetables and whole grains.

CASE STUDY SOLUTION Problems with Milk Intake

1. Myeshia suspected that she was sensitive to milk because she developed bloating and gas when she consumed milk during one meal.
2. The disaccharide, lactose, is the most likely cause of Myeshia's symptoms.
3. When the enzyme lactase is missing or in short supply in the small intestine, undigested lactose travels to the large intestine and is fermented by bacteria there. The fermentation of lactose by bacteria produces abdominal gas, bloating, and pain. Diarrhea can also occur because the presence of lactose in the colon draws water from the blood vessels into the large intestine.
4. Lactose maldigestion is a normal pattern of physiology that often begins to develop at about ages 3 to 5 years in populations that do not rely on milk and other dairy foods as a main food source. When significant symptoms develop after lactose intake, it is then called lactose intolerance.
5. The primary form of lactose maldigestion is estimated to occur in about 75% of the world's population. In the United States, most lactose maldigesters are African-Americans (like Myeshia), Asian-Americans, and Latino/Hispanic Americans. The occurrence of lactose maldigestion also increases as people age.
6. The active bacteria cultures in yogurt digest the lactose when these bacteria are broken apart in the small intestine and release their lactase.
7. Many products, such as low-lactose milk and lactase pills, are available to enable lactose maldigesters to tolerate milk.
8. Many lactose maldigesters can consume moderate amounts of lactose with minimal or no intestinal discomfort because of the eventual lactose breakdown by bacteria in the large intestine.
9. In North America and Western Europe, milk and other dairy products are important sources of calcium and vitamin D. These are important for maintaining bone health.
10. Secondary lactose maldigestion occurs when a condition (e.g., long-standing diarrhea from a viral infection) results in a significant decline in lactase production.

Further Readings

1. American Diabetes Association: Nutrition recommendations and interventions for diabetes: A position statement of the American Diabetes Association. *Diabetes Care* 31 (suppl 1):S61, 2008.

2. American Diabetes Association and Academy of Nutrition and Dietetics: *Choose your foods:* Food lists for diabetes. 2014.

3. American Diabetes Association: Standards of Medical Care in Diabetes-2014: *Diabetes Care* 37:S14, 2014.

4. Centers for Disease Control and Prevention: *National Diabetes Fact Sheet: National Estimates and General Information on Diabetes and Prediabetes in the United States, 2011.* Atlanta, GA: U.S. Department of Health and Human Services, Centers for Disease Control and Prevention, 2011.

5. Franz MJ and others: The evidence for medical nutrition therapy for type 1 and type 2 diabetes in adults. *Journal of the American Dietetic Association* 110:1852, 2010.

6. Getz L: Deciphering whole grain food labels: Separating fact from fiction. *Today's Dietitian* 14(6):44, 2012.

7. Hall M: Fiber facts about cereal. *Today's Dietitian* 14(12):30, 2012.

8. Hayes C and Kriska A: Role of physical activity in diabetes management and prevention. *Journal of the American Dietetic Association* 108:S19, 2008.

9. Johnson RK and others: Dietary sugars intake and cardiovascular health: A scientific statement from the American Heart Association. *Circulation* 120(11):1011, 2009.

10. Position of the American Dietetic Association: Health implications of dietary fiber. *Journal of the American Dietetic Association* 108:1716, 2008.

11. Position of the Academy of Nutrition and Dietetics: Use of nutritive and nonnutritive sweeteners. *Journal of the Academy of Nutrition and Dietetics* 112:739, 2012.

12. Rippe JM and Angelopoulos TJ: Sucrose, high-fructose corn syrup, and fructose, their metabolism and potential health effects: What do we really know? *Advances in Nutrition* 4:236, 2013.

13. Schaeffer J: Boosting whole grain consumption. *Today's Dietitian* 15 (2):33, 2013.

14. Slavin J and others: How fiber affects weight regulation. *Food Technology,* p. 34, February 2008.

15. Webb D: Whole grain goodness. *Today's Dietitian* 14(9):56, 2012.

16. Welsh JA and others: Consumption of added sugars is decreasing in the United States. *American Journal of Clinical Nutrition* 94:726, 2011.

 **To get the most out of your study of nutrition, visit McGraw-Hill Connect at www.mcgrawhillconnect.com where you will find NutritionCalc Plus, LearnSmart, and many other dynamic tools.**

Rate Your Plate

Estimate Your Fiber Intake

Review the sample menus shown in Table 4-6. The first menu contains 1600 kcal and 25 grams of fiber (AI for women); the second menu contains 2100 kcal and 38 grams of fiber (AI for men).

TABLE 4-6 ▶ **Sample Menus Containing 1600 kcal with 25 grams of Fiber and 2000 kcal with 38 grams of Fiber***

Menu	25 grams of Fiber			38 grams of Fiber		
	Serving Size	Carbohydrate Content (g)	Fiber Content (g)	Serving Size	Carbohydrate Content (g)	Fiber Content (g)
Breakfast						
Muesli cereal	1 cup	60	6	1 cup	60	6
Raspberries	½ cup	11	2	½ cup	11	2
Whole-wheat toast	1 slice	13	2	2 slices	26	4
Margarine	1 tsp	0	0	1 tsp	0	0
Orange juice	1 cup	28	0	1 cup	28	0
1% milk	1 cup	24	0	1 cup	24	0
Coffee	1 cup	0	0	1 cup	0	0
Lunch						
Bean and vegetable burrito	2 small	50	4.5	3 small	75	7
Guacamole	¼ cup	5	4	¼ cup	5	4
Monterey Jack cheese	1 oz	0	0	1 oz	0	0
Pear (with skin)	1	25	4	1	25	4
Carrot sticks	—	—	—	¾ cup	6	3
Sparkling water	2 cups	0	0	2 cups	0	0
Dinner						
Grilled chicken (no skin)	3 oz	0	0	3 oz	0	0
Salad						
Red cabbage	½ cup	7	3	½ cup	19	6
Romaine	½ cup			½		
Peach slices	¼ cup			1 cup		
Toasted almonds	—	—	—	½ oz	3	2
Fat-free salad dressing	2 tbsp	0	0	2 tbsp	0	0
1% milk	1 cup	24	0	1 cup	24	0
Total		247	25		306	38

*The overall diet is based on MyPlate breakdown of approximate energy content: carbohydrate, 58%; protein, 12%; and fat, 30%.

To roughly estimate your daily fiber consumption, determine the number of servings that you ate yesterday from each food category listed here. If you are not meeting your needs, how could you do so? Multiply the serving amount by the value listed and then add up the total amount of fiber.

Food	Servings	Grams
Vegetables (serving size: 1 cup raw leafy greens or 1/2 cup other vegetables)	_____ × 2	_____
Fruits (serving size: 1 whole fruit, 1/2 grapefruit, 1/2 cup berries or cubed fruit, and 1/4 cup dried fruit)	_____ × 2.5	_____
Beans, lentils, split peas (serving size: 1/2 cup cooked)	_____ × 7	_____
Nuts, seeds (serving size: 1/4 cup; 2 tbsp peanut butter)	_____ × 2.5	_____
Whole grains (serving size: 1 slice whole-wheat bread; 1/2 cup whole-wheat pasta, brown rice, or other whole grain; and 1/2 each bran or whole-grain muffin)	_____ × 2.5	_____
Refined grains (serving size: 1 slice bread, 1/2 cup pasta, rice, or other processed grains; and 1/2 each refined bagels or muffins)	_____ × 1	_____
Breakfast cereals (serving size: check package for serving size and amount of fiber per serving)	_____ × grams of fiber per serving	_____
Total Grams of Fiber =		_____

Source: Adapted from Fiber: Strands of protection. *Consumer Reports on Health,* p. 1, August 1999.

How does your total fiber intake for yesterday compare with the general recommendation of 25 to 38 grams of fiber per day for women and men, respectively?

If you are not meeting your needs, how could you do so?

▲ This dessert is an excellent source of fiber from 2 slices of whole-wheat banana bread (7 grams) with ½ cup of berries (1.8 grams) and 2 ounces of yogurt for a total of 8.8 grams of fiber.

Student Learning Outcomes

Chapter 5 is designed to allow you to:

5.1 Understand the common properties of lipids.

5.2 List three structural forms of lipids (fats) and the role of each in our food supply. Distinguish between fatty acids and triglycerides. Discuss the importance of the essential fatty acids, as well as the balance between omega-3 and omega-6 fatty acids, with respect to health.

5.3 Differentiate among food sources of saturated, monounsaturated, and polyunsaturated fatty acids and cholesterol.

5.4 Explain how lipids are digested and absorbed.

5.5 Name the classes of lipoproteins and classify them according to their functions.

5.6 List the function of lipids, including the roles of phospholipids and the functions of cholesterol in the body.

Chapter 5
Lipids

What Would You Choose?

Several of your friends are getting together for a cook-out, and you have volunteered to be the grill master. Hamburgers will be the main "protein" portion of your meal. (Remember the size of that group on MyPlate?) You have heard that eating red meat is bad for your heart. Is all red meat bad for your heart? In the meat section of the grocery store, you see several varieties of ground beef. Which of the following types of ground beef would you choose for heart health?

- **a** Ground round
- **b** Ground chuck
- **c** Ground sirloin
- **d** Ground beef

connect NUTRITION Think about your choice as you read Chapter 5, then see What the Dietitian Chose at the end of the chapter. To learn more about ground beef choices, check out the Connect site: www.mcgrawhillconnect.com.

5.7 Explain the recommendations for fat intake.

5.8 Characterize the symptoms of cardiovascular disease and highlight some known risk factors.

Lipids are energy-dense: fatty acids from triglycerides contain more than twice the calories per gram (on average, 9 kcal) as proteins and carbohydrates (on average, 4 kcal each). Consumption of common solid fats also contributes to the risk of cardiovascular disease. For these reasons, some concern about excessive consumption of certain lipids is warranted. Lipids do play vital roles, however, both in the body and in foods. Their presence in the diet is essential to good health. In fact, lipids found in seafood, such as the salmon pictured here, have been linked to *lower* risk for several chronic diseases. In general, lipids should comprise 20% to 35% of an adult's total calorie intake.

Humans need very little fat in their diet to maintain health. In fact, the body's need for the essential fatty acids can be met by daily consumption of only about 2 to 4 tablespoons of plant oil incorporated into foods and consumption of fatty fish such as salmon or tuna at least twice weekly. The Food and Nutrition Board suggests that fat intake can be as high as 35% of calories consumed for an adult. Some experts suggest that an intake as high as 40% of calories is appropriate as long as the predominant type of fat is a healthy one, such as olive oil. After learning more about lipids—fats, oils, and related compounds—in this chapter, you can decide for yourself how much fat you want to consume.

Let us look at lipids in detail—their forms, functions, metabolism, and food sources. This chapter will also discuss the link between various lipids and the major "killer" disease in North America: cardiovascular disease, which involves the heart, including the coronary arteries (coronary heart disease), as well as other arteries in the body.

5.1 Lipids: Common Properties

Lipids are a diverse group of chemical compounds. They share one main characteristic: They do not readily dissolve in water. Think of an oil-and-vinegar salad dressing. The oil is not soluble in the water-based vinegar; on standing, the two separate into distinct layers, with oil on the top and vinegar on the bottom.

✅ CONCEPT CHECK 5.1

1. What is the common property that all lipid compounds share?

5.2 Lipids: Triglycerides, Phospholipids, and Sterols

The chemical structures of lipids are diverse. Lipids (mostly fats and oils) are composed primarily of the elements carbon and hydrogen; they contain fewer oxygen **atoms** than do carbohydrates. Lipids yield more calories per gram than do carbohydrates—on the average, 9 kcal per gram—because of this difference in composition. **Triglycerides** are the most common type of lipid found in the body and in foods. Each triglyceride molecule consists of three fatty acids bonded to **glycerol. Phospholipids** and **sterols,** including **cholesterol,** are also classified as lipids, although their structures can be quite different from the structure of triglycerides. All of these lipid compounds are described in this chapter.

Food experts, such as chefs, call lipids that are solid at room temperature *fats* and lipids that are liquid *oils.* Most people use the word *fat* to refer to all lipids because they do not realize that there is a difference. However, *lipid* is a generic term that includes triglycerides and many other substances. To simplify our discussion, this chapter primarily uses the term *fat.* When necessary for clarity, the name of a specific lipid, such as cholesterol, will be used. This word use is consistent with the way many people use these terms.

FATTY ACIDS AND TRIGLYCERIDES

Fatty Acids: The Simplest Form of Lipids. Triglycerides are the primary form of lipids in the body and in foods. Fatty acids are found in triglycerides. A fatty acid is basically a long chain of carbons bonded together and flanked by hydrogens. At one end of the molecule (the alpha end) is an **acid group.** At the other end (the omega end) is a **methyl group** (Fig. 5-1).

Fats in foods are not composed of a single type of fatty acid. Rather, each dietary fat, or triglyceride, is a complex mixture of many different fatty acids, the combination of which provides each food its unique taste and smell.

Fatty acids can be saturated or unsaturated with hydrogen. Chemically speaking, a carbon atom can form four bonds. Within the carbon chain of a fatty acid, each carbon bonds to two other carbons and to hydrogens. The carbons that make up the chain of a **saturated fatty acid** are all connected to each other by single bonds. This allows the maximum number of hydrogens to be bound. Just as a sponge can be saturated (full) with water, a saturated fatty acid, such as stearic acid, is saturated with hydrogen (Fig. 5-1a). The saturated fatty acids are very straight and linear and therefore can pack very close together. This close packing or stacking of saturated fat molecules makes them solidify at room temperature.

Animal fats are high in saturated fatty acids and remain solid at room temperature. A good example is the solid fat surrounding a piece of uncooked steak. Chicken fat, semisolid at room temperature, contains less saturated fat than beef fat.

triglyceride The major form of lipid in the body and in food. It is composed of three fatty acids bonded to glycerol, an alcohol.

glycerol A three-carbon alcohol used to form triglycerides.

phospholipid Any of a class of fat-related substances that contain phosphorus, fatty acids, and a nitrogen-containing base. The phospholipids are an essential part of every cell.

sterol A compound containing a multi-ring (steroid) structure and a hydroxyl group (–OH). Cholesterol is a typical example.

cholesterol A waxy lipid found in all body cells. It has a structure containing multiple chemical rings that is found only in foods that contain animal products.

saturated fatty acid A fatty acid containing no carbon–carbon double bonds.

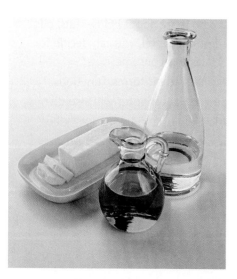

▲ Saturated fats, such as butter, are solid at room temperature, whereas unsaturated fats, such as olive and corn oil, are liquid at room temperature.

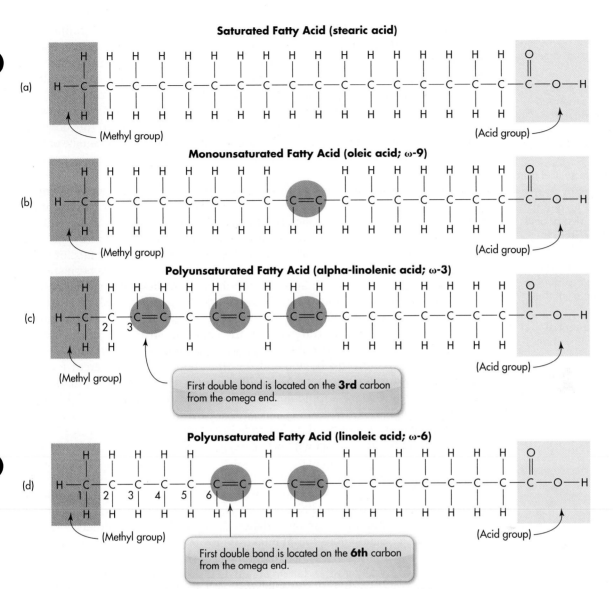

FIGURE 5-1 ▲ (a-d) Chemical forms of saturated, monounsaturated, and polyunsaturated fatty acids. Each of the depicted fatty acids contains 18 carbons, but they differ from each other in the number and location of double bonds. The double bonds are shaded. The linear shape of saturated fatty acids, as shown in (a), allows them to pack tightly together and so form a solid at room temperature. In contrast, unsaturated fatty acids (b-d) have "kinks" where double bonds interrupt the carbon chain (see Figure 5-2). Thus, unsaturated fatty acids pack together only loosely and are usually liquid at room temperature.

monounsaturated fatty acid A fatty acid containing one carbon–carbon double bond.

polyunsaturated fatty acid A fatty acid containing two or more carbon–carbon double bonds.

However, in some foods, saturated fats are suspended in liquid, such as the butterfat in whole milk, so the solid nature of these fats at room temperature is less apparent.

If the carbon chain of a fatty acid contains a double bond, those carbons in the chain have fewer bonds to share with hydrogen, and the chain is said to be *unsaturated*. A fatty acid with only one double bond is **monounsaturated** (Fig. 5-1b). Canola and olive oils contain a high percentage of monounsaturated fatty acids. Likewise, if two or more of the bonds between the carbons are double bonds, the fatty acid is even less saturated with hydrogens, and so it is **polyunsaturated** (Fig. 5-1c, d). The double bonds in unsaturated fatty acids create kinks in their structure that keep them from packing closely together, so they are liquid at room temperature. Corn, soybean, sunflower, and safflower oils are rich in polyunsaturated fatty acids.

Unsaturated fatty acids, with their double bonds, can exist in two different structural forms, the *cis* and *trans* forms. In their natural form, monounsaturated and

▲ Avocados are a rich source of monounsaturated fat.

cis **fatty acid** A form of an unsaturated fatty acid that has the hydrogens lying on the same side of the carbon-carbon double bond.

trans **fatty acid** A form of an unsaturated fatty acid, usually a monounsaturated one when found in food, in which the hydrogens on both carbons forming the double bond lie on opposite sides of that bond.

long-chain fatty acid A fatty acid that contains 12 or more carbons.

omega-3 (ω-3) fatty acid An unsaturated fatty acid with the first double bond on the third carbon from the methyl end (–CH₃).

omega-6 (ω-6) fatty acid An unsaturated fatty acid with the first double bond on the sixth carbon from the methyl end (–CH₃).

alpha-linolenic acid An essential omega-3 fatty acid with 18 carbons and 3 double bonds.

linoleic acid An essential omega-6 fatty acid with 18 carbons and 2 double bonds.

polyunsaturated fatty acids usually are in the *cis* form (Fig. 5-2). By definition, the resulting ***cis* fatty acid** has the hydrogens on the same side of the carbon-carbon double bond. During certain types of food processing (discussed in Section 5.3), some hydrogens are transferred to opposite sides of the carbon–carbon double bond, creating the *trans* form, or a ***trans* fatty acid.** As seen in Figure 5-2, the *cis* bond causes the fatty acid backbone to bend, whereas the *trans* bond allows the backbone to remain straighter. This makes it similar to the shape and functions of a saturated fatty acid. The Food and Nutrition Board suggests limiting intake of *trans* fatty acids (also referred to as *trans* fats) in processed foods as much as possible. Later you will see why.

You may be surprised to learn that some *trans* fatty acids, known as conjugated linoleic acid (CLA), occur naturally. CLA is a family of derivatives of the fatty acid linoleic acid. The bacteria that live in the rumens of some animals (cows, sheep, and goats, for example) produce *trans* fatty acids from the polyunsaturated fats in the grass the animals are fed. These natural *trans* fats, or CLA, eventually appear in foods such as beef, milk, and butter, and have been studied extensively. Results of clinical and epidemiological studies suggest that natural *trans* fats improve insulin levels in diabetics and decrease the risk of heart disease, cancer, and obesity, the same diseases that industrial *trans* fats have been shown to increase. A slight difference in the chemical structures of the natural and industrial *trans* fats cause these significant differences in their effects. CLA contains both *cis* and *trans* bonds, and the *trans* bond is in a different location compared to industrial *trans* fats. About 20% of *trans* fatty acids in our diets come from this source. Dietary supplements of CLA are available but are highly variable in their quality.

An important aspect of the double bonds in polyunsaturated fatty acids is their chemical susceptibility to react with active oxygen molecules. These oxidation reactions can take place in food and in the body. Some of the random oxidation can happen rapidly in the body and cause inflammation and tissue damage thought to be a cause of aging, atherosclerosis, and cancer. Fortunately, antioxidants such as vitamin E stop the chain of oxidation. Many features of the dietary recommendations for fat intake are related to decreasing the likelihood of these harmful oxidation reactions.

Overall, a fat or an oil is classified as saturated, monounsaturated, or polyunsaturated based on the type of fatty acids present in the greatest concentration (Fig. 5-3). Fats in foods that contain primarily saturated fatty acids are solid at room temperature, especially if the fatty acids have long carbon chains (i.e., a **long-chain fatty acid**), as opposed to shorter versions. In contrast, fats containing primarily polyunsaturated or monounsaturated fatty acids (long chain or shorter) are usually liquid at room temperature. Almost all fatty acids in the body and in foods are long-chain varieties.

The location of the first double bond on an unsaturated fatty acid is an important feature and will be three, six, or nine carbons from the methyl end. If the first double bond starts three carbons from the methyl (omega) end of the fatty acid, it is an **omega-3 (ω-3) fatty acid** (review Fig. 5-1c). If it is located six carbons from the omega end, it is an **omega-6 (ω-6) fatty acid** (review Fig. 5-1d). An omega-9 fatty acid has its first double bond starting at the ninth carbon from the methyl end (review Fig. 5-1b). In foods, **alpha-linolenic acid** is the major omega-3 fatty acid, and **linoleic acid** is the major omega-6 fatty acid. These are also the **essential fatty acids** we need to consume (more on this in Section 5.6 about putting lipids to work in the body). **Oleic acid** is the major omega-9 fatty acid.

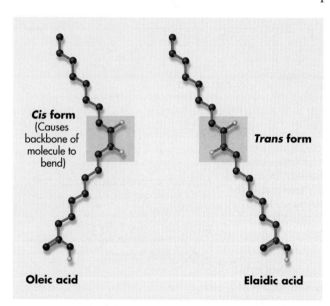

Cis form
(Causes backbone of molecule to bend)

Trans form

Oleic acid

Elaidic acid

FIGURE 5-2 ▲ *Cis* and *trans* fatty acids. In the *cis* form at carbon-carbon double bonds in a fatty acid, the hydrogens (in white) lie on the same side of the double bond. This causes a "kink" at that point in the fatty acid, typical of unsaturated fatty acids in foods. In contrast, in the *trans* form at carbon-carbon double bonds in a fatty acid, the hydrogens lie across from each other at the double bond. This causes the fatty acid to exist in a linear form, like a saturated fatty acid. *Cis* fatty acids are much more common in foods than *trans* fatty acids. The latter are primarily found in foods containing partially hydrogenated fats, notably stick margarine, shortening, and deep-fat fried foods.

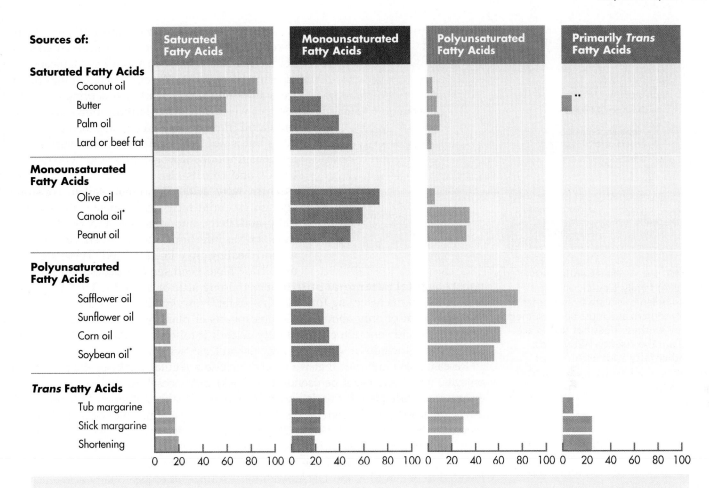

Sources of:	Saturated Fatty Acids	Monounsaturated Fatty Acids	Polyunsaturated Fatty Acids	Primarily *Trans* Fatty Acids

*Rich source of the omega-3 fatty acid alpha-linolenic acid (7% and 12% of total fatty acid content for soybean oil and canola oil, respectively).

**The natural *trans* fatty acids in butter are not harmful and may even have health-promoting properties, such as preventing certain forms of cancer.

FIGURE 5-3 ▲ Saturated, monounsaturated, polyunsaturated, and *trans* fatty acid composition of common fats and oils (expressed as percent of all fatty acids in the product).

Essential Fatty Acids. The various classes of lipids have diverse functions in the body and are necessary for health. Of all the classes of lipids, however, only certain polyunsaturated fatty acids are essential parts of a diet. All others can be made by the body and therefore are not needed in our diet. Linoleic acid (an omega-6 fatty acid) and alpha-linolenic acid (an omega-3 fatty acid) are called *essential* fatty acids because we must obtain them from foods to maintain health (Fig. 5-4). These omega-6 and omega-3 fatty acids form parts of vital body structures, perform important roles in immune system function and vision, help form cell membranes, and produce **eicosanoids,** which are involved in practically all important functions in the body. Omega-6 and omega-3 fatty acids must be obtained through the diet because human cells lack the enzymes needed to produce these fatty acids. Only plants can make the omega-3 and omega- 6 parent fatty acids.

The essential fats start with the dietary omega-3 and omega-6 parents. Human enzymes can then convert these vital fatty acids to long-chain polyunsaturated fatty acids such as **eicosapentaenoic acid (EPA)** and **docosahexaenoic acid (DHA),** which are particularly important for the brain and nervous system. In addition to its role in brain function, EPA is needed for concentration and vision and is converted into a powerful anti-inflammatory agent. Because of its role in brain structure, DHA is especially important during pregnancy for the fetal brain and nervous system development. Other fatty acids, such as omega-9 fatty acids, can be synthesized in the body and therefore are not essential components of the diet.

essential fatty acids Fatty acids that must be supplied by the diet to maintain health. Currently, only linoleic acid and alpha-linolenic acid are classified as essential.

oleic acid An omega-9 fatty acid with 18 carbons and 1 double bond.

eicosanoids A class of hormone compounds, including the prostaglandins, derived from polyunsaturated fatty acids such as arachidonic acid and involved in the cellular activity affecting practically all important functions in the body.

eicosapentaenoic acid (EPA) An omega-3 fatty acid with 20 carbons and 5 carbon-carbon double bonds. It is present in large amounts in fatty fish and is slowly synthesized in the body from alpha-linolenic acid.

docosahexaenoic acid (DHA) An omega-3 fatty acid with 22 carbons and 6 carbon-carbon double bonds. It is present in large amounts in fatty fish and is slowly synthesized in the body from alpha-linolenic acid. DHA is especially present in the retina and brain.

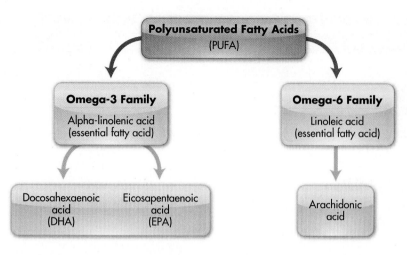

FIGURE 5-4 ▲ The essential fatty acid (EFA) family. Linoleic acid and alpha-linolenic acid are available from dietary sources and must be consumed as body synthesis does not take place. These are the essential fatty acids. The other fatty acids in this figure can be synthesized from the essential fatty acids.

total parenteral nutrition The intravenous feeding of all necessary nutrients, including the most basic forms of protein, carbohydrates, lipids, vitamins, minerals, and electrolytes.

arachidonic acid An omega-6 fatty acid made from linoleic acid with 20 carbon atoms and 4 carbon-carbon double bonds.

hemorrhagic stroke Damage to part of the brain resulting from rupture of a blood vessel and subsequent bleeding within or over the internal surface of the brain.

Omega-3 Fatty Acids in Fish (grams per 3 ounce serving)

Atlantic salmon	1.8
Anchovy	1.7
Sardines	1.4
Rainbow trout	1.0
Coho salmon	0.9
Bluefish	0.8
Striped bass	0.8
Tuna, white, canned	0.7
Halibut	0.4
Catfish, channel	0.2

Recommended omega-3 fatty acid (alpha-linolenic acid) intake per day:

Men	1.6 grams
Women	1.1 grams

Nevertheless, we need to consume only about 5% of our total calories per day from essential fatty acids. That corresponds to about 2 to 4 tablespoons of plant oil each day. We can easily get that much from nuts and seeds, and two servings of fatty fish per week. The best plant oil source would be canola oil (flax seed oil has greater omega-3 fatty acids but is not as versatile in the kitchen). Regular consumption of vegetables and whole-grain breads and cereals also helps to supply enough essential fatty acids. If humans fail to consume enough essential fatty acids, their skin becomes flaky and itchy, and diarrhea and other symptoms such as infections often are seen. Growth and wound healing may be restricted. These signs of deficiency have been seen in people fed intravenously by **total parenteral nutrition** containing little or no fat for 2 to 3 weeks, as well as in infants receiving formulas low in fat. However, because our bodies need the equivalent of only about 2 to 4 tablespoons of plant oils a day, even a low-fat diet will provide enough essential fatty acids if it follows a balanced plan such as MyPlate and includes a serving of fatty fish at least twice a week.

Research also indicates that we should include a regular intake of the preformed omega-3 fatty acids, eicosapentaenoic acid (EPA) and docosahexaenoic acid (DHA), to ensure levels adequate for brain and cardiovascular health. Although EPA and DHA can be made in the body from the essential fatty acid, alpha-linolenic acid, conversion is thought to be inadequate. EPA and DHA are naturally high in fatty fish such as salmon, tuna, sardines, anchovies, striped bass, catfish, herring, mackerel, trout, and halibut. Consumption of a serving of one or more of these fish at least twice a week is recommended to obtain EPA and DHA. Additional sources of omega-3 fatty acids include canola and soybean oils, walnuts, chia seeds, and flax seeds (see Further Reading 5). An EPA/DHA supplement is recommended if these dietary sources are not consumed.

The recommendation to consume omega-3 fatty acids stems from the observation that compounds made from omega-3 fatty acids tend to decrease blood clotting and inflammatory processes in the body. The omega-6 parent fat, linoleic acid, is converted by the body into the long-chain fatty acid **arachidonic acid (AA),** which is needed by the body but also acts as a pro-inflammatory agent. When omega-6 fatty acids are consumed in excess, the arachidonic acid made from linoleic acid and directly consumed from animal sources can contribute to inflammatory conditions such as arthritis, clotting, and other cardiovascular problems. Some studies show that people who eat fish at least twice a week (total weekly intake: 8 ounces) run lower risks for heart attack than do people who rarely eat fish. In these cases, the omega-3 fatty acids in fish oil are probably acting to reduce blood clotting. As will be covered in detail in the Nutrition and Your Health section at the end of this chapter, blood clots are part of the heart attack process. In addition, these omega-3 fatty acids have a favorable effect on heart rhythm. Consequently, the risk of heart attack decreases with the consumption of omega-3 fatty acids from fish, especially for people already at high risk.

We need to remember, however, that blood clotting is a normal body process. Certain groups of people, such as Eskimos in Greenland, eat so much seafood that their normal blood-clotting ability can be impaired. An excess of omega-3 fatty acid intake can allow uncontrolled bleeding and may cause **hemorrhagic stroke.** However, no increase in risk of stroke has been observed in studies using moderate amounts of omega-3 fatty acids. The excess long-chain fatty acids in seafood can also lower immunity, resulting in more infections.

Studies also have shown that large amounts of omega-3 fatty acids from fish (2 to 4 grams per day) can lower blood triglycerides in people with high triglyceride concentrations. In addition, these omega-3 fatty acids are thought to be helpful in managing the pain of inflammation associated with rheumatoid arthritis by suppressing immune system responses. The use of omega-3 fatty acids has also improved symptoms of certain behavioral disorders and cases of mild depression (see Further Reading 11).

Triglycerides. Fats and oils in foods are mostly in the form of triglycerides. The same is true for fats found in body structures. Although some fatty acids are transported in the bloodstream attached to proteins, most fatty acids are formed into triglycerides by cells in the body.

Triglycerides contain a simple three-carbon alcohol, glycerol, that serves as a backbone for the three attached fatty acids (Fig. 5-5a). Removing one fatty acid from a triglyceride forms a **diglyceride.** Removing two fatty acids from a triglyceride forms a **monoglyceride.** Later you will see that before most dietary fats are absorbed, the two

diglyceride A breakdown product of a triglyceride consisting of two fatty acids bonded to a glycerol backbone.

monoglyceride A breakdown product of a triglyceride consisting of one fatty acid attached to a glycerol backbone.

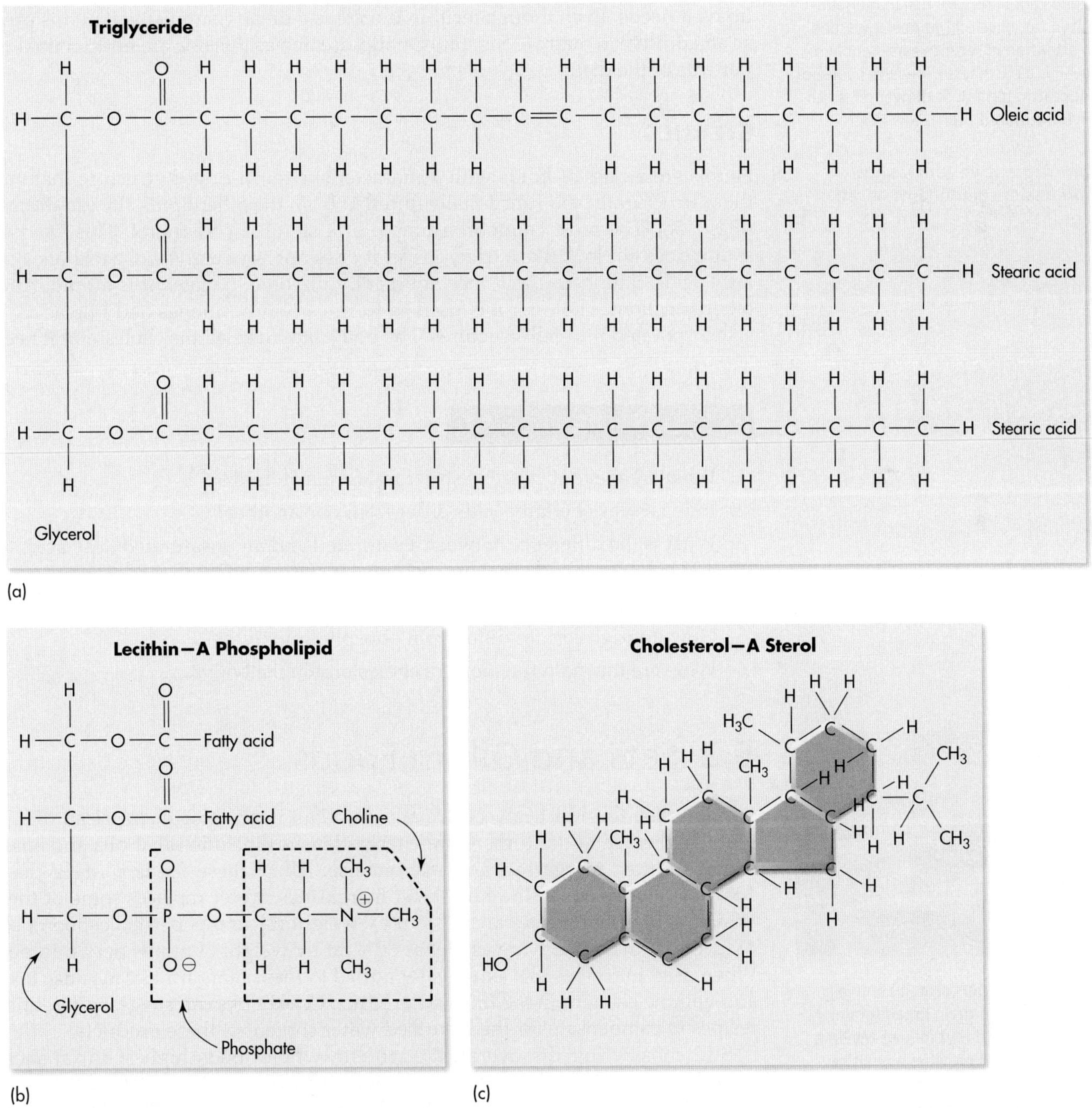

FIGURE 5-5 ▲ Chemical forms of common lipids: (a) triglyceride, (b) phospholipid (in this case, lecithin), and (c) sterol (in this case, cholesterol).

▲ Plant oils vary in their content of specific fatty acids. Oils similar in appearance may vary significantly in fatty acid composition. Olive and canola oils are rich in monounsaturated fat; olive oil has been awarded much attention in recent years. Canola oil, however, is a much less expensive choice of monounsaturated fat. Safflower oil is rich in polyunsaturated fat.

lecithin A group of compounds that are major components of cell membranes.

outer fatty acids are typically removed from the triglyceride during digestion in the small intestine. This produces a mixture of fatty acids and monoglycerides, absorbed into the intestinal cells. After absorption, the fatty acids and monoglycerides are mostly reformed into triglycerides.

PHOSPHOLIPIDS

Phospholipids are another class of lipid. Like triglycerides, they are built on a backbone of glycerol. However, at least one fatty acid is replaced with a compound containing phosphorus (and often other elements, such as nitrogen) (Fig. 5-5b). Many types of phospholipids exist in the body, especially in the brain. They are an important part of cell membranes. **Lecithin** is a common example of a phospholipid. Various forms of phospholipids are found in body cells, and they participate in fat digestion, absorption, and transport. The body is able to produce all the phospholipids it needs. Even though lecithin is sold as a dietary supplement and is present as an additive in many foods, phospholipids such as this one are not essential components of the diet.

STEROLS

Sterols are a class of lipids with a characteristic multi-ringed structure that makes them structurally and functionally different from the other lipids already discussed (Fig. 5-5c). The most common example of a sterol is cholesterol. This waxy substance does not look like a triglyceride: it does not have a glycerol backbone or any fatty acids. Still, because it does not readily dissolve in water, it is a lipid. Among other functions, cholesterol is used to form certain hormones and bile acids, and is incorporated into cell structures. The body can make all the cholesterol it needs.

✔ CONCEPT CHECK 5.2

1. What are the four possible structural forms of lipid?
2. In what ways do fatty acids differ from one another?
3. What is the difference between a saturated and an unsaturated fatty acid, and between an omega-3 and an omega-6 fatty acid?
4. Which fatty acids are essential?
5. How do triglycerides differ from phospholipids?
6. What are the main functions of cholesterol in the body?

5.3 Fats and Oils in Foods

Lipids in the form of triglycerides are abundant in the North American diet. The foods highest in fat (and therefore energy density) include salad oils and spreads such as butter, margarine, and mayonnaise. All of these foods contain close to 100% of calories as fat. In reduced-fat margarines, water replaces some of the fat. While regular margarines are 80% fat by weight (11 grams per tablespoon), some reduced-fat margarines are as low as 30% fat by weight (4 grams per tablespoon). When used in recipes, the extra water added to these margarines can cause texture and volume changes in the finished product. Cookbooks can suggest alterations in recipes to compensate for the increased water content of these products.

Still considering the overall fat content, whole foods highest in fat include nuts, bologna, avocados, and bacon, which have about 80% of calories as fat (Fig. 5-6). Next, peanut butter and cheddar cheese have about 75%. Marbled steak

▲ This peanut butter (**protein**) and jelly sandwich on whole-grain bread (**grain**) with low-fat milk (**dairy**) follows **MyPlate** guidelines well and provides a healthy source of plant fat from the peanut butter. Which sections are we missing?

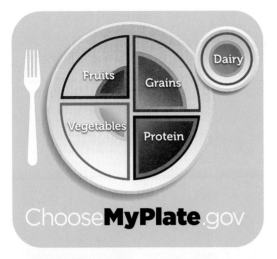

MyPlate:
Sources of Fats

Grains
- Crackers
- Pasta dishes with added fat

0–18 grams per serving

Vegetables
- French fried potatoes
- Deep-fried vegetables
- Vegetables roasted with oil

0–27 grams per serving

Fruits
- Fruit pies
- Avocados

0–11 grams per serving

Dairy
- Whole milk
- Low-fat milk (1%, 2%)
- Some yogurts
- Many cheeses
- Premium ice cream

0–10 grams per serving

Protein
- Marbled meat
- Bacon
- Poultry (skin)
- Deep-fat-fried meat
- Nuts

7–17 grams per serving

FIGURE 5-6 ▲ Sources of fats from MyPlate. The fill of the background color (none, 1/3, 2/3, or completely covered) within each group in the plate indicates the average nutrient density for fat in that group. The fruit group and vegetable groups are generally low in fat. In the other groups, both high-fat and low-fat choices are available. Careful reading of food labels can help you choose lower-fat versions of some foods. In general, any type of frying adds significant amounts of fat to a product, as with French fries and fried chicken.

and hamburgers (ground chuck) have about 60%, and chocolate bars, ice cream, doughnuts, and whole milk have about 50% of calories as fat. Eggs, pumpkin pie, and cupcakes have 35%, as do lean cuts of meat, such as top round (and ground round) and sirloin. Bread contains about 15%. Finally, foods such as cornflakes, sugar, and fat-free milk have essentially no fat. Figure 5-7 shows examples of food sources of fat. Label reading is necessary to determine the true fat content of food. When considering the contribution that meat makes to the fat calories of our diets, it is important to remember to limit the serving size of meats to just 3 ounces.

The type of fat in food is important to consider along with the total amount of fat. Animal fats are the chief contributors of saturated fatty acids to the North American diet. About 40% to 60% of total fat in dairy and meat products is in the form of saturated fatty acids. Fatty fish provide about 50% of the calories from fat but are a significant source of unsaturated omega-3 fatty acids. Plant oils contain mostly unsaturated fatty acids, ranging from 73% to 94% of total fat. A moderate to high proportion of total fat (49% to 77%) is supplied by monounsaturated fatty acids in canola oil, olive oil, and peanut oil. Some animal fats are also good sources of monounsaturated fatty acids (30% to 47%) (review Fig. 5-3). Corn, cottonseed, sunflower, soybean, and safflower oils contain mostly polyunsaturated fatty acids

▲ Dairy products are a primary contributor of saturated fat to our diets.

FIGURE 5-7 ▶ Food sources of fat compared to the American Heart Association (AHA) recommendation of 70 grams per day, or 30% of calories from fat for a 2100-kcal diet.

Mercury in Fish

Consumption of fatty fish at least twice a week is recommended as a good source of the omega-3 fatty acids. Some fish can be a source of mercury, which is toxic in high amounts, especially swordfish, shark, king mackerel, and tile fish (see Chapter 13). Albacore tuna also is a potential source, while other forms of tuna are much lower in mercury. Those fish low in mercury are salmon, sardines, and herring. For others, varying your choices rather than always eating the same species of fish and limiting overall intake to 12 ounces per week (on average two to three meals of fish or shellfish per week) is recommended to reduce mercury exposure, especially for pregnant women and children. Research indicates that the benefits of fish intake, especially in reducing the risk of cardiovascular disease, outweigh the possible risks of mercury contamination.

Food Sources of Fat

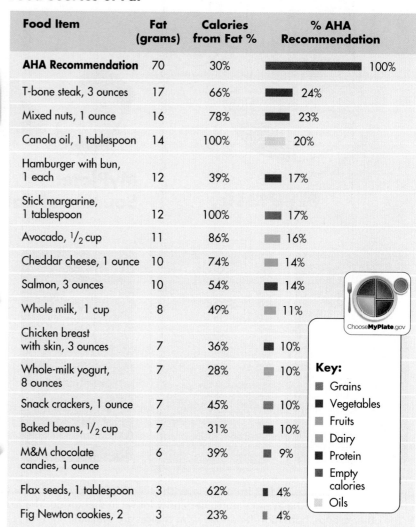

Food Item	Fat (grams)	Calories from Fat %	% AHA Recommendation
AHA Recommendation	70	30%	100%
T-bone steak, 3 ounces	17	66%	24%
Mixed nuts, 1 ounce	16	78%	23%
Canola oil, 1 tablespoon	14	100%	20%
Hamburger with bun, 1 each	12	39%	17%
Stick margarine, 1 tablespoon	12	100%	17%
Avocado, ½ cup	11	86%	16%
Cheddar cheese, 1 ounce	10	74%	14%
Salmon, 3 ounces	10	54%	14%
Whole milk, 1 cup	8	49%	11%
Chicken breast with skin, 3 ounces	7	36%	10%
Whole-milk yogurt, 8 ounces	7	28%	10%
Snack crackers, 1 ounce	7	45%	10%
Baked beans, ½ cup	7	31%	10%
M&M chocolate candies, 1 ounce	6	39%	9%
Flax seeds, 1 tablespoon	3	62%	4%
Fig Newton cookies, 2	3	23%	4%

ChooseMyPlate.gov

Key:
- ■ Grains
- ■ Vegetables
- ■ Fruits
- ■ Dairy
- ■ Protein
- ■ Empty calories
- ■ Oils

emulsifier A compound that can suspend fat in water by isolating individual fat droplets, using a shell of water molecules or other substances to prevent the fat from coalescing.

▲ Peanuts are a source of lecithins, as are wheat germ and egg yolks.

(54% to 77%). These plant oils are commonly consumed and are the primary sources of the omega-6 linoleic acid in our diets. It is important to balance this supply of omega-6 with rich sources of the omega-3 alpha-linolenic acid. Of the plant oils, canola and soybean oils supply the most alpha-linolenic acid. Additional sources of omega-3 fatty acids include fatty fish, chia seeds, walnuts, and flax seeds.

Wheat germ, peanuts, egg yolk, soy beans, and organ meats are rich sources of phospholipids. Phospholipids such as lecithin, a component of egg yolks, are often added to salad dressing. Lecithin is used as an **emulsifier** in these and other products because of its ability to keep mixtures of lipids and water from separating (see Fig. 5-8). Emulsifiers are added to salad dressings to keep the vegetable oil suspended in water. Eggs added to cake batters likewise emulsify the fat with the milk.

Cholesterol is found only in animal foods. An egg yolk contains about 210 milligrams of cholesterol. Eggs are our main dietary source of cholesterol, along with meats and whole milk. Manufacturers who advertise their brand of peanut butter, vegetable shortening, margarines, and vegetable oils as "cholesterol-free" are taking advantage of uninformed consumers: all of these products are naturally cholesterol-free. Some plants contain other sterols similar to cholesterol, but they do not pose the heart health risks associated with cholesterol. In fact, some plant sterols have blood cholesterol-lowering properties (see this chapter's Nutrition and Your Health section on medical interventions to lower blood lipids).

Emulsifiers and Agitation in Salad Dressing

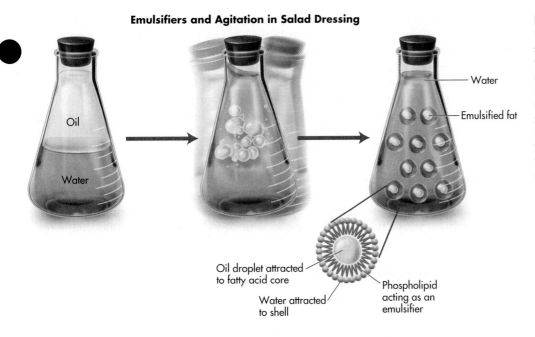

Oil droplet attracted
to fatty acid core

Water attracted
to shell

Phospholipid
acting as an
emulsifier

FIGURE 5-8 ◄ Emulsifiers in action. Emulsifiers prevent many brands of salad dressings and other condiments from separating into layers of water and fat. Emulsifiers attract fatty acids inside and have a water-attracting group on the outside. Add them to salad dressing, shake well, and they hold the oil in the dressing away from the water. Emulsification is important in both food production and fat digestion/absorption.

FAT IS HIDDEN IN SOME FOODS

Some fat discussed so far is obvious: butter on bread, mayonnaise in potato salad, and marbling in raw meat. Fat is harder to detect in other foods that also contribute significant amounts of fat to our diets. These foods that contain hidden fat include whole milk, pastries, cookies, cake, cheese, hot dogs, crackers, French fries, and ice cream. When we try to cut down on fat intake, hidden fats need to be considered, along with the more obvious sources.

A place to begin searching for hidden fat and learn more about the fat content of the foods you eat is on the Nutrition Facts labels of foods you buy (Fig. 5-9). Some signals from the ingredients list that can alert you to the presence of fat are animal fats, such as bacon, beef, ham, lamb, pork, chicken, and turkey fats; lard; dairy fats, such as butter and cream; egg and egg-yolk solids; nuts; vegetable oils; and partially hydrogenated shortening or vegetable oil. Conveniently, the label lists ingredients by order of weight in the product. If fat is one of the first ingredients listed, you are probably looking at a high-fat product.

▲ The North American diet contains many high-fat foods—including typical pastry choices. Portion control with these foods is thus important, especially if one is trying to control calorie intake.

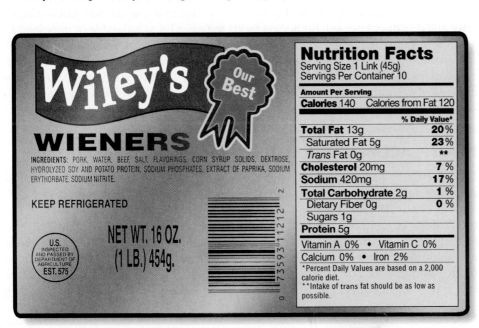

FIGURE 5-9 ◄ Reading labels helps locate hidden fat. Who would think that wieners (hot dogs) can contain about 85% of food calories as fat? Looking at the hot dog does not suggest that almost all of its food calories come from fat, but the label shows otherwise. Let us do the math: 13 grams total fat × 9 kcal per gram of fat = 120 kcal from fat; 120 kcal/140 kcal per link = 0.86 or 86% kcal from fat.

Definitions for Nutrient Claims About Fat and Cholesterol on Food Labels

Fat

- **Fat free:** less than 0.5 gram of fat per serving

- **Saturated fat free:** less than 0.5 gram per serving, and the level of *trans* fatty acids does not exceed 0.5 gram per serving

- **Low fat:** 3 grams or less per serving and, if the serving is 30 grams or less or 2 tablespoons or less, per 50 grams of the food; 2% milk can no longer be labeled low fat, as it exceeds 3 gram per serving. *Reduced fat* will be the term used instead.

- **Low saturated fat:** 1 gram or less per serving and not more than 15% of kcal from saturated fatty acids

- **Reduced or less fat:** at least 25% less per serving than reference food

- **Reduced or less saturated fat:** at least 25% less per serving than reference food

Cholesterol

- **Cholesterol free:** less than 2 milligrams (mg) of cholesterol and 2 gram or less of saturated fat per serving

- **Low cholesterol:** 20 milligrams or less cholesterol and 2 grams or less of saturated fat per serving and, if the serving is 30 grams or less or 2 tablespoons or less, per 50 grams of the food

- **Reduced or less cholesterol:** at least 25% less cholesterol and 2 grams or less of saturated fat per serving than reference food

▲ Fat replacements such as gum fiber are typically seen in soft serve ice cream.

The definitions for various fat descriptors on food labels, such as "low fat," "fat free," and "reduced fat," were listed in Table 2-10 in Chapter 2 and are reprinted in the margin here. Recall that "low fat" indicates, in most cases, that a product contains no more than 3 grams of fat per serving. Products marketed as "fat free" must have less than one-half of a gram of fat per serving. A claim of "reduced fat" means the product has at least 25% less fat than is usually found in that type of food. When there is no Nutrition Facts label to inspect, such as on restaurant menus, controlling portion size is a good way to control fat intake.

The healthiest way to create a low-fat diet is to focus primarily on fruits, vegetables, and whole-grain breads and cereals, rather than on the reduced-fat versions of pastries, cookies, and cakes that many are choosing. Whether to choose a fat-rich food should depend on how much fat you have eaten or will eat during that particular day. So, if you plan to eat high-fat foods at your evening meal, you should reduce your fat intake at a previous meal to balance overall fat intake for the day.

FAT IN FOOD PROVIDES SOME SATIETY, FLAVOR, AND TEXTURE

There is much controversy about whether high-fat foods increase satiety. Certain research suggests that some types of fat may have an effect on satiety, whereas others may not. Other studies indicate that protein and carbohydrate probably lead to the most satiety. What everyone knows for sure is that fats contain more than twice the calories of carbohydrates and proteins. Therefore, a high-fat meal is likely to be a high-calorie meal.

Various fats play important roles in foods, so much ingenuity must go into the production of reduced-fat products to preserve flavor and texture. In some cases, "fat-free" also means tasteless. Fat components in foods provide important textures and carry flavors. If you have ever eaten a high-fat yellow cheese or cream cheese, you probably agree that fat melting on the tongue feels good. The fat in reduced-fat and whole milk also gives body, which fat-free milk lacks. The most tender cuts of meat are high in fat, visible as the marbling of meat. In addition, many flavorings dissolve in fat. Heating spices in oil intensifies the flavors of an Indian curry or a Mexican dish.

LOW-FAT DIETS

A person who has been following a typical North American diet will probably need some time to adjust to the taste of a lower-fat diet. Emphasizing flavorful fruits, vegetables, and whole grains will help one to adapt to a low-fat diet. Interestingly, after an adjustment period, higher-fat foods may not be as palatable or may lead to gastrointestinal discomfort. For example, after switching from whole to 1% low-fat milk for a few weeks, whole milk begins to taste more like cream than milk. It is certainly possible to make the change from a higher-fat diet to a lower-fat diet. The benefits of weight control and reduced risk for several chronic diseases make the adjustment worth the effort.

FAT-REPLACEMENT STRATEGIES FOR REDUCED-FAT FOODS

Manufacturers have introduced reduced-fat versions of numerous food products. The fat content of these alternatives ranges from 0% in fat-free Fig Newtons to about 75% of the original fat content in other products. However, the total calorie content of most fat-reduced products is not substantially lower than that of their conventional versions. Generally, when fat is removed from a product, something must be added—commonly, sugars—in its place. It is difficult to reduce both the fat and sugar contents of a product at the same time and maintain flavor and texture. For this reason, many reduced-fat products (e.g., cakes and cookies) are still energy dense. Use the Nutrition Facts label to choose the portion size that fits into your daily calorie needs.

To help consumers trim their fat intake and still enjoy the mouthfeel fat provides, food companies offer low-fat versions of many foods. To lower the fat in foods, manufacturers may replace some of the fat with water, protein (Simplesse®, Dairy-Lo®), or forms of carbohydrates such as starch derivatives (Z-trim®), fiber (Maltrin®, Stellar™, Oatrim), and gums. Manufacturers also may use engineered fats, such as olestra (Olean®) and salatrim (Benefat®), that are made with fat and sucrose (table sugar) but that provide few or no calories because they cannot be digested and/or absorbed well. The main problem with the fat replacer olestra is that it can bind the fat-soluble vitamins and reduce their absorption.

So far, fat replacements have had little impact on our diets, partly because the currently approved forms are either not very versatile or not used extensively by manufacturers. In addition, fat replacements are not practical for use in the foods that provide the most fat in our diets: beef, cheese, whole milk, and pastries.

FAT RANCIDITY LIMITS SHELF LIFE OF FOODS

Decomposing oils emit a disagreeable odor and taste sour and stale. Stale potato chips are a good example. The double bonds in unsaturated fatty acids break down, producing **rancid** by-products. Ultraviolet light, oxygen, and heat (as in deep-fat frying) can break double bonds and, in turn, destroy the structure of polyunsaturated fatty acids. Saturated fats and *trans* fats can more readily resist these effects because they contain fewer carbon-carbon double bonds.

Rancidity is not a major problem for consumers because the odor and taste generally discourage us from eating enough to become sick. However, rancidity is a problem for the food industry and restaurant business because it reduces a product's shelf life. To increase shelf life, manufacturers often add partially-hydrogenated plant oils to products. Foods most likely to become rancid are deep-fried foods and foods with a large amount of exposed surface. While polyunsaturated oils are a good choice for products such as salad dressings, more saturated fats are best for high-heat cooking. The fat in fish is also susceptible to rancidity because it is highly polyunsaturated.

Antioxidants such as vitamin E help protect foods against rancidity by guarding against fat breakdown. The vitamin E naturally occurring in plant oils reduces the breakdown of double bonds in fatty acids. When food manufacturers want to prevent rancidity in polyunsaturated fats, they often add the synthetic antioxidants **BHA** and **BHT** or vitamin C to products that contain fat, such as salad dressings and cake mixes. Manufacturers also tightly seal products and use other methods to reduce oxygen levels inside packages.

HYDROGENATION OF FATTY ACIDS IN FOOD PRODUCTION INCREASES *TRANS* FATTY ACID CONTENT

As mentioned previously, most fats with long-chain saturated fatty acids are solid at room temperature, and those with unsaturated fatty acids are liquid at room temperature. In the production of some foods, solid fats work better than liquid oils. In pie crust, for example, solid fats yield a flaky product, whereas crusts made with liquid oils tend to be greasy and more crumbly. If oils with unsaturated fatty acids are used to replace solid fats, they often must be made more saturated (with hydrogen), as this solidifies the vegetable oils into shortenings and margarines. Hydrogen is added by bubbling hydrogen gas under pressure into liquid vegetable oils in a process called **hydrogenation** (Fig. 5-10). The fatty acids are not fully hydrogenated to the saturated fatty acid form, as this would make the product too hard

Canada has not approved the use of olestra in food products; the United States is the sole country that permits the use of this fat substitute in foods.

rancid Containing products of decomposed fatty acids that have an unpleasant flavor and odor.

CRITICAL THINKING

Allison has decided to start eating a low-fat diet. Allison has mentioned to you that all she needs to do is add less butter, oil, or margarine to her foods and she will dramatically lower her fat intake. How can you explain to Allison that she needs to be aware of the hidden fats in her diet as well?

BHA, BHT Butylated hydroxyanisole and butylated hydroxytoluene—two common synthetic antioxidants added to foods.

hydrogenation The addition of hydrogen to a carbon-carbon double bond, producing a single carbon-carbon bond with two hydrogens attached to each carbon.

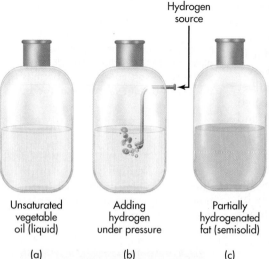

Hydrogen source

Unsaturated vegetable oil (liquid)	Adding hydrogen under pressure	Partially hydrogenated fat (semisolid)
(a)	(b)	(c)

FIGURE 5-10 ▲ How liquid oils become solid fats. (a) Unsaturated fatty acids are present in liquid form in large metal tanks. (b) Hydrogens are added (hydrogenation), changing some carbon-carbon double bonds to single bonds and producing some *trans* fatty acids. (c) The partially hydrogenated product is likely to be used in margarine, shortening, or for deep-fat frying.

and brittle. Partial hydrogenation—leaving some monounsaturated fatty acids—creates a semi-solid product.

The process of **hydrogenation** produces *trans* fatty acids, which were described in Section 5.2. Most natural monounsaturated and polyunsaturated fatty acids exist in the *cis* form, causing a bend in the carbon chain, whereas the straighter carbon forms of *trans* fat more closely resemble saturated fatty acids. This may be the mechanism whereby *trans* fat increases the risk for heart disease. Studies also indicate that *trans* fats increase overall inflammation in the body, which is not healthful. Thus, people should avoid intake of partially hydrogenated fat and thus *trans* fat. This may not be such a concern for the average person, as long as *trans* fat intake is not excessive and the diet is adequate in polyunsaturated fat. However, because *trans* fatty acids serve no particular role in maintaining body health, the latest Dietary Guidelines for Americans, the American Heart Association, and the Food and Nutrition Board each recommend minimal *trans* fat intake.

Partially hydrogenated soybean oil—rich in *trans* fat—became the major fat in food processing when manufacturers eliminated the tropical oils rich in saturated fat (palm, palm olein, and coconut). While foods made with partially hydrogenated oil were becoming popular, there was very little known about the *trans* fats they contained and their harmful effects. It was not until the 1990s that scientists began identifying the adverse health effects of *trans* fats. Currently, *trans* fat intake in North America is estimated to contribute about 3% to 4% of total calories, amounting to 10 grams per day, on average. Table 5-1 includes typical sources of *trans* fat.

The American Heart Association strongly advises that we limit *trans* fat intake to less than 1% of total daily calories. This means only 2 grams of *trans* fats a day, or 20 calories, if you need 2000 calories a day. FDA has taken action to make consumers more aware of the amounts of *trans* fat in foods. Since 2006, federal regulations have required the disclosure of *trans* fat content on food packages (review Fig. 5-9). The food labels in Canada also must list *trans* fat content. North American companies are already responding to this issue by creating products free of *trans* fat. For example, Promise, Smart Beat, and some Fleischmann's margarines are lower in or free of *trans* fat (less than 0.5 grams per serving) compared to typical margarines. In 2013, FDA proposed a change to the way *trans* fats are regulated in the United States. If approved, food manufacturers would no longer be able add *trans* fats to food products.

What about foods that do not bear a Nutrition Facts panel? When dining out, it can be difficult to know which foods are high in *trans* fat because restaurant foods are not currently covered by federal nutrition labeling requirements. One state (California) and several localities (e.g. New York City) have passed legislation to ban *trans* fat use in food service establishments (see Further Reading 4). FDA's recently proposed legislation would prevent restaurants from purchasing partially hydrogenated oils with *trans* fats.

To minimize *trans* fat intake, a general guideline is to limit consumption of fried (especially deep-fat fried) food items, any pastries or flaky bread products (such as pie crusts, crackers, croissants, and biscuits), and cookies. Most importantly, use little or no stick margarine or shortening. Instead, substitute vegetable oils and softer tub margarines (whose labels list vegetable oil or water as the first ingredient). Avoid deep-fat frying any food in shortening. Substitute baking, panfrying, broiling, steaming, grilling, or deep-fat frying in unhydrogenated

▲ Fried foods are a rich source of fat and *trans* fats. Reducing the intake of these foods can help lower blood lipid levels.

TABLE 5-1 ▶ Main Sources of Fatty Acids and Their State at Room Temperature

Type and Health Effects	Main Sources	State at Room Temperature
Saturated Fatty Acids Increase blood levels of cholesterol		
Long chain	Lard; fat in beef, pork, and lamb	Solid
Medium and short chain	Milk fat (butter), coconut oil, palm oil, palm kernel oil	Soft or liquid
Monounsaturated Fatty Acids Decrease blood levels of cholesterol	Olive oil, canola oil, peanut oil	Liquid
Polyunsaturated Fatty Acids Decrease blood levels of cholesterol	Sunflower oil, corn oil, safflower oil, fish oil	Liquid
Essential Fatty Acids Omega-3: alpha–linolenic acid Reduces inflammation responses, blood clotting, and plasma triglycerides	Cold-water fish (salmon, tuna, sardines, mackerel), walnuts, flaxseed, hemp oil, canola oil, soybean oil, chia seeds, and perilla oil	Liquid
Omega-6: linoleic acid Regulates blood pressure and increases blood clotting	Beef, poultry, safflower oil, sunflower oil, corn oil	Solid to liquid
Trans Fatty Acids Increase blood cholesterol more than saturated fat	Margarine (squeeze, tub, stick), shortening	Soft to very solid

Nut Butters: Are They "Good" Fat?

Nut butters are plant products, so they are naturally cholesterol free and provide 2 or 3 grams of fiber, which can lower blood cholesterol. Nuts are also a good source of unsaturated fats. Two tablespoons of peanut butter provide about 12 grams of unsaturated fats and only 3 grams of saturated fat. As you've learned, monounsaturated and polyunsaturated fats tend to lower blood cholesterol levels. To steer clear of *trans* fats, be on the lookout for partially hydrogenated oils in the list of ingredients. These have been used in some products to increase shelf life, although food manufacturers are gradually phasing them out. Lastly, check the sugar content. Nut butters naturally contain 1 or 2 grams of sugar, but processed varieties (e.g., cinnamon swirl) may contain up to 9 grams of sugar. Reap the most heart-health benefits by choosing natural nut butters and controlling the size of your portion.

oils with high "smoke points," meaning they do not break down at deep-frying temperatures. Peanut, safflower, sunflower, and canola oils are good choices. Replace nondairy creamers with reduced-fat or fat-free milk, since most nondairy creamers are rich in partially hydrogenated vegetable oils. Finally, read the ingredients on food labels, using the previous tips to estimate *trans* fat content. If partially hydrogenated vegetable oil is one of the first three ingredients on the label, you can assume that there is a significant amount of *trans* fat in the product.

✓ CONCEPT CHECK 5.3

1. Which foods are the most fat-dense (>60% total calories as fat)?
2. Where is cholesterol found in the food supply?
3. Which types of fat are used as emulsifiers, and what is their function in food?
4. What are some strategies used to produce reduced-fat foods?
5. How do fats become rancid, and how can this be prevented?
6. What happens to unsaturated fats during the process of hydrogenation?
7. What regulations are in place to educate and protect the public from *trans* fats in foods?

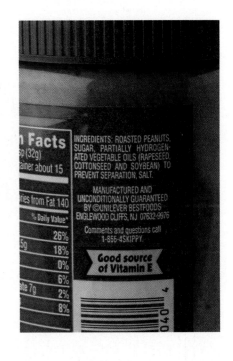

5.4 Making Lipids Available for Body Use

It is no secret that fats and oils make foods more appealing. Their presence in foods adds flavor, moisture, and texture. What happens to lipids once they are eaten? Let us take a closer look at the digestion, absorption, and physiological roles of lipids in the body.

DIGESTION

In the first phase of fat digestion, the stomach (and salivary glands to some extent) secretes the enzyme **lipase.** Lipase acts primarily on triglycerides that have fatty acids with short chain lengths, such as those found in butterfat. The action of salivary and stomach lipase, however, is usually dwarfed by that of the lipase enzyme released from the pancreas and active in the small intestine. Triglycerides and other lipids found in common vegetable oils and meats have longer chain lengths and are generally not digested until they reach the small intestine (Fig. 5-11).

In the small intestine, triglycerides are broken down by lipase into smaller products, namely monoglycerides (glycerol backbones with a single fatty acid attached) and fatty acids. Under the right circumstances, digestion is rapid and thorough. The "right" circumstances include the presence of bile from the gallbladder. Acids present in the bile act as emulsifiers on the digestive products of lipase action, suspending

lipase Fat-digesting enzyme produced by the salivary glands, stomach, and pancreas.

FIGURE 5-11 ▶ A summary of fat digestion and absorption. Chapter 3 covered general aspects of this process.

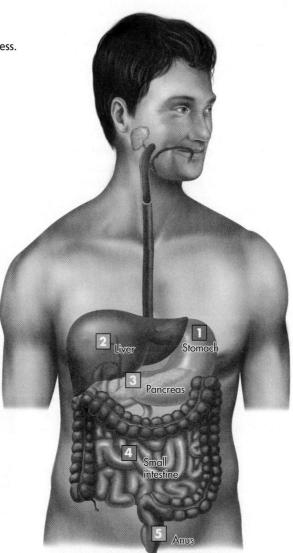

Fat Digestion and Absorption

1 Stomach: Only minor digestion of fat takes place in the stomach through the action of lipase enzymes.

2 Liver: The liver produces bile, stored in the gallbladder and released through the bile duct into the small intestine. Bile aids in fat digestion and absorption by emulsifying lipids in the digestive juices.

3 Pancreas: The pancreas secretes a mixture of enzymes, including lipase, into the small intestine.

4 Small intestine: The small intestine is the primary site for digestion and absorption of lipids. Once absorbed, long-chain fatty acids are packaged for transport through the lymph and bloodstream. (Shorter-chain fatty acids are absorbed directly into portal circulation.)

5 Large intestine: Less than 5% of ingested fat is normally excreted in the feces.

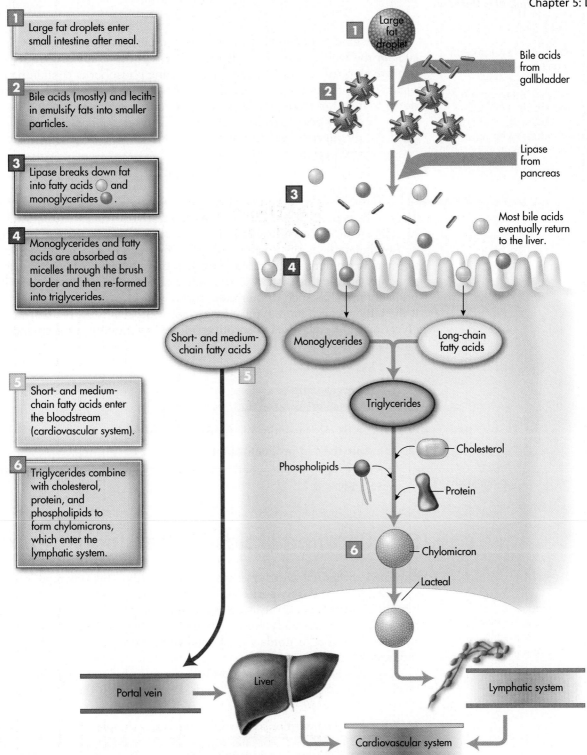

1 Large fat droplets enter small intestine after meal.

2 Bile acids (mostly) and lecithin emulsify fats into smaller particles.

3 Lipase breaks down fat into fatty acids ⚪ and monoglycerides ⚫ .

4 Monoglycerides and fatty acids are absorbed as micelles through the brush border and then re-formed into triglycerides.

5 Short- and medium-chain fatty acids enter the bloodstream (cardiovascular system).

6 Triglycerides combine with cholesterol, protein, and phospholipids to form chylomicrons, which enter the lymphatic system.

1 Large fat droplet

Bile acids from gallbladder

Lipase from pancreas

Most bile acids eventually return to the liver.

Short- and medium-chain fatty acids

Monoglycerides

Long-chain fatty acids

Triglycerides

Cholesterol

Phospholipids

Protein

Chylomicron

Lacteal

Portal vein

Liver

Lymphatic system

Cardiovascular system

FIGURE 5-12 ▲ Bile acids mix with fats to form small droplets that facilitate the absorption of monoglycerides and fatty acids into the mucosal cells of the small intestine.

the monoglycerides and fatty acids in the watery digestive juices. This emulsification improves digestion and absorption because as large fat globules are broken down into smaller ones, the total surface area for lipase action increases (Fig. 5-12). During meals, bile acids circulate in a path that begins in the liver, goes on to the gallbladder, and then moves to the small intestine. After participating in fat digestion, most bile acids are absorbed and end up back at the liver. Approximately 98% of the bile acids are recycled. Only 1% to 2% ends up in the large intestine to be eliminated in the feces.

If the gallbladder is surgically removed (e.g., in cases of gallstone formation), bile will enter the small intestine directly from the liver. Moderate intakes of fat can still be digested adequately, but some people experience loose stools after eating high-fat meals because some fat reaches the large intestine unabsorbed.

With regard to phospholipid digestion, certain enzymes from the pancreas and cells in the wall of the small intestine digest phospholipids. The eventual products are glycerol, fatty acids, and the remaining phosphorus-containing parts. With regard to cholesterol digestion, any cholesterol with a fatty acid attached is broken down to free cholesterol and fatty acids by certain enzymes released from the pancreas. Any fatty acids that are part of these structures could be broken down to yield energy but contribution to overall calorie intake is miniscule compared to the energy stored in triglycerides.

ABSORPTION

The products of fat digestion in the small intestine are fatty acids and monoglycerides. These products diffuse into the absorptive cells of the small intestine. About 95% of dietary fat is absorbed in this way. The chain length of fatty acids affects the ultimate fate of fatty acids and monoglycerides after absorption. If the chain length of a fatty acid is less than 12 carbon atoms, it is water soluble and will therefore probably travel as such through the portal vein that connects directly to the liver. If the fatty acid is a more typical long-chain variety, it must be reformed into a triglyceride in the intestinal absorptive cell and eventually enter circulation via the lymphatic system.

CONCEPT CHECK 5.4

1. What enzyme is responsible for digestion of triglycerides?
2. What are the end products of fat digestion?
3. What are the differences between the absorption of long- versus short-chain fatty acids?

5.5 Carrying Lipids in the Bloodstream

As noted earlier, fat and water do not mix easily. This incompatibility presents a challenge for the transport of fats through the watery media of the blood and lymph. **Lipoproteins** serve as vehicles for transport of lipids from the small intestine and liver to the body tissues (Table 5-2). They are found in the bloodstream and contain a core of lipids with a shell composed of protein, phospholipid, and cholesterol (Fig. 5-13).

Lipoproteins are classified into four groups—chylomicrons, VLDL, LDL, and HDL—based on their densities. Lipids are less dense than proteins. Therefore,

lipoprotein A compound found in the bloodstream containing a core of lipids with a shell composed of protein, phospholipid, and cholesterol.

TABLE 5-2 ▶ Composition and Roles of the Major Lipoproteins in the Blood

Lipoprotein	Primary Component	Key Role
Chylomicron	Triglyceride	Carries dietary fat from the small intestine to cells
VLDL	Triglyceride	Carries lipids made and taken up by the liver to cells
LDL	Cholesterol	Carries cholesterol made by the liver and from other sources to cells
HDL	Protein	Contributes to cholesterol removal from cells and, in turn, excretion of it from the body

lipoproteins that contain a large percentage of lipids in comparison to protein are less dense than those depleted of lipids (Fig. 5-14).

DIETARY FATS ARE CARRIED BY CHYLOMICRONS

As you learned in the previous section, digestion of dietary fats results in a mixture of glycerol, monoglycerides, and fatty acids. Once these products are absorbed

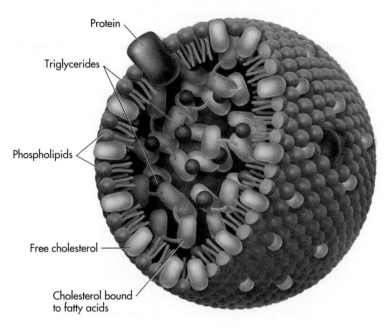

FIGURE 5-13 ▲ The structure of a lipoprotein, in this case an LDL. This structure allows fats to circulate in the water-based bloodstream. Various lipoproteins are found in the bloodstream. The primary component of LDL is cholesterol.

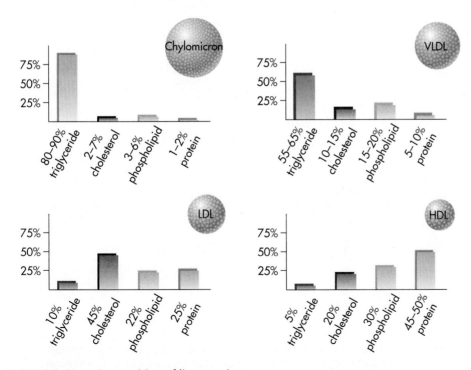

FIGURE 5-14 ▲ Composition of lipoproteins.

chylomicron Lipoprotein made of dietary fats surrounded by a shell of cholesterol, phospholipids, and protein. Chylomicrons are formed in the absorptive cells of the small intestine after fat absorption and travel through the lymphatic system to the bloodstream.

lipoprotein lipase An enzyme attached to the cells that form the inner lining of blood vessels; it breaks down triglycerides into free fatty acids and glycerol.

by the cells of the small intestine, they are reassembled into triglycerides. Then, the intestinal cells package the triglycerides into **chylomicrons,** which enter the lymphatic system and eventually the bloodstream. Chylomicrons are the largest lipoproteins and contain dietary fat and originate only from the intestinal cells. Like the other lipoproteins, chylomicrons are composed of large droplets of lipid surrounded by a thin, water-soluble shell of phospholipids, cholesterol, and protein (Fig. 5-14). The water-soluble shell around a chylomicron allows the lipid to float freely in the water-based blood. Some of the proteins present may also help other cells identify the lipoprotein as a chylomicron.

Once a chylomicron enters the bloodstream, the triglycerides in its core are broken down into fatty acids and glycerol by an enzyme called **lipoprotein lipase,** attached to the inside walls of the blood vessels (Fig. 5-15). As soon as the fatty acids are released to the bloodstream, they are absorbed by nearby cells, while much of the glycerol circulates back to the liver. Muscle cells can immediately use the absorbed fatty acids for fuel. Adipose cells, on the other hand, tend to re-form the fatty acids into triglycerides for storage. After triglycerides have been removed, a chylomicron remnant remains. Chylomicron remnants are removed from circulation by the liver, and their components are recycled to make other lipoproteins and bile acids.

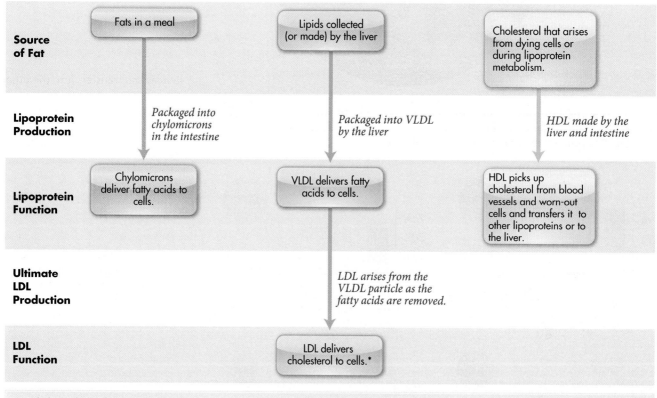

* = Cholesterol not taken up by body cells can be taken up by scavenger cells in the arteries. The eventual cholesterol buildup leads to atherosclerosis.

VLDL = very-low-density lipoprotein

LDL = low-density lipoprotein

HDL = high-density lipoprotein

FIGURE 5-15 ▲ Lipoprotein production and function. Chylomicrons carry absorbed fat to body cells. VLDL carries fat taken up from the bloodstream by the liver, as well as any fat made by the liver, to body cells. LDL arises from VLDL and carries mostly cholesterol to cells. HDL arises mostly from the liver and intestine. HDL carries cholesterol from cells to other lipoproteins and to the liver for excretion.

OTHER LIPOPROTEINS TRANSPORT LIPIDS FROM THE LIVER TO THE BODY CELLS

The liver takes up various lipids from the blood. The liver also is the manufacturing site for lipids and cholesterol. The raw materials for lipid and cholesterol synthesis include free fatty acids taken up from the bloodstream, as well as carbon and hydrogen derived from carbohydrates, protein, and alcohol. The liver then must package the lipids it makes into lipoproteins for transport in the blood to body tissues.

First in our discussion of lipoproteins made by the liver are **very-low-density lipoproteins (VLDL).** These particles are composed of cholesterol and triglycerides surrounded by a water-soluble shell. VLDLs are rich in triglycerides and thus are very low in density. Once in the bloodstream, lipoprotein lipase on the inner surface of the blood vessels breaks down the triglyceride in the VLDL into fatty acids and glycerol. Fatty acids and glycerol are released into the bloodstream and taken up by the body cells.

As its triglycerides are released, the VLDL becomes proportionately denser. Much of what eventually remains of the VLDL fraction is then called **low-density lipoprotein (LDL);** this is composed primarily of the remaining cholesterol. The primary function of LDL is to transport cholesterol to tissues. LDL particles are taken up from the bloodstream by specific receptors on cells, especially liver cells, and are then broken down. The cholesterol and protein components of LDL provide some of the building blocks necessary for cell growth and development, such as synthesis of cell membranes and hormones.

The final group of lipoproteins, **high-density lipoproteins (HDL),** is a critical and beneficial participant in this process of lipid transport. Its high proportion of protein makes it the densest lipoprotein. The liver and intestine produce most of the HDL in the blood. It roams the bloodstream, picking up cholesterol from dying cells and other sources. HDL donates the cholesterol primarily to other lipoproteins for transport back to the liver to be excreted. Some HDL travels directly back to the liver.

very-low-density lipoprotein (VLDL) The lipoprotein created in the liver that carries cholesterol and lipids that have been taken up or newly synthesized by the liver.

low-density lipoprotein (LDL) The lipoprotein in the blood containing primarily cholesterol; elevated LDL is strongly linked to cardiovascular disease risk.

high-density lipoprotein (HDL) The lipoprotein in the blood that picks up cholesterol from dying cells and other sources and transfers it to the other lipoproteins in the bloodstream, as well as directly to the liver; low HDL increases the risk for cardiovascular disease.

"GOOD" AND "BAD" CHOLESTEROL IN THE BLOODSTREAM

HDL and LDL are often described as "good" and "bad" cholesterol, respectively. Many studies demonstrate that the amount of HDL in the bloodstream can closely predict the risk for cardiovascular disease. Risk increases with low HDL because little cholesterol is transported back to the liver and excreted. Women tend to have high amounts of HDL, especially before **menopause,** compared to men. High amounts of HDL slow the development of cardiovascular disease, so any cholesterol carried by HDL can be considered "good" cholesterol.

On the other hand, LDL is sometimes considered "bad" cholesterol. In our discussion of LDL, you learned that LDL is taken up by receptors on various cells. If LDL is not readily cleared from the bloodstream, **scavenger cells** in the arteries take up the lipoprotein, leading to a buildup of cholesterol in the blood vessels. This buildup, known as **atherosclerosis,** greatly increases the risk for cardiovascular disease. Scientists believe that atherosclerosis starts because the innermost layer of the artery becomes damaged. Because of the damage, over time, fats, cholesterol, platelets, cellular debris, and calcium are deposited in the artery wall. The longer LDL cholesterol is in the bloodstream, the more likely it is to be oxidized and cleared by scavenger cells, which "swallow" the cholesterol, become **"foam cells,"** and embed in the artery wall (see the Nutrition and Your Health section in this chapter). LDL is only a problem when it is too high in the bloodstream because low amounts are needed as part of routine body functions.

The cholesterol in foods is not designated as "good" or "bad." It is only after cholesterol has been made or processed by the liver that it shows up in the bloodstream as LDL or HDL. Dietary patterns can affect the metabolism of cholesterol, however.

menopause The cessation of the menstrual cycle in women, usually beginning at about 50 years of age.

scavenger cells Specific form of white blood cells that can bury themselves in the artery wall and accumulate LDL. As these cells take up LDL, they contribute to the development of atherosclerosis.

atherosclerosis A buildup of fatty material (plaque) in the arteries, including those surrounding the heart.

foam cells Lipid-loaded white blood cells that have surrounded large amounts of a fatty substance, usually cholesterol, on the blood vessel walls.

It appears that saturated fatty acids promote an increase in the amount of free cholesterol (not attached to fatty acids) in the liver, whereas unsaturated fatty acids do the opposite. As free cholesterol in the liver increases, it causes the liver to reduce cholesterol uptake from the bloodstream, contributing to elevated LDL in the blood. (*Trans* fatty acids are thought to act in the same ways as saturated fatty acids.)

Diets low in saturated fat, *trans* fat, and cholesterol encourage the uptake of LDL by the liver, thereby removing LDL from the bloodstream and decreasing the ability of scavenger cells to form atherosclerotic plaques in the blood vessels. Likewise, diets high in saturated fat, *trans* fat, and cholesterol reduce the uptake of LDL by the liver, increasing cholesterol in the blood and the risk for cardiovascular disease. Consider what foods in your diet are high in saturated fat, *trans* fat, or cholesterol.

✓ CONCEPT CHECK 5.5

1. How do lipids move through the bloodstream?
2. How are dietary fats packaged in the small intestine and transported?
3. Where are VLDLs made and what do they contain?
4. Where do the LDLs originate and what is their destination?
5. Why are HDLs considered "good" cholesterol?

5.6 Broader Roles for Lipids in the Body

Many key functions of fat in the body require the use of fatty acids in the form of triglycerides. Triglycerides are used for energy storage, insulation, and transportation of fat-soluble vitamins.

PROVIDING ENERGY

Triglycerides contained in the diet and stored in adipose tissue provide the fatty acids that are the main fuel for muscles while at rest and during light activity. A mix of fat and carbohydrates is burned for fuel during endurance exercise, but muscles rely mainly on carbohydrates during short, intense exercise. As the duration of exercise increases, the percent of fatty acids used for exercise rises as the percent of carbohydrates falls. Other body tissues also use fatty acids for energy needs. Overall, about half of the energy used by the entire body at rest and during light activity comes from fatty acids. When considering the whole body, the use of fatty acids by skeletal and heart muscle is balanced by the use of glucose by the nervous system and red blood cells. Recall from Chapter 4 that cells need a supply of carbohydrate to efficiently process fatty acids for fuel. The details about how we burn fat as a fuel will be discussed in Chapters 7 and 12.

STORING ENERGY FOR LATER USE

We store energy mainly in the form of triglycerides. The body's ability to store fat is essentially limitless. Its fat storage sites, adipose cells, can increase about 50 times in weight. If the amount of fat to be stored exceeds the ability of the existing cells to expand, the body can form new adipose cells.

An important advantage of using triglycerides to store energy in the body is that they are energy dense. Recall that these yield, on average, 9 kcal per gram, whereas proteins and carbohydrates yield only about 4 kcal per gram. In addition, triglycerides are chemically stable, so they are not likely to react with other cell constituents, making them a safe form for storing energy. Finally, when we store triglycerides in adipose cells, we store little else, especially water. Adipose cells contain about 80% lipid and only 20% water and protein. In contrast, imagine if we were to store energy as muscle tissue, which is about 73% water. Body weight linked to energy storage would increase dramatically. The same would be true if we stored energy primarily as glycogen, as about 3 grams of water are stored for every gram of glycogen.

▲ When at rest or during light activity, the body uses mostly fatty acids for fuel.

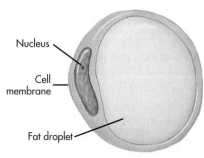

Nucleus

Cell membrane

Fat droplet

Adipose cell

INSULATING AND PROTECTING THE BODY

The insulating layer of fat just beneath the skin is made mostly of triglycerides. Fat tissue also surrounds and protects some organs—kidneys, for example—from injury. We usually do not notice the important insulating function of fat tissue, because we wear clothes and add more as needed. A layer of insulating fat is important in animals living in cold climates. Polar bears, walruses, and whales all build a thick layer of fat tissue around themselves to insulate against cold-weather environments. The extra fat also provides energy storage for times when food is scarce.

TRANSPORTING FAT-SOLUBLE VITAMINS

Triglycerides and other fats in food carry fat-soluble vitamins to the small intestine and aid their absorption. People who absorb fat poorly, such as those with the disease cystic fibrosis, are at risk for deficiencies of fat-soluble vitamins, especially vitamin K. A similar risk comes from taking mineral oil as a laxative at mealtimes. The body cannot digest or absorb mineral oil, so the undigested oil carries the fat-soluble vitamins from the meal into the feces, where they are eliminated. Unabsorbed fatty acids can bind minerals, such as calcium and magnesium, and draw them into the stool for elimination. This can harm mineral status (see Chapter 9). Recall that the main problem with the fat replacer, olestra, is that it can bind the fat-soluble vitamins and reduce their absorption.

PHOSPHOLIPIDS IN THE BODY

Many types of phospholipids exist in the body, especially in the brain. They form important parts of cell membranes. Phospholipids are found in body cells, and they participate in fat digestion in the intestine. Recall that the various forms of lecithin (discussed in Section 5.2) are common examples of phospholipids (review Fig. 5-5b).

Cell membranes are composed primarily of phospholipids. A cell membrane looks much like a sea of phospholipids with protein "islands" (Fig. 5-16). The proteins form receptors for hormones, function as enzymes, and act as transporters for nutrients. The fatty acids on the phospholipids serve as a source of essential fatty acids for the cell. Some cholesterol is also present in the membrane.

By breaking fat globules into small droplets, emulsifiers enable a fat to be suspended in water. They act as bridges between the oil and water that in turn lead to the formation of tiny oil droplets surrounded by thin shells of water. Phospholipids in the shells of lipoproteins suspend fats in blood in much the same way lecithin suspends vinegar in oil in salad dressing. Sphingomyelin is a phospholipid that protects nerve cells.

The body's main emulsifiers are the lecithins and bile acids, produced by the liver and released into the small intestine via the gallbladder during digestion.

CHOLESTEROL IN THE BODY

Cholesterol plays many vital roles in the body. It forms part of some important hormones, such as estrogen, testosterone, and a precursor of the active vitamin D hormone. Cholesterol is an essential structural component of cells and the outer layer of the lipoprotein particles that transport lipids in the blood. The cholesterol content of the heart, liver, kidney, and brain is high, reflecting its critical role in these organs. Cholesterol is also the building block of bile acids, needed for fat digestion. As discussed earlier, about 98% of the bile acids are absorbed and end up back at the liver after participating in fat digestion. Using medicines that block some of this reabsorption of bile acids is one way to treat high blood cholesterol. The liver takes cholesterol from the bloodstream to form replacement bile acids. Viscous fiber in the diet can also bind to bile acids to produce the same effect (see this chapter's Nutrition and Your Health section on medical interventions related to cardiovascular disease for details).

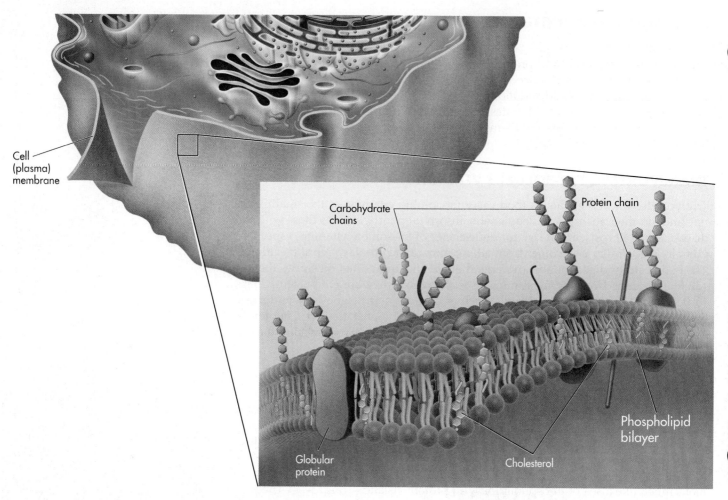

Cell
(plasma)
membrane

Carbohydrate
chains

Protein chain

Phospholipid
bilayer

Globular
protein

Cholesterol

FIGURE 5-16 ▲ Phospholipids are the main components of cell membranes, forming a double layer (bilayer) of lipid.

About two-thirds of the cholesterol circulating through your body is made by body cells; the remaining one-third is consumed in the diet. Except in cases of genetically related high blood cholesterol, cholesterol synthesis by the body is regulated. A higher intake of dietary cholesterol leads to an overall decrease in body production, and a lower intake from food results in more cholesterol made in the body. Generally, each day, our cells produce approximately 875 milligrams of cholesterol. Of the 875 milligrams of cholesterol made by the body, about 400 milligrams are used to make new bile acids to replenish those lost in the feces, and about 50 milligrams are used to make hormones. In addition to all the cholesterol cells make, we consume about 180 to 325 milligrams of cholesterol per day from animal-derived food products, with men consuming the higher amount compared to women. Absorption of cholesterol from food ranges from about 40% to 65%. The effect of blood cholesterol, especially LDL cholesterol, on cardiovascular disease risk will be discussed in the Nutrition and Your Health section in this chapter.

▲ Trimming the fat off meats can help reduce saturated fat intake, but you cannot remove the marbling (streaks of fat running through the meat). Limit portion size to 3 ounces and choose fatty cuts of meat less often to help keep your blood cholesterol levels in check.

✓ CONCEPT CHECK 5.6

1. What are the functions of triglycerides in the body?

2. Where are phospholipids found in the body?

3. What are some compounds that are made from cholesterol in the body?

5.7 Recommendations for Fat Intake

There is no RDA for total fat intake for adults, although there is an Adequate Intake set for total fat for infants (see Chapter 15). The 2010 Dietary Guidelines for Americans recommendation and the Acceptable Macronutrient Distribution Range is that total fat intake should be 20% to 35% of total calories, which equates to 44 to 78 grams per day for a person who consumes 2000 kcal daily. The most specific recommendations for fat intake come from the American Heart Association (AHA) (see Further Reading 6). Many North Americans are at risk for developing cardiovascular disease, so AHA promotes dietary and lifestyle goals aimed at reducing this risk. The AHA diet and lifestyle goals for cardiovascular disease risk reduction for the general public include aiming for an overall healthy eating pattern; appropriate body weight; and a desirable blood cholesterol profile, blood pressure, and blood glucose level. In Table 5-3, a more detailed list of recommendations is provided for those who currently are at high risk or have cardiovascular disease.

To reduce risk for cardiovascular disease, the AHA recommends that no more than 7% of total calories come from saturated fat and no more than 1% from *trans* fat. These are the primary fatty acids that raise LDL. In addition, cholesterol should amount to a maximum of 300 mg per day. Table 5-4 lists the cholesterol content of some foods. This often happens along with the reduction in saturated fat and *trans* fat intake. Table 5-5 is an example of a diet that adheres to 20% or 30% of calories as fat. Compare these recommendations to the actual dietary intake patterns of these fats by North Americans: 33% of calories from total fat, about 13% of calories from saturated fat, and 180 to 320 milligrams of cholesterol each day.

The regular consumption of fatty fish is advised and thought to have greater benefits and be safer than using fish oil supplements. Fish is not only a rich source of omega-3 fatty acids but also a valuable source of protein and trace elements that may also provide protective effects for the cardiovascular system. Broiled or baked fish is recommended rather than fried fish because frying may increase the ratio of omega-6 to omega-3 fatty acids and may produce *trans* fatty acids and oxidized lipid products that may increase cardiovascular disease risk.

One goal of *Healthy People 2020* is to reduce consumption of saturated fat in the population ages 2 years and older to 9.5% of total calorie intake.

▲ The American Heart Association recommends eating fatty fish such as salmon at least twice a week. As a source of omega-3 fatty acids, fish is a heart-healthy alternative to other animal sources of protein, which can be high in saturated fat and cholesterol.

Compare the advice regarding fat intake from the 2010 Dietary Guidelines for the general population to AHA's recommendations in Table 5-3.

- Consume less than 10% of calories from saturated fatty acids by replacing them with monounsaturated and polyunsaturated fatty acids.
- Consume less than 300 milligrams per day of dietary cholesterol.
- Keep *trans* fatty acid consumption as low as possible by limiting foods that contain synthetic sources of *trans* fats, such as partially hydrogenated oils, and by limiting other solid fats.
- Reduce the intake of calories from solid fats and added sugars.
- Use monounsaturated oils (olive, canola, and peanut) to replace solid fats where possible.
- Limit the consumption of foods that contain refined grains, especially refined grain foods that contain solid fats, added sugars, and sodium.
- Replace protein foods that are higher in solid fats with choices that are lower in solid fats (e.g., beans, nuts, and seeds) and calories and/or are sources of oils.
- Increase the amount and variety of seafood in place of some meat and poultry.
- Increase intake of fat-free or low-fat milk and milk products, such as milk, yogurt, cheese, or fortified soy beverages.

TABLE 5-3 ► **American Heart Association 2006 Diet and Lifestyle Recommendations for Cardiovascular Disease Risk Reduction**

- Balance calorie intake and physical activity to achieve or maintain a healthy body weight.
- Consume a diet rich in vegetables and fruits.
- Choose whole-grain, high-fiber foods.
- Consume fish, especially oily fish, at least twice a week.
- Limit your intake of saturated fat to less than 7% of energy, *trans* fat to less than 1% of energy, and cholesterol to less than 300 milligrams per day by
  — choosing lean meats and vegetable alternatives;
  — selecting fat-free (skim), 1%-fat, and low-fat dairy products; and
  — minimizing intake of partially hydrogenated fats.
- Minimize your intake of beverages and foods with added sugars.
- Choose and prepare foods with little or no salt.
- If you consume alcohol, do so in moderation.
- When you eat food prepared outside of the home, follow the AHA Diet and Lifestyle Recommendations.

Source: Lichtenstein AH and others: Diet and lifestyle recommendations revision 2006. A scientific statement from the American Heart Association Nutrition Committee. *Circulation* 114:82, 2006.

▲ **Whole grains** (shredded wheat, whole-wheat bread, oatmeal cookies, popcorn), **fruits** (orange juice, apple, banana, raisins), **vegetables** (carrots, lettuce, tomato), **lean meats** (roast beef, turkey, chicken), and **fat-free milk** are the primary components of the low-fat menus in Table 5-5.

TABLE 5-4 ▶ Cholesterol Content of Foods

3 oz beef brains	2635 mg
3 oz beef liver	337 mg
1 large egg yolk*	209 mg
3 oz shrimp	166 mg
3 oz beef*	75 mg
3 oz pork	75 mg
3 oz chicken or turkey (white meat)*	75 mg
1 cup ice cream	63 mg
3 oz trout	60 mg
3 oz tuna	45 mg
3 oz hot dog	38 mg
1 oz cheddar cheese*	30 mg
1 cup whole milk*	24 mg
1 cup 1% milk	12 mg
1 cup fat-free milk	5 mg
1 large egg white	0 mg

*Leading dietary sources of cholesterol in American diets.

TABLE 5-5 ▶ Daily Menu Examples Containing 2000 kcal and 30% or 20% of Calories as Fat

30% of Calories as Fat		20% of Calories as Fat	
Food	Fat (grams)	Food	Fat (grams)
Breakfast			
Orange juice, 1 cup	0.5	Same	0.5
Shredded wheat, ¾ cup	0.5	Shredded wheat, 1 cup	0.7
Toasted whole-grain bagel	1.1	Same	1.1
Peanut butter, 3 teaspoons	8.0	Same	8.0
1% low-fat milk, 1 cup	2.5	Fat-free milk, 1 cup	0.6
Lunch			
Whole-wheat bread, 2 slices	2.4	Same	2.4
Roast beef, 2 ounces	4.9	Light turkey roll, 2 ounces	0.9
Mustard, 3 teaspoons	0.6	Same	0.6
Swiss cheese, 2 slices	15.6	Swiss cheese, 1 slice	7.8
Lettuce	—	Same	
Tomato	—	Same	—
Oatmeal cookie, 1	3.3	Oatmeal cookie, 2	6.6
Snack			
Apple	—	Same	—
Dinner			
Chicken tenders frozen meal	18.0	Fat-free chicken tenders	—
Carrots, ½ cup	—	Same	—
Dinner roll, 1	2.0	Same	2.0
Olive oil, 1.5 teaspoon	6.8	Same	6.8
Banana	0.6	Same	0.6
1% low-fat milk, 1 cup	2.5	Fat-free milk, 1 cup	0.6
Snack			
Raisins, 2 teaspoons	—	Raisins, ½ cup	—
Air-popped popcorn, 3 cups	1.0	Air-popped popcorn, 6 cups	2.0
Parmesan cheese, 2 tablespoons	2.8	Same	2.8
Totals	**73.1**		**44.0**

Although consuming fish is thought to have greater benefits than using fish oil supplements, in some instances, fish oil capsules can be safely substituted for fish if a person does not like fish. Generally, about 1 gram of omega-3 fatty acids (about three capsules) from fish oil per day is recommended, especially for people with evidence of cardiovascular disease. (Freezing fish oil capsules before consumption, using enteric coated capsules, and using capsules with vitamin E will reduce the fishy aftertaste.) The American Heart Association also recently suggested that fish oil supplements (providing 2 to 4 grams of omega-3 fatty acids per day) could be employed to treat elevated blood triglycerides. However, fish oil capsules should be limited for individuals who have bleeding disorders, take anticoagulant medications, or anticipate surgery, because they may increase risk of uncontrollable bleeding and hemorrhagic stroke. Thus, for fish oil capsules, as well as other dietary supplements, it is important to follow a physician's recommendations. Remember that fish oil supplements are not regulated by FDA. The quality of these supplements, therefore, is not standardized, and contaminants naturally present in the fish oil may not have been removed.

Flax seeds and walnuts are readily available, rich plant sources of the omega-3 alpha-linolenic acid. About 2 tablespoons of flax seed per day is typically recommended if used as an omega-3 fatty acid source. Flax seeds can be purchased in many natural food stores rather inexpensively. These need to be chewed thoroughly, or they will pass through the GI tract undigested. Many people find it easier to grind them in a coffee grinder before eating them. Flax seed oil is also available, but it turns **rancid** very quickly, especially if not refrigerated. Compared to other nuts and seeds, walnuts are one of the richest sources of alpha-linolenic acid (2.6 grams per 1-ounce serving or 14 walnut halves). The AI for alpha-linolenic acid is 1.6 grams per day for men and 1.1 grams per day for women. In addition, walnuts are a rich source of plant sterols known to inhibit intestinal absorption of cholesterol.

Regarding essential fatty acids, the Food and Nutrition Board has issued recommendations for both omega-6 and omega-3 fatty acids. The amounts listed in Table 5-6 work out to about 5% of calorie intake for the total of both essential fatty acids. Infants and children have lower needs (see Chapter 15). Consumption of fish at least twice a week is one step toward meeting requirements for essential fatty acids.

The typical North American diet derives about 7% of calories from polyunsaturated fatty acids and thus meets essential fatty acid needs. An upper limit of 10% of calorie intake as polyunsaturated fatty acids is often recommended, in part because the breakdown (oxidation) of those present in lipoproteins is linked to increased cholesterol deposition in the arteries (see the Nutrition and Your Health section in this chapter). Depression of immune function is also suspected to be caused by an excessive intake of long-chain polyunsaturated fatty acids (EPA/DHA).

Mediterranean Diet. In recent years, the Mediterranean diet (see Further Readings 3 and 13, and Newsworthy Nutrition on page 186) has attracted a lot of attention as a result of lower rates of chronic diseases seen in people following such a diet plan. Reduction of cardiovascular disease has been one of the most consistent results of the Mediterranean diet. The major sources of fat in the Mediterranean diet include liberal amounts of olive oil compared to a small amount of animal fat (from animal flesh, eggs, and dairy products). In contrast, major sources of fat in the typical North American diet include animal flesh, whole milk, pastries, cheese, margarine, and mayonnaise. While dietary fat sources definitely play a role in prevention of chronic disease, it is important to remember that other aspects of one's lifestyle also contribute to disease risk. People who follow a Mediterranean diet also tend to consume moderate alcohol (usually in the form of red wine, which contains many antioxidants), eat plenty of whole grains and few refined carbohydrates, and are also more physically active than typical North Americans (see Further Reading 10).

Ornish Diet. An alternative plan for reduction of cardiovascular disease is Dr. Dean Ornish's purely vegetarian (**vegan**) diet plan (see Further Reading 9). This diet is very low in fat, including only a scant quantity of vegetable oil used in cooking and the small amount of oils present in plant foods. Individuals restricting fat intake to 20% of calories should be monitored by a physician, as the resulting increase in carbohydrate intake can increase blood triglycerides in some people, which is not a healthful change. Over time, however, the initial problem

▲ Walnuts are one of the richest plant sources of the omega-3 fatty acid, alpha-linolenic acid, and are a good source of plant sterols.

The advice to consume 20% to 35% of calories as fat does not apply to infants and toddlers below the age of 2 years. These youngsters are forming new tissue that requires fat, especially in the brain, so their intake of fat and cholesterol should not be greatly restricted.

vegan A person who eats only plant foods.

TABLE 5-6 ▶ Food and Nutrition Board Recommendations for Omega-6 and Omega-3 Fatty Acids per Day

	Men (grams per day)	Women (grams per day)
Linoleic acid (omega-6)	17 (3.6 tsp)	12 (2.6 tsp)
Alpha-linolenic acid (omega-3)	1.6 (0.35 tsp)	1.1 (0.24 tsp)

NEWSWORTHY NUTRITION

Mediterranean diet decreases risk of cardiovascular disease

Based on earlier observations of an inverse relationship between adherence to the Mediterranean diet and risk of cardiovascular disease, the effect of the Mediterranean diet pattern on the primary prevention of cardiovascular disease events was studied in Spain, using a randomized multicenter trial design. Participants (7447 persons, 55 to 80 years, 57% women) were at high cardiovascular risk but had no cardiovascular disease at the beginning of the study and were randomly assigned to either a Mediterranean diet supplemented with extra-virgin olive oil; a Mediterranean diet supplemented with mixed nuts; or a control diet with advice to reduce dietary fat. Participants were followed for a median of 4.8 years and assessed for the occurrence of the following major cardiovascular events: myocardial infarction, stroke, or death from cardiovascular causes. There was good adherence to the interventions for the two Mediterranean-diet groups. At least one major cardiovascular event occurred in 288 participants. This included 96 events for the group assigned to a Mediterranean diet with extra-virgin olive oil and 83 events for the group assigned to a Mediterranean diet with nuts versus 109 events for the control group. The authors concluded that a Mediterranean diet, unrestricted in calories but supplemented with extra-virgin olive oil or nuts, resulted in a substantial reduction in the incidence of major cardiovascular events among persons at high cardiovascular risk. These results, therefore, support the benefits of the Mediterranean diet for the prevention of cardiovascular disease.

Source: Estruch R and others: Primary prevention of cardiovascular disease with a Mediterranean diet. *New England Journal of Medicine* 368:1279–1290, 2013.

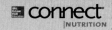 Check out the Connect site **www.mcgrawhillconnect.com** to further explore the implications of various eating patterns on risk for cardiovascular disease.

of high blood triglycerides on a low-fat diet may self-correct. Among people following the Ornish plan, blood triglycerides initially increased but within a year fell to normal values as long as the individuals emphasized high-fiber carbohydrate sources, controlled (or improved) body weight, and followed a regular exercise program.

In summary, the general consensus among nutrition experts suggests that limitation of saturated fat, cholesterol, and *trans* fat intake should be the primary focus and that the diet needs to contain a balance of omega-3 and omega-6 fatty acids (Table 5-7). Furthermore, if fat intake exceeds 30% of total calories, the extra fat should come from monounsaturated fat, primarily olive oil. The Academy of Nutrition and Dietetics, the Dietitians of Canada (see Further Reading 1), the National Cholesterol Education Program (NCEP), and the Food and Nutrition Board are in agreement with the advice of the AHA. The 2010 Dietary Guidelines also support this advice. In addition to fat intake, controlling total calorie intake is also significant, as weight control is a vital component of cardiovascular disease prevention.

▲ If you are looking to decrease the amount of saturated and *trans* fats in your diet, it is a good idea to opt for lower-fat substitutes for some of your current high-fat food choices. How do you think this meal compares with the fried meal on p. 172? How does it compare to MyPlate recommendations?

✓ CONCEPT CHECK 5.7

1. How does the percent of calories as fat in the North American diet compare to recommendations?
2. What is the limit for cholesterol intake?
3. What are the characteristics of the Mediterranean diet?

TABLE 5-7 ▶ Tips for Avoiding Too Much Fat, Saturated Fat, Cholesterol, and *Trans* Fat

	Eat Less of These Foods	Eat More of These Foods
Grains	• Pasta dishes with cheese or cream sauces • Croissants • Pastries • Doughnuts • Pie crust	• Whole-grain breads • Whole-grain pasta • Brown rice • Air-popped popcorn
Vegetables	• French fries • Potato chips • Vegetables cooked in butter, cheese, or cream sauces	• Fresh, frozen, baked, or steamed vegetables
Fruit	• Fruit pies	• Fresh, frozen, or canned fruits
Dairy	• Whole milk • Ice cream • High-fat cheese • Cheesecake	• Fat-free and reduced-fat milk • Low-fat frozen desserts (e.g., yogurt, sherbet, and ice milk) • Reduced-fat/part-skim cheese
Protein	• Bacon • Sausage • Organ meats (e.g., liver) • Egg yolks	• Fish • Skinless poultry • Lean cuts of meat (with fat trimmed away) • Soy products • Egg whites/egg substitutes

CASE STUDY Planning a Heart-Healthy Diet

Jackie is a 21-year-old health-conscious individual majoring in business. She recently learned that a diet high in saturated fat can contribute to high blood cholesterol and that exercise is beneficial for the heart. Jackie now takes a brisk 30-minute walk each morning before going to class, and she has started to cut as much fat out of her diet as she can, replacing it mostly with carbohydrates. A typical day for Jackie now begins with a 2-cup bowl of Fruity Pebbles with 1 cup of skim milk and ½ cup of apple juice. For lunch, she might pack a turkey sandwich on white bread with lettuce, tomato, and mustard; a 1-ounce package of fat-free pretzels; and five reduced-fat vanilla wafers. Dinner could be a 2-cup portion of pasta with some olive oil and garlic mixed in, and a small iceberg lettuce salad with lemon juice squeezed over it. Her snacks are usually baked chips, low-fat cookies, fat-free frozen yogurt, or fat-free pretzels. She drinks five diet soft drinks throughout the day as her main beverage.

Answer the following questions, and check your response at the end of this chapter.

1. Has Jackie made the best diet changes with regard to lowering blood cholesterol and maintaining heart health?
2. Is there much fat left in Jackie's new diet plan? Is it necessary for her to drastically lower her fat intake?
3. What types of fat should Jackie try to consume? Why are these types of fat the most desirable?
4. What types of foods has Jackie used to replace the fat in her diet?
5. What food groups are missing from her new diet plan? How many servings should she be including from these food groups?
6. Is Jackie's new exercise routine appropriate?

▲ Are there important food groups missing from Jackie's new diet plan?

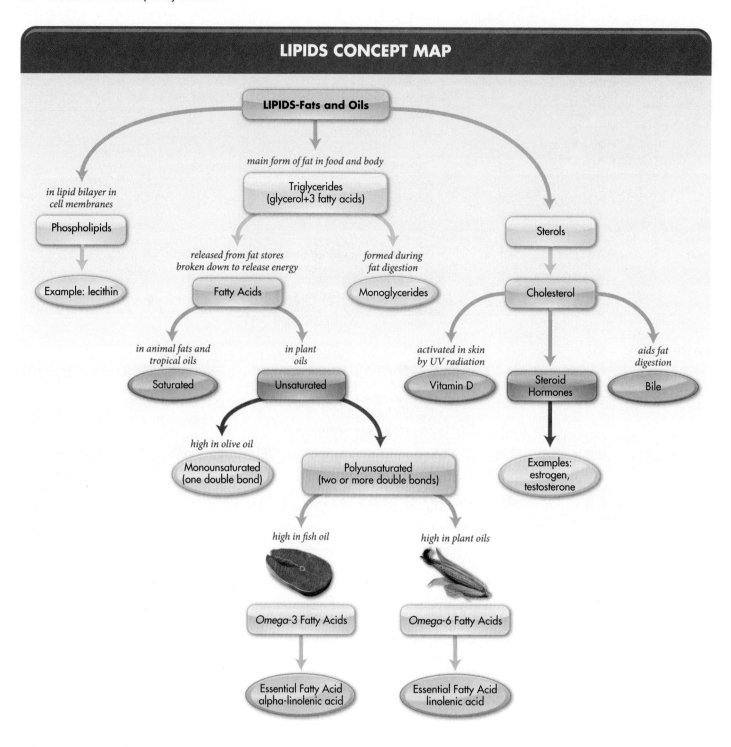

LIPIDS CONCEPT MAP

LIPIDS-Fats and Oils

in lipid bilayer in cell membranes

Phospholipids

Example: lecithin

main form of fat in food and body

Triglycerides (glycerol+3 fatty acids)

released from fat stores broken down to release energy

Fatty Acids

formed during fat digestion

Monoglycerides

in animal fats and tropical oils

Saturated

in plant oils

Unsaturated

high in olive oil

Monounsaturated (one double bond)

Polyunsaturated (two or more double bonds)

high in fish oil

high in plant oils

Omega-3 Fatty Acids

Omega-6 Fatty Acids

Essential Fatty Acid alpha-linolenic acid

Essential Fatty Acid linolenic acid

Sterols

Cholesterol

activated in skin by UV radiation

Vitamin D

Steroid Hormones

aids fat digestion

Bile

Examples: estrogen, testosterone

Nutrition and Your Health
Lipids and Cardiovascular Disease

The typical forms of cardiovascular disease—coronary heart disease and strokes—are associated with inadequate blood circulation in the heart and brain related to buildup of this plaque. Blood supplies the heart muscle, brain, and other body organs with oxygen and nutrients. When blood flow via the coronary arteries surrounding the heart is interrupted, the heart muscle can be damaged. A heart attack, or **myocardial infarction,** may result (review Fig. 5-17). This may cause the heart to beat irregularly or to stop. About 25% of people do not survive their first heart attack. If blood flow to parts of the brain is interrupted long enough, part of the brain dies, causing a **cerebrovascular accident,** or stroke.

A heart attack can strike with the sudden force of a sledgehammer, with pain radiating up the neck or down the arm. It can sneak up at night, masquerading as indigestion, with slight pain or pressure in the chest. Crushing chest pain is a more common symptom in men (see Further Reading 12). Many times, the symptoms are so subtle in women that death occurs before she or the health professional realizes that a heart attack is taking place. If there is any suspicion at all that a heart attack is taking place, the person should first call 911

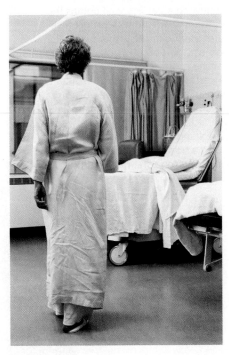

▲ Cardiovascular disease kills more women than any other disease.

Cardiovascular disease is the major killer of North Americans. It typically involves the coronary arteries and, thus, frequently the term *coronary heart disease (CHD)* or *coronary artery disease (CAD)* is used. Each year, about 600,000 people die of coronary heart disease in the United States. About 715,000 people in the United States have a heart attack each year. Women generally lag about 10 years behind men in developing the disease. Still, it eventually kills more women than any other disease. The cost of coronary heart disease alone is $108.9 billion each year, which includes the costs of health care services, medications, and lost productivity.

Development of Cardiovascular Disease

The symptoms of cardiovascular disease develop over many years and often do not become obvious until old age. Nonetheless, autopsies of young adults under 20 years of age have shown that many of them had atherosclerotic **plaque** in their arteries (Fig. 5-17). This finding indicates that plaque buildup can begin in childhood and continue throughout life, although it usually goes undetected for some time (see Further Reading 12).

plaque A cholesterol-rich substance deposited in the blood vessels; it contains various white blood cells, smooth muscle cells, various proteins, cholesterol and other lipids, and eventually calcium.

myocardial infarction Death of part of the heart muscle. Also termed a *heart attack.*

cerebrovascular accident (CVA) Death of part of the brain tissue due typically to a blood clot. Also termed a *stroke.*

and then chew an aspirin (325 milligrams) thoroughly. Aspirin helps reduce the blood clotting that leads to a heart attack.

Continuous formation and breakdown of blood clots in the blood vessels is a normal process. However, in areas where plaques build up, blood clots are more likely to remain intact and then lead to a blockage, cutting off or diminishing the supply of blood to the heart (via the coronary arteries) or brain (via the carotid arteries). More than 95% of heart attacks are caused by total blockage of the coronary arteries due to a blood clot forming in an area of the artery already partially blocked by plaque. Disruption of the plaque may even lead to eventual clot formation.

Atherosclerosis probably first develops to repair damage in a vessel lining. Ongoing inflammation in the blood vessel is suspected of causing blood vessel damage. (A laboratory test for c-reactive protein in the blood is used to detect inflammation; see Further Reading 8). The damage that starts this process can also be caused by smoking, diabetes, hypertension, viral and bacterial infection, and LDL. In addition to causing damage to arteries, LDL then enters the damaged areas of the artery walls and begins to accumulate. Atherosclerosis can be seen in arteries throughout the body. The damage develops especially at points where an artery branches into two smaller

vessels. A great deal of stress is placed on the vessel walls at these points due to changes in blood flow.

Once blood vessel damage has occurred, the next step in the development of atherosclerosis is the progression phase. This step is characterized by deposition of plaque at the site of initial damage. The rate of plaque buildup during the progression phase is directly related to the amount of LDL in the blood. The form of LDL that contributes to atherosclerosis has been changed by free radicals into **oxidized** LDL. This form is preferentially taken up by scavenger cells in the arterial wall. Nutrients and phytochemicals that have **antioxidant** properties may reduce LDL oxidation. Fruits and vegetables are particularly rich in these compounds. Eating fruits and vegetables regularly is one positive step we can make to reduce plaque

oxidize In the most basic sense, an electron has been lost or an oxygen has been gained by a chemical substance. This change typically alters the shape and/or function of the substance.

antioxidant Generally a compound that stops the damaging effects of reactive substances seeking an electron (i.e., oxidizing agents). This prevents breakdown (oxidizing) of substances in foods or the body, particularly lipids.

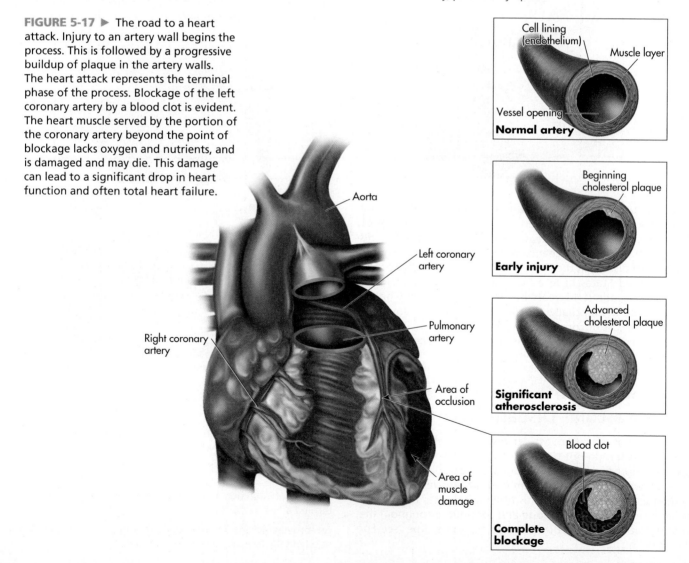

FIGURE 5-17 ▶ The road to a heart attack. Injury to an artery wall begins the process. This is followed by a progressive buildup of plaque in the artery walls. The heart attack represents the terminal phase of the process. Blockage of the left coronary artery by a blood clot is evident. The heart muscle served by the portion of the coronary artery beyond the point of blockage lacks oxygen and nutrients, and is damaged and may die. This damage can lead to a significant drop in heart function and often total heart failure.

Typical warning signs of a heart attack are:

- Intense, prolonged chest pain or pressure, sometimes radiating to other parts of the upper body (men and women)
- Shortness of breath (men and women)
- Sweating (men and women)
- Nausea and vomiting (especially women)
- Dizziness (especially women)
- Weakness (men and women)
- Jaw, neck, and shoulder pain (especially women)
- Irregular heartbeat (men and women)

buildup and slow the progression of cardiovascular disease. Some fruits and vegetables particularly helpful in this regard include legumes (beans), nuts, dried plums (prunes), raisins, berries, plums, apples, cherries, oranges, grapes, spinach, broccoli, red bell peppers, and onions. Tea, coffee, and dark chocolate are also sources of antioxidants. The American Heart Association does not support use of antioxidant supplements (such as vitamin E) to reduce cardiovascular disease risk. This is because large-scale studies have shown no decrease in cardiovascular disease risk with use of antioxidant supplements.

The body also sends white blood cells called macrophages to the location of the cholesterol accumulation on the blood vessel wall. In an attempt to destroy it, the macrophage surrounds the fatty deposit and produces lipid-loaded foam cells. Over years, as cholesterol plaques form and grow inside arteries, they eventually can begin to block off blood flow. The LDL-rich center of the plaque grows in a slow, controlled way and seldom causes heart attacks but arteries harden, narrow, and lose their elasticity. Affected arteries become further damaged as blood pumps through them and pressure increases. Some plaques can become unstable and tear away from the artery. If they rupture, a blood clot will form inside the artery, and within minutes, blood flow is cut off, resulting in a heart attack or stroke.

Factors that typically bring on a heart attack in a person already at risk include dehydration; acute emotional stress (such as firing an employee); strenuous physical activity when not otherwise physically fit (shoveling snow, for example); waking during the night or getting up in the morning (linked to an abrupt increase in stress); and consuming large, high-fat meals (increases blood clotting).

Risk Factors for Cardiovascular Disease

Many of us are free of the risk factors that contribute to rapid development of atherosclerosis. If so, the advice of health experts is to consume a balanced diet, perform regular physical activity, have a complete fasting lipoprotein analysis performed at age 20 or beyond, and reevaluate risk factors every 5 years.

For most people, however, the most likely risk factors are:

- **Total blood cholesterol over 200 milligrams per 100 milliliters of blood** (mg/dl; dl is short for deciliter or 100 milliliters). Risk is especially high when total cholesterol is at or over 240 mg/dl and LDL-cholesterol readings are over

Healthy People 2020 has set a goal of reducing death from coronary heart disease by 20%, compared with today's incidence.

130 to 160 mg/dl. (The terms *LDL-cholesterol* and *HDL-cholesterol* are used when expressing the blood concentration because it is the cholesterol content of these lipoproteins that is measured.)

- **Smoking.** Smoking is the main cause of about 20% of cardiovascular disease deaths and generally negates the female advantage of later occurrence of the disease. A combination of smoking and oral contraceptive use increases the risk of cardiovascular disease in women even more. Smoking greatly increases the expression of a person's genetically linked risk for cardiovascular disease, even if one's blood lipids are low. Smoking also makes blood more likely to clot. Even secondhand smoke has been implicated as a risk factor.
- **Hypertension. Systolic blood pressure** over 139 (millimeters of mercury) and **diastolic blood pressure** over 89 indicate hypertension. Healthy blood pressure values are less than 120 and 80, respectively. (Treatment of hypertension is reviewed in Chapter 9.)
- **Diabetes.** Diabetes virtually guarantees development of cardiovascular disease and so puts a person with diabetes in the high-risk group. Insulin increases cholesterol synthesis in the liver, in turn increasing LDL in the bloodstream. This disease negates any female advantage.

Together, the previous four risk factors explain most cases of cardiovascular disease.

Other risk factors to consider:

- **HDL-cholesterol** under 40 mg/dl, especially when the ratio of total cholesterol to HDL-cholesterol is greater than 4:1 (3.5:1 or less is optimal). Women often have high values for HDL-cholesterol; therefore, it is important for this to be measured in women to establish cardiovascular disease risk. A value of 60 mg/dl or more is especially protective. Exercising for at least 45 minutes four times a week can increase HDL by about 5 mg/dl. Losing excess weight (especially around the waist) and avoiding smoking and overeating also help maintain or raise HDL, as does moderate alcohol consumption.
- **Age.** Men over 45 years and women over 55 years.
- **Family history** of cardiovascular disease, especially before age 50.
- **Blood triglycerides** 200 mg/dl or greater in the fasting state (less than 100 mg/dl is optimal).
- **Obesity** (especially fat accumulation in the waist). Typical weight gain seen in adults is a chief contributor to the increase in LDL seen with aging. Obesity also typically leads to insulin resistance, creating a diabetes-like state, and ultimately the disease itself. It also increases overall inflammation throughout the body.

systolic blood pressure The pressure in the arterial blood vessels associated with the pumping of blood from the heart.

diastolic blood pressure The pressure in the arterial blood vessels when the heart is between beats.

▲ The buildup of plaque within artery walls is called atherosclerosis. The body sends white blood cells called macrophages to the location of the plaque in an attempt to destroy it. The macrophage surrounds the fatty deposit and produces lipid-loaded foam cells, seen here as a foamy appearance on the artery wall.

- **Inactivity.** Exercise conditions the arteries to adapt to physical stress. Regular exercise also improves insulin action in the body. The corresponding reduction in insulin output leads to a reduction in lipoprotein synthesis in the liver. Both regular aerobic exercise and resistance exercise are recommended. A person with existing cardiovascular disease should seek physician approval before starting such a program, as should older adults.

The term *risk factor* is not equivalent to cause of disease; nevertheless, the more of these risk factors one has, the greater the chances of ultimately developing cardiovascular disease. A good example is the **metabolic syndrome,** discussed in Chapter 4. A person with the metabolic syndrome would have abdominal obesity, high blood triglycerides, low HDL-cholesterol, hypertension, poor blood glucose regulation (i.e., high fasting blood glucose), and increased blood clotting. This profile raises the risk for cardiovascular disease considerably. On a positive note, cardiovascular disease is rare in populations that have low LDL-cholesterol, normal blood pressure, and do not smoke or have diabetes. By minimizing these risk factors, as well as following the dietary recommendations of the American Heart Association on page 183 and staying physically active, one will most likely reduce many of the other controllable risk factors listed. In other words, develop and follow a total lifestyle plan. Medications may also be added to lower blood lipids, as discussed next. Finally, if a person has a family history of cardiovascular disease but the usual risk factors are not present, a rarer defect might be the cause. In this case, having a detailed physical examination for other potential causes is advised.

Medications to Lower Blood Lipids

For some people, diet and lifestyle changes are simply not enough to lower blood cholesterol. Fortunately, medications offer a more aggressive approach to treating high cholesterol. Populations that may need cholesterol-lowering medications include individuals who

- Have clinical evidence of atherosclerotic cardiovascular disease (Fig. 5-17),
- Have LDL levels of 190 milligrams per deciliter or higher,
- Are ages 40 – 75 and have LDL levels 70 - 189 milligrams per deciliter *and* diabetes, or

Medicine Cabinet

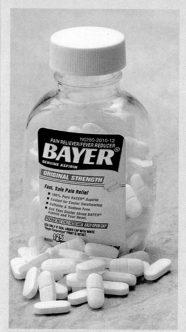

▲ At the first sign of a possible heart attack, the person should first call 911 and then thoroughly chew an aspirin. Aspirin in small doses reduces blood clotting. It is often used under a physician's guidance to treat people at risk for heart attack or stroke, especially if one has already occurred. About 80 to 160 milligrams per day is needed for such benefits. Individuals who may especially benefit from aspirin therapy are men over 40, smokers, postmenopausal women, and people with diabetes, hypertension, or a family history of cardiovascular disease.

Aspirin. Most common medication recommended to prevent heart attacks in people with coronary artery disease by helping to maintain blood flow to the heart.

Statins. Most frequently prescribed cholesterol-lowering drugs that block a liver enzyme involved in cholesterol synthesis and thus reduce the amount of cholesterol in the blood. Examples:
- Atorvastatin (Lipitor®)
- Fluvastatin (Lescol, Lescol XL®)
- Lovastatin (Altoprev, Mevacor®)
- Pravastatin (Pravachol®)
- Simvastatin (Zocor®)
- Rosuvastatin (Crestor®)
- Simvastatin + ezetimibe (Vytorin®)

Bile acid sequestering agents (Resins). These resins bind to bile acids in the intestine and are excreted in the feces, reducing their supply. This stimulates the liver to produce more bile acids, which uses more cholesterol and causes a decrease in blood cholesterol levels. Examples:
- Cholestyramine (Questran®)
- Colestipol (Colestid®)
- Colesevelam (Welchol®)

- Are ages 40 – 75 and have LDL levels 70 - 189 milligrams per deciliter *and* greater than 7.5% risk for atherosclerotic cardiovascular disease within the next 10 years (determined by a global risk assessment tool, see Table 5-8).

For more information on cardiovascular disease, see the website of the American Heart Association at **www.americanheart.org** or the heart disease section of Healthfinder at **www.healthfinder.gov/tours/heart.htm**. This is a site created by the U.S. government for consumers. In addition, visit the website **www.nhlbi.nih.gov/**.

Healthy People 2020 has set a goal of reducing total blood cholesterol among adults from the average of 198 mg/dl to 178 mg/dl, as well as reducing the percentage of adults with high blood cholesterol from 15% to 13.5%.

Two approaches have been shown to cause reversal of atherosclerosis in the body. One employs a vegan diet and other lifestyle changes that are part of the Dr. Dean Ornish program. The other employs aggressive LDL lowering with medications.

The global risk assessment tool is based on data from several large observational studies of the development and progression of cardiovascular disease. It is meant to assess non-Hispanic Black and non-Hispanic white men and women between the ages of 40 and 79 who do not have existing cardiovascular disease. Alternative calculators are available for individuals of other ages and races.

Medications work to lower blood cholesterol in several ways. The statins are a group of medications that reduce cholesterol synthesis in the liver. Examples are listed in the Medicine Cabinet feature on page 192. These medications may lead to adverse effects, especially on liver function, so physician monitoring is required. The cost of treatment with one of these drugs can vary widely – from as little as $12 per month to more than $500 (see Further Reading 2).

A second group of medications works to lower cholesterol by binding to bile acids in the small intestine, thus preventing their reabsorption and allowing them to be excreted as part of the feces. Recall from Chapter 3 that bile (which contains cholesterol) is secreted from the liver (via the gallbladder) into the small intestine to help digest fats. Usually, about 98% of bile acids is reabsorbed and recycled. When these medications bind to bile acids, they cannot be reabsorbed and recycled, so the liver must synthesize new bile acids. The liver removes LDL from the blood to do this. Medications that block cholesterol absorption are not as effective at lowering cholesterol as statins. Also, some of these medications taste gritty and therefore are not popular.

A third group of drugs can be used to lower blood triglycerides by decreasing the triglyceride production of the liver. These include gemfibrozil (Lopid) and megadoses of the vitamin nicotinic acid. The use of nicotinic acid does result in side effects, however, but these are typically manageable.

Some pharmaceutical companies combine medications with different mechanisms of action. A statin drug (simvastatin) has been combined with another drug (ezetimibe) and is marketed as Vytorin®, a drug that will treat the two sources of cholesterol, "food and family." While the statin reduces the cholesterol made by the liver, the ezetimibe helps to block the absorption of cholesterol from food.

TABLE 5-8 ▶ Factors Used in Interpreting Risk for Atherosclerotic Cardiovascular Disease*

Risk Factors	Interpretation
Gender	Compared to women, men are at higher risk of developing atherosclerotic cardiovascular disease.
Age	As age increases, risk also increases.
Race	African-Americans are at higher risk than Whites.
Total cholesterol	Risk progressively increases as total cholesterol rises above 170 milligrams per deciliter.
HDL cholesterol	Risk progressively increases as HDL cholesterol decreases below 50 milligrams per deciliter.
Systolic blood pressure	Risk progressively increases as systolic blood pressure rises above 110 mmHg.
Taking medications to treat hypertension	Risk is higher for a person with a history of hypertension.
Diabetes	Risk is higher for a person with a history of diabetes.
Smoking	Compared to nonsmokers, risk is higher for smokers.

*Goff DC and others: 2013 ACC/AHA guideline on the assessment of cardiovascular risk: a report of the American College of Cardiology/American Heart Association Task Force on Practice Guidelines. *Journal of the American College of Cardiology* 63: 2935, 2014.

▲ Dark chocolate has a high percentage of cocoa solids that contain phytonutrients called flavonols known to have antioxidant and anti-inflammatory properties. Eating dark chocolate has been found to decrease LDL cholesterol and increase HDL cholesterol.

The intensity of therapy depends on the severity of the patient's condition and his or her tolerance of the medication. For example, a patient who has already suffered a heart attack would be treated with a high-dose statin, with the aim of reducing LDL cholesterol by 50%. However, if that patient suffered adverse effects from the high-dose medication (e.g., impaired liver function), the physician would either lower the dose or try a different medication.

Plant Sterols for Cardiovascular Disease

CoroWise® is a leading brand of natural cholesterol reducers called plant sterols. Plant sterols have been clinically shown to reduce LDL (bad) cholesterol, and products that contain these natural cholesterol reducers are backed by the following FDA-approved health claim: *Foods containing at least 0.4 grams per serving of plant sterols, eaten twice a day with meals for a daily total intake of at least 0.8 grams, as part of a diet low in saturated fat and cholesterol, may reduce the risk of heart disease.* Products such as Smart Balance® margarines and Minute Maid HeartWlse® orange juice contain these plant sterols. The plant stanols/sterols, also called phytosterols, work by reducing cholesterol absorption in the small intestine and lowering its return to the liver. The liver responds by taking up more cholesterol from the blood so it can continue to make bile acids. The studies done on the cholesterol-lowering effect of these plant compounds have found that 2 to 5 grams of plant stanols/sterols per day reduces total blood cholesterol by 8% to 10% and LDL-cholesterol by 9% to 14% (similar to what is seen with some cholesterol-lowering drugs; see Further Reading 7).

In people who have borderline high total blood cholesterol (between 200 and 239 mg/dl), these plant compounds can be helpful in avoiding future drug therapy. Plant stanols/sterols have been made available in pill form as well. Remember that plant sterols are naturally present in nuts in high concentrations. Wheat germ, sesame seeds, pistachios, and sunflower seeds are some of the richest sources.

Surgical Treatment for Cardiovascular Disease

The two most common surgical treatments for coronary artery blockage are percutaneous transluminal coronary angioplasty (PTCA) and coronary artery bypass graft (CABG). PTCA involves the insertion of a balloon catheter into an artery. Once it is advanced to the area of the lesion, the balloon is expanded to crush the lesion. This method works best when only one vessel is blocked, and it may be held open with metal mesh, called a stent. CABG involves the removal and use of a saphenous vein (a large vein in the leg) or use of a mammary artery. The relocated vein is sewn to the main heart vessel (aorta) and then used to bypass the blocked artery. The procedure can be performed on one or more blockages.

Summary (Numbers refer to numbered sections in the chapter.)

5.1 Lipids are a group of compounds that do not dissolve in water. Fatty acids are the simplest form of lipid. There are three fatty acids on every triglyceride, the most common type of lipid found in the body and foods. Phospholipids and sterols are two other classes of lipids in food and our bodies.

5.2 Saturated fatty acids contain no carbon-carbon double bonds, monounsaturated fatty acids contain one carbon–carbon double bond, and polyunsaturated fatty acids contain two or more carbon-carbon double bonds in the carbon chain. In omega-3 polyunsaturated fatty acids, the first of the carbon-carbon double bonds is located three carbons from the methyl end of the carbon chain. In omega-6 polyunsaturated fatty acids, the first carbon-carbon double bond counting from the methyl end occurs at the sixth carbon. The essential fatty acids are linoleic acid (an omega-6 fatty acid) and alpha-linolenic acid (an omega-3 fatty acid). These must be included in the diet to maintain health. The hormone-like compounds produced from omega-3 fatty acids tend to reduce blood clotting, blood pressure, and inflammatory responses in the body. Those produced from omega-6 fatty acids tend to increase blood clotting.

Triglycerides are formed from a glycerol backbone with three fatty acids. Triglycerides rich in long-chain saturated fatty acids tend to be solid at room temperature, whereas those rich in monounsaturated and polyunsaturated fatty acids are liquid at room temperature. Triglyceride is the major form of fat in both food and the body. It allows for efficient energy storage, protects certain organs, transports fat-soluble vitamins, and helps insulate the body.

5.3 Foods rich in fat include salad oils, butter, margarine, and mayonnaise. Nuts, bologna, avocados, and bacon are also high in fat, as are peanut butter and cheddar cheese. Steak and hamburger are moderate in fat content, as is whole milk. Many grain products and fruits and vegetables in general are low in fat.

Fats and oils have several functions as components of foods. Fats add flavor and texture to foods and provide some satiety after meals. Some phospholipids are used in foods as emulsifiers, which suspend fat in water. When fatty acids break down, food becomes rancid, resulting in a foul odor and unpleasant flavor.

Hydrogenation is the process of converting carbon–carbon double bonds into single bonds by adding hydrogen at the point of unsaturation. The partial hydrogenation of fatty acids in vegetable oils changes the oils to semisolid fats and helps in food formulation and reduces rancidity. Hydrogenation also increases *trans* fatty acid content. High amounts of *trans* fat in the diet are discouraged, as these increase LDL and reduce HDL.

5.4 Fat digestion takes place primarily in the small intestine. Lipase enzyme released from the pancreas digests long-chain triglycerides into monoglycerides (glycerol backbones with single fatty acids attached) and fatty acids. The breakdown products are then taken up by the absorptive cells of the small intestine. These products are mostly remade into triglycerides inside intestinal cells and eventually enter the lymphatic system, in turn passing into the bloodstream.

5.5 Lipids are carried in the bloodstream by various lipoproteins, which consist of a central triglyceride core encased in a shell of protein, cholesterol, and phospholipid. Chylomicrons are released from intestinal cells and carry lipids arising from dietary intake. Very-low-density lipoprotein (VLDL) and low-density lipoprotein (LDL) carry lipids both taken up by and synthesized in the liver. High-density lipoprotein (HDL) picks up cholesterol from cells and facilitates its transport back to the liver.

5.6 Triglycerides are used for energy storage, insulation, and transportation of fat-soluble vitamins. Phospholipids are derivatives of triglycerides in which one or two of the fatty acids are replaced by phosphorus-containing compounds. Phospholipids are important parts of cell membranes, and some act as efficient emulsifiers.

Cholesterol forms vital biological compounds, such as hormones, components of cell membranes, and bile acids. Cells in the body make cholesterol whether we eat it or not. It is not a necessary part of an adult's diet.

5.7 There is currently no RDA for fat for adults. Plant oils should contribute about 5% of total calories to achieve the Adequate Intakes proposed for essential fatty acids (linoleic acid and alpha-linolenic acid). Fatty fish are a rich source of omega-3 fatty acids and should be consumed at least twice a week.

Many health agencies and scientific groups suggest a fat intake of no more than 30% to 35% of total calories. Some health experts advocate an even further reduction to 20% of calorie intake for some people to maintain a normal LDL value, but such a diet requires professional guidance. Medications such as "statins" may be added also to lower LDL. If fat intake exceeds 30% of total calories, the diet should emphasize monounsaturated fat. The typical North American diet contains about 33% of total calories as fat.

NAYH In the blood, elevated amounts of LDL and low amounts of HDL are strong predictors of risk for cardiovascular disease. Additional risk factors for the disease are smoking, hypertension, diabetes, obesity, and inactivity.

Check Your Knowledge (Answers to the following questions are below.)

1. Margarine usually is made by a process called _____, in which hydrogen atoms are added to carbon-carbon double bonds in the polyunsaturated fatty acids found in vegetable oils.
   a. saturation
   b. esterification
   c. isomerization
   d. hydrogenation

2. Essential fatty acids that cause a decrease in blood clotting are
   a. omega-3.
   b. omega-6.
   c. omega-9.
   d. prostacyclins.

3. Cholesterol is
   a. a dietary essential; the human body cannot synthesize it.
   b. found in foods of plant origin.
   c. an important part of human cell membranes and necessary to make some hormones.
   d. All of the above.

4. Which of the following groups of foods would be important sources of saturated fatty acids?
   a. olive oil, peanut oil, canola oil
   b. palm oil, palm kernel oil, coconut oil
   c. safflower oil, corn oil, soybean oil
   d. All of the above.

5. Lipoproteins are important for
   a. transport of fats in the blood and lymphatic system.
   b. synthesis of triglycerides.
   c. synthesis of adipose tissue.
   d. enzyme production.

6. Which of the following foods is the best source of omega-3 fatty acids?
   a. fatty fish
   b. peanut butter and jelly
   c. lard and shortenings
   d. beef and other red meats

7. Immediately after a meal, newly digested and absorbed dietary fats appear in the lymph and then the blood as part of which of the following?
   a. LDL
   b. HDL
   c. chylomicrons
   d. cholesterol

8. High blood concentrations of _____ decrease the risk for cardiovascular disease.
   a. low-density lipoproteins
   b. chylomicrons
   c. high-density lipoproteins
   d. cholesterol

9. Phospholipids such as lecithin are used extensively in food preparation because they
   a. provide the agreeable feel of fat melting on the tongue.
   b. are excellent emulsifiers.
   c. provide important textural features.
   d. impart delicate flavors.

10. The main form of lipid found in the food we eat is
    a. cholesterol.
    b. phospholipids.
    c. triglycerides.
    d. plant sterols.

Answer Key: 1. d (LO 5.3), 2. a (LO 5.2), 3. c (LO 5.6), 4. b (LO 5.3), 5. a (LO 5.5), 6. a (LO 5.3), 7. c (LO 5.5), 8. c (LO 5.5), 9. b (LO 5.6), 10. c (LO 5.2)

Study Questions (Numbers refer to Learning Outcomes)

1. Describe the chemical structures of saturated and polyunsaturated fatty acids and their different effects in both food and the human body. **(LO 5.2)**

2. Relate the need for omega-3 fatty acids in the diet to the recommendation to consume fatty fish at least twice a week. **(LO 5.7)**

3. Describe the structures, origins, and roles of the four major blood lipoproteins. **(LO 5.5)**

4. What are the recommendations from various health care organizations regarding fat intake? What does this mean in terms of food choices? **(LO 5.7)**

5. What are two important attributes of fat in food? How are these different from the general functions of lipids in the human body? **(LO 5.3)**

6. Describe the significance of and possible uses for reduced-fat foods. **(LO 5.6)**

7. Does the total cholesterol concentration in the bloodstream tell the whole story with respect to cardiovascular disease risk? **(LO 5.8)**

8. List the four main risk factors for the development of cardiovascular disease. **(LO 5.8)**

9. What three lifestyle factors decrease the risk of cardiovascular disease development? **(LO 5.8)**

10. When are medications most needed in cardiovascular disease therapy, and how in general do the various classes of medications operate to reduce risk? **(LO 5.8)**

What the Dietitian Chose

With heart health in mind, you will want to choose the ground meat with the lowest fat, saturated fat, and cholesterol. Check the Nutrition Facts panel, if available, to compare the lipid content of various products. Look for cuts of meat with "round " or "loin" in the name for lowest fat content.

USDA allows up to 30% fat (by weight) in raw ground beef, so the cut of beef makes a difference. Regular ground beef typically contains the most fat (about 20% to 30% fat). Next comes ground chuck (about 15% fat), followed by ground round (about 10% fat), and ground sirloin (about 3% fat). Table 5-9 shows a further breakdown of the fat content of varieties of ground beef. Based on this information, you can see that the ground sirloin will give you the leanest burgers.

The ground sirloin, however, will also be the most expensive variety of ground beef. You may, therefore, want to take advantage of the fact that ground beef loses a lot of fat during cooking—as much as 50% for the highest fat products. Typically, regular ground beef is the least expensive product. As the percent lean increases, so does price.

For foods in which the ground beef will be shaped, then cooked (e.g., hamburger patties, meatballs, or meatloaf), bake, grill, or broil the ground meat on a rack so that fat will drain from the product as it cooks, then let the cooked product rest on paper towels for 1 minute after cooking. For recipes that incorporate browned ground beef into a mixed dish (e.g., casseroles or spaghetti sauce), brown ground beef, crumbling as you cook, then blot with paper towels or rinse under warm water to achieve a final cooked product with nearly the same fat content as the ground round. Using the rinsing method, 100 grams (about 3.5 ounces) of ground beef yields a final product with just 4 grams of fat. Because a greater percentage of the starting product is lost during cooking, you will end up with less meat in the final product. However, this is not such a big deal because most Americans consume two to three times as much protein as they need. To further enhance the heart-healthiness of mixed dishes, replace some of that lost product with beans. Chili or tacos would be excellent recipes to try with this method.

▲ Whereas the more expensive ground round and ground sirloin will give you lowest-fat burgers to start, the less expensive regular ground beef or ground chuck burgers will lose fat while cooking. You will therefore want to consider both fat content and cost when making your choice.

TABLE 5-9 ▶ **Calorie, Fat, and Cholesterol Content of Types of Ground Beef**

	Energy (kcal)	Total Fat(g)	Saturated Fat (g)	Cholesterol (mg)
Regular ground beef, 3.5 oz cooked	273	18	7	82
Ground chuck, 3.5 oz cooked	232	14	5	86
Ground round, 3.5 oz cooked	204	11	4	82
Ground sirloin, 3.5 oz cooked	164	6	3	76

CASE STUDY SOLUTION Planning a Heart-Healthy Diet

1. Jackie's approach to lowering blood cholesterol does not incorporate the best choices. She has excluded a great deal of fat in her diet, perhaps more than necessary, and has failed to include some key food groups that have cholesterol-lowering components.
2. Lowering fat as drastically as she has is not really necessary, especially for a physically active 21-year-old female.
3. Jackie could allow a more liberal amount of fat in her diet by including more monounsaturated fats. Canola oil and olive oil, as well as fats found in nuts and avocados, are rich sources of monounsaturated fats. These fats do not increase blood cholesterol. In addition, she should include good sources of omega-3 fatty acids such as fatty fish, walnuts, flaxseeds, or soybean oil.

One option is to use a canola oil-and-vinegar dressing on her salad, rather than lemon juice.
4. She has excluded a great deal of fat in her diet by merely replacing it with refined carbohydrates.
5. To make a shift to a more heart-healthy diet, Jackie would need to include at least 2 cups of fruit and 3 cups of vegetables a day, along with more whole-grain products. She should use whole-wheat bread instead of white bread for her sandwiches and eat a breakfast cereal that has at least 3 grams of fiber per serving.
6. Taking a brisk walk each morning is an excellent way to guarantee at least 30 minutes of activity on most days of the week. Exercise has been shown to increase concentration of HDL-cholesterol in the blood and decrease risk of cardiovascular disease.

Further Readings

1. ADA Reports: Position of the American Dietetic Association and Dietitians of Canada: Dietary fatty acids. *Journal of the American Dietetic Association* 107:1599, 2007.
2. Consumers Union: Evaluating statin drugs to treat high cholesterol and heart disease: Comparing effectiveness, safety, and price. *Consumer Reports Health Best Buy Drugs.* Consumers Union of United States, Inc., **http://www.consumerreports.org/health/resources/pdf/best-buy-drugs/StatinsUpdate-FINAL.pdf 2012.** Accessed March 10, 2013.
3. Estruch R and others: Primary prevention of cardiovascular disease with a Mediterranean diet. *New England Journal of Medicine* 368:1279–1290, 2013.
4. Getz L: A burger and fries (hold the trans fats). *Today's Dietitian* 11(2):35, 2009.
5. Kris-Etherton PM and Hill AM: Omega-3 fatty acids: Food or supplements? *Journal of the American Dietetic Association* 108:1125, 2008.
6. Lichtenstein AH and others: Diet and lifestyle recommendations revision 2006. A scientific statement from the American Heart Association Nutrition Committee. *Circulation* 114:82, 2006.
7. Micallef MA and Garg ML: The lipid-lowering effects of phytosterols and (omega-3) polyunsaturated fatty acids are synergistic and complementary in hyperlipidemic men and women. *Journal of Nutrition* 138:1086, 2008.
8. Myers GL and others: National Academy of Clinical Biochemistry laboratory medicine practice guidelines: Emerging biomarkers for primary prevention of cardiovascular disease. *Clinical Chemistry* 55:378, 2009.
9. Palmer S: Fighting heart disease the Dean Ornish way. *Today's Dietitian* 2:48, 2009.
10. Palmer S: The Mediterranean diet: A practical guide to shopping, menu ideas, and recipes. *Today's Dietitian* 14(5):30, 2012.
11. Sarris J and others: Omega-3 for bipolar disorder: Meta-analyses of use in mania and bipolar depression. *Journal of Clinical Psychiatry* 73:81, 2012.
12. Schaeffer J: Prevent heart disease: How to dispel the five common heart-health myths with clients and set the record straight. *Today's Dietitian* 15(2); 20, 2013.
13. Shai I and others: Weight loss with a low-carbohydrate, Mediterranean, or low-fat diet. *New England Journal of Medicine* 359:229, 2008.

 **connect** | NUTRITION
To get the most out of your study of nutrition, visit McGraw-Hill Connect at www.mcgrawhillconnect.com where you will find NutritionCalc Plus, LearnSmart, and many other dynamic tools.

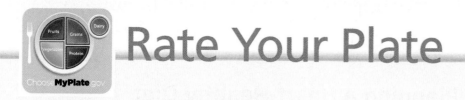

Rate Your Plate

I. Choosing Foods for Heart Health

Instructions: The foods listed in column A tend to be high in saturated fat, *trans* fatty acids, cholesterol, and total fat. Fill in column B with alternative foods that are generally low in these dietary components. After filling in your ideas, see our suggestions below for column B for foods that will reduce your risk of cardiovascular disease.

Column A			Column B
Bacon and eggs		or	
Doughnut or sweet roll		or	
Breakfast sausage		or	
Whole milk		or	
Cheeseburger		or	
French fries		or	
Ground chuck		or	
Soup with cream base		or	
Macaroni and cheese		or	
Cream/fruit pie		or	
Cream-filled cookies		or	
Ice cream		or	
Butter or stick margarine		or	

Ready-to-eat whole-grain breakfast cereal
Whole-wheat roll, bagel, or bread
Fruit
Reduced-fat, low-fat, or fat-free milk
Turkey sandwich, no cheese
Plain baked potato with salsa
Ground round
Soup with broth base
Macaroni with marinara sauce
Graham crackers
Granola bar
Frozen yogurt, sherbet, or reduced-fat ice cream
Vegetable oils or soft margarine in a tub

198

II. Applying the Nutrition Facts Label to Your Daily Food Choices

Imagine that you are at the supermarket looking for a quick snack to help you keep your energy up during afternoons. In the snack section, you settle on two choices (see labels a and b). Evaluate the products using the table on the left.

Compare the nutrients in each product by completing this list. For each serving, which product is lower in each of the following?

Calories	(a)	(b)	no difference
Calories from Fat	(a)	(b)	no difference
Total Fat	(a)	(b)	no difference
Saturated Fat	(a)	(b)	no difference
Trans Fat	(a)	(b)	no difference
Cholesterol	(a)	(b)	no difference
Sodium	(a)	(b)	no difference
Total Carbohydrates	(a)	(b)	no difference
Dietary Fiber	(a)	(b)	no difference
Sugars	(a)	(b)	no difference
Protein	(a)	(b)	no difference
Iron	(a)	(b)	no difference

Which package has more servings per container?

(a)	(b)	no difference

(a)

Nutrition Facts

Serving Size: 2 bars (42g)
Servings Per Container: 6

Amount Per Serving

Calories 180   Calories from Fat 50

	% Daily Value*
Total Fat 6g	9%
Saturated Fat 0.5g	3%
Trans fat 0g	**
Cholesterol 0mg	0%
Sodium 160mg	7%
Total Carbohydrates 29g	10%
Dietary Fiber 2g	8%
Sugars 11g	
Protein 4g	
Iron	6%

Not a significant source of Vitamin A, Vitamin C, and calcium.

** Intake of *trans* fat should be as low as possible.

* Daily values are based on a 2,000 calorie diet. Your daily values may be higher or lower depending on your calorie needs:

** Intake should be as low as possible.	Calories	2,000	2,500
Total Fat	Less than	65g	80g
Saturated Fat	Less than	20g	25g
Cholesterol	Less than	300mg	300mg
Sodium	Less than	2,400mg	2,400mg
Total Carbohydrates		300g	375g
Dietary Fiber		25g	30g

INGREDIENTS: WHOLE GRAIN ROLLED OATS, SUGAR, CANOLA OIL, CRISP RICE WITH SOY PROTEIN (RICE FLOUR, SOY PROTEIN CONCENTRATE, SUGAR, MALT, SALT), HONEY, BROWN SUGAR SYRUP, HIGH FRUCTOSE CORN SYRUP, SALT, SOY LECITHIN, BAKING SODA, NATURAL FLAVOR, PEANUT FLOUR, ALMOND FLOUR, HAZELNUT FLOUR, WALNUT FLOUR, PECAN FLOUR.

(b)

Nutrition Facts

Serving Size: 2 cookies (38g)
Servings Per Container: about 12

Amount Per Serving

Calories 180   Calories from Fat 70

	% Daily Value*
Total Fat 7g	11%
Saturated Fat 2g	10%
Trans fat 2g	**
Cholesterol 0mg	0%
Sodium 100mg	4%
Total Carbohydrate 26g	9%
Dietary Fiber 1g	4%
Sugars 12g	
Protein 2g	

Vitamin A 0%	•	Vitamin C 0%
Calcium 0%	•	Iron 2%

** Intake of *trans* fat should be as low as possible.

* Daily values are based on a 2,000 calorie diet. Your daily values may be higher or lower depending on your calorie needs:

** Intake should be as low as possible.	Calories	2,000	2,500
Total Fat	Less than	65g	80g
Saturated Fat	Less than	20g	25g
Cholesterol	Less than	300mg	300mg
Sodium	Less than	2,400mg	2,400mg
Total Carbohydrate		300g	375g
Dietary Fiber		25g	30g

Calories per gram: • Fat 9 • Carbohydrate 4
• Protein 4

INGREDIENTS: ENRICHED FLOUR (WHEAT FLOUR, NIACIN, REDUCED IRON, THIAMINE MONONITRATE, RIBOFLAVIN, FOLIC ACID), SUGAR, VEGETABLE OIL SHORTENING (PARTIALLY HYDROGENATED SOYBEAN, COCONUT, COTTONSEED, CORN AND/OR SAFFLOWER AND/OR CANOLA OIL), CORN SYRUP, HIGH FRUCTOSE CORN SYRUP, WHEY (A MILK INGREDIENT), CORN STARCH, SALT, SKIM MILK, LEAVENING (BAKING SODA, AMMONIUM BICARBONATE), ARTIFICIAL FLAVOR, SOYBEAN LECITHIN, COLOR (CONTAINING FD&C YELLOW #5 LAKE).

1. Which of the two brands would you choose?

_____

_____

2. What information on the Nutrition Facts labels contributed to your decision?

_____

_____

Student Learning Outcomes

Chapter 6 is designed to allow you to:

6.1 Distinguish between essential and nonessential amino acids and explain why adequate amounts of each of the essential amino acids are required for protein synthesis.

6.2 Describe how amino acids form proteins.

6.3 Identify food sources of protein, distinguish between high-quality and low-quality proteins, and describe the concept of complementary proteins.

6.4 Describe how protein is digested, absorbed, and metabolized in the body.

6.5 List the primary functions of protein in the body.

Chapter 6
Proteins

What Would You Choose?

About 3 weeks ago, you started lifting weights at the student recreation center. You are disappointed that you have not seen the results you were anticipating. You can lift more now than you could when you started, but you were hoping to tone and define the muscles in your arms, back, and abs. Perhaps you need more protein. Bodybuilding magazines have numerous advertisements for protein and amino-acid supplements, but they are expensive. What would you choose as optimal nutrition to support your weight-training regimen?

a Take individual amino-acid supplements.

b Take whey protein supplements.

c Increase your consumption of animal protein.

d Consume a diet that provides 10% to 35% of calories from a variety of sources of protein.

connect **NUTRITION** Think about your choice as you read Chapter 6, then see **What the Dietitian Chose** at the end of the chapter. To learn more about supplements and diet, check out the Connect site: www.mcgrawhillconnect.com.

Consuming enough protein is vital for maintaining health. Proteins form important structures in the body, make up a key part of the blood, help regulate many body functions, and can fuel body cells.

North Americans generally eat more protein than is needed to maintain health. Our daily protein intake comes mostly from animal sources, such as meat, poultry, fish, eggs, milk, and cheese. In contrast, in the developing world, diets can be deficient in protein.

Diets that are mostly vegetarian still predominate in much of Asia and areas of Africa, and some North Americans are currently adopting the practice. Plant sources of protein are worthy of more attention from North Americans. In the early 1900s, plant sources of proteins—nuts, seeds, and legumes—were consumed just as often as animal proteins. Over the years, though, plant proteins have been sidelined by meats. During this time, nuts were viewed as high-fat foods, and beans had the inferior reputation of "the poor man's meat." Contrary to these popular misconceptions, sources of plant proteins offer a wealth of nutritional benefits—from lowering blood cholesterol to preventing certain forms of cancer.

We could benefit from eating more plant sources of proteins, but it takes some knowledge to do so. It is possible—and desirable—to enjoy the benefits of animal *and* plant protein as we work toward the goal of meeting protein needs. This chapter takes a close look at protein, including the benefits of plant proteins in a diet. It will also examine the potential risks and benefits of vegetarian diets. Let's see why a detailed study of protein is worth your attention.

6.6 Apply current recommendations for protein intake to determine protein needs for healthy adults.

6.7 Describe the harmful effects of a high-protein diet and describe what is meant by positive protein balance, negative protein balance, and protein equilibrium.

6.8 Describe how protein-calorie malnutrition eventually can lead to disease in the body.

6.9 Develop vegetarian diet plans that meet the body's nutritional needs.

6.1 Amino Acids—Building Blocks of Proteins

protein Food and body compounds made of amino acids; proteins contain carbon, hydrogen, oxygen, nitrogen, and sometimes other atoms in a specific configuration. Proteins contain the form of nitrogen most easily used by the human body.

amino acid The building block for proteins containing a central carbon atom with nitrogen and other atoms attached.

Diets in the developed parts of the world, such as the United States and Canada, are typically rich in **protein,** and therefore, a specific focus on eating enough protein is generally not needed. In the developing world, however, it is important to focus on protein in diet planning because diets in those areas of the world can be deficient in protein.

Thousands of substances in the body are made of proteins. Aside from water, proteins form the major part of lean body tissue, totaling about 17% of body weight. **Amino acids**—the building blocks for proteins—are unique in that they contain nitrogen along with carbon, oxygen, and hydrogen. Plants combine nitrogen from the soil with carbon and other elements to form amino acids. They then link these amino acids together to make proteins. We get the nitrogen we need by consuming dietary proteins. Proteins are thus an essential part of a diet because they supply nitrogen in a form we can readily use, namely, amino acids. Using simpler forms of nitrogen is, for the most part, impossible for humans.

Proteins are crucial to the *regulation* and *maintenance* of the body. Body functions such as blood clotting, fluid balance, hormone and enzyme production, visual processes, transport of many substances in the bloodstream, and cell repair require specific proteins. The body makes proteins in many configurations and sizes so that they can serve these greatly varied functions. Formation of these body proteins begins with amino acids from both the protein-containing foods we eat and those synthesized from other compounds within the body. Proteins can also be broken down to *supply energy* for the body—on average, 4 kcal per gram.

If you fail to consume an adequate amount of protein for weeks at a time, many metabolic processes slow down. This is because the body does not have enough amino acids available to build the proteins it needs. For example, the immune system no longer functions efficiently when it lacks key proteins, thereby increasing the risk of infections, disease, and death.

Amino acids are formed mostly of carbon, hydrogen, oxygen, and nitrogen. Figure 6-1 shows the structure of a generic amino acid and two examples of specific amino acids. The amino acids are slight variations of the generic amino acid pictured but all have different chemical makeups (see Appendix D). Each amino acid has an "acid" group, an "amino" group, and a "side" or R group specific to the amino acid.

The R group on some amino acids has a branched shape, like a tree. These so-called **branched-chain amino acids** are leucine, isoleucine, and valine. The branched-chain amino acids are the primary amino acids used by muscles for energy needs. Whey protein (from milk) is popular among strength-training athletes because it is particularly rich in branched-chain amino acids (see Chapter 10).

Your body uses 20 different amino acids to function (Table 6-1). Although all these commonly found amino acids are important, 11 (alanine, arginine, asparagine, aspartic acid, cysteine, glutamic acid, glutamine, glycine, proline, serine, and tyrosine) are considered **nonessential** with respect to our diets. All tissues have

TABLE 6-1 ▶ Classification of Amino Acids

Essential Amino Acids	Nonessential Amino Acids
Histidine	Alanine
Isoleucine*	Arginine
Leucine*	Asparagine
Lysine	Aspartic acid
Methionine	Cysteine
Phenylalanine	Glutamic acid
Threonine	Glutamine
Tryptophan	Glycine
Valine*	Proline
	Serine
	Tyrosine

*A branched-chain amino acid.

branched-chain amino acids Amino acids with a branching carbon backbone; these are leucine, isoleucine, and valine. All are essential amino acids.

nonessential amino acids Amino acids that can be synthesized by a healthy body in sufficient amounts; there are 11 nonessential amino acids. These are also called *dispensable amino acids*.

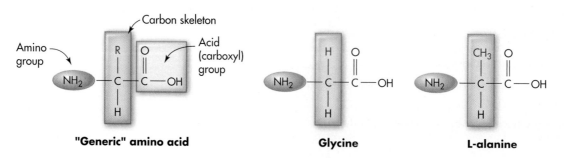

FIGURE 6-1 ▲ Amino acid structure. The side chain (R) differentiates glycine (H) and alanine (CH₃).

some ability for synthesis of the nonessential amino acids as long as the right ingredients are present—the key factor being nitrogen that is already part of another amino acid. Therefore, it is not essential that these amino acids be in our diet.

ESSENTIAL AMINO ACIDS

The nine amino acids (histidine, isoleucine, leucine, lysine, methionine, phenylalanine, threonine, tryptophan, and valine) the body cannot make in sufficient amounts or at all are known as **essential** because they must be obtained from foods. This is because body cells cannot make the needed carbon-based foundation of the amino acid, cannot put a nitrogen group on the needed carbon-based foundation, or just cannot do the whole process fast enough to meet body needs.

Eating a balanced diet can supply us with both the essential and nonessential amino-acid building blocks needed to maintain good health. Both nonessential and essential amino acids are present in foods that contain protein. If you do not eat enough essential amino acids, your body first struggles to conserve what essential amino acids it can. However, eventually your body slows production of new proteins until at some point, you will break protein down faster than you can make it. When that happens, health deteriorates.

The essential amino acid in smallest supply in a food or diet in relation to body needs becomes the limiting factor (called the **limiting amino acid**) because it limits the amount of protein the body can synthesize. Adults need only about 11% of their total protein requirement to be supplied by essential amino acids. Typically, 50% of the amino acids in dietary proteins are essential.

Because of the needs of rapid growth and development, the estimated requirements for essential amino acids for infants and preschool children are greater (40% of total protein intake); however, in later childhood, the need drops to 20%. Diets designed for infants and young children need to take this into account to make sure enough proteins are present to yield sufficient high-quality protein. Including some animal products in the diet, such as human milk or formula for infants, or cow's milk for children, helps ensure this. A major health risk for infants and children occurs in famine situations in which only one type of cereal grain is available, increasing the probability that one or more of the nine essential amino acids are lacking in the total diet. This is discussed further in Section 6.8 on protein–calorie malnutrition (PCM).

Conditionally Essential Amino Acids. Some of the nonessential amino acids, which are usually synthesized in the body, can become essential during times of rapid growth, disease, or metabolic stress. For example, patients recovering from surgery or burns have very high needs for amino acids to heal their injuries, so the activity of enzymes that synthesize nonessential amino acids cannot keep up with demands. Arginine and glutamine are two examples of **conditionally essential amino acids.**

The disease phenylketonuria (PKU) illustrates how a nonessential amino acid can become conditionally essential as a result of a genetic disease. A person with PKU has a limited ability to metabolize the essential amino acid phenylalanine. Normally, the body uses an enzyme to convert much of our dietary phenylalanine intake into tyrosine. In PKU-diagnosed persons, the activity of the enzyme used in processing phenylalanine to tyrosine is insufficient. The results are that (1) tyrosine becomes essential (it must be obtained from the diet), whereas (2) phenylalanine builds up to toxic levels in the blood. Elevated phenylalanine disrupts brain function, leading to mental retardation. PKU is treated with a special diet that limits phenylalanine.

▲ Soy products such as soy milk, tofu, soy bread, and soybeans themselves provide a plant source of all the essential amino acids.

essential amino acids The amino acids that cannot be synthesized by humans in sufficient amounts or at all and therefore must be included in the diet; there are nine essential amino acids. These are also called *indispensable amino acids*.

limiting amino acid The essential amino acid in lowest concentration in a food or diet relative to body needs.

conditionally essential amino acids Amino acids that must be made from essential amino acids if insufficient amounts are eaten.

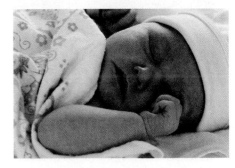

▲ Within the first few days of life, all newborns are tested for phenylketonuria.

✔ CONCEPT CHECK 6.1

1. What is the basic structure of an amino acid?
2. What is the difference between the essential and nonessestial amino acids?

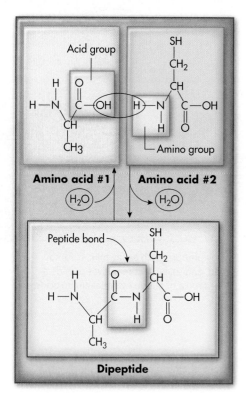

Amino acid #1   **Amino acid #2**

Dipeptide

FIGURE 6-2 ▲ Peptide bonds link amino acids. The synthesis of a peptide bond is a dehydration reaction in which a molecule of water is removed. When peptide bonds are broken (as in digestion), a molecule is added (hydrolysis).

peptide bond A chemical bond formed between amino acids in a protein.

polypeptide A group of amino acids bonded together, from 50 to 2000 or more.

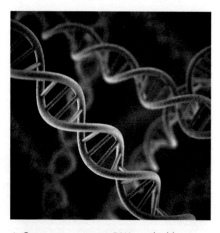

▲ Genes are present on DNA—a double-stranded helix. The cell nucleus contains most of the DNA in the body. DNA can be compared to a "cookbook" that contains the "recipes" (genes) for the synthesis of all proteins that cells need to make. Because the cookbook cannot leave the nucleus to get to the "cooks" (the ribosomes), the recipes for proteins must be transcribed into a form (mRNA) that can leave the nucleus. The amino acids that form a protein can be equated to the "ingredients" required for the recipe.

6.2 Protein Synthesis and Organization

Within body cells, amino acids are linked together by chemical bonds—technically called **peptide bonds**—to form proteins (Fig. 6-2). Peptide bonds form between the amino group of one amino acid and the acid (carboxyl) group of another. Through peptide bonding of amino acids, cells can synthesize dipeptides (joining of two amino acids), tripeptides (joining of three amino acids), oligopeptides (joining of four to nine amino acids), and **polypeptides** (joining of 10 or more amino acids). Most proteins are polypeptides ranging from about 50 to 2000 amino acids. These bonds are difficult to break, but heat, acids, enzymes, and other agents are able to do so during cooking and chemical digestion.

The body can synthesize many different proteins by linking together the 20 common types of amino acids with peptide bonds.

PROTEIN SYNTHESIS

Our discussion of protein synthesis begins with DNA. DNA is present in the nucleus of the cell and contains coded instructions for protein synthesis (i.e., which specific amino acids are to be placed in a protein and in which order). Recall from Chapter 3 that DNA is a double-stranded molecule.

Protein synthesis in a cell, however, takes place in the cytoplasm, not in the nucleus. Thus, the DNA code used for synthesis of a specific protein must be transferred from the nucleus to the cytoplasm to allow for protein synthesis. This transfer is the job of messenger RNA (mRNA). Enzymes in the nucleus read the code (a gene) on a DNA base sequence that encodes one or more proteins and *transcribe* that information into a single-stranded mRNA molecule (Fig. 6-3) that is ready to leave the nucleus.

Once in the cytoplasm, mRNA travels to the ribosomes. The ribosomes read the mRNA code and *translate* those instructions to produce a specific protein. Amino acids are added one at a time to the growing polypeptide chain according to the instructions on the mRNA. Another key participant in protein synthesis, transfer RNA (tRNA), is responsible for bringing the specific amino acids to the ribosomes as needed during protein synthesis (review Fig. 6-3). Energy input is required to add each amino acid to the chain, making protein synthesis "costly" in terms of calorie use.

Once synthesis of a polypeptide is complete, it twists and folds into the appropriate three-dimensional structure of the intended protein. These structural changes occur based on specific interactions between the amino acids that make up the polypeptide chain. Some polypeptides, such as the hormone insulin, also undergo further changes in the cell before they are functional.

PROTEIN ORGANIZATION

By bonding together various combinations of the 20 common types of amino acids, the body synthesizes thousands of different proteins. The sequential order of the amino acids then ultimately determines the protein's shape. The main point is that only correctly positioned amino acids can interact and fold properly to form the intended shape for the protein. The resulting unique, three-dimensional form, such as that shown for the protein hemoglobin in Figure 6-4, goes on to dictate the function of each particular protein. If it lacks the proper structure, a protein cannot function.

The relationship between DNA and the proteins eventually produced by a cell is very important. If the DNA code contains errors, an incorrect mRNA will be produced. The ribosomes will then read this incorrect message and an incorrect amino acid will be added and an incorrect polypeptide chain will be produced. Genetic engineering may ultimately be able to correct many gene defects in humans by placing the correct DNA code in the nucleus, so that the correct protein can be made by the ribosomes.

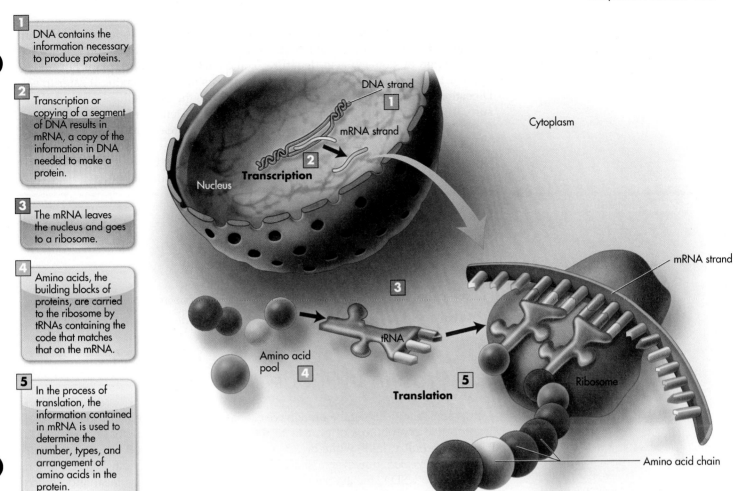

1. DNA contains the information necessary to produce proteins.

2. Transcription or copying of a segment of DNA results in mRNA, a copy of the information in DNA needed to make a protein.

3. The mRNA leaves the nucleus and goes to a ribosome.

4. Amino acids, the building blocks of proteins, are carried to the ribosome by tRNAs containing the code that matches that on the mRNA.

5. In the process of translation, the information contained in mRNA is used to determine the number, types, and arrangement of amino acids in the protein.

FIGURE 6-3 ▲ Protein synthesis (simplified). Once the mRNA is fully read, the amino acids have been connected into the polypeptide, which is released into the cytoplasm. It generally is then processed further to become a cell protein.

Sickle cell disease (also called **sickle cell anemia**) is one example of what happens when amino acids are out of order on a protein (Fig. 6-5). North Americans of African descent are especially prone to this genetic disease. Sickle cell anemia is not a nutritional disease but is caused by a mutation in the genetic code for hemoglobin, the protein depicted in Figure 6-4 that carries oxygen in red blood cells. The mutation causes the amino acid glutamic acid to be replaced with the amino acid valine. This error produces a profound change in hemoglobin structure. It can no longer form the shape needed to carry oxygen efficiently inside the red blood cell. Instead of forming normal circular disks, the red blood cells collapse into crescent (or sickle) shapes (Fig. 6-5). Sickle red blood cells become hard and sticky, which causes them to clog blood flow and break apart. This can cause severe bone and joint pain, abdominal pain, headache, convulsions, paralysis, and even death due to the lack of oxygen.

These life-threatening symptoms are caused by a minute, but critical, error in amino-acid order. Why does this error happen? It results from a defect in a person's genetic blueprint, DNA, inherited from one's parents. A defect in the DNA can dictate that a wrong amino acid will be built into the sequence of the body proteins. Many diseases, including cancer, stem from errors in the DNA code.

DENATURATION OF PROTEINS

Exposure to acid or alkaline substances, heat, or agitation (e.g., whipping egg whites) can alter a protein's structure, leaving it uncoiled or otherwise deformed. This process of altering the three-dimensional structure of a protein is called **denaturation** (see Fig. 6-11 on page 210). Changing a protein's shape often destroys its ability to function normally, such that it loses its biological activity.

sickle cell disease (sickle cell anemia) An illness that results from a malformation of the red blood cell because of an incorrect structure in part of its hemoglobin protein chains.

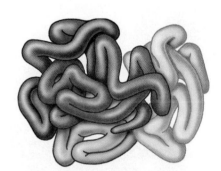

FIGURE 6-4 ▲ Protein organization. Proteins often form a coiled shape, as shown by this drawing of the blood protein hemoglobin. This shape is dictated by the order of the amino acids in the protein chain. To get an idea of its size, consider that each teaspoon (5 milliliters) of blood contains about 10^{18} hemoglobin molecules. (One billion is 10^9.)

denaturation Alteration of a protein's three-dimensional structure, usually because of treatment by heat, enzymes, acid or alkaline solutions, or agitation.

Denaturation of proteins is useful for some body processes, especially digestion. The heat produced during cooking denatures some proteins. After food is ingested, the secretion of stomach acid denatures some bacterial proteins, plant hormones, many active enzymes, and other forms of proteins in foods, making it safer to eat. Digestion is also enhanced by denaturation because the unraveling increases exposure of the polypeptide chain to digestive enzymes. Denaturing proteins in some foods can also reduce their tendencies to cause allergic reactions.

Recognize that we need the essential amino acids that the proteins in the diet supply—not the proteins themselves. We dismantle ingested dietary proteins and use the amino-acid building blocks to assemble the proteins we need.

✓ CONCEPT CHECK 6.2

1. What is the role of DNA in protein synthesis?
2. What are the steps of protein synthesis?
3. Why is the amino acid order within a protein important?
4. What are some of the ways a protein can become denatured?

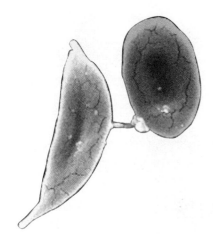

FIGURE 6-5 ▲ An example of the consequences of errors in DNA coding of proteins. A normal circular red blood cell is shown along with an abnormal sickle-shaped red blood cell.

6.3 Protein in Foods

For decades, about 70% of the protein we eat has come from animal sources (Fig. 6-6). The top five contributors of protein to the North American diet have traditionally been beef, poultry, milk, white bread, and cheese (Fig. 6-7). Food consumption surveys have revealed, however, a sharp drop in U.S. meat and poultry consumption. According to USDA, the average American consumed 12.2% less meat and poultry in 2012 than in 2007. More specifically, beef consumption has dropped steadily over the past two decades. In 2011, U.S. beef consumption per person was 57.4 pounds, which was down 13% from 10 years ago and down about 25% from 1980. Decreases in chicken and pork consumption have also occurred over the past 5 years. The downward trend also includes eggs and milk, and seems to be a result of the rising cost of these foods as well as conscious decisions to eat less meat for health, environmental, and social reasons.

While the consumption of animal products has declined in the United States, world meat and dairy consumption doubled between 1950 and 2009, and is set to increase by an additional factor of four over the next 40 years (see Further Reading 7). It is particularly interesting that the demand for meat, eggs, and dairy products has increased in developing nations at a dramatic rate. The United Nations Food and Agriculture Organization (FAO) reports that in developing countries, milk consumption per person has almost doubled between 1980 and 2005, meat consumption more than tripled, and egg consumption increased fivefold. The greatest increases in consumption are occurring in East and Southeast Asia. For example, China's per capita milk consumption increased tenfold (2.3 kilograms to 23.3 kilograms between 1980 and 2005), while per capita meat consumption quadrupled. India has also boosted its milk production and consumption and is now the largest milk producer in the world. Overall, these increases in demand and consumption have had a substantial impact on agriculture, because of the significant increase (23%) in the farm animal population over the past three decades. The "factory farms" that have developed have a significant environmental impact because of the large amount of land and water used, and the high level of waste produced. These negative effects are in addition to the adverse health implications of consuming a diet high in animal products (see Further Reading 8). It is predicted that the increasing intake of meat and dairy will have profound impacts on our agricultural

▲ This grilled chicken (protein) sandwich with lettuce and tomato (vegetables) on foccacia bread (grains) completes three section of MyPlate. Which sections are missing?

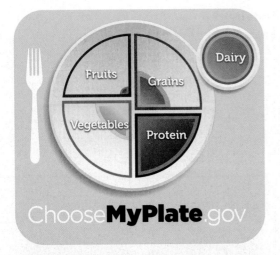

MyPlate:
Sources of Protein

Grains

- Bread
- Breakfast cereals
- Rice
- Noodles

2–3 grams per serving

Vegetables

- Carrots
- Corn
- Broccoli

2–3 grams per serving

Fruits

- Apples
- Oranges
- Bananas

<1 gram per serving

Dairy

- Milk
- Yogurt
- Cheese

8–10 grams per serving

Protein

- Meat
- Eggs
- Fish
- Dry beans
- Nuts

7 grams per serving

FIGURE 6-6 ▲ Sources of protein from MyPlate. The fill of the background color (none, 1/3, 2/3, or completely covered) within each group on the plate indicates the average nutrient density for protein in that group. Overall, the dairy group and the protein group contain many foods that are nutrient-dense sources of protein. Based on serving sizes listed for MyPlate, the fruits group provides little or no protein (less than 1 gram per serving). Food choices from the vegetables group and grains group provide moderate amounts of protein (2 to 3 grams per serving). The dairy group provides much protein (8 to 10 grams per serving), as does the protein group (7 grams per serving).

and ecological resources, and our ability to provide food in the future. Because the Mediterranean diet is recognized as being low in meat, rich in fresh fruit and vegetables, and low in added sugar and saturated fatty acids, it has been recommended as a sustainable alternative dietary pattern. It has been advocated by the FAO and could be a starting point for discussing the creation of policies to support sustainable food production (see Further Reading 2).

PROTEIN QUALITY OF FOODS

Animal and plant proteins can differ greatly in their proportions of essential and nonessential amino acids. Animal proteins contain ample amounts of all nine essential amino acids. (Gelatin—made from the animal protein collagen—is an exception because it loses one essential amino acid during processing and is low in other essential amino acids.) With the exceptions of soy protein and quinoa seed, plant proteins do not match our need for essential amino acids as precisely as animal proteins. Many plant proteins, especially those found in grains, are low in one or more of the nine essential amino acids.

As you might expect, humans are able to use proteins from any single animal source more efficiently to support growth and maintenance than from any

▲ Small amounts of animal protein in a meal easily add up to meet daily protein needs when mixed with grains and vegetables.

FIGURE 6-7 ▶ Food sources of protein compared to the RDA of 56 grams for a 70-kilogram man.

Food Sources of Protein

Food Item and Amount	Protein (grams)	% RDA
RDA	56*	100%
Canned tuna, 3 ounces	21.6	38.6%
Broiled chicken, 3 ounces	21.3	38%
Roast beef, 3 ounces	15.3	27%
Yogurt, 1 cup	10.6	19%
Kidney beans, ½ cup	8.1	14.5%
1% low-fat milk, 1 cup	8.0	14%
Peanuts, 1 ounce	7.3	13%
Cheddar cheese, 1 ounce	7.0	12.5%
Egg, 1	5.5	10%
Cooked corn, ½ cup	2.7	5%
Seven-grain bread, 1 slice	2.6	4.6%
White rice, ½ cup	2.1	4%
Pasta, 1 ounce	1.2	2%
Banana, 1	1.2	2%

* for 70 kilogram man

Key:
- Grains
- Vegetables
- Fruits
- Dairy
- Protein

ChooseMyPlate.gov

high-quality (complete) proteins Dietary proteins that contain ample amounts of all nine essential amino acids.

lower-quality (incomplete) proteins Dietary proteins that are low in or lack one or more essential amino acids.

CRITICAL THINKING

Evan, a vegetarian, has heard of the "all-or-none principle" of protein synthesis but does not understand how this principle applies to protein synthesis in the body. He asks you, "How important is this nutritional concept for diet planning?" How would you answer his question?

complementary proteins Two food protein sources that make up for each other's inadequate supply of specific essential amino acids; together, they yield a sufficient amount of all nine and so provide high-quality (complete) protein for the diet.

single plant source. For this reason, animal proteins (except gelatin) are considered **high-quality** (also called **complete**) **proteins,** which contain sufficient amounts of the nine essential amino acids. Individual plant sources of proteins (except for soy beans and quinoa seed) are considered **lower-quality** (also called **incomplete**) **proteins** because their amino-acid patterns can be quite different from ours. Thus, a single plant protein source, such as corn alone, cannot easily support body growth and maintenance. To obtain a sufficient amount of essential amino acids, a variety of plant proteins needs to be consumed because each plant protein lacks adequate amounts of one or more essential amino acids.

When only lower-quality protein foods are consumed, an adequate amount of the essential amino acids needed for protein synthesis may not be obtained. Therefore, when compared to high-quality proteins, a greater amount of lower-quality protein is needed to meet the demands of protein synthesis. Moreover, once any of the nine essential amino acids in the plant protein we have eaten is used up, further protein synthesis becomes impossible. Because the depletion of just one of the essential amino acids prevents protein synthesis, the process illustrates the *all-or-none principle:* Either all essential amino acids are available, or none can be used. The remaining amino acids would then be used for energy needs or converted in to carbohydrate or fat.

When two or more proteins combine to compensate for deficiencies in essential amino acid content in each protein, the proteins are called **complementary proteins.** Mixed diets generally provide high-quality protein because a complementary protein pattern results. Many legumes, for example, are deficient in the essential amino acid methionine, whereas grains are limited in lysine. Eating a combination of legumes and grains, such as beans and rice, will supply the body with adequate amounts of all essential amino acids (Fig. 6-8). Likewise vegetables, which are limited in methionine, can be combined with nuts, which are limited in lysine. Therefore, healthy adults should have little concern about balancing foods to obtain enough of all nine essential amino acids. Even on plant-based diets, complementary proteins need not be consumed at the same meal by adults. Meeting amino-acid needs over the course of a day is a reasonable goal because there is a ready supply of amino acids from those present in body cells and in the blood (see Fig. 6-14 on page 215).

In general, plant sources of protein deserve more attention and use than they currently receive from many North Americans. Plant foods contribute fewer calories to the diet than most animal products, and they supply an ample amount of protein (Fig. 6-9). Vegetable sources of proteins, especially legumes and nuts, are a heart-healthy alternative to animal proteins because they contain very little cholesterol and saturated fat, aside from that added during processing or cooking.

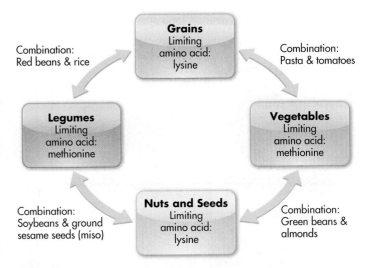

FIGURE 6-8 ▲ Plant group combinations in which the proteins complement each other in a meal based on their limiting amino acids.

A CLOSER LOOK AT PLANT SOURCES OF PROTEINS

Per gram of protein, plant foods provide more magnesium, fiber, folate, vitamin E, iron (absorption is increased by the vitamin C also present), zinc, and some calcium than animal sources of protein. In addition, phytochemicals from these foods are implicated in prevention of a wide variety of chronic diseases.

Legumes are a plant family with pods that contain a single row of seeds. Examples include garden and black-eyed peas, kidney beans, great northern beans, lentils, soy beans, and peanuts. Dried varieties of the mature legume seeds—what we know as beans—also make an impressive contribution to the protein, vitamin, mineral, and fiber content of a meal. A ½ cup serving of legumes provides 100 to 150 kcal, 5 to 10 grams of protein, less than 1 gram of fat, and about 5 grams of fiber. Recall from Chapter 4 that consumption of beans can lead to intestinal gas because our bodies lack the enzymes to break down certain carbohydrates that beans contain. It is helpful to soak dry beans in water, which leaches the indigestible carbohydrates into the water so they can be disposed. However, intestinal gas is not harmful. In fact, fermentation products of ingestible carbohydrates promote the health of your colon.

Nuts and seeds are also excellent sources of plant protein. Commonly consumed nuts include almonds, pistachios, walnuts, and pecans. The defining characteristic of a nut is that it grows on a tree. Remember that peanuts, because they grow underground, are legumes. Seeds, including pumpkin, sesame, and sunflower seeds, are similar to nuts in nutrient composition. A 1-ounce serving of nuts or seeds generally supplies 160 to 190 kcal, 6 to 10 grams of protein, and 14 to 19 grams of fat. Although they are a dense source of calories, nuts and seeds make a powerful contribution to health when consumed in moderation.

In sum, plant proteins are a nutritious alternative to animal proteins. They are inexpensive, versatile, tasty, add color to your plate, and benefit health beyond their contribution of protein to the diet. Learning to substitute plant proteins in place of less healthy foods is one way to reduce your risk for many diseases. The impact of plant proteins on health will be discussed in the Nutrition and Your Health section at the end of this chapter.

FIGURE 6-9 ▲ Legumes are rich sources of protein. These kidney beans provide a significant amount of protein per serving: 7 grams per 1/2 cup.

FIGURE 6-10 ▼ Most common food allergens.

Peanut/tree nuts

Milk products

Soy

Wheat

Eggs

Fish/shellfish

FOOD PROTEIN ALLERGIES

Allergies occur when the immune system reacts to what it thinks is a foreign protein. In the case of food allergies, the immune system mistakes a food protein for a harmful invader. For some of us, food allergies to soy, peanuts and tree nuts (e.g., almonds and walnuts), and wheat are a concern. Overall, food allergies occur in up to 8% of children 4 years of age or younger and in up to 2% of adults. Eight foods account for 90% of food-related allergies; soy, peanuts, tree nuts, and wheat are four of these foods. (The other foods are milk, eggs, fish, and shellfish; Fig. 6-10.) The allergic reactions can range from a mild intolerance to fatal allergic reactions. Introducing allergenic foods such as peanut butter and eggs to infants as young as 4 to 6 months old is a new approach to combating food allergies that will be discussed further in Chapter 15.

✔ CONCEPT CHECK 6.3

1. What types of foods contain high-quality proteins?
2. Why are complementary proteins important when pairing plant food sources?

pepsin A protein-digesting enzyme produced by the stomach.

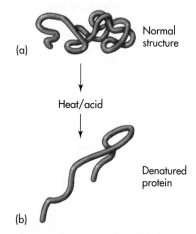

FIGURE 6-11 ▲ Denaturation.
(a) Protein showing typical coiled state.
(b) Protein is now partly uncoiled. This uncoiling can reduce biological activity and allow digestive enzymes to act on peptide bonds.

6.4 Protein Digestion and Absorption

As with carbohydrate digestion, protein digestion begins with the cooking of food. Cooking unfolds (denatures) proteins (Fig. 6-11) and softens tough connective tissue in meat. Cooking also makes many protein-rich foods easier to chew and swallow, and facilitates their breakdown during later digestion and absorption. Cooking also makes many protein-rich foods, such as meats, eggs, fish, and poultry, much safer to eat.

DIGESTION

The enzymatic digestion of protein begins in the stomach (Fig. 6-12). Proteins are first denatured by stomach acid. **Pepsin,** a major stomach enzyme for digesting proteins, then goes to work on the unraveled polypeptide chains. Pepsin breaks the polypeptide into shorter chains of amino acids because it can break only a few of the many peptide bonds found in these large molecules. The release of pepsin is controlled by the hormone gastrin. Thinking about food or chewing food stimulates gastrin release in the stomach. Gastrin also strongly stimulates the stomach to produce acid.

The partially digested proteins move from the stomach into the small intestine along with the rest of the nutrients and other substances in a meal (chyme). Once

FIGURE 6-12 ▶ A summary of protein digestion and absorption. Enzymatic protein digestion begins in the stomach and ends in the absorptive cells of the small intestine, where any remaining short groupings of amino acids are broken down into single amino acids. Stomach acid and enzymes contribute to protein digestion. Absorption from the intestinal lumen into the absorptive cells requires energy input.

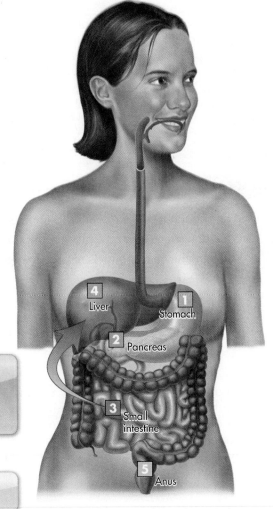

Protein Digestion and Absorption

1 Partial protein digestion by the enzyme pepsin and stomach acid.

2 Pancreas releases enzymes that will further digest polypeptides in the small intestine.

3 Final digestion of peptides into single amino acids takes place mostly inside cells of the small intestine. Once in the small intestine, the partially digested proteins (and fats) trigger the release of the hormone cholecystokinin (CCK) from the walls of the small intestine. CCK causes the pancreas to release protein-splitting enzymes, such as trypsin.

4 Liver

1 Stomach

2 Pancreas

3 Small intestine

5 Anus

4 Amino acids absorbed into the portal vein and transported to the liver. From there, they enter the general bloodstream.

5 Little dietary protein is present in feces.

in the small intestine, the partially digested proteins (and any fats accompanying them) trigger the release of the hormone cholecystokinin (CCK) from the walls of the small intestine. CCK, in turn, travels through the bloodstream to the pancreas where it causes the pancreas to release protein-splitting enzymes, such as **trypsin.** These digestive enzymes further divide the chains of amino acids into segments of two to three amino acids and some individual amino acids. Eventually, this mixture is digested into amino acids, using other enzymes from the lining of the small intestine and enzymes present in the absorptive cells themselves.

trypsin A protein-digesting enzyme secreted by the pancreas to act in the small intestine.

ABSORPTION

The short chains of amino acids and any individual amino acids in the small intestine are taken up by active transport into the absorptive cells lining the small intestine. Any remaining peptide bonds are broken inside intestinal cells to yield individual amino acids. They are water-soluble, so the amino acids travel to the liver via the portal vein, which drains absorbed nutrients from the intestinal tract. In the liver, individual amino acids can undergo several modifications, depending on the needs of various body tissues. Individual amino acids may be combined into the proteins needed by specific cells; broken down for energy needs; released into the bloodstream; or converted into nonessential amino acids, glucose, or fat. With excess protein intake, amino acids are converted into fat as a last resort.

Gluten Sensitivity

Gluten is a type of protein found in certain grains: wheat, rye, and barley. Protein-digesting enzymes in the GI tract break down some of the peptide bonds in gluten, but digestion is incomplete, leaving some small peptides in addition to individual amino acids. The small peptides that arise from digestion can be absorbed into the cells lining the small intestine. When people with a genetic predisposition for celiac disease are exposed to these small peptides from gluten, they experience an inflammatory reaction. Although many people think celiac disease is a food allergy, it is actually an autoimmune response: the immune system attacks and destroys its own cells. Development of celiac disease depends on two factors: a genetic predisposition and dietary exposure to a protein called gluten. Strict dietary avoidance of food products containing wheat, rye, and barley is the only proven way to manage the disease. (See Further Reading 3 and recall the discussion of celiac disease in Chapter 3. You will learn more about food allergies in Chapter 15.)

protein turnover The process by which cells break down old proteins and resynthesize new proteins. In this way, the cell will have the proteins it needs to function at that time.

▲ Protein contributes to the structure and function of muscle.

Except during infancy, it is uncommon for intact proteins to be absorbed from the digestive tract. In infants up to 4 to 5 months of age, the gastrointestinal tract is somewhat permeable to small proteins, so some whole proteins can be absorbed. Because proteins from some foods (e.g., cow's milk and egg whites) may predispose an infant to food allergies, experts recommend waiting until an infant reaches 4 to 6 months of age to introduce solid foods (see Further Reading 5 and Chapter 15 for details).

✔ CONCEPT CHECK 6.4

1. Where and how does protein digestion begin?
2. What digestion steps take place in the stomach and small intestine?
3. What are the final products of protein digestion, and where do they go after absorption?

6.5 Putting Proteins to Work in the Body

Proteins function in many crucial ways in human metabolism and in the formation of body structures. We rely on foods to supply the amino acids needed to form these proteins. However, only when we also eat enough carbohydrate and fat can food proteins be used most efficiently. If we do not consume enough calories to meet needs, proteins are broken down, allowing cells to harness the energy stored in proteins and rendering the amino acids unavailable to build body proteins.

PRODUCING VITAL BODY STRUCTURES

The amino-acid pool in a cell can be used to form body proteins, as well as a variety of other possible products. Every cell contains protein. Muscles, connective tissue, mucus, blood-clotting factors, transport proteins in the bloodstream, lipoproteins, enzymes, immune antibodies, some hormones, visual pigments, and the support structure inside bones are made of protein. Excess protein in the diet does not enhance the synthesis of these body components, but eating too little protein can prevent it.

Most vital body proteins are in a constant state of breakdown, rebuilding, and repair. For example, the intestinal tract lining is constantly sloughed off. The digestive tract treats sloughed cells just like food particles, digesting them and absorbing their amino acids. In fact, most of the amino acids released throughout the body can be recycled to become part of the pool of amino acids available for the synthesis of future proteins. Overall, **protein turnover** is a process by which a cell can respond to its changing environment by producing proteins that are needed and disassembling proteins that are not needed.

During any day, an adult makes and degrades about 250 grams of protein, recycling many of the amino acids. Relative to the 65 to 100 grams of protein typically consumed by adults in North America, recycled amino acids make an important contribution to total protein metabolism.

If a person's diet is low in protein for a long period, the processes of rebuilding and repairing body proteins will slow down. Over time, skeletal muscles; blood proteins; and vital organs, such as the heart and liver, will decrease in size or volume. Only the brain resists protein breakdown.

Arterial end of a capillary bed

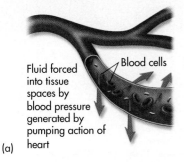

Fluid forced into tissue spaces by blood pressure generated by pumping action of heart

Blood cells

(a)

Venous end of a capillary bed

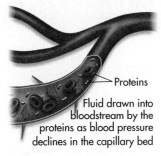

Proteins

Fluid drawn into bloodstream by the proteins as blood pressure declines in the capillary bed

Normal tissue

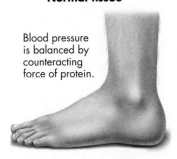

Blood pressure is balanced by counteracting force of protein.

Swollen tissue (edema)

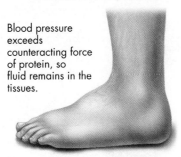

Blood pressure exceeds counteracting force of protein, so fluid remains in the tissues.

(b)

FIGURE 6-13 ◄ Proteins help to maintain fluid balance. As blood is pumped through the circulatory system, some fluid leaks out of the vessels. Normally, proteins in the blood draw water back into the blood vessels. Without sufficient protein in the bloodstream, fluid remains in the tissues and edema develops.

MAINTAINING FLUID BALANCE

Blood proteins help maintain body fluid balance. Normal blood pressure in the arteries forces blood into capillary beds. The blood fluid then moves from the **capillary beds** into the spaces between nearby cells (**extracellular spaces**) to provide nutrients to those cells (Fig. 6-13). Proteins in the bloodstream are too large however, to move out of the capillary beds into the tissues. The presence of these proteins in the capillary beds attracts the proper amount of fluid back to the blood, partially counteracting the force of blood pressure.

With an inadequate consumption of protein, the concentration of proteins in the bloodstream drops below normal. Excessive fluid then builds up in the surrounding tissues because the counteracting force produced by the smaller amount of blood proteins is too weak to pull enough of the fluid back from the tissues into the bloodstream. As fluids accumulate in the tissues, the tissues swell, causing **edema.** Edema may be a symptom of a variety of medical problems, so its cause must be identified. An important step in diagnosing the cause is to measure the concentration of blood proteins.

capillary bed Network of one-cell-thick vessels that create a junction between arterial and venous circulation. It is here that gas and nutrient exchange occurs between body cells and the blood.

extracellular space The space outside cells; represents one-third of body fluid.

edema The buildup of excess fluid in extracellular spaces.

CONTRIBUTING TO ACID–BASE BALANCE

Proteins help regulate acid–base balance in the blood. Proteins located in cell membranes pump chemical ions in and out of cells. The ion concentration that results from the pumping action, among other factors, keeps the blood slightly alkaline. In addition, some blood proteins are especially good **buffers** for the body. Buffers are compounds that maintain acid–base conditions within a narrow range.

buffers Compounds that cause a solution to resist changes in acid–base conditions.

FORMING HORMONES AND ENZYMES

Amino acids are required for the synthesis of many hormones—our internal body messengers. Some hormones, such as the thyroid hormones, are made from only one type of amino acid, tyrosine. Insulin, on the other hand, is a hormone composed of 51 amino acids. Almost all enzymes are proteins or have a protein component.

Neurotransmitters, released by nerve endings, are often derivatives of amino acids. This is true for dopamine and norepinephrine (both synthesized from the amino acid tyrosine) and serotonin (synthesized from the amino acid tryptophan).

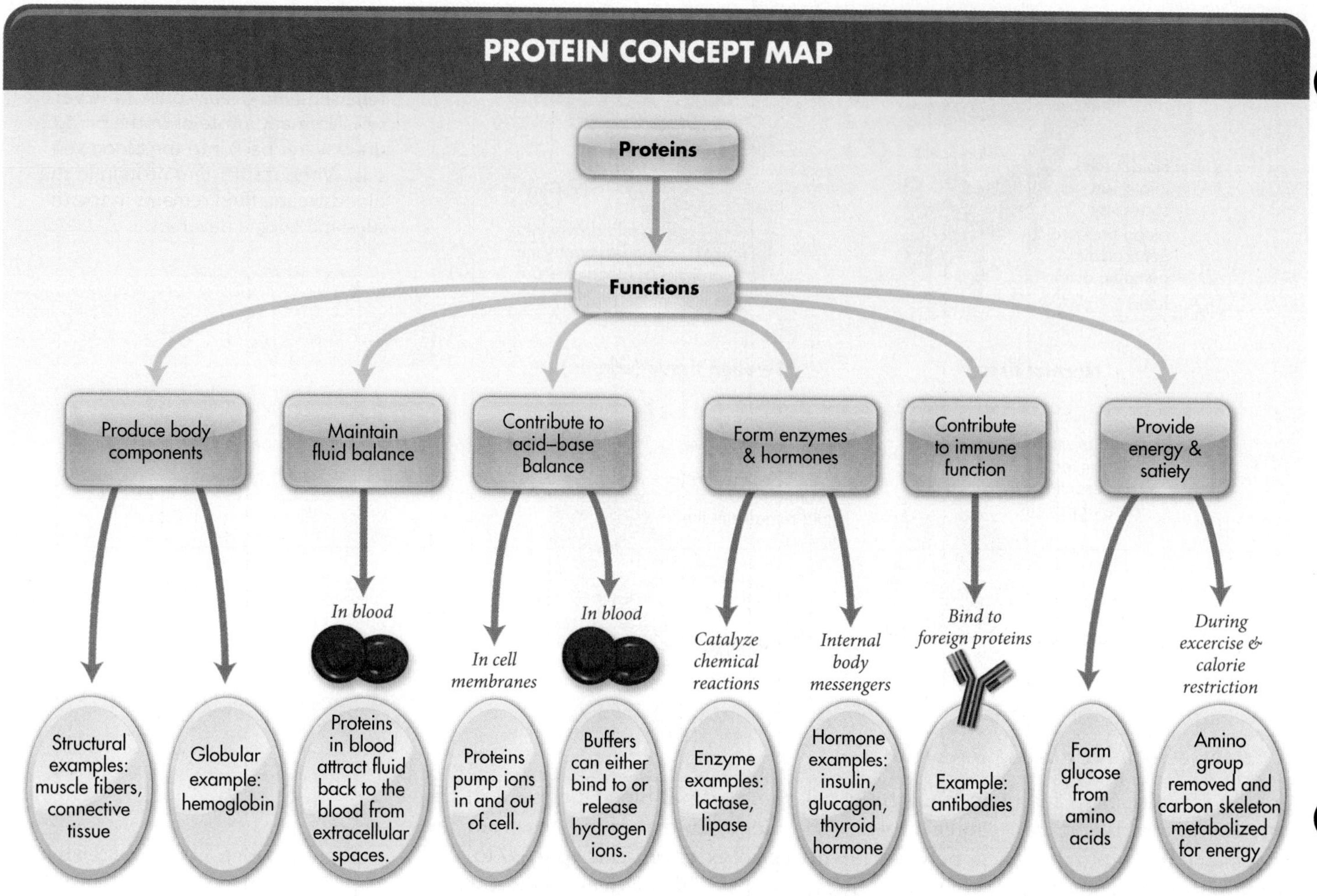

PROTEIN CONCEPT MAP

```
                              ┌──────────┐
                              │ Proteins │
                              └────┬─────┘
                                   │
                              ┌────▼─────┐
                              │Functions │
                              └──────────┘
```

Produce body components	Maintain fluid balance	Contribute to acid–base Balance	Form enzymes & hormones	Contribute to immune function	Provide energy & satiety

In blood

In cell membranes

In blood

Catalyze chemical reactions

Internal body messengers

Bind to foreign proteins

During excercise & calorie restriction

Structural examples: muscle fibers, connective tissue	Globular example: hemoglobin	Proteins in blood attract fluid back to the blood from extracellular spaces.	Proteins pump ions in and out of cell.	Buffers can either bind to or release hydrogen ions.	Enzyme examples: lactase, lipase	Hormone examples: insulin, glucagon, thyroid hormone	Example: antibodies	Form glucose from amino acids	Amino group removed and carbon skeleton metabolized for energy

CONTRIBUTING TO IMMUNE FUNCTION

Proteins are a key component of the cells within the immune system. An example is the antibodies, proteins produced by one type of white blood cell. These antibodies can bind to foreign proteins in the bloodstream an important step in removing invaders from the body. Without sufficient dietary protein, the immune system lacks the materials needed to function properly. For example, a low-protein status can turn measles into a fatal disease for a malnourished child.

FORMING GLUCOSE

The vitamin niacin can be made from the amino acid tryptophan, illustrating another role of proteins.

In Chapter 4, you learned that a fairly constant concentration of glucose must be maintained in the blood to supply energy for the brain, red blood cells, and nervous tissue. At rest, the brain uses about 19% of the body's energy requirements, and it gets most of that energy from glucose. If you do not consume enough carbohydrate to supply the glucose, your liver (and kidneys, to a lesser extent) will be forced to make glucose from amino acids present in body tissues (Fig. 6-14).

Making some glucose from amino acids is normal. For example, when you skip breakfast and have not eaten since 7 P.M. the preceding evening, glucose must be manufactured. In an extreme situation, however, such as in starvation, amino acids from muscle tissue are converted into glucose, which wastes muscle tissue and can produce edema.

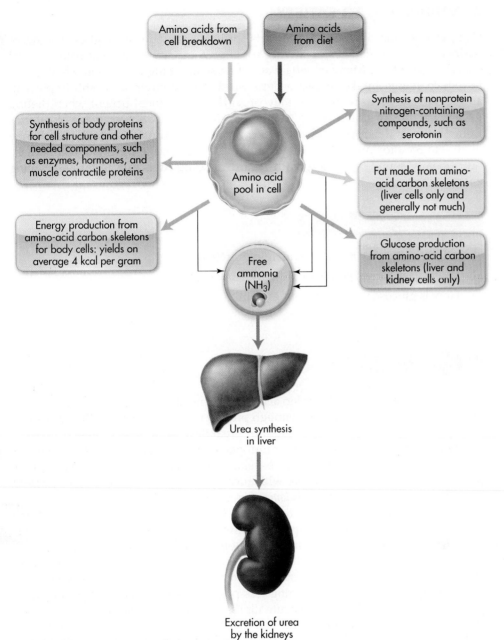

FIGURE 6-14 ◄ Amino-acid metabolism. The amino-acid **pool** in a cell can be used to form body proteins, as well as a variety of other possible products. When the **carbon skeletons** of amino acids are metabolized to produce glucose or fat, ammonia (NH_3) is a resulting waste product. The ammonia is converted in to **urea** and excreted in the urine.

pool The amount of a nutrient stored within the body that can be mobilized when needed.

carbon skeleton Amino-acid structure that remains after the amino group (—NH_2) has been removed.

PROVIDING ENERGY

Proteins supply little energy for a weight-stable person. Two situations in which a person does use protein to meet energy needs are during prolonged exercise and during calorie restriction, as with a weight-loss diet. In these cases, the amino group (—NH_2) from the amino acid is removed, and the remaining carbon skeleton is metabolized for energy needs (Fig. 6-14). When the carbon skeletons of amino acids are metabolized to produce glucose or fat, ammonia (NH_3) is a resulting waste product. The ammonia is converted into **urea** and excreted in the urine. Under most conditions, cells primarily use fats and carbohydrates for energy needs. Although proteins contain the same amount of calories (on average, 4 kcal per gram) as carbohydrates, proteins are a costly source of calories, considering the amount of processing the liver and kidneys must perform to use this calorie source.

urea Nitrogenous waste product of protein metabolism; major source of nitrogen in the urine,

$$\text{chemically } NH_2-\overset{\displaystyle O}{\overset{\displaystyle \|}{C}}-NH_2$$

satiety A state in which there is no longer a desire to eat; a feeling of satisfaction.

protein equilibrium A state in which protein intake is equal to related protein losses; the person is said to be in protein balance.

positive protein balance A state in which protein intake exceeds related protein losses, as is needed during times of growth.

negative protein balance A state in which protein intake is less than related protein losses, as is often seen during acute illness.

The 2010 Dietary Guidlines for Americans in "**Food and Nutrients to Increase**" provide the following recommendations regarding protein intake as part of a healthy eating pattern and while staying within their calorie needs:

- Increase intake of fat-free or low-fat milk and milk products, such as milk, yogurt, cheese, or fortified soy beverages.
- Choose a variety of protein foods, which include seafood, lean meat and poultry, eggs, beans and peas, soy products, and unsalted nuts and seeds.
- Increase the amount and variety of seafood consumed by choosing seafood in place of some meat and poultry.
- Replace protein foods that are higher in fats with choices that are lower in solid fats and calories and/or are sources of oils.

CONTRIBUTING TO SATIETY

Compared to the other macronutrients, proteins provide the highest feeling of **satiety** after a meal. Thus, including some protein with each meal helps control overall food intake. Many experts warn against skimping on protein when trying to reduce energy intake to lose weight. Meeting protein needs is still important, and exceeding needs somewhat may provide an additional benefit when dieting to lose weight. Several effective weight-loss diets include a percentage of calories from protein at the upper end of the Acceptable Macronutrient Distribution Range of 10% to 35% for protein. So in general, these diets are appropriate if otherwise nutritionally sound, especially with regard to being moderate in fat and having enough fiber.

✔ CONCEPT CHECK 6.5

1. Which body constituents are mainly proteins?
2. What is the turnover rate of protein each day?
3. How is glucose produced from protein?

6.6 Protein Needs

How much protein (actually amino acids) do we need to eat each day? People who are not growing need to eat only enough protein to match whatever they lose daily from protein breakdown. The amount of breakdown can be determined by measuring the amount of urea and other nitrogen-containing compounds in the urine, as well as losses of protein from feces, skin, hair, nails, and so on. In short, people need to balance protein intake with such losses to maintain a state of **protein equilibrium,** also called *protein balance* (Fig. 6-15).

When a body is growing or recovering from an illness or injury, it needs a **positive protein balance** to supply the raw materials required to build new tissues. To achieve this, a person must eat more protein daily than he or she loses. In addition, the hormones insulin, growth hormone, and testosterone all stimulate positive protein balance. Resistance exercise (weight training) also enhances positive protein balance. Consuming less protein than needed leads to **negative protein balance,** such as when acute illness reduces the desire to eat and so one loses more protein than consumed.

For healthy people, the amount of dietary protein needed to maintain protein equilibrium (wherein intake equals losses) can be determined by increasing protein intake until it equals losses of protein and its related breakdown products (e.g., urea). Calorie needs must also be met so that amino acids are not diverted for such use.

Today, the best estimate for the amount of protein required for nearly all adults to maintain protein equilibrium is 0.8 grams of protein per kilogram of healthy body weight. This is the RDA for protein. Requirements are higher during periods of growth, such as pregnancy and infancy. Healthy weight is used as a reference in the determination of protein needs because excess fat storage does not contribute much to protein needs (the concept of healthy weight will be discussed in Chapter 7). Calculations using this RDA are shown in the margin and estimate a requirement of about 56 grams of protein daily for a typical 70-kilogram (154-pound) man and about 46 grams of protein daily for a typical 57-kilogram (125-pound) woman.

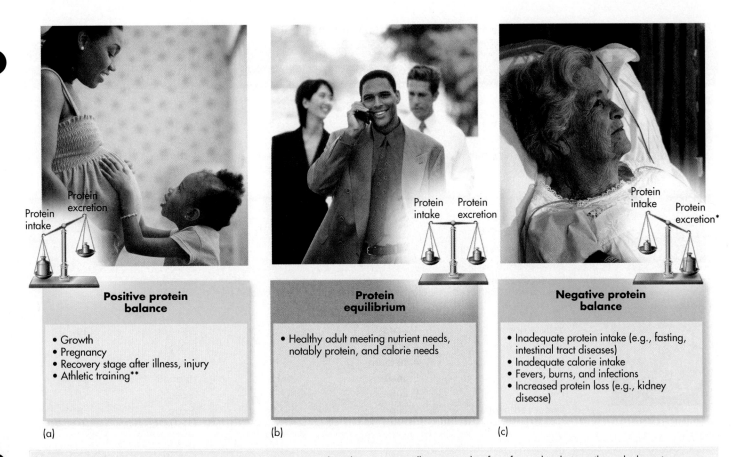

*Based on losses of urea and other nitrogen-containing compounds in the urine, as well as protein lost from feces, skin, hair, nails, and other minor routes.
**Only when additional lean body mass is being gained. Nevertheless, the athlete is probably already eating enough protein to support this extra protein synthesis; protein supplements are not needed.

FIGURE 6-15 ▲ Protein balance in practical terms: (a) positive protein balance, (b) protein equilibrium, and (c) negative protein balance.

The RDA for protein translates into about 10% of total calories. Many experts recommend up to 15% of total calories to provide more flexibility in diet planning and to allow for the variety of protein-rich foods North Americans typically consume. Certain protein-rich foods, such as beans, nuts, seeds, fat-free milk, and seafood, are also part of the "Food and Nutrients to Increase" in the 2010 Dietary Guidelines. As noted earlier, the Food and Nutrition Board has set an upper range for protein intake at 35% of calories consumed. It is easy to meet currently suggested daily protein needs, as given in Table 6-2. On a daily basis, typical North American protein intakes are about 100 grams of protein for men and 65 grams for women. Thus, most of us consume much more protein than the RDA recommends because we like many high-protein foods and can afford to buy them. Our bodies cannot store excess protein once it is consumed, so the excess amino acids are stripped of the nitrogen-containing amino group and may be turned into glucose or fat and then stored as fat or metabolized for energy needs (review Fig. 6-14).

Mental stress, physical labor, and recreational weekend sports activities do not require an increase in the protein RDA. For some highly trained athletes, such as those participating in endurance or strength training, protein consumption may need to exceed the RDA. This is an area of debate in sports nutrition: the Food and Nutrition Board does not support an increased need, but some experts suggest an increase to about 1.7 grams per kilogram. Many North Americans already consume that much protein, especially men.

TABLE 6-2 ▶ Protein Content of Sample Menus Containing 1600 and 2000 kcal

Menu	1600 kcal		2000 kcal	
	Serving Size	Protein (g)	Serving Size	Protein (g)
Breakfast				
Low-fat granola	⅔ cup	5	⅔ cup	5
Blueberries	1 cup	1	1 cup	1
Fat-free (skim) milk	1 cup	8.5	1 cup	8.5
Coffee	1 cup	0	1 cup	0
Lunch				
Broiled chicken breast	3 oz	25	4 oz	33
Salad greens	3 cups	5	3 cups	5
Baked taco shell strips	½ cup	2	½ cup	2
Low-fat salad dressing	2 tbsp	0	2 tbsp	0
Fat-free (skim) milk	1 cup	8.5	1 cup	8.5
Dinner				
Yellow rice	1¼ cups	5	2½ cups	10
Shrimp	4 large	5	6 large	7
Mussels	4 medium	8	6 medium	12
Clams	5 small	12	10 small	24
Peas	¼ cup	2	½ cup	4
Sweet red pepper	¼ cup	0	½ cup	0
Snack				
Muffin	1 small	4	1 small	4
Swiss cheese	1 oz	7.5	1 oz	7.5
Banana	½ small	0.5	½ small	0.5
Total		101		132

✓ CONCEPT CHECK 6.6

1. During what situations is the body in positive protein balance?
2. What is the RDA for protein for a 70-kilogram person?
3. How much protein does the average American consume?

6.7 Does Eating a High-Protein Diet Harm You?

People frequently ask about the potential harm of protein intakes in excess of the RDA. Problems with diets that are high in protein foods primarily stem from the fact that they typically are based on animal sources of protein. Diets rich in animal products will most likely be simultaneously low in the beneficial substances found in plant sources, including fiber, some vitamins (e.g., folate), some minerals (e.g., magnesium), as well as phytochemicals, and high in substances such as saturated fat and cholesterol. A high animal-protein diet, therefore, is unlikely to follow the recommendations of the Dietary Guidelines for Americans or the American Heart Association in terms of reducing risk for cardiovascular disease.

Some, but not all, studies show that high-protein diets can increase calcium losses in urine. For people with adequate calcium intakes, little concern about this relationship is warranted, but keep in mind that calcium is commonly deficient in North American diets.

Meat is one of the richest sources of protein. Excessive intake of red meat, however, especially processed forms, is linked to colon cancer in population studies. Furthermore, an association between red meat consumption and total mortality, as well as premature mortality caused by cardiovascular disease and cancer, has been found in men and women (see Further Readings 6 and 10, and Newsworthy Nutrition below). There are several possible explanations for this connection. The curing agents used to process meats such as ham and salami may cause cancer. Substances that form during cooking of muscle meats at high temperatures may also cause cancer. The excessive fat or low-fiber contents of diets rich in red meat may also be a contributing factor. Because of these concerns, some nutrition experts suggest we focus more on poultry, fish, nuts, legumes, and seeds to meet protein needs. In addition, any red or other type of meat should be trimmed of all visible fat before cooking, especially grilling.

Some researchers have expressed concern that a high-protein intake may overburden the kidneys by forcing them to excrete the extra nitrogen as urea. Additionally, animal proteins may contribute to kidney stone formation in certain individuals. There is some support for limiting protein intake for people in

---

Convert weight from pounds to kilograms:

$$\frac{154 \text{ pounds}}{2.2 \text{ pounds/kilogram}} = 70 \text{ kilograms}$$

$$\frac{125 \text{ pounds}}{2.2 \text{ pounds/kilogram}} = 57 \text{ kilograms}$$

Calculate protein RDA:

$$70 \text{ kilograms} \times \frac{0.8 \text{ grams protein}}{\text{kilogram body weight}} = 56 \text{ grams}$$

$$57 \text{ kilograms} \times \frac{0.8 \text{ grams protein}}{\text{kilogram body weight}} = 46 \text{ grams}$$

Newsworthy Nutrition

Red meat consumption linked to increased premature mortality

Because red meat is a major source of protein and fat in most diets and because consumption of red meat has been associated with an increased risk of several chronic diseases, researchers hypothesized that red meat consumption would be associated with increased premature mortality. The study design was an assessment of meat consumption and causes of death from two large prospective studies that included 37,698 men from the Health Professionals Follow-up Study (1986–2008) and 83,644 women from the Nurses' Health Study (1980–2008) who were free of cardiovascular disease (CVD) and cancer at baseline. During the study, there were 23,926 documented deaths, including 5910 caused by CVD and 9464 caused by cancer. Analysis of red meat intakes revealed that consumption of both unprocessed and processed red meat was associated with an increased risk of total, CVD, and cancer mortality in men and women. The authors were also able to estimate that substitutions of one serving per day of other healthy protein sources (including fish, poultry, nuts, legumes, low-fat dairy, and whole grains) for one serving per day of red meat were associated with a 7% to 19% lower mortality risk. They also estimated that 9.3% of deaths in men and 7.6% in women in these cohorts could be prevented at the end of the study if all the individuals consumed less than one-half serving per day (approximately 42 grams per day) of red meat. These results support the hypothesis that red meat consumption is associated with an increased risk of total mortality and premature mortality from CVD and cancer.

Source: Pan A and others: Red meat consumption and mortality: Results from two prospective cohort studies. *Archives of Internal Medicine* 172:555, 2012 (see Further Reading 10).

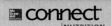 Check out the Connect site **www.mcgrawhillconnect.com** to further explore red meat consumption and heart disease and cancer.

▲ Animal protein foods, such as the roast beef (2 ounces) and Swiss cheese (1 ounce) on this bagel (3.5 inches), are typically our main sources of protein in the North American diet. This sandwich provides 332 kcal and 31 grams of protein.

the early stages of kidney disease because low-protein diets somewhat slow the decline in kidney function. Laboratory animal studies show that moderate protein intakes that just meet nutritional needs preserve kidney function over time better than high-protein diets. Preserving kidney function is especially important for those who have diabetes, early signs of kidney disease, or only one functioning kidney, so a high-protein diet is not recommended for these people. High-protein diets increase urine output, in turn posing a risk for dehydration. This is a special concern for athletes.

AMINO-ACID SUPPLEMENTS

Protein and amino-acid supplements are used primarily by athletes and dieters. Athletes use them hoping they will help build muscle. The branched-chain amino acids described earlier are especially popular with athletes looking to enhance their performance. Dieters have turned to these supplements hoping that they will increase their weight loss. Although the right amount of protein in the diet will aid athletic performance and help in weight control, consuming this protein in the form of amino-acid supplements cannot be considered safe.

Earlier in this chapter, you learned that the body's gastrointestinal system is adapted to handle whole proteins as a dietary source of amino acids. When individual amino-acid supplements are taken, they can overwhelm the absorptive mechanisms in the small intestine, triggering amino-acid imbalances in the intestinal tract. Imbalances occur because groups of chemically similar amino acids compete for absorption sites in the absorptive cells. For example, lysine and arginine are absorbed by the same transporter, so an excess of lysine can impair absorption of arginine. The amino acids most likely to cause toxicity when consumed in large amounts are methionine, cysteine, and histidine. Due to the potential for imbalances and toxicities of individual amino acids, the best advice to ensure adequacy is to stick to whole foods rather than supplements as sources of amino acids. Amino acids as such also have a disagreeable odor and flavor and are much more expensive than food protein. In Canada, the sale of individual amino acids to consumers is banned. Read more about this issue in What the Dietitian Chose at the end of this chapter.

✔ CONCEPT CHECK 6.7

1. What are some of the harmful effects of regularly consuming a high-protein diet?

6.8 Protein–Calorie Malnutrition

Protein deficiency is rarely an isolated condition and usually accompanies a deficiency of calories and other nutrients resulting from insufficient food intake. In the developed world, alcoholism can lead to cases of protein deficiency because of the low protein content of alcoholic beverages that make up a high percent of calories. Protein and calorie malnutrition is a significant problem in hospitals worldwide, affecting patients from infancy through older adulthood. Malnutrition can be caused by the illnesses or injuries for which the patients are admitted to the hospital and by the hospitalization itself (see Further Reading 4). In developing areas of the world, people often have diets low in calories and protein. This state of undernutrition stunts the growth of children and makes them more susceptible to disease throughout life. (Undernutrition is a main focus of Chapter 12.) People who consume too few of their calories

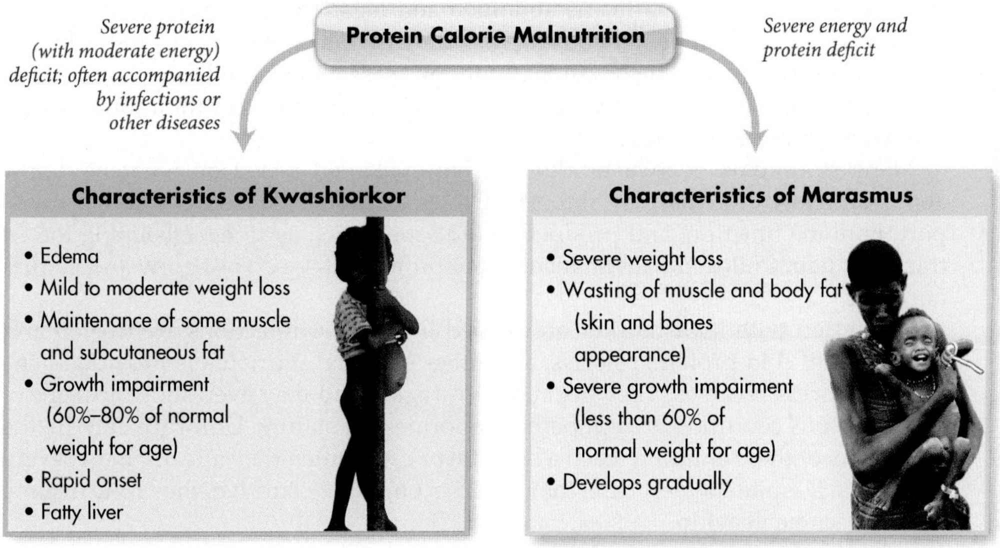

FIGURE 6-16 ◄ Classification of undernutrition in children.

as protein can eventually develop **protein-calorie malnutrition (PCM),** also referred to as *protein–energy malnutrition (PEM)*. In its milder form, it is difficult to tell if a person with PCM is consuming too little calories or protein, or both. When an inadequate intake of nutrients, including protein, is combined with an existing disease, especially an infection, a form of malnutrition called **kwashiorkor** can develop. But if the nutrient deficiency—especially for calories—becomes severe, a deficiency disease called **marasmus** can result. Both conditions are seen primarily in children but also may develop in adults, even in those hospitalized in North America. These two conditions form the tip of the iceberg with respect to states of undernutrition, and symptoms of these two conditions can even be present in the same person (Fig. 6-16).

KWASHIORKOR

Kwashiorkor is a word from Ghana that means "the disease that the first child gets when the new child comes." From birth, an infant in developing areas of the world is usually breastfed. Often by the time the child reaches 1 to 1.5 years of age, the mother is pregnant or has already given birth again, and the newborn infant gets preference for breastfeeding. The older child's diet then abruptly changes from nutritious human milk to starchy roots and **gruels.** These foods have low protein densities relative to their calorie content. Additionally, the foods are usually full of plant fibers, often bulky, making it difficult for the child to consume enough to meet calorie needs. The child generally also has infections, which acutely raise calorie and protein needs. For these reasons, calorie needs of these children are met just barely, at best, and their protein consumption is grossly inadequate, especially in view of the increased amount needed to combat infections. Many vitamin and mineral needs are also far from being fulfilled. Famine victims face similar problems.

The major symptoms of kwashiorkor are apathy, diarrhea, listlessness, failure to grow and gain weight, and withdrawal from the environment. These symptoms complicate other diseases present. For example, a condition such as measles, a disease that normally makes a well-nourished child ill for only a week or so, can become severely debilitating and even fatal. Further symptoms of kwashiorkor are changes in hair color, potassium deficiency, flaky skin, fatty liver, reduced muscle

protein-calorie malnutrition (PCM) A condition resulting from regularly consuming insufficient amounts of calories and protein. The deficiency eventually results in body wasting, primarily of lean tissue, and an increased susceptibility to infections. Also known as *protein–energy malnutrition*.

kwashiorkor A disease occurring primarily in young children who have an existing disease and consume a marginal amount of calories and insufficient protein in relation to needs. The child generally suffers from infections and exhibits edema, poor growth, weakness, and an increased susceptibility to further illness.

marasmus A disease resulting from consuming a grossly insufficient amount of protein and calories; one of the diseases classed as protein–calorie malnutrition. Victims have little or no fat stores, little muscle mass, and poor strength. Death from infections is common.

gruels A thin mixture of grains or legumes in milk or water.

▲ Unsafe water supplies in developing countries contribute to the incidence of marasmus, particularly in bottle-fed infants.

preterm An infant born before 37 weeks of gestation; also referred to as premature.

mass, and massive edema in the abdomen and legs. The presence of edema in a child who has some subcutaneous fat (i.e., directly under the skin) is the hallmark of kwashiorkor (review Fig. 6-16). In addition, these children seldom move. If you pick them up, they do not cry. When you hold them, you feel the plumpness of edema, not muscle and fat tissue.

Many symptoms of kwashiorkor can be explained based on what we know about proteins. Proteins play important roles in fluid balance, lipoprotein transport, immune function, and production of tissues, such as skin, cells lining the GI tract, and hair. Children with an insufficient protein intake do not grow and mature normally.

If children with kwashiorkor are helped in time—if infections are treated and a diet plentiful in protein, calories, and other essential nutrients is provided—the disease process reverses. They begin to grow again and may even show no signs of their previous condition, except perhaps shortness of stature. Unfortunately, by the time many of these children reach a hospital or care center, they already have severe infections. Despite the best care, they still die. Or, if they survive, they return home only to become ill again.

MARASMUS

Marasmus typically occurs as an infant slowly starves to death. It is caused by diets containing minimal amounts of calories, as well as too little protein and other nutrients. As previously noted, the condition is also commonly referred to as *protein-calorie malnutrition,* especially when experienced by older children and adults. The word *marasmus* means "to waste away," in Greek. Victims have a "skin-and-bones" appearance, with little or no subcutaneous fat (review Fig. 6-16).

Marasmus commonly develops in infants who either are not breastfed or have stopped breastfeeding in the early months. When people are poor and sanitation is lacking, bottle feeding often leads to marasmus. Often the weaning formula used is improperly prepared because of unsafe water and because the parents cannot afford sufficient infant formula for the child's needs. The latter problem may lead the parents to dilute the formula to provide more feedings, not realizing that this provides only more water for the infant.

Marasmus in infants commonly occurs in the large cities of poverty-stricken countries. In the cities, bottle feeding is often necessary because the infant must be cared for by others when the mother is working or away from home. An infant with marasmus requires large amounts of calories and protein—like a **preterm** infant—and, unless the child receives them, full recovery from the disease may never occur. The majority of brain growth occurs between conception and the child's first birthday. In fact, the brain is growing at its highest rate after birth. If the diet does not support brain growth during the first months of life, the brain may not grow to its full adult size. This reduced or retarded brain growth may lead to diminished intellectual function. Both kwashiorkor and marasmus plague infants and children; mortality rates in developing countries are often 10 to 20 times higher than in North America.

✔ CONCEPT CHECK 6.8

1. What are the characteristics of kwashiorkor and marasmus?
2. Why are bottle-fed infants at high risk for marasmus in poverty-striken countries?

Nutrition and Your Health
• Vegetarian and Plant-Based Diets

who choose a vegetarian diet can meet their nutritional needs by following a few basic rules and knowledgeably planning their diets.

There are many documented health benefits of following a vegetarian diet. Studies show that death rates from some chronic diseases, such as certain forms of cardiovascular disease, hypertension, many forms of cancer, type 2 diabetes, and obesity, are lower for vegetarians than for non-vegetarians. Vegetarians often live longer, as shown in religious groups that practice vegetarianism. Other factors of healthful lifestyles, such as not smoking, abstaining from alcohol and drugs, and regular physical activity, are typical of vegetarians and probably partially account for the lower risks of chronic disease and longer lives seen in this population.

As you learned in Chapter 2, MyPlate and the 2010 Dietary Guidelines for Americans emphasize a plant-based diet of whole-grain breads and cereals, fruits, and vegetables. In addition, the American Institute for Cancer Research promotes "The New American Plate," which includes plant-based foods covering two-thirds (or more) of the plate and meat, fish, poultry, or low-fat dairy covering only one-third (or less) of the

Vegetarianism has evolved over the centuries from a necessity into an option. It is difficult to know exactly how many Americans follow a vegetarian or plant-based diet. A recent Gallup Poll found that 5% of U.S. adults are self-proclaimed vegetarians, whereas a 2012 study from *Vegetarian Times* found 4% of U.S. adults (9 million people) to be vegetarian with only 2 million being strictly vegan (see Further Reading 11). Vegetarianism is popular among college students. According to a 2009–2010 survey, 12% of college students consider themselves vegetarian.

Vegetarian diets have evolved to include many soy-based meat alternatives. In addition, cookbooks that feature the use of a variety of fruits, vegetables, and seasonings are enhancing food selection for vegetarians of all degrees.

The growing popularity of vegetarian diets has prompted many changes in the marketplace. Many restaurants offer vegetarian meals in response to the growing number of customers who want a vegetarian option when they eat out (see Further Reading 11). Campus dining services offer vegetarian options at every meal.

As nutrition science has grown, new information has enabled the design of nutritionally adequate vegetarian diets. It is important for vegetarians to take advantage of this information because a diet of only plant-based foods has the potential to promote various nutrient deficiencies and substantial growth retardation in infants and children. People

▲ Meatless Monday is a nonprofit initiative that began in 2003 in response to the Healthy People 2010 goal to reduce dietary saturated fat by 15%. The Meatless Monday campaign recommends that we cut meat from our diet on Monday and thus encourages us to increase our consumption of fruits, vegetables, whole grains, and legumes. Ideally, this eating pattern will flow into other days of the week and translate into healthier eating habits (see Further Reading 9).

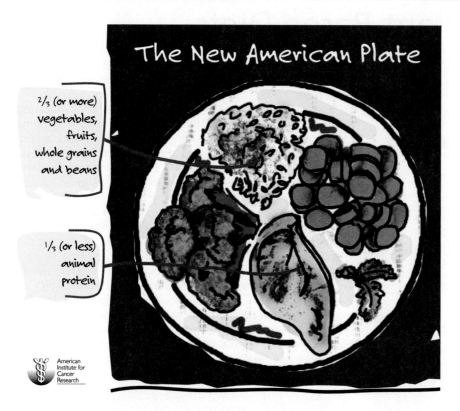

The New American Plate

2/3 (or more) vegetables, fruits, whole grains and beans

1/3 (or less) animal protein

American Institute for Cancer Research

American Institute for Cancer Research Recommendations for Cancer Prevention

1. Be as lean as possible without becoming underweight.

2. Be physically active for at least 30 minutes every day.

3. Avoid sugary drinks. Limit consumption of energy-dense foods.

4. Eat more of a variety of vegetables, fruits, whole grains and legumes such as beans.

5. Limit consumption of red meats (such as beef, pork and lamb) and avoid processed meats.

6. If consumed at all, limit alcoholic drinks to 2 for men and 1 for women a day.

7. Limit consumption of salty foods and foods processed with salt (sodium).

8. Don't use supplements to protect against cancer.

And always remember . . .

Do not smoke or chew tobacco.

"The New American Plate" published by the Amercian Institute for Cancer Research was a forerunner of and embodies the same concepts as the MyPlate program.

Source: **http://www.aicr.org/reduce-your-cancer-risk/recommendations-for-cancer-prevention/** Reprinted with permission from the American Institute for Cancer Research.

plate. Although these recommendations do allow the inclusion of animal products, they are definitely more "vegetarian-like" than typical North American diets.

Why Do People Become Vegetarians?

People choose vegetarianism for a variety of reasons including ethics, religion, economics, and health. Some believe that killing animals for food is unethical. Hindus and Trappist monks eat vegetarian meals as a practice of their religion. In North America, many Seventh Day Adventists base their practice of vegetarianism on biblical texts and believe it is a more healthful way to live.

Some advocates of vegetarianism base their food preference upon the inefficient use of animals as a source of protein. In the United States, nearly 70% of the grain crop is used for animal feed, and globally 35% of the grain harvest is used to produce animal protein. Although animals that humans eat sometimes eat grasses that humans cannot digest, many also eat grains that humans can eat. According to the United Nations Environment Programme, it takes approximately 3 kilograms of grain to produce 1 kilogram of animal protein in cereal-fed animals (see Further Reading 7). An advantage of grass-fed livestock that eat from a pasture is that they convert grass into protein more efficiently than those raised on grains.

People might also practice vegetarianism because it limits saturated fat and cholesterol intake, while encouraging a high intake of complex carbohydrates; vitamins A, E, and C; carotenoids; magnesium; and fiber.

Good for Disease Prevention

Plant sources of proteins can positively impact heart health in several ways. First, the plant foods we eat contain no cholesterol nor *trans* fat and little saturated fat. As noted in Chapter 5, a vegan diet coupled with regular exercise and other lifestyle changes can lead to a reversal of atherosclerotic plaque in various arteries in the body. The major type of fats in plant foods are monounsaturated and polyunsaturated fats. Nuts in particular are high in monounsaturated fat, which helps to keep blood cholesterol low.

Beans and nuts contain soluble fiber, which binds to cholesterol in the small intestine and prevents it from being absorbed by the intestinal cells. Also, due to the activity of some phytochemicals, foods made from soybeans can lower production of cholesterol by the liver. The effect is modest (about a 2% to 6% drop). Since 1999, Food and Drug Administration has allowed health claims for the cholesterol-lowering properties of soy foods, and the American Heart Association has recommended inclusion of some soy protein in the diets of people with high blood cholesterol. As noted in Chapter 2, to list a health claim for soy on the label, a food product must have at least 6.25 grams of soy protein and less than 3 grams of fat, 1 gram of saturated fat, and 20 milligrams of cholesterol per serving.

There are several other heart-protective compounds in plant foods. Some of the phytochemicals may help to prevent blood clots and relax the blood vessels. Nuts are an especially good source of nutrients implicated in heart health, including vitamin E, folate, magnesium, and copper. Frequent consumption of nuts (about 1 ounce of nuts five times per week) is associated with a decreased risk of cardiovascular disease. Recall from Chapter 2 that FDA allows a provisional health claim on food labels to link nuts with a reduced risk of developing cardiovascular disease.

The numerous phytochemicals in plant foods are also thought to aid in preventing cancers of the breast, prostate, and colon. Many of the proposed anticancer effects of foods containing plant protein are through antioxidant mechanisms.

Consumption of plant sources of proteins can aid in prevention of cardiovascular disease and cancer, but there are also other areas for future study. Plants may be particularly good sources of protein for people with diabetes or impaired glucose tolerance because the high fiber content of plant foods leads to a slower increase in blood glucose. Frequent nut consumption may even reduce the risk of developing gallstones, obesity, and type 2 diabetes.

Increasing Plant Proteins in Your Diet

The following are some suggestions for including plant proteins in your diet.

- At your next cookout, try a veggie burger instead of a hamburger. These are usually made from beans and are available in the frozen foods section of the grocery store and come in a variety of delicious flavors. Many restaurants have added veggie burgers to their menus.
- Sprinkle sunflower seeds or chopped almonds on top of your salad to add taste and texture.
- Mix chopped walnuts into the batter of your banana bread to boost your intake of monounsaturated fats.
- Eat soy nuts (oil-roasted soybeans) or edamame as a great snack.
- Spread some peanut butter on your bagel instead of butter or cream cheese.
- Instead of having beef or chicken tacos for dinner, heat up a can of great northern beans in your skillet with one half of a packet of taco seasoning and chopped tomatoes. Use this as a filling in a tortilla shell.
- Consider using soy milk, especially if you have lactose malabsorption or lactose intolerance. Look for varieties fortified with calcium.

Food Planning for Vegetarians

Of the estimated 5% of American adults who call themselves vegetarians, less than 2% are total vegetarians, or **vegans,** who eat only plant foods (and do not use animal products for other purposes, such as leather shoes or feather pillows). **Fruitarians** primarily eat fruits, nuts, honey, and vegetable oils. This plan is not recommended because it can lead to nutrient deficiencies in people of all ages. **Lactovegetarians** are a bit more liberal than vegans; they allow dairy products in their plant-based diet. **Lactoovovegetarians** modify the diet even further and eat dairy products and eggs, as well as plant foods. Including these

▲ Plant proteins, like those in walnuts, can be incorporated into one's diet in numerous ways, such as adding them to banana nut muffins.

animal products makes food planning easier because the dairy and eggs are rich in some nutrients that are missing or minimal in plants, such as vitamin B-12 and calcium. The more variety in the diet, the easier it is to meet nutritional needs. Thus, the practice of eating no animal sources of food significantly separates the vegans and fruitarians from all other semivegetarian styles.

Vegan Diet Planning

Planning a vegan diet requires knowledge and creativity to yield high-quality protein and other key nutrients without animal products. In Section 6.3, you learned about complementary

▲ Keep in mind that amino acids in vegetables are best used when a combination of sources is consumed.

vegan A person who eats only plant foods.

fruitarian A person who primarily eats fruits, nuts, honey, and vegetable oils.

lactovegetarian A person who consumes plant products and dairy products.

lactoovovegetarian A person who consumes plant products, dairy products, and eggs.

▲ A salad containing numerous types of vegetables and legumes is a healthy vegetarian choice.

proteins, whereby the essential amino acids deficient in one protein source are supplied by those of another consumed at the same meal or the next (Fig. 6-8). Many legumes are deficient in the essential amino-acid methionine, whereas grains are limited in lysine. Eating a combination of legumes and grains such as beans and rice, will supply the body with adequate amounts of all essential amino acids (Fig. 6-8). As for any diet, variety is an especially important characteristic of a nutritious vegan diet. Table 6-3 exemplifies vegetarian food plans, which emphasize grains, legumes, nuts, and seeds to help meet protein needs.

Aside from amino acids, low intakes of certain micronutrients can be a problem for the vegan. At the forefront of

nutritional concerns are riboflavin, vitamins D and B-12, iron, zinc, iodide, and calcium. Although use of a balanced multivitamin and mineral supplement can help, the following dietary advice should be implemented.

Riboflavin can be obtained from green leafy vegetables, whole grains, yeast, and legumes—components of most vegan diets. Alternate sources of vitamin D include fortified foods (e.g., margarine), some mushroom varieties, as well as regular sun exposure.

Vitamin B-12 only occurs naturally in animal foods. Plants can contain soil or microbial contaminants that provide trace amounts of vitamin B-12, but these are negligible sources of the vitamin. Because the liver can store vitamin B-12 for about 4 years, it may take a long time after removal of animal foods from the diet for a vitamin B-12 deficiency to surface. If dietary B-12 inadequacy persists, deficiency can lead to anemia, nerve damage, and mental dysfunction. These deficiency consequences have been noted in the infants of vegetarian mothers whose breast milk was low in vitamin B-12. Vegans can prevent a vitamin B-12 deficiency by finding a reliable source of vitamin B-12, such as fortified soybean milk, ready-to-eat breakfast cereals, and special yeast grown on media rich in vitamin B-12.

For iron, the vegan can consume whole grains and ready-to-eat breakfast cereals, dried fruits and nuts, and legumes. The iron in these foods is not absorbed as well as iron in animal foods but consuming these foods with a good source of vitamin C can enhance iron absorption. Cooking in iron pots and skillets can also add iron to the diet.

The vegan can find zinc in whole grains (especially ready-to-eat breakfast cereals), nuts, and legumes, but phytic acid and other substances in these foods limit zinc absorption. Breads are a good source of zinc because the leavening process (rising of the bread dough) reduces the influence of

TABLE 6-3 ▶ **Food Plan for Vegetarians Based on MyPlate**

Food Group	MyPlate Servings		Key Nutrients Supplied ‡
	Lactovegetarian*	Vegan†	
Grains	6–11	8–11	Protein, thiamin, niacin, folate, vitamin E, zinc, magnesium, iron, and fiber
Beans and other legumes	2–3	3	Protein, vitamin B-6, zinc, magnesium, and fiber
Nuts, seeds	2–3	3	Protein, vitamin E, and magnesium
Vegetables	3–5 (include 1 dark-green or leafy variety daily)	4–6 (include 1 dark-green or leafy variety daily)	Vitamin A, vitamin C, folate, vitamin K, potassium, and magnesium
Fruits	2–4	4	Vitamin A, vitamin C, and folate
Dairy	3	____	Protein, riboflavin, vitamin D, vitamin B-12, and calcium
Fortified soy milk	____	3	Protein, riboflavin, vitamin D, vitamin B-12, and calcium

*This plan contains about 75 grams of protein in 1650 kcal.

†This plan contains about 79 grams of protein in 1800 kcal.

‡One serving of vitamin- and mineral-enriched ready-to-eat breakfast cereal is recommended to meet possible nutrient gaps. Alternatively, a balanced multivitamin and mineral supplement can be used. Vegans also may benefit from the use of fortified soy milk to provide calcium, vitamin D, and vitamin B-12.

phytic acid. Iodized salt is a reliable source of iodide. It should be used instead of plain salt, both of which are found in U.S. supermarkets.

Of all nutrients, calcium and vitamin D are the most difficult to consume in sufficient quantities for vegans. Fortified foods including fortified soy milk, fortified orange juice, calcium-rich tofu (check the label), and certain ready-to-eat breakfast cereals and snacks are the vegan's best option for obtaining these nutrients. Green leafy vegetables and nuts also contain calcium, but the mineral is either not well absorbed or not very plentiful from these sources. Dietary supplements are another option. Special diet planning is always required because even a multivitamin and mineral supplement will not supply enough calcium to meet the needs for bone health.

Consuming adequate quantities of omega-3 fatty acids is yet another nutritional concern for vegetarians, especially vegans. Fish and fish oils, abundant sources of these heart-healthy fats, are omitted from many types of vegetarian diets. Alternative plant sources of omega-3 fatty acids include canola oil, soybean oil, seaweed, microalgae, flax seeds, chia seeds, and walnuts.

Special Concerns for Infants and Children

Infants and children, notoriously picky eaters in the first place, are at highest risk for nutrient deficiencies as a result of improperly planned vegetarian and vegan diets. With the use of complementary proteins and good sources of the problem nutrients just discussed, the calorie, protein, vitamin, and mineral needs of vegetarian and vegan infants and children can be met (see Further Reading 1). The most common nutritional concerns for

▲ Children can safely enjoy vegetarian and vegan diets as long as certain adjustments are made to meet their age-specific nutritional needs.

infants and children following vegetarian and vegan diets are deficiencies of iron, vitamin B-12, vitamin D, and calcium.

Vegetarian and vegan diets tend to be high in bulky, high-fiber, low-calorie foods that cause a feeling of fullness. While this is a welcome advantage for most adults, children have small stomach capacity and relatively high nutrient needs compared to their size and therefore may feel full before their calorie needs are met. For this reason, the fiber content of a child's diet may need to be decreased by replacing high-fiber sources with some refined grain products, fruit juices, and peeled fruit. Other concentrated sources of calories for vegetarian and vegan children include fortified soy milk, nuts, dried fruits, and avocados.

CASE STUDY Planning a Vegetarian Diet

Jordan is a freshman in college. He lives in a campus residence hall and teaches martial arts in the afternoon. He eats two or three meals a day at the residence hall cafeteria and snacks between meals. Jordan and his roommate both decided to become vegetarians because they recently read an article on a fitness website describing the health benefits of a vegetarian diet. Yesterday Jordan's vegetarian diet consisted of a Danish pastry for breakfast and a tomato-rice dish (no meat) with pretzels and a diet soft drink for lunch. In the afternoon, after his martial arts class, he had a milk shake and two cookies. At dinnertime, he had a vegetarian sub sandwich consisting of lettuce, sprouts, tomatoes, cucumbers, and cheese, with two glasses of fruit punch. In the evening, he had a bowl of popcorn.

Answer the following questions and check your response at the end of this chapter.

1. What type of health benefits can Jordan expect from following a well-planned vegetarian diet?

2. What is missing from Jordan's current diet plan in terms of foods that should be emphasized in a vegetarian diet?

3. Which nutrients are missing in this current diet plan?

4. Are there any food components in the current diet plan that should be minimized or avoided?

5. How could he improve his new diet at each meal and snack to meet his nutritional needs and avoid undesirable food components?

▲ Has Jordan planned a healthy and nutritious vegetarian diet?

Summary (Numbers refer to numbered sections in the chapter.)

6.1 Amino acids, the building blocks of proteins, contain a very usable form of nitrogen for humans. Of the 20 common types of amino acids found in food, nine must be consumed in food (essential) and the rest can be synthesized by the body (nonessential).

6.2 Individual amino acids are bonded together to form proteins. The sequential order of amino acids determines the protein's ultimate shape and function. This order is directed by DNA in the cell nucleus. Diseases such as sickle cell anemia can occur if the amino acids are incorrect on a polypeptide chain. When the three-dimensional shape of a protein is unfolded—denatured—by treatment with heat, acid or alkaline solutions, or other processes, the protein also loses its biological activity.

6.3 Almost all animal products are nutrient-dense sources of protein. The high quality of these proteins means that they can be easily converted into body proteins. Rich plant sources of protein, such as beans, are also available.

High-quality (complete) protein foods contain ample amounts of all nine essential amino acids. Lower-quality (incomplete) protein foods lack sufficient amounts of one or more essential amino acids. This is typical of plant foods, especially cereal grains. Different types of plant foods eaten together often complement each other's amino-acid deficits, thereby providing high-quality protein in the diet.

6.4 Protein digestion begins in the stomach, where stomach acid and pepsin break down proteins into shorter polypeptide chains of amino acids. In the small intestine, these polypeptide chains eventually separate into amino acids in the absorptive cells. The free amino acids then travel via the portal vein that connects to the liver. Some then enter the bloodstream.

6.5 Important body components—such as muscles, connective tissue, transport proteins in the bloodstream, visual pigments, enzymes, some hormones, and immune cells—are made of proteins. These proteins are in a state of constant turnover. The carbon chains of proteins may be used to produce glucose (or fat) when necessary.

6.6 The protein RDA for adults is 0.8 grams per kilogram of healthy body weight. For a typical 70-kilogram (154-pound) person, this corresponds to 56 grams of protein daily; for a 57-kilogram (125-pound) person, this corresponds to 46 grams per day. The North American diet generally supplies plenty of protein. Men typically consume about 100 grams of protein daily, and women consume closer to 65 grams. These usual protein intakes are also of sufficient quality to support body functions. This is even true for well-balanced vegetarian diets.

6.7 People need to balance protein intake with losses to maintain a state of protein equilibrium, also called *protein balance.*

When a body is growing or recovering from an illness or injury, it needs a positive protein balance to supply the raw materials required to build new tissues. To achieve this, a person must eat more protein daily than he or she loses. Consuming less protein than needed leads to negative protein balance, such as when acute illness reduces the desire to eat and so one loses more protein than consumed.

6.8 Undernutrition can lead to protein–calorie malnutrition in the form of kwashiorkor or marasmus. Kwashiorkor results primarily from an inadequate protein intake in comparison with body needs, which often increase with concurrent disease and infection. Kwashiorkor often occurs when a child is weaned from human milk and fed mostly starchy gruels. Marasmus results from extreme starvation—a negligible intake of both protein and calories. Marasmus commonly occurs during famine, especially in infants.

NAYH Consumption of vegetarian and other plant-based diets provides many health benefits, including lower risks of such chronic diseases as cardiovascular disease, diabetes, and certain cancers. The benefits associated with the plant-based diets appear to stem from the lower content of saturated fat and cholesterol and the higher amount of fiber, vitamins, minerals, and phytochemicals.

Check Your Knowledge (Answers to the following questions are on the next page.)

1. The "instructions" for making proteins are located in the
   a. cell membrane.
   b. cell nucleus.
   c. cytoplasm.
   d. lysosome.

2. A nutrient that could easily be deficient in the diet of a vegan would be
   a. vitamin C.
   b. folic acid.
   c. calcium.
   d. All of the above.

3. An example of protein complementation used in vegetarian diet planning would be the combination of
   a. cereal and milk.
   b. bacon and eggs.
   c. rice and beans.
   d. macaroni and cheese.

4. If an essential amino acid is unavailable for protein synthesis, the
   a. cell will make the amino acid.
   b. synthesis of the protein will stop.
   c. cell will continue to attach amino acids to the protein.
   d. partially completed protein will be stored for later completion.

5. An individual who eats only plant food is referred to as a
   a. planetarium.
   b. vegan.
   c. lactovegetarian.
   d. lactoovovegetarian.

6. Which of the following groups accounts for the differences among amino acids?
   a. amine group
   b. side chain
   c. acid group
   d. keto group

7. Absorption of amino acids primarily takes place in the
   a. stomach.
   b. liver.
   c. small intestine.
   d. large intestine.

8. Jack is not an athlete and weighs 176 pounds (80 kilograms). His RDA for protein is _____ grams.
   a. 32
   b. 40
   c. 64
   d. 80

9. The basic building block of a protein is called a(n)
   a. fatty acid.
   b. monosaccharide.
   c. amino acid.
   d. gene.

10. Which of the following is true about protein intake of people in the United States?
    a. Most do not consume enough protein.
    b. Most consume the amount needed to balance losses.
    c. Athletes do not get enough protein without supplementation.
    d. Most consume more than is needed.

Answers Key: 1. b (LO 6.2), 2. c (LO 6.3), 3. c (LO 6.2), 4. b (LO 6.9), 5. b (LO 6.9), 6. b (LO 6.1), 7. c (LO 6.4), 8. c (LO 6.6), 9. c (LO 6.1), 10. d (LO 6.7)

Study Questions (Numbers refer to Learning Outcomes)

1. Discuss the relative importance of essential and nonessential amino acids in the diet. Why is it important for essential amino acids lost from the body to be replaced in the diet? **(LO 6.1)**

2. What is the role of cholecystokinin (CCK) in protein digestion? **(LO 6.4)**

3. What is a limiting amino acid? Explain why this concept is a concern in a vegetarian diet. How can a vegetarian compensate for limiting amino acids in specific foods? **(LO 6.8)**

4. Briefly describe the organization of proteins. How can this organization be altered or damaged? What might be a result of damaged protein organization? **(LO 6.2)**

5. Describe four functions of proteins. Provide an example of how the structure of a protein relates to its function. **(LO 6.5)**

6. How are DNA and protein synthesis related? **(LO 6.2)**

7. What would be one health benefit of reducing high-protein intake(s) to RDA amounts for some people? **(LO 6.6)**

8. Which eight foods are the major sources of proteins that cause food allergies? **(LO 6.3)**

9. Outline the major differences between kwashiorkor and marasmus. **(LO 6.8)**

10. What are the possible long-term effects of an inadequate intake of dietary protein among children between the ages of 6 months and 4 years? **(LO 6.8)**

What the Dietitian Chose

Consuming adequate protein is important for muscle repair and synthesis, but consuming excessive protein from any source provides no advantage. Excess protein is not stored in the body but may be metabolized as fuel or stored as fat, not muscle. In fact, some excess subcutaneous fat is probably the reason you have not seen your toned muscles yet. Cutting back on calories or including more endurance exercise in your routine can help to reduce body weight and body fat as you tone your muscles. Animal protein, in particular, is often a source of excess total fat, saturated fat, and cholesterol. Not only will this hinder efforts at weight loss, but it is detrimental to heart health.

Americans typically consume two to three times as much protein as they actually need. The average American intake of 100 grams per day for men or 65 grams per day for women easily meets even the highest protein recommendations for resistance exercise (1.7 grams of protein per kilogram). Also, omnivorous diets and even appropriately planned vegan diets can supply ample amino acids to support muscle recovery and synthesis.

It is not a good idea to take individual amino-acid supplements. Your digestive tract is adapted to handle whole proteins: hydrochloric acid and enzymes break polypeptides into amino acids that can be absorbed by the cells of the small intestine. Taking large doses of one amino acid can impair the absorption and/or metabolism of other amino acids.

Whey protein, a by-product of cheese production from cow's milk is considered to be a high-quality protein because it is easily digested and contains all of the essential amino acids. It is a source of the BCAAs valine, leucine, and isoleucine. BCAAs can be used for fuel by exercising muscles and are particularly important for synthesizing muscle tissue. The rationale is that whey protein supports muscle recovery and anabolism after exercise. Protein and amino-acid supplements are expensive and may be detrimental to health. It is safer and more economical to focus on a diet that provides 10% to 35% of calories from a variety of lean sources of protein.

▲ This 4-ounce grilled chicken breast is an excellent choice, providing 38 grams of high-quality yet inexpensive lean protein.

CASE STUDY SOLUTION Planning a Vegetarian Diet

1. Studies show that the incidence of and the death rates from some chronic diseases, such as certain cardiovascular diseases, hypertension, many forms of cancer, type 2 diabetes, and obesity, are lower for vegetarians than for nonvegetarians.
2. Many components of a healthy vegetarian diet—whole grains, nuts, soy products, beans, two to four servings of fruits, and three to five servings of vegetables per day—are missing. With so few fruits and vegetables, his diet is also low in the many phytochemicals under study for numerous health benefits.
3. Protein appears to be low in the diet because Jordan has not replaced meat with a good plant source of protein. Vitamin B-12, iron, and zinc are also missing from the foods he is choosing. His diet seems to be adequate in calcium and riboflavin because he is including dairy products (milk shake and cheese).
4. Some of Jordan's food choices are high in fat (Danish pastry, milk shake, and cookies) and sugar (fruit punch).
5. A healthy vegetarian diet includes whole grains, nuts, soy products, beans, two to four servings of fruits, and three to five servings of vegetables per day. It is apparent that he has not yet learned to implement the concept of complementary proteins, so the quality of protein in his diet is low. Meals that combine legumes or vegetables with grains or nuts (see Fig. 6-8) will provide the needed amounts of all amino acids.

Further Readings

1. ADA Reports: Position of the American Dietetic Association and Dietitians of Canada: Vegetarian diets. *Journal of the American Dietetic Association* 109:1266, 2009.

2. Burlingame B and Dernini S: Sustainable diets: The Mediterranean diet as an example. *Public Health Nutrition* 14:2285, 2011.

3. Cooper CC: Gluten free and healthy. *Today's Dietitian* 14(5):24, 2012.

4. Fessler TA: Malnutrition: A serious concern for hospitalized patients. *Today's Dietitian* 10(7):44, 2008.

5. Fleischer DM and others: Primary prevention of allergic disease through nutritional interventions. *Journal of Allergy and Clinical Immunology: In Practice* 1:29, 2013.

6. Micha R and others: Red and processed meat consumption and risk of incident coronary heart disease, stroke, and diabetes mellitus: A systemic review and meta-analysis. *Circulation* 121:2271, 2010.

7. Moomaw WT and others: The critical role of global food consumption patterns in achieving sustainable food systems and food for all.

A UNEP Discussion Paper, United Nations Environment Programme, Division of Technology, Industry and Economics, Paris, France, 2012. **http://www.humanmedia.org/dcc/pdf/unep_food_report_2012 .pdf.** Accessed 4/12/2013.

8. Nierenberg D and Reynolds L: Farm animal populations continue to grow. *Vital Signs,* Worldwatch Institute, March 23, 2012. **http://vital-signs.worldwatch.org/vs-trend/farm-animal-populations-continue-grow.** Accessed 4/12/2013.

9. Palmer S: Meatless Monday. *Today's Dietitian* 15(1):38, 2013.

10. Pan A and others: Red meat consumption and mortality: Results from two prospective cohort studies. *Archives of Internal Medicine* 172:555, 2012.

11. Stahler C: How often do Americans eat vegetarian meals? And how many adults in the U.S. are vegetarian? *Vegetarian Journal* 31:12, 2012.

12. Williams CD and other: Associations of red meat, fat, and protein intake with distal colorectal cancer risk. *Nutrition and Cancer* 62(6):701, 2010.

 **|NUTRITION**

To get the most out of your study of nutrition, visit McGraw-Hill Connect at www.mcgrawhillconnect.com where you will find NutritionCalc Plus, LearnSmart, and many other dynamic tools.

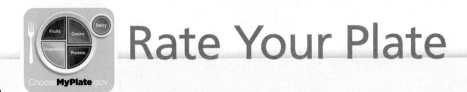

Rate Your Plate

Protein and the Vegetarian

Alana is excited about all the health benefits that might accompany a vegetarian diet. However, she is concerned that she will not consume enough protein to meet her needs. She is also concerned about possible vitamin and mineral deficiencies. Use NutritionCalc Plus or a food composition table to calculate her protein intake and see if her concerns are valid.

	Protein (grams)
Breakfast Calcium-fortified orange juice, 1 cup Soy milk, 1 cup Fortified bran flakes, 1 cup Banana, medium	
Snack Calcium-enriched granola bar	
Lunch GardenBurger, 4 oz Whole-wheat bun Mustard, 1 tbsp Soy cheese, 1 oz Apple, medium Green leaf lettuce, 1 ½ cups Peanuts, 1 oz Sunflower seeds, 1/4 cup Tomato slices, 2 Mushrooms, 3 Vinaigrette salad dressing, 2 tbsp Iced tea	
Dinner Kidney beans, ½ cup Brown rice, 3/4 cup Fortified margarine, 2 tbsp Mixed vegetables, 1/4 cup Hot tea	
Dessert Strawberries, ½ cup Angel food cake, 1 small slice Soy milk, ½ cup	
	Total Protein (grams) _____

Alana's diet contained 2150 kcal, with _____ grams (you fill in) of protein (Is this plenty for her?), 360 grams of carbohydrate, 57 grams of total dietary fat (only 9 grams of which came from saturated fat), and 50 grams of fiber. Her vitamin and mineral intake with respect to those of concern to vegetarians—vitamin B-12, vitamin D, calcium, iron, and zinc—met her needs.

 **connect** NUTRITION   **Find more Rate Your Plate activities for this chapter in Connect at www.mcgrawhillconnect.com.**

Student Learning Outcomes

Chapter 7 is designed to allow you to:

7.1 Describe energy balance and the uses of energy by the body.

7.2 Compare methods to determine energy use by the body.

7.3 Discuss methods for assessing body composition and determining whether body weight and composition are healthy.

7.4 Explain factors associated with the development of obesity and outline the risks to health posed by being overweight or obese.

7.5 List and discuss characteristics of a sound weight-loss program.

7.6 Describe why reduced calorie intake is the main key to weight loss and maintenance.

7.7 Discuss why physical activity is a key to weight loss and especially important for later weight maintenance.

Chapter 7
Energy Balance and Weight Control

7.8 Describe why and how behavior modification fits into a weight-loss program.

7.9 Outline the benefits and hazards of various weight-loss methods for severe obesity.

7.10 Discuss the causes and treatment of being underweight.

7.11 Evaluate popular weight-reduction diets and determine which are safe and successful.

In the last 25 years (the years since most of you were born), there has been a dramatic increase in the percentage of individuals who are overweight or obese. The ranks of the obese are growing in North America and worldwide. Recall from Chapter 1 that an estimated 1 billion people in the world are overweight. This problem is increasing not only in the United States but also globally among affluent peoples and in developing countries where Westernized diets (high-fat, high-sugar) are increasing in popularity. Excess weight increases the likelihood of many health problems, such as cardiovascular disease, cancer, hypertension, strokes, certain bone and joint disorders, and type 2 diabetes, especially if a person performs minimal physical activity.

Currently, most weight-reduction efforts fail before dieters reach a healthy weight range. Typical popular ("fad") diets are generally monotonous and restrictive. The more restrictive diets may even endanger some populations, such as children, teenagers, pregnant women, and people with various health disorders. A more logical "total diet" approach to weight loss is straightforward: (1) eat less; (2) increase physical activity; and (3) change problematic eating behaviors (see Further Reading 1).

A variety of groups, including government agencies, the food industry, health professionals, and communities, have begun to address the growing weight problem in North America. It is expected that, without this national effort to promote effective new approaches to maintaining healthy weight, the current trends will continue (Fig. 7-1; see Futher Readings 4, 14, and 20). This chapter will lead you to an understanding of obesity's causes, consequences, and potential treatments.

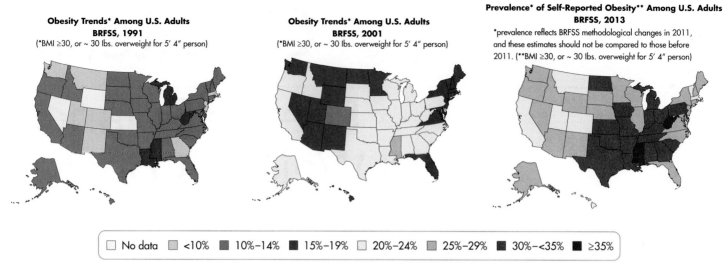

| □ No data | ■ <10% | ■ 10%–14% | ■ 15%–19% | □ 20%–24% | ■ 25%–29% | ■ 30%–<35% | ■ ≥35% |

FIGURE 7-1 ▲ Obesity trends among U.S. adults: 1991, 2001, 2013.
Source: CDC Behavioral Risk Factor Surveillance System (see Further Reading 3).

"How many calories do you burn by downloading diet apps?"

Reprinted with permission of Marty Bucella.

▲ What components make up a successful diet plan? What constitutes a "fad" diet? Why are overweight and obesity growing problems worldwide? What might be the future consequences of this trend? This chapter provides some answers.

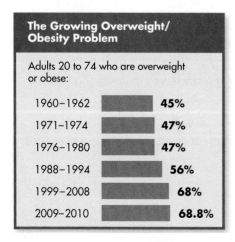

The Growing Overweight/Obesity Problem	
Adults 20 to 74 who are overweight or obese:	
1960–1962	45%
1971–1974	47%
1976–1980	47%
1988–1994	56%
1999–2008	68%
2009–2010	68.8%

7.1 Energy Balance

We begin this chapter with some good news and some bad news. The good news is that if you stay at a healthy body weight, you increase your chances of living a long and healthy life. The bad news is that currently 68.8% of all North American adults are overweight, significantly more than in the 1980s. Of those, about 50% (34% of the total population) are obese. There is a good chance that any of us could become part of those statistics if we do not pay attention to the prevention of significant weight gain in adulthood. Gaining more than 10 pounds or 2 inches in waist circumference are signals that a reevaluation of diet and lifestyle is in order.

The prevalence and trends in obesity rates in the United States have been analyzed in several recent studies. One study estimated the prevalence of adult obesity and severe obesity through the year 2030 using statistical models of CDC obesity data from 1990 to 2008. The study estimated a 33% increase in obesity prevalence and a 130% increase in severe obesity prevalence over the next two decades, with 51% of the population forecasted to be obese by 2030 if current trends continue. If obesity rates stop climbing and remain at current levels, the savings in medical expenditures over the next two decades would be over $500 billion (see Further Reading 5).

There is good news regarding our obesity trends as seen in two other studies. In the first, the prevalence of adult obesity in the United States was estimated from the 2009–2010 National Health and Nutrition Examination Survey (NHANES) and compared to adult obesity and body mass index (BMI) from earlier years of NHANES data. In 2009–2010, obesity rates were 35.5% among adult men and 35.8% among adult women, which was not a significant change since 2003–2008. These results suggest a slowing or leveling off of the obesity and BMI trends compared to increases observed between 1980 and 1999 (see Further Reading 6).

Because energy intake is a key determinant of weight, another study examined trends in energy intake using NHANES surveys from 1971–1975 to 2009–2010. They found that while energy intake increased from 1955 kcal per day during 1971–1975 to 2269 kcal per day during 2003–2004, it has declined more recently to 2195 kcal per day during 2009–2010. In fact, from 1999–2000 to 2009–2010, a downward trend in energy intake was observed (see Further Reading 7). These recent results suggest that the decrease in energy intakes over the past 10 years or more has had a minimal effect on slowing the obesity trend. There is no quick cure for overweight, despite what the advertisements claim. Successful weight loss comes from hard work and commitment. A combination of decreased calorie intake, increased physical activity, and behavior modification is considered to be the most reliable treatment for the overweight condition. And without a doubt, the prevention of the overweight condition in the first place is the most successful approach.

POSITIVE AND NEGATIVE ENERGY BALANCE

A healthy weight can result from paying more attention to the important concept of **energy balance** (Fig. 7-2). Think of energy balance as an equation:

$$\begin{array}{ll} \text{Energy input} & \text{Energy output} \\ \text{(calories from food intake)} \;=\; & \text{(metabolism; digestion, absorption, and transport} \\ & \text{of nutrients; physical activity)} \end{array}$$

energy balance The state in which energy intake, in the form of food and beverages, matches the energy expended, primarily through basal metabolism and physical activity.

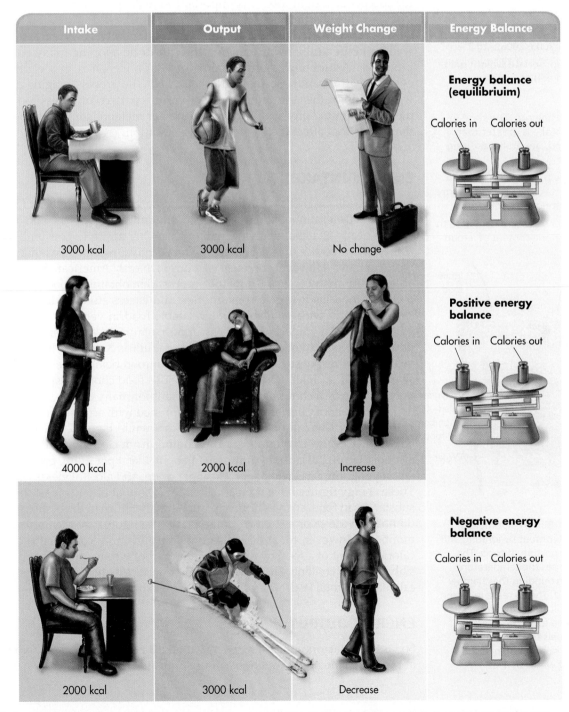

Intake	Output	Weight Change	Energy Balance
3000 kcal	3000 kcal	No change	**Energy balance (equilibriuim)** — Calories in, Calories out
4000 kcal	2000 kcal	Increase	**Positive energy balance** — Calories in, Calories out
2000 kcal	3000 kcal	Decrease	**Negative energy balance** — Calories in, Calories out

FIGURE 7-2 ▲ A model for energy balance: Intake versus output. This figure depicts energy balance in practical terms.

positive energy balance The state in which energy intake is greater than energy expended, generally resulting in weight gain.

negative energy balance The state in which energy intake is less than energy expended, resulting in weight loss.

bomb calorimeter An instrument used to determine the calorie content of a food.

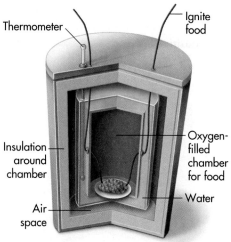

Thermometer

Ignite food

Insulation around chamber

Oxygen-filled chamber for food

Air space

Water

FIGURE 7-3 ▲ Bomb calorimeters measure calorie content by igniting and burning a dried portion of food. The burning food raises the temperature of the water surrounding the chamber holding the food. The increase in water temperature indicates the number of kilocalories in the food because 1 kilocalorie equals the amount of heat needed to raise the temperature of 1 kilogram of water by 1°C.

The balance of calories (measured in kcals) on the two sides of this equation can influence energy stores, especially the amount of triglyceride stored in adipose tissue. When energy input is greater than energy output, the result is **positive energy balance.** The excess calories consumed are stored, which results in weight gain. There are some situations in which positive energy balance is normal and healthy. During pregnancy, a surplus of calories supports the developing fetus. Infants and children require a positive energy balance for growth and development. In adults, however, even a small positive energy balance is usually in the form of fat storage rather than muscle and bone and, over time, can cause body weight to climb.

On the other hand, if energy input is less than energy output, there is a calorie deficit and **negative energy balance** results. A negative energy balance is necessary for successful weight loss. It is important to realize that when we lose weight, we lose some lean tissue in addition to adipose tissue.

The maintenance of energy balance substantially contributes to health and well-being in adults by minimizing the risk of developing many common health problems. Adulthood is often a time of subtle increases in weight gain, which eventually turns into obesity if left unchecked. The process of aging does not cause weight gain; rather the problem stems from a pattern of excess food intake coupled with limited physical activity and slower metabolism. Let us look in detail at the factors that affect the energy balance equation.

ENERGY INTAKE

Energy needs are met by food intake, represented by the number of calories eaten each day. Determining the appropriate amount and type of food to match our energy needs is a challenge for many of us. Our desire to consume food and the ability of our bodies to use it efficiently are survival mechanisms that have evolved with humans. However, because of current North American food supplies and accessibility, many of us are now too successful in obtaining food energy. The abundance of food leads to overconsumption and excess stores of body fat. Given the cheap cost and wide availability of palatable food in vending machines, drive-up windows, social gatherings, and fast-food restaurants—combined with *supersized* portions—it is no wonder that the average adult is 8 pounds heavier than just 10 years ago. In response to this cultural trend of food being widely available, "defensive eating" (i.e., making careful and conscious food choices, especially in regard to portion size) on a continual basis is important for many of us.

The number of calories in a food is determined with an instrument called a **bomb calorimeter.** This calorie determination is shown in Figure 7-3. The bomb calorimeter measures the amount of calories coming from carbohydrate, fat, protein, and alcohol. Recall that carbohydrates yield about 4 kcal per gram, proteins yield about 4 kcal per gram, fats yield about 9 kcal per gram, and alcohol yields 7 kcal per gram. These energy figures have been adjusted for (1) our ability to digest the food and (2) substances in food, such as fibrous plant parts that burn in the bomb calorimeter but do not provide calories to the human body. The figures are then rounded to whole numbers. However, today it is also possible and more common to determine the calorie content of a food by quantifying its carbohydrate, protein, and fat (and possibly alcohol) content. Then the kcal per gram factors listed previously are used to calculate the total kcals. (Recall that Chapter 1 showed how to do this calculation.)

ENERGY OUTPUT

So far, some factors concerning energy intake have been discussed. Now let us look at the other side of the equation: energy output.

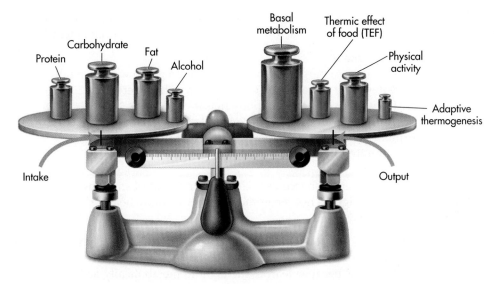

FIGURE 7-4 ▲ The components of energy intake and expenditure. This figure incorporates the major variables that influence energy balance. *Remember that alcohol is an additional source of energy for some of us.* The size of each component shows the relative contribution of that component to energy balance.

CRITICAL THINKING

As she gets closer to age 30, a 28-year-old classmate of yours has been thinking about the process of aging. One of the things she fears most as she gets older is gaining weight. How would you suggest that she adjust her energy balance as she gets older?

The body uses energy for three general purposes: basal metabolism; physical activity; and digestion, absorption, and processing of ingested nutrients. A fourth minor form of energy output, known as adaptive thermogenesis, refers to energy expended during fidgeting or shivering in response to cold (Fig. 7-4).

Basal Metabolism. Basal metabolism is expressed as basal metabolic rate (BMR) and represents the minimal amount of calories expended in a fasting state to keep a resting, awake body alive in a warm, quiet environment. For a sedentary person, basal metabolism accounts for about 60% to 75% of total energy use by the body. Some of the processes involved include the beating of the heart, respiration by the lungs, and the activity of other organs such as the liver, brain, and kidney. It does not include energy used for physical activity or digestion, absorption, and processing of recently consumed nutrients. If the person is not fasting or completely rested, the term **resting metabolism** is used and expressed as resting metabolic rate (RMR). An individual's RMR is typically higher than his or her BMR.

To see how basal metabolism contributes to energy needs, consider a 130-pound woman. First, knowing that there are 2.2 pounds for every kilogram, convert her weight into metric units:

$$130 \text{ lbs} \div 2.2 \text{ lbs/kilograms} = 59 \text{ kilograms}$$

Then, using a rough estimate of BMR of 0.9 kcal per kilogram per hour for an average female (1.0 kcal per kilogram per hour is used for an average male), calculate her BMR:

$$59 \text{ kg} \times 0.9 \text{ kcal/kg} = 53 \text{ kcal per hour.}$$

Finally, use this hourly BMR to find her BMR for an entire day:

$$53 \text{ kcal/hr} \times 24 \text{ hrs} = 1272 \text{ kcal.}$$

basal metabolism The minimal amount of calories the body uses to support itself in a fasting state when resting and awake in a warm, quiet environment. It amounts to roughly 1 kcal per kilogram per hour for men and 0.9 kcal per kilogram per hour for women; these values are often referred to as *basal metabolic rate (BMR)*.

resting metabolism The amount of calories the body uses when the person has not eaten in 4 hours and is resting (e.g., 15 to 30 minutes) and awake in a warm, quiet environment. It is usually slightly higher than basal metabolism due to the less strict criteria for the test; often referred to as *resting metabolic rate (RMR)*.

lean body mass Body weight minus fat storage weight equals lean body mass. This includes organs such as the brain, muscles, and liver, as well as bone and blood and other body fluids.

These calculations give only an estimate of basal metabolism, as it can vary 25% to 30% among individuals. Factors that increase basal metabolism include:

- Greater **lean body mass**
- Larger body surface area per body volume (e.g., for a given weight, a tall person has greater body surface area compared to a short person)
- Male gender (caused by greater lean body mass)
- Body temperature (fever or cold environmental conditions)
- Thyroid hormones (elevated, or hyperthyroid, condition)
- Stress (release of norepinephrine)
  - Pregnancy
  - Caffeine and tobacco use (Using the practice of smoking to control body weight is not recommended as too many health risks are increased.)

Of those factors, the amount of lean body mass a person has is the most important one. Persons with higher amounts of lean body mass have a higher BMR because lean tissue is more metabolically active than fat. The lean tissue, therefore, requires more energy to support its activity. Although overweight and obese persons have an increased amount of body fat, they also typically have a high amount of lean body mass and therefore a high BMR to go along with it (Fig. 7-5).

In contrast to factors that increase basal metabolism, a low calorie intake, such as an extreme diet regime, decreases basal metabolism by about 10% to 20% (about 150 to 300 kcal per day) as the body shifts into a conservation mode. This is a barrier to sustained weight loss during dieting that involves extremely low calorie diets. In addition, the effects of aging make weight maintenance a challenge. As lean body mass slowly and steadily decreases, basal metabolism declines 1% to 2% for each decade past the age of 30. However, because physical activity aids in maintenance of lean body mass, remaining active as one ages helps to preserve a high basal metabolism and, in turn, aids in weight control.

FIGURE 7-5 ▲ Lean body mass (LBM), the most important determinant of the basal metabolic rate, varies greatly between individuals. Persons of the same body weight can have different amounts of LBM and body fat.

Energy for Physical Activity. Physical activity increases energy expenditure above and beyond basal energy needs by as much as 15% to 35%. In choosing to be active or inactive, we determine much of our total calorie expenditure for a day. Calorie expenditure from physical activity varies widely among people. For example, climbing stairs rather than riding the elevator, walking rather than driving to the store, and standing in a bus rather than sitting increase physical activity and, hence, energy use. The alarming incidence of and recent increase in obesity in North America are partially the result of our inactivity. Jobs demand less physical activity, and leisure time is often spent in front of a television or computer.

While a person (130-pound woman) is resting, the percentage of total energy use and corresponding energy use by various organs are approximately as follows:

Brain	19%	242 kcal/day
Skeletal muscle	18%	229 kcal/day
Liver	27%	343 kcal/day
Kidney	10%	127 kcal/day
Heart	7%	89 kcal/day
Other	19%	242 kcal/day
Total	100%	1272 kcal/day

Thermic Effect of Food (TEF). In addition to basal metabolism and physical activity, the body uses energy to digest food, and absorb and further process the nutrients recently consumed. Energy used for these tasks is referred to as the **thermic effect of food (TEF).** TEF is similar to a sales tax: it is like being charged about 5% to 10% for the total amount of calories you eat to cover the cost of processing that food eaten. We may even recognize this increase in metabolism as a warming of the body during and right after a meal. Because of this "tax," you must eat between 5 to 10 kcal extra for every 100 kcal needed for basal metabolism and physical activity. If your daily calorie intake was 3000 kcal, TEF would account for 150 to 300 kcal. As with other components of energy output, the total amount can vary somewhat among individuals.

thermic effect of food (TEF) The increase in metabolism that occurs during the digestion, absorption, and metabolism of energy-yielding nutrients. This represents 5% to 10% of calories consumed.

Food composition influences TEF. For example, the TEF value for a protein-rich meal is 20% to 30% of the calories consumed and is higher than that of a carbohydrate-rich (5% to 10%) or fat-rich (0% to 3%) meal. This is because it takes more energy to metabolize amino acids into fat than to convert glucose into glycogen or transfer absorbed fat into adipose stores. In addition, large meals result in higher TEF values than the same amount of food eaten over many hours. The TEF value for alcohol is 20%.

Adaptive Thermogenesis. Adaptive thermogenesis represents the increase in nonvoluntary physical activity triggered by cold conditions or overeating. Some examples of nonvoluntary activities include shivering when cold, fidgeting, maintenance of muscle tone, and maintaining body posture when not lying down. Studies have shown that some people are able to resist weight gain from overfeeding by inducing thermogenesis, while others are not able to do so to a great extent.

The contribution of thermogenesis to overall calorie output is fairly small. The combination of basal metabolism and TEF accounts for 70% to 85% of energy used by a sedentary person. The remaining 15% to 30% is used mostly for physical activity, with a small amount used for thermogenesis.

Brown adipose tissue is a specialized form of adipose tissue that participates in thermogenesis. It is found in small amounts in infants. The brown appearance results from its greater number of mitochondria. Brown adipose tissue contributes to thermogenesis by releasing some of the energy from energy-yielding nutrients into the environment as heat instead of producing ATP. In infants, brown adipose tissue contributes as much as 5% of body weight and is thought to be important for heat regulation. Hibernating animals also use brown adipose tissue to generate heat to withstand a long winter. Adults have very little brown adipose tissue, and its role in adulthood is unknown.

✓ CONCEPT CHECK 7.1

1. What are the main components of energy balance?
2. How is the energy content of food determined and expressed?
3. What are the four main purposes for which the body uses energy? Approximately how much does each component contribute to overall energy use by the body?

7.2 Determination of Energy Use by the Body

The amount of energy a body uses can be measured by both direct and indirect calorimetry or can be estimated based on height, weight, degree of physical activity, and age.

DIRECT AND INDIRECT CALORIMETRY

Direct calorimetry measures the amount of body heat released by a person. The person is put into an insulated chamber, often the size of a small bedroom, and body heat released raises the temperature of a layer of water surrounding the chamber. A kcal, as you recall, is related to the amount of heat required to raise the temperature of water. By measuring the water temperature in the direct calorimeter before and after the body releases heat, the energy expended can be determined. Direct calorimetry works because almost all the energy used by the body eventually leaves as heat. However, mostly because of its expense and complexity, direct calorimetry is rarely used.

The most commonly used method of **indirect calorimetry** measures the respiratory gas exchange, which is the amount of oxygen a person consumes and the amount of carbon dioxide he or she expels (Fig. 7-6). A predictable relationship

▲ A few foods, such as celery, use more calories for TEF than they contain, making them negative calorie foods. Lean protein foods such as chicken breast, egg whites, and white fish have the highest thermic effect at almost 30%. This means that if you eat 100 calories of chicken breast, almost 30 of those calories are burned off just to digest it.

adaptive thermogenesis This term encompasses the ability of humans to regulate body temperature within narrow limits (thermoregulation). Two visible examples of thermogenesis are fidgeting and shivering when cold.

brown adipose tissue A specialized form of adipose tissue that produces large amounts of heat by metabolizing energy-yielding nutrients without synthesizing much useful energy for the body. The unused energy is released as heat.

▲ The cultural trend of serving large quantities of high-calorie foods can easily lead to positive energy balance. Sharing your meal with another person is a good way to avoid overeating when served large portions.

direct calorimetry A method of determining a body's energy use by measuring heat released from the body. An insulated chamber is usually used.

indirect calorimetry A method to measure energy use by the body by measuring oxygen uptake and carbon dioxide output. Formulas are then used to convert this gas exchange value into energy use, estimating the proportion of energy nutrients that are being oxidized for energy in the fuel mix.

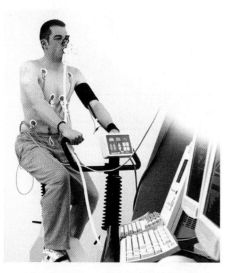

FIGURE 7-6 ▲ Indirect calorimetry measures oxygen intake and carbon dioxide output to determine energy expended during daily activities.

exists between the body's use of energy and oxygen. For example, when metabolizing a typical mixed diet of the energy-yielding nutrients, carbohydrate, fat, and protein, the human body uses 1 liter of oxygen to yield about 4.85 kcal of energy.

Instruments to measure oxygen consumption for indirect calorimetry are widely used. They can be mounted on carts (metabolic carts) and rolled up to a hospital bed or carried in a backpack while a person plays tennis or jogs to measure how many kcals are burned during these activities. There are even handheld instruments (BodyGem). Tables presenting energy costs of various forms of exercises rely on information gained from indirect calorimetry studies. You will also see an estimation of calories burned during a workout on most exercise equipment. The accuracy of these values is discussed on page 255.

ESTIMATES OF ENERGY NEEDS

As covered in Chapter 2, the Food and Nutrition Board has published a number of formulas to estimate energy needs, called Estimated Energy Requirements (EERs). Those for adults are shown here. As you calculate your own EER, remember to do multiplication and division before addition and subtraction! (Specific information regarding energy needs for pregnant women, lactating women, and children is covered in Chapters 14 and 15.) The calories used for basal metabolism are already factored into these formulas.

The variables in the formulas correspond to the following:

$$EER = \text{Estimated Energy Requirement}$$
$$AGE = \text{age in years}$$
$$PA = \text{physical activity estimate (see following table)}$$
$$WT = \text{weight in kilograms (pounds} \div 2.2)$$
$$HT = \text{height in meters (inches} \div 39.4)$$

Estimated Energy Requirement Calculation for Men 19 Years and Older

$$EER = 662 - (9.53 \times AGE) + PA \times (15.91 \times WT + 539.6 \times HT)$$

Estimated Energy Requirement Calculation for Women 19 Years and Older

$$EER = 354 - (6.91 \times AGE) + PA \times (9.36 \times WT + 726 \times HT)$$

Track Your Energy Needs

Use the forms in Appendix C to track your energy expenditure and estimate your energy needs. List all of your activities for a 24-hour period. Record the number of minutes spent in each activity to total 1440 minutes (24 hours). Record the energy cost, in kcal, for each activity and multiply the energy cost by the minutes to obtain the energy expended for each activity. Total all the kcal values for your estimated energy expenditure for the day.

Physical Activity (PA) Estimates

Activity Level	PA (Men)	PA (Women)
Sedentary (e.g., no exercise)	1.00	1.00
Low activity (e.g., walks the equivalent of 2 miles per day at 3 to 4 mph)	1.11	1.12
Active (e.g., walks the equivalent of 7 miles per day at 3 to 4 mph)	1.25	1.27
Very active (e.g., walks the equivalent of 17 miles per day at 3 to 4 mph)	1.48	1.45

The following is a sample calculation for a man who is 25 years old, 5 feet, 9 inches (1.75 meters), 154 pounds (70 kilograms), and has an active lifestyle. His EER is as follows:

$$EER = 662 - (9.53 \times 25) + 1.25 \times (15.91 \times 70 + 539.6 \times 1.75) = 2997 \text{ kcal}$$

The next equation is a sample calculation for a woman who is 25 years old, 5 feet, 4 inches (1.62 meters), 120 pounds (54.5 kilograms), and has an active lifestyle. Her EER is as follows:

$$EER = 354 - (6.91 \times 25) + 1.27 \times (9.36 \times 54.5 + 726 \times 1.62) = 2323 \text{ kcal}$$

You have determined the man's EER to be about 3000 kcal and the woman's EER to be about 2300 kcal per day. Remember that this is only an estimate; many other factors, such as genetics and hormones, can affect actual energy needs.

The **www.ChooseMyPlate.gov** website provides an interactive tool to estimate your calorie needs called Daily Food Plan. Figure 7-7 shows the range of activity levels and calorie recommendations for age and gender groups.

MyPlate Calorie Guidelines		
Children	**Sedentary** ⟶	**Active**
2–3 years	1000 ⟶	1400
Females	**Sedentary** ⟶	**Active**
4–8 years	1200 ⟶	1800
9–13	1400 ⟶	2200
14–18	1800 ⟶	2400
19–30	1800 ⟶	2400
31–50	1800 ⟶	2200
51+	1600 ⟶	2200
Males	**Sedentary** ⟶	**Active**
4–8 years	1200 ⟶	2000
9–13	1600 ⟶	2600
14–18	2000 ⟶	3200
19–30	2400 ⟶	3000
31–50	2200 ⟶	3000
51+	2000 ⟶	2800

FIGURE 7-7 ▲ MyPlate Calorie Guidelines for age and gender groups.

CONCEPT CHECK 7.2

1. What methods can be used to measure energy use by the body?
2. Estimated Energy Requirement can be calculated based on which five factors?

7.3 Assessing Healthy Body Weight

Numerous methods are used to establish what body weight should be, typically called *healthy weight*. Healthy weight is currently the preferred term to use for weight recommendations. Older terms, such as *ideal weight* and *desirable weight*, are subjective and are no longer used in medical literature. Several tables exist, generally based on weight for height. These tables arise from studies of large population groups. When applied to a population, they provide good estimates of weight associated with health and longevity. However, they do not necessarily indicate the healthiest body weight for each individual. For example, *athletes with large, lean body mass but low fat content will have higher body weights than sedentary individuals.*

Listening to the body for hunger cues, regularly eating a healthy diet, and remaining physically active eventually helps one maintain an appropriate height/weight value. The clearest idea regarding a healthy weight is that it is personal. Weight has to be considered in terms of health, not a mathematical calculation. Overall, the individual, under a physician's guidance, should establish a "personal" healthy weight (or need for weight reduction) based on weight history, fat distribution patterns, family history of weight-related disease, and current health status. Indications that your weight is not healthy would include the following weight-related conditions:

- Hypertension
- Elevated LDL-cholesterol
- Family history of obesity, cardiovascular disease, or certain forms of cancer (e.g., uterus, colon)
- Pattern of upper-body fat distribution
- Elevated blood glucose

This assessment points out how well the person is tolerating any existing excess weight. Thus, current height/weight standards are only a rough guide. On a more practical note, other questions can be pertinent: What is the least one has weighed as an adult for at least a year? What is the largest size clothing one would be happy with? What weight has one been able to maintain during previous diets without feeling constantly hungry? Furthermore, a healthy lifestyle may make a more important contribution to a person's health status than the number on the scale. Fit and overweight are not necessarily mutually exclusive, and neither is thin synonymous with healthy if the person is not also physically active.

▲ Physical activity, such as walking, is an important component of our energy expenditure.

Women   Men

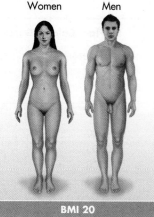

BMI 20

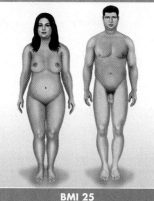

BMI 25

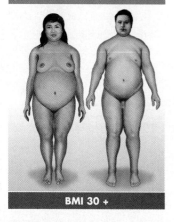

BMI 30 +

FIGURE 7-8 ▲ Estimates of body shapes at different BMI values.

body mass index (BMI) Weight (in kilograms) divided by height (in meters) squared; a value of 25 and above indicates overweight, and a value of 30 and above indicates obesity.

◄ A high BMI may not reflect overweight or fatness. Extra muscle tissue can result in a BMI greater than 25.

BODY MASS INDEX (BMI)

Currently, **body mass index (BMI)** is the preferred weight-for-height standard because it is the clinical measurement most closely related to body fat content (Fig. 7-8).

$$\text{Body mass index is calculated as } \frac{\text{body weight (in kilograms)}}{\text{height}^2 \text{ (in meters)}}$$

$$\text{An alternate method for calculating BMI is } \frac{\text{weight (pounds)} \times 703}{\text{height}^2 \text{ (inches)}}$$

BMI weight categories are shown in Table 7-1. A healthy weight for height is a BMI between 18.5 and 24.9. Health risks from excess weight may begin when the BMI is 25 or more. What is your BMI? How much would your weight need to change to yield a BMI of 25? 30? These are general cutoff values for the presence of overweight and obesity, respectively. Figure 7-9 lists the BMI for various heights and weights.

The concept of BMI is convenient to use because the values apply to both men and women (i.e., gender neutral). However, any weight-for-height standard is a crude measure. Keep in mind, also, that a BMI of 25 to 29.9 is a marker of *overweight* (compared to a standard population) and not necessarily a marker of *overfat*. Many men (especially athletes) have a BMI greater than 25 because of extra muscle tissue. Also, very short adults (under 5 feet tall) may have high BMIs that may not reflect overweight or fatness. For this reason, BMI should be used only as a screening test for overweight or obesity. Even agreed-upon weight standards for BMI are not for everyone. Adult BMIs should not be applied to children, still growing adolescents, frail older people, pregnant and lactating women, and highly muscular individuals. Pregnant women and children have unique BMI standards (see Chapters 14 and 15).

Still, overfat and overweight conditions generally appear together. The focus is on BMI in clinical settings mainly because this is easier to measure than total body fat.

ESTIMATING BODY FAT CONTENT AND DIAGNOSING OBESITY

If calorie intake exceeds output over time, overweight (and often obesity) is a likely result. Often, health problems eventually follow (Table 7-2) (see Further Reading 16). As just discussed, BMI values can be used as a convenient clinical tool to screen for overweight (BMI ≥ 25), obesity (BMI ≥ 30), and severe obesity (BMI ≥ 40) in individuals older than 20 years of age. Medical experts, however, recommend that an individual's diagnosis of obesity should not be based primarily on body weight but rather on the total amount of fat in the body, the location of body fat, and the presence or absence of weight-related medical problems.

Body fat varies widely among individuals. Good to acceptable amounts of body fat are about 11% to 20% for men and 16% to 30% for women. Men with over 24% body fat and women with over 37% body fat are considered obese. The higher range

TABLE 7-1 ► **BMI categories. BMI is a useful measure of weight-for-height and an estimate of body fat. It is a good gauge of your risk for diseases associated with more body fat.**

Category	BMI
Underweight	<18.5
Healthy weight	18.5–24.9
Overweight	25–29.9
Obese	30–39.9
Severely (morbidly) obese	>40

TABLE 7-2 ▶ Health Problems Associated with Excess Body Fat

Health Problem	Partially Attributable To
Surgical risk	Increased anesthesia needs, as well as greater risk of wound infections (the latter is linked to a decrease in immune function)
Pulmonary disease and sleep disorders	Excess weight over lungs and pharynx
Type 2 diabetes	Enlarged adipose cells, which poorly bind insulin and poorly respond to the message insulin sends to the cell; less synthesis of factors that aid insulin action and greater synthesis of factors by adipose cells that lessen insulin action
Hypertension	Increased miles of blood vessels found in the adipose tissue, increased blood volume, and increased resistance to blood flow related to hormones made by adipose cells
Cardiovascular disease (e.g., coronary heart disease and stroke)	Increases in LDL cholesterol and triglyceride values, low HDL cholesterol, decreased physical activity, and increased synthesis of blood clotting and inflammatory factors by enlarged adipose cells. A greater risk for heart failure is also seen, due in part to altered heart rhythm.
Bone and joint disorders (including gout)	Excess pressure put on knee, ankle, and hip joints
Gallstones	Increased cholesterol content of bile
Skin disorders	Trapping of moisture and microorganisms in tissue folds
Various cancers, such as in the kidney, gallbladder, colon and rectum, uterus (women), and prostate gland (men)	Estrogen production by adipose cells; animal studies suggest excess calorie intake encourages tumor development
Shorter stature (in some forms of obesity)	Earlier onset of puberty
Pregnancy risks	More difficult delivery, increased number of birth defects, and increased needs for anesthesia
Reduced physical agility and increased risk of accidents and falls	Excess weight that impairs movement
Menstrual irregularities and infertility	Hormones produced by adipose cells, such as estrogen
Vision problems	Cataracts and other eye disorders are more often present
Premature death	A variety of risk factors for disease listed in this table
Infections	Reduced immune system activity
Liver damage and eventual failure	Excess fat accumulation in the liver
Erectile dysfunction in men	Low-grade inflammation caused by excess fat mass and reduced function of the cells lining the blood vessels associated with being overweight

The greater the degree of obesity, the more likely and the more serious these health problems generally become. They are much more likely to appear in people who show an upper-body fat distribution pattern and/or are greater than twice their healthy body weight.

of body fat percentage for women is needed physiologically to maintain reproductive functions, including estrogen production.

To measure body fat content accurately using typical methods, both body weight and body volume of the person are used to calculate body density. Body weight is easy to measure on a conventional scale. Of the typical methods used to estimate body volume, **underwater weighing** is the most accurate. This technique determines body volume using the difference between conventional body weight and body weight under water, along with the relative densities of fat tissue and lean tissue, and a specific mathematical formula. This procedure requires that an individual be totally submerged in a tank of water, with a trained technician directing the procedure (Fig. 7-10). **Air displacement** is another method of determining body volume. Body volume is quantified by measuring the space a person takes up inside a measurement chamber, such as the BodPod (Fig. 7-11).

underwater weighing A method of estimating total body fat by weighing the individual on a standard scale and then weighing him or her again submerged in water. The difference between the two weights is used to estimate total body volume.

air displacement A method for estimating body composition that makes use of the volume of space taken up by a body inside a small chamber.

Weight in pounds

Height	120	130	140	150	160	170	180	190	200	210	220	230	240	250
4'6"	29	31	34	36	39	41	43	46	48	51	53	56	58	60
4'8"	27	29	31	34	36	38	40	43	45	47	49	52	51	56
4'10"	25	27	29	31	34	36	38	40	42	44	46	48	50	52
5'0"	23	25	27	29	31	33	35	37	39	41	43	45	47	49
5'2"	22	24	26	27	29	31	33	35	37	38	40	42	44	46
5'4"	21	22	24	26	28	29	31	33	34	36	38	40	41	43
5'6"	19	21	23	24	26	27	29	31	32	34	36	37	39	40
5'8"	18	20	21	23	24	26	27	29	30	32	34	35	37	38
5'10"	17	19	20	22	23	24	26	27	29	30	32	33	35	36
6'0"	16	18	19	20	22	23	24	26	27	28	30	31	33	34
6'2"	15	17	18	19	21	22	23	24	26	27	28	30	31	32
6'4"	15	16	17	18	20	21	22	23	24	26	27	28	29	30
6'6"	14	15	16	17	19	20	21	22	23	24	25	27	28	29
6'8"	13	14	15	17	18	19	20	21	22	23	24	26	26	28

(Height in feet and inches)

☐ Healthy weight   ☐ Overweight   ■ Obese

Developed by the National Center for Health Statistics in collaboration with the National Center for Chronic Disease Prevention and Health Promotion

FIGURE 7-9 ▲ Convenient height/weight table based on BMI. A healthy weight for height generally falls within a BMI range of 18.5 to 24.9 kilograms/meters^2.

bioelectrical impedance The method to estimate total body fat that uses a low-energy electrical current. The more fat storage a person has, the more impedance (resistance) to electrical flow will be exhibited.

dual energy X-ray absorptiometry (DEXA) A highly accurate method of measuring body composition and bone mass and density using multiple low-energy X rays.

The total cost attributable to obesity-related disease is about $147 billion annually in the United States. This is double what it was nearly a decade ago. Half of this cost is borne by the taxpayers, who fund Medicare and Medicaid.

upper-body obesity The type of obesity in which fat is stored primarily in the abdominal area; defined as a waist circumference more than 40 inches (102 centimeters) in men and more than 35 inches (89 centimeters) in women. Also known as *android obesity*.

Once body volume is known, it can be used along with body weight in the following equation to calculate body density. Then using body density, body fat content finally can be determined.

$$\text{Body density} = \frac{\text{body weight}}{\text{body volume}}$$

$$\% \text{ body fat} = (495 \div \text{body density}) - 450$$

For example, assume that the individual in the underwater weighing tank in Figure 7-10 has a body density of 1.06 grams per centimeter^3. We can use the second formula to calculate that he has 17% body fat ([495 ÷ 1.06] − 450 = 17).

Skinfold thickness is also a common anthropometric method to estimate total body fat content, although there are some limits to its accuracy. Clinicians use calipers to measure the fat layer directly under the skin at multiple sites and then plug these values into a mathematical formula (Fig. 7-12).

The technique of **bioelectrical impedance** is also used to estimate body fat content. The instrument sends a painless, low-energy electrical current to and from the body via wires and electrode patches to estimate body fat. This estimation is based on the assumption that adipose tissue resists electrical flow more than lean tissue because it has a lower electrolyte and water content than lean tissue. More adipose tissue therefore means proportionately greater electrical resistance. Within a few seconds, bioelectrical impedance analyzers convert body electrical resistance into an approximate estimate of total body fat, as long as body hydration status is normal (Fig. 7-13). Body composition monitors, better known as body fat calculators, which use bioelectric impedence are now available for home use. These machines are similar in shape and use to bathroom scales, but their main purpose is to measure body fat. A current passes easily through conductive foot pads and/or handheld electrodes. These in-home devices will hopefully encourage people to be less concerned with what they weigh than whether their weight comes from fat or muscle.

A more advanced determination of body fat content can be made using **dual energy X-ray absorptiometry (DEXA).** DEXA is considered the most accurate way to determine body fat, but the equipment is expensive and not widely available for this use. This X-ray system allows the clinician to separate body weight into three separate components: fat, fat-free soft tissue, and bone mineral. The usual whole-body scan requires about 5 to 20 minutes, and the dose of radiation is less than a chest X-ray. An assessment of bone mineral density and the risk of osteoporosis can also be made using this method (Fig. 7-14).

There are a few other methods of assessing body composition, but these are the five you will encounter most often in health clinics, fitness centers, and in the research articles referenced in this book. In the hands of a skilled clinician, these assessments provide valuable information about body fat in addition to simple measures of height and weight.

USING BODY FAT DISTRIBUTION TO FURTHER EVALUATE OBESITY

In addition to the amount of fat we store, the location of that body fat is an important predictor of health risks. Some people store fat in upper-body areas, whereas others store fat lower on the body. **Upper-body obesity,** characterized by a large abdomen, is more often called *abdominal* or *central* obesity and is related to insulin resistance and fatty liver leading to diabetes, high blood lipids, and heart disease. Because men more typically develop upper-body obesity, it is also known as *android*

FIGURE 7-10 ◀ Underwater weighing. In this technique, the subject exhales as much air as possible and then holds his or her breath and bends over at the waist. Once the subject is totally submerged, the underwater weight is recorded. Using this value, body volume can be calculated.

FIGURE 7-11 ▶ BodPod. This device determines body volume based on the volume of displaced air, measured as a person sits in a sealed chamber for a few minutes.

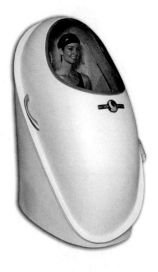

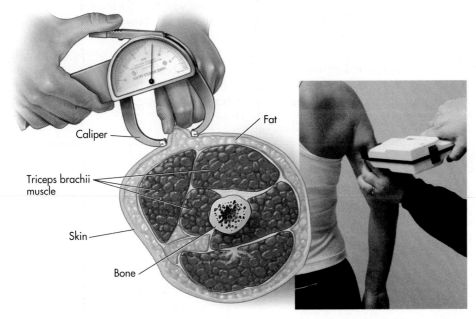

Fat

Caliper

Triceps brachii muscle

Skin

Bone

FIGURE 7-12 ◀ Skinfold measurements. With proper technique and calibrated equipment, skinfold measurements around the body can be used to predict body fat content in about 10 minutes. Measurements are made at several locations, including the triceps (photo and drawing) skinfolds.

FIGURE 7-13 ▲ Bioelectrical impedance estimates total body fat in less than 5 minutes and is based on the principle that body fat resists the flow of electricity, since it is low in water and electrolytes. The degree of resistance to electrical flow is used to estimate body fatness. This handheld device sends an electrical current through the body and gives a percentage of body fat when it has completed its process.

FIGURE 7-14 ▶ Dual energy X-ray absorptiometry (DEXA). This method measures body fat by passing small doses of radiation through the body. The radiation reacts differently with fat, lean tissue, or bone, allowing these components to be quantified. The scanner arm moves from head to toe and in doing so can determine body fat and bone density. DEXA is currently considered the most accurate method for determining body fat (as long as the person can fit under the arm of the instrument). The radiation dose is minimal.

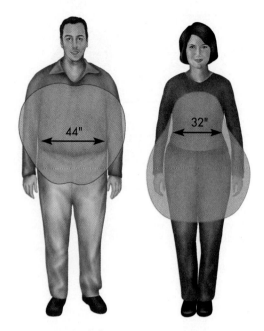

Upper-body fat
distribution
(android: apple shape)

Lower-body fat
distribution
(gynoid: pear shape)

FIGURE 7-15 ▲ Body fat stored primarily in the upper-body (android) brings higher risks of ill health associated with obesity than does lower-body (gynoid) fat. The woman's waist circumference of 32 inches and the man's waist circumference of 44 inches indicate that the man has upper-body fat distribution but the woman does not, based on a cutoff of 35 inches for women and 40 inches for men.

lower-body obesity The type of obesity in which fat storage is primarily located in the buttocks and thigh area. Also known as *gynoid* or *gynecoid obesity*.

obesity. While other adipose cells empty fat into general blood circulation, the fat released from abdominal adipose cells goes directly to the liver, by way of the portal vein. This influx of fat interferes with the liver's ability to use insulin and negatively affects lipoprotein metabolism by the liver. These abdominal adipose cells are not just storage depots but are metabolically active and release many hormones and other peptides called adipokines involved in long-term energy regulation. When they fill with excess fat, the cells become dysfunctional and release altered secretions, resulting in inflammation, insulin resistance, and other adverse health conditions leading to chronic disease.

High blood testosterone levels apparently encourage upper-body obesity, as does alcohol intake, and smoking. This pattern of fat storage is commonly known as the apple shape (large abdomen and small buttocks and thighs). Upper-body obesity is assessed by measuring the circumference of the abdomen at the natural waist (in between the lowest rib and the top of the hip bone), or at the umbilicus (belly button). A waist circumference more than 40 inches (102 centimeters) in men and more than 35 inches (89 centimeters) in women indicates upper-body obesity (Fig. 7-15). If BMI is also 25 or more, health risks are significantly increased.

Estrogen and progesterone encourage **lower-body** (*gynecoid* or *gynoid*) **obesity**—the typical female pattern. The small abdomen and much larger buttocks and thighs give a pearlike appearance. Fat deposited in the lower body is not mobilized as easily as the other type and often resists being shed. After menopause, blood estrogen falls, encouraging upper-body fat distribution and raising the risk of chronic disease dramatically for postmenopausal females.

✓ CONCEPT CHECK 7.3

1. How is the body mass index determined?
2. What are the BMI, body fat percentage, and waist circumference values for men and women that are associated with increased risk of health problems related to being overweight?
3. What are five methods by which body fat content can be estimated?
4. Obesity leads to an increased risk of which diseases?

7.4 Why Some People Are Obese—Nature Versus Nurture

The energy imbalance that promotes obesity stems from cultural, economic, and social factors (see Further Reading 6). Many studies of obesity attribute the increasing trends to the growth of the global food system, including advancements in food processing and marketing, and the resulting availability of affordable food.

Both genetic (nature) and environmental (nurture) factors can increase the risk for obesity (Table 7-3). The eventual location of fat storage is strongly influenced by genetics, particularly during pregnancy when gene expression is being imprinted on the fetus. For example, research studies have found that offspring born to obese mothers are at heightened risk of obesity later in life. Consider the possibility that obesity is nurture allowing nature to express itself. Some obese people begin life with a slower basal metabolism; maintain an inactive lifestyle; and consume highly refined, calorie-dense diets. These people are nurtured into gaining weight, promoting their natural tendency toward obesity. Even with a genetic tendency toward obesity, individuals can attain a healthier body weight with increased physical activity and decreased calorie consumption.

▲ Waist circumference is an important measure of weight-related health risk.

TABLE 7-3 ▶ What Encourages Excess Body Fat Stores and Obesity?

Factor	How Fat Storage Is Affected
Age	Excess body fat is more common in adults and middle-age individuals.
Menopause	Increase in abdominal fat deposition is typical.
Gender	Females have more fat.
Positive energy balance	Over a long period, positive energy balance promotes storage of fat.
Composition of diet	Excess calorie intake from fat, alcohol, and calorie-dense foods contributes to obesity.
Physical activity	Low physical activity ("couch potato") leads to positive energy balance and body fat storage.
Basal metabolism	A low BMR due to factors such as thyroid problems or energy restriction is linked to weight gain.
Thermic effect of food	Some obese individuals metabolize nutrients more efficiently.
Increased hunger sensations	Some people have excessive trouble resisting the abundant availability of food, which is likely linked to the activity of various brain chemicals in food reward pathways.
Ratio of fat to lean tissue	A high ratio of fat mass to lean body mass is correlated with weight gain.
Fat uptake by adipose tissue	Efficiency of fat storage is high in some obese individuals and remains high (perhaps even increases) with weight loss.
Variety of social and behavioral factors	Obesity is associated with socioeconomic status; familial conditions; network of friends; busy lifestyles; binge eating; availability of inexpensive, "supersized" high-fat food; pattern of leisure activities; screen time; smoking cessation; excessive alcohol intake; and meals eaten away from home.
Undetermined genetic characteristics	These affect energy expenditure, the deposition of the energy surplus as adipose tissue or as lean tissue, and the relative proportion of fat and carbohydrate used by the body.
Ethnicity	In some groups, higher body weight may be more acceptable, which may promote overeating.
Certain medications	Increased hunger/appetite can be a side effect.
Childbearing	A pattern of weight gain during the childbearing years can occur if the mother does not return to her approximate prepregnancy weight after the pregnancy. Fat stored during pregnancy to support subsequent lactation may not be lost in women who do not breastfeed.
National region	Regional differences, such as high-fat diets and sedentary lifestyles in the Midwest and areas of the South, lead to higher rates of obesity compared to other regions.

HOW DOES NATURE CONTRIBUTE TO OBESITY?

Studies in pairs of **identical twins** give us some insight into the contribution of nature to obesity. Even when identical twins are raised apart, they tend to show similar weight gain patterns, both in overall weight and body fat distribution. It appears that nurture—eating habits and nutrition, which varies between twins raised apart—has less to do with obesity than nature does. In fact, research suggests that genes account for up to 70% of weight differences between people. A child with no obese parent has only a 10% chance of becoming obese. When a child has one obese parent (common in our society), that risk advances up to 40%, and with two obese parents, it soars to 80%. Our genes help determine metabolic rate, fuel use, and differences in brain chemistry—all of which affect body weight.

We also inherit specific body types. Tall, thin people appear to have an inherently easier time maintaining healthy body weight. This is probably because basal metabolism increases as body surface increases, and therefore, taller people use more calories than shorter people, even at rest.

Humans have inherited a so-called "thrifty" metabolism that enables us to store fat readily. In early human history, our genes adapted to an environment where food was sometimes scarce; thus, a metabolism that efficiently stored fat would have been a safeguard against starvation in lean times. Now, with a constant abundance of

identical twins Two offspring that develop from a single ovum and sperm and, consequently, have the same genetic makeup.

▲ Studies in identical twins give us insight into the genetic contribution to obesity.

set point Often refers to the close regulation of body weight. It is not known what cells control this set point or how it functions in weight regulation. There is evidence, however, that mechanisms exist that help regulate weight.

hypothalamus A region at the base of the brain that contains cells that play a role in the regulation of energy intake, respiration, body temperature, and other body functions.

leptin A hormone made by adipose tissue in proportion to total fat stores in the body that influences long-term regulation of fat mass. Leptin also influences release of the hormone insulin.

food, we require wise food choices and regular physical activity for energy balance. Depending on genetic traits we inherit from our parents, some of us are more prone to weight gain in the modern food environment than others.

Does the Body Have a Set Point for Weight? The **set-point** theory of weight maintenance proposes that humans have a genetically predetermined body weight or body fat content, which the body closely regulates. Several physiological changes that occur during calorie reduction and weight loss support this theory. For example, research suggests that the **hypothalamus** monitors the amount of body fat in humans and tries to keep that amount constant over time. The release and circulation of the hormone, **leptin,** from adipose, promotes a sense of fullness, and thus tends to reduce appetite. As adipose cells increase in size and number, overall production of leptin increases, which should suppress appetite. If adipose mass is reduced, leptin levels are reduced, so appetite should be increased. This system, however, is not foolproof. Research has shown that overweight persons have large amounts of leptin coming from the excess body fat, but their brains seem to be "leptin resistant" and are not getting the signal to stop eating.

Thyroid hormone levels change, too. When calorie intake is reduced, the blood concentration of thyroid hormones falls, which slows basal metabolism. Also, the calorie cost of weight-bearing activity decreases, so that an activity that burned 100 kcal before weight loss may only burn 80 kcal after weight loss. Furthermore, with weight loss, the body becomes more efficient at storing fat by increasing the activity of the enzyme lipoprotein lipase, which takes fat into cells. All of these changes protect the body from losing weight.

If a person overeats, in the short run, basal metabolism tends to increase. This causes some resistance to weight gain. However, in the long run, resistance to weight gain is much less than resistance to weight loss. When a person gains weight and stays at that weight for a while, the body tends to establish energy balance at a new set point.

Opponents of the set-point theory argue that weight does not remain constant throughout adulthood: the average person gains weight slowly, at least until old age. Also, if an individual is placed in a different social, emotional, or physical environment, weight can be altered and maintained markedly higher or lower. These arguments suggest that humans, rather than having a set point determined by genetics or the number of adipose cells, settle into a particular stable weight based on their circumstances, often regarded as a "settling point."

The size-acceptance nondiet movement, "Health at Every Size," indirectly refers to a set point for weight by defining healthy weight as the natural weight the body adopts, given a healthy diet and meaningful levels of physical activity. Overall, the set point is weaker in preventing weight gain than in preventing weight loss. Even with a set point helping us, the odds are in favor of eventual weight gain unless we commit to a healthy lifestyle.

DOES NURTURE HAVE A ROLE?

Environmental factors, such as high-fat diets and inactivity, literally shape us. This seems likely when we consider that our gene pool has not changed at all in the past 50 years, whereas according to the U.S. Centers for Disease Control and Prevention, the ranks of obese people have grown to epidemic proportions over the last 25 years.

Some would argue that body weight similarities between family members stem more from learned behaviors than genetic similarities. Even couples, who have no genetic link, may behave similarly toward food and eventually assume similar degrees of leanness or fatness. Adult obesity in women is often rooted in childhood obesity. In addition, relative inactivity and periods of stress or boredom, as well as excess weight gain during pregnancy, contribute to female obesity. (Chapter 14 notes that breastfeeding one's infant contributes to loss of some of the excess fat associated with pregnancy.) These patterns suggest both social and genetic links. Male obesity, however, is not strongly linked to childhood obesity and, instead, tends to appear after age 30. This powerful and prevalent pattern suggests a primary role of nurture in obesity, with less genetic influence.

Is poverty associated with obesity? Ironically, in developed nations, the answer is *yes*. North Americans of lower socioeconomic status, especially minorities, are more likely to be obese than those of higher socioeconomic status. Several social and behavioral factors promote fat storage and support the link between socioeconomic status and obesity. These factors include lower socioeconomic status, overweight friends and family, a cultural/ethnic group that prefers higher body weight, a lifestyle that discourages healthy meals and adequate exercise, easy availability of inexpensive high-calorie food, limited access to fresh fruits and vegetables, excessive television viewing, smoking cessation, lack of adequate sleep, emotional stress, and meals frequently eaten away from home. Weight gain associated with cessation of smoking has been shown to be a result of increased calorie consumption, with 96% of the weight gain as fat.

▲ Body weight is influenced by many factors related to both nature and nurture. We resemble our parents because of the genes we have inherited, as well as the lifestyle habits, including diet, that we have learned from them.

✔ CONCEPT CHECK 7.4

1. Explain how body weight is influenced by nature.
2. What role does nurture play in determining body weight?

7.5 Treatment of Overweight and Obesity

Treatment of overweight and obesity should be long-term, similar to that for any chronic disease. Treatments require long-term lifestyle changes, rather than a quick fix promoted by many popular (also called fad) diet books. We often view a "diet" as something one goes on temporarily, only to resume prior (typically poor) habits once satisfactory results have been achieved. This is a big reason that so many people regain lost weight. Instead, an emphasis on healthy, active living with acceptable dietary and behavior modifications will promote weight loss and later weight maintenance. Maintenance of a healthy weight requires lifelong changes in habits, not a short-term weight-loss period.

LOSING BODY FAT

One pound of weight loss includes both adipose tissue plus supporting lean tissues and represents approximately 3300 kcal per pound (about 7.2 kcal per gram). Since there are approximately 3500 calories in a pound of fat, the past 50 years of weight-loss advice has centered on the notion that a deficit of approximately 500 kcal per day is required to lose 1 pound of adipose tissue per week. This fairly simple "3500-kcal rule," however, is now regarded as an inaccurate predictor of weight change, resulting in unrealistic expectations that may undermine efforts to lose extra pounds. Over the years, energy balance research has shown that weight loss occurs more gradually than would be predicted by the 3500-kcal rule because of several factors, including the loss of muscle along with fat during weight loss.

Experts have developed new weight-loss prediction formulas that predict a much slower and realistic pattern of weight loss. For a typical overweight adult, the basic formula calculates that a calorie deficit of 10 kcal per day will result in a 1-pound weight loss over 3 years. Researchers emphasize that weight change is not linear; it occurs most rapidly during the first year after a change in energy balance, but tapers off over the next 2 years. In addition, changes in energy intake and physical activity must be strictly maintained over time. For example, reducing calorie intake by 500 kcal per day would result in a 25-pound weight loss in 1 year, with continued loss of another 22 pounds by the end of three years. In contrast, the outdated 3500-kcal rule would have predicted a 52-pound weight loss in just one year.

▲ Student life is often full of physical activity. This is not necessarily true for a person's later working life; hence, weight gain is a strong possibility.

You can read more about the new weight-loss prediction formulas in Further Reading 8. A web-based "body weight simulator" can more accurately predict expected weight loss over time. The body weight simulator, which can be found

As you read brochures, blogs, articles, or research reports about specific diet plans, look beyond the initial weight loss to see if the reported weight loss was maintained. If the weight loss was not maintained, then the program was not successful.

The 2010 Dietary Guidelines for Americans provide the following recommendations regarding "Balancing Calories to Manage Weight":

• Prevent and/or reduce overweight and obesity through improved eating and physical activity behaviors.
• Control total calorie intake to manage body weight. For people who are overweight or obese, this will mean consuming fewer calories from foods and beverages.
• Increase physical activity and reduce time spent in sedentary behaviors.
• Maintain appropriate calorie balance during each stage of life—childhood, adolescence, adulthood, pregnancy and breastfeeding, and older age.

at **http://bwsimulator.niddk.nih.gov,** projects weight loss over time based on an individual's height, weight, age, current calorie intake, calorie reduction, and activity level. As always, the new guidelines emphasize that the daily calorie deficit can come from decreased calorie intake, increased physical activity, or a combination of both.

WHAT TO LOOK FOR IN A SOUND WEIGHT-LOSS PLAN

A dieter can develop a plan of action by seeking advice from a health professional, such as a registered dietitian. Interactive tools such as **www.ChooseMyPlate.gov** can also aid in weight loss. A sound weight-loss program (Fig. 7-16) should especially include these components:

RATE OF LOSS

☐ Encourages slow and steady weight loss, rather than rapid weight loss, to promote lasting weight management
☐ Sets goal of 1 pound of fat loss per week
☐ Includes a period of weight maintenance for a few months after 10% of body weight is lost
☐ Evaluates need for further dieting before more weight loss begins

FLEXIBILITY

☐ Supports participation in normal activities (e.g., parties, and restaurants)
☐ Adapts to individual habits and tastes

INTAKE

☐ Meets nutrient needs (except for energy needs)
☐ Includes common foods, with no foods being promoted as magical or special
☐ Recommends a fortified ready-to-eat breakfast cereal or balanced multivitamin/mineral supplement, especially when intake is less than 1600 kcal per day
☐ Uses MyPlate or a comparable food guide as a pattern for food choices

BEHAVIOR MODIFICATION

☐ Focuses on maintenance of healthy lifestyle (and weight) for a lifetime
☐ Promotes reasonable changes that can be maintained
☐ Encourages social support
☐ Includes plans for relapse, so that one does not quit after a setback
☐ Promotes changes that control problem eating behaviors
☐ Promotes self-monitoring pratices such as keeping food diaries and setting goals

OVERALL HEALTH

☐ Requires screening by a physician for people with existing health problems, those over 40 (men) to 50 (women) years of age who plan to increase physical activity substantially, and those who plan to lose weight rapidly
☐ Encourages regular physical activity, sufficient sleep, stress reduction, and other healthy changes in lifestyle
☐ Addresses underlying psychological weight issues, such as depression or marital stress

▲ Slow, steady weight loss is one of the characteristics of a sound weight-loss program.

FIGURE 7-16 ▲ Characteristics of a sound weight-loss plan. Use this checklist to evaluate any new diet plan before putting it into practice.

1. **Control of calorie intake.** One recommendation is to decrease calorie intake by 500 kcal per day. This should allow for slow and steady weight loss.
2. **Increased physical activity** to the equivalent of more than 150 minutes (probably 300 minutes) of moderate-intensity aerobic activity each week.
3. **Behavior modification.**

A one-sided approach that focuses only on restricting calories is a difficult plan of action. Instead, adding physical activity and an appropriate psychological component will contribute to success in weight loss and eventual weight maintenance (Fig. 7-17). The 2010 Dietary Guidelines for Americans recommendations regarding "Balancing Calories to Manage Weight" are listed in the margin and focus on healthier dietary choices translating to reduced calorie intake and increased physical activity.

WEIGHT LOSS IN PERSPECTIVE

These principles point to the importance of preventing obesity. This concept has wide support because conquering the disorder is so difficult. Public health strategies to address the current obesity problem must speak to all age groups. There is a particular need to focus on children and adolescents because patterns of excess weight and sedentary lifestyle developed during youth may form the basis for a lifetime of weight-related illness and increased mortality. In the adult population, attention should be directed toward weight maintenance by making healthy dietary choices and increasing physical activity.

For more information on weight control, obesity, and nutrition, visit the Weight-Control Information Network (WIN) at **http://win.niddk.nih.gov/index.htm** or call 800-WIN-8098. Complete guidelines for weight management are available at **www.nhlbi.nih.gov/guidelines/index.htm**. Other websites include **www.caloriecontrol.org, www.weight.com, www.obesity.org,** and **www.cyberdiet.com.**

Control calorie intake

Control problem behaviors

Perform regular physical activity

FIGURE 7-17 ◀ Weight-loss triad. The key to weight loss and maintenance can be thought of as a triad, which consists of three parts: (1) controlling calorie intake, (2) performing regular physical activity, and (3) controlling problem behaviors. The three parts of the triad support each other in that without one part of the triad, weight loss and later maintenance become unlikely.

✔ CONCEPT CHECK 7.5

1. What are the characteristics of an appropriate weight-loss program?
2. What are the Dietary Guidelines related to Balancing Calories to Manage Weight?

Kilocalorie or Calorie? Kilocalorie (kcal) is a unit that describes the energy content of food. Specifically, a kilocalorie (kcal) is the heat energy needed to raise the temperature of 1000 grams (1 liter) of water 1° Celsius. Remember that although kcal refers to a 1000-calorie unit of measurement, it is commonly referred to as "calories". Calories is a familiar term for the energy content of a food, so we use it in this book.

Newsworthy Nutrition

Small changes result in large weight change over the years

Subjects in a large 20-year study of lifestyle factors and weight change gained an average of 20 pounds in 20 years. Less exercise led to weight gain, but the kinds of foods people ate had a larger effect than changes in physical activity. Increased consumption of French fries alone led to an average weight gain of 3.4 pounds every 4 years. Greater intake of fruits, vegetables, whole grains, yogurt, and nuts resulted in weight loss or no gain. The results of this study show how small changes in eating, exercise, and other habits can result in large changes in body weight over the years.

Source: Mozaffarian D and others: Changes in diet and lifestyle and long-term weight gain in women and men. *New England Journal of Medicine* 364:2392, 2011.

7.6 Control of Calorie Intake is Essential for Weight Management

Per day, the average woman requires 1800 to 2400 calories and the average male requires 2200 to 3000 calories. A goal of losing 1 pound or so of stored fat per week may require limiting calorie intake to 1200 kcal per day for women and 1500 kcal per day for men. While the number of calories allowed could be higher for very active people, decreasing calorie intake is necessary in our sedentary society. Keep in mind that this calorie intake is very restrictive, especially for those who have been eating in excess of their calorie requirements. In addition, any calorie reduction will promote the loss of lean along with fat tissue. With regard to consuming fewer calories, some experts suggest consuming less fat (especially saturated fat and *trans* fat), while others suggest consuming less carbohydrate, especially refined (high glycemic load) carbohydrate sources. Protein intakes in excess of what is typically needed by adults are also receiving attention. Using any of these approaches is fine. At this time, the low-fat, high-fiber approaches have been the most successful in long-term studies. A recent report (see Newsworthy Nutrition) confirms that the kinds of food we eat have a large effect on weight gain over the years. Finding what works for an individual is a process of trial and error. The notion that any type of diet promotes significantly greater calorie use by the body is unfounded (see Further Readings 17, 18, and 19).

Portion control is another challenge that influences our calorie intake and requires a change to our approach to eating. The concept of energy density can help dieters choose more foods that have fewer calories per gram. With this technique, we can fill our plates with larger portions of low-energy density foods, which are low in calories but high in volume. These low-energy density foods enable us to eat fewer calories without eating less food and help us feel full and satisfied while we are losing weight. Fruits and vegetables are great examples of low-energy density foods. The Volumetrics Diet (see Further Reading 7 in Chapter 2) is based on this energy-density approach to eating. While no foods are off limits, dieters are encouraged to eat foods similar to the ones they are craving such as crunchy carrots and hummus instead of chips and dip.

One way for a dieter to monitor calorie intake at the start of a weight-loss program is by reading labels. Label reading is important, because many foods are more energy dense than people suppose (Fig. 7-18). Another method is to write down food intake for 24 hours (Appendix C) and then calculate calorie intake from the food table in the textbook supplement or by using your diet-analysis software. With knowledge of current calorie intake, future food choices can be adjusted as needed. People often underestimate portion size when recording food intake, so measuring cups and a food scale can help.

Table 7-4 shows how to start reducing calorie intake. As you should realize, it is best to consider healthy eating a lifestyle change, rather than a weight-loss plan. Also, liquids deserve attention because liquid calories do not stimulate satiety mechanisms to the same extent as solid foods. The corresponding advice from experts is to use beverages that have few or no calories and limit sugar-sweetened beverages.

CONTROLLING HUNGER

A challenge to most weight-loss programs is to regulate hunger while eating less and exercising more. Separating true hunger from the eating you do out of habit or emotion is the first step toward controlling the hunger that can sabotage a diet. Hormones and the nervous system help you know when you are hungry. The hormone ghrelin in your blood, along with an empty stomach, signals the brain that you are hungry. Likewise, nerves in the stomach signal the brain when you are full, but it can take up to 20 minutes for these signals to reach the brain. The

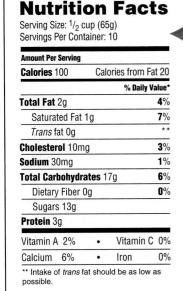

Nutrition Facts

Serving Size: ¹/₂ cup (65g)
Servings Per Container: 10

Amount Per Serving

Calories 100	Calories from Fat 20

	% Daily Value*
Total Fat 2g	**4%**
Saturated Fat 1g	**7%**
Trans fat 0g	* *
Cholesterol 10mg	**3%**
Sodium 30mg	**1%**
Total Carbohydrates 17g	**6%**
Dietary Fiber 0g	**0%**
Sugars 13g	
Protein 3g	

Vitamin A 2%	•	Vitamin C 0%
Calcium 6%	•	Iron 0%

** Intake of *trans* fat should be as low as possible.

Nutrition Facts

Serving Size: ¹/₂ cup (106g)
Servings Per Container: 10

Amount Per Serving

Calories 270	Calories from Fat 150

	% Daily Value*
Total Fat 17g	**25%**
Saturated Fat 11g	**54%**
Trans fat 0g	* *
Cholesterol 120mg	**40%**
Sodium 85mg	**4%**
Total Carbohydrates 20g	**7%**
Dietary Fiber 0g	**0%**
Sugars 20g	
Protein 5g	

Vitamin A 10%	•	Vitamin C 0%
Calcium 15%	•	Iron 0%

** Intake of *trans* fat should be as low as possible.

goal is to be hungry at meal time but not so ravenous that you are tempted to binge, and then to stop eating when you are comfortably full. If hunger strikes between meals, determine if you are feeling true hunger, and if you are, then choose a small, high-fiber snack to hold you over until the next meal. Drinking a glass of water can also help decrease hunger pangs between meals. Including lean protein (nuts, low-fat dairy, soy protein, or lean meat, fish, or chicken) in meals and snacks will also keep hunger at bay longer. Eating high-volume foods that are rich in water and fiber will provide bulk with fewer calories, fill up your stomach, and send the satiety signal to the brain. Becoming more mindful of what you eat will go a long way to controlling hunger and overall calorie intake (read more about mindful eating in Section 7.8). When you slow down, you will not only enjoy each mouthful but also give your stomach time to signal your brain that you are full.

FIGURE 7-18 ▲ Reading labels helps you choose foods with fewer calories. Which of these frozen desserts is the best choice, per ½ cup serving, for a person on a weight-loss diet? The % Daily Values are based on a 2000-kcal diet. Read more about these dessert choices in What the Dietitian Chose.

CONQUERING THE WEIGHT-LOSS PLATEAU

It is important for anyone on a weight-loss program to know that healthy weight loss is slow, sometimes erratic, and that it is normal to reach a weight-loss plateau. After losing pounds for weeks, suddenly the weight ceases to budge. Fortunately, there are some strategies to overcome these plateaus and start getting results again. There are several reasons why weight loss may stall. During the first part of a weight-loss program, dieters are typically losing fluid in addition to fat, causing a weight loss larger than the expected 1 to 2 pounds per week. Because a healthy weight-loss program is designed to cause fat loss rather than loss of muscle or fluid,

TABLE 7-4 ▶ **Saving Calories: Ideas for Getting Started**

Save This Much	By Choosing This	Instead of This
45 kcal	1 cup 1% milk	1 cup whole milk
50 kcal	12 oz light beer	12 oz regular beer
60 kcal	1 cup cornflakes	1 cup sugar-coated cornflakes
65 kcal	½ cup boiled potatoes	½ cup fried potatoes
135 kcal	1 cup plain popcorn	1 oz potato chips
140 kcal	3 oz lean beef	3 oz well-marbled beef
140 kcal	1 cup raw vegetables	½ cup potato salad
150 kcal	2 tbsp low-calorie salad dressing	2 tbsp regular salad dressing
150 kcal	1 English muffin	1 Danish pastry
150 kcal	6 oz wine cooler made with sparkling water	6 oz gin and tonic
150 kcal	12 oz sugar-free soft drink	12 oz regular soft drink
175 kcal	½ broiled chicken	½ batter-fried chicken
185 kcal	1 slice angel food cake	1 slice white iced cake
210 kcal	3 oz lean roast beef	½ cup beef stroganoff
310 kcal	1 apple	1 slice apple pie

your weight loss will begin to slow down after the first week or so. Also the level of calorie deprivation needed to lose weight is hard to maintain, and you may begin to eat more calories than your diet recommends. This "calorie creep" can contribute to the weight-loss plateau and eventually lead to weight gain. When this happens, it is important to go back to tracking your calories by weighing and measuring your food to understand proper portions. Another possible reason for the weight-loss plateau is that your metabolism is adjusting to your lower calorie intake. In this case, it may be time to reduce your calories somewhat. Your metabolism may also be adapting to your physical activity routine. Changing up the intensity of your workout routine, therefore, will help your muscles burn more calories and get you past the weight plateau. Strength training, along with the calorie-burning cardio exercises, is important to build muscle mass, which will ultimately use more calories for its metabolism.

✔ CONCEPT CHECK 7.6

1. How are hormones involved in hunger control?
2. What are "calorie creep" and the weight-loss plateau and how can they lead to weight gain?

7.7 Regular Physical Activity Promotes Weight Loss and Maintenance of a Healthy Weight

Regular physical activity is important for everyone, especially those trying to lose weight or maintain a lower body weight. Calorie burning is enhanced both during and after physical activity. Therefore, activity greatly complements a reduction in calorie intake for weight loss. Many of us rarely do more than sit, stand, and sleep. More calories are used during physical activity than at rest. Expending only 100 to 300 extra kcal per day above and beyond normal daily activity, while controlling calorie intake, can lead to a steady weight loss. Furthermore, physical activity has so many other benefits, including a boost for overall self-esteem. A Key Recommendation from the 2010 Dietary Guidelines is to increase physical activity and reduce time spent in sedentary behaviors. The Dietary Guidelines point to the 2008 Physical Activity Guidelines for Americans for specific recommendations. Weight management, as well as other health outcomes including diseases and risk factors for disease, was considered in developing the Physical Activity Guidelines. Although some adults will need a higher level of physical activity than others, it is recommended that adults should perform a minimum of 150 minutes of moderate-intensity aerobic activity each week to achieve and maintain a healthy body weight. Some may need more than the equivalent of 300 minutes per week of moderate-intensity activity.

Adding any of the activities in Table 7-5 to one's lifestyle can increase calorie use. Duration and regular performance, rather than intensity, are the keys to success with this approach to weight loss. One should search for activities that can be continued over time. In this regard, walking vigorously 3 miles per day can be as helpful as aerobic dancing or jogging, if it is maintained. Moreover, activities of lighter intensity are less likely to lead to injuries. Some resistance exercises (weight training) should also be added to retain and increase lean body mass and, in turn, fat use (see Chapter 10). As lean muscle mass increases, so will one's overall metabolic rate. Keep in mind that aerobic activity, including brisk walking, jogging, or cycling, burns more fat than resistance activity. An added benefit of including exercise in a weight-reduction program is maintenance of bone health.

Unfortunately, opportunities to expend calories in our daily lives are diminishing as technology systematically eliminates almost every reason to move our muscles.

Calorie estimation on exercise machines

The control panel of your exercise machine will typically display your time, speed, distance covered, and calories burned. While time, speed, and distance are generally accurate values, calories burned is a rough estimate based on the weight you enter before you start your workout. The calories burned are estimated based on formulas derived from testing on volunteers using the machine. They are not completely accurate because they do not consider factors other than weight, such as body fat percentage, fitness level, form, and running efficiency.

Treadmills and other cardio machines have been shown to overestimate calories burned by up to 15% to 20%. A heart rate monitor is usually more accurate at estimating the number of calories you burn during a workout than are calorie estimators on cardio machines.

▲ Physical activity complements any diet plan.

TABLE 7-5 ▶ **Approximate Calorie Costs of Various Activities and Specific Calorie Costs Projected for a 150-Pound (68-Kilogram) Person**

Activity	kcal per Kilogram per Hour	Total kcal per Hour	Activity	kcal per Kilogram per Hour	Total kcal per Hour
Aerobics—heavy	8.0	544	Horseback riding	5.1	346
Aerobics—medium	5.0	340	Ice skating (10 mph)	5.8	394
Aerobics—light	3.0	204	Jogging—medium	9.0	612
Backpacking	9.0	612	Jogging—slow	7.0	476
Basketball—vigorous	10.0	680	Lying—at ease	1.3	89
Bowling	3.9	265	Racquetball—social	8.0	544
Calisthenics—heavy	8.0	544	Roller skating	5.1	346
Calisthenics—light	4.0	272	Running or jogging (10 mph)	13.2	897
Canoeing (2.5 mph)	3.3	224	Skiing downhill (10 mph)	8.8	598
Cleaning (female)	3.7	253	Sleeping	1.2	80
Cleaning (male)	3.5	236	Swimming (0.25 mph)	4.4	299
Cooking	2.8	190	Tennis	6.1	414
Cycling (13 mph)	9.7	659	Volleyball	5.1	346
Cycling (5.5 mph)	3.0	204	Walking (3.75 mph)	4.4	299
Dressing/showering	1.6	106	Walking (2.5 mph)	3.0	204
Driving	1.7	117	Water skiing	7.0	476
Eating (sitting)	1.4	93	Weight lifting—heavy	9.0	612
Food shopping	3.6	245	Weight lifting—light	4.0	272
Football—touch	7.0	476	Window cleaning	3.5	240
Golf (using power cart)	3.6	244	Writing (sitting)	1.7	118

The values in Table 7-5 refer to total energy expenditure, including that needed to perform the physical activity, plus that needed for basal metabolism, the thermic effect of food, and thermogenesis. You can find the calorie costs of additional activities using the Physical Activity Calorie Counter at **http://www.acefitness.org/acefit/health_living_tools_content.aspx?id=9.**

The easiest way to increase physical activity is to make it an enjoyable part of a daily routine. To start, one might pack a pair of athletic shoes and walk around the parking lot before coming home after school or work every day. Other ideas are avoiding elevators in favor of stairs and parking the car farther away from the shopping mall.

A pedometer is an inexpensive device that monitors activity as steps. A recommended goal for activity is to take at least 10,000 steps per day; typically, we take half that many or less. A pedometer tracks this activity. Calorie counters, such as the Bodybugg, are new devices that track calorie expenditures throughout the day. The counters calculate calories by measuring heart rate, sweat rate, or heat loss and production. Like pedometers, calorie counters can motivate users to do more activity.

✔ CONCEPT CHECK 7.7

1. What should dieters remember about physical activity as part of their weight loss plan?

7.8 Behavior Modification Strategies for Weight Management

Setting realistic goals should be the first behavior modification step toward weight loss and management. Realistic weight-loss goals will keep you focused and motivated, and help to guarantee success. A successful weight-loss program is all about transitioning to a healthier lifestyle. The most helpful goals will focus on changes to

The motivation to lose weight and keep it off generally comes with a proverbial "flip of the switch," in which the desire to lose weight finally becomes more important than the desire to overeat.

your behavior, such as exercising three times a week or eating five servings of vegetables a day, rather than reaching a certain target weight. Making lifestyle changes means finding the problem behaviors that led to weight gain in the first place. Controlling calorie intake means modifying those *problem* behaviors. Only the dieter can decide what behaviors are preventing calorie control. What events cause us to start (or stop) eating? What factors influence food choices?

The 2010 Dietary Guidelines identify the following behaviors as having the strongest evidence related to body weight:

- Focus on the total number of calories consumed.
- Monitor food intake.
- When eating out, choose smaller portions or lower-calorie options.
- Prepare, serve, and consume smaller portions of foods and beverages, especially those high in calories.
- Eat a nutrient-dense breakfast.
- Limit screen time.

MINDFUL EATING

▲ Fruit is a great snack—high in nutrients and low in calories.

As we mentioned in the previous section on calorie control, becoming more mindful of what you eat will help control hunger and overall calorie intake. The concept of eating mindfully has become increasingly popular since publication in 2006 of the best-selling book *Mindless Eating: Why We Eat More Than We Think* by an expert on eating behavior, Dr. Brian Wansink of Cornell University. Making the break from mindless eating focuses on making changes to avoid triggers that may tempt you to eat less healthy food or to eat too much, or both. In contrast, the principles of mindful eating include being aware of the entire eating experience from food preparation to consumption, including recognizing and respecting the physical hunger and satiety cues, and being wise in eating decisions. Today, there are many publications, websites, and programs associated with the art of eating mindfully. The "Mindful Eating Cycle" from the book *Eat What You Love, Love What You Eat* by Michelle May includes the following key questions that can help you discern if you are eating for reasons other than physical hunger (see Further Reading 9). The answers to these questions will be instrumental in changing problem eating behavior.

- Why do I eat?
- When do I want to eat?
- What do I eat?
- How do I eat?
- How much do I eat?
- Where does the energy go?

A key component of mindful eating is a nonjudgmental awareness of eating patterns that will hopefully lead to a healthier relationship between you and food.

OTHER BEHAVIOR MODIFICATION STRATEGIES

Chain-breaking, stimulus control, cognitive restructuring, contingency management, and *self-monitoring* are behavior modification strategies used by psychologists that help place the problem in perspective and organize the intervention into manageable steps.

chain-breaking Breaking the link between two or more behaviors that encourage overeating, such as snacking while watching television.

stimulus control Altering the environment to minimize the stimuli for eating; for example, removing foods from sight and storing them in kitchen cabinets.

cognitive restructuring Changing one's frame of mind regarding eating; for example, instead of using a difficult day as an excuse to overeat, substituting other pleasures for rewards, such as a relaxing walk with a friend.

Chain-breaking separates behaviors that tend to occur together, for example, snacking on chips while watching television. Although these activities do not have to occur together, they often do. Dieters may need to break the chain reaction.

Stimulus control puts us in charge of temptations. Options include pushing tempting food to the back of the refrigerator, removing fat-laden snacks from the kitchen counter, and avoiding the path by the vending machines. Provide a positive stimulus by keeping low-fat snacks available to satisfy hunger/appetite.

Cognitive restructuring changes our frame of mind. For example, after a hard day, avoid using alcohol or comfort foods as quick relief for stress. Instead, plan for healthful, relaxing activities for stress reduction. For example, take a walk around the neighborhood or have a satisfying talk with a friend.

Labeling some foods as "off limits" sets up an internal struggle to resist the urge to eat that food. This hopeless battle can keep us feeling deprived. We lose the fight. Managing food choices with the principle of moderation is best. If a favorite food becomes troublesome, place it off limits only temporarily, until it can be enjoyed in moderation.

Contingency management prepares one for situations that may trigger overeating (e.g., when snacks are served at a party) or hinder physical activity (e.g., rain).

Self-monitoring can reveal problem eating behaviors—such as unconscious overeating—that may lead to weight gain. Records of dietary and physical activity behaviors can encourage new habits that will counteract unwanted behaviors. Obesity experts note that this is the key behavioral tool to use in any weight-loss program (see Further Reading 10). See the margin for a list of free online tools available for self-monitoring. Several of these online tools and others are also available as apps for mobile devices and smartphones (see Further Reading 4 for a review of smartphone apps for heart-healthy living).

Overall, it's important to address specific problems, such as snacking, compulsive eating, and mealtime overeating. Behavior modification principles (Table 7-6) are critical components of weight reduction and maintenance. Without them, it is difficult to make lifelong lifestyle changes needed to meet weight-control goals. Sign on to Connect to put these behavior modification strategies into action.

RELAPSE PREVENTION IS IMPORTANT

Preventing relapse is thought to be the hardest part of weight control—even harder than losing weight. A dieter needs to plan for lapses, not overreact, and take charge immediately. Change responses such as "I ate that cookie; I'm a failure" to "I ate that cookie, but I did well to stop after only one!" When dieters lapse from their diet plan, newly learned food habits should steer them back toward the plan. Without a strong behavioral program for **relapse prevention** in place, a lapse frequently turns into a relapse and a potential collapse. Once a pattern of poor food choices begins, dieters may feel failure and stray farther from the plan. As the relapse lengthens, the diet plan collapses, and falls short of the weight-loss goal. Losing weight is difficult. Overall, maintenance of weight loss is fostered by the "3 *Ms*": motivation, movement, and monitoring.

SOCIAL SUPPORT AIDS BEHAVIORAL CHANGE

Healthy social support is helpful in weight control. Helping others understand how they can be supportive can make weight control easier. Family and friends can provide praise and encouragement. Unfortunately, your social network can also sabotage your dieting efforts, so be aware of whom you can rely on for support. A registered dietitian or other weight-control professional can keep dieters accountable and help them learn from difficult situations. Long-term contact with a professional can be helpful for later weight maintenance. Groups of individuals attempting to lose weight or maintain losses can provide empathetic support.

SOCIETAL EFFORTS TO REDUCE OBESITY

The incidence of obesity in the United States is now considered an epidemic. An epidemic is a public health problem, and public health problems call for collective action. In fact, improvement in the health of our nation requires an approach that includes many sectors. Although we ultimately make our own choices at an individual level, partnerships, programs, and policies that support healthy eating and active living must be coordinated. The 2010 Dietary Guidelines' Call to Action includes three guiding principles:

1. Ensure that all Americans have access to nutritious foods and opportunities for physical activity.
2. Facilitate individual behavior change through environmental strategies that make it easier for individuals to make healthier choices.
3. Set the stage for lifelong healthy eating, physical activity, and weight-management behaviors.

contingency management Forming a plan of action to respond to a situation in which overeating is likely, such as when snacks are within arm's reach at a party.

self-monitoring Tracking foods eaten and conditions affecting eating; actions are usually recorded in a diary, along with location, time, and state of mind. This is a tool to help people understand more about their eating habits.

Food and Activity Tracking

Here are some websites where you can record your food and physical activity online for free:

http://www.myfitnesspal.com/
https://www.supertracker.usda.gov/
www.fitday.com
http://www.livestrong.com
http://nutritiondata.self.com
www.sparkpeople.com

relapse prevention A series of strategies used to help prevent and cope with weight-control lapses, such as recognizing high-risk situations and deciding beforehand on appropriate responses.

▲ Individuals who successfully maintain their weight loss employ a variety of strategies, such as yoga, to cope with the stresses and challenges of changing problem behaviors.

ChooseMyPlate.gov

▲ Large portions of food, such as this steak, provide us with many opportunities to overeat. It takes much perseverance to eat sensibly. How do the portion sizes shown here compare to those recommended on MyPlate?

TABLE 7-6 ▶ Behavior Modification Principles for Weight Loss

Shopping
1. Shop for food after eating; buy nutritious foods.
2. Shop from a list; limit purchases of irresistible "problem" foods. Shopping for fresh foods around the perimeter of the store first helps.
3. Avoid ready-to-eat foods.
4. Put off food shopping until absolutely necessary.

Plans
1. Plan to limit food intake as needed.
2. Substitute periods of physical activity for snacking.
3. Eat meals and snacks at scheduled times; don't skip meals.

Activities
1. Store food out of sight, preferably in the freezer, to discourage impulsive eating.
2. Eat all food in a "dining" area.
3. Keep serving dishes off the table, especially dishes of sauces and gravies.
4. Use smaller dishes and utensils.

Holidays and Parties
1. Drink fewer alcoholic beverages.
2. Plan eating behavior before parties.
3. Eat a low-calorie snack before parties.
4. Practice polite ways to decline food.
5. Don't get discouraged by an occasional setback.

Eating Behavior
1. Put the fork down between mouthfuls.
2. Chew thoroughly before taking the next bite.
3. Leave some food on the plate.
4. Pause in the middle of the meal.
5. Do nothing else while eating (for example, reading and watching television).

Reward
1. Plan specific rewards for specific behavior (behavioral contracts).
2. Solicit help from family and friends and suggest how they can help you. Encourage family and friends to provide this help in the form of praise and material rewards.
3. Use self-monitoring records as basis for rewards.

Self-Monitoring
1. Note the time and place of eating.
2. List the type and amount of food eaten.
3. Record who is present and how you feel.
4. Use the diet diary to identify problem areas.
5. Use online or mobile/smartphone apps to track your progress, including your new nutrition and health goals and habits (see Further Reading 4).

Cognitive Restructuring
1. Avoid setting unreasonable goals.
2. Think about progress, not shortcomings.
3. Avoid imperatives such as *always* and *never*.
4. Counter negative thoughts with positive restatements.

Portion Control
1. Make substitutions, such as a regular hamburger instead of a "quarter pounder" or cucumbers instead of croutons in salads.
2. Think small. Order the entrée and share it with another person. Order a cup of soup instead of a bowl or an appetizer in place of an entrée.
3. Use a doggie bag. Ask your server to put half the entrée in a doggie bag before bringing it to the table.

As we said at the start of this chapter, many of us need to become "defensive eaters." Know when to refuse food after satiety registers and reduce portion sizes.

Successful weight losers and maintainers from the National Weight Control Registry:

- Eat a low-fat, high-carbohydrate diet (on average 25% of calorie intake as fat).
- Eat breakfast almost every day.
- Self-monitor by regularly weighing oneself and keeping a food journal.
- Exercise for about 1 hour per day.
- Eat at restaurants only once or twice per week.

Other recent studies support this approach, especially the last four characteristics.

Public, private, and nonprofit organizations have begun to work together to address and reverse this public health crisis. For example, the U.S. Food and Drug Administration has brought together leaders from industry, government, academia, and the public health community to seek solutions to the obesity epidemic by making changes in foods eaten outside the home (restaurant and carry-out foods). These groups have collaborated and made recommendations to support the consumer's ability to manage calorie intake. Recommendations include "social marketing" programs that promote healthy eating and active living.

✔ CONCEPT CHECK 7.8

1. What behavior modification steps are helpful in changing problem eating behaviors to increase weight-loss success?
2. How does mindful eating help control hunger while dieting?

7.9 Professional Help for Weight Loss

The first professional to see for advice about a weight-loss program is one's family physician. Doctors are best equipped to assess overall health and the appropriateness of weight loss by examining health parameters such as blood pressure, blood lipids, and blood glucose that can be altered because of excess weight. The physician may then recommend a registered dietitian for a specific weight-loss plan and answers to diet-related questions. Registered dietitians are uniquely qualified to help design a weight-loss plan because they understand both food composition and the psychological importance of food. Exercise physiologists can provide advice about programs to increase physical activity. The expense for such professional interventions is tax deductible in the United States in some cases (see a tax advisor) and often covered by health insurance plans if prescribed by a physician.

Many communities have a variety of weight-loss organizations. These include self-help groups, such as Take Off Pounds Sensibly and Weight Watchers, which offer social support. Other programs, such as Jenny Craig and Physicians' Weight Loss Center, are less desirable for the average dieter because they require food purchases and don't encourage healthy food preparation. Often, the employees are not registered dietitians or other appropriately trained health professionals. These programs also tend to be expensive because of their requirements for intense counseling or mandatory diet foods and supplements. These commercial diet products as well as diet programs that require product purchases promote weight cycling, sometimes called yo-yo dieting, ultimately resulting in rebound weight gain at a higher level. In addition, the Federal Trade Commission has charged these and other commercial diet-program companies with misleading consumers through unsubstantiated weight-loss claims and deceptive testimonials.

MEDICATIONS FOR WEIGHT LOSS

Candidates for medications for obesity include those with a BMI of 30 or more or a BMI of 27 to 29.9 with weight-related health conditions, such as type 2 diabetes, cardiovascular disease, hypertension, or excess waist circumference; those with no contraindications to use the medication; and those ready to undertake lifestyle change. Drug therapy alone has not been found to be successful. Success with medications has been shown only in those who also modify their behavior, decrease calorie intake, and increase physical activity. For a drug to be considered effective in treating obesity, it must pass FDA guidelines and prove to be relatively safe.

Three main classes of medications have been used. An **amphetamine**-like medication (phenteramine [Fastin or Ionamin]) prolongs the activity of epinephrine and norepinephrine in the brain. This therapy is effective for some people in the short run but has not yet been proven effective in the long run. Most state medical boards limit use of

▲ All weight-loss programs should begin with a visit to your family physician.

amphetamine A group of medications that induce stimulation of the central nervous system and have other effects in the body. Abuse is linked to physical and psychological dependence.

Fat Digestion Without Orlistat

Enzymes called lipases help digest fat.

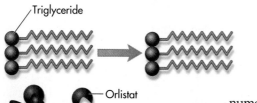

Lipase enzyme
Free fatty acid
Triglyceride
Lipase enzyme
Monoglyceride
Free fatty acid

Monoglycerides and free fatty acids are ready to be absorbed.

Fat Digestion With Orlistat

Orlistat attaches to lipases and blocks them from digesting some of the fat.

Triglyceride
Orlistat
Lipase enzyme

Intact triglycerides are not absorbed.

FIGURE 7-19 ▲ Orlistat is a weight-loss drug that works in the digestive system to block digestion of about one-third of the fat in the food we eat. A low-dose form of this drug (alli™) is now available without a prescription.

▲ alli™ is an over-the-counter weight-loss drug that helps block fat digestion and absorption.

this drug to 12 weeks unless the person is participating in a medical study using the product. The medication should not be used in pregnant or nursing women or those under 18 years of age.

The second class of medication approved by FDA for weight loss is orlistat (Xenical). This medication reduces fat digestion by about 30% by inhibiting lipase enzyme action in the small intestine (Fig. 7-19). This cuts absorption of dietary fat by one-third for about 2 hours when taken along with a meal containing fat. This malabsorbed fat is deposited in the feces. *Fat intake has to be controlled,* however, because large amounts of fat in the feces cause numerous side effects, such as gas, bloating, and oily discharge. Interestingly, orlistat use can remind the person to follow a fat-controlled diet, as the symptoms resulting from consuming a high-fat meal are unpleasant and develop quickly. Orlistat is taken with each meal containing fat. The malabsorbed fat also carries fat-soluble vitamins into the feces, so the person taking orlistat must take a multivitamin and mineral supplement at bedtime. In this way, any micronutrients not absorbed during the day can be replaced; fat malabsorption from the dinner meal will not greatly influence micronutrient absorption in the late evening. A low-dose form of orlistat (alli™) is now available over the counter without a prescription.

Three recently-approved weight loss medications work in various ways to curb appetite. Lorcaserin hydrochloride (Belviq®) suppresses appetite by altering serotonin receptors found within the feeding center of the hypothalamus in the brain (see Further Reading 13). It works similarly to the banned drugs fenfluramine (Fen-Phen) and sibutramine (Meridia), which caused various heart problems (see Further Reading 11). However, because lorcaserin is more selective, none of the heart issues have been documented in clinical tests. Weight loss with lorcaserin can also help people with type 2 diabetes to control blood sugar (see the Newsworthy Nutrition on the next page).

Qsymia® is a combination of phentermine and the seizure/migraine drug topiramate. Topiramate is effective in causing weight loss because it helps you feel full, makes foods taste less appealing, and helps you burn more calories. Qsymia® has serious side effects, however, including the potential to cause birth defects if taken during pregnancy.

The most recent drug approved for weight loss is Contrave. Contrave combines bupropion (the active ingredient in the antidepressant Wellbutrin) and naltrexone (an anti-addiction drug). The drug includes a warning about increased risk of suicidal thoughts or seizures. All of these new drugs continue to be tested.

Sometimes, physicians may prescribe medications that are not approved for weight loss but have weight loss as a side effect. Such an application is termed *off-label.* Over-the-counter medications and supplements are widely marketed as miracle cures for obesity, but in some cases, they do more harm than good. Today more than ever, let the buyer beware concerning any purported weight-loss aid not prescribed by a physician.

Overall, in skilled hands, prescription medications can aid weight loss in some instances. However, they do not replace the need for reducing calorie intake, modifying "problem" behaviors, and increasing physical activity, both during and after therapy. Often, any weight loss during drug treatment can be attributed mostly to the individual's hard work at balancing calorie intake with calorie output.

TREATMENT OF SEVERE OBESITY

Severe (morbid) obesity—having a BMI greater than or equal to 40 or weighing at least 100 pounds over healthy body weight (or twice one's healthy body weight)—requires professional treatment. Because of the serious health implications of severe

Newsworthy Nutrition

New weight-loss drug approved

The new obesity drug, lorcaserin, has been studied in five clinical trials and was approved by FDA in June 2012 for overweight and obesity management. Approval was based in part on the results of these studies (Behavioral Modification and Lorcaserin for Overweight and Obesity Management [BLOOM] and BLOOM in Diabetes Management [DM]). The hypothesis of the BLOOM-DM study was that use of lorcaserin would result in weight loss and improvement in glycemic control in patients with type 2 diabetes. The study was a randomized, placebo-controlled trial of 604 patients (18 to 65 years old, BMI 27 to 45 kg/m^2, and HbA_{1c} 7% to 10%) who received either the placebo, lorcaserin 10 milligrams once daily, or lorcaserin 10 milligrams twice daily, and received diet and exercise counseling. Safety monitoring included serial echocardiograms. Adverse events included headache, back pain, nasopharyngitis, and nausea. The use of lorcaserin for up to 1 year in these obese and overweight patients with type 2 diabetes was associated with statistically significant and clinically meaningful weight loss. More patients lost 5% or more of their body weight with lorcaserin twice a day (37.5%) or lorcaserin once a day (44.7%), compared to those taking a placebo (16.1%). Lorcaserin was also associated with improvement in measures of glycemic control, with decreases in HbA_{1c} and fasting glucose. Based on the significant improvements in glycemic control, the authors concluded that lorcaserin could be considered a useful weight management tool for overweight and obese type 2 diabetic patients.

Source: O'Neil PM and others: Randomized placebo-controlled clinical trial of lorcaserin for weight loss in type 2 diabetes mellitus: The BLOOM-DM Study. *Obesity* 20(7):1426, 2012.

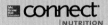 **connect** |NUTRITION   Check out the Connect site **www.mcgrawhillconnect.com** to further explore long-term weight-loss medications.

obesity, drastic measures may be necessary. Such treatments are recommended only when traditional diets and medications fail. Drastic weight-loss procedures are not without side effects, both physical and psychological, making careful monitoring by a physician a necessity.

Very-Low-Calorie Diets. If more traditional diet changes have failed, treating severe obesity with a **very-low-calorie diet (VLCD)** is possible, especially if the person has obesity-related diseases that are not well controlled (e.g., hypertension and type 2 diabetes). Some researchers believe that people with body weight greater than 30% above their healthy weight are appropriate candidates. The VLCD is dangerous because of its lack of fat, an essential nutrient. Major health risks include heart problems and gallstones. VLCD programs are offered almost exclusively by medical centers or clinics since careful monitoring by a physician is crucial throughout this very restrictive form of weight loss.

Optifast is one such commercial program. In general, the diet allows a person to consume only 400 to 800 kcal per day, often in liquid form. (These diets were previously known as protein-sparing modified fasts.) Of this amount, about 30 to 120 grams (120 to 480 kcal) is carbohydrate. The rest is high-quality protein, in the amount of about 70 to 100 grams per day (280 to 400 kcal). This low carbohydrate intake often causes ketosis, which may decrease hunger. However, the main reasons for weight loss are the minimal energy consumption and the absence of food choice. About 3 to 4 pounds can be lost per week; men tend to lose at a faster rate than women. When physical activity and resistance training augment this diet, a greater loss of adipose tissue occurs.

Weight regain remains a nagging problem, especially without a behavioral and physical activity component. If behavioral therapy and physical activity supplement a long-term support program, maintenance of the weight loss is more likely but still difficult. Any program under consideration should include a maintenance plan. Today, antiobesity medications may also be included in this phase of the program.

Bariatric Surgery. Bariatrics is the medical specialty focusing on the treatment of obesity. Bariatric surgery is only considered for people with severe obesity and includes operations aimed at promoting weight loss. Two types of bariatric operations are now common and effective (see Further Reading 12). Both procedures can be performed using an open (8- to 10-inch) incision in the middle of the abdomen or a laparoscopic approach in which several smaller (½- to 2-inch) incisions are used

very-low-calorie diet (VLCD) Known also as *protein-sparing modified fast (PSMF),* this diet allows a person 400 to 800 kcal per day, often in liquid form. Of this, 120 to 480 kcal is carbohydrate, and the rest is mostly high-quality protein.

bariatrics The medical specialty focusing on the treatment of obesity.

adjustable gastric banding A restrictive procedure in which the opening from the esophagus to the stomach is reduced by a hollow gastric band.

gastroplasty Gastric bypass surgery performed on the stomach to limit its volume to approximately 30 milliliters. Also referred to as stomach stapling.

Sleeve gastrectomy Surgical reduction of stomach by about 75% (to about the size of a banana). Also known as the gastric sleeve.

that allow cameras and instruments to enter the abdomen. **Adjustable gastric banding** (also known as the lap-band procedure) is a restrictive procedure in which the opening from the esophagus to the stomach is reduced by a hollow gastric band. This creates a small pouch and a narrow passage into the rest of the stomach and thus decreases the amount of food that can be eaten comfortably. The band can be inflated or deflated via an access port placed just under the skin. Studies have shown that adjustable gastric banding is more effective long term than a very low-calorie diet (400 to 800 kcal) for people who are about 50 pounds overweight.

Gastric bypass (also called **gastroplasty** or stomach stapling) is another bariatric surgical procedure used for treating severe obesity. The most common and effective approach (the Roux-en-Y gastric bypass procedure) works by reducing the stomach capacity to about 30 milliliters (the volume of one egg or shot glass) and bypassing a short segment of the upper small intestine. **Sleeve gastrectomy,** or gastric sleeve, reduces the size of the stomach by about 75% (to about the size of a banana) by removal of a large portion of the stomach following the major curve (Fig. 7-20). With both of these procedures, weight loss is promoted mainly because overeating of solid foods is now less likely due to rapid satiety and discomfort or vomiting after overeating.

Patient selection criteria for bariatric surgery include:

- BMI should be greater than 40.
- BMI between 35 and 40 is considered when there are serious obesity-related health concerns.
- Obesity must be present for a minimum of 5 years, with several nonsurgical attempts to lose weight.
- There should be no history of alcoholism or major untreated psychiatric disorders.

The risks of bariatric surgery include death and both early and late postoperative complications, such as bleeding, blood clots, hernias, and severe infections. Risk of death from these demanding surgeries can be as high as 2% (less risk with experienced surgeons). These risks depend on many factors related to the surgeon and facility, the patient, and the procedure. The procedures that are simply restrictive (e.g., adjustable gastric banding and sleeve gastrectomy) do not cause malabsorption and rarely affect bowel function. However, for those procedures that induce malabsorption (e.g., Roux-en-Y gastric bypass), nutrient deficiencies are of greater concern if the person is not adequately treated in the years following the surgery. Anemia and bone loss might then be the result.

The person also must consider that the surgery is costly and may not be covered by medical insurance. The average cost for gastric bypass produres is $18,000 to $35,000 and for adjustable gastric banding is $17,000 to $30,000. In addition, follow-up surgery is often needed after weight loss to correct stretched skin, previously filled with fat. Furthermore, the surgery necessitates major lifestyle changes, such as the need to plan frequent, small meals. Therefore, the dieter who has chosen this drastic approach to weight loss faces months of difficult adjustments.

Despite potential adverse effects, the benefits of bariatric surgery usually outweigh the risks. In terms of long-term weight-loss success, bariatric surgeries have the best results of any intervention. Weight loss statistics vary by surgical method, but on average, about 75% of people with severe obesity eventually lose *and keep off* 50% or more of excess body weight. In addition, many patients experience significant improvements in blood sugar, cholesterol, and blood pressure. By no means is bariatric surgery a quick and easy fix for obesity, but with a serious commitment to permanent lifestyle changes and long-term follow-up with a health professional, these procedures can positively impact both quality and quantity of life.

Lipectomy. Spot reducing by using diet and physical activity is not possible. "Problem" local fat deposits can be reduced in size, however, using suction lipectomy. Lipectomy or liposuction means surgical removal of fat. A pencil-thin tube is inserted into an incision in the skin, and the fat tissue, such as that in the buttocks and thigh area, is suctioned. This procedure carries some risks, such as infection; lasting depressions in

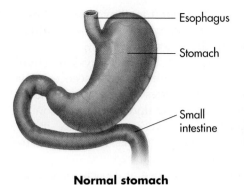

Normal stomach

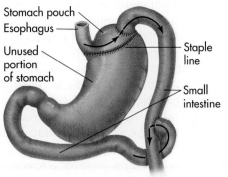

Gastric bypass

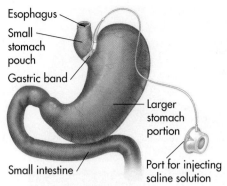

Gastric banding (e.g., LAP-BAND® procedure)

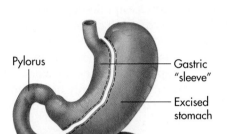

Sleeve gastrectomy

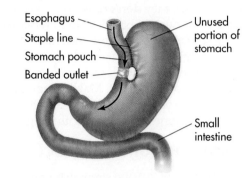

Vertical-banded gastroplasty

FIGURE 7-20 ▲ The normal stomach and the most common forms of gastroplasty for treating severe obesity. The gastric bypass is the most effective method. In banded gastroplasty, the band prevents expansion of the outlet for the stomach pouch.

the skin; and blood clots, which can lead to kidney failure and sometimes death. The procedure is designed to help a person lose about 4 to 8 pounds per treatment. Cost is about $1800 per site; total costs range from $2600 to $9000.

✔ CONCEPT CHECK 7.9

1. What are the surgical options for obese people who have failed to lose weight with other weight-loss strategies?
2. How restrictive is a very-low-calorie diet plan? Why is monitoring by a qualified health professional important?

7.10 Treatment of Underweight

Underweight is defined by a BMI less than 18.5 and can be caused by a variety of factors, such as cancer, infectious disease (e.g., tuberculosis), digestive tract disorders (e.g., chronic inflammatory bowel disease), and excessive dieting or physical activity. Genetic background may also lead to a higher RMR, a slight body frame, or both. Health problems associated with underweight include the loss of menstrual function, low bone mass, complications with pregnancy and surgery, and slow recovery after illness. Significant underweight is also associated with increased death rates, especially when combined with cigarette smoking. We frequently hear about the risks of obesity but seldom of underweight. In our culture, being underweight is much more socially acceptable than being obese.

Sometimes being underweight requires medical intervention. A physician should be consulted first to rule out hormonal imbalances; depression; cancer; infectious disease; digestive tract disorders; excessive physical activity; and other hidden disease, such as a serious eating disorder (see Chapter 11 for a detailed discussion of eating disorders).

underweight A body mass index below 18.5. The cutoff is less precise than for obesity because this condition has been less studied.

▲ Underweight people should increase their consumption of calorie-dense foods, such as smoothies, that are also loaded with nutrients.

The causes of underweight are not altogether different from the causes of obesity. Internal and external satiety-signal irregularities, the rate of metabolism, hereditary tendencies, and psychological traits can all contribute to underweight.

In growing children, the high demand for calories to support physical activity and growth can cause underweight. During growth spurts in adolescence, active children may not take the time to consume enough calories to support their needs. Moreover, gaining weight can be a formidable task for an underweight person. An extra 500 kcal per day may be required to gain weight, even at a slow pace, in part because of the increased expenditure of energy in thermogenesis. In contrast to the weight loser, the weight gainer may need to increase portion sizes.

When underweight requires a specific intervention, one approach for treating adults is to gradually increase their consumption of calorie-dense foods, especially those high in vegetable fat. Nuts and granola can be good calorie sources with low saturated fat content. Dried fruits and bananas are good fruit choices. If eaten at the end of a meal, they don't cause early satiety. The same advice applies to salads and soups. Underweight people should replace such foods as diet soft drinks with good calorie sources, such as fruit juices and smoothies.

Encouraging a regular meal and snack schedule also aids in weight gain and maintenance. Sometimes underweight people have experienced stress at work or have been too busy to eat. Making regular meals a priority may not only help them attain an appropriate weight but also help with digestive disorders, such as constipation, sometimes associated with irregular eating times.

GAINING WEIGHT AS MUSCLE, NOT FAT

A combination of diet and strength training is needed to gain weight as muscle. Strength training slows muscle loss that comes with dieting and age, builds the strength of your muscles and connective tissues, and increases bone density. It is especially important for dieters. When weight is lost, up to a quarter of the loss may come from muscle, which can slow metabolism. Strength training helps rebuild any muscle lost by dieting—or keeps it from being lost in the first place. The best bet when starting a strength-training program is to get one-on-one help from a qualified fitness trainer who can address personal goals and limitations and can help with alignment and execution of each exercise.

There are several things to consider when designing the diet to accompany training. During a workout, the body goes into a state of breaking down muscle. Once you're done lifting, you want to put it back into a state of building muscle again. It is very important to get something in the body after a workout to start that building phase. Ideally, a serving of protein should be eaten both before and after a workout to optimize performance and build lean muscle mass. Whey protein can be added to shakes, peanut butter, oatmeal, and just about anything else. It is also important to have some carbohydrate along with protein to increase the protein absorption and to provide fuel for the workout. While a quick protein bar or shake is great when you're at the gym, it should not be the only source of protein. During meals, lean protein sources such as lean chicken, lean beef, soy, and beans are the healthiest. Those who work out but eat nothing but food high in fat and calories will gain fat on top of their muscle. To gain lean muscle mass, one needs a balanced diet of protein and carbohydrates, including lots of fruits, vegetables, and whole grains, and a low level of fat. The number of calories you require each day varies greatly and will depend on your weight, activity level, age, and muscle mass. If you are working out 3 days a week, you can eat about 15 kcal per pound of body weight. If you who work out 5 days a week, you can up that kcal count to 20 per pound.

☑ CONCEPT CHECK 7.10

1. How is underweight defined and what are some of its primary causes?
2. What are the components necessary to gain weight as muscle and not as fat?

Nutrition and Your Health

Popular Diets—Cause for Concern

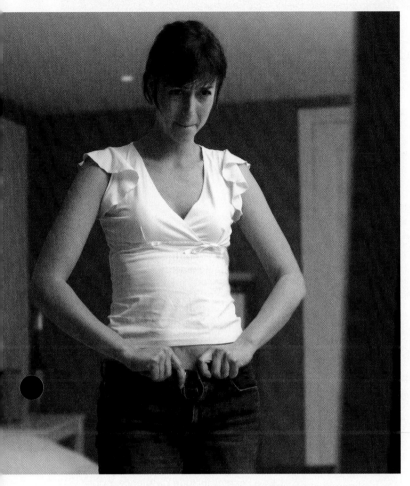

Many overweight people try to help themselves by using the latest popular (also called fad) diet book. But, as you will see, most of these diets do not help, and some can actually harm those who follow them (Table 7-7). Research has shown that early dieting and other unhealthful weight control practices in adolescents lead to an increased risk of weight gain, overweight, and eating disorders.

Recently, weight-loss experts came together at the request of the USDA to evaluate weight-loss diets. They came to this conclusion: Forget these fads when it comes to dieting. Most of the popular diets are nutritionally inadequate and include certain foods that people would not normally choose to consume in large amounts. The experts stated that eating less of one's favorite foods and becoming more physically active can be much more effective when trying to implement a weight-loss diet. People need a plan they can live with in the long run so that a healthy weight becomes permanent. The goal should be weight control over a lifetime, not immediate weight loss. Every popular diet leads to some immediate weight loss simply because daily intake is monitored and monotonous food choices are typically part of the plan. A well-known example of the effectiveness of monotony contributing to weight loss is the experience of Jared Fogle. He ate primarily Subway sandwiches for 11 months and lost 245 pounds. He notes, however, that this is not a miracle diet; it

takes a lot of hard work to lead to the success he experienced. There are also many other examples where diet monotony has led to weight loss. Overall, a traditional moderate diet coupled with regular physical activity is adequate for weight loss.

People on diets often fall within a healthy BMI of 18.5 to 25. Rather than worrying about weight loss, these individuals should be focusing on a healthy lifestyle that allows for weight maintenance. Incorporating necessary lifestyle changes and learning to accept one's particular body characteristics should be the overriding goals.

The dieting mania can be viewed as mostly a social problem, stemming from unrealistic weight expectations (especially for women) and lack of appreciation for the natural variety in body shape and weight. Not every woman can look like a fashion model, nor can every man look like a Greek god, but all of us can strive for good health and, if physically possible, an active lifestyle.

The size-acceptance nondiet movement, "Health at Every Size," has attempted to shift the paradigm away from the use of "popular" weight loss diets. The goals of the movement are all independent of body weight and include improvement of self-image, normalization of eating behavior, and increase in physical activity.

How to Recognize an Unreliable Diet

The criteria for evaluating weight-loss programs with regard to their safety and effectiveness were discussed in Section 7.5 (review Fig. 7-16). In contrast, unreliable diets typically share some common characteristics:

1. They promote quick weight loss. This is the primary temptation that attracts the dieter. As mentioned, this initial weight loss primarily results from water loss and lean muscle mass depletion.
2. They limit food selections and dictate specific rituals, such as eating only fruit for breakfast or cabbage soup every day.
3. They use testimonials from famous people and tie the diet to well-known cities, such as Beverly Hills and South Beach.
4. They bill themselves as cure-alls. These diets claim to work for everyone, whatever the type of obesity or the person's specific strengths and weaknesses.
5. They often recommend expensive supplements.
6. No attempts are made to change eating habits permanently. Dieters follow the diet until the desired weight is reached and then revert to old eating habits; they are told, for example, to eat rice for a month, lose weight, and then return to old habits.
7. They are generally critical of and skeptical about the scientific community. The lack of a quick fix from medical and dietetic professionals has led some of the public to seek advice from those who appear to have the answer.
8. They claim that there is no need to exercise.

Probably the cruelest characteristic of these diets is that they essentially guarantee failure for the dieter. The diets are not designed for permanent weight loss. Habits are not changed, and the food selection is so limited that the person cannot

TABLE 7-7 ▶ **Summary of Popular Diet Approaches to Weight Control**

Approach	Examples*	Characteristics	Dietitian's Review
Moderate calorie restriction	• *Dieting for Dummies* (2003) • *Dieting with the Duchess* (2000) • *Dr. Phil's Ultimate Weight Solution* (2003, 2005) • Flat Belly Diet (2008) • Jenny Craig (1980s) • *Jumpstart to Skinny* (2013) • *Picture Perfect Weight Loss* (2003) • Slim-fast (1980s) • *Sonoma Diet* (2005) • *Ultimate Volumetrics Diet* (2012) • *Wedding Dress Diet* (2000) • Weight Watchers (1960s) • *You on a Diet* (2006)	• Generally 1200 to 1800 kcal per day • Moderate fat intake • Reasonable balance of macronutrients • Encourage exercise • May use behavioral approach	These diets are acceptable if a multivitamin and mineral supplement is used and permission of family physician is granted.
Carbohydrate focused	• *Carbohydrate Addicts Diet* (1993, 2001) • *Dr. Atkin's Diet Revolution* (1973, 2002) • *Dr. Gott's No Flour, No Sugar Diet* (2006) • *Eat, Drink & Weigh Less* (2006) • *G.I. (Glycemic Index) Diet* (2003) • *Healthy for Life* (2005) • *New Glucose Revolution* (2002) • Nutrisystem (2003) • *South Beach Diet* (especially initial phases) (2003) • *Sugar Busters Diet* (1998, 2003) • *Zone Diet* (1995)	• Restricted carbohydrate diets generally advise consumption of fewer than 100 grams of carbohydrate per day. • Some plans focus on carbohydrate choices (e.g., choosing low rather than high glycemic index foods).	Selecting high-fiber, whole-grain sources of carbohydrates is an advisable practice for weight control and prevention of several chronic diseases. However, severe carbohydrate restriction may lead to ketosis, reduced exercise capacity (due to poor glycogen stores in the muscles), excessive animal fat intake, constipation, headaches, halitosis (bad breath), and muscle cramps. Severe carbohydrate restriction is not a nutritionally sound, long-term weight-loss solution.
Low fat	• *20/30 Fat and Fiber Diet Plan* (2000) • *Complete Hip and Thigh Diet* (1989, 1999) • *Eat More, Weigh Less* (1993, 2001) • *Fit or Fat* (1977, 2005) • *Foods That Cause You to Lose Weight* (1992, 2003) • *McDougall Program* (1983, 1995) • Pritikin Diet (1984, 1995) • *Rice Diet Solution* (2005) • *T-Factor Diet* (1989, 2001) • *Okinawa Program* (2002)	• Generally less than 20% of calories from fat • Limited (or elimination of) animal protein sources; also limited plant oils, nuts, and seeds	Low-fat diet plans are not necessarily to be avoided, but certain aspects may be unacceptable. Some potentially negative outcomes include flatulence, poor mineral absorption (from excess fiber intake), and a sense of deprivation (due to limited food choices).
Novelty diets	• *17-Day Diet* (2011) • *3-Hour Diet* (2005) • *Alkaline Cure* (2014) • *Beverly Hills Diet* (1981, 1996) • Cabbage-Soup Diet (2004) • *Eat Right 4 Your Type* (1996) • Fast Metabolism Diet (2013) • *Fat Smash Diet* (2006) • *Fit for Life* (1987, 2001) • *Metabolic Typing Diet* (2002) • Paleo Diet (2010) *New Hilton Head Metabolism Diet* (1983, 1996) • *Ultrametabolism* (2006) • *Weigh Down Diet* (2002) • *Wheat Belly* (2011)	• Promote certain nutrients, foods, or combinations of foods as having unique, magical, or previously undiscovered qualities	Novelty diets are usually not nutritionally balanced, thus malnutrition is a possible result. Also, failure to make long-term changes may lead to relapse, and unrealistic food choices lead to possible bingeing.

*Dates listed are original release date followed by most recent release date, if applicable.

follow the diet in the long run. Although dieters assume that they have lost fat, they have lost mostly muscle and body fluids. As soon as they begin eating normally again, much of the lost tissue is replaced. In a matter of weeks, most of the lost weight is back. The dieter appears to have failed, when actually the *diet* has failed. The gain and loss cycle is called weight cycling or "yo-yo" dieting. This whole scenario can add more blame and guilt, challenging the self-worth of the dieter. It can also come with some health costs, such as increased upper-body fat deposition. If someone needs help losing weight, professional help is advised. It is unfortunate that current trends suggest that people are spending more time and money on "quick fixes" than on such professional help.

The hCG diet is an fairly recent example of an unsafe diet that fits these characteristics. Human chorionic gonadotropin (hCG) is a hormone found in the urine of pregnant women and has been associated with dieting because, theoretically, it is supposed to suppress hunger and trigger your body's use of fat for fuel. The truth is that clinical studies have shown hCG to be ineffective as a weight-loss aid. Not only has FDA not approved any hCG weight-loss products, but FDA and Federal Trade Commission have issued warning letters to companies marketing over-the-counter hCG products labeled as "homeopathic" for weight loss. Companies have been warned that they are violating federal law by selling drugs that have not been approved and by making unsupported claims for the products. It is true that individuals will lose weight on the hCG diets, but only because most hCG diets limit dieters to only 500 calories per day in addition to the hCG injections or serum drops under the tongue. These extremely low-calorie diets are virtually impossible to maintain and can lead to nutrient deficiencies and harmful side effects from the injections.

Types of Popular Diets

HIGH-PROTEIN, LOW-CARBOHYDRATE APPROACHES

High-protein, low-carbohydrate diets are a popular approach to losing weight. These diets typically recommend 30% to 50% of their total calories from protein and drastically restrict carbohydrate. Low-carbohydrate intake leads to less glycogen synthesis and therefore less water in the body (about 3 grams of water are stored per gram of glycogen). As discussed in Chapter 4, a very-low-carbohydrate intake also forces the liver to produce

▲ In time, the very-low-carbohydrate, high-protein diets typically leave a person wanting more variety in meals, and so the diets are abandoned. Dropout rates are high on these diets.

needed glucose. The source of carbons for this glucose is mostly proteins from tissues such as muscle, resulting in loss of protein tissue, about 72% water. Essential ions such as potassium are also lost in the urine. In the initial stages of a low-carbohydrate diet, losses of glycogen stores, lean tissue, and water cause rapid weight loss. When a normal diet is resumed, the protein tissue is rebuilt and the weight is regained.

In addition, restricting carbohydrate causes your body to burn fat instead of carbohydrate for fuel. In theory, this burning of excess fat stores makes a lot of sense for weight loss. Remember, however, that when we burn fat without carbohydrate, it causes the body to go into the metabolic state called ketosis. Although ketosis has the benefit of making you feel less hungry, risks of chronic, diet-induced ketosis have not been adequately assessed in long-term studies.

For some dieters, a low-carb plan is such a major change from normal habits that it is difficult to maintain. However, research indicates that low-carbohydrate diets may be an effective long-term alternative to low-fat diets for some people. In a recent 12-month study, otherwise healthy, overweight or obese adults who followed a low-carbohydrate diet were able to lose more weight than adults on a low-fat diet. In addition, despite concerns that a high-protein diet may promote cardiovascular disease, the low-carb dieters experienced greater improvements in blood lipids as they lost weight (see Further Reading 2). In a recent 2-year study published in the *New England Journal of Medicine,* moderately obese adults on a low-carbohydrate diet lost and kept off about 12 pounds, compared to 10 pounds for those following the traditional Mediterranean diet and 7 pounds for those on a restricted-fat plan. This study suggested that the ability of the dieter to stick with moderate calorie restriction over the long term is a better determinant of success than the exact macronutrient composition of the diet (see Further Reading 17).

The most popular diet using a low-carbohydrate approach is Dr. Atkins' New Diet Revolution. More moderate approaches are found in the various Zone diets (40% of calorie intake as carbohydrate), Sugar Busters diet, and the South Beach diet (especially initial phases).

CARBOHYDRATE-FOCUSED DIETS

Several recent diets, including Sugar Busters, the Glucose Revolution, and Eat, Drink, and Weigh Less, do not restrict carbohydrates but rather emphasize the "good" carbohydrates in place of the "bad" or "harmful" ones. These diets recommend eating plenty of fruits, vegetables, whole grains, and cutting out simple sugars and processed grains. The carbohydrate-focused diets rely largely on low glycemic index foods. In theory, these foods will cause a slow, steady rise and fall in blood sugar after a meal, which will help control hunger.

LOW-FAT APPROACHES

The very-low-fat diets contain approximately 5% to 10% of calories as fat and are very high in carbohydrates. The most notable are the Pritikin Diet and the Dr. Dean Ornish "Eat More, Weigh Less" diet plans. If followed consistently, these approaches lead to weight loss and they may be helpful for reducing heart disease risks. However, they are difficult to follow. People are quickly bored with this type of diet because they cannot eat many of their favorite foods. These dieters eat primarily grains, fruits, and vegetables, which most people cannot do for very long. Eventually, the person wants some foods higher in fat or protein. Furthermore, such diets may be too much carbohydrate for some people who have diabetes in their family history.

NOVELTY DIETS

A variety of diets are built on gimmicks. Some novelty diets emphasize one food or food group and exclude almost all others. A rice diet was designed in the 1940s to lower blood pressure; now it has resurfaced as a weight-loss diet. The first phase consists of eating only rice and fruit. On the Beverly Hills Diet, you eat mostly fruit. The Paleo Diet is designed to mimic the diets of cavemen and includes meats, seafood, vegetables, fruits, and nuts.

The most questionable of the novelty diets propose that "food gets stuck in your body." Fit for Life, the Beverly Hills Diet, and Eat Great, Lose Weight are examples. The supposition is that food gets stuck in the intestine, putrefies, and creates toxins, which invade the blood and cause disease. In response, recommendations are to not consume meat with potatoes or to consume fruits only after noon. These recommendations make no physiological sense and promote weight cycling.

MEAL REPLACEMENTS

Meal replacements come in many forms, including beverages or formulas, frozen or shelf-stable entrees, and meal or snack bars.

Most meal replacements are fortified with vitamins and minerals and are appropriate to replace one or two regular meals or snacks per day. Although they are not a "magic bullet" for weight loss, they have been shown to help some people lose weight. Advantages of these convenient products are that they provide portion- and calorie-controlled foods that can serve as a visual education on appropriate portion sizes. A disadvantage is that when dieters rely on foods selected and prepared by someone else, they do not learn to select and prepare healthy foods on their own.

QUACKERY IS CHARACTERISTIC OF MANY POPULAR DIETS

Many popular diets fall under the category of quackery—people taking advantage of others. They usually involve a product or service that costs a considerable amount of money. Often, those offering the product or service don't realize that they are promoting quackery because they were victims themselves. For example, they tried the product and by pure coincidence it worked for them, so they wish to sell it to all their friends and relatives.

Numerous other gimmicks for weight loss have come and gone and are likely to resurface. If in the future an important aid for weight loss is discovered, you can feel confident that major journals, such as the *Journal of the Academy of Nutrition and Dietetics, Journal of the American Medical Association,* or *New England Journal of Medicine,* will report it.

CASE STUDY Choosing a Weight-Loss Program

Joe has a hectic schedule. During the day, he works full-time at a warehouse distribution center filling orders. At night, three times a week, he attends class at the local community college in pursuit of computer certification. On weekends, he likes to watch sports on TV, spend time with family and friends, and study. Joe has little time to think about what he eats—convenience rules. He stops for coffee and a pastry on his way to work, has a burger or pizza for lunch at a quick service restaurant, and for dinner picks up fried chicken or fish at the drive-through on his way to class. Unfortunately, over the past few years, Joe's weight has been climbing. He is 5 feet, 10 inches tall and weighs 200 pounds. Lately, he has frequently been out of breath during his shift at work. Watching a game on television a few nights ago, he saw an infomercial for a weight-loss supplement that promises to increase his energy level and allow him to continue to eat large portions of tasty foods but not gain weight. A famous actor supports the claim that this product allows one to eat at will and not gain weight. This claim is tempting to Joe.

Answer the following questions, and check your response at the end of the chapter.

▲ What changes can Joe make in his daily routine and diet to stop his weight gain?

1. Has Joe been experiencing positive or negative energy balance over the past few years? What is his current BMI?
2. What aspects of Joe's lifestyle (other than diet) are causing this effect on his energy balance? What changes could Joe make in his habits to promote weight loss or maintenance?
3. What changes could Joe make in his diet that would promote weight loss or maintenance?
4. Why should Joe be skeptical of the claims he heard about the weight-loss product in the infomercial?
5. Referring back to the characteristics of unreliable diets, what advice can you offer Joe for evaluating weight-loss programs?

Summary (Numbers refer to numbered sections in the chapter.)

7.1 Energy balance considers energy intake and energy output. Negative energy balance occurs when energy output surpasses energy intake, resulting in weight loss. Positive energy balance occurs when calorie intake is greater than output, resulting in weight gain.

Basal metabolism, the thermic effect of food, physical activity, and adaptive thermogenesis account for total energy use by the body. Basal metabolism, which represents the minimum amount of calories required to keep the resting, awake body alive, is primarily affected by lean body mass, surface area, and thyroid hormone concentrations. Physical activity is energy use above that expended at rest. The thermic effect of food describes the increase in metabolism that facilitates digestion, absorption, and processing of the nutrients recently consumed. Adaptive thermogenesis includes nonvoluntary activities, such as shivering and fidgeting, that increase energy use and may counter extra calories from overeating. In a sedentary person, about 70% to 85% of energy use is accounted for by basal metabolism and the thermic effect of food.

7.2 Energy use by the body can be measured as heat given off by direct calorimetry or as oxygen used by indirect calorimetry. A person's Estimated Energy Requirement can be calculated based on the following factors: gender, height, weight, age, and amount of physical activity.

7.3 A body mass index (weight in kilograms ÷ height² in meters) of 18.5 to 24.9 is one measure of healthy weight. A healthy weight is best determined in conjunction with a thorough health evaluation by a physician. A body mass index of 25 to 29.9 represents overweight. Obesity is defined as a total body fat percentage over 25% (men) or 35% (women) or a body mass index of 30 or more.

Fat distribution greatly determines health risks from obesity. Upper-body fat storage, as measured by a waist circumference greater than 40 inches (102 centimeters) for men or 35 inches (88 centimeters) for women typically results in higher risks of hypertension, cardiovascular disease, and type 2 diabetes than does lower-body fat storage.

7.4 Both genetic (nature) and environmental (nurture) factors can increase the risk of obesity. The set-point theory proposes that we have a genetically predetermined body weight or body fat content, which the body regulates.

7.5 A sound weight-loss program meets the dieter's nutritional needs by emphasizing a wide variety of low-calorie, bulky foods; adapts to the dieter's habits; consists of readily obtainable foods; strives to change poor eating habits; stresses regular physical activity; and stipulates the supervision by a physician if weight is to be lost rapidly or if the person is over the age of 40 (men) or 50 (women) and plans to perform substantially greater physical activity than usual.

7.6 New weight-loss formulas predict a much slower and realistic pattern of weight loss, with a calorie deficit of 10 kcal per day resulting in a 1-pound weight loss over 3 years. Weight change is not linear; it occurs most rapidly during the first year after a change in energy balance, but tapers off over the next 2 years if changes in energy intake and physical activity are strictly maintained over time.

7.7 Physical activity as part of a weight-loss program should be focused on duration rather than intensity. Ideally, about 60 minutes of moderate-intensity physical activity should be part of each day to prevent adult weight gain.

7.8 Behavior modification is a vital part of a weight-loss program because the dieter may have many habits that discourage weight maintenance. Specific behavior modification techniques, such as stimulus control and self-monitoring, can be used to help change problem behavior.

7.9 Medications to blunt appetite, such as phentermine (Fastin) and lorcaserin (Belviq), can aid weight loss. Orlistat (Xenical) reduces fat absorption from a meal when taken with the meal. Weight-loss drugs are reserved for those who are obese or have weight-related problems, and they must be administered under close physician supervision.

The treatment of severe obesity may include very-low-calorie diets containing 400 to 800 kcal per day or bariatric surgery to reduce stomach volume to approximately 30 milliliters (1 ounce). Both of these measures should be reserved for people who have failed at more conservative approaches to weight loss. They also require close medical supervision.

7.10 Underweight can be caused by a variety of factors, such as excessive physical activity and genetic background. Sometimes being underweight requires medical attention. A physician should be consulted first to rule out underlying disease. The underweight person may need to increase portion sizes and learn to like calorie-dense foods. In addition, encouraging a regular meal and snack schedule aids in weight gain, as well as weight maintenance.

NAYH Many overweight people try popular diets that most often are not helpful and may actually be harmful. Unreliable diets typically share some common characteristics, including promoting quick weight loss, limiting food selections, using testimonials as proof, and requiring no exercise.

Check Your Knowledge (Answers to the following questions are below.)

1. An energy deficit of 100 kcal per day would result in a total weight loss of about 10 pounds over a _____ period.
   a. 1-week
   b. 4-week
   c. 1-year
   d. 3-year

2. Thermic effect of food represents the energy cost of
   a. chewing food.
   b. peristalsis.
   c. basal metabolism.
   d. digesting, absorbing, and packaging nutrients.

3. A well-designed weight-loss diet should
   a. increase physical activity.
   b. alter problem behaviors.
   c. reduce energy intake.
   d. All of the above.

4. All of the following factors are associated with a higher basal metabolic rate *except*
   a. stress.
   b. low calorie intake.
   c. fever.
   d. pregnancy.

5. The intent of bariatric surgery is to
   a. limit stomach volume.
   b. slow transit time.
   c. surgically remove adipose tissue.
   d. prevent snacking.

6. Basal metabolism
   a. represents about 30% of total energy expenditure.
   b. is energy used to maintain heartbeat, respiration, other basic functions, and physical activities.
   c. represents about 60% to 75% of total calories used by the body during a day.
   d. includes energy to digest food.

7. It is recommended that adults should do the equivalent of _____ minutes of moderate-intensity aerobic activity each week for weight loss and to achieve and maintain a healthy body weight.
   a. 60
   b. 90
   c. 150
   d. 300

8. Probably the most important contributing factor for obesity rates today in the United States is
   a. food advertising.
   b. snacking practices.
   c. inactivity.
   d. eating French fries.

9. The major goal for weight reduction in the treatment of obesity is the loss of
   a. weight.
   b. body fat.
   c. body water.
   d. body protein.

10. For most adults, the greatest portion of their energy expenditure is for
    a. physical activity.
    b. sleeping.
    c. basal metabolism.
    d. the thermic effect of food.

Answer Key: 1.d (LO 7.6), 2.d (LO 7.2), 3.d (LO 7.6), 4.b (LO 7.1), 5.a (LO 7.10), 6.c (LO 7.1), 7.d (LO 7.8), 8.c (LO 7.5), 9.b (LO 7.6), 10.c (LO 7.1)

Study Questions (Numbers refer to Learning Outcomes)

1. Explain how nuture and nature can contribute to the development of obesity. What are the two most convincing pieces of evidence that both genetic and environmental factors play significant roles in the development of obesity? **(LO 7.4)**

2. How does energy imbalance, including the role of physical activity, lead to weight gain and obesity? **(LO 7.1)**

3. Define a healthy weight in a way that makes the most sense to you. **(LO 7.3)**

4. Describe a practical method to define obesity in a clinical setting. **(LO 7.3)**

5. List three health problems that obese people typically face and a reason that each problem arises. **(LO 7.3)**

6. What are three key characteristics of a sound weight-loss program? **(LO 7.5)**

7. Why is the claim for quick, effortless weight loss by any method always misleading? **(LO 7.5)**

8. Define the term *behavior modification*. Relate it to the terms *stimulus control, self-monitoring, chain-breaking, relapse prevention,* and *cognitive restructuring*. Give examples of each. **(LO 7.8)**

9. Why should obesity treatment be viewed as a lifelong commitment rather than a short episode of weight loss? **(LO 7.5)**

10. What steps are important to remember when an underweight person wants to gain muscle but not fat? **(LO 7.10)**

What the Dietitian Chose

In this case, the dietitian starts with the Nutrition Facts on the package when making a dessert choice. Research has shown that reading labels plays a role in decreasing obesity, especially in women, who were found to be more likely to read nutrition labels (see Further Reading 13). Regular varieties of ice cream contain at least 10% milk fat by weight and are therefore a source of fat, saturated fat, and cholesterol. The milk fat gives the product the smooth, creamy texture for which ice cream is famous. You can see that the regular ice cream has 180 kcal, 10 grams of fat, 6 grams of saturated fat, and 65 mg of cholesterol. The sugar content is 19 grams; some of this comes from natural milk sugar (lactose), and some sugar is added for flavor.

Premium ice cream varieties, such as those you would find in an ice cream shop, may have up to 300 kcal, 18 grams of fat, and 30 grams of sugar per ½-cup serving! That's equivalent to 4 teaspoons of butter and 7 teaspoons of sugar. Such decadent treats can be enjoyed in moderation but are best reserved for special occasions.

Reduced-fat, low-fat, light, or fat-free varieties of ice cream have less than 10% milk fat. This can be accomplished by starting with lower-fat milk, using gelatin instead of eggs, using fat replacers, incorporating more air into the product (e.g., "churned"), or any combination thereof. "Reduced fat" means the product has at least 25% less fat than the original product. "Low fat" sets the standard at 3 grams of fat or less per ½-cup serving. "Light" ice cream has at least 50% less fat than the original product. "Fat free" signifies 0.5 grams of fat or less per ½-cup serving. The brand shown here is "light": it contains lower fat and sugar than the regular ice cream. It has 110 kcal, 3 grams of fat, 2 grams of saturated fat, 10 mg cholesterol, and 14 grams of sugar per serving. For weight-management purposes, this saves you 70 kcals, 7 grams of fat, and 5 grams of sugar per serving compared to regular ice cream.

Reducing the fat content of ice cream might also reduce the creamy texture, but for most reduced-fat, low-fat, and light products, the change is barely noticeable.

Frozen yogurt is not always a lower-calorie option than ice cream. It does not have to comply with the same 10% milk fat standard as ice cream does, so there may be a wide range of fat and sugar contents. This brand of frozen yogurt has 180 kcal, 3 grams of fat, 1 gram of saturated fat, and 45 mg cholesterol. The product is marketed as "low fat"—and it does only contain 3 grams of fat per ½-cup serving—but there are still 22 grams of sugar. This is a great example of how low fat does not necessarily mean low calorie.

Serving size is the most important part of the Nutrition Facts panel when it comes to calorie control. No matter how low the calorie, fat, and sugar contents are, if you consume multiple servings of any product, you are likely to take in too many calories. Likewise, you can control your calorie intake by choosing a smaller portion of premium ice cream.

Nutrition facts
Kroger Private Selection—Country Made Vanilla Ice Cream

Serving Size: 1/2 cup

Calories	180	Sodium	45 mg
Total Fat	10 g	Potassium	0 g
Saturated	6 g	Total Carbs	19 g
Polyunsaturated	0 g	Dietary Fiber	0 g
Monounsaturated	0 g	Sugars	19 g
Trans	0 g	Protein	6 g
Cholesterol	65 mg		
Vitamin A	8%	Calcium	10%
Vitamin C	2%	Iron	2%

*Percent Daily Values are based on a 2000 calorie diet. Your Daily Values may be higher or lower depending on your calories needs.

Nutrition facts
Kroger Deluxe—Vanilla Bean Light Ice Cream

Serving Size: 1/2 cup

Calories	110	Sodium	50 mg
Total Fat	3 g	Potassium	0 g
Saturated	2 g	Total Carbs	17 g
Polyunsaturated	0 g	Dietary Fiber	0 g
Monounsaturated	0 g	Sugars	14 g
Trans	0 g	Protein	3 g
Cholesterol	10 mg		
Vitamin A	6%	Calcium	10%
Vitamin C	2%	Iron	0%

*Percent Daily Values are based on a 2000 calorie diet. Your Daily Values may be higher or lower depending on your calories needs.

Nutrition facts
Haagen-Dazs—Yogurt Frozen Low Fat Vanilla

Serving Size: 1/2 cup

Calories	180	Sodium	45 mg
Total Fat	3 g	Potassium	0 g
Saturated	1 g	Total Carbs	30 g
Polyunsaturated	0 g	Dietary Fiber	0 g
Monounsaturated	0 g	Sugars	22 g
Trans	0 g	Protein	9 g
Cholesterol	45 mg		
Vitamin A	2%	Calcium	20%
Vitamin C	0%	Iron	0%

*Percent Daily Values are based on a 2000 calorie diet. Your Daily Values may be higher or lower depending on your calories needs.

CASE STUDY SOLUTION Choosing a Weight-Loss Program

1. Joe's weight has been climbing, so he has been experiencing positive energy balance.
2. When Joe is not working, his activity level is low. His time watching TV and studying does not use much energy. Joe should try to find time for physical activity after work and on weekends.
3. Balancing calorie intake with energy expenditure is the main key to weight loss and weight maintenance. Changes that Joe could make to his diet include a decrease in high-fat and high-calorie items in his take-out meals. He should include more fruits, vegetables, and whole grains in his meals. A whole-grain cereal with fruit would be a healthy and convenient substitute for his pastry for breakfast.

   A grilled chicken sandwich or a taco salad would be a nutritious change for lunch. Grilled chicken or chili with a baked potato or salad would be healthier choices for dinner. Table 7-4 includes many ideas for food substitutions that will lower calories.
4. Joe will be wasting his money if he buys the product seen in the infomercial. Unfortunately, regulation of the supplement industry is woefully lacking. In the future, if there is a meaningful breakthrough in weight loss and weight control, health authorities such as the Surgeon General's Office or the National Institutes of Health will make North Americans aware of that fact.
5. The characteristics of a sound weight-loss diet are shown in Figure 7-16. These characteristics include a slow and steady rate of loss, flexibility relative to your habits and tastes, nutritional adequacy, behavior modification, physical activity, and maintenance of overall health.

Further Readings

1. Academy of Nutrition and Dietetics. Position of the Academy of Nutrition and Dietetics: Total Diet Approach to Healthy Eating. *Journal of the Academy of Nutrition and Dietetics* 113:307, 2013.

2. Bazzano LA and others: Effects of low-carbohydrate and low-fat diets: A randomized trial. *Annals of Internal Medicine* 161(5): 309, 2014.

3. Centers for Disease Control and Prevention (CDC). *Overweight and Obesity* September 5, 2014. **http://www.cdc.gov/obesity/data/prevalence-maps.html.** Accessed September 9, 2014.

4. Dyczkowski CT and Seher CL: Smartphone apps for heart-healthy living—clients can track diet and exercise habits at their fingertips. *Today's Dietitian* 14(8):18, 2012.

5. Finkelstein EA and others: Obesity and severe obesity forecasts through 2030. *American Journal of Preventive Medicine* 42(6): 563, 2012.

6. Flegal KM and others: Prevalence of obesity and trends in the distribution of body mass index among U.S. adults, 1999–2010. *Journal of the American Medical Association* 307(5):491, 2012.

7. Ford ES and Dietz WH: Trends in energy intake among adults in the United States: Findings from NHANES. *American Journal of Clinical Nutrition* 97(4): 848, 2013.

8. Hall KD and others: Energy balance and its components: Implications for body weight regulation. *American Journal of Clinical Nutrition* 95:989, 2012.

9. Harris C: Mindful eating—studies show this concept can help clients lose weight and better manage chronic disease. *Today's Dietitian* 15(3): 42, 2013.

10. Hollis JF and others: Weight loss during the intensive intervention phase of the weight-loss maintenance trial. *American Journal of Preventive Medicine* 35:118, 2008.

11. James WPT and others: Effect of sibutramine on cardiovascular outcomes in overweight and obese subjects. *New England Journal of Medicine* 363(10):905, 2010.

12. Keidar A: Bariatric surgery for type 2 diabetes reversal: The risks. *Diabetes Care* 34:S361, 2011.

13. Loureiro ML and others: The effects of nutritional labels on obesity. *Agricultural Economics* 43:333, 2012.

14. Miller LE: Lorcaserin for weight loss: Insights into U.S. Food and Drug Administration approval. *Journal of the Academy of Nutrition and Dietetics* 113:25, 2013.

15. National Center for Chronic Disease Prevention and Health Promotion: *Obesity: Halting the Epidemic by Making Health Easier.* Centers for Disease Control and Prevention, Department of Health and Human Services, 2009.

16. See R and others: The association of differing measures of overweight and obesity with prevalent atherosclerosis: The Dallas Heart Study. *Journal of the American College of Cardiology* 50:752, 2007.

17. Shai I and others: Weight loss with a low-carbohydrate, Mediterranean, or low-fat diet. *New England Journal of Medicine* 359:229, 2008.

18. Slavin J and others: How fiber affects weight regulation. *Food Technology* 62:34, 2008.

19. Svetkey LP and others: Comparison of strategies for sustaining weight loss: The weight-loss maintenance randomized controlled trial. *Journal of the American Medical Association* 299:1139, 2008.

20. Trust for America's Health: F as in Fat: How Obesity Threatens America's Future 2012. Robert Wood Johnson Foundation, (2012). Available at **http://healthyamericans.org/report/100/.** Accessed April 5, 2013.

 **connect** | NUTRITION

To get the most out of your study of nutrition, visit McGraw-Hill Connect at www.mcgrawhillconnect.com where you will find NutritionCalc Plus, LearnSmart, and many other dynamic tools.

Rate Your Plate

A Close Look at Your Weight Status

Determine the following two indices of your body status: body mass index and waist circumference.

Body Mass Index (BMI)

Record your weight in pounds: _____ pounds

Divide your weight in pounds by 2.2 to determine your weight in kilograms: _____ kilograms

Record your height in inches: _____ in

Divide your height in inches by 39.4 to determine your height in meters: _____ meters

Calculate your BMI using the following formula:

$$BMI = \text{weight (kilograms)}/\text{height}^2 \text{ (meters)}$$

BMI = _____ kg/ _____ m^2 = _____

Waist Circumference

Use a tape measure to measure the circumference of your waist (at the umbilicus with stomach muscles relaxed). Circumference of waist (umbilicus) = _____ in

Interpretation

1. When BMI is greater than 25, health risks from obesity often begin. It is especially advisable to consider weight loss if your BMI exceeds 30. Does yours exceed 25 (or 30)?

   Yes _____ No _____

2. When a person has a BMI greater than 25 and a waist circumference of more than 40 inches (102 centimeters) in men or 35 inches (88 centimeters) in women, there is an increased risk of cardiovascular disease, hypertension, and type 2 diabetes. Does your waist circumference exceed the standard for your gender?

   Yes _____ No _____

3. Do you feel you need to pursue a program of weight loss?

   Yes _____ No _____

Application

From what you have learned in Chapter 7, list three specific actions you can take to move toward (or maintain) a healthy weight.

_____

_____

_____

connect NUTRITION   Find more Rate Your Plate activities for this chapter on Connect at www.mcgrawhillconnect.com.

Student Learning Outcomes

Chapter 8 is designed to allow you to:

8.1 Describe the general characteristics of the fat-soluble or water-soluble groups of vitamins, the general process of vitamin absorption and storage, the dangers of vitamin toxicity, and the preservation of vitamins in foods.

8.2 Describe the functions of vitamin A and carotenoids in vision, growth, development, reproduction, and cancer prevention, as well as the signs and symptoms of deficiency and toxicity, and understand its dietary requirements and how to get enough vitamin A in the diet and avoid too much from dietary supplements.

8.3 Explain the functions and sources of vitamin D, including its synthesis by the action of ultraviolet radiation; understand its dietary requirements; and describe the signs and symptoms of vitamin D deficiency and toxicity.

8.4 Summarize the functions of vitamin E, understand its dietary requirements, and describe its food sources and deficiency and toxicity symptoms.

8.5 Describe the role of vitamin K in blood clotting and the signs and symptoms of deficiency and toxicity, understand its dietary requirements, and identify its food sources.

8.6 Describe the characteristics of the water-soluble vitamins and the roles of the B vitamins in energy metabolism.

8.7 Describe the functions of thiamin in carbohydrate metabolism, and list its sources, dietary requirements, and deficiency symptoms.

Chapter 8
Vitamins

What Would You Choose?

Since being accepted into the business program, your courses are getting tougher and you are staying up late doing homework. You also need work experience, so you are putting in 20 hours a week at the bookstore. Between classes, homework, work, and hanging out with friends, you are only getting about 4 to 5 hours of sleep a night and barely have time to eat. You are feeling exhausted and looking for a good, quick energy source. Which of the following would you choose to get energized? (Supplement Facts for these products are shown at the end of the chapter.)

a 2-ounce 5-Hour Energy® shot

b Nature Made® B-Complex with Vitamin C Dietary Supplement

c Centrum® Adults Multivitamin and Mineral Supplement

d Several small meals throughout the day and 7 to 9 hours of sleep each night

CONNECT | NUTRITION — Think about your choice as you read Chapter 8, then see **What the Dietitian Chose** at the end of the chapter. To learn more about nutrients and energy metabolism, check out the Connect site: www.mcgrawhillconnect.com.

Although the vitamins are essential nutrients, the amount of vitamins we need to prevent deficiency is small. Some people believe that consuming vitamins far in excess of their needs provides them with extra energy, protection from disease, and prolonged youth. They seem to think that if a little is good, then more must be better. More than half of the U.S. adult population have taken vitamin and/or mineral supplements on a regular basis, some at unsafe levels.

Vitamins are found in plants and animals. Plants synthesize all the vitamins they need and are a healthy source of vitamins for animals. Animals vary in their ability to synthesize vitamins. For example, guinea pigs and humans are two of the few organisms unable to make their own supply of vitamin C.

Every major public health authority recommends that we increase our intake of fruits and vegetables. Which vitamins are especially found in fruits and vegetables? What are some other health-related attributes of fruits and vegetables in general? Which chronic diseases are associated with a poor intake of fruits and vegetables? Should we take a daily vitamin supplement if we do not include fruits and vegetables in our diet? This chapter provides some answers.

8.8 Describe the functions of riboflavin in energy metabolism and understand the dietary requirements and sources of riboflavin that will prevent deficiency symptoms.

8.9 Describe the functions of niacin in energy metabolism and list its sources, dietary requirements, and deficiency and toxicity symptoms.

8.10 Describe the functions of vitamin B-6 in amino acid metabolism and understand how to get enough vitamin B-6 in the diet to avoid a deficiency and prevent toxicity from dietary supplements.

8.11 Describe the functions of pantothenic acid and biotin in energy metabolism, understand their dietary requirements, and list their food sources and deficiency signs and symptoms.

8.12 Summarize the functions of folate in cell metabolism, understand the dietary requirements needed to prevent megaloblastic anemia, and identify its food sources.

8.13 Understand the process of vitamin B-12 absorption and its role in folate metabolism, identify the dietary requirements for vitamin B-12 needed to avoid deficiency, and identify its food sources.

8.14 Describe the functions of vitamin C as well as its dietary requirements, sources, and deficiency and toxicity signs and symptoms.

8.15 Understand the functions and sources of choline and other vitamin-like substances.

8.16 Evaluate the use of dietary supplements with respect to their potential benefits and hazards to the body.

8.17 Describe how calorie and fat intakes contribute to cancer risk and understand the role other food constituents play in inhibiting cancer.

vitamin An essential organic (carbon-containing) compound needed in small amounts in the diet to help regulate and support chemical reactions and processes in the body.

fat-soluble vitamins Vitamins that dissolve in fat and the substances such as ether and benzene but not readily in water. These vitamins are A, D, E, and K.

water-soluble vitamins Vitamins that dissolve in water. These vitamins are the B vitamins and vitamin C.

scurvy The vitamin C–deficiency disease characterized by weakness, fatigue, slow wound healing, opening of previously healed wounds, bone pain, fractures, sore and bleeding gums, diarrhea, and pinpoint hemorrhages on the skin.

rickets A disease characterized by poor mineralization of newly synthesized bones because of low calcium content. Arising in infants and children, this deficiency is caused by insufficient amounts of the vitamin D hormone in the body.

8.1 Vitamins: Vital Dietary Components

By definition, **vitamins** are essential organic (carbon-containing) substances needed in small amounts in the diet for normal function, growth, and maintenance of the body. In general, humans require a total of about 1 ounce (28 grams) of vitamins for every 150 pounds (70 kilograms) of food consumed. However slight its requirement is, each vitamin is essential for one or more functions in the body (Fig. 8-1). Vitamins can be divided into two broad classes based on solubility: vitamins A, D, E, and K are **fat-soluble vitamins,** whereas the B vitamins and vitamin C are **water-soluble vitamins.** The B vitamins include thiamin, riboflavin, niacin, pantothenic acid, biotin, vitamin B-6, folate, and vitamin B-12. Choline is a related nutrient but is not classified as a vitamin.

Vitamins are essential in human diets because they cannot be synthesized in the human body or because their synthesis can be decreased by environmental factors. Notable exceptions to having a strict dietary need for a vitamin are vitamin A, which we can synthesize from certain pigments in plants; vitamin D, synthesized in the body if the skin is exposed to adequate sunlight; niacin, synthesized from the amino acid tryptophan; and vitamin K and biotin, synthesized to some extent by the bacteria in the intestinal tract.

To be classified as a vitamin, a compound must meet the following criteria: (1) the body is unable to synthesize enough of the compound to maintain health and (2) absence of the compound from the diet for a defined period produces deficiency symptoms that, if caught in time, are quickly cured when the compound is resupplied. A compound does not qualify as a vitamin merely because the body cannot make it. Evidence must suggest that health declines when the substance is not consumed.

As scientists began to identify various vitamins, related deficiency diseases such as **scurvy** and **rickets** were dramatically cured. For the most part, as the vitamins were discovered, they were named alphabetically: A, B, C, D, E, and so on. Later, many substances originally classified as vitamins were found not to be essential for humans and were dropped from the list. Other vitamins, thought at first to be only one chemical, turned out to be several chemicals, so the alphabetical names had to be broken down by numbers (B-6, B-12, and so on).

In addition to their use in correcting deficiency diseases, a few vitamins have also proved useful in treating several nondeficiency diseases. These medical applications require administration of **megadoses,** well above typical human needs for the vitamins. For example, megadoses of a form of niacin can be used as part of blood cholesterol-lowering treatment for certain individuals. Still, any claimed benefits from use of vitamin supplements, especially intakes in excess of the Upper Level (if set), should be viewed critically because unproved claims are common. Remember, whenever you take a supplement at high doses, you are taking it at a pharmacological dose—that of a drug. Expect side effects as you would from any drug.

Vitamins isolated from foods or synthesized in the laboratory are the same chemical compounds and work equally well in the body. Contrary to claims in the health-food literature, "natural" vitamins isolated from foods are, with few exceptions, no more healthful or effective than those

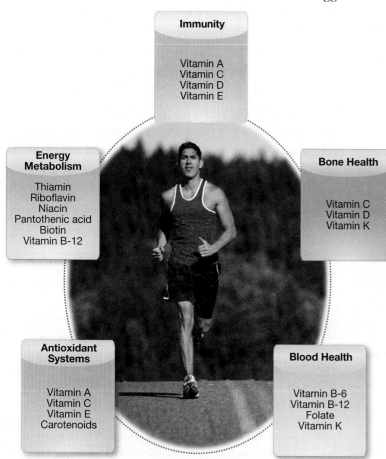

Immunity

Vitamin A
Vitamin C
Vitamin D
Vitamin E

Energy Metabolism

Thiamin
Riboflavin
Niacin
Pantothenic acid
Biotin
Vitamin B-12

Bone Health

Vitamin C
Vitamin D
Vitamin K

Antioxidant Systems

Vitamin A
Vitamin C
Vitamin E
Carotenoids

Blood Health

Vitamin B-6
Vitamin B-12
Folate
Vitamin K

FIGURE 8-1 ▲ Vitamins contribute to many functions in the body.

synthesized in a laboratory. Of note, the natural form of vitamin E is much more potent than the synthetic form. In contrast, synthetic folic acid, the form of the vitamin added to ready-to-eat breakfast cereals and flour, is 1.7 times more potent than the natural vitamin form (see Further Reading 5).

ABSORPTION AND STORAGE OF VITAMINS IN THE BODY

The fat-soluble vitamins (A, D, E, and K) are absorbed along with dietary fat. These vitamins then travel with dietary fats as part of chylomicrons through the bloodstream to reach body cells. Special carriers in the bloodstream help distribute some of these vitamins. Fat-soluble vitamins are stored mostly in the liver and fatty tissues.

When fat absorption is efficient, about 40% to 90% of the fat-soluble vitamins are absorbed. Anything that interferes with normal digestion and absorption of fats, however, also interferes with fat-soluble vitamin absorption. For example, people with cystic fibrosis, a disease that often hampers fat absorption, may develop deficiencies of fat-soluble vitamins. Some medications, such as the weight-loss drug orlistat (Alli), discussed in Chapter 7, also interfere with fat absorption. Unabsorbed fat carries these vitamins to the large intestine, and they are excreted in the feces. People with fat-malabsorption conditions are especially susceptible to vitamin K deficiency because body stores of vitamin K are lower than those of the other fat-soluble vitamins. Vitamin supplements, taken under a physician's guidance, are part of the treatment for preventing a vitamin deficiency associated with fat malabsorption. Finally, people who use mineral oil as a laxative at mealtimes risk fat-soluble vitamin deficiencies. Fat-soluble vitamins dissolve in the mineral oil, but the intestine does not absorb mineral oil. Hence, the fat-soluble vitamins are eliminated with the mineral oil in the feces.

Water-soluble vitamins are handled much differently than fat-soluble vitamins. After being ingested, the B vitamins from food are first broken down from their active **coenzyme** forms into free vitamins in the stomach and small intestine. The vitamins are then absorbed, primarily in the small intestine. Typically, about 50% to 90% of the water-soluble vitamins in the diet are absorbed, which means they have relatively high **bioavailability.** Water-soluble vitamins are transported to the liver via the hepatic portal vein and are distributed to body tissues. Once inside cells, the active coenzyme forms are resynthesized. Although some supplement manufacturers sell vitamins in their coenzyme forms, there is no benefit of consuming the coenzyme forms, as these are broken down during digestion and activated inside cells when needed.

Excretion of vitamins varies primarily on their solubility. Except for vitamin K, fat-soluble vitamins are not readily excreted from the body. Hence, toxicity can be an issue. Water-soluble vitamins are excreted based on **tissue saturation,** the degree to which the tissue vitamin stores are full. Tissue storage capacity is limited. As the tissues become saturated, the rate of excretion via the kidney increases sharply, preventing potential toxicity. Unlike other water-soluble vitamins, B-6 and B-12 are stored in the liver and not easily excreted in the urine.

In light of the limits of tissue saturation for many water-soluble vitamins, these vitamins should be consumed in the diet daily. However, an occasional lapse in the intake of even water-soluble vitamins causes no harm. Symptoms of a vitamin deficiency occur only when that vitamin is lacking in the diet and the body stores are essentially exhausted. For example, for an average person, the diet must be devoid of thiamin for 10 days or lacking in vitamin C for 20 to 40 days before the first symptoms of deficiencies of these vitamins develop.

VITAMIN TOXICITY

For most water-soluble vitamins, when you consume more than the RDA or AI, the kidneys efficiently filter the excess from the blood and excrete these compounds in urine. Notable exceptions are vitamin B-6 and vitamin B-12, which are stored in the liver. Although they are water-soluble, these two B vitamins may accumulate to toxic levels.

megadose Intake of a nutrient beyond estimates of needs to prevent a deficiency or what would be found in a balanced diet; 2 to 10 times human needs is a starting point for such a dosage.

coenzyme A compound (e.g., water-soluble vitamin) that combines with an inactive enzyme to form a catalytically active form. In this manner, coenzymes aid in enzyme function.

bioavailability The degree to which an ingested nutrient is digested and absorbed and thus is available to the body.

tissue saturation The limited storage capacity of water-soluble vitamins in the tissues.

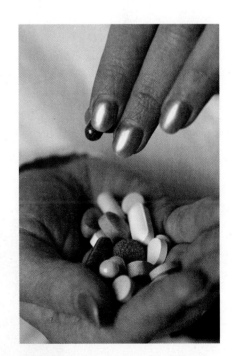

▲ Vitamins are not likely to be toxic unless taken in large amounts as supplements.

In contrast to the water-soluble vitamins, fat-soluble vitamins are not readily excreted, so some can easily accumulate in the body and cause toxic effects. Although a toxic effect from an excessive intake of any vitamin is theoretically possible, toxicity of the fat-soluble vitamin A is the most frequently observed. Vitamin A causes toxicity at intakes as little as two times the RDA. Vitamin E and the water-soluble vitamins niacin, vitamin B-6, and vitamin C can also cause toxic effects but only when consumed in very large amounts (15 to 100 times human needs or more). Overall, vitamins are unlikely to cause toxic effects unless taken in supplement (pill) form.

Some people believe that consuming vitamins far in excess of their needs provides them with extra energy, protection from disease, and prolonged youth. They seem to think that if a little is good, then more must be better. A "one-a-day" type of multivitamin and mineral supplement usually contains less than two times the Daily Values of its components, so daily use of these products is unlikely to cause toxic effects in men and nonpregnant women. However, consuming many vitamin pills, especially potent sources of vitamin A, can cause problems. See Section 8.16 to learn more about appropriate uses of dietary supplements.

PRESERVATION OF VITAMINS IN FOODS

Good sources of vitamins can be found in all food groups, especially fruits and vegetables (Fig. 8-2). However, storage time and several environmental factors can affect vitamin content of foods. The riper the food is, the more vitamins it will contain, but substantial amounts of vitamins can be lost from the time a fruit or vegetable is picked until it is eaten. Therefore, it is best to eat fresh produce as soon as possible after harvest. Food cooperatives, **community-supported agriculture (CSA),** and farmers' markets are great sources of freshly harvested fruits and vegetables. The water-soluble vitamins, particularly thiamin, vitamin C, and folate, can be destroyed with improper storage and excessive cooking. Heat, light, exposure to the air, cooking in water, and alkalinity are factors that can destroy vitamins.

There are several steps you can take to preserve nutrients when you are purchasing, storing, and preparing fruits and vegetables (Table 8-1). Frozen vegetables and fruits are often as nutrient-rich as freshly picked ones because fruits and vegetables are typically frozen immediately after harvesting. As part of the freezing process, vegetables are quickly blanched in boiling water. Blanching destroys the enzymes that would otherwise degrade the vitamins. If a food is not to be eaten within a few days of harvest, freezing is the best preservation method to retain nutrients.

Community-supported agriculture (CSA) Farms that are supported by a community of growers and consumers who provide mutual support and share the risks and benefits of food production, usually including a system of weekly delivery or pickup of vegetables and fruit, and sometimes dairy products and meat.

FIGURE 8-2 ▼ Certain food groups on MyPlate are especially rich sources of various vitamins and choline. This is true for those listed. Each may be also found in other MyPlate groups but in lower amounts. In addition to those vitamins listed here, pantothenic acid is present in moderate amounts in many groups, and vitamin E is abundant in plant oils.

Source: United States Department of Agriculture, **ChooseMyPlate.Gov**

**MyPlate:
Sources of Vitamins
and Choline**

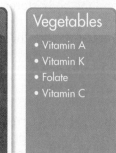

Grains
- Thiamin
- Riboflavin
- Niacin
- Folic acid

Vegetables
- Vitamin A
- Vitamin K
- Folate
- Vitamin C

Fruits
- Vitamin A
- Vitamin C

Dairy
- Vitamin D
- Riboflavin
- Vitamin B-12
- Choline

Protein
- Thiamin
- Riboflavin
- Niacin
- Biotin
- Vitamin B-6
- Vitamin B-12
- Choline

TABLE 8-1 ▶ **Tips for Preserving Vitamins in Fruits and Vegetables**

Preservation Methods	Why?
Keep fruits and vegetables cool until eaten.	Enzymes in fruits and vegetables begin to degrade vitamins once they are harvested. Chilling limits this process.
Refrigerate fruits and vegetables (except bananas, onions, potatoes, and tomatoes) in moisture-proof, airtight containers or in the vegetables drawer.	Nutrients keep best at temperatures near freezing, at high humidity, and away from air.
Trim, peel, and cut fruits and vegetables minimally—just enough to remove inedible parts.	Oxygen breaks down vitamins faster when more of the food surface is exposed. Whenever possible, cook fruits and vegetables in their skins.
Microwave, steam, or stir-fry vegetables.	More nutrients are retained when there is less contact with water and shorter cooking time.
Minimize cooking time.	Prolonged cooking (slow simmering) and reheating reduce vitamin content.
Avoid adding fats to vegetables during cooking if you plan to discard the liquid.	Fat-soluble vitamins will be lost in discarded fat. If you want to add fats, do so after vegetables are fully cooked and drained.
Do not add baking soda to vegetables to enhance the green color.	Alkalinity destroys vitamin D, thiamin, and other vitamins.
Store canned and frozen fruits and vegetables carefully.	To protect canned foods, store them in a cool, dry location. To protect frozen foods, store them at 0°F (−32°C) or colder. Eat within 12 months.

▲ It is time for a healthy snack! Store fresh produce in the refrigerator to preserve vitamin content and reduce spoilage.

✔ CONCEPT CHECK 8.1

1. Coenzyme Q (CoQ) is an organic compound that is required for the electron transport chain and has some antioxidant functions. This compound is synthesized within cells, and under most circumstances, the body synthesizes enough CoQ to meet its needs. Is CoQ a vitamin? Why or why not?

2. What is a megadose? Are there any negative consequences of consuming megadoses of vitamins? Are there any situations in which megadoses of vitamins are useful?

3. List at least three differences between fat-soluble and water-soluble vitamins.

4. List three ways to preserve vitamin content when storing, preparing, or cooking foods.

8.2 Vitamin A (Retinoids) and Carotenoids

Vitamin A was the first fat-soluble vitamin to be recognized as an important component of food essential for human health. Almost all (90%) of vitamin A is stored in the liver; the remaining 10% is in adipose tissue, kidneys, and the lungs. Either a deficiency or toxicity can cause severe problems, and there is a narrow range of optimal intakes between these two states.

Vitamin A is in a group of compounds known as **retinoids.** There are three active forms of vitamin A: **retinol, retinal,** and **retinoic acid.** These are often called preformed vitamin A. They exist only in animal products. When retinol is stored, it

retinoids Chemical forms of preformed vitamin A; one source is animal foods.

retinol Alcohol form of vitamin A.

retinal Aldehyde form of vitamin A.

retinoic acid Acid form of vitamin A.

retinyl Storage form of vitamin A.

carotenoids Precursors of vitamin A found in plant foods.

provitamin A A substance that can be converted into vitamin A.

is esterified (joined to a fatty acid) and becomes **retinyl.** In supplements, you will often find vitamin A listed as retinyl acetate or retinyl palmitate.

Plants contain pigments called **carotenoids.** Carotenoids are phytochemicals, that is, chemicals within plants that have health-promoting properties for humans. Carotenoids, which are precursors of vitamin A, have profound antioxidant properties. Because carotenoids can be turned into vitamin A, they are termed **provitamin A.** Beta-carotene, the orange-yellow pigment in carrots, is the only carotenoid that can be sufficiently absorbed and converted into retinol to play a significant role as a source of vitamin A and as a powerful antioxidant. The other two carotenoids that can be converted into vitamin A (though not very effectively) are alpha-carotene and beta-cryptoxanthin. Beta-carotene is split in half by cells in the body to form two molecules of retinal. Neither the conversion of the provitamin A carotenoids into vitamin A nor the absorption of carotenoids is an efficient process. Other carotenoids that may play a role in human health but are not vitamin A precursors include lycopene, zeaxanthin, and lutein.

FUNCTIONS OF VITAMIN A AND CAROTENOIDS

Health of Epithelial Cells and Immune Function. Vitamin A maintains the health of epithelial cells, which line internal and external surfaces of the lungs, intestines, stomach, vagina, urinary tract, and bladder, as well as those of the eyes and skin. Retinoic acid is required for immature epithelial cells to develop into mature, functional epithelial cells. Without vitamin A, mucus-forming cells, such as those in the intestines and lungs, deteriorate and lose function. For the eye, this can lead to blindness (see next section). Hyperkeratosis, also a result of vitamin A deficiency, is a condition in which skin cells produce too much keratin, blocking the hair follicles and causing "gooseflesh" or "toadskin" appearance. The excessive keratin in these skin cells causes the skin to be hard and dry.

The epithelial tissues described earlier serve as important barriers to infection. Vitamin A also supports the activity of certain immune system cells, specifically, the T-lymphocytes, or T-cells. Vitamin A–deficient animals and humans have an increased infection rate, but when they are supplemented with vitamin A, the immune response improves. Indeed, vitamin A is sometimes called the "anti-infection" vitamin.

night blindness Vitamin A–deficiency disorder that results in loss of the ability to see under low-light conditions.

retina A light-sensitive lining in the back of the eye. It contains retinal.

macular degeneration A painless condition leading to disruption of the central part of the retina (in the eye) and, in turn, blurred vision.

Vision. The link between vitamin A and night vision has been known since ancient Egyptians used juice extracted from liver to cure **night blindness.** Vitamin A performs important functions in light-dark vision and, to a lesser extent, color vision. Light entering the eye reaches a lining called the **retina.** The retina consists of rods, cones, and nerve cells. Rods detect black and white, and are responsible for night vision. Cones are responsible for color vision. Rods and cones require vitamin A for normal function. One form of vitamin A (retinal) allows certain cells in the eye to adjust to dim light (such as after seeing the headlights of an oncoming car; Fig. 8-3).

Some provitamin A carotenoids are also important for vision. The macula (also known as the macula lutea, meaning yellow spot) is in the central area of the retina and is responsible for the most detailed central vision. It contains the carotenoids lutein and zeaxanthin in high enough concentrations to impart a yellow color. Age-related **macular degeneration** (Fig. 8-4), the leading cause of blindness among older adults in North America, occurs because of changes in this macular area of the retina. In one study of older adults, the higher the total number of carotenoids (beta-carotene, lutein, and zeaxanthin) consumed in the diet was, the lower was the risk for age-related macular degeneration. The richest sources of lutein and zeaxanthin are green, leafy vegetables (Table 8-2).

These carotenoids may also decrease the risk of cataracts in the eyes. Research studies point toward the actual fruit and vegetables with their high carotenoid content as contributing to reduced risk for eye disorders. Consuming just carotenoids

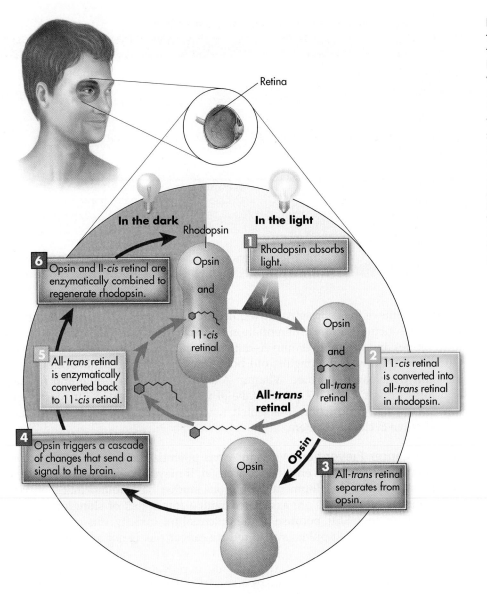

Retina

In the dark

In the light

Rhodopsin

Opsin

and

11-*cis*
retinal

**All-*trans*
retinal**

Opsin

Opsin
and
all-*trans*
retinal

Opsin

1 Rhodopsin absorbs
light.

2 11-*cis* retinal
is converted into
all-*trans* retinal
in rhodopsin.

3 All-*trans* retinal
separates from
opsin.

4 Opsin triggers a cascade
of changes that send a
signal to the brain.

5 All-*trans* retinal
is enzymatically
converted back
to 11-*cis* retinal.

6 Opsin and ll-*cis* retinal are
enzymatically combined to
regenerate rhodopsin.

FIGURE 8-3 ◄ Vitamin A functions to maintain vision. Light enters the eye through the cornea and lens, and then hits the retina. The light reacts with vitamin A–containing rhodopsin, which is stored in the rod cells of the retina. Rod cells allow us to see in black and white. When light reacts with rhodopsin, retinal is cleaved from rhodopsin (bleaching), a process that stimulates an electrical impulse to the brain. A new molecule of vitamin A then combines with opsin to regenerate rhodopsin. The yellow background indicates the bleaching events that occur in the light; the gray background indicates the regenerative events that can occur in either light or dark conditions.

as supplements for this purpose is not recommended. Multivitamin and mineral supplements formulated for older adults (e.g., Centrum Silver™) are being marketed as a source of lutein. Consumers must be aware that studies on carotenoids are generally conducted with food rather than supplements.

Cardiovascular Disease Prevention. Carotenoids may play a role in preventing cardiovascular disease in persons at high risk. This role may be linked to carotenoids' ability to inhibit the oxidation of low-density lipoproteins (LDLs). Until definitive studies are complete, many scientists recommend that we consume a total of at least five servings of a combination of fruits and vegetables per day as part of an overall effort to reduce the risk of cardiovascular disease.

Growth, Development, and Reproduction. Vitamin A participates in the processes of growth, development, and reproduction in several ways. At the genetic level, vitamin A binds to receptors on DNA to increase synthesis of a variety of

FIGURE 8-4 ▲ The blurry center of the image simulates the vision of a person with macular degeneration.

Vitamin A
RDA
    Men: 900 micrograms RAE
    Women: 700 micrograms RAE
DV: 1000 micrograms (5000 IU)
UL: 3000 micrograms RAE

TABLE 8-2 ▸ Vegetables Rich in Lutein and Zeaxanthin

Vegetables (serving size)	Lutein and Zeaxanthin (mg)
Kale (1 cup cooked)	23.8
Spinach (1 cup cooked)	20.4
Swiss chard (1 cup cooked)	19.2
Collard greens (1 cup cooked)	14.6
Spinach (2 cups raw)	7.4
Peas (1 cup cooked)	3.8
Broccoli (1 cup cooked)	2.4
Romaine lettuce (2 cups raw)	2.2
Brussels sprouts (1 cup cooked)	2.0
Zucchini (1 cup cooked)	2.0

▲ Inuits long knew and explorers soon learned to avoid eating the liver of polar bears. Just 4 ounces (120 grams) of polar bear liver will deliver a toxic dose of 1.36 million RAE of vitamin A. That is 136 times the RDA!

▲ The carotenoid lycopene is the red pigment found in tomatoes, watermelon, guava, and pink grapefruit. Lycopene seems to protect against prostate cancer.

prostate gland A solid, chestnut-shaped organ surrounding the first part of the urinary tract in the male. The prostate gland secretes substances into the semen.

proteins. Some of these proteins are required for growth. During early fetal growth, vitamin A functions in the differentiation and maturation of cells, which ultimately form tissues and organs. Vitamin A–deficient children experience stunted growth. For bones to grow and elongate, old bone must be remodeled (broken down) so that new bone can be formed. Vitamin A assists with breakdown and formation of healthy bone tissue. Adequate intake of vitamin A is needed for reproduction; it aids in sperm production (associated with its epithelial role) and in a normal reproductive cycle for women.

Possible Cancer Prevention. Vitamin A has potential benefits but also potential dangers where cancer prevention is concerned. It plays a role in cellular differentiation and embryonic development. Numerous studies have found that diets rich in provitamin A carotenoids are associated with a lower risk of skin, lung, bladder, and breast cancers. Still, because of the potential for toxicity, unsupervised use of megadose vitamin A supplements to reduce cancer risk is not advised and can be potentially dangerous.

Carotenoids by themselves may play a role in preventing cancer, as a by-product of their antioxidant activity. Population studies show that regular consumption of foods rich in carotenoids decreases the risk of lung and oral cancers. The carotenoid lycopene may decrease skin cancer risk. In contrast, recall from Chapter 1 that studies from the United States and Finland failed to show a reduction in lung cancer in male smokers and nonsmokers given supplements of the carotenoid β-carotene for 5 or more years. In fact, β-carotene use in male smokers *increased* the number of lung cancer cases compared with control groups. No comparable studies have been conducted with women. Although further research continues, most researchers are convinced that β-carotene supplementation offers no protection against cancer. Again, the best advice is to rely on food sources for this or any other carotenoids.

Cancer of the **prostate gland** is one of the most common cancers among North American men. The dietary carotenoid lycopene (the red pigment found in tomatoes, watermelon, pink grapefruit, and guava) seems to protect against this type of cancer. The proposed biological role of lycopene again appears to be that of an antioxidant. Some food companies have even marketed their tomato products as important sources of lycopene.

VITAMIN A DEFICIENCY

Without sufficient dietary vitamin A, the cells in the eye cannot quickly readjust to dim light, causing night blindness.

If vitamin A deficiency progresses, the cells that line the cornea of the eye (the clear window of the eye) lose the ability to produce mucus. The eye then becomes dry. This disease is called **xerophthalmia,** which means dry eye. Eventually, dirt particles scratch the dry surface of the eye, leading to blindness. Xerophthalmia can progress to the stage where there is an accumulation of dead cells and secretions on the surface of the eye. This condition is called Bitot's spots (Fig. 8-5).

Vitamin A deficiency is the leading cause of blindness worldwide. North Americans are at low risk because typical American diets contain plentiful sources of preformed vitamin A, such as fortified milk and eggs. However, poor vitamin A intakes, low fat intakes that do not allow for sufficient vitamin A absorption, and low stores of vitamin A lessen the ability of children to meet high needs during periods of rapid growth. Worldwide, about one-third of children suffer from vitamin A deficiency. Hundreds of thousands of children in developing nations, especially Southeast Asia and Africa, become blind each year because of vitamin A deficiency. Some of these children ultimately die from infections. Worldwide, attempts to reduce this problem have included promoting breastfeeding, giving large doses of vitamin A twice yearly, and fortifying sugar and margarine with vitamin A. These food vehicles are used because they are commonly consumed by the population of less-developed nations. This effort has proved effective in some countries.

Another effort is to treat maternal night blindness in these same areas of the world. Among pregnant women, night blindness is a marker of vitamin A deficiency that will likely lead to pregnancy-related deaths, malnutrition, anemia, and infant mortality. Screening and supplementing populations of pregnant women can be effective in treating and preventing this public health problem.

GETTING ENOUGH VITAMIN A AND CAROTENOIDS

Preformed vitamin A (e.g., retinol, retinal, and retinoic acid) is found in liver, fish, fish oils, fortified milk, butter, yogurt, and eggs (Fig. 8-6). Margarine and spreads are also fortified with vitamin A.

About 65% of the vitamin A in the typical North American diet comes from preformed vitamin A sources, whereas provitamin A (carotenoids) dominates in the diet among poor people in other parts of the world. The provitamin A carotenoids are mainly found in dark green and yellow-orange vegetables and some fruits. Carrots, spinach and other greens, winter squash, sweet potatoes, broccoli, mangoes, cantaloupe, peaches, and apricots are examples of such sources. Beta-carotene accounts for some of the orange color of carrots. Green vegetables also contain provitamin A. The yellow-orange beta-carotene is masked by dark-green chlorophyll pigments. Green, leafy vegetables, such as spinach and kale, have high concentrations of lutein and zeaxanthin. Tomato products contain significant amounts of lycopene. Cooking food improves the bioavailability of carotenoids. In raw fruits and vegetables, carotenoids are bound to proteins. Cooking disrupts this protein bond and frees the carotenoid for better absorption.

The RDA for vitamin A (see margin) is expressed in retinol activity equivalents (RAE). These RAE units consider the activity of both preformed vitamin A and the carotenoids that are synthesized into vitamin A in humans. There is no separate DRI for beta-carotene or any of the other carotenoids. The total RAE value for a food is calculated by adding the concentration of preformed vitamin A to the amount of provitamin A carotenoids in the food that will be converted into vitamin A.

You may have noticed that the RDA for vitamin A is given in micrograms, whereas the vitamin A content on Supplement Facts labels is given in **international units (IU).** Legally, supplement manufacturers still must provide amounts of vitamin A in IU, which is a measure of the biological activity of a nutrient rather than its absolute quantity (see margin for conversion factors).

The diets of North American adults typically contain adequate vitamin A. Most adults in North America have liver reserves of vitamin A three to five times higher

FIGURE 8-5 ▲ Vitamin A deficiency leads to accumulation of dead cells and eventually leads to blindness. Note the severe effects on this eye. This problem is commonly seen today in Southeast Asia.

xerophthalmia Hardening of the cornea and drying of the surface of the eye, which can result in blindness.

▲ Provitamin A carotenoids from foods (not supplements) are the safest way to meet vitamin A needs. A serving of these carrots would be an excellent addition to the vegetable section of MyPlate. Cooking food sources of carotenoids actually makes them more bioavailable.

international unit (IU) A crude measure of vitamin activity, often based on the growth rate of animals in response to the vitamin. Today IUs have largely been replaced by more precise milligram or microgram measures.

FIGURE 8-6 ◄ **Food sources of vitamin A and carotenoids.** (a) The fill of the background color (none, 1/3, 2/3, or completely covered) within each food group on MyPlate indicates the average nutrient density for vitamin A and provitamin A carotenoids in that group. (b) The bar graph shows the vitamin A content of several foods from each food group compared to the RDA for adult males and females. Overall, the fruits and vegetables groups provide many rich sources of carotenoids, whereas fortified dairy products and certain choices in the protein group are good sources of preformed vitamin A. The grains group also contains some foods that are nutrient dense because they are fortified with vitamin A. *Nutrition data from USDA National Nutrient Database for Standard Reference, Release 26.*

(a)

(b)

Food Item and Amount	Vitamin A (micrograms RAE*)	% RDA for Adult Males (900 micrograms RAE)	% RDA for Adult Females (700 micrograms RAE)
Grains Cream of Wheat®, Cooked, ½ cup	280	31%	40%
Kellogg's® Raisin Bran® cereal, 1 cup	228	25%	33%
Corn muffin, 1 medium	59	7%	8%
Vegetables Sweet potato, baked, 1 large	1730	192%	247%
Spinach, cooked, 1 cup	943	105%	135%
Kale, cooked, 1 cup	885	98%	126%
Fruits Cantaloupe, 1 cup	300	33%	43%
Apricots, dried, ½ cup	117	13%	17%
Mango, 1 cup	90	10%	13%
Dairy Milk, fat-free, 1 cup	150	17%	21%
Soymilk (fortified), 1 cup	134	15%	19%
Cheese, cheddar, 1.5 ounces	113	13%	16%
Protein Beef liver, pan-fried, 3 ounces	6273	697%	896%
Tuna, bluefin, broiled, 3 ounces	643	71%	92%
Egg, hard-boiled, 1 large	74	8%	11%
Fats & Oils Butter, 1 teaspoon	34	4%	5%
Margarine, 1 teaspoon	0	0%	0%
Olive oil, 1 teaspoon	0	0%	0%

* Retinol activity equivalents.

1 RAE =
- 1 microgram of retinol (or retinal or retinoic acid: all the same)
- 12 micrograms of β-carotene from food or supplements
- 24 micrograms of other carotenoids from food

Converting IU to RAE
- IU ÷ 3.3 if the supplement is preformed vitamin A (such as retinyl acetate)
- IU ÷ 6.6 if the supplement is carotenoids (usually β-carotene)

than needed to provide good health. Thus, the use of vitamin A supplements by most people is unnecessary. Populations in North America that may be at risk for vitamin A deficiency include those with low vegetable intakes (e.g., some children, older adults, and the urban poor); people with alcoholism or liver disease; or people with severe fat malabsorption.

AVOIDING TOO MUCH VITAMIN A AND CAROTENOIDS

Intakes in excess of the UL for vitamin A are linked to birth defects and liver toxicity. Other possible side effects include an increased risk of hip fracture and poor pregnancy outcomes.

During the early months of pregnancy, a high intake of preformed vitamin A is especially dangerous because it may cause fetal malformations and spontaneous abortions. This is because vitamin A binds to DNA and thus influences cell development. The FDA recommends that women of child-bearing age limit their overall intake of preformed vitamin A from diet plus supplements to a total of about 100% of the Daily Value. It is also important to limit consumption of rich food sources, such as liver. These precautions also apply to women who may possibly become pregnant; vitamin A is stored in the body for long periods, so women who ingest large amounts during the months before pregnancy place their **fetus** at risk.

In contrast, ingesting large amounts of vitamin A–yielding carotenoids does not cause toxic effects. A high carotenoid concentration in the blood (called hypercarotenemia) can occur if someone routinely consumes large amounts of carrots or takes pills containing beta-carotene (more than 30 milligrams daily) or if infants eat a great deal of squash. The skin turns yellow-orange, particularly the palms of the hands and soles of the feet. It differs from jaundice, a sign of liver failure. In jaundice, the yellow discoloration extends to the sclera (whites) of the eye, whereas in hypercarotenemia, it does not. Hypercarotenemia does not appear to cause harm and disappears when carotenoid intake decreases. Dietary carotenoids do not produce toxic effects because (1) their rate of conversion into vitamin A is relatively slow and regulated and (2) the efficiency of carotenoid absorption from the small intestine decreases markedly as oral intake increases.

Medicine Cabinet

Two derivatives of vitamin A are used to treat moderate to severe acne. Tretinoin (Retin-A) is used topically (applied to the skin), and isotretinoin (Roaccutane) is taken orally. These drugs appear to work by altering genes expressed by the skin cells. However, taking vitamin A supplements or making a paste from supplements and applying it to your skin will have no effect on acne. In fact, high doses of vitamin A can induce toxic symptoms, including birth defects. Isotretinoin's label clearly advises against the use of the medication during pregnancy, and its use is strictly monitored by the Food and Drug Administration (FDA). In order for women to receive this medication, they must have two negative pregnancy tests; sign a patient information/consent form; agree to use two effective forms of birth control; register, along with their physicians and pharmacists, with iPLEDGE (**https://www.ipledgeprogram.com/**); and agree to follow all instructions of the program.

fetus The developing human life form from 8 weeks after conception until birth.

As little as four to five large carrots daily can produce hypercarotenemia. Taking a break from carrots, squash, and pumpkins will turn the skin back to normal.

✔ CONCEPT CHECK 8.2

1. What are the names of four carotenoids known to have antioxidant functions?
2. How are the carotenoids related to vitamin A?
3. What are the consequences of vitamin A deficiency?
4. What are some rich food sources of carotenoids?

8.3 Vitamin D (Calciferol or Calcitriol)

Vitamin D is a fat-soluble vitamin with two unique qualities. First, vitamin D is the only nutrient that is also a hormone. A hormone is a compound manufactured by one organ or tissue in the body that enters the bloodstream and has a physiological effect on another organ or tissue. The cells that participate in the synthesis of the active vitamin D hormone (skin, liver, and kidney) are different from the cells that respond to vitamin D (e.g., bone and intestine); therefore, vitamin D is a hormone.

Second, vitamin D is the only nutrient that can be produced in the skin upon exposure to ultraviolet light. The human production of vitamin D begins when the ultraviolet B (UVB) rays of the sun convert a cholesterol precursor of vitamin D **(7-dehydrocholesterol)** found in the skin into an inactive form of **vitamin D (cholecalciferol).** As illustrated in Figure 8-7, this compound must be activated to **25-hydroxyvitamin D₃ (calcidiol)** in the liver and to **1,25-dihydroxyvitamin D₃ (calcitriol)** in the kidney before it can function as the vitamin D hormone.

Our ability to absorb UVB rays and synthesize vitamin D is affected by many factors. Dark skin pigmentation, geographic latitude, time of day, season of the year, weather conditions, and amount of body surface covered with clothing or sunscreen

7-dehydrocholesterol Precursor of vitamin D found in the skin.

vitamin D₃ (cholecalciferol) Previtamin form found naturally in some animal sources, including fish and egg yolks.

25-hydroxyvitamin D₃ (calcidiol or calcifediol) Form found in blood. Sometimes shortened to 25(OH)D₃.

1,25-dihydroxyvitamin D₃ (calcitriol) Biologically active form of vitamin D.

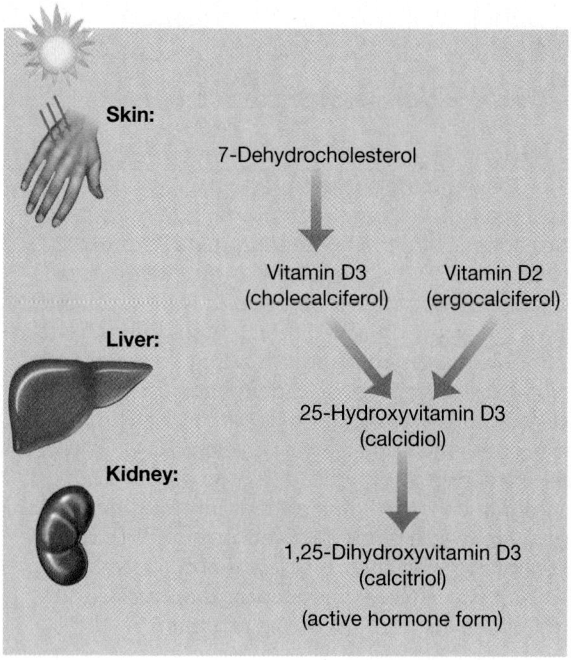

Skin:

7-Dehydrocholesterol

Vitamin D3   Vitamin D2
(cholecalciferol)   (ergocalciferol)

Liver:

25-Hydroxyvitamin D3
(calcidiol)

Kidney:

1,25-Dihydroxyvitamin D3
(calcitriol)

(active hormone form)

FIGURE 8-7 ▲ A precursor to vitamin D is synthesized when skin is exposed to sunlight. Previtamin D must be further modified by the liver and kidney for maximal activity.

affect the skin's exposure to UVB rays and therefore influence vitamin D synthesis. Something as simple as complete cloud cover or severe pollution can reduce UVB rays by about 50%. In addition, UVB rays will not penetrate glass. Aging reduces our ability to synthesize vitamin D—as much as 70% by age 70! Exposure of the hands, face, and arms for about 15 minutes daily will support adequate vitamin D synthesis for most healthy children and young adults. However, older adults and individuals with dark skin pigmentation require about three to five times this amount of sun exposure to synthesize an equivalent amount of vitamin D. Table 8-3 summarizes the various factors that can inhibit vitamin D synthesis, absorption, and activation.

FUNCTIONS OF VITAMIN D

Blood Calcium Regulation. The main function of vitamin D (calcitriol) is to maintain the normal range of calcium and phosphorus in the blood. Together with the hormones parathyroid hormone (PTH) and calcitonin, vitamin D closely maintains blood calcium within a narrow range. This tight regulation of blood calcium ensures that an appropriate amount of calcium is available to all cells. Vitamin D regulates blood calcium in three ways: (1) it influences the absorption of calcium and phosphorus from the small intestine; (2) in combination with PTH and calcitonin, it regulates calcium excretion via the kidney; and (3) it affects the deposition or withdrawal of minerals from the bones (Fig. 8-8).

TABLE 8-3 ▶ Factors That Impair Vitamin D Status

Factor	Description
Inadequate sun exposure • Northern latitudes • Excess clothing (e.g., robes/veils) • Air pollution (i.e., smog) • Sunscreen with SPF >8 • Excessive time spent indoors (e.g., due to health, work, or environmental conditions)	Limited exposure to UVB reduces the skin's ability to synthesize vitamin D.
Age	Vitamin D synthesis by the skin decreases. Vitamin D activation by the kidneys decreases.
Dark skin pigmentation	Melanin reduces the skin's ability to produce vitamin D, particularly for older adults and especially among women.
Inadequate dietary intake	Dietary intake of vitamin D is unable to compensate for inadequate skin synthesis of vitamin D.
Exclusive breastfeeding or low consumption of infant formula	Infants typically have limited sun exposure. Breast milk is a poor source of vitamin D. Infant formula contains vitamin D, but young infants may not consume adequate quantities to meet needs.
Fat malabsorption • Liver disease • Cystic fibrosis • Weight-loss medications	Poor absorption of dietary fat limits absorption of vitamin D from the small intestine.
Obesity	Release of vitamin D stored in subcutaneous fat is inefficient.
Liver diseases	Vitamin D activation by the liver decreases.
Kidney diseases	Vitamin D activation by the kidneys decreases.

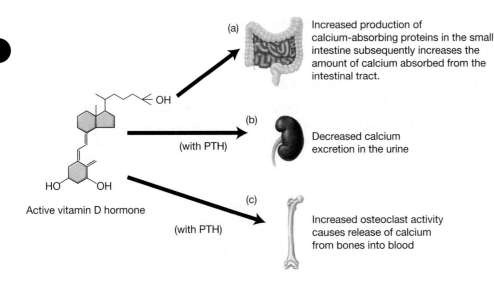

(a) Increased production of calcium-absorbing proteins in the small intestine subsequently increases the amount of calcium absorbed from the intestinal tract.

(b) (with PTH) Decreased calcium excretion in the urine

Active vitamin D hormone

(c) (with PTH) Increased osteoclast activity causes release of calcium from bones into blood

FIGURE 8-8 ◄ Vitamin D regulates blood calcium. When blood levels of calcium begin to drop from the normal range, PTH stimulates the synthesis of the most active form of vitamin D (calcitriol) by the kidney. Calcitriol acts at three different sites to increase blood calcium: (a) small intestine, (b) kidney, and (c) bone. When blood calcium levels increase above the normal range, PTH release is inhibited and calcitonin is released, which has the opposite effects of PTH.

Gene Expression and Cell Growth. It is now accepted that the biological effects of vitamin D extend far beyond its roles in calcium regulation and bone health. Vitamin D is involved in gene expression and cell growth; it binds to and subsequently affects cells of the immune system, brain and nervous system, parathyroid gland, pancreas, skin, muscles, and reproductive organs. In fact, vitamin D is considered one of the most potent regulators of cell growth, capable of influencing normal development of some cells (e.g., skin, colon, prostate, and breast), in turn reducing cancer risk in these sites (see Further Reading 21). Besides its role in influencing cancer risk, evidence suggests that vitamin D has a role in the prevention of several other chronic diseases, such as cardiovascular disease, diabetes, and hypertension.

The pigment that imparts color to skin, melanin, is a potent natural sunscreen. Anthropologists believe that dark skin is protective against UVB rays because it is found in populations that have lived near the equator for many generations.

osteomalacia Adult form of rickets. The bones have low mineral density and subsequently are at risk for fracture.

VITAMIN D DEFICIENCY

When vitamin D levels are adequate, about 30% to 40% of dietary calcium is absorbed by the small intestine. If blood levels of vitamin D are low, the small intestine is able to absorb only about 10% to 15% of calcium from the diet, which is not enough to maintain the calcium requirements for bone health and other functions. Vitamin D deficiency can occur at any time, but when it occurs during infancy and early childhood, the resulting disease is known as rickets. The skeletal abnormalities of rickets include bowed legs, thick wrists and ankles, curvature of the spine, a pigeon chest (chest protrudes above the sternum), skull malformations, and pelvic deformities (Fig. 8-9). Studies show that vitamin D deficiency is prevalent among children and adolescents in the United States. It was estimated that 9% (7.6 million) of children in the United States were vitamin D deficient in 2009 (see Further Reading 7).

Osteomalacia, which means soft bone, is an adult disease comparable to rickets. It can result from inadequate calcium intake, inefficient calcium absorption in the intestine, or poor conservation of calcium by the kidneys. It occurs most commonly in people with kidney, stomach, gallbladder, or intestinal disease (especially when most of the intestine has been removed) and in people with cirrhosis of the liver. These diseases affect both vitamin D activation and calcium absorption, leading to a decrease in bone mineral density. Bones become porous and weak, and break easily. Research shows that treatment with 10 to 20 micrograms (400 to 800 IU) per day of vitamin D, in conjunction with adequate dietary calcium, can reduce fracture risk among older adults. Thus, vitamin D is just as important as calcium when it comes to bone health.

Individuals who are most at risk of a vitamin D deficiency are older than the age of 60, live in northern latitudes, have dark skin, are rarely outdoors or always wear sunscreen while outdoors, have chronic kidney disease, take medications that

FIGURE 8-9 ▲ Vitamin D deficiency causes rickets, in which the bones and teeth do not develop normally.

vitamin D₂ (ergocalciferol) Form found in nonanimal sources, such as in some mushrooms.

Vitamin D

RDA: 15 micrograms (600 IU)
DV: 10 micrograms (400 IU)
UL: 100 micrograms (4000 IU)

Many tanning machines produce ultraviolet light of the appropriate wavelength to cause vitamin D synthesis in the skin. Relying on tanning machines, however, is not recommended by FDA and the American Academy of Dermatology because of the potential health hazards (burns, damage to eyes, and skin cancer) associated with this practice.

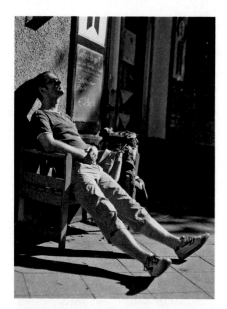

▲ Solar radiation (UVB rays) on the skin is the most reliable way to maintain vitamin D status and provides about 80% to 100% of the vitamin D humans use. In the absence of adequate sun exposure, the body must rely on dietary sources of vitamin D to meet needs. To get the vitamin D value of 10 minutes' exposure to sunlight, you would have to eat 30 servings of fortified cereal or 30 cups of fortified orange juice.

interfere with vitamin D absorption, have a condition that prevents fat absorption, or have had part of their stomach or intestine removed. A combination of sun exposure, dietary vitamin D intake, and vitamin D supplementation can prevent deficiency.

GETTING ENOUGH VITAMIN D

There are two forms of vitamin D: vitamin D₂ and vitamin D₃. **Vitamin D₂ (ergocalciferol)** is a synthetic product derived from the irradiation of plant sterols (ergosterol) and is used in some supplements. Vitamin D₃ (cholecalciferol)—the form synthesized in the human body—is more commonly used in supplements and fortified foods. Both forms of vitamin D must be modified by chemical reactions that occur in the kidney and liver (Fig. 8-7) before they can be active in the body.

Sunlight is the best source of vitamin D. Unlike with supplements, you can never receive a toxic dose. It is estimated that we synthesize roughly 3000 IU of vitamin D with 10 minutes' exposure to arms and legs. To some extent, the vitamin D synthesized on sunny days can be stored "for a rainy day" in the liver and adipose cells. However, most people now limit sun exposure to decrease their risk of skin cancer. Unless you live in a year-round sunny climate and find yourself outside most days between 10 A.M. and 3 P.M., you are not meeting your vitamin D needs from sun exposure alone. Therefore, people residing in northern climates and with limited sun exposure in general should find alternative (dietary) sources, especially in the winter months (Fig. 8-10). Overall, anyone who does not receive enough direct exposure to sunshine to synthesize an adequate amount of vitamin D must have a dietary source of the vitamin. The RDA for vitamin D (see margin) is based on a daily intake that is sufficient to maintain bone health and normal calcium metabolism, assuming minimal sun exposure.

Dietary sources of vitamin D are limited, with very few foods being naturally high in vitamin D (Fig. 8-11). Fatty fish are considered the richest natural sources of vitamin D. Wild-caught salmon tops the list with around 600 to 1000 IU of vitamin D in a 3.5-ounce serving. Farmed salmon contains less at between 100 to 250 IU per 3.5-ounce serving. Tuna is a bit farther down the list with 6 ounces of canned light tuna providing about 300 IU. Eggs are another natural source with 1 large egg yolk delivering 41 IU. Although butter, liver, and a few brands of margarine contain some vitamin D, large servings must be eaten to obtain an appreciable amount of the vitamin; therefore, these foods are not considered significant sources.

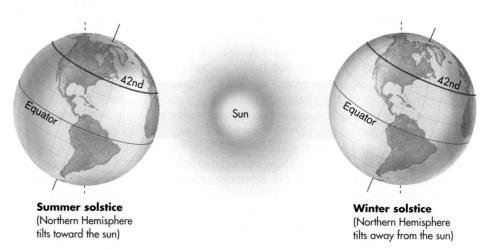

Summer solstice
(Northern Hemisphere tilts toward the sun)

Winter solstice
(Northern Hemisphere tilts away from the sun)

FIGURE 8-10 ▲ Seasonal variations to sunlight intensity. If you live north of 42° N latitude, the angle of the winter sun is such that the sun's rays must pass through more of the atmosphere than at other times of the year. As a result, skin forms less pro-hormone vitamin D in the winter. The 42nd parallel crosses North America at the northern border of California on the West Coast and Boston on the East Coast. In the far North (e.g., Alaska), this effect can last up to 6 months. Living below the 34th parallel (i.e., south of Los Angeles to Columbia, South Carolina), 10 minutes of UVB exposure per day is adequate to meet vitamin D needs year-round.

FIGURE 8-11 ◄ Food sources of vitamin D. (a) The fill of the background color (none, 1/3, 2/3, or completely covered) within each food group on MyPlate indicates the average nutrient density for vitamin D in that group. (b) The bar graph shows the vitamin D content of several foods compared to the RDA for adult males and females. Overall, the richest sources of vitamin D are fish, fortified dairy products, and fortified breakfast cereals. The vegetables group is not included in the bar graph because, aside from a select variety of mushrooms, vegetables are not a source of vitamin D. *Nutrition data from USDA National Nutrient Database for Standard Reference, Release 26.*

(a) (b)

Food Item and Amount	Vitamin D (micrograms)	% RDA for Adult Males and Females (15 micrograms)
Total® Raisin Bran cereal, 1 cup	2.5	17%
Cheerios® cereal, 1 cup	1.0	7%
White bread, 1 slice	0.0	0%
Orange juice (fortified), 1 cup	2.5	17%
Blackberries, 1 cup	0.0	0%
Kiwi, 1 cup	0.0	0%
Milk, whole, 1 cup	3.2	21%
Milk, fat-free, 1 cup	2.9	19%
Almond milk, sweetened, 1 cup	2.4	16%
Salmon, baked, 3 ounces	11.0	73%
Egg, hard-boiled, 1 large	1.1	7%
Cod, baked, 3 ounces	1.0	7%
Margarine (fortified), 1 teaspoon	0.5	3%
Butter, 1 teaspoon	0.1	1%
Olive oil, 1 teaspoon	0.0	0.0

Group labels: Grains, Fruits, Dairy, Protein, Fats & Oils

Fortified foods and supplements are effective ways to add vitamin D to your diet. Use of vitamin D–fortified milk, which began in the 1930s, effectively wiped out rickets in the United States. Because vitamin D is a fat-soluble vitamin, there is slightly more in a cup of whole milk (124 IU) compared to 1% (120 IU) or non-fat (115 IU). Ready-to-eat breakfast cereals are also fortified with vitamin D and other vitamins and minerals. A few provide up to 100 IU per 1-cup serving, but most offer closer to 40 IU per serving. You will almost double your intake of vitamin D if you add milk on top of your fortified cereal. Some brands of orange juice are now fortified with vitamin D (about 140 IU of vitamin D per 1-cup serving). You can count on foods fortified with vitamin D containing not much more than 100 IU per serving because U.S. government regulations cap the added amount at that level.

It takes some planning to meet daily vitamin D requirements from food sources. If you work indoors, live in a northern location, have a family history of skin cancer, and are not eating wild salmon every day, you are going to need a supplement. In supplements and fortified foods, vitamin D is available as vitamin D_2 and vitamin D_3; recent research shows that vitamin D_3 is most effective at raising blood levels of vitamin D and reducing fracture risk.

During times of growth (i.e., from infancy through adolescence), consuming adequate vitamin D supports optimal bone mineralization. The American Academy of

▲ The newest addition to foods naturally high in vitamin D is mushrooms. Mushrooms are the only vegetable to contain ergosterol, the precursor of vitamin D. Similar to humans, mushrooms also have the ability to make vitamin D when exposed to ultraviolet light. Although mushrooms are typically grown in the dark and do not contain the vitamin, some brands are now grown in ultraviolet light to stimulate vitamin D production. A vitamin D–rich portobello mushroom that provides 400 IUs of vitamin D per 3-ounce serving (about 1 cup of diced mushrooms) is already available in stores.

▲ Milk is usually fortified with vitamin D as well as vitamin A. Note that the DV for vitamin D is lower than the RDA for adults.

The 2010 Dietary Guidelines for Americans recommend that all children over age 9 and adults should choose more foods that provide vitamin D, which is a nutrient of concern in American diets (see Further Reading 2).

Pediatrics recommends that all infants, children, and adolescents consume a minimum of 400 IU of vitamin D daily. Until such intake can be obtained from foods, vitamin D supplementation is endorsed (under a physician's guidance). This recommendation includes all infants (exclusively breastfed, partially breastfed, or formula-fed). As you will learn in Chapter 14, breast milk is a poor source of vitamin D, and exclusively breastfed infants with limited sun exposure are at risk for developing rickets. Even though infant formula contains vitamin D (60 IU per 100 kcal), the total intake of formula among young infants may not provide adequate vitamin D to meet needs. Evidence from clinical trials and historical precedence support this dosage of vitamin D for infants and children (see Further Reading 3). Keep in mind, however, that supplements need to be used carefully to avoid vitamin D toxicity in the infant.

For adults over the age of 70, the RDA increases because of the reduced ability to absorb vitamin D from the intestine and the decreased ability to synthesize it in the skin. A number of experts suggest older adults, especially those over age 70 who have limited sun exposure or dark skin, receive about 1000 IU (25 micrograms) from a combination of vitamin–D–fortified foods and a multivitamin and mineral supplement, with an individual supplement of vitamin D added if needed (see Further Reading 14).

Other population groups that have difficulty meeting vitamin D needs include vegans and people with milk allergies or lactose intolerance. Vitamin D supplements or vitamin D–fortified soy milk or fruit juices are options for people who do not consume dairy products.

AVOIDING TOO MUCH VITAMIN D

Too much vitamin D taken regularly can create serious health consequences in infants and children. Due to the role of vitamin D in calcium absorption, excretion, and release of calcium from bone, supplementation with high doses of vitamin D can cause calcium levels in the blood to increase above the normal range. The UL (see margin on page 288) is based on the risk of overabsorption of calcium and eventual calcium deposits in the kidneys and other organs. Calcium deposits in organs can cause metabolic disturbances and cell death. Toxicity symptoms also include weakness, loss of appetite, diarrhea, vomiting, mental confusion, and increased urine output. Please note that vitamin D toxicity does not result from excessive exposure to the sun because the body regulates the amount made in the skin (i.e., as exposure to sunlight increases, vitamin D synthesis decreases).

✅ CONCEPT CHECK 8.3

1. Why is vitamin D sometimes not considered an "essential" nutrient?
2. How is vitamin D activated in the body?
3. How does vitamin D work to maintain blood calcium levels?
4. What are some rich food sources of vitamin D?
5. Can vitamin D be toxic?

8.4 Vitamin E (Tocopherols)

In the 1920s, a fat-soluble compound was found to be essential for fertility in rats. This compound was named tocopherol from the Greek words *tokos*, meaning birth, and *phero*, meaning to bring forth. Later, this essential nutrient was named vitamin E. Vitamin E is a family of four tocopherols and four tocotrienols called alpha, beta, gamma, and delta. They differ in that tocopherols have a saturated side chain, whereas the tocotrienols have an unsaturated side chain. Tocotrienols have not been as extensively studied as tocopherols, but recent research explores their potential roles in prevention of cancer, diabetes, and cardiovascular diseases. Of these eight forms of vitamin E, alpha (α)-tocopherol is the most biologically active and the most potent.

FUNCTIONS OF VITAMIN E

Antioxidant. The principal function of vitamin E in humans is as an anti-oxidant. Vitamin E is a fat-soluble vitamin found primarily in adipose tissue and in the lipid bilayers of cell membranes (Fig. 8-12). Many of the lipids within these membranes are polyunsaturated fatty acids (PUFA), which are particularly susceptible to oxidative attack by free radicals. The formation of free radicals may destabilize the cell membrane, which may ultimately alter the ability of the cell to function properly. Vitamin E can donate electrons or hydrogen to free radicals found in membranes, thereby making them more stable. The antioxidant function of vitamin E appears to be critical in cells continually exposed to high levels of oxygen, particularly red blood cells and the cells lining the lungs.

Increasing vitamin E intake has been suggested as a way to prevent several chronic diseases that are linked to oxidative damage. For example, oxidized LDL cholesterol is a major component of the plaque that develops in arteries, which leads to atherosclerosis. Vitamin E is thought to attenuate the development of atherogenic plaque due to its ability to prevent or reduce the formation of oxidized LDL cholesterol (see Further Reading 4). In addition, oxidative damage to proteins in the eye leads to the development of cataracts. Oxidized proteins combine and precipitate in the lens, causing cloudiness and decreasing visual acuity. Insufficient consumption of antioxidants from foods increases one's risk of these diseases.

Experts do not know whether supplementation with megadoses of vitamin E can confer any significant protection against diseases linked to oxidative damage. The consensus among the scientific community is that the established benefits of lifestyle choices have a far greater effect than any proposed benefits of antioxidant supplementation. The position of scientific research groups (e.g., American Heart Association, U.S. Preventive Services Task Force) is that it is premature to recommend vitamin E supplements to the general population, based on current knowledge and the failure of large clinical trials to show any consistent benefit. This conclusion is in agreement with the latest report on vitamin E by the Food and Nutrition Board of the National Academy of Sciences. In addition, FDA has denied the request of the dietary supplement industry to make a health claim that vitamin E supplements reduce the risk of cardiovascular disease and cancer.

Other Roles of Vitamin E. Although vitamin E is essential for fertility in many animal species, it does not appear to serve this role in humans. It is, however, important for the formation of muscles and the central nervous system in early human development. Vitamin E has been shown to improve vitamin A absorption if the dietary intake of vitamin A is low. It also functions in the metabolism of iron within cells, and it helps maintain nervous tissue and immune function.

VITAMIN E DEFICIENCY

Specific population groups are especially susceptible to developing marginal vitamin E status. Preterm infants tend to have low vitamin E stores because this vitamin is transferred from mother to baby during the late stages of pregnancy. Hence, the potential for oxidative damage, which could cause the cell membranes of red blood cells to break (hemolysis), is of particular concern for preterm infants. The rapid growth of preterm infants, coupled with the high oxygen needs of their immature lungs, greatly increases the stress on red blood cells. Special vitamin E–fortified formulas and supplements designed for preterm infants compensate for lack of vitamin E. Smokers are another group at high risk for vitamin E deficiency, as smoking readily destroys vitamin E in the lungs. One study showed that megadosing will not correct this vitamin E destruction by smokers. Others at risk of vitamin E deficiency include adults on very low-fat diets (< 15% total fat) or those with fat malabsorption.

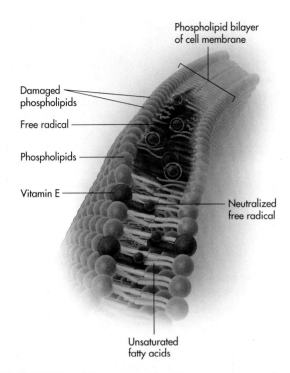

Phospholipid bilayer of cell membrane

Damaged phospholipids

Free radical

Phospholipids

Vitamin E

Neutralized free radical

Unsaturated fatty acids

FIGURE 8-12 ▲ Fat-soluble vitamin E can insert itself into cell membranes, where it helps stop free-radical chain reactions. If not interrupted, these reactions cause extensive oxidative damage to cells and, ultimately, cell death.

▲ The avocado in guacamole and the oil in the tortilla chip are good sources of vitamin E.

▲ Olive oil, leafy greens such as spinach, and egg yolks are sources of vitamin E.

Vitamin E

RDA: 15 milligrams

DV: 30 milligrams

UL: 1000 milligrams

(a)

GETTING ENOUGH VITAMIN E

Because vitamin E is only synthesized by plants, plant products (especially the oils) are the best sources. In the North American diet, nearly two-thirds of vitamin E is supplied by salad oils, margarines, spreads (low-fat margarine), and shortening (Fig. 8-13). Breakfast cereals fortified with vitamin E are good sources, but other than wheat germ, few other grain products provide much vitamin E. Milling of grains removes the germ, which contains the oils (mostly PUFAs) and vitamin E. By removing the germ, the resulting grain product has less chance of spoiling (i.e., rancidity of the PUFAs) and thus a longer shelf life. Other good sources of vitamin E are nuts and seeds.

Because plant oils contain mostly unsaturated fatty acids, the relatively high amount of vitamin E in plant oils naturally protects these unsaturated lipids from oxidation. Animal products (meat, dairy, and eggs) and fish oils, on the other hand, contain almost no vitamin E (Fig. 8-13). Vitamin E is susceptible to destruction by oxygen, metals, light, and heat, especially when oil is repeatedly reused in deep-fat frying; thus, the vitamin E content of a food depends on how it is harvested, processed, stored, and cooked.

The RDA of vitamin E for adults is 15 milligrams per day of alpha-tocopherol, the most active, natural form of vitamin E (d isomer). This amount equals 22.4 milligrams of the less active, synthetic source (dl isomer). Typically, North American adults consume about two-thirds of the RDA for vitamin E from food sources. The Daily Value used on food and supplement labels is 30 milligrams.

FIGURE 8-13 ◄ Food sources of vitamin E. (a) The fill of the background color (none, 1/3, 2/3, or completely covered) within each food group on MyPlate indicates the average nutrient density for vitamin E in that group. (b) The bar graph shows the vitamin E content of several foods compared to the RDA for adult males and females. Overall, the richest sources of vitamin E are nuts, seeds, plant oils, and fortified breakfast cereals. The dairy group is not pictured in the bar graph because, except for plant-based dairy alternatives, foods in this group yield no vitamin E. *Nutrition data from USDA National Nutrient Database for Standard Reference, Release 26.*

(b)

	Food Item and Amount	Vitamin E (milligrams)	Vitamin E (IU)	% RDA for Adult Males and Females (15 milligrams)
Grains	Total® Raisin Bran cereal, 1 cup	13.5	20.1	90%
Grains	Whole-wheat bread, 1 slice	0.9	1.3	6%
Grains	Quinoa, cooked, ½ cup	0.6	0.9	4%
Vegetables	Spinach, cooked, 1 cup	3.7	5.5	25%
Vegetables	Asparagus, cooked, 1 cup	2.7	4.0	18%
Vegetables	Sweet potato, baked, 1 cup	1.4	2.1	9%
Fruits	Blackberries, 1 cup	1.7	2.5	11%
Fruits	Mango, 1 cup	1.5	2.2	10%
Fruits	Olives, 5 large	0.7	1.0	5%
Protein	Sunflower seeds, dry roasted, 1 ounce	7.4	11.0	49%
Protein	Shrimp, cooked, 3 ounces	1.9	2.8	12%
Protein	Egg, hard-boiled, 1 large	0.5	0.8	3%
Fats & Oils	Sunflower oil, 1 teaspoon	1.9	2.8	12%
Fats & Oils	Salad dressing, Italian, 2 tablespoons	0.6	1.0	4%
Fats & Oils	Butter, 1 teaspoon	0.1	0.2	1%

As with vitamin A, the vitamin E content of dietary supplements is given in international units rather than simple metric measurements. International units reflect the different biological activity of natural vitamin E compared with synthetic preparations of vitamin E. Supplements may contain any of the variety of chemical forms and stereoisomers of vitamin E. Typical synthetic preparations of vitamin E contain a mixture of equal parts of d and l stereoisomers, only half of which are biologically active. Therefore, a supplement containing synthetic vitamin E has lower biological activity (i.e., lower IU) than the same quantity of natural vitamin E. The more expensive natural vitamin E supplements contain only the biologically active d-form. See the margin for conversion factors used to calculate the vitamin E content of dietary supplements.

AVOIDING TOO MUCH VITAMIN E

Unlike other fat-soluble vitamins, vitamin E is not stored in the liver. It is stored in adipose tissue throughout the body. The UL for vitamin E is 1000 milligrams per day of supplemental alpha-tocopherol. Excessive intake of vitamin E can interfere with vitamin K's role in the clotting mechanism, leading to hemorrhage. The risk of insufficient blood clotting is especially high if vitamin E is taken in conjunction with anticoagulant medications (e.g., Coumadin or heavy aspirin use). Always be cautious about using dietary supplements. In addition to the significant risk of drug interference and prolonged bleeding, vitamin E supplements can produce nausea, gastrointestinal distress, and diarrhea.

✓ CONCEPT CHECK 8.4

1. How does vitamin E work to prevent oxidative damage?
2. What are some rich food sources of vitamin E?
3. Why are preterm infants, smokers, and people with fat malabsorption particularly susceptible to oxidative damage to cell membranes?
4. What are the possible results of vitamin E toxicity?

8.5 Vitamin K (Quinone)

A family of compounds known collectively as vitamin K is found in plants, plant oils, fish oils, and animal products. Vitamin K is also synthesized by bacteria in the human colon, which normally fulfills approximately 10% of human requirements. Vitamin K has three forms: phylloquinone, the most abundant form of vitamin K, is synthesized by green plants; gut bacteria synthesize menaquinone; and menadione is the synthetic form found in supplements. Interestingly, the synthetic menadione form of vitamin K is twice as biologically available as the other two!

FUNCTIONS OF VITAMIN K

Vitamin K serves as a cofactor in chemical reactions that add CO_2 molecules to various proteins, thus enabling these proteins to bind calcium. This is the biochemical basis for vitamin K's role in the life-and-death process of blood clotting. In the clotting cascade (Fig. 8-14), vitamin K imparts calcium-binding ability to seven different proteins, eventually leading to the conversion of soluble fibrinogen into insoluble fibrin (i.e., the clot). The "K" stands for *koagulation* in the language spoken by the Danish researchers who first noted the relationship between vitamin K and blood clotting.

Calculating Vitamin E Content of Dietary Supplements

To find milligrams of *natural* vitamin E (d-alpha-tocopherol), multiply IU by 0.67. Conversely, to calculate IU from milligrams, multiply milligrams by 1.49.

Example: The RDA for vitamin E is 15 milligrams per day. This is equal to how many IU of natural vitamin E?

15 mg × 1.49 IU/mg = 22.35 IU

To find milligrams of *synthetic* vitamin E (dl-alpha-tocopherol), multiply IU by 0.45. Conversely, to calculate IU from milligrams, multiply milligrams by 2.22.

Example: If a vitamin E supplement contains 400 IU of dl-alpha-tocopherol, then how many milligrams of vitamin E does it contain?

400 IU × 0.45 mg/IU = 180 mg

Medicine Cabinet

People who are prone to develop blood clots may take anticoagulants or "blood thinners." One example is Plavix (clopidogrel), which works by inhibiting the activity of platelets. Another commonly prescribed anticoagulant is Coumadin (warfarin). This medication inhibits vitamin K-dependent coagulation factors. When taking Coumadin or similar drugs, it is important to keep vitamin K intake consistent from day to day (see Further Reading 12).

Vitamin K

AI

Men: 120 micrograms

Women: 90 micrograms

DV: 80 micrograms

UL: None

Besides its role in blood clotting, vitamin K is also important for bone health. Three calcium-binding proteins (e.g., osteocalcin) in the bone depend upon vitamin K for their function in bone mineralization.

VITAMIN K DEFICIENCY

At birth, a newborn has a "sterile gut"—an intestinal tract with an insufficient amount of bacteria. Thus, the newborn cannot produce enough vitamin K to allow for effective blood clotting if the infant is injured or needs surgery. Therefore, vitamin K is routinely administered by injection shortly after birth. In adults, deficiencies of vitamin K have occurred when a person takes antibiotics for an extended time (this destroys the bacteria that normally produce some of the vitamin K that is absorbed and used by the body) or when fat absorption is limited.

GETTING ENOUGH VITAMIN K

Major food sources of the phylloquinone form of vitamin K are green, leafy vegetables, broccoli, asparagus, and peas (Fig. 8-15). The menaquinone form of vitamin K is found in some meats, eggs, and dairy products, and is the form synthesized by bacteria. Compared to that of plant sources, the nutrient density of vitamin K in foods of animal origin is rather low. Vitamin K is resistant to cooking losses.

As with other fat-soluble vitamins, absorption of vitamin K requires dietary fat and adequate liver and pancreatic secretions. Unlike other fat-soluble vitamins, though, not much vitamin K is stored in the body and excesses can be excreted via urine. Thus, a deficiency could develop rather quickly if dietary intake of this nutrient is poor, which can be a problem among older adults whose diets lack vegetables. However, because vitamin K is fairly widespread in foods and some can be synthesized by bacteria in the colon, deficiencies of this vitamin rarely occur. No reports of toxicity have been published.

Table 8-4 reviews what we have covered so far regarding the fat-soluble vitamins.

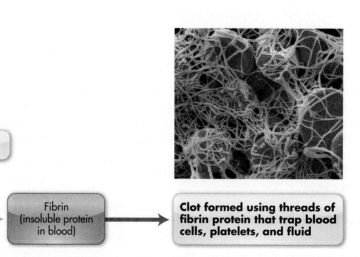

FIGURE 8-14 ▲ Vitamin K works to activate clotting factors, which are then able to bind to calcium. The binding of calcium to clotting factors is necessary for clot formation.

Clot formed using threads of fibrin protein that trap blood cells, platelets, and fluid

✓ CONCEPT CHECK 8.5

1. What is the role of vitamin K in blood clotting?
2. Why is it important for people who take Coumadin to monitor their dietary intake of vitamin K?

FIGURE 8-15 ◄ Food sources of vitamin K. (a) The fill of the background color (none, 1/3, 2/3, or completely covered) within each food group on MyPlate indicates the average nutrient density for vitamin K in that group. (b) The bar graph shows the vitamin K content of several foods compared to the AI for adult males and females. Overall, the richest sources of vitamin K are green, leafy vegetables. The grains group is not shown in the bar graph because grains provide very little vitamin K. *Nutrition data from USDA National Nutrient Database for Standard Reference, Release 26. and Elder SJ and others: Vitamin K contents of meat, dairy, and fast food in the U.S. diet.* Journal of Agricultural and Food Chemistry *54: 463, 2006.*

(a)

(b)

	Food Item and Amount	Vitamin K (micrograms)	% AI for Adult Males (120 micrograms)	% AI for Adult Females (90 micrograms)
Vegetables	Spinach, cooked, 1 cup	889	741%	987%
	Broccoli, cooked, 1 cup	220	183%	244%
	Brussels sprouts, cooked, 1 cup	219	183%	243%
Fruits	Rhubarb, cooked, 1 cup	51	42%	56%
	Blueberries, 1 cup	29	24%	32%
	Pomegranates, 1 cup arils	29	24%	32%
Dairy	Cheese, cheddar, 1.5 ounces	6*	5%	7%
	Soymilk, 1 cup	4	3%	4%
	Milk 2%, 1 cup	1*	1%	1%
Protein	Tofu, firm, cooked, 1 cup	6*	5%	7%
	Egg, hard-boiled, 1 large	4*	3%	4%
	Beef liver, pan-fried, 3 ounces	3	3%	4%
Fats & Oils	Soybean oil, 1 teaspoon	8	7%	9%
	Margarine, 1 teaspoon	4	4%	5%
	Butter, 1 teaspoon	0	0%	0%

* Value includes phylloquinone and menaquinone.

TABLE 8-4 ► **Summary of the Fat-Soluble Vitamins**

Vitamin	Major Functions	RDA or AI	Dietary Sources	Deficiency Symptoms	Toxicity Symptoms
Vitamin A (preformed vitamin A and provitamin A)	• Promotes vision: night and color • Promotes growth • Prevents drying of skin and eyes • Promotes resistance to bacterial infection and overall immune system function	*Men:* 900 micrograms RAE (3000 IU preformed vitamin A) *Women:* 700 micrograms RAE (2300 IU preformed vitamin A)	Preformed vitamin A: • Liver • Fortified milk • Fortified breakfast cereals Provitamin A: • Sweet potatoes • Spinach • Greens • Carrots • Cantaloupe • Apricots • Broccoli	• Night blindness • Xerophthalmia • Poor growth • Dry skin	• Fetal malformations • Hair loss • Skin changes • Bone pain • Fractures Upper Level is 3000 micrograms (10,000 IU) of preformed vitamin A based on the risk of birth defects and liver toxicity.

(continued)

TABLE 8-4 ▶ **Summary of the Fat-Soluble Vitamins** *(continued)*

Vitamin	Major Functions	RDA or AI	Dietary Sources	Deficiency Symptoms	Toxicity Symptoms
Vitamin D	• Increases absorption of calcium and phosphorus • Maintains optimal blood calcium and calcification of bone • Regulation of cell development	15 micrograms (600 IU)	• Fortified milk • Fortified breakfast cereals • Fish oils • Sardines • Salmon	• Rickets in children • Osteomalacia in adults	• Growth retardation • Kidney damage • Calcium deposits in soft tissue Upper Level is 100 micrograms (4000 IU) based on the risk of elevated blood calcium.
Vitamin E	• Antioxidant; prevents breakdown of vitamin A and unsaturated fatty acids	15 milligrams alpha-tocopherol (22 IU natural form, 33 IU synthetic form)	• Plant oils • Products made from plant oils • Some greens • Some fruits • Nuts and seeds • Fortified breakfast cereals	• Hemolysis of red blood cells • Nerve degeneration	• Muscle weakness • Headaches • Nausea • Inhibition of vitamin K metabolism Upper Level is 1000 milligrams (1100 IU synthetic form, 1500 IU natural form) based on the risk of hemorrhage.
Vitamin K	• Activation of blood-clotting factors • Activation of proteins involved in bone metabolism	*Men:* 120 micrograms *Women:* 90 micrograms	• Green vegetables • Liver • Some plant oils • Some calcium supplements	• Hemorrhage • Fractures	No Upper Level has been set.

Abbreviations: RAE = retinol activity equivalents; IU = international units.

8.6 The Water-Soluble Vitamins and Choline

Regular consumption of good sources of the water-soluble vitamins is important. Most water-soluble vitamins are readily excreted from the body with any excess generally ending up in the urine or stool and very little being stored. They dissolve in water, so large amounts of these vitamins can be lost during food processing and preparation. Vitamin content is best preserved by light cooking methods, such as stir-frying, steaming, and microwaving (review Table 8-1).

The B vitamins are thiamin, riboflavin, niacin, pantothenic acid, biotin, vitamin B-6, folate, and vitamin B-12. Choline is a related nutrient, but currently is not classified as a vitamin. Vitamin C is also a water-soluble vitamin.

The B vitamins often occur together in the same foods, so a lack of one B vitamin may mean other B vitamins are also low in a diet. The B vitamins function as coenzymes, small molecules that interact with enzymes to enable the enzymes to function. In essence, the coenzymes contribute to enzyme activity (Fig. 8-16).

As coenzymes, the B vitamins play many key roles in metabolism. The metabolic pathways used by carbohydrates, fats, and amino acids all require input from B vitamins. Because of their role in energy metabolism, needs for many B vitamins increase somewhat as energy expenditure increases. Still, this is not a major concern because this increase in energy expenditure is usually accompanied by a corresponding increase in food intake, which contributes more B vitamins to a diet. Many B vitamins are interdependent because they participate in the same processes (Fig. 8-17). B vitamin–deficiency symptoms typically occur in the brain and nervous system, skin, and GI tract. Cells in these tissues are metabolically active, and those in the skin and GI tract are also constantly being replaced.

After being ingested, the B vitamins are first broken down from their active coenzyme forms into free vitamins in the stomach and small intestine. The vitamins are then absorbed, primarily in the small intestine. Typically, about 50% to 90% of the B vitamins in the diet are absorbed, which means that they have relatively high bioavailability. Once inside cells, the active coenzyme forms are resynthesized. There is no need to consume the coenzyme forms themselves. Some vitamins are sold in their coenzyme forms, but these are broken down during digestion and we activate them when needed.

B VITAMIN INTAKES OF NORTH AMERICANS

The nutritional health of most North Americans with regard to the B vitamins is good. Typical diets contain plentiful and varied natural sources of these vitamins. In addition, many common foods, such as ready-to-eat breakfast cereals, are fortified with one or more of the B vitamins. In some developing countries, however, deficiencies of the B vitamins are more common, and the resulting deficiency diseases pose significant health problems. (We will discuss these worldwide nutritional deficiencies in more detail in Chapter 12.)

Because B vitamins are water soluble, very little is stored and excess ends up in the urine or stool. About 10% to 25% of these vitamins are lost from food during food processing and preparation because they dissolve in water. Light cooking methods, such as stir-frying, steaming, and microwaving, best preserve vitamin content (review Table 8-1).

Despite good B vitamin status of North Americans, marginal deficiencies of these vitamins may occur in some cases, especially among older adults who eat little food and in other people with poor dietary patterns. In the short run, a marginal deficiency likely leads only to fatigue or other unspecified physical effects. Although the long-term effects of such marginal deficiencies are yet unknown, increased risks of cardiovascular disease, cancer, and cataracts of the eye are suspected. With rare exceptions, healthy adults do not develop the more serious B–vitamin–deficiency diseases from dietary inadequacy alone. The main exceptions are people with alcoholism. The combination of extremely unbalanced diets and alcohol-induced alterations of vitamin absorption and metabolism creates significant risks for serious nutrient deficiencies among people with alcoholism.

B VITAMINS IN GRAINS

The production of refined grains, such as white flour from wheat, leads to the loss of B vitamins as well as other vitamins and minerals. In milling grains to make refined products, seeds are crushed and the germ, bran, and husk layers are discarded, leaving just the starch-containing endosperm in the refined grains. This starch is used to make white flour, bread, and cereal products. Unfortunately, many nutrients are lost along with the discarded germ, bran, and husk materials. To counteract these losses, in the United States, bread and cereal products made from milled grains are enriched with four B vitamins (thiamin, riboflavin, niacin, and folic acid) and with the mineral iron.

Food enrichment was initiated by federal legislation in the 1930s to help combat nutrient deficiencies such as pellagra (niacin deficiency) and iron-deficiency **anemia.** Federal regulations added folate to the list of nutrients required to be added to refined grain products in 1998. Not all nutrients lost in milling are added back through enrichment; these products remain lower in vitamins E and B-6,

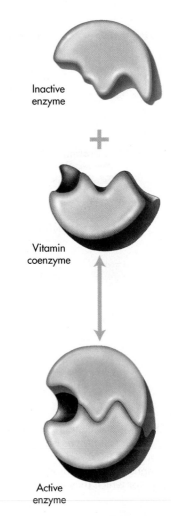

FIGURE 8-16 ▲ Coenzymes, such as those formed from B vitamins, aid in the function of various enzymes. Without the coenzyme, the enzyme cannot function, and deficiency symptoms associated with the missing vitamin eventually appear. Health-food stores sell the coenzyme forms of some vitamins. These more expensive forms of vitamins are unnecessary. The body makes all the coenzymes it needs from vitamin precursors.

▲ Rapid cooking of vegetables in minimal fluids aids in preserving vitamin content. Steaming is one effective method.

anemia A decreased oxygen-carrying capacity of the blood. This can be caused by many factors, such as iron deficiency or blood loss.

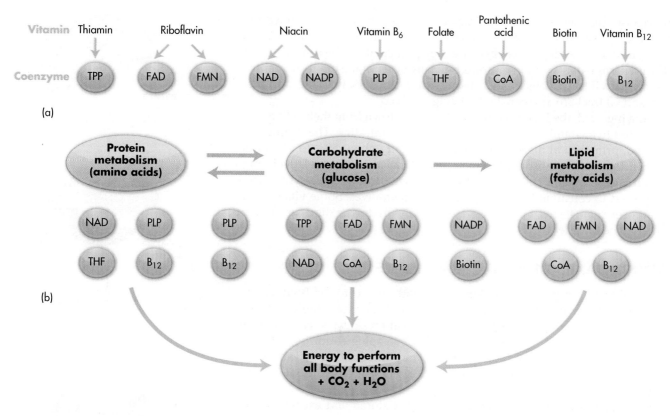

FIGURE 8-17 ▲ B vitamins are essential components of many coenzymes involved in energy metabolism. (a) The B vitamins and their coenzymes (shown as their commonly used abbreviation): TPP, thiamin pyrophosphate; FAD, flavin adenine dinucleotide; FMN, flavin mononucleotide; NAD, nicotinamide adenine dinucleotide; NADP, nicotinamide adenine dinucleotide phosphate; PLP, pyridoxal phosphate; CoA, coenzyme A. (b) The coenzymes are shown associated with the metabolic pathways for which they are essential.

potassium, magnesium, fiber, and other nutrients than the whole grains. This lower nutrient density is why nutrition experts and the Dietary Guidelines advocate daily consumption of whole-grain products, such as whole-wheat bread and brown rice, rather than refined grain products (Fig. 8-18).

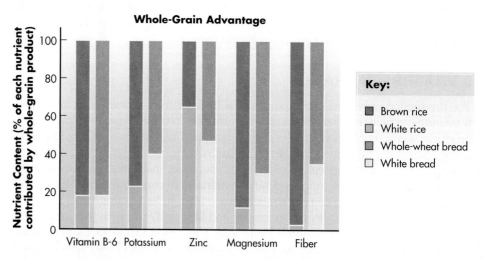

FIGURE 8-18 ▲ Compare the relative nutrient contents of refined versus whole grains. Nutrients are expressed as a percentage of the nutrient contribution of the whole-grain product.

✔ CONCEPT CHECK 8.6

1. What are the coenzymes associated with each of the B vitamins involved in energy metabolism?
2. What body organs or tissues are most likely to show symptoms if there is a deficiency of B vitamins?
3. Why are B vitamins lost when foods are cooked in water?
4. What group of people is at very high risk of deficiency of B vitamins?
5. What happens during the refining of grains that causes a decrease in nutrient density?
6. What is added during the "enrichment" of grain products, and why is it important?

8.7 Thiamin (Vitamin B-1)

FUNCTIONS OF THIAMIN

Thiamin was the first water-soluble vitamin to be discovered. One of its primary functions is to help release energy from carbohydrate. Its coenzyme form, thiamin pyrophosphate (TPP), participates in reactions in which carbon dioxide (CO_2) is released. Such reactions are particularly important in the body's ATP-producing energy pathways, which involve the breakdown of carbohydrates and certain amino acids (see Fig. 8-17). Thiamin also functions in chemical reactions that make RNA, DNA, and neurotransmitters.

THIAMIN DEFICIENCY

The thiamin-deficiency disease is called **beriberi,** a word that means "I can't, I can't" in the Sri Lankan language of Sinhalese. This disease was described long before thiamin was discovered to be a vitamin in 1910. The symptoms include weakness, loss of appetite, irritability, nervous tingling throughout the body, poor arm and leg coordination, and deep muscle pain in the calves. A person with beriberi often develops an enlarged heart and sometimes severe edema.

Beriberi is seen in areas where rice is a staple and polished (white) rice is consumed rather than brown (whole-grain) rice. In most parts of the world, even poor countries, white rice is preferred and is made by removing the bran and germ layer from brown rice. White rice is a poor source of thiamin, except for the enriched variety sold in the United States.

Beriberi results when glucose, the primary fuel for brain and nerve cells, cannot be metabolized to release energy because of the lack of thiamin. Because the thiamin coenzyme participates in glucose metabolism, problems with functions that depend on glucose such as brain and nerve action are the first signs of a thiamin deficiency. Symptoms can develop in just 10 days on a thiamin-free diet.

Alcohol abuse increases risk for thiamin deficiency. Absorption and use of thiamin are profoundly diminished and excretion is increased by consumption of alcohol. The low-quality diet that often accompanies severe alcoholism makes matters worse. There is limited storage in the body; therefore, an alcoholic binge lasting 1 to 2 weeks may quickly deplete already diminished amounts of the vitamin and result in deficiency symptoms. The beriberi associated with alcoholism is also called Wernicke-Korsakoff syndrome.

GETTING ENOUGH THIAMIN

Average daily intakes of thiamin for men exceed the DV by 50% or more, and women generally meet the RDA. Adults with low incomes and older people may

beriberi The thiamin-deficiency disorder characterized by muscle weakness, loss of appetite, nerve degeneration, and sometimes edema.

▲ Pork is an excellent source of thiamin.

FIGURE 8-19 ◄ Food sources of thiamin. (a) The fill of the background color (none, 1/3, 2/3, or completely covered) within each food group on MyPlate indicates the average nutrient density for thiamin in that group. (b) The bar graph shows the thiamin content of several foods in each food group compared to the RDA for adult males and females. Overall, the richest sources of thiamin are meats (especially pork), whole grains, and fortified breakfast cereals. *Nutrition data from USDA National Nutrient Database for Standard Reference, Release 26.*

(a)

(b)

	Food Item and Amount	Thiamin (milligrams)	% RDA for Adult Males (1.2 milligrams)	% RDA for Adult Females (1.1 milligrams)
Grains	Cheerios® cereal, 1 cup	0.4	31%	34%
	Wheat germ, 2 tablespoons	0.3	25%	27%
	Flour tortilla, 8″	0.2	17%	18%
Vegetables	Green peas, cooked, 1 cup	0.4	35%	38%
	Acorn squash, cooked, 1 cup	0.3	29%	31%
	Corn, cooked, 1 cup	0.1	12%	13%
Fruits	Orange juice, fresh, 1 cup	0.2	17%	18%
	Raisins, ½ cup	0.1	6%	7%
	Watermelon, 1 cup	0.1	4%	5%
Dairy	Yogurt, plain, non-fat, 1 cup	0.1	10%	11%
	Milk, fat-free, 1 cup	0.1	9%	10%
	Soymilk, 1 cup	0.1	6%	6%
Protein	Ham, canned, 3 ounces	0.8	68%	74%
	Kidney beans, cooked, ½ cup	0.1	12%	13%
	Chicken breast, roasted, 3 ounces	0.1	5%	5%

Thiamin

RDA
  Men: 1.2 milligrams
  Women: 1.1 milligrams
DV: 1.5 milligrams
UL: none

barely meet their needs for thiamin. Potential contributors to thiamin deficiency are diets dominated by highly processed and unenriched foods, sugar, and fat; and heavy alcohol intake combined with a poor diet. Oral thiamin supplements are typically nontoxic because thiamin is rapidly lost in the urine. Thus, no UL has been set for thiamin.

Major sources of thiamin include pork products, whole grains (wheat germ), ready-to-eat breakfast cereals, enriched grains and flour, green beans, milk, orange juice, organ meats, peanuts, dried beans, and seeds (Fig. 8-19). When considering the sections of MyPlate, the protein and grains groups contain the most foods that are nutrient-dense sources of thiamin.

✔ CONCEPT CHECK 8.7

1. How is thiamin involved in energy metabolism?
2. What body organs or tissues are most likely to show symptoms if there is a deficiency of thiamin?
3. What group of people is at very high risk of deficiency of thiamin?
4. What are some excellent sources of thiamin?

8.8 Riboflavin (Vitamin B-2)

FUNCTIONS OF RIBOFLAVIN

Riboflavin derives its name from its yellow color (*flavus* means yellow in Latin). The coenzyme forms of riboflavin, flavin dinucleotide (FAD) and flavin mononucleotide (FMN), participate in many energy-yielding metabolic pathways, such as the breakdown of fatty acids (see Fig. 8-17). Some metabolism of vitamins and minerals also requires riboflavin. Indirectly, riboflavin also has an antioxidant role in the body through its support of the enzyme glutathione peroxidase.

RIBOFLAVIN DEFICIENCY

Symptoms associated with riboflavin deficiency (**ariboflavinosis**) include inflammation of the mouth and tongue, dermatitis, cracking of tissue around the corners of the mouth (called **cheilosis**), various eye disorders, sensitivity to the sun, and confusion (Fig. 8-20). Such symptoms develop after approximately 2 months on a riboflavin-poor diet. Riboflavin deficiency typically would occur jointly with deficiencies of niacin, thiamin, and vitamin B-6 because these nutrients often occur in the same foods.

GETTING ENOUGH RIBOFLAVIN

On average, daily intakes of riboflavin are slightly above the RDA. As with thiamin, people with alcoholism risk riboflavin deficiency because they eat nutrient-poor diets. No specific symptoms indicate that riboflavin taken in megadoses is toxic, so no UL has been set. Riboflavin supplementation, including the riboflavin found in multivitamin supplements or in a heavily fortified breakfast cereal, can cause the urine to become bright yellow.

The grains, dairy, and protein groups of MyPlate contain the most nutrient-dense sources of riboflavin (Fig. 8-21). Major sources of riboflavin are ready-to-eat breakfast cereals, milk and milk products, enriched grains, meat, and eggs. Vegetables such as asparagus, broccoli, and various greens (e.g., spinach) are also good sources. Riboflavin is a relatively stable water-soluble vitamin; however, it is destroyed by light. Milk is sold in paper or opaque plastic containers rather than clear glass to protect the riboflavin. In the United States, many meet the riboflavin recommendation by consuming three servings of dairy products each day.

✔ CONCEPT CHECK 8.8

1. How is riboflavin involved in energy metabolism?
2. What body organs or tissues are most likely to show symptoms if there is a deficiency of riboflavin?
3. What types of foods are the best sources of riboflavin?

8.9 Niacin (Vitamin B-3)

FUNCTIONS OF NIACIN

Niacin functions in the body as one of two related compounds: nicotinic acid and nicotinamide. The coenzyme forms of niacin function in many cellular metabolic pathways. When you are generating energy (ATP) by burning carbohydrate and fat, a niacin coenzyme, nicotinamide adenine dinucleotide (NAD) or nicotinamide adenine dinucleotide phosphate (NADP), is used. Anabolic pathways in the cell—those that make new compounds—also often use a niacin coenzyme. This is especially true for fatty-acid synthesis (see Fig. 8-17).

FIGURE 8-20 ▲ Angular cheilitis, also called cheilosis or angular stomatitis, is result of a riboflavin deficiency. It causes painful cracks at the corners of the mouth. Angular cheilitis can be caused by other medical conditions; thus, further evaluation is required before diagnosing a nutrient deficiency.

Riboflavin
RDA
 Men: 1.3 milligrams
 Women: 1.1 milligrams
DV: 1.7 milligrams
UL: None

ariboflavinosis Riboflavin deficiency disease resulting in irritation of the skin, mouth, and throat; usually accompanied by low overall intakes of calories and protein.

cheilosis Inflammation of the skin at one or both corners of the mouth; may be a nonspecific symptom of a nutrient deficiency or just opportunistic infection; also called *angular cheilitis*.

FIGURE 8-21 ◀ Food sources of riboflavin. (a) The fill of the background color (none, 1/3, 2/3, or completely covered) within each food group on MyPlate indicates the average nutrient density for riboflavin in that group. (b) The bar graph shows the riboflavin content of several foods in each food group compared to the RDA for adult males and females. Overall, the richest sources of riboflavin are meats (especially liver), dairy products, and fortified breakfast cereals. Fruits (not shown) are not particularly good sources of riboflavin. *Nutrition data from USDA National Nutrient Database for Standard Reference, Release 26.*

(a)

(b)

	Food Item and Amount	Riboflavin (milligrams)	% RDA for Adult Males (1.3 milligrams)	% RDA for Adult Females (1.1 milligrams)
Grains	Kellogg's All Bran® cereal, 1 cup	0.8	62%	73%
Grains	Egg noodles, cooked, ½ cup	0.1	8%	9%
Grains	Bread, white, 1 slice	0.1	8%	9%
Vegetables	Mushrooms, raw, 5 medium	0.4	31%	36%
Vegetables	Spinach, cooked, 1 cup	0.4	31%	36%
Vegetables	Asparagus, cooked, 1 cup	0.3	23%	27%
Dairy	Yogurt, plain, non-fat, 1 cup	0.5	38%	45%
Dairy	Milk, fat-free, 1 cup	0.4	31%	36%
Dairy	Cheese, feta, 1.5 ounces	0.4	31%	36%
Protein	Beef liver, pan-fried, 3 ounces	2.8	215%	255%
Protein	Egg, hard-boiled, 1 large	0.3	23%	27%
Protein	Black beans, cooked, ½ cup	0.1	4%	5%

Nicotinic acid has been promoted as a natural method to lower blood lipids including LDL cholesterol; however, due to potential adverse side effects, its use is discouraged.

NIACIN DEFICIENCY

Because niacin coenzymes function in over 200 enzymatic reactions, niacin deficiency causes widespread problems in the body. Early symptoms include poor appetite, weight loss, and weakness. The distinct group of niacin-deficiency symptoms is known as **pellagra,** which means rough or painful skin (Fig. 8-22). The symptoms of the disease are **dementia,** diarrhea, and dermatitis (especially on areas of skin exposed to the sun). Left untreated, death often results.

Pellagra is the only dietary deficiency disease ever to reach epidemic proportions in the United States. It became a major problem in the southeastern United States in the late 1800s and persisted until the 1930s, when standards of living and diets improved. Pellagra was particularly prevalent in populations that consumed corn as a major part of their diet. Niacin in corn is bound by a protein that inhibits its absorption, making it less bioavailable. Soaking corn in an alkaline solution, such as lime water (water with calcium hydroxide), releases bound niacin and renders it more bioavailable. Hispanic people in North America traditionally soak corn in lime water before making tortillas. This treatment is one of the reasons why this Hispanic population never experienced much pellagra. Today, pellagra is rare in Western societies but can be seen in the developing world.

pellagra Niacin-deficiency disease characterized by dementia, diarrhea, and dermatitis, and possibly leading to death.

dementia A general loss or decrease in mental function.

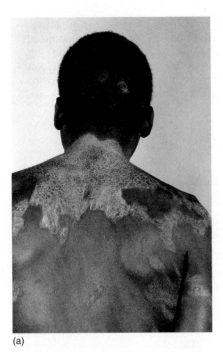

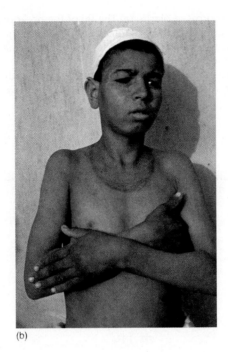

(a)   (b)

FIGURE 8-22 ◄ The dermatitis of pellagra. (a) Dermatitis on both sides (bilateral) of the body is a typical symptom of pellagra. Sun exposure worsens the condition. (b) The rough skin around the neck is referred to as Casal's necklace.

Niacin

RDA

   Men: 16 milligrams

   Women: 14 milligrams

DV: 20 milligrams

UL: 35 milligrams (nicotinic acid form)

GETTING ENOUGH NIACIN

The RDA for niacin is expressed as niacin equivalents (NE) to account for niacin received intact from the diet, as well as that synthesized from tryptophan: 60 milligrams of tryptophan in a diet yield about 1 milligram of niacin. Intakes of niacin by adults are about double the RDA, without considering the contribution from tryptophan. (Tables of food composition values also ignore tryptophan contribution.) Although pellagra was once common, today it is typically only seen associated with chronic alcoholism, in conjunction with poverty and malnutrition, and in those with rare disorders of tryptophan metabolism (for example, Hartnup disease).

The best food sources of niacin are found in the protein group on MyPlate (Fig. 8-23). Major sources of niacin are tuna, poultry, peanuts, fish, ready-to-eat cereals, beef, and asparagus. Coffee and tea also contribute some niacin to the diet. Niacin is heat stable; little is lost in cooking. Besides the preformed niacin found in protein foods, we can synthesize niacin from the amino acid tryptophan. In this manner, we synthesize about 50% of the niacin required each day. This reaction requires two other vitamins (riboflavin and vitamin B-6) to function as coenzymes in this chemical conversion.

AVOIDING TOO MUCH NIACIN

The UL for niacin pertains only to the nicotinic acid form (found in supplements). Side effects of niacin toxicity include headache; itching; and increased blood flow to the skin, because of blood vessel dilation or flushing in various parts of the body. These symptoms are especially seen when intakes are above 100 milligrams per day. In the long run, GI tract and liver damage are possible, so any use of megadoses, including large doses recommended for treatment for cardiovascular disease, requires close medical monitoring.

▲ Corn is treated in an alkaline solution to release protein-bound niacin so it can be available in corn products such as tortillas, taco shells, tortilla chips, and corn flour.

✓ CONCEPT CHECK 8.9

1. How is niacin involved in energy metabolism?
2. What are the three distinct signs of a niacin deficiency?
3. What are some excellent sources of niacin?
4. What is the relationship between tryptophan and niacin?

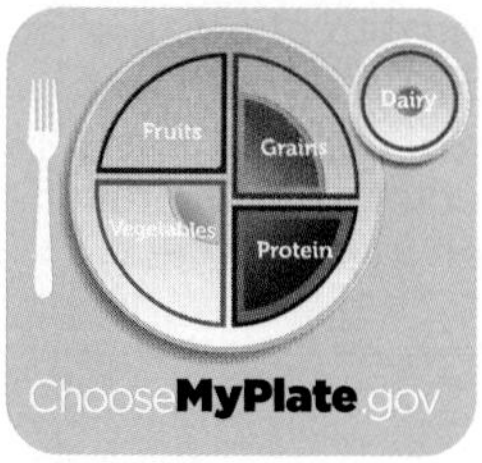

FIGURE 8-23 ◄ Food sources of niacin. (a) The fill of the background color (none, 1/3, 2/3, or completely covered) within each food group on MyPlate indicates the average nutrient density for niacin in that group. (b) The bar graph shows the niacin content of several foods compared to the RDA for adult males and females. Overall, the richest sources of niacin are foods in the protein group and fortified breakfast cereals. Foods in the dairy group (not shown) contain very little niacin, but the tryptophan in dairy foods can be converted into niacin. *Nutrition data from USDA National Nutrient Database for Standard Reference, Release 26.*

(a)

(b)

Food Item and Amount	Niacin (milligrams)	% RDA for Adult Males (16 milligrams)	% RDA for Adult Females (14 milligrams)
Total ® Raisin Bran cereal, 1 cup	20	125%	143%
Tortilla, flour, 8″	1.8	11%	13%
Bread, whole wheat, 1 slice	1.4	9%	10%
Raw mushrooms, 5	4.7	29%	34%
Potato, 1	2.1	13%	15%
Asparagus, cooked, 1 cup	2.0	13%	14%
Orange juice, fresh, 1 cup	1.0	6%	7%
Banana, 1 medium	0.8	5%	6%
Blueberries, 1 cup	0.6	4%	4%
Tuna, yellow fin, grilled, 3 ounces	18.8	118%	134%
Chicken breast, roasted, 3 ounces	11.8	74%	84%
Peanut butter, 2 tablespoons	4.2	26%	30%

8.10 Vitamin B-6 (Pyridoxine)

This vitamin is known by its number, rather than its general name. Vitamin B-6 is a family of three structurally similar compounds. All can be converted into the active vitamin B-6 coenzyme, pyridoxal phosphate (PLP).

FUNCTIONS OF VITAMIN B-6

The coenzymes of vitamin B-6 are needed for the activity of numerous enzymes involved in carbohydrate, protein, and lipid metabolism.

One of the primary functions is as a coenzyme in over 100 chemical reactions that involve the metabolism of amino acids and protein (see Fig. 8-17). The B-6 coenzyme, PLP, participates in reactions that allow the synthesis of nonessential (dispensable) amino acids by helping split the nitrogen group ($-NH_2$) from an amino acid and making it available to another amino acid. Vitamin B-6 also plays a role in the metabolism of the nonprotein amino acid **homocysteine** (see Newsworthy Nutrition and Further Reading 10).

Among the other important functions of vitamin B-6 are the synthesis of neurotransmitters such as serotonin and gamma aminobutyric acid (GABA); conversion of tryptophan to niacin; breakdown of stored glycogen to glucose; and synthesis of hemoglobin and white blood cells. Vitamin B-6 is also important for the synthesis of the heme portion of **hemoglobin.**

homocysteine An amino acid that arises from the metabolism of methionine. Vitamin B-6, folate, vitamin B-12, and choline are required for its metabolism. Elevated levels are associated with an increased risk of cardiovascular disease.

hemoglobin The iron-containing part of the red blood cell that carries oxygen to the cells and carbon dioxide away from the cells. The heme iron portion is also responsible for the red color of blood.

VITAMIN B-6 DEFICIENCY

Because of the role of vitamin B-6 in hemoglobin synthesis, a deficiency in vitamin B-6 would affect multiple body systems, including the cardiovascular, immune, and nervous systems, as well as overall energy metabolism. Vitamin B-6 deficiency also results in widespread symptoms, including depression, vomiting, skin disorders, irritation of the nerves, anemia, and impaired immune response.

People with alcoholism are susceptible to a vitamin B-6 deficiency. A metabolite formed in alcohol metabolism can displace the coenzyme form of B-6, increasing its tendency to be destroyed. In addition, alcohol decreases the absorption of vitamin B-6 and decreases the synthesis of its coenzyme form. Cirrhosis and hepatitis (both can accompany alcoholism) also destroy healthy liver tissue. Thus, a cirrhotic liver cannot adequately metabolize vitamin B-6 or synthesize its coenzyme form.

GETTING ENOUGH VITAMIN B-6

Major sources of vitamin B-6 are animal products and fortified ready-to-eat breakfast cereals (Fig. 8-24). Other sources are vegetables and fruits such as potatoes, spinach, bananas, and cantaloupes. Overall, the protein group of MyPlate offers many rich food sources of vitamin B-6. Animal sources and fortified grain products are the most reliable food sources because the vitamin B-6 they contain is more absorbable than that in plant foods. Vitamin B-6 is rather unstable; heating and freezing can easily destroy it.

With their ample consumption of animal products, North Americans have an average daily consumption of vitamin B-6 that is greater than the RDA (see margin). There is some research to indicate that athletes may need slightly more vitamin B-6 than sedentary adults. The athlete's body processes large quantities of glycogen and protein, and the metabolism of these compounds requires vitamin B-6. However, unless athletes restrict their food intake, they are likely to consume plenty of this B vitamin.

AVOIDING TOO MUCH VITAMIN B-6

The UL for vitamin B-6 (see margin) is based on the risk of developing nerve damage. Studies have shown that intakes of 2 to 6 grams of vitamin B-6 per day for 2 or more months can lead to irreversible nerve damage. Symptoms of vitamin B-6 toxicity include walking difficulties, and hand and foot tingling and numbness. Some nerve damage in individual sensory neurons is probably reversible, but damage

▲ Bananas are a plant source of vitamin B-6.

Vitamin B-6
RDA
  Men: 1.7 milligrams
  Women: 1.3 milligrams
DV: 2 milligrams
UL: 100 milligrams

Newsworthy Nutrition

Vitamin B-6 and Decreased Risk of Colorectal Cancer

Several chronic diseases, including cancer, have been linked to elevated blood levels of homocysteine. Vitamin B-6 is involved in more than 100 enzymatic reactions, including those that metabolize homocysteine. Evidence indicates that vitamin B-6 may reduce the risk of colorectal cancer and that its intake may be inadequate in older adults. This study was a meta-analysis of research studies that measured serum vitamin B-6 (pyridoxal 5' phosphate) in comparison to cancers of the colon and rectum. The authors found that colorectal cancer risk is decreased in subjects with high blood levels of vitamin B-6.

Source: Larsson SC and others: Vitamin B-6 and risk of colorectal cancer: A meta-analysis of prospective studies. *Journal of the American Medical Association* 303:1077, 2010.

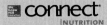 To further explore the link between vitamin B-6 and colorectal cancer, check out the Connect site: **www.mcgrawhillconnect.com**.

FIGURE 8-24 ◄ Food sources of vitamin B-6. (a) The fill of the background color (none, 1/3, 2/3, or completely covered) within each food group on MyPlate indicates the average nutrient density for vitamin B-6 in that group. (b) The bar graph shows the vitamin B-6 content of several foods compared to the RDA for adult males and females. Overall, the richest and most bioavailable sources of vitamin B-6 are animal sources of protein and fortified breakfast cereals. However, dairy foods (not shown) are not a particularly good source of vitamin B-6. *Nutrition data from USDA National Nutrient Database for Standard Reference, Release 26.*

(a)

(b)

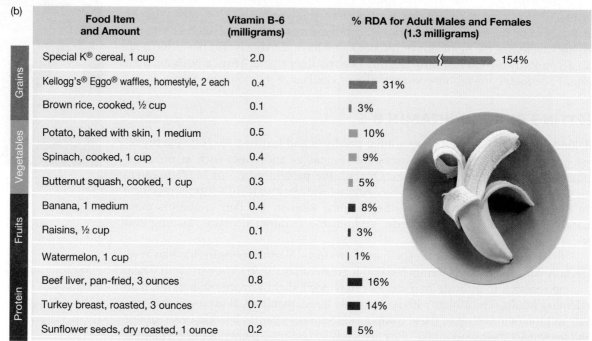

	Food Item and Amount	Vitamin B-6 (milligrams)	% RDA for Adult Males and Females (1.3 milligrams)
Grains	Special K® cereal, 1 cup	2.0	154%
	Kellogg's® Eggo® waffles, homestyle, 2 each	0.4	31%
	Brown rice, cooked, ½ cup	0.1	3%
Vegetables	Potato, baked with skin, 1 medium	0.5	10%
	Spinach, cooked, 1 cup	0.4	9%
	Butternut squash, cooked, 1 cup	0.3	5%
Fruits	Banana, 1 medium	0.4	8%
	Raisins, ½ cup	0.1	3%
	Watermelon, 1 cup	0.1	1%
Protein	Beef liver, pan-fried, 3 ounces	0.8	16%
	Turkey breast, roasted, 3 ounces	0.7	14%
	Sunflower seeds, dry roasted, 1 ounce	0.2	5%

to the ganglia (where many nerve fibers converge) appears to be permanent. With 500-milligram tablets of vitamin B-6 available in health-food stores, taking a toxic dose is easy (see Further Reading 15).

✓ CONCEPT CHECK 8.10

1. What is the role of vitamin B-6 in energy metabolism and other body functions?
2. What are the primary sources of vitamin B-6?
3. Are vitamin B-6 supplements safe?

8.11 Pantothenic Acid (Vitamin B-5) and Biotin (Vitamin B-7)

PANTOTHENIC ACID

Pantothenic acid is required for the synthesis of coenzyme A (CoA), a coenzyme in chemical reactions that allow the release of energy from carbohydrates, lipids, and protein. It also activates fatty acids so they yield energy (see Fig. 8-17) and is used in the initial steps of fatty-acid synthesis. Pantothenic acid is so widespread in foods that a nutritional deficiency among healthy people who eat varied diets is unlikely. *Pantothen* means "from every side" in Greek. A deficiency of pantothenic acid might occur in alcoholism along with a nutrient-deficient diet. However, the symptoms would probably be hidden among deficiencies of thiamin,

(a)

(b)

FIGURE 8-25 ◄ Food sources of pantothenic acid. (a) The fill of the background color (none, 1/3, 2/3, or completely covered) within each food group on MyPlate indicates the average nutrient density for pantothenic acid in that group. (b) The bar graph shows the pantothenic acid content of several foods in each food group compared to the AI for adult males and females. Overall, fortified foods and foods rich in protein are the best sources of pantothenic acid. *Nutrition data from USDA National Nutrient Database for Standard Reference, Release 26.*

	Food Item and Amount	Pantothenic Acid (milligrams)	% AI for Adult Males and Females (5 milligrams)
Grains	Whole Grain Total® cereal, 1 cup	10.0	200%
	Brown rice, cooked, ½ cup	0.4	8%
	White bread, 1 slice	0.1	2%
Vegetables	Mushrooms, raw, 5 medium	1.3	26%
	Acorn squash, cooked, 1 cup	1.0	20%
	Broccoli, cooked, 1 cup	1.0	20%
Fruits	Orange juice, fresh, 1 cup	0.5	10%
	Blueberries, 1 cup	0.2	4%
	Apple, Fuji, 1 medium	0.1	2%
Dairy	Yogurt, plain, non-fat, 1 cup	1.6	32%
	Milk, fat-free, 1 cup	0.9	18%
	Soymilk, 1 cup	0.3	6%
Protein	Power bar, 1 each	10.8	216%
	Beef liver, pan-fried, 3 ounces	5.6	112%
	Sunflower seeds, dry roasted, ¼ cup	2.6	52%

riboflavin, vitamin B-6, and folate, so the pantothenic acid deficiency might be unrecognizable. No toxicity is known for pantothenic acid, so no UL has been set.

Getting Enough Pantothenic Acid. The Adequate Intake (AI) set for pantothenic acid is 5 milligrams per day for adults. Average consumption is well in excess of this amount. The DV on food and supplement labels is 10 milligrams. Rich sources of pantothenic acid are sunflower seeds, mushrooms, peanuts, and eggs (Fig. 8-25). Other rich sources are meat, milk, and many vegetables.

BIOTIN

In its coenzyme form, biotin aids in dozens of chemical reactions. Biotin assists in the addition of carbon dioxide to other compounds, a reaction critical in synthesizing glucose and fatty acids, as well as breaking down certain amino acids. Symptoms of biotin deficiency include a scaly inflammation of the skin, changes in the tongue and lips, decreased appetite, nausea, vomiting, a form of anemia, depression, muscle pain and weakness, and poor growth.

Getting Enough Biotin. Foods rich in protein, such as egg yolks, peanuts, and cheese, are good sources of biotin (Fig. 8-26). The biotin content of food is typically not measured and therefore often not available in food composition tables or nutrient databases. Because intestinal bacteria synthesize some biotin that you can absorb, a biotin deficiency is unlikely. Scientists are not sure how much of the

Pantothenic Acid
AI: 5 milligrams
DV: 10 milligrams
UL: none set

Biotin
AI: 30 micrograms
DV: 300 micrograms
UL: none set

(a)

FIGURE 8-26 ◀ Food sources of biotin. (a) The fill of the background color (none, 1/3, 2/3, or completely covered) within each food group on MyPlate indicates the average nutrient density for biotin in that group. (b) The bar graph shows the biotin content of several foods compared to the AI for adult males and females. Overall, foods rich in protein are the best sources of biotin. Grains (even fortified varieties) contain very little biotin, so they are not shown in the bar graph. *Nutrition data from Staggs CG and others: Determination of the biotin content of select foods using accurate and sensitive HPLC/avidin binding.* J Food Compost Anal *2004;17:767–776.*

(b)

	Food Item and Amount	Biotin (micrograms)	% AI for Adult Males and Females (30 micrograms)
Vegetables	Sweet potato, baked, 1 cup	2.9	10%
	Mushrooms, canned, ½ cup	1.7	6%
	Carrots, cooked, 1 cup	0.9	3%
Fruits	Strawberries, 1 cup	2.3	8%
	Orange juice, from concentrate, 1 cup	1.0	3%
	Raisins, ½ cup	0.3	1%
Dairy	Cheese, American, 2 ounces	1.7	6%
	Cheese, cheddar, 1.5 ounces	0.6	2%
	Milk, non-fat, 1 cup	0.3	1%
Protein	Beef liver, pan-fried, 3 ounces	35.0	106%
	Egg, hard-boiled, 1 large	10.0	33%
	Peanuts, dry roasted, 1 ounce	4.9	16%

bacteria-synthesized biotin in our intestines is absorbed, so we still need to consume some in our diet. If bacterial synthesis in the intestines is not sufficient, as in people who are missing a large part of the colon or who take antibiotics for many months, special attention must be paid to meeting biotin needs.

Biotin's bioavailability varies significantly among foods based on the food's biotin-protein complex. In raw egg whites, biotin is bound to avidin, which inhibits absorption of the vitamin. Consuming many raw egg whites can eventually lead to biotin-deficiency disease. Cooking, however, denatures the protein avidin in eggs so it cannot bind biotin. In addition to food safety concerns (see Chapter 13), this is an important reasons to avoid consuming raw eggs.

Our food supply is thought to provide 40 to 60 micrograms per person per day. The DV used on food and supplement labels is 10 times the AI; the DV for biotin was set years before the vitamin had an AI level set and is based on outdated recommendations. However, biotin is relatively nontoxic. Large doses (up to 1 milligram per day) have been given over an extended period without harmful side effects to children who exhibit defects in biotin metabolism. Thus, no UL for biotin has been set.

✓ CONCEPT CHECK 8.11

1. What is the role of pantothenic acid in energy metabolism?
2. What are some rich sources of pantothenic acid?
3. What is the role of biotin in energy metabolism?
4. What are the signs and symptoms of biotin deficiency?
5. What are the best sources of biotin?
6. Why does consumption of raw eggs lead to biotin deficiency?

8.12 Folate (Vitamin B-9)

The term *folate* is used to describe a variety of forms of this B vitamin found in foods and in the body. Folic acid is the synthetic form added to fortified foods and present in supplements.

FUNCTIONS OF FOLATE

A key role of the folate coenzyme is to supply or accept single carbon compounds. In this role, folate coenzymes help form DNA and metabolize amino acids and their derivatives, such as homocysteine.

Along with vitamins B-6 and B-12, folate is required for the metabolism of homocysteine (see Further Reading 23). Blood homocysteine levels have been associated with cardiovascular risk; elevated homocysteine is considered an independent risk factor for atherosclerosis. Research is ongoing to examine the impact of folate (and other B vitamins) on cardiovascular disease risk.

Research is also underway on the link between folate and cancer protection. Folate aids in DNA synthesis, so it is hypothesized that even mild folate deficiency contributes to abnormal DNA integrity, which in turn affects certain cancer-producing genes. Meeting the RDA for folate may be one way to reduce cancer risk.

A final function of folate is the formation of neurotransmitters in the brain. Meeting folate needs can improve some cases of depression

FOLATE DEFICIENCY

One major result of a folate deficiency is that in the early phases of red blood cell synthesis, immature cells cannot divide because they cannot form new DNA. The cells grow progressively larger because they can still synthesize enough protein and other cell parts to make new cells. When the time comes for the cells to divide, the amount of DNA is insufficient to form two nuclei. The cells then remain in a large immature form, known as a **megaloblast** (Fig. 8-27).

Few mature red blood cells arrive in the bloodstream because the bone marrow of a folate-deficient person produces mostly immature megaloblast cells. When fewer mature red blood cells are present, the blood's capacity to carry oxygen decreases, causing a condition known as **megaloblastic anemia** (also called **macrocytic** or large-cell anemia).

Clinicians focus on red blood cells as an indicator of folate status because they are easy to collect and examine. Folate deficiency, however, disrupts cell division throughout the entire body. Other symptoms of folate deficiency are inflammation of the tongue, diarrhea, poor growth, mental confusion, depression, and problems in nerve function.

Maternal folate deficiency (along with a genetic abnormality related to folate metabolism) has been linked to the development of **neural tube defects** in the fetus. These defects include **spina bifida** (spinal cord or spinal fluid bulge through the back) and **anencephaly** (absence of a brain). Adequate folate status is crucial for all women of childbearing age because the neural tube closes within the first 28 days of pregnancy, a time when many women are not even aware that they are pregnant. Refer to Chapter 14 for more information on nutrition and birth defects.

Older people may be at risk for folate deficiency due to a combination of inadequate folate intake and decreased absorption. Perhaps these people fail to consume sufficient amounts of fruits and vegetables because of poverty or

megaloblast A large, immature red blood cell that results from the inability of the cell to divide normally (*megalo* = large; *blast* = primitive or immature).

megaloblastic (macrocytic) anemia Anemia characterized by the presence of abnormally large red blood cells.

neural tube defect A defect in the formation of the neural tube occurring during early fetal development. This type of defect results in various nervous system disorders, such as spina bifida. Folate deficiency in the pregnant woman increases the risk that the fetus will develop this disorder.

spina bifida Birth defect resulting from improper closure of the neural tube during embryonic development. The spinal cord or fluid may bulge outside the spinal column.

anencephaly Birth defect characterized by the absence of some or all of the brain and skull.

Medicine Cabinet

Methotrexate is a cancer drug that closely resembles a form of folate but cannot act in its place. Taken in high doses, methotrexate hampers folate metabolism. Consequently, DNA synthesis is retarded and cell division decreases. Cancer cells are among the most rapidly dividing cells in the body, so they are among those first affected. However, other rapidly dividing cells, such as intestinal cells and skin cells, are also affected. Not surprisingly, typical side effects of methotrexate therapy—diarrhea, vomiting, and hair loss—are also symptoms of folate deficiency. Today, when patients are given methotrexate, they need to consume a folate-rich diet or take folic acid supplements in order to reduce the toxic side effects of the drug. High supplemental doses have little or no influence on methotrexate's effectiveness as a cancer therapy.

FIGURE 8-27 ► Macrocytic (megalo-blastic) anemia occurs when red blood cells are unable to divide, leaving large, immature red blood cells. Either a folate or vitamin B-12 deficiency may cause this condition. Measurements of blood concentrations of both vitamins are taken to help determine the cause of the anemia.

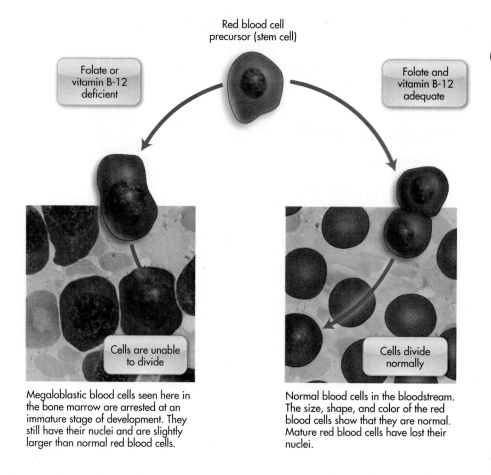

Megaloblastic blood cells seen here in the bone marrow are arrested at an immature stage of development. They still have their nuclei and are slightly larger than normal red blood cells.

Normal blood cells in the bloodstream. The size, shape, and color of the red blood cells show that they are normal. Mature red blood cells have lost their nuclei.

physical problems, such as poor dental health. In addition, folate deficiencies often occur with alcoholism, mostly to poor intake and absorption. Symptoms of a folate-related anemia can alert a physician to the possibility of alcoholism.

GETTING ENOUGH FOLATE

The name of the vitamin is derived from the Latin word *folium,* which means foliage or leaves. Quite predictably, the richest sources of folate are green, leafy vegetables. In addition, other vegetables, orange juice, dried beans, and organ meats are excellent sources of folate (Fig. 8-28). Fortified ready-to-eat breakfast cereals, bread, and milk are important sources of folic acid for many adults.

Folate is susceptible to destruction by heat and oxygen. The vitamin C present in some food sources of folate, such as orange juice, helps to reduce folate destruction, but food processing and preparation destroy 50% to 90% of the folate in food. This underscores the importance of regularly eating fresh fruits and raw or lightly cooked vegetables.

Folate recommendations for all but women of childbearing age are based on dietary folate equivalents (DFE). Synthetic folic acid, found in supplements and fortified foods, is more bioavailable than the folate that naturally occurs in food. The DFE unit considers these differences in bioavailability (see margin).

Pregnant women need extra folate (a total of 600 micrograms DFE) to accommodate the increased rates of cell division and DNA synthesis in their bodies and in the developing fetus. A healthy diet can supply this much. Still, prenatal care often includes a specially formulated multivitamin and mineral supplement enriched with folic acid to meet the higher RDA during pregnancy.

Prior to 1998, average daily folate intakes in the United States were approximately 320 micrograms for men and 220 micrograms for women. In 1998, FDA

Folate
RDA: 400 micrograms
DV: 400 micrograms
UL: 1000 micrograms (synthetic only)

Dietary Folate Equivalents
1 DFE = 1 microgram folate from food
= 0.6 microgram folic acid from food
= 0.5 microgram folic acid from a supplement on an empty stomach

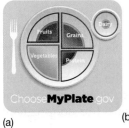

FIGURE 8-28 ◄ Food sources of folate. (a) The fill of the background color (none, 1/3, 2/3, or completely covered) within each food group on MyPlate indicates the average nutrient density for folate in that group. (b) The bar graph shows the folate content of several foods in each food group compared to the RDA for adult males and females. Overall, the richest sources of folate are green, leafy vegetables and fortified grains. *Nutrition data from USDA National Nutrient Database for Standard Reference, Release 26.*

(a)   (b)

	Food Item and Amount	Folate (micrograms DFE)	% RDA for Adult Males and Females (400 micrograms DFE)
Grains	Cheerios® cereal, 1 cup	336	84%
Grains	Wheat germ, 2 tablespoons	150	38%
Grains	Flour tortilla, 8″	98	25%
Vegetables	Asparagus, cooked, 1 cup	268	67%
Vegetables	Spinach, cooked, 1 cup	263	66%
Vegetables	Broccoli, cooked, 1 cup	168	42%
Fruits	Orange juice, fresh, 1 cup	74	19%
Fruits	Cantaloupe, 1 cup	37	9%
Fruits	Strawberries, raw, 1 cup	36	9%
Dairy	Yogurt, plain, 1 cup	29	7%
Dairy	Cottage cheese, low-fat, 1 cup	23	6%
Dairy	Soymilk, 1 cup	22	6%
Protein	Beef liver, pan-fried, 3 ounces	211	53%
Protein	Lentils, cooked, ½ cup	179	45%
Protein	Sunflower seeds, dry roasted, ¼ cup	76	19%

mandated the fortification of grain products with folate with the aim of reducing birth defects of the spine. With this mandate, average intakes have increased by about 200 micrograms per day. Factors other than folate deficiency (e.g., genetics and environment) play a role in the development of neural tube defects, but studies have shown that fortification of grain products with folic acid has decreased the rates of neural tube defects in infants by an estimated 15% to 30% in the United States and up to 50% in other countries with higher background rates of neural tube defects (see Further Reading 8).

The mandated folic acid enrichment of grains has also been accompanied by a noticeable decline in cardiovascular risk, especially risk for stroke, owing to a drop in blood homocysteine levels among U.S. adults. Supplements of folic acid, B-12, and B-6 have been promoted to help lower homocysteine and decrease cardiac and stroke risks. This is only likely to be effective for individuals who start out with elevated homocysteine levels. Taking pharmacological doses of folate is not likely to benefit people who have blood homocysteine levels in the normal range. In addition, some research indicates that too much folic acid could promote tumor development. Thus, even though folic acid fortification has been a public health success story for prevention of neural tube defects, there are concerns about the appropriate dose for the entire population (see Further Reading 8).

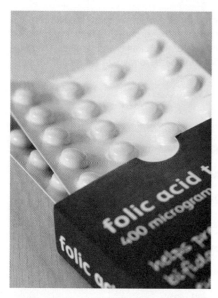

▲ An objective of Healthy People 2020 is to increase by 10% the proportion of women of childbearing potential who take supplements containing folic acid prior to pregnancy.

R-proteins Proteins produced by the salivary glands that bind to free vitamin B-12 in the stomach and protect it from stomach acid.

intrinsic factor A proteinlike compound produced by the stomach that enhances vitamin B-12 absorption in the ileum.

▲ With age, absorption of vitamin B-12 from food becomes less efficient, usually owing to the decreases in stomach acid production.

AVOIDING TOO MUCH FOLATE

The UL for folate (see margin) only refers to folic acid. This is because folate, the natural form in food, has limited absorption. Large doses of folic acid can hide the signs of vitamin B-12 deficiency and therefore complicate its diagnosis. Specifically, regular consumption of large amounts of folate can prevent the appearance of an early warning sign of vitamin B-12 deficiency: enlarged red blood cell size. For this reason, FDA limits the amount of folic acid in supplements (for nonpregnant adults) to 400 micrograms.

✔ CONCEPT CHECK 8.12

1. Explain why macrocytic (megaloblastic) anemia occurs.
2. Why do daily folate needs increase from 400 to 600 micrograms for pregnant women?

8.13 Vitamin B-12 (Cobalamin or Cyanocobalamin)

Vitamin B-12 is quite unique among the water-soluble vitamins. It is the only vitamin that contains a mineral as part of its structure. Its structure is the largest of all the vitamins. Unlike most water-soluble vitamins, B-12 can be stored to a significant extent in the liver, so it takes many months on a diet devoid of vitamin B-12 for a deficiency to surface. Vitamin B-12 is only naturally found in foods of animal origin. Finally, the means by which the body absorbs vitamin B-12 is complex; a problem at any one of several steps could impair absorption and lead to deficiency. To illustrate the multistep process by which vitamin B-12 is absorbed, we will trace the path of a meal containing vitamin B-12 through the digestive tract (Fig. 8-29).

In food, much of the vitamin B-12 is bound to protein and therefore cannot be absorbed. When food enters the mouth, **R-proteins** are secreted by the salivary glands. The bolus of food, including the R-proteins, travels down the esophagus to the stomach. Acid and enzymes present in the stomach release vitamin B-12 from food proteins, and the free vitamin B-12 then binds to R-protein. While food is in the stomach, the stomach cells release a proteinlike compound called the **intrinsic factor.** When the chyme reaches the duodenum, pancreatic enzymes release vitamin B-12 from R-proteins. The free vitamin B-12 then combines with intrinsic factor. The vitamin B-12–intrinsic factor complex travels the length of the small intestine to the ileum, where vitamin B-12 is finally absorbed.

If any of these steps fails or is altered, absorption can drop to 1% to 2%. In these cases, the person usually takes monthly injections of vitamin B-12, uses nasal gels of the vitamin to bypass the need for absorption, or takes megadoses of a supplemental form (300 times the RDA). In this latter case, the vitamin B-12 absorption defect is overcome by providing enough of the vitamin via simple diffusion across the intestinal tract.

About 95% of cases of vitamin B-12 deficiencies in healthy people result from defective absorption, rather than from inadequate intakes. This is especially true for older people. As we age, stomach acid production declines and our stomachs have a decreased ability to synthesize the intrinsic factor needed for vitamin B-12 absorption.

FUNCTIONS OF VITAMIN B-12

Vitamin B-12 participates in a variety of cellular processes. The most important function is folate metabolism. B-12 is required to convert folate coenzymes into the active forms needed for metabolic reactions, such as DNA synthesis. Without vitamin B-12, reactions that require certain active forms of folate do not take place in the cell. Thus,

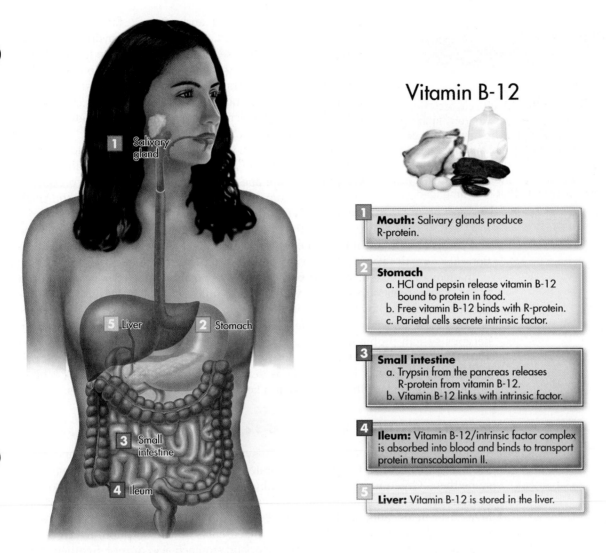

Vitamin B-12

1. **Mouth:** Salivary glands produce R-protein.

2. **Stomach**
   a. HCl and pepsin release vitamin B-12 bound to protein in food.
   b. Free vitamin B-12 binds with R-protein.
   c. Parietal cells secrete intrinsic factor.

3. **Small intestine**
   a. Trypsin from the pancreas releases R-protein from vitamin B-12.
   b. Vitamin B-12 links with intrinsic factor.

4. **Ileum:** Vitamin B-12/intrinsic factor complex is absorbed into blood and binds to transport protein transcobalamin II.

5. **Liver:** Vitamin B-12 is stored in the liver.

FIGURE 8-29 ▲ The absorption of vitamin B-12 requires several compounds produced in the mouth, stomach, and small intestine. Defects arising in the stomach or small intestine can interfere with absorption and result in vitamin B-12 deficiency.

a deficiency of vitamin B-12 can result in symptoms of a folate deficiency, including elevated homocysteine levels and macrocytic anemia (see Further Reading 23).

Another vital function of vitamin B-12 is maintaining the myelin sheath that insulates neurons. Initial neurological symptoms of B-12 deficiencies include irregular muscular actions and impaired reflexes. Eventual destruction of the myelin sheath causes paralysis and, perhaps, even death. In the past, vitamin B-12 deficiencies eventually led to death, mainly due to the destruction of nerves.

VITAMIN B-12 DEFICIENCY

The fatal consequences of vitamin B-12 deficiency can be observed in a disease known as **pernicious anemia.** The word *pernicious* means "leading to death." Indeed, before the discovery that consuming large doses of raw liver—a rich source of vitamin B-12—could be used to treat this particular disease, many people did die from it. Pernicious anemia is characterized by macrocytic anemia (with all the usual signs of anemia), sore mouth, depression, back pain, apathy, and severe nerve degeneration that can lead to tingling in the extremities, weakness, paralysis, and, eventually, death from heart failure.

People with pernicious anemia usually do not lack vitamin B-12 in their diets. Instead, they suffer from an autoimmune disease that destroys the stomach cells

pernicious anemia The anemia that results from a lack of vitamin B-12 absorption; it is *pernicious* because of associated nerve degeneration that can result in eventual paralysis and death.

Medicine Cabinet

Some medications may limit vitamin B-12 absorption. Antacids or other medications used to inhibit acid secretions will increase the pH within the stomach, thereby limiting release of B-12 from protein. People who have ulcers or reflux may take these drugs. Metformin, a popular medication for controlling diabetes, may reduce B-12 absorption. With any of these medications, you should check with your physician to see if supplemental B-12 is recommended.

that produce gastric acid and intrinsic factor. The resulting malabsorption of vitamin B-12 is responsible for all the symptoms of the disease. Because we are able to store some vitamin B-12, symptoms of nerve destruction do not develop until after about 3 years from the onset of the disease. Unfortunately, substantial nerve destruction often occurs before clinical signs of deficiency, such as anemia, are detected. The nerve destruction is irreversible.

Pernicious anemia, which affects about 2% of older adults, is the most common cause of vitamin B-12 malabsorption. Other causes are age-related atrophy of the acid-producing cells of the stomach and bacterial overgrowth in the small intestine. When acid production is low, bacteria normally present in the large intestine may colonize the small intestine and compete with our intestinal cells for vitamin B-12 absorption. Certain medications also impair vitamin B-12 absorption (see Medicine Cabinet).

Inadequate intake of vitamin B-12 is rarely responsible for deficiency, but it can occur. Vegan diets supply little vitamin B-12 unless they include vitamin B–12-enriched food (e.g., soy milk) or supplements. Infants breastfed by vegetarian mothers are at risk for vitamin B-12 deficiency accompanied by anemia and long-term nervous system problems, such as diminished brain growth, degeneration of the spinal cord, and poor intellectual development. The problems may have their origins during pregnancy if the mother is deficient in vitamin B-12. Certainly, achieving an adequate vitamin B-12 intake is a key diet-planning goal for vegans.

GETTING ENOUGH VITAMIN B-12

Vitamin B-12 compounds are originally synthesized by bacteria, fungi, and other lower organisms, then become incorporated into animal tissues when animals consume them. Organ meats (e.g., liver, kidneys, and heart) are especially rich sources of vitamin B-12. Other major sources of vitamin B-12 include meat, seafood, fortified breakfast cereals, milk, and eggs (Fig. 8-30). Adults over age 50 are encouraged to seek a synthetic vitamin B-12 source to increase absolute absorption, which can be limited due to both reduced intrinsic factor and stomach acid output. Synthetic vitamin B-12 is not food-bound, so it does not need stomach acid to release it from foods.

▲ Salmon, rainbow trout, and other types of seafood are excellent sources of B-12.

Newsworthy Nutrition

Folic acid and vitamin B-12 may prevent cognitive decline

Nine hundred and nine older adults (aged 60 to 74 years) received either a placebo or a dietary supplement containing 400 micrograms folic acid and 100 micrograms vitamin B-12 for 24 months as part of a randomized controlled trial to assess the effects of these B vitamins on depression. A secondary analysis revealed that subjects who received folic acid and vitamin B-12 performed better on tests of short- and long-term recall, which was also related to smaller increases in plasma homocysteine levels over time. Supplementation with these B vitamins may help older adults maintain cognitive functioning as they age.

Source: Walker JG and others: Oral folic acid and vitamin B-12 supplementation to prevent cognitive decline in community-dwelling older adults with depressive symptoms—the Beyond Ageing Project: A randomized controlled trial. *American Journal of Clinical Nutrition* 95:194, 2012.

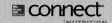 To further explore the link between B vitamins and cognitive function, check out the Connect site: **www.mcgrawhillconnect.com**.

It will be more readily absorbed than the form found in food (see Further Reading 16). Fortified breakfast cereals and dietary supplements are two possible synthetic sources.

On average, adults consume two times the RDA or more (see margin). This high intake provides the average meat-eating person with 2 to 3 years' storage of vitamin B-12 in the liver. A person would have to consume a diet essentially free of vitamin B-12 for approximately 20 years before exhibiting nerve destruction caused by a dietary deficiency. (Pernicious anemia develops more rapidly because of the reduced ability to reabsorb vitamin B-12 excreted into the GI tract during digestion, coupled with reduced absorption of dietary sources.) Still, vegans, who eat no animal products, should find a reliable source of vitamin B-12, such as fortified soy or rice milk, ready-to-eat breakfast cereals, and a form of yeast grown on media rich in vitamin B-12. Use of a multivitamin and mineral supplement containing vitamin B-12 is another option. Vitamin B-12 supplements are essentially nontoxic, so no UL has been set.

Vitamin B-12
RDA: 2.4 micrograms
DV: 6 micrograms
UL: None

Following the success of folate fortification, some experts advocate widespread fortification of the food supply with vitamin B-12 (see Further Reading 6).

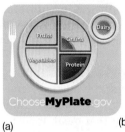

FIGURE 8-30 ◄ Food sources of vitamin B-12. (a) The fill of the background color (none, 1/3, 2/3, or completely covered) within each food group on MyPlate indicates the average nutrient density for vitamin B-12 in that group. (b) The bar graph shows the vitamin B-12 content of several foods compared to the RDA for adult males and females. Overall, foods of animal origin and fortified grains are the richest sources of vitamin B-12. Except for fortified grains, foods of plant origin (not shown) do not contain vitamin B-12. *Nutrition data from USDA National Nutrient Database for Standard Reference, Release 26.*

Food Item and Amount	Vitamin B-12 (micrograms)	% RDA for Adult Males and Females (2.4 micrograms)
Cheerios® cereal, 1 cup	1.9	79%
Egg noodles, cooked, ½ cup	0	0%
Quinoa, cooked, ½ cup	0	0%
Soymilk (fortified), 1 cup	2.1	86%
Yogurt, plain, non-fat, 1 cup	1.5	63%
Milk, fat-free, 1 cup	1.2	50%
Claims, baked, 3 ounces	84.1	3504%
Beef liver, pan-fried, 3 ounces	67.3	2804%
Kidney beans, cooked, ½ cup	0	0%

✔ CONCEPT CHECK 8.13

1. Explain the role of specific factors required for vitamin B-12 absorption.
2. Which two nutrient deficiencies could be responsible for macrocytic anemia?
3. Identify two population groups that are at risk for vitamin B-12 deficiency. Explain why these people are at risk.

8.14 Vitamin C (Ascorbic Acid)

FUNCTIONS OF VITAMIN C

Formation of Collagen. The best understood function of vitamin C (also known as *ascorbic acid* or *ascorbate*) is its role in the synthesis of collagen. This protein is highly concentrated in connective tissue, bone, teeth, tendons, and blood vessels. The important function of vitamin C in the formation of connective tissue is

FIGURE 8-31 ▲ Pinpoint hemorrhages of the skin—an early symptom of scurvy. The spots on the skin are caused by slight bleeding. The person may experience poor wound healing. These are signs of defective collagen synthesis.

exemplified in the early symptoms of a deficiency: pinpoint hemorrhages under the skin (Fig. 8-31), bleeding gums, and joint pain. Vitamin C is very important for wound healing; it strengthens structural tissues by increasing the cross-connections between amino acids found in collagen.

Formation of Other Compounds. Vitamin C has a specific function in the synthesis of numerous other compounds in the body. It is required for the synthesis of carnitine, a compound that transports fatty acids into the mitochondria. In addition, it takes part in the formation of two neurotransmitters, serotonin and norepinephrine.

Antioxidant. Vitamin C also has a more general function as an antioxidant because it can readily accept and donate electrons. These antioxidant properties have been postulated to reduce the formation of cancer-causing nitrosamines in the stomach. Vitamin C also aids in the reactivation of vitamin E after it has donated an electron to a free radical. Population studies suggest that the antioxidant properties of vitamin C may be effective in the prevention of certain cancers (esophagus, mouth, and stomach) and cataracts. The extent to which vitamin C functions in the reduction of diseases is debatable based on the scientific studies to this point.

Absorption of Iron. Vitamin C enhances iron absorption by keeping iron in its most absorbable form, especially as the mineral travels through the alkaline environment of the small intestine. Consuming 75 milligrams or more of vitamin C at a meal significantly increases absorption of the iron consumed at that meal. Increasing intake of vitamin–C–rich foods is beneficial for those with poor iron status or for those who choose to limit iron-rich food sources. Iron-deficiency anemia is common in the United States and is the number–one nutritional deficiency worldwide.

Immune Function. Last, but certainly not least, vitamin C is vital for the proper function of the immune system. Not only does vitamin C protect the immune cells from being degraded by the powerful oxidant reactions they use to kill pathogens, but it also promotes the proliferation of white blood cells. Can taking vitamin C fend off the common cold? Numerous well-designed, double-blind studies have failed to show that vitamin C prevents colds. Nevertheless, vitamin C does appear to reduce the duration of symptoms by a day or so and to lessen the severity of the symptoms. The key to success is to start the supplement as soon as symptoms appear. Once the cold has taken hold, it is too late!

VITAMIN C DEFICIENCY

On long sea voyages before the mid-eighteenth century, half or more of sailing crews died due to scurvy, the vitamin C–deficiency disease. The symptoms of scurvy, which include bleeding gums, tooth loss, bruising, and scaly skin, illustrate the important function of vitamin C in the formation of connective tissue. Without vitamin C, the skin and blood vessels weaken and wounds will not heal. In 1740, the Englishman Dr. James Lind first showed that citrus fruits—two oranges and one lemon a day—could prevent the development of scurvy. Fifty years after Lind's discovery, daily rations for British sailors included limes (thus their nickname, limeys). Even after this discovery, scurvy continued to affect many people as thousands died during the American Civil War owing to inadequate intake of vitamin C.

GETTING ENOUGH VITAMIN C

Major sources of vitamin C are citrus fruits, strawberries, green peppers, cauliflower, broccoli, cabbage, papayas, and romaine lettuce (Fig. 8-32). The brighter the fruit or vegetable is, the higher it tends to be in vitamin C. Fresh, ripe fruits and vegetables are loaded with vitamin C. Allowing them to "ripen" further in the grocery store or on your countertop at home will not increase vitamin C and actually can decrease it.

Ready-to-eat breakfast cereals, potatoes, and fortified fruit drinks are also good sources of vitamin C (Fig. 8-32). The five to nine servings of fruit and vegetables suggested by MyPlate's Daily Food Plan can easily provide enough vitamin C. Vitamin C is rapidly lost in processing and cooking as it is unstable in the presence of heat, iron, copper, or oxygen and is water soluble. Boiling fruits and vegetables for an extended time can destroy much of the vitamin C or cause it to leach out of the food.

Average daily consumption of vitamin C in the United States is 70 to 100 milligrams. Despite a generally poor intake of fresh fruits and vegetables, Americans still achieve dietary adequacy because of the widespread use of vitamin C as a food additive (see Chapter 13). Absorption efficiency is about 80% to 90% for moderate intakes (i.e., 60 to 100 milligrams per day). Tobacco users, however, need to add an extra 35 milligrams per day to the RDA. The toxic by-products of cigarette smoke and the oxidizing agents found in tobacco products increase the need for the antioxidant action of vitamin C.

▲ Citrus fruits are good sources of vitamin C.

AVOIDING TOO MUCH VITAMIN C

Note that when vitamin C is consumed in large doses, the amount in excess of daily needs mostly ends up in the feces or urine. The kidneys start rapidly excreting vitamin C when intakes exceed 100 milligrams per day. As the amount ingested increases, absorption efficiency decreases precipitously—to approximately 50% with intake of 1000 milligrams per day and to 20% with intakes of 6000 milligrams daily. Regular consumption of more than 2000 milligrams per day may cause stomach inflammation and diarrhea. Even 1000-mg supplement pills can cause some nausea and GI distress. A recent study found that men who said they took a vitamin C supplement regularly were at higher risk of developing kidney stones; therefore, ingesting large amounts of vitamin C supplements is discouraged in people predisposed to kidney stones. Because vitamin C enhances the absorption of iron, vitamin C supplements are also not recommended for those who overabsorb iron or have excessive iron stores. High doses of vitamin C may interfere with medical tests for

Vitamin C
RDA
  Men: 90 milligrams
  Women: 75 milligrams
DV: 60 milligrams
UL: 2000 milligrams

FIGURE 8-32 ◄ Food sources of vitamin C. (a) The fill of the background color (none, 1/3, 2/3, or completely covered) within each food group on MyPlate indicates the average nutrient density for vitamin C in that group. (b) The bar graph shows the vitamin C content of several foods compared to the RDA for adult males and females. Overall, fruits and vegetables are the richest sources of vitamin C. Foods in the dairy and protein groups (not shown) are poor sources of vitamin C. *Nutrition data from USDA National Nutrient Database for Standard Reference, Release 26.*

(a)

(b)

Food Item and Amount	Vitamin C (milligrams)	% RDA for Adult Males (90 milligrams)	% RDA for Adult Females (75 milligrams)
Whole Grain Total® cereal, 1 cup	60	67%	80%
Tortellini with cheese filling, ½ cup	0	0%	0%
Brown rice, cooked, ½ cup	0	0%	0%
Red peppers, raw, 1 cup	190	211%	253%
Brussels sprouts, cooked, 1 cup	97	108%	129%
Spinach, raw, 2 cups	17	19%	23%
Strawberries, 1 cup	89	99%	119%
Orange, 1 medium	83	92%	111%
Grapefruit juice, canned, 1 cup	72	80%	96%

diabetes or blood in the feces. If you take vitamin C supplements at any dose, be sure to inform your doctor. Physicians may misdiagnose conditions if they do not realize the influence of large doses of vitamin on your medical test results.

✓ CONCEPT CHECK 8.14

1. How does vitamin C function as an antioxidant and regenerate vitamin E?
2. How do the signs of vitamin C deficiency relate to the many roles of the vitamin discussed in this chapter?
3. Why are fresh foods the best sources of vitamin C?

8.15 Choline and Other Vitamin-Like Compounds

The dietary component choline is the latest addition to the list of essential nutrients. In 1998, the Institute of Medicine recognized choline as an essential nutrient. When the Dietary Reference Intakes were released in 2000, only limited research on the dietary requirements for choline existed. One study of male volunteers showed decreased choline stores and liver damage when they were fed choline-deficient intravenous nutrition solutions. Based on this human study, plus laboratory animal studies, choline has been deemed essential, but it is not yet classified as a vitamin.

FUNCTIONS OF CHOLINE

Despite its lack of vitamin status, choline is needed by all cells and plays several important roles in the body.

Cell Membrane Structure. Choline is a precursor for several phospholipids. Phosphatidylcholine (also known as lecithin) accounts for about half of the phospholipids in cell membranes. Recall from Chapter 5 that phospholipids contribute to the flexibility of cell membranes and allow for the presence of both water- and fat-soluble compounds in cell membranes. With its role in cell membrane structure, choline is important for the health of every cell and particularly for the health of brain tissue, where it is present in high levels.

▲ Choline is important for proper development of the fetal brain. Milk and other dairy products supply some choline.

Single-Carbon Metabolism. Choline is a precursor for betaine, a compound that participates in many chemical reactions that involve the transfer of single-carbon groups in metabolism. Important examples of metabolic pathways that involve the transfer of single-carbon groups include the synthesis of neurotransmitters, modifications of DNA during embryonic development, and the metabolism of homocysteine. As you learned in Sections 8.10, 8.12, and 8.13, high levels of homocysteine in the blood are related to increased risk of heart disease. Betaine and the B vitamin folate both donate single-carbon groups to convert homocysteine into another compound, thus reducing levels of homocysteine in the blood.

Recent research points to a role of adequate choline for the prevention of birth defects (see Further Reading 22). Choline's purported role in prevention of birth defects is similar to that of folate. Both folate and choline are involved in the formation of DNA during embryonic development. As you will read in Chapter 14, problems with DNA formation lead to birth defects. Indeed, animal studies show that maternal choline supplementation during critical stages of embryonic development can improve learning and memory in the offspring. In humans, as well, studies show that babies born to women with low choline intakes have four times higher rates of birth defects than babies born to women with high choline intakes.

Nerve Function and Brain Development. Choline is part of acetylcholine, a neurotransmitter associated with attention, learning, memory, muscle control, and many other functions. Sphingomyelin, a choline-containing phospholipid, is part of the myelin sheath that insulates nerve cells. As already mentioned, brain tissue is particularly high in choline. During pregnancy, the concentration of choline in amniotic fluid is high, supplying choline to the developing brain of the fetus. Animal studies demonstrate that choline deficiencies during pregnancy lead to poor brain development, learning ability, and memory. The AI for choline is increased during pregnancy and breastfeeding to assist in proper brain development.

Lipid Transport. As part of phospholipids, choline is a component of lipoproteins, which carry lipids through the blood. Choline deficiencies in animals and humans lead to decreased production of lipid transport proteins, such as very low density lipoproteins (VLDL). The inability of the liver to export fat to the rest of the body leads to the buildup of fat in the liver. A small amount of fat in the liver is normal, but excess fat leads to scarring of the liver tissue and eventual dysfunction. Fatty liver is a common cause of cirrhosis.

The roles of choline in lipid transport and homocysteine metabolism have implicated the nutrient in the prevention of cardiovascular disease. However, research has also raised some concerns about possible negative effects of choline on cardiovascular disease risk. Recent findings have shown that metabolism of dietary phosphatidylcholine, from foods such as eggs, by the intestinal microbiota produces an atherosclerosis-promoting compound that has been associated with an increased risk of major adverse cardiovascular events (see Further Reading 19).

GETTING ENOUGH CHOLINE

Choline is widely distributed in foods (Fig. 8-33). Soybeans, egg yolks, beef, cauliflower, almonds, and peanuts are good sources. In addition to natural food sources, lecithin is often added to food products as an emulsifier during processing, so many other foods are sources of choline.

Choline can exist in foods as free choline or as part of other compounds, such as phospholipids. Pancreatic enzymes break down some of the phospholipid forms prior to absorption. Free choline is water soluble and can be absorbed from the small intestine into portal circulation for transport to the liver. Choline that is part of phospholipids, on the other hand, is fat soluble and gets absorbed into the lymphatic system.

To some extent, choline also can be synthesized in the body by a process that involves other nutrients, such as folate and the amino acid methionine. If the body must synthesize choline to meet its needs, functional deficiencies of folate could result.

Eggs (with yolks) are by far the most nutrient-dense source of choline. One whole egg supplies about ¼ of the daily choline needs in a 70-kcal package. However, dietary advice to limit saturated fat and cholesterol discourages frequent intake of egg yolks. Choline researchers suggest that an average of one egg per day would assist in achieving the AI for choline while still supplying less than the 300-milligram-per-day limit for cholesterol.

An AI has been set for adults (see margin), but is unknown whether a dietary supply is essential for infants or children. As already noted, some choline can be synthesized in the body, but recent research indicates that synthesis by the body is not sufficient to meet the body's needs for choline. Nutrition surveys show that fewer than 10% of Americans meet the AI for choline. In addition, the AIs do not reflect wide genetic variation in individual choline requirements. Research suggests that at least half the population has genetic variations that increase dietary requirements for nutrients that serve in single-carbon metabolism, including choline and folate. Thus, even meeting the AI may not provide enough choline to support the body's needs for some people (see Further Reading 20).

▲ Dairy products, soy, almonds, and peanuts are natural sources of choline.

FIGURE 8-33 ◀ Food sources of choline. (a) The fill of the background color (none, 1/3, 2/3, or completely covered) within each food group on MyPlate indicates the average nutrient density for choline in that group. (b) The bar graph shows the choline content of several foods compared to the RDA for adult males and females. Overall, foods that are rich sources of protein are good sources of choline. Grains and fruits are not included in the bar graph because, in general, these are poor sources of choline. *Nutrition data from USDA National Nutrient Database for Standard Reference, Release 26.*

(a)

(b)

	Food Item and Amount	Choline (milligrams)	% AI for Adult Males (550 milligrams)	% AI for Adult Females (425 milligrams)
Vegetables	Swiss chard, cooked, 1 cup	50	9%	12%
Vegetables	Cauliflower, cooked , 1 cup	49	9%	12%
Vegetables	Mushrooms, shiitake, 4 each	30	5%	7%
Dairy	Milk, fat-free, 1 cup	38	7%	9%
Dairy	Yogurt, plain, non-fat, 1 cup	37	7%	9%
Dairy	Cottage cheese, low-fat, ½ cup	18	3%	4%
Protein	Beef liver, pan-fried , 3 ounces	339	62%	78%
Protein	Egg, hard-boiled, 1 large	147	27%	35%
Protein	Cod, baked, 3 ounces	71	13%	17%

Choline

AI

   Men: 550 milligrams

   Women: 425 milligrams

DV: not established

UL: 3.5 grams

The AI for choline increases during pregnancy (to 450 milligrams per day) and breastfeeding (to 550 milligrams per day) to support the brain development of the fetus or infant. Prenatal vitamins do not contain choline. Therefore, consumption of rich dietary sources of choline, such as eggs, is important for pregnant and breastfeeding women.

AVOIDING TOO MUCH CHOLINE

The UL for adults is set at 3.5 grams per day. Routinely exceeding the UL will result in a fishy body odor and low blood pressure.

OTHER VITAMIN-LIKE COMPOUNDS

A variety of vitamin-like compounds are found in the body. These include the following:

- Carnitine, needed to transport fatty acids into cell mitochondria
- Inositol, part of cell membranes
- Taurine, part of bile acids
- Lipoic acid, which participates in carbohydrate metabolism and acts as an antioxidant

These vitamin-like compounds can be synthesized by cells using common building blocks, such as amino acids and glucose. Our diets are also a source. In disease states or periods of active growth, the synthesis of vitamin-like compounds may not meet needs, so dietary intake can be crucial. The needs for vitamin-like compounds in certain groups of individuals, such as for preterm infants, are being investigated. Although promoted and sold by health-food stores, these vitamin-like compounds need not be included in the diet of the average healthy adult.

Table 8-5 summarizes much of what we know about the water-soluble vitamins. Now that you have studied the vitamins, review MyPlate, and note how each food group can make an important vitamin contribution (Fig. 8-2).

TABLE 8-5 ▶ Summary of the Water-Soluble Vitamins and Choline

Vitamin	Major Functions	RDA or AI	Dietary Sources*	Deficiency Symptoms	Toxicity Symptoms
Thiamin	• Coenzyme of carbohydrate metabolism • Nerve function	*Men:* 1.2 milligrams *Women:* 1.1 milligrams	• Sunflower seeds • Pork • Whole and enriched grains • Dried beans • Peas	*Beriberi* • Nervous tingling • Poor coordination • Edema • Heart changes • Weakness	None
Riboflavin†	• Coenzyme of carbohydrate metabolism	*Men:* 1.3 milligrams *Women:* 1.1 milligrams	• Milk • Mushrooms • Spinach • Liver • Enriched grains	• Inflammation of the mouth and tongue • Cracks at the corners of the mouth • Eye disorders	None
Niacin	• Coenzyme of energy metabolism • Coenzyme of fat synthesis	*Men:* 16 milligrams (NE) *Women:* 14 milligrams (NE)	• Mushrooms • Bran • Tuna • Salmon • Chicken • Beef • Liver • Peanuts • Enriched grains	*Pellagra* • Diarrhea • Dermatitis • Dementia • Death	• Headache • Itching • Flushing of skin • Damage to the GI tract or liver Upper Level is 35 milligrams from supplements, based on flushing of skin.
Pantothenic acid	• Coenzyme of energy metabolism • Coenzyme of fat synthesis	5 milligrams	• Mushrooms • Liver • Broccoli • Eggs *Most foods have some.*	No natural deficiency disease or symptoms	None
Biotin	• Coenzyme of glucose production • Coenzyme of fat synthesis	30 micrograms	• Cheese • Egg yolks • Cauliflower • Peanut butter • Liver	• Dermatitis • Tongue soreness • Anemia • Depression	Unknown
Vitamin B-6†	• Coenzyme of energy metabolism, especially protein • Neurotransmitter synthesis • Red blood cell synthesis *Many other functions*	*Men (up to 50 years):* 1.3 milligrams *Women (up to 50 years):* 1.3 milligrams	• Animal protein foods • Spinach • Broccoli • Bananas • Salmon • Sunflower seeds	• Headache • Anemia • Convulsions • Nausea • Vomiting • Flaky skin • Sore tongue	• Difficulty walking • Numbness or tingling in hands or feet Upper Level is 100 milligrams, based on nerve destruction.
Folate (folic acid)†	• Coenzyme involved in DNA synthesis *Many other functions*	400 micrograms (DFE)	• Green, leafy vegetables • Orange juice • Organ meats • Sprouts • Sunflower seeds	• Macrocytic • Inflammation of tongue • Diarrhea • Poor growth • Depression	None likely Upper Level for adults is set at 1000 micrograms for synthetic folic acid (exclusive of food folate), based on masking of B-12 deficiency.

(continued)

TABLE 8-5 ▶ **Summary of the Water-Soluble Vitamins and Choline** *(continued)*

Vitamin	Major Functions	RDA or AI	Dietary Sources*	Deficiency Symptoms	Toxicity Symptoms
Vitamin B-12[†]	• Coenzyme of folate metabolism • Nerve function *Many other functions*	2.4 micrograms *Older adults and vegans should use fortified foods or supplements.*	• Animal foods (not natural in plants) • Organ meats • Oysters • Clams • Fortified, ready-to-eat breakfast cereals	• Macrocytic anemia • Poor nerve function	None
Vitamin C	• Connective tissue synthesis • Hormone synthesis • Neurotransmitter synthesis • Possible antioxidant activity	*Men:* 90 milligrams *Women:* 75 milligrams *Smokers should add 35 milligrams.*	• Citrus fruits • Strawberries • Broccoli • Greens	• Scurvy • Poor wound healing • Pinpoint hemorrhages • Bleeding gums	• GI distress Upper Level is 2 grams, based on development of diarrhea. • Can also alter some diagnostic tests
Choline[†]	• Neurotransmitter synthesis • Phospholipid synthesis	*Men:* 550 milligrams *Women:* 425 milligrams	• Liver • Eggs • Milk • Peanuts • Soybeans • Wheat germ *Also synthesized by the body*	No natural deficiency	• Hypotension • Fishy body odor Upper Level is 3.5 grams per day, based on development of fishy body odor and reduced blood pressure.

*Fortified, ready-to-eat breakfast cereals are good sources for most of these vitamins and a common source of B vitamins for many of us.

[†]29 These nutrients also participate in homocysteine metabolism; meeting the RDA or AI may reduce the risk of developing cardiovascular disease.

Abbreviations: NE = niacin equivalents; DFE = dietary folate equivalents.

✓ CONCEPT CHECK 8.15

1. Describe three functions of choline in the human body.
2. List three ways to incorporate more choline into the diet.
3. Is it necessary to take dietary supplements of vitamin-like compounds, such as carnitine and taurine? Why or why not?

8.16 Dietary Supplements—Who Needs Them?

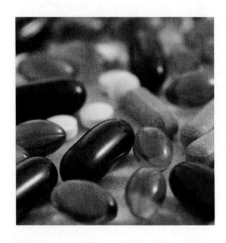

The phrase *multivitamin* and *mineral supplement* has been mentioned many times so far in this textbook. Often, these and other supplements are marketed as cures for anything and everything. This cure-all approach is promoted by the supplement industry and countless health-food stores, pharmacies, and supermarkets.

According to the Dietary Supplement Health and Education Act of 1994, a supplement in the United States is a product intended to supplement the diet that bears or contains one or more of the following ingredients:

• A vitamin
• A mineral
• An herb or another botanical

- An amino acid
- A dietary substance to supplement the diet, which could be an extract or a combination of the first four ingredients in this list

The definition is broad and covers a wide variety of nutritional substances. The use of dietary supplements is a common practice among North Americans and generates about $35 to $36 billion annually for the industry in the United States (Fig. 8-34). Supplements can be sold without proof that they are safe and effective. Unless FDA has evidence that a supplement is inherently dangerous or marketed with an illegal claim, it will not regulate such products closely. (The vitamin folate is an exception.) FDA has limited resources to police supplement manufacturers and has to act against these manufacturers one at a time. Thus, we cannot rely on FDA to protect us from vitamin and mineral supplement overuse and misuse. We bear that responsibility ourselves, with the help of professional advice from a physician or registered dietitian.

The supplement makers can make broad claims about their products under the "structure or function" provision of the law. The products, however, cannot claim to prevent, treat, or cure a disease. Menopause in women and aging are not diseases per se, so products alleging to treat symptoms of these conditions can be marketed without FDA approval. For example, a product that claims to treat hot flashes arising during menopause can be sold without any evidence to prove that the product works, but a product that claims to decrease the risk of cardiovascular disease by reducing blood cholesterol must have results from scientific studies that justify the claim.

Why do people take supplements? Frequently given reasons include:

- Reducing susceptibility to health problems (e.g., colds)
- Preventing heart attacks
- Preventing cancer
- Reducing stress
- Increasing "energy"

CRITICALTHINKING

Believing that supplements provide the nutrition her body needs, Janice regularly takes numerous supplements while paying relatively little attention to daily food choices. How would you explain to her that this practice may lead to health problems?

SHOULD YOU TAKE A SUPPLEMENT?

Multivitamin and mineral supplements (MVMs) are popularly regarded as a simple back-up plan or an insurance policy, even for people who consciously try to consume a balanced diet. Users aim to prevent nutrient deficiencies or chronic diseases by filling any gaps between dietary intake and nutrient needs. However, evidence to support the widespread use of MVMs is mixed. While there is little risk of harm from consuming a balanced MVM that supplies no more than 100% of the Daily Value for the nutrients it contains, most studies indicate no discernible advantage. The National Institutes of Health, in its State-of-the-Science report, concluded that the present evidence is insufficient to recommend either for or against the use of MVMs by Americans to prevent chronic disease (see Further Readings 11, 13, 17, and 18).

FIGURE 8-34 ▼ The dietary supplement industry is a growing multibillion-dollar business.

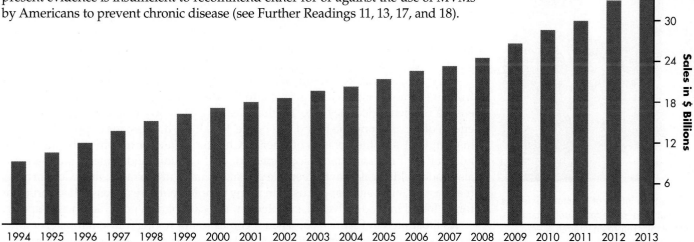

TABLE 8-6 ▶ **People Who Are Most Likely to Benefit from Dietary Supplements**

Type of Supplement	Who May Benefit
MVM	• People who consume <1200 kilocalories per day (e.g., some women and older adults) • People with unbalanced or inadequate diets (e.g., in cases of food insecurity or children who are "picky" eaters) • People with malabsorptive diseases • People who take medications that interfere with nutrient absorption or metabolism
Various B vitamins	• People who abuse alcohol
Folic acid	• Women of childbearing age (especially during pregnancy and breastfeeding)
Vitamin B-12	• Older adults • Strict vegans
Vitamin C	• People who smoke
Vitamin D	• People with limited milk intake (due to allergies or lactose intolerance) • People with limited exposure to sunlight (e.g., all infants, many African-Americans, and some older adults) • Strict vegans
Vitamin E	• People who follow diets low in fat (especially low in plant oils)
Vitamin K	• Newborns (given by injection shortly after birth)
Calcium	• Strict vegans • Older adults with bone loss
Fluoride	• Some older infants and children (as directed by a dentist)
Iron	• Women with excessive bleeding during menstruation • Women who are pregnant • Strict vegans
Zinc	• Strict vegans

▲ Long-term intake of just three times the Daily Value for some fat-soluble vitamins—particularly preformed vitamin A—can cause toxic effects. Know what you are taking if you use supplements.

Do specific vitamin or mineral supplements provide any benefit? The experts found that only a few studies of vitamin and mineral supplements demonstrate beneficial effects for the prevention of deficiencies or chronic diseases. For example, post-menopausal women may benefit from taking calcium and vitamin D supplements to increase bone mineral density and decrease fracture risk. Table 8-6 outlines the population groups that are most likely to benefit from taking dietary supplements.

While there may be moderate benefits of consuming dietary supplements, uninformed use of supplements can be risky. Indeed, most cases of nutrient toxicity are a result of supplement use. High doses of one nutrient can affect absorption or metabolism of other nutrients. For example, excessive zinc intake can inhibit copper absorption, and large amounts of folate can mask signs and symptoms of a vitamin B-12 deficiency. In addition, some supplements can interfere with medications. For instance, high intakes of vitamin K or vitamin E alter the action of anticlotting medications, vitamin B-6 can offset the action of L-dopa (used in treating Parkinson's disease), and large doses of vitamin C might interfere with certain cancer therapy regimens.

For most Americans, finding ways to incorporate the recommended servings of fruits, vegetables, and whole grains into the diet is the safest and healthiest way to ensure nutrient adequacy (see Further Reading 1). Many of the health-promoting effects of foods cannot be found in a bottle. Recall the discussions of phytochemicals in Chapter 1 and the benefits of fiber in Chapter 4. Few or no phytochemicals and no fiber are present in most supplements. Multivitamin and mineral supplements also contain little calcium to keep the pill size small. Furthermore, the oxide forms of magnesium, zinc, and copper used in many supplements are not as well absorbed as forms found in foods. Overall, supplement use cannot fix a poor diet in all respects.

As illustrated in Figure 8-35, when it comes to improving nutrient intake, emphasize foods before considering dietary supplements. First, you should assess your current dietary habits. The Dietary Guidelines for Americans and MyPlate are two tools consumers can use to plan a healthy diet (see Chapter 2). If nutrient gaps still remain, identify food sources that can help. For example, fortified, ready-to-eat breakfast cereals supply a variety of micronutrients, including vitamin E, folic acid, vitamin B-6, and highly absorbable forms of vitamin B-12. Other fortified foods, such as calcium-fortified orange juice, can also be useful. Be mindful of portion sizes of highly fortified foods, however, as multiple servings could lead to excessive intakes of some nutrients, such as vitamin A, iron, and synthetic folic acid. Lastly, if supplement use is desired, educate yourself and discuss this choice with a physician or registered dietitian.

WHICH SUPPLEMENT SHOULD YOU CHOOSE?

If you decide to take a multivitamin and mineral supplement, start by choosing a nationally recognized brand (from a supermarket or pharmacy) that contains about 100% of the Daily Values for the nutrients present. A multivitamin and mineral supplement should generally be taken with or just after meals to maximize absorption. Make sure also that intake from the total of this supplement, any other supplements used, and highly fortified foods (such as ready-to-eat breakfast cereals) provides no more than the Upper Level for each vitamin and mineral. (See the inside cover of this textbook for Upper Levels.) This is especially important with regard to preformed vitamin A intake. Two exceptions are: (1) both men and older women should make sure any product used is low in iron or iron-free to avoid possible iron overload (see Chapter 9 for details) and (2) somewhat exceeding the Upper Level for vitamin D is likely a safe practice for adults. Read the labels carefully to be sure of what is being taken (Fig. 8-36). Because research on a variety of nutrient supplements has revealed a lack of product quality, FDA now requires supplement makers to test the identity, purity, strength, and composition of all their products. As an extra protection, select supplements that bear the logo of the United States Pharmacopeial Convention (USP). The USP is an independent, nonprofit group of scientists that reviews products for strength, quality, purity, packaging, labeling, speed of dissolution, and shelf-stability. The USP designation on a supplement label indicates that the product has been evaluated and meets professionally accepted standards of supplement quality.

Another consideration in choosing a supplement is avoiding superfluous ingredients, such as para-aminobenzoic acid (PABA), hesperidin complex, inositol, bee pollen, and lecithins. These are not needed in our diets. They are especially common in expensive supplements sold in health-food stores and online. In addition, use of l-tryptophan and high doses of beta-carotene or fish oils is discouraged.

A few websites to help you evaluate ongoing claims and evaluate safety of supplements are:

http://acsh.org/
www.quackwatch.com
www.ncahf.org
http://ods.od.nih.gov/
www.eatright.org
www.usp.org/dietary-supplements/overview

These sites are maintained by groups or individuals committed to providing reasoned and authoritative nutrition and health advice to consumers.

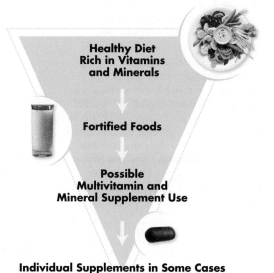

FIGURE 8-35 ▲ Supplement savvy—an approach to the use of nutrient supplements. Emphasizing a healthy diet rich in vitamins and minerals is always the first option.

▲ Look for the USP symbol on your vitamin or mineral supplement.

✔ CONCEPT CHECK 8.16

1. Name four types of ingredients that are classified as dietary supplements by the Dietary Supplement Health and Education Act.
2. Identify three potential risks from use of dietary supplements.
3. Describe three situations in which use of dietary supplements is necessary.

FIGURE 8-36 ▶ Nutrient supplements display a nutrition label different from that of foods. This Supplement Facts label must list the ingredient(s), amount(s) per serving, serving size, suggested use, and % Daily Value if one has been established. In addition, this label includes structure/function claims, which are not mandatory elements of the supplement label. When structure/function claims are made, however, the label also must include the FDA warning that these claims have not been evaluated by the agency.

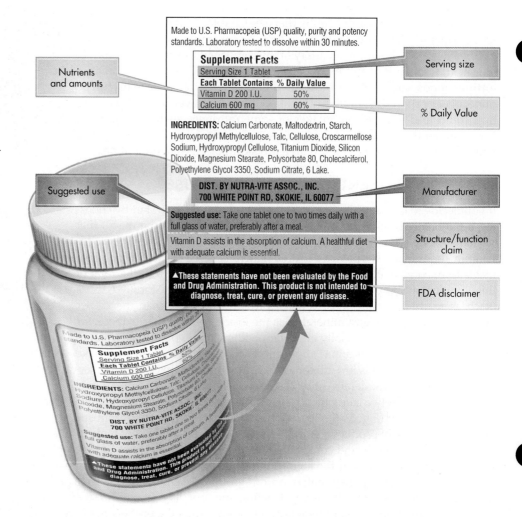

Made to U.S. Pharmacopeia (USP) quality, purity and potency standards. Laboratory tested to dissolve within 30 minutes.

Nutrients and amounts

Serving size

% Daily Value

Supplement Facts

Serving Size 1 Tablet

Each Tablet Contains	% Daily Value
Vitamin D 200 I.U.	50%
Calcium 600 mg	60%

INGREDIENTS: Calcium Carbonate, Maltodextrin, Starch, Hydroxypropyl Methylcellulose, Talc, Cellulose, Croscarmellose Sodium, Hydroxypropyl Cellulose, Titanium Dioxide, Silicon Dioxide, Magnesium Stearate, Polysorbate 80, Cholecalciferol, Polyethylene Glycol 3350, Sodium Citrate, 6 Lake.

DIST. BY NUTRA-VITE ASSOC., INC. 700 WHITE POINT RD, SKOKIE, IL 60077

Manufacturer

Suggested use: Take one tablet one to two times daily with a full glass of water, preferably after a meal.

Suggested use

Vitamin D assists in the absorption of calcium. A healthful diet with adequate calcium is essential.

Structure/function claim

▲These statements have not been evaluated by the Food and Drug Administration. This product is not intended to diagnose, treat, cure, or prevent any disease.

FDA disclaimer

CASE STUDY Getting the Most Nutrition from Your Food

In the dietary supplements aisle of the grocery store, the choices are endless—and expensive. Julie, a college sophomore, just read the Academy of Nutrition and Dietetics' position paper on nutrient supplementation for her class. She learned that dietary supplements, such as a balanced multivitamin and mineral supplement, can be a good back-up plan to ensure adequate nutrition, but the jury is still out when it comes to demonstrating a benefit of dietary supplements for long-term health. About one-third of Americans regularly take nutrient supplements, but it is usually the people who already consume a healthy diet who take them. Getting more than the recommended amount of a nutrient does not confer additional health benefits. In fact, too much of some vitamins and minerals can lead to toxicity.

Julie decides she would rather focus on getting her nutrients from foods. How can she get the most vitamins and minerals out of the foods she eats? Answer the following questions and check your responses at the end of the chapter.

1. What factors can damage or reduce vitamins in food?
2. To maximize vitamin content, what should Julie keep in mind as she selects fresh produce for purchase?
3. How does food processing affect vitamin and mineral content? Does it make a difference if Julie chooses products with whole grains or refined grains?
4. When storing fruits and vegetables in her apartment, what steps can Julie take to minimize nutrient losses?
5. Which cooking methods are best for preserving vitamin content?

Nutrition and Your Health

● Nutrition and Cancer

Cancer is the second leading cause of death for North American adults. It is estimated that more than 1600 people die each day of cancer in the United States. Cancer-related expenses exceed $200 billion each year. The top four cancers, causing more than 50% of cancer deaths, are lung, colorectal, breast, and prostate cancers (Fig. 8-37).

Cancer is many diseases; these differ in the types of cells affected and, in some cases, in the factors contributing to cancer development. For example, the factors leading to skin cancer differ from those leading to breast cancer. Similarly, the treatments for the different types of cancer often vary.

Cancer Terminology

Cancer essentially represents abnormal and uncontrollable division of cells that results from mutations in DNA. This initiates the cancer process. The cells then go through a variety of steps (called promotion and progression). A cancer cell is the result. Without effective treatment, cancer then typically leads to death. Most cancers take the form of tumors, although not all tumors are cancers. A **tumor** is spontaneous new tissue growth that serves no physiological purpose. It can be **benign,** like a wart, or **malignant,** like most lung cancers. The terms *malignant tumor* and *malignant neoplasm* are synonymous with cancer.

While benign tumors are dangerous only if their presence interferes with normal body functions, malignant (cancerous) tumors are capable of invading surrounding structures, including blood vessels, the lymph system, and nervous tissue. Cancer can also spread, or **metastasize,** to distant sites via the blood and lymphatic circulation, thereby producing invasive tumors in almost any part of the body. Metastasis then makes the cancer much more difficult to treat. That cancer can spread explains why early detection of cancer is so important. Cancers that can be diagnosed in the early stages are those in the colon, breast, and cervix.

Detecting Cancer

If cancer is left untreated, it can spread quickly throughout the body. When this happens, it is much more likely to lead to death. Thus, early detection is critical. The CAUTION acronym (see below) is a useful aid for remembering early warning signs for cancer. Unexplained weight loss can be an additional warning sign.

- **C**hange in bowel or bladder habits
- **A** sore that does not heal
- **U**nusual bleeding or discharge
- **T**hickening or lump in the breast or elsewhere
- **I**ndigestion or difficulty in swallowing
- **O**bvious change in a wart or mole
- **N**agging cough or hoarseness

Routine screenings are important for early detection of cancer. Colonoscopy examinations for middle-aged and older adults, prostate-specific antigen (PSA) tests for middle-aged and older men, and Papanicolaou tests (Pap smears) and regular breast examinations (and mammograms starting about age 40) for women are recommended by the American Cancer Society.

▲ Cruciferous vegetables such as cabbage and cauliflower are rich in cancer-preventing phytochemicals.

tumor Mass of cells; may be cancerous (malignant) or noncancerous (benign).

benign Noncancerous; tumors that do not spread.

malignant Malicious; in reference to a tumor, the property of spreading locally and to distant sites.

metastasize The spreading of disease from one part of the body to another, even to parts of the body that are remote from the site of the original tumor. Cancer cells can spread via blood vessels, the lymphatic system, or direct growth of the tumor.

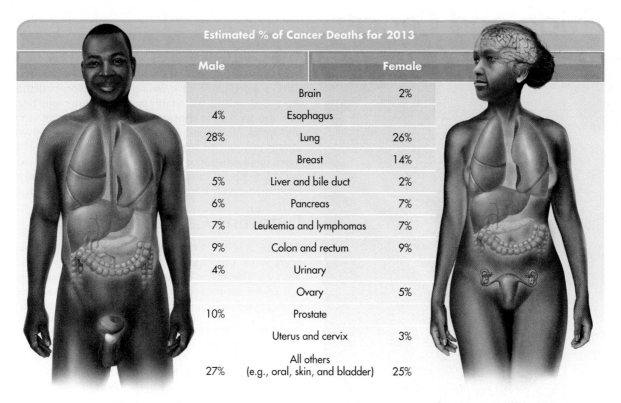

Estimated % of Cancer Deaths for 2013		
Male		**Female**
	Brain	2%
4%	Esophagus	
28%	Lung	26%
	Breast	14%
5%	Liver and bile duct	2%
6%	Pancreas	7%
7%	Leukemia and lymphomas	7%
9%	Colon and rectum	9%
4%	Urinary	
	Ovary	5%
10%	Prostate	
	Uterus and cervix	3%
27%	All others (e.g., oral, skin, and bladder)	25%

FIGURE 8-37 ▲ Cancer is many diseases. Numerous types of cells and organs are its target. About one-third of all cancers arise from smoking (primarily lung cancer).

Factors That Influence Development of Cancer

Genetics, environment, and lifestyle are potent forces that influence the risk for developing cancer. A genetic predisposition is especially important in development of colon cancer, some types of breast cancer (e.g., mutated BRCA1 or BRCA2 gene), and prostate cancer (35%, 27%, and 42% of all cases, respectively). About 30 cancer-susceptibility genes have been identified. However, experts estimate that only 5% of most cancers can be explained by the inheritance of a cancer gene. Overall, lifestyle and environmental exposures are also critical factors in most forms of cancer, as evidenced by the variation in cancer rates from country to country. In fact, diet likely accounts for 30% to 40% or more of all cancers.

Although we have little control over our genetic risks for cancer, we have tremendous influence when it comes to making decisions about lifestyle risks, especially with regard to smoking, alcohol abuse, physical activity, and nutrient intake (food choice). It is well established that one-third of all cancers in North America are due directly to tobacco use. About half of the cancers of the mouth, pharynx, and larynx are associated with heavy use of alcohol. A combination of alcohol use and smoking increases cancer risks even more.

A Closer Look at the Influence of Diet on Cancer

Some food constituents may contribute to cancer development, whereas others have a protective effect (Table 8-7). First, the association between fat/calorie intake and cancer risk is discussed, and then some of the food constituents that may reduce the risk for cancer are presented (see Further Reading 9).

CONTRIBUTION OF CALORIE AND FAT INTAKES TO CANCER RISK

An excess calorie intake, leading eventually to obesity, is responsible for an estimated 14% of cancer deaths among men and as many as 20% among women. This is the main diet-cancer risk factor. This includes cancer of the breast (especially in postmenopausal women), pancreas, kidney, gallbladder, colon, **endometrium,** and prostate gland. There are several ways excess adipose tissue can influence cancer risk. First, adipose tissue secretes estrogens and several other proteins that tend to promote the development of cancer. Excess insulin output resulting from an obese, insulin-resistant state is also implicated. In addition, higher levels of inflammation and oxidative stress typical among overweight and obese individuals contribute to cancer risk.

The National Cancer Institute (NCI) believes there is sufficient evidence for a link between dietary fat and cancer to encourage North Americans to reduce fat intake. It recommends initially decreasing dietary fat to about 30% of total calorie intake and eventually to 20% or less of total calories if the person is at high risk and can follow such a dietary pattern.

Some scientists, however, believe that the NCI has overreacted to the fat and cancer issue. Although epidemiological

endometrium The membrane that lines the inside of the uterus. It increases in thickness during the menstrual cycle until ovulation occurs. The surface layers are shed during menstruation if conception does not take place.

evidence does link fat and certain forms of cancer, the evidence is not strong. A stronger link exists between cancer and excess calories in the diet. In animal experiments, restricting total calorie intake to about 70% of usual intake results in about a 40% reduction in tumor development, regardless of the amount of fat in the diet. Calorie restriction is the most effective technique for preventing cancer in laboratory animals.

Unfortunately, it is difficult for humans to reduce dietary calories to 70% of usual intake. Therefore, while the data obtained from laboratory animal studies are interesting, nutritionists do not see any practical way to make recommendations on the basis of these studies. In addition, once cancer is present, calorie restriction is no longer helpful.

CANCER-INHIBITING FOOD CONSTITUENTS

Many single nutrients may have cancer-inhibiting properties. These anticarcinogens include antioxidants and certain phytochemicals (review Table 8-7).

The antioxidant activity of vitamin C and vitamin E helps to prevent formation of **nitrosamines** in the GI tract, thus preventing formation of a potent carcinogen. Vitamin E also helps protect unsaturated fatty acids from damage by free radicals. Overall, carotenoids, vitamin E, vitamin C, and selenium contribute to antioxidant protection for the body. Some of these antioxidant systems help prevent DNA mutations by electron-seeking compounds, the main way that cancer develops in the first place.

In addition, phytochemicals from fruits and vegetables, and even tea, block cancer development in some types of cells. Numerous studies suggest that fruit and vegetable intake reduces the risk of nearly all types of cancer. These foods are normally rich in carotenoids, vitamin C, and vitamin E. Adequate vitamin D intake is suspected of reducing breast, colon, prostate, and other forms of cancer. Calcium is also linked to a decreased risk for developing colon cancer. In sum, a diet that conforms to MyPlate, so that fruits, vegetables, whole grains, low-fat and fat-free dairy products, and some plant oils are eaten daily, is aimed at cancer prevention. It is likely that all of these foods have a "cocktail" effect and as strong a claim can be made for no one particular food (see Further Reading 9).

Nutrition Concerns During Cancer Treatment

Nutrition concerns during cancer treatment may vary depending on the site of the cancer, but the overall goals of medical nutrition therapy are to minimize weight loss and prevent nutrient deficiencies. Weight loss, particularly loss of muscle mass, is a major concern during cancer treatment because poor nutrition status can limit recovery from illness. Common effects of cancer and/or cancer treatments include fatigue, mouth sores, dry mouth, taste abnormalities, nausea, and diarrhea—all of which can lead to poor food intake.

During cancer treatment, the best food to eat is any food the patient can tolerate. Food choices vary based on each patient's individual symptoms, but cool, non-acidic liquids and soft, mildly flavored foods are generally well–accepted. Small, frequent meals and foods with high nutrient density

and energy density should be emphasized to meet calorie and protein needs. Often, liquid nutritional supplements are needed. Because immunity may be suppressed during cancer treatment, safe food handling practices also are extremely important (see Chapter 13).

Guidance for Cancer Prevention

Given the devastating toll of cancer treatment and lack of a definitive cure, efforts at prevention are of prime importance. Several health organizations have issued their own sets of diet and lifestyle guidelines for cancer prevention. Here, we present the recommendations of the American Institute for Cancer Research, which are representative of the advice of other organizations.

AMERICAN INSTITUTE FOR CANCER RESEARCH RECOMMENDATIONS FOR CANCER PREVENTION

1. *Be as lean as possible without becoming underweight.* Excess body fat is linked to increased risk of many chronic diseases, including cancer. As already described, hormones and other proteins secreted by adipose tissue combined with high levels of insulin tend to promote development of cancer.
2. *Be physically active for at least 30 minutes every day.* Physical activity assists with weight management, improves insulin sensitivity, and strengthens the immune system.
3. *Avoid sugary drinks. Limit consumption of energy-dense foods.* These dietary strategies assist efforts at weight management.
4. *Eat more of a variety of vegetables, fruits, whole grains, and legumes.* Plant-based diets provide plenty of fiber and antioxidant nutrients, and can promote weight management.
5. *Limit consumption of red meats (e.g., beef, pork, and lamb) and avoid processed meats.* Specifically, limit red meat consumption to 18 ounces (cooked) per week. High consumption of red and processed meats has been linked to increased risk for colon cancer. This may be due to fat content or the formation of carcinogens (e.g., nitrosamines or heterocyclic amines).
6. *If consumed at all, limit alcoholic drinks to two per day (men) or one per day (women).* Alcoholic drinks especially increase the risk of cancer of the organs of the digestive system. This may be due to damaging effects of alcohol on DNA. For cancer prevention, it is best to avoid alcohol completely, but alcohol does afford some protection against cardiovascular disease.
7. *Limit consumption of salty foods and foods processed with salt.* High salt intake is especially linked to increased risk for stomach cancer.
8. *Do not use supplements to protect against cancer.* The phytochemicals present in whole foods have cancer-protective effects beyond those of isolated nutrients. In

---

nitrosamine A carcinogen formed from nitrates and breakdown products of amino acids; can lead to stomach cancer.

TABLE 8-7 ▶ Some Food Constituents Suspected of Having a Role in Cancer

Constituent	Dietary Sources	Action
Possibly Protective*		
Vitamin A	Liver, fortified milk, fruits, vegetables	Encourages normal cell development
Vitamin D	Fortified milk, fatty fish	Increases production of a protein that suppresses cell growth, such as in the colon
Vitamin E	Whole grains, vegetable oils, green, leafy vegetables	Prevents formation of nitrosamines; general antioxidant properties
Vitamin C	Fruits, vegetables	Can block conversion of nitrites and nitrates to potent carcinogens; likely has general antioxidant properties
Folate	Fruits, vegetables, whole grains	Encourages normal cell development; especially reduces the risk of colon cancer
Selenium	Meats, whole grains	Part of antioxidant system that inhibits tumor growth and kills developing cancer cells
Carotenoids, such as lycopene	Fruits, vegetables	Likely act as antioxidants; some of these possibly influence cell metabolism. Lycopene in particular may reduce the risk of prostate cancer.
Flavonoids, indoles, phenols, and other phytochemicals	Vegetables, especially cabbage, cauliflower, broccoli, Brussels sprouts, garlic, onions, tea	May reduce cancer in the stomach and other organs
Calcium	Milk products, green vegetables	Slows cell division in the colon and binds bile acids and free fatty acids, thus reducing colon cancer risk
Omega-3 fatty acids	Cold-water fish, such as salmon and tuna	May inhibit tumor growth
Soy products	Tofu, soy milk, tempeh, soy nuts	Phytic acid present possibly binds carcinogens in the intestinal tract; the genistein component possibly reduces growth and metastasis of malignant cells.
Conjugated linoleic acid	Milk products, meats	May inhibit tumor development and act as an antioxidant
Fiber-rich foods	Fruits, vegetables, whole-grain breads and cereals, beans, nuts	Colon and rectal cancer risk may be decreased by accelerating intestinal transit or binding carcinogens such that they are excreted.
Possibly Carcinogenic		
Excessive calorie intake	All macronutrients can contribute.	Excess fat mass leading to obesity; linked to increased synthesis of estrogen and other sex hormones, which in excess may themselves increase the risk for cancer. Resulting excess insulin output from creation of an insulin-resistant state is also implicated.
Total fat	Meats, high-fat milk and milk products, animal fats and vegetable oils	The strongest evidence is for excessive saturated and polyunsaturated fat intake. Saturated fat is linked to an increased risk of prostate cancer.
High glycemic index carbohydrates	Cookies, cakes, sugared soft drinks, candy	Insulin surges associated with these foods may increase tumor growth, such as in the colon.
Alcohol	Beer, wine, liquor	Contributes to cancers of the throat, liver, bladder, breast, and colon (especially if the person does not consume enough folate)
Nitrites, nitrates	Cured meats, especially ham, bacon, and sausages	Under very high temperatures, they will bind to amino acid derivatives to form nitrosamines, potent carcinogens
Multi-ring compounds: aflatoxin	Formed when mold is present on peanuts or grains	May alter DNA structure and inhibit its ability to properly respond to physiologic controls; aflatoxin in particular is linked to liver cancer.
Benzo(a)pyrene and other heterocyclic amines	Charcoal-broiled foods, especially meats	Linked to stomach and colon cancer. To limit this risk, trim fat from meat before cooking, cut barbecuing time by partially cooking meat (such as in a microwave oven) prior to grilling, and do not consume blackened parts.

*Many of the actions listed for these possibly protective agents are speculative and have been verified only by experimental animal studies. The best evidence supports obtaining these nutrients and other food constituents from foods. The U.S. Preventive Services Task Force supports this statement, noting there is no clear evidence that nutrient supplements provide the same benefits.

addition, high intakes of certain food components, such as beta-carotene, have been shown in large clinical trials to actually increase rates of cancer.

9. *It is best for mothers to breastfeed exclusively for up to 6 months and then add other liquids and foods.* Breast-feeding reduces the risk of breast and other reproductive cancers for the mother and lowers the risk of future obesity in the infant.

10. *After treatment, cancer survivors should follow the recommendations for cancer prevention.* During cancer treatment, follow the advice of your health care specialist. After cancer treatment, following these guidelines may help to prevent recurrence of cancer.

And always remember . . . Do not smoke or chew tobacco.

Diet and lifestyle changes can exert a powerful influence on the risk for developing cancer, but it is important to note that they are no substitute for preventive screening and appropriate medical care. Once cancer has developed, diet and lifestyle changes will not be adequate to prevent growth or metastasis.

To learn more about cancer, review these sources of credible cancer information on the Internet:

American Cancer Society
www.cancer.org
National Cancer Institute
www.cancer.gov
American Institute for Cancer Research
www.aicr.org
Abramson Cancer Center's Oncolink
www.oncolink.org
Harvard School of Public Health's Disease Risk Index
www.diseaseriskindex.harvard.edu

Summary (Numbers refer to numbered sections in the chapter.)

8.1 Vitamins are organic substances required in small amounts in the diet for growth, function, and body maintenance. These can be categorized as fat soluble (vitamins A, D, E, and K) or water soluble (B vitamins and vitamin C). Vitamins cannot be synthesized by the body in adequate amounts to support health, and absence of a vitamin from the diet leads to the development of a deficiency disease. Fat-soluble vitamins require dietary fat for absorption and are carried by lipoproteins in the blood. Vitamin toxicity is most likely to occur from megadoses of fat-soluble vitamins because they are readily stored in the body. Intakes of water-soluble vitamins that exceed the storage ability of tissues are typically excreted in urine. Some vitamins are susceptible to destruction by light, heat, air, or alkalinity, or may be lost from foods in cooking water or fats.

8.2 Vitamin A, sometimes called the "anti-infection vitamin," maintains the health of epithelial tissues and is responsible for the function of mucus-secreting cells. Vitamin A is found in meats, fortified dairy products, fish, eggs, and a variety of red, orange, or leafy green vegetables. Carotenoids are phytochemicals that can be converted into vitamin A in the body. Three forms of carotenoids can yield vitamin A in humans: beta-carotene, alpha-carotene, and beta-cryptoxanthin. Although carotenoids are not essential nutrients, some have health-promoting qualities for humans. In addition to their contribution to vitamin A intake, carotenoids are powerful antioxidants. The antioxidant abilities of several carotenoids are linked to prevention of macular degeneration, cataracts,

cardiovascular disease, and cancer. Carotenoids are plentiful in dark-green and orange vegetables.

8.3 Vitamin D is both a hormone and a vitamin. Human skin synthesizes it using sunshine and a cholesterol-like substance. If we do not spend enough time in the sun, foods such as fish and fortified milk can supply the vitamin. The active hormone form of vitamin D helps regulate blood calcium in part by increasing calcium absorption from the intestine. Infants and children who do not get enough vitamin D may develop rickets, and adults with inadequate amounts in the body develop osteomalacia. Older people and infants often need a supplemental source. Toxicity may lead to calcification of soft tissue, weakness, and gastrointestinal disturbances.

8.4 Vitamin E functions primarily as an antioxidant and is found in plant oils. By donating electrons to electron-seeking free-radical (oxidizing) compounds, it neutralizes them. This effect shields cell membranes and red blood cells from breakdown. Claims are made about the curative powers of vitamin E, but more information is needed before megadose vitamin E recommendations for healthy adults can be made with certainty. The Upper Level is set at about 50 times adult needs.

8.5 Vitamin K is essential for blood clotting and imparts calcium-binding ability to various proteins, including those in bone. Some vitamin K absorbed each day comes from bacterial synthesis in the intestine, but most comes from foods, primarily green, leafy vegetables.

8.6 The B vitamins yield no energy directly, but they contribute to energy-yielding chemical reactions in the body by virtue of their coenzyme functions. B vitamins are highly bioavailable. North American diets are typically adequate in B vitamins except in cases of poverty, metabolic disorders, or alcoholism. Whole grains are more nutrient-dense sources of B vitamins (as well as other nutrients) than refined grains. Several B vitamins function as coenzymes in energy metabolism.

8.7 Thiamin's coenzyme form is involved in the metabolism of carbohydrates and proteins as well as the synthesis of RNA, DNA, and neurotransmitters. Rich food sources of thiamin include pork, enriched or fortified grain products, and milk. Beri-beri, the thiamin-deficiency disease, leads to muscle weakness and nerve damage. Thiamin toxicity is unknown, and no UL has been set.

8.8 The coenzymes of riboflavin participate in the catabolism of fatty acids, metabolism of other vitamins and minerals, and antioxidant activity of glutathione peroxidase. Dairy products, enriched and fortified grain products, meat, and eggs are rich food sources of riboflavin. Symptoms of ariboflavinosis include glossitis and angular cheilitis. There is no evidence of toxicity with high doses of riboflavin; no UL has been set.

8.9 Niacin's coenzymes function in many synthetic reactions, especially fatty-acid synthesis. Rich food sources include seafood, poultry, meats, peanuts, and enriched or fortified grains. Pellagra, the disease of niacin deficiency, results in dermatitis, diarrhea, dementia, and, eventually, death. Megadoses of niacin have been used to lower blood lipids, but they cause side effects, such as flushing of the skin.

8.10 Vitamin B-6 coenzymes activate many enzymes of carbohydrate, lipid, and, especially, protein metabolism. They also help synthesize neurotransmitters and participate in homocysteine metabolism. Rich food sources include animal products and enriched or fortified grain products, as well as some fruits and vegetables. A deficiency of vitamin B-6 leads to headaches, depression, gastrointestinal symptoms, skin disorders, nerve problems, anemia, and impaired immunity. Vitamin B-6 toxicity can result in nerve damage.

8.11 Pantothenic acid functions as a coenzyme in reactions that yield energy from carbohydrates, lipids, and protein, as well as fatty–acid synthesis. It is widely distributed among foods, with sunflower seeds, mushrooms, peanuts, and eggs among the richest sources. A deficiency of pantothenic acid is unlikely, but symptoms would be similar to those seen with deficiencies of other B vitamins. There is no known toxicity and no UL for pantothenic acid. Biotin's coenzyme form aids in reactions that synthesize glucose and fatty acids and in the metabolism of amino acids. Egg yolks, peanuts, and cheese provide dietary biotin, but this vitamin is also synthesized by bacteria in the intestines. Consuming raw egg whites may lead to a biotin deficiency because avidin in egg whites binds biotin and reduces its bioavailability. Biotin deficiency can lead to inflammation of the skin and mouth, gastrointestinal symptoms, muscle pain and weakness, poor growth, and anemia. No UL has been set for biotin as no toxicity has ever been observed.

8.12 Folate plays an important role in DNA synthesis and homocysteine metabolism. Symptoms of a deficiency include generally poor cell division in various areas of the body, megaloblastic anemia, tongue inflammation, diarrhea, and poor growth. Pregnancy puts high demands for folate on the body; deficiency during the first month of pregnancy can result in neural tube defects in offspring. A deficiency can also occur in people with alcoholism. Food sources are leafy vegetables, organ meats, and orange juice.

8.13 Vitamin B-12 is needed to metabolize folate and homocysteine, and to maintain the insulation surrounding nerves. Absorption of vitamin B-12 is a complex process that requires a salivary protein, adequate stomach acid production, and an intrinsic factor produced by the stomach. A deficiency, which results in anemia and nerve degeneration, most likely results from poor absorption of vitamin B-12 rather than poor dietary intake. Pernicious anemia is one condition that can impair vitamin B-12 absorption. Vitamin B-12 is found in foods of animal origin, fortified foods, and supplements.

8.14 Vitamin C is a potent antioxidant and functions in the synthesis of collagen, carnitine, and neurotransmitters. A vitamin C deficiency results in scurvy, evidenced by pinpoint hemorrhages in the skin, bleeding gums, and joint pain. Vitamin C also modestly enhances iron absorption. Fresh fruits and vegetables, especially citrus fruits, are good sources. A great amount of vitamin C is lost in storage and cooking; therefore, the diet should emphasize fresh or minimally cooked fruits and vegetables. Deficiencies can occur in people with alcoholism and those whose diets lack sufficient fruits and vegetables. Smoking increases the possibility of vitamin C deficiency. The Upper Level is set at about 20 times adult needs.

8.15 Choline is an essential nutrient but has not yet been classified as a vitamin. As a component of phospholipids, it is important for cell membrane structure, myelination of nerves, and lipid transport. Like folate, choline plays a role in single-carbon metabolism, which has implications for prevention of birth defects, cancer, and heart disease. Egg yolks, meats, dairy products, soybeans, and nuts are good food sources of choline.

8.16 To meet nutrient needs and prevent chronic disease, foods should be emphasized, but dietary supplements can be useful in some circumstances. For example, women of childbearing age, older adults, vegans, and people with malabsorptive diseases are most likely to benefit from dietary supplements. Consumers should educate themselves about possible benefits and risks.

NAYH Given the toll of cancer treatment and lack of a definitive cure, efforts at prevention are key. A variety of dietary changes will reduce your risk for cancer. Start by making sure that your diet is moderate in calorie and fat content and that you consume many fruits and vegetables, whole-grain breads and cereals, beans, some fish, and low-fat or fat-free milk products. In addition, remain physically active; avoid obesity; consume alcohol in moderation (if at all); and limit intake of animal fat and salt-cured, smoked, and nitrate-cured foods.

Check Your Knowledge (Answers to the following questions are below.)

1. Vitamins are classified as
   a. organic and inorganic.
   b. fat soluble and water soluble.
   c. essential and nonessential.
   d. elements and compounds.

2. A vitamin synthesized by bacteria in the intestine is
   a. A.
   b. D.
   c. E.
   d. K.

3. A deficiency of vitamin A can lead to the disease called
   a. xerophthalmia.
   b. osteomalacia.
   c. scurvy.
   d. pellagra.

4. Vitamin D is called the sunshine vitamin because
   a. it is available in orange juice.
   b. exposure to sunlight converts a precursor into vitamin D.
   c. it can be destroyed by exposure to sunlight.
   d. it is an ingredient in sunscreen.

5. Vitamin E functions as
   a. a coenzyme.
   b. a hormone.
   c. an antioxidant.
   d. a peroxide.

6. Bowed legs, an enlarged and misshapen head, and enlarged knee joints in children are all symptoms of
   a. rickets.
   b. xerophthalmia.
   c. osteoporosis.
   d. vitamin D toxicity.

7. A deficient intake of _____ has been shown to increase the risk of having a baby with a neural tube defect such as spina bifida.
   a. vitamin A
   b. vitamin C
   c. vitamin E
   d. folate

8. Vitamin C is necessary for the production of
   a. stomach acid.
   b. collagen.
   c. insulin.
   d. clotting factors.

9. B vitamins, including thiamin, riboflavin, and niacin, are called the "energy" vitamins because they
   a. can be broken down to provide energy.
   b. are ingredients in energy drinks such as Powerade.
   c. are part of coenzymes needed for release of energy from carbohydrates, fats, and proteins.
   d. are needed in large amounts by competitive athletes.

10. Noodles, spaghetti, and bread are made from wheat flour that is enriched with all of the following nutrients except
   a. vitamin B-6.
   b. thiamin.
   c. niacin.
   e. riboflavin.

11. Which of the B vitamins is sensitive to and can be degraded by light?
   a. riboflavin
   b. niacin
   c. thiamin
   d. pantothenic acid

12. Niacin can be synthesized in the body from the amino acid
   a. tyrosine.
   b. tryptophan.
   c. phenylalanine.
   d. glutamine.

13. Avidin, a component of raw egg whites, may decrease the absorption of
   a. biotin.
   b. thiamin.
   c. iron.
   d. riboflavin.

14. Choline is an important component of
   a. cholesterol.
   b. an antioxidant.
   c. a phospholipid.
   d. proteins.

15. Which of the following meals is most compatible with American Institute for Cancer Research guidelines for cancer prevention?
   a. Flame-broiled chicken breast, baked potato, and mixed vegetables
   b. Poached salmon, steamed broccoli, and corn on the cob
   c. Baked ham, sweet potato casserole, and spinach salad
   d. Cheese pizza and bread sticks with marinara sauce

Answer Key: 1. b (LO 8.1), 2. d (LO 8.5), 3. a (LO 8.2), 4. b (LO 8.3), 5. c (LO 8.4), 6. a (LO 8.3), 7. d (LO 8.12), 8. b (LO 8.14), 9. c (LO 8.6), 10. a (LO 8.6), 11. a (LO 8.8), 12. b (LO 8.6), 13. a (LO 8.11), 14. c (LO 8.15), 15. b (LO 8.17)

Study Questions (Numbers refer to Learning Outcomes)

1. Why is the risk of toxicity greater with the fat-soluble vitamins A and D than with water-soluble vitamins in general? **(LO 8.1)**

2. How would you determine which fruits and vegetables displayed in the produce section of your supermarket are likely to provide plenty of carotenoids? **(LO 8.2)**

3. What is the primary function of the vitamin D hormone? Which groups of people likely need to supplement their diets with vitamin D, and on what do you base your answer? **(LO 8.3)**

4. Describe how vitamin E functions as an antioxidant. **(LO 8.4)**

5. Milling (refining) grains removes which vitamins and minerals? Which of these are replaced during processing? **(LO 8.6)**

6. What are the best food sources for thiamin? **(LO 8.7)**

7. What are the signs of a riboflavin deficiency? **(LO 8.8)**

8. Describe the three signs of the niacin deficiency pellagra. **(LO 8.9)**

9. Describe how the RDA, DV, and UL for vitamin B-6 should be used in everyday life. **(LO 8.10)**

10. How is pantothenic acid involved in energy metabolism? **(LO 8.11)**

11. Why does the consumption of raw eggs lead to a biotin deficiency? **(LO 8.11)**

12. Why does FDA limit the amount of folate that may be included in supplements and fortified foods? **(LO 8.12)**

13. Is it necessary for North Americans to consume a great excess of vitamin C to avoid the possibility of a deficiency? Do vitamin C intakes well above the RDA have any negative consequences? **(LO 8.14)**

14. Why is choline not considered a vitamin? **(LO 8.15)**

CASE STUDY SOLUTION — Getting the Most Nutrition from Your Food

1. Exposure to light, heat, oxygen, or alkalinity leads to destruction of some vitamins, particularly the water-soluble vitamins. In addition, some vitamins can be leached into cooking water or fat during food preparation.

2. Because light, heat, and oxygen can destroy some vitamins, prolonged storage of produce can reduce vitamin content. Imported (i.e., not in season) produce in the grocery store may have been harvested several days or even weeks before it reaches the shopping cart. If Julie focuses on buying fruits and vegetables that are in season for her area and have been grown locally, the total time from harvest to purchase is likely to be less. Finding fresh, local produce at a farmers' market also decreases the number of steps from the field to Julie's plate, so vitamin content will be higher.

3. Food processing dramatically decreases the nutrient content of grains. Discarding the bran and germ parts of the grain reduces vitamin E, the B vitamins, and minerals such as magnesium, iron, and zinc. The enrichment of grains adds back some B-vitamins and iron, but refined grains are still lower than whole grains in many micronutrients. Choosing to make half your grains whole will help in meeting micronutrient needs without dietary supplements.

4. Decreasing exposure to light, heat, and air will limit destruction of vitamins. Most produce should be stored in closed containers in the refrigerator. Julie should not peel or chop her fruits and vegetables until just before consuming them to limit exposure to air. Foods that will not be eaten within a few days should be frozen to maintain nutrient content.

5. Steaming, microwaving, and stir-frying are the best cooking methods for preserving vitamin content of foods. During food preparation, some vitamins may be destroyed by heat or leached into cooking water or fat, thus reducing nutrient availability. Boiling vegetables, for example, results in losses of B vitamins, especially folate. Higher temperatures and longer cooking times increase vitamin losses. Using quick cooking methods, such as microwaving, and avoiding overcooking vegetables can help to retain nutrients. Using a lid on the cooking pot can reduce cooking time, helping to preserve vitamin content. When vegetables are steamed, fewer water-soluble vitamins are lost in cooking water. Fat-soluble vitamins, such as vitamin A and carotenoids, can be lost in fat used for cooking, so use minimal fat during cooking or add any fats for flavoring after food is cooked.

What the Dietitian Chose

Who does not occasionally yearn for more energy? In our fast-paced society, it is not surprising that energy supplements are the fastest-growing segment of the supplement industry. Products marketed as energy boosters typically contain at least 100% of the DV of the B vitamins that are involved in energy metabolism. As you learned in Chapter 8, the B vitamins function as coenzymes in chemical reactions that yield usable energy from carbohydrates, fats, and protein. However, taking B vitamins *alone* will not give you any energy. To obtain energy, you must also take in adequate calories. Furthermore, supplemental B vitamins will only enhance energy metabolism if they compensate for a pre-existing deficiency. If your current B vitamin status is adequate, you will not gain any extra energy. How is your current B vitamin status? How much, if any, should you take? Are there other ingredients that can help to boost your energy level? Let us take a closer look at our options.

The 5-Hour Energy shot catches your eye because it promises a quick, convenient boost of energy with fewer calories (it is artificially sweetened) than full-size energy drinks. This product provides 150% of the DV of niacin, 2000% of the DV for vitamin B-6, and 8333% of the DV for vitamin B-12! In addition, 5-Hour Energy contains a proprietary "energy blend," which includes several stimulant ingredients (e.g., caffeine and citicholine) and amino acids that are purported to enhance physical and mental performance.

Do you need that much niacin, vitamin B-6, and vitamin B-12? If your current B vitamin status is adequate, taking these B vitamins will offer no boost of energy. On the other hand, if your usual intake of B vitamins is poor, this product could optimize your B vitamin status and thereby enhance energy metabolism. Beyond your body's needs, however, excess water-soluble vitamins will be excreted in urine. In addition, the B vitamins are highly interdependent. Consuming unbalanced quantities of B vitamins could prevent them from working together properly in energy-yielding metabolic processes. Recognize that the high dose of niacin in this product has been known to cause uncomfortable symptoms of flushing in sensitive individuals.

Experts agree that caffeine provides most of the kick from an energy shot. Presently, energy shot makers are not required to disclose their products' caffeine content on the supplement facts label, but on their website, 5-Hour Energy

volunteers that each shot contains around 200 milligrams of caffeine. You should be cautious if using energy shots alongside coffee or other caffeinated drinks. Excess caffeine can cause nervousness, trouble sleeping, nausea, vomiting, rapid heartbeats, and higher blood pressure. The mix of other ingredients in energy shots, such as the taurine, tyrosine, and phenylalanine in 5-Hour Energy, are cause for concern; there is currently not enough research on how they react together.

Finally, note that 5-Hour Energy provides no energy! The metabolic reactions that involve the coenzyme form of these B vitamins require some carbohydrates, fats, or proteins to generate energy. Therefore, even though this small shot has as much caffeine as a 12-ounce cup of coffee and 20 times the Daily Value of vitamin B-6, it still does not have the necessary source of sustained energy: calories.

Leaving behind the proprietary energy blend of the energy shot, let us take a look at a B-complex. The product label for Nature Made® B-Complex with Vitamin C claims that it *"helps convert food into energy."* The supplement facts panel reveals that it contains thiamin, riboflavin, niacin, and vitamin B-6 in concentrations 2.5 to 10 times the Daily Value. Like the 5-Hour Energy Shot, the B vitamins in this supplement can assist the chemical reactions that generate energy, but only in the presence of adequate carbohydrates, fats, and/or proteins. The B-complex will not enhance energy levels in a person who already has adequate B vitamin status. The quantities of B vitamins in this product

Nature Made® B-Complex w/Vitamin C Dietary Supplement

Supplement Facts
Serving Size 1 Caplet

Amount Per Caplet	% Daily Value
Vitamin C 300 mg	500%
Thiamin 15 mg	1,000%
Riboflavin 10.2 mg	600%
Niacin 50 mg	250%
Vitamin B6 5 mg	250%
Pantothenic Acid 10 mg	100%

Supplement Facts
Serving Size 1.93 fl. oz.

Amount Per Serving	% Daily Value
Niacin (as Niacinamide) 30 mg	150%
Vitamin B6 (as Pyridoxine Hydrochloride) 40 mg	2,000%
Folic Acid 400 mcg	100%
Vitamin B12 (as Cyanocobalamin) 500 mcg	8333%
Sodium 18 mg	<1%
Energy Blend 1870 mg	‡
Taurine, Glucuronic acid (as or from glucuronolactone), Malic Acid, N-Acetyl L-Tyrosine, L-Phenylalanine, Caffeine, Citicoline	

‡ Daily value not established.

Supplement Facts
Serving Size 1 Tablets

Each Tablet Contains	% Daily Value
Vitamin A 3,500 IU (29% as Beta-Carotene)	70%
Vitamin C 60 mg	100%
Vitamin D 400 IU	100%
Vitamin E 30 IU	100%
Vitamin K 25 mcg	31%
Thiamin 1.5 mg	100%
Riboflavin 1.7 mg	100%
Niacin 20 mg	100%
Vitamin B6 2 mg	100%
Folic Acid 400 mcg	100%
Vitamin B12 6 mcg	100%
Biotin 30 mcg	10%
Pantothenic Acid 10 mg	100%
Calcium 200 mg	20%
Iron 18 mg	100%
Phosphorus 20 mg	2%
Iodine 150 mcg	100%
Magnesium 50 mg	13%
Zinc 11 mg	73%
Selenium 55 mcg	79%

exceed human needs; much of it will be excreted. Unless there is a specific medical need, it is not necessary to supplement with more than 100% of the DV for any nutrient. Bluntly stated, you will be flushing your hard-earned cash down the toilet!

These megadoses of B vitamins are not necessary, but what if you would like a simple backup plan for days when your dietary intake is less than adequate? In this case, a balanced multivitamin-mineral supplement providing no more than 100% of the DV for nutrients is a prudent idea. The Centrum® product shown here claims to support energy and metabolism by providing B vitamins and iron (see Chapter 9), plus it contains a full array of essential vitamins and minerals needed for other aspects of overall health. With this product, you can meet (but not exceed) your needs for nutrients involved in energy metabolism. Keep in mind, for the majority of the healthy population, studies show that there is no overall benefit of taking multivitamin-mineral supplements (see Further Reading 3). Such products are most valuable for people who do not meet their nutrient requirements from food sources.

Although a deficiency of B vitamins theoretically can affect your energy levels, typical American diets supply ample B vitamins. Good animal sources are fish, poultry, meat, eggs, and dairy products. Leafy green vegetables, lentils, beans, peas, and whole grains are rich plant sources of B vitamins. Recall that grain products are enriched with thiamin, niacin, riboflavin, and folate. Fortified breakfast cereals are especially rich sources of B vitamins and many other micronutrients. If you consume a diet consistent with the Dietary Guidelines, you will likely meet your daily needs for B vitamins.

Try to gain perspective on some other factors that may be draining you of energy. Lack of sleep is a likely culprit for busy college students who try to balance school, work, and a social life. Besides B vitamin status, other common dietary causes for fatigue include inadequate calories and poor iron status. High levels of stress and a sedentary lifestyle could also leave you feeling tired. To boost energy levels naturally:

- Eat several small meals throughout the day rather than three large ones.
- Avoid large amounts of sugar and fat.
- Do not skip meals, especially breakfast.
- Take a quick exercise break, such as a short walk.
- Exercise regularly.
- Reduce stress.
- Get 7 to 9 hours sleep each night.

Further Readings

1. ADA Reports: Position of the American Dietetic Association: Nutrient supplementation. *Journal of the American Dietetic Association* 109:2073, 2009.

2. Adams JS and Hewison M: Update in vitamin D. *Journal of Clinical Endocrinology and Metabolism* 95:471, 2010.

3. Casey CF and others: Vitamin D supplementation in infants, children, and adolescents. *American Family Physician* 81(6):745, 2010.

4. Cordero Z and others: Vitamin E and risk of cardiovascular diseases: A review of epidemiologic and clinical trial studies. *Critical Reviews in Food Science and Nutrition* 50:420, 2010.

5. Fulgoni VL 3rd and others: Foods, fortificants, and supplements: Where do Americans get their nutrients? *Journal of Nutrition* 141:1847, 2011.

6. Green R: Is it time for vitamin B-12 fortification? What are the questions? *American Journal of Clinical Nutrition* 89:712S, 2009.

7. Keller M: Vitamin D deficiency in children. *Today's Dietitian* 14 (12):26, 2012.

8. Krider KS, Bailey LB, and Berry RJ: Folic acid food fortification—its history, effect, concerns and future directions. *Nutrients* (3):370, 2011.

9. Kushi LH and others: American Cancer Society guidelines on nutrition and physical activity for cancer prevention: Reducing the risk of cancer with healthy food choices and physical activity. *CA: A Cancer Journal for Clinicians* 62:30, 2012.

10. Larsson SC and others: Vitamin B-6 and risk of colorectal cancer: A meta-analysis of prospective studies. *Journal of the American Medical Association* 303:1077, 2010.

11. Martini LA and others: Role of vitamins and minerals in prevention and management of type 2 diabetes mellitus. *Nutrition Reviews* 68:341, 2010.

12. National Institutes of Health Clinical Center's Drug-Nutrient Interaction Task Force: Important information to know when you are taking: Coumadin and vitamin K. Updated 9/5/2012. Available at: http://www.cc.nih.gov/ccc/patient_education/drug_nutrient/coumadin1.pdf.

13. National Institutes of Health State-of-the-Science Conference Statement on Multivitamin/Mineral Supplements and Chronic Disease Prevention. *NIH Consensus State-of-the-Science Statements* 23(2):1, 2006.

14. Office of Dietary Supplements, National Institutes of Health: Dietary supplement fact sheet: Vitamin D. Updated 6/24/11. Available at: http://ods.od.nih.gov/factsheets/vitamind.

15. Office of Dietary Supplements, National Institutes of Health: Dietary supplement fact sheet: vitamin B-6 fact sheet. Updated 9/15/2011. http://ods.od.nih.gov/factsheets/vitaminb6.

16. Office of Dietary Supplements, National Institutes of Health: Dietary supplement fact sheet: Vitamin B-12 fact sheet. Updated 6/24/2011. http://ods.od.nih.gov/factsheets/vitaminb12.

17. Palmer S: Sorting out the science on multivitamins and minerals. *Today's Dietitian* 11:38, 2009.

18. Sesso HD and others: Multivitamins in the prevention of cardiovascular disease in men: The Physician's Health Study II randomized controlled trial. *Journal of the American Medical Association* 308:1751, 2012.

19. Tang WH and others: Intestinal microbial metabolism of phosphatidylcholine and cardiovascular risk. *New England Journal of Medicine* 368(17):1575, 2013.

20. Webb D: Choline—rethinking the dietary requirement. *Today's Dietitian* 13:38, 2011.

21. Webb D: Vitamin D and cancer—evidence suggests this vital nutrient may cut risk. *Today's Dietitian* 14(10): 58, 2012.

22. Zeisel SH and da Costa K: Choline: An essential nutrient for public health. *Nutrition Reviews* 67:615, 2009.

23. Zhang SM and others: Effect of combined folic acid, vitamin B-6, and vitamin B-12 on cancer risk in women: A randomized trial. *Journal of the American Medical Association* 300:2012, 2008.

 **connect** NUTRITION

To get the most out of your study of nutrition, visit McGraw-Hill Connect at www.mcgrawhillconnect.com where you will find NutritionCalc Plus, LearnSmart, and many other dynamic tools.

Rate Your Plate

Throughout this book, you see breakfast cereals described as enriched, fortified, cold, or ready to eat. What do these terms mean?

The cereal you pour out of the box, add milk to, and enjoy is called ready-to-eat or cold cereal. Any addition of nutrients to a food product is called fortification. In the United States, FDA strictly defines enrichment, which is a special type of fortification of refined grain products. Per pound of flour, all refined grains must contain at least 2.9 milligrams of thiamin, 1.8 milligrams of riboflavin, 24 milligrams of niacin, 0.7 milligrams of folic acid, and 20 milligrams of iron. Enrichment replaces some (but not all) of the nutrients lost during the milling of grains.

You can tell if a product is fortified by reading the Nutrition Facts Panel. See for yourself in the following table. **Cereal A** has been enriched to levels mandated by FDA. **Cereal B,** the enriched or fortified choice, contains 100% of the DV for select vitamins and minerals, especially the B vitamins and iron. While all refined grain products are enriched with thiamin, riboflavin, niacin, folic acid, and iron, many food manufacturers choose to fortify their products with additional nutrients. In this book, when we refer to fortified breakfast cereals, we are referring to this latter group of food products.

Nutrient	Cereal A (1 ounce) % DV	Cereal B (1 ounce) % DV	Your Favorite Cereal % DV	Nutrient	Cereal A (1 ounce) % DV	Cereal B (1 ounce) % DV	Your Favorite Cereal % DV
Vitamin A	25	15		Niacin	25	100	
Vitamin C	25	100		Vitamin B-6	25	100	
Calcium	0	0		Folic acid	25	100	
Iron	50	100		Pantothenic acid	—*	100	
Vitamin D	10	10		Phosphorus	4	4	
Vitamin E	25	100		Magnesium	—*	4	
Thiamin	25	100		Zinc	—*	100	
Riboflavin	25	100					

*Information not provided on label.

Locate a nutrition facts panel for one of your favorite breakfast cereals. In the blank spaces in the table, record the % DV for each nutrient provided by one serving. If some data are missing, fill in the blanks with an asterisk. Based on your observations, answer the following questions.

1. Is your favorite cereal fortified? How can you tell?

_____

_____

2. Examine the calorie, fat, and sugar content listed on your Nutrition Facts panel. How does the cereal compare to the recommendations of the Dietary Guidelines for Americans and other health authorities to limit calories, fat, and added sugars?

_____

_____

3. Which special population groups can especially benefit from daily use of a fortified breakfast cereal?

_____

_____

4. What could you eat with your breakfast cereal to build a meal that resembles MyPlate?

_____

_____

Student Learning Outcomes

Chapter 9 is designed to allow you to:

9.1 Understand the functions of water in the body, the regulation of fluid balance, and the health consequences of fluid imbalance (i.e., dehydration); list recommended intakes and sources of water.

9.2 Describe the general characteristics of the major and trace minerals, the general process of mineral absorption and storage, the dangers of mineral toxicities, and ways to preserve minerals in foods.

9.3 Describe the roles of sodium in controlling fluid balance, acid-base balance, and nerve impulse transmission and list its dietary sources and requirements, as well as the dangers of exceeding sodium recommendations.

9.4 List the functions, dietary sources, and requirements of potassium, as well as the dangers of getting too much potassium.

9.5 List the functions of chloride, as well as its dietary sources and requirements.

9.6 List the dietary sources and requirements of calcium and describe its role in bone growth, maintenance and repair, as well as the process of osteoporosis development and prevention.

9.7 Describe the functions, dietary requirements, and sources of phosphorus, as well as the benefit of avoiding too much phosphorus.

Chapter 9
Water and Minerals

What Would You Choose?

Water is the best choice for everyday hydration: it quenches thirst without calories. There are currently many "water" choices available. Do you need water that contains vitamins and minerals? What's the difference between spring water and mineral water? Is bottled water safer to drink than tap water? As you peruse the convenience store shelves, which is the healthiest choice to take along to the student recreation center for your workout?

a Reusable bottle filled with tap water

b Aquafina® Pure Water

c San Pellegrino Mineral Water

d Vitaminwater®

connect NUTRITION **Think about your choice as you read Chapter 9, then see What the Dietitian Chose at the end of the chapter. To learn more about water choices, check out the Connect site: www.mcgrawhillconnect.com.**

Water (H_2O)—the most abundant molecule in the human body and the most versatile medium for a variety of chemical reactions—constitutes the major portion of the human body. Without water, biological processes necessary to life would cease in a matter of days. We must replenish water regularly because the body does not store it *per se*. Fluids must be consumed daily to replenish what is lost through respiration (lungs), perspiration (skin), and excretion (urine and feces). We recognize this constant demand for water as thirst, but the body has several intricate mechanisms to ensure fluid conservation. Maintenance of fluid balance relies on strict control of levels of dissolved minerals inside and outside the cells. These dissolved minerals—sodium, chloride, potassium, and phosphorus—are called electrolytes. Not only do they regulate the distribution of water throughout the body, they are also involved in maintenance of acid-base balance and in the conduction of nerve impulses. In this chapter, you will learn about the importance of water and how the electrolytes work together to regulate fluid balance, acid-base balance, and nerve function.

Like water, many minerals are vital to health. They are considered inorganic because they are typically not bonded to carbon atoms. In addition to water balance, minerals are key participants in body metabolism, muscle movement, body growth, and other wide-ranging processes (Fig. 9-1). We also know that some mineral deficiencies can cause severe health problems.

9.8 List magnesium functions and sources, as well as the dangers of magnesium toxicity from nonfood sources.

9.9 Describe the functions of iron in the maintenance of blood health and list its sources, dietary requirements, and deficiency and toxicity problems.

9.10 Describe the functions of zinc, including its role in immune function, as well as its food sources, dietary requirements, and signs and symptoms of deficiency and toxicity.

9.11 Summarize the functions of selenium and describe its dietary requirements, food sources, and symptoms of deficiency, as well as the dangers of toxicity from supplements.

9.12 Describe the functions of iodine in thyroid metabolism and list its sources, dietary requirements, and problems of deficiency and toxicity.

9.13 Describe the functions of copper and its potential for toxicity and list its sources, dietary requirements, and conditions that can lead to a deficiency.

9.14 Describe the role of fluoride in tooth development and bone health and its primary sources, as well as recommendations for use and problems of toxicity.

9.15 Describe the functions of chromium in glucose metabolism, and list its sources, dietary requirements, and deficiency symptoms.

9.16 Describe the functions of manganese and molybdenum, and understand how to get enough from dietary sources.

9.17 Describe factors that can contribute to the development of hypertension.

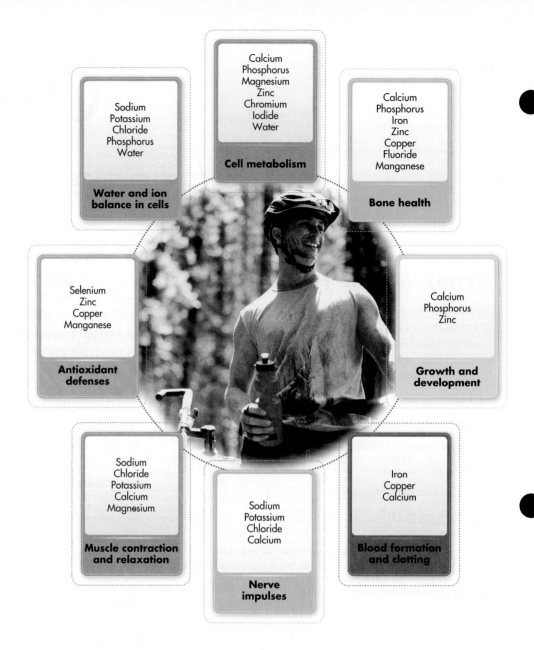

FIGURE 9-1 ▶ Water and minerals are involved in many processes in the body.

Calcium
Phosphorus
Magnesium
Zinc
Chromium
Iodide
Water

Cell metabolism

Sodium
Potassium
Chloride
Phosphorus
Water

Water and ion balance in cells

Calcium
Phosphorus
Iron
Zinc
Copper
Fluoride
Manganese

Bone health

Selenium
Zinc
Copper
Manganese

Antioxidant defenses

Calcium
Phosphorus
Zinc

Growth and development

Sodium
Chloride
Potassium
Calcium
Magnesium

Muscle contraction and relaxation

Sodium
Potassium
Chloride
Calcium

Nerve impulses

Iron
Copper
Calcium

Blood formation and clotting

9.1 Water

Life as we know it could not exist without water. Every cell, tissue, and organ contains some water. Overall, water comprises 50% to 70% of the human body (Fig. 9-2). Indeed, water is essential for life. While humans can live for several weeks without food, we cannot survive for more than a few days without water. The simple H_2O molecule has some amazing properties: it is a versatile solvent, a heat sink, and a lubricant. As detailed in the next few sections, these properties make water uniquely suited to carry out a number of essential roles in the human body.

WATER IN THE BODY—INTRACELLULAR AND EXTRACELLULAR FLUID

Water flows in and out of body cells through cell membranes. Water inside cells forms part of the **intracellular fluid.** When water is outside cells or in the bloodstream, it is part of the **extracellular fluid** (Fig. 9-3). Cell membranes are permeable to water, so water shifts freely in and out of cells. For example, if blood volume decreases, water can move into the bloodstream from the areas inside and around cells to increase blood volume. On the other hand, if blood volume increases, water

intracellular fluid Fluid contained within a cell; it represents about two-thirds of body fluid.

extracellular fluid Fluid present outside the cells; it represents about one-third of body fluid.

340

can shift out of the bloodstream into cells and the surrounding areas, leading to edema (review Fig. 6-13 in Chapter 6).

The body controls the amount of water in the intracellular and extracellular compartments mainly by controlling *ion* movement and concentrations. Ions are minerals that have electrical charges and so are called **electrolytes.** Water is attracted to ions, such as sodium, potassium, chloride, phosphate, magnesium, and calcium. The body maintains the appropriate amount of water in each compartment by controlling the movements of ions into and out of the cellular compartments in a process called **osmosis.** Overall, where ions go, water follows (Fig. 9-4).

Positive ions, such as sodium and potassium, end up pairing with negative ions, such as chloride and phosphate. Maintenance of intracellular water volume depends primarily on intracellular potassium and phosphate concentrations. Extracellular water volume depends primarily on the extracellular sodium and chloride concentrations.

Healthy Man (170 pounds)	Healthy Woman (130 pounds)
1 pound of glycogen (<1%)	1 pound of glycogen (<1%)
10 pounds of minerals (6%)	7 pounds of minerals (5%)
27 pounds of protein (16%)	17 pounds of protein (13%)
27 pounds of fat (16%)	32 pounds of fat (25%)
105 pounds of water (62%)	74 pounds of water (57%)

FIGURE 9-2 ▲ Although the percentages vary for men and women, the main constituent of the body is water.

electrolytes Substances that separate into ions in water and, in turn, are able to conduct an electrical current. These include sodium, chloride, and potassium.

osmosis The passage of water through a semipermeable membrane from a less concentrated compartment to a more concentrated compartment.

Roles of Water in the Body
• Distribution of nutrients
• Removal of waste products
• Temperature regulation
• Lubrication
• Cushioning
• Medium for and participant in chemical reactions
• Acid-base balance

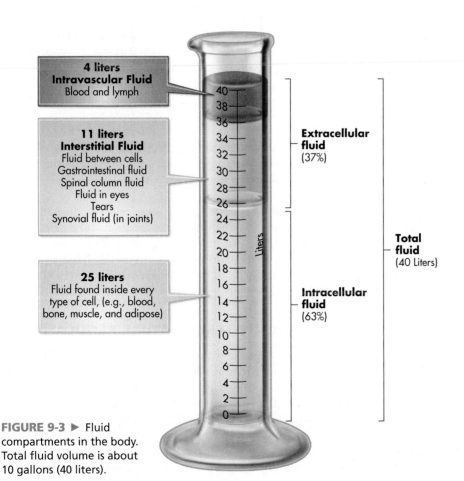

**4 liters
Intravascular Fluid**
Blood and lymph

**11 liters
Interstitial Fluid**
Fluid between cells
Gastrointestinal fluid
Spinal column fluid
Fluid in eyes
Tears
Synovial fluid (in joints)

25 liters
Fluid found inside every type of cell, (e.g., blood, bone, muscle, and adipose)

Extracellular fluid (37%)

Intracellular fluid (63%)

Total fluid (40 Liters)

FIGURE 9-3 ▶ Fluid compartments in the body. Total fluid volume is about 10 gallons (40 liters).

FIGURE 9-4 ▶ Effects of various ion concentrations in a fluid on red blood cells. This shows the process of osmosis. Fluid is shifting in and out of the red blood cells in response to changing ion concentrations in the flasks.

Red blood cell

H_2O

Dilute solution   Normal concentration   Concentrated solution

(a) A dilute solution with a low ion concentration results in swelling (*black arrows*) and subsequent rupture (*puff of red in the lower left part of the cell*) of a red blood cell placed into the solution.

(b) A normal concentration (a concentration of ions outside the cell equal to that inside the cell) results in a typically shaped red blood cell. Water moves into and out of the cell in equilibrium (*black arrows*), but there is no net water movement.

(c) A concentrated solution, with a high ion concentration, causes shrinkage of the red blood cell as water moves out of the cell and into the concentrated solution (*black arrows*).

The percentage of water varies tremendously between tissues. For example, muscle is 73% water, adipose tissue is 10% to 20% water, and bone contains approximately 20% water. As the fat content of the body increases, the percentage of lean decreases, and subsequently, the percentage of body water decreases. When body composition measurements are performed on extremely lean athletes, the percentage of body water will be around 70%.

solvent A liquid substance in which other substances dissolve.

metabolic water Water formed as a by-product of carbohydrate, lipid, and protein metabolism.

Should you feed a fever? Fever is an increase in body temperature caused by an immune response. When you have a fever, you do need more energy. For every increase of 1°F (0.5°C) in internal temperature above normal, your basal metabolic rate (BMR) increases by 7%. Although BMR is elevated for a person with a fever, physical activity will likely be lower than normal, which lowers overall energy needs. Thus, having a fever is not a reason to overeat.

WATER IS THE UNIVERSAL SOLVENT

Water is often called the "universal **solvent**" because so many different solutes can be dissolved in it. This property of water makes it (1) an ideal transport vehicle for nutrients and wastes and (2) a medium for many chemical reactions of human metabolism.

Water Transports Nutrients and Wastes. The majority of the nutrients we consume—carbohydrates, proteins, minerals, and many vitamins—are water soluble. Lipids, although they are not soluble in water, can be surrounded by a layer of water-soluble protein so that they can be dispersed throughout the water-based environment within and around cells and tissues (recall Chapter 5). As the primary component of blood and lymph, water acts to transport nutrients to all the cells of the body.

The metabolism of nutrients generates some waste products, most of which can dissolve in water and exit the body as part of urine. For example, when proteins are broken down as fuel, the nitrogen portion of amino acids cannot be used for energy production. Recall from Chapter 6 that the liver converts these nitrogenous waste products into urea. In addition, when we consume more than we need of some nutrients, such as sodium, the excess can be dissolved in water and excreted in urine. Typical urine output is about 1 liter per day, depending mostly on our intakes of fluid, protein, and sodium.

Water Is a Medium for Chemical Reactions. Because so many compounds dissolve in water, it provides a medium in which chemical reactions take place in the body. Furthermore, water itself (H_2O) is an important participant in many chemical reactions. When carbohydrates, lipids, and proteins are metabolized as sources of energy, water is one of the by-products. In fact, this **metabolic water** (1 cup or more per day) contributes to the maintenance of fluid balance in the body.

WATER CONTRIBUTES TO BODY TEMPERATURE REGULATION

Water temperature changes slowly because water has a great ability to hold heat. It takes much more energy to heat water than it does to heat air. Water molecules are polar (charged), so they are attracted to each other. This attractive force is strong, and energy is required to separate the water molecules. Because the human body is 50% to 70% water (see Fig. 9-2), it takes a lot of energy to change body temperature.

When overheated, the body secretes fluids in the form of perspiration, which evaporates through skin pores. As water evaporates from the skin, heat energy is released. So, as perspiration evaporates, heat energy is removed from the skin, cooling the body in the process (Fig. 9-5). In response to an increased body temperature, blood vessels in the skin become larger, allowing greater water loss through perspiration. Each quart (approximately 1 liter or 2 pounds) of perspiration that evaporates represents approximately 600 kilocalories of energy lost from the skin and surrounding tissues.

When carbohydrates, lipids, and proteins are used by cells in the body, energy is released in the form of heat. About 60% of the chemical energy in food is turned into body heat; the other 40% is converted into forms of energy that cells can use (principally adenosine triphosphate [ATP]). Almost all of that energy eventually leaves the body in the form of heat. If this heat could not be dissipated, the body temperature would rise enough to prevent enzyme systems from functioning efficiently, ultimately leading to death. Perspiration is the primary way to prevent this rise in body temperature.

WATER MOISTENS, LUBRICATES, AND CUSHIONS

The body secretes many fluids that are primarily water. Water-based secretions are produced by the digestive tract, respiratory tract, urogenital tract, eyes, and skin. Saliva acts as a lubricant, allowing food to pass through the esophagus to the stomach. Mucus provides a protective fluid coating throughout the digestive tract. The lungs are coated with a layer of mucus that provides an important immunologic function. Water helps form the lubricant found in knees and other joints of the body. The spinal cord and brain are cushioned by cerebrospinal fluid. Water is also the basis of amniotic fluid, which functions as a shock absorber surrounding the growing fetus in the mother's womb. Without adequate availability of water, the ability of the body to produce these critical secretions will be limited.

THE WATER BALANCING ACT

You have just learned about the many functions of water in the human body. Despite its critical importance for human survival, however, water is not stored in the body. It is continually lost through respiration (lungs), perspiration (skin), and excretion (urine and feces). Through mechanisms that monitor blood pressure and the concentration of solutes in body fluids, the nervous, endocrine, digestive, and urinary systems work elegantly together to maintain fluid balance and support life.

▲ Tears are one example of water's role as a lubricant. This fluid allows the eyeball to move smoothly in its socket and helps to flush away foreign particles. If an infant cries without producing tears, he or she may be dehydrated.

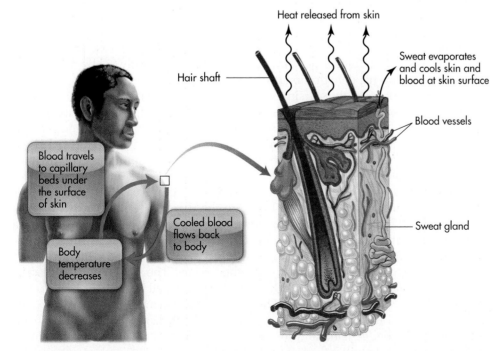

Heat released from skin

Hair shaft

Sweat evaporates and cools skin and blood at skin surface

Blood vessels

Sweat gland

Blood travels to capillary beds under the surface of skin

Cooled blood flows back to body

Body temperature decreases

FIGURE 9-5 ◄ Body temperature is reduced when heat is transported from the body through the bloodstream to the surface of the skin. As perspiration evaporates from the surface of the skin, heat is dissipated. This cools the blood, which circulates back to the body, reducing body temperature.

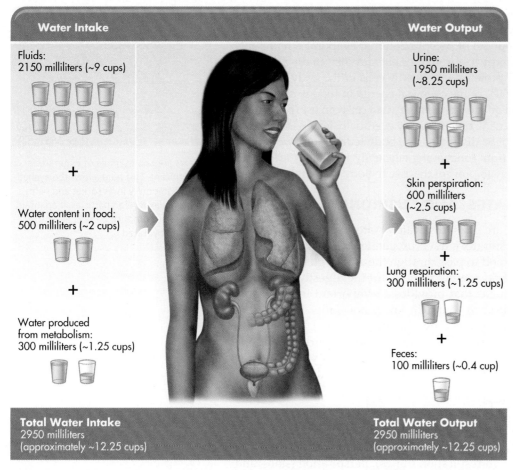

Water Intake

Fluids:
2150 milliliters (~9 cups)

+

Water content in food:
500 milliliters (~2 cups)

+

Water produced
from metabolism:
300 milliliters (~1.25 cups)

Water Output

Urine:
1950 milliliters
(~8.25 cups)

+

Skin perspiration:
600 milliliters
(~2.5 cups)

+

Lung respiration:
300 milliliters (~1.25 cups)

+

Feces:
100 milliliters (~0.4 cup)

Total Water Intake
2950 milliliters
(approximately ~12.25 cups)

Total Water Output
2950 milliliters
(approximately ~12.25 cups)

FIGURE 9-6 ▲ Estimate of water balance—intake versus output—in a woman. We primarily maintain body fluids at an optimum amount by adjusting water output to intake. As you can see with this woman, most water comes from the liquids we consume. Some comes from the moisture in foods, and the remainder is manufactured during metabolism. Water output includes that lost via urine, skin, lungs, and feces.

Water Intake. The Adequate Intake (AI) for *total* water is 2.7 liters (11 cups) for adult women and 3.7 liters (15 cups) for adult men. This amount is based primarily on average water intake from fluids and foods. For *fluid* alone, this corresponds to about 2.2 liters (9 cups) for women and about 3 liters (13 cups) for men.

Fluid intake—including water, fruit juice, coffee, tea, soft drinks, milk, and even alcoholic beverages—makes the biggest contribution to our total water needs. For the woman in Figure 9-6, fluid intake adds up to about 9 cups. In addition, nearly all foods contain water. Many fruits and vegetables are more than 80% water, and many meats contain at least 50% water (Fig. 9-7). In Figure 9-7, water from foods supplies another 2 cups. As mentioned previously, the body produces 250 to 350 milliliters (1 to 1½ cups) of water each day as a by-product of the chemical reactions used to metabolize energy. For the woman in Figure 9-6, metabolic water amounts to 1.25 cups. However, the amount of metabolic water produced can double in physically active people.

Water Output. Usually, urinary excretion of water accounts for the greatest source of output. The daily amount of urine produced may vary based on intake of fluids, protein, and sodium, but average urinary water loss per day is approximately 1650 milliliters (7½ cups). Removal of waste products requires at least 500 milliliters (2 cups) of urine production per day. Urine output consistently below this level is often a sign of chronic dehydration due to low fluid intake.

Water is lost through the skin in the form of perspiration. On days of low physical activity, these losses amount to about 1 liter. Under hot, humid conditions or with strenuous physical activity, losses can be much greater than 1 liter per day. Some water is also lost from the lungs in the form of water vapor in exhaled air. Together with perspiration, the fluid lost through lung respiration is sometimes called "insensible" water loss because it is difficult to measure.

A relatively small amount of water is lost daily in the feces. When we consider the large amount of water used to lubricate the digestive tract, the loss of only 100 milliliters (½ cup) of water each day through the feces is remarkable. Besides the variable amount of water ingested in the form of foods and fluids, about 8000 milliliters (34 cups) of water enters the digestive tract daily through secretions from the mouth, stomach, intestine, pancreas, and other organs. The small intestine absorbs most of this water, while the colon takes up a lesser but still important amount. The kidneys also greatly conserve water. They can reabsorb as much as 97% of the water filtered each day. The volumes of water intake and output shown in Figure 9-6 are estimates. Altitude, caffeine and alcohol intake, ambient temperature, humidity, and physical activity will influence water loss.

**MyPlate:
Sources of water**

FIGURE 9-7 ◄ Sources of water from MyPlate. The fill of the background color (none, 1/3, 2/3, or completely covered) within each group on the plate indicates the average nutrient density for water in that group. Overall, the vegetables, fruits, dairy, and protein groups contain many foods that are nutrient-dense sources of water. Although not depicted on MyPlate, all beverages are nearly 100% water. Fats and oils, on the other hand, have almost no water.

Source: United States Department of Agriculture, **ChooseMyPlate.Gov.**

Grains	Vegetables	Fruits	Dairy	Protein
• Bread • Pasta (cooked) • Rice (cooked)	• Potatoes • Lettuce • Tomatoes • Squash • Green beans • Vegetable juices	• Oranges • Apples • Bananas • Melons • Plums • Fruit juices	• Milk • Yogurt	• Beef • Poultry • Seafood • Beans

Fluid Conservation. The blood pressure and the concentration of solutes in the blood are closely monitored by receptors in the kidneys, blood vessels, and brain. Once the body registers a shortage of available water, it increases fluid conservation (Fig. 9-8). Hormones that participate in this process are **antidiuretic hormone (ADH),**

antidiuretic hormone A hormone secreted by the pituitary gland when blood concentration of solutes is high. It causes the kidneys to decrease water excretion, which increases blood volume.

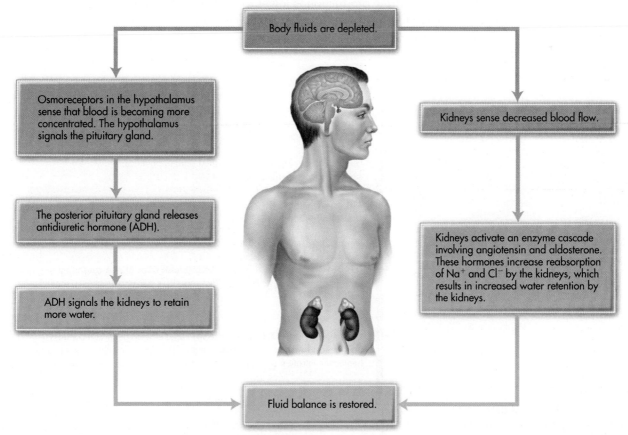

FIGURE 9-8 ▲ When body fluids are depleted, hormonal signals from the pituitary gland and the kidneys work together to increase fluid retention by the kidneys and thereby restore fluid balance.

angiotensin A hormone produced by the liver and activated by enzymes from the kidneys. It signals the adrenal glands to produce aldosterone and also directs the kidneys to conserve sodium (and therefore water). Both of these actions have the effect of increasing blood volume.

aldosterone A hormone produced by the adrenal glands when blood volume is low. It acts on the kidneys to conserve sodium (and therefore water) to increase blood volume.

dehydration Inadequate intake of water to replace losses.

angiotensin, and **aldosterone.** The pituitary gland, located in the brain, senses the concentration of solutes in the blood. When blood concentration of solutes is high, the pituitary gland releases ADH. The kidneys respond to ADH by reducing urine production and output. ADH also causes blood vessel constriction, which acts to raise blood pressure. Meanwhile, the kidneys possess receptors that monitor blood pressure. Low blood pressure triggers the release of an enzyme that activates angiotensin and, eventually, aldosterone, two hormones that signal the kidneys to retain more sodium and, in turn, more water via osmosis. As sodium and water are retained, blood pressure increases back to normal.

Dehydration. Despite mechanisms that work to conserve water, fluid continues to be lost via the feces, skin, and lungs. Those losses must be replaced. In addition, there is a limit to how concentrated urine can become. Eventually, if fluid is not consumed, the body becomes dehydrated and suffers ill effects.

By the time a person loses 1% to 2% of body weight in fluids, he or she will be thirsty (Fig. 9-9). Even this small water deficit can cause one to feel tired and dizzy and to experience headaches. At a 4% loss of body weight, muscles lose significant strength and endurance, and central nervous system function is negatively affected (e.g., memory and reaction time are compromised and one becomes impatient). By the time body weight is reduced by 10%, heat tolerance is decreased and weakness results. Ultimately, **dehydration** will lead to kidney failure, coma, and death. Dehydration is a contributing factor to the development of heatstroke, a very serious condition. Performing strenuous physical activity in hot, humid conditions can lead to dehydration and the inability to control body temperature. Heart rate is increased and the skin becomes dry. Unassisted, the individual will become unconscious and die. Adequate fluid intake and, if possible, avoidance of physical activity in hot, humid conditions are the best steps to prevent heat illness.

Another potential consequence of inadequate fluid intake is kidney stones. When urine production is lower than about 500 milliliters per day, the kidneys are forced to form concentrated urine. Heavy ion concentration, in turn, increases the

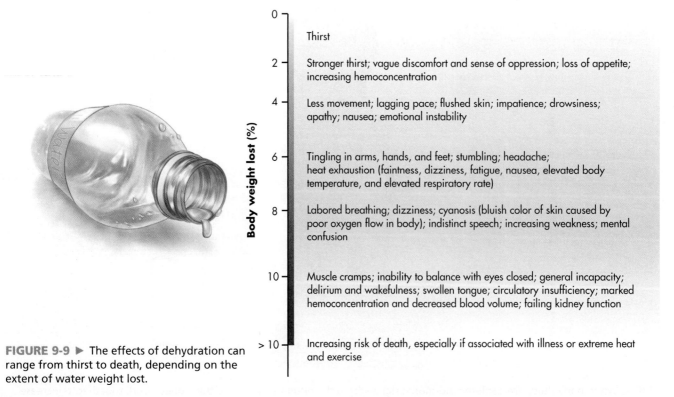

FIGURE 9-9 ▶ The effects of dehydration can range from thirst to death, depending on the extent of water weight lost.

risk of kidney stone formation in susceptible people (generally men). Kidney stones form from minerals and other substances that have precipitated out of the urine and accumulate in the kidney.

The simplest way to determine if water intake is adequate is to observe the urine color (Fig. 9-10). If hydration is adequate, urine should be clear or pale yellow (like pale lemonade or the color of straw); concentrated urine is dark yellow (like apple juice). Urine color can be influenced by consuming supplements (especially some B vitamins), medications, and food. Lots of carrot juice, pumpkin, or winter squash can tint your urine orange. Too many fava beans or too much rhubarb will turn it dark brown. A reddish or pinkish urine results from eating too many beets or blackberries. In the spring, too much asparagus not only makes your urine smell funny but can also turn it a bit green.

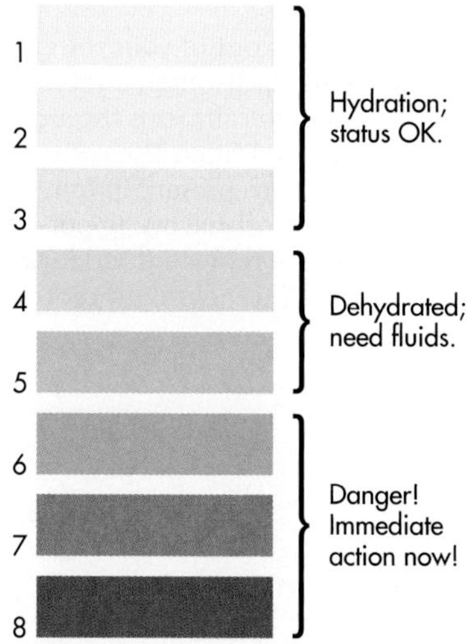

Is Thirst a Good Indicator of Hydration Status? If you do not drink enough water, your body soon lets you know by signaling thirst. Your brain is communicating the need to drink. In most cases, drinking fluids in response to the thirst sensation will result in adequate hydration. However, the thirst mechanism can lag behind actual water loss during prolonged exercise and illness. The thirst sensation decreases with age, as well. Sick children, especially those with fever, vomiting, diarrhea, and increased perspiration, and older persons often need to be reminded to drink plenty of fluids.

As you will learn in Chapter 10, athletes need to monitor fluid status. They should weigh themselves before and after training sessions to determine their rate of water loss and, thus, their water needs. The old saying, "A pint's a pound the world around" does not apply here. True, 1 pound of water is 1 pint (2 cups) of water. However, the body can absorb only about 60% of the water consumed. So athletes should drink about 50% more than what they lose through sweat in a workout. Their goal is to consume 2 to 3 cups of fluid for every pound lost.

Can a Person Consume Too Much Water? Even though the kidneys of a healthy person can process up to 15 liters of urine per day, it is possible to drink too much water. As water intake increases above what is needed, kidneys process the excess fluid and excrete dilute urine. If water intake far exceeds the kidneys' processing ability, overhydration and sodium dilution in the blood result. This condition is commonly known as **water intoxication,** or more accurately, **hyponatremia.** Water intoxication can happen to healthy people when they drink a great deal of water in a very short period of time. Endurance athletes exercising for prolonged times, drinking large volumes of water to replace sweat losses, are especially at risk. Rapid dilution of the blood causes tissues to swell. The heartbeat becomes irregular, allowing fluid to enter the lungs; the brain and nerves swell, causing severe headaches, confusion, seizure, and coma. Unless water is restricted and a concentrated salt solution administered under close medical monitoring, the person will die. For endurance athletes and those who labor outdoors for many hours per day, sports drinks can be used to replace electrolytes along with fluids (see Chapter 10).

SOURCES OF WATER

What Is the Difference Between Hard and Soft Water? Water can be classified as either hard or soft. **Hard water** contains relatively high levels of the minerals calcium and magnesium. In North America, 89% of homes have hard water. Naturally occurring **soft water,** which can be high in sodium, is found in the Pacific North and Northwest, New England, South Atlantic-Gulf, and Hawaii. Hard water can be converted into soft water through the use of a commercial water softener. As water travels through the water softener, calcium and magnesium exchange with sodium found in the water softener device. The water that exits the softener has a low calcium and magnesium content; however, the

FIGURE 9-10 ▲ Monitoring the color of urine is a good gauge of hydration.

water intoxication Potentially fatal condition that occurs with a high intake of water, which results in severe dilution of the blood and other fluid compartments.

hyponatremia Dangerously low blood sodium level.

hard water Water that contains high levels of calcium, magnesium, and sometimes iron.

soft water Water that contains little or no calcium or magnesium. Soft water may contain sodium and other minerals.

hypertension High blood pressure. The pressure inside the blood vessels exceeds 140/90 millimeters of mercury.

Water that comes out of your refrigerator dispenser or through a filter on your faucet typically runs through a charcoal (carbon) filter. The carbon attracts compounds present in tap water to remove off flavors. Importantly, it does not remove fluoride, a mineral essential to help fight tooth decay.

To find a list of brands of bottled water that contain fluoride, check out the website of the International Bottled Water Association at **http://www.bottledwater.org/fluoride.**

Since the 1930s, BPA has been used to make clear plastic bottles. Unfortunately, this organic compound can leach into a bottle's contents when the plastic is exposed to acidic or hot conditions. BPA is considered an endocrine disrupter. At low doses, it can mimic the body's own hormones. Thus, there is concern about chronic exposure for infants and young children. BPA has been banned for use in Canada and many European countries. In the United States, FDA has banned use of BPA in the manufacture of baby bottles and sippy cups. Many manufacturers of reusable water bottles are switching to "BPA free" plastics.

sodium content is increased. The additional intake of sodium from softened water (about 12.5 milligrams per cup) is small compared to typical American intakes of the mineral but could be undesirable for people restricting sodium intake due to **hypertension.** The additional intake of calcium and magnesium afforded by consuming hard water would be more beneficial than increasing sodium intake through the use of softened water.

Is Bottled Water Healthier than Tap Water? Bottled water is a popular alternative to tap water; the latest statistics show that Americans consume 30.8 gallons of bottled water per capita per year. Many people choose bottled water because they believe that it is less likely to be contaminated with pathogens or impurities than tap water. There are some differences in water treatment methods: rather than using chlorine to disinfect water, most bottled water is treated with ozone, which does not impart a flavor to the water. However, the standards for quality and contaminant levels are identical for bottled and tap water. The Environmental Protection Agency regulates and monitors public water supplies, whereas the Food and Drug Administration (FDA) regulates bottled water. Truth be told, much of the bottled water produced in the United States is actually processed municipal tap water.

Beyond the level of contaminants, there are some definite differences between bottled and tap water. As you'll learn in Section 9.14, a small amount of fluoride is added to many municipal water supplies to prevent dental caries. Very few bottled water manufacturers add the mineral fluoride to the water. People who drink primarily bottled water should include regular tap water throughout the day in order to receive the benefits of fluoride. Trips to the drinking fountain or making coffee or tea with tap water should suffice.

Another big difference between bottled water and tap water is the way it is delivered. Tap water travels through pipes into your home, but bottled water requires extra packaging and more costly methods of transport and storage. The large amount of plastic used to package the more than 9 billion gallons of bottled water consumed in America each year creates huge energy use, recycling, and solid waste disposal concerns.

Beyond its impact on the environment, the use of plastics may pose additional threats to human health. Drinking water from a freshly washed or newly opened bottle is fine. But plastic, like the food we eat, has a shelf life. Over time, the chemicals that make up plastic break down and can leach into the liquid inside the container. Temperature, age of the bottle, acidity of the contents, and type of plastic (recycling code) all make a difference. Age of the consumer makes a difference, too, as babies and young children are more susceptible to problems than adults. Table 9-1 lists several guidelines to ensure the safety of the bottled water you drink.

TABLE 9-1 ▶ Guidelines for Safe Use of Water Bottles

1. For repeated use, choose stainless steel bottles.

2. If you choose plastic, look for recycling codes 2 and 4 (see Further Reading 8). When these types of plastics break down, they are the least harmful.

3. Avoid bottles with recycling codes 3 and 7 unless they have "BPA free" embedded in the plastic. These hard plastics may contain bisphenyl A (BPA, see margin). If your plastic bottle contains BPA, avoid cleaning it with harsh (i.e., laboratory) detergents or filling it with acidic (e.g., lemonade and fruit juices) or hot beverages (see Further Reading 20).

4. Choose a reusable bottle with a wide mouth so you can easily clean it. Wash and soak it in a sanitizer, such as ¼ teaspoon (4 milliliters) bleach in 1 quart (1 liter) water.

5. When your plastic bottle becomes scratched or cracked, throw it away. Bacteria can grow inside.

6. Do not store your containers of water in a hot garage or in the back of your hot car. The heat quickly breaks down the plastic. If a bottle has been exposed to heat, pitch the water and recycle the bottle.

✔ **CONCEPT CHECK 9.1**

1. Why is it significant that water is the "universal solvent"?
2. Describe how water regulates body temperature.
3. Provide two examples of water's role as a lubricant.
4. List the components of water intake and water output.
5. What is the AI for fluid intake for adult men? For adult women?
6. Examine Figure 9-8. In your own words, describe the hormonal regulation of water balance.
7. List two situations in which thirst is *not* a reliable indicator of fluid needs.
8. What is water intoxication?

9.2 Minerals: Essential Elements for Health

While vitamins are compounds consisting of many elements (e.g., carbon, oxygen, and hydrogen), **minerals** are individual chemical elements. They cannot be broken down further. The mineral content of foods is sometimes called "ash" because it is all that remains after the whole food has been destroyed by high temperatures or chemical degradation. In humans, minerals make up about 4% of adult body weight (Fig. 9-11). A mineral is essential for humans if a dietary inadequacy results in a physiological or structural abnormality and its addition to the diet prevents such illness or reinstates normal health. Sixteen minerals are known to be essential in the diet.

Minerals are categorized based on the amount we need per day. Recall from Chapter 1 that if we require greater than 100 milligrams (1/50 of a teaspoon) of a mineral per day, it is considered a **major mineral.** These include calcium, phosphorus, magnesium, sulfur, sodium, potassium, and chloride. **Trace minerals** are required at levels less than 100 milligrams per day. Nine essential trace minerals

mineral Element used in the body to promote chemical reactions and form body structures.

major mineral Vital to health, a mineral required in the diet in amounts greater than 100 milligrams per day.

trace mineral Vital to health, a mineral required in the diet in amounts less than 100 milligrams per day.

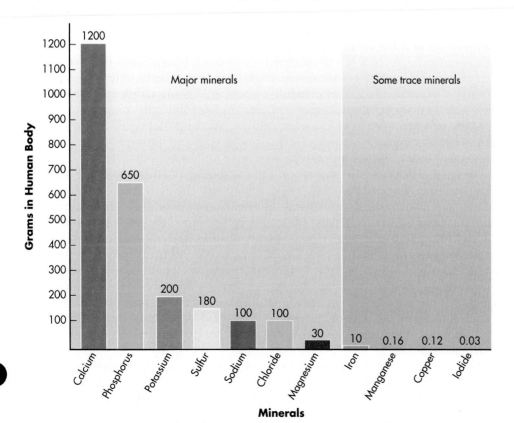

FIGURE 9-11 ◄ Approximate amounts of various minerals present in the average human body. Other trace minerals of nutritional importance not listed include chromium, fluoride, molybdenum, selenium, and zinc.

ultratrace mineral A mineral present in the human diet in trace amounts but that has not been shown to be essential to human health.

phytic acid (phytate) A constituent of plant fibers that binds positive ions to its multiple phosphate groups.

oxalic acid (oxalate) An organic acid found in spinach, rhubarb, and sweet potatoes that can depress the absorption of certain minerals present in the food, such as calcium.

Sulfur is unique among the major minerals because it has no known dietary requirement. Proteins supply all the sulfur we need. It is found in many important compounds in the body, such as some amino acids (e.g., methionine) and vitamins (e.g., biotin and thiamin). Sulfur helps to maintain acid-base balance in the body and is an important part of the liver's drug-detoxifying pathways.

▲ The zinc deficiencies found among some Middle Eastern populations are attributed partly to their consumption of unleavened breads, such as matzo, resulting in low bioavailability of dietary zinc. If grains are leavened with yeast, as they are in bread making, enzymes produced by the yeast can break some of the bonds between phytic acid and minerals. This increases mineral absorption.

(iron, zinc, copper, iodide, selenium, molybdenum, fluoride, manganese, and chromium) have been identified for humans.

Information about trace minerals is perhaps the most rapidly expanding area of knowledge in nutrition. With the exceptions of iron and iodide, the importance of trace minerals to humans has been recognized only within the last 50 years or so. Although we need 100 milligrams or less of each trace mineral daily, they are as essential to good health as are major minerals.

In some cases, discovering the importance of a trace mineral reads like a detective story, and the evidence is still unfolding. In 1961, researchers linked dwarfism in Middle Eastern villagers to a zinc deficiency. Other scientists recognized that a rare form of heart disease in an isolated area of China was linked to a selenium deficiency. In North America, some trace mineral deficiencies were first observed in the late 1960s and early 1970s, when the minerals were not added to synthetic formulas used for intravenous feeding.

It is difficult to define precisely our trace mineral needs because we need only minute amounts. Highly sophisticated technology is required to measure such small amounts in both food and body tissues.

There are several additional trace minerals (sometimes called **ultratrace minerals**) found in the human body, but many of them have no known requirements. These include arsenic, boron, nickel, silicon, and vanadium.

ABSORPTION AND STORAGE OF MINERALS IN THE BODY

Foods offer us a plentiful supply of many minerals, but the ability of our bodies to absorb and use them varies. The bioavailability of minerals depends on many factors, including many nonmineral components of foods. Age, gender, genetic variables, nutritional status, and diet will affect mineral absorption and bioavailability. Numerous prescription drugs also adversely affect mineral absorption. The mineral content listed in a food composition table is a starting point for estimating the contribution the food will make to our mineral needs.

Components of fiber, such as **phytic acid (phytate)** and **oxalic acid (oxalate),** can limit absorption of some minerals by binding to them. Spinach, for example, contains plenty of calcium, but only about 5% (compared to the average 25% bioavailability of calcium from foods) of it can be absorbed because of the vegetable's high concentration of oxalic acid, which binds calcium. High-fiber diets—particularly those in excess of current recommendations of 25 (adult women) to 38 grams (adult men) of fiber per day—can decrease the absorption of iron, zinc, and possibly other minerals.

Many minerals, such as magnesium, calcium, iron, and copper, are of similar sizes and electrical charges ($+2$ charge). Having similar sizes and the same electrical charge causes these minerals to compete with each other for absorption; therefore, an excess of one mineral decreases the absorption and metabolism of other minerals. For example, a large intake of zinc decreases copper absorption. Therefore, people should avoid taking individual mineral supplements unless a dietary deficiency or medical condition specifically warrants it. Food sources, however, pose little risk for these mineral interactions, giving us another reason to emphasize foods in meeting nutrient needs.

Several beneficial vitamin-mineral interactions occur during nutrient absorption and metabolism. When consumed in conjunction with vitamin C, absorption of certain forms of iron—such as that in plant products—improves. The active form of vitamin D hormone improves calcium absorption. Many vitamins require specific minerals to act as components in their structure and as cofactors for their function. For example, without magnesium or manganese, the thiamin coenzyme cannot function efficiently.

The average North American diet derives minerals from both plant and animal sources. Overall, minerals from animal products are better absorbed than those from plants because binders such as fiber are not present to hinder absorption. Also, the mineral content of plants greatly depends on mineral concentrations of the soil in which they are grown. Vegans must be aware of the potentially poor mineral content of some plant foods and choose some concentrated sources of minerals. Soil conditions have less of an influence on the mineral content of animal products because livestock usually consume a variety of plant products grown from soils of differing mineral contents.

Like vitamins, the majority of the minerals are absorbed in the small intestine. Minor amounts may be absorbed in the stomach, and some sodium and potassium are absorbed in the large intestine. After minerals are absorbed, some travel freely in the bloodstream, but many are carried by specific transport proteins to their sites of action or storage. Calcium is one example of a mineral that can travel as an ion in the blood or bound to a blood protein called albumin. Iron, on the other hand, has damaging effects in its unbound form, so it is transported bound to proteins, such as transferrin.

Minerals are stored in various tissues throughout the body. Some minerals must remain in the bloodstream to maintain fluid balance and supply body functions. Others, such as calcium, phosphorus, magnesium, and fluoride, are stored mainly in bones. Iron, copper, zinc, and many trace minerals are stored in the liver. Still others are stored in muscle tissue, organs, or glands.

MINERAL TOXICITIES

Excessive mineral intake, especially of trace minerals such as iron and copper, can have toxic results. For many trace minerals, the gap between just enough and too much is small. Taking minerals in supplement form poses the biggest threat for mineral toxicity, whereas food sources are unlikely culprits. Mineral supplements exceeding current standards for mineral needs—especially those that supply more than 100% of the Daily Values on supplement labels—should be taken only under a physician's supervision. The Daily Values are for the most part higher than our current standards (e.g., Recommended Dietary Allowances [RDA]) for mineral needs. Without close monitoring, doses of minerals should not exceed any Upper Level set on a long-term basis.

The potential for toxicity is not the only reason to carefully consider the use of mineral supplements. Harmful interactions with other nutrients are possible. Also, contamination of mineral supplements—with lead, for example—is a possibility. Use of brands approved by the United States Pharmacopeial Convention (USP) lessens this risk (see Chapter 8). Even with the best intentions, people can harm themselves using mineral supplements.

PRESERVATION OF MINERALS IN FOODS

Minerals are found in plant and animal foods (Fig. 9-12), but as you previously read, the **bioavailability** of minerals varies widely. Minerals are not typically lost from animal sources during processing, storage, or cooking; but for plant sources, significant amounts may be lost during food processing. When grains are refined, the final products have lost the majority of their vitamin E, many B vitamins, and trace minerals. The more refined a plant food, as in the case of white flour, the lower its mineral content. During the enrichment of refined grain products, iron is the only mineral added, whereas the selenium, zinc, copper, and other minerals lost during refinement are not replaced. Following the recommendation of the 2010 Dietary Guidelines for Americans to "make half your grains whole" will effectively preserve the mineral content of foods.

bioavailability The degree to which a consumed nutrient is absorbed and used by the body.

CRITICAL THINKING

Gwen follows a vegetarian diet. What factors in foods of plant origin may limit the bioavailability of minerals, such as calcium and zinc, in her diet?

✔ CONCEPT CHECK 9.2

1. Define bioavailability.
2. Are ultratrace minerals essential in the human diet? List three examples of ultratrace minerals.
3. Should people take individual mineral supplements? Why or why not?
4. Where are minerals stored in the body?

FIGURE 9-12 ▶ Certain groups of MyPlate are especially rich sources of various minerals. This is true for the minerals listed. Each mineral may also be found in other groups but in lower amounts. Other trace minerals are also present in moderate amounts in many groups. With regard to the grains group, whole-grain varieties are the richest sources of most trace minerals listed.

**MyPlate:
Sources of Minerals**

Grains

- Sodium chloride
- Calcium (fortified products)
- Phosphorus
- Magnesium
- Iron
- Zinc
- Copper
- Selenium
- Chromium

Vegetables

- Potassium
- Magnesium

Fruits

- Potassium
- Boron

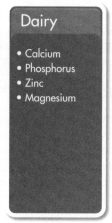

Dairy

- Calcium
- Phosphorus
- Zinc
- Magnesium

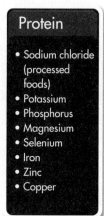

Protein

- Sodium chloride (processed foods)
- Potassium
- Phosphorus
- Magnesium
- Selenium
- Iron
- Zinc
- Copper

9.3 Sodium (Na)

Are you "worth your salt"? That shaker of salt on your kitchen table contains a nutritionally important and once highly valued chemical compound! Historically, salt was treasured because it enabled people to preserve foods. Salt was traded and used as a form of payment. In fact, the word *salary* comes from the Latin *salarium*, which is linked to the use of salt to pay Roman soldiers. Salt is 40% sodium and 60% chloride by weight. (The chemical symbol Na represents the Latin term *natrium*.) One teaspoon of salt contains 2400 milligrams of sodium. What was once difficult to find, however, is now abundant in our food supply. Indeed, nearly all Americans exceed dietary requirements for sodium by so much that reducing sodium in our diets is the focus of major public health campaigns.

FUNCTIONS OF SODIUM

The digestive tract absorbs nearly all consumed sodium. When sodium chloride (NaCl) is dissolved in water, the chemical bond holding the two atoms together breaks and the charged ions Na^+ and Cl^- are released. These electrolytes, as well as others, attract water. The concentration of intracellular and extracellular water is controlled by the concentration of the electrolytes. Fluid balance is maintained by moving or actively pumping sodium ions where more water is needed. Sodium ions also function in nerve impulse conduction and absorption of some nutrients (e.g., glucose).

Unless purposely controlled for health reasons, sodium consumption varies tremendously from day to day, even meal to meal. Yet, our blood levels vary only slightly. Kidneys function as a filter. If blood sodium is low, as blood flows through the kidney, sodium is secreted back into the blood, resulting in a decreased urine output. Conversely, if our blood sodium levels are too high, the sodium is filtered out by the kidneys and excreted into the urine. When this excess sodium is

Is sea salt a healthier choice than common table salt? Sea salt is made by evaporating sea water. It is minimally processed, which gives it a coarser texture, and it may contain traces of magnesium, calcium, and potassium. In contrast, table salt is typically mined from the earth and processed into fine grains. Many consumers prefer the taste and texture of sea salt over table salt, but when it comes to heart health, there is no significant difference in sodium or chloride content. Furthermore, as you will learn in Section 9.12, most of the table salt sold in North America is fortified with iodide, an essential nutrient for thyroid function. Next time you purchase salt, compare the labels of several products to assess any nutritional differences for yourself.

352

removed, water follows, resulting in greater urine output. Without drinking extra water, dehydration can result. Fortunately, high-sodium (salty) foods make us thirsty and drive us to drink more fluids.

SODIUM DEFICIENCY

A diet low in sodium, coupled with excessive perspiration and persistent vomiting or diarrhea, has the ability to deplete the body of sodium. This state can lead to muscle cramps, nausea, vomiting, dizziness, and later shock and coma. The likelihood of this occurring is low because the kidney is very efficient at conserving sodium under conditions of low-sodium status.

When weight loss from perspiration exceeds 2% to 3% of total body weight (or about 5 to 6 pounds), sodium losses should raise concern. Even then, merely salting food or selecting some salty foods such as soup or crackers is sufficient to restore body sodium for most people. Athletes who perspire for hours during endurance activities need to consume electrolyte-replacement drinks during exercise to avoid depletion of sodium, which can lead to hyponatremia. Perspiration contains about two-thirds the sodium concentration found in blood or about 1 gram of sodium per liter.

GETTING ENOUGH SODIUM

About 77% of the sodium we consume is added in the form of salt during food manufacturing and food preparation at restaurants. Sodium added while cooking or at the table at home provides about 11% of our intake, and naturally occurring sodium in foods provides the remaining 12% (Fig. 9-13). Most unprocessed foods are relatively low in sodium; milk is one exception (about 120 milligrams per cup).

The more processed and restaurant food consumed, the higher one's sodium intake. Conversely, the more home-cooked meals prepared, the more control a person has over sodium in the diet. Major contributors of sodium in the adult diet are white bread and rolls; hot dogs and lunch meat; cheese; soups; and foods with tomato sauce—partly because these foods are consumed so often. Other foods that can be major contributors to sodium intake include French fries, pretzels, potato chips, sauces, and gravies.

If we ate only unprocessed foods and added no salt, we would consume about 500 milligrams of sodium per day. The AI for sodium is 1500 milligrams for adults through age 50. The recommendation is lower for older adults (see margin). If we compare 500 milligrams of sodium from a diet of unprocessed foods with the 2300 to 4700 milligrams or more typically consumed by adults, it is clear that food processing, dining out, and salt added while cooking are the major contributors to sodium intake. When dietary sodium must be restricted, attention to food labels is valuable to monitor sodium intake.

Most people can adapt to wide variations in dietary sodium intakes; today's sodium intake is found in tomorrow's urine. However, approximately 10% to 15% of adults are *sodium sensitive;* that is, sodium intake has a direct effect on their blood pressure. As their sodium intake increases, so does their blood pressure. Among these people, lower-sodium diets (about 2000 milligrams daily) often decrease blood pressure. Groups that appear to be especially affected are African-Americans, Asian-Americans, and people who have diabetes and/or are overweight (see Newsworthy Nutrition in the margin and the Nutrition and Your Health section on minerals and hypertension at the end of the chapter). Lifestyle factors such as being overweight and inactive are the major contributors to the development of hypertension.

The medical community suggests that adults should reduce salt and sodium consumption to limit the risk of developing hypertension later in life (see Further Reading 3). It is also a good idea to have your blood pressure checked regularly. If you are diagnosed with hypertension, you should reduce sodium intake as

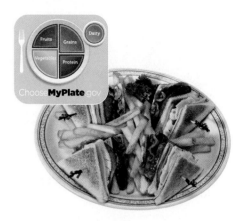

▲ How does this meal of a turkey club sandwich and French fries compare to MyPlate? How could you reduce the sodium content of this meal?

Sodium
AI
  9–50 years: 1500 milligrams
  51–70 years: 1300 milligrams
  >70 years: 1200 milligrams
DV: 2400 milligrams
UL: 2300 milligrams

Newsworthy Nutrition

Lower sodium and higher potassium intakes reduce risk of cardiovascular disease

Increasing potassium intake appears effective in lowering blood pressure and is most effective at reducing risk of cardiovascular disease when combined with lower intakes of sodium. The sodium-to-potassium ratio in urine is a good predictor of cardiovascular disease.

Source: Cook NR and others: Joint effects of sodium and potassium intake on subsequent cardiovascular disease. *Archives of Internal Medicine* 169:32, 2009.

connect
NUTRITION

Check out the Connect site **www.mcgrawhillconnect.com** to further explore the roles of sodium and potassium in cardiovascular health.

Sodium in food

Added during food processing or in restaurant meals

Added during cooking at home

Added at the table

Present naturally

77%

5%

6%

12%

FIGURE 9-13 ▲ Salt arrives in the American diet largely as a result of food processing.

you follow a comprehensive plan to treat this disease. Reducing sodium intake may also help maintain a healthy calcium status, as sodium intake greater than about 2000 milligrams per day may increase urinary calcium loss along with the sodium excreted. The effect of this increased urinary calcium loss on bone health is controversial.

Adopting a reduced-salt diet is a significant lifestyle change for most people because many typical food choices will have to be limited (Fig. 9-14). At first, foods may taste bland, but eventually, you will perceive more flavor as the taste receptors in the tongue become more sensitive to the salt content of foods. It takes 6 to 8 weeks to retrain your taste buds to sense sodium at a lower level. Slowly reducing sodium intake by substituting lemon juice, herbs, and spices will allow you to become accustomed to a diet that contains minimal amounts of salt. Many cookbooks and online sources offer excellent recipes for flavorful dishes.

AVOIDING TOO MUCH SODIUM

The Upper Level (UL) for sodium for adults is 2300 milligrams (2.3 grams), approximately 1 teaspoon. Intakes exceeding this amount typically increase blood pressure. Recent research also links excessive sodium consumption to overweight and obesity. As salt intake increases, fluid intake also increases; if calorie-laden beverages are chosen, weight gain may ensue (see Further Reading 7). About 95% of

FIGURE 9-14 ◄ Food sources of sodium. (a) The fill of the background color (none, 1/3, 2/3, or completely covered) within each food group on MyPlate indicates the average nutrient density for sodium for *natural, unprocessed foods* in that group. (b) The bar graph shows the sodium content of several natural and processed foods from each food group compared to the AI for adult males and females. Overall, the dairy group is the only food group that provides much sodium in its natural form. Food processing adds significant sodium to foods such as canned vegetables and cured meats. *Nutrition data from USDA National Nutrient Database for Standard Reference, Release 26.*

(a)

(b)

Food Item and Amount	Sodium (milligrams)	% AI for Adult Males and Females (1500 milligrams)
Grains Chicken noodle soup, canned, 1 cup	870	58%
Kellogg's Corn Flakes®, 1 cup	200	13%
Whole-wheat bread, 1 slice	146	10%
Vegetables Tomatoes, stewed (canned), 1 cup	564	38%
Green beans, cooked (from canned), 1 cup	461	31%
Green beans, cooked (from frozen), 1 cup	3	0%
Fruits Olives, green, bottled, 5 medium	210	14%
Cantaloupe, 1 cup	25	2%
Fruit cocktail, 1 cup	9	1%
Dairy Cottage cheese, low-fat, 1 cup	746	18%
Cheddar cheese, 1.5 ounces	264	50%
Milk, non-fat, 1 cup	103	7%
Protein Pepperoni pizza, 2 slices	1538	103%
Ham steak, cooked, 3 ounces	1079	72%
Chicken breast, roasted, 3 ounces	64	4%

North American adults have sodium intakes that exceed that UL. It must be noted that the Daily Value (DV) of 2400 milligrams exceeds the UL for sodium. A healthier goal is to aim for the AI of 1500 milligrams.

✓ CONCEPT CHECK 9.3

1. List three food sources of sodium in your diet.
2. Which organ regulates the amount of sodium in your blood?
3. Define *sodium sensitivity.*
4. List three specific changes you could make to your diet to decrease your sodium intake.

9.4 Potassium (K)

FUNCTIONS OF POTASSIUM

Potassium performs many of the same functions as sodium, such as water balance and nerve impulse transmission. (The chemical symbol K represents the Latin term *kalium.*) All membranes contain an energy-dependent pump that can transfer sodium from inside to outside the cell. When sodium (Na^+) is actively pumped out of the cell, potassium (K^+) enters the cell in an attempt to balance the loss of the positively charged sodium ions. That makes potassium the principal positively charged ion inside cells. Intracellular fluids contain 95% of the potassium in the body. Higher potassium intake is associated with *lower* rather than higher blood pressure values.

POTASSIUM DEFICIENCY

Low blood potassium, also known as *hypokalemia,* is a life-threatening problem. Symptoms often include a loss of appetite, muscle cramps, confusion, and constipation. Eventually, the heart beats irregularly, decreasing its capacity to pump blood.

Hypokalemia can result from continually deficient food intake, but it is most commonly seen with chronic diarrhea or vomiting, or as a side effect of medications, including laxatives and some diuretics (see Medicine Cabinet). Vulnerable populations include people with certain eating disorders (see Chapter 11) or alcohol use disorders (see Chapter 16). Other populations at increased risk for potassium deficiency include people on very low-calorie diets and athletes who exercise for prolonged periods. These people should compensate for potentially low body potassium by consuming potassium-rich foods.

GETTING ENOUGH POTASSIUM

Unprocessed foods are rich sources of potassium, including fruits, vegetables, milk, whole grains, dried beans, and meats (Fig. 9-15). Here is an easy guide: the more processed your food, the higher it is in sodium and the lower it is in potassium. Major contributors of potassium to the adult diet include milk, potatoes, beef, coffee, tomatoes, and orange juice.

Approximately 90% of the potassium consumed is absorbed, but diets are more likely to be lower in potassium than sodium because we add salt to our food, not potassium. In fact, North Americans typically consume only 2000 to 3000 milligrams of potassium per day, much less than the AI (see margin). Thus, many of us need to increase potassium intake, preferably by increasing fruit and vegetable intake.

The 2010 Dietary Guidelines for Americans recommend less than 2300 milligrams of sodium (approximately 1 teaspoon of salt) per day for the general population. African-Americans; middle-aged or older adults; and individuals with hypertension, diabetes, or chronic kidney disease should consume no more than 1500 milligrams of sodium per day. The American Heart Association takes a stronger position, advising no more than 1500 milligrams of sodium per day for all Americans.

▲ Fruits and vegetables are rich sources of potassium.

Potassium
AI: 4700 milligrams
DV: 3500 milligrams
UL: none

diuretic A substance that increases urinary fluid excretion.

Medicine Cabinet

Some people take **diuretics** to lower their blood pressure. Diuretics cause the kidneys to excrete more urine but at the same time may increase urinary excretion of minerals. This is nutritionally relevant for regulation of blood levels of potassium, magnesium, and zinc. People who take potassium-wasting diuretics need to carefully monitor their dietary intake of this mineral. Increased intake of fruits and vegetables or potassium chloride supplements are prescribed by physicians.

Examples:

Hydrochlorothiazide (Microzide)

Furosemide (Lasix)

FIGURE 9-15 ◄ Food sources of potassium. (a) The fill of the background color (none, 1/3, 2/3, or completely covered) within each food group on MyPlate indicates the average nutrient density for potassium in that group. (b) The bar graph shows the potassium content of several foods in each food group compared to the AI for adult males and females. Overall, the richest sources of potassium are unprocessed foods of plant origin, such as fruits, vegetables, and beans. *Nutrition data from USDA National Nutrient Database for Standard Reference, Release 26.*

(a)   (b)

	Food Item and Amount	Potassium (milligrams)	% AI for Adult Males and Females (4700 milligrams)
Grains	Raisin bran cereal, 1 cup	352	7%
	Quinoa, cooked, ½ cup	159	3%
	White bread, 1 slice	29	1%
Vegetables	Swiss chard, cooked, 1 cup	961	20%
	Potato, baked (with skin), 1 medium	926	20%
	Acorn squash, cooked, 1 cup	896	19%
Fruits	Orange juice, fresh, 1 cup	496	11%
	Cantaloupe, 1 cup	426	9%
	Banana, 1 medium	422	9%
Dairy	Yogurt, plain, non-fat, 1 cup	625	13%
	Milk, fat-free, 1 cup	382	8%
	Soymilk, 1 cup	296	6%
Protein	Kidney beans, cooked, ½ cup	358	8%
	Sirloin steak, grilled, 3 ounces	286	6%
	Pistachios, dry roasted, 1 ounce	285	6%

AVOIDING TOO MUCH POTASSIUM

If the kidneys function normally, potassium from dietary sources will not lead to toxicity. Thus, no Upper Level for potassium has been set. When the kidneys function poorly, potassium builds in the blood, inhibiting heart function and leading to a slowed heartbeat. If left untreated, the heart eventually stops beating, resulting in a cardiac arrest and death. Therefore, in cases of kidney failure or kidney disease, close monitoring of blood levels of potassium and potassium intake becomes critical.

☑ CONCEPT CHECK 9.4

1. List two functions of potassium in the body.
2. How is potassium intake related to blood pressure?
3. List three specific changes you could make to your diet to increase your potassium intake.

9.5 Chloride (Cl)

FUNCTIONS OF CHLORIDE

Chloride (Cl⁻) is a negative ion found primarily in the extracellular fluid. Along with sodium and potassium, chloride helps to regulate fluid balance in the body. In fact, chloride itself may be partially responsible for increases in blood pressure that accompany high-salt diets.

Chloride ions are also a component of the acid produced in the stomach (hydrochloric acid) and are important for overall maintenance of acid-base balance in the body. This electrolyte is used during immune responses as white blood cells attack foreign cells. In addition, nervous system function relies on the presence of chloride.

CHLORIDE DEFICIENCY

Low levels of chloride in the blood can lead to a disturbance of the body's acid-base balance. A chloride deficiency is unlikely, however, because our dietary salt intake is so high. Frequent and lengthy bouts of vomiting, if coupled with a nutrient-poor diet, can contribute to a deficiency because stomach secretions contain a lot of chloride. Individuals with bulimia or severe cases of gastroenteritis are at risk for chloride deficiency. In addition, low chloride levels could occur as a side effect of some medications, such as diuretics or laxatives.

GETTING ENOUGH CHLORIDE

When it comes to sources of chloride, it is important to make the distinction between the chloride ion, which is vital for body functions, and chlorine (Cl_2), which is a poisonous gas. *Chlorine* is used to disinfect municipal water supplies. A small amount of chlorine may remain in tap water, but the substance evaporates quickly. Municipal and well water supplies usually contain some *chloride* (leached from the earth) as well, but water does not represent a significant source of chloride in the diet.

A few fruits and vegetables, such as seaweed, celery, tomatoes, and olives, are naturally good sources of chloride. Most of our dietary chloride, however, comes from salt added to foods. Knowing a food's salt content allows for a close prediction of its chloride content. Salt is 60% chloride by weight.

Like sodium, nearly all the chloride consumed is efficiently absorbed. The AI for chloride (see margin) is based on the 40:60 ratio of sodium to chloride in salt (1500 milligrams of sodium: 2300 milligrams of chloride). If the average adult consumes about 9 grams of salt daily, that yields 5.4 grams (5400 milligrams) of chloride. The principal route of excretion is the kidneys, although some chloride is lost in perspiration.

AVOIDING TOO MUCH CHLORIDE

The average adult typically consumes an excess of this mineral. Because chloride has a role in raising blood pressure, it is important that aging adults consciously control salt intake to decrease risk of developing hypertension. Learning at a young age to select lower-salt foods is the best way to start.

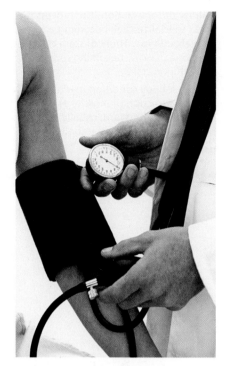

▲ Chloride is likely part of the blood-pressure-raising property of sodium chloride (salt).

Chloride
AI: 2300 milligrams
DV: 3400 milligrams
UL: 3600 milligrams

✔ CONCEPT CHECK 9.5

1. List two functions of chloride in the body.
2. How is chloride intake related to blood pressure?

9.6 Calcium (Ca)

FUNCTIONS OF CALCIUM

Calcium represents 40% of all the minerals present in the body and equals about 2.5 pounds (1200 grams) in the average person. All cells require calcium to function; however, more than 99% of the calcium in the body is used for growth, development, and maintenance of bones. Calcium is the main component of hydroxyapatite, the crystalline compound responsible for the structure and hardness of bone.

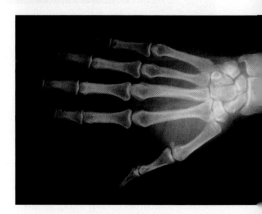

▲ Ninety-nine percent of calcium in the body is in bones.

cellular differentiation The process of a less specialized cell becoming a more specialized type. Think of stem cells in the bone marrow becoming red and white blood cells.

tetany A body condition marked by sharp contraction of muscles and failure to relax afterward; usually caused by abnormal calcium metabolism.

osteoporosis The presence of a stress-induced fracture or a T-score of −2.5 or lower. The bones are porous and fragile due to low mineral density.

Beyond its prominent role in bone health, calcium is critical for many other processes, as well.

Calcium is essential for formation of a blood clot. Muscle contraction is activated by calcium release and the flow of calcium along the surface of the muscle cell. In nerve transmission, calcium assists in the release of neurotransmitters and permits the flow of ions in and out of nerve cells. Calcium helps regulate cellular metabolism by influencing the activities of various enzymes and hormonal responses. Calcium also functions in the maintenance of cell membrane integrity, normal blood pressure, regulation of glucose concentration, and **cellular differentiation.** It is the tight regulation of the concentration of calcium in the blood that keeps these processes going, even if a person fails to consume enough dietary calcium from day to day.

Other Possible Health Benefits of Calcium. Researchers have been examining links between calcium intake and risks for a wide array of diseases. An adequate calcium intake can reduce the risk of colon cancer, especially in people who consume a high-fat diet (see Further Reading 17). A decreased risk of some forms of kidney stones and reduced lead absorption are other possible benefits when calcium is part of a meal. Calcium intakes of 800 to 1200 milligrams per day can also decrease blood pressure, compared with intakes of 400 milligrams per day or less. Calcium intakes of 1200 milligrams per day in combination with a low-fat, low-cholesterol diet can help people with elevated LDL improve their blood lipid profiles. For women, an adequate calcium intake might also reduce the risk of premenstrual syndrome and high blood pressure that can develop during pregnancy. Overall, the benefits of a diet providing adequate calcium extend far beyond bone health.

CALCIUM DEFICIENCY

The body tightly regulates blood calcium concentration within a narrow range, regardless of dietary intake. If dietary calcium intake is inadequate and blood calcium concentration begins to decrease, three hormonally controlled actions are stimulated to reestablish calcium blood levels: (1) bones release calcium, (2) intestines absorb more calcium, and (3) the kidneys retain more calcium in the blood. (Recall the role of vitamin D in blood calcium regulation, discussed in Chapter 8.) Because of this tight hormonal regulation, poor dietary intake of calcium is not likely to result in low blood calcium. Rather, kidney diseases, hormonal abnormalities, or medications are the likely culprits. If blood calcium does fall below a critical point, muscles cannot relax after contraction and nerve function is disrupted. The result is a condition called **tetany,** in which muscles become stiff or twitch involuntarily.

You can see that the skeleton does more than simply provide the framework for the body; it also functions as a bank from which calcium can be added or withdrawn. Only about 1% of the calcium in bone is available at any time for this purpose. Over time, however, bone loss due to inadequate calcium intake and/or absorption may occur, though slowly. Clinical symptoms of the calcium loss from bones show after many years. By not meeting calcium needs, some people, especially women, are setting the stage for osteoporosis and future bone fractures.

Osteoporosis. *Healthy People 2020* identified the prevention of the bone disease **osteoporosis** as one of its major focus areas. About 12 million Americans over the age of 50 currently have osteoporosis. Of these, about 9.6 million are women. An additional 40 million Americans have low bone density, placing them at risk for developing osteoporosis. By 2020, these numbers are expected to increase to 10.5 million women and 3.3 million men with osteoporosis, and another 48 million with low bone mass. Consistently, osteoporosis has led to approximately 2 million bone fractures per year in the United States, including nearly 300,000 broken hips. The health care costs associated with these fractures are projected to reach more than $25 billion by the year 2025, with the most rapid increases among minority populations (see Further Reading 6).

▲ Osteoporosis leads to millions of bone fractures each year, including debilitating fractures of the hip and spine. These fractures frequently result in the need for assistance with activities of daily living, as we see with this woman who needs a walker to help her navigate.

The personal cost of a hip fracture surpasses any financial cost. Hip fractures are regarded as devastating. They result in loss of mobility and need for long-term care. The average age for hip fracture is 82 years. Afterward, the person may require long-term care for the remainder of his or her life. Only 40% of people with hip fractures regain their earlier level of independence. More specifically, it is estimated that a year after fracturing a hip, 90% of those who needed no assistance climbing stairs before the fracture will not be able to climb five stairs; 66% will need help to get on or off a toilet; 50% will not be able to raise themselves from a chair; 31% will need assistance to get out of bed; and 20% will not be able to put on a pair of pants by themselves.

Hip fracture is associated with significant mortality. A recent study estimates that 20% to 30% of the 300,000 Americans 65 or older who fracture a hip each year will die within 12 months (see Further Reading 5). Other types of fractures or even fear of fracture due to osteoporosis can affect quality of life. Vertebral fractures, especially if there are several of them, cause significant pain, reduced lung function, loss of height, and a curved spine. Movement can be restricted and gait altered, increasing the risk and fear of falls and/or more fractures.

Women tend to lose 1% to 3% of their bone mass each year after menopause. Men also lose bone mass as they age, but the loss is more gradual. When the bone-demineralization activities of bone-resorbing cells exceed the bone-building activities of bone-building cells, bone mass declines. This is a normal part of aging and does not always lead to unhealthy bones. However, if bone mass is low when this process begins, even moderate bone demineralization can lead to a condition called **osteopenia.** As more bone is lost, the entire matrix of the bone tissue also begins to break down. When this occurs, osteoporosis is likely to result. About 25% of women older than age 50 develop or have osteoporosis. Among people older than age 80, osteoporosis becomes the rule, not the exception. To young adults, it may sound benign to have osteoporosis. To older adults, the diagnosis presents the reality of life permanently changed.

There are two types of osteoporosis. **Type 1 osteoporosis,** also called postmenopausal osteoporosis, typically occurs in women between 50 and 60 years of age. This type of osteoporosis is directly linked to decreased estrogen concentrations that occur at menopause. Type 1 osteoporosis most dramatically affects **trabecular bone,** as this type of bone undergoes faster remodeling than cortical bone (Fig. 9-16). A woman can lose 20% to 30% of trabecular bone and 5% to 10% of **cortical bone** between ages 50 and 60, unless intervention occurs (Fig. 9-17).

Trabecular bone has more bone-building and bone-resorbing cells than cortical bone. The bone-building cells require estrogen for maximal activity. After menopause, the rate of bone synthesis declines, whereas the rate of bone resorption remains high, leading to net bone loss. Minerals are released but not reincorporated into bone, leaving weak areas and open spaces within the bone matrix. The bones at greatest risk for osteoporotic fractures are the trabecular-rich bones of the pelvis (7% of fractures), vertebrae of the spine (27%) (Fig. 9-18), and portions of long bones such as the wrist (19%).

Type 2 osteoporosis tends to be diagnosed later in life (70 to 75 years of age). Type 2 osteoporosis is a result of breakdown of both cortical and trabecular bone. It is due to a combination of dietary and age-related factors; low dietary intake of bone-building nutrients compounds the problems associated with decreased ability to absorb or metabolize nutrients.

osteopenia A bone disease defined by low mineral density.

type 1 osteoporosis Porous trabecular bone characterized by rapid bone demineralization following menopause.

trabecular bone The less dense, more open structure bone found in the inner layer of bones.

cortical bone The compact or dense bone found on the outer surfaces of bone.

type 2 osteoporosis Porous trabecular and cortical bone observed in men and women after the age of 70.

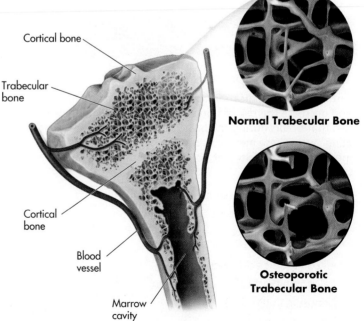

Cortical bone

Trabecular bone

Cortical bone

Blood vessel

Marrow cavity

Normal Trabecular Bone

Osteoporotic Trabecular Bone

FIGURE 9-16 ▲ Cortical and trabecular bone. Cortical bone forms the shafts of bones and their outer mineral covering. Trabecular bone supports the outer shell of cortical bone in various bones of the body. Note how the osteoporotic bone has much less trabecular bone. This leads to a more fragile bone and is not reversible to any major extent with current therapies.

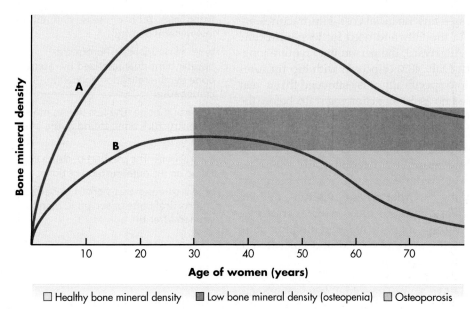

• **Woman A** had developed a high peak bone mass by age 30. Her bone loss was slow and steady between ages 30 and 50 and sped up somewhat after age 50 because of the effects of menopause. At age 75, the woman had a healthy bone mineral density value and did not show evidence of osteoporosis.

• **Woman B** with low peak bone mass experienced the same rate of bone loss as Woman A. By age 50, she already had low bone mineral density and by age 70, kyphosis and spinal fractures had occurred.

□ Healthy bone mineral density   ■ Low bone mineral density (osteopenia)   ▨ Osteoporosis

FIGURE 9-17 ▲ The relationship between peak bone mass and the ultimate risk of developing osteoporosis and related bone fractures.

Source: Based on data from R. Burge, "Incidence and economic burden of osteoporosisrelated fractures in the United States, 2005–2025," Journal of Bone and Mineral Research, 22, 2007, 465.

kyphosis Abnormally increased bending of the spine.

bisphosphonates Drugs that bind minerals and prevent osteoclast breakdown of bone. Examples are alendronate (Fosamax) and risedronate (Actonel).

Medicine Cabinet

Types of Osteoporosis Medications

Antiresorptive medications are used to prevent bone loss and decrease the risk of bone fractures. They slow bone loss that occurs during the breakdown phase of the remodeling cycle. These medicines slow bone loss without affecting bone synthesis so that bone density can increase.

- Bisphosphonates
  - Alendronate (Fosamax® and Fosamax Plus D®)
  - Ibandronate (Boniva®)
  - Risedronate (Actonel®, Actonel® with Calcium, and Atelvia™)
  - Zoledronic acid (Reclast®)
- Calcitonin (Fortical® and Miacalcin®)
- Denosumab (Prolia®)
- Estrogen therapy
- Estrogen agonists/antagonists (Evista®)

Anabolic drugs are used to increase the rate of bone formation and decrease the risk of fractures.

- Teriparatide (Forteo®) is currently the only osteoporosis medicine approved by FDA that rebuilds bone.

People with either form of osteoporosis can lose significant height and experience severe pain, especially in the vertebrae. A woman may lose an inch or more in height as the bone is demineralized (Fig. 9-18). Both men and women can develop a curvature in the upper spine called **kyphosis** or dowager's hump. Kyphosis is a major concern because the bending of the spine may decrease the volume of the chest cavity, resulting in difficulty breathing, abdominal pain, decreased appetite, and premature satiety. As already discussed, osteoporotic bone is also more susceptible to fracture following a fall.

According to the National Osteoporosis Foundation 2010 guidelines (see Further Reading 12), all men and women diagnosed with osteoporosis should first be counseled on risk factor reduction. Daily habits of calcium and vitamin D consumption as well as exercise should be stressed. Clear guidelines are provided to physicians for patients to be considered for pharmacological (drug) intervention. These drugs are indicated for postmenopausal women and men over 50 who meet specific standards for level of risk of future fracture as well as medical history (see Further Reading 10). Current medication options for men and women are **bisphosphonates** and **parathyroid hormone.** Women may also use calcium and hormone replacement therapy (see Medicine Cabinet in the margin for a list of osteoporosis medications).

The incidence of osteoporosis increases as a person ages. Its importance as a personal and public health problem is intensifying as the U.S. population ages. It appears, however, that a large percentage of the cases of osteoporosis can be prevented. The key to prevention is to build dense bones during the first 30 years of life and then limit the amount of bone loss in adulthood. People with higher peak bone mass have more calcium to lose before bones become weak and fracture easily. A higher peak bone mass is also the reason males experience osteoporosis less often than females do; they have more bone mass to lose.

CASE STUDY Worried About Grandma

Grace, a 23-year-old woman of Korean descent, is in her final year of nursing school in Boston. She also works 20 hours per week at a local pharmacy. Grace is worried about a phone call she just received from her mom. While out on her walk yesterday, Grandmother Kyon caught her toe on an uneven section of the sidewalk, fell, and broke her hip. The doctor diagnosed Grandmother Kyon with osteoporosis, and the family is worried about the long recovery ahead. As a nursing student, Grace knows how devastating a hip fracture can be. She also knows that osteoporosis can run in families. Grace resolves to do whatever she can to learn about osteoporosis and strengthen her bones now.

She starts by searching the Internet for a website that can help her determine if she is at high risk for osteoporosis. The New York State Department of Health Osteoporosis Education Prevention Program, **www.health.ny.gov/diseases/conditions/osteoporosis/index.htm**, offers a list of risk factors for osteoporosis. For herself, Grace sees a few risk factors that she cannot change, but there are a few that are glaringly obvious. She uses this website to "bone up" on ways to promote her bone health while she is still young.

Grace discovers she is probably pretty low in vitamin D intake. She also does not have much free time for exercise. She cooks traditional Korean dishes for half of her meals but eats fast food or sandwiches for the rest. Daily, she takes a multivitamin/mineral pill plus one calcium carbonate pill. She generally takes them when she brushes her teeth at night before bed. On weekends, she gets out with friends or her boyfriend and has a glass or two of wine. Grace does not smoke.

Answer the following questions about Grace's situation and check your responses at the end of the chapter.

1. Go to the URL that Grace visited and click on the link for Risk Factors for Osteoporosis. What risk factors do you see for Grace? What additional questions would you ask Grace to assess her risk for osteoporosis?
2. Grace went to the section "Keeping Your Bones and Teeth Strong for Life" to determine key actions she can take to strengthen her bones now. Which ones does she already do? Which lifestyle actions could Grace integrate into her daily life right now?
3. What are the environmental conditions that inhibit Grace from adequately synthesizing vitamin D? How can she overcome some of them?
4. What are some calcium and vitamin D sources in traditional Korean dishes?
5. Give Grace some tips to increase her calcium and vitamin D intake at fast-food restaurants without significantly increasing costs or calories.
6. Would you recommend any changes to her supplement regimen?

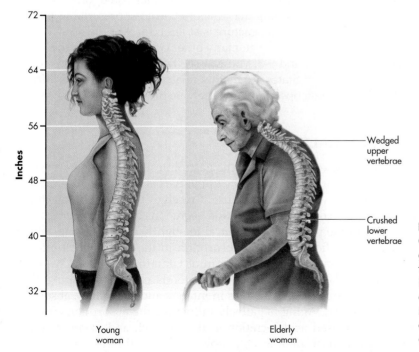

Young woman

Elderly woman

parathyroid hormone (PTH) A hormone made by the parathyroid gland that increases synthesis of the active form of vitamin D. This hormone works with vitamin D to increase blood levels of calcium.

FIGURE 9-18 ◄ Normal and osteoporotic woman. Osteoporotic bones have less substance, so osteoporosis generally leads to loss of height, distorted body shape, fractures, and possibly loss of teeth. Monitoring changes in adult height is one way to detect early evidence of osteoporosis. Kyphosis, or curvature of the upper spine, results from demineralization of the vertebrae. This can lead to both physical and emotional pain. Kyphosis occurs in both men and women.

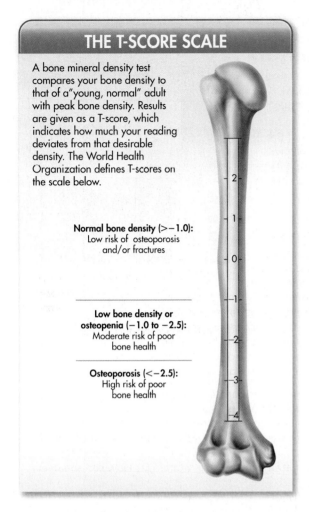

THE T-SCORE SCALE

A bone mineral density test compares your bone density to that of a"young, normal" adult with peak bone density. Results are given as a T-score, which indicates how much your reading deviates from that desirable density. The World Health Organization defines T-scores on the scale below.

Normal bone density (>−1.0):
Low risk of osteoporosis and/or fractures

Low bone density or osteopenia (−1.0 to −2.5):
Moderate risk of poor bone health

Osteoporosis (<−2.5):
High risk of poor bone health

FIGURE 9-19 ▲ The classification for diagnosing osteoporosis is used by the World Health Organization. A T-score of −1.0 or higher is normal; a T-score from −1.0 to −2.5 indicates low bone mass and is known as osteopenia, and a T-score less than −2.5 indicates osteoporosis. Severe osteoporosis is diagnosed with a T-score of less than −2.5 and a personal history of fragility fracture.

TABLE 9-2 ▶ Biological Factors Associated with Bone Status

Biological Factors	Effect on Bone Status
Sex	Women have lower bone mass and density than men.
Age	Bone loss occurs after age 30.
Ethnicity	Individuals of Caucasian or Asian heritage are at greater risk for poor bone health than individuals of African descent.
Frame size	People with "small bones" have a lower bone mass.

Bone Health Assessment. Today, we have tools that can quantitate bone mass and bone density, and, subsequently, the likelihood of a person's developing bone disease. The most accurate test for assessing bone density is the central **dual energy X-ray absorptiometry (DEXA)** measurement of the hip and spine. The central DEXA procedure is simple, painless, safe, noninvasive, and generally takes less than 15 minutes. The hip and spine are measured because these sites are commonly affected by osteoporosis and are likely to result in more serious injuries. The ability of the bone to block the path of a low-level X ray is used as a measure of bone mineral density. A very low dose of radiation is used for the DEXA—about one-tenth of the exposure from a chest X ray.

From the DEXA measurement of bone density, a T-score is generated, which compares the observed bone density to that of a person at peak bone mass (e.g., age 30). The T-score is interpreted as shown in Figure 9-19.

The National Osteoporosis Foundation 2010 recommendations call for DEXA testing for the following groups of people.

- All women age 65 and older and men age 70 and older;
- Younger postmenopausal women and men (ages 50 to 69) who have risk factors;
- Women going through perimenopause (transitioning into menopause) who have low body weight, have prior low-trauma fracture, or take high-risk medications, such as steroids
- Adults with fracture after age 50;
- Adults with a health condition for which they take steroids for a prolonged period (e.g., rheumatoid arthritis, Crohn's disease, asthma);
- Anyone being considered for medication for osteoporosis or receiving therapy for osteoporosis.

Whether or not you have had your bone mineral density tested, you can also use the online Fracture Risk Assessment Tool (FRAX) to estimate your risk of fracture. You can calculate your FRAX score at **http://www.shef.ac.uk/FRAX**. The National Osteoporosis Foundation Guide recommends that doctors consider prescribing medication if your risk of a hip fracture is at least 3% over the next 10 years. The incidence of osteoporosis is far less in many other countries than in the United States. In reviewing Tables 9-2 and 9-3, you can see that the contributions for the greater incidence in the United States are mainly related to diet and other lifestyle behaviors. That means that osteoporosis can largely be prevented.

GETTING ENOUGH CALCIUM

Calcium requires an acidic environment in the gastrointestinal tract to be absorbed efficiently. Absorption occurs primarily in the upper part of the small intestine. This area tends to remain somewhat acidic because it receives the acidic stomach contents. After the first section of the small intestine, secretions from the pancreas enter the small intestine and the pH becomes neutral to slightly basic, causing calcium absorption to decrease. Efficient calcium absorption in the upper small intestine also depends on the presence of the active form of vitamin D (review Chapter 8). Adults absorb about 30% of the calcium in the foods eaten, but during times when the body needs extra calcium, such as in infancy and pregnancy, absorption increases to as high as 60%. Aging negatively influences the absorption efficiency of calcium; due to decreased acid secretion in the stomach and lower synthesis, absorption, and activation of vitamin D, people over age 40 have a harder time meeting their needs for calcium.

TABLE 9-3 ▶ Modifiable Lifestyle Factors Associated with Bone Status

Lifestyle Factors	Call to Action
Adequate diet containing an appropriate amount of nutrients	• Follow MyPlate with special emphasis on adequate amounts of fruits, vegetables, and low-fat and fat-free dairy products. • Consider use of fortified foods (or supplements) to make up for specific nutrient shortfalls, such as vitamin D and calcium.
Healthy body weight	• Maintain a healthy body weight (BMI of 18.5–24.9) to support bone health.
Normal menses	• During childbearing years, seek medical advice if menses cease (such as in cases of anorexia nervosa or extreme athletic training). • Women at menopause and beyond should consider use of current medical therapies to reduce bone loss linked to the fall in estrogen output.
Weight-bearing physical activity	• Perform weight-bearing activity as this contributes to bone maintenance, whereas bed rest and a sedentary lifestyle lead to bone loss. Strength training, especially upper body, is helpful to bone maintenance.
Smoking	• Smoking lowers estrogen synthesis in women. Cessation is advised. Passive exposure is a risk.
Medications	• Some medications (e.g., thyroid hormone, cortisol, and diuretics) stimulate urinary calcium excretion. • Some medications (e.g., alcohol, diuretics, and cancer medications) stimulate urinary excretion of magnesium.
Excessive intake of protein, phosphorus, sodium, caffeine, wheat bran, or alcohol	• Moderate intake of these dietary constituents is recommended. Problems primarily arise when excessive intakes of these nutrients are combined with inadequate calcium consumption. • Excessive soft drink consumption is especially discouraged.
Inadequate UV-B exposure	• If sunlight exposure is limited (<10–15 minutes per day without sunscreen), focus on food or supplements to meet current RDA for vitamin D.

dual energy X-ray absorptiometry (DEXA) A scientific tool used to measure bone mineral density.

The Recommended Dietary Allowance for calcium is 1000 milligrams per day for adults through 50 years of age (see margin). For women older than age 50 and for both men and women over age 70, the RDA increases to 1200 milligrams per day. The RDA is based on the amount of calcium needed each day to offset calcium losses in urine, feces, and other routes. The RDA for young people ages 9 to 18 (1300 milligrams per day) includes an additional amount to allow for increases in bone mass during growth and development.

The calcium intakes of many Americans fall short of meeting the RDA. During late childhood and adolescence—a critical time for accretion of bone mass—many youngsters are choosing sugar- and caffeine-laden beverages instead of milk or dairy alternatives. This calcium deficit sets the stage for osteoporosis later in life. Into adulthood, average daily calcium intakes are approximately 800 milligrams for women and 1000 milligrams for men. Approximately half of adult women in the United States consume less than 60% of the recommended intake of calcium (Fig. 9-20). Some of this is due to their perception that dairy products are high in calories (although many reduced-fat dairy products are available). A growing number of Americans are choosing plant-based diets, including a vegan diet, which eliminates dairy products. Other adults simply lose their taste for milk as they age. In addition, lactose intolerance becomes more prevalent as people age.

Calcium is found in both plant and animal foods. Overall, dairy products provide about 75% of the calcium in North American diets. In fact, fat-free milk is the most nutrient-dense (milligrams per kcal) source of this bone-building nutrient. Calcium tends to be highly bioavailable from dairy products because they contain vitamin D and lactose, which enhance calcium absorption. An exception is cottage cheese; much of its calcium is bound and unavailable for absorption.

Bread, rolls, crackers, and foods made with milk products also make a significant contribution to North Americans' calcium intake. Other sources are leafy greens (such as kale), almonds, some legumes, sardines, and canned salmon (Fig. 9-21).

Through the production of more products fortified with calcium, food and beverage companies are responding to consumers' desire to increase consumption of

▲ Weight-bearing exercise such as walking or running is associated with increased bone density. Female athletes, however, must maintain an adequate energy intake to maintain estrogen levels, which stimulate bone formation.

Calcium
RDA: 1000 milligrams
DV: 1000 milligrams
UL: 2500 milligrams

FIGURE 9-20 ▶ The gap between recommended dietary intake of calcium and daily consumption of calcium in males and females in the United States. Female calcium intake is low throughout the life span, whereas males' intakes drop below recommendations after age 50.

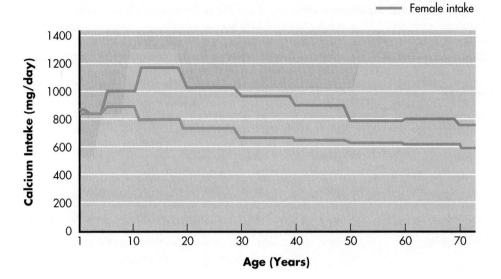

FIGURE 9-21 ◀ Food sources of calcium. (a) The fill of the background color (none, 1/3, 2/3, or completely covered) within each food group on MyPlate indicates the average nutrient density for calcium in that group. (b) The bar graph shows the calcium content of several foods in each food group compared to the RDA for adult males and females. Overall, the richest sources of calcium are dairy foods (and dairy alternatives), legumes, green leafy vegetables, and fortified foods. *Nutrition data from USDA National Nutrient Database for Standard Reference, Release 26.*

(a)

(b)

Food Item and Amount	Calcium (milligrams)	% RDA for Adult Males and Females (1000 milligrams)
Grains		
Total Raisin Bran cereal, 1 cup	1000	100%
English muffin, whole-wheat, 1 each	175	18%
Pancake, 4"	83	8%
Vegetables		
Spinach, cooked, 1 cup	245	25%
Butternut squash, cooked, 1 cup	84	8%
Broccoli, cooked, 1 cup	62	6%
Fruits		
Orange juice (fortified), 1 cup	349	35%
Figs, dried, ½ cup	121	12%
Raisins, ½ cup	36	4%
Dairy		
Yogurt, plain, non-fat, 1 cup	488	49%
Milk, fat-free, 1 cup	299	30%
Soymilk, (fortified), 1 cup	299	30%
Protein		
Tofu (with calcium sulfate), 3 ounces	574	57%
Salmon, canned with bones, 3 ounces	203	20%
Almonds, dry roasted, 1 ounce	76	8%

calcium. Calcium-fortified foods such as orange juice, breakfast cereals, breakfast bars, waffles, and soy products provide considerable amounts of calcium. In fact, an 8-ounce glass of calcium-fortified orange juice can provide up to 350 milligrams of calcium, whereas a serving of milk has 300. Another source of calcium is soybean curd (tofu) if it is made with calcium (check the label). It is easy to assess the calcium content of foods because it is among those nutrients that are required to be listed on the Nutrition Facts panel on food labels. The DV for calcium used for food and supplement labels is 1000 milligrams.

To estimate your calcium intake, use the rule of 300s. Count 300 milligrams for the calcium provided by foods scattered throughout the diet. Add another 300 milligrams to that for every cup of milk or yogurt or 1.5 ounces of cheese. If you eat a lot of tofu, almonds, or sardines or drink calcium-fortified beverages, use Figure 9-21 or diet-analysis software to get a more accurate calculation of calcium intake.

Keep in mind, however, that many factors can influence the bioavailability of calcium. Calcium absorption can be reduced by the presence of oxalates, tannins, and phytic acid. These compounds chelate (chemically bind) calcium in the digestive tract. Oxalates are found in sweet potatoes, collard greens, spinach, and rhubarb. It is estimated that a person would have to consume over eight servings (8 cups) of spinach to absorb the same amount of calcium present in one serving (1 cup) of milk! Oxalates bind only the calcium in the food they are in; oxalate-containing foods do not affect the calcium availability from other foods. This does not hold true for phytates or tannins. Tea and some legumes are rich sources of tannins. Phytates are found in whole grains, raw beans, and nuts. Diets high in dietary fiber reduce mineral absorption. With this in mind, vegans should take extra care to include good sources of calcium in their diets.

AVOIDING TOO MUCH CALCIUM

The Upper Level (UL) for calcium intake is 2500 milligrams per day for young adults, based on the observation that greater intakes increase the risk for some forms of kidney stones. Excessive calcium intakes by some people can also cause high blood and urinary calcium concentrations, irritability, headache, kidney failure, soft tissue calcification, and decreased absorption of other minerals, as noted previously.

Dairy Alternatives. Many dairy alternatives are available in the grocery store today. These are products that look, feel, and taste like typical dairy foods but are not made from the milk of animals. Most of the milk alternatives are healthy and delicious, but not exact nutritional replicas of cow's milk (Table 9-4). *Soy milk* contains the highest quantity of protein of all the dairy alternatives with 6 to 10 grams per cup and also the best quality in that it contains all of the essential amino acids. It also has about the same proportion of protein (3.5%), fat (2%), and carbohydrate (3%) as cow's milk. It contains omega-3 fatty acids, fiber, magnesium, and manganese, and most brands are fortified with calcium, riboflavin, and vitamins A, D, and B-12. Most brands contain

Many factors *enhance* calcium absorption, including:
- High need (e.g., during growth, pregnancy, lactation)
- Increased blood levels of **parathyroid hormone** and vitamin D
- The presence of lactose in the diet
- Decreased flow rate (motility) of digestive contents through the intestine
- Acidic environment of the stomach

Many factors *inhibit* calcium absorption, including:
- Large amounts of phytic acid and fiber from grains (if fiber intake exceeds 30 grams per day)
- Oxalates found in specific foods— but only calcium in the food itself; they do not inhibit calcium absorption from other foods.
- Great excess of phosphorus, magnesium, sodium, and zinc in the diet if calcium intake is very low
- Tannins (polyphenols) in tea and some legumes (e.g., soy)
- A vitamin D deficiency
- Diarrhea
- Old age
- Some medications (anticonvulsants, cortisone, antacids)

▲ Calcium is found in plant sources such as green leafy vegetables but is much more bioavailable from dairy foods.

Determining the amount of calcium in food

Convert the percent Daily Value of calcium on food labels to milligrams by adding a "0" to the percentage. For example, if a food label states a serving supplies 30% of the DV for calcium, convert this into milligrams by adding a 0, so 30% is the same as 300 milligrams.

TABLE 9-4 ▶ Comparing the Nutrient Content (per cup) of Dairy Alternatives to Cow's Milk

Nutrients	Whole Milk	Soy Milk	Rice Milk	Almond Milk	Coconut Milk	Flax Milk
Calories	149	90	120	60	90	50
Fat (grams)	7.7	3.5	2.5	2.5	5	2.5
Protein (grams)	8	6	1	1	1	0
Calcium (% Daily Value)	28	45	30	45	45	30

▲ Almonds are a natural source of calcium. One ounce contains 80 milligrams.

▲ Freshly pressed coconut milk is loaded with calories because of its high fat content and contains very little calcium and vitamin D. Coconut milk beverages sold alongside cow's milk have been altered to reduce fat content and increase bone-building nutrients.

▲ Chickpeas (garbanzo beans) are a good plant source of calcium at 80 milligrams per cup. A cup of hummus made from chickpeas provides 93.5 milligrams of calcium.

450 milligrams of calcium (45% DV). Soy milk is a popular product for vegans because it is plant based and for people with lactose intolerance because it does not contain lactose. *Rice milk* is naturally sweeter than cow's milk, much higher in carbohydrates (24 grams per cup), and has significantly less protein (1 gram per cup) and no lactose. Most brands are fortified with calcium, iron, riboflavin, and vitamins A, D, and B-12 and contain about 300 milligrams of calcium per cup (30% DV). *Almond milk* naturally contains a significant amount of calcium and vitamin D and is an excellent source of the antioxidant vitamin E. A cup of almond milk contains about 200 mg of calcium (20% DV), 100 IU of vitamin D (25% DV), and 10 mg of vitamin E (50% DV). Almond milk is much lower in calories than cow's milk and contains very little protein (1 gram per cup), carbohydrate (2 grams per cup), fat (3 grams per cup), and fiber (1 gram per cup). Because almond milk is low in fat and calories and contains some essential vitamins and minerals, it can be used as a milk substitute for those who would like to lose weight. Many find the taste of almond milk more acceptable compared to other dairy substitutes. *Coconut milk* has a high oil content and therefore is much higher in calories, fat, and saturated fat than other milks and milk substitutes. A 1-cup serving of canned coconut milk contains 445 calories and 48 grams of fat, of which 43 grams are saturated fat. It contains very little calcium or vitamin D but does contain about 40% DV of iron. There are also coconut milk beverages available that compare closely to soy, rice, and almond milks. These beverages are a mixture of coconut cream and water, and per cup have only 5 grams of fat, 1 gram of protein, 80 calories, and 450 milligrams of calcium (45% DV). *Flax milk* is cold-pressed flax oil mixed with filtered water. One cup has 50 calories, 1200 mg of omega-3s, and 30% DV of calcium but no protein.

In summary, many dairy alternatives are available. Not all of these substitutes, however, have the same nutritional profiles as the original milk-based product. Thus, it is very important to know what nutrients you are looking for and then to read labels carefully to compare products.

Calcium Supplementation. There has been much debate over the effectiveness of calcium and vitamin D supplementation for maintaining bone health. Recent studies have ignited controversy over the potential health risks of consuming large amounts of these nutrients, especially for older adults. Although achieving the RDAs for calcium and vitamin D through dietary modifications has been found to improve bone mineral density and reduce rate of fractures, use of supplements to meet RDAs may actually impair skeletal health and worsen cardiovascular health. A recent study found that a dietary intake of calcium near the current recommendations was not related to higher bone mineral density of the hip or lumbar spine compared with lower intakes of calcium in elderly men and women in the United States. In addition, calcium intake beyond the RDA for elderly women and men, usually achieved by calcium supplements, did not provide any benefit for hip or lumbar spine bone mineral density in older adults (see Further Reading 2). Other large studies, including the large Women's Health Initiative trial, have also shown a small but statistically significant increase in the rate of heart attacks among older adults taking calcium supplements with or without vitamin D (see Further Reading 4). Although the observed relationships between calcium supplements and heart attack risk are small, the regular use of calcium supplements by so many older adults could translate into a large public health problem.

On the positive side, a more recent report from the Women's Health Initiative study indicates that long-term use of a daily calcium and vitamin D supplement that is close to the RDA results in a substantial reduction in the risk of hip fracture among postmenopausal women. These authors also reported that the level of calcium and vitamin D supplementation did not result in an increase in other chronic diseases, including heart disease. The more positive effects of supplementation appear to happen when the level of total calcium and vitamin D intake is kept very close to the RDA (see Newsworthy Nutrition on the next page). Therefore, taking 1000 milligrams of calcium carbonate or calcium citrate daily as a supplement in divided doses (about 500 milligrams per tablet) is probably safe in many instances.

So which is better: calcium from food or supplements? The National Osteoporosis Foundation continues to encourage all individuals to consume the recommended amounts of calcium and vitamin D to protect bone health. Experts agree that we should strive to meet our calcium and vitamin D needs from foods first and that more research is needed to better comprehend the benefits and risks associated with calcium and vitamin D supplementation. Modification of eating habits to include foods that are good sources of calcium is a better plan of action and appears to be the safest means to prevent osteoporosis without jeopardizing heart health. In addition to this important mineral, foods that contain calcium also supply other vitamins, minerals, phytochemicals, and fats needed to support health. Problems associated with excessive consumption of calcium, such as constipation, are not likely when foods are the primary sources of calcium.

In an effort to provide guidance for the public, the U.S. Preventive Services Task Force recently reviewed current research studies on the use of vitamin D and calcium supplements to prevent fractures and issued recommendations in its *Vitamin D and Calcium Supplementation to Prevent Fractures* report in February 2013. The recommendations apply to adult men and women who live at home. They do not apply to those living in assisted living or skilled nursing facilities or who have been diagnosed with osteoporosis or vitamin D deficiency. The Task Force conclusions are listed in the margin and are based on what is known about the potential benefits and harms of using these supplements.

Vitamin D and Calcium Supplementation to Prevent Fractures **Task Force Conclusions:**

(1) There is not enough evidence to determine whether vitamin D and calcium supplements can prevent fractures in men and in women who have not yet gone through menopause.

(2) There is not enough evidence to determine whether vitamin D and calcium supplements with greater than 400 IU of vitamin D and greater than 1000 mg of calcium can prevent fractures in noninstitutionalized postmenopausal women.

(3) There is enough evidence to recommend against daily supplementation with 400 IU or less of vitamin D and 1000 milligrams or less of calcium for the primary prevention of fractures in noninstitutionalized postmenopausal women because lower doses of vitamin D and calcium supplements do not prevent fractures in older women and may increase the risk of kidney stones.

The full Task Force statement can be viewed at **http://www.uspreventiveservicestaskforce.org/uspstf12-vitamind/vitdfact.pdf.**

Newsworthy Nutrition

Calcium supplements decrease risk of hip fracture in women

The Women's Health Initiative (WHI) clinical trial randomly assigned 36,282 postmenopausal women in the United States to a placebo or 1000 milligram supplement of calcium carbonate plus 400 IU of vitamin D daily for 7.0 years. The hypothesis of this study was that calcium plus vitamin D supplementation would reduce hip fracture. This study also examined the health benefits and risks of calcium and vitamin D supplementation on total fractures, cardiovascular disease, cancer, and total mortality. Women in both groups were also allowed to take personal calcium and vitamin D supplements during the study, which was accounted for during the analysis of the data. Regarding bone health, the women who took the assigned calcium and vitamin D supplement but took no personal calcium supplements had a 75 percent lower risk of hip fracture than women who got the placebo and took no personal calcium supplements. The results also showed that women assigned to take calcium and vitamin D had no higher risk of heart disease, heart attacks, stroke, colorectal cancer, or total mortality. The supplement group did have a 17 percent increased risk of kidney stones. The authors concluded that long-term use of calcium and vitamin D appears to confer a substantial reduction in the risk of hip fracture among postmenopausal women. Because the risk reduction was significant only in women not taking any additional calcium supplements, it is recommended that women strive to consume the RDA for calcium (1200 milligrams per day) and vitamin D (600 IU/per day up to age 70 and 800 IU per day over 70) from food and supplements combined.

Source: RL Prentice, "Health risks and benefits from calcium and vitamin D supplementation: Women's Health Initiative clinical trial and cohort study," Osteoporosis International, 24, 2, 2013, 567. Copyright © 2013 Springer-Verlag London LTD. All rights reserved. Used with permission.

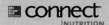 **connect** |NUTRITION   Check out Connect at **www.mcgrawhillconnect.com** to further explore calcium and vitamin D supplements.

TABLE 9-5 ▶ **Calcium Supplement Comparisons**

Supplement form
Calcium carbonate (40% calcium) • Forms: tablets, chewable tablets, soft chews • Most common form • Least expensive • Needs acid environment in stomach, so take with acid food or take with meals
Calcium citrate (21% calcium) • Forms: pills, liquid • Best absorbed • Most expensive • *Does not* need acid environment to be absorbed • Pills can be quite large • Liquid (colloidal) form sometimes easier to tolerate

Keeping in mind the recommendations just discussed, increasing calcium intake through the use of a calcium supplement is beneficial if you have a milk allergy; do not like milk; are ovovegetarian, vegan, or lactose intolerant; or cannot incorporate enough calcium-containing foods into your diet (see Further Reading 13). Always look for a supplement with added vitamin D, as it enhances calcium uptake. This additional vitamin D typically does not add to the cost of the supplement. Table 9-5 compares the two most common forms of calcium supplements. Calcium carbonate should be taken with meals because it requires an acid environment in the stomach to dissolve and maximize calcium absorption. Calcium citrate is indicated for people who cannot remember to take calcium carbonate with meals and for those who have low-acid stomach conditions, such as people who take acid-reducing medications for ulcers or reflux, or have had surgery for obesity reduction.

Calcium supplements have side effects, including gas, bloating, or constipation. Distributing small-dose supplements throughout the day, taking it with meals, or even changing the brand of supplement may alleviate some problems. Intake of calcium from supplements and/or food above 500 milligrams at any one time significantly reduces the percent absorbed.

With calcium supplements, interactions with other minerals are a concern. There is evidence that calcium supplements may decrease zinc, iron, and other mineral absorption. An effect of calcium supplementation on iron absorption is possible; however, this appears to be small over the long term. To be safe, people using a calcium supplement on a regular basis should notify their physician of the practice. Calcium supplements can also interfere with the body's ability to absorb certain antibiotics. If your doctor prescribes antibiotics, especially tetracycline, be sure to talk with your pharmacist about timing of your supplement, medication, and meals.

Tablet or liquid calcium supplements with the United States Pharmacopeia (USP) symbol are considered the safest. FDA has cautioned the public on the use of calcium supplements from dolomite, bone meal, coral, or oyster shell because of the potential for unhealthy levels of environmental contaminants.

✔ **CONCEPT CHECK 9.6**

1. What percent of calcium in the body is found in bone and teeth?
2. What are the two types of osteoporosis and how do they differ?
3. What type of bone is most affected by osteoporosis?
4. Why is the achievement of peak bone mass as a young adult so important in preventing osteoporosis?
5. What are some current treatments for osteoporosis?
6. What is the most accurate test for bone density and how is it done?
7. According to the National Osteoporosis Foundation, who should have their bone density checked?
8. Beyond its role in bone health, what are some other critical functions of calcium?
9. What role does vitamin D play in calcium metabolism?
10. What factors reduce calcium absorption?

9.7 Phosphorus (P)

FUNCTIONS OF PHOSPHORUS

Phosphorus is the second most abundant mineral in the body. Approximately 85% of phosphorus is found as a component of hydroxyapatite crystals that provide the functional component of bone and teeth. The remaining 15% of phosphorus is in the soft tissues, blood, and extracellular fluid. Phosphorus is part of DNA and RNA, the genetic material present in every cell. Therefore, phosphorus is critical for cellular replication and growth because DNA and RNA are responsible for mitosis and protein synthesis. Phosphorus is also a primary component of adenosine triphosphate (ATP), the energy molecule that fuels body functions. Phosphorus is essential for the activation and deactivation of many enzymes and many of the B vitamins are functional only when a phosphate group is attached.

A major class of lipids contains phosphorus: phospholipids are the principal structural component of cell membranes, making up approximately 60% of membranes. These phospholipid membranes regulate the transport of nutrients and waste products into and out of cells. Phosphorus also serves as a buffer to maintain blood pH. Lastly, phosphorus (in the form of the phosphate ion) is the principal negatively charged ion in intracellular fluid and thus is essential for maintenance of fluid balance.

PHOSPHORUS DEFICIENCY

Phosphorus deficiency is uncommon because the mineral is widespread in our food supply. However, if starvation (e.g., anorexia nervosa), hormonal imbalance, or medications lead to low levels of phosphate in the blood, bones can become demineralized. Other symptoms of low phosphate levels include poor appetite, weakness, and difficulty walking.

Marginal phosphorus status can be found in preterm infants, vegans, people with alcoholism, older people on nutrient-poor diets, and people with long-term bouts of diarrhea. A 2004 review of older women undergoing osteoporosis treatment identified a small but significant group at high risk for phosphorus deficiency. These women, estimated to comprise 10% to 15% of all women ages 60 years and older, tend to live alone and eat a poor-quality diet that is below 70% of the RDA for phosphorus. To control their osteoporosis, these women are taking calcium supplements and antiresorptive medications to inhibit further bone loss. To maintain bone hydroxyapatite, however, both calcium and phosphorus must be available. High calcium carbonate or citrate supplements can bind phosphorus in the intestine, thus inhibiting its absorption. These women are at high risk for phosphorus deficiency, as their intake is low and their absorption is inhibited. Attempts must be made to enhance dietary intake or calcium supplements should be switched to calcium phosphate, all under close medical supervision.

GETTING ENOUGH PHOSPHORUS

In contrast to calcium, phosphorus is naturally abundant in many foods. Milk, cheese, meat, and bread provide most of the phosphorus in the adult diet. Nuts, fish, breakfast cereals, bran, and eggs are also good sources (Fig. 9-22). About 20% to 30% of dietary phosphorus comes from food additives, especially in baked goods, cheeses, processed meats, and many soft drinks (about 75 milligrams per 12 ounces). As a food additive, phosphorus is considered a GRAS (generally recognized as safe) substance, and its function is to increase water binding and taste. Phosphoric acid, which gives a tangy, sour taste, will also significantly lower the pH of a food or beverage (pH of a soft drink is less than 3). Absorption of phosphorus is generally high, ranging from 55% to 80%. Phosphorus absorption from grains, however, is reduced because of the high phytic acid content. Vitamin D enhances phosphorus absorption.

The RDA for phosphorus is 700 milligrams for adult men and women (see margin). The recommendation is higher (1250 milligrams per day) for young people ages 9 to 18 to support growth and development. Average daily adult consumption is about 1000 to 1600 milligrams. As a general rule, deficiencies of phosphorus are unlikely in healthy adults, especially because it is so efficiently absorbed.

Phosphorus
RDA: 700 milligrams
DV: 1000 milligrams
UL: 4000 milligrams

FIGURE 9-22 ◄ Food sources of phosphorus. (a) The fill of the background color (none, 1/3, 2/3, or completely covered) within each food group on MyPlate indicates the average nutrient density for phosphorus in that group. (b) The bar graph shows the phosphorus content of several foods compared to the RDA for adult males and females. Overall, the richest sources of phosphorus are dairy products and protein foods. Foods from the fruits group (not shown) are poor sources of phosphorus. *Nutrition data from USDA National Nutrient Database for Standard Reference, Release 26.*

(a)

(b)

Food Item and Amount	Phosphorus (milligrams)	% RDA for Adult Males and Females (700 milligrams)
Grains Trail mix (with chocolate, nuts, seeds), ½ cup	283	40%
Raisin bran cereal, 1 cup	205	29%
Oatmeal, prepared with water, ½ cup	90	13%
Vegetables Potato, baked (with skin), 1 medium	121	17%
Asparagus, cooked, 1 cup	97	14%
Mushrooms, white, raw, 1 cup	60	9%
Dairy Yogurt, plain, non-fat, 1 cup	385	55%
Milk, fat-free, 1 cup	247	35%
Soymilk, 1 cup	104	15%
Protein Salmon, baked, 3 ounces	218	31%
Great northern beans, cooked, ½ cup	146	21%
Almonds, dry roasted, 1 ounce	134	19%

AVOIDING TOO MUCH PHOSPHORUS

The UL for phosphorus intake is 3 to 4 grams per day. Intakes greater than this can result in mineralization of soft tissues. Phosphorus levels in the blood are regulated primarily by the kidneys, and these organs are particularly sensitive to phosphorus toxicity. High intakes can lead to serious problems in people with certain kidney diseases. In addition, a high phosphorus intake coupled with a low calcium intake can cause a chronic imbalance in the calcium-to-phosphorus ratio in the diet and contribute to bone loss. This situation most likely arises when the RDA for calcium is not met, as can occur in adolescents and adults who regularly substitute soft drinks for milk or otherwise underconsume calcium.

✓ CONCEPT CHECK 9.7

1. What are the key functions of phosphorus beyond bone health?
2. Is phosphorus deficiency very common, or is the RDA met by most North Americans?
3. What are the primary food sources of phosphorus?
4. What are the risks of excess intake of phosphorus?
5. What effect does vitamin D have on phosphorus absorption?

9.8 Magnesium (Mg)

FUNCTIONS OF MAGNESIUM

Magnesium is important for nerve and heart function and aids in many enzyme reactions. Magnesium is similar to calcium and phosphorus in that most of the magnesium in the body is found in bones. Bone contains 60% of the body's magnesium. Magnesium serves a structural role in bones to help provide rigidity, and it functions as a storage site drawn upon by other tissues when dietary intake is inadequate to meet the body's needs. Magnesium functions to relax muscles after contraction. It promotes resistance to tooth decay by stabilizing calcium in tooth enamel. Over 300 enzymes use magnesium, and many energy-yielding compounds in cells require magnesium to function properly (e.g., ATP). Magnesium plays a critical role in the synthesis of DNA and protein. Another notable function of magnesium as it relates to bone health is that magnesium is required for the synthesis of vitamin D in the liver.

Other possible benefits of magnesium in relation to cardiovascular disease include decreasing blood pressure by dilating arteries and preventing heart abnormalities. People with cardiovascular disease should closely monitor magnesium intake, especially because they are often on medications such as diuretics that reduce magnesium status. A diet rich in food sources of magnesium and calcium is associated with lower risk of type 2 diabetes in some populations.

MAGNESIUM DEFICIENCY

In humans, low blood magnesium causes an irregular heartbeat, sometimes accompanied by weakness, muscle pain, disorientation, and seizures. In terms of bone health, low magnesium disrupts the hormonal regulation of blood calcium by parathyroid hormone and affects the activity of vitamin D. You might expect that magnesium deficiency would result in diminished bone mass, but to date, this has only been observed in animals. (There is some evidence, however, that magnesium supplementation may improve bone density in postmenopausal women.)

Magnesium deficiency develops very slowly. Not only is the mineral present in foods of both plant and animal origin but the kidneys also are very efficient at retaining magnesium. Thus, it is uncommon for healthy people to suffer a clinically relevant magnesium deficiency due to dietary inadequacy alone. Poor magnesium status is most commonly found among people with abnormal kidney function, whether as a result of kidney disease or as a side effect of certain diuretics. Alcohol use disorders also can increase the risk of deficiency because dietary intake may be poor and because alcohol increases magnesium excretion in the urine. The disorientation and weakness associated with alcohol use disorders closely resemble the behavior of people with low blood magnesium. In addition, people with malabsorptive diseases (e.g., Crohn's disease), heavy perspiration, or prolonged bouts of diarrhea or vomiting are susceptible to low blood levels of magnesium.

GETTING ENOUGH MAGNESIUM

Magnesium is found in the plant pigment chlorophyll, so rich sources for magnesium are plant products, such as squash, whole grains (such as wheat bran), beans, nuts, seeds, and broccoli (Fig. 9-23). Animal products (e.g., milk and meats) and chocolate supply some magnesium, although not as much as foods of plant origin. Two other sources of magnesium are hard tap water, which contains a high mineral content, and coffee (espresso, not brewed).

The adult RDA for magnesium (see margin) is based on the amount needed to offset daily losses. Adult men consume on average 320 milligrams daily, whereas women consume closer to 220 milligrams daily, suggesting that many of us should improve our intakes of magnesium-rich foods, such as whole-grain breads and

Magnesium
RDA
  Adult men: 400 milligrams
  Adult women: 310 milligrams
DV: 400 milligrams
UL: 350 milligrams (nonfood sources only)

FIGURE 9-23 ◄ Food sources of magnesium. (a) The fill of the background color (none, 1/3, 2/3, or completely covered) within each food group on MyPlate indicates the average nutrient density for magnesium in that group. (b) The bar graph shows the magnesium content of several foods in each food group compared to the RDA for adult males and females. Overall, the richest sources of magnesium are vegetables and whole grains. *Nutrition data from USDA National Nutrient Database for Standard Reference, Release 26.*

(a)

(b)

	Food Item and Amount	Magnesium (milligrams)	% RDA for Adult Males (400 milligrams)	% RDA for Adult Females (310 milligrams)
Grains	Raisin bran cereal, 1 cup	72	18%	23%
Grains	Whole-wheat bread, 1 slice	24	6%	8%
Grains	White bread, 1 slice	7	2%	2%
Vegetables	Spinach, cooked, 1 cup	157	39%	51%
Vegetables	Okra, cooked, 1 cup	58	15%	19%
Vegetables	Sweet potato, baked, 1 cup	54	14%	17%
Fruits	Banana, 1 medium	32	8%	10%
Fruits	Blackberries, 1 cup	29	7%	9%
Fruits	Pineapple, 1 cup	20	5%	6%
Dairy	Yogurt, plain, non-fat, 1 cup	47	12%	15%
Dairy	Soymilk, 1 cup	36	9%	12%
Dairy	Milk, fat-free, 1 cup	27	7%	9%
Protein	Peanut butter, 2 tablespoons	49	12%	16%
Protein	Navy beans, cooked, ½ cup	48	12%	15%
Protein	Chicken breast, roasted, 3 ounces	25	6%	8%

cereals. The refined grain products that dominate the diets of many North Americans are poor sources of this mineral, as refining reduces the magnesium content by as much as 80%. This low value also reflects poor intake of green and other brightly colored vegetables. If dietary intake of magnesium is inadequate, a balanced multivitamin and mineral supplement containing approximately 100 milligrams of magnesium can help close the gap between intake and needs.

Nutrient-nutrient interactions can reduce magnesium absorption. Diets very high in phosphorus or fiber (phytate) limit intestinal absorption, as do diets too low in protein.

AVOIDING TOO MUCH MAGNESIUM

The UL for magnesium intake is 350 milligrams per day, based on the risk of developing diarrhea. This guideline refers only to nonfood sources such as antacids, laxatives, or supplements (see Further Reading 13). Food sources are not known to cause toxicity. Magnesium toxicity especially occurs in people who have kidney failure or who overuse over-the-counter medications that contain magnesium, such as certain antacids and laxatives (e.g., milk of magnesia). Older people are at particular risk, as kidney function may be compromised.

Table 9-6 summarizes much of what we have covered regarding the major minerals.

TABLE 9-6 Summary of the Major Minerals

Mineral	Major Functions	RDA, or AI	Dietary Sources	Deficiency Symptoms	Toxicity Symptoms
Sodium	• Major positive ion of the extracellular fluid • Aids nerve impulse transmission • Water balance	*Age 19–50 years:* 1500 milligrams *Age 51–70 years:* 1300 milligrams *Age > 70 years:* 1200 milligrams	• Table salt • Processed foods • Condiments • Sauces • Soups • Chips	• Muscle cramps	• Contributes to hypertension in susceptible individuals • Increases calcium loss in urine Upper Level is 2300 milligrams.
Potassium	• Major positive ion of intracellular fluid • Aids nerve impulse transmission • Water balance	4700 milligrams	• Spinach • Squash • Bananas • Orange juice • Milk • Meat • Legumes • Whole grains	• Irregular heartbeat • Loss of appetite • Muscle cramps	• Slowing of the heartbeat, as seen in kidney failure
Chloride	• Major negative ion of extracellular fluid • Participates in acid production in stomach • Aids nerve impulse transmission • Water balance	2300 milligrams	• Table salt • Some vegetables • Processed foods	• Convulsions in infants	• Linked to hypertension in susceptible people when combined with sodium Upper Level is 3600 milligrams.
Calcium	• Bone and tooth structure • Blood clotting • Aids in nerve impulse transmission • Muscle contractions • Other cell functions	*Age 9–18 years:* 1300 milligrams *Age > 18 years:* 1000–1200 milligrams	• Dairy products • Canned fish • Leafy vegetables • Tofu • Fortified orange juice (and other fortified foods)	• Increased risk of osteoporosis	• May cause kidney stones and other problems in susceptible people Upper Level is 2500 milligrams.
Phosphorus	• Major negative ion of intracellular fluid • Bone and tooth strength • Part of various metabolic compounds • Acid/base balance	*Age 9–18 years:* 1250 milligrams *Age >18 years:* 700 milligrams	• Dairy products • Processed foods • Fish • Soft drinks • Bakery products • Meats	• Possibility of poor bone maintenance	• Impairs bone health in people with kidney failure • Poor bone mineralization if calcium intakes are low Upper Level is 3 to 4 grams.
Magnesium	• Bone formation • Aids enzyme function • Aids nerve and heart function	*Men:* 400–420 milligrams *Women:* 310–320 milligrams	• Wheat bran • Green vegetables • Nuts • Chocolate • Legumes	• Weakness • Muscle pain • Poor heart function	• Causes diarrhea and weakness in people with kidney failure Upper Level is 350 milligrams but refers to nonfood sources (e.g., supplements) only.

1. What are the key functions of magnesium other than bone health?
2. What are the primary food sources of magnesium?
3. Who is at greatest risk of developing a magnesium deficiency?
4. When is magnesium toxicity most likely to occur?

9.9 Iron (Fe)

Iron is the trace mineral present in the largest amount in the body (Fig. 9-11). Although the importance of dietary iron has been recognized for many years, iron deficiency is the most common nutrient deficiency worldwide. About 30% of the world's population is anemic; half of these cases are caused by iron deficiency. Iron is the only nutrient for which young women have a greater RDA than do adult men.

FUNCTIONS OF IRON

Iron is part of the hemoglobin in red blood cells and myoglobin in muscle cells. Hemoglobin molecules in red blood cells transport oxygen (O_2) from the lungs to cells and then transport carbon dioxide (CO_2) from cells to the lungs for excretion. In addition, iron is used as part of many enzymes, some proteins, and compounds that cells use in energy production. Iron also is needed for brain and immune function, drug detoxification in the liver, and synthesis of collagen for bone health.

IRON DEFICIENCY

hematocrit The percentage of blood made up of red blood cells.

If neither the diet nor body stores can supply the iron needed for hemoglobin synthesis, the concentration of hemoglobin in red blood cells decreases. Medical professionals use both the percentage of blood that consists of red blood cells **(hematocrit)** and the hemoglobin concentration to assess iron status. Other measures indicative of poor iron status include the concentration of iron and iron-containing proteins in blood (serum iron, ferritin, or transferrin).

When hematocrit and hemoglobin fall, an iron deficiency is suspected. In severe deficiency, hemoglobin and hematocrit fall so low that the amount of oxygen carried in the bloodstream is decreased. This condition is called iron-deficiency anemia.

Iron deficiency can be categorized into three stages:

- **Stage 1:** Iron stores become depleted, but no physiological impairment is observed.
- **Stage 2:** The amount of iron in transferrin is depleted; some physiological impairment occurs. Heme production is decreased, and activities of enzymes that require iron as a cofactor are limited.
- **Stage 3 (iron-deficiency anemia):** Red blood cells are small (microcytic), pale (hypochromic), and reduced in number; oxygen-carrying capacity of red blood cells declines.

Clinical symptoms of iron-deficiency anemia are associated with the lack of oxygen getting to the tissues. One experiences pale skin, fatigue upon exertion, poor temperature regulation (always cold, especially toes and fingers), loss of appetite, and apathy. Poor iron stores may decrease learning ability, attention span, work performance, and immune status even before a person is anemic. Children with chronic anemia have abnormal cognitive development (see Further Reading 21).

It is important to note that many more North Americans have an iron deficiency without anemia (stages 1 or 2) than have iron-deficiency anemia (stage 3). Their blood hemoglobin values are still normal, but they have no stores to draw from in times of pregnancy or illness, and basic functioning may be marginally impaired.

▲ In the United States, minority populations are more likely than whites to have iron-deficiency anemia. Compared to non-Hispanic white women, iron-deficiency anemia is at least two times more common among Hispanic and black women (see Further Reading 12).

That could mean anything from too little energy to perform everyday tasks in an efficient manner to difficulties staying mentally alert.

There are many conditions that lead to an anemic state; iron-deficiency anemia is the most prevalent worldwide. Probably about 10% of North Americans in high risk categories have iron-deficiency anemia. This appears most often in infancy, the preschool years, and at puberty for both males and females. Growth—with accompanying expansion of blood volume and muscle mass—increases iron needs, making it difficult to consume enough iron. Women are vulnerable to anemia during childbearing years due to menstrual blood loss. Anemia is also found in pregnant women because blood volume expands during pregnancy and extra iron is needed to synthesize red blood cells for the mother and the fetus. Iron-deficiency anemia in adult men is usually caused by blood loss from ulcers, colon cancer, or hemorrhoids. Athletes can have increased iron requirements due to increased blood loss in feces and urine and chronic lysis of red blood cells in the feet due to the trauma of running.

To cure iron-deficiency anemia, a person needs to take iron supplements (see Further Reading 14). A physician should also find the cause so that the anemia does not reoccur. Changes in diet may *prevent* iron-deficiency anemia, but supplemental iron is the only reliable *cure* once it has developed. Supplements must be taken for 3 to 6 months or perhaps longer. Hemoglobin levels respond quickly to dietary changes and supplementation, but terminating supplements too soon means that iron stores (blood, bone marrow, etc.) will not be replenished. Remember, it takes longer than 1 month to become anemic, so it will take longer than 1 month to cure it.

ABSORPTION AND DISTRIBUTION OF IRON

Overall, iron absorption depends on the following factors: (1) the person's iron status, (2) its form in food, (3) the acidity of the GI tract, and (4) other dietary components consumed with iron-containing foods. Controlling iron levels in the body is important because, when it comes to iron, there is a narrow gap between just enough and too much. As you've learned, too little iron can impair oxygen transport. To prevent deficiency, the human body highly conserves iron. Except for bleeding associated with menstruation, injury, or childbirth, body loss of iron is minimal. Approximately 90% is recovered and reused every day. On the other hand, too much iron in the body is also extremely damaging. It can accumulate in organs and promote oxidative damage. To avoid toxicity, iron absorption from the small intestine is tightly regulated.

The most important factor influencing iron absorption is body need. Iron needs are increased during pregnancy and growth. At high altitudes, the lower oxygen concentration of the air causes an increase in the hemoglobin concentration of blood and thus an increase in iron needs.

The principal mechanism to regulate iron content in the body is tight control of absorption. High doses of iron can still be toxic, but absorption is carefully regulated under most conditions. When iron stores are inadequate or needs are high due to growth or pregnancy, the main protein that carries iron (transferrin) more readily binds iron, shifting it from intestinal cells into the bloodstream. Absorption efficiency in times of need can be as high as 50%. On the other hand, if iron stores are adequate and the iron-binding protein in the blood is fully saturated with iron, absorption from the intestinal cells is minimal—as low as 2%. The iron remains in the intestinal cells, and it will be excreted in the feces when those intestinal cells slough off, which occurs every 5 to 6 days. In general, healthy people with adequate iron stores absorb between 5% and 15% of dietary iron, which is quite low compared to other nutrients.

Another major influence on iron absorption is the form of iron in the food. **Heme iron,** derived from hemoglobin and **myoglobin,** comprises 40% of the iron in meat, fish, and poultry (MFP). Absorption of heme iron ranges from about 15% to 35%. Almost nothing affects its absorption. **Nonheme iron,** on the other hand, is subject to many conditions that can either enhance or inhibit its absorption, which ranges from

heme iron Iron provided from animal tissues in the form of hemoglobin and myoglobin. Approximately 40% of the iron in meat, fish, and poultry is heme iron; it is readily absorbed.

myoglobin Iron-containing protein that binds oxygen in muscle tissue.

nonheme iron Iron provided from plant sources, supplements, and animal tissues other than in the forms of hemoglobin and myoglobin. Nonheme iron is less efficiently absorbed than heme iron; absorption is closely dependent on body needs.

▲ Red meat is a major source of iron in the North American diet. Heme iron, which makes up about 40% of the iron in foods of animal origin, is better absorbed than nonheme iron.

Iron
RDA
  Men: 8 milligrams
  Women: 18 milligrams (8 milligrams after menopause)
DV: 18 milligrams
UL: 45 milligrams

The iron added to foods and in most dietary supplements is nonheme iron.

2% to 8%. Table 9-7 summarizes dietary factors that affect bioavailability of nonheme iron. Nonheme iron makes up 60% of iron in MFP and 100% of the iron found in dairy, eggs, fruit, vegetables, grains, fortified foods, and supplements. Because most of our dietary iron is nonheme iron, our overall dietary iron absorption is 5% to 15%.

Acidity also affects iron absorption: an acidic environment solubilizes iron and keeps it in a form that can be readily absorbed. Therefore, any medication or health condition that lowers acid production of the stomach can decrease iron absorption. For example, acid-reducing medications that people take to control heartburn or ulcers can impair iron absorption. Also, as people age, gastric acid secretion may decline. This puts older adults at risk for iron-deficiency anemia.

Lastly, other micronutrients affect iron absorption and availability. Megadoses of zinc or calcium compete with iron for absorption in the small intestine. In contrast, adequate copper is important for iron metabolism, and vitamin C is a powerful enhancer of iron absorption. Doses of 75 milligrams of vitamin C can increase nonheme absorption by 4%—a lot for nonheme. If you want to get the most iron out of your dietary supplement, take it with a glass of orange juice.

GETTING ENOUGH IRON

Animal sources contain approximately 40% heme iron, the most bioavailable form. The major iron sources in the adult diet are ready-to-eat breakfast cereals, beans, and animal products (Fig. 9-24). Iron is added to flour during the enrichment process. Other iron sources are peas and legumes, but the absorption of nonheme iron found in these products is relatively low. Milk and eggs are poor sources of iron. A common cause of iron-deficiency anemia in children is high consumption of milk coupled with insufficient meat intake. Vegans are particularly susceptible to iron-deficiency anemia because of their lack of dietary heme iron.

The adult RDA is based on a 10% absorption rate to cover average losses of about 0.8 milligrams per day. For women of reproductive age, average menstrual losses are about 1 gram of additional iron per day. Thus, iron is the only nutrient for which women have higher requirements than men. Most women do not consume the recommended 18 milligrams of iron daily. The average daily amount consumed by women is closer to 13 milligrams, while in men, it is about 18 milligrams per day. Women of reproductive age can close this gap between average daily intakes and needs by seeking out iron-fortified foods, such as ready-to-eat breakfast cereals that contain at least 50% of the DV. Use of a balanced multivitamin and mineral supplement containing up to 100% of the DV for iron is another option. Consuming more than that much iron is not advised unless recommended by a physician.

AVOIDING TOO MUCH IRON

The UL for iron is 45 milligrams per day. Higher amounts can lead to stomach irritation. Although iron overload is not as common as iron deficiency, the consequences can be dire. Even a large single dose of 60 milligrams of iron can be life

▲ During both pregnancy and childhood, demands for iron are high. Young children and women of reproductive age are at risk for iron-deficiency anemia.

TABLE 9-7 ▶ Dietary Iron Enhancers and Inhibitors

Nonheme Enhancers	Nonheme Inhibitors

• Vitamin C
  • Add marinara sauce to your spaghetti noodles.
• MFP (meat, fish, poultry) meat protein
  • Add some tuna to your snack of crackers.
	• Tannins (found in tea)
	  • Can lower absorption up to 60%, so drink tea between meals. Does not apply to herbal "tea," which contains no tea leaves.
	• Oxalates (spinach, rhubarb, and chard)
	• Phytates (whole grains, bran, and soybean)
	• Megadoses of zinc, calcium, or copper

FIGURE 9-24 ◀ Food sources of iron. (a) The fill of the background color (none, 1/3, 2/3, or completely covered) within each food group on MyPlate indicates the average nutrient density for iron in that group. (b) The bar graph shows the iron content of several foods in each food group compared to the RDA for adult males and females. Overall, the richest sources of iron are meats, legumes, and fortified grain products. *Nutrition data from USDA National Nutrient Database for Standard Reference, Release 26.*

(a)

(b)

	Food Item and Amount	Iron (milligrams)	% RDA for Adult Males (8 milligrams)	% RDA for Adult Females (18 milligrams)
Grains	Frosted mini wheats, 24 biscuits	17.6	220%	98%
Grains	Flour tortilla, 8"	1.0	13%	6%
Grains	Whole-wheat bread, 1 slice	0.8	10%	4%
Vegetables	Spinach, cooked, 1 cup	6.4	80%	36%
Vegetables	Parsley, raw, 1 cup	3.7	47%	21%
Vegetables	Potato, baked (with skin), 1 medium	1.9	23%	10%
Fruits	Prune juice, 1 cup	3.0	38%	17%
Fruits	Raisins, ½ cup	1.6	19%	9%
Fruits	Orange, navel, 1 medium	0.2	2%	1%
Dairy	Soymilk, 1 cup	1.0	13%	6%
Dairy	Feta cheese, 1.5 ounces	0.3	4%	2%
Dairy	Milk, fat-free, 1 cup	0.1	1%	0%
Protein	Oysters, steamed, 3 ounces*	7.8	98%	44%
Protein	Kidney beans, cooked, ½ cup	2.0	25%	11%
Protein	Sirloin steak, grilled, 3 ounces*	1.5	18%	8%

*Contains heme iron

threatening to a 1-year-old. Children are frequently victims of iron poisoning (acute toxicity) because supplements, which look a lot like candy, may be easily accessible on kitchen tables and from cabinets. FDA requires that all iron supplements carry a warning about toxicity. Furthermore, supplements with 30 milligrams of iron or more per tablet must be individually wrapped.

Iron toxicity accompanies the genetic disease hereditary **hemochromatosis.** The disease is associated with a substantial increase in iron absorption from both food and supplements. The harshest effects are seen in iron-storing organs such as the liver and heart. Some iron is deposited in the pancreas and muscles. Blood levels remain high, too, which increases the likelihood of infections and may promote cardiovascular disease.

Development of hereditary hemochromatosis requires that a person carry two defective copies of a particular gene. People with one defective gene and one normal gene (i.e., carriers) may also absorb too much dietary iron but not to the same extent as those with two defective genes. About 5% to 10% of North Americans of Northern European descent are carriers of hemochromatosis. Approximately one in 250 North Americans has both hemochromatosis genes. These numbers are high, considering that many physicians regard hemochromatosis as a rare disease and therefore do not routinely test for it.

hemochromatosis A disorder of iron metabolism characterized by increased iron absorption and deposition in the liver and heart. This eventually poisons the cells in those organs.

therapeutic phlebotomy Periodic blood removal, as a blood donation, for the purpose of ridding the body of excess iron.

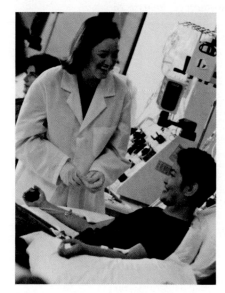

▲ Red blood cells contain about two-thirds of the body's total iron supply. Each time a person donates blood, he or she sacrifices about 10% of his or her total blood volume, which removes about 7% of the body's iron supply. Over the next few weeks, the red blood cells will be replaced, so healthy people can usually donate blood two to four times a year without harmful consequences. As a precaution, blood banks first screen potential donors' blood for the presence of anemia.

Anyone who has a blood relative (including uncles, aunts, and cousins) who has hemochromatosis or is a carrier should be screened for iron overload. At your next visit to a physician, ask for a transferrin saturation test to assess iron stores. A ferritin test may also be added to assess your stores. Hemochromatosis can go undetected until a person is in his or her 50s or 60s, so some experts recommend screening for anyone over the age of 20.

If the disease goes untreated, iron accumulates and serious health problems may result: darkening of skin, arthritis, heart disease, diabetes, liver disease, and cancer (colon). Even with iron overload, the person may have anemia due to damage to the bone marrow or liver. Treatment of hemochromatosis is relatively easy, but it must be monitored consistently. **Therapeutic phlebotomy** to remove excess iron is essential. One must be very careful about the diet. Few sources of heme iron should be eaten, and supplements with iron or vitamin C should be avoided. Highly fortified breakfast cereals must also be avoided.

☑ CONCEPT CHECK 9.9

1. List three symptoms of iron deficiency. How do these symptoms relate to the roles of iron in the body?

2. What are heme and nonheme iron? What can you do to enhance your absorption of nonheme iron?

3. What is hemochromatosis?

9.10 Zinc (Zn)

Zinc deficiency was first recognized in the early 1960s in Egypt and Iran, where it was linked to growth retardation and poor sexual development (Fig. 9-25). Even though the zinc content of the diets of people in these areas was fairly high, absorption of the mineral was limited by the phytic acid in unleavened bread. Parasite infestation and the practice of eating clay and other parts of soil also contributed to the severe zinc deficiency.

FUNCTIONS OF ZINC

Approximately 200 enzymes require zinc as a cofactor for activity. Adequate zinc intake is necessary to support many physiological functions:

- DNA synthesis and function
- Protein metabolism, wound healing, and growth
- Development of bones and reproductive organs
- Storage, release, and function of insulin
- Cell membrane structure and function
- Component of superoxide dismutase (SOD), an enzyme that aids in the prevention of oxidative damage to cells (zinc, therefore, has an indirect antioxidant function)
- White blood cell formation

It is important to note that although zinc is important for immune function, intakes in excess of the RDA do not provide any extra benefit for immune function. In fact, chronic excessive intakes of zinc can actually depress immune function. Zinc supplementation may be useful to slow the progression of macular degeneration of the eye and reduce the risk for developing certain forms of cancer.

ZINC DEFICIENCY

Symptoms of adult zinc deficiency include an acnelike rash, diarrhea, lack of appetite, delayed wound healing, impaired immunity, reduced sense of taste (metallic-like) and smell, and hair loss. In children and adolescents with zinc deficiency, growth, sexual development, and learning ability may also be hampered.

GETTING ENOUGH ZINC

Protein-rich diets, especially those that include many animal sources of protein, are high in zinc. The average North American consumes 10 to 14 milligrams of zinc per day, about 80% of which is provided by meat, fish, poultry, fortified cereal, and dairy products (Fig. 9-26). There are no indications of moderate or severe zinc deficiencies in an otherwise healthy adult population. It is likely, however, that some North Americans—especially some poor children, vegans, and older people with alcohol use disorders—have marginal zinc status (see Further Reading 9). These and other people who show deterioration in taste sensation, recurring infections, poor growth, or depressed wound healing should have their zinc status checked.

Overall, about 40% of dietary zinc is absorbed. Absorption efficiency depends on the body's need for zinc and the form of the mineral in foods. When zinc status is poor, absorption of the mineral increases. The zinc found in animal foods is better absorbed than that found in plants. Worldwide, however, most people rely on unfortified cereal grains (low in zinc) as their source of protein, calories, and zinc. As described above, phytic acid in plant foods binds to zinc and limits its availability. Adding yeast to grains (leavening) breaks down phytic acid, increasing zinc bioavailability from leavened grain products. In populations that consume mainly unleavened bread, zinc deficiency can be a problem.

The form generally used in multivitamin and mineral supplements (zinc oxide) is not as well-absorbed as zinc found naturally in foods but still contributes to meeting zinc needs. High-dose calcium supplementation decreases zinc availability if taken too close to mealtime. Finally, zinc competes with copper and iron for absorption, and vice versa, when supplemental sources are consumed. Supplements with more than 100% of the Daily Value for individual minerals are not advised without medical supervision.

AVOIDING TOO MUCH ZINC

Excessive zinc intake over time can lead to problems by interfering with copper metabolism. The interference with copper metabolism is the basis for setting the UL (see margin). Zinc toxicity can occur from zinc supplements and over-consumption of zinc-fortified foods. A person using megadose supplementation should be under close medical supervision and take a supplement containing copper (2 milligrams per day). Zinc intakes over 100 milligrams result in diarrhea, cramps, nausea, vomiting, and loss of appetite. Intakes consistently over 2000 milligrams per day can lead to depressed immune function and decreased high density lipoproteins (HDLs).

FIGURE 9-25 ▲ Low intakes of zinc limit growth in people world-wide. On the right, an Egyptian farm boy, 16 years old and 49 inches tall, experienced limited growth and sexual development associated with zinc deficiency.

Zinc
RDA:
  Men: 11 milligrams
  Women: 8 milligrams
DV: 15 milligrams
UL: 40 milligrams

✔ CONCEPT CHECK 9.10

1. List three good sources of zinc.
2. What are the consequences of zinc deficiency?

(a)

FIGURE 9-26 ◄ Food sources of zinc. (a) The fill of the background color (none, 1/3, 2/3, or completely covered) within each food group on MyPlate indicates the average nutrient density for zinc in that group. (b) The bar graph shows the zinc content of several foods compared to the RDA for adult males and females. Overall, the richest sources of zinc are in the protein group. Foods in the fruits group (not shown) provide very little zinc. *Nutrition data from USDA National Nutrient Database for Standard Reference, Release 26.*

(b)

Food Item and Amount	Zinc (milligrams)	% RDA for Adult Males (11 milligrams)	% RDA for Adult Females (8 milligrams)
White rice, enriched, ½ cup	1.1	10%	14%
Quinoa, cooked, ½ cup	1.0	9%	13%
Special K cereal, 1 cup	0.4	4%	5%
Asparagus, cooked, 1 cup	1.1	10%	14%
Zucchini, cooked, 1 cup	0.6	5%	7%
Spinach, raw, 2 cups	0.3	3%	4%
Mozzarella cheese, part skim, 1.5 ounces	1.2	11%	15%
Milk, fat-free, 1 cup	1.0	9%	13%
Soymilk, 1 cup	0.6	6%	8%
Oysters, steamed, 3 ounces	66.8	607%	835%
Beef pot roast, 3 ounces	7.9	71%	98%
Peanuts, dry roasted, 1 ounce	0.9	9%	12%

9.11 Selenium (Se)

FUNCTIONS OF SELENIUM

Selenium is a trace mineral that exists in many readily absorbed chemical forms. Selenium's best-understood role is aiding the activity of one of the body's natural antioxidant enzymes, glutathione peroxidase. Glutathione peroxidase chemically converts potentially damaging peroxides (hydrogen peroxide, for example) into water. In functioning as part of our natural antioxidant enzyme system, selenium spares vitamin E and indirectly helps maintain cell-membrane integrity. Selenium is also a component of an enzyme essential for the activation of thyroid hormone (T3).

SELENIUM DEFICIENCY

Selenium content of foods is strongly dependent on the selenium content of the soil where plants are raised or animals graze. Worldwide, only one region—the Keshan province in China—has such low soil selenium levels that selenium deficiencies result. Such deficiencies were first reported by Chinese scientists in 1979. (Of note, communication and relations between the West and China were closed from 1949 until 1972. In the mid- and late 1970s, medical dialogue opened significantly.)

Selenium deficiency symptoms in humans include muscle pain and wasting, and a certain form of heart damage. Also, due to its role in thyroid hormone metabolism, selenium deficiency may impair thyroid function, thereby limiting growth. In China's

Newsworthy Nutrition

Selenium and vitamin E supplements offer no benefit for cancer prevention

In the late 1990s and early 2000s, secondary results of two large trials of nutrient supplementation in cancer prevention hinted at possible roles for micronutrients in the prevention of prostate cancer. As a result, the 12-year Selenium and Vitamin E Cancer Prevention Trial (SELECT) was designed to further explore the roles of selenium and vitamin E for the prevention of prostate cancer. After just 7 years, the trial was stopped due to lack of evidence of any beneficial effect of the supplements on endpoints related to cancer. Follow-up of prostate cancer development in the men in this study showed a higher number of prostate cancer cases in the groups receiving vitamin E, selenium, or a combination of both, when compared to those receiving the placebo. The authors concluded that dietary supplementation with vitamin E significantly increased the risk of prostate cancer among healthy men. This trial demonstrated that neither selenium (200 micrograms per day of selenomethionine) nor vitamin E (400 IU per day of dl-alpha-tocopherol), either alone or in combination, reduced risk for prostate cancer. This is an example of why replication of research results is so important. The results of one study are not sufficient to inform clinical practice.

Source: Klein EA and others: Vitamin E and the risk of prostate cancer: The Selenium and Vitamin E Cancer Prevention Trial (SELECT). *Journal of the American Medical Association* 306:1549, 2011.

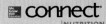 **connect** NUTRITION Check out the Connect site **www.mcgrawhillconnect.com** to further explore nutrition research on cancer prevention.

Keshan province, unless they receive supplements, children and adults develop characteristic muscle and heart disorders associated with inadequate selenium intake.

Low blood levels of selenium have been linked with an increased incidence of some forms of cancer, specifically prostate cancer. Although selenium could prove to have a role in prevention of cancers in those with low or marginal selenium stores, it is premature to recommend selenium supplementation for this purpose. Animal studies in this area are conflicting. Current studies examine the interaction of selenium and vitamin E on **gene expression** in some cancers.

GETTING ENOUGH SELENIUM

Fish, meat (especially organ meats), shellfish, and eggs are good animal sources of selenium (Fig. 9-27). Brazil nuts, and grains and seeds grown in soils containing selenium are good plant sources. Major selenium contributors to the adult diet are animal and grain products. Some geographic regions identified with low-selenium soil in North America include the Northeast, Pacific, Southwest, and coastal plain of the Southeast in the United States; and the north central and eastern regions in Canada. We eat a varied diet of foods supplied from many geographic areas, so it is unlikely that local areas with low soil selenium will mean inadequate selenium in our diets.

The RDA for selenium is 55 micrograms per day for adults. This intake maximizes the activity of selenium-dependent enzymes. The Daily Value used on food and supplement labels is 70 micrograms. Adults meet the RDA, consuming on average 105 micrograms each day.

AVOIDING TOO MUCH SELENIUM

High concentrations of selenium are rarely found in food, with the exception of Brazil nuts. Therefore, selenium toxicity has not been reported from eating food.

gene expression Use of information from a gene to synthesize a functional gene product. In the case of DNA, these products are proteins.

Selenium
RDA: 55 micrograms
DV: 70 micrograms
UL: 400 micrograms

▲ Be careful with Brazil nuts. This portion of 10 nuts contains 960 micrograms of selenium, which exceeds the Upper Limit for selenium of 400 micrograms per day.

Choose **MyPlate**.gov

(a)

(b)

FIGURE 9-27 ◄ Food sources of selenium. (a) The fill of the background color (none, 1/3, 2/3, or completely covered) within each food group on MyPlate indicates the average nutrient density for selenium in that group. (b) The bar graph shows the selenium content of several foods compared to the RDA for adult males and females. Overall, the richest sources of selenium are found in the protein foods and grains groups. Foods from the fruits group (not shown) provide very little selenium. *Nutrition data from USDA National Nutrient Database for Standard Reference, Release 26.*

Food Item and Amount	Selenium (micrograms)	% RDA for Adult Males and Females (55 micrograms)
Egg noodles, cooked, ½ cup	19	35%
Puffed wheat cereal, 1 cup	19	34%
White rice, cooked, ½ cup	7	13%
Mushrooms, crimini, raw, 1 cup	19	35%
Asparagus, cooked, 1 cup	11	20%
Spinach, cooked, 1 cup	3	5%
Swiss cheese, 1.5 ounces	8	14%
Milk, fat-free, 1 cup	8	14%
Soymilk, 1 cup	6	10%
Brazil nuts, 2 each	192	349%
Tuna, grilled, 3 ounces	92	167%
Oysters, steamed, 3 ounces	34	62%

Excessive supplementation for an extended period has been shown to be toxic. The UL for selenium is 400 micrograms per day for adults. This is based on overt signs of selenium toxicity, such as hair loss, weakness, nausea, vomiting, and cirrhosis. Because Brazil nuts are such a concentrated source of selenium, it is recommended to avoid consuming them daily, thus preventing an inadvertent overload.

✔ **CONCEPT CHECK 9.11**

1. How does selenium play an antioxidant role?
2. What other functions does selenium play in the body?
3. What are the signs of a selenium deficiency?
4. What food groups are the best sources of selenium?
5. What are the signs of selenium toxicity?

9.12 Iodide (I)

FUNCTIONS OF IODIDE

The thyroid gland actively accumulates and traps iodide from the bloodstream to support thyroid hormone synthesis. Thyroid hormones are synthesized using iodide and the amino acid tyrosine. Because these hormones help regulate metabolic rate and promote growth and development throughout the body, iodide adequacy is important for overall energy metabolism.

Chapter 9: Water and Minerals **383**

IODIDE DEFICIENCY

During World War I, a link was discovered between a deficiency of iodide and an enlarged thyroid gland, a condition called **goiter.** Men drafted into the military from areas such as the Great Lakes Region of the United States had a much higher rate of goiter than did men from other areas of the country. The soils in these areas have low iodide contents. In the 1920s, a researcher in Ohio found that low doses of iodide given to children over a 4-year period could prevent goiter. That finding led to the addition of iodide to salt beginning in the 1920s, the first time a nutrient was purposely added to food to prevent a disease.

Today, many nations, such as Canada, require iodide fortification of salt. In the United States, salt can be purchased either iodized or plain. Check for this on the label when you purchase salt. Some areas of Europe, such as northern Italy, have very low soil levels of iodide but have yet to adopt an iodide-fortification program. People in these areas, especially women, still suffer from goiter, as do people in areas of Latin America, the Indian subcontinent, Southeast Asia, and Africa. About 2 billion people worldwide are at risk of iodide deficiency, and approximately 800 million of these people have suffered the widespread effects of such a deficiency. Eradication of iodide deficiency is a goal of many health-related organizations worldwide.

If a person's iodide intake is insufficient, the thyroid gland enlarges as it attempts to take up more iodide from the bloodstream. This eventually leads to goiter. Simple goiter is a painless condition but, if uncorrected, can lead to pressure on the trachea (windpipe), which may cause difficulty in breathing. Although iodide can prevent goiter formation, it does not significantly shrink a goiter once it has formed. Surgical removal may be required in severe cases.

If a woman has an iodide-deficient diet during the early months of her pregnancy, the fetus suffers iodide deficiency because the mother's body uses up the available iodide. The infant then may be born with short length and develop intellectual delays. Collectively, the stunted growth and developmental delays that result are known as **congenital hypothyroidism** (formerly called cretinism). This deficiency disease appeared in North America before iodide fortification of table salt began. Today, congenital hypothyroidism still appears in Europe, Africa, Latin America, and Asia.

GETTING ENOUGH IODIDE

The RDA for iodide (see margin) was set to support thyroid gland function. This is the same as the DV used on food and supplement labels. A half teaspoon of iodide-fortified salt (about 2 grams) supplies that amount. Most North American adults consume more iodide than the RDA—an estimated 190 to 300 micrograms daily, not including that from use of iodized salt at the table. This extra amount adds up because dairies use it as a sterilizing agent, bakeries use it as a dough conditioner, food producers use it as part of food colorants, and it is added to salt. There is concern, however, that vegans may not consume enough unless iodized salt is used. Iodized salt, dairy products, and grain products contain various forms of iodide (Fig. 9-28). Sea salt and kosher salt, however, are not typically iodized.

AVOIDING TOO MUCH IODIDE

The UL for iodide is 1.1 milligrams per day. When high amounts of iodide are consumed, thyroid hormone synthesis is inhibited, as in a deficiency. This can appear in people who eat a lot of seaweed, because some seaweeds contain as much as 1% iodide by weight. Total iodide intake then can add up to 60 to 130 times the RDA.

goiter An enlargement of the thyroid gland; this is often caused by insufficient iodide in the diet.

Iodide
RDA: 150 micrograms
DV: 150 micrograms
UL: 1.1 milligrams

congenital hypothyroidism The stunting of body growth and poor development in the offspring that result from inadequate maternal intake of iodide during pregnancy (formerly called cretinism).

▲ This woman has an enlargement of the thyroid gland (also known as a goiter) caused by insufficient iodide in the diet.

FIGURE 9-28 ◄ Food sources of iodide. (a) The fill of the background color (none, 1/3, 2/3, or completely covered) within each food group on MyPlate indicates the average nutrient density for iodide in that group. (b) The bar graph shows the iodide content of several foods compared to the RDA for adult males and females. Overall, the richest sources of iodide are iodized salt (added to foods in any group), seafood and seaweed, and dairy products. Fruits and vegetables (not shown) are poor sources of iodide. *Nutrition data from FDA/CFSAN Total Diet Study Statistics on Elements Results—2006–2008.*

(a) (b)

	Food Item and Amount	Iodide (micrograms)	% RDA for Adult Males and Females (150 micrograms)
	Iodized salt, ½ tsp	195	130%
Grains	White bread, 1 slice	29	19%
Grains	Corn muffin, 1 small	21	14%
Grains	Whole wheat bread, 1 slice	8	5%
Dairy	Yogurt, plain, non-fat, 1 cup	106	71%
Dairy	Milk, fat-free, 1 cup	103	69%
Dairy	Cheese, cheddar, 1.5 ounces	22	15%
Protein	Pepperoni pizza, 1 slice	51	34%
Protein	Eggs, hard-boiled, 1 large	32	21%
Protein	Tuna, grilled, 3 ounces	13	9%

✓ CONCEPT CHECK 9.12

1. What is the role of iodide in thyroid metabolism?
2. What are the effects of iodide deficiency? What is a goiter?
3. Is salt always a good source of iodide?

9.13 Copper (Cu)

Copper and iron are similar in terms of food sources, absorption, and functions. Copper is a component of blood. In the body, it is found in highest concentration in the liver, brain, heart, kidneys, and muscles. **Ceruloplasmin** is the name of the protein that carries most of the body's copper in the blood.

FUNCTIONS OF COPPER

Copper is a cofactor for many enzymes, including some involved in the body's antioxidant defenses. Copper serves as a cofactor for superoxide dismutase, an enzyme that defends the body against free-radical damage. Copper also has a role in the function of enzymes that create cross-links in connective tissue proteins, such as the collagen in bone. Another very important role of copper is as a cofactor in the electron transport chain, the last stage of cellular respiration, which converts the energy stored in carbohydrates, fats, and proteins into ATP.

Copper is also involved in blood health because one of the roles of copper is to transport iron, which is needed to form red blood cells. Copper is part of three different enzymes that assist in the transport of iron out of intestinal cells, through the blood, and to the bone marrow, where iron is incorporated into hemoglobin. The discoveries of these iron-transporting proteins are relatively recent; there is still much to learn about the significance of these copper-containing proteins to iron metabolism.

ceruloplasmin Copper-containing protein in the blood; functions in the transport of iron.

A genetic disease called **Menkes syndrome** decreases the amount of copper available to the brain and nervous system. Babies born with Menkes syndrome suffer from nervous system disorders, weak muscle tone, and delays in physical and cognitive development related to the lack of copper-containing enzymes that help to form nervous tissue and synthesize neurotransmitters. They usually do not live past the age of 3.

In addition, copper is needed by enzymes involved in nerve myelination, neurotransmitter synthesis, immune system function, blood clotting, and blood lipoprotein metabolism.

Copper
RDA: 900 micrograms
DV: 2 milligrams
UL: 10 milligrams

COPPER DEFICIENCY

Considering the myriad roles of copper discussed above, it is not surprising that copper deficiency affects so many different body systems. Symptoms of copper deficiency include a form of anemia, low white blood cell count, bone loss, poor growth, and some forms of cardiovascular disease.

The groups most likely to develop copper deficiencies are preterm infants recovering from semistarvation on a milk-dominated diet (a poor source of copper) and people recovering from intestinal surgery. A copper deficiency can also result from overzealous supplementation of zinc, because zinc and copper compete with each other for absorption.

GETTING ENOUGH COPPER

Rich sources of copper include liver, legumes, seeds, whole-grain breads and cereals, and cocoa (Fig. 9-29). Milk and dairy products, fruit, and vegetables are generally poor sources of copper. Also, the form of copper typically found in multivitamin

FIGURE 9-29 ◀ Food sources of copper. (a) The fill of the background color (none, 1/3, 2/3, or completely covered) within each food group on MyPlate indicates the average nutrient density for copper in that group. (b) The bar graph shows the copper content of several foods in each food group compared to the RDA for adult males and females. Overall, the richest sources of copper are found in the protein foods and grains groups. *Nutrition data from USDA National Nutrient Database for Standard Reference, Release 26.*

(a)

(b)

	Food Item and Amount	Copper (micrograms)	% RDA for Adult Males and Females (900 micrograms)
Grains	Whole Grain Total cereal, 1 cup	140	16%
Grains	Whole-wheat bread, 1 slice	73	8%
Grains	White rice, cooked, ½ cup	35	4%
Vegetables	Spinach, cooked, 1 cup	313	35%
Vegetables	Potato, baked (with skin) medium	185	21%
Vegetables	Butternut squash, cooked, 1 cup	133	15%
Fruits	Blackberries, fresh, 1 cup	238	26%
Fruits	Red seedless grapes, 1 cup	192	21%
Fruits	Orange juice, fresh, 1 cup	109	12%
Dairy	Soymilk, 1 cup	401	45%
Dairy	Cottage cheese, 1 cup	68	8%
Dairy	Milk, fat-free, 1 cup	32	4%
Protein	Beef liver, pan-fried, 3 ounces	11816	1313%
Protein	Walnuts, 1 ounce	386	43%
Protein	Kidney beans, cooked, ½ cup	191	21%

▲ Seafood is one source of copper in the diet. How does this meal of grilled tuna steak with peach salsa compare to MyPlate?

Wilson's disease A genetic disorder that results in accumulation of copper in the tissues; characterized by damage to the liver, nervous system, and other organs.

and mineral supplements (copper oxide) is not readily absorbed. It is best to rely on food sources to meet copper needs (see margin).

Copper absorption is highly variable, with higher intakes associated with lower absorption efficiency. Absorption takes place in the stomach and upper small intestine. Excess copper is not stored to a great extent, so when intake exceeds needs, the liver incorporates it into bile, which is excreted as part of the feces. Phytates, fiber, and excessive zinc and iron supplements may all interfere with copper absorption.

The copper status of adults appears to be good: the average adult intake is about 1 milligram for women and 1.6 milligrams for men per day. However, sensitive laboratory tests to determine copper status arc lacking.

AVOIDING TOO MUCH COPPER

A single dose of copper greater than 10 milligrams can cause toxicity. Consequences of copper toxicity include GI distress, vomiting blood, tarry feces, and damage to the liver and kidneys. Toxicity cannot occur with food, just supplements or excessive exposure to copper salts used in agriculture.

Wilson's disease is a genetic disease in which the liver cannot synthesize ceruloplasmin. In turn, copper accumulates in tissues, such as lungs and liver. People with Wilson's disease suffer damage to the liver and nervous system. A primary treatment for Wilson's disease is a vegan diet, as fruits and vegetables are low in copper. Researchers are currently interested in how excess copper in the blood may influence the development of Alzheimer's disease and Parkinson's disease.

✔ CONCEPT CHECK 9.13

1. List three functions of copper.
2. Describe some interactions among iron, zinc, and copper in the body.
3. What dietary changes will be required for a person with Wilson's disease?

9.14 Fluoride (F)

The fluoride ion (F^-) is the form of this trace mineral essential for human health. Nearly all (about 95%) of the fluoride in the body is found in the teeth and skeleton. Dentists in the early 1900s noticed a lower rate of dental caries (cavities) in the southwestern United States. These areas contained high amounts of fluoride in the water. The amounts of fluoride were sometimes so high that small spots developed on the teeth (mottling). Even though mottled teeth were discolored, they contained few dental caries. Experiments in the early 1940s showed that fluoride in the water decreased the incidence of dental caries by 20% to 80% in children. Fluoridation of public water supplies in many parts of the United States has since been instituted (see Further Reading 19).

FUNCTIONS OF FLUORIDE

Fluoride functions in the following ways to prevent dental caries: (1) fluoride is incorporated into the tooth structure, causing it to be stronger and more resistant to acid degradation from bacteria found in plaque; (2) stimulation of remineralization of enamel and inhibition of tooth demineralization; and (3) antibacterial effect on acid-producing microorganisms found in plaque.

Fluoride has been shown to increase protein synthesis in the bone-building cells and subsequently have an effect on the production of new bone. The combination of calcium and fluoride supplementation has been investigated regarding increasing bone mass; however, to date, the studies have been inconclusive.

Except for increased risk of dental caries, there is no fluoride deficiency disease.

GETTING ENOUGH FLUORIDE

The list of foods that are good sources of fluoride is rather short: marine fish, clams, lobster, crab, shrimp, tea, and seaweed. Most of our fluoride actually comes from oral hygiene products and the water supply. Numerous products are available to apply fluoride to teeth topically. These include gels applied at a dentist's office or toothpaste and mouth rinses for everyday use. Fluoride is also available in supplement form, although use should be directed by a dentist or pediatrician. The most economical method of distributing fluoride is to add the mineral to the community's drinking water.

In a few areas of the world, the fluoride content of groundwater is naturally high, but most groundwater supplies contain low levels of fluoride. In the 1950s, after researchers established a connection between fluoride and rates of dental caries, communities in the United States began adding fluoride to the municipal water supply to achieve a fluoride level of 0.7 to 1.2 milligrams per liter. (The lower levels are for communities in hotter climates, where total water consumption is higher.) About two-thirds of North Americans currently consume fluoridated water; these policies are made by individual municipalities. Because most people now have ample access to oral hygiene products with fluoride, in 2011, the U.S. Department of Health and Human Services and the Environmental Protection Agency issued new recommendations to lower the level of water fluoridation to just 0.7 milligrams per liter.

The AI for fluoride for adults is 3.1 to 3.8 milligrams per day. This range of intake provides the benefit of resistance to dental caries without causing ill effects. As described above, typical fluoridated water contains about 1 milligram per liter, which works out to about 0.25 milligrams per cup. In communities without fluoridated water (e.g., those that rely on private well water), use of fluoride-containing oral hygiene products or dietary supplements is of heightened importance for combating dental decay. Note that fluoride is generally not added to bottled water. Frequent use of bottled water or a household reverse osmosis water purification system significantly restricts fluoride intake. A refrigerator or Brita filter does not remove fluoride. When water fluoridation and fluoridated oral hygiene products are used in combination, the reductions in dental caries are additive.

▲ *Healthy People 2020* objectives for the nation set a target goal of 79.6% of the population using piped water to have that water optimally fluoridated. As of 2010, the benefits of community water fluoridation reached nearly 74% of the U.S. population served by community water systems.

AVOIDING TOO MUCH FLUORIDE

The UL for fluoride is set at 1.3 to 2.2 milligrams per day for young children and 10 milligrams per day for children over 9 years of age and adults, based on skeletal and tooth damage seen with higher doses. Children may swallow large amounts of fluoride toothpaste as part of daily tooth care and develop **fluorosis.** During tooth development (first decade of life), fluorosis permanently damages teeth. Fluorosis leads to stained and pitted teeth. Not swallowing toothpaste and limiting the amount used to "pea" size are the best ways to prevent this problem. In addition, children under 6 years should have tooth brushing supervised by an adult and should never use fluoride mouthwash. In adults, fluorosis is associated with hip fractures, weak or stiff joints, and chronic stomach inflammation.

There have been opponents to the fluoridation of public water supplies. Some people argue that water fluoridation standards were set at a time when much of the population did not have adequate access to fluoride-containing oral hygiene products, and that addition of fluoride to the water supply is no longer necessary. Other critics claim that chronic exposure to fluoridated water is linked to a variety of health ailments affecting the skeletal, nervous, or endocrine systems. At this time, there is little scientific evidence to support claims that water fluoridation at current levels has adverse health effects other than dental fluorosis, but this is definitely an area for continued study. The updated recommendations for the level of water fluoridation aim to take advantage of the oral health benefits of fluoride while limiting unwanted health effects, including fluorosis (see Further Reading 15).

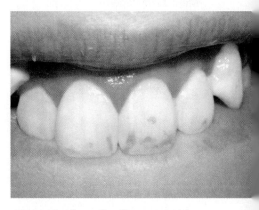

▲ Example of mottling (brown spots) in a tooth caused by overexposure to fluoride.

Fluoride

AI:	3.1 to 3.8 milligrams
UL	
Young children:	1.3 to 2.2 milligrams
>9 years:	10 milligrams

1. When is fluoride incorporated into teeth?
2. How does fluoride help to reduce the development of dental caries?
3. What are our primary sources of fluoride?
4. What are the risks of excessive fluoride intake?

9.15 Chromium (Cr)

FUNCTIONS OF CHROMIUM

Chromium enhances the function of insulin, so it is required for glucose uptake into cells. The mineral is involved in the metabolism of lipids and proteins as well, although the exact mechanisms are not known. Chromium supplements have been promoted for building muscle mass and for weight loss, but there is not much evidence to support these claims.

CHROMIUM DEFICIENCY

A chromium deficiency is characterized by impaired blood glucose control and elevated blood cholesterol and triglycerides. Low or marginal chromium intakes may contribute to an increased risk for developing type 2 diabetes, but opinions are mixed on the true degree of this effect. Chromium deficiency appears in people maintained on intravenous nutrition solutions not supplemented with chromium and in children with malnutrition. Marginal deficiencies may go undetected because sensitive measures of chromium status are not available.

GETTING ENOUGH CHROMIUM

Specific data regarding the chromium content of various foods are scant, and most food-composition tables do not include values for this trace mineral. Because of two major limits, we really cannot accurately determine the amount of chromium in food: (1) the content is significantly affected by agricultural and manufacturing processes; and (2) when foods are analyzed, they may be "contaminated" by the chromium in the instruments themselves. Meat, whole-grain products, eggs, mushrooms, nuts, beer, and spices are relatively good sources of chromium. Brewer's yeast is also a very good source.

Chromium absorption is quite low: only 0.4% to 2.5% of the amount consumed. Absorption is enhanced by vitamin C and niacin. Any unabsorbed chromium is excreted in the feces. Once absorbed, it is stored in the liver, spleen, soft tissue, and bone, and excreted via urine. Certain conditions can enhance urinary excretion of chromium: diets high in simple sugars (more than 35% of total calories), significant infection, acute prolonged exercise, pregnancy and lactation, and major physical trauma. If chromium intakes are already low, these states potentially can lead to deficiency.

The Adequate Intake (AI) for chromium is 25 to 35 micrograms per day, based on the amount present in a balanced diet. The DV used on food and supplement labels is 120 micrograms. Average adult intakes in North America are estimated at about 30 micrograms per day but could be somewhat higher.

No UL for chromium has been set because toxicity from food sources has not been observed. Chromium toxicity, however, has been reported in people exposed to industrial waste and in painters who use art supplies with high chromium content. Liver damage and lung cancer can result. Use of any supplement should normally not exceed the DV unless supervised by a physician because of the risk of toxicity.

▲ Mushrooms are a good source of chromium.

Chromium
AI
  Men: 35 micrograms
  Women: 25 micrograms
DV: 120 micrograms
UL: none

1. How is chromium involved in carbohydrate metabolism?
2. Why is it hard to know the chromium content of foods?
3. Which foods are considered the best sources of chromium?

9.16 Other Trace Minerals

MANGANESE (Mn)

The mineral manganese is easily confused with magnesium (Mg). Not only are their names similar, but they also often substitute for each other in metabolic processes. As a participant in energy metabolism, manganese is required as a cofactor for synthesis of glucose and metabolism of some amino acids. Manganese is also needed by some enzymes, such as those used in free-radical metabolism (via superoxide dismutase). Manganese is also important in bone formation.

Manganese deficiency does not develop in humans unless the mineral is purposely removed from the diet. Animals on manganese-deficient diets suffer alterations in brain function, bone formation, and reproduction. If human diets were low in manganese, these symptoms would probably appear as well. As it happens, our need for manganese is very low, and our diets tend to be adequate in this trace mineral.

The AI for manganese is 1.8 to 2.3 milligrams to offset daily losses. Average intakes fall within this range. The DV used on food and supplement labels is 2 milligrams. Good food sources of manganese are nuts, rice, oats, and other whole grains, beans, and leafy vegetables. Manganese is toxic at high doses. Supplements are not recommended, as large doses can decrease absorption of other minerals. People with low iron stores must avoid manganese supplements or risk worsening anemia. The UL is 11 milligrams per day. This value is based on the development of nerve damage. Miners who have inhaled dust fumes high in manganese experience symptoms that mimic Parkinson's disease, including cognitive and muscular dysfunction.

Manganese
AI
  Men: 2.3 milligrams
  Women: 1.8 milligrams
DV: 2 milligrams
UL: 11 milligrams

▲ Nuts are sources of manganese and molybdenum.

MOLYBDENUM (Mo)

Several human enzymes use molybdenum, including some involved in metabolism of amino acids that contain sulfur. No molybdenum deficiency has been reported in people who consume diets orally. Deficiency symptoms have appeared in people maintained on intravenous nutrition devoid of this trace mineral. Symptoms include increased heart and respiratory rates, night blindness, mental confusion, edema, and weakness.

Good food sources of molybdenum include milk and dairy products, beans, whole grains, and nuts. The RDA for molybdenum is 45 micrograms to offset daily losses. The DV used on food and supplement labels is 75 micrograms. Our daily intakes average 76 micrograms (for women) and 109 micrograms (for men). The Upper Level for molybdenum is 2 milligrams per day. When consumed in high doses, molybdenum causes toxicity in laboratory animals, resulting in weight loss and decreased growth. Toxicity risk in humans is quite low.

Table 9-8 summarizes much of what we know about the trace minerals.

Molybdenum
RDA: 45 micrograms
DV: 75 micrograms
UL: 2 milligrams

1. What are the primary functions of manganese and molybdenum in metabolism of nutrients?
2. What are the good food sources of manganese and molybdenum?
3. Why is a deficiency of manganese or molybdenum possible when patients are fed intravenously?

TABLE 9-8 ▶ Summary of Key Trace Minerals

Mineral	Major Functions	RDA or AI	Dietary Sources	Deficiency Symptoms	Toxicity Symptoms
Iron	• Component of hemoglobin and other key compounds used in respiration • Immune function • Cognitive development	*Men* and *Postmenopausal Women:* 8 milligrams *Premenopausal Women:* 18 milligrams	• Meats • Seafood • Broccoli • Peas • Bran • Enriched breads	• Fatigue • Anemia	• Liver and heart damage (extreme cases) • GI upset Upper Level is 45 milligrams.
Zinc	• Required for nearly 200 enzymes • Growth • Immunity • Alcohol metabolism • Sexual development • Reproduction • Antioxidant protection	*Men:* 11 milligrams *Women:* 8 milligrams	• Seafood • Meats • Greens • Whole grains	• Skin rash • Diarrhea • Decreased appetite and sense of taste • Hair loss • Poor growth and development • Poor wound healing	• Reduced copper absorption • Diarrhea • Cramps • Depressed immune function Upper Level is 40 milligrams.
Selenium	• Part of an antioxidant system	55 micrograms	• Meats • Eggs • Fish • Seafood • Whole grains	• Muscle pain • Weakness • Form of heart disease	• Nausea • Vomiting • Hair loss • Weakness • Liver disease Upper Level is 400 micrograms.
Iodide	• Component of thyroid hormones	150 micrograms	• Iodized salt • White bread • Saltwater fish • Dairy products	• Goiter • Mental retardation • Poor growth in infancy when mother is iodide deficient during pregnancy	• Inhibition of thyroid gland function Upper Level is 1.1 milligrams.
Copper	• Aids in iron metabolism • Works with many antioxidant enzymes • Involved with enzymes of protein metabolism and hormone synthesis	900 micrograms	• Liver • Cocoa • Beans • Nuts • Whole grains • Dried fruits	• Anemia • Low white blood cell count • Poor growth	• Vomiting • Nervous system disorders Upper Level is 8–10 milligrams.
Fluoride	• Increases resistance of tooth enamel to dental caries	*Men:* 3.8 milligrams *Women:* 3.1 milligrams	• Fluoridated water • Toothpaste • Tea • Seaweed • Dental treatments	• Increased risk of dental caries	• Stomach upset • Mottling (staining) of teeth during development • Bone pain Upper Level is 10 milligrams for adults.
Chromium	• Enhances insulin action	*Men (up to 50 years):* 35 micrograms *Women (up to 50 years):* 25 micrograms	• Egg yolks • Whole grains • Pork • Nuts • Mushrooms • Beer	• High blood glucose after eating	Caused by industrial contamination, not dietary excesses, so no Upper Level has been set.
Manganese	• Cofactor of some enzymes, such as those involved in carbohydrate metabolism • Works with some antioxidant systems	*Men:* 2.3 milligrams *Women:* 1.8 milligrams	• Nuts • Oats • Beans • Tea	None observed in humans	• Nervous system disorders Upper Level is 11 milligrams.
Molybdenum	• Aids in action of some enzymes	45 micrograms	• Beans • Grains • Nuts	None observed in healthy humans	• Poor growth in laboratory animals Upper Level is 2 milligrams.

Nutrition and Your Health

● Minerals and Hypertension

Among North Americans, an estimated one in five adults has hypertension. Over the age of 65, the number rises to one in every two adults. Only about half of cases are being treated. Blood pressure is expressed by two numbers. The higher number represents systolic blood pressure, the pressure in the arteries when the heart muscle is contracting and pumping blood into the arteries. Optimal systolic blood pressure is 120 millimeters of mercury (mm Hg) or less. The second value is diastolic blood pressure, the artery pressure when the heart is relaxed. Optimal diastolic blood pressure is 80 mm Hg or less. Elevations in both systolic and diastolic blood pressure are strong predictors of disease (Fig. 9-30).

Hypertension is defined as sustained systolic pressure exceeding 139 mm Hg or diastolic blood pressure exceeding 89 mm Hg. Most cases of hypertension (about 95%) have no clear-cut cause. Such cases are classified as **primary** or essential **hypertension.** Kidney disease, sleep-disordered breathing (sleep apnea), and other causes often lead to the other 5% of cases, classified as **secondary hypertension.**

primary hypertension Blood pressure of 140/90 mm Hg or higher with no identified cause; also called *essential hypertension.*

secondary hypertension Blood pressure of 140/90 mm Hg or higher as a result of disease (e.g., kidney dysfunction or sleep apnea) or drug use.

African-Americans and Asian-Americans are more likely than Caucasians to develop hypertension and to do so earlier in life.

Unless blood pressure is periodically measured, the development of hypertension is easily overlooked. Because it usually does not cause symptoms, it is described as a silent disorder.

Why Control Blood Pressure?

Blood pressure needs to be controlled mainly to prevent cardiovascular disease, kidney disease, strokes and related declines in brain function, poor blood circulation in the legs, problems with vision, and sudden death. These conditions are much more likely to be found in individuals with hypertension than in people with normal blood pressure. Smoking and elevated blood lipoproteins make these diseases even more likely. Individuals with hypertension need to be diagnosed and treated as soon as possible, as the condition generally progresses to a more serious stage over time and even resists therapy if it persists for years.

Contributors to Hypertension

Because we do not know the cause of 95% of the cases of hypertension, we can identify only risk factors that contribute to its development. A family history of hypertension is a risk factor, especially if both parents have (or had) the problem. In addition, blood pressure can increase as a person ages. Some increase is caused by atherosclerosis. As plaque builds up in the arteries, the arteries become less flexible and cannot expand. When vessels remain rigid, blood pressure remains high. Eventually, the plaque begins to decrease the blood supply to the kidneys, decreasing their ability to control blood volume and, in turn, blood pressure.

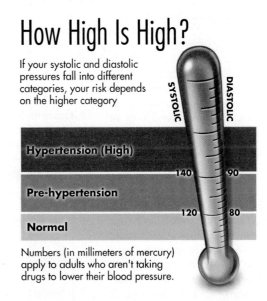

How High Is High?

If your systolic and diastolic pressures fall into different categories, your risk depends on the higher category

SYSTOLIC   DIASTOLIC

Hypertension (High)

140   90

Pre-hypertension

120   80

Normal

Numbers (in millimeters of mercury) apply to adults who aren't taking drugs to lower their blood pressure.

FIGURE 9-30 ▲ The cutoff for hypertension is 140/90 mm Hg, but the risk of heart attacks and stroke precedes the rise in blood pressure.

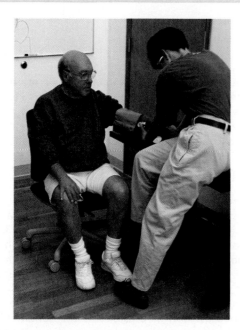

▲ Older adults are particularly at risk of hypertension.

Overweight people have six times greater risk of having hypertension than lean people. Overall, obesity is considered the number one lifestyle factor related to hypertension. This is especially the case in minority populations. Additional blood vessels develop to support excess tissue in overweight and obese individuals, and these extra miles of associated blood vessels increase work by the heart and also blood pressure. Hypertension is linked to obesity if elevated blood insulin levels result from insulin-resistant adipose cells. This increased insulin level augments sodium retention in the body and accelerates atherosclerosis. In such cases, a weight loss of as little as 10 to 15 pounds often can help treat hypertension.

Inactivity is considered the number two lifestyle factor related to hypertension. If an obese person can engage in regular physical activity (at least 5 days per week for 30 to 60 minutes) and lose weight, blood pressure often returns to normal.

Third, excess alcohol intake is responsible for about 10% of all cases of hypertension, especially in middle-aged males and among African-Americans in general. When hypertension is caused by excessive alcohol intake, it is usually reversible. A sensible alcohol intake for people with hypertension is two or fewer drinks per day for men and one or no drinks per day for women and all older adults. Did you recognize that this is the same recommendation given in the 2010 Dietary Guidelines for Americans? Some studies suggest that such a minimal alcohol intake may reduce the risk of ischemic stroke. These data, however, should not be used to encourage alcohol use.

In some people, particularly African-Americans and older overweight persons, blood pressure is especially sensitive to sodium. In these people, excess salt leads to fluid retention by the kidney and a corresponding increase in blood volume, resulting in increased blood pressure. It is not clear whether sodium or chloride is more responsible for the effect. Still, as reviewed in this chapter, if one reduces sodium intake, chloride intake naturally falls. For the most part, a recommendation to consume less sodium is equivalent to a call for less salt in the diet. Only some North Americans are susceptible to increases in blood pressure from salt intake, so it is only the number four lifestyle factor related to hypertension. It is unfortunate that salt intake receives the major portion of public attention with regard to hypertension. Efforts to prevent hypertension should also focus on obesity, inactivity, and excessive consumption of alcohol.

Other Minerals and Blood Pressure

Minerals such as calcium, potassium, and magnesium also deserve attention when it comes to prevention and treatment of hypertension (see Further Reading 22). Studies show that a diet rich in these minerals and low in salt can decrease blood pressure within days of beginning this type of diet, especially among African-Americans. The response is even similar to that seen with commonly used medications. The diet is called the Dietary Approaches to Stop Hypertension (DASH) diet (Table 9-9). The diet is rich in calcium, potassium, and magnesium and low in salt. It takes a standard MyPlate Daily Food Plan; adds one to two extra vegetables and fruits servings; and emphasizes consumption of nuts, seeds, or legumes (beans) 4 to 5 days of the week. In DASH studies, participants also consumed no more than 3 grams of sodium and no more than one to two alcoholic drinks per day. A DASH 2 diet trial tested three daily sodium intakes (3300, 2400, and 1500 milligrams). People showed a steady decline in blood pressure on the DASH diet as sodium intake declined. Overall, the DASH diet is seen as a total dietary approach to treating hypertension. It is not clear which of the many healthful practices of this diet are responsible for the fall in blood pressure.

Other studies also show a reduction in stroke risk among people who consume a diet rich in fruits, vegetables, and vitamin C (recall that fruits and vegetables are rich in vitamin C). Overall, a diet low in salt and rich in low-fat and fat-free dairy products, fruits, vegetables, whole grains, and

TABLE 9-9 ▶ What Is the DASH Diet?
The DASH diet is low in fat and sodium and rich in fruits, vegetables, and low-fat dairy products. Here is the breakdown:

Per Day	Per Week
6–8 servings of grains and grain products	4–5 servings of nuts, seeds, or legumes
4–5 servings of fruit	5 servings of sweets and added sugars
4–5 servings of vegetables	
2–3 servings of low-fat or fat-free dairy products	
2 or fewer servings of meats, poultry, and fish	
2–3 servings of fats/oils	

CASE STUDY Focus on the Positive

Jerome enjoys listening to his nutrition professor's experiences with successes and failures in advising people on diet change. The instructor says that when people focus on doing something positive in their diet, they automatically start to do less of the negative—like substituting one behavior with another. While at the pharmacy the other day, Jerome stopped at the station to check his blood pressure. He did not like the results. He went to the student health center for a recheck, as advised by the American Heart Association. He was diagnosed as prehypertensive, with a blood pressure of 129/85. Jerome was sad but not shocked. He is African-American and has a family history of hypertension. He exercises four times per week, controls stress through playing piano, is close to his family, enjoys friends, and does not drink or smoke. Jerome lives in an apartment with three friends. His mom taught him to cook, which he does three nights per week. Jerome does not skip meals (eats breakfast) and packs a lunch to eat on campus. However, too often dinner is late and he grabs packaged snack foods to ward off hunger until he can fix dinner.

   Jerome talks with his nutrition professor about his dilemma. She wants him to stay positive and asks him to focus on potassium. She tells him that if he focuses on including high-potassium foods, he will receive some added benefits.

   Answer the following questions about Jerome's situation and check your responses at the end of the chapter.

1. Since Jerome is an African-American, why is it prudent for him to do something about his blood pressure right now?
2. In what other nutrients are potassium-rich foods high?
3. How might these foods be beneficial for Jerome in his desire to control his blood pressure?
4. What are some potassium-rich foods that Jerome could have readily available at home to help him ward off hunger until he is ready to eat his dinner? Can he pack any of these for a quick snack?

some nuts can substantially reduce hypertension and stroke risk in many people, especially those with hypertension (see Further Reading 1).

Medications to Treat Hypertension

Diuretic medications are one class of drugs used to treat hypertension. These "water pills" work to reduce blood volume (and therefore blood pressure) by increasing fluid output in the urine. Other medications act by slowing heart rate or by causing relaxation of the small muscles lining the blood vessels. A combination of two or more medications is commonly required to treat hypertension that does not respond to diet and lifestyle therapy.

Prevention of Hypertension

Many of the risk factors for hypertension and stroke are controllable, and appropriate lifestyle changes can reduce a person's risk (Fig. 9-31), depending on the severity of the hypertension. Experts also recommend that those with hypertension lower blood pressure through diet and lifestyle changes before resorting to blood pressure medications.

Advice	Details	Drop in Systolic Blood Pressure
Lose excess weight	For every 20 pounds you lose (if BMI > 25)	5 to 20 points
Follow a DASH diet	Eat a lower-fat diet rich in vegetables, fruits, and low-fat dairy foods	8 to 14 points
Exercise daily	Get 30 minutes a day of aerobic activity (such as brisk walking)	4 to 9 points
Limit sodium	Eat no more than 2400 mg per day (1500 mg per day is better)	2 to 8 points
Limit alcohol	Have no more than 2 drinks per day for men, 1 drink per day for women (1 drink = 12 oz beer, 5 oz wine, or 1.5 oz 80-proof whiskey)	2 to 4 points

FIGURE 9-31 ▶ What works? If your blood pressure is high, here's how much lifestyle changes should lower it.

Source: The Seventh Report of the Joint National Committee on Prevention, Detection, Evaluation, and Treatment of High Blood Pressure (**www.nhlbi.nih.gov/guidelines/hypertension**).

Summary (Numbers refer to numbered sections in the chapter.)

9.1 Water constitutes 50% to 70% of the human body. Its unique chemical properties enable it to dissolve substances as well as serve as a medium for chemical reactions, temperature regulation, and lubrication. Water also helps regulate the acid-base balance in the body.

For adults, daily water needs are estimated at 9 cups (women) to 13 cups (men) per day; fluid intake contributes to meeting this need.

Overall, the United States enjoys a safe water supply. However, people with poor immune status should boil water used for drinking and cooking to avoid waterborne illness. Bottled water can be used, but reliance on it to meet fluid needs is costly to the environment and our personal budgets.

9.2 Minerals are categorized based on the amount we need per day. If we require greater than 100 milligrams of a mineral per day, it is considered a major mineral; otherwise it is considered a trace mineral. Many minerals are vital for sustaining life. For humans, animal products are the most bioavailable sources of most minerals. Supplements of minerals exceeding 100% of the Daily Values listed on the label should be taken only under a physician's supervision. Toxicity and nutrient interactions are especially likely if the Upper Level (when set) is exceeded on a long-term basis.

9.3 Sodium, the major positive ion found outside cells, is vital in fluid balance and nerve impulse transmission. The North American diet provides abundant sodium through processed foods and table salt. About 10% to 15% of the adult population, such as overweight people, is especially sodium-sensitive and at risk for developing hypertension from consuming excessive sodium.

9.4 Potassium, the major positive ion found inside cells, has a similar function to sodium. Milk, fruits, and vegetables are good sources.

9.5 Chloride is the major negative ion found outside cells. It is important in digestion as part of stomach acid and in immune and nerve functions. Table salt supplies most of the chloride in our diets.

9.6 Calcium forms a part of bone structure and plays a role in blood clotting, muscle contraction, nerve transmission, and cell metabolism. Calcium absorption is enhanced by stomach acid and the active vitamin D hormone. Dairy products are important calcium sources. Deficient calcium intake decreases bone mineralization, ultimately leading to osteopenia and osteoporosis. Women are particularly at risk for developing osteoporosis as they age. Numerous lifestyle and medical options can help reduce this risk, including an adequate intake of calcium and many other minerals.

9.7 Phosphorus aids enzyme function and forms part of key metabolic compounds, cell membranes, and bone. It is efficiently absorbed, and deficiencies are rare, although there is concern about possible poor intake by some older women. Good food sources are dairy products, bakery products, and meats.

9.8 Magnesium is a mineral found mostly in plant food sources. It is important for nerve and heart function and as an activator for many enzymes. Whole-grain breads and cereals (bran portion), vegetables, nuts, seeds, milk, and meats are good food sources.

9.9 Iron absorption depends mainly on the form of iron present and the body's need for it. Heme iron from animal sources is better absorbed than the nonheme iron obtained primarily from plant sources. Consuming vitamin C or meat simultaneously with nonheme iron increases absorption. Iron operates mainly in synthesizing hemoglobin and myoglobin and in the action of the immune system. Women are at highest risk for developing iron deficiency, which decreases blood hemoglobin and hematocrit. When this condition is severe, iron-deficiency anemia develops. This decreases the amount of oxygen carried in the blood. Iron toxicity usually results from a genetic disorder called hemochromatosis. This disease causes overabsorption and accumulation of iron, which can result in severe liver and heart damage.

9.10 Zinc aids in the action of up to 200 enzymes important for growth, development, cell membrane structure and function, immune function, antioxidant protection, wound healing, and taste. A zinc deficiency results in poor growth, loss of appetite, reduced sense of taste and smell, hair loss, and a persistent rash. Zinc is best absorbed from animal sources. The richest sources of zinc are oysters, shrimp, crab, and beef. Good plant sources are whole grains, peanuts, and beans.

9.11 An important role of selenium is decreasing the action of free-radical (oxidizing) compounds. In this way, selenium acts along with vitamin E in providing antioxidant protection. Muscle pain, muscle wasting, and a form of heart damage may result from a selenium deficiency. Meats, eggs, fish, and shellfish are good animal sources of selenium. Good plant sources include grains and seeds.

9.12 Iodide forms part of the thyroid hormones. A lack of dietary iodide results in the development of an enlarged thyroid gland or goiter. Iodized salt is a major food source.

9.13 Copper is important for iron metabolism, cross-linking of connective tissue, and other functions, such as enzymes that provide antioxidant protection. A copper deficiency can result in a form of anemia. Copper is found mainly in liver, seafood, cocoa, legumes, and whole grains.

9.14 Fluoride as part of regular dietary intake or toothpaste use makes teeth resistant to dental caries. Most North Americans receive the bulk of their fluoride from fluoridated water and toothpaste.

9.15 Chromium aids in the action of the hormone insulin. Chromium deficiency results in impaired blood glucose control. Egg yolks, meats, and whole grains are good sources of chromium.

9.16 Manganese and molybdenum are used by various enzymes. One enzyme that uses manganese provides antioxidant protection. Clear deficiencies in otherwise healthy people are rarely seen for these nutrients. Human needs for other trace minerals are so low that deficiencies are uncommon

NAYH Controlling weight and alcohol intake; exercising regularly; decreasing salt intake; and ensuring adequate potassium, magnesium, and calcium in the diet all can play a part in controlling high blood pressure.

Check Your Knowledge (Answers to the following questions are below.)

1. Dietary heme iron is derived from
   a. elemental iron in food.
   b. animal flesh.
   c. breakfast cereal.
   d. vegetables.

2. Chloride is
   a. a component of hydrochloric acid.
   b. an intracellular fluid ion.
   c. a positively charged ion.
   d. converted to chlorine in the intestinal tract.

3. Minerals involved in fluid balance are
   a. calcium and magnesium.
   b. copper and iron.
   c. calcium and phosphorus.
   d. sodium and potassium.

4. In a situation where there is an insufficient intake of dietary iodide, the thyroid-stimulating hormone promotes the enlargement of the thyroid gland. This condition is called
   a. Graves' disease.
   b. goiter.
   c. hyperparathyroidism.
   d. congenital hypothyroidism.

5. Ninety-nine percent of the calcium in the body is found in
   a. intracellular fluid.
   b. bones and teeth.
   c. nerve cells.
   d. the liver.

6. At the end of long bones, inside the spinal vertebrae, and inside the flat bones of the pelvis is a spongy type of bone known as _____ bone.
   a. cortical
   b. osteoporotic
   c. trabecular
   d. compact

7. Which compartment contains the greatest amount of body fluid?
   a. intracellular
   b. extracellular
   c. They contain the same amount.

8. The primary function of sodium is to maintain
   a. bone mineral content.
   b. hemoglobin concentration.
   c. immune function.
   d. fluid distribution.

9. Hypertension is defined as a blood pressure greater than
   a. 110/60.
   b. 120/65.
   c. 140/90.
   d. 190/80.

10. Which of the following individuals are most likely to develop osteoporosis?
    a. premenopausal women athletes
    b. women taking estrogen replacement therapy
    c. slender, inactive women who smoke
    d. women who eat a lot of high-fat dairy products

Answer Key: 1. b (LO 9.9), 2. a (LO 9.5), 3. d (LO 9.3), 4. b (LO 9.12), 5. b (LO 9.6), 6. c (LO 9.12), 7. a (LO 9.1), 8. d (LO 9.3), 9. c (LO 9.17), 10. c (LO 9.6)

Study Questions (Numbers refer to Learning Outcomes)

1. Approximately how much water do you need each day to stay healthy? Identify at least two situations that increase the need for water. Then list three sources of water in the average person's diet. **(LO 9.1)**

2. Identify four factors that influence the bioavailability of minerals from food. **(LO 9.2)**

3. What is the relationship between sodium and water balance, and how is that relationship monitored as well as maintained in the body? **(LO 9.3)**

4. List three sources of dietary calcium. Identify two factors that negatively influence the absorption of calcium. Identify two factors that positively influence the absorption of calcium. **(LO 9.6)**

5. Describe two methods that can be used to assess bone density. What demographic groups should have bone density measured? **(LO 9.6)**

6. List three roles of magnesium in the body. Identify two chronic diseases that may be affected by magnesium status. **(LO 9.8)**

7. Describe the symptoms of iron-deficiency anemia, and explain possible reasons they occur. **(LO 9.9)**

8. What is the relationship between iodine, the thyroid gland, and energy metabolism? **(LO 9.12)**

9. Describe the functions of fluoride in the body. List three sources of fluoride. **(LO 9.14)**

10. Explain the function of chromium in carbohydrate metabolism. **(LO 9.15)**

11. List three dietary strategies to lower blood pressure. **(LO 9.17)**

What the Dietitian Chose

Bottled water has become very popular. In 2013, each American consumed more than 31 gallons of bottled water, adding up to about 10.1 billion gallons for the nation. The public perception that bottled water is safer and healthier than tap water has stimulated the current bottle-toting habit. The truth is that bottled and tap water are both regulated: bottled water by FDA and tap water by the Environmental Protection Agency (EPA). FDA requirements for bottled water mimic those of the EPA for water quality, but in neither does the water have to be contaminant-free.

"Pure" is an advertising term and means nothing about the quality of the water. Recent controversy over misleading advertising by major companies such as Pepsico and Coca-Cola that use public source water in their respective products, Aquafina and Dasani, has led to the printing of "Public Source Water" or "PSW" on labels.

Mineral water must contain consistent levels of natural elements from an underground source; no minerals may be added to the water. The precise mineral content varies by source; but some common minerals found in water include calcium, magnesium, potassium, sodium, sulfur, iron, fluoride, zinc, and some ultratrace minerals. The minerals impart some taste to the water but are typically a minor contributor to overall mineral intake.

There is no legal definition for "vitamin water." Manufacturers of vitamin water (i.e., soft-drink companies) usually use filtered or distilled water and add sweeteners (e.g., high-fructose corn syrup) and citric acid as flavoring agents, plus several vitamins (mostly vitamin C and an assortment of B vitamins). In 2010, the Coca-Cola Bottling Company was sued by the Center for Science in the Public Interest for false marketing of its popular vitaminwater® brand as "nutritious." These products are marketed as healthy because they contain added vitamins, but most of them also supply a surprising amount of sugar. Recall that the American Heart Association recommends that men and women get no more than 150 or 100 kcal per day, respectively, from added sugars. A 20-ounce bottle of vitaminwater® has 33 grams of sugar,

providing 132 kcal from added sugar. This is about half the sugar in a regular soft drink, but you can see how this can add up to excess. Furthermore, most North Americans consume adequate amounts of vitamin C and the B vitamins without the aid of these beverages.

Consumers should also be aware of FDA definitions for other types of bottled water.

- Artesian water must come from a confined aquifer.

- Springwater must flow naturally to the surface.

- Purified water is produced through an approved process such as distillation or reverse osmosis.

Relying on bottled water is an expensive habit, personally and environmentally. If you use bottled water to meet your recommended fluid needs, you will spend close to $1500 per year compared with about 50 cents for the same volume of tap water. On a larger scale, it is estimated that close to 90% of water bottles end up in the trash, clogging our landfills or being shipped to other countries for recycling. It is well known that America has some of the cleanest, safest tap water in the world. While following the recommendations to drink more water, hydrate with tap or home-filtered water in a reusable container (see Further Reading 16).

CASE STUDY SOLUTION Worried About Grandma

1. Grace's known risk factors:
   a. Race: Asian
   b. Relative with osteoporosis
   c. Little exercise
   What we don't know:
   a. Body weight: Is she under 127 pounds?
   b. What is her menstrual status? Has she had times of amenorrhea or irregular periods?
   c. Has she had any chronic illnesses, or does she take any medications that may affect her bone mineralization or absorption of calcium?
   d. What are the details about her calcium intake?

2. What Grace does right:
   a. Limits alcohol intake
   b. Does not smoke
   c. Has access to health care professionals (e.g., pharmacy, nursing instructors, and nursing training experience) for advice
   d. Is aware of the issue of calcium and vitamin D. Already takes a calcium supplement.
   Lifestyle changes Grace could make right now:
   a. Exercise. Grace needs to get moving! Because her time is limited, she can accomplish this objective through work. If she takes the bus to work, she could get off a few stops early and walk the remaining blocks. She could talk with her boss/supervisor to schedule her to stock shelves twice a week, allowing her to "lift weights" at work.

3. As a nursing student and employee in a pharmacy, Grace is inside most of the day; she cannot rely on sunlight to synthesize vitamin D. Boston is 42° north of the equator, a location considered to have inadequate sunlight from November through March for vitamin D synthesis. Because Grace is Korean, her skin will take longer to synthesize vitamin D than if she were fair-skinned. To overcome these limitations, Grace must work hard to expose her arms and face to sunlight (between the hours of 11 A.M. to 2 P.M. would be optimal) for 20 minutes, 3 to 5 times per week during the sunny months. She must also work to consciously consume foods with vitamin D or take a vitamin D supplement of 400 IU per day.

4. Korean dishes vary by province, just as food in the United States varies by region. Rice, noodles, tofu, vegetables, and meat are commonly used in Korean dishes. Family meals usually include soup, rice, three to four side dishes, and kimchi. Food is usually steamed, not fried. Tea is usually drunk with meals. A sweet dessert may follow. The tofu (if fortified with calcium) and some vegetables would be sources of calcium. Egg yolks and fish (especially salmon and tuna) would be excellent sources of vitamin D. Dairy foods are not common in Korean dishes.

5. If Grace goes to a sub or other sandwich restaurant, she could ask for low-fat cheese to be added to her sandwiches for added calcium and vitamin D. Asking for tuna salad would also increase her vitamin D intake. At a fast-food restaurant, drinking a carton of skim milk in place of diet soda will add only 90 kcal. Substituting a salad for the fries will reduce any added calories from the choice of milk.

6. Grace should take her calcium supplement with meals rather than on an empty stomach. The meal will increase the acidity of the stomach, thus enhancing calcium absorption.

CASE STUDY SOLUTION Focus on the Positive

1. African-Americans are at higher risk for hypertension than any other racial or ethnic group. It is more common, happens at an earlier age, and is more severe. Approximately 40% of African-Americans have hypertension. The longer it is left untreated, the more severe its complications become. These complications include stroke, heart attack, kidney disease, and heart failure.

2. Fruits and vegetables that are high in potassium tend also to be high in vitamins A (especially beta-carotene) and C. The only exceptions are two white foods: bananas are high in potassium but low in vitamins A and C; potatoes are high in potassium and vitamin C but low in vitamin A. In addition, fruits and vegetables provide plenty of soluble fiber. Whole grains are higher in potassium than refined grains. With whole grains, there is more insoluble fiber. Beans (black, red, and pinto) and peas (chickpeas, split peas, and lentils) are high not only in potassium but also in proteins and soluble fiber. They are naturally low in fat. Low-fat dairy products provide plenty of potassium, calcium, and vitamin D.

3. Eating fruits, vegetables, whole grains, beans, and peas will provide plenty of fiber. Fiber helps with weight management by contributing to satiety. Additionally, soluble fiber helps to lower cholesterol. Beans provide Jerome with low-cost, low-fat sources of

protein, so he can substitute them for meat at meals. When he includes high-potassium foods in his diet, he will naturally begin to increase his intake of fiber and vitamins A and C, while simultaneously lowering his intake of fat and sodium.

4. If Jerome has a time-bake feature on his oven at home, before he leaves for classes, he could set the oven to have a baked potato ready for him the minute he walks through the door. He could also learn to use a Crock-Pot. By slow-cooking food while he is gone, Jerome would come home to a place filled with delicious aromas and healthy food. Carrots, peas, potatoes, chicken, canned tomatoes, celery, spices, and chickpeas can be combined the night before, refrigerated, poured into the Crock-Pot in the morning, and set on low to cook in his absence. Jerome will come home to a meal high that is high in potassium, vitamins A and C, soluble fiber, and protein, and low in fat and sodium. If he adds some crunchy whole-grain bread, he'll have a feast. For a heart-healthy snack, he could pack a mixture of dried fruit, nuts, and cereal, and buy a carton of low-fat milk.

Further Readings

1. Aburto NJ and others: Effect of lower sodium intake on health: Systematic review and meta-analyses. *British Medical Journal* 346:f1326, 2013.

2. Anderson JJ and others: Calcium intakes and femoral and lumbar bone density of elderly U.S. men and women: National Health and Nutrition Examination Survey 2005–2006 analysis. *Journal of Clinical Endocrinology and Metabolism* 97:4531, 2012.

3. Appel LJ and others: The importance of population-wide sodium reduction as a means to prevent cardiovascular disease and stroke: A call to action from the American Heart Association. *Circulation* 123:1138, 2011.

4. a. Bolland MJ and others: Effect of calcium supplements on risk of myocardial infarction and cardiovascular events: Meta-analysis. *British Medical Journal* 341:c3691, 2010, 3. b. Bolland MJ and others: Calcium supplements with or without vitamin D and risk of cardiovascular events: Reanalysis of the Women's Health Initiative limited access dataset and meta-analysis. *British Medical Journal* 342:d2040, 2011.

5. Brauer CAC and others: Incidence and mortality of hip fractures in the United States. *Journal of the American Medical Association* 302:1573, 2009.

6. Burge R and others: Incidence and economic burden of osteoporosis-related fractures in the United States, 2005–2025. *Journal of Bone and Mineral Research* 22:465, 2007.

7. Grimes CA and others: Dietary salt intake, sugar-sweetened beverage consumption, and obesity risk. *Pediatrics* 131:14, 2013.

8. Howard BC: What do recycling symbols on plastics mean? *The Daily Green:* March 2008. Available at: **http://www.thedailygreen.com/green-homes/latest/recycling-symbols-plastics-460321**.

9. King JC: Zinc: An essential but elusive nutrient. *American Journal of Clinical Nutrition* 94:679S, 2011.

10. Lewiecki EM: Current and emerging pharmacological therapies for the management of postmenopausal osteoporosis. *Journal of Women's Health* 19(10):1615, 2009.

11. Mei Z and others: Assessment of iron status in US pregnant women from the National Health and Nutrition Examination Survey (NHANES), 1999-2006. *American Journal of Clinical Nutrition* 93:1312, 2011.

12. National Osteoporosis Foundation: 2013 *Clinician's guide to prevention and treatment of osteoporosis.* Washington, DC: National Osteoporosis Foundation. Available at: **www.nof.org/hcp**.

13. Office of Dietary Supplements, National Institutes of Health: Magnesium fact sheet. Updated 11/4/2013. Available at: **http://ods.od.nih.gov/factsheets/magnesium;** and Calcium fact sheet. Updated 11/21/2013. Available at: **http://ods.od.nih.gov/factsheets/calcium**.

14. Office of Dietary Supplements, National Institutes of Health: Dietary supplement fact sheet: Iron. Updated 4/8/2014. Available at: **http://ods.od.nih.gov/factsheets/iron**.

15. Palmer CA and Gilbert JA: Position of the Academy of Nutrition and Dietetics: The impact of fluoride on health. *Journal of the Academy of Nutrition and Dietetics* 112:1443, 2012.

16. Palmer S: Busting bottled water. *Today's Dietitian* 9(12):60, 2007.

17. Park Y and others: Dairy food, calcium, and risk of cancer in the NIH-AARP diet and health study. *Archives of Internal Medicine* 169:391, 2009.

18. Popkin BM and others: A new proposed guidance system for beverage consumption in the United States. *American Journal of Clinical Nutrition* 83:529, 2006.

19. Rugg-Gunn AJ and Do L: Effectiveness of water fluoridation in caries prevention. *Community Dentistry and Oral Epidemiology* 240 (Suppl. 2): 55, 2012.

20. U.S. Food and Drug Administration. FDA continues to study BPA. *FDA Consumer Update:* March 2012. Available at: **http://www.fda.gov/downloads/ForConsumers/ConsumerUpdates/UCM297971.pdf**.

21. Yadav D and Chandra J: Iron deficiency: Beyond anemia. *Indian Journal of Pediatrics* 78:65, 2011.

22. Yang Q and others: Sodium and potassium intake and mortality among US adults. *Archives of Internal Medicine* 171:1183, 2011.

Rate Your Plate

I. Rate Your Beverage Choices

Water is the key component in the guidelines that have been developed by the Beverage Guidance Panel. (Table 9-10). These recommendations give guidance on the health and nutritional benefits as well as the risks of various beverage categories. The basis of the Beverage Guidance System is that fluids should not provide a significant amount of the energy nutrients in a healthy diet. More specifically, the system recommends that beverages provide less than 10% of total calories consumed for a 2200 kcal diet (see Further Reading 18).

TABLE 9-10 ▶ The Beverage Guidance System

Level	Category*	Recommended Servings per Day
1	Water	50 fluid ounces (1.7 liters)
2	Tea or coffee, unsweetened	0 to 40 fluid ounces (0 to 1.4 liters)
3	Low-fat and skim milk and soy beverages	0 to 16 fluid ounces (0 to 0.5 liter)
4	Noncalorically sweetened beverages (diet drinks)	0 to 32 fluid ounces (0 to 1 liter)
5	Calorie beverages with some nutrients (100% fruit juices, alcoholic beverages, whole milk, and sports drinks)	0 to 8 fluid ounces 100% fruit juices (0 to 0.25 liter) 0 to 1 alcoholic drink for women 0 to 2 alcoholic drinks for men
6	Calorically sweetened beverages (regular soft drinks)	0 to 8 fluid ounces (0 to 0.25 liter)

*Categories established based on their possible health benefits or risks.

Adapted from Popkin and others. A new proposed guidance system for beverage consumption in the United States. *American Journal of Clinical Nutrition* 83:529, 2006.

1. Think of all the beverages you drank yesterday, from the time you woke up to the time you went to bed. Do your best to recall the types and amounts of your fluid intake in the space below.

   _____

   _____

   _____

   _____

2. Next to each beverage choice you listed for question 1, indicate its category from the Beverage Guidance System.

3. Did you come close to the recommendation to consume 50 fluid ounces (1.7 liters) of fluid in the form of water yesterday?

   _____

   _____

4. How many fluid ounces of your beverages fit into categories 2, 3, 4, 5, or 6? Did you exceed the recommended servings per day for these categories?

   _____

   _____

5. What changes could you make in your beverage choices to follow the Beverage Guidance System more closely? How does following the Beverage Guidance System contribute to good health?

   _____

   _____

II. Working for Denser Bones

Osteoporosis and related low bone mass affect many adults in North America, especially older women. One-third of all women experience fractures because of this disease, amounting to about 2 million bone fractures per year.

Osteoporosis is a disease you can do something about. Some risk factors cannot be changed, but others, such as poor calcium intake, can. Is this true for you? To find out, complete this tool for estimating your current calcium intake. For all the following foods, write the number of servings you eat in a day. Total the number of servings in each category and then multiply the total number of servings by the amount of calcium for each category. Finally, add the total amount for each category to estimate your calcium intake for that day.

Does your intake meet your RDA set for calcium?

Food	Serving Size	Number of Servings	Calcium (mg)	Total Calcium (mg)
Plain low-fat yogurt	1 cup	_____		
Fat-free dry milk powder	½ cup	_____		
Total servings		_____	× 400	= _____ mg
Canned sardines (with bones)	3 ounces	_____		
Fruit-flavored yogurt	1 cup	_____		
Milk: fat-free, reduced-fat, whole, chocolate, buttermilk	1 cup			
Calcium-fortified soy, rice, or almond milk (e.g., Silk®)	1 cup	_____		
Parmesan cheese (grated)	¼ cup	_____		
Swiss cheese	1 ounce	_____		
Total servings		_____	× 300	= _____ mg
Cheese (all other hard cheese)	1 ounce	_____		
Pancakes	3	_____		
Total servings		_____	× 200	= _____ mg
Canned pink salmon	3 ounces	_____		
Tofu (processed with calcium)	4 ounces	_____		
Total servings		_____	× 150	= _____ mg
Collards or turnip greens, cooked	½ cup	_____		
Ice cream or ice milk	½ cup	_____		
Almonds	1 ounce	_____	× 75	= _____ mg
Total servings		_____		
Chard, cooked	½ cup	_____		
Cottage cheese	½ cup	_____		
Corn tortilla	1 medium	_____		
Orange	1 medium	_____		
Total servings		_____	× 50	= _____ mg
Kidney, lima, or navy beans, cooked	½ cup	_____		
Broccoli	½ cup	_____		
Carrot, raw	1 medium	_____		
Dates or raisins	¼ cup	_____		
Egg	1 large	_____		
Whole-wheat bread	1 slice	_____		
Peanut butter	2 tablespoons	_____		
Total servings		_____	× 25	= _____ mg

Food	Serving Size	Number of Servings	Calcium (mg)	Total Calcium (mg)
Calcium-fortified orange juice	6 ounces	_____		
Calcium-fortified snack bars	1 each	_____		
Calcium-fortified breakfast bars	½ bar	_____		
	Total servings	_____	× 200	= _____ mg
Calcium-fortified chocolate candies	1 each	_____		
Calcium supplements*	1 each	_____	× 500	= _____ mg
	Total servings	_____	Total calcium intake	= _____ mg

Other calcium sources to consider include many breakfast cereals (100–250 mg per cup) and some vitamin/mineral supplements (200–500 mg or more per tablet).

*Amount varies, so check the label for the amount in a specific product and then adjust the calculation as needed.

Source: Adapted from *Topics in Clinical Nutrition,* "Putting Calcium into Perspective for Your Clients," G. Wardlaw and N. Weese, 11:1, © 1995 Aspen Publishers, Inc.

▲ Milk is a rich as well as convenient source of calcium.

Student Learning Outcomes

Chapter 10 is designed to allow you to:

10.1 List five positive health-related outcomes of a physically active lifestyle.

10.2 Enumerate three key elements of a sound fitness regimen.

10.3 Describe the use of carbohydrates, fat, and protein to meet energy needs during different activities.

10.4 Differentiate between anaerobic and aerobic uses of glucose and identify advantages and disadvantages of each.

10.5 Show how muscles and related organs adapt to an increase in physical activity.

10.6 Outline how to estimate an athlete's calorie needs and discuss the general principles for meeting overall nutrient requirements in the training diet.

10.7 Examine problems associated with weight loss by dehydration and outline the importance of fluids during exercise.

Chapter 10
Nutrition: Fitness and Sports

What Would You Choose?

You are gearing up to run your first half-marathon in a few weeks. From your long training runs, you know that physical and mental fatigue sets in a few miles before the finish. You'd like to try one of the sports nutrition products you've seen at the grocery store. What would you choose to take along during the race to help you make it to the finish line?

a Clif Shot Turbo energy gel with 100 milligrams caffeine, 22 grams of carbohydrate, and electrolytes

b PowerBar ProteinPlus energy bar with 23 grams of protein

c Powerade sports drink with electrolytes, vitamins, and 14 grams of carbohydrate

d Essential Amino Energy drink with branched-chain amino acids

connect NUTRITION Think about your choice as you read Chapter 10, then see **What the Dietitian Chose** at the end of the chapter. To learn more about store-bought and homemade sports nutrition products, check out the Connect site: www.mcgrawhillconnect.com.

Are you a sedentary individual looking to start a personal exercise program? Do you participate in recreational sports, such as intramural volleyball? Or are you a competitive athlete who would like to take your performance to the next level? One thing is certain: at every level of fitness, sensible nutrition and physical activity complement each other in the pursuit of wellness.

Some athletes may fall into the trap of thinking that participation in sports alleviates the need to pay close attention to eating habits. Sports nutritionists and athletic trainers frequently remind their clients that "you can't out-train a poor diet." Indeed, as you will learn in this chapter, adequate intakes of overall calories and specific nutrients can have a big impact on physical performance.

On the other hand, good eating habits cannot substitute for physical training and genetic endowment. Most athletes are on the lookout for any advantage, whether real or perceived, that might enhance performance and give them the winning edge. For this reason, athletes are likely targets for nutrition marketing and misinformation. As you try to sort fact from fiction, be sure of this: no dietary practice or supplement can increase speed or promote gains in muscle mass without proper physical training.

Overall, making informed choices about foods, beverages, and dietary supplements can optimize many aspects of physical performance, from preventing fatigue to gaining muscle to recovering from workouts. In this chapter, you will discover how physical fitness benefits the entire body and how nutrition relates to physical performance. This information is important for the everyday, recreational athlete who exercises to manage weight or just to have fun . . . but it is crucial for elite athletes for whom a fraction of a second can make the difference between first and second place.

10.8 Understand how athletes can optimize performance by consuming foods and fluids before, during, and after exercise.

10.9 List several ergogenic aids and describe their effects, if any, on an athlete's performance.

10.1 An Introduction to Physical Fitness

Healthy People 2020 objectives focus on reducing the proportion of adults who engage in no leisure time physical activity and increasing the proportion of adults who meet the 2008 Physical Activity Guidelines for aerobic and muscle-strengthening physical activity.

physical activity Any movement of skeletal muscles that requires energy.

Ravi and Chandra used to chat over a cup of coffee after their 8:00 a.m. nutrition class, but now they meet at the court for a game of racquetball. Terry listens to recorded political science lectures as he jogs around the track. Claudia takes the stairs instead of the elevator to her dorm room on the fourth floor. Marlie reviews her biology notes while she works out on the elliptical machine. What do these students have in common? They each have found inventive ways to incorporate physical activity into their crazy-busy college schedules.

Without a doubt, the benefits of **physical activity** outweigh the risks for most Americans. The potential benefits of regular physical activity include enhanced heart function, less injury, better sleep habits, and improvement in body composition. Physical activity also can reduce stress and positively affect blood pressure, blood cholesterol, blood glucose regulation, and immune function. In addition, physical activity aids in weight control, both by raising resting energy expenditure for a short time after exercise and by increasing overall energy expenditure. In fact, as physical fitness improves, so does your ability to mobilize fat as a source of energy. See Figure 10-1 and Further Reading 5 for a closer look at these and other benefits of a physically active lifestyle.

It is important to note that physical activity and exercise are not synonymous. Physical activity refers to any movement of skeletal muscles that requires energy. It includes exercise, sports, as well as all the simple, unplanned activities of daily living, such as raking the yard, walking up and down the stairs, and carrying

FIGURE 10-1 ▶ The benefits of regular, moderate physical activity and exercise.

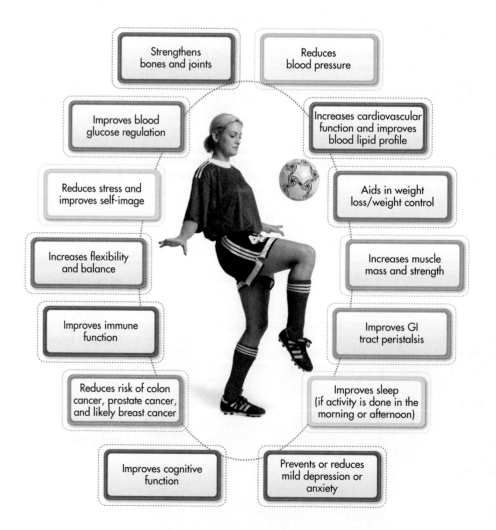

- Strengthens bones and joints
- Reduces blood pressure
- Improves blood glucose regulation
- Increases cardiovascular function and improves blood lipid profile
- Reduces stress and improves self-image
- Aids in weight loss/weight control
- Increases flexibility and balance
- Increases muscle mass and strength
- Improves immune function
- Improves GI tract peristalsis
- Reduces risk of colon cancer, prostate cancer, and likely breast cancer
- Improves sleep (if activity is done in the morning or afternoon)
- Improves cognitive function
- Prevents or reduces mild depression or anxiety

bags of groceries into the house. **Exercise** specifically refers to those physical activities that are planned, repetitive, and intended to improve **physical fitness.** Examples of exercise include walking, biking, swimming, participating in team sports, and running.

The 2008 Physical Activity Guidelines for Americans recommended that all adults should avoid inactivity. Some physical activity is better than none, and adults who participate in any amount of physical activity gain some health benefits. The Physical Activity Guidelines set specific time goals for adults (review Chapter 2):

- For substantial health benefits, adults should do at least 150 minutes per week of **moderate-intensity,** or 75 minutes per week of **vigorous-intensity aerobic physical activity,** or an equivalent combination of moderate- and vigorous-intensity aerobic activity. Activity should be performed in episodes of at least 10 minutes, and preferably, it should be spread throughout the week.
- For additional and more extensive health benefits, adults should increase their physical activity to 300 minutes per week of moderate-intensity, or 150 minutes per week of vigorous-intensity aerobic physical activity, or an equivalent combination of moderate- and vigorous-intensity activity. Additional health benefits are gained by engaging in physical activity beyond this amount.
- Adults should also include **muscle-strengthening activities** that involve all major muscle groups on 2 or more days a week.

These recommendations are consistent with guidance from the American College of Sports Medicine (ACSM; see Further Reading 5).

Complying with the Physical Activity Guidelines is the first step toward managing weight, reducing risks for chronic diseases, and improving physical fitness. Losing excess weight and maintaining weight loss will require more than 150 minutes of physical activity per week.

Unfortunately, as noted in Chapter 7, many North American adults lead sedentary lives. A recent report from the Department of Health and Human Services revealed that more than 80% of American adults fail to achieve levels of physical activity set forth by the Physical Activity Guidelines. When New Year's Day rolls around each year, many of us resolve to make exercise a priority in our lives, but statistics show that about half of all adults quit an exercise program within 3 months of initiation.

exercise Physical activities that are planned, repetitive, and intended to improve physical fitness.

physical fitness The ability to perform moderate to vigorous physical activity without undue fatigue.

moderate-intensity aerobic physical activity Aerobic activity that increases a person's heart rate and breathing to some extent (4–6 on RPE scale). Examples include brisk walking, dancing, swimming, or bicycling on level terrain.

vigorous-intensity aerobic physical activity Aerobic activity that greatly increases a person's heart rate and breathing (7–8 on RPE scale). Examples include jogging, singles tennis, swimming continuous laps, or bicycling uphill.

muscle-strengthening activity Physical activity that increases skeletal muscle strength, power, endurance, and mass. Examples include lifting weights, using weight machines, and calisthenics (e.g., push-ups).

✓ CONCEPT CHECK 10.1

1. Differentiate between physical activity and exercise.
2. How much physical activity is recommended by the Physical Activity Guidelines for Americans to reduce risks for chronic diseases?
3. List the five benefits of regular physical activity that are most important to you.

10.2 Achieving and Maintaining Physical Fitness

Much of the nutrition advice that appears later in this chapter focuses on enhancing the performance of highly competitive athletes. However, only a few students would describe themselves as highly competitive athletes. Furthermore, if you enter a health profession, most of your clients will be at a beginner or intermediate level of fitness. Certainly, proper nutrition supports physical performance at all levels. Perhaps more importantly, nutrition and physical activity complement each other when it comes to improving health outcomes. This section outlines how to get started with a plan to achieve physical fitness.

ASSESS YOUR CURRENT LEVEL OF FITNESS

Start by assessing your current level of fitness. In some cases, it is beneficial to seek medical advice before getting started. Men aged 40 years or older and women aged 50 years or older, anyone who has been inactive for many years, or those who have an existing health problem should discuss their fitness goals with their physician before increasing activity. Health problems that require medical evaluation before beginning an exercise program include obesity, cardiovascular disease (or family history of it), hypertension, diabetes (or family history), chest pains or shortness of breath after mild exertion, and arthritis. Even if you do not have pre-existing medical problems, enlisting the aid of a fitness professional (e.g., a qualified personal trainer) can help you to determine a safe starting point and establish realistic goals.

SET A GOAL

For any behavior change, goal setting will enhance your success. Taking into account your current level of fitness, choose a goal that is challenging yet attainable. For someone who has never exercised regularly, running a mile without stopping might be a worthwhile goal. An elite runner might strive to run that mile in under 4 minutes. Whatever goal you set, write it down and post it where you and everyone else can see it! Based on your long-term goal, also incorporate several smaller, short-term goals into your plan. Measurable, attainable, and realistic goals provide focus and motivation for any endeavor.

PLAN YOUR PROGRAM

What activities should you include in your fitness program? Although your goal may focus on just one aspect of fitness, such as being able to bench press your body weight, a balanced fitness program will include three key types of activities: aerobic exercise, strength training, and flexibility.

Many fitness experts use the FITT principle to design a fitness program. FITT stands for *frequency, intensity, time,* and *type* of exercise. Frequency describes the number of days per week you engage in a particular exercise. Intensity describes how hard you work when you exercise: how much you increase your heart rate or how much resistance you use when you lift weights. Type (also called *mode*) is your choice of activity, such as walking or running. Table 10-1 summarizes ACSM's guidance for planning a general fitness program (see Further Reading 5).

Aerobic Exercise Enhances Heart and Lung Function. As you learned in Chapter 3, *aerobic* means "with oxygen." Aerobic exercises use large muscle groups in a rhythmic fashion and aim to increase your heart rate. The ability to perform aerobic

TABLE 10-1 ▶ Elements of a Well-Rounded Fitness Program

	Aerobic Fitness	Muscular Fitness	Flexibility
Frequency	5 days per week	2 to 3 days per week	2 to 3 days per week
Intensity	55% to 85% of MHR or RPE of 4 or higher (see Fig. 10-3)	40% to 80% of 1 RM (lower to build endurance and higher to build strength)	To the point of tension
Time	20 to 60 minutes per day	1 to 3 sets of 8 to 12 repetitions of 8 to 10 different exercises	2 to 4 repetitions of 8 to 10 different exercises, held for 15 to 30 seconds each
Type (examples)	Brisk walking, running, cycling, swimming, basketball, tennis, and soccer	Bench press, squat, biceps curl, and abdominal crunch	Hamstring stretch, shoulder reach, and side bend

exercise depends on the fitness of your heart and lungs—those organ systems that provide oxygen to the cells of the body.

Aerobic activities usually form the backbone of a fitness program. Indeed, many of the benefits of fitness mentioned in the previous section are direct effects of aerobic training. The ACSM recommends including at least 30 minutes per day of moderate-intensity aerobic activities on 5 days per week. Alternatively, less time spent in aerobic exercises of vigorous intensity can yield similar results. Walking, running, dancing, and biking are examples of aerobic activities.

There are a few ways to determine the intensity of aerobic exercise. A popular and simple method is to use a percentage of your age-predicted maximum heart rate (MHR) (Fig. 10-2). To find your MHR, subtract your age from 220. For a 20-year-old person, MHR equals 200 beats per minute (220 − 20 = 200).

At the initiation of an aerobic exercise program, aim for about 50% to 65% of MHR. Thus, (200 × 0.5) and (200 × 0.65) yields a target training zone of 100 to 130 beats per minute. As you progress and become more physically fit, you can work up to a higher heart rate. For an intermediate fitness enthusiast, 60% to 75% of MHR is recommended. For more experienced exercisers, 70% to 85% MHR is suitable.

Remember that heart rate is just an estimate of workout intensity. Medications, such as those for hypertension and other health conditions, may impact heart rate. If you have health concerns, a physician can help to personalize your target zone.

Another way of determining the intensity of exercise is the Rating of Perceived Exertion (RPE) scale. One version includes a range of 1 to 10, with each number corresponding to a subjective feeling of exertion. For example, the number 0 is "nothing at all" (e.g., sitting at a table), and the number 10 is considered close to maximal effort or "very, very strong" (e.g., all-out sprint; Fig. 10-3).

When using the 10-point RPE scale, the goal is to aim for the number 4, which corresponds to the beginning of "somewhat strong." This is the point at which you begin to see significant fitness results. You should be working hard but still be able to talk to an exercise partner (sometimes called the "talk test").

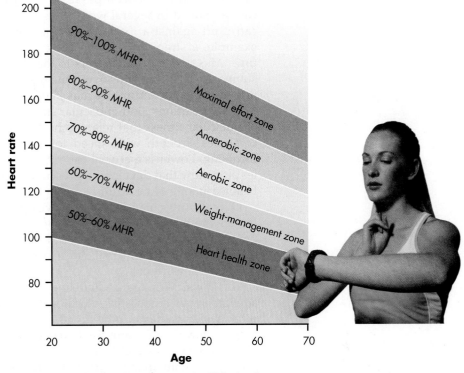

*MHR (maximum heart rate) = 220 - Age in years

FIGURE 10-2 ▲ Heart rate training chart. This chart shows the number of heart beats per minute that corresponds to various exercise intensities.

Measuring heart rate (pulse) is easy: Stop and count your pulse for 10 seconds and then multiply that number by 6 to determine your heart rate for 1 minute. Some exercise equipment and high-tech devices contain heart rate monitors.

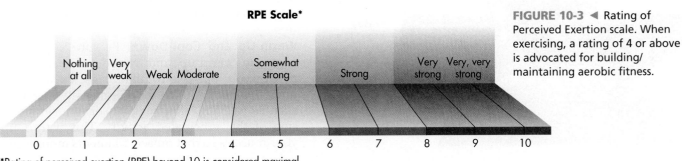

*Rating of perceived exertion (RPE) beyond 10 is considered maximal.

FIGURE 10-3 ◄ Rating of Perceived Exertion scale. When exercising, a rating of 4 or above is advocated for building/maintaining aerobic fitness.

Muscular Fitness Encompasses Strength, Endurance, and Power. Muscles can be trained in several ways. *Muscular strength* is the maximal force a muscle can exert against a load at one time (one repetition maximum or 1 RM). For example, a strength-trained athlete may be able to dead-lift 400 pounds one time. *Muscular endurance* refers to the ability of the muscle to perform repeated, submaximal contractions over time without becoming fatigued. An athlete training for muscular endurance may bench press 80 to 100 pounds for several sets of 8 to 12 repetitions. Both muscular strength and endurance are important aspects of muscular fitness that relate to health for athletes of all levels. *Muscular power* combines strength with speed for explosive movements such as jumping or throwing. Power is a crucial aspect of muscular fitness for many athletes. Studies also show that developing muscular power can help to improve function and balance among older adults.

Overall, muscular fitness is developed by performing resistance exercises for all the major muscle groups of the body, including the arms and shoulders, back, abdominals, and legs. This resistance may come from free weights (e.g., barbells), weight machines (e.g., leg press), or the weight of your own body (e.g., push-ups).

The Physical Activity Guidelines and ACSM recommend including muscle-strengthening activities in your fitness program on 2 to 3 nonconsecutive days per week. (Taking a day or more to rest between bouts of resistance exercise is important to allow time for muscles to recover and increase in size.) Performing regular resistance exercises will help you become stronger and better able to handle the activities of daily living. It promotes increases in lean mass, which assist efforts at weight management. Compared to aerobic exercise alone, resistance exercise can also provide further reductions in risks for cardiovascular diseases, osteoporosis, and type 2 diabetes.

Flexibility Exercises Enhance Balance and Stability. Flexibility is an often-overlooked aspect of physical fitness, and it tends to decline with age. It refers to the ability to move a joint through its full range of motion. Poor flexibility is often linked to chronic pain, especially in the lower back. Contrary to popular belief, research studies do not clearly support a role of flexibility exercises in preventing injury or muscle soreness from aerobic or strength-training activities. However, gains in flexibility can improve balance and stability, thereby reducing risks of falls and injuries, especially among older adults.

ACSM recommends performing flexibility exercises at least 2 to 3 days per week. It is best to perform stretches when the muscles have already been warmed up—after a few minutes of light-intensity aerobic activity. Examples of flexibility exercises include hamstring stretch, side bends, and shoulder reach. Some forms of exercise, such as Pilates or tai chi, combine stretching with strength moves.

Warm-Up and Cool-Down. Be sure to plan for adequate warm-up and cool-down periods as part of your exercise routine. Begin by warming up with 5 to 10 minutes of low-intensity exercises, such as walking, slow jogging, or any slow version of the anticipated activity. This warms up your muscles so that muscle filaments more easily

Josie's Fitness Plan

Monday
Spinning class (45 min)
Stretching routine (20 min)

Tuesday
Walking (30 min)
Weight machines (30 min)

Wednesday
Spinning class (45 min)
Stretching routine (20 min)

Thursday
Walking (30 min)
Weight machines (30 min)

Friday
Yoga (30 min)

Saturday
Trail hiking (60 min)

Sunday
Rest

slide over one another to increase range of motion and decreases the risk of injury. It is also thought to lower cardiovascular risks, particularly among people who are not accustomed to regular exercise. During cool-down, slow down for 5 to 10 minutes of low-intensity activity followed by 5 to 10 minutes of stretching. The same exercises performed during warm-up are appropriate. Although the cool-down does not actually prevent muscle soreness, it does reduce the dizziness or light-headedness that can occur with an abrupt end to a vigorous workout.

Get Started. For sedentary people who are otherwise healthy, gradual **progression** toward a goal of regular physical activity is recommended. During the first phase of a fitness program to promote health, you should begin to incorporate short periods of physical activity into your daily routine. This includes walking, taking the stairs instead of the elevator, house cleaning, gardening, and other activities that cause you to "huff and puff" a bit. A sensible goal is a total of 30 minutes of this moderate type of physical activity on most (and preferably all) days of the week. If necessary, this can be broken up into increments lasting at least 10 minutes. Experts suggest starting with short intervals, building up to a total of 30 minutes of activity incorporated into each day's tasks. If there is not much time for activity, you can get similar benefits from performing shorter sessions of increased intensity, such as running instead of walking or biking on hilly instead of flat terrain.

Once you can perform physical activity for 30 minutes per day, turn your attention to more specific goals, such as increasing muscle mass and strength, to reap even more benefits.

✓ CONCEPT CHECK 10.2

1. Why is goal-setting important for the success of an exercise program?
2. What is the FITT principle? Demonstrate how you would use the FITT principle to make a plan for aerobic fitness.
3. Differentiate between muscular strength, muscular endurance, and muscular power.
4. List two evidence-supported benefits of flexibility exercises.
5. Provide three tips for a person who needs help with long-term maintenance of his or her fitness program.

10.3 Energy Sources for Exercising Muscles

Like other cells, muscle cells cannot directly use the energy released from breaking down glucose or triglycerides. Muscle cells need a specific form of energy for contraction. Body cells must first convert food energy (i.e., calories) to **adenosine triphosphate (ATP).**

The chemical bonds between phosphates in ATP and related molecules are high-energy bonds. Using the energy obtained from foodstuffs, cells make ATP from its breakdown product **adenosine diphosphate (ADP)** and a phosphate group (abbreviated P_i). Conversely, to release energy from ATP, cells partially break the compound down into ADP and P_i. The released energy is used for many cell functions (see margin).

progression Incremental increase in frequency, intensity, and time spent in each type of physical activity over several weeks or months.

adenosine triphosphate (ATP) The main energy currency for cells. ATP energy is used to promote ion pumping, enzyme activity, and muscular contraction.

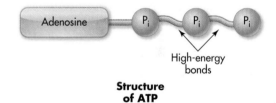

Structure of ATP

adenosine diphosphate (ADP) A breakdown product of ATP. ADP is synthesized into ATP using energy from foodstuffs and a phosphate group (abbreviated P_i).

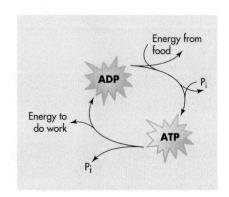

phosphocreatine (PCr) A high-energy compound that can be used to re-form ATP. It is used primarily during bursts of activity, such as lifting and jumping.

creatine An organic (i.e., carbon-containing) molecule in muscle cells that serves as a part of a high-energy compound (termed creatine phosphate or phosphocreatine) capable of synthesizing ATP from ADP.

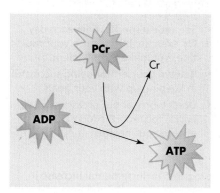

pyruvic acid A three-carbon compound formed during glucose metabolism; also called *pyruvate.*

lactic acid A three-carbon acid formed during anaerobic cell metabolism; a partial breakdown product of glucose; also called *lactate.*

anaerobic Not requiring oxygen.

aerobic Requiring oxygen.

ANAEROBIC METABOLISM SUPPLIES ENERGY FOR SHORT BURSTS OF INTENSE ACTIVITY

Stored ATP. Essentially, ATP is the immediate source of energy for body functions (Table 10-2). The primary goal in the use of any fuel, whether carbohydrate, fat, or protein, is to make ATP. A resting muscle cell contains only a small amount of ATP that can be used immediately. This amount of ATP could keep the muscle working maximally for only about 2 to 4 seconds if no resupply of ATP were possible. Fortunately, the cells have various mechanisms to resupply ATP. Overall, cells must constantly and repeatedly use and then re-form ATP, using a variety of energy sources.

Phosphocreatine. As soon as ATP stored in muscle cells begins to be used, another high-energy compound, **phosphocreatine (PCr),** is used to resupply ATP. An enzyme in the muscle cell is activated to split PCr into phosphate and **creatine.** This releases energy that can be used to re-form ATP from its breakdown products. If no other source of energy for ATP resupply were available, PCr could probably maintain maximal muscle contractions for about 10 seconds.

The main advantage of PCr is that it can be activated instantly and can replenish ATP at rates fast enough to meet the energy demands of the fastest and most powerful actions, including jumping, lifting, throwing, and sprinting. The disadvantage of PCr is that not much of it is made and stored in the muscles. Strength-training athletes sometimes use creatine supplements in an effort to increase PCr in muscles (see Nutrition and Your Health at the end of this chapter).

Anaerobic Glucose Breakdown. Carbohydrates are an important fuel for muscles. The most useful form of carbohydrate fuel is the simple sugar glucose, available to all cells from the bloodstream. As you will recall from Chapter 4, glucose is stored as glycogen in the liver and muscle cells. Blood glucose is maintained by the breakdown of liver glycogen. Breakdown of glycogen stored in a specific muscle also helps meet the carbohydrate demand of that muscle, but the actual amount of glycogen stored in muscle is limited (about 350 grams for all the muscles in the body, which would yield about 1400 kcal).

When oxygen supply in the muscle is limited (anaerobic conditions), glucose is broken down into a three-carbon compound called **pyruvic acid.** The pyruvic acid accumulates in the muscle and is then converted to **lactic acid.** Only

TABLE 10-2 ▶ Energy Sources Used by Resting and Working Muscle Cells

Energy Source*	When in Use	Activity
ATP	At all times	All types
Phosphocreatine (PCr)	All exercise initially; short bursts of exercise thereafter	Shotput, high jump, bench press
Carbohydrate (anaerobic)	High-intensity exercise, especially lasting 30 seconds to 2 minutes	200-yard (about 200 meters) sprint
Carbohydrate (aerobic)	Exercise lasting 2 minutes to several hours; the higher the intensity (for example, running a 6-minute mile), the greater the use	Basketball, swimming, jogging, power walking, soccer, tennis
Fat (aerobic)	Exercise lasting more than a few minutes; greater amounts are used at lower exercise intensities	Long-distance running, long-distance cycling; much of the fuel used in a 30-minute brisk walk is fat
Protein (aerobic)	Low amount during all exercise; slightly more in endurance exercise, especially when carbohydrate fuel is depleted	Long-distance running

*At any given time, more than one source is used. The relative amount of use differs during various activities.

about 5% of the total amount of ATP that could be formed from complete breakdown of glucose is released through this anaerobic process (Fig. 10-4).

The advantage of anaerobic glucose breakdown is that it is the fastest way to resupply ATP, other than PCr breakdown. It therefore provides most of the energy needed for events that require a quick burst of energy, ranging from about 30 seconds to 2 minutes. Examples of activities that primarily rely on anaerobic glucose breakdown include sprinting 400 meters or swimming 100 meters.

The two major disadvantages of the anaerobic process are that (1) the high rate of ATP production cannot be sustained for long periods and (2) the rapid accumulation of lactic acid increases the acidity of the muscle. Normally, the pH of the muscle is about 7.1. Intense exercise that relies on anaerobic glucose breakdown can decrease the pH of muscle tissue to about 6.5. Acidity inhibits the activities of key enzymes in the muscle cells, slowing anaerobic ATP production and causing short-term fatigue. In addition, the acidity leads to a net potassium loss from muscle cells, which also contributes to fatigue. During exercise, you need to choose a pace that you can sustain according to the goals of the activity.

Before long, the muscle cells release the accumulating lactic acid into the bloodstream. The liver (and the kidneys, to some extent) takes up the lactic acid and resynthesizes it into glucose. Glucose can then reenter the bloodstream, where it is available for cell uptake and breakdown. Individuals vary in their ability to clear lactic acid from the muscles and recycle it. Physical training may improve the ability of the body to remove and recycle lactic acid. The Nutrition and Your Health section at the end of this chapter describes an ergogenic aid to neutralize lactic acid (see Further Reading 3).

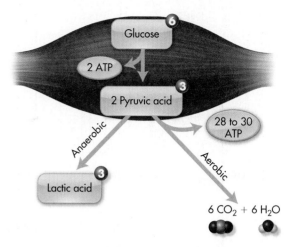

FIGURE 10-4 ▲ ATP yield from aerobic versus anaerobic glucose use. Encircled numbers indicate the number of carbons in each molecule.

AEROBIC METABOLISM FUELS PROLONGED, LOWER-INTENSITY ACTIVITY

Carbohydrates. If plenty of oxygen is available in the muscle (aerobic conditions), such as when the exercise is of low to moderate intensity, the bulk of the three-carbon pyruvic acid is shuttled to the mitochondria of the cell, where it is fully metabolized into carbon dioxide (CO_2) and water (H_2O) (Fig. 10-5). This aerobic breakdown of glucose yields approximately 95% of the ATP made from complete glucose metabolism (glucose → → CO_2 + H_2O).

Aerobic glucose breakdown supplies more ATP than does the anaerobic process, but it releases the energy more slowly. This slower rate of aerobic energy supply can be sustained for hours. One reason is that the products are carbon dioxide and water, not lactic acid. Aerobic glucose breakdown makes a major energy contribution to activities that last anywhere from 2 minutes to several hours. Examples of such activities include jogging or distance swimming (review Table 10-2).

Endurance athletes sometimes reach a point in an event at which extreme physical and mental fatigue sets in; it feels impossible to stand up, let alone continue competing. Long-distance runners call this phenomenon "hitting the wall," and cyclists refer to it as "bonking." This occurs because muscle glycogen has been depleted and blood glucose has begun to decline during exercise, leading to deterioration of both physical and mental function. As you will learn in Section 10.5, maximizing glycogen storage before exercise, supplying carbohydrates during activity, and replenishing glycogen stores between events can help athletes avoid the game-stopping effects of glycogen depletion.

▲ Bursts of muscle activity, such as the 100-meter freestyle, use a variety of energy sources, including ATP, PCr, and glucose.

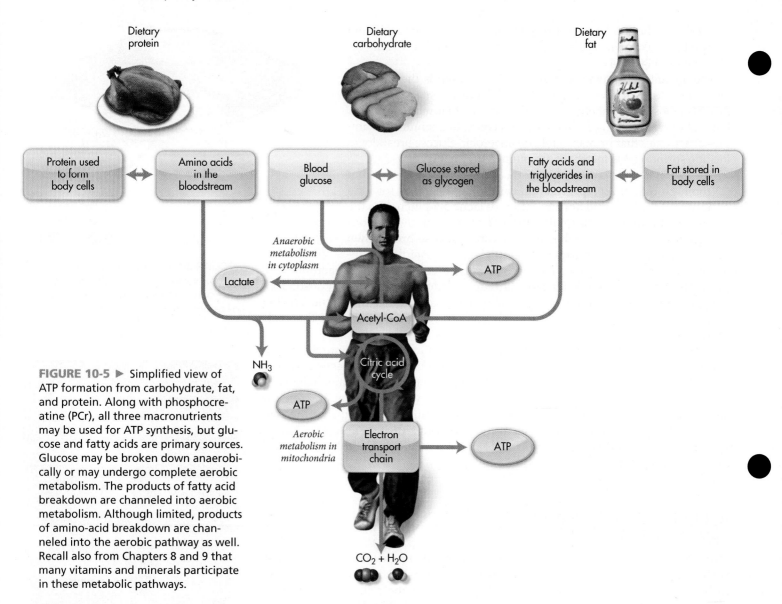

Dietary protein

Dietary carbohydrate

Dietary fat

| Protein used to form body cells | ⟷ | Amino acids in the bloodstream | | Blood glucose | ⟷ | Glucose stored as glycogen | | Fatty acids and triglycerides in the bloodstream | ⟷ | Fat stored in body cells |

Anaerobic metabolism in cytoplasm

Lactate

ATP

Acetyl-CoA

NH_3

Citric acid cycle

ATP

Aerobic metabolism in mitochondria

Electron transport chain

ATP

$CO_2 + H_2O$

FIGURE 10-5 ▶ Simplified view of ATP formation from carbohydrate, fat, and protein. Along with phosphocreatine (PCr), all three macronutrients may be used for ATP synthesis, but glucose and fatty acids are primary sources. Glucose may be broken down anaerobically or may undergo complete aerobic metabolism. The products of fatty acid breakdown are channeled into aerobic metabolism. Although limited, products of amino-acid breakdown are channeled into the aerobic pathway as well. Recall also from Chapters 8 and 9 that many vitamins and minerals participate in these metabolic pathways.

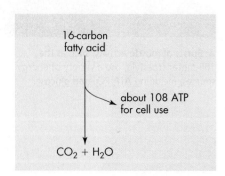

16-carbon fatty acid

about 108 ATP for cell use

$CO_2 + H_2O$

▲ ATP yield from aerobic fatty acid utilization.

Fat. When fat stores in body tissues begin to be broken down for energy, each triglyceride first yields three fatty acids and a glycerol. The majority of the stored energy is found in the fatty acids. During physical activity, fatty acids are released from various adipose tissue depots into the bloodstream and travel to the muscles, where they are taken into each cell and broken down aerobically to carbon dioxide and water. Some of the fat stored in muscles (intramuscular triglycerides) also is used, especially as activity increases from a low to a moderate pace (see Further Reading 4).

Fat is an advantageous fuel for muscles: we generally have plenty of it stored, and it is a concentrated source of energy. For a given weight of fuel, fat supplies more than twice as much energy as carbohydrate does. However, the ability of muscles to use fat for fuel depends on the intensity of exercise. During intense, brief exercise, muscles may not be able to use much fat. The reason for this is that some of the steps involved in fat breakdown cannot occur fast enough to meet the ATP demands of short-duration, high-intensity exercise. However, fat becomes a progressively more important energy source as duration increases, especially when exercise remains at a low or moderate (aerobic) rate for more than 20 minutes (Fig. 10-6).

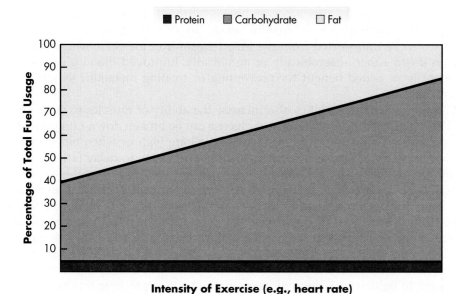

FIGURE 10-6 ▲ Generalized relationship between fuel use and exercise intensity. At rest or during light activity, nearly equal amounts of carbohydrate and fat are used to generate ATP. As activity intensifies (e.g., sprinting), anaerobic processes supply quick fuel, so the relative proportion of carbohydrates used for fuel increases. Except during endurance exercise, very little protein is used for fuel.

For lengthy activities at a moderate pace (for example, hiking) or even sitting at a desk for 8 hours a day, fat supplies about 70% to 90% of the energy required. Carbohydrate use is much less. As intensity increases, carbohydrate use goes up and fat use decreases. During a 5-mile run at a moderate pace, muscles use about a 50:50 ratio of fat to carbohydrate. In comparison, for a sprint, the contribution of fat to resupply ATP is minimal. To summarize, remember that the only fast-paced (anaerobic) fuel we eat is carbohydrate; slow and steady (aerobic) activity uses fat in addition to carbohydrate.

Protein. Although amino acids derived from protein can be used to fuel muscles, their contribution is relatively small, compared with that of carbohydrate and fat. Most protein is reserved for building and repairing body tissues and for synthesizing important enzymes, hormones, and transporters. As a rough guide, only about 5% of the body's energy comes from the metabolism of amino acids.

During endurance exercise, proteins can contribute importantly to energy needs, perhaps as much as 10% to 15%, especially as glycogen stores in the muscle are exhausted. Most of the energy supplied from protein comes from metabolism of the branched-chain amino acids: leucine, isoleucine, and valine. A normal diet provides ample branched-chain amino acids to supply this amount of fuel, so protein or amino-acid supplements are not typically needed.

CAN PHYSICAL TRAINING AFFECT FUEL USE?

As people start exercising regularly (e.g., a minimum of 150 minutes per week of moderate-intensity aerobic activity), they experience a "training effect." Initially, these individuals might be able to exercise for 20 minutes before tiring. Months later, exercise can be extended to an hour before they become fatigued. The training effect results from changes in the ability of exercising cells to use food fuel to generate ATP.

▲ Fatty acids can come from all over the body, not necessarily from fat stored near the active muscles. *This is why spot reducing does not work.* Exercise can tone the muscles near adipose tissue but does not preferentially use those stores.

▲ The calories needed to perform come from carbohydrate, fat, and protein. The relative mix depends on the pace.

Almost immediately after a person begins a fitness program, both aerobic and strength-training improve the insulin sensitivity of cells. In other words, more glucose can be transported from the bloodstream into the cells, where it can be broken down either anaerobically or aerobically. Improved blood glucose management is an added benefit for preventing or treating metabolic syndrome or type 2 diabetes.

Endurance aerobic activities also increase the ability of muscles to store glycogen. This highly branched polymer of glucose can be broken down into individual glucose units when the energy needs of the cell are high or when blood glucose levels start to drop. Increasing glycogen storage will help to delay fatigue during prolonged exercise.

Training induces elevations in muscle triglyceride content. This enhances the ability of muscles to use triglycerides to meet energy needs, particularly during low- or moderate-intensity endurance exercise. As you just learned, fat is a concentrated source of calories; complete oxidation of a long-chain fatty acid yields about three times as much ATP as metabolism of glucose. Many endurance athletes attempt to train their muscles to more readily use fat for fuel and thereby conserve muscle glycogen. In Section 10.5, you will learn more about this technique, known as **fat adaptation.**

fat adaptation Manipulating the diet and physical training regimen so that muscles become more efficient at metabolizing fat as fuel during aerobic activity.

Protein use becomes more efficient with training, too. Endurance training increases the ability of muscle cells to use branched-chain amino acids for fuel during prolonged activity. However, the ability to use carbohydrates and fats for fuel is also increased; as long as the diet is adequate in carbohydrates and fat, most protein is spared for muscle synthesis and repair.

In addition, training increases the number of mitochondria within muscle cells. Recall from Chapter 3 that mitochondria are the powerhouses of the cells; this is where glucose and fat are broken down aerobically to generate ATP. With more mitochondria, muscle cells can use carbohydrates and fat more efficiently.

Overall, the cardiovascular and respiratory systems become more efficient at providing oxygen to the cells of the body. Plasma volume increases shortly after a training program is started, and red blood cell volume eventually increases as well. The heart pumps more blood with each contraction. Training also increases the number of capillaries in muscles, which increases oxygen supply to the muscles. Meanwhile, lung capacity increases, so more oxygen is available. The increased supply of oxygen translates into more efficient aerobic metabolism of carbohydrates and fats. Thus, lactic acid production from anaerobic glucose metabolism decreases. Lactic acid contributes to short-term muscle fatigue, so the less lactic acid produced, the longer the exercise can be sustained.

Through all these adaptations, physical training improves the ability of cells to convert food energy into fuel for exercise.

CRITICAL THINKING

Marty started going to the gym about 8 weeks ago. At first, he noticed that he began "huffing and puffing" about 7 minutes into his aerobic workout. Now, however, he can work out for about 25 minutes without tiring. What is a possible explanation for this ability to work out longer?

✓ CONCEPT CHECK 10.3

1. Describe one process used to resupply ATP during a short, intense burst of activity, such as a 100-meter sprint.

2. How does the ATP yield of anaerobic breakdown of glucose compare to that of aerobic breakdown of glucose?

3. Why is fat a useful source of energy during exercise? Name three types of activity during which fat supplies 50% or more of fuel.

4. Is protein a useful source of energy during exercise? Why or why not?

10.4 Tailoring Nutrient Recommendations for Athletes

Athletic training and genetic makeup are two important determinants of athletic performance. A good diet won't substitute for either factor, but diet can help to enhance and maximize an athlete's potential. On the other hand, a poor diet can seriously reduce performance (see Further Readings 13 and 19).

CALORIES

The daily calorie needs of athletes are highly individualized; genetics, hormones, age, sex, body size and composition, and training volume influence energy expenditure. A small, female gymnast may need only 1800 kcal daily to sustain her training regimen without losing body weight, whereas a large, muscular football player may need 4000 kcal per day. Because athletes are such a heterogeneous group, there is no perfect equation to estimate their daily calorie needs. Even for nonathletes, the Estimated Energy Requirement (EER) equation (see Chapter 7) provides only a rough approximation. However, you can use the EER equation as a starting point and individualize recommendations based on trial and error.

An estimate of the calories required to sustain moderate activity is 5 to 8 kcal per minute. The calories required for sports training or competition then have to be added to those used to carry on normal activities. For example, consider a 135-pound young woman who requires 2200 kcal per day to fuel her normal activities. If she starts teaching two 45-minute Zumba classes each day, she will need about 500 extra kcal (a total of 2700 kcal per day) to maintain her current body weight. If an athlete experiences daily fatigue, the first consideration should be whether he or she is consuming enough food. Up to six meals per day may be needed, including one before each workout.

How can we know if an athlete is getting enough calories? Estimating daily intake from a food diary kept by the athlete is one way. Another step is to estimate the athlete's body fat percentage via skinfold measurements, bioelectrical impedance, or underwater weighing (review Chapter 7). Body fat should be the typical amount found for athletes in the specific sport practiced. This corresponds to 5% to 18% for most male athletes and 17% to 28% for most female athletes. The next step is to monitor body weight changes on a daily or weekly basis. If body weight starts to fall, calories should be increased; if weight rises and it is because of increases in body fat, the athlete should eat less.

If the body composition test shows that an athlete has too much body fat, the athlete should lower food intake by about 200 to 500 kcal per day, while maintaining a regular exercise program, until the desirable fat percentage is achieved. On the other hand, if an athlete needs to gain weight, increasing food intake by 500 to 700 kcal per day will eventually lead to the needed weight gain. A mix of carbohydrate, fat, and protein is advised, coupled with exercise to make sure this gain is mostly in the form of lean tissue and not fat stores.

Historically, athletes who competed in sports with weight classes (e.g., wrestlers, boxers, judoists, and oarsmen) would try to lose weight before a competition. Being certified to compete in a lower weight class could help an athlete gain a mechanical advantage over an opponent of smaller stature. Many of the methods used to cut weight are unhealthy and dangerous. For example, an athlete could lose up to 22 pounds (10 kilograms) of body weight as water in 1 day by sitting in a sauna, exercising in a plastic sweat suit, or taking diuretic drugs, which speed water loss via the kidneys. As you learned in Chapter 9, losing as little as 2% of body weight by dehydration can adversely affect physical and mental performance, especially in hot weather. A pattern of repeated weight loss or gain of more than 5% of body

▲ Intense athletic training may require thousands of additional calories. This increased food intake should easily provide ample protein and other nutrients to support activity.

Review Table 7-5 in Chapter 7, which listed the energy costs of typical forms of physical activity.

weight by dehydration carries risk of kidney malfunction and heat-related illness. Death is also a possibility.

To discourage such unhealthy practices and prevent future deaths, the National Collegiate Athletic Association and many states have authorized physicians or athletic trainers to set safe weight and body fat content minimums (e.g., 7% or more of total body weight for male athletes and 12% or more for females) in weight-class sports. Under current guidelines, athletes are assigned to weight classes at the beginning of the season and are not allowed to "cut weight" to gain a competitive advantage. In addition, urine specific gravity is measured prior to events to check for hydration status. This practice deters athletes from dehydrating themselves before weigh-in. Weight gain in the days after a competition (reflecting regain of body water) can now be no greater than 2 pounds. If athletes, such as wrestlers, wish to compete in a lower-bodyweight class and have enough extra fat stores, they should begin a gradual, sustained reduction in calorie intake long before the competitive season starts.

CARBOHYDRATES

Anyone who exercises vigorously, especially for more than 1 hour per day on a regular basis, needs to consume a diet that includes moderate to high amounts of carbohydrates. Numerous servings of varied grains, starchy vegetables, and fruits provide enough carbohydrate to maintain adequate liver and muscle glycogen stores, especially for replacing glycogen losses from workouts on the previous day.

▲ High-carbohydrate foods should form the basis of the diet for athletes.

Depletion of carbohydrate ranks just behind depletion of fluid and electrolytes as a major cause of fatigue. To prevent chronic fatigue and load the muscles and liver with glycogen, carbohydrate intake should be at least 6 grams per kilogram of body weight. As the duration of aerobic exercise increases, carbohydrate needs can increase to as much as 10 grams per kilogram of body weight. Most sports nutrition experts recommend that carbohydrates should make up around 60% of total kilocalories. For endurance athletes who train for several hours per day, carbohydrate intake may be 600 grams per day (or more). Attention to carbohydrate intake is especially important when performing multiple training bouts in a day (e.g., 2-a-day swim practices) or heavy training on successive days (e.g., cross-country running). During short events (e.g., 30 minutes or so), carbohydrate intake during exercise is not as important because the muscles do not take up much blood glucose during short-term exercise, relying instead primarily on their glycogen stores for fuel.

Table 10-3 gives sample menus, based on MyPlate's Daily Food Plan recommendations, for diets providing food energy ranging from 1500 to 5000 kcal per day. In addition, the *Choose Your Foods* system designed for people with diabetes (see Appendix B) is a very useful tool for planning all types of diets, including high-carbohydrate diets for athletes. As noted, athletes should obtain at least 60% of their total energy needs from carbohydrate (rather than the 50% typical of most North American diets), especially if exercise duration is expected to exceed 2 hours and total caloric intake is about 3000 kcal per day or less. Diets providing 4000 to 5000 kcal per day can be as low as 50% carbohydrate, as these will still provide sufficient carbohydrate (e.g., 500 to 600 grams or so per day).

One does not have to give up any specific food when planning a high-carbohydrate diet. The focus is to include more high-carbohydrate foods while moderating concentrated fat sources. Sports nutritionists emphasize the difference between a high-carbohydrate meal and a high-carbohydrate/high-fat meal. Before endurance events, such as marathons or triathlons, some athletes seek to increase their carbohydrate reserves by eating foods such as potato chips, French fries, banana cream pie, and pastries. Although such foods provide carbohydrate,

TABLE 10-3 ▶ **Sample Daily Menus Based on MyPlate's Daily Food Plan***

1500 kcal	2000 kcal	3000 kcal	4000 kcal	5000 kcal
Breakfast Fat-free milk, 1 cup Cheerios, 1/2 cup Bagel, 1/2 Cherry jam, 2 tsp Margarine, 1 tsp	**Breakfast** Fat-free milk, 1 cup Cheerios, 1 cup Bagel, 1/2 Cherry jam, 1 tbsp Margarine, 1 tsp	**Breakfast** Fat-free milk, 1 cup Cheerios, 2 cups Bagel, 1 Cherry jam, 2 tsp Margarine, 1 tsp Bran muffins, 2	**Breakfast** Fat-free milk, 1 cup Cheerios, 2 cups Orange, 1 Bran muffins, 2	**Breakfast** Fat-reduced milk, 1 cup Cheerios, 2 cups Bran muffins, 2 Orange, 1
			Snack Chopped dates, 3/4 cup	**Snack** Low-fat yogurt, 1 cup Chopped dates, 1 cup

Lunch
Chicken breast (roasted),
  2 oz
Figs, 1
Fat-free milk, 1/2 cup
Banana, 1
	Lunch
	Chicken breast (roasted),
	  2 oz
	Wheat bread, 2 slices
	Mayonnaise, 1 tsp
	Raisins, 1/4 cup
	Cranberry juice, 1 1/2
	  cups
	Banana, 1
		Lunch
		Chicken breast (roasted),
		  2 oz
		Wheat bread, 2 slices
		Provolone cheese, 1 oz
		Mayonnaise, 1 tsp
		Raisins, 1/3 cup
		Cranberry juice,
		  1 1/2 cups
		Low-fat fruit yogurt,
		  1 cup
			Lunch
			Romaine lettuce, 1 cup
			Garbanzo beans, 1 cup
			Grated carrots, 1/2 cup
			French dressing, 2 tbsp
			Macaroni and cheese,
			  3 cups
			Apple juice, 1 cup
				Lunch
				Chicken enchilada, 1
				Romaine lettuce, 1 cup
				Garbanzo beans, 1 cup
				Shredded carrots, 3/4 cup
				Chopped celery, 1/2 cup
				Seasoned croutons, 1 oz
				French dressing, 2 tbsp
				Wheat bread, 2 slices
				Margarine, 1 tbsp
				Apple juice, 1 cup

Snack
Oatmeal-raisin cookie, 1
Low-fat fruit yogurt, 1
  cup
	Snack
	Oatmeal-raisin cookies, 3
	Low-fat fruit yogurt,
	  1 cup
		Snack
		Banana, 1
		Oatmeal-raisin cookies, 3
			Snack
			Wheat bread, 2 slices
			Margarine, 1 tsp
			Jam, 2 tbsp
				Snack
				Banana, 1
				Bagel, 1
				Cream cheese, 1 tbsp

Dinner
Spaghetti w/ meatballs,
  1 cup
Romaine lettuce, 1 cup
Italian dressing, 2 tsp
Green beans, 1/2 cup
Cranberry juice,
  1 1/2 cups
	Dinner
	Broiled beef sirloin, 3 oz
	Romaine lettuce, 1 cup
	Italian dressing, 2 tsp
	Green beans, 1 cup
	Fat-free milk, 1/2 cup
		Dinner
		Broiled beef sirloin, 3 oz
		Romaine lettuce, 1 cup
		Garbanzo beans, 1 cup
		Italian dressing, 2 tsp
		Spinach pasta noodles,
		  1 1/2 cups
		Margarine, 1 tsp
		Green beans, 1 cup
		Fat-free milk, 1/2 cup
			Dinner
			Skinless turkey breast,
			  2 oz
			Mashed potatoes, 2 cups
			Peas and onions, 1 cup
			Banana, 1
			Fat-free milk, 1 cup
				Dinner
				Beef sirloin, 5 oz
				Mashed potatoes, 2 cups
				Spinach pasta noodles,
				  1 1/2 cups
				Grated parmesan cheese,
				  2 tbsp
				Green beans, 1 cup
				Oatmeal-raisin cookies, 3
				Fat-reduced milk, 1 cup

			Snack Pasta, 1 cup cooked Margarine, 2 tsp Parmesan cheese, 2 tbsp Cranberry juice, 1 cup	**Snack** Air-popped popcorn, 4 cups Raisins, 1/3 cup Cranberry juice, 2 cups
18% protein (68 grams) 64% carbohydrate (240 grams) 19% fat (32 grams)	17% protein (85 grams) 63% carbohydrate (315 grams) 20% fat (44 grams)	17% protein (128 grams) 62% carbohydrate (465 grams) 21% fat (70 grams)	14% protein (140 grams) 61% carbohydrate (610 grams) 26% fat (116 grams)	14% protein (175 grams) 63% carbohydrate (813 grams) 24% fat (136 grams)

* Daily Food Plans are available online for diets supplying 1000 to 3200 kcal/day (see **www.ChooseMyPlate.gov**). Additional servings from each food group have been added to supply increased calories while maintaining healthy proportions of protein, carbohydrate, and fat.

they also contain a lot of fat. Better high-carbohydrate food choices include pasta, rice, potatoes, bread, fruit and fruit juices, and many breakfast cereals (check the label for carbohydrate content) (Table 10-4). Sports drinks appropriate for carbohydrate loading, such as GatorLode and UltraFuel, can also help. Consuming a moderate (rather than high) amount of fiber during the final day of training is a good precaution to reduce the chances of bloating and intestinal gas during the next day's event.

▲ Fruits provide a good source of carbohydrate, especially starch and natural sugars, for athletes.

TABLE 10-4 ▸ **Grams of Carbohydrate Based on Serving Size of Typical Carbohydrate-Rich Foods**

Starches—15 Grams Carbohydrate per Serving (80 kcal)

One Serving

Dry breakfast cereal*, 1/2–3/4 cup	Baked potato, 1/4 large
Cooked breakfast cereal, 1/2 cup	Bagel, 1/4 (of 4 ounces)
Cooked grits, 1/2 cup	English muffin, 1/2
Cooked rice, 1/3 cup	Bread, 1 slice
Cooked pasta, 1/3 cup	Pretzels, 3/4 ounce
Baked beans, 1/3 cup	Saltine crackers, 6
Cooked corn, 1/2 cup	Pancake, 4 inches in diameter, 1
Cooked dry beans, 1/2 cup	Taco shells, 2 (add 45 kcal)

Vegetables—5 Grams Carbohydrate per Serving (25 kcal)

One Serving
Cooked vegetables, 1/2 cup
Raw vegetables, 1 cup
Vegetable juice, 1/2 cup
Examples: carrots, green beans, broccoli, cauliflower, onions, spinach, tomatoes, and vegetable juice

Fruits—15 Grams Carbohydrate per Serving (60 kcal)

One Serving

Canned fruit or berries, 1/2 cup	Grapes (small), 17
Fruit juice, 1/2 cup	Grapefruit, 1/2
Figs (dried), 1 1/2	Dates, 3
Apple or orange, 1 small	Peach, 1
Apricots (dried), 8	Watermelon cubes, 1 1/4 cups
Banana, 1 small	

Milk—12 Grams Carbohydrate per Serving

One Serving	Soymilk, 1 cup
Milk, 1 cup	
Plain low-fat yogurt, 2/3 cup	

Sweets—15 Grams Carbohydrate per Serving (variable calories)

One Serving	
Cake, 2-inch square	Ice cream, 1/2 cup
Cookies, 2 small	Sherbet, 1/2 cup

*The carbohydrate content of dry cereal varies widely. Check the labels of the ones you choose and adjust the serving size accordingly.

Source: Modified from *Choose Your Foods: Food Lists for Diabetes* by the American Diabetes Association and Academy of Nutrition and Dietetics, 2014.

FAT

A diet containing up to 35% of calories from fat is generally recommended for athletes. Rich sources of monounsaturated fat, such as canola oil, should be emphasized, and saturated fat and *trans* fat intake should be limited.

PROTEIN

For athletes, the American College of Sports Medicine, Academy of Nutrition and Dietetics, and Dietitians of Canada recommend protein intake within the range of 0.8 to 1.7 grams of protein per kilogram of body weight. The International Society of Sports Nutrition recommends up to 2.0 grams of protein per-kilogram (see Further Reading 9). Estimates of protein needs vary by the type of physical activity (Table 10-5).

Some of the protein recommendations listed in Table 10-5 are considerably higher than the RDA of 0.8 grams per kilogram of body weight recommended by the Food and Nutrition Board for all adults. Sports nutrition experts contend that the RDAs have been set to prevent deficiency among the general population, not to optimize physical performance among athletes (see Further Reading 13). What are some reasons why athletes require more protein than sedentary adults?

TABLE 10-5 ▸ Estimated Protein Needs of Athletes Based on Kilograms Body Weight[1]

Activity Level	Protein Needs (g/kg) Males	Protein Needs (g/kg) Females	Amount (g/d) for a 70-kilogram (154-pound) Man
Sedentary adults[2]	0.8	0.8	56
Recreational endurance athletes[3]	0.8–1.0	0.8–0.9	56–70
Moderate-intensity endurance athletes[4]	1.2	1.0–1.1	84
Elite endurance athletes	1.6	1.3–1.4	112
Football, power sports	1.4–1.7	1.1–1.5	98–119
Resistance athletes (early training)	1.5–1.7	1.2–1.5	105–119
Resistance athletes (steady state)	1.0–1.2	0.8–1.1	70–84

Source: Adapted from Burke L, Deakin V: *Clinical Sports Nutrition,* 4th ed., McGraw-Hill, Australia, 2009.
[1]Calculate kilograms by dividing pounds by 2.2.
[2]RDA, as recommended by the Food and Nutrition Board.
[3]Exercising four to five times per week for 30 minutes.
[4]Exercising four to five times per week for 45–60 minutes.

- The overall calorie needs of athletes are increased to meet the demands of physical activity, so more amino acids will be metabolized as fuel. For endurance athletes, in particular, protein may provide up to 15% of overall energy needs.
- Athletes need additional amino acids for repair of damaged muscle tissue and synthesis of new muscle protein.
- Beyond their use as fuel or building blocks, certain amino acids also act as chemical signals that regulate protein synthesis and other metabolic processes.

For athletes beginning a strength-training program, some experts recommend up to 2.0 grams of protein per kilogram of body weight. That is more than twice the RDA for protein. To date, the value of protein intake above 1.7 grams per kilogram of body weight has not been supported by sufficient research. Protein intakes above this amount result in an increased use of amino acids for energy needs; no further increase in muscle protein synthesis is seen. Once the desired muscle mass is achieved, protein intake need not exceed 1.2 grams per kilogram of body weight.

Is there an optimal dose of protein? Some research suggests that a dose of 20 to 25 grams of protein per meal is ideal to promote muscle synthesis (see Further Reading 13).

Unless an athlete follows a low-calorie diet, the recommended ranges of protein intake listed in Table 10-5 can be met by eating a variety of foods (review Table 10-3). To illustrate, a 123-pound (53-kilogram) woman performing moderate-intensity endurance activity can consume 58 grams of protein (53 × 1.1) during a single day by including 3 ounces of chicken (one chicken breast), 3 ounces of beef (a small, lean hamburger), and two glasses of milk in her diet. Similarly, a 180-pound (77-kilogram) man who aims to gain muscle mass through strength training needs to consume only 6 ounces of chicken (a large chicken breast), $^1/_2$ cup of cooked beans, a 6-ounce can of tuna, and three glasses of milk to achieve an intake of 130 grams of protein (77 × 1.7) in a day. And, for both athletes, these calculations do not even include the protein present in grains or vegetables they will also eat. By simply meeting their calorie needs, many athletes consume more protein than is required.

Despite marketing claims, protein supplements are usually an expensive and unnecessary part of a fitness plan. However, many athletes choose to use protein powders (e.g., whey, casein, or soy) as a lean and convenient way to add protein to their diet. Whey protein is a popular product, especially among strength-trained athletes. Whey is an easily digested, high-quality protein derived from cow's milk. It is particularly rich in leucine, an essential branched-chain amino acid that has been shown to stimulate gains in muscle mass during strength training. Except in cases of milk protein allergies and kidney disease, whey protein is safe to use at the doses recommended on product labels. However, these supplements tend to be expensive. It may be more economical to rely on food sources, such as low-fat or fat-free dairy products.

▲ Weight-restricted athletes especially should make sure they are consuming enough protein as well as other essential nutrients.

Consuming excessive amounts of protein has drawbacks. As noted in Chapter 6, it increases calcium loss somewhat in the urine. It also leads to increased urine production, possibly compromising body hydration. Excess animal protein also may lead to kidney stones in people with a history of this or other kidney problems. Finally, enough carbohydrate fuel may not be consumed on such a diet, leading to fatigue. Athletes who either feel they must significantly limit their calorie intake or are vegetarians should specifically determine how much protein they eat. They should make sure to follow a diet that provides at least 1.2 grams of protein per kilogram of body weight per day, the upper recommendation for most athletes.

VITAMINS AND MINERALS

Vitamin and mineral needs are the same or slightly higher for athletes, compared with those of sedentary adults. At this time, there are not enough data to support separate DRIs specific to athletes for any of the micronutrients, but research continues. Athletes usually have high calorie intakes, so they tend to consume plenty of vitamins and minerals. An exception is athletes consuming low-calorie diets (about 1200 kcal or less), as seen with some female athletes participating in events in which maintaining a low body weight is crucial. These diets may not meet B-vitamin and other micronutrient needs. Vegetarian athletes are also a concern. In these cases, consuming fortified foods, such as ready-to-eat breakfast cereals, or a balanced multivitamin and mineral supplement is recommended.

B Vitamins Support Energy Metabolism and Red Blood Cell Health. As you learned in Chapter 8, coenzyme forms of B vitamins facilitate chemical reactions that generate ATP from carbohydrates, proteins, and fats. Some B vitamins are involved in biosynthetic reactions, such as synthesis of glycogen from glucose, as well as reactions that metabolize amino acids. Compared to the metabolism of sedentary adults, the higher volume of energy metabolism of athletes will increase demands for these coenzymes. Although no separate DRIs have been set, athletes may need more than the current RDA for some B vitamins, such as riboflavin and vitamin B-6.

Furthermore, physical performance is highly dependent on the availability of oxygen to exercising muscles. Folate, vitamin B-6, and vitamin B-12 are involved in the formation of healthy red blood cells, which transport oxygen to all body tissues (see Chapter 8).

As you can imagine, an inadequate supply of B vitamins could impair an athlete's physical performance. Certainly, deficiency symptoms such as muscle weakness, nervous system dysfunction, and anemia are not compatible with participation in competitive sports! In reality, deficiencies of B vitamins are not very common. As athletes consume greater quantities of food to meet their increased calorie needs, they typically consume enough B vitamins from food sources to support energy metabolism and red blood cell health. Taking more than the RDA for B vitamins is not likely to enhance performance.

On the other hand, for a person who is deficient in one or more of the B vitamins, supplementation could improve athletic performance. At-risk populations include vegan or senior athletes (vitamin B-12), female athletes of childbearing age (folate), and any athlete who restricts dietary intake to control body weight (a variety of micronutrients). In these cases, fortified foods or a balanced multivitamin and mineral supplement would be beneficial to overall health and athletic performance (see Further Reading 20).

Antioxidant Nutrients May Prevent Oxidative Damage. Exercise leads to increased production of free radicals. A low level of free radicals in muscle tissue is actually beneficial for muscle contraction and adaptation to exercise. However, excessive free radicals can lead to fatigue and cell damage.

Athletes' needs for antioxidants such as vitamin E and vitamin C may be somewhat greater because of the potential protection these nutrients provide. However, there is evidence that antioxidant systems in the body increase in activity as exercise training progresses. The use of large doses of vitamin E and vitamin C requires more study and is not currently an accepted part of the dietary guidance for athletes. Experts suggest consuming a diet containing foods rich in antioxidants, such as fruits, vegetables, whole-grain breads and cereals, and vegetable oils (see Further Reading 10).

Iron Deficiency Impairs Performance. Iron is involved in red blood cell production, oxygen transport, and energy production, so a deficiency of this mineral can noticeably detract from optimal athletic performance. Some of the consequences of iron deficiency include weakness, fatigue, and decreased work capacity. The potential causes for iron deficiency in athletes vary (see Further Reading 14). As in the general population, female athletes are most susceptible to low iron status due to monthly menstrual losses. Special diets followed by athletes, such as low-calorie and vegetarian (especially vegan) diets, are likely to be low in iron. Distance runners should pay special attention to iron intake because their intense workouts may lead to gastrointestinal bleeding.

> Recall from Chapter 8 that nutrients in dietary supplements should not exceed any Upper Levels set over the long term. As well, men should be cautious about any use of supplements containing iron.

Another concern is *sports anemia*—which occurs because exercise causes blood plasma volume to expand, particularly at the start of a training regimen before the synthesis of red blood cells increases. This results in dilution of the blood; even if iron stores are adequate, blood iron tests may appear low. Sports anemia is not detrimental to performance, but it is hard to differentiate between sports anemia and true anemia. If iron status is low and not replenished, iron-deficiency anemia can markedly impair endurance performance.

> "Blood doping" is the injection of red blood cells, naturally containing iron, to enhance aerobic capacity. This is an illegal practice under Olympic guidelines.

Although true iron-deficiency anemia (a depressed blood hemoglobin level) is not that common among athletes, some studies suggest that iron deficiency *without* anemia may have a negative impact on physical activity and performance. Recall from Chapter 9 that iron deficiency occurs long before anemia is detected clinically. As body stores of iron are depleted, body processes that use iron, such as energy-yielding reactions, are impaired.

It is a good idea for athletes (especially adult women) to have their iron status checked at the beginning of a training season and at least once midseason. Current evidence suggests that as many as half of female athletes are iron deficient. To identify iron deficiency without anemia among athletes, many experts advocate serum ferritin testing. Ferritin is an iron transport protein; low serum ferritin levels indicate low iron stores even before red blood cell health is affected.

Any blood test indicating low iron status—sports anemia or not—warrants follow-up. A physician will need to determine the cause of iron depletion: is it due to poor dietary intake, heavy menstrual losses, or a serious medical condition, such as internal bleeding? Whatever the cause, once depleted, iron stores can take months to replenish. Dietary sources are not enough to correct iron deficiency anemia; supplementation (under physician supervision) is required. Athletes must be especially careful to meet iron needs because preventing iron deficiency is a lot simpler than treating it.

Knowing that iron is required for red blood cell synthesis, athletes may be tempted to self-prescribe iron supplements in an attempt to boost the oxygen-carrying capacity of the blood. However, indiscriminate use of iron supplements for people with normal hemoglobin and serum ferritin levels is NOT advised (see Further Reading 6). Research does not clearly support a benefit of iron supplementation on athletic performance for athletes with normal iron status. Furthermore, liver damage and increased rates of heart disease and some forms of cancer are possible consequences of iron toxicity. A safer alternative would be to have iron status checked periodically. In addition, monitor dietary patterns to become aware of usual iron intakes. If dietary iron intake is low, incorporate more food sources of heme iron and pair nonheme sources with vitamin C to enhance absorption. Beyond using a balanced multivitamin and mineral supplement, the decision to use an iron supplement is best left to a physician.

Calcium Intake Is Important, Especially in Women. Athletes, especially women trying to lose weight by restricting their intake of dairy products, can have marginal or low dietary intakes of calcium. This practice compromises optimal bone health. Of still greater concern are women athletes who have stopped menstruating because their arduous training and low body fat interferes with the normal secretion of reproductive hormones. Disturbing reports show that female athletes who do not menstruate regularly have spinal bones far less dense than those of both nonathletes and female athletes who menstruate regularly. They are more likely to suffer **stress fractures** during training and will be susceptible to bone injuries throughout life. The negative impacts of low dietary calcium intake and irregular menses in female athletes outweigh the benefits of weight-bearing exercise on bone density. Increasing energy intake to restore body weight and body fat stores is important to correct hormonal imbalances and prevent further bone loss. This topic is discussed further in Chapter 11, with respect to the female athlete triad, and in Chapter 9, where osteoporosis was reviewed in detail.

Female athletes whose menstrual cycles become irregular should consult a physician to determine the cause. Decreasing the amount of training or increasing energy intake and body weight often restores regular menstrual cycles. If irregular menstrual cycles persist, severe bone loss (much of which is not reversible) and osteoporosis can result. Extra calcium in the diet does not necessarily compensate for these damaging effects of menstrual irregularities, but inadequate dietary calcium can make matters worse.

FLUID

Fluid needs for an average adult are about 9 cups per day for women and 13 cups per day for men. Athletes generally need even more water to regulate body temperature. Heat production in contracting muscles can rise 15 to 20 times above that of resting muscles. Unless this heat is quickly dissipated, heat exhaustion, heat cramps, and potentially fatal heatstroke may ensue.

Fluid intake during exercise, when possible, should be adequate to minimize body weight loss. Remember this advice even when sweating can go unnoticed, such as when swimming or during the winter. Fluid and electrolyte needs vary widely, based on differences in body mass, environmental conditions, level of training, event duration, and even genetics. Because fluid needs are highly individualized and dynamic, it is not appropriate to make general recommendations for fluid replacement (see Further Reading 11). Rather, an athlete should aim to replace the total amount of fluid lost during exercise. This can be accomplished by knowing the body's hourly sweat rate, which can be calculated from the weight lost during exercise per hour plus the fluid consumed during exercise per hour.

The American College of Sports Medicine recommends losing no more than 2% of body weight during exercise, especially in hot weather. Athletes should first calculate 2% of their body weight and then by trial and error determine how much fluid they must take in to avoid excessive losses. Monitoring pre- and post-workout body weight is the most straightforward method of determining fluid losses. Much of this fluid replacement will have to take place after exercise because it is difficult to consume enough fluid during exercise to prevent weight loss. If weight change cannot be monitored, urine color is another measure of hydration status (see Chapter 9, Fig. 9-10). Urine color should be no more yellow than lemonade.

Thirst is a late sign of dehydration. An athlete who drinks only when thirsty may take 48 hours to replenish fluid losses. After several days of training, an athlete

---

stress fracture A fracture that occurs from repeated jarring of a bone. Common sites include bones of the foot.

Before the Event

Drink 5 to 7 milliliters per kilogram of body weight of water or sports drink at least 4 hours before exercise.

During the Event

Consume fluid to prevent dehydration (i.e., losses of more than 2% body weight).

After the Event

Drink 2 to 3 cups of fluid per pound of body weight lost within 4 to 6 hours after exercise.

relying on thirst can build up a fluid debt that will impair performance. The following fluid-replacement approach can meet athletes' fluid needs in most cases:

- Freely drink beverages (e.g., water, diluted fruit juice, and sports drinks) during the 24-hour period before an event, even if not particularly thirsty.
- Drink 5 to 7 milliliters per kilogram of body weight (about 1.5 to 2 cups for a 150-pound male) of water or sports drink at least 4 hours before exercise. This allows time for both adequate hydration and excretion of excess fluid.
- During events lasting more than 30 minutes, athletes should consume fluid to prevent dehydration (i.e., losses of >2% body weight). Research in marathon runners suggests about 1 ½ to 3 ½ cups (400 to 800 milliliters) per hour to prevent dehydration. Football players wearing equipment for two-a-day practices during the heat of August may need even more than 800 milliliters per hour to prevent dehydration. The best plan is to determine individual rate of fluid losses during training and plan accordingly. In many cases, athletes, especially children and teenagers, need to be reminded to consume fluids during exercise.
- Within 4 to 6 hours after exercise, about 2 to 3 cups of fluid should be consumed for every pound lost. It is important that weight be restored before the next exercise period. Skipping fluids before or during events will almost certainly impair performance.

The popularity of caffeine-containing energy drinks has surged in recent years (see Newsworthy Nutrition). Some studies show that caffeine improves athletic performance during endurance events (e.g., cycling) or sports that require a high level of mental alertness (e.g., archery). However, excessive caffeine consumption can lead to shakiness, nervousness, anxiety, nausea, and insomnia. In addition, the diuretic effect of caffeine may not support optimal hydration, particularly for athletes who are not accustomed to caffeine. Compare the caffeine and kilocalorie contents of several top-selling energy drinks (Table 10-6).

As environmental temperature rises above 95°F (35°C), virtually all body heat is lost through the evaporation of sweat from the skin. Sweat rates during prolonged exercise range from 3 to 8 cups (750 to 2000 milliliters) per hour.

Wearing football equipment in hot weather can lead to a loss of 2% of body weight in 30 minutes. Marathon runners have been shown to lose 6% to 10% of body weight during a race. Dehydration reduces endurance, strength, and overall performance. As humidity rises, especially above 75%, evaporation slows and sweating is insufficient to cool the body. The result is rapid fatigue, increased work for the heart, and difficulty with prolonged exertion. Heat-related injuries—heat exhaustion, heat cramps, and heatstroke—can be deadly (Table 10-7). To decrease the risk of developing heat-related injuries, watch for rapid body-weight changes (2% or more of body weight), replace lost fluids, and avoid exercising under extremely hot, humid conditions.

▲ Dehydration, which can lead to illness and death, is a problem that must be avoided during physical activity in hot, humid environments.

TABLE 10-6 ▶ **Caffeine, Kilocalorie, and Sugar Content of Popular Energy Drinks**

Beverage	Container Size (fl oz)	Caffeine (mg)	Energy (kcal)	Sugars (g)
5-Hour Energy	1.93	215*	4	0
Amp	16	142	220	58
Full Throttle	16	197	220	58
Monster	16	160	200	54
NOS	16	260	210	54
Red Bull	8.4	80	110	27
Red Bull Sugarfree	8.4	80	10	0
Red Rain	8.4	80	180	47
Rockstar	16	160	280	62
SoBe Energize	20	160	88	23

*Determined by independent lab analysis reported by *Consumer Reports,* December 2012. Caffeine content is not specified by the manufacturer but rather as part of a total amount of "energy blend." The labels state that this product has approximately the same amount of caffeine as a strong cup of coffee (typical range: 80–175 milligrams).

TABLE 10-7 ▶ Heat-Related Illnesses

Heat-Related Illness	Symptoms	Recommended Treatment
Heat exhaustion is the first stage of heat-related illness that occurs because of depletion of blood volume from fluid loss by the body.	• Profuse sweating • Headache • Dizziness • Nausea • Muscle weakness • Visual disturbances • Flushed skin • Hyperthermia • Heat cramps (see below)	• Move to cool environment. • Remove excess clothing. • Cool the skin with ice packs or cold water. • Replenish lost fluids and electrolytes.
Heat cramps are a frequent complication of heat exhaustion. They usually occur in people who have experienced large sweat losses from exercising for several hours in a hot climate and have consumed a large volume of water without replacing electrolytes.	• Painful skeletal muscle cramps	• Replenish lost fluids and electrolytes.
Heatstroke can occur when internal body temperature reaches 104°F. Sweating generally ceases if left untreated, and blood circulation is greatly reduced. Nervous system damage may ensue, and death results in 10% of cases.	• Hyperthermia • Hot, dry skin • Nausea • Confusion • Irritability • Poor coordination • Fainting • Seizures • Coma	• Cool the skin with ice packs or cold water. *SEEK PROFESSIONAL MEDICAL ATTENTION*

Sports Drinks. Is it better to rehydrate with water or a sports drink during a workout? For sports that require less than 60 minutes of exertion or when total weight loss is less than 5 to 6 pounds, the primary concern is replacing the water lost in sweat, because losses of carbohydrate stores and electrolytes (sodium, chloride,

Newsworthy Nutrition

Sports drinks and energy drinks are not interchangeable

The value of electrolyte- and carbohydrate-containing *sports* drinks (e.g., Gatorade) is well documented for replenishing fluids during physical activity lasting more than 1 hour. However, an assortment of *energy* drinks are now marketed alongside sports drinks to young people, especially male athletes. Experts caution athletes against using energy drinks, such as the top-selling Red Bull, Rockstar, Monster, and Throttle, to meet fluid needs during exercise. The sugar content is too high, and a multitude of additives, including caffeine, taurine, ginseng, guarana, and others, have limited (if any) documented benefits for athletes. In fact, binge consumption of energy drinks has led to illness and at least four deaths. Combining energy drinks with alcohol can mask the symptoms of intoxication and result in hazardous behavior and accidents. Experts recommend limiting use of energy beverages to 1 can (500 milliliters) per day, avoiding combinations with alcohol, and, for people with cardiac problems, consulting a physician before use. Athletes should use a sports drink rather than an energy drink for the purpose of rehydration during physical activity.

Source: Higgins JP and others: Energy beverages: Content and safety. *Mayo Clinic Proceedings* 85:1033, 2010.

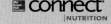 Check out the Connect site **www.mcgrawhillconnect.com** to further explore sports and energy drinks.

potassium, and other minerals) are not usually very great. (Sweat is about 99% water and only 1% electrolytes and other substances.) When exercise extends beyond 60 minutes, electrolyte (especially sodium) and carbohydrate replacement becomes increasingly important.

Use of sports drinks (Fig. 10-7) during long bouts of exercise—especially in hot weather—offers several advantages:

- *Water* increases blood volume to allow for efficient cooling and transport of fuels and waste products to and from cells.
- *Carbohydrates* supply glucose to muscles as they become depleted of glycogen and also add flavor, which encourages athletes to drink.
- *Electrolytes* in sports drinks help to maintain blood volume, enhance the absorption of water and carbohydrate from the intestine, and stimulate thirst.

Overall, the decision to use a sports drink hinges primarily on the duration of the activity. As the projected duration of continuous activity approaches 60 minutes or longer, the advantages of the use of a sports drink over plain water clearly emerge. However, athletes should first experiment with sports drinks during practice, instead of trying them for the first time during competition.

Water Intoxication. It is also possible for some athletes to drink too much water and thereby develop water intoxication (hyponatremia). Endurance athletes (especially novices) may compete at relatively low exercise intensities for prolonged periods and therefore may not sweat as much as one might predict. Thus, water losses are not high. Drinking less fluid, choosing a sports drink containing sodium (usually in the form of sodium chloride), and not gaining weight during the activity can help prevent this problem. (A drop in blood sodium can occur in both hot and cold weather.)

✅ CONCEPT CHECK 10.4

1. What does it mean to "cut weight" before a competition? How may this affect physical performance?
2. Greta, a point guard on the women's basketball team, complains of chronic fatigue. Describe three nutritional concerns you would investigate.
3. During one day of preseason training for football, David loses 7 pounds as a result of sweat losses. How much fluid should he drink to rehydrate after practice?

10.5 Specialized Advice for Endurance, Strength, and Power Athletes

In Section 10.4, you learned about ways in which nutrient needs can be universally affected by participation in sports. The definition of sports, however, is broad and each athlete is unique. Endurance athletes, who need to fuel activity that lasts several hours, should take a different approach to nutrition than strength and power athletes, who focus on gains in muscle mass. Below, we present specific nutrition strategies for endurance, strength, and power athletes.

ENDURANCE ATHLETES: STRATEGIES TO DELAY OR PREVENT FATIGUE

The overarching goal for endurance athletes is to consume adequate carbohydrates and fluids. *Before* the event, endurance athletes should focus on maximizing muscle and liver glycogen stores, which will later be used to fuel muscles and maintain

FIGURE 10-7 ▲ Sports drinks for fluid and electrolyte replacement typically contain a form of simple carbohydrate plus sodium and potassium. The various sugars in this product total 14 grams per 1 cup (240 milliliters) serving. Sports drinks typically contain about 6% to 8% sugar. This provides ample glucose and other monosaccharides to fuel working muscles, and it is well tolerated. Drinks with a sugar content above 10%, such as soft drinks or fruit juices, may cause gastrointestinal distress and are not recommended.

▲ Carbohydrate loading is appropriate only for endurance activities such as a long-distance race.

carbohydrate loading A process in which a high-carbohydrate diet is consumed for several days before an athletic event while tapering exercise duration in an attempt to increase muscle glycogen stores.

Carbohydrate Loading May Be Beneficial for These Activities
- Marathons
- Long-distance swimming
- Cross-country skiing
- 30-kilometer runs
- Triathlons
- Tournament-play basketball
- Soccer
- Cycling time trials
- Long-distance canoe racing

Carbohydrate Loading Is Not Beneficial for These Activities
- American football games
- 10-kilometer or shorter runs
- Walking and hiking
- Most swimming events
- Single basketball games
- Weight lifting
- Most track and field events

TABLE 10-8 ▶ Carbohydrate Loading Regimen

Days Before Competition	6	5	4	3	2	1
Exercise time (minutes)	60	40	40	20	20	Rest
Carbohydrate (grams)	450	450	450	600	600	600

blood glucose. *During* an event, the goal is to prevent dehydration and glycogen depletion, as both of these conditions lead to fatigue and detract from physical performance. *After* an event, muscle glycogen stores need to be replenished, damaged muscle tissue must be repaired, and hydration should be restored.

Maximize Glycogen Stores Before the Event. For athletes who compete in continuous, intense aerobic events lasting more than 90 minutes (or in shorter events taking place more than once within a 24-hour period), a **carbohydrate-loading** regimen can help to maximize the amount of energy stored in the form of muscle glycogen for the event. In one possible regimen, during the week prior to the event, the athlete gradually reduces the intensity and duration of exercise ("tapering") while simultaneously increasing the percentage of total calories supplied by carbohydrates. Shorter carbohydrate-loading regimens (e.g., 1 or 2 days before an event) may also be effective.

For example, consider the carbohydrate-loading schedule of a 25-year-old man preparing for a marathon. His typical calorie needs are about 3500 kcal per day. Six days before competition, he completes a final, hard workout of 60 minutes. On that day, carbohydrates contribute 45% to 50% of his total calorie intake. As he goes through the rest of the week, the duration of his workouts decreases to 40 minutes and then to about 20 minutes by the end of the week. Meanwhile, he increases the amount of carbohydrate in his diet to reach 70% to 80% of total calorie intake as the week continues (Table 10-8). Total calorie intake should decrease as exercise time decreases throughout the week. On the final day before competition, he rests while maintaining the high carbohydrate intake.

This carbohydrate-loading technique usually increases muscle glycogen stores by 50% to 85% over typical conditions (that is, when dietary carbohydrate constitutes only about 50% of total calorie intake). A potential disadvantage of carbohydrate loading is that additional water (about 3 grams) is incorporated into the muscles along with each gram of glycogen. Although the additional water aids in maintaining hydration, for some individuals this additional water weight and related muscle stiffness detract from their sports performance.

Athletes considering a carbohydrate loading regimen should try it during training (and well before an important competition) to experience its effects on performance. They can then determine whether it is worth the effort. Currently, expert advice is shifting away from such regimented carbohydrate loading in favor of supplying carbohydrates during the event (along with a daily diet high in carbohydrate).

Even if an endurance athlete chooses not to practice a strict carbohydrate loading regimen, a light meal supplying up to 1000 kcal should be eaten about 2 to 4 hours before an endurance event to top off muscle and liver glycogen stores, prevent hunger during the event, and provide extra fluid. The longer the period before an event, the larger the meal can be, because there will be more time available for digestion. A pre-event meal should consist primarily of carbohydrate (about 200 grams), contain little fat or fiber, and include a moderate amount of protein (Table 10-9). Anything consumed 1 hour or so before an event should be blended or liquid to promote rapid stomach emptying. Examples are low-fat smoothies and sports drinks.

Carbohydrate-rich food choices for a pre-event meal include spaghetti, muffins, bagels, pancakes with fresh fruit topping, oatmeal with fruit, a baked potato topped with a small amount of sour cream, toasted bread with jam, bananas, or low-sugar breakfast cereals with reduced-fat or fat-free milk. Liquid meal-replacement formulas, such as Carnation Instant Breakfast, also can be used. Foods especially rich in fiber should be eaten the previous day to help empty the colon before an event, but they should not be eaten the night before or on the morning of the event. Avoid fatty or fried foods, such as sausage, bacon, sauces, and gravies. Some foods (e.g., dairy products) may cause gastrointestinal upset. Athletes should experiment with the size, timing, and composition of pre-event meals during training to determine what will be tolerated.

Emerging Research on Fat Adaptation. Traditionally, high-carbohydrate diets have been the nutrition norm for endurance athletes. However, an alternative approach in training known as fat adaptation is becoming more popular among endurance athletes.

Of all the energy-yielding nutrients, carbohydrates are utilized most rapidly to fuel exercising muscles. When athletes consume a high-carbohydrate diet or practice carbohydrate loading before an endurance event, they ensure that muscle and liver glycogen will be available to muscles throughout the race. Even after carbohydrate loading, however, the total amount of energy available from muscle glycogen is limited to about 2500 kcal.

In comparison, the supply of energy from triglycerides stored in the muscle and adipose tissue is virtually limitless. Recall from Figure 10-6 that, depending on intensity, about half of the energy for endurance events comes from fat. The metabolism of fat for energy occurs more slowly, but it provides more than twice as many kilocalories per gram as carbohydrates or protein.

With fat adaptation, rather than following a traditional high-carbohydrate diet (about 65% of kilocalories from carbohydrates and only about 20% from fat) during the days leading up to an event, endurance athletes replace much of the carbohydrates with fat. For example, a high-fat training diet might consist of just 25% of kilocalories from carbohydrate with a whopping 60% of kilocalories from fat. The rationale is that high-carbohydrate diets, especially those with many simple sugars and refined grains, boost insulin secretion, which inhibits the breakdown of fat. By lowering carbohydrates and increasing the fat content of the diet, the cells will adapt to greater use of fat for fuel. If the athlete uses more fat for fuel during an endurance event, muscle glycogen might be spared, so that those stored carbohydrates would be available for a burst of speed at the end of the race.

Research comparing the effects of high-carbohydrate or high-fat training diets on athletic performance has yielded mixed results. A possible explanation is that the muscles of "fat-adapted" athletes are able to break down more fat for fuel during exercise at low or moderate intensities, but the low carbohydrate intake depletes glycogen stores, so higher-intensity activity is impaired.

A compromise between these two approaches might be the best strategy. Some studies indicate that several weeks on a high-fat diet followed by 2 or 3 days of carbohydrate loading just before the event may enhance the ability of muscles to use fat for fuel, yet still facilitate adequate glycogen storage to fuel the sprint to the finish line.

TABLE 10-9 ▶ Convenient High-Carbohydrate Pre-Event Meals

Breakfast	
Cheerios, ¾ cup Reduced-fat milk, 1 cup Blueberry muffin, 1 Orange juice, 4 ounces or	450 kcal 92 grams (82%) carbohydrate
Low-fat fruit yogurt, 1 cup Plain bagel, ½ Apple juice, 4 ounces Peanut butter (for bagel), 1 tbsp	482 kcal 84 grams (68%) carbohydrate
Lunch or Dinner	
Broiled pork chop, 3 ounces White rice, 1½ cups Steamed zucchini, 1 cup Chocolate milk, 1 cup Jello, ½ cup or	839 kcal 120 grams (57%) carbohydrate
Spaghetti noodles, 2 cups Spaghetti sauce, 1 cup Reduced-fat milk, 1½ cups Green beans, 1 cup	761 kcal 129 grams (66%) carbohydrate

With regard to the timing of pre-activity meals, a general guide is to allow 4 hours for a big meal (about 1200 kcal), 3 hours for a moderate meal (about 800 to 900 kcal), 2 hours for a light meal (about 400 to 600 kcal), and an hour or less for a snack (about 300 kcal).

▲ Pre-event meals may require a higher proportion of grains than suggested by MyPlate to boost carbohydrate content. Choose starchy vegetables and grain-based snacks to help top off glycogen stores. How does this pasta meal compare to MyPlate?

General Guide for Approximate Pre-Event Carbohydrate Intake

Hours Before	Grams per kilogram Body Weight	For a 70-kilogram Person
1	1	70
2	2	140
3	3	210
4	4	280

▲ Elite athletes are subject to all the same nutritional challenges as the general public: overreliance on convenience foods, abundance of nutrition misinformation, temptations to eat out of boredom or for emotional comfort. Furthermore, they must adapt to the seasonal demands of their sports and maintain exhausting training and travel schedules. In her role as a sports dietitian for the United States Olympic Committee, Jennifer Gibson, M.Sc., R.D., CSSD carefully tailors nutrition recommendations for each of her athletes based on nutritional status, sport, season, position, and body composition goals.

At this time, there is not enough research evidence to support a recommendation for high-fat diets for endurance athletes. Evidence *does* clearly support a performance-enhancing effect of carbohydrate ingestion before and during physical activity. This will continue to be an area of active research, so stay on the lookout for more information on fat adaptation.

Replenish Fuel During the Event. We have already established the importance of consuming adequate fluids during endurance exercise. For sporting events longer than 60 minutes, consumption of carbohydrate during activity can also improve athletic performance. Prolonged exercise depletes muscle glycogen stores and may transiently lower blood glucose, leading to physical and mental fatigue. One way to avoid "hitting the wall" is to maintain normal blood glucose concentrations by carbohydrate feedings during exercise.

A general guideline for endurance events is to consume 30 to 60 grams of carbohydrate per hour. A current trend in sports nutrition is to use multiple sources of carbohydrates (e.g., glucose, fructose, and maltodextrin) with different routes and rates of absorption to maximize the supply of glucose to cells and lessen the risk of gastrointestinal distress (see Further Reading 1).

Some experts suggest that consuming protein with carbohydrate during exercise provides added benefit. In particular, branched-chain amino acids, which can be used for fuel, are thought to delay fatigue by supplying energy and altering the production of certain neurotransmitters (see Further Reading 2). This is an area of ongoing research, but there is not enough evidence at this time to support a clear recommendation for ingesting protein during exercise. Some products formulated for consumption during exercise do contain amino acids, so you can try various formulations to see what works best for you.

What about fat? Compared to carbohydrates, fat is more slowly digested, absorbed, and metabolized. Thus, although fat serves as fuel during prolonged aerobic activity, consumption of fat during activity is not likely to improve athletic performance and is likely to cause gastrointestinal distress.

Sports drinks are a good source of carbohydrate calories during endurance events. Sports drinks, described in detail in the previous section, usually contain about 14 grams of carbohydrate per 8-oz serving (Table 10-7). They supply the necessary fluid, electrolytes, and carbohydrates to keep an athlete performing at his or her best.

As an alternative to sports drinks, some athletes use carbohydrate gels or chews. Gels and chews are formulated with one or more sugars or starches to rapidly supply about 25 grams of carbohydrates per serving (Table 10-10). In addition, they provide electrolytes to replenish those lost in sweat. Some of these products also

TABLE 10-10 ▶ Energy and Macronutrient Content of Popular Energy Bars, Gels, and Chews

Energy Bars	Serving (oz)	Energy (kcal)	Carbohydrates (g)	Fiber (g)	Protein (g)	Fat (g)
Clif Bar (chocolate chip)	2.4	240	44	5	10	4.5
LUNA Bar (nutz over chocolate)	1.69	180	25	4	9	6
Met-Rx Protein Plus (chocolate fudge deluxe)	3.0	310	32	2	32	9
PowerBar Performance (peanut butter)	2.3	240	44	1	9	4
Snickers Marathon Protein Bar (caramel nut rush)	2.82	290	40	10	20	10
Energy Gels & Chews						
Carb BOOM! Energy Gel (strawberry kiwi)	1.4	110	27	0	0	0
Clif Shot (vanilla)	1.1	100	24	0	0	0
GU Energy Gel (lemon sublime)	1.1	100	25	0	0	0
Jelly Belly Sport Beans (berry)	1.0	100	24	0	0	0
PowerBar Gel (strawberry banana)	1.44	110	27	0	0	0

Overall, choosing energy bars is preferable to choosing candy bars and packaged cakes. When used in sports situations, energy bars can be handy. Better yet, however, is to eat a variety of wholesome foods; these offer more health-protective compounds. This is also a less expensive choice, especially for day-to-day snacking. An additional concern is that micronutrient toxicity might occur if numerous bars are eaten in a day, as many are highly fortified. Vitamin A and iron are two nutrients of special concern in this regard.

may contain certain amino acids, vitamins, caffeine (see the Nutrition and Your Health section at the end of this chapter), or herbal ingredients. An advantage of gels compared to energy bars or sports drinks is that they are so convenient to carry.

Popular energy bars (e.g., PowerBar) typically provide about 180 to 250 kcal and anywhere from 2 to 45 grams of carbohydrate. The wide range of carbohydrate content in energy bars is due to a variety of marketing trends in the sports supplement industry. Overall, endurance athletes should choose a bar with about 40 grams of carbohydrate and no more than 10 grams of protein, 4 grams of fat, and 5 grams of fiber. The bars are fortified with vitamins and minerals in amounts ranging from about 25% to 100% of typical human needs. Outside of sporting events, some people use energy bars as a quick and convenient meal or snack.

Check the label on these products to gauge the amount of gel or bar that provides 30 to 60 grams of carbohydrate per hour. In addition, remember that any carbohydrate-containing food must be accompanied by fluid to ensure adequate hydration.

At minimum, one serving of any of these will cost at least $1, and some brands with all natural or organic ingredients cost as much as $5. Are sports drinks, energy bars, and gels worth the price? It depends how much you are willing to pay for convenient packaging. Critics suggest that these products are essentially the nutritional equivalent of a cup of low-fat yogurt and piece of fruit (see Table 10-3). For an athlete on a tight budget, a small bag of graham crackers or jelly beans could just as easily provide a quick shot of glucose during a race. With a little bit of time and an Internet connection, you can even find recipes to make your own sports drinks and energy bars at home for a fraction of the cost of name-brand products.

After Exercise, Replenish Glycogen and Fluid. After prolonged aerobic exercise, muscle and liver glycogen stores will be depleted. An athlete's need to pay special attention to nutrition during recovery from exercise depends on the type of workout completed and the timing of the next workout. For example, multiple events in the same day will require rapid restoration of glycogen stores. However, if an athlete will be able to rest for 1 or 2 days before the next exercise session, immediate consumption of a post-workout meal is not as crucial.

To rapidly restore glycogen stores, carbohydrate-rich foods providing 1 to 1.5 grams of carbohydrate per kilogram body weight should be consumed within 30 minutes after extended (endurance) exercise. Immediately after exercise is when glycogen synthesis is greatest because the muscles are insulin-sensitive at this point. This process should then be repeated every 2 hours for the next 4 to 6 hours. Foods with high a glycemic index, such as fruit, fruit juice, bread, short-grain white rice, or a sports drink, contribute to rapid restoration of glycogen stores. Recall that Table 4-3 gives the glycemic index of various foods.

Although carbohydrate intake is the most important factor for replenishing glycogen after endurance exercise, adding an appropriate amount of high-quality protein during recovery can be helpful for stimulating glycogen synthesis and repairing damaged muscle tissue. Read more about recovery meals in the next section on nutrition strategies for strength and power athletes.

Fluid and electrolyte (i.e., sodium and potassium) intake is another essential component of recovery for an endurance athlete, especially if two workouts a day are performed or if the environment is hot and humid. Specialized recovery drinks containing carbohydrates, amino acids, and electrolytes are available, but if food and fluid intake is sufficient to restore weight loss, it generally will also supply enough electrolytes to meet needs during recovery from endurance activities.

STRENGTH AND POWER ATHLETES: STRATEGIES TO ENHANCE MUSCLE GAIN

Strength training—improving the maximal force that can be exerted by the muscles—should be part of any well-rounded fitness program. Resistance exercises may utilize free weights, specialized weight machines, or the weight of one's own body.

There is evidence that the ergogenic effect of carbohydrate ingestion during exercise is due not only to the increased availability of fuel but also to changes in chemical signals that relay information about carbohydrate ingestion from the gastrointestinal tract to the brain. Some studies among athletes participating in events lasting less than 1 hour suggest that simply rinsing the mouth with a carbohydrate-containing fluid can decrease fatigue and enhance performance without risking gastrointestinal distress.

▲ Do strength-trained athletes need to ingest tuna, chicken, and lean beef at every meal to build muscle? Patrik Baboumian is strongman competitor and former bodybuilder who follows a vegan diet. The plant-based protein sources that built his 250-pound physique include beans, peas, lentils, nuts, seeds, and protein shakes made with soy protein powder. Baboumian is the current world record-holder for the log lift in his weight class and has been named Germany's strongest man.

▲ A special nutrition issue that concerns strength-trained athletes is muscle dysmorphia, informally called "bigorexia." Chapter 11 will touch upon this form of body dysmorphic disorder in which an individual sees himself or herself as being too thin, even though he or she is more muscular than average. People who suffer from muscle dysmorphia may practice disordered eating behaviors or use steroids to achieve high levels of muscularity.

Many power athletes utilize a training technique called **periodization,** in which physical stresses on the body change throughout the year:

- Early in the training season, athletes work on building aerobic endurance.
- After gains in aerobic capacity have been achieved, the focus shifts to building strength, power, and sport-specific skills.
- During the competitive season, daily workouts are scaled back, but activity is intense and of long duration on game days.
- In the off-season, athletes continue to work out to stay in shape, but the volume is certainly lower than it was in season.

Athletes taking part in periodized training will use the full spectrum of energy systems we have discussed. Nutrition recommendations should also be periodized to match such dynamic training plans.

periodization Cycling the volume, intensity, and activities of workouts throughout the training season.

A workout typically includes 8 to 12 different exercises that target all the major muscle groups of the body. Most of the people you see lifting weights in the gym are probably performing several sets of 8 to 10 repetitions each, lifting about 50% of the maximum weight they could lift (1 RM). This type of workout improves muscular *endurance,* which is an important part of muscular fitness for overall health. However, to truly build muscular *strength,* athletes need to work against greater resistance (around 80% of 1 RM) over just a few repetitions (2 to 5 repetitions). For a few athletes, such as those who participate in weight-lifting or body-building competitions, muscular strength is the focus of training.

Muscular power combines strength with speed, improving the ability to apply force quickly. Examples of power sports include middle-distance running, gridiron football, rowing, and swimming. In reality, many sports and everyday activities involve muscular power: jumping for a rebound in basketball, delivering a round-house kick to an opponent in martial arts, or driving the ball down the fairway in a round of golf are examples of muscular power in sports.

For strength and power athletes, a few changes are needed to the general nutrition advice for athletes. First, calorie needs will be high due to the additional lean mass and high-volume training routines of these athletes. Recall that the primary types of fuel for strength and power moves are phosphocreatine (PCr) and carbohydrates for the brief bursts of activity, with fat providing energy during the resting stages. Very little protein is used as fuel during resistance exercises (review Fig. 10-6). Second, there will be some extra emphasis on protein intake in the recovery phase (see Further Readings 17 and 18).

Strength and power athletes tend to be extremely focused on consuming adequate protein to support muscle protein synthesis. Strength-training athletes in the early phases of training do have the highest estimated protein needs of any athletes (Table 10-5). Once desired muscle mass has been achieved, protein requirements for maintenance of muscular strength decrease slightly. Meeting these recommendations for protein intake optimizes muscle protein synthesis, but consuming more than the recommended range of protein intake does not appear to offer any advantage. Recall that excess amino acids are used as fuel or stored as fat; they do not directly translate into increased muscular strength.

Before and During Strength and Power Training, Focus on Calories, Carbohydrates, and Fluids. Adequate hydration supports optimal athletic performance; strength and power athletes are no exception. Checking urine color or urine-specific gravity is a good indication of fluid status. If the athlete is poorly hydrated before an event, water or a sports drink will be sufficient to restore hydration.

Similar to nutrition strategies for endurance athletes, adequate carbohydrate ingestion in the days leading up to and hours immediately before exercise has been shown to enhance performance for strength and power events, too. Athletes who perform many repetitions with moderate resistance will use more of their muscle glycogen stores than athletes who perform fewer repetitions with high resistance. Overall, research has shown that consuming 4 to 7 grams of carbohydrates per kilogram of body weight per day is appropriate for strength and power training. The optimal rate of carbohydrate ingestion before and during resistance activities has not yet been established, but some research indicates that 1 to 4 grams of carbohydrate per kilogram of body weight in a pre-event meal or beverage will enhance work capacity during resistance workouts.

In strength and power sports, many athletes also use creatine supplements to increase levels of phosphocreatine in muscles. Recall that phosphocreatine is used to resupply ATP during short, intense bursts of activity. When phosphocreatine stores are increased, muscle glycogen may be preserved (see Further Reading 17). The Nutrition and Your Health section at the end of this chapter provides more information on creatine and other ergogenic aids.

Focus on achieving adequate hydration and maximizing muscle glycogen before activities because there may not be an opportunity to replenish fluids and carbohydrates during strength or power competitions. During extended training sessions, however, supplying fluids and carbohydrates will enhance both physical and mental performance.

While a few sports nutrition experts advocate ingesting protein before or during a resistance workout in an effort to promote muscle protein synthesis, the bulk of evidence points to emphasizing protein in the recovery meal.

For athletes (and adults in general), fat intake should fall into the range of 20% to 35% of overall calorie intake. Dietary surveys show that usual fat intake of resistance-trained athletes is slightly higher than recommended, probably because too many of their choices of protein-rich foods (e.g., meats and dairy products) are also rich sources of fat. If fat intake is above 35% of total calories, replacing the excess fat with carbohydrates would have a favorable effect on protein balance. This is because insulin, secreted in response to glucose in the blood, triggers uptake of amino acids by cells, which provides materials for protein synthesis within the cells.

After Strength and Power Activities, Consuming Carbohydrates and Protein Promotes Recovery. Those first few hours after resistance exercise, according to many researchers, are the best time to provide carbohydrates and protein to replenish muscle glycogen and promote muscle repair and synthesis. Right after exercise, the cells are insulin sensitive, so they rapidly take up glucose from the blood and store it as glycogen. General ACSM guidelines are to consume 1.0 to 1.5 grams of carbohydrate per kilogram of body weight to restore muscle glycogen. To promote muscle protein synthesis along with glycogen restoration, many sports nutrition experts recommend intakes at the upper end of that range (e.g., 1.2 to 1.5 grams of carbohydrates per kilogram of body weight) shortly after exercise (see Further Reading 17). The presence of certain amino acids further stimulates insulin secretion to enhance the uptake of glucose and synthesis of glycogen.

To promote gains in muscle mass, most sports nutrition experts recommend at least 20 grams of high-quality protein within the first 1 or 2 hours after exercise to maximize protein synthesis. Novice strength-training athletes who are seeking to gain muscle mass have the highest requirements for protein. With advanced training, the rate of protein turnover during exercise decreases. Therefore, well-trained strength athletes require less protein to repair and maintain muscles than their untrained counterparts. Some amino acids (e.g., leucine) may stimulate the metabolic pathways that lead to synthesis of muscle protein. The process of muscle protein synthesis not only requires amino acids as building blocks, of course, but also depends on carbohydrates as a source of energy.

Overall, recovery from resistance exercise requires a combination (3:1 ratio) of carbohydrates and high-quality protein. For a 154-pound (70-kilogram) athlete, this corresponds to about 70 grams of carbohydrate and 25 grams of protein in each 2-hour interval. Table 10-11 describes sample meals of this composition.

CONCLUDING REMARKS

Nutritional strategies have the potential to optimize athletic performance. Here, we have presented several generalized guidelines to plan nutritionally adequate diets that optimize energy stores, ensure hydration, and give athletes a competitive edge. We have stressed the importance of carbohydrates and fluids before, during, and after exercise, as well as protein for muscle recovery. Above all, recognize that each athlete is unique. Remember that genetics can impact nutrition requirements. Each type of exercise demands its own set of energy sources. Sports vary in training regimens, duration, and opportunities to acquire nourishment before, during, and between events. Even within a particular sport, each player's position has its own physical demands, which can alter nutritional needs (see Further

▲ Chocolate milk is the go-to recovery drink for many athletes. This 2-cup serving of low-fat chocolate milk is a tasty vehicle for 52 grams of carbohydrate and 16 grams of protein.

For more information on sports nutrition, check out the following websites:
Gatorade Sports Science Institute
www.gssiweb.com
The Physician and Sportsmedicine
https://physsportsmed.org
American College of Sports Medicine
www.acsm.org
CDC's Division of Nutrition, Physical Activity and Obesity
www.cdc.gov/nccdphp/dnpao
American Council on Exercise
www.acefitness.org

TABLE 10-11 ▸ Sample Recovery Meals

Option 1
1 regular bagel
1 oz deli turkey
1 oz Swiss cheese
1 cup fat-free milk
509 kcal, 71 grams carbohydrate, 33 grams protein, and 10 grams fat

Option 2
16 oz flavored, low-fat Greek yogurt
1 medium banana
533 kcal, 86 grams carbohydrate, 31 grams protein, and 8 grams fat

Option 3
1 carton Gatorade Recover Post-Game Recovery Beverage (8 fl oz)
1 carton Gatorade Recover Protein Shake (11 fl oz)
380 kcal, 65 grams carbohydrate, 28 grams protein, and 1.5 grams fat

Reading 7). Finally, personal taste preferences and gastrointestinal tolerance will dictate adherence to any nutrition plan. When working with athletes, start with your solid foundation of knowledge about nutrient needs, but be attentive to individual concerns, be adaptable, and always continue to learn.

✓ CONCEPT CHECK 10.5

1. Which nutrient(s) should be emphasized in a pre-event meal for an endurance athlete? Provide an example of a suitable pre-event meal for a long-distance cyclist.
2. What is carbohydrate loading? List three sports for which carbohydrate loading could enhance performance.
3. Why is a combination of carbohydrate and protein recommended for recovery after resistance exercise? When should a recovery meal be provided? Suggest a suitable recovery meal.

CASE STUDY Planning a Training Diet

Michael is training for a 10K run coming up in 3 weeks. He has read a lot about sports nutrition and especially about the importance of eating a high-carbohydrate diet while in training. He also has been struggling to keep his weight in a range that he feels contributes to better speed and endurance. Consequently, he is also trying to eat as little fat as possible. Unfortunately, over the past week, his workouts in the afternoon have not met his expectations. His run times are slower, and he shows signs of fatigue after just 20 minutes into his training program.

His breakfast yesterday was a large bagel, a small amount of cream cheese, and orange juice. For lunch, he had a small salad with fat-free dressing, a large plate of pasta with marinara sauce and broccoli, and a diet soft drink. For dinner, he had a small broiled chicken breast, a cup of rice, some carrots, and iced tea. Later, he snacked on fat-free pretzels.

Answer the following questions, and check your responses at the end of this chapter.

1. Is the high-carbohydrate diet a good idea during Michael's training?
2. Are there any important components missing in Michael's diet? Are missing components contributing to his fatigue?
3. Describe some changes that should be made in Michael's diet, including some specific foods that should be emphasized.
4. How should fluid needs be met during workouts?
5. Should Michael focus on fueling his body before, during, or after workouts?

Nutrition and Your Health
● Ergogenic Aids and Athletic Performance

Extreme diet manipulation to improve athletic performance is not a recent innovation. Today's athletes are as likely as their predecessors to experiment with any substance that promises a competitive advantage. In 2011, sports nutrition supplements accounted for about 12% ($3.6 billion) of the $30 billion in dietary supplement sales. Artichoke hearts, bee pollen, dried adrenal glands from cattle, seaweed, freeze-dried liver flakes, gelatin, and ginseng are just some of the ineffective substances used by athletes in hopes of gaining an **ergogenic** (work-producing) edge.

Based on what is known at this time, today's athletes can benefit from scientific evidence documenting the ergogenic properties of a few dietary substances. These ergogenic aids include sufficient water and electrolytes, adequate carbohydrates, and a balanced and varied diet consistent with MyPlate. Protein and amino acid supplements are not among those aids because athletes can easily meet protein needs from foods, as Table 10-3 demonstrated. In general, nutrient supplements should only be used to meet a specific dietary shortcoming, such as an inadequate iron intake. Dietary supplements rumored to enhance athletic performance must be given close scrutiny before use. The risk-benefit ratio of any ergogenic aid merits careful evaluation (see Further Readings 8 and 15).

As summarized in Table 10-12 very little scientific evidence supports the effectiveness of many substances touted as performance-enhancing aids. Many are useless; some are dangerous enough to cause organ damage. The liver is particularly susceptible to damage because it is responsible for detoxifying harmful compounds (see Further Reading 12). Athletes should be skeptical of any substance until its ergogenic effect is scientifically validated. FDA has a limited ability to regulate these dietary supplements (review Chapter 8), and the manufacturing processes for dietary supplements are not as tightly regulated by FDA as they are for prescription drugs.

Some supplements may contain substances that will cause athletes to "test positive" for various banned substances. This was demonstrated in the 2008 Summer Olympics, when swimmer Jessica Hardy tested positive for a banned substance that she inadvertently consumed as part of a dietary supplement. Studies also show that many supplements do not contain the substance and/or the amount listed on the label. Not only must the athlete determine whether there is evidence that a dietary supplement is safe and effective but he or she must also question if the dietary supplement contains what it is supposed to contain.

Even substances whose ergogenic effects have been supported by systematic scientific studies should be used with caution, as the testing conditions may not match those of the intended use. Caution should be exercised when it comes to using the appropriate dose of supplements or using multiple types of supplements concurrently.

Rather than waiting for a magic bullet to enhance performance, athletes are advised to concentrate their efforts on improving their training routines and sport techniques, while consuming well-balanced diets as described in this chapter.

NCAA and Supplements

The NCAA's Committee on Competitive Safeguards and Medical Aspects of Sports has developed lists of permissible and nonpermissible supplements for athletic departments to provide to student athletes. The NCAA advises students to discuss their use of *any* dietary supplement with their team medical staff to avoid unknowingly ingesting banned substances. Following are a few key examples:

Permissible	Nonpermissible
Vitamins and minerals	Amino acids
"Energy" bars (if no more than 30% protein)	Creatine
	Glycerol
Sports drinks	HMB
Meal replacement drinks such as Ensure Plus or Boost	L-carnitine
	Protein powders

For a complete explanation of NCAA's rules regarding dietary supplements, see **www.ncaa.org**.

ergogenic Work-producing. An ergogenic aid is a mechanical, nutritional, psychological, pharmacological, or physiological substance or treatment intended to directly improve exercise performance.

TABLE 10-12 ▶ **An Evaluation of Popular Ergogenic Aids**

Substance/Practice	Rationale	Reality
Useful in Some Circumstances		
Caffeine	Decrease perception of fatigue, promote mental alertness, and improve vigilance	Drinking two to three 5-ounce cups of coffee (equivalent to 3 to 9 milligrams of caffeine per kilogram of body weight) about 1 hour before events lasting about 5 minutes or longer is useful for some athletes; benefits are less apparent in those who have ample stores of glycogen, are highly trained, or habitually consume caffeine; intake of more than about 600 milligrams (6 to 8 cups of coffee) may elicit a urine concentration illegal under NCAA rules (greater than 15 micrograms per milliliter). Possible side effects include shakiness, nervousness, nausea, anxiety, and insomnia. Cost: $0.08 per 300 milligrams.
Creatine	Increase phosphocreatine (PCr) in muscles to keep ATP concentration high	Use of 20 grams per day for 5 to 6 days and then a maintenance dose of 2 grams per day may improve performance in those who undertake repeated bursts of activity, such as in sprinting and weight lifting. Vegetarian athletes benefit the most because their diets are low in or devoid of creatine. Some of the muscle weight gain noted with use results from water contained in muscles. Endurance athletes do not benefit from use. Little is known about the safety of long-term creatine use. Continual use of high doses has led to kidney damage in a few cases. Cost: $25 to $65 per month.
Sodium bicarbonate (baking soda)	Neutralize acidic compounds that contribute to muscle fatigue	Partially effective in some circumstances (when lactic acid is rapidly produced), such as wrestling, but can cause nausea and diarrhea. The dose used is 300 milligrams/kilogram, given 1 to 3 hours before exercise. Cost: nil.
Beta-alanine	Increase muscle carnosine, a protein that neutralizes acidic compounds that contribute to muscle fatigue during exercise at high intensity	Supplementation with beta-alanine may improve performance in strength and power sports, such as sprinting and lifting. In research studies, doses vary from 3 to 6 grams per day. Higher doses may cause tingling of the skin. Cost: $25 per month.
Possibly Useful, Still Under Study		
Beta-hydroxy-beta methylbutyric acid (HMB)	Decrease protein catabolism, causing a net growth-promoting effect	Some research suggests that supplementation with this substance may increase muscle mass by 0.5 to 1 kilogram beyond unsupplemented diets when taken during initial phases of weight training. Still, safety and effectiveness of long-term HMB use in humans are unknown. Cost: $100 per month.
Branched-chain amino acids (BCAA) (leucine, isoleucine, and valine)	Increase gains in muscle mass during resistance training Important energy source, especially when carbohydrate stores are depleted	Several studies point to positive effects of BCAA on increases in muscle mass when taken before or after resistance training. This effect is most dramatic among untrained athletes who are just beginning a resistance-training program. Protein-rich foods (especially dairy proteins) are also rich in BCAA. For endurance athletes, supplementation of BCAA (10 to 30 grams per day) during exercise can increase BCAA in the blood when it has been depleted due to exercise, although there is no consistent evidence of improved endurance performance. Carbohydrate feeding, by delaying use of BCAA as fuel, may negate the need for BCAA supplementation in endurance activities. Cost: $20 per month.

(continued)

TABLE 10-12 ▶ An Evaluation of Popular Ergogenic Aids *(concluded)*

Substance/Practice	Rationale	Reality
Glutamine (an amino acid)	Enhance immune function and preserve lean body mass	Some preliminary studies show decreased occurrence of upper respiratory tract infections in athletes with use. It also may promote muscle growth, but long-term studies are lacking. Protein foods are a rich source of glutamine. Cost: $10 to $20 per month for 1 to 2 grams per day.
Dangerous or Illegal Substances/Practices		
Anabolic steroids (and related substances, such as androstenedione and tetrahydrogestrinone)	Increase muscle mass and strength	Although effective for increasing protein synthesis, anabolic steroids are illegal in the United States unless prescribed by a physician. They have numerous potential side effects, such as premature closure of growth plates in bones (thus possibly limiting the adult height of a teenage athlete), bloody cysts in the liver, increased risk of cardiovascular disease, increased blood pressure, and reproductive dysfunction. Possible psychological consequences include increased aggressiveness, drug dependence (addiction), withdrawal symptoms (such as depression), sleep disturbances, and mood swings (known as "roid rage"). Use of needles for injectable forms adds further health risk. Banned by the International Olympic Committee.
Blood doping	To enhance aerobic capacity by injecting red blood cells harvested previously from the athlete, or alternately the athlete may use the hormone erythropoietin (Epogen) to increase red blood cell number	May offer aerobic benefit; very serious health consequences are possible, including thickening of the blood, which puts extra strain on the heart; is an illegal practice under Olympic guidelines.
Ephedrine (stimulant)	Increase muscle strength and power, promote mental alertness, decrease fatigue and weight loss	A limited number of studies demonstrate modest improvements in anaerobic tasks, but most studies do not support the use of ephedrine as an ergogenic aid. Use has been linked to a variety of adverse effects, such as heart palpitations, anxiety, and even death. Ephedrine is banned by the International Olympic Committee and the National Collegiate Athletic Association.
Gamma hydroxybutyric acid (GHB)	Promoted as a steroid alternative for bodybuilding	FDA has never approved it for sale as a medical product; it is illegal to produce or sell GHB in the United States. GHB-related symptoms include vomiting, dizziness, tremors, and seizures. Many victims have required hospitalization, and some have died. Clandestine laboratories produced virtually all of the chemical accounting for GHB abuse. FDA is working with the U.S. Attorney's office to arrest, indict, and convict individuals responsible for the illegal operations.
Growth hormone	Increase muscle mass	At critical ages, may increase height; may also cause uncontrolled growth of the heart and other internal organs and even death; potentially dangerous; requires careful monitoring by a physician. Use of needles for injections adds further health risk. Banned by the International Olympic Committee.

Summary

10.1 A gradual increase in regular physical activity is recommended for all healthy persons. Benefits include improvements in cardiovascular health, gastrointestinal function, blood glucose regulation, and sleep; reduced risk of certain cancers; and enhanced muscle and bone strength.

10.2 The Physical Activity Guidelines for Americans advise adults to do at least 150 minutes of moderate-intensity or 75 minutes of vigorous-intensity aerobic physical activity per week. In addition, adults should perform muscle-strengthening activities and flexibility exercises at least twice per week. Workouts should allow time for warm-up exercises to increase blood flow and warm the muscles and then end with cooldown exercises, including stretching.

10.3 Human metabolic pathways extract chemical energy from carbohydrate, fat, and protein to yield ATP. Phosphocreatine is high-energy compound that can be used to resupply ATP during short, intense activities. The mix of macronutrients used for fuel depends on the intensity and duration of activity: short-term, intense exercises primarily use carbohydrate for fuel, whereas low- or moderate-intensity endurance exercises use more fat for fuel. Protein makes a minor contribution as a fuel source.

10.4 To support physical activity, athletes require 5 to 8 kcal per minute of activity above energy needs for a sedentary person. Monitoring weight changes over time is a good way to assess the adequacy of energy intake. Athletes should obtain energy from a varied diet that includes sources of carbohydrates (6 to 10 grams of carbohydrate per kilogram of body weight; usually 60% of total energy), protein (0.8 to 2.0 grams of protein per kilogram, depending on the type of training), and fat (up to 35% of energy, focusing on vegetable oils instead of solid fats). The increased overall food intake of athletes typically furnishes adequate vitamins and minerals. Some micronutrients of concern are iron and calcium, especially for women. Athletes should drink fluid before, during (to minimize loss of body weight), and after exercise (2 to 3 cups per pound lost). Sports drinks help replace fluid, electrolytes, and carbohydrates lost during workouts that last beyond 60 minutes.

10.5 Endurance athletes can delay or prevent fatigue by consuming enough fluids, electrolytes, and carbohydrates before, during, and after events. In addition, protein in the post-exercise period will aid muscle recovery. In addition to these strategies to maintain hydration and muscle glycogen stores, athletes who train to develop strength or power should place special emphasis on protein during the recovery period.

NAYH Athletes can benefit from ergogenic properties of sufficient water, electrolytes, and carbohydrates, and a balanced and varied diet consistent with the Dietary Guidelines and MyPlate. Under some circumstances, creatine, sodium bicarbonate, and caffeine are useful for enhancing athletic performance. Protein and amino-acid supplements are not necessary because athletes meet protein needs from foods.

Check Your Knowledge

1. An energy-rich compound, phosphocreatine (PCr), is found in _____ tissue.
   a. adipose
   b. muscle
   c. liver
   d. kidney

2. A fitness program should include
   a. aerobic exercises 5 days per week.
   b. strength-training exercises 2 to 3 days per week.
   c. stretching exercises 2 to 3 days per week.
   d. All of the above.

3. During muscle-building regimens, athletes should consume _____ grams of protein per kilogram body weight.
   a. 0.5 to 0.7
   b. 0.8
   c. 1.5 to 1.7
   d. 2 to 2.5

4. Which of these foods is the best choice for carbohydrate loading before endurance events?
   a. potato chips
   b. French fries
   c. All-Bran (high-fiber) cereal
   d. rice

5. As the body adapts to regular exercise, the "training effect" results in
   a. decreased blood flow to muscles.
   b. increased lactic acid production.
   c. decreased muscle triglyceride content.
   d. decreased resting heart rate.

6. A physically active lifestyle leads to
   a. increased bone strength.
   b. decreased risk of colon cancer.
   c. reduced anxiety and depression.
   d. All of the above.

7. How many cups of fluid are required to replace each pound of weight lost during an athletic event or workout?
   a. 0.5 to 0.75
   b. 1 to 1.5
   c. 2 to 3
   d. 4 to 5

8. The benefit of a "sports" drink is to provide
   a. water to hydrate.
   b. electrolytes to enhance water absorption in the intestine and maintain blood volume.
   c. carbohydrate for energy.
   d. All of the above.

9. Compared to anaerobic glucose metabolism, aerobic glucose metabolism produces more
   a. lactic acid.
   c. phosphocreatine.
   b. ATP.
   d. fatty acids.

10. Caffeine is used as an ergogenic aid by some athletes because it is thought to
    a. decrease fatigue.
    b. decrease the buildup of lactic acid.
    c. serve as an energy source.
    d. increase muscle mass and strength.

Answers Key: 1. b (LO 10.3), 2. d (LO 10.2), 3. c (LO 10.6), 4. d (LO 10.8), 5. d (LO 10.5), 6. d (LO 10.1), 7. c (LO 10.7), 8. d (LO 10.7), 9. b (LO 10.4), 10. a (LO 10.9)

Study Questions (Numbers refer to Learning Outcomes)

1. How does greater physical fitness contribute to better overall health? Explain the process. **(LO 10.1)**

2. You have set a goal to increase muscle mass and decrease body fat. Plan a weekly fitness regimen using the FITT principle. **(LO 10.2)**

3. How are carbohydrates, fat, and protein used to supply energy during a 100-meter sprint? During a weight-lifting session? During a 3-mile walk? **(LO 10.3)**

4. What is the difference between anaerobic and aerobic exercise? Explain why aerobic metabolism is increased by a regular exercise routine. **(LO 10.4)**

5. Is fat from adipose tissue used as an energy source during exercise? If so, when? **(LO 10.5)**

6. What are some typical measures used to assess whether an athlete's calorie intake is adequate? **(LO 10.6)**

7. List five specific nutrients that athletes need and the appropriate food sources from which these nutrients can be obtained. **(LO 10.6)**

8. You plan to participate in a half-marathon. Plan your menu for the day of the event, being sure to include appropriate levels of macronutrients and fluids before, during, and after the athletic event. **(LO 10.8)**

9. Your neighbor is planning to run a 5-kilometer race. Summarize for her what you have learned about fluid intake before, during, and after the event. **(LO 10.7)**

10. Should competitive athletes take amino acid supplements? Why or why not? **(LO 10.9)**

What the Dietitian Chose

In this chapter, you have learned that it's important to replenish carbohydrates, fluids, and electrolytes for optimal performance during endurance events, like this half-marathon. General guidelines are to consume ½ to 1 ½ cups of fluid every 15 minutes and 30 to 60 grams of carbohydrate per hour. Choosing a sports nutrition product is something you should do during training. With a few weeks to go, you have time to experiment with a product that can help you avoid "hitting the wall."

The PowerBar ProteinPlus energy bar provides 300 kcal, 6 grams of fat, 39 grams of carbohydrate, and 23 grams of protein. This might be a tasty and convenient meal replacement during your busy weeks of training. Alternatively, it could serve as a recovery meal to help replenish your muscle glycogen stores after the race. However, during the race, its high fat and protein content would slow digestion, possibly leading to abdominal cramps. Furthermore, this product would not contribute to fluid replenishment.

The Essential Amino Energy drink provides 2 grams of carbohydrate and 5 grams of essential amino acids in free form, including branched-chain amino acids. The drink may help meet fluid needs, but the carbohydrate content is not sufficient to maintain blood glucose during the race. It is true that BCAAs are utilized for fuel during endurance activities and that they can enhance muscle building during weight training, but there is no consistent evidence that supplying BCAAs during endurance exercise enhances performance.

A sports drink, such as Powerade, is superior to water for events lasting more than 60 minutes. The carbohydrates help to supply energy to the brain and muscles. The added flavor encourages consumption. The electrolytes replace those lost in sweating, increase fluid absorption, and stimulate thirst. A 32-ounce bottle of Powerade (4 servings) contains a total of 200 kcal and 56 grams of carbohydrate in 4 cups of fluid. One bottle of Powerade per hour while you compete would meet your fluid and carbohydrate needs but may be inconvenient to tote while running.

The energy gel is small enough to take along during the race. The label says that it has 110 kcal, 1.5 grams of fat, 22 grams of carbohydrate, 12 grams of sugar, and 100 milligrams of caffeine. A Clif Shot during the race could help replenish blood glucose and electrolytes, but it definitely won't meet your fluid requirements while racing. However, there will be water available at numerous stops along the course.

Some energy gels are caffeine-free, but this one contains as much caffeine as a strong cup of coffee. Research shows

that caffeine can decrease feelings of fatigue and increase mental alertness among people who are not habitual caffeine consumers. On the other hand, negative effects of excessive caffeine can include dehydration, heart palpitations, and gastrointestinal discomfort, so moderation is necessary.

In summary, both the Powerade and the Clif Shot would replenish carbohydrates and electrolytes. The Powerade would also fulfill your fluid needs, whereas you'd need to grab water along the course if you choose the energy gel. The energy gel has the advantage of being light and portable, plus the caffeine may have an ergogenic effect. Give both the Powerade and the Clif Shot energy gel a trial run during your training sessions and see which one helps you most.

CASE STUDY SOLUTION Planning a Training Diet

1. Michael is correct in following a high-carbohydrate diet.
2. In his effort to minimize his fat intake, he is probably not consuming enough calories, protein, iron, and calcium to support his training routine. He has fallen into the bagel, pasta, and pretzel routine that sports nutritionists warn is not conducive to peak performance. Low protein, iron, and overall calories may contribute to fatigue.
3. Michael's performance would improve if he also had a source of protein at each meal. He could include milk with breakfast and possibly some low-fat yogurt or low-fat cheese at lunch. He should have a carbohydrate-and-protein snack before his workout, such as half a sandwich with fruit and some water. The

sandwich and fruit will help provide him with fuel to support his vigorous training. In the evenings, he could add some reduced-fat cheese or hummus to his snack of pretzels to improve protein intake.
4. During his workouts, he could consume a sports drink to meet fluid needs and supply some carbohydrate, or he could consume water, along with a few graham crackers or other high-carbohydrate food.
5. Overall, it is important for Michael to fuel his body before, during, and after workouts. Before and during exercise, carbohydrates and fluid will improve athletic performance. After exercise, carbohydrates coupled with protein will enhance muscle recovery.

Further Readings

1. Burd NA and others: A-Z of nutritional supplements: Dietary supplements, sports nutrition foods, and ergogenic aids for health and performance—Part 26. *British Journal of Sports Medicine* 45:1163, 2011.
2. Burke LM and others: BJSM reviews: A-Z of nutritional supplements: Dietary supplements, sports nutrition foods and ergogenic aids for health and performance—Part 4. *British Journal of Sports Medicine* 43:1088, 2009.
3. Castell LM and others: BJSM review: A-Z of nutritional supplements: Dietary supplements, sports nutrition foods and ergogenic aids for health and performance—Part 5. *British Journal of Sports Medicine* 44:77, 2010.
4. DeJonge L, Smith MR: Macronutrients and exercise. *Obesity Management* February:11, 2008.
5. Garber CE and others: American College of Sports Medicine Position Stand: Quantity and quality of exercise for developing and maintaining cardiorespiratory, musculoskeletal, and neuromotor fitness in apparently healthy adults: Guidance for prescribing exercise. *Medicine and Science in Sports and Exercise* 43:1334, 2011.
6. Goodman C and others: A to Z of nutritional supplements: Dietary supplements, sports nutrition foods and ergogenic aids for health and performance—Part 21. *British Journal of Sports Medicine* 45:677, 2011.
7. Holway FE and Spriet LL: Sport-specific nutrition: Practical strategies for team sports. *Journal of Sports Sciences* 29:S115, 2011.
8. Jenkinson DM, Harbert AJ. Supplements and sports. *American Family Physician* 78:1039, 2008.

9. Kreider RB and others: ISSN exercise and sport nutrition review: Research and recommendations. *Journal of the International Society of Sports Nutrition* 7:7, 2010.

10. Margaritis I, Rousseau AS: Does physical exercise modify antioxidant requirements? *Nutrition Research Reviews* 21:3, 2008.

11. Maughan RJ, Shirreffs SM: Development of individual hydration strategies for athletes. *International Journal of Sport and Nutrition and Exercise Metabolism* 18:457, 2008. Angie/April: This article is epub ahead of print. I was not sure how to cite it. See **http://www.ncbi.nlm. nih.gov/pubmed/25043597**

12. Navarro VJ and others: Liver injury from herbals and dietary supplements in the U.S. Drug-Induced Livery Injury Network. *Hepatology*, 2014.

13. Phillips SM and Van Loon LJC: Dietary protein for athletes: From requirements to optimum adaptation. *Journal of Sports Sciences* 29:S29, 2011.

14. Rodriguez NR and others: Position of the American Dietetic Association, Dietitians of Canada, and the American College of Sports Medicine: Nutrition and athletic performance. *Journal of the American Dietetic Association* 109:509, 2009.

15. Rosenbloom C and Rosbruck M. Popular dietary supplements used in sports. *Nutrition Today* 43:60, 2008.

16. Sawka MN and others: American College of Sports Medicine Position Stand: Exercise and fluid replacement. *Medicine and Science in Sports and Exercise* 39:377, 2007.

17. Slater G and Phillips SM: Nutrition guidelines for strength sports: Sprinting, weightlifting, throwing events, and bodybuilding. *Journal of Sports Sciences* 29:S67, 2011.

18. Stellingwerff T, Maughan RJ, and Burke LM: Nutrition for power sports: Middle-distance running, track cycling, rowing, canoeing/ kayaking, and swimming. *Journal of Sports Sciences* 29:S79, 2011.

19. Williams MH, Anderson DE, and Rawson ES: *Nutrition for health, fitness, and sport.* 10th ed. Boston: McGraw-Hill, 2013.

20. Woolf K and Manore MM: B-vitamins and exercise: Does exercise alter requirements? *International Journal of Sport Nutrition and Exercise-Metabolism* 16:453, 2006.

 **NUTRITION**

To get the most out of your study of nutrition, visit McGraw-Hill Connect at www.mcgrawhillconnect.com where you will find NutritionCalc Plus, LearnSmart, and many other dynamic tools.

Rate Your Plate

I. Evaluating Protein Intake—A Case Study

Mark is a college student who has been lifting weights at the student recreation center. The trainer at the center recommended a protein drink to help Mark build muscle mass. Answer the following questions about Mark's current food intake and determine whether a protein drink is needed to supplement Mark's diet.

The following is a tally of yesterday's intake.

Breakfast	Frosted Mini-Wheats cereal, 2 oz 1% milk, 1½ cups Orange juice, chilled, 6 oz Glazed yeast doughnut, 1 Brewed coffee, 1 cup
Lunch	Double hamburger with condiments, 1 French fries, 30 Cola, 12 oz Medium apple, 1
Dinner	Frozen lasagna w/meat, 2 pieces 1% milk, 1 cup Looseleaf lettuce, chopped, 1 cup Creamy Italian salad dressing, 2 tsp Medium tomato, ½ Whole carrot, raw, 1
Evening snack	Low-fat vanilla ice cream, 1 cup Hot fudge chocolate topping, 2 tsp Soft chocolate chip cookies, 2

1. Evaluate Mark's diet using NutritionCalc Plus. Is he meeting the minimum recommendations of MyPlate's Daily Food Plan for his calorie needs?

   _____

2. Mark's weight has been stable at 70 kilograms (154 pounds). Determine his protein needs based on the RDA (0.8 grams per kilogram).
   a. Mark's estimated protein RDA: _____

   b. What are the maximum recommendations for protein intake for strength-training athletes (see Table 10-5)?
   _____

   c. Calculate the maximum protein recommendation for Mark. _____
   _____

3. An analysis of the total calorie and protein content of Mark's current diet is 3470 kcal, 125 grams of protein (14% of total calories supplied by protein). This diet is representative of the food choices and amounts of food that Mark chooses on a regular basis.
   a. What is the difference between Mark's estimated protein needs as an athlete (from number 2) and the amount of protein that his current diet provides? _____
   b. Is his current protein intake inadequate, adequate, or excessive? _____

4. Mark takes his trainer's advice and goes to the supermarket to purchase a protein drink to add to his diet. Four products are available; they contain the following label information.

	Amino Fuel	Joe Weider's Sugar-Free 90% Plus Protein	Joe Weider's Dynamic Muscle Builder	Victory Super Mega Mass 2000
Serving size	3 tbsp	3 tbsp	3 tbsp	¼ scoop
Kcal	104	110	103	104
Protein (grams)	15	24	10	5

The trainer recommends adding the supplement to Mark's diet two times a day. Mark chooses the Muscle Builder protein drink.

   a. How much protein would be added to Mark's diet daily from two servings of the supplement alone (prior to mixing it with a beverage)?

   b. Mark mixes the powder with the milk he already consumes at breakfast and dinner. How much protein total would Mark now consume in 1 day? (Add the protein amount from the nutrition analysis to the value from question 4a.)

   c. What is the difference between Mark's estimated protein needs as an athlete and his total intake?

5. What is your conclusion—does Mark need the protein supplement?

 **Find more Rate Your Plate activities for this chapter on Connect at www.mcgrawhillconnect.com.**

Student Learning Outcomes

Chapter 11 is designed to allow you to:

11.1 Contrast healthy attitudes toward uses of food with behavior patterns that could lead to unhealthy uses of food.

11.2 Describe current hypotheses about the origins of eating disorders.

11.3 Enumerate physical and mental characteristics of anorexia nervosa, and outline current best practices for its treatment.

11.4 List physical and mental characteristics of bulimia nervosa and outline current best practices for its treatment.

11.5 Enumerate physical and mental characteristics of binge-eating disorder and outline current best practices for its treatment.

Chapter 11
Eating Disorders

What Would You Choose?

Your college roommate has been acting strangely lately, and you are starting to suspect she has an eating disorder. Over the first few months at college, she gained the "freshman 15." She got kind of depressed about it when her jeans started to fit too snugly and declared she was going on a diet. Her weight has not really decreased much since then, but her eating behaviors certainly have changed. She stopped joining your group of friends for meals in the dining hall, opting instead to eat by herself in your dorm room. She has been spending at least 2 hours a day in the campus recreation center, sometimes going in the morning and evening. A few days ago, you thought you heard her throwing up in the restroom, but when she came out of the stall, she said she was fine. What would you choose to help your roommate?

a Tell her she has bulimia and needs to get some treatment.

b Give her some weight-management advice that you learned in Chapter 7.

c Express your concern and ask her if you can go with her to the student health center to talk to someone about it.

d Ignore her behavior; she is just trying to draw attention to herself, and this phase will pass in a few weeks.

connect **NUTRITION** Think about your choice as you read Chapter 11, then see **What the Dietitian Chose** at the end of the chapter. To learn more about warning signs of eating disorders, check out the Connect site: www.mcgrawhillconnect.com.

Many of us occasionally eat until we're stuffed and uncomfortable, such as at Thanksgiving dinner. Faced with savory and tempting foods, we find that we cannot easily stop eating. Usually we forgive ourselves, vowing not to overeat the next time. Nevertheless, many of us have problems controlling our food intake and body weight. The combination of too many instances of simple overeating, and too little physical activity eventually leads to progressive weight gain.

The obesity epidemic tends to upstage all other nutrition-related problems. However, the consequences of eating disorders are just as serious as obesity; indeed, if left untreated, eating disorders can be fatal. What is most alarming about these disorders—anorexia nervosa, bulimia nervosa, and binge-eating disorder—is the increasing number of cases reported each year.

Eating disorders arise as a result of interactions among brain biology, personality, and environmental influences. Successful treatment of eating disorders, therefore, is complex and must go beyond nutritional therapy. Keep in mind that eating disorders are not restricted to any socioeconomic class or ethnicity. They can also strike at any age in either females or males. Let us examine the causes, effects, and treatments of these conditions in detail because eating disorders touch many of our lives.

11.6 Describe other forms of eating disorders, including night eating syndrome and the female athlete triad.

11.7 Describe methods to reduce the development of eating disorders, including the use of warning signs to identify early cases.

▲ Maintaining an ultraslim body type is an all too common goal in today's culture. The media and the fashion world bombard us with body images that are unrealistic for most people.

endorphins Natural body tranquilizers that may be involved in the feeding response and function in pain reduction.

disordered eating Mild and short-term changes in eating patterns that occur in relation to a stressful event, an illness, or a desire to modify one's diet for a variety of health and personal appearance reasons.

11.1 From Ordered to Disordered Eating Habits

Eating—a completely instinctive behavior for animals—serves an extraordinary number of psychological, social, and cultural purposes for humans. Eating practices may take on religious meanings; signify bonds within families and ethnic groups; and provide a means to express hostility, affection, prestige, or class values. Within the family, supplying, preparing, and even withholding food may be a means of expressing love, hatred, or power.

We are bombarded daily with images of our society's portrayal of the "ideal" body. Dieting is promoted to achieve this ideal body—eternally young and admired. Television programs, billboard and Internet advertisements, magazine pictures, movies, and newspapers suggest that an ultraslim body will bring happiness, love, and ultimately, success. This fantasy notion is contradictory to the fact that the rate of obesity in our society increases with each passing year. In response to this social pressure, some of us take an extreme approach: the pathological pursuit of weight control or weight loss.

Early in life, we develop images of "acceptable" and "unacceptable" body types. Of the attributes that constitute attractiveness, many people view body weight as the most important, partly because we can control our weight somewhat. Fatness is the most dreaded deviation from our cultural ideals of body image, the one most derided and shunned. In a study about weight bias conducted by researchers at Yale University, nearly half of survey respondents said they would rather give up 1 year of life than live with obesity.

It is difficult to resist comparing ourselves to the "ideal" body, but not everyone can look like a fashion model. For some people, the disparity between their own body image and the perceived ideal may be enough to trigger an eating disorder.

FOOD: MORE THAN JUST A SOURCE OF NUTRIENTS

From birth, we link food with personal and emotional experiences. As infants, we associate milk with security and warmth, so the breast or bottle becomes a source of comfort and food. As noted in Chapter 1, even when older, most people continue to derive comfort and great pleasure from food. This is both a biological and a psychological phenomenon. Food can be a symbol of comfort, but eating can also stimulate the release of certain neurotransmitters (e.g., serotonin) and *natural opioids* (including **endorphins**), which produce a sense of calm and euphoria in the human body. Thus, in times of great stress, some people turn to food for a druglike, calming effect.

Food is also used as a reward or a bribe. You have probably heard or spoken something similar to the following comments.

You can have your dessert if you eat five more bites of your vegetables.
You cannot play until you clean your plate.
I will eat the broccoli if you let me watch TV.
If you love me, you will eat your dinner.

On the surface, using food as a reward or bribe seems harmless. Eventually, however, this practice encourages both caregivers and children to use food to achieve goals other than satisfying hunger and nutrient needs. Food may then become much more than a source of nutrients. Regularly using food as a bargaining chip can contribute to abnormal eating patterns. Carried to the extreme, these patterns can lead to **disordered eating.**

Disordered eating can be defined as mild and short-term changes in eating patterns that occur in response to a stressful event, an illness, or even a desire to modify the diet for a variety of health and personal appearance reasons. The problem may be no more than a bad habit, a style of eating adapted from friends or family members, or an aspect of preparing for athletic competition. While disordered eating can lead to weight loss or weight gain, as well as certain nutritional problems, it rarely requires in-depth professional attention. If, however, disordered eating becomes sustained, distressing, or starts to interfere with everyday activities and is linked to physiological changes, it may require professional intervention.

ORIGINS OF EATING DISORDERS

Given the common practice of dieting in North America, it can sometimes be difficult to draw a definitive line between disordered eating and an **eating disorder.** Indeed, many eating disorders start with a simple diet. Eating disorders then go on to involve physiological changes associated with food restriction, binge eating, purging, and fluctuations in weight. They also involve a number of emotional and cognitive changes that affect the way a person perceives and experiences his or her body, such as feelings of distress or extreme concern about body shape or weight. Eating disorders are not due to a failure of will or behavior; rather, they are real, treatable medical illnesses in which certain maladaptive patterns of eating take on a life of their own.

People who suffer from eating disorders can experience a wide range of health complications, including heart conditions and kidney failure, which may even lead to death. Recognition of eating disorders as important and treatable diseases, therefore, is critical. The three main types of eating disorders are **anorexia nervosa, bulimia nervosa,** and **binge-eating disorder.** Although it is convenient to label patients with a clear-cut diagnosis, the various types of eating disorders have more similarities than distinctions, particularly in the biological and psychological processes that underlie them.

Specific criteria from the recently revised *Diagnostic and Statistical Manual of Mental Disorders, 5th edition* (*DSM-5*) are used by clinicians to diagnose eating disorders. As you read about the different eating disorders in this chapter, look for the relevant diagnostic criteria in Tables 11-1, 11-2, and 11-3. Keep in mind that individuals may exhibit a few symptoms of eating disorders but not enough to warrant a formal diagnosis. These people may be classified as having "subthreshold" eating disorders. Also, some people show characteristics of more than one eating disorder or may migrate from one disorder to another over time. Indeed, about half of women diagnosed as having anorexia nervosa eventually develop bulimic symptoms. Still, appreciating the differences between the disorders helps us to understand the various approaches to prevention and treatment.

Over the years, researchers have theorized that dysfunctional family interactions, especially between parents and adolescents, precipitate eating disorders. While unhealthy family relationships may lead to emotional distress, there is little scientific evidence that family functioning is a primary cause for eating disorders. In fact, insinuating that the family has caused a person's eating disorder leads to feelings of guilt and shame within the family that may actually hinder efforts at treatment (see Further Reading 10).

Scientists now recognize that genes bear much of the blame for eating disorders (see Further Reading 9). Twin studies have shown that identical twins (who share the same DNA) have a higher likelihood of sharing eating disorders than do fraternal twins. Genetic background accounts for an estimated 50% to 83% of the overall risk for developing an eating disorder.

A variety of genes could be involved in the development of eating disorders, including those responsible for the expression of hormones and neurotransmitters involved in weight regulation and eating behaviors (review Chapter 1). In about 80% of cases, eating disorders co-occur with other psychological disorders, such as anxiety disorders, major depression, and substance abuse disorders. It appears that genes influence the brain biology that determines how we perceive ourselves and respond to food stimuli or stress. When they are unable to tolerate negative feelings such as sadness, anger, or guilt, genetically predisposed individuals may resort to self-destructive coping mechanisms, including disordered eating.

Because stressful life events may precipitate an eating disorder in genetically predisposed individuals, there is a strong association between eating disorders

eating disorder Severe alterations in eating patterns linked to physiological changes. The alterations are associated with food restriction, binge eating, inappropriate compensatory behaviors, and fluctuations in weight. They also involve a number of emotional and cognitive changes that affect the way a person perceives and experiences his or her body.

anorexia nervosa An eating disorder characterized by extreme restriction of energy intake relative to requirements leading to significantly low body weight.

bulimia nervosa An eating disorder characterized by recurrent episodes of binge eating followed by inappropriate compensatory behaviors to prevent weight gain.

binge-eating disorder An eating disorder characterized by recurrent episodes of binge eating that are associated with marked distress and lack of control over behavior but not followed by inappropriate compensatory behaviors to prevent weight gain.

Progression from Ordered to Disordered Eating

Attention to hunger and satiety signals; limitation of calorie intake to restore weight to a healthful level

↓

Some disordered eating habits begin as weight loss is attempted, such as very restricted eating

↓

Clinically evident eating disorder recognized

epigenetics Changes in gene function that are independent of DNA sequence. For example, malnutrition during pregnancy may modify gene expression in the fetus and affect long-term body weight regulation in the offspring.

and a history of abuse. In fact, physical and sexual abuse are about twice as common among people with eating disorders compared to the population as a whole (see Further Reading 7). Other stressful life events, such as wartime military service, the death of a loved one, or constant social pressures to achieve thinness, are potential triggers for eating disorders.

In addition to the genetic code itself, researchers are studying **epigenetics** (the interaction between genes and environmental influences), as it relates to the development of eating disorders (see Further Reading 3). Some studies suggest that maternal stress or hormone levels during pregnancy can predict the future development of an eating disorder in their children.

Overall, genes may set the stage for development of an eating disorder, but environmental factors also play a role. Identifying genes linked to eating disorders eventually could help in tailoring prevention and treatment efforts for at-risk individuals. However, the psychiatric counseling that is part of current therapy will still be of value.

THE CHANGING FACE OF EATING DISORDERS

If you were asked to paint a picture of a person with an eating disorder, who would you depict? The predominant stereotype is that eating disorders only affect young, white females of middle or upper socioeconomic status. However, the face of eating disorders is changing.

When it comes to anorexia and bulimia, women do outnumber men by about 9:1. Perhaps social pressures can account for part of this disparity: in the media, women are held to standards of unnatural thinness, whereas the image conveyed for men is big and muscular. However, men are affected by eating disorders as well. Among men, exercise status and sexual orientation are factors that particularly influence development of anorexia and bulimia. Male athletes are more prone than nonathletes to develop these eating disorders, especially those who participate in sports that require weight classes, (e.g., boxers, wrestlers, and jockeys) or where judging is partly based on aesthetics (e.g., swimming, diving, or dancing) (see Further Reading 12). With reference to sexual orientation, the prevalence of eating disorders in gay men is two to three times that in heterosexual men (see Further Reading 4). With the release of *DSM-5* in 2013, binge-eating disorder is now recognized as a distinct eating disorder. While women outnumber men by about 5 to 1 for cases of anorexia and bulimia, men account for about 40% of cases of binge-eating disorder.

Eating disorders typically develop during adolescence or young adulthood. Adolescence is a period of turbulent sexual and social tensions. At this time of life, teenagers establish their own identities. While declaring independence, they seek acceptance and support from peers and parents and react strongly to how they think others perceive them. At the same time, their bodies are changing, and much of the change is beyond their control. This is a time when extreme dieting practices may take root. It is alarming that eating disorders are being diagnosed at earlier ages (see Further Reading 1). It is important to note that calorie restriction is not always evidenced as weight loss; stunting (failure to grow in height) and delayed sexual maturation also could be signs of eating disorders among children and adolescents.

While much of the focus has been on youth, middle-aged and older adults are not immune to the devastating effects of eating disorders. In North America, aging baby boomers are the largest demographic group. Although eating disorders rarely make their first appearance late in adulthood, it may not be until later in adulthood that people who have suffered from eating disorders for years finally seek treatment. Furthermore, some recovered adults may relapse into former disordered eating practices. Research reveals that disordered eating behaviors and dissatisfaction with body weight and shape are quite common among older women. In one study, binge eating was reported by 3.5% of a community-based sample of women over age 50; purging behaviors, such as excessive exercise, were employed by 7.8%

▼ Individuals who participate in sports with weight classes, such as wrestling, may practice disordered eating behaviors in order to gain a competitive advantage over other athletes in a lower weight class.

of the women (see Further Reading 5). Negative body image can have a dramatic impact on self-esteem and overall quality of life at any age.

Until recently, most researchers have reported that eating disorders primarily affect middle- and upper-class Caucasian women. Now, studies show greater similarities in the rates of body dissatisfaction and disordered eating behaviors across ethnic and cultural groups. Perhaps minorities with eating disorders have been less likely to seek help in the past due to fear of shame or stigma, lack of resources, or language barriers. It seems more likely, however, that health care workers have been less likely to diagnose non-Caucasians as having eating disorders. Previously, it seemed that non-Caucasian cultures were more accepting of larger body shapes, but mainstream pressures for thinness now cut across cultural lines.

Do you know someone who is at risk for an eating disorder? If so, suggest that the person seek a professional evaluation because the sooner treatment begins, the better the chances for recovery. However, do not try to diagnose eating disorders in your friends or family members. Only a professional can exclude other possible diseases and correctly diagnose an eating disorder. Once an eating disorder is diagnosed, immediate treatment is advisable. As a friend, the best you can do is to encourage an affected person to seek professional help. Such help is commonly available at student health centers and student guidance/counseling facilities on college campuses.

> Our passion for thinness may have its roots in the Victorian era of the nineteenth century, which specialized in denying "unpleasant" physical realities, such as appetite and sexual desire. Flappers of the 1920s set the standard for the twentieth and twenty-first-century trend for thinness. Since 1922, the BMI values of Miss America winners have steadily decreased; during the last three decades, most winners had a BMI in the "underweight" range (less than 18.5).

☑ CONCEPT CHECK 11.1

1. Differentiate between disordered eating and an eating disorder.
2. Describe how genetics and environment interact in the development of eating disorders.
3. Why are eating disorders more common among adolescents than other age groups?

11.2 Anorexia Nervosa

Anorexia nervosa was first described in early medical literature in 1689 and is an eating disorder characterized by extreme weight loss, a distorted body image, and an irrational fear of weight gain and obesity. These three criteria are outlined in Table 11-1 and described in detail below. Anorexia nervosa affects an estimated 0.8% of American women (see Further Reading 17).

First, people with anorexia nervosa severely restrict energy intake relative to requirements. The term *anorexia* implies a loss of appetite; however, a denial of one's appetite more accurately describes the behavior of people with anorexia nervosa. Low energy intake leads to a body weight that is significantly less than expected when compared to others of the same age, sex, stage of physical development, and

TABLE 11-1 ▶ Diagnostic Criteria for Anorexia Nervosa

A. Restriction of energy intake relative to requirements, leading to a significantly low body weight in the context of age, sex, developmental trajectory, and physical health.
B. Intense fear of gaining weight or becoming fat, or persistent behavior that interferes with weight gain, even though at a significantly low weight.
C. Disturbance in the way in which one's body weight or shape is experienced, undue influence of body weight or shape on self-evaluation, or persistent lack of recognition of the seriousness of the current low body weight.

From *Diagnostic and Statistical Manual for Mental Disorders, 5th edition.* Copyright 2013 by the American Psychiatric Association.

▲ Self-image is an important part of adolescence. For people with eating disorders, the difference between the real and desired body images may be too difficult to accept. See the website **womenshealth.gov/body-image**.

activity level. While low body weight (i.e., less than 85% of that expected for a given age and sex or BMI of less than 17) may indicate anorexia nervosa, a variety of other medical conditions could also result in low body weight. The next two diagnostic criteria set anorexia nervosa apart from other problems related to poor food intake or low body weight.

The second key criterion for diagnosis of anorexia nervosa involves an intense fear of gaining weight or becoming obese. Some individuals with eating disorders may deny a fear of weight gain, so persistent behaviors that interfere with weight gain are also included in this criterion. To be diagnosed with anorexia nervosa, an individual experiences fear of weight gain or practices behaviors to prevent weight gain at least 75% of the days in the last three months.

Third, as depicted in the margin, individuals with anorexia nervosa have a very distorted body image. The term *nervosa* refers to an attitude of disgust with one's body. Anorexic individuals irrationally believe they are fat, even though others constantly comment on their thin physique. Some anorexics realize they are thin but continue to be haunted by certain areas of their bodies that they believe to be fat (such as thighs, buttocks, and stomach). Even though extremely low body weight results in severe health effects, as described below, a person with anorexia does not acknowledge the problem. He or she may persist in efforts at weight loss and try to thwart efforts of family members and medical professionals to increase their body weight to a healthy level.

COMMON BEHAVIORS OF ANOREXIA NERVOSA

Individuals who develop anorexia share some common personality traits. Take a young woman, for example, who is described by parents and teachers as responsible, meticulous, and obedient. She holds herself to high standards of performance and appearance. She is competitive and often obsessive. At home, she may not allow clutter in her bedroom. Physicians note that after a physical examination, she may fold her examination gown very carefully and clean up the examination room before leaving. As mentioned previously, genes dictate the brain biology that determines how we perceive ourselves and respond to stress. The personality traits described above are probably related to the same brain biology that predicts development of eating disorders.

Anorexia nervosa may begin as a simple attempt to lose weight. A comment from a well-meaning friend, relative, or coach suggesting that the person seems to be gaining weight or is too fat may be all that is needed. The stress of having to maintain a certain weight to look attractive or competent on a job can also lead to disordered eating. Abusive experiences, a difficult break-up, or the stress of leaving home for college are examples of triggers for extreme dieting. Changing one's appearance might be viewed as a way to avoid future conflict or ensure success in a new situation.

Still, looking "good" does not necessarily help people deal with anger, depression, low self-esteem, or past experiences with sexual abuse. If these issues are behind the disorder and are not resolved as weight is lost, the individual may intensify efforts to lose weight "to look even better," rather than work through unresolved psychological concerns. At first, dieting becomes the life focus. The person may think, "The only thing I am good at is dieting. I can't do anything else." This innocent beginning often leads to very abnormal self-perceptions and eating habits, such as cutting a pea in half before eating it. Other habits include hiding and storing food or spreading food around a plate to make it look as if much has been eaten. An anorexic person may cook a large meal and watch others eat it while refusing to eat anything, or he or she may insist on having different meals from the rest of the family. Frequent weighing—multiple times per day—is common (see Further Reading 19).

▲ Extreme dietary rules severely limit nutritional intake among people with anorexia nervosa.

Extreme dieting is the most important predictor of an eating disorder. (Adolescents expressing concern about their weight should be advised to focus on exercise, which does not appear to impart a risk for subsequent problems.) Once dieting begins, a person developing anorexia nervosa does not stop. The result is a long period of rigidly self-enforced semistarvation, practiced almost with a vengeance, in a relentless pursuit of control.

Among some people with anorexia nervosa, disordered eating behaviors may eventually include bingeing on large amounts of food in a short time, followed by inappropriate behaviors to compensate for the large number of calories consumed. Covered further in Section 11.3, compensatory behaviors (sometimes called purging) include vomiting, laxatives, diuretics, and excessive exercise. Thus, a person with anorexia nervosa may exist in a state of continuous semistarvation or may alternate between periods of starvation with periods of bingeing and purging.

As the disorder progresses, the range of foods eaten may narrow; the list of "safe foods" shortens, whereas the list of "unsafe foods" gets longer each day. For people developing anorexia nervosa, these practices say "I am in control." These people may be hungry, but they deny it, driven by the belief that good things will happen by just becoming thin enough. It becomes a question of willpower.

People with anorexia become irritable and hostile, and begin to withdraw from family and friends. School performance generally crumbles. They refuse to eat out with family and friends, thinking, "I won't be able to have the foods I want to eat," or "I won't be able to throw up afterward."

Anorexic persons see themselves as rational and others as irrational. They also tend to be excessively critical of themselves and others. Nothing is good enough. Because it cannot be perfect, life appears meaningless and hopeless. A sense of joylessness pervades everything.

Ultimately, an anorexic person eats very little food; 300 to 600 kcal daily is not unusual. In place of food, the person may consume up to 20 cans of diet soft drinks and chew many pieces of sugarless gum each day.

PHYSICAL EFFECTS OF ANOREXIA NERVOSA

The state of semistarvation disturbs many body systems as it forces the body to conserve energy stores as much as possible (Fig. 11-1). Many of the complications listed below can be reversed by returning to a healthy weight, provided the duration of the semistarvation has not been too long.

- Lowered body temperature and cold intolerance from loss of insulating fat layer.
- Slowed metabolic rate from decreased synthesis of thyroid hormones.
- Decreased heart rate as metabolism slows, leading to premature fatigue, fainting, and an overwhelming need for sleep. Other changes in heart function may occur as well, including loss of heart tissue and irregular heart rhythm.
- Iron-deficiency anemia, which leads to further weakness.
- Rough, dry, scaly, and cold skin, which may show multiple bruises because of the loss of the protective fat layer normally present under the skin.
- Low white blood cell count, which increases the risk of infection—one cause of death in people with anorexia nervosa.
- Abnormal feeling of fullness or bloating, which can last for several hours after eating.
- Loss of hair.
- Appearance of **lanugo**—downy hairs on the body that trap air to partially counteract heat loss that occurs with loss of fat tissue.
- Constipation due to deterioration of the gastrointestinal tract and abuse of laxatives. In extreme cases, impaired motility can cause rupture of the GI tract, leading to infection and even death.
- Low blood potassium, worsened by potassium losses during vomiting and use of some types of diuretics. This increases the risk of heart rhythm disturbances, another leading cause of death in anorexic people.
- Loss of menstrual periods because of low body weight, low body fat content, and the stress of the disease. Accompanying hormonal changes contribute to bone loss.
- Changes in brain size, blood flow to the brain, and neurotransmitter function, all of which contribute to depression and complicate treatment attempts.

▲ Eating disorders are commonly seen in people who must maintain low body weight, such as ballet dancers.

lanugo Downlike hair that appears after a person has lost much body fat through semi-starvation. The hair stands erect and traps air, acting as insulation for the body to compensate for the relative lack of body fat, which usually functions as insulation.

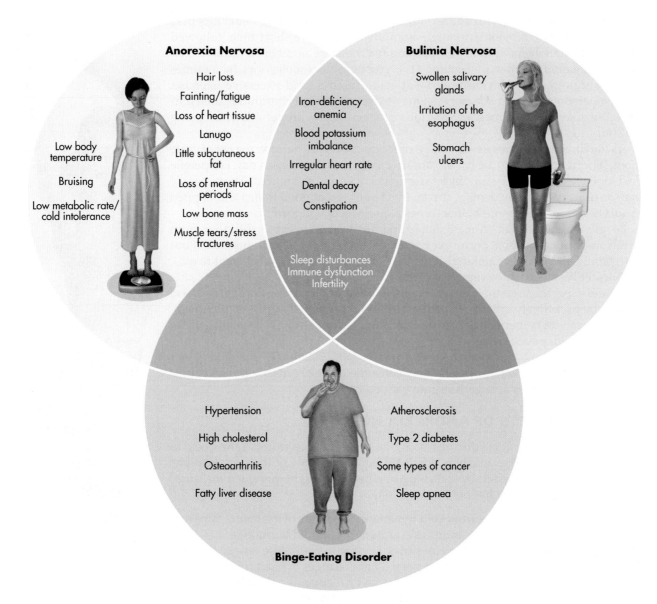

Anorexia Nervosa

Low body temperature

Bruising

Low metabolic rate/ cold intolerance

Hair loss

Fainting/fatigue

Loss of heart tissue

Lanugo

Little subcutaneous fat

Loss of menstrual periods

Low bone mass

Muscle tears/stress fractures

Iron-deficiency anemia

Blood potassium imbalance

Irregular heart rate

Dental decay

Constipation

Bulimia Nervosa

Swollen salivary glands

Irritation of the esophagus

Stomach ulcers

Sleep disturbances Immune dysfunction Infertility

Hypertension

High cholesterol

Osteoarthritis

Fatty liver disease

Atherosclerosis

Type 2 diabetes

Some types of cancer

Sleep apnea

Binge-Eating Disorder

FIGURE 11-1 ▲ Physical effects of eating disorders. This figure contains many but is not an exhaustive list of all potential consequences. These physical effects can also serve as warning signs that a problem exists. Professional evaluation is then indicated.

- Osteopenia (i.e., low bone mass, evident in 90% of adult women with anorexia nervosa) and osteoporosis (evident in at least one site in 40% of adult women with anorexia nervosa). Bone loss is due to decreased body weight and lean mass, several related hormonal changes, and prolonged use of some antidepressant medications.
- Eventual loss of teeth caused by acid erosion of tooth enamel if frequent vomiting occurs. Until vomiting ceases, one way to reduce this effect on teeth is to rinse the mouth with water right away and brush the teeth as soon as possible. Loss of teeth (along with low bone mass) can be lasting signs of the disease, even if the other physical and mental problems are resolved.
- Muscle tears and stress fractures in athletes because of decreased bone and muscle mass.
- Sleep disturbances and depression.

Many of the psychological and physical problems associated with anorexia nervosa arise from insufficient calorie intake, as well as deficiencies of nutrients, such as thiamin, calcium, and iron. A person with this disorder is psychologically and physically ill and needs help.

About one-quarter of those with anorexia nervosa recover within 6 years, whereas the rest simply exist with the disease or go on to develop another form of eating disorder. Among all psychiatric diseases, anorexia nervosa has the highest mortality rate: about 3% of people with anorexia nervosa eventually die from the disease—from suicide, heart ailments, and infections. The longer someone suffers from this eating disorder, the poorer the chances for complete recovery. A young patient with a brief episode and a cooperative family has a better outlook than an older patient with a long history of disordered eating and no family support. Overall, prompt and vigorous treatment, and close long-term follow-up improve the chances for success.

TREATMENT FOR THE PERSON WITH ANOREXIA NERVOSA

People with anorexia often sink into shells of isolation and fear. They deny that a problem exists. Frequently, their friends and family members meet with them to confront the problem in a loving way. This is called an *intervention*. They present evidence of the problem and encourage immediate treatment. Treatment then requires a multidisciplinary team of experienced physicians, registered dietitians, psychologists, and other health professionals working together. An ideal setting is an eating disorders clinic in a medical center. Outpatient therapy generally begins first. This may be extended to 3 to 5 days per week. Day hospitalization (6 to 12 hours per day) is another option. Total hospitalization is necessary once a person falls below 75% of expected weight, experiences acute medical problems, and/or exhibits severe psychological problems or suicidal risk. Still, even in the most skilled hands at the finest facilities, efforts may fail. This tells us that the prevention of anorexia nervosa is of utmost importance.

Once a medical team has gained the cooperation and trust of an anorexic person, the team works together to restore a sense of balance, purpose, and future possibilities. However, the anorexic person who has been barely existing in a state of semistarvation cannot focus on much besides food. Dreams and even morbid thoughts about food will interfere with therapy until sufficient weight is regained. Currently, the average time for recovery from anorexia nervosa is 7 years. Many insurance companies cover only a fraction of the estimated $150,000 cost of treatment.

Nutrition Therapy. The first goal of nutrition therapy is to gain the person's cooperation and trust, with the ultimate objective of increasing oral food intake. Ideally, weight gain must be enough to raise the metabolic rate to normal and reverse as many physical signs of the disease as possible. Food intake is designed first to minimize or stop any further weight loss. Then, the focus shifts to restoring appropriate food habits. After this, the expectation can be switched to slowly gaining weight. A gain of 2 to 3 pounds per week is appropriate. Tube and/or intravenous (IV) feeding is used only if immediate renourishment is required, as this can frighten the person and cause him or her to distrust medical staff.

Persons with anorexia nervosa need considerable reassurance during the refeeding process because of uncomfortable and unfamiliar effects—such as bloating, increase in body heat, and increase in body fat. This is a frightening process because these changes can symbolize a loss of control. Rapid changes in electrolytes and minerals in the blood associated with refeeding, especially potassium, phosphorus, and magnesium, can be dangerous. Therefore, monitoring blood levels of these minerals is of critical importance during the process of incorporating more food into the diet.

In addition to helping persons with anorexia nervosa reach and maintain adequate nutritional status, the registered dietitian on the medical team provides accurate nutrition information throughout the treatment, promotes a healthy attitude toward food, and helps the person learn to eat in response to natural hunger and satiety cues. The focus then turns to identifying healthy and adequate food choices that promote weight gain to achieve and maintain a clinically estimated goal weight (e.g., BMI of 20 or more).

▲ In 2010, the death of the French fashion model Isabelle Caro increased awareness of the serious nature of anorexia nervosa. In 2006, the 5-foot-4-inch model was hospitalized when she slipped into a coma at her lowest weight of 55 pounds. After surviving the coma, she resolved to speak out against dieting in the fashion industry and posed for a controversial billboard ad under the words "No Anorexia." She also authored a book about her 15-year struggle with an eating disorder. Her weight had increased to at least 80 pounds by the filming of her interview for National Geographic's *Taboo: Beauty* documentary, but she died, at age 28, before it aired.

Links exist between disordered eating behaviors and substance abuse, particularly alcohol abuse. Some people may use food restriction or purging to offset the calories consumed during drinking binges. Popularly called "drunkorexia," this practice can lead to severe nutrient deficiencies and dehydration.

CASE STUDY Eating Disorders—Steps to Recovery

At age 16, Sarah suddenly became self-conscious about her body when her peers teased her about being overweight. She began exercising to an aerobics video for an hour each day and found that she had success in losing weight; this was the beginning of her obsession to be thin. Next, Sarah turned to eating less food to lose even more weight and began eliminating certain foods from her diet, such as candy and meat. She increased her water and vegetable intake and chewed sugarless gum to curb her appetite. Once she began dieting, it was impossible for her to stop. She enjoyed having a high degree of self-control over her body. Still, she was literally obsessed with food, even staring at others while they were eating a meal. She occasionally cooked large meals and then refused to eat all but a few bites. By the time Sarah was 19 years old and 5 feet 6 inches tall, her weight had dropped from 150 to 105 pounds. Her family was concerned about her weight status, demanding that she go to a physician for an evaluation. Sarah was not happy about this idea because she worried the doctors would force her to eat and gain back unwanted weight, but she believed that her family would stop pestering her if she went. Sarah did not think she had a problem; she thought she was still grotesquely overweight. She did notice, however, that she always felt cold and was concerned that she had not menstruated in a year.

Answer the following questions and check your responses at the end of this chapter.

1. Sarah appears to have an eating disorder. Which eating disorder best describes her behavior?
2. List the behaviors that Sarah developed between ages 16 and 19 that are signs of the development of this eating disorder.
3. What physical symptoms of this disorder does Sarah have (review Fig. 11-1)?
4. Outline the therapies you think the physician will prescribe for Sarah. Where could she go for the therapy she needs? Which types of professionals would be involved?
5. Do you think Sarah has developed any vitamin or mineral deficiencies? Which ones would be most likely? How could these deficiencies be best treated?
6. What is the likelihood that she will fully recover from her condition?

A disturbing Internet trend is the attempt to promote eating disorders as a way of life. Some anorexic individuals have personified their illness into a role model named "Ana," who tells them what to eat and mocks them when they don't lose weight. Similarly, pro-"Mia" sites provide tips and encouragement for people with bulimia (e.g., how to induce vomiting and cover up evidence of compensatory behaviors). Pro-Ana and pro-Mia websites reject the serious health risks of eating disorders and instead dispense unsafe "thinspiration" to vulnerable individuals (see Further Reading 2).

As noted, nutrient deficiencies are commonly seen in anorexic persons. A multivitamin and mineral supplement will be added, as well as enough calcium to raise intake to about 1500 milligrams per day. Bone loss will not likely be completely restored, even if nutrition is adequate. Particularly in adolescent patients, a critical time for accrual of bone mass or gains in height may have been missed. However, supplementation with calcium and vitamin D is still necessary to prevent further bone loss.

Excessive physical activity prevents weight gain, so professionals must help anorexic persons regulate their activity. At many treatment centers, moderate bed rest is used in the early stages of treatment to help promote weight gain.

Experienced professional help is the key (see Further Reading 13). An anorexic person may be on the verge of suicide and starvation. In addition, anorexic people are often very resistant to therapy. They may try to hide weight loss by wearing many layers of clothes, putting coins in their pockets, and drinking numerous glasses of water before stepping on the scale.

Psychological Therapy. Once the immediate physical problems of anorexia nervosa are addressed, the treatment focus shifts to the underlying emotional problems that led to excessive dieting and other symptoms of the disorder. To heal, these anorexic persons must reject the sense of accomplishment they have associated with an emaciated body and begin to accept themselves at a healthy body weight. If therapists can discover the psychological conflicts that triggered the disorder, they can develop more effective treatment strategies. Education about the medical consequences of semistarvation is also helpful. A key aspect of psychological treatment is showing affected individuals how to regain control of other facets of their lives and cope with difficult situations. As eating evolves into a normal routine, they can turn to previously neglected activities.

Family-based therapy (usually 6 to 12 months) is the preferred method of psychological treatment for anorexia nervosa among younger individuals who still live with their families (see Further Reading 19). It focuses on the role of the illness among family members, the reactions of individual family members and ways in which their subconscious behavior might contribute to the abnormal eating patterns. Frequently, a therapist finds family struggles at the heart of the problem. As the disorder resolves, patients must relate to family members in new ways to gain the attention that was needed and previously tied to the disease. For example, the family may need to help the young person ease into adulthood and accept its responsibilities along with its advantages.

At this time, there is not a lot of conclusive evidence to support one particular type of psychological therapy over another for adults with anorexia nervosa. Therapists may use **cognitive behavioral therapy,** which involves helping the person confront and change irrational beliefs about body image, eating, relationships, and weight. However, when anorexic individuals have been subsisting in a starved state, their brain chemistry is so altered that attempts at cognitive restructuring are usually not effective in the early stages of illness. Underlying issues that may have triggered the eating disorder, such as sexual abuse, also must be identified and addressed by the therapist (see Further Reading 8). Guided self-help groups for anorexic (and bulimic) people, as well as their families and friends, represent additional nonthreatening first steps into treatment.

Pharmacological Therapy. There are no FDA-approved medications specifically for the treatment of eating disorders. Rather, food is the drug of choice in the treatment of anorexic patients. Generally, medications are not effective in management of the primary symptoms of anorexia nervosa. Fluoxetine (Prozac®) and related medications may stabilize recovery once 85% of expected body weight has been attained. These medications work by prolonging serotonin activity in the brain, which in turn regulates mood and feelings of satiety. A variety of other pharmacological agents, such as olanzapine (Zyprexa®), may have some role in treating mood changes, anxiety, or psychotic symptoms associated with anorexia nervosa, but they have limited value unless weight gain is also achieved.

With professional help, many people with anorexia nervosa can lead normal lives. Although they may not be totally cured, recovering anorexic individuals no longer depend on unusual eating habits to cope with daily problems. They recover a sense of normalcy in their lives. Longer follow-up—sometimes several years—is associated with better outcomes. Recovery rates are around 20% to 30% for short-term therapy but increase to 70% to 80% with 8 years of follow-up. No universal approach exists, because each case is unique. Establishing a strong relationship with either a therapist or another supportive person is especially important to recovery. Once anorexic persons feel understood and accepted by another person, they can begin to build a sense of self and exercise some autonomy. As they learn alternative coping mechanisms, they can relinquish their dysfunctional relationships with food and instead develop healthy personal relationships.

✔ CONCEPT CHECK 11.2

1. Identify the three diagnostic criteria for anorexia nervosa.
2. List five physical effects of anorexia nervosa.
3. Describe elements of nutrition, psychological, and pharmacological therapy for anorexia nervosa.

CRITICAL THINKING

Jennifer is an attractive 13-year-old. However, she is very compulsive. Everything has to be perfect—her hair, clothes, even her room. Her body is beginning to mature, so she is obsessed with having perfect physical features as well. Her parents are worried about her behavior. Given her behavioral traits, what symptoms of an eating disorder should they watch for? What should Jennifer's parents do if they suspect an eating disorder?

cognitive behavioral therapy Psychological therapy in which the person's assumptions about dieting, body weight, and related issues are confronted. New ways of thinking are explored and then practiced by the person. In this way, an individual can learn new ways to control disordered eating behaviors and related life stress.

▲ Early treatment for an eating disorder, such as anorexia nervosa, improves chances of success.

11.3 Bulimia Nervosa

Literally translated, *bulimia* means ravenous (oxlike) hunger. This eating disorder is characterized by recurrent episodes of binge eating followed by some type of compensatory behavior to prevent weight gain (see Table 11-2). Also, as with anorexia nervosa, bulimic individuals overvalue body weight and shape.

Binge eating is defined as consuming an abnormally large amount of food within a short time period (e.g., 2 hours). These binges are characterized by a lack of control over the food consumed. **Compensatory behaviors** (also known as purging) used to rid the body of excess calories consumed during a binge may include vomiting; misuse of laxatives, diuretics, or enemas; or excessive exercise. For a diagnosis of bulimia nervosa, binge eating followed by inappropriate compensatory behaviors must take place at least four times per month over a period of 3 months or more.

It is likely that many people with bulimic behavior are never diagnosed. People with bulimia nervosa lead secret lives, hiding their abnormal eating habits. Moreover, it can be difficult to recognize the disorder based on appearance, because people with bulimia nervosa are usually at or slightly above normal weight. By rough estimate, among North American women, approximately 2.6% will develop bulimia nervosa by the age of 20. An estimated 4% of college-age women have bulimia nervosa. About 10% of cases occur in men. However, most diagnoses of bulimia nervosa rely on self-reports, so the disorder may be much more widespread than commonly thought.

COMMON BEHAVIORS OF BULIMIA NERVOSA

Bulimia nervosa involves episodes of binge eating followed by various means to rid the body of excess calories. Susceptible people often have genetic factors and lifestyle patterns that predispose them to becoming overweight, and many try frequent weight-reduction diets as teenagers. The person with bulimia nervosa may think of food constantly. However, unlike the anorexic person, who turns away from food when faced with problems, the bulimic person turns toward food in critical situations. Also, unlike those with anorexia nervosa, people with bulimia nervosa recognize their behavior as abnormal.

binge eating Consuming an abnormally large amount of food within a short time period (e.g., 2 hours).

compensatory behaviors Actions taken to rid the body of excess calories and/or to alleviate guilt or anxiety associated with a binge; examples include vomiting, misuse of laxatives, or excessive exercise.

Bingeing and purging (via vomiting) was evident in pre-Christian Roman times but was practiced in a group setting. The eating disorder bulimia nervosa is generally practiced in private. It was first described in the medical literature in 1979.

TABLE 11-2 ▶ Diagnostic Criteria for Bulimia Nervosa

A. Recurrent episodes of binge eating. An episode of binge eating is characterized by both of the following:
1. Eating, in a discrete period of time (i.e., within any 2-hour period), an amount of food that is definitely larger than most people would eat during a similar period of time and under similar circumstances.
2. A sense of lack of control over eating during the episode (e.g., a feeling that one cannot stop eating or control what or how much one is eating).

B. Recurrent inappropriate compensatory behavior in order to prevent weight gain, such as self-induced vomiting; misuse of laxatives, diuretics, enemas, or other medications; fasting; or excessive exercise.

C. The binge eating and inappropriate compensatory behaviors both occur, on average, at least once a week for 3 months.

D. Self-evaluation is unduly influenced by body shape and weight.

E. The disturbance does not occur exclusively during episodes of anorexia nervosa.

From *Diagnostic and Statistical Manual for Mental Disorders, 5th edition.* Copyright 2013 by the American Psychiatric Association.

Bulimic people tend to be impulsive, which may be expressed in behaviors such as stealing, increased sexual activity, drug and alcohol abuse, self-mutilation, or attempted suicide. Some experts have suggested that part of the problem may actually arise from an inability to control responses to impulse and desire. Approximately half of the people with bulimia nervosa have major depression. Lingering effects of child abuse may be one reason for these feelings. Many bulimic persons report that they have been sexually abused. They appear competent to outsiders, while they actually feel out of control, ashamed, and frustrated.

For intake to qualify as a binge, an atypically large amount of food must be consumed in a short time and the person must exhibit a lack of control over his or her behavior. Among sufferers of bulimia nervosa, bingeing often alternates with attempts to rigidly restrict food intake. Elaborate food rules are common, such as avoiding all sweets. Thus, eating just one cookie or doughnut may cause individuals with this disorder to feel as though they have broken a rule. At that point, in the mind of a bulimic person, the objectionable food must be eliminated. Usually this leads to further overeating, partly because it is easier to regurgitate a large amount of food than a small amount.

Binge-compensate cycles may be practiced daily, weekly, or across longer intervals. A specific time often is set aside. Most binge eating occurs at night, when other people are less likely to interrupt and usually lasts from ½ to 2 hours. A binge can be triggered by stress, boredom, loneliness, depression, or any combination thereof. It often follows a period of strict dieting and thus can be linked to intense hunger. The binge is not at all like normal eating; once begun, it seems to propel itself. The person not only loses control but generally does not even taste or enjoy food during a binge. This separates the practice from overeating.

Most commonly, bulimic people consume cakes, cookies, ice cream, and other high-carbohydrate convenience foods during binges because these foods can be purged relatively easily and comfortably by vomiting. In a single binge, foods supplying 3000 kcal or more may be eaten. Compensating behaviors follow in hopes that no weight will be gained. However, even when vomiting follows the binge, 33% to 75% of the calories taken in are still absorbed, inevitably causing some weight gain. When laxatives or enemas are used, about 90% of the calories are absorbed, as laxatives act in the large intestine, beyond the point of most nutrient absorption. The belief that purging soon after bingeing will prevent excessive calorie absorption and weight gain is a misconception.

At the onset of bulimia nervosa, sufferers often induce vomiting by placing their fingers deep into the mouth. They may bite down on these fingers inadvertently, resulting in bite marks and scars around the knuckles, a characteristic sign of this disorder. Once the disease is established, however, a person may be able to vomit simply by contracting the abdominal muscles. Vomiting may also occur spontaneously.

Another way bulimic persons attempt to compensate for a binge is by engaging in excessive exercise to expend a large amount of calories. In this practice, referred to as "debting," bulimic individuals try to estimate the amount of calories eaten during a binge and then exercise to counteract the excess.

People with bulimia nervosa are not proud of their behavior. After a binge, they usually feel guilty and depressed. Over time, they experience low self-esteem, feel hopeless about their situation, and are caught in a vicious cycle of obsession (Fig. 11-2). Compulsive lying, shoplifting to obtain food, and drug abuse can further intensify these feelings. Bulimic people discovered in the act of bingeing by a friend or family member may order the intruder to "get out" and "go away." Sufferers gradually distance themselves from others, spending more time preoccupied by and engaging in bingeing and compensating.

Because individuals with bulimia nervosa attempt to hide their behaviors, it may be difficult to identify them early in the disease process, when treatment is likely to be most effective. An early warning sign of bulimia is frequent trips to

▲ Excessive exercise can be one component of bulimia if it is used as a way to offset the calorie intake from a binge. Exercise is considered excessive when it is done at inappropriate times or settings, or when a person does it despite injury or other medical complications.

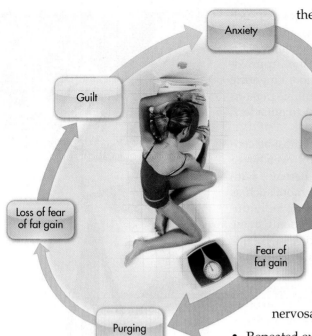

FIGURE 11-2 ▲ Bulimia nervosa's vicious cycle of obsession.

the bathroom during or after meals. To cover the sounds of vomiting, the individual may run the bathroom fan or turn on the shower. Despite efforts to disguise their behavior with air fresheners, mouthwash, or breath mints, there may be a lingering odor of vomit. Be suspicious of packages or receipts for laxatives, diuretics, diet pills, or enemas. A person who uses exercise to compensate for binges is usually preoccupied with their workout schedule or might seem extremely stressed when they are unable to exercise. If you suspect someone is falling victim to bulimia nervosa, encourage them to get help. Early intervention can prevent some of the serious physical health effects described in the next section.

PHYSICAL EFFECTS OF BULIMIA NERVOSA

Repeated vomiting is a physically destructive method of purging. Indeed, the majority of health problems associated with bulimia nervosa, as noted here, arise from vomiting:

- Repeated exposure of teeth to stomach acid causes demineralization (Fig. 11-3), making the teeth painful and sensitive to heat, cold, and acids. Eventually, the teeth may decay severely, erode away from fillings, and finally fall out. Dental professionals are sometimes the first health professionals to notice signs of bulimia nervosa.

- Blood potassium can drop significantly with regular vomiting or the use of certain diuretics. This can disturb the heart's rhythm and even produce sudden death.

- Salivary glands may swell as a result of infection and irritation from persistent vomiting.

- Stomach ulcers and bleeding and tears in the esophagus develop in some cases.

- Constipation may result as a complication of frequent laxative use.

- Syrup of ipecac, formerly used as a home remedy to induce vomiting in cases of accidental poisoning, has been abused by people with bulimia nervosa. Major health agencies (e.g., the American Academy of Pediatrics) now advise against its use, and most pharmacies no longer sell it because ipecac is now known to be toxic to the heart, liver, and kidneys.

The mortality rate for patients with bulimia nervosa is estimated around 0.4%. Patients with anorexia are about five times more likely to die from their disease than those with bulimia nervosa. Nonetheless, bulimia is a serious and potentially debilitating disorder that can lead to death, usually from suicide, cardiac arrest, or overwhelming infections.

TREATMENT FOR THE PERSON WITH BULIMIA NERVOSA

Therapy for bulimia nervosa, as for anorexia nervosa, requires a team of experienced clinicians (see Further Reading 18). Bulimic individuals are less likely than those with anorexia to enter treatment in a state of semistarvation. However, if a bulimic person has lost significant weight, this must be treated before psychological treatment begins. Although clinicians have yet to agree on the best therapy for bulimia nervosa, they generally agree that treatment should last at least 16 weeks. Hospitalization may be indicated in cases of extreme laxative abuse, regular vomiting, substance abuse, and depression, especially if physical harm is evident.

FIGURE 11-3 ▼ Dental erosion resulting from frequent episodes of self-induced vomiting by a patient with bulimia nervosa.

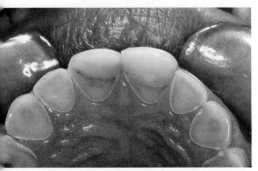

Nutritional Therapy. Nutritional counseling by a registered dietitian has two main goals: re-establishing regular eating habits and correcting misconceptions about food.

In general, the focus is not on stopping bingeing and purging *per se* but on developing regular eating habits. Once this is achieved, the binge-purge cycle should start to break down. Initially, the registered dietitian must help the person to decrease the amount of food in a binge. This will decrease the risk of esophageal tears from related purging by vomiting. A decrease in the frequency of this type of purging will also decrease damage to the teeth. Next, to develop a normal eating pattern, some specialists encourage their patients to develop daily meal plans and keep a food diary in which they record food intake, internal sensations of hunger, environmental factors that precipitate binges, and thoughts and feelings that accompany binge-compensate cycles.

Avoiding binge foods and not constantly stepping on a scale may be recommended early in treatment. Setting time limits for the completion of meals and snacks can also be important for people with eating disorders. Many bulimic persons eat quickly, reflecting their difficulties with satiety. Suggesting that the person put his or her utensil down after each bite is a behavioral technique that a therapist might try during recovery for bulimia. (In comparison, many anorexic persons eat in an excessively slow manner; for example, taking 1 hour to eat a muffin cut into tiny, bite-size pieces.)

Providing accurate information about bulimia nervosa and its consequences can help the affected individual to see the need to change his or her behavior. Over the long term, individuals are discouraged from following strict rules about healthy food choices because this mimics the typical obsessive attitudes associated with bulimia nervosa. Rather, encouraging a mature perspective on nutrient intake—that is, regular consumption of moderate amounts of a variety of foods from each food group—helps one overcome this disorder.

▲ Developing regular eating habits helps a person with bulimia nervosa to stop the binge-purge cycle.

Psychological Therapy. The primary aim of psychotherapy is to improve a person's self-acceptance and help him or her to be less concerned about body weight. Cognitive behavioral therapy is generally used (see Further Reading 8). Psychotherapy helps correct the all-or-none thinking that is typical of bulimic persons: "If I eat one cookie, I'm a failure and might as well binge." The premise of this psychotherapy is that if abnormal attitudes and beliefs can be altered, normal eating will follow. In addition, the therapist guides the person in establishing food habits that will minimize bingeing: avoiding fasting, eating regular meals, and using alternative methods—other than eating—to cope with stressful situations. Another goal of therapy is to help bulimic persons accept some depression and self-doubt as normal. Group therapy is often useful to foster strong social support.

The binge-purge cycle can create an initial state of euphoria in the person. Giving up this euphoria has been equated to giving up an addiction.

Pharmacological Therapy. Although pharmacological agents should not be used as the sole treatment for bulimia nervosa, studies indicate that some medications may be beneficial in conjunction with other therapies. Fluoxetine (Prozac®) is the only antidepressant that has been approved by FDA for use in the treatment of bulimia nervosa (in combination with psychological therapy). It can help to increase feelings of satiety after eating and thereby reduce the frequency of binges. Physicians also may prescribe other similar forms of antidepressants, other classes of psychiatric medications, or certain antiseizure medications (e.g., topiramate [Topamax®]).

Medicine Cabinet

Various antidepressant medications are prescribed to help people with eating disorders, but only fluoxetine (Prozac®) is approved by FDA for treatment of bulimia nervosa. Studies show that this medication can help to decrease the frequency of binge eating and purging. Women of childbearing age should be cautious because fluoxetine has been linked to increased rates of miscarriages and other complications of pregnancy.

People with bulimia nervosa must recognize that they have a serious disorder that can have grave medical complications if not treated. Relapse is likely, so therapy should be long-term. As seen with anorexia, earlier intervention and longer duration of treatment are associated with better outcomes. After just 1 year of therapy, about one-fourth of bulimic individuals experience recovery, but most bulimics require many years of therapy. Those with bulimia nervosa need professional help because they can be very depressed and are at a high risk for suicide. About 50% of people with bulimia nervosa recover completely from the disorder. Others continue to struggle with it to varying degrees for the rest of their lives. Such a difficult course of treatment underscores the need for prevention.

✔ CONCEPT CHECK 11.3

1. Describe the eating behaviors of bulimia nervosa. What triggers them? What distinguishes them from simple overeating at a holiday meal?
2. What is a compensatory behavior? List at least three examples of inappropriate compensatory behaviors.
3. Describe at least three physical effects of bulimia nervosa.
4. Outline the basic components of nutritional therapy for people with bulimia nervosa. How do psychological and pharmacological therapies contribute to recovery?

11.4 Binge-Eating Disorder

First officially described in 1994, binge-eating disorder is a growing, complex, and potentially serious problem. With the 2013 publication of *DSM-5*, binge-eating disorder is now recognized as a distinct eating disorder alongside anorexia nervosa and bulimia nervosa. Its inclusion as an official diagnosis will improve the availability of treatments and success of outcomes for people who suffer with the disorder.

Generally, binge-eating disorder can be defined as binge-eating episodes not accompanied by compensatory behaviors (as seen in bulimia nervosa) at least one time per week, on average, for at least 3 months. The diagnostic criteria for binge-eating disorder are listed in Table 11-3.

Like anorexia and bulimia, there seems to be a genetic predisposition for binge-eating disorder. While anorexia and bulimia most often affect females, about 40% of people with binge-eating disorder are males. For adults, the lifetime prevalence of binge-eating disorder is estimated to be about 3.5% for women and about 2.0% for men. Among the general U.S. population, about 4 million have this disorder. However, many more people in the general population are likely to have less severe forms of the disorder that do not meet all the criteria described in Table 11-3. The number of cases of binge-eating disorder is far greater than that of either anorexia nervosa or bulimia nervosa. This disorder is also more common among the severely obese and those with a long history of frequent restrictive dieting, although obesity is not a criterion for having binge-eating disorder. Compared to anorexia and bulimia, binge-eating disorder tends to be diagnosed later in life, usually in the late 40s or early 50s.

COMMON BEHAVIORS OF BINGE-EATING DISORDER

As previously described, a binge refers to the uncontrolled consumption of an unusually large amount of food within a discrete period of time. A binge can include any food, but most often consists of foods that carry the social stigma of

Up to 25% of college-age women exhibit some degree of binge eating. Such behavior may have a negative impact on physical appearance, health, social life, and academics.

▲ Binge-eating disorder is seen in both men and women.

TABLE 11-3 ▶ Diagnostic Criteria for Binge-Eating Disorder

A. Recurrent episodes of binge eating. An episode of binge eating is characterized by both of the following:
   1. Eating, in a discrete period (e.g., within any 2-hour period), an amount of food that is definitely larger than most people would eat during a similar period of time under similar circumstances.
   2. A sense of lack of control over eating during the episode (e.g., a feeling that one cannot stop eating or control what or how much one is eating).

B. The binge-eating episodes are associated with three (or more) of the following:
   1. Eating much more rapidly than normal.
   2. Eating until feeling uncomfortably full.
   3. Eating large amounts of food when not feeling physically hungry.
   4. Eating alone because of being embarrassed by how much one is eating.
   5. Feeling disgusted with oneself, depressed, or very guilty afterward.

C. Marked distress regarding binge eating is present.

D. The binge eating occurs, on average, at least once a week for 3 months.

E. The binge eating is not associated with the recurrent use of inappropriate compensatory behavior as in bulimia nervosa and does not occur exclusively during the course of bulimia nervosa or anorexia nervosa.

From *Diagnostic and Statistical Manual for Mental Disorders, 5th edition.* Copyright 2013 by the American Psychiatric Association.

"junk" or "bad" foods—ice cream, cookies, sweets, potato chips, and similar snack foods. During binges, food is eaten without regard to biological need and often in a recurrent, ritualized fashion. Unlike those with bulimia nervosa, people with binge-eating disorder do not attempt to purge the excess calories.

Binge eating is usually triggered by negative emotions, such as stress, anxiety, loneliness, grief, or anger (see Further Reading 15). In fact, almost half of those with severe binge-eating disorder exhibit clinical depression. Often, people with binge-eating disorder have not learned to express or appropriately deal with their feelings, so they turn to food to cope with stress or meet emotional needs. Those who regularly practice binge eating may grow up nurturing others instead of themselves, avoiding their own feelings and taking little time for themselves. Unfortunately, unresolved conflicts and unmet emotional needs will resurface. To make matters worse, binge eating itself brings added feelings of guilt, embarrassment, and shame.

Typical binge eaters isolate themselves and eat large quantities of a favorite food. For example, a person might find comfort in consuming large quantities of food—such as a whole pizza in one sitting—when an emotional setback occurs. Other people with this disorder eat food continually over an extended period, called grazing. For instance, someone with a stressful or frustrating job might come home every night and graze until bedtime.

Many people with binge-eating disorder have struggled to lose weight throughout their lives. As noted in Chapter 7, overly restrictive diets can lead to hunger and a sense of deprivation that triggers binge eating. People with binge-eating disorder tend to perceive themselves as hungry more often than normal. During periods when little food is eaten, they get very hungry and obsessive about food. Restricting favorite foods, such as chocolate, leads to feelings of deprivation. When these individuals finally give themselves permission to eat a forbidden food or loosen up a rigid meal plan, they feel driven to eat in a compulsive, uncontrolled way. People with binge-eating disorder usually began this cycle of strict dieting alternating with binge eating during adolescence or in their early twenties and have had little success with traditional weight-control programs.

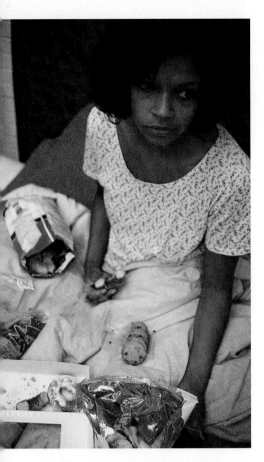

▲ Some experts describe binge-eating disorder as an addiction to food, involving psychological dependence. The person becomes attached to the behavior and is driven to continue it, senses only limited control over it, and persists at it despite negative consequences.

Mindfulness meditation is another strategy being studied to address the psychological issues that lead to binge-ing. Mindful eating programs train a person to become aware of, accept, and acknowledge the daily experiences that can be triggers for binge eating but not to respond with emotional eating. Early studies show beneficial effects on body weight, stress, and symptoms of depres-sion (review the discussion of mindful eating in Chapter 7 and see Further Reading 16).

PHYSICAL EFFECTS OF BINGE-EATING DISORDER

Although obesity is not among the criteria for diagnosis of binge-eating disorder, about 70% of those with the disorder are obese. The physical effects of the disor-der reflect the comorbid conditions of obesity, which were extensively described in Chapter 7 (see Table 7-2). The most deadly physical effects are listed here.

- Hypertension from excess body weight and high sodium intake.
- Elevated cholesterol levels, which contribute to atherosclerosis. Binge eating may have a more severe effect on blood lipids than simply overeating by graz-ing throughout the day because large meals are linked to high insulin and high triglycerides.
- Cardiovascular disease, which contributes to deaths from heart attacks and strokes.
- Type 2 diabetes is strongly linked to obesity. An estimated one in five women with type 2 diabetes exhibits binge-eating behaviors.

TREATMENT FOR THE PERSON WITH BINGE-EATING DISORDER

Approximately 30% of subjects in organized weight-control programs have binge-eating disorder. This means registered dietitians and others who lead weight man-agement programs may be the first to identify binge-eating disorder among their clients. As with anorexia and bulimia, binge-eating disorder truly is a psychological problem with nutritional consequences. The success of traditional weight-loss ther-apies for people with binge-eating disorder has been poor because these approaches fail to address the underlying psychological causes of the disorder. True, people with binge-eating disorder will likely benefit from increased nutrition knowledge, but unless they find a way to manage or overcome negative emotions that underlie eating behaviors, the success of attempts to improve food choices and incorporate physical activity into the lifestyle will be short-lived. Thus, first addressing the psy-chological needs of individuals with binge-eating disorder is essential to successful treatment.

Psychological Therapy. Similar to treatments for bulimia nervosa, there is grow-ing evidence that cognitive behavioral therapy techniques are useful for overcom-ing binge-eating disorder. Many people with binge-eating disorder may experience difficulty in identifying personal emotional needs and expressing emotions. This problem is a common predisposing factor in binge eating, so communication issues should be addressed during treatment. Binge eaters often must be helped to recog-nize their buried emotions in anxiety-producing situations and then encouraged to share them with their therapist or therapy group. Learning simple but appropriate phrases to say to oneself can help stop bingeing when the desire is strong. Even if negative situations cannot be changed, people must learn how to adapt to and bear with them effectively—not through self-destructive binge-eating behaviors.

Cognitive behavioral therapy can be delivered in many forms. One-on-one sessions with a therapist and group therapy are the most common methods. Self-help groups such as Overeaters Anonymous can also help. Their treatment philosophy, which par-allels that of Alcoholics Anonymous, is to create an environment of encouragement and accountability to overcome this eating disorder. Recently, even web-based adap-tations of cognitive behavioral therapy have proved to be useful.

Although psychological therapy has been valuable for correcting binge-eating behavior, it is not always successful at inducing weight loss among a population of overweight and obese individuals who may have one or more other health issues related to obesity. Thus, nutrition therapy is an important part of treatment for binge-eating disorder.

Nutrition Therapy. Once effective coping mechanisms are learned, the registered dietitian can educate the patient on developing normal eating patterns and making healthful food choices. First, those with binge-eating disorder must learn to eat in response to hunger—a biological signal—rather than in response to emotional needs or external factors (such as the time of day, boredom, or the simple presence of food). Counselors often direct binge eaters to record their perceptions of physical hunger throughout the day and at the beginning and end of every meal. These people must learn to respond to a prescribed amount of fullness at each meal.

Individuals recovering from binge-eating disorder should initially avoid weight-loss diets because feelings of food deprivation can lead to more disruptive emotions and a greater sense of unmet needs. Restrictive "diets" are likely to encourage more intense problems, such as extreme hunger, that can trigger binge eating. Even if exposure to favorite binge foods is limited in the early stages of treatment, many experts feel that learning to eat all foods—but in moderation—is an effective long-term goal for people with binge-eating disorder. This practice can prevent the feelings of desperation and deprivation that come from limiting particular foods.

Pharmacological Therapy. Psychological and nutritional therapies are useful tools to treat binge-eating disorder, but they are not 100% effective. Thus, there is growing interest in drug therapy. No medications are currently approved by FDA specifically for treatment of binge-eating disorder, but some antidepressants (e.g., fluoxetine [Prozac®] and duloxetine [Cymbalta®]) and antiseizure medications (e.g., topamirate [Topamax®]) have demonstrated success in reducing binge eating and decreasing the depression that triggers disordered eating. Despite their usefulness for reducing binge eating, they still may not induce significant weight loss. Orlistat (Xenical®) and phentermine (Adipex-P®), discussed in Chapter 7, can assist with weight-loss efforts after binge eating is under control. The newest weight-loss medication, lorcaserin (Belviq®), has yet to be extensively tested for treatment of binge-eating disorder.

Newsworthy Nutrition

Strong connection between eating disorders and binge drinking

A survey of 480 college-age women at high risk for eating disorders found a positive correlation between binge drinking (consuming four or more drinks in one sitting) and disordered eating behaviors. At baseline, 67% of the women in the study engaged in binge drinking at least once per month and 30% did so frequently (three or more times per month). Additional research also supports this interplay between substance abuse and disordered eating behavior—purging, skipping meals, or fasting to compensate for calories consumed during drinking episodes—particularly among women. These studies exemplify how disordered eating behaviors and abuse of substances such as alcohol are used as inappropriate coping mechanisms. The stress of college and the influence of peers in this age group demand university-based intervention programs to educate students on the dangers of both eating disorders and substance abuse and to help students develop effective coping mechanisms.

Sources: Khaylis A and others: Binge drinking in women at risk for developing eating disorders. *International Journal of Eating Disorders* 42:409, 2009.

Kelly-Weeder S: Binge drinking and disordered eating in college students. *Journal of the American Academy of Nurse Practitioners* 23:33, 2011.

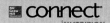 Check out the Connect site **www.mcgrawhillconnect.com** to further explore eating disorders and binge drinking.

Given the similarities between binge-eating and other addictive behaviors, medications used to treat substance abuse (e.g., naltrexone) are being studied for use among patients with binge-eating disorder. Other novel therapies that require additional research include stimulants (e.g., Ritalin®, routinely used to treat attention deficit hyperactivity disorder) and glutamate-modulating agents (Namenda®, used in patients with Alzheimer's disease).

Overall, people who have binge-eating disorder are usually unsuccessful in controlling it on their own. Furthermore, unrecognized binge-eating disorder will undermine the success of weight-loss therapies among many overweight and obese individuals who do seek professional help. By asking questions about eating patterns, feelings of loss of control over eating behaviors, and feelings of guilt after eating, health professionals can screen weight-loss clients for binge-eating disorder and refer affected individuals to appropriate treatments.

✔ CONCEPT CHECK 11.4

1. What differentiates binge-eating disorder from bulimia nervosa?
2. List at least three health effects that may result from binge-eating disorder.
3. Why do people with binge-eating disorder have little success in traditional weight-loss programs?

11.5 Other Eating Disorders

Besides those already covered, there are several other types of eating disorders. Some of these, such as **pica,** are distinct eating disorders with a complete set of diagnostic criteria. In *DSM-5,* the category of Other Specified Feeding or Eating Disorders encompasses several disorders that do not quite meet all of the criteria for diagnosis of anorexia, bulimia, or binge-eating disorder. These include several categories of subthreshold eating disorders, **purging disorder,** and **night eating syndrome.** Two additional distinct eating disorders (rumination disorder and avoidant/restrictive food intake disorder) are most prevalent in infants and young children and, therefore, will be discussed in Chapter 15.

PICA

Pica, which will be mentioned again in Chapter 14 as it relates to pregnancy, is a disorder in which a person persistently eats nonnutritive, nonfood substances over a period of at least 1 month. A few examples of nonfood substances ingested by people with pica are clay, dirt, ice, chalk, or wood. Pica could lead to some serious health consequences, including microbial infections, poisoning from toxins present in the nonfood material, gastrointestinal blockages, or nutrient deficiencies (in cases when nonfood substances displace nutritive foods in the diet). Pica tends to co-occur with other mental disorders, such as autism or obsessive-compulsive disorder.

SUBTHRESHOLD EATING DISORDERS

Individuals who meet some but not all of the criteria for diagnosis with anorexia nervosa, bulimia nervosa, or binge-eating disorder may fall into one of five subthreshold classifications. For example, **atypical anorexia nervosa** describes a person who meets most of the criteria for diagnosis of anorexia nervosa but whose weight is still within a normal range. This could occur if an overweight person

purging disorder An eating disorder characterized by repeated purging (e.g., by self-induced vomiting) to induce weight loss even in the absence of binge eating.

night eating syndrome Eating a lot of food in the late evening and nocturnal awakenings with ingestion of food.

atypical anorexia nervosa A subthreshold eating disorder in which a person meets most of the criteria for diagnosis of anorexia nervosa, except weight is within a normal range.

has just begun severely restricting calories. Despite significant weight loss, BMI may still fall within the healthy range of 18.5 to 24.9. Other diagnoses in this category include cases of bulimia nervosa or binge-eating disorder in which episodes of binge-eating occur less than once per week (e.g., bulimia nervosa of low frequency) or have taken place for less than 3 months (e.g., binge-eating disorder of limited duration).

PURGING DISORDER

Purging disorder is the name given to the behavior of people who repeatedly purge (i.e., vomit) to promote weight loss even in the absence of binge eating. This disorder is related to body dissatisfaction, anxiety, and depression. The physical effects of purging disorder are the same as those of bulimia nervosa and include dental problems, mouth sores, damage to the esophagus, constipation, dehydration, electrolyte imbalances, and overall malnutrition.

NIGHT EATING SYNDROME

Night eating syndrome is characterized by recurrent episodes of night eating, manifested by eating after awakening from sleep or by excessive food consumption after the evening meal (see Further Reading 11). With night eating syndrome, the person is fully aware of and able to recall the behavior, which results in marked distress. Although night eating syndrome was first observed among obese patients, it also occurs among nonobese persons. It has been estimated to occur in 1.5% of the general population and in 8.9% of persons treated in obesity clinics. Some typical signs and symptoms of night eating syndrome include:

▲ Night eating syndrome is characterized by waking at least once during the night and needing to eat to be able to fall asleep again.

- Not feeling hungry in the morning and delaying the first meal until several hours after waking.
- Overeating in the evening with more than 25% of daily food intake consumed after dinner.
- Difficulty falling asleep and a need to eat something to help fall asleep faster.
- Waking at least once during the night with a need to eat to be able to fall asleep again.
- Eating produces feelings of guilt and shame.
- Feeling depressed, especially at night.

Research shows that the circadian rhythm (your body's 24-hour clock) of food intake appears to be disturbed in night eating syndrome. Studies have also shown that night eating syndrome is prevalent among outpatients with sleep apnea, restless leg syndrome, or other psychiatric conditions. Behavioral changes, such as establishing and monitoring the sleep-wake schedule and performing regular exercise, can help. Symptoms are significantly improved with use of the antidepressant sertraline (Zoloft®).

✔ CONCEPT CHECK 11.5

1. List three ways pica could harm health.
2. Describe three different cases in which a person would be diagnosed with subthreshold eating disorders.
3. How is purging disorder similar to bulimia nervosa? How do these two disorders differ?
4. List three characteristics of night eating syndrome.

11.6 Additional Disordered Eating Patterns

There is a growing number of other patterns of disordered eating that have not yet been classified by *DSM-5* but are worthy of mention. Some of those mentioned below, such as **female athlete triad,** have been the subject of much research. Others have been discussed mainly in popular media, yet health practitioners report encountering them regularly. Continued research is required to establish a set of diagnostic criteria, describe the course of the disorder, and establish evidence-based treatments.

FEMALE ATHLETE TRIAD

Mentioned in Chapter 10, females participating in appearance-based and endurance sports are at risk of developing an eating disorder. One study of college-age female athletes found that 15% of swimmers, 62% of gymnasts, and 32% of varsity athletes exhibited disordered eating patterns.

Food restriction and stress can precipitate irregular menstruation among female athletes. When body fat is low, estrogen production decreases. Remember from Chapter 9, estrogen stimulates osteoblasts (bone-forming cells). In this way, low body fat impairs bone health. Some of these young women have bone mass values equivalent to those of 50- to 60-year-olds, increasing their risk for bone fractures during sports and general activities.

The American College of Sports Medicine (ACSM) has named this syndrome "female athlete triad" because it consists of three parts: disordered eating, lack of menstrual periods, and osteoporosis (Fig. 11-4). Many coaches, trainers, and even some health professionals mistakenly believe that loss of menstrual periods is a normal consequence of a high level of physical activity. To combat misinformation, the ACSM has issued a plea to teachers, coaches, health professionals, and parents to educate female athletes about the triad and its health consequences.

Those exhibiting symptoms of the female athlete triad should seek treatment from a multidisciplinary team of health professionals. Involving the coach or trainer in therapy is usually a key factor in the success of the treatment plan. Suggestions for treatment are as follows:

- Reduce preoccupation with food, weight, and body fat.
- Gradually increase meals and snacks to an appropriate amount.
- Achieve an appropriate weight for height.
- Establish regular menstrual periods.
- Decrease training time and/or intensity by 10% to 20%.

Increasing caloric intake will increase body weight and body fat. Estrogen production then increases and should stabilize bone mass. During therapy, a physician may prescribe a multivitamin and mineral supplement as well as calcium supplements as needed to maintain a daily calcium intake of 1200 to 1500 milligrams. Even with intervention, however, much of the bone loss that has already occurred is irreversible.

MUSCLE DYSMORPHIA

Muscle dysmorphia has gained attention as a psychological condition with many similarities to other eating disorders. First identified among male bodybuilders in the 1990s, muscle dysmorphia was initially termed "reverse anorexia." Men (and some women) with this disorder perceive themselves as *too thin,* rather than too fat, and are preoccupied with strict weightlifting and diet regimens to achieve a high level of muscularity. Among high school and college-aged men, in particular, body dissatisfaction may lead to disordered eating and exercise practices.

In muscle dysmorphia, numerous hours are devoted to working out at the gym; planning and eating low-carbohydrate, high-protein meals; and keeping meticulous

female athlete triad A condition characterized by disordered eating, lack of menstrual periods (amenorrhea), and osteoporosis.

The tragic case of Christy Henrich illustrates why anyone at risk for the female athlete triad should seek professional help. As a young teenager, Christy weighed 95 pounds and was 4 feet 11 inches tall. She showed promise as a gymnast but was told that she was too fat to excel in gymnastics. Christy continued her training but often starved herself, some days consuming just an apple and frequently purging by vomiting. Christy's eating disorder eventually rendered her too weak to compete. She was hospitalized after her weight had dropped to 47 pounds. She died, just after her 22nd birthday, from the effects of long-term semistarvation.

FIGURE 11-4 ▼ The female athlete triad occurs when the athlete has disordered eating, lack of menstrual periods, and osteoporosis. Stress fractures and chronic fatigue also occur. This triad often is seen in appearance-related sports, such as gymnastics. Long-term health is at risk; thus, prevention and early treatment are crucial.

records of exercise, body measurements, and food intake. The dietary practices (e.g., high-volume eating, protein intakes of up to 5 grams per kilogram) and use of unproven ergogenic aids or anabolic steroids can result in physical impairment. Such exercise and eating routines also interfere with social, occupational, and recreational activities. People with muscle dysmorphia may avoid social contact, eating in restaurants, and being seen without clothes because of distress over appearing too thin. Considering its many similarities with anorexia nervosa and a tendency for people with muscle dysmorphia to cross over to other forms of eating disorders over time, some experts support recognition of muscle dysmorphia as an eating disorder.

DIABULIMIA

Adolescents with diabetes are at heightened risk of eating disorders (see Further Reading 14). As you learned in Chapter 4, when insulin is absent or when cells are insulin resistant, cells are unable to use glucose for energy. Weight loss occurs because these carbohydrate calories essentially are wasted. After a person starts insulin or insulin-sensitizing therapy, cells are able to utilize glucose and weight gain is a common side effect. About one of three teens with type 1 diabetes admits to intentionally skipping doses of insulin to induce weight loss. This practice, so-called "diabulimia," can lead to severe hyperglycemia and its myriad consequences, which include eye damage, kidney damage, diabetic coma, or death. On the other hand, about one of five teens with type 1 diabetes uses overdoses of insulin to compensate for episodes of binge eating, a practice that could lead to dangerously low levels of blood sugar. Among adolescents with type 2 diabetes, up to 25% show signs of binge-eating disorder. For youth with diabetes, evidence of poor blood sugar control, frequently missed clinic visits, and the presence of depression can be warning signs of eating disorders.

ORTHOREXIA

As you have learned about nutrition in this course, you may have changed some of your own food choices. Obviously, a focus on healthy eating can help with weight management and prevention of disease, but when strict food rules begin to interfere with everyday life, they can be pathological. Some people worry excessively about the availability of food they permit themselves to eat, such as low-fat, sugar-free, or organic food choices. Although not currently classified as an eating disorder, orthorexia describes an emerging condition in which healthful eating becomes an obsession (see Further Reading 6). The term comes from Greek words meaning straight or proper appetite. Unlike cases of anorexia or bulimia, orthorexia does not usually originate in the drive for thinness. Rather, a need for perfection or purity lies at the heart of orthorexia. Such extreme dietary perfectionism may be related to obsessive-compulsive disorder.

▲ Orthorexia describes a condition in which a person is unduly concerned with eating only healthful foods.

☑ CONCEPT CHECK 11.6

1. List the three components of the female athlete triad.
2. How is muscle dysmorphia similar to anorexia nervosa? How is it different?
3. What is diabulimia? List some of the harmful physical effects of such practices.
4. How would you distinguish between healthy eating and orthorexia?

11.7 Prevention of Eating Disorders

A key to developing and maintaining healthful eating behavior is to realize that some concern about diet, health, and weight is normal. It is also normal to experience variation in what we eat, how we feel, and even how much we weigh. For example, it is common to experience some minimal weight change (up to 2 to 3 pounds) throughout the day and even more over the course of a week. A large weight fluctuation or ongoing weight gain or weight loss is more likely to indicate a problem. If you notice

CRITICAL THINKING

Mr. Thomas, a high school teacher, is concerned about eating disorders. He wants to try to prevent young adults from falling into the discouraging traps of anorexia nervosa and bulimia nervosa. What are some of the topics and issues he should discuss with students in his health classes?

a large change in your eating habits, how you feel, or your body weight, it is a good idea to consult your physician. Treating physical and emotional problems early helps lead you to peace of mind and good health.

With a view on society as a whole, many people begin to form opinions about food, nutrition, health, weight, and body image prior to or during puberty. Parents, friends, and professionals working with young adults should consider the following advice for preventing eating disorders:

- Discourage restrictive dieting and meal skipping. Fasting is also discouraged (except for religious occasions).
- Provide information about normal changes that occur during puberty.
- Correct misconceptions about nutrition, healthy body weight, and approaches to weight loss.
- Carefully phrase any weight-related recommendations and comments.
- Do not overemphasize numbers on a scale. Instead, teach the basics of proper nutrition and regular physical activity in school and at home.
- Encourage normal expression of disruptive emotions.
- Encourage children to eat only when they are hungry.
- Provide adolescents with an appropriate, but not unlimited, degree of independence, choice, responsibility, and self-accountability for their actions.
- Increase self-acceptance and appreciation of the power and pleasure emerging from one's body.
- Enhance tolerance for diversity in body weight and shape.
- Build respectful environments and supportive relationships.
- Encourage coaches to be sensitive to weight and body-image issues among athletes.
- Emphasize that thinness is not necessarily associated with better athletic performance.
- Support programs for eating disorder screening and prevention at high schools and colleges.

Our society as a whole can benefit from a fresh focus on nutritious food practices and a healthful outlook toward food and body weight. Not only is treatment of eating disorders far more difficult than prevention, these disorders also have devastating effects on the entire family. For this reason, caregivers and health care professionals must emphasize the importance of an overall healthful diet that focuses on moderation, as opposed to restriction and perfection.

Overall, the challenge facing many North Americans is achieving a healthy body weight without excessive dieting. A growing number of health professionals support a nondiet approach to weight management. This means adopting and maintaining sensible eating habits, a physically active lifestyle, and realistic and positive attitudes and emotions while practicing creative ways to handle stress. Certain cultural ideals of beauty can trigger eating disorders; changing these values might reduce the pressures predisposing some people to various types of disordered eating behavior. The "size acceptance" approach, discussed in Chapter 7, fights discrimination against overweight and obese people and asserts that each person should be free to find his or her natural weight. Women who combine careers and motherhood are saying that they have more important things to worry about; some fashion leaders are tolerating more curves; exercise programs are encouraging regular brisk walking, rather than high-intensity activities such as jogging. We cannot change the genes that predispose people to eating disorders (at least not yet!), but we can make a difference in the environmental triggers that set these devastating disorders in motion.

For further information on eating disorders, contact:
- Academy for Eating Disorders, **www.aedweb.org**
- The National Eating Disorders Association, **www.nationaleatingdisorders.org**
- The National Institute of Mental Health has published a concise review of eating disorders (**www.nimh.nih.gov/health/publications/eating-disorders/eating-disorders.pdf**).

✔ CONCEPT CHECK 11.7

1. Why is prevention of eating disorders so important?
2. Imagine you are a coach for a high school girls' softball team. What could you do to foster an environment that prevents eating disorders among this group of adolescent girls?

Nutrition and Your Health
● Eating Disorder Reflections

Thoughts of an Anorexic Woman

It was the spring of my freshman year of high school, and I had just turned 15. I wanted to get a leading role in the upcoming high school musical *West Side Story*. I thought I should lose some weight to look more attractive to the student director Shawn, so I decided to give up "junk" food. The next day my friend Sandra looked at my lunch, spread out neatly on a napkin before me, and squawked, "Dill pickles?! Who brings dill pickles for lunch in a zip-lock baggie?" The other girls at the table fell into a fit of hysterics. "Casting for *West Side Story* is coming up," I said, "and I gave up 'junk' food to try to lose a few pounds." One of my friends thought it would be funny to give me an M&M—just to smell. Ha, ha. I put it in a little Tupperware container and kept it for days in my backpack as a reminder. Every once in a while, I did smell it.

For the next few weeks, there were times when I would find myself cracking open the refrigerator door and just staring down at what I knew to be a deliciously crunchy, crisp, and cold Kit Kat bar in the dairy bin. I didn't eat it though. At the mall with my friends (since at 15, that's about all my parents allowed me to do), Bridgette and Nora wanted to stop and

get a cinnamon bun. They chided me, but I didn't budge. The cinnamon bun smelled so good. But as I sat opposite them in the food court and watched them overdramatize its ooey-gooey goodness, I felt a sense of pride that I could make a decision and stick with it. I could see that they were jealous of my willpower.

When Easter came around, I took a look at the contents of the Easter basket my mom insisted on preparing and turned up my nose at it. I had proven to myself that I could resist temptation . . . why stop now?

I was looking so much better as the pounds kept coming off. I'm pleased to say the thinner, new me got a part in the musical! I couldn't rest, though. I had to stick with it or I might regain the weight that I'd lost!

Also at this time, I started running, and my friend Laura became my running partner. She was getting in shape for the next season of field hockey. After school, we met in the locker room, changed out of our school clothes, and out we went. The running helped. Every morning, just after going to the bathroom and before getting any breakfast, I would pop onto the scale in my mom's bathroom. One hundred and fifteen pounds and still going. At 5 feet 7 inches, that wasn't too bad.

Cheese and butter had made it to the "no" list by the time I was 16 and down to 105 pounds. Fat-free was my mantra. For my sixteenth birthday, my friends threw a little surprise party for me. Nora, knowing I would put up a fight, made me a cake. "It's your birthday! You can have a piece of cake!" I politely said no, that I would cut it for everyone else, but I really didn't want any. They pestered me, and Nora started to feel offended, so finally I took a few bites so she wouldn't burst into tears. It had been so long since I'd had so much sugar. I felt bloated and sick. I ate nothing for the rest of the day, and only six saltines, one apple, and two stalks of celery the next day. Those foods were on the "yes" list. Salads also were okay but only with salt and vinegar. I told my parents that the dissections in biology class had given me a distaste for meat, but really, I just didn't want all those calories. For a while, I craved food day and night, but I was getting better and better at holding my ground.

By my senior year, I was skipping lunches altogether, opting instead to hang out in the library and read over my AP bio text. "Where were you at lunch today?" Bridgette would ask later. "Oh, I had some reading to do. The AP exam is going to be tough." At 100 pounds, I was getting closer to finding out what "tough" really meant.

Even though Laura moved out of state, I didn't give up on exercising. Now, my mom's stair stepper in the basement was my favorite. I'd take my biology notes, prop them up in front of me, and step-step-step until I had burned 400 kcal. I felt so efficient knowing that I could multitask. Sometimes I'd go twice a day. As senior year wore on, though, it got harder and harder to get up in the morning and put on my tennis shoes. And then one morning, in the shower, I just collapsed under the stream of hot water.

I ended up in this hospital bed with an IV tube in my arm. At 92 pounds, my body was starving. As it turns out, if you don't give your body any fuel, you start to cannibalize yourself, in a sense. My body had been so hungry, my muscles had been wasting away, and the episode in the shower was due to a problem with my heart. It's a problem I have created . . . not my parents or my distant group of friends . . . just me. My mom was there, next to me, caressing the arm with the IV tube, putting her whole life on hold because of me. Isn't this what I'd wanted—to be in control of my own destiny?

Where do I go from here?

Thoughts of a Bulimic Woman

From Hall L, Cohn L: *Bulimia—A Guide to Recovery.* Gurze Books: Carlsbad, CA, 1992. Reprinted with permission.

I am wide awake and immediately out of bed. I think back to the night before, when I made a new list of what I wanted to get done and how I wanted to be. My husband is not far behind me on his way into the bathroom to get ready for work. Maybe I can sneak onto the scale to see what I weigh this morning before he notices me. I am already in my private world. I feel overjoyed when the scale says that I stayed the same weight as I was the night before, and I can feel that slightly hungry feeling. Maybe it will stop today; maybe today everything will change. What were the projects I was going to get done?

We eat the same breakfast, except that I take no butter on my toast, no cream in my coffee, and never take seconds (until Doug gets out the door). Today I am going to be really good, and that means eating certain predetermined portions of food and not taking one more bite than I think I am allowed. I am very careful to see that I don't take more than Doug. I judge myself by his body. I can feel the tension building. I wish Doug would hurry up and leave so I can get going!

As soon as he shuts the door, I try to get involved with one of the myriad responsibilities on my list. I hate them all! I just want to crawl into a hole. I don't want to do anything. I'd rather eat. I am alone; I am nervous; I am no good; I always do everything wrong anyway; I am not in control; and I can't make it through the day, I know it. It has been the same for so long. I remember the starchy cereal I ate for breakfast. I am into the bathroom and onto the scale. It measures the same, but I don't want to stay the same! I want to be thinner! I look into the mirror. I think my thighs are ugly and deformed looking. I see a lumpy, clumsy, pear-shaped wimp. There is always something wrong with what I see. I feel frustrated, trapped in this body, and I don't know what to do about it.

I float to the refrigerator knowing exactly what is there. I begin with last night's brownies. I always begin with the sweets. At first, I try to make it look like nothing is missing,

▲ Bulimic episodes add to the despair felt in this disorder.

but my appetite is huge and I resolve to make another batch of brownies. I know there is half of a bag of cookies in the bathroom, thrown out the night before, and I polish them off immediately. I take some milk so my vomiting will be smoother. I like the full feeling I get after downing a big glass. I get out six pieces of bread and toast one side of each in the broiler, turn them over and load them with pats of butter, and put them under the broiler again until they are bubbling. I take all six pieces on a plate to the television and go back for a bowl of cereal and a banana to have along with them. Before the last piece of toast is finished, I am already preparing the next batch of six more pieces. Maybe another brownie or five and a couple of large bowls full of ice cream, yogurt, or cottage cheese.

My stomach is stretched into a huge ball below my rib cage. I know I'll have to go into the bathroom soon, but I want to postpone it. I am in never-never land. I am waiting, feeling the pressure, pacing the floor in and out of the rooms. Time is passing. Time is passing. It is getting to be time. I wander aimlessly through each of the rooms again, tidying, making the whole house neat and put back together. I finally make the turn into the bathroom. I brace my feet, pull my hair back and stick my finger down my throat, stroking twice, and get up a huge pile of food. Three times, four times, and another pile of food. I can see everything come back. I am so glad to see those brownies because they are so fattening. The rhythm of the emptying is broken and my head is beginning to hurt. I stand up feeling dizzy, empty, and weak. The whole episode has taken about an hour.

Summary (Numbers refer to numbered sections in the chapter.)

11.1 Disordered eating encompasses mild and short-term changes in eating patterns that occur as a result of life stress, illness, or a desire to change body weight. When carried to the extreme, disordered eating may progress to an eating disorder, in which severe changes in eating patterns have lasting and detrimental effects. Current research on the origins of eating disorders indicates that genetic factors dictate brain biology, which affects how certain individuals perceive their bodies and respond to life stresses. Thus, a person who is genetically predisposed to eating disorders may use disordered eating behaviors to cope with feelings of depression, anger, or guilt. The three main types of eating disorders are anorexia nervosa, bulimia nervosa, and binge-eating disorder.

11.2 Anorexia nervosa is characterized by extreme weight loss (e.g., BMI of less than 17), a distorted body image, and an irrational fear of weight gain and obesity. Weight loss is achieved primarily by restricting food intake. Physical consequences include a profound decrease in body weight and body fat, heart irregularities, iron-deficiency anemia, impaired immunity, digestive dysfunction, and loss of menstrual periods. Treatment of anorexia nervosa includes increasing food intake to support gradual weight gain. Family-based psychological counseling can help anorexic individuals establish healthy eating behaviors and body image.

11.3 Similar to anorexia nervosa, bulimia is characterized by overvaluation of body weight and shape. However, the disordered eating patterns of bulimic individuals involve recurrent binge eating followed by compensatory behaviors. Binge eating is consuming an abnormally large amount of food within a short time period. A person with bulimia nervosa has lack of control over bingeing behaviors and feels extremely distressed after a binge. Inappropriate compensatory behaviors used to rid the body of excess calories include vomiting or misusing laxatives, diuretics, or enemas. Alternately, fasting and excessive exercise may be used. Vomiting as a means of purging is especially destructive to the body; it can cause severe tooth decay, stomach ulcers, irritation of the esophagus, low blood potassium, and other problems. Treatment of bulimia nervosa includes psychological and nutritional counseling. Certain medications, such as antidepressants, can aid recovery.

11.4 Binge-eating disorder is the most widespread eating disorder. It tends to be diagnosed in middle age and affects men and women nearly equally. It is characterized by recurrent episodes of binge eating, which cause marked distress, but are not followed by compensatory behaviors. About 70% of people with binge-eating disorder are obese. The health effects that stem from binge-eating disorder are comorbid conditions of obesity, including hypertension, high blood cholesterol, cardiovascular disease, and type 2 diabetes. Treatment involves cognitive behavioral therapy and nutrition counseling. Antidepressants and other medications may enhance treatment success.

11.5 Pica is an eating disorder in which a person persistently ingests nonnutritive, nonfood items such as clay, dirt, or ice. Subthreshold eating disorders (e.g., atypical anorexia nervosa, bulimia nervosa of low frequency or limited duration, or binge-eating disorder of low frequency or limited duration) capture those individuals who meet some but not all of the criteria for diagnosis of anorexia nervosa, bulimia nervosa, or binge-eating disorder. People with purging disorder practice purging behaviors to achieve weight loss, but they do not exhibit the binge-eating behaviors typical of bulimia nervosa. Individuals with night eating syndrome consume more than 25% of daily food intake after dinner, may have difficulty falling asleep without eating, and wake up at least once during the night to consume food.

11.6 Several additional disordered eating patterns that are not classified as eating disorders include the female athlete triad, muscle dysmorphia, diabulimia, and orthorexia. The female athlete triad consists of disordered eating, loss of menstrual periods, and osteoporosis. Muscle dysmorphia could be described as "reverse anorexia"; a person views him- or herself as less muscular than desired, which results in patterns of disordered eating and obsessive exercise to achieve a muscular body shape. Diabulimia refers to the misuse of diabetes medication to regulate body weight. Some type 1 diabetic patients will skip doses of insulin to induce weight loss but then suffer the consequences of hyperglycemia. Alternatively, overdoses of insulin or glucose-lowering medications could be used to counter the effects of a binge. This practice could result in hypoglycemia. Orthorexia refers to an obsession with healthy eating such that overly restrictive food choices interfere with other aspects of life, such as social interactions.

11.7 Prevention of eating disorders is crucial because treatments are expensive, lengthy, and not 100% effective. Encouraging healthy attitudes about eating and exercise from a young age will aid in preventing the development of eating disorders. Those who work closely with children and young adults should encourage acceptance of diversity in body sizes and carefully phrase comments about body weight. Helping children to develop healthy ways to cope with emotions is also important.

Check Your Knowledge (Answers to the following questions are below.)

1. For 3 weeks leading up to her friend's wedding, Teresa skipped meals and restricted her food intake to 800 kcal per day so that she could fit into her bridesmaid dress. After the wedding, she resumed eating 2200 kcal per day. This is an example of
   a. disordered eating.
   b. an eating disorder.
   c. size acceptance.
   d. muscle dysmorphia.

2. Factors that contribute to development of eating disorders include
   a. genetics.
   b. social pressures to be thin.
   c. sexual abuse.
   d. All of the above.

3. Anorexia nervosa can be defined as
   a. compulsive eating.
   b. hyperactivity.
   c. denial of appetite.
   d. purging.

4. The most likely long-term health consequence of anorexia nervosa could be
   a. fractures resulting from bone loss.
   b. atherosclerotic heart disease.
   c. esophageal ulcers.
   d. cancer.

5. Bulimia is most frequently first recognized by a
   a. dietitian.
   b. physician.
   c. dentist.
   d. physical therapist.

6. The *most life-threatening* health risk from frequent vomiting due to bulimia nervosa is
   a. a drop in blood potassium.
   b. constipation.
   c. weight gain.
   d. swollen salivary glands.

7. Binge-eating disorder can be characterized as
   a. bingeing accompanied by purging.
   b. secretive eating.
   c. eating to avoid feeling and dealing with emotional pain.
   d. the early phase of bulimia nervosa.

8. Night eating syndrome is characterized by
   a. eating dinner but no breakfast or lunch.
   b. the need to eat to fall asleep.
   c. waking at night to purge by vomiting.
   d. consuming all of the daily calories at night.

9. Female athlete triad consists of
   a. anorexia nervosa, lack of family support, and overtraining.
   b. disordered eating, overtraining, and lack of menstrual periods.
   c. osteoporosis, lack of menstrual periods, and disordered eating.
   d. osteoporosis, lack of sleep, and disordered eating.

10. If you were assigned to speak to a group of middle school 4-H students about healthy eating, which message would be best?
    a. Ask children to sort various snack ideas into "good" or "bad" groups.
    b. Illustrate how many minutes of exercise are needed to burn the calories in various snacks.
    c. Advise kids to restrict favorite treats (e.g., ice cream), except as a reward for reaching a goal, such as getting a good grade on a test.
    d. Emphasize that children should eat when they are hungry and stop eating when they are full.

Answer Key: 1.a (LO 11.1), 2.d (LO 11.2), 3.c (LO 11.3), 4.a (LO 11.3), 5.c (LO 11.4), 6.a (LO 11.4), 7.c (LO 11.5), 8.b (LO 11.6), 9.c (LO 11.6), 10.d (LO 11.7)

Study Questions (Numbers refer to Learning Outcomes)

1. What are the typical characteristics of a person with anorexia nervosa? What may influence a person to begin rigid, self-imposed dietary patterns? **(LO 11.3)**

2. List the detrimental physical effects of bulimia nervosa. Describe important goals of the psychological and nutrition therapy used to treat bulimic patients. **(LO 11.4)**

3. What is the current thinking concerning use of medications for treatment of anorexia nervosa, bulimia nervosa, and binge-eating disorder? **(LOs 11.3, 11.4, and 11.5)**

4. Explain the role of excessive exercise in eating disorders. **(LO 11.1)**

5. How might parents or other role models contribute to the development of an eating disorder? Suggest an attitude that a parent or an adult friend of yours displayed that may not have been conducive to developing a healthy relationship to food. **(LO 11.7)**

6. Based on your knowledge of good nutrition and sound dietary habits, answer the following questions:
   a. How can repeated bingeing and purging lead to significant nutrient deficiencies?
   b. How can significant nutrient deficiencies contribute to major health problems in later life?
   c. A friend asks you, the nutrition expert, if it is okay to "cleanse" the body by eating only grapefruit for a week. What is your response? **(LO 11.1)**

7. Provide an example of the way society contributes to development of eating disorders. **(LO 11.2)**

8. List the three components of the female athlete triad. What is the major health risk associated with loss of menstrual periods in the female athlete? **(LO 11.6)**

9. How does binge-eating disorder differ from bulimia nervosa? Describe the factors that contribute to the development of binge-eating disorder. **(LO 11.5)**

10. Provide two recommendations to reduce the problem of eating disorders in our society. **(LO 11.7)**

What the Dietitian Chose

It can be uncomfortable to confront a friend about an unhealthy behavior. You may worry about compromising your friendship. The eating and excessive exercise behaviors you have witnessed may be short-term disordered eating and may pass in a few weeks. On the other hand, they might just be the tip of the iceberg. Ignoring her behavior will not make it go away. As you have learned in this chapter, eating disorders can have very serious consequences, and the earlier they are treated, the better the outcome. If you feel unable to address your concerns one on one with your roommate, you can seek help from a resident assistant in your dorm, a school counselor, or a nurse in the student health center. You will not be betraying your friendship; you will be showing how much you care about her by taking action.

Diagnosing an eating disorder is a complex task. People with eating disorders may deny the problem, hide their behaviors, and be irrational and uncooperative with those who try to help them. Do not try to diagnose your friend, but do try to keep the lines of communication open. Tell her what you have observed and ask her how you can help. She may open up to you and be glad she is not alone in her struggle. Then again, she may get angry and withdraw from you. Try to share what you know about eating disorders without lecturing or admonishing her. Also, be aware that your friend is going through a difficult emotional struggle that has probably been years in the making. Do not assume you know what she is going through.

Discussing healthy weight-management techniques with your roommate is not likely to help her if she has fallen prey to an eating disorder. She is already critical of her appearance and weight; your advice, although well intended and backed by the nutrition knowledge you have picked up in this course, may be construed as judgmental. Eating disorders are not about food. Rather, they involve issues of self-concept and control. It is important to be supportive but not critical. Taking the focus off food and body weight is important. Do not try to force her to eat or slow down on her exercise. If you can help get your roommate to seek professional treatment, psychiatric counseling will help to address her unhealthy relationship with food; eventually, nutrition counseling will help her to establish better eating habits.

Get prepared before you confront your roommate. Find out what information and resources are available on campus so you can share them with her. Recognize that you are not going to cure her disorder—she needs professional help. You can be supportive and patient as she deals with her inner turmoil. Listen and offer to go with her to seek care. Someday, you may be a health professional who is able to diagnose and counsel victims of eating disorders. For now, leave it to a qualified professional.

CASE STUDY SOLUTION Eating Disorders—Steps to Recovery

1. Sarah has the characteristics to be diagnosed with anorexia nervosa: she restricts food intake to maintain a weight below 85% of that expected (BMI = 17), has a distorted view of her appearance, and fears weight gain.
2. Sarah's disordered eating habits began when she became self-conscious about her body when her peers teased her about being overweight. She began exercising each day and was successful at losing weight. Her disordered eating then began with restricting her food intake more. She enjoyed the self-control she had over her body.
3. The case study specifically mentions that Sarah's menstrual periods have ceased and she is intolerant to cold temperatures.
4. Most of her treatment will involve outpatient therapy. A team of health professionals, most likely consisting of a physician, registered dietitian, and psychologist, would provide therapy. She will need to reduce her physical activity and consume enough calories (e.g., 1000 to 1600 kcal per day initially) to promote weight gain. Her food intake will be increased by increments of about 100 to 200 kcal every few days until an acceptable rate of weight gain is achieved.

The goal is to achieve a BMI that falls within the normal range (e.g., 20). Family-based psychological therapy will help Sarah to correct misperceptions about food and body image and develop healthy coping strategies. After her body weight has stabilized, antidepressant medications may assist her recovery.
5. Several nutrient deficiencies are likely. Those with the most salient health effects are deficiencies of calcium and iron. The physician would likely prescribe a multivitamin and mineral supplement, along with an additional supplement of calcium as needed to make sure intake is in the range of 1200 to 1500 milligrams per day. Additional iron may be needed if Sarah is anemic. This overall practice will correct vitamin and mineral deficiencies. The calcium in particular will contribute to bone maintenance.
6. Sarah's outlook for recovery is not good unless she realizes she has a problem. Even if she is willing to accept the therapy and counseling, relapse is common. Sarah's disordered eating habits have been in place for about 3 years, so her problem is deep-rooted. The chances of recovery are greater if intervention is early in the course of the disease and if follow-up is continued for several years.

Further Readings

1. American Academy of Pediatrics: Clinical Report—Identification and management of eating disorders in children and adolescents. *Pediatrics* 126:1240, 2010.

2. Borzekowski DLG and others: e-Ana and e-Mia: A content analysis of pro-eating disorder websites. *American Journal of Public Health* 100:1526, 2010.

3. Campbell IC and others: Eating disorders, gene-environment interactions, and epigenetics. *Neuroscience and Behavioral Reviews* 35:784, 2011.

4. Feldman MB and Meyer IH: Eating disorders in diverse lesbian, gay, and bisexual populations. *International Journal of Eating Disorders* 40:218, 2007.

5. Gagne DA and others: Eating disorder symptoms and weight and shape concerns in a large web-based convenience sample of women ages 50 and above: Results of the Gender and Body Image (GABI) Study. *International Journal of Eating Disorders* 45:832, 2012.

6. Getz L: Orthorexia: When eating healthy becomes an unhealthy obsession. *Today's Dietitian* 11:40, 2009.

7. Harrop EN and Marlatt GA: The comorbidity of substance use disorders and eating disorders in women: Prevalence, etiology, and treatment. *Addictive Behaviors* 35:392, 2010.

8. Hay P: A systematic review of evidence for psychological treatments in eating disorders: 2005 – 2012. *International Journal of Eating Disorders* 46:462, 2013.

9. Klump K and others: Academy for Eating Disorders position paper: Eating disorders are serious mental illnesses. *International Journal of Eating Disorders* 42:97, 2009.

10. Le Grange D and others: Academy of Eating Disorders Position Paper: The role of the family in eating disorders. *International Journal of Eating Disorders* 43:1, 2010.

11. Leman C: Night eating syndrome. *Today's Dietitian* 12:8, 2010.

12. McFarland MB and Kaminski PL: Men, muscles, and mood: The relationship between self-concept, dysphoria, and body image disturbances. *Eating Behaviors* 10:68, 2009.

13. Ozier AD and Henry BW: Position of the American Dietetic Association: Nutrition intervention in the treatment of eating disorders. *Journal of the American Dietetic Association* 111:1236, 2011.

14. Pinhas-Hamiel O and Levy-Shraga Y: Eating disorders in adolescents with type 2 and type 1 diabetes. *Current Diabetes Reports* 13:289, 2013.

15. Seher CL: Binge-eating disorder—learning about this condition can help RDs counsel patients more effectively. *Today's Dietitian* 14:34, 2012.

16. Sojcher R and others: Evidence and potential mechanisms for mindfulness practices and energy psychology for obesity and binge-eating disorder. *Explore* 8:271, 2012.

17. Stice E and others: Prevalence, incidence, impairment, and course of the proposed DSM-5 eating disorder diagnoses in an 8-year prospective community study of young women. *Journal of Abnormal Psychology* 122:445, 2013.

18. Waterhous T and others: Practice paper of the American Dietetic Association: Nutrition intervention in the treatment of eating disorders. *Journal of the American Dietetic Association* 111:1261, 2011.

19. Yeo M and Hughes E: Eating disorders: Early identification in general practice. *Australian Family Physician* 40:108, 2011.

 **NUTRITION**

To get the most out of your study of nutrition, visit McGraw-Hill Connect at www.mcgrawhillconnect.com where you will find NutritionCalc Plus, LearnSmart, and many other dynamic tools.

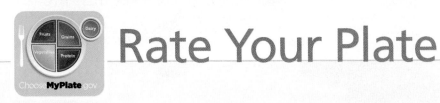

Rate Your Plate

I. Assessing Risk of Developing an Eating Disorder

British investigators have developed a five-question screening tool called the SCOFF Questionnaire for recognizing eating disorders:[†]

1. Do you make yourself **Sick** because you feel full?

2. Do you lose **Control** over how much you eat?

3. Have you lost more than **One** stone (about 13 pounds) recently?

4. Do you believe yourself to be **Fat** when others say you are thin?

5. Does **Food** dominate your life?

Two or more positive responses suggest an eating disorder.

1. After completing this questionnaire, do you feel that you might have an eating disorder or the potential to develop one?

   _____

   _____

2. Do you think any of your friends might have an eating disorder?

   _____

   _____

3. What counseling and education resources exist in your area or on your campus to help with a potential eating disorder?

   _____

   _____

4. If a friend has an eating disorder, what do you think is the best way to assist him or her in getting help?

   _____

   _____

[†]Morgan JF and others: The SCOFF Questionnaire, *British Medical Journal* 319:1467, 1999.

II. Helping Prevent Eating Disorders

You have been asked to speak to a junior high school class about eating disorders. What are four major points that you would make to help prevent disordered eating in this population?

1. _____

2. _____

3. _____

4. _____

Here are points you may consider:

1. Extreme thinness is oversold in the media. Extremely low weight (i.e., BMI of less than 17.5) is generally not healthy.

2. Self-induced vomiting is dangerous. Damage to the teeth, stomach, and esophagus often results.

3. Loss of menstrual periods is a sign of illness. It is important to see a physician about this. Bone deterioration is a common result.

4. The treatment of eating disorders in early phases aids success. These diseases are difficult to treat once firmly established.

Student Learning Outcomes

Chapter 12 is designed to allow you to:

12.1 Define and characterize the terms *hunger, malnutrition,* and *undernutrition.*

12.2 Examine undernutrition in the United States and highlight several programs established to combat this problem.

12.3 Examine undernutrition in the developing world and evaluate the major obstacles that hinder a solution.

12.4 Outline some possible solutions to undernutrition in the developing world.

12.5 Evaluate the consequences of undernutrition during critical periods in a person's life.

Chapter 12

Undernutrition Throughout the World

What Would You Choose?

"You'd better clean your plate! There are starving children in Africa who would love to eat that dinner!" Mom was right: we often overlook the fact that we live in a rich nation with access to plentiful food and clean water, whereas something as simple as clean drinking water is merely a dream in many developing nations. In fact, the problems of hunger and food insecurity exist in your own community. It is probably not feasible (or sanitary) to send your rejected green beans overseas, but there are some simple steps you can take to support local, national, and international agencies seeking to provide clean water and nutritious food for people in need. After considering the benefits of direct aid versus empowering others to help themselves, what would you choose to help put an end to world hunger?

a Organize a group of friends to prepare and hand out peanut butter and jelly sandwiches to homeless people downtown.

b Donate $10 per month to an international aid organization that supplies food to starving children in Africa.

c Participate in a 5K run/walk event that raises funds and awareness to combat world hunger.

d Buy locally grown fruits and vegetables at your community's farmers market.

connect |NUTRITION **Think about your choice as you read Chapter 12, then see What the Dietitian Chose at the end of the chapter. To learn more about combating hunger, check out the Connect site: www.mcgrawhillconnect.com.**

Today, nearly one in seven people worldwide is chronically undernourished—too hungry to lead a productive, active life. The problems of poverty and undernutrition are widespread—even though there is enough food available to sufficiently feed all of us. Some progress has been made in decreasing the amount of poverty, and there is a push to make great strides by 2015.

Within the developing world, over 900 million people are malnourished, with Asia having the largest number. This level of malnutrition is the main cause of lowered resistance to disease, infection, and death, with children under 5 years being the most susceptible.

This chapter examines the problem of undernutrition, the conditions that create it, and some solutions that are in progress. If we are to eradicate undernutrition, we all have to understand the problem. We must begin to assume responsibility today, not tomorrow, to work on solutions to hunger close to home and in faraway nations. It is important to recognize that many political, economic, and social factors worldwide, such as rising food prices, war, environmental catastrophes, and the global threat of AIDS contribute to the hunger problem. The United Nations Millennium Development Goals (MDGs) are eight goals that address poverty, hunger, disease, illiteracy, environmental degradation, and discrimination against women. A global action plan was adopted in 2010 to achieve the MDGs by 2015. Why must we act today to stem this tide of undernutrition? This chapter provides some answers.

food insecure Condition in which the quality, variety, and/or desirability of the diet is reduced and there is difficulty at times providing enough food for everyone in the household.

nutrition security Secure access to a nutritious diet coupled with a sanitary environment and adequate health services and care.

▲ Malnutrition is associated with overpopulation in some developing countries.

FIGURE 12-1 ▶ Undernutrition in 2012–2014 by region (in millions). Regions have differed in their rates of progress toward reducing hunger. South-Eastern Asia, Eastern Asia, Latin America, and the Caribbean have seen the greatest decline, whereas the rates have increased in Western Asia and sub-Saharan Africa.
Source: FAO.

12.1 World Hunger: A Crisis of Nutrition Security

Uncertainty regarding the source of one's next meal remains a daily experience for close to 1 billion people around the world. This situation is troubling, considering that agriculture worldwide produces more than enough food to meet the energy requirements of each of the planet's 7 billion persons. Even with this abundance, 805 million people—11.3% of the world population—were still unable to access enough food in 2014 to lead active, healthy lives; that is, they were **food insecure.** This statistic comes from the United Nations Food and Agriculture Organization (FAO), which measures undernutrition around the world. Availability, access, utilization, and stability are the four pillars of food security, which exists when people have physical, social, and economic access to sufficient, safe, and nutritious food to meet their dietary needs and food preferences for an active and healthy life.

Food security is actually part of a bigger concept called **nutrition security.** The FAO defines nutrition security as secure access to an appropriately nutritious diet (i.e., protein, carbohydrate, fat, vitamins, minerals, and water), coupled with a sanitary environment and adequate health services and care, in order to ensure a healthy and active life for all household members (see Further Reading 10). The combined term *food and nutrition security* can be used to emphasize both food and health requirements (see Further Reading 1).

The serious problems of food insecurity and malnutrition exist in virtually every nation (Fig. 12-1). They are most common in regions of the developing world, with Asia and the Pacific having the largest number (520 million), along with sub-Saharan Africa (214 million), and Latin America and the Caribbean (37 million). Nearly all people suffering from food insecurity, hunger, or malnutrition are poor. There has been global interest in the poor and hungry for decades, and new development goals have been set forth by the United Nations (UN) since the beginning of the 2000 millennium. In 2010, the UN Millennium Development Goals (MDGs) of 2000 were reaffirmed and a global action plan was adopted to achieve them by 2015. The MDGs are eight goals that address poverty, hunger, disease, illiteracy, environmental degradation, and discrimination against

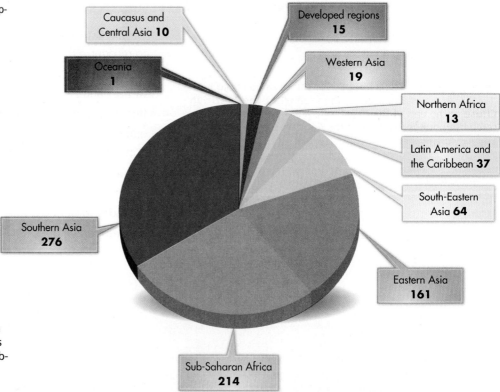

Total = 805 million

women. The likelihood of reaching these goals is optimistic in that 193 UN member states have agreed to make progress to achieve them by the year 2015. The MDGs range from cutting extreme poverty rates by half to stopping the spread of HIV/AIDS and have been agreed upon by all countries and leading development institutions of the world. The MDGs and their target outcomes are listed in Table 12-1 and will be discussed throughout this chapter (see Further Reading 16). We will begin our look at the problem of world hunger and malnutrition by defining some key terms.

TABLE 12-1 ▶ Millennium Development Goals by 2015

Millennium Development Goals	Targets
Goal 1: Eradicate extreme poverty and hunger	Target 1A: Halve, between 1990 and 2015, the proportion of people whose income is less than $1.25 a day
	Target 1B: Achieve full and productive employment and decent work for all, including women and young people
	Target 1C: Halve, between 1990 and 2015, the proportion of people who suffer from hunger
Goal 2: Achieve universal primary education	Target 2A: Ensure that, by 2015, children everywhere, boys and girls alike, will be able to complete a full course of primary schooling
Goal 3: Promote gender equality and empower women	Target 3A: Eliminate gender disparity in primary and secondary education, preferably by 2005, and in all levels of education no later than 2015
Goal 4: Reduce child mortality	Target 4A: Reduce by two-thirds, between 1990 and 2015, the under 5 mortality rate
Goal 5: Improve maternal health	Target 5A: Reduce by three quarters, between 1990 and 2015, the maternal mortality ratio (number of women who die during pregnancy and childbirth, per 100,000 live births)
	Target 5B: Achieve universal access to reproductive health
Goal 6: Combat HIV/AIDS, malaria and other diseases	Target 6A: Have halted, by 2015, and begun to reverse the spread of HIV/AIDS
	Target 6B: Achieve, by 2010, universal access to treatment for HIV/AIDS for all those who need it
	Target 6C: Have halted, by 2015, and begun to reverse the incidence of malaria and other major diseases
Goal 7: Ensure environmental sustainability	Target 7A: Integrate the principles of sustainable development into country policies and programs and reverse the loss of environmental resources
	Target 7B: Reduce biodiversity loss, achieving, by 2010, a significant reduction in the rate of loss
	Target 7C: Halve, by 2015, the proportion of people without sustainable access to safe drinking water
	Target 7D: Achieve, by 2020, a significant improvement in the lives of at least 100 million slum dwellers
Goal 8: Global partnership for development	Target 8A: Develop further an open, rule-based, predictable, nondiscriminatory trading and financial system
	Target 8B: Address the special needs of the least developed countries
	Target 8C: Address the special needs of landlocked countries and small island developing states
	Target 8D: Deal comprehensively with the debt problems of developing countries
	Target 8E: In cooperation with pharmaceutical companies, provide access to affordable essential drugs in developing countries
	Target 8F: In cooperation with the private sector, make available benefits of new technologies, especially information and communications

hunger The primarily physiological (internal) drive to find and eat food.

food insecurity A condition of anxiety regarding running out of either food or money to buy more food.

malnutrition Failing health that results from longstanding dietary practices that do not coincide with nutritional needs.

undernutrition Failing health that results from a longstanding dietary intake that is not enough to meet nutritional needs.

▲ Minimal intakes of protein and zinc limit the growth of children worldwide. About 25% of children in developing countries show evidence of poor growth rates.

HUNGER

The physiological state that results when not enough food is eaten to meet energy needs is **hunger.** It also describes an uneasiness, discomfort, weakness, or pain caused by lack of food. The medical and social costs of the undernutrition that can result from hunger are high: preterm births, mental disabilities, inadequate growth and development in childhood, poor school performance, decreased work output in adulthood, and chronic disease. Although malnutrition does occur in North America, it is not due to extreme poverty over a large section of the population. Instead, there are usually specific causes such as an eating disorder, alcoholism, problems in skilled nursing facilities (nursing homes), or homelessness. There is also some degree of moderate malnutrition in some of the poorer segments of North American society (i.e., those earning less than the current poverty level). Fortunately, there are resources such as food pantries and food stamps, though sometimes there are bureaucratic obstacles to getting these resources to the people who need them. In addition, there is the problem known as **food insecurity,** which describes a state of anxiety about running out of food or running out of money to buy more food. In 2011, 14.9% of households in the United States and 12.2% of Canadian households reported experiencing food insecurity.

Around the world, poverty is a primary cause of malnutrition. In the United States, child poverty reached a record high level in 2013, with 16.7 million children, 20% of all children under the age of 18, in families living below the federal poverty level. Poverty is defined as a family of four having an income less than of $23,850 a year in 2014 (more information available at **http://aspe.hhs.gov/poverty/13poverty.cfm**). Research, however, estimates that families need an income of about twice that level to cover basic expenses. Living in poverty impairs a child's ability to learn and contributes to behavioral problems and poor health. The risks of poverty are greatest for young children. Fortunately, the United States does have food assistance programs for low-income families, and therefore, most children in the United States are shielded from hunger.

MALNUTRITION AND MICRONUTRIENT DEFICIENCES

A condition of impaired development or function caused by either a long-term deficiency or excess in calorie and/or nutrient intake is **malnutrition.** When food supplies are low and the population is large, **undernutrition** is common, leading to nutritional deficiency diseases, such as goiter (from an iodide deficiency) and xerophthalmia (eye problems caused by poor vitamin A intake). However, when the food supply is ample or overabundant, incorrect food choices coupled with an excessive intake can lead to overnutrition and obesity and its related chronic diseases, such as type 2 diabetes.

Undernutrition is the most common form of malnutrition among the poor in both developing and developed countries. Undernutrition is also the primary cause of specific nutrient deficiencies that can result in muscle wasting, blindness, scurvy, pellagra, beriberi, anemia, rickets, goiter, and a host of other problems (Table 12-2).

The most critical micronutrients missing from diets worldwide (see Fig. 12-2) are iron, vitamin A, iodide, zinc, and various B vitamins (e.g., folate), as well as selenium and vitamin C (see Further Readings 4 and 5). About 1 billion people, mostly in the developing world, are affected by iron deficiency. The same is true for zinc deficiencies. With poor iron status, cognitive development will likely be impaired, particularly if prolonged deficiency occurs during early infancy. An estimated 50 million people worldwide also suffer brain damage from preventable maternal iodide deficiency. In addition, new studies from the United Kingdom and Australia suggest that pregnant women who do not consume enough iodide may put their children at risk of lower intelligence and reading ability (see Newsworthy Nutrition).

It is estimated that in vitamin A-deficient areas, a large percentage of pregnant women and at least 250 million preschool children are vitamin A deficient. Although severe vitamin A deficiency, which causes blindness, is on the decline, up

TABLE 12-2 ▶ **Nutrient-Deficiency Diseases That Commonly Accompany Undernutrition**

Disease and Key Nutrient Involved*	Typical Effects	Foods Rich in Deficient Nutrient	Target Populations for Intervention
Xerophthalmia Vitamin A	Blindness from chronic eye infections, restricted growth, dryness, and keratinization of epithelial tissues	Fortified milk, sweet potatoes, spinach, greens, carrots, cantaloupe, and apricots	Asia and Africa
Rickets Vitamin D	Poorly calcified bones, bowed legs, and other bone deformities	Fortified milk, fish oils, and sun exposure	Asia, Africa, and parts of the world where religious dress codes prevent women and children from receiving adequate sun exposure; older adults in developed nations
Beriberi Thiamin	Nerve degeneration, altered muscle coordination, and cardiovascular problems	Sunflower seeds, pork, whole and enriched grains, and dried beans	Victims of famine in Africa and alcoholics
Ariboflavinosis Riboflavin	Inflammation of tongue, mouth, face, and oral cavity; nervous system disorders	Milk, mushrooms, spinach, liver, and enriched grains	Victims of famine in Africa
Pellagra Niacin	Diarrhea, dermatitis, and dementia	Mushrooms, bran, tuna, chicken, beef, peanuts, and whole and enriched grains	Victims of famine in Africa and survivors of war-torn Eastern Europe
Megaloblastic anemia Folate	Enlarged red blood cells, fatigue, and weakness	Green leafy vegetables, legumes, oranges, and liver	Asia and Africa
Scurvy Vitamin C	Delayed wound healing, internal bleeding, and abnormal formation of bones and teeth	Citrus fruits, strawberries, and broccoli	Victims of famine in Africa
Iron-deficiency anemia Iron	Reduced work output, retarded growth, and increased health risk in pregnancy	Meats, seafood, broccoli, peas, bran, whole-grain, and enriched breads	Worldwide
Goiter Iodide	Enlarged thyroid gland in teenagers and adults, possible mental retardation, and congenital hypothyroidism	Iodized salt and saltwater fish	South America, Eastern Europe, and Africa

*Although the nutrients are listed separately to illustrate the important role of each one, often two or more nutrition-deficiency diseases are found in an undernourished person in the developing world.

to 500,000 preschool-age children are still blinded by it each year, with half of them likely dying within 12 months of losing their sight. The United Nations Children's Fund (UNICEF) reports that the lives of 1 to 3 children could be saved annually in the developing world if vitamin A supplements were provided a few times each year. The annual cost per child would be about 6 cents.

Of the 7.1 billion people in the world, about 2 billion may experience episodes of food shortages and be affected by some form of micronutrient malnutrition. Death and disease from infections, particularly those causing acute and prolonged diarrhea or respiratory disease, increase dramatically when the infections occur during a state of chronic undernutrition. Chronic undernutrition leaves many people in the developing world in a continual state of depressed immune function, in turn greatly increasing the risk of death, especially in childhood.

The Irish potato famine of 1840 to 1850 caused an estimated 2 million deaths and resulted in nearly as many people emigrating to other countries, such as the United States and Canada. More than 3 million people may have perished in the great famine of 1943 in Bengal, India. China suffered a famine from 1959 to 1961: estimates of mortality range from 16 to 64 million. In 1974, another 1.5 million starved in the country of Bangladesh.

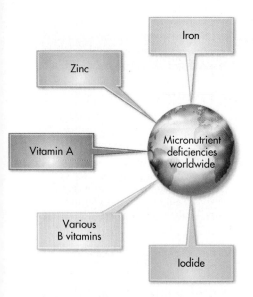

FIGURE 12-2 ▲ Critical micronutrient deficiencies worldwide.

famine An extreme shortage of food, which leads to massive starvation in a population; often associated with crop failures, war, and political unrest.

The effects of hunger are diverse and widespread:

- Reduced energy and strength
- Diminished concentration
- Impaired ability to learn
- Lowered productivity
- Worsening of chronic health conditions
- Increased susceptibility to infectious diseases
- Deterioration of mood
- Slowed recovery from illness and injury
- Decreased birthweights
- Decreased growth in infants and children

Protein-calorie malnutrition (PCM) is a form of undernutrition caused by an extremely deficient intake of calories or protein and is generally accompanied by an illness. The dramatic results of PCM—kwashiorkor and marasmus—were described in Chapter 6. This chapter focuses on the more subtle effects of a chronic lack of food.

FAMINE

The extreme form of chronic hunger is **famine.** Periods of famine are characterized by large-scale loss of life, social disruption, and economic chaos that slows food production. As a result of these extreme events, the affected community experiences a downward spiral characterized by human distress; sales of land, livestock, and other farm assets; migration; division and impoverishment of the poorest families; crime; and humanitarian crises, as seen in Sudan. In the midst of all this, undernutrition rates soar; infectious diseases, such as cholera, spread; and many people die.

Special efforts are needed to eradicate the fundamental causes of famine. Causes vary by region and decade, but the most common is crop failure. The most obvious reasons for crop failure are extreme weather conditions such as floods or drought, war, and civil strife. War deserves a special focus and will be specifically addressed in Section 12.3.

GENERAL EFFECTS OF SEMISTARVATION

In the initial stages, the results of undernutrition from semistarvation are often so mild that physical symptoms are absent and blood tests do not usually detect the slight metabolic changes. Even in the absence of clinical symptoms, however, undernourishment may affect the ability to work, learn, reproduce, and recover from illnesses or injuries. Recall from Chapter 2 that as tissues continue to be depleted of nutrients, blood tests eventually detect biochemical changes, such as a drop in blood hemoglobin concentration. Physical symptoms, such as body

Newsworthy Nutrition

Maternal iodine (iodide) deficiency could lead to lower educational outcomes

The hypothesis of this study was that children born to mothers with mild iodine deficiency during pregnancy will have poorer educational outcomes in primary school than peers whose mothers did not have iodine deficiency during pregnancy. The study was a longitudinal design with follow-up when the children were 9 years old. The study participants were children born in Australia between 1999 and 2001. The mild iodine deficiency during pregnancy was followed with the children subsequently growing up in an iodine-adequate environment. Children whose mothers had mild iodine deficiency had reductions of 10.0% in spelling, 7.6% in grammar, and 5.7% in English-literacy performance compared with children whose mothers were not iodine deficient. The authors conclude that even mild iodine deficiency during pregnancy can have long-term adverse impacts on fetal brain development that will affect cognition and are not improved by iodine sufficiency during childhood.

Source: Hynes KL and others: Mild iodine deficiency during pregnancy is associated with reduced educational outcomes in the offspring: 9-Year follow-up of the gestational iodine cohort. *Journal of Clinical Endocrinology and Metabolism* 98(5):1954, 2013.

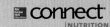

Check out Connect **www.mcgrawhillconnect.com** to further explore the effects of maternal gestational nutrient deficiencies on long-term health outcomes.

weakness, appear with further depletion. Finally, the full-blown symptoms of the deficiency are recognizable, such as when blindness accompanies a vitamin A deficiency.

When a few people in a population develop a severe deficiency, this may represent only the "tip of the iceberg." Typically, a much greater number have milder degrees of undernutrition. These deficiencies should not, therefore, be dismissed as trivial, especially in the developing world. In many low- and even middle-income countries, it is typical for a combination of micronutrient deficiencies to occur. These are caused by any number of factors, including diets of poor nutritional quality related to seasonal variation in food availability, low bioavailability of nutrients from plant sources, cultural food practices, and poverty. It is becoming clear that combined deficiencies of specific vitamins and the minerals iron and zinc can seriously reduce work performance, even when they do not cause obvious physical symptoms. This resulting state of ill health, in turn, diminishes the ability of individuals, communities, and even whole countries to perform at peak levels of physical and mental capacity (Fig. 12-3). Because nutritional requirements are high during periods of rapid growth, pregnant women, infants, and children are especially vulnerable to the effects of undernutrition. Studies of pregnant women and children in developing countries have shown the negative impact of micronutrient deficiencies on birth size, length of gestation, growth, and intellectual development (see Newsworthy Nutrition).

Strategies to address the multiple nutrient deficiencies in developing countries have included supplementation and fortification of ready-to-use foods. While supplementation has been the most widely practiced intervention, public health experts believe that fortification of a commonly consumed food could be a single, cost-effective intervention strategy that would target a larger population. Studies of the use of micronutrient supplementation during pregnancy have shown an increase in birth weight and a reduction in low birth weights but no impact on preterm births or perinatal mortality. In children, micronutrient supplementation with three or more micronutrients has resulted in increases in height and weight (see Further Reading 4).

Added to their lack of nourishment, the inhabitants of poorer countries must also contend with recurrent infections, poor sanitation, extreme weather conditions,

Historical Research on Undernutrition

In the 1940s, a group of researchers led by Dr. Ansel Keys examined the general effects of undernutrition on adults. Previously healthy men were fed a diet averaging about 1800 kcal daily for 6 months. During this time, the men lost an average of 24% of their body weight. After about 3 months, the participants complained of fatigue, muscle soreness, irritability, intolerance to cold, and hunger pains. They exhibited a lack of ambition, self-discipline, and concentration, and were often moody, apathetic, and depressed. Their heart rate and muscle tone decreased and they developed edema. When the men were permitted to eat normally again, feelings of recurrent hunger and fatigue persisted even after 12 weeks of rehabilitation. Full recovery required about 8 months. This study helps us understand the general state of undernourished adults worldwide.

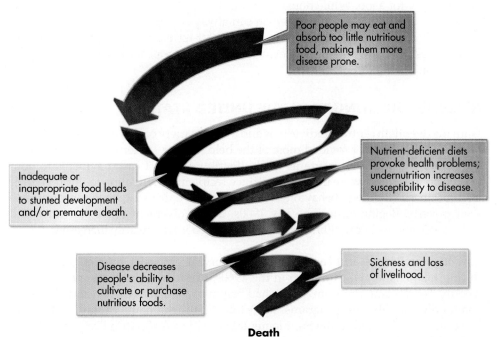

Poor people may eat and absorb too little nutritious food, making them more disease prone.

Nutrient-deficient diets provoke health problems; undernutrition increases susceptibility to disease.

Inadequate or inappropriate food leads to stunted development and/or premature death.

Sickness and loss of livelihood.

Disease decreases people's ability to cultivate or purchase nutritious foods.

Death

FIGURE 12-3 ◄ The downward spiral of poverty and illness can ultimately end in death (based on World Food Program graphic).

and regular exposure to infectious diseases. They require greater amounts of certain nutrients—especially iron—to combat rampant parasite and other infections. Deficiencies of both iron and zinc can lead to reduced immune function and thereby increase the risk of diseases, such as diarrhea and pneumonia.

✔ CONCEPT CHECK 12.1

1. What are the characteristics of hunger?
2. What is the definition of food security?
3. Define the term nutrition security.
4. What are the results of chronic hunger?
5. What is the primary cause of undernutrition?
6. At which stages of the life cycle is undernutrition especially damaging?
7. What are the effects of chronic undernutrition?

12.2 Undernutrition in the United States

In 2013, 45.3 million (14.5%) people in the United States were living at or below the poverty level, estimated at about $23,830 annually for a family of four (see Further Reading 7). After steady increases between 2006 and 2010, the fact that the overall percent of Americans living below the poverty line held steady in 2013 is good news. For children under age 18, the poverty rate declined from 21.8% in 2012 to 19.9% in 2013. Between 2012 and 2013, the number of people in poverty held at 9.5% for people aged 65 and older. In 2008, the percentage of the population living in food-insecure households in Canada is lower (9.4%) than in the United States (see Further Reading 14). In 2013 the national poverty rate was 27.2% for Blacks, 23.5% for Hispanics, 10.5% for Asians, and 9.6% for whites.

The poor often face difficult choices: whether to buy groceries for the family or pay this month's rent; whether to have dental work done or pay the current utility bill; whether to replace clothes the children have outgrown or pay for transportation to apply for a job. While housing and utility costs, medical care, and transportation fares are nonnegotiable, a person can always eat less. Food is one of the few flexible items in a poor person's budget. The short-term consequences of eating less may be less dramatic than getting evicted, but the long-term cumulative effects are significant.

HELPING THE HUNGRY IN THE UNITED STATES

Until the twentieth century, individuals and a wide variety of charitable, often church-related organizations, provided most of the help to poor, undernourished people in the United States. Early programs rarely distributed direct cash payments to poor people because these were thought to reduce recipients' motivation to improve their circumstances or change behaviors, such as excessive drinking, that contributed to their poverty. Beginning in the early 1900s, the involvement of local, county, and state governments in providing assistance to the poor has steadily increased.

Low-Income People and Families. After observing extensive hunger and poverty during his presidential campaign in the 1960s, President John F. Kennedy revitalized the Food Stamp Program, which had begun two decades earlier, and expanded commodity distribution programs. In October 2008, the name of the Food Stamp Program was changed to the Supplemental Nutrition Assistance Program (SNAP) (Table 12-3). The new name reflects the program's focus on nutrition and putting

Hispanics were the only major race and ethnic group to have a statistically significant decline in poverty rate (fell from 25.6% to 23.5%) and number of people in poverty (fell from 13.6 million to 12.7 million).

▲ This man is hoping to find work to fill his basic need of food. There are many federal food assistance programs for those in need.

TABLE 12-3 ▶ **Some Federally Subsidized Programs That Supply Food for People in the United States**

Program	Eligibility	Description
Supplemental Nutrition Assistance Program (formerly Food Stamp Program)*	Low-income families	Electronic benefit transfer (debit) cards are given to purchase food at grocery stores; the amount is based on size of household and income.
The Emergency Food Assistance Program (TEFAP)*	Low-income families	Provides nutrition assistance to needy Americans through distribution of USDA food commodities.
Commodity Supplemental Food Program	Certain low-income populations, such as pregnant women, children up to the age of 6 years, and seniors	USDA surplus foods are distributed by county agencies; not found in all states; may be based on nutritional risk.
Special Supplemental Nutrition Program for Women, Infants, and Children (WIC)*	Low-income pregnant/lactating women, infants, and children less than 5 years old at nutritional risk	Coupons are given to purchase milk, cheese, fruit juice, cereal, infant formula, and other specific food items at grocery stores; includes nutrition education component. Includes new Farmers' Market Nutrition Program (FMNP).
National School Lunch Program*	Low-income children of school age	Free or reduced-price lunch is distributed by the school; meal follows USDA pattern based on MyPlate; cost for the child depends on family income. For students who do not participate in the lunch program, special milk program may be available.
School Breakfast Program	Low-income children of school age	Free or reduced-price breakfast is distributed by the school; meal follows USDA pattern; and cost for the child depends on family income.
Child and Adult Care Food Program	Children enrolled in organized child-care programs and seniors in adult-care programs; income guidelines are the same as those for the National School Lunch Program	Reimbursement is given for meals supplied to children at the site; meals must follow USDA guidelines based on MyPlate.
Congregate Meals for the Elderly	Age 60 or over (no income guidelines)	Free noon meal is furnished at a site; meal follows specific pattern based on one-third of nutrient needs.
Home-Delivered Meals	Age 60 or over, homebound	Noon meal is delivered at no cost or for a donation at least 5 days a week. Sometimes additional meals for later consumption are delivered at the same time; often referred to as "Meals on Wheels."
Summer Food Service Program	Residence in a low-income neighborhood or participation in a program	Free, nutritious meals and snacks are given to children in a low-income area at a central site, such as a school or a community center during long school vacations.
Food Distribution Program on Indian Reservations*	Low-income American Indian and non-Indian households on reservations; members of federally recognized tribes	Alternative to Supplemental Nutrition Assistance Program, distributes monthly food packages; includes nutrition education component.
Fresh Fruit and Vegetable Program	Low-income elementary schools	Provides free fresh fruits and vegetables to increase their consumption and combat childhood obesity.

*Benefits increased through the American Recovery and Reinvestment Act of 2009.

healthy food within reach for low-income households. SNAP helps low-income people and families buy food they need for good health. The program allows recipients to use an Electronic Benefit Transfer (EBT) card to purchase food and garden seeds—but not tobacco, cleaning items, alcoholic beverages, and nonedible products—at stores authorized to accept them. In 2014, the maximum dollar amount of SNAP benefits a household of four can receive each month was $632, depending on their income and expenses. A record high of 46.7 million Americans—about one in five adults—participated in this program in June 2012. This was a spike of 51% since the start of the economic crisis in October 2008. The number of Americans using SNAP is estimated to be about the same as the number living in poverty.

School Breakfast and Lunch Programs. The U.S. Congress established the School Breakfast Program in 1965 as politicians became aware of the number of hungry children coming to school. School breakfast and lunch programs still enable low-income students—12.8 million for breakfast and 31.6 million for lunch in 2012—to receive meals free or at reduced cost if certain income guidelines are met (under $29,965 for free and $42,643 for reduced cost for annual income of a family of four).

Elderly Nutrition Services. The Older Americans Act authorizes the elderly nutrition services program to provide funds ($797 million in 2014) for congregate (group noon-time meals) and home-delivered meals for all citizens aged 60 and older. These meals are available regardless of income, but donations are requested. The program not only addresses problems of food insecurity through nutrition and nutrition-related services but also promotes socialization and the health and well-being of older persons. Approximately 228 million meals were served to about 2.5 million people in 2011, the most recent year for which data are available; 61% were served to older people living at home and 39% were served in congregate settings. Both remain active programs, serving about 1 million meals each day, but they still do not reach all who need help. The number of home-delivered meals served decreased from 2008 to 2011, whereas the number of congregate meals served increased. The increase in the number of congregate meals stemmed from an additional $100 million in Recovery Act funding that was awarded in 2009 to support senior nutrition programs. This allowed an approximate 14 million additional meals to be provided through community senior nutrition programs.

Women, Infants, and Children. In 1972, the Special Supplemental Nutrition Program for Women, Infants, and Children (WIC) was authorized. This program provides food vouchers and nutrition education to low-income, nutritionally at-risk pregnant and lactating women and their young children. It served 8.6 million women and children each month in 2013. Read more about child and adolescent nutrition assistance programs in Further Reading 3.

Sometimes, severe undernutrition due to involuntary hunger does occur in the United States. More often, though, Americans experience periodic episodes of hunger and food insecurity. Unemployment, medical and housing expenses, and even occasional holiday shopping can cause a household to be hungry or food insecure.

Government food assistance programs are like a "safety net": they are strong, yet porous. The Recovery Act, also known as the "stimulus" or the "stimulus package," that was signed by President Barack Obama in February 2009 was an unprecedented effort to increase a variety of federal benefits and services, including those offered by the federal food and nutrition programs (Table 12-3). The enhancement of these programs helped those workers and families hardest hit by the economic crisis. Privately funded programs have also stepped in to add to state and federal efforts to combat hunger and related food insecurity in the United States. There are more than 150,000 charitable food providers (such as food banks and food pantries) helping to cope with this problem. Many low-income U.S. households rely on food pantries. A USDA survey of food security found that 5.1 percent of all U.S. households (6.1 million) accessed emergency food from a food pantry one or more times in 2011. Statistics also show that most people requesting emergency food assistance are members of families—children and their parents. This is not surprising considering that households with children report a higher rate of food insecurity than households without

▲ Food insecurity is part of the North American landscape. A "safety net" of programs exists, but it is "porous."

children, and the rate is even higher for households with children headed by single parents.

Undernutrition in North America is a much more subtle problem than in developing countries. To the untrained eye, undernourished children may just seem skinny, when, in fact, their growth is being stunted by insufficient nutrients. More likely, though, children from food-insecure households are prone to be overweight. This may be the result of considerable reliance on convenience foods that provide mostly fat and sugar. The availability of cooking facilities also affects nutrient intake among the poor. Without cooking facilities, people may buy expensive convenience foods that require no preparation. These are typically highly processed snack foods, which provide calories but are often lacking in nutrients.

▲ Food pantries and soup kitchens are important sources of nutrients for a growing number of people in the United States. Consider volunteering some of your time to a local program.

SOCIOECONOMIC FACTORS RELATED TO UNDERNUTRITION

In the United States, persistent hunger and food insecurity are largely associated with two interrelated conditions: poverty and homelessness. Thus, the economic, social, and political changes that lead to an increase in the number of poor or homeless people also tend to intensify the problem of undernutrition.

Poverty. Poverty is determined in the United States using poverty thresholds that are issued each year by the Census Bureau. Persons are designated as poor if they have income less than that deemed sufficient to purchase the basic needs of food, shelter, clothing, and other essentials. Poverty, however, is complex and does not mean the same thing to all people. Poverty can usually be characterized as *situational* or *generational*. Households can fall into situational poverty as the result of dire or unexpected circumstances, such as the head of the household getting laid off from a job or a family member needing expensive medical treatment. Some have linked situational poverty to the seven D's: divorce, death, disease, downsized, disabled, disasters, and debt. Situational poverty may affect entire communities as a result of economic downturns and may be alleviated by temporary social service programs. In contrast, generational poverty refers to a culture of persistent poverty passed from parents to children, with two or more generations living in poverty. Generational poverty is more complex because the cultures and traditions of previous generations are passed onto future generations and often include a lack of value placed on the benefits of education and striving for a better way of life.

A major driving force behind poverty is underemployment. This relationship increased significantly with the economic recession that began late in 2007 and caused unemployment to soar. Between December 2007 and September 2009, 7.6 million jobs were lost. These job losses occurred across all major private-sector industries, including manufacturing, retail, leisure and hospitality, financial, transportation, and warehousing. The impact of unemployment has been enhanced because, during the economic downturn, many states cut their welfare assistance. This combination of events has resulted in more people living in poverty and in need of food assistance, at risk of homelessness, and without sufficient health care. Fortunately, the economy has more recently experienced some recovery. After cresting at 10% in October 2009, the unemployment rate has slowly declined as the labor market has grown slowly and steadily. In September 2013, the number of unemployed persons was 11.3 million, resulting in an unemployment rate of 7.2%.

Access to Healthy Food. Access to affordable and nutritious foods from supermarkets, grocery stores or other retailers is also a challenge to many Americans, making it harder for them to eat a healthy diet (see Further Reading 6). In 2011, 23.5 million Americans, including 6.5 million children, lived in low-income areas lacking stores likely to sell affordable healthy foods such as fresh fruits and vegetables.

food desert An area where 33% or 500 people, whichever is less, live more than a mile from a grocery store in an urban area or more than 10 miles away in a rural area.

These impoverished areas with little access to healthy foods have been named "**food deserts**." The USDA originally defined a food desert as a low-income area where a significant number of residents (33% or 500 people, whichever was less), lived far from a supermarket. "Far" was more than a mile in an urban area or more than 10 miles in a rural area. In a new tool called the Food Access Research Atlas, low-income census tracts where a substantial number or share of people are far from supermarkets can be viewed and mapped. New measures include alternative distance markers with ½-mile and 1-mile demarcations to the nearest supermarket used for urban areas and 10-mile and 20-mile demarcations used for rural areas. Household vehicle availability is also measured since access to a vehicle is an important factor for food access (see Further Reading 9).

First Lady Michelle Obama has made the availability of nutritious and affordable food from grocery stores, small retailers, corner markets, and farmers markets in communities with limited access an important part of the Healthy Communities aspect of her Let's Move! initiative (**www.letsmove.gov/**). In 2011, several national retail stores, including Wal-Mart, Walgreens, and SuperValu, as well as some regional companies, agreed to bring more nutritious and fresh food to underserved communities by opening or expanding more than 1500 stores. These changes were aimed at serving about 9.5 million people and creating tens of thousands of jobs.

Homelessness. The growing shortage of affordable rental housing and a simultaneous increase in poverty are the two trends largely responsible for the rise in homelessness over the past 25 years. The more recent foreclosure and economic crises have added to this situation and created a significant increase in homelessness and the number of families at risk of homelessness across the country. In 2009, data from major cities indicated increases of homelessness by as much as 20% since the foreclosure crisis began in 2007. The homelessness created by the gap between the number of affordable housing units and the number of people needing them impacts the health and well-being of all individuals. Homeless children are most vulnerable and are twice as likely to experience hunger compared to children with a home. Data collected by the U.S. Department of Housing and Urban Development (HUD) in January 2012 revealed that 633,782 people in the United States experience homelessness on any given night. People in families account for 239,403 of homeless, and 394,379 are individuals. Single mothers in their late 20s with approximately two children are the most common heads of homeless families. Families most commonly become homeless as a result of an unexpected financial crisis that prevents them from retaining their housing. Events similar to those that cause situational poverty—a medical emergency, death in the family, loss of a job—can lead to homelessness. Poor people must make difficult choices when limited resources cover only some necessities such as housing, food, child care, health care, and education. Because housing absorbs a high percentage of income, it is often given up. The National Alliance to End Homelessness (**www.endhomelessness.org/**) indicates that most homeless families are able to quickly rebound from homelessness, requiring only short-term public assistance before returning to independence and stability.

As the recession gripped America from 2007 to 2010, the number of people using shelters or transitional housing in suburban and rural areas increased 57 percent while at the same time the use of shelters in urban areas decreased. Regrettably, veterans make up about one-third of the homeless population, often due to war-related disabilities that affect their ability to work.

Although the "chronically homeless" are often the most common image of homelessness, they account for slightly less than 16% of the homeless population. Based on the survey taken every two years, on a single night in 2012, there were 99,894 adults in America experiencing chronic homelessness. Chronic homelessness is defined as involving either long-term and/or repeated occurrences of

▲ Three-quarters, or about 75,000, of the chronically homeless are men with the average age approaching 50.

homelessness combined with physical and/or mental disability. The chronically homeless are most likely to live in shelters and consume the majority of the homeless assistance resources. Three-quarters, or about 75,000, of this population are men with the average age approaching 50. Fortunately, progress has been made in addressing chronic homelessness in the last decade, resulting in a 19.3% decline in this population since 2007.

POSSIBLE SOLUTIONS TO POVERTY AND HUNGER IN THE UNITED STATES

For many years, government-funded food assistance programs have helped to alleviate some problems of undernutrition in the United States (see Further Reading 2).

It is documented that an increasing number of people who are experiencing poverty are accessing the Supplemental Nutrition Assistance Program benefits and other federal programs. As a result of the American Recovery and Reinvestment Act of 2009, more federal funds were infused into assistance programs, but these additional funds have recently begun to expire.

Obtaining food to survive is a tremendous challenge, especially when homeless people are forced to live outside. Although many people assume that food pantries and soup kitchens are abundant and accessible for every needy person, there are many obstacles to these resources. Food pantries are often ineffective because they can give only one box of food to each family per month, which falls short of meeting their needs. Furthermore, homeless people lack the cooking facilities necessary to prepare the food. Food availability through soup kitchens is also limited in many cities. An additional challenge is that cities, including Houston, New York City, and Philadelphia, have used ordinances and policies to discourage or prohibit the sharing of food with poor or homeless persons by individuals and groups. These laws are troubling, considering that most cities do not have adequate food resources to meet the needs of their poor and homeless. Advocacy groups have raised questions about individual rights and freedoms and have challenged these practices in court cases. The American Civil Liberties Union of Pennsylvania filed a lawsuit against the city of Philadelphia's ban. The case resulted in a federal judge issuing an injunction that prevents Philadelphia from enforcing the ban and ordered the city to remove any signage related to the ban. The success in Philadelphia is just one example. As a result of the backlash, many cities, including Chicago and Seattle, have scrapped their bans against feeding the homeless. Advocates and food providers appear eager to work with cities and other government agencies to help address the problems of hunger and homelessness by improving access to federal food benefits and other food resources. State governments have also begun moving to protect the right of individuals and groups to share food with others. New Jersey recently passed a law that strengthens the state's Good Samaritan Law by allowing "universities to donate food to charitable organizations without fear of lawsuit[s]."

Despite even the highest motivation, the outlook is bleak for many people who attempt to gain independence from assistance programs. Teen pregnancy may have cut short the education or vocational training of one or both parents, thwarting efforts to earn adequate income. Often, the expense of reliable and safe child care far exceeds the meager income from a minimum-wage job. Illness of either the parents or children may prevent adults from holding steady employment. Poor communication skills, inability to relocate, and a lack of economic reserves also complicate financial independence. Regardless of how wasteful government assistance appears to some people, it will probably always be necessary to some extent.

▲ The Healthy Food Financing Initiative was established to bring grocery stores, corner stores, and farmers markets to underserved communities in America so that children like this girl and their families can have access to healthy foods.

Two objectives of *Healthy People 2020* are to eliminate very low food security among children from 1.3% of households in 2008 to 0.2% in 2020 and to reduce hunger by reducing food insecurity from 14.6% of households in 2008 to 6% in 2020.

Looking beyond food assistance programs, access to healthy food has become a recent target in the efforts to eradicate hunger in the United States. Of the 23.5 million people living in low-income areas that are more than a mile from a supermarket, 11.5 million are in households with incomes at or below 200% of the poverty line. In low-income rural areas, 1.1 million low-income people live more than 10 miles from a supermarket. A new government program has been designed to improve access to healthy food. In February 2010, the Healthy Food Financing Initiative (HFFI) was launched by the Obama Administration to bring grocery stores, small retailers, corner stores, and farmers markets to underserved urban and rural communities across America. This new initiative is targeted toward achieving long-term progress and the eventual relief of poverty and hunger.

Through these recent efforts, lack of access to healthy food is now recognized as a solvable problem. The HFFI is a partnership between the U.S. Departments of Treasury, Agriculture, and Health and Human Services. The multiyear HFFI engages the private sector and aims to eliminate food deserts across the country within 7 years. HFFI attracts investment in underserved communities by providing financing tools such as low-cost loans, grants, tax credits, and technical assistance to healthy food retailers. From 2010 through 2012, over $100 million has been allocated for HFFI, and the president's 2013 budget requested another $285 million. The HFFI investments have not only expanded healthy food options into food deserts but also provided business development opportunities and created thousands of jobs in urban and rural low-income communities.

The HFFI appears to be an effective and sustainable solution to the limited access to healthy foods. Its impact can be far-reaching in terms of improving the nutritional status and overall health of families and children and reducing health disparities while at the same time creating jobs and stimulating the local economy in low-income areas. Start-up funds are helping healthy food retailers overcome the high initial costs of entering underserved communities. Grants are also used for the needed renovation and expansion of existing stores looking to provide the healthy foods their communities need. Farmers markets are also being encouraged as an innovative yet simple approach to improving access to fresh produce for those with limited or fixed incomes. Farmers markets also participate in the senior benefits program, WIC, SNAP, and the SNAP Double Dollar program, where shoppers using their SNAP EBT card will receive matching funds for SNAP purchases at that market.

Finally, several U.S. legislators have sponsored a variety of bills such as the "Let's Grow Act" that would expand the use of SNAP benefits to purchase produce at farmers markets, encourage food production on vacant land, and connect farmers with preschool programs so that young children would have access to a healthy meal. Another bill, the Local Food, Farms, and Jobs Act of 2013, encourages sustainable agriculture and the growth of local food markets. This type of legislation is aimed at counteracting recent cuts to SNAP and other programs that promote health and create opportunities for local and minority farmers.

CONCEPT CHECK 12.2

1. What are the primary food assistance programs established in the United States during the 1960s?
2. What factors influence the presence of poverty, homelessness, and undernutrition?
3. What programs have been launched recently to provide long-term solutions to hunger?

12.3 Undernutrition in the Developing World

Undernutrition in the developing world is also tied to poverty, and any true solution must address this problem. However, these countries have a multitude of problems so complex and interrelated that they cannot be treated separately. Programs that have proved helpful in the United States (and throughout the rest of North America) are only a starting point in this context. The major obstacles challenging those seeking a solution are illustrated in Figure 12-4 and deserve individual consideration.

▲ Poverty aggravates the problem of hunger in the developing world

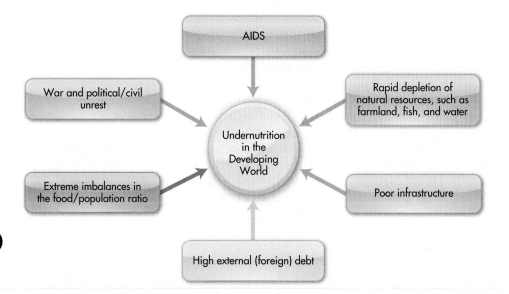

FIGURE 12-4 ▲ Many factors contribute to undernutrition in the developing world. Any solutions to the problem must take these factors into consideration.

FOOD/POPULATION RATIO

The world has just over 7 billion inhabitants. Population growth exceeds economic growth in much of the developing world, and as a result, poverty is increasing. This disrupts the balance in the food/population ratio, tipping it toward food shortages. If we want to ensure a decent life for a widening segment of humanity, many experts suggest that the growth in the earth's most vulnerable populations should slow. If not, by 2050, the world may have 1 to 3 billion more people than it does today—most of them in countries where the average person earns less than $2 per day. In developing countries, nearly 1.3 billion people live on less than $1.25 per day. Unless a catastrophe occurs, more than 9 of 10 infants in the next generation will be born in the poorest parts of the world.

More than three-quarters of people in the world live in developing countries, and more than half live in Asia. A United Nations report on worldwide hunger revealed that almost two-thirds of the world's undernourished live in Asia and the Pacific Rim. The world's food supplies also are not distributed equally among consumers. Gross disparities exist between developed and developing countries, among the rich and the poor within countries, and even within families (i.e., males may be fed before females).

Still, economists estimate that world food production will continue to increase more rapidly than the world population in the near future, allowing the food/

infrastructure The basic framework of a system of organization. For a society, this includes roads, bridges, telephones, and other basic technologies.

population ratio to increase through the year 2020. This will come at a high cost, however, in terms of the water, fertilizer, and pesticides needed to allow for this production. Overall, in the short run, the primary problem appears not to be food production but distribution and use, especially in poverty-stricken areas of developing nations. A major factor is that many people in the world do not have enough income to purchase land to grow enough food.

Eventually, food production will begin to lag behind population growth. Most good farmland in the world is already in use, and because of poor farming practices or competing land-use demands, the number of farmable acres worldwide decreases annually. For many reasons, *sustainable* world food output—an amount that does not deplete the earth's resources—is now running well behind food consumption. This discrepancy suggests that food production in less-developed countries will barely keep up with population growth and will soon lag behind.

▲ This young Maasai woman, with her baby on her back, lives in Kenya, which is in a region with low contraceptive rates and the highest rates of maternal death.

Birth control programs, an obvious brake on population expansion, have been effective in developed countries but relatively ineffective in many developing countries that could really benefit from them. Family planning, if available to all who wanted it, can contribute to the reduction of poverty *and* hunger, and *decrease* the number of maternal and childhood deaths. However, women (and men) in many developing countries still lack adequate access to contraceptives. Organizations such as Population Services International (**www.psi.org**) are trying to keep distribution costs low and make the products available to as many people as possible by subsidizing condoms and oral contraceptives to areas such as Bangladesh. Reports from the United Nations and the World Bank suggest that between 1990 and 2010, family planning contributed to halving the number of maternal deaths worldwide. East Asia, with a contraceptive prevalence rate of 84%, has made the greatest progress in preventing maternal deaths, whereas sub-Saharan Africa, with a contraceptive prevalence rate of 22%, has the highest rates of maternal death.

Promoting breastfeeding also contributes to the goal of birth control. Although it is not a completely reliable method of contraception, exclusively breastfeeding an infant delays ovulation after childbirth, thereby lowering the likelihood of fertilization, for an average of 6 months. (Women who do not breastfeed generally begin to ovulate within a month or so after giving birth.) When childbirths are more widely spaced, not only do fewer total births occur, but the mother has a longer chance to recover from pregnancy, and the infant receives feeding priority for a longer time. One possible exception to the healthful nature of breastfeeding occurs, however, when mothers are infected with the **human immunodeficiency virus (HIV).** The risk of transferring the virus through human milk is about 10%. Depending on the circumstances, this may outweigh the benefits of breastfeeding.

▲ Women in developing countries are forced to spend large parts of their day fetching water.

human immunodeficiency virus (HIV) The virus that leads to **acquired immune deficiency syndrome (AIDS).**

acquired immune deficiency syndrome (AIDS) A disorder in which a virus (human immunodeficiency virus [HIV]) infects specific types of immune system cells. This leaves the person with reduced immune function and, in turn, defenseless against numerous infectious agents; typically contributes to the person's death.

Experience with family planning programs in developing countries and historical changes in birth rates in many developed countries suggest an important conclusion. Generally, only when people have enough to eat and are financially secure do they feel confident that having fewer children will still result in enough surviving children to provide for their care in later years. Increasing per capita income and improving education, especially for women in developing nations, are considered to be the most likely long-term solutions to excessive population growth. This effort has led to a decline in family size in Brazil, Egypt, India, and Mexico. A major concern is whether there are enough resources worldwide to raise per capita income and provide enough education to slow population growth.

WAR AND POLITICAL/CIVIL UNREST

The Millennium Summit of the United Nations in September 2000 resulted in the Millennium Declaration that included the resolution to "spare no effort to free our peoples from the scourge of war." Against that background stands the reality that worldwide military spending was reported to be over $1.75 trillion in annual expenditure in 2013 and has increased 50% since 2001. The entire budget of the United Nations is only a fraction of the world's military expenditure—approximately 1.8%.

Aside from the economic impact of military spending, civil disruptions and wars are setting back the progress of the poor and contributing to massive undernutrition. Many regions of Africa and Asia have been involved in numerous civil wars and conflicts in recent years. These include internal conflicts in Burma, Afghanistan, and Sudan, and civil wars in Somalia and Syria. These conflicts have resulted in millions of deaths as well as millions of refugees and internally displaced people. While war has raged, health, education, and public services have declined for African people, and poverty has increased in sub-Saharan Africa. War-related famine affects millions of people in southern and northeastern Africa. In 2011, the worst drought in 60 years devastated communities throughout the Horn of Africa. Famine, war, and drought are the three major crises that have led to devastation in the Horn of Africa, leaving more than 13 million people in a state of crisis. The drought led to an outbreak of famine in Somalia, where 20 years of war and violence have limited humanitarian efforts and destroyed the country's ability to respond. Millions suffered from the effects of the famine, war, and drought. Many walked more than 100 miles to reach refugee camps in Kenya and Ethiopia. Overall, most people in war-torn areas are without sufficient shelter, clothing, food, or means of obtaining them.

Even when food is available, political divisions may impede its distribution to the point that undernutrition will plague many people for years. Especially during emergencies, programs designed to help the poor have been undermined by unstable administration, corruption, and political influence. During such political chaos, relief agencies are often caught between warring factions and those they are trying to help. This has been the case in Sudan where nongovernmental relief organizations, such as the Children's Hunger Relief Fund, were expelled in 2009. This caused deterioration in the Darfur region, and hundreds of thousands of refugees were left without the help they depended on from relief organizations.

During the 1960s and 1970s, the problem of undernutrition in developing countries was perceived as a technical one: how to produce enough food for the growing world population. The problem is now seen as largely political: how to achieve cooperation among and within nations, so that gains in food production and infrastructure are not wiped out by war. The best answer lies in a combination of approaches: finding technical solutions to help with the problems of chronic hunger and poverty, and resolving political crises that have pushed developing nations into a state of acute hunger and chaos.

AGRICULTURE AND THE RAPID DEPLETION OF NATURAL RESOURCES

As we quickly deplete the earth's resources, population control grows increasingly critical. Agriculture production is approaching its limits in many areas worldwide. Environmentally unsustainable farming methods have been undermining food production, especially in developing countries.

The **green revolution** was a phenomenon that began in the 1960s when crop yields rose dramatically in some countries, such as the Philippines, India, and Mexico (countries in Africa did not benefit because climates were not compatible with the crops used). The increased use of fertilizers, irrigation, and the development of superior crops through careful plant breeding made this boost in agricultural production possible. The green revolution was intended as a stopgap measure until world leaders could control population growth. Many of the technologies associated with the green revolution have now achieved their potential. Additional gains in productivity have been a challenge to accomplish because of the lack or loss of productive farmland. The movement toward sustainable agricultural practices and

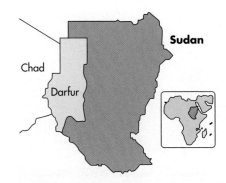

▲ The conflicts in the Darfur region of Sudan have led to high rates of death and undernutrition among people displaced by war.

▲ This mother, holding her 27-month-old malnourished child at a camp for internally displaced persons in North Darfur, Sudan, is dependent on assistance from relief organizations that are often not welcome in the war-torn country.

green revolution This refers to increases in crop yields that accompanied the introduction of new agricultural technologies in less-developed countries, beginning in the 1960s. The key technologies were high-yielding, disease-resistant strains of rice, wheat, and corn; greater use of fertilizer and water; and improved cultivation practices.

▲ Rich agricultural resources in North America, seen in the wheat fields above, in contrast to the gardens being tended by farmer on a floating island in Myanmar (formerly Burma).

the development of new crops through food biotechnology have begun to improve the crop yields on the shrinking amount of farmable land. Crops that are pest or chemical resistant or biofortified have been produced as a result of biotechnology. Controversy still surrounds the use of this technology, and more research is warranted on the effects of these crops on human health and the environment (see Section 12.4 on use of biotechnology).

Areas of the world that remain uncultivated or ungrazed are mostly too rocky, steep, infertile, dry, wet, or inaccessible to sustain farming. With agriculture being the largest user of freshwater resources, nearly all irrigation water available worldwide is being used, and groundwater supplies are becoming depleted at rapid rates in many regions. An eventual water shortage is projected to increase war and civil unrest in arid areas of the world, such as Northern Africa and the Middle East. China, which has more than 20% of the world's irrigated land, is also plagued with a growing scarcity of fresh water. In the future, billions of people will face ongoing water shortages.

The prospects of obtaining substantially more food from the oceans are also poor. In recent years, the amount of fish caught worldwide has leveled off. Fish was once considered the poor person's protein, but this is not likely to continue because farming of fish does not come close to compensating for the degree of reduction in wild fish populations.

Clearly, we can exploit the earth's resources only so far: the world population probably cannot continue to expand as it does today without the potential for serious famine and death. The importance of adequate resources for agriculture is suggested in one of the key messages from the Food and Agriculture Organization *State of Food Insecurity In the World 2014* report: "Hunger reduction requires an integrated approach, which would include: public and private investments to raise agricultural productivity; better access to inputs, land, services, technologies and markets; measures to promote rural development; social protection for the most vulnerable, including strengthening their resilience to conflicts and natural disasters; and specific nutrition programmes, especially to address micronutrient deficiencies in mothers and children under five." Thus, if food production is to keep up with the expanding population, immediate action is needed to protect the earth's already deteriorated environment from further destruction (see Further Reading 10). Agricultural systems should sustain the environment while providing adequate, healthy food.

INADEQUATE SHELTER AND SANITATION

When people die from undernutrition in developing countries, other factors, such as inadequate shelter and sanitation, almost always contribute. None of us can survive very long without water, nor can we stay healthy for any length of time without sanitation or shelter. Regrettably, many more people suffer from the effects of poor sanitation and an unreliable water supply than are affected by war and political/civil unrest. Without water, sanitation, and hygiene (WASH), sustainable development is impossible.

The tremendous movement of people to urban settings has caused a population redistribution that has challenged the capacity for shelter and sanitation in the urban setting. Between 1995 and 2005, urban populations of developing countries increased by an average of 1.2 million people per week (around 170,000 people every day). In developing countries, the poor make up most of the urban population, and their needs for housing and community services often go beyond available governmental resources. It is expected that in the next 20 years, 9 of the world's 10 largest cities will be in poor countries, and in the next 30 years, all urban population growth will be in cities of developing countries, reaching almost 5.2 billion in 2050. Such a skewed population distribution will result in more poverty. The urban explosion is the result of both high birth rates and continuing migration of people to the cities from rural areas. People go to the cities to find employment and resources the countryside can no longer provide. Pollution is another result of the increased urbanization. The World Bank reports that 16 of the 20 most polluted cities in the world are in China.

Most of these urban poor live in overcrowded, self-made shelters, which lack a safe and adequate water supply and are only partially served by public utilities. The WHO/UNICEF Joint Monitoring Programme for Water Supply and Sanitation (JMP) has estimated that as of 2013, 2.5 billion people—36% percent of the world's population—lack improved sanitation facilities, and 768 million people still use unsafe drinking water sources. The shantytowns and ghettos of the developing world are often worse than the rural areas the people left behind. The urban poor need cash to purchase food, so they often subsist on diets even more meager than the homegrown rural fare. Making matters worse, haphazard shelters often lack facilities to protect food from spoilage or damage by insects and rodents. This inability to protect food supplies in some developing countries leads to the loss of as much as 40% of all perishable foods.

The shift from rural to urban life takes its greatest toll on infants and children. WHO/UNICEF reports that inadequate access to safe water and sanitation services, combined with poor hygiene practices, kills and sickens thousands of children every day. For example, many of the 6.9 million deaths of children under 5 years old that occurred in developing countries in 2012 could have been prevented by improving the standards of environmental hygiene. Contributing to this issue is the fact that infants are often weaned early from breast milk to infant formula, partly because the mother must find employment. Mothers may also be influenced by advertisements depicting images of sophisticated, formula-feeding women. Unfortunately, because infant formulas are relatively expensive, poor parents may try to conserve the formula by either overdiluting the mixture or using too little to meet the infant's needs. The water supply may not be safe, so the prepared formula is also likely to be contaminated with bacteria. Human milk, in contrast, is much more hygienic, readily available, and nutritious. It also provides infants with immunity to some ailments. Promoting breastfeeding is important when it is safe for the infant.

Poor sanitation also creates a critical public health problem and, along with undernutrition, particularly raises the risk of infection. Inadequate sanitation is another example of the inferior infrastructure in the developing world. Human urine and feces are two of the most dangerous substances encountered in routine daily living. In addition, rotting garbage and associated insect and rodent infestations are potent sources of disease-causing organisms commonly seen in urban areas of the developing world. The inability to dispose of the massive numbers of dead people (and dead animals) resulting from civil wars causes additional sanitation problems. In some developing countries, diarrheal diseases account for as many as one-third of all deaths in children younger than 5 years of age. Additional repercussions of poor sanitation are that children are denied access to education because their schools lack private and decent sanitation facilities.

Fortunately, organizations such as UNICEF are working in more than 90 countries across the globe to improve water supplies and sanitation facilities in schools and communities and to promote safe hygiene practices. The UNICEF WASH programs are designed to support the Millennium Development Goal for water and sanitation, which is to halve, by 2015, the proportion of people without sustainable access to safe water and basic sanitation. Read more about the sanitation crisis in Further Reading 17.

THE IMPACT OF AIDS WORLDWIDE

Nutrition security is impacted greatly in developing counties by the high prevalence of HIV infection and AIDS. HIV/AIDS impairs absorption of nutrients, increases nutrient requirements, and decreases the capacity to work. About 35 million people around the world were known to be infected with HIV or had gone on to develop AIDS from the infection at the end of 2013. The epidemic varies considerably among

Blood loss caused by intestinal and blood-borne parasite infections is a common cause of anemia among poor populations, especially when people do not wear shoes. Parasites, such as hookworms, can easily penetrate the soles of the feet and legs and enter the bloodstream. Although hookworm disease has been largely eradicated in the United States and other industrialized nations through improved sanitation, it continues to plague more than one-eighth of the world's population, mostly in tropical regions.

▲ Inadequate sanitation facilities and the consumption of contaminated water cause the majority of all diseases. WHO/UNICEF estimated that as of 2013, 2.5 billion people lack improved sanitation facilities, and 768 million people still use unsafe drinking water sources.

Newsworthy Nutrition

Should HIV-positive women choose breastfeeding for their infants?

Exclusive breastfeeding is recommended for infants of HIV-positive women during the first months of life if available replacement feeding methods are not acceptable, feasible, affordable, sustainable, or safe for their circumstances. This study of South African HIV-positive women found that exclusive breastfeeding was successful when the women had a strong belief in the benefits of breast-feeding, a supportive home environment, when they could recall key messages on mother-to-child transmission risks and mixed feeding, and could resist pressure from the family to introduce other fluids.

Source: Doherty T and others: A longitudinal qualitative study of infant-feeding decision making and practices among HIV-positive women in South Africa. *Journal of Nutrition* 136:2421, 2006.

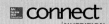 Check out the Connect site **www.mcgrawhillconnect.com** to further explore HIV and breastfeeding.

countries and regions. Sub-Saharan Africa continues to be most severely affected, with 24.7 million (nearly 5% of its adults) living with HIV, which accounts for 71% of the people living with HIV worldwide. After sub-Saharan Africa, Asia and the Pacific (4.8 million), and the Caribbean (250,000), were the regions of the world most heavily affected in 2013 (see Further Reading 12).

An individual can be infected with HIV through contact with bodily fluids including blood, semen, vaginal secretions, and human milk. Thus, the virus can be transmitted through sexual contact; through blood-to-blood contact; and from a mother to an infant during pregnancy, delivery, or breastfeeding. The virus has a very limited ability to exist outside the body. Read more about HIV and breastfeeding in Newsworthy Nutrition and Further Reading 13.

Once infected with HIV, the individual is said to be HIV-positive. If untreated, the viral disease progresses over the next few years, and the individual develops opportunistic infections with symptoms such as diarrhea, lung disease, weight loss, and a form of cancer. Once the individual has developed these symptoms, they are said to have AIDS. Without treatment, an individual will likely die from AIDS within 4 to 5 years.

Global efforts at preventing HIV infection have begun to have significant effects. The Millennium Development Goal 6 is to combat HIV/AIDS, malaria, and other diseases with a target to halt and begin to reverse the spread of HIV/AIDS by 2015. The HIV epidemic worldwide has stabilized, and the rate of new HIV infections has decreased in several countries (see Further Reading 12). It is encouraging that between 2011 and 2013, new HIV infections have decreased by 13%. Globally, the number of adults and children newly infected with HIV in 2013 (2.1 million) was 38% lower than in 2001 (3.4 million). The sharpest decline in the number of new HIV cases since 2001 has occurred in the Caribbean (42%). Sub-Saharan Africa also experienced a significant 33% drop in the number of people acquiring HIV infection between 2005 and 2013. Despite this progress, HIV infection in sub-Saharan Africa still accounted for 71% of all the new cases worldwide in 2013. This high infection rate and the fact that AIDS continues to reduce life expectancy in sub-Saharan Africa reinforces the need to enhance HIV prevention efforts in this region. Another region of concern is Indonesia, where the number of people living with HIV has increased by 48% and the country's

TABLE 12-4 ▶ Regional Statistics for Adults and Children Living with HIV for 2001 and 2011. Arrows Indicate Increases or Decreases Between 2001 and 2013.

Region		Adults and Children Living with HIV	Adults and Children Newly Infected with HIV
Sub-Saharan Africa	2013	24.7 million ↑	1.5 million ↓
	2001	20.9 million	2.4 million
Middle East and North Africa	2013	230,000 ↑	25,000 ↓
	2001	210,000	27,000
Asia and the Pacific	2013	4.8 million ↑	350,000 ↓
	2001	4.1 million	449,000
Latin America	2013	1.6 million ↑	94,000 ↑
	2001	1.2 million	93,000
Caribbean	2013	250,000 ↑	12,000 ↓
	2001	240,000	22,000
Eastern Europe and Central Asia	2013	1.1 million ↑	110,000 ↓
	2001	970,000	130,000
Western and Central Europe and North America	2013	2.3 million ↑	88,000 ↑
	2001	1.7 million	79,000
Global	2013	35.0 million ↑	2.1 million ↓
	2001	29.4 million	3.2 million

Source: UNAIDS *Report on the Global AIDS Epidemic,* 2014.

share of new HIV infections in the South and South-East Asia region reached 23% in 2013, second only to India.

AIDS leads to millions of deaths worldwide and creates orphans in many countries. Although its death toll has declined worldwide since 2001, AIDS killed 1.7 million adults and children in 2011. In 2009, 2.5 million children under 15 years of age were living with HIV/AIDS globally. Most of those children, almost 9 in 10, live in sub-Saharan Africa. It is estimated that more than 16 million children worldwide have been orphaned by AIDS, including 14.8 million in sub-Saharan Africa alone.

One of the specific targets of the UNAIDS 2011 Political Declaration on HIV/AIDS is to eliminate new infections among children and substantially reduce the number of mothers dying from AIDS-related causes. Reductions in the number of children newly infected with HIV have begun. In 2013, there were 240,000 new cases of HIV infection in children, which was a 58% decline since 2002, the year when the highest number (580,000) children became newly infected with HIV. Furthermore, two-thirds of the total drop in new HIV infections between 2009 and 2011 was in newborns. The significant reduction in infected newborns is largely attributed to provision of antiretroviral drugs to HIV-positive mothers during pregnancy, delivery, and breastfeeding. Progress, however, has not been universal. Again, more than 90% of the children who acquired HIV infection in 2011 live in sub-Saharan Africa. Several countries have seen only modest declines in the number of children newly infected with HIV, and the Middle East and North Africa is the only region that has seen no reduction in the number of newly infected children (Table 12-4). Reductions in the number of adults that are newly infected with HIV, as well as the direct use of preventive antiretroviral medications, appear to be the reason for the

▲ A particularly sad consequence of AIDS in Africa is the number of AIDS orphans, children whose parents have both died of AIDS. AIDS orphans then become responsible for the care of their siblings and other family members.

declines in the number of children acquiring HIV. Antiretroviral prophylaxis is now credited for preventing more than 900,000 new HIV infections among children since 2009.

North America also has an AIDS problem. In the United States, it is estimated that more than 1.1 million people are infected with HIV, with one in five (18.1%) unaware of their infections. It is estimated that 15,529 people with AIDS died in 2010 and that about 636,000 people in the United States have died from the disease since it surfaced in the early 1980s. Over 49,000 new cases of HIV infection were reported in the United States in 2011. HIV affects a large percentage of blacks in the United States, and gay and bisexual men are most seriously affected by HIV. In 2011, the areas with the highest rates were in the South, a few states in the Northeast (Massachusetts, New Jersey, and New York), Illinois, Puerto Rico, and the U.S. Virgin Islands. Overall, from 2008 through 2011, the annual estimated number and rate of diagnoses of HIV infection remained stable in the United States.

Toward Universal Treatment. Although no vaccine is available to prevent AIDS, the latest antiretroviral drugs can significantly slow the progression of the disease. Providing AIDS drugs to pregnant women is also an effective preventive measure. If a woman begins taking AIDS drugs such as zidovudine (AZT) by the fourteenth week of pregnancy, the risk of transferring the virus to her offspring is greatly reduced. Providing the drug immediately before birth also helps (see Further Reading 8). However, there are many barriers to the use of these drugs, especially in the developing world. For example, the newest therapies require a person to take at least three different drugs in the form of about 14 pills each day. A few missed doses can significantly reduce the effectiveness of the drugs and result in faster disease progression. Another barrier is economic: a typical drug regimen can cost approximately $14,000 per year, not including unforeseen hospital stays.

Fortunately, drug companies and governments have worked successfully to lower the cost of AIDS drugs for developing nations. Most countries that aspire to expand access to treatment set a goal of providing antiretroviral treatment to around 80% of those in need. It is estimated that 8 million people are currently on the life-saving antiretroviral therapy. Although this is an increase of 60% over the past two years, global treatment is still reaching only 54% of those that need it, meaning there are another 7 million people waiting to be treated. The goal is to achieve universal access to treatment, care, and prevention for HIV and AIDS by 2015. The international community appears committed to achieving this goal.

Nutrition and AIDS. Although adequate nutrition cannot prevent or cure HIV infection or AIDS, nutritional status can affect the progression of the disease. A diet adequate in energy, protein, and micronutrients can help to lessen the impact of infections associated with AIDS. Poor nutritional status, such as a low status of vitamin A and vitamin E, contributes to a quicker onset of symptoms such as body wasting and fever, and a more rapid demise. Overall, maintenance of nutritional status should be an integral part of the treatment for AIDS. The daily use of a balanced vitamin and mineral supplement has also been shown to slow health declines in people with HIV/AIDS. Studies in developing countries have found that adults living with HIV need 10% to 30% more energy per day than persons who are not HIV positive. For a person infected with HIV but showing no symptoms, a 10% increase in a healthy adult's 2100 kcal energy requirements would mean a total of 2310 kcal per day. HIV-positive children have also been shown to require a 10% increase in daily energy intake compared to children without HIV (see Further Reading 11). To learn more about AIDS, check out the website **www.unaids.org.**

☑ **CONCEPT CHECK 12.3**

1. Why does undernutrition continue despite adequate food resources?

2. What are some solutions to limiting population growth, especially in developing countries where birth rates are high?

3. How have war and declines in natural resources contributed to undernutrition in developing countries?

4. What effects have overpopulation and increased unbanization had on nutrition and disease risk?

5. How many people globally are currently infected with HIV?

6. How is HIV transmitted?

7. What have been the trends in terms of numbers of new cases of HIV infection since 2001?

12.4 Reducing Undernutrition in the Developing World

As you have probably guessed, greatly reducing undernutrition in the developing world is complicated and takes considerable time to accomplish. It has been a common practice for the more affluent nations to supply famine areas with direct food aid. However, direct food aid is not a long-term solution. Although it reduces the number of deaths from famine, it can also reduce incentives for local production by driving down food prices. In addition, the affected countries may have little or no means of transporting the food to those who need it most, and the donated foods may not be culturally acceptable.

In the short run, there is no choice: aid must be given because people are starving (see Further Reading 1). Still, improving the infrastructure for poor people, especially rural people, needs to be the long-term focus. This future-minded approach is the focus of the strategies outlined for the Millennium Development Goals.

The United Nations Millennium Declaration, which contains the eight MDGs, is an international agreement that is already proving to be effective because of its overarching design. Nutrition security is a critical component for success of the MDGs. MDG 1 is to eradicate poverty and hunger with targets of halving, between 1990 and 2015, the proportion of people whose income is less than $1.25 a day and the proportion of people who suffer from hunger. By 2015, the global poverty rate is expected to be reduced to 15%, which would achieve MDG 1. Despite the overall progress being made, however, poverty is still a significant problem, primarily in rural areas. It is estimated that 80% of hungry people live in rural areas and work mainly on small farms. In addition, certain regions, particularly those affected by political conflicts and instability, are still falling behind and have yet to achieve any MDGs.

Development Tailored to Local Conditions Is Important. Although world food supplies have grown faster than the population in recent years, an increase in undernutrition has been caused by an increase in the number of people cutoff from their share of this supply. Millions of farmers are losing access to resources they need to be self-reliant. There is a growing realization that unless economic opportunities can be created as part of a plan for sustainable development, rural people who own no land will flock to the overcrowded cities. In response, careful, small-scale regional development is one option.

For the most part, the solution lies in helping people meet their own needs and directing them to resources and employment opportunities, rather than giving them resources. Experience has shown that the provision of credit—along with training,

▲ Food security is fostered by communities raising and distributing locally grown food.

food storage facilities, and marketing support—allows rural people to actively participate in their development, which will benefit their families and communities.

One U.S. program, the Peace Corps, has helped improve conditions in developing nations for over 50 years by providing education, distributing food and medical supplies, and building structures for local use. The aim of the Peace Corps is to help create independent, self-sustaining economies around the world. The MDGs are now a key component of the educational curriculum for Peace Corps volunteers. Becoming a Peace Corps volunteer is a significant way you can make a difference around the world. Learn more about this program at **www.peacecorps.gov.**

Impoverished women are a special concern. In addition to working longer hours than men, they grow most of the food for family consumption and make up three-fourths of the labor force in the informal sector of the economy and an increasing proportion in the formal sector. Economic opportunities for women and education regarding family planning must be augmented, and thus, MDG 3 aims to promote gender equality and empower women. A specific target of MDG 3 is to eliminate gender disparity in all levels of education no later than 2015. Of the 3 billion people in the world living on less than $2 a day, 70% are women. Thus, an important means of propelling nations out of poverty is to end the cycle of female neglect. MDG 3 is not only a goal in itself but also recognized as essential to the achievement of all other MDGs. Empowerment of women is critical to improve the level of nutrition security, increase the production and distribution of food and other agricultural products, and enhance living conditions in general. The **gender and development (GAD) approach** works toward improving the status of women through the active participation of both men and women. By increasing women's access to education, information and communication technologies, economic resources, and governance, the GAD approach reduces poverty, promotes development, achieves gender equality, protects women's human rights, and eliminates violence against women. Progress on achieving MDG 3 has been slow, but the proportion of girls in primary education rose to 96 for every 100 boys in 2008. It is possible that gender disparity at all education levels may be eliminated by 2015.

Suitable technologies for processing, preserving, marketing, and distributing nutritious local staples also need to be encouraged, so that small farmers can flourish. Education on how to use these foods to create healthful diets, such as preparing vitamin A-rich vegetables, adds further benefit. Supplementing indigenous foods with nutrients that are in short supply, such as iron, various B vitamins, zinc, and iodide, also deserves consideration. One program involves adding iron to sugar in various parts of the world. The next section examines the role of biotechnology in improving nutrient quality and other plant and animal characteristics, another possible positive step in lessening undernutrition. In addition, advances in water purification need to be employed.

MDG 7 aims to ensure environmental sustainability, which may include promoting extensive land ownership and thus increasing the availability of food. If food resources are concentrated among a minority of people, as often happens with unequal land ownership, food is not likely to be equally distributed unless efficient transportation systems are in place.

Raising the economic status of impoverished people by employing them is as important as expanding the food supply. If an increase in food supply is achieved without an accompanying rise in employment, there may be no long-term change in the number of undernourished people. Although food prices may fall with increased mechanization, use of fertilizers, and other modern technologies, it needs to be realized that these advances can also displace people from jobs, a result that harms rather than helps the population.

With the adoption of the Millennium Declaration in 2000 by all United Nations member states, world leaders committed to tackle the many dimensions of extreme poverty and create a better life for those in need. The commitment was articulated

---

CRITICAL THINKING

Stan has read about various relief efforts to help undernourished people in developing countries, especially the emergency food aid programs for famine-ravaged areas. Many of these efforts appear to be only temporary, and he wonders what long-range approaches might help alleviate the problem of undernutrition. What suggestions would you give Stan about possible long-term solutions for undernutrition in developing countries?

gender and development (GAD) approach Understanding the roles and responsibilities of both men and women in the process of sustainable development.

▲ This school teacher and her students are in a classroom in a village in Senegal, Africa. Eliminating gender disparity by including more girls at all levels of education no later than 2015 is a target of Millennium Development Goal 3.

as the MDGs, which provide a framework of measurable goals and targets that address extreme poverty and hunger, education, women's empowerment and gender equality, health, environmental sustainability, and global partnership.

SUSTAINABLE AGRICULTURE

Over the years, changes in agricultural practices have had many positive results on the availability of food around the world. Along with the positive effects on farming, however, there have been significant negative impacts. Most significant among these are depletion of topsoil, contamination of groundwater, the decline of family farms, neglect of living and working conditions for farm laborers, increasing costs of production, and the lack of integration of economic and social conditions in rural communities.

The concept of **sustainable development** has been embraced as economic growth that will reduce poverty while at the same time protect the environment and preserve natural capital. The United Nations cites economic development, social development, and environmental protection as the "reinforcing pillars" of sustainable development. The role of the agriculture industry in promoting practices that contribute to environmental and social problems has been at the center of discussions of sustainable development. As a result, interest in alternative farming practices has grown and a movement toward **sustainable agriculture** has emerged. Sustainable agriculture describes farming systems that can indefinitely maintain their productivity and usefulness to society. Sustainable agriculture flows directly from MDG 7 and depends on the integration of several goals, including environmental health, economic profitability, and social and economic equity. It addresses many environmental and social concerns and offers innovative and economically viable opportunities for many in the food system, including growers, laborers, consumers, and policymakers. Sustainable agriculture is gaining support and acceptance from conventional farmers in many countries.

Sustainable agriculture involves maintaining or enhancing the land and natural resources for use long into the future. In addition, the social responsibilities of human resources, including working and living conditions of laborers, the needs of rural communities, and consumer health and safety, must be considered both in the present and in the future. The potential of sustainability is best understood when the consequences of farming practices on both human communities and the environment are considered.

Farmers around the globe are transitioning to sustainable agriculture by taking small realistic steps based on their personal goals and family economics. Reaching the goal of worldwide sustainable agriculture requires participation by all stakeholders, including farmers, laborers, retailers, consumers, researchers, and policymakers, as articulated by MDG 8 for global partnership for development.

The ability of humans to manipulate nature has enabled us to improve the production and yield of many important foods. Traditional **biotechnology** is almost as old as agriculture. The first farmer to improve stock by selectively breeding the best bull with the best cows was implementing biotechnology in a simple sense. The first baker to use yeast to make bread rise took advantage of biotechnology.

By the 1930s, biotechnology made possible the selective breeding of better plant hybrids. As a result, corn production in the United States quickly doubled. Through similar methods, agricultural wheat was crossed with wild grasses to confer more desirable properties, such as greater yield, increased resistance to mildew and bacterial diseases, and tolerance to salt or adverse climatic conditions.

Another type of biotechnology uses hormones rather than breeding. In the last decade, Canadian salmon have been treated with a hormone that allows them to mature three times faster than normal—without changing the fish in any other way. In general terms, biotechnology can be understood as the use of living things—plants, animals, bacteria—to manufacture products.

sustainable development Economic growth that will simultaneously reduce poverty, protect the environment, and preserve natural capital.

sustainable agriculture Agricultural system that provides a secure living for farm families; maintains the natural environment and resources; supports the rural community; and offers respect and fair treatment to all involved, from farm workers to consumers to the animals raised for food.

biotechnology A collection of processes that involves the use of biological systems for altering and, ideally, improving the characteristics of plants, animals, and other forms of life.

BIOTECHNOLOGY

The new biotechnology used in agriculture includes several methods that directly modify products. It differs from traditional methods because it more directly changes some of the genetic material (DNA) of organisms to improve characteristics. Cross-breeding plants or animals is no longer the only tool. Development of the new process, called **genetic engineering,** began in the 1970s. The field now features a wide range of cell and subcell techniques for the synthesis and placement of genetic material in organisms (Fig. 12-5).

This process of **recombinant DNA technology** allows access to a wider gene pool, and it permits faster and more accurate production of new and more useful microbial, plant, and animal species. In comparison, conventional breeding is inefficient and has inconsistent results; biotechnology uses genetic material more precisely. Scientists select the traits they want and genetically engineer or introduce the gene that produces the desired trait into plants or animals (now called a **genetically modified organism [GMO]** or **transgenic organism**). Genetic engineering has not replaced conventional breeding practices; both work together.

Biotechnology has been used to enhance crops in three main categories. The primary category is the addition of a unique characteristic called an *input trait* to a crop. These enhanced input traits include herbicide (weed killer) tolerance, insect and virus protection, and tolerance to environmental stressors such as drought. Other categories are value-added *output traits*, such as plant oils with increased levels of omega-3 fatty acids, and crops that produce pharmaceuticals. Scientists have engineered plants that grow with the use of less pesticides and new forms of potatoes that can be stored longer without preservatives. In addition, biotechnology allows scientists to create fruits and grains with greater amounts of nutrients such as beta-carotene (e.g., "golden rice") and vitamins E and C. Biotechnology is being used cautiously and conservatively, so benefits of the new biotechnology will strike us as only subtly different. The ultimate benefits, however, could be important if foods eaten by people in the developing world can be so enhanced.

Few consumers in the United States realize that as of July 2013, 90% of corn and 93% of soybeans produced in the United States have been genetically engineered to either resist certain insects, thereby reducing pesticide use, and/or survive when sprayed with herbicides that kill surrounding weeds. Papaya and sugar beet plants have been genetically engineered for viral resistance.

Corn has been genetically altered by inserting a gene from the bacterium *Bacillus thuringiensis*, usually referred to as the Bt gene, into the corn DNA (Fig. 12-5). The gene allows the corn plant to make a protein lethal to certain caterpillars that destroy the plant. The Bt protein in the corn, however, is present in the plant in low concentrations and has no effect on humans: it is digested along with the other proteins in corn. In fact, for many years organic farmers have used the Bt bacteria as a spray on plants to destroy pests without changing the DNA of the plant.

FDA is confident that approved varieties of genetically engineered foods are safe to consume. Food manufacturers, therefore, are not required to disclose the genetically modified ingredient content on food labels. FDA does not believe labeling of GMO products is needed because they pose no health risk. Recently, the debate about whether products containing GMOs are dangerous to human health has resurfaced in several states. A grass-roots coalition, Right to Know GMO, has been active, and GMO labeling bills

genetic engineering Manipulation of the genetic makeup of any organism with recombinant DNA technology.

recombinant DNA technology A test tube technology that rearranges DNA sequences in an organism by cutting the DNA, adding or deleting a DNA sequence, and rejoining DNA molecules with a series of enzymes.

genetically modified organism (GMO) Any organism created by genetic engineering.

transgenic organism Organism that contains genes originally present in another organism.

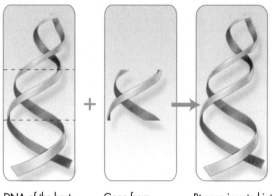

DNA of the host plant, corn

Gene from bacteria (Bt gene) that produces a protein toxic to the European corn borer

Bt gene inserted into DNA of corn plant. Now the corn plant is genetically modified. It makes the Bt toxin and so is resistant to the European corn borer.

FIGURE 12-5 ▲ Biotechnology involves various techniques for transferring foreign DNA into an organism. In this diagram, a sample of DNA is cleaved out of a larger DNA fragment and inserted into the DNA of a host cell. Thus, the host cell contains new genetic information, with the potential of providing the cell with new capabilities. For corn, this could mean resistance to the European corn borer. The corn plant is now referred to as a genetically modified organism (GMO). In another application, bacteria can be engineered to produce the human form of the hormone insulin.

or referenda have been introduced in many states, including California, Washington, and Vermont. In June 2013, Connecticut became the first state to pass a law requiring a package label indicating that a food is made from GMOs. A bill has also been introduced at the federal level that would direct FDA to "clearly label" genetically engineered foods. Some legislators recommend that the United States should follow the 64 other nations, including those of the European Union, that require labeling for genetically engineered foods. Discussions are also occurring about the "probable impacts" of having differing labeling requirements passed by states rather than a having a federal standard. The companies that supply seeds for genetically engineered crops to farmers are frequent targets of protests against GMOs. Their argument against requiring labels on genetically engineered foods is that this type of labeling would imply that those foods are unsafe.

Public response to use of GMO biotechnology, however, has been mixed. Even the scientific community has conflicting opinions about this technology, with supporters as convinced about the benefits as opponents are of the risks. The biggest debate in the United States surrounds the potential environmental hazards of introducing genes from one species to another. Some challengers even question the reduction in pesticide use that accompanies the cultivation of genetically modified crops. Although the use of genetically modified crops may reduce the need for environmentally harmful activities, such as spraying crops with pesticides, critics point out that seeds produced with additional insecticide, such as the Bt protein, potentially will lead to rapid insect resistance because the insecticides are continuously emitted. Use of traditional pesticides involves prudent application, in part to avoid insect resistance. In addition, accidental release of genetically modified animals, such as fish, may go on to harm wild varieties.

Although the risks of biotechnology may appear to be momentarily negligible, they may be cumulative and therefore of concern in the long run. FDA carefully examines all products developed using this technology and will enforce labeling of potential allergens that may be newly present in food altered by biotechnology.

The public has long been opposed to processes perceived as harmful to the environment, such as producing unnatural products. Food reserves are high in the United States, Canada, and Europe, so some question the need to increase food production. Skepticism surrounds unnatural products, as exemplified by Western Europe's ban of hormones used in beef and milk production, and of almost all genetically modified foods.

▲ Both traditional plant breeding and biotechnology have produced high-yielding and disease-resistant plant varieties, including new varieties of corn.

ROLE OF THE NEW BIOTECHNOLOGY IN THE DEVELOPING WORLD

Twenty-seven countries planted 433 million acres of crops enhanced through biotechnology in 2013. These crops included improved varieties of soybeans, corn, cotton, canola, papaya, and squash. The United States maintained the largest growth of biotech crops with 173 million acres of soybeans, corn, cotton, canola, sugar beet, alfalfa, papaya, and squash. Brazil was second with 100 million acres. More than half (52%) of the global biotech crop, equivalent to 233 million acres, were grown in 19 developing countries in 2013. Of the 18 million farmers who grew biotech crops in 2013, 16.5 million or 90% were small resource-poor farmers from developing countries. These included 7 million small farmers in China and another 7 million small farmers in India, who collectively planted a record 37 million acres of biotech crops.

Whether applications of genetic engineering will help to significantly reduce undernutrition in the developing world remains to be seen. Unless price cuts accompany the increased production, only landowners and suppliers of biotechnology will enjoy the benefits. Small farmers may benefit if they can afford to purchase the genetically modified seeds. This point deserves emphasis: The person who cannot afford to buy enough food today will still face that same predicament in the future.

As with most innovations, the more successful farmers—often those with larger farms—will adopt the new biotechnology first. Because of this, the present trend toward fewer and larger farms will continue in the developing world, a movement that undermines the solution to one of the most pressing undernutrition issues there. Furthermore, biotechnology does not promise dramatic increases in the production of most grains and cassava, the primary food resources in developing parts of the world.

With the introduction of drought- and pest-resistant, as well as self-fertilizing crops, agricultural biotechnology may help to lessen world hunger. Perhaps the most promising potential of genetically modified foods today lies within the realm of plant breeding for micronutrients. Developing countries will have a tool to treat and prevent nutrient deficiencies among their populations if they have access to farming resources to increase the micronutrient composition of crops. In addition, greater yields for indigenous plants, such as tomatoes that tolerate high soil salinity, are another hopeful outcome. Biotechnology will likely be a useful tool against the complex scourge of world undernutrition. Improved crops produced by this technology, together with political and other efforts, can contribute to success in the battle against worldwide undernutrition.

SOME CONCLUDING THOUGHTS

The economic loss from undernutrition is staggering, and the amount of human pain and suffering is incalculable. With all the international relief efforts and assistance from governments and private organizations combined, we are still battling undernutrition. Read more about the progress toward the Millennium Development Goals and the lack of progress on hunger in several regions of the world, even as income poverty has decreased, in Further Reading 16.

Ultimately, the depletion of world resources, the massive debt incurred by poorer countries, the threat of danger to more prosperous countries nearby, and the toll taken in human lives affect the world economy and well-being. The resulting instability can go on to affect the developing world, as has been apparent in recent years. Life is not necessarily fair, but the aim of civilization should be to make it more so. The world has both enough food and the technical expertise to end hunger. With 38 countries affirming the UN Millennium Declaration of 2000 and working to achieve its Millennium Development Goals, we have seen a promising start to a coordinated political effort that is making progress toward lasting reductions in hunger.

▲ Soybeans are a common GMO food in the marketplace. Over 90% of the soybeans grown in the United States are genetically modified.

✔ CONCEPT CHECK 12.4

1. Why is development tailored to local conditions important in combating world hunger?
2. What is the definition of sustainable agriculture?
3. What are three traits for which biotechnology is used to enhance crops?

Nutrition and Your Health
Undernutrition at Critical Life Stages

PREGNANCY

Undernutrition poses the greatest health risk during pregnancy. About 500,000 women worldwide die each year from complications of pregnancy and childbirth. A pregnant woman needs extra nutrients to meet both her own needs and those of her developing offspring. Nourishing the fetus may deplete maternal stores of nutrients. Maternal iron deficiency anemia is one possible consequence (Chapter 14).

The birth rates in Africa are the highest in the world. In Niger, for example, a woman gives birth to an average of nearly eight children. Close behind are countries such as Uganda, Mali, and Somalia where an average of six to seven children are born to each woman. Coupled with undernutrition, a woman's risk of dying increases with every pregnancy and birth. This results in a strong connection between a high fertility rate and a high maternal mortality rate. Although most of us consider pregnancy and childbirth as very natural parts of life, for women living in developing countries without access to health care, pregnancy and childbirth complications are among the leading causes of death. The cumulative effect of successive pregnancies too close together does not allow the mother to recover nutrients such as iron and folate lost during pregnancy and breastfeeding. Progress toward meeting the maternal mortality goal (MDG 5) to reduce the maternal mortality ratio by three quarters between 1990 and 2015 has been lagging. Improvements will need to double to meet this goal by 2015.

Critical Life Stages When Undernutrition Is Particularly Devastating

Prolonged undernutrition is detrimental to many aspects of human health, resulting in increased maternal, infant, and child mortality; loss of parents (especially linked to AIDS); exploitation of women; reduced work capacity; reduced intellectual and social development; and overall human suffering, especially in a famine. It is particularly damaging during some periods of growth and old age. A conceptual framework of malnutrition developed by UNICEF lists the immediate causes of malnutrition as inadequate dietary intake and unsatisfactory health. The underlying causes relate to families and include inadequate access to food, inadequate care for women and children, and insufficient health services. Basic causes relate to human and economic resources and affect communities and nations. Effective strategies to combat malnutrition should embrace a life cycle approach. Public health programs should be complementary and comprehensive across vulnerable periods of the whole reproductive cycle (see Further Reading 15).

▲ This Lobi tribeswoman, standing with her six children in a grassy field in Burkina Faso, Africa, is an illustration of the typical birth rate in this region. Maternal nutrient levels become depleted with so many successive pregnancies followed by breastfeeding.

FETAL AND INFANT STAGES

The fetus faces major health risks from undernutrition during gestation. To support growth and development of the brain and other body tissues, a growing fetus requires a rich supply of protein, vitamins, and minerals. When these needs are not met, the infant is often born before 37 weeks of gestation, well before the ideal 40 weeks of gestation. The consequences of this preterm birth include reduced lung function and a weakened immune system. These conditions not only compromise health but also increase the likelihood of premature death. If the infant survives, long-term problems in growth and development can result. In extreme cases, low-birth-weight infants (about 5.5 pounds [2.5 kilograms] or less) face 5 to 10 times the normal risk of dying before the age of 1 year, primarily because of reduced lung development. When low birth weight is accompanied by other physical abnormalities, medical intervention can cost $200,000 or more. These costs can be met only in developed countries.

Worldwide, the incidence of infants born with low birth weight is 15% of all births. Much higher rates, however, occur in developing countries, with the highest rates in Mauritania (34%), Pakistan (32%), India (28%), Nanru (27%), Niger (27%), and Haiti (25%). Undernutrition is a major contributor to these high rates of low birth weight in developing countries. About 8% of infants born in the United States and 6% in Canada have low birth weights. In the United States, low birth weight accounts for more than half of all infant deaths and 75% of deaths of infants younger than 1 month old. Pregnancy during the teenage years, while a girl's body is still growing, contributes to low birth weight worldwide.

CHILDHOOD

Early childhood, when growth is rapid, is another period when undernutrition is extremely risky. The greatest impact of undernutrition occurs in the first 1000 days of life from conception to a child's second birthday. Irreversible lifelong damage is the result if the right nutrition is not available at this critical time. The central nervous system—including the brain—continues to be vulnerable because of rapid growth through early childhood. After the preschool years, brain growth and development slow dramatically until maturity. Nutritional deprivation, especially in early infancy, can lead to permanent brain impairment. Without an effective intervention, it is projected that ongoing undernutrition could leave more than 1 billion children with mental impairment by 2020.

In general, poor children are at the greatest risk for nutritional deprivation and subsequent illness. Stunted growth, which stops children's bodies from developing properly, is an obvious effect. The WHO reports that nearly 165 million suffer from stunting. In addition, iron-deficiency anemia is much more common among low-income children than children from less deprived families. This deficiency can lead to fatigue upon exertion, reduced stamina, stunted growth, impaired motor development, and learning problems. Undernutrition in childhood can also weaken resistance to infection because immune function decreases when nutrients such as protein, vitamin A, and zinc are low in a diet. Clearly, undernutrition and illness have a cyclical relationship. Not only does undernutrition lead to illness, but illness, particularly diarrhea and infectious diseases, worsens undernutrition. For this reason, many children in developing countries are dying from the combination of malnutrition and infection. Conversely, when missing nutrients such as vitamin A and zinc are restored to children's diets, improvements in health can be obvious.

Reducing the under-5 child mortality rate by two-thirds between 1990 and 2015 is MDG 4. Improvement has been seen with a narrowing of the gap in child mortality between the top and bottom countries. Death rates have already been reduced from 171 per 1000 live births in 1990 to 107 per 1000 live births in 2011. Despite this progress, 6.9 million children under 5 years of age died in 2011. Furthermore, the progress made in recent decades has been unequally distributed across regions and countries and within countries. In 2011, nearly 80 percent of the under-5 deaths occurred in only 25 countries, with India (24 percent) and Nigeria (11 percent) together accounting for more than a third of these deaths worldwide. These statistics indicate that important challenges remain for the child mortality goal to be achieved globally. Undernutrition is one of the most important challenges to overcome because it is estimated to be an underlying cause in as many as 45% of the deaths among children under age 5.

LATER YEARS

The WHO predicts a significant increase in the number of people aged 65 years or older from an estimate of 524 million in 2010 to almost 1.5 billion in 2050. Most of this increase in global aging will be in developing countries. Older adults, especially older women living alone in poverty, are also at risk for undernutrition. Older adults in general require nutrient-dense foods, in amounts dependent on their state of health and degree of physical activity. Many of them have fixed incomes and incur significant medical costs, so food often becomes a low-priority item. In addition, depression, social isolation, and declining physical and mental health can compound the problem of undernutrition in older adults.

▲ This senior Lobi tribeswoman in Burkina Faso, Africa, with an ivory piercing in her upper lip, reminds us that the population of people aged 65 years or older will increase the most in developing countries.

CASE STUDY Undernutrition During Childhood

Jamal traveled to the Philippines with his church group last summer. During their stay, they helped build shelters for people in a village where, a few weeks before, a storm had destroyed several houses. Jamal noticed that many of the children were very short, much shorter than the children in his neighborhood in the United States. His group worked in a remote, low-elevation area where the storm and subsequent flooding had caused the most damage. On several occasions, he noticed young mothers crouched on curbs or in doorways, holding their children. These children rarely moved: they appeared pale and listless. In contrast to the children Jamal's group had met at a church in the capital city, most of the children in this village were not active and lively. One evening, a nurse from the local clinic came to speak to Jamal's group. She said that many children in this area do not get enough to eat and that health problems were rampant. She considered the recent storm a blessing in disguise, hoping it would spur the Philippine government to send supplies to the village, particularly food and medicines. Jamal is shocked by such a degree of suffering. He wonders why children in the Philippines can be starving to death while many children in his hometown in the United States are overweight.

   Answer the following questions, and check your responses at the end of this chapter.

1. Should Jamal have been surprised by widespread disease and general listlessness in the Philippine children?
2. Which nutrients are likely to be deficient in the diets of these children?
3. Which nutrient deficiencies contribute to poor growth or "stunting" of growth?
4. Which nutrient deficiencies may be causing diarrhea and illness?
5. Are these children likely to be consuming adequate calories? What effect does consumption of inadequate calories have on growth?
6. Why might the recent storm be a blessing for this small village?
7. List some reasons why the children in Jamal's neighborhood at home are more likely to be overweight.

Summary (Numbers refer to numbered sections in the chapter.)

12.1 Poverty is commonly linked to chronic or periodic undernutrition. Malnutrition can occur when the food supply is either scarce or abundant. The resulting deficiency conditions and degenerative diseases contribute to poor health.

   Undernutrition is the most common form of malnutrition in developing countries. It results from inadequate intake, absorption, or use of nutrients or food energy. Many deficiency conditions consequently appear, and infectious diseases thrive because the immune system cannot function properly.

   Undernutrition diminishes both physical and mental capabilities. In poor countries, this is worsened by recurrent infections, unsanitary conditions, extreme weather, inadequate shelter, and exposure to diseases.

12.2 In North America, famine is not seen, but food insecurity and undernutrition remain problems, due in large part to poverty. Single parents and their children are likely to live in poverty. Soup kitchens, food stamps, the school lunch and breakfast programs, and the Special Supplemental Nutrition Program for Women, Infants, and Children have focused on improving the nutritional health of poor and at-risk people. When adequately funded, these programs have proved effective in reducing undernutrition.

12.3 Multiple factors contribute to the problem of undernutrition in the developing world. In densely populated countries, food resources, as well as the means for distributing food, may be inadequate. Environmentally unsustainable farming methods hamper future efforts to grow food. Limited water availability hinders food production. Naturally occurring devastation from droughts, excessive rainfall, fire, crop infestation, and human causes—such as urbanization, war and civil unrest, poor sanitation, and AIDS—all contribute to the major problem of undernutrition.

12.4 Proposed solutions to world undernutrition must consider multiple interacting factors, many thoroughly embedded in cultural traditions. Family planning efforts, for example, may not succeed until life expectancy increases. Through education, efforts should be made to upgrade farming methods, improve crops, limit pregnancies, encourage breastfeeding when it is safe to do so, and improve sanitation and hygiene.

   Direct food aid is only a short-term solution. In what may appear to be a step backward, many experts recommend more sustainable subsistence-level farming. Small-scale industrial development is another way to create meaningful employment and purchasing power for vast numbers of the rural poor. Various biotechnology applications may also prove beneficial. Since the year 2000, the United Nations member states have combined efforts to achieve goals related to eliminating poverty and hunger.

NAYH The greatest risk of undernutrition occurs during critical periods of growth and development: gestation, infancy, and childhood. Low birth weight is a leading cause of infant deaths worldwide. Many developmental problems are caused by nutritional deprivation during critical periods of brain growth. People in their later years are also at great risk.

Check Your Knowledge (Answers to the following questions are below.)

1. There are an estimated _____ chronically undernourished people in the world.
   a. 14 million
   b. 1 billion
   c. 3 billion
   d. 6 billion

2. The number one killer of children in developing countries is
   a. xerophthalmia.
   b. iron-deficiency anemia.
   c. iodide deficiency.
   d. diarrhea.

3. The human organism is particularly susceptible to the effects of undernutrition during
   a. pregnancy.
   b. infancy.
   c. childhood.
   d. all of the above.

4. A barrier to solving undernutrition in the developing world is NOT
   a. external debt.
   b. poor infrastructure.
   c. a lack of manpower.
   d. expanding population.

5. Many of the child deaths each year in developing countries could be prevented if
   a. technology were improved.
   b. doctors were more specialized.
   c. mothers would learn more about nutrition.
   d. sanitation and hygiene were improved.

6. The Supplemental Nutrition Assistance Program allows
   a. low-income families to buy surplus food at government stores with government-issued electronic benefit transfer cards.
   b. low-income people to purchase food, cleaning supplies, alcoholic beverages, and anything else sold in supermarkets with electronic benefit transfer cards.
   c. low-income people to turn in government-issued electronic benefit transfer cards for cash to buy food.
   d. low-income people to purchase food and seeds with government-issued electronic benefit transfer cards.

7. A long-term solution to world hunger is
   a. the green revolution.
   b. cash crops.
   c. jobs and self-sufficiency.
   d. government and private aid.

8. Bt corn has been genetically modified to make
   a. the bacterium *Bacillus thuringiensis.*
   b. a protein toxic to caterpillars that can destroy the corn plant.
   c. a sugar that makes the corn sweeter.
   d. a fat that makes corn oil healthier.

9. Genetically modified soybeans make up approximately _____% of the soybeans grown in the United States.
   a. 50
   b. 75
   c. 80
   d. 90

10. FDA requires the statement "contains genetically modified ingredients" on the label of all foods containing genetically modified ingredients.
    a. True
    b. False

Answer Key: 1. b (LO 12.1), 2. d (LO 12.3), 3. d (LO 12.5), 4. c (LO 12.3), 5. d (LO 12.3), 6. d (LO 12.2), 7. c (LO 12.4), 8. b (LO 12.4), 9. d (LO 12.4), 10. b (LO 12.4)

Study Questions (Numbers refer to Learning Outcomes)

1. Describe the difference between malnutrition and undernutrition. **(LO 12.1)**

2. Describe in a short paragraph any evidence of undernutrition that you saw while you were growing up, such as on television. What are or were the likely roots of these problems? **(LO 12.1)**

3. What do you believe are the major factors contributing to undernutrition in wealthy nations, such as the United States? What are some solutions to this problem? **(LO 12.2)**

4. List three long-term consequences of undernutrition during fetal development or infancy. **(LO 12.5)**

5. What federal programs are available to address the problem of undernutrition in the United States? **(LO 12.2)**

6. Outline how war and civil unrest in developing countries have worsened problems of chronic hunger over the past few years. **(LO 12.3)**

7. How important is population control in addressing the problem of world hunger now and in the future? Support your answer with three main points. **(LO 12.3)**

8. Why is solving the problem of undernutrition a key factor in the ability of developing countries to reach their full potential? **(LO 12.3)**

9. Name three nutrients often lacking in the diets of undernourished people. What effects can be expected with each deficiency? **(LO 12.1)**

10. Describe how sustainable agriculture and biotechnology can improve food availability worldwide. **(LO 12.4)**

What the Dietitian Chose

The correct response for the question posed at the beginning of this chapter is *all of the above*! Do what comes naturally to you—whether it is donating your spare change or volunteering your spare time—to move thought into action against world hunger. Indeed, the need is great—both globally and locally—and opportunities to get involved abound.

As you learned in this chapter, there are basically two ways to address the problem of hunger: direct food aid or empowering others to help themselves. Direct aid (e.g., donating money each month) is a short-term solution—one that necessarily fills an immediate gap between resources and requirements. However, in the long run, solving the problem of world hunger will take bottom-up changes that enable struggling communities to provide their own food and clean water.

It may be easier and more comfortable to visualize the hunger problem in a faraway place, such as in a developing country. In reality, however, poverty and hunger lurk in your own community. You can help local efforts by donating nonperishable items to local food pantries, volunteering at Summer Food Service Programs for school-age children, or simply by shopping for locally grown foods direct from farmers in your neighborhood.

In your quest to get involved, look for organizations whose focus is to help set up sustainable solutions in communities: empowering local citizens to grow their own food, maintain their own clean water supplies, or build their own local economies.

Action Against Hunger:
www.actionagainsthunger.org
Alliance to End Hunger:
www.alliancetoendhunger.org

▲ Volunteer your time at a local food pantry or soup kitchen. Check out **www.serve.gov/endhunger.asp** to find a volunteer opportunity near you.

Blood: Water Mission:
www.bloodwatermission.com
Bread for the World: **www.bread.org**
The Hunger Project: **www.thp.org**
The International Food Policy Research Institute:
www.ifpri.org
MAZON: A Jewish Response to Hunger:
http://mazon.org
Stop Hunger Now:
www.stophungernow.org

There are many creative ideas for raising funds and awareness to end world hunger. Check out the Rate Your Plate section at the end of this chapter for more ideas and put your own talents and passions to work for a good cause!

CASE STUDY SOLUTION   Undernutrition During Childhood

1. Jamal should not be surprised that the children in the village were often sick and listless. The living conditions and recent natural disaster have added to the already poor conditions in this village. We know that children are especially susceptible to the effects of poverty and undernutrition.
2. Based on the reports that the children do not get enough to eat and on the outward signs of the children's health, protein, vitamin A, iron, iodide, and zinc are likely to be deficient in the diets of these children.
3. Protein, vitamin A, iron, iodide, and zinc deficiencies contribute to poor growth and stunting because of their essential functions in the body. One or more of these deficiencies is likely present in many children in the village.

4. Protein, vitamin A, iron, iodide, and zinc deficiencies may also contribute to depressed immune function, which can cause diarrhea and illness. These deficiencies are particularly risky in children.
5. The diets of these children are most likely also marginal in caloric content. A low calorie intake during childhood will further depress the growth and overall health compromised by the protein and micronutrient deficiencies. Recall from Chapter 6 that protein-calorie malnutrition (PCM) is a form of undernutrition caused by an extremely deficient intake of calories or protein and generally is accompanied by an illness.
6. The recent storm will hopefully cause the Philippine government to send food and medicines to the village,

which will be a short-term solution to some of the residents' nutrition and health needs.

7. Poverty and undernutrition occur at much lower rates in the United States and other developed countries. Childhood nutrition concerns in developed countries are more likely to be the increasing incidence of overweight and obesity. As we discussed in Chapter 7, maintaining a healthy weight requires the proper balance of calories in and calories out. In Chapter 15, we will learn that overweight children are becoming more common because the typical number of calories they consume per day has increased at the same time that calorie expenditure in physical activity has decreased.

Further Readings

1. Academy of Nutrition and Dietetics: Position of the Academy of Nutrition and Dietetics: Nutrition security in developing nations: Sustainable food, water, and health. *Journal of the Academy of Nutrition and Dietetics* 113:581, 2013.

2. American Dietetic Association: Position of the American Dietetic Association: Food insecurity in the United States. *Journal of the American Dietetic Association* 110:1368, 2010.

3. American Dietetic Association: Position of the American Dietetic Association: Child and adolescent nutrition assistance programs. *Journal of the American Dietetic Association* 110:791, 2010.

4. Bhutta ZA and others: Meeting the challenges of micronutrient malnutrition in the developing world. *British Medical Bulletin* 106 (1): 7, 2013.

5. Christian P, Tielsch JM: Evidence for multiple micronutrient effects based on randomized controlled trials and meta-analyses in developing countries. *Journal of Nutrition* 142: 173S, 2012.

6. Dammann KW, Smith C: Race, homelessness, and other environmental factors associated with the food-purchasing behavior of low-income women. *Journal of the American Dietetic Association* 110:1351, 2010.

7. DeNavas-Walt C and Proctor BD. *Income and Poverty in the United States: 2013 Current Population Reports.* United States Census Bureau. September 2014.

8. Dworkin SL, Ehrhardt AA: Going beyond "ABC" to include "GEM": Critical reflections on progress in the HIV/AIDS epidemic. *American Journal of Public Health* 97:13, 2007.

9. Economic Research Service, U.S. Department of Agriculture: Food Access Research Atlas, **www.ers.usda.gov/data-products/food-access-research-atlas.aspx. Last updated:** Friday, March 01, 2013. Accessed June 20, 2013.

10. Food and Agriculture Organization, International Fund for Agricultural Development, and World Food Programme: *The State of Food Insecurity in the World 2014. Strengthening the enabling environment for food security and nutrition.* FAO, Rome, Italy, 2014.

11. Janssens B and others: Effectiveness of highly active antiretroviral therapy in HIV-positive children: Evaluation at 12 months in a routine program in Cambodia. *Pediatrics,* 120 (5):1134, 2007.

12. Joint United Nations Programme on HIV/AIDS (UNAIDS) and World Health Organization. UNAIDS Report on the Global AIDS Epidemic 2013.

13. Kuhn L and others: Breastfeeding and AIDS in the developing world. *Current Opinion in Pediatrics* 21:83, 2009.

14. Nord M, Hopwood H: A comparison of household food security in Canada and the United States. *Economic Research Report* 67, December 2008. **www.ers.usda.gov/publications/err67/.**

15. Stoltzfus RJ: Iron interventions for women and children in low-income countries. *Journal of Nutrition* 141:756S, 2011.

16. United Nations: *The Millennium Development Goals Report 2012.* United Nations, New York, 2012.

17. United Nations Childrens Fund: *Water, Sanitation, and Hygiene.* **www.unicef.org/wash/.** Accessed June 20, 2013.

 **connect** | NUTRITION

To get the most out of your study of nutrition, visit McGraw-Hill Connect at www.mcgrawhillconnect.com where you will find NutritionCalc Plus, LearnSmart, and many other dynamic tools.

Rate Your Plate

I. Fighting World Undernutrition on a Personal Level

If you want to do something about world and domestic undernutrition, consider the following activities. It is a noble act to try to make a difference, even if you make only one small step. As with any change in behavior, do not try to do too many things at once. Try one or two activities that represent your commitment to solving this problem.

1. Volunteer at a local soup kitchen or homeless shelter for a time (1 month, for example). What insights did you gain?

2. Coordinate the efforts of a campus organization to donate money to an antihunger agency such as the following:

Bread for the World	Catholic Relief Services	Oxfam America	Feeding America
50 F Street NW, Suite 500	228 W. Lexington St	226 Causeway St., 5th Floor	35 E. Wacker Dr., #2000
Washington, DC 20001	Baltimore, MD 21201	Boston, MA 02114	Chicago, IL 60601
www.bread.org/	**www.crs.org**	**www.oxfamamerica.org/**	**http://feedingamerica.org**

CARE USA	EarthSave International	Save the Children
151 Ellis Street, NE	PO Box 96	Foundation
Atlanta, GA 30303	New York, NY 10108	54 Wilton Rd.
www.care.org/	**www.earthsave.org/**	Westport, CT 06880
		www.savethechildren.org/

3. Make a contribution of nonperishable foods to the ongoing offering at a place of worship near you. If such an offering does not exist, start one.

4. Get on a food recovery program's mailing list, read its newsletters for information on upcoming fund-raisers and other activities, and become involved.

5. Utilize your love of good food and healthy cooking skills to organize a local meal for the hungry or a benefit dinner to raise money for local food banks. Resources to set up projects may be available from mission organizations. For example, with Blood: Water Mission's Lemon: Aid project, students or families can use something as simple as the sale of a refreshing drink to fund potable drinking water in rural communities in Africa.

6. Participate in food drives organized by local grocery stores by contributing food or services. Food-drive organizers may need volunteers to transport the donations to a food pantry. Pay attention to events around World Food Day, October 16.

7. Point, click, and fight hunger. Internet users can find information on hunger at several sites, including the following:

- Someone somewhere dies of hunger every 3.6 seconds. You can help stop the clock: go to **www.thehungersite.com** and click on Donate Free Food to send a meal to a needy someone. This site is affiliated with the UN World Food Program, which tracks the number of clicks and then sends a bill to one of its corporate or nonprofit sponsors.

- The Food and Agriculture Organization of the United Nations has worked to alleviate poverty and hunger by promoting agricultural development, improved nutrition, and the pursuit of food security. This website will keep you up-to-date on recent issues and provides an extensive list of publications related to food security. **www.fao.org**

- America's Second Harvest, the largest domestic hunger-relief organization, shows you how to help online and has information about the latest updates. **www.secondharvest.org**

- Bread for the World is a nationwide Christian citizens' movement seeking justice for the world's hungry people by lobbying our nation's decision makers. **www.bread.org**

- CARE is one of the world's largest private international relief and development organizations, with the goal of saving lives, building opportunities, and bringing hope to people in need. **www.care.org**

- Turn your personal fitness goals into a campaign to raise funds and awareness for projects that provide food and clean water to communities around the world. Water Walks (see **www.bloodwatermission.com**) or Crop Walks (see **www.churchworldservice.org**) are great ways to walk a mile in someone else's shoes.

II. Joining the Battle Against Undernutrition

Imagine that you recently spent your summer vacation in a developing country and saw evidence of undernutrition and hunger. Then imagine that you are now asking a large corporation to support your efforts to ease hunger and suffering in this area. Develop a two-paragraph statement outlining why addressing hunger issues in this area is important. Include how the corporation could assist you.

Student Learning Outcomes

Chapter 13 is designed to allow you to:

13.1 List some of the types and common sources of viruses, bacteria, fungi, and parasites that can make their way into food.

13.2 Compare and contrast food-preservation methods.

13.3 Understand the foodborne illnesses caused by bacteria, viruses, and parasites.

13.4 Describe the main reasons for using chemical additives in foods, the general classes of additives, and the functions of each class.

13.5 Identify sources of toxic environmental contaminants in food and the consequences of their ingestion.

Chapter 13
Safety of Our Food Supply

What Would You Choose?

By now, you have learned the importance of including more fruits and vegetables in your diet. You are choosing fresh or frozen produce more often than canned to cut back on sodium and other preservatives. You are steaming or stir-frying your vegetables to retain nutrients and avoid adding too much fat during preparation. In the grocery store, you have seen the growing selection of organic products, including produce. Do they have health benefits that make them worth the extra cost? If cost was not a concern, what type of fruits and vegetables would you choose from the following list to satisfy your desire for increased nutritional value, lower risk for foodborne illness, and decreased exposure to preservatives and pesticides?

a Canned low-sodium

b Organically grown

c Frozen

d Fresh

connect | **NUTRITION** Think about your choice as you read Chapter 13, then see **What the Dietitian Chose** at the end of the chapter. To learn more about organic foods, check out the Connect site: www.mcgrawhillconnect.com.

13.6 Understand the reasons behind pesticide use, the possible long-term health complications, and the safety limits set for their use.

13.7 Understand the effects of conventional and sustainable agriculture on our food choices.

13.8 Describe the procedures that can be used to limit the risk of foodborne illness.

The first Food and Drug Act in the United States was passed over 100 years ago, in 1906, as a result of public pressure to improve food preparation standards. Today, warnings about the safety of food and water appear everywhere. Attention has turned to more contemporary concerns, such as microbial and chemical contamination. While we are told to eat more fruits, vegetables, fish, and poultry and to drink more water, we are also warned that these may contain dangerous substances. Therefore, we still must ask, "How safe is our food and water?"

Scientists and health authorities agree that North Americans enjoy a relatively safe food supply, especially if foods are stored and prepared properly. Over the past 100 or so years, tremendous progress has been made to allow for this. Nonetheless, microorganisms and certain chemicals in foods still can pose a health risk. Thus, the nutritional and health benefits of food must be balanced against any food-related hazards. This chapter focuses on these hazards: how real they are and how you can minimize their effect on your life. You bear some responsibility for this; government agencies and industry can only do so much. The 2010 Dietary Guidelines for Americans encourage us to prepare and store foods safely. Which foods pose the greatest risk for foodborne illness? Is any food safe after being stored in the refrigerator for 6 months? Are food additives and pesticides an even greater day-to-day concern? This chapter provides some answers.

13.1 Food Safety: Setting the Stage

pasteurizing The process of heating food products to kill pathogenic microorganisms and reduce the total number of bacteria.

virus The smallest known type of infectious agent, many of which cause disease in humans. A virus is essentially a piece of genetic material surrounded by a coat of protein. They do not metabolize, grow, or move by themselves. They reproduce only with the aid of a living cellular host.

bacteria Single-cell microorganisms; some produce poisonous substances, which cause illness in humans. Bacteria can be carried by water, animals, and people. They survive on skin, clothes, and hair, and thrive in foods at room temperature. Some can live without oxygen and survive by means of **spore** formation.

spores Dormant reproductive cells capable of turning into adult organisms without the help of another cell. Various bacteria and fungi form spores.

fungi Simple parasitic life forms, including molds, mildews, yeasts, and mushrooms. They live on dead or decaying organic matter. Fungi can grow as single cells, like yeast, or as a multicellular colony, as seen with molds.

parasite An organism that lives in or on another organism and derives nourishment from it.

foodborne illness Sickness caused by the ingestion of food containing harmful substances.

toxins Poisonous compounds produced by an organism that can cause disease.

During the early stages of urbanization in North America, contaminated water and food—notably, milk—were responsible for many large outbreaks of typhoid fever, septic sore throat, scarlet fever, diphtheria, and other devastating human diseases. These experiences led to the development of processes for purifying water, treating sewage, and **pasteurizing** milk. Since that time, safe water and milk have become universally available, with only occasional problems from either.

The greatest health risk from food today is contamination by **viruses** and **bacteria** and, to a lesser extent, by various forms of **fungi** and **parasites.** These microorganisms can all cause **foodborne illness.** In 2013, 113 people became ill and 50 were hospitalized in seven states from an organic blend of frozen berries that was contaminated with the hepatitis A virus. In the United States, *Escherichia coli* O157:H7 has caused several multistate outbreaks in the past few years. In December 2012, 33 persons from five states were infected with *E. coli* O157:H7 from prepackaged leafy greens. Of these ill persons, 13 were hospitalized and two developed kidney failure, but no deaths were reported. In a more widespread outbreak in 2006, 199 persons in 26 states were infected with *E. coli* O157:H7 from fresh spinach. Of these ill persons, 102 were hospitalized, 31 developed kidney failure, and 22 were children under 5 years. Three deaths were associated with the outbreak. A deadlier outbreak of *Escherichia coli* O104:H4 occurred in Germany during the summer of 2011, resulting in over 3000 infections, hundreds of cases of kidney failure, and a death toll of at least 36. Read more about this outbreak that was linked to sprouts in Table 13.1.

Although microbial contamination is the cause of most incidents of foodborne illness, North Americans are also concerned about health risks from chemicals in foods. In the long run, this concern has some merit. On a day-to-day basis, however, food additives cause only about 4% of all cases of foodborne illness in North America. Microbial contamination of food is by far the more important issue for our short-term health, so it will be discussed first. This chapter will then cover chemical food safety hazards, including the use and safety of food additives, and discuss the risks of pesticides in foods.

EFFECTS OF FOODBORNE ILLNESS

According to the U.S. Centers for Disease Control and Prevention, foodborne illness causes 48 million illnesses, 128,000 hospitalizations, and 3000 deaths in the United States each year (see Further Readings 7 and 9). Some people are particularly susceptible to foodborne illness, including the following:

- Infants and children
- Older adults
- Those with liver disease, diabetes, HIV infection (and AIDS), or cancer
- Postsurgical patients
- Pregnant women
- People taking immunosuppressant agents (e.g., transplant patients)

Some bouts of foodborne illness, especially when coupled with ongoing health problems, are lengthy and lead to food allergies, seizures, blood poisoning (from **toxins** or microorganisms in the bloodstream), or other illnesses. Foodborne illnesses often result from the unsafe handling of food at home, so we each bear some responsibility for preventing them (see Further Readings 6 and 10). You cannot usually tell that a particular food contains harmful microorganisms by taste, smell, or sight; therefore, you might not even suspect that food has caused your distress. In fact, your last case of diarrhea may have been caused by foodborne illness (Table 13-1). In response to the significant public health burden of foodborne illness that is largely preventable, the FDA Food Safety Modernization Act was signed into law by President Barack Obama on January 4, 2011. This new law strengthens the food safety system, enabling FDA to better protect public health.

TABLE 13-1 ▸ **Some Examples of Cases of Foodborne Illness. We Generally Have a Safe Food Supply, but There Are Occasional Instances of Foodborne Illnesses, Such as Those Listed**

Viruses

- *Norovirus:* In 2012, there were 16 outbreaks on international cruise ships. Outbreaks occurred on seven Princess cruises, affecting 288 passengers and 75 crew members on one ship and 364 passengers and 32 crew members on another. In December 2012, a new Australian strain of Norovirus, Gil 4 Sydney, resulted in 204 passengers and 16 crew members becoming ill on the *Queen Mary 2.*

- *Hepatitis A:* In 2013, 113 people were sickened and 50 were hospitalized in seven states by an organic blend of frozen berries sold at Costco that was contaminated with hepatitis A. In 2003, over 500 adults in the United States contracted hepatitis A after eating raw green onions in a Mexican restaurant. These were contaminated during growth in Mexico and not properly washed by food service workers.

Bacteria

- *Salmonella:* In 2012, 224 persons in 34 states were infected with a strain of *Salmonella*. This outbreak was linked to contact with chicks, ducklings, and other live baby poultry purchased from multiple feed stores. In 2008–2009, two outbreaks of salmonella cost the food industry millions of dollars. The first outbreak was traced to imported jalapeno and serrano peppers from one Mexican farm. Illnesses led to 282 hospitalizations and 2 deaths. The second outbreak, traced to peanut butter from a processing plant in Georgia, may have caused at least eight deaths.
- *Shigella:* In 1994, more than 600 people on a cruise ship developed shigellosis and one person died.
- *Listeria:* In 2012, ricotta salata cheese imported from Italy caused 22 cases of foodborne illness in 14 states and resulted in 4 deaths. In 2011, 147 persons in 28 states were infected with *Listeria* from cantaloupes. Most of those affected were older than 60 years; 143 were hospitalized and 33 died.
- *E. coli:* In June 2013, an outbreak of *E. coli* O157:H7 associated with food from Los Burritos restaurant in Lombard, Illinois resulted in 11 confirmed and 10 more probable cases of *E. coli* poisoning. The world's deadliest outbreak of *E. coli* occurred during the summer of 2011 with 3332 persons infected, more than 600 in intensive care, and a death toll of 36. The outbreak occurred mainly in Germany and involved the rare *E.coli* O104:H4 strain. Several hundred also contracted its potentially fatal kidney complication. The *E.coli* infection was caused by tainted vegetable sprouts from a small, rather traditional organic sprout farm. The sprout seeds were mostly imported

from overseas, and the *E.coli* bacteria were antibiotic resistant. In the United States, the largest *E. coli* O157:H7 outbreak on record infected more than 1000 people in upstate New York at a county fair. The bacterium was found in infected well water. It killed a 79-year-old man and a 4-year-old girl, and it caused 10 other children to undergo kidney dialysis. Six adults and a 2-year-old child were killed after an *E. coli* outbreak from contaminated drinking water in Canada. The bacteria entered the water supply from animal manure after flooding from a heavy storm. In an outbreak of 199 persons infected with *E. coli* from fresh spinach, *E. coli* O157:H7 was isolated from 13 packages of fresh spinach in 10 states.
- *Clostridium botulinum:* A man in Arkansas developed botulism after eating stew that was cooked and then kept at room temperature for 3 days. He spent 42 days on mechanical ventilation.
- *Vibrio:* Since 1992, 17 people in Florida have died of *Vibrio vulnificus* infections after eating raw oysters.
- *Bacillus cereus:* A teenage boy and his father experienced abdominal pain, vomiting, and diarrhea within 30 minutes of eating 4-day-old homemade pesto that had been reheated and left out a number of times and was apparently contaminated with *B. cereus*. The boy died of liver failure.
- *Campylobacter:* In June 2013, many people became ill with *Campylobacter* infections after attending a Burger and Beer Bash outdoor festival at the Kenisco Dam in Valhalla, New York. The bacteria was identified from samples taken from people who became ill after attending the festival.

Parasites

- *Cryptosporidium:* In March 2012, a cryptosporidiosis outbreak involved 20 firefighters who responded to a fire in a barn housing calves infected with *Cryptosporidium*. The firefighters experienced gastrointestinal illness that caused missed work

days and one hospitalization. Previous outbreaks were associated with exposure to young livestock as well as water parks and contaminated municipal water sources.

Risks from Seafood

- *Ciguatera:* An outbreak of ciguatera fish poisoning involved 17 crew members of a cargo ship that caught, cooked, and ate a barracuda in the Bahamas. All became ill with nausea, vomiting, abdominal cramps, and diarrhea within hours of

eating the fish. Within 2 days, all suffered from muscle pain and weakness; dizziness; and numb or itchy feet, hands, and mouth.

It allows FDA to focus on prevention of food safety problems before they occur. The law also provides new tools for inspection and compliance and for holding imported foods to the same standards as domestic foods. As a result of this legislation, the National Integration of Food Safety Initiative supports states, institutions, and disciplines that work together in food safety research, extension, and education activities. The law also directed FDA to build a national food safety system that is integrated and in partnership with state and local authorities. Although government agencies are at work on problems regarding food safety, this does not substitute for individual safety efforts (Table 13-2) (see Further Reading 1).

TABLE 13-2 ▶ Agencies Responsible for Monitoring the Food Supply in the United States

Agency Name	Responsibilities	Methods	How to Contact
United States Department of Agriculture (USDA)	• Enforces wholesomeness and quality standards for grains and produce (while in the field), meat, poultry, milk, eggs, and egg products	• Inspection • Grading • "Safe Handling Label"	**www.fsis.usda.gov**
Bureau of Alcohol, Tobacco, Firearms and Explosives (ATF)	• Enforces laws on alcoholic beverages	• Inspection	**www.atf.gov**
Environmental Protection Agency (EPA)	• Regulates pesticides • Establishes water quality standards	• Approval required for all U.S. pesticides • Sets pesticide residue limits in food	**www.epa.gov**
Food and Drug Administration (FDA)	• Ensures safety and wholesomeness of all foods in interstate commerce (except meat, poultry, and processed egg products) • Regulates seafood • Controls product labels	• Inspection • Food sample studies • Sets standards for specific foods	**www.fda.gov** or call 1-800-FDA-4010 For the FDA Center for Food Safety and Applied Nutrition (CFSAN) information, call 1-888-SAFEFOOD.
Centers for Disease Control and Prevention (CDC)	• Promotes food safety	• Responds to emergencies concerning foodborne illness • Surveys and studies environmental health problems • Directs and enforces quarantines • National programs for prevention and control of foodborne and other diseases	**www.cdc.gov**
National Marine Fisheries Service or NOAA Fisheries	• Domestic and international conservation and management of living marine resources	• Voluntary seafood inspection program • Can use official mark to show federal inspection	**www.nmfs.noaa.gov**
State and local governments	• Milk safety • Monitors food industry within their borders	• Inspection of food-related establishments	Government pages of telephone book Visit **www.FoodSafety.gov** for food safety information.

Government agencies responsible for monitoring food safety in Canada and the specific laws followed can be found at **http://www.inspection.gc.ca/**.

WHY IS FOODBORNE ILLNESS SO COMMON?

Foodborne illness is carried or transmitted to people by food. Most foodborne illnesses are transmitted through food in which microorganisms are able to grow rapidly. These foods are generally moist, rich in protein, and have a neutral or slightly acidic pH. Unfortunately, this describes many of the foods we eat every day, such as meats, eggs, and dairy products.

Our food industry tries whenever possible to increase the shelf life of food products; however, a longer shelf life allows more time for bacteria in foods to multiply. Some bacteria even grow at refrigeration temperatures. Partially cooked—and some fully cooked—products pose a particular risk because refrigerated storage may only slow, not prevent, bacterial growth. The risk of contracting foodborne illness also is high because of consumer trends. First, there is greater consumer interest in eating raw or undercooked animal products. In addition, more people receive medication that suppresses their ability to combat foodborne infectious agents. Another factor is the continuing increase in the number of older adults in the population.

The risk of illness from foodborne microorganisms increases as more of our foods are prepared in kitchens outside the home. With the high number of two-income families, many people look for convenient, easy-to-prepare, nutritious foods. Supermarkets have become an alternative to cooking at home by offering a variety of prepared foods from specialty meat departments, salad bars, and bakeries. Supermarkets offer entrées that can be served immediately or reheated. The foods are usually prepared in central kitchens or processing plants and shipped to individual stores.

This centralization of food production by the food-processing and restaurant industry enhances the risk of foodborne illness. If a food product is contaminated in a central processing plant, consumers over a wide area can suffer foodborne illness. For example, a malfunction in an ice cream plant in Minnesota resulted in 224,000 suspected cases of *Salmonella* bacterial infections, which were linked to use of contaminated ice cream mix. At least 4 people died and 700 became ill in Washington and surrounding western states after eating at a chain of fast-food restaurants. The source of the problem was undercooked hamburger contaminated with the bacterium *E. coli* 0157:H7. Restaurants are inspected by health departments only about every 6 months, so we must rely on each restaurant to handle foods safely.

Greater consumption of ready-to-eat foods imported from foreign countries is still another cause of increased foodborne illness in North America. In the past, food imports were mostly raw products processed here under strict sanitation standards. Now, however, we import more ready-to-eat processed foods—such as berries from Guatemala and shellfish from Asia—some of which are contaminated. U.S. authorities are re-examining inspection procedures for these imports. For example, a 2012 outbreak of *Listeria* was caused by contaminated ricotta cheese imported from Italy.

The use of antibiotics in animal feeds is also increasing the severity of cases of foodborne illness. FDA estimates that antibiotic use in livestock increased from about 18 million pounds in 1999 to almost 30 million pounds in 2011. Fortunately, the animals pass the antibiotics through their systems long before they are slaughtered. In addition, animal products to be used for human consumption are tested for antibiotics. The potentially harmful effect of antibiotic use is that it encourages the development of antibiotic-resistant strains of bacteria, which can grow even if exposed to typical antibiotic medicines. This issue is receiving considerable attention by scientists in the field.

Finally, more cases of foodborne disease are reported now because scientists are more aware of the roles of various players in the process. Every decade, the list of microorganisms suspected of causing foodborne illness lengthens. In addition, physicians are more likely to suspect foodborne contaminants as a cause of illness. Furthermore, we now know that food, besides serving as a good growth medium for some microorganisms, transmits many others as well. Seafood is receiving greater scrutiny and surveillance by FDA as a cause of foodborne illness. FDA's Center for Food Safety and Applied Nutrition is a good source of information about food safety.

▲ Restaurants that choose to serve raw or undercooked foods of animal origin, such as eggs or meats, are required to have a consumer advisory on the menu warning customers of the health risks of ordering such foods.

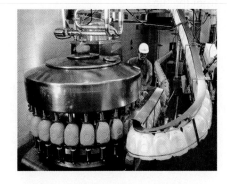

▲ Food contaminated in a central plant can go on to produce illness in people across the nation. In the case of juices, pasteurization is an effective method of reducing the risk of foodborne illness.

✓ CONCEPT CHECK 13.1

1. What are the effects of foodborne illnesses in the United States each year in terms of hospitalizations and deaths?
2. What lifestyle changes have made foodborne illness so common today?
3. What agencies are responsible for monitoring the safety of our food supply?

13.2 Food Preservation—Past, Present, and Future

For centuries, salt, sugar, smoke, fermentation, and drying have been used to preserve food. Ancient Romans used sulfites to disinfect wine containers and preserve wine. In the age of exploration, European adventurers traveling to the New World preserved their meat by salting it. Most preserving methods work on the principle of decreasing water content. Bacteria need abundant stores of water to grow; yeasts and molds can grow with less water, but some is still necessary. Adding sugar or salt binds water and so decreases the water available to these microbes. The process of drying evaporates free water.

Decreasing the water content of some high-moisture foods, however, would cause them to lose essential characteristics. To preserve such foods—cucumber pickles, sauerkraut, milk (yogurt), and wine—fermentation has been a traditional alternative. Selected bacteria or yeast are used to ferment or pickle foods. The fermenting bacteria or yeast makes acids and alcohol, which minimize the growth of other bacteria and yeast.

Today, we can add pasteurization, sterilization, refrigeration, freezing, food **irradiation,** canning, and chemical preservation to the list of food preservation techniques. An additional method of food preservation—**aseptic processing**—simultaneously sterilizes the food and package separately before the food enters the package. Liquid foods, such as fruit juices, are especially easy to process in this manner. With aseptic packaging, boxes of sterile milk and juices can remain unrefrigerated on supermarket shelves, free of microbial growth, for many years.

Food irradiation uses minimal doses of radiation to control pathogens such as *E. coli* O157:H7 and *Salmonella.* The **radiation** energy used does not make the food radioactive. The energy essentially passes through the food, as in microwave cooking, and no radioactive residues are left behind. However, the energy is strong enough to break chemical bonds, destroy cell walls and cell membranes, break down DNA, and link proteins together. Irradiation thereby controls growth of insects, bacteria, fungi, and parasites in foods.

FDA approved the use of irradiation for raw red meat to reduce risk of *E. coli* and other infectious microorganisms. Other additions to the approved list are shell eggs and seeds. Prior to this, the only animal products so treated were pork and chicken. Irradiation also extends the shelf life of spices, dry vegetable seasonings, other meats, and fresh fruits and vegetables.

Irradiated food, except for dried seasonings, must be labeled with the international food irradiation symbol, the Radura, and a statement that the product has been treated by irradiation. Foods treated in this way are safe in the opinion of FDA and many other health authorities, including the American Academy of Pediatrics. Although the demand for irradiated foods is still low in the United States, other countries, including Canada, Japan, Italy, and Mexico, all use food irradiation technology widely. Certain consumer groups continually try to block its use in the United States, claiming that irradiation diminishes the nutritional value of food and that it can lead to the formation of harmful compounds, such as carcinogens. The safety of irradiated food has been evaluated and confirmed by FDA for over 30 years. It has also been shown that irradiation does not compromise nutritional quality or noticeably change the taste, texture, or appearance of food. Harmful compounds are not found in irradiated food at any higher level than in foods that are not irradiated. Keep in mind that even when foods, especially meats, have been irradiated, it is still important to follow basic food-safety procedures, as later contamination during food preparation is possible.

irradiation A process in which **radiation** energy is applied to foods, creating compounds (free radicals) within the food that destroy cell membranes, break down DNA, link proteins together, limit enzyme activity, and alter a variety of other proteins and cell functions of microorganisms that can lead to food spoilage. This process does not make the food radioactive.

radiation Literally, energy that is emitted from a center in all directions. Various forms of radiation energy include X-rays and ultra-violet rays from the sun.

aseptic processing A method by which food and container are separately and simultaneously sterilized; it allows manufacturers to produce boxes of milk that can be stored at room temperature.

▲ This is the Radura, the international label denoting prior irradiation of the food product.

✔ CONCEPT CHECK 13.2

1. What food preservation techniques have been used for centuries?
2. Why is irradiation considered a safe technique to preserve food?

13.3 Foodborne Illness Caused by Microorganisms

Most cases of foodborne illness are caused by specific viruses, bacteria, and other fungi. Prions—proteins involved in maintaining nerve cell function—can also turn infectious and lead to diseases such as bovine spongiform encephalopathy, better known as mad cow disease. Bacteria specifically cause health problems either directly by invading the intestinal wall and producing an *infection* via a toxin contained in the organism or indirectly by producing a toxin secreted into the food, which later harms us (called an *intoxication*). The main way to distinguish an infectious route from an intoxication is time: If symptoms appear in 4 hours or less, it is an intoxication.

BACTERIA

Bacteria are single-cell organisms found in the food we eat, the water we drink, and the air we breathe. Many types of bacteria cause foodborne illness, including *Bacillus, Campylobacter, Clostridium, Escherichia, Listeria, Vibrio, Salmonella,* and *Staphylococcus* (Table 13-3). Bacteria are everywhere: each teaspoon of soil contains about 2 billion bacteria. Luckily, only a small number of all bacteria pose a threat.

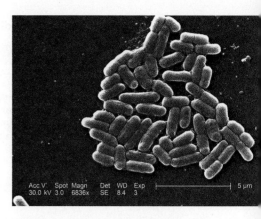

▲ This is an electron micrograph of *Escherichia coli* bacteria, strain O157:H7, magnified 6836x. Although most strains of *E. coli* are harmless and live in the intestines of healthy humans and animals, this strain produces a toxin that causes severe illness. The first recognized outbreak of illness from *E. coli* O157:H7 occurred in 1982 from contaminated hamburgers and caused bloody diarrhea. Since then, most infections have been linked to eating undercooked ground beef.

TABLE 13-3 ▶ Bacterial Causes of Foodborne Illness

Bacteria	Typical Food Sources	Illness	Additional Information
Salmonella species 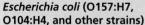	Raw and undercooked meats, poultry, eggs, and fish; produce, especially raw sprouts; peanut butter; unpasteurized milk (see Further Reading 2)	Onset: 12–72 hours; Symptoms: nausea, fever, headache, abdominal cramps, diarrhea, and vomiting; can be fatal in infants, the elderly, and those with impaired immune systems; Duration: 4–7 days	Estimated 1 million infections/year; bacteria live in the intestines of animals and humans; food is contaminated by infected water and feces; about 2000 strains of *Salmonella* bacteria can cause disease but 3 strains account for almost 50% of cases; *Salmonella enteritidis* infects the ovaries of healthy hens and contaminates eggs; almost 20% of cases are from eating undercooked eggs or egg-containing dishes; reptiles, such as turtles, also spread the disease
Campylobacter jejuni 	Raw and undercooked meat and poultry (more than half of raw poultry in the United States is contaminated), unpasteurized milk, and contaminated water	Onset: 2–5 days; Symptoms: muscle pain, abdominal cramping, diarrhea (sometimes bloody), and fever; Duration: 2–7 days	Estimated 845,000 infections/year; produces a toxin that destroys intestinal mucosal surfaces; can cause Guillain-Barré syndrome, a rare neurological disorder that causes paralysis
Escherichia coli (O157:H7, O104:H4, and other strains)	Undercooked ground beef; produce (e.g., lettuce, spinach, and sprouts); unpasteurized juice and milk	Onset: 1–8 days; Symptoms: bloody diarrhea and abdominal cramps; in children under age 5 and the elderly, hemolytic uremic syndrome (HUS) is a serious complication; red blood cells are destroyed and kidneys fail; can be fatal; Duration: 5–10 days	Leading cause of bloody diarrhea in the United States; estimated 73,000 cases/year; lives in the intestine of healthy cattle; cattle and cattle manure are chief sources; illness caused by a powerful toxin made by the bacteria; petting zoos, lakes, and swimming pools can contain pathogenic *E. coli*

(continued)

TABLE 13-3 **Bacterial Causes of Foodborne Illness** *(continued)*

Bacteria	Typical Food Sources	Illness	Additional Information
Shigella species	Fecal/oral transmission; water supplies, produce, and other foods contaminated by infected food handlers with poor hygiene	Onset: 1–3 days; Symptoms: abdominal cramps, fever, and diarrhea (often bloody); Duration: 5–7 days	Estimated 448,000 cases/year; humans and primates are the only sources; common in day-care centers and custodial institutions from poor hygiene; traveler's diarrhea often caused by *Shigella dysenteriae*
Staphylococcus aureus	Ham, poultry, egg salads, cream-filled pastries, custards, and whipped cream	Onset: 1–6 hours; Symptoms: diarrhea, vomiting, nausea, and abdominal cramps; Duration: 1–3 days	Estimated 241,148 cases/year; bacteria live on skin and nasal passages of up to 25% of people; can be passed to foods; multiplies rapidly when contaminated foods are held for extended time at room temperature; illness caused by a heat-resistant toxin that cannot be destroyed by cooking
Clostridium perfringens	Beef, poultry, gravy, and Mexican food	Onset: 8–24 hours; Symptoms: abdominal pain and diarrhea, usually mild; can be more serious in elderly or ill persons; Duration: lasts 1 day or less	Estimated 966,000 cases/year; anaerobic bacteria widespread in soil and water; multiplies rapidly in prepared foods, such as meats, casseroles, and gravies, held for extended time at room temperature
Listeria monocytogenes	Unpasteurized milk and soft cheeses, raw meats, uncooked vegetables, ready-to-eat deli meats and hotdogs, and refrigerated smoked fish	Onset: 9–48 hours for early symptoms, 14–42 days for severe symptoms; Symptoms: fever, muscle aches, headache, vomiting; can spread to nervous system, resulting in stiff neck, confusion, loss of balance, or convulsion; can cause premature birth and stillbirth	Estimated 1600 cases with 255 fatalities/year; widespread in soil and water and can be carried in healthy animals; grows at refrigeration temperatures; about one-third of cases occur during pregnancy; high-risk persons should avoid uncooked deli meats, soft cheeses (e.g., feta, Brie, and Camembert), blue-veined cheeses, Mexican-style cheeses (e.g., queso blanco made from unpasteurized milk), refrigerated meat spreads or pates, uncooked refrigerated smoked fish
Clostridium botulinum	Incorrectly home-canned vegetables, meats, and fish; incorrectly canned commercial foods; herb-infused oils; bottled garlic; potatoes baked in foil and held at room temperature; honey	Onset: 18–36 hours but can be 6 hours to 10 days; Symptoms: neurological symptoms—double and blurred vision, drooping eyelids, slurred speech, difficulty swallowing, muscle weakness, and paralysis of face, arms, respiratory muscles, trunk, and legs; can be fatal; Duration: lasts days to weeks	Estimated 100 cases/year; caused by a neurotoxin; *C. botulinum* grows only in the absence of air in nonacidic foods; incorrect home canning causes most botulism, but in 2007, commercially canned chili sauce caused an outbreak; honey can contain botulism spores and should not be given to infants younger than 1 year of age

(continued)

TABLE 13-3 Bacterial Causes of Foodborne Illness *(continued)*

Bacteria	Typical Food Sources	Illness	Additional Information
Vibrio 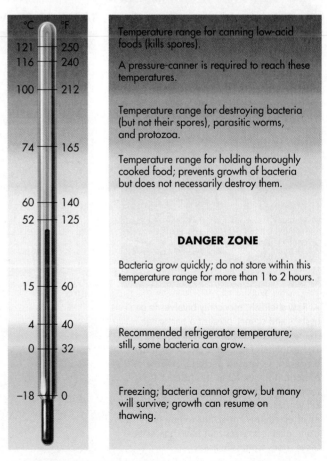	*V. parahemolyticus:* raw and undercooked shellfish, especially oysters	Onset: 24 hours; Symptoms: watery diarrhea, nausea, vomiting, fever, and chills; Duration: lasts 3 days	Found in coastal waters; more infections in summer; number of infections hard to determine because it is difficult to isolate in the laboratory
	V. vulnificus: raw and undercooked shellfish, especially oysters	Onset: 1–2 days; Symptoms: vomiting, diarrhea, and abdominal pain; in more severe cases, bloodstream infection with fever, chills, decreased blood pressure, and blistering skin lesions; Duration: lasts 3 or more days	Estimated 95 cases/year; found in coastal waters; more infections in summer; those with impaired immune systems and liver disease at higher risk of infection; fatality rate of 50% with bloodstream infection
	V. cholerae: contaminated water and food, human carriers	Onset: 2–3 days; Symptoms: severe, dehydrating diarrhea, and vomiting; dehydration, cardiovascular collapse, and death can occur	Occurs mainly in countries without adequate water purification and sewage treatment
Yersinia enterocolitica	Raw or undercooked pork, particularly pork intestines (chitterlings); tofu; water; unpasteurized milk	Onset: 4–7 days; Symptoms: fever, abdominal pain, and diarrhea (often bloody); Duration: lasts 1–3 weeks or longer	Yersinosis most common in children under age 5 years; relatively rare; bacteria live mainly in pigs but can be found in other animals

Some foodborne bacteria cause infections, whereas others cause intoxications. *Salmonella,* for example, causes an infection because the bacteria cause the illness. *Clostridium botulinum, Staphylococcus aureus,* and *Bacillus cereus* produce toxins and therefore cause illness from intoxication. In addition, whereas most strains of *E. coli* are harmless, *E. coli* O157:H7 and O104:H4 produce a toxin that can cause severe illness, including severe bloody diarrhea and hemolytic uremic syndrome (HUS). Bacterial foodborne illnesses typically cause gastrointestinal symptoms such as vomiting, diarrhea, and abdominal cramps. *Salmonella, Listeria, E. coli* O157:H7 and O104:H4, and *Campylobacter* are the bacterial foodborne illnesses of particular interest because they are the ones most often associated with death. *E. coli* O157:H7 and O104:H4 have caused deaths when HUS has developed. Of the 1651 cases of listeriosis in the United States from 2009 through 2011, 292 were fatal. Listeriosis is of particular concern for pregnant women because they are about 20 times more likely to get this infection than other healthy adults, and listeriosis can cause spontaneous abortion or stillbirth because the *Listeria* bacteria can cross the placenta and infect the fetus.

To proliferate, bacteria require nutrients, water, and warmth. Most grow best in *danger zone* temperatures of 40° to 140°F (4° to 60°C) (Fig. 13-1). Pathogenic bacteria typically do not multiply when food is held at temperatures above 140°F (60°C) or stored at safe refrigeration temperatures, 32° to 40°F (0° to 4.4°C). One important exception is *Listeria* bacteria, which can multiply at refrigeration temperatures. Also note that high temperatures can kill toxin-producing bacteria, but any toxin produced in the food will not be inactivated by high temperatures. Most pathogenic bacteria also require oxygen

FIGURE 13-1 ▲ Effects of temperature on microbes that cause foodborne illness.

for growth, but *Clostridium botulinum* and *Clostridium perfringens* grow only in anaerobic (oxygen-free) environments, such as those found in tightly sealed cans and jars. Food acidity can affect bacterial growth, too. Although most bacteria do not grow well in acidic environments, some, such as disease-causing *E. coli*, can grow in acidic foods, such as fruit juice.

VIRUSES

Viruses, like bacteria, are widely dispersed in nature. Unlike bacteria, however, viruses can reproduce only after invading body cells, such as those that line the intestines. Experts speculate that about 70% of foodborne illness cases go undiagnosed because they result from viral causes, and there is no easy way to test for these pathogens. Table 13-4 describes the two most common viral causes of foodborne illness and typical food sources and symptoms of the illnesses they cause. Norovirus is the number one pathogen contributing to domestically acquired foodborne illnessses. It causes an illness commonly misdiagnosed as the "stomach flu." Norovirus infection has a sudden onset and usually a short duration of only one to two days. Noroviruses cause almost all (over 90%) of the highly publicized diarrheal disease outbreaks on cruise ships. They are hardy and survive freezing, relatively high temperatures, and chlorination up to 10 parts per million. The most commonly reported Norovirus outbreaks are at long-term care facilities (Fig. 13-2).

PARASITES

Parasites live in or on another organism, known as the host, from which they absorb nutrients. Humans may serve as hosts to parasites. These tiny ravagers rob millions of people around the globe of their health and, in some cases, their lives. Those

▲ Raw shellfish, especially bivalves (e.g., oysters and clams), present a particular risk related to foodborne viral disease. These animals filter feed, a process that concentrates viruses, bacteria, and toxins present in the water as it is filtered for food. Adequate cooking of shellfish will kill viruses and bacteria, but toxins may not be affected. It is important to buy shellfish from reliable sources that have harvested these foods from safe areas.

TABLE 13-4 ▶ Viral Causes of Foodborne Illness

Viruses	Typical Food Sources	Illness	Additional Information
Norovirus (Norwalk and Norwalk-like viruses), human rotavirus	Foods prepared by infected food handlers; shellfish from contaminated waters; vegetables and fruits contaminated during growing, harvesting, and processing (see Further Reading 9)	Onset: 1–2 days; Symptoms: "stomach flu"—severe diarrhea, nausea, vomiting, stomach cramping, low-grade fever, chills, and muscle aches; Duration: lasts 1–2 days or longer	Estimated to cause over 20 million cases of gastroenteritis, 70,000 hospitalizations, and 800 deaths per year. Viruses found in stool and vomit of infected persons; food handlers can contaminate foods or work surfaces; noroviruses are very infectious—as few as 10–100 particles can lead to infection; workers with norovirus symptoms should not work until 2 or 3 days after they feel better.
Hepatitis A virus	Foods prepared by infected food handlers, especially uncooked foods or those handled after cooking, such as sandwiches, pastries, and salads; shellfish from contaminated waters; vegetables and fruits contaminated during growing, harvesting, and processing	Onset: 15–50 days; Symptoms: anorexia, diarrhea, fever, jaundice, dark urine, and fatigue; may cause liver damage and death; Duration: lasts several weeks up to 6 months	Infected food handlers contaminate food and transmit the disease to dozens of persons; children and young adults are more susceptible; a vaccine is available, decreasing the number of infections dramatically; immunoglobulin given within 1 week to those exposed to hepatitis A virus can also decrease infection.

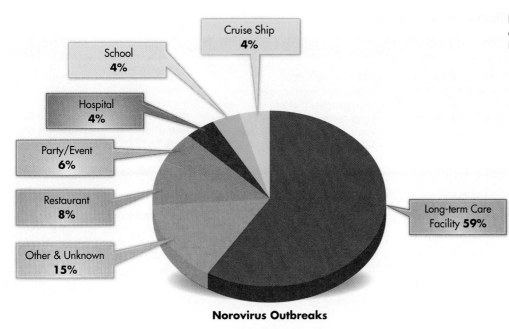

Norovirus Outbreaks

hardest hit live in tropical countries where poor sanitation fosters the growth of parasites.

The more than 80 foodborne parasites known to affect humans include mainly **protozoa** (one-celled animals), such as *Cryptosporidium* and *Cyclospora,* and **helminths,** such as tapeworms and the roundworm *Trichinella spiralis.* Table 13-5 describes common parasites and typical food sources and symptoms of the illnesses they cause. Parasitic infections spread via person-to-person contact and contaminated food, water, and soil.

protozoa One-celled animals that are more complex than bacteria. Disease-causing protozoa can be spread through food and water.

helminth Parasitic worm that can contaminate food, water, feces, animals, and other substances.

TABLE 13-5 ▶ **Parasitic Causes of Foodborne Illness**

Parasite	Typical Food Sources	Illness	Additional Information
Trichinella spiralis	Pork, wild game	Onset: weeks to months; Symptoms: GI symptoms followed by muscle weakness, fluid retention in the face, fever, and flulike symptoms	The number of trichinosis infections has decreased greatly because pigs are now less likely to harbor this parasite; cooking pork to 160°F (72°C) will kill *trichinella,* as will freezing it for 3 days at −4°F (−20°C).
Anisakis	Raw or undercooked fish	Onset: 12 hours or less; Symptoms: violent stomach pain, nausea, and vomiting	Caused by eating the larvae of roundworms; the infection is more common where raw fish is routinely consumed
Tapeworms	Raw beef, pork, and fish	Symptoms: Abdominal discomfort and diarrhea	Tapeworm larvae can get into the muscles of their hosts and cause infection when raw or undercooked meat from an infected animal is eaten.

(continued)

TABLE 13-5 Parasitic Causes of Foodborne Illness *(continued)*

Parasite	Typical Food Sources	Illness	Additional Information
Toxoplasma gondii	Raw or undercooked meat, unwashed fruits and vegetables	Onset: 5–20 days; Symptoms: most people are asymptomatic; those with symptoms have fever, headache, sore muscles, and diarrhea; can be fatal to the fetus of pregnant women	Parasite is spread to humans from animals, including cats, the main reservoir of the disease; humans acquire the disease from ingesting contaminated meat or from fecal contamination from handling cat litter.
Cyclospora cayetanensis	Water, contaminated food	Onset: 1 week; Symptoms: watery diarrhea, vomiting, muscle aches, fatigue, anorexia, and weight loss; Duration: lasts 10–12 weeks	Most common in tropical and subtropical areas, but since 1990, about a dozen outbreaks, affecting 3600 people, have occurred in the United States and Canada.
Cryptosporidium	Water, contaminated food	Onset: 2–10 days; Symptoms: watery diarrhea, abdominal pain, fever, nausea, vomiting, and weight loss; those with impaired immune systems become more ill; Duration: lasts 1–2 weeks in otherwise healthy persons	Outbreaks occur worldwide; the largest U.S. outbreak was in 1993 in Milwaukee, with more than 443,000 persons affected; also can be spread in water parks and community swimming pools.

☑ CONCEPT CHECK 13.3

1. What is the temperature "danger zone"?
2. What type of microorganisms pose the greatest risk for foodborne illness?

13.4 Food Additives

By the time you see a food on the market shelf, it usually contains substances added to make it more palatable or increase its nutrient content or shelf life. Manufacturers also add some substances to foods to make them easier to process. Other substances may have accidentally found their way into the foods you buy. All of these extraneous substances are known as **additives,** and, although some may be beneficial, others, such as sulfites, may be harmful for some people. All purposefully added substances must be evaluated by FDA.

WHY ARE FOOD ADDITIVES USED?

Most additives are used to limit food spoilage. Common food additives serve the general function of **preservatives,** including acidic or alkaline agents, antioxidants, antimicrobial agents, curing and pickling agents, and **sequestrants.** Table 13-6 helps you to understand exactly why these are used and to learn more about the specific substances used. Food additives, such as potassium sorbate, are used to maintain the safety and acceptability of foods by retarding the growth of microbes implicated in foodborne illness.

additives Substances added to foods, either intentionally or incidentally.

preservatives Compounds that extend the shelf life of foods by inhibiting microbial growth or minimizing the destructive effect of oxygen and metals.

sequestrants Compounds that bind free metal ions. By so doing, they reduce the ability of ions to cause rancidity in foods containing fat.

TABLE 13-6 ▶ Types of Food Additives—Sources and Related Health Concerns

Food Additive Class	Attributes	Health Risks
Acidic or alkaline agents, such as citric acid, calcium lactate, and sodium hydroxide	Acids impart a tart taste to soft drinks, sherbets, and cheese spreads; inhibit mold growth; and lessen discoloration and rancidity. They also reduce the risk of botulism in naturally low-acid vegetables, such as canned green beans. Alkaline agents neutralize acids produced during fermentation and so improve flavor.	No known health risks when used properly
Alternative low-calorie sweeteners, such as saccharin, sucralose, ascesulfame potassium, aspartame, neotame, and tagatose	Sweeten foods without adding more than a few calories	Moderate use of these alternative sweeteners is considered safe (except for use of aspartame by people with the disease PKU)
Anticaking agents, such as calcium silicate, magnesium stearate, and silicon dioxide	Absorb moisture to keep table salt, baking powder, or powdered sugar and powdered food products free-flowing and prevent caking and lumping	No known health risks when used properly
Antimicrobial agents, such as salt, sodium benzoate, sorbic acid, and calcium propionate	Inhibit mold and fungal growth	Salt increases the risk of developing hypertension, especially in sodium-sensitive individuals. No known health risks from other agents when used properly
Antioxidants, such as BHA (butylated hydroxyanisole), BHT (butylated hydroxytoluene), alpha-tocopherol (vitamin E), ascorbic acid (vitamin C), and sulfites	Delay food discolorations from oxygen exposure; reduce rancidity from the breakdown of fats; maintain the color of luncheon meats; prevent the formation of cancer-causing nitrosamines	Sulfites can cause an allergic reaction in about 1 in every 100 people. Symptoms include difficulty breathing, wheezing, hives, diarrhea, abdominal pain, cramps, and dizziness. Salad bars, dried fruit, and wine are typical sources of sulfites.
Color additives, such as tartrazine	Make foods more appealing	Tartrazine (FD&C yellow number 5) can cause allergic symptoms such as hives and nasal discharge in some people, especially those allergic to aspirin. FDA requires manufacturers to list all forms of synthetic colors on the labels of foods that contain them.
Curing and pickling agents, such as salt, nitrates, and nitrites	Nitrates and nitrites act as preservatives, especially to prevent the growth of *Clostridium botulinium;* often used in conjunction with salt.	Salt increases the risk of developing hypertension, especially in sodium-sensitive individuals. Nitrate and nitrite consumption from both cured foods and that found naturally in some vegetables has been associated with synthesis of nitrosamines. (An adequate vitamin C intake may reduce this synthesis.) Some nitrosamines are cancer-causing agents, particularly for the stomach, esophagus, and colon, but the risk is low. The National Cancer Institute advises consuming these foods in moderation.
Emulsifiers, such as monoglycerides and lecithins	Suspend fat in water to improve uniformity, smoothness, and body of foods, such as baked goods, ice cream, and mayonnaise	No known health risks when used properly
Fat replacements, such as Paselli SA2, Dur-Low, Oatrim, Sta-Slim 143, Stellar, and Olean	Limit calorie content of foods by reducing some of the fat content	Generally no known health risks when used properly; possible loss of fat-soluble vitamins and GI side effects if used in excess
Flavor and flavoring agents (such as natural and artificial flavors), sugar, and corn syrup	Impart more or improve flavor of foods	Sugar and corn syrup can increase risk for dental caries. Generally no known health risks for flavoring agents when used properly; possible weight gain from excess calories
Flavor enhancers, such as monosodium glutamate (MSG) and salt	Help bring out the natural flavor of foods, such as meats	Some people (especially infants) are sensitive to the glutamate portion of MSG and after exposure experience flushing, chest pain, facial pressure, dizziness, sweating, rapid heart rate, nausea, vomiting, increase in blood pressure, and headache. Those so affected should look for the word *glutamate* on food labels, especially on isolated protein, yeast extract, bouillon, and soup stock. Salt increases the risk of developing hypertension, especially in sodium-sensitive people.

(continued)

Food Additive Class	Attributes	Health Risks
Humectants, such as glycerol, propylene glycol, and sorbitol	Retain more moisture, texture, and fresh flavor in foods such as candies, shredded coconut, and marshmallows	No known health risk when used properly
Leavening agents, such as yeast, baking powder, and baking soda	Introduce carbon dioxide into food products	No known health risk when used properly
Maturing and bleaching agents, such as bromates, peroxides, and ammonium chloride	Shorten the time needed for maturation of flour to become usable for baking products	No known health risk when used properly
Nutrient supplements, such as vitamin A, vitamin D, and potassium iodide	Enhance the nutrient content of foods such as margarine, milk, and ready-to-eat breakfast cereals	No known health risk if intake from such supplemental sources combined with other natural food sources of a nutrient does not exceed the Upper Level set for a particular nutrient
Stabilizers and thickeners, such as pectins, gums, gelatins, and agars	Impart a smooth texture and uniform color and flavor to candies, ice cream and other frozen desserts, chocolate milk, and beverages containing alternative sweeteners; prevent evaporation and deterioration of flavorings used in cakes, puddings, and gelatin mixes	No known health risk when used properly
Sequestrants, such as EDTA and citric acid	Bind free ions, helping preserve food quality by reducing ability of ions to cause rancidity in products containing fat	No known health risk when used properly

▲ Emulsifiers improve the texture of foods such as ice cream, baked goods, and cookies.

intentional food additives Additives knowingly (directly) incorporated into food products by manufacturers.

incidental food additives Additives that appear in food products indirectly, from environmental contamination of food ingredients or during the manufacturing process.

generally recognized as safe (GRAS) A list of food additives that in 1958 were considered safe for consumption. Manufacturers were allowed to continue to use these additives, without special clearance, when needed for food products. FDA bears responsibility for proving they are not safe but can remove unsafe products from the list.

Additives are also used to reduce the activity of some enzymes that leads to undesirable changes in color and flavor in foods but does not cause anything as serious as foodborne illness. This second type of food spoilage occurs when enzymes in a food react to oxygen; for example, when apple and peach slices darken or turn rust color as they are exposed to air. Antioxidants are a type of preservative that slows the action of oxygen-requiring enzymes on food surfaces. These preservatives are not necessarily novel chemicals. They include vitamins E and C and a variety of sulfites.

Without the use of some food additives, it would be impossible to produce massive quantities of foods and safely distribute them nationwide or worldwide, as is now done. Despite consumer concerns about the safety of food additives, many have been extensively studied and proven safe when FDA guidelines for their use are followed.

INTENTIONAL VERSUS INCIDENTAL FOOD ADDITIVES

Food additives are classified into two types: **intentional food additives** (directly added to foods) and **incidental food additives** (indirectly added as contaminants). Both types of agents are regulated by FDA in the United States. Currently, more than 2800 different substances are intentionally added to foods. As many as 10,000 other substances enter foods as contaminants. This includes substances that may reasonably be expected to enter food through surface contact with processing equipment or packaging materials.

THE GRAS LIST

In 1958, all food additives used in the United States and considered safe at that time were put on a **generally recognized as safe (GRAS)** list. The U.S. Congress established the GRAS list because it believed manufacturers did not need to prove the safety of substances that had been used for a long time and were already generally recognized as safe. Since that time, FDA has been responsible for proving that a substance does *not* belong on the GRAS list. Substances may be added to the GRAS list if data and information about the use of the substance are known and accepted widely by qualified experts and establish that the substance is safe under the conditions of its intended use.

Since 1958, some substances on the list have been reviewed. A few, such as cyclamates, failed the review process and were removed from the list. The additive red dye #3 was removed because it is linked to cancer. Many chemicals on the GRAS list have not yet been rigorously tested, primarily because of expense. These chemicals have received a low priority for testing, mostly because they have long histories of use without evidence of toxicity or their chemical characteristics do not suggest that they are potential health hazards.

The American Heart Association (AHA) and other experts have recently questioned the appropriateness of the GRAS listing for salt. They have suggested that sodium, one of the two components of salt, "has negative health consequences" and therefore does not meet the "safe" requirement of the GRAS. The AHA would like to see FDA amend the GRAS listing for sodium chloride in an effort to reduce sodium content in processed foods (see Further Reading 5).

ARE SYNTHETIC CHEMICALS ALWAYS HARMFUL?

Nothing about a natural product makes it inherently safer than a synthetic product. Many synthetic products are laboratory copies of chemicals that also occur in nature. Moreover, although human endeavors contribute some toxins to foods, such as synthetic pesticides and industrial chemicals, nature's poisons are often even more potent and widespread. Some cancer researchers suggest that we ingest at least 10,000 times more (by weight) natural toxins produced by plants than we do synthetic pesticide residues. (Plants produce these toxins to protect themselves from predators and disease-causing organisms.) This comparison does not make synthetic chemicals any less toxic, but it does put them in a more accurate perspective.

Lastly, toxicity is related to dosage. Consider vitamin E, often added to food to prevent rancidity of fats. This chemical is safe when used within certain limits. However, high doses have been associated with health problems, such as interfering with vitamin K activity in the body (review Chapter 8). Thus, even well-known chemicals we are comfortable using can be toxic in some circumstances and at some concentrations.

TESTS OF FOOD ADDITIVES FOR SAFETY

Food additives are tested by FDA for safety on at least two animal species, usually rats and mice. Scientists determine the highest dose of the additive that produces *no observable effects* in the animals. These doses are proportionately much higher than humans are ever exposed to. The maximum dosage that produced no observable effects is then divided by at least 100 to establish a margin of safety for human use. This 100-fold margin is used because it is assumed that we are at least 10 times more sensitive to food additives than are laboratory animals and that any one person might be 10 times more sensitive than another. This broad margin essentially ensures that the food additive in question will cause no harmful health effects in humans. In fact, many synthetic chemicals are probably less dangerous at these low doses than some of the natural compounds in common foods such as apples or celery.

One important exception applies to the schema for testing intentional food additives: If an additive is shown to cause cancer, even though only in high doses, no margin of safety is allowed. The food additive cannot be used because it would violate the **Delaney Clause** in the 1958 Food Additives Amendment. This clause prohibits intentionally adding to foods a compound introduced after 1958 and that causes cancer at any level of exposure. Evidence for cancer could come from either laboratory animal or human studies. Few exceptions to this clause are allowed; exceptions are discussed regarding curing and pickling agents in Table 13-6.

Incidental food additives are another matter. FDA cannot ban various industrial chemicals, pesticide residues, and mold toxins from foods, even though some of these contaminants can cause cancer. These products are not purposely added to foods. FDA sets an acceptable level for these substances. An incidental substance

---

Some important terms used by toxicologists:

toxicology	Scientific study of harmful substances
safety	Relative certainty that a substance will not cause injury
hazard	Chance that injury will result from use of a substance
toxicity	Capacity of a substance to produce injury or illness at some dosage

---

Sugar, salt, corn syrup, and citric acid constitute 98% of all additives (by weight) used in food processing.

▲ Color additives make some foods more desirable.

Delaney Clause A clause to the 1958 Food Additives Amendment of the Pure Food and Drug Act in the United States that prevents the intentional (direct) addition to foods of a compound shown to cause cancer in laboratory animals or humans.

found in a food cannot contribute to more than one cancer case during the lifetimes of 1 million people. If a higher risk exists, the amount of the compound in a food must be reduced until the guideline is met.

In general, if you consume a variety of foods in moderation, the chances of food additives jeopardizing your health are minimal. Pay attention to your body. If you suspect an intolerance or a sensitivity, consult your physician for further evaluation. Remember that in the short run, you are more likely to suffer either from foodborne illness due to poor food-handling practices that allow viral and bacterial contamination in food or from the consumption of raw animal foods than from consuming additives. Excess calories, saturated fat, cholesterol, *trans* fat, salt, and other potential "problem" nutrients in our diets pose the greatest long-term health risk.

APPROVAL FOR A NEW FOOD ADDITIVE

Before a new food additive can be added to foods, FDA must approve its use. Besides rigorously testing an additive to establish its safety margins, manufacturers must give FDA information that (1) identifies the new additive, (2) gives its chemical composition, (3) states how it is manufactured, and (4) specifies laboratory methods used to measure its presence in the food supply at the amount of intended use.

Manufacturers must also offer proof that the additive will accomplish its intended purpose in a food, that it is safe, and that it is to be used in no higher amount than needed. Additives cannot be used to hide defective food ingredients, such as rancid oils; to deceive customers; or replace good manufacturing practices. A manufacturer must establish that the ingredient is necessary for producing a specific food product.

Despite these guidelines, many activists and public health watchdogs are not satisfied with FDA procedures for regulating and monitoring the safety of food additives. These groups have urged FDA and food manufacturers to stop the use of various chemicals until their safety can be more fully determined. A major incentive to these requests is the fact that many of the chemicals used in the United States are illegal to use as food additives in the countries of the European Union and others such as Brazil, Canada, India, and Japan.

If you are bewildered or concerned about all the additives in your diet, you can easily avoid most of them by consuming unprocessed whole foods. However, no evidence shows that this will necessarily make you healthier, nor can you avoid all additives, because some, such as pesticides, are used even on whole foods. It amounts to a personal decision. Do you have confidence that FDA and food manufacturers are adequately protecting your health and welfare, or do you want to take more personal control by minimizing your intake of compounds not naturally found in foods?

▲ Your choice to consume fresh rather than processed foods will lower your intake of food additives. For most people, however, food additives are not a threat to health.

✓ CONCEPT CHECK 13.4

1. What are some of the functions of the additives used in food?
2. What is the difference between intentional and incidental additives?
3. How do scientists determine safe limits for the amount of additives that are allowed in foods?
4. What is the purpose of the Delaney Clause?

13.5 Substances That Occur Naturally in Foods and Can Cause Illness

Foods contain a variety of naturally occurring substances that can cause illness. Here are some of the more important examples:

- *Safrole*—found in sassafras, mace, and nutmeg; causes cancer when consumed in high doses

- *Solanine*—found in potato shoots and throughout the potato when it has been stressed by exposure to light or pests and indicated by green spots on potato skins; inhibits the action of neurotransmitters
- *Mushroom toxins*—found in some species of mushrooms such as aminita; can cause stomach upset, dizziness, hallucinations, and other neurological symptoms. The more lethal varieties can cause liver and kidney failure, coma, and even death. FDA regulates commercially grown and harvested mushrooms. These are cultivated in concrete buildings or caves. However, there are no systematic controls on individual gatherers harvesting wild species, except in Illinois and Michigan.
- *Avidin*—found in raw egg whites (cooking destroys avidin); binds the vitamin biotin in a way that prevents its absorption, so a biotin deficiency may ultimately develop over the long term
- *Thiaminase*—found in raw fish, clams, and mussels; destroys the vitamin thiamin
- *Tetrodotoxin*—found in puffer fish; causes respiratory paralysis
- *Oxalic acid*—found in spinach, strawberries, sesame seeds, and other foods; binds calcium and iron in the foods and so limits absorption of these nutrients
- *Herbal teas* containing senna or comfrey—can cause diarrhea and liver damage

People have coexisted for centuries with these naturally occurring substances and have learned to avoid some of them and limit intake of others. They pose little health risk. Farmers know potatoes must be stored in the dark, so that solanine will not synthesized. Furthermore, we have developed cooking and food-preparation methods to limit the potency of other substances, such as thiaminase. Spices are used in such small amounts that health risks do not result. Nevertheless, it is important to understand that some potentially harmful chemicals in foods occur naturally.

▲ Many varieties of wild mushrooms contain deadly toxins.

IS CAFFEINE A CAUSE FOR CONCERN?

Why all the controversy over a cup of coffee? Researchers have spent a great deal of time on the study of caffeine, the substance of greatest concern in the favorite beverage of many of us. So why do caffeine recommendations change from year to year?

Caffeine is a stimulant found as a natural or added ingredient in many beverages and chocolate. On average, we consume 75% of our caffeine intake as coffee, 15% as tea, 10% as soft drinks, and 2% as chocolate (Table 13-7). (For teenagers and young adults, this ratio is often relatively higher for soft drinks and lower for coffee.)

Caffeine is not often consumed by itself. With the popularity of trendy coffee shops that serve everything from mocha java to flavored lattes, it is difficult to separate caffeine intake from cream, sugar, alternative sweeteners, and flavorings. Although a 6-ounce cup of black coffee contains just 7 calories, adding cream and sugar increases the calorie count significantly. Adding half-and-half will give you 46 extra calories; liquid nondairy creamer adds 48 calories; and a teaspoon of sugar adds about 23 calories. So what is the conscientious coffee drinker to think? Let us explore the myths and facts of caffeine intake.

Caffeine does not accumulate in the body and is normally excreted within several hours following consumption. Caffeine can cause anxiety, increased heart rate, insomnia, increased urination (possibly resulting in dehydration), diarrhea, and gastrointestinal upset in high doses. In addition, those already suffering from ulcers may experience irritation due to increased acid production; those who have anxiety or panic attacks may find that caffeine worsens their symptoms; and those prone to heartburn may find that caffeine worsens this symptom because it relaxes sphincter muscles in the esophagus. Some people need little caffeine to feel such effects, and the threshold for children is likely even lower than that for adults.

Withdrawal symptoms are also real. Former coffee drinkers may experience headache, nausea, and depression for a short time after discontinuing use. These symptoms can be expected to peak at 20 to 48 hours following the last intake of caffeine. Symptoms hold true even for those trying to quit as little as one cup of coffee per day. Slow tapering of use over a few days is recommended to avoid these problems.

TABLE 13-7 ► Caffeine Content of Common Sources

Item	Milligrams of Caffeine	
	Typical	Range*
Coffee (8 fl oz)		
Brewed, drip method	85	65–120
Brewed, percolator	75	60–85
Decaffeinated, brewed	3	2–4
Espresso (1 fl oz serving)	40	30–50
Teas (8 fl oz)		
Brewed, black tea	40	20–90
Brewed, green tea	20	8–30
Iced	25	9–50
Instant	28	24–31
Some soft drinks (8 fl oz)	24	20–40
"Energy drinks" such as Red Bull (8.3 fl oz)	80	0–80
Cocoa beverage (8 fl oz)	6	3–32
Chocolate milk beverage (8 fl oz)	5	2–7
Milk chocolate (1 oz)	6	1–15
Dark chocolate, semisweet (1 oz)	20	5–35
Baker's chocolate (1 oz)	26	26
Chocolate-flavored syrup (1 fl oz)	4	4

*For the coffee and tea products, the range varies due to brewing method, plant variety, brand of product, and so on.

For a more extensive list of caffeine content of food and drugs, visit **www.cspinet.org/new/cafchart.htm.**

Are there more serious consequences of consuming caffeine regularly? It has been hypothesized that caffeine consumption can lead to certain types of cancer, such as pancreatic and bladder cancers. The association of caffeine with cancer, however, has not been supported in recent literature. In fact, regular coffee consumption has been linked to a decreased risk of colon cancer.

Negative press has dwindled with regard to a link between cardiovascular disease and moderate coffee consumption. Heavy use does increase blood pressure for a short period of time. Coffee consumption also has been linked to increased LDL-cholesterol and triglycerides in the blood. This association was found to be caused by cafestol and kahweol, two oils in ground coffee. However, filtered and instant coffees do not contain the harmful oils. It is prudent, though, to limit the amount of coffee in general, especially from French coffee presses and from espresso, as these beverages are not filtered.

Women are thought to be at higher risk for a variety of deleterious effects with caffeine consumption, including miscarriages, osteoporosis, and birth defects in their offspring. Heavy caffeine use does mildly increase the amount of calcium excreted in urine. For this reason, it is important that heavy coffee drinkers check their diets for adequate calcium sources. Some studies do show a higher likelihood for miscarriages in women consuming more than 500 milligrams of caffeine per day (about five 8-ounce cups of coffee). FDA warns women to consume caffeine in moderation (no more than the equivalent of one to two 8-ounce cups of coffee per day).

In contrast to these potential harmful effects of caffeine, coffee consumption has been linked to reducing the risk of developing basal cell carcinoma, depression, and cardiovascular disease. Swedish scientists recently found that women ages 40 to 83 who consumed more than a cup of coffee per day for 10 years had a 22% to 25% lower risk of stroke. Coffee consumption has also been shown to decrease a person's risk of type 2 diabetes and Parkinson's disease. Though some women attest to the idea that caffeine improves premenstrual symptoms, no study proves this theory. Some older formulations of weight-loss drugs contained caffeine, under the assumption that it made the drugs more effective. FDA has since banned this use as it was found to be ineffective. Some newer research findings suggest caffeine may reduce the risk of developing headaches, cirrhosis of the liver, some forms of kidney stones, gallbladder stones, some nerve-related diseases, and type 2 diabetes. You may have heard that caffeine can improve physical performance. This has been shown in highly trained athletes; recall that use of large amounts of caffeine is banned by the NCAA (review Chapter 10). For those below professional status, though, no benefit has been shown. Also keep in mind that coffee will not "sober up" a person who is drunk.

Though the debate over caffeine will likely continue as long as North Americans drink coffee, research does not support many old misconceptions about caffeine. These studies are reinforcing the idea of moderation—the equivalent of about two to three 8-ounce cups of coffee per day. A prudent dose of caffeine is 200 to 300 milligrams per day. Review Table 13-3 concerning the caffeine content of typical sources.

▲ Coffee is a common source of caffeine for many adults.

✔ CONCEPT CHECK 13.5

1. What are some examples of naturally occurring substances that can cause illness?
2. What is the typical caffeine content of 8-ounce cups of coffee and tea?
3. What are some of the negative effects of caffeine on the body?
4. Coffee consumption has been linked to a decrease risk of which disorders?

13.6 Environmental Contaminants in Food

A variety of environmental contaminants can be found in foods. Aside from pesticide residues, other potential contaminants that deserve attention are listed in Table 13-8. A general program to minimize exposure to environmental contaminants includes knowing which foods pose greater risks and consuming a wide variety of foods in moderation.

PESTICIDES IN FOOD

Pesticides used in food production produce both beneficial and unwanted effects. Most health authorities believe that the benefits outweigh the risks. Pesticides help ensure a safe and adequate food supply and help make foods available at reasonable cost. Consumers have come to assume that synthetic is dangerous and organic is safe. Some researchers believe that this sentiment is grounded in fear and fueled by unbalanced reports. Other researchers say concern about pesticides is valid and overdue.

Most concern about pesticide residues in food appropriately focuses on long-term rather than acute toxicity because the amounts of residue present, if any, are extremely small. These low concentrations found in foods are not known to produce adverse effects in the short term, although harm has been caused by the high amounts that occasionally result from accidents or misuse. For humans, pesticides pose a danger mainly in their cumulative effects, so their threats to health are difficult to determine. However, problems of the contamination of underground water supplies and destruction of wildlife habitats, indicate that North Americans would probably be better off if we could reduce our use of pesticides. Both the U.S. federal government and many farmers are working toward that end. Chapter 12 discussed the lastest use of biotechnology to reduce pesticide use.

Genetic alteration of foods such as corn and soybeans has created concern, especially in Europe. FDA considers genetically altered products safe if approval for human use has been granted (see Chapter 12 for details).

One of the problems with pesticides is that they create new pests because they destroy the predators (spiders, wasps, and beetles) that naturally keep most plantfeeding insect populations in check. The brown plant hopper, which has plagued Indonesian rice fields, was not a serious problem before heavy pesticide use began to kill its predators in the early 1970s. In the United States, such major pests as spider mites and the cotton bollworm were merely nuisances until pesticides decimated their predators.

TABLE 13-8 ▶ Potential Environmental and Other Contaminants in Our Food Supply

Chemical Substance	Sources	Toxic Effects	Preventive Measures
Acrylamide	Fried foods rich in carbohydrate cooked at high temperatures for extended periods, such as French fries and potato chips	Potential neurotoxin and carcinogen. Known carcinogen for laboratory animals; however, studies have not clearly proven the relationship between acrylamide ingestion and the development of cancer in humans	Limit intake of deep-fat fried foods rich in carbohydrate.
Cadmium	Plants in general if much cadmium is in the soil Clams, shellfish, tobacco smoke Occupational exposure in some cases	Kidney disease Liver disease Prostate cancer (debatable) Bone deformities Lung disease (when inhaled)	Consume a wide variety of foods, including seafood sources.
Dioxin	Trash-burning incinerators Bottom-feeding fish from the Great Lakes Animal fats from animals exposed to such contamination via water or soil	Abnormal reproduction and fetal/infant development Immune suppression Cancer (to date only clearly shown in laboratory animals)	Pay attention to warnings of dioxin risks from local fish; if risk exists, limit intake as suggested on the fishing license. Consume a variety of fish from local waters rather than mostly one specific species.
Lead	Lead-based paint chips and related dust in older homes Occupational exposure (e.g., radiator repair) Lead caps on wine bottles Fruit juices and pickled vegetables stored in galvanized or tin containers or leaded glass Some types of solder used in joining copper pipes (mostly in older homes) Mexican pottery dishes Koo Soo herbal remedies Leaded glass containers	Anemia Kidney disease Nervous system damage (tiredness and changes in behavior are symptoms) Reduced learning capacity in childhood (even from mild lead exposure)	Avoid paint chips and related dust in older homes; regular cleaning of these homes is also important (see www.hud.gov/offices/lead). Meet iron and calcium needs to reduce lead absorption. Wipe the inside and outside neck of wine bottles before use if the bottle has a lead cap. Store fruit juices and pickled vegetables in glass or plastic or waxed paper containers. Let water run 1 minute or so if off for more than 2 hours, and use only cold water for cooking; do not soften drinking water. Do not store alcoholic beverages in leaded glass containers.
Mercury	Swordfish, shark, king mackerel, and tilefish. Fresh and canned albacore tuna is also a possible source. (In contrast, the more typical light chunk tuna is very low in mercury.)	Reduced fetal/child development and birth defects; toxic to nervous system	Consume these sources no more than once per week, (no more than two times per week for albacore tuna). Pregnant women should avoid these species of fish, but some albacore tuna consumption is fine. Two to three fish meals per week is appropriate for pregnant (and nursing) women if different types of fish are eaten.
Polychlorinated biphenyls (PCBs)	Fish from the Great Lakes and Hudson River Valley (e.g., coho salmon) Farmed salmon are a possible source, but less so	Cancer (to date only clearly shown in laboratory animals), as well as a potential for liver, immune, and reproductive disorders	Pay attention to warnings of PCB contamination from local fish; if risk exists, limit intake as suggested on the fishing license or on state advisories. Vary the type of fish eaten during a specific week.
Urethane	Alcoholic beverages such as sherry, bourbon, sake, and fruit brandies	Cancer (to date only clearly shown in laboratory animals)	Avoid generous amounts of typical sources.

WHAT IS A PESTICIDE?

Federal law defines a pesticide as any substance or mixture of substances intended to prevent, destroy, repel, or mitigate any pest. The built-in toxic properties of pesticides lead to the possibility that other, nontarget organisms, including humans, might also be harmed. The term *pesticide* tends to be used as a generic reference to many types of products, including insecticides, herbicides, fungicides, and rodenticides. A pesticide product may be chemical or bacterial, natural or synthetic. For agriculture, EPA allows about 10,000 pesticides to be used, containing some 300 active ingredients. About 5.1 billion pounds of pesticides are used each year in the United States, much of which is applied to agricultural crops.

Once a pesticide is applied, it can turn up in a number of unintended and unwanted places. It may be carried in the air and dust by wind currents, remain in soil attached to soil particles, be taken up by organisms in the soil, decompose to other compounds, be taken up by plant roots, enter groundwater, or invade aquatic habitats. Each is a route to the food chain; some are more direct than others.

WHY USE PESTICIDES?

The primary reason for using pesticides is economic: the use of agricultural chemicals increases production and lowers the cost of food, at least in the short run. Many farmers believe that it would be impossible to stay in business without pesticides, which help protect farmers from ruinous losses. It is estimated that for every $1 that is spent on pesticides to increase crops yields, $4 are saved in crops.

Consumer demands also have changed over the years. At one time, we would not have thought twice about buying an apple with a worm hole. Today, consumers find worm holes less acceptable, so farmers rely more on pesticides to produce cosmetically attractive fruits and vegetables. On the practical side, pesticides can protect against the rotting and decay of fresh fruits and vegetables. Also, food grown without pesticides can contain naturally occurring organisms that produce carcinogens at concentrations far above current standards for pesticide residues. For example, fungicides help prevent the carcinogen aflatoxin (caused by growth of a fungus) from forming on some crops. Thus, although some pesticides may do little more than improve the appearance of food products, others help keep foods fresher and safer to eat.

REGULATION OF PESTICIDES

The responsibility for ensuring that residues of pesticides in foods are below amounts that pose a danger to health is shared by FDA, EPA, and the Food Safety and Inspection Service of USDA in the United States. Table 13-2 listed the roles of various food protection agencies. FDA is responsible for enforcing pesticide tolerances in all foods except meat, poultry, and certain egg products, which are monitored by USDA. A newly proposed pesticide is exhaustively tested, perhaps over 10 years or more, before it is approved for use. EPA must decide that the pesticide causes no unreasonable adverse effects on people and the environment and that benefits of use outweigh the risks of using it. FDA tests thousands of raw products each year for pesticide residues. (A pesticide is considered illegal in this case if it is not approved for use on the crop in question or if the amount used exceeds the allowed tolerance.) The latest FDA studies show no residues in about 60% of samples. Less than 1% of domestic and about 3% of import samples have residues continually over tolerance. These findings continue to support previous FDA studies over the past 10 years that pesticide residues in food are generally well below EPA tolerances, and they confirm the safety of the food supply relative to pesticide residues.

▲ There are risks and benefits associated with pesticide use. The greatest short-term risk is in rural communities, where exposure is more direct. These crops are being dusted from the air with pesticide.

▲ Fruits and vegetables grown without use of pesticides are available and may bear an "organic" label (see Table 2-10 for rules regarding the use of the term *organic* on food labels). These products generally are more expensive than those grown using pesticides. Consumers need to decide if the potential benefits of the products are worth the extra cost.

HOW SAFE ARE PESTICIDES?

Dangers from exposure to pesticides through food depend on how potent the chemical toxin is, how concentrated it is in the food, how much and how frequently it is eaten, and the consumer's resistance or susceptibility to the substance. Accumulating information links pesticide use to increased cancer rates in farm communities. For rural counties in the United States, the incidence of lymph, genital, brain, and digestive tract cancers increases with higher-than-average pesticide use. Respiratory cancer cases increase with greater insecticide use. In tests using laboratory animals, scientists have found that some of the chemicals present in pesticide residues cause birth defects, sterility, tumors, organ damage, and injury to the central nervous system. Some pesticides persist in the environment for years.

Still, some researchers argue that the cancer risk from pesticide residues is hundreds of times less than the risk from eating such common foods as peanut butter, brown mustard, and basil. Plants manufacture toxic substances to defend themselves against insects, birds, and grazing animals (including humans). When plants are stressed or damaged, they produce even more of these toxins. Because of this, many foods contain naturally occurring chemicals considered toxic, and some are even carcinogenic. Other scientists argue that if natural carcinogens are already in the food supply, then we should reduce the number of added carcinogens whenever possible. In other words, we should do what we can to decrease our overall exposure.

PERSONAL ACTION

Every year, the Environmental Working Group (EWG) publishes a *Shopper's Guide to Pesticides*. The EWG is an environmental health research and advocacy organization whose mission is to serve as a watchdog to help Americans get factual information so they can make healthier choices and enjoy a cleaner environment. One of its goals is to help ensure that food products are free of harmful chemicals. Its *Shopper's Guide* includes a list of fruits and vegetables called "The Dirty Dozen™" because they are most likely to contain the highest amounts of pesticide residue (Table 13-9). In 2014, apples topped the list, followed by strawberries, grapes, celery, and peaches. The Dirty Dozen™ Plus category includes crops that did not meet traditional Dirty Dozen™ criteria but were commonly contaminated with pesticides exceptionally toxic to the nervous system. The *Shopper's Guide* also features a list of fruits and vegetables called "The Clean 15" because they are least likely to contain pesticide residues. To avoid pesticide exposure, consumers are encouraged to choose organic produce when buying anything on "The Dirty Dozen™" list. In addition, the EWG says that buying nonorganic produce from the "The Clean 15" group is relatively safe. The EWG reminds consumers that "the health benefits of a diet rich in fruits and vegetables outweigh the risks of pesticide exposure." Additional advice for limiting exposure to pesticides is found in Table 13-10.

FDA and other scientific organizations believe that the hazards are comparatively low and in the short run are less dangerous than the hazards of foodborne illness created in our kitchens. We can also encourage farmers to use fewer pesticides to reduce exposure to our foods and water supplies, but we will have to settle for produce that is not perfect in appearance or that has been grown with the aid of biotechnology.

ENVIRONMENTAL CONTAMINANTS IN FISH

The presence of the environmental contaminants mercury and polychlorinated biphenyls (PCBs) in fish has caused some confusion regarding the risks and benefits of fish consumption. In our previous discussion of the benefits of omega-3 fatty acids, it was recommended that we include cold-water fatty fish, such as salmon or tuna, in our diet about twice a week. Conversely, you may have heard

▲ FDA's yearly evaluation of a "market basket" of typical foods shows that pesticide content is minimal in most foods.

TABLE 13-9 ▶ The 2013 Dirty Dozen™ Fruits and Vegetables to Avoid (Unless Organically Grown) and the Clean 15 That Are Relatively Safe

Dirty Dozen™	Clean 15
Apples	Asparagus
Celery	Avocados
Cherry tomatoes	Cabbage
Cucumbers	Cantaloupe (domestic)
	Cauliflower
Grapes	Sweet corn
	Eggplant
Nectarines (imported)	Grapefruit
Peaches	Kiwi
Potatoes	Mangoes
Snap peas (imported)	
Spinach	
Strawberries	Onions
Sweet bell peppers	Papayas
Dirty Dozen™ Plus	Pineapples
Hot peppers	Sweet peas (frozen)
Blueberries (domestic)	Sweet potatoes

Source: Environmental Working Group, *Shopper's Guide to Pesticides.*

▲ This fruit salad is a healthy choice because cantaloupe, pineapple, and kiwi are some of the fruits found on the Clean 15 list.

recommendations to eat less fish because they are a source of environmental contaminants. Balancing the benefits and risks of consuming fish is tricky, and not all experts agree (see Further Reading 8). For example, advice from FDA and EPA indicates that salmon is safe to eat, even during pregnancy, because it is low in mercury. In contrast, the Environmental Defense Fund (EDF) advocacy group recommends that all adults, not just pregnant women, limit consumption of wild salmon (except Alaskan) to only one serving per month and farmed salmon to no more than two servings a month (based on PCB contamination).

TABLE 13-10 ▶ What You Can Do to Reduce Exposure to Pesticides

WASH: Wash and scrub all fresh fruits and vegetables thoroughly under running water. Running water has an abrasive effect that soaking does not have. This will help remove bacteria and traces of chemicals from the surface of fruits and vegetables and dirt from crevices. Not all pesticide residues can be removed by washing.

PEEL AND TRIM: Peel fruits and vegetables when possible to reduce dirt, bacteria, and pesticides. Discard outer leaves of leafy vegetables. Trim fat from meat and skin from poultry and fish because some pesticide residues collect in fat.

SELECT A VARIETY OF FOODS: Eat a variety of foods from a variety of sources. This will give you a better mix of nutrients and reduce your likelihood of exposure to a single pesticide.

CHOOSE ORGANICALLY GROWN FOODS: Some consumers are looking to organically grown and produced foods as a way to reduce exposure to synthetic pesticides.

USE INSECT REPELLENTS SAFELY: Read the label for pesticide safety information and apply insect repellents safely. See **www.epa.gov/pesticides/factsheets/pest_ti.htm** for more pesticide safety tips.

Source: **www.epa.gov/pesticides/food/tips.htm.**

▲ Wash fresh fruits and vegetables under running water to remove bacteria and soil. Special antibacterial washing products are not necessary.

▲ This wild Alaskan salmon on organic green beans is a top choice among types of fish based on its high nutritional value and its low mercury levels.

Mercury and PCBs are by-products of industrial processes and accumulate in fish tissue. PCBs were banned from use in 1979, but environmental levels have been decreasing very slowly and therefore still persist in our food supply, especially seafood. The contaminants become more concentrated in bigger fish as they eat smaller, contaminated fish. Fish are of primary concern because they are the only predators we eat regularly. The Institute of Medicine (IOM), Food and Agriculture Organization (FAO), FDA, and EPA have issued similar guidelines for fish consumption. These groups advise pregnant women to eat up to 12 ounces of low-mercury fish per week and to avoid the four highest-mercury fish, which are swordfish, shark, tilefish, and king mackerel. For other adults, the basic recommendation is to "eat fish" but do not always eat the same type so you are not exposed to the same contaminants over and over.

As a nation, Americans do not eat enough fish to cause concern about high intakes of environmental contaminants. On average, we consume only about 4 ounces of seafood per week. Around 80% of that is shrimp, canned tuna, salmon, and white fish, which are relatively low in environmental contaminants. Most Americans would benefit from eating more fish—a rich source of omega-3 fatty acids. Research shows that the risk of dying from heart disease is about 50% greater among people who do not eat fish compared to those who eat one or two servings of fatty fish each week. Overall, it appears that the benefits of consuming fish twice per week outweigh the potential risks discussed above. Pregnant women should follow the FDA/EPA guidelines, and the rest of us should eat a variety of types of fish, focusing on the smaller, fatty fish at the bottom of the food chain.

✔ CONCEPT CHECK 13.6

1. What is a pesticide?
2. What are the benefits of pesticide use?
3. What agencies regulate the use of pesticides?
4. What foods are on the "Dirty Dozen™" list?
5. What environmental contaminants can be found in fish, and which fish are most likely to contain these toxins?

13.7 Food Production Choices

Agriculture—the production of food and livestock—has supplied humans with food for millennia. At one time, nearly everyone was involved in food production. Today, only about one in three people around the globe (less than 1% in the United States) is now involved in farming. Numerous advances in agricultural sciences are affecting our food supply; of particular note are organic food production and sustainable agriculture. Many of these new developments in agriculture are aimed at reducing the overall carbon footprint (carbon dioxide and methane emissions) generated from the farm to the fork.

ORGANIC FOODS

biological pest management Control of agricultural pests by using natural predators, parasites, or pathogens. For example, ladybugs can be used to control an aphid infestation.

The term **organic** refers to the way agricultural products are produced. Organic production relies on farming practices such as **biological pest management,** composting, manure applications, and crop rotation to maintain healthy soil, water, crops, and animals. Synthetic pesticides, fertilizers, and hormones; antibiotics;

sewage sludge (used as fertilizer); genetic engineering; and irradiation are not permitted in the production of organic foods. Additionally, organic meat, poultry, eggs, and dairy products must come from animals allowed to graze outdoors and consume only organic feed.

Organic foods are increasingly available in supermarkets, specialty stores, farmers markets, and restaurants (see Further Readings 11 and 12). Consumers can select organic fruits, vegetables, grains, dairy products, meats, eggs, and many processed foods, including sauces and condiments, breakfast cereals, cookies, and snack chips. Interest in personal and environmental health has contributed to the increasing availability and sales of organic foods. According to the Organic Trade Association, U.S. sales of organic foods exceeded $35 billion in 2013. This reflects an increase of 11.5% over the previous year, even in the face of tough economic times. Canada's organic market has also tripled since 2006, reaching $3.7 billion in 2012. Despite this rapid growth, only 4.2% of foods sold are organic. Organic foods, because they often cost more to grow and produce, are typically more expensive than comparable conventional foods.

The Organic Foods Production Act of 1990 established standards for the production of foods that bear the USDA organic seal. Foods labeled and marketed as organic must be grown on farms that are certified by the USDA as following all of the rules established in the 1990 act. Foods made from multiple ingredients (e.g., breakfast cereal) labeled as organic must have at least 95% of their ingredients (by weight) meet organic standards. The term *made with organic* can be used if at least 70% of the ingredients are organic. Small organic producers and farmers with sales less than $5000 per year are exempt from the certification regulation. Some farmers use organic production methods but choose not to be USDA certified. Their foods cannot be labeled as organic, but many of these farmers market and sell to those seeking organic foods.

The organic food market received a boost in 2009 when USDA offered $50 million in new funding to encourage greater production of organic food in the United States. This funding has encouraged farmers to use organic practices and increased the U.S. production of organic food to meet growing consumer demand. With the creation of opportunities for farmers and ranchers, the number of certified organic operations grew by 1109, or more than 6%, between 2009 and 2011. Because most stores now offer organic products, consumers have the opportunity to shop around. Increased availability and use of coupons, the proliferation of private label and store brands, and better-value products offered by major organic brands all have contributed to increased sales.

▲ October 2012 marked the tenth anniversary of the USDA organic seal, which identifies organic foods grown on USDA-certified organic farms.

Organic Foods and Health. Consumers may choose to eat organic foods to reduce their pesticide intake, to protect the environment, and out of the belief that they will improve the nutritional quality of their diets. Those who consume organic produce do ingest lesser amounts of pesticides (only 1 in 4 organically grown fruits and vegetables contains pesticides and in lower amounts than conventional produce), but it is still not known whether or how this affects the health of most consumers. However, organic foods may be a wise choice for young children because pesticide residues may pose a greater risk to them. Consumers also may opt for organic foods to encourage environmentally friendly **sustainable agriculture** practices.

Most studies do not show that organic foods have higher amounts of vitamins and minerals (see Further Reading 3). However, researchers have found that, in some cases, organic fruits and vegetables contain more vitamin C and antioxidants that help protect cells against damage. At this point, it is not possible to recommend organic foods over conventional foods based on nutrient content: both can meet nutritional needs. A healthy dose of common sense also is important; an "organic" label does not change a less healthy food into a more healthy food. For example, organic potato chips have the same calorie and fat content as conventional potato chips.

sustainable agriculture Agricultural system that provides a secure living for farm families; maintains the natural environment and resources; supports the rural community; and offers respect and fair treatment to all involved, from farm workers to consumers to the animals raised for food.

One concern raised about organic foods is that food safety may be jeopardized because animal manures used for fertilizers might cause more pathogen contamination of food. A major outbreak of foodborne illness linked to organically grown sprouts in Europe confirmed that an organic label does not guarantee good health. However, research does not show that certified organic food has higher contamination with bacterial pathogens. Consumers should wash or scrub all produce—organic and conventional—under running water.

Unlike the term *organic*, the term *natural* is not regulated by any federal agency. Products labeled as "natural" are generally those derived from natural ingredients, such as a plant source, which retain their natural properties in the finished product. Meat or poultry labeled "natural" is expected to be minimally processed and contain no artificial flavoring, coloring, chemical preservative, or other artificial or synthetic ingredients. Generally, however, no one is checking these products, and there is some debate over what constitutes "minimally processed." Although all organic products fit this definition of natural, not all natural products are necessarily organic. Legislation has recently been proposed (the Food Labeling Modernization Act) that seeks to define and regulate use of the term "natural" on food labels.

SUSTAINABLE AGRICULTURE

Conventional agriculture focuses on maximizing production through the use of large acreages, powerful machines, chemicals to control pests, and synthetic fertilizers to boost growth. In contrast, sustainable agriculture is an integrated system of plant and animal production that will, over the long-term, satisfy human food needs, enhance environmental quality, efficiently use nonrenewable resources, sustain the economic viability of farm operations, and enhance the quality of life for farmers and society as a whole. A culture of sustainability has emerged, including a clear trend for sustainable food choices manufactured in an environmentally responsible way.

Sustainable practices, such as crop rotation, intercropping (growing two or more crops in proximity), and step farming (also known as terrace farming), have increased productivity manyfold in countries in South America. Crop rotation helps protect the soil by not depleting a particular nutrient. Intercropping helps different species of plants take advantage of the different soil characteristics. Step farming increases productivity by enabling planting on hillsides by terracing slopes to hold water for a long duration and retain the top soil more effectively. The food industry has responded with a move toward "green" initiatives that should be sustainable for the long term. A new demographic term, LOHAS (Lifestyle of Health and Sustainability), describes a growing demographic group focused on sustainable living. An increasing number of today's college students are joining this market segment and developing behaviors associated with social responsibility. These consumers are driving changes in many areas, including the food industry. Slow Food Nation is an example of a nonprofit group dedicated to creating a framework for a deeper environmental connection to our food and aiming to inspire and empower Americans to build a food system that is sustainable, healthy, and delicious.

Sustainable Seafood. The recommended fish consumption increased from 3.5 to 8 ounces per week with the 2010 Dietary Guidelines. The challenge of increasing our fish consumption, while also being wary of environmental contaminants in seafood, was discussed in Section 13.6. Our seafood choices become even more complex when we consider the issue of overfishing and protecting endangered species of fish (see Further Reading 4). An "overfished" species is a population that is below a prescribed threshold, which is jeopardizing its survival, due to being harvested at a rate that is too high for the stock to replenish itself. The good news is that fish production, whether farmed or wild-caught, happens at a lower environmental cost compared to the production of meats. Fewer greenhouse gases are emitted, fewer chemicals and antibiotics are used, and fewer pounds of protein in feed are used than in beef, pork, or poultry production.

Fortunately, the United States has rigorous standards and closely monitors its fishing and aquaculture operations. The National Oceanic and Atmospheric Administration (NOAA) Fisheries establishes fishing catch levels in U.S. waters, enforces these catch limits on fishing vessels, docks, and ports, and oversees, monitors, and enforces the standards for U.S. aquaculture (fish farming). As a result, when we buy U.S. wild-caught or farmed fish, we are making a sustainable choice. Look for the words "U.S. seafood product," which indicate that the fish or shellfish has been sustainably harvested.

The sustainable solution, however, is not that simple at the fish market or restaurant. The challenge persists because the vast majority (85%) of the seafood we eat in the United States comes from an international source. About half of this seafood is from Southeast Asia. As we pointed out earlier, Americans have a limited seafood palate, with only 10 species making up 90% of the seafood we eat. Fortunately, of the 10 most popular fish, most of the shrimp, salmon, tilapia, and pangasius are raised by certified and sustainable aquaculture operations. Farmed salmon, for example, is necessary to supply two-thirds of the 350 million tons of salmon that Americans consume annually. Many consumers perceive that wild-caught fish are the more environmentally friendly variety and that aquaculture is damaging or polluting the environment. The fact is that many unregulated wild-caught fishing harvests have reached their peak capacity and are threatening future global seafood supplies.

While sustainable seafood sourcing is complex, consumers need to know that both farmed and wild-caught fish can be healthy, sustainable, and economical choices. Groups such as the New England Aquarium (**http://www.neaq.org/index.php**) and Monterey Bay Aquarium (**http://www.montereybayaquarium.org/cr/seafoodwatch.aspx**) regularly update lists of acceptable fish. Conservation experts from the New England Aquarium have bluefish, Pacific halibut, and sardines on their list of ocean-friendly wild-caught seafood. They also recommend farm-raised bay scallops, mussels, oysters, barramundi, and rainbow trout from the United States and Arctic char from Iceland.

locavore Someone who eats food grown or produced locally or within a certain radius such as 50, 100, or 500 miles.

LOCALLY GROWN FOODS

With people everywhere more interested in the origins of their food, more grocery store shelves are devoted to "locally grown" products. Consumers are demanding increased transparency with the food supply, and local food helps answer questions about where food comes from and how it was grown. Retailers are using the "locally grown" label to respond to consumer desires for fresh, safe products that also support small, local farmers and help the environment. Local products provide fresher options, do not have the added costs of long transportation, and thus use less fossil fuel. Foodservice establishments are giving greater emphasis to local producers, focusing on where food was grown and how it was handled.

Farmers markets are the most obvious way that consumers have access to locally grown, farm-fresh produce. Farmers markets are also an integral part of the way that urban communities are linked to farms and continue to gain popularity. In 2013, there were 8144 farmers markets operating throughout the United States, nearly double what was available just 5 years ago.

The interest in "local" foods has become such a phenomenon that the term *locavore* was the 2007 Word of the Year in the *New Oxford American Dictionary*. **Locavore** is defined as someone who eats food grown or produced locally or within a certain

▲ The locavore movement is based on the assumption that local products are more nutritious and taste better and encourages consumers to buy from farmers markets or produce their own food.

Newsworthy Nutrition

German organic sprouts cause deadly *E. coli* outbreak

The world's deadliest outbreak of *Escherichia coli* occurred during the summer of 2011. The outbreak occurred mainly in Germany and involved a rare entero-hemorrhagic strain of *E. coli* known as O104:H4. In June 2011, 3950 persons had been infected, more than 600 were in intensive care, and the death toll from the outbreak reached 53. Several hundred also contracted its potentially fatal kidney complication, known as hemolytic uremic syndrome. This German case-control study concluded that the *E. coli* infection was caused by tainted vegetable sprouts from a small, rather traditional organic sprout farm. The sprout seeds were mostly imported from overseas, and the *E. coli* bacteria were antibiotic resistant. Experts say these types of outbreaks are becoming more common due to large-scale industrial farming and the widespread use of antibiotics.

Source: Buchholz U and others: German outbreak of *Escherichia coli* O104:H4 associated with sprouts. *New England Journal of Medicine* 365:1763, 2011.

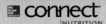 Check out the Connect site **www.mcgrawhillconnect.com** to further explore issues concerning food safety and food production.

radius, such as 50, 100, or 150 miles. The locavore movement has gained prominence due to food-safety concerns by consumers and the search for local, sustainable foods. It encourages consumers to buy from farmers markets or produce their own food, with the argument that fresh, local products are more nutritious and taste better.

There is no evidence, however, that locally grown products are safer. Although many small producers have good food-safety practices, they often lack the expensive food-safety audits more common among big producers. Food-safety auditors determine such things as whether or not there is evidence of insects on produce and whether or not producers have enough bathrooms for workers. In addition, undetected foodborne illness outbreaks are more likely with "local" products delivered in small quantities and sold in a small area. Local products are not necessarily pesticide-free and may not be cheaper, given that smaller growers lack the economic advantages of bigger growers. Read about the outbreak of foodborne illness linked to a local organic farm in Newsworthy Nutrition.

Unlike organic products, there are no regulations specifying the meaning of "locally grown." Whole Foods Market, Inc., is the biggest retailer of natural and organic food and probably the best-known for buying and selling locally grown produce. Whole Foods considers local to be anything produced within 7 hours of one of its stores, with most local producers located within 200 miles of a store. Wal-Mart, the world's largest retailer, has also become a large buyer of locally grown fruits and vegetables and considers anything local if it is grown in the same state as it is sold. Searchable databases and mapping resources such as MarketMaker (**http://national.marketmaker.uiuc.edu/**) are available to connect growers with buyers, restaurants with distributors, and consumers with local farmers markets. These tools make it easier for people to find and sell locally grown foods. Positive attitudes toward organic, local, and sustainable food production practices are on the rise and appear to be increasing the quality of diets. A recent study of college students in Minnesota showed that college students who put a high importance on alternative food production methods had a higher-quality diet. They consumed more fruits and vegetables and dietary fiber, fewer added sugars and sugar-sweetened drinks, and less fat.

COMMUNITY SUPPORTED AGRICULTURE

Consumers are not only taking comfort in knowing where their food comes from but also starting to have interest in community connections with local/regional farmers. Stemming from the interest in locally grown food, there is growing national support for local food collaboratives and community supported agriculture. Community supported agriculture programs (CSA) involve a partnership between local food producers and local consumers. During each growing season, CSA farmers offer a share of foods to individuals, families, or companies that have pledged support to the CSA either financially and/or by working for the CSA.

Another example of a farm-community partnership is the National Farm to School Network, a nonprofit effort to connect farmers with nearby school (K-12) cafeterias. The objectives of this program are to serve healthy meals in school cafeterias; improve student nutrition; provide agriculture, health and nutrition education opportunities; and support local and regional farmers. Between 1997 and 2012, this program grew from only six local programs to programs in all 50 states, resulting in 12,429 schools incorporating the local bounty into their menus. Administrators of the program have found that if kids can meet the farmer who actually grew the food, they are much more likely to eat it.

▲ Farmers participating in community supported agriculture (CSA) offer a share of foods from each growing season to individuals, families, or companies that support the CSA financially and/or by working for the CSA.

Visit: **www.localharvest.org** to find farmers markets, family farms, and other sources of sustainably grown food in your area.

Look up **www.farmtoschool.org** to find a Farm to School Program near you.

✔ CONCEPT CHECK 13.7

1. What are the requirements for foods to be labeled as organic?
2. What is the definition of sustainable agriculture?
3. What is the locavore movement, and what are its advantages?
4. What is a community supported agriculture program (CSA)?

Nutrition and Your Health
Preventing Foodborne Illness

You can greatly reduce the risk of foodborne illness by following some important rules (see Further Readings 6 and 10). It is a long list because many risky habits need to be addressed.

Purchasing Food

- When shopping, select frozen foods and perishable foods, such as meat, poultry, or fish, last. Always have these products put in separate plastic bags, so that drippings do not contaminate other foods in the shopping cart. Do not let groceries sit in a warm car; this allows bacteria to grow. Get the perishable foods such as meat, egg, and dairy products home and promptly refrigerate or freeze them.
- Do not buy or use food from damaged containers that leak, bulge, or are severely dented or from jars that are cracked or have loose or bulging lids. Do not taste or use food that has a foul odor or spurts liquid when the can is opened; the deadly *Clostridium botulinum* toxin may be present.
- Purchase only pasteurized milk and cheese (check the label). This is especially important for pregnant women

because highly toxic bacteria and viruses that can harm the fetus thrive in unpasteurized milk.
- Purchase only the amount of produce needed for a week's time. The longer you keep fruits and vegetables, the more time is available for bacteria to grow.
- When purchasing precut produce or bagged salad greens, avoid those that look slimy, brownish, or dry; these are signs of improper holding temperatures.
- Observe sell-by and expiration dates on food labels and do not buy products that are near or past these dates.
- Follow food recalls because a Class I recall means that there is a "reasonable probability" that consuming the food will cause serious health consequences or death.

Preparing Food

- Thoroughly wash your hands for 20 seconds with hot, soapy water before and after handling food. This practice is especially important when handling raw meat, fish, poultry, and eggs; after using the bathroom; after playing with pets; or after changing diapers.
- Make sure counters, cutting boards, dishes, and other equipment are thoroughly sanitized and rinsed before use. Be especially careful to use hot, soapy water to wash surfaces and equipment that have come in contact with raw meat, fish, poultry, and eggs as soon as possible to remove *Salmonella* bacteria that may be present. Otherwise, bacteria on the surfaces will infect the next foods that come in contact with the surface, a process called cross-contamination. In addition, replace sponges and wash kitchen towels frequently. (Microwaving sponges for 30 to 60 seconds also helps rid them of live bacteria.)
- If possible, cut foods to be eaten raw on a clean cutting board reserved for that purpose. Then clean this cutting board using hot, soapy water. If the same board must be used for both meat and other foods, cut any potentially contaminated items, such as meat, last. After cutting the meat, wash the cutting board thoroughly.
- Ignore the 5-second rule of picking up food that has fallen on the floor because food picks up bacteria immediately upon contact.

FDA recommends cutting boards with unmarred surfaces made of easy-to-clean, nonporous materials, such as plastic, marble, or glass. If you prefer a wooden board, make sure it is made of a nonabsorbent hardwood, such as oak or maple, and has no obvious seams or cracks. Then reserve it for a specific

One tool in the battle against foodborne illness is Hazard Analysis Critical Control Point (HACCP). By applying the principles of HACCP, food handlers critically analyze how they approach food preparation and what conditions may exist that might allow pathogenic microorganisms to enter and thrive in the food system. Once specific hazards and critical control points (potential problems) are identified, preventive measures can be used to reduce specific sources of contamination.

purpose; for example, set it aside for cutting raw meat and poultry. Keep a separate wooden cutting board for chopping produce and slicing bread to prevent these products from picking up bacteria from raw meat. Many foods are served raw, so any bacteria clinging to them are not destroyed.

Furthermore, FDA recommends that all cutting boards be replaced when they become streaked with hard-to-clean grooves or cuts, which may harbor bacteria. In addition, cutting boards should be sanitized once a week in a dilute bleach solution. Flood the board with the solution, let it sit for a few minutes, then rinse thoroughly.

- When thawing foods, do so in the refrigerator, under cold potable running water, or in a microwave oven. Also, cook foods immediately after thawing under cold water or in the microwave. Never let frozen foods thaw unrefrigerated all day or night. Also, marinate food in the refrigerator.
- Avoid coughing or sneezing over foods, even when you are healthy. Cover cuts on hands with a sterile bandage. This helps stop *Staphylococcus* bacteria from entering food.
- Carefully wash fresh fruit and vegetables under running water to remove dirt and bacteria clinging to the surface, using a vegetable brush if the skin is to be eaten. People have become ill from *Salmonella* introduced from melons used in making a fruit salad and from oranges used for fresh-squeezed orange juice. The bacteria were on the outside of the melons and oranges.
- Completely remove moldy portions of food or do not eat the food. If a food is covered in mold, throw the food out.

The World Health Organization's Golden Rules for Safe Food Preparation

1. Choose foods processed for safety.
2. Cook food thoroughly.
3. Eat cooked foods immediately.
4. Store cooked foods carefully.
5. Reheat cooked foods thoroughly.
6. Avoid contact between raw and cooked foods.
7. Wash hands repeatedly.
8. Keep all kitchen surfaces meticulously clean.
9. Protect foods from insects, rodents, and other animals.
10. Use pure water.

▲ Washing hands thoroughly (for at least 20 to 30 seconds) with hot water and soap should be the first step in food preparation. The 4 "F's" of food contamination are fingers, foods, feces, and flies. Handwashing especially combats the finger and fecal routes.

The USDA simplified these rules into four actions as a part of its food safety program (check out **www.foodsafety.gov**):

1. Clean. Wash hands and surfaces often.
2. Separate. Don't cross-contaminate.
3. Cook. Cook to proper temperatures.
4. Chill. Refrigerate promptly.

The 2010 Dietary Guidelines for Americans also stress the importance of these four actions.

▲ Food safety logo of USDA.

Mold growth is prevented by properly storing food at cold temperatures and using the food promptly. Also discard soft foods with high moisture content such as bread, yogurt, soft cheeses, and deli meats if there are spots of mold on them. It is safe to trim off any moldy spots of dense foods such as hard cheeses or firm fruits and vegetables.

- Use refrigerated ground meat and patties in 1 to 2 days and frozen meat and patties within 3 to 4 months.

Cooking Food

- Cook food thoroughly and use a bimetallic thermometer to check for doneness, especially for fresh beef and fish (145°F [63°C]), pork (145°F [63°C]), and poultry (165°F [74°C]) (Fig. 13-3). Eggs should be cooked until the yolk and white are hard. FDA does not recommend that eggs be prepared sunny-side-up. Alfalfa sprouts and other types of sprouts should be cooked until they are steaming. Cooking is by far the most reliable way to destroy foodborne viruses and bacteria, such as Norovirus and toxic strains of *E. coli.* Freezing only halts viral and bacterial growth.

As noted, many restaurants now include an advisory on menus stating that an increased risk of foodborne illness is associated with eating undercooked eggs. As long as restaurants provide this warning on their menus, however, they are allowed to cook eggs to any temperature requested by the consumer. FDA warns us not to consume homemade ice cream, eggnog, and mayonnaise if made with unpasteurized, raw

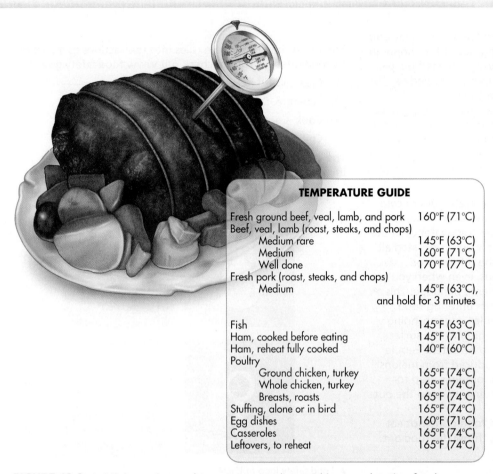

TEMPERATURE GUIDE

Fresh ground beef, veal, lamb, and pork	160°F (71°C)
Beef, veal, lamb (roast, steaks, and chops)	
Medium rare	145°F (63°C)
Medium	160°F (71°C)
Well done	170°F (77°C)
Fresh pork (roast, steaks, and chops)	
Medium	145°F (63°C), and hold for 3 minutes
Fish	145°F (63°C)
Ham, cooked before eating	145°F (71°C)
Ham, reheat fully cooked	140°F (60°C)
Poultry	
Ground chicken, turkey	165°F (74°C)
Whole chicken, turkey	165°F (74°C)
Breasts, roasts	165°F (74°C)
Stuffing, alone or in bird	165°F (74°C)
Egg dishes	160°F (71°C)
Casseroles	165°F (74°C)
Leftovers, to reheat	165°F (74°C)

FIGURE 13-3 ▲ Minimum internal temperatures when cooking or reheating foods.

Source: USDA Food Safety and Inspection Service, Safe Minimum Internal Temperature Chart at **http://www.fsis.usda.gov/**

eggs because of the risk of *Salmonella* foodborne illness. It is safer to use eggs or egg products that have been pasteurized, which kills *Salmonella* bacteria. Overall, a good general precaution is to eat no raw animal products.

Seafood also poses a risk of foodborne illness, especially oysters. Properly cooked seafood should flake easily and/or be opaque or dull and firm. If it is translucent or shiny, it is not done.

- Cook stuffing separately from poultry (or stuff immediately before cooking, and then transfer the stuffing to a clean bowl immediately after cooking). Make sure the stuffing reaches 165°F (74°C). *Salmonella* is the major concern with poultry.
- Once a food is cooked, consume it right away, or cool it to 40°F (4°C) within 2 hours. If it is not to be eaten immediately, in hot weather (80°F and above) make sure that this cooling is done within 1 hour. Do this by separating the food into as many shallow pans as needed to provide a large surface area for cooling. Be careful not to recontaminate cooked food by contact with raw meat or juices from hands, cutting boards, dirty utensils, or in other ways.
- Serve meat, poultry, and fish on a clean plate—never the same plate used to hold the raw product. For example, when grilling hamburgers, do not put cooked items on the same plate used to carry the raw product out to the grill.
- For outdoor cooking, cook food completely at the picnic site, with no partial cooking in advance.

Storing and Reheating Cooked Food

- Keep foods out of the "danger zone" (Fig.13-1) by keeping hot foods hot and cold foods cold. Hold food below 40°F (4°C) or above 140°F (60°C). Foodborne microorganisms thrive in more moderate temperatures (60°F to 110°F [16°C to 43°C]). Some microorganisms can even grow in the refrigerator. Again, do not leave cooked or refrigerated foods, such as meats and salads, at room temperature for more than 2 hours (or 1 hour in hot weather) because that gives microorganisms an opportunity to grow. Store dry food at 60°F to 70°F (16°C to 21°C).
- Reheat leftovers to 165°F (74°C); reheat gravy to a rolling boil to kill *Clostridium perfringens* bacteria, which may be present. Merely reheating to a good eating temperature is not sufficient to kill harmful bacteria.
- Store peeled or cut-up produce, such as melon balls, in the refrigerator.
- Keep leftovers in the refrigerator only for the recommended length of time (Fig. 13-4).
- Make sure the refrigerator stays below 40°F (4°C). Either use a refrigerator thermometer or keep it as cold as possible without freezing milk and lettuce.
- When the power goes out, keep the freezer and refrigerator doors closed as much as possible. Food can stay cold in an unopened refrigerator for about 4 hours; after 4 hours without power, discard perishable foods such as milk, meat, leftovers, and deli meats. Unopened freezers will keep food frozen for 2 days if full and 1 day if half full. Meat, poultry, and seafood can be refrozen if the freezer has not risen above 40°F.

Cross-contamination is not only a threat

▲ Sushi, like all raw fish or meat dishes, is a high-risk food. For maximum protection from foodborne illness, animal foods should be cooked thoroughly before eating.

When in doubt, throw it out!

Food	Refrigerator Storage Time (days)
Meats	
Cooked ground beef/turkey	3–4
Deli meat	2–3
Cooked pork	3–4
Cooked poultry	3–4
Cooked beef, bison, lamb	3–4
Seafood	
Raw (e.g. sushi/sashimi)	Must consume on day of purchase
Cooked	2
Other Entrees	
Pizza	1–2
Pasta/rice	1–2
Casserole	3–4
Soups and Chili	
Chili with meat	2–3
Chili without meat	3–4
Soup/stew	3–4
Side Dishes	
Fresh salad	1–2
Fresh vegetables	1–2
Pasta or potato salad	2–3
Deviled egg	2–3
Hard boiled egg	7
Potato (any style)	3–4
Cooked vegetables	3–4
Dessert	
Cream pie	2–3
Fruit pie	2–3
Pastries	7
Cake	7
Cheesecake	7

FIGURE 13-4 ▲ Length of time to keep leftovers in the refrigerator. Visit **www.homefoodsafety.org** for more tips on keeping your food safe.

To reduce the risk of bacteria surviving during microwave cooking,

- Cover food with glass or ceramic when possible to decrease evaporation and heat the surface.
- Stir and rotate food at least once or twice for even cooking. Then, allow microwaved food to stand, covered, after heating is completed to help cook the exterior and equalize the temperature throughout.
- Use the oven temperature probe or a meat thermometer to check that food is done. Insert it at several spots.
- If thawing meat in the microwave, use the oven's defrost setting. Ice crystals in frozen foods are not heated well by the microwave oven and can create cold spots, which later cook more slowly.

during food preparation; it can also become a problem during food storage. Make sure all foods, including leftovers, are contained and covered in the refrigerator to prevent drippings from uncooked and potentially hazardous foods from tainting other foods. It is a good idea to store foods likely to pose risk of foodborne illness on lower shelves of the refrigerator, beneath other foods to be eaten raw.

Raw fish dishes, such as sushi, can be safe for most people to eat if they are made with very fresh fish that has been commercially frozen and then thawed. The freezing is important to eliminate potential health risks from parasites. FDA recommends that the fish be frozen to an internal temperature of −10°F for 7 days. If you choose to eat uncooked fish, purchase the fish from reputable establishments that have high standards for quality and sanitation. If you are at high risk for foodborne illness, it is wise to avoid raw fish products.

In summary, thoroughly cook all meat and poultry to reduce the risk of foodborne illness from *E. coli* and *Salmonella*. In addition, always separate raw meats and poultry products from cooked foods. To prevent foodborne intoxication from *Staphylococcus* organisms, cover cuts on hands and avoid sneezing on foods. To avoid intoxication from *Clostridium perfringens,* rapidly cool leftover foods and thoroughly reheat them. To avoid intoxication from *Clostridium botulinum,* carefully examine canned foods. Overall, do not allow cooked food to stand for more than 1 to 2 hours at room temperature. For other causes of foodborne illness, precautions already mentioned generally apply as well. In addition, thoroughly cook fish and other seafood; consume only pasteurized dairy products; wash all fruits and vegetables; and thoroughly wash your hands with soap and water before and after preparing food and after using the bathroom.

Safe Handling Instructions for Eggs

To prevent illness from bacteria: keep eggs refrigerated, cook eggs until yolks are firm, and cook foods containing eggs thoroughly.

Regularly cleaning food-preparation surfaces and equipment with a dilute bleach solution (1:10) is helpful in reducing the risk of cross–contamination of foods.

CASE STUDY   Preventing Foodborne Illness at Gatherings

Nicole attended a gathering of her coworkers on a warm Saturday in July. The theme of the party was international dining. Nicole and her husband brought an Argentinian dish, potato and beef empanadas. They followed the recipe and cooking time carefully, removing the dish from the oven at 1 P.M. and keeping it warm by wrapping the pan in a towel. They traveled in their car to the party and set the dish out on the buffet table at 3 P.M. Dinner was to be served at 4 P.M. However, the guests were enjoying themselves so much lounging around the host's pool and drinking ginger beer (also on the menu) that no one began to eat until 6 P.M. Nicole made sure she sampled the empanadas that she and her husband made, while her husband did not. She also had some salad, garlic bread, and a sweet dessert made with coconut.

The couple returned home at 11 P.M. and went to bed. At about 2 A.M., Nicole knew something was wrong. She had severe abdominal pain and had to make a dash to the bathroom. She spent most of the next 3 hours in the bathroom with severe diarrhea. By dawn, the diarrhea subsided and she started to feel better. After a few cups of tea and a light breakfast, she was feeling like herself by noon.

Answer the following questions and check your responses at the end of the chapter.

1. Based on her symptoms, what type of foodborne illness did Nicole contract?
2. Why is the beef the most likely vehicle for this type of foodborne illness?
3. Why is consuming food at large gatherings risky?
4. What precautions for avoiding foodborne illness were ignored by Nicole and the rest of the people at the party?
5. How could this scenario be rewritten to substantially reduce the risk of foodborne illness?

Summary (Numbers refer to numbered sections in the chapter.)

13.1 Infants, children, older adults, postsurgical patients, immunosuppressed individuals, and pregnant women are most susceptible to foodborne illness. The risk of foodborne illness has increased as more of our foods are prepared outside of the home.

13.2 In the past, salt, sugar, smoke, fermentation, and drying were used to protect against foodborne illness. Today, careful cooking, pasteurization, irradiation, keeping hot foods hot and cold foods cold, and thorough handwashing provide additional insurance.

13.3 Viruses, bacteria, and other microorganisms in food pose the greatest risk for foodborne illness. Major causes of foodborne illness are Norovirus and the bacteria *Campylobacter jejuni, Salmonella, Staphylococcus aureus*, and *Clostridium perfringens*. In addition, such bacteria as *Clostridium botulinum, Listeria monocytogenes*, and *Escherichia coli* have been found to cause illness.

13.4 Food additives are used primarily to extend shelf life by preventing microbial growth and the destruction of food components by oxygen, metals, and other substances. Food additives are classified as those intentionally added to foods and those that incidentally appear in foods. An intentional additive is limited to no more than one-one-hundredth of the greatest amount that causes no observed symptoms in animals. Under its jurisdiction in the United States, the Delaney Clause allows FDA to ban the use of any intentional food additive that causes cancer.

13.5 Toxic substances occur naturally in a variety of foods, such as green potatoes, raw fish, mushrooms, and raw egg whites. Cooking foods limits their toxic effects in some cases; others are best to avoid altogether, such as toxic mushroom species and the green parts of potatoes.

13.6 A variety of environmental contaminants and pesticide residues can be found in foods. It is helpful to know which foods pose the greatest risks and act accordingly to reduce exposure, such as washing fruits and vegetables before use.

13.7 While conventional agriculture emphasizes large yields and low costs, the more recent trend toward sustainability considers the long-term environmental impact of agricultural practices. Consumers are driving up demand for organic and locally grown products.

NAYH Safe food handling can be summed up in four easy steps: (1) *Clean* hands and surfaces often; (2) *Separate* raw and ready-to-eat foods to prevent cross-contamination; (3) *Cook* (and reheat) potentially hazardous foods thoroughly, measuring temperature with a food thermometer; and (4) *Chill* foods by refrigerating promptly after eating. Two of these practices focus on food temperature; foods should not be held in the "danger zone" (40°F to 140°F) for more than 2 hours.

Check Your Knowledge (Answers to the following questions are below.)

1. Nitrite prevents the growth of
   a. *Clostridium botulinum.*
   b. *Escherichia coli.*
   c. *Staphylococcus aureus.*
   d. yeasts.

2. Substances used to preserve foods by lowering the pH are
   a. smoke and irradiation.
   b. baking powder and soda.
   c. salt and sugar.
   d. vinegar and citric acid.

3. Food additives widely used for many years without apparent ill effects are on the _____ list.
   a. FDA
   b. GRAS
   c. USDA
   d. Delaney

4. The foodborne illness organism often associated with small cuts and boils is
   a. *Listeria.*
   b. *Staphylococcus.*
   c. *C. botulinum.*
   d. *Salmonella.*

5. *Salmonella* bacteria are usually spread via
   a. raw meats, poultry, and eggs.
   b. pickled vegetables.
   c. home-canned vegetables.
   d. raw vegetables.

6. It is unwise to thaw meats or poultry
   a. in a microwave oven.
   b. in the refrigerator.
   c. under cool running water.
   d. at room temperature.

7. Milk that can remain on supermarket shelves, free of microbial growth, for many years has been processed by which of the following methods?
   a. use of humectants
   b. using antibiotics in animal feed
   c. use of sequestrants
   d. aseptic processing

8. Those at greatest risk for foodborne illness include
   a. pregnant women.
   b. infants and children.
   c. immunosuppressed individuals.
   d. All of the above.

9. Pasteurization involves the
   a. exposure of food to high temperatures for short periods to destroy harmful microorganisms.
   b. exposure of food to heat to inactivate enzymes that cause undesirable effects in foods during storage.
   c. fortification of foods with vitamins A and D.
   d. use of irradiation to destroy certain pathogens in foods.

10. Food can be kept for long periods by adding salt or sugar because these substances
    a. make the food too acidic for spoilage to occur.
    b. bind to water, thereby making it unavailable to the microorganisms.
    c. effectively kill microorganisms.
    d. dissolve the cell walls in plant foods.

Answer Key: 1. a (LO 13.4), 2. d (LO 13.2), 3. b (LO 13.4), 4. b (LO 13.1), 5. a (LO 13.1), 6. d (LO 13.8), 7. d (LO 13.2), 8. d (LO 13.3), 9. a (LO 13.2), 10. b (LO 13.4)

Study Questions (Numbers refer to Learning Outcomes)

1. What three trends in food purchasing and production have led to a greater number of cases of foodborne illness? (LO 13.1)

2. Which types of foods are most likely to be involved in foodborne illness? Why are they targets for contamination? (LO 13.2)

3. Identify three major classes of microorganisms responsible for foodborne illness. (LO 13.3)

4. Define the term *food additive* and give examples of four intentional food additives. What are their specific functions in foods? What is their relationship to the GRAS list? (LO 13.4)

5. Describe the federal process that governs the use of food additives, including the Delaney Clause. (LO 13.4)

6. Put into perspective the benefits and risks of using additives in food. Point out an easy way to reduce the consumption of food additives. Do you think this is worth the effort in terms of maintaining health? Why or why not? (LO 13.4)

7. Name some substances that occur naturally in foods but may cause illness. (LO 13.5)

8. Describe four recommendations for reducing the risk of toxicity from environmental contaminants. (LO 13.6)

9. Describe some of the advances in agricultural science that are positively affecting our food supply. (LO 13.7)

10. List four techniques other than thorough cooking that are important in preventing foodborne illness. (LO 13.8)

What the Dietitian Chose

Some organic fruits and vegetables do have higher levels of vitamin C, iron, phosphorus, magnesium, and phytochemicals. Exposure to environmental stressors may cause plants to produce more phytochemicals that have a positive impact on human health. However, current research is insufficient to recommend organic over conventional produce on the basis of nutrient content. Canned and frozen fruits and vegetables have greater nutrient content compared to fresh or organic products that are purchased several days or weeks after harvest.

Products labeled organic must comply with

▲ Exposure to pesticides is lower from organic foods.

standards regarding use of fertilizers, pesticides, hormones, antibiotics, genetic engineering, and irradiation. Organic food producers may use natural preservatives. Most preservatives are used to prevent food spoilage and are not linked to negative health effects. Buying fresh or frozen produce and preparing meals at home are the best ways to avoid excessive preservative intake.

Organic produce is no less likely to be contaminated with microorganisms than conventionally grown foods. Although manure is used as organic fertilizer, statistics show similar levels of foodborne illness organisms from either type of food. It is still important to follow food-safety advice, such as washing all fresh produce before eating it.

Exposure to pesticides makes organically grown produce stand apart. Because of strict production standards, we can expect a dramatically lower intake of pesticides from organic produce compared to conventionally grown produce. The health benefits of lower pesticide intake is greatest for children.

CASE STUDY SOLUTION Preventing Foodborne Illnesses at Gatherings

1. Nicole likely contracted *Clostridium perfringens,* based on the fact that she had diarrhea but did not vomit and the symptoms occurred about 8 hours after consuming the contaminated food (review Table 13-3).

2. Spores of *Clostridium perfringens* are typically present in meat. Thorough cooking will kill any of the live bacteria present, but the product may still contain spores. These can later develop into bacteria if the food is kept in a warm setting for a few hours. The beef in the empanadas was likely the bearer of the spores, and they likely germinated and produced a toxin as the beef sat in the car and on the buffet table.

3. Consuming food at large gatherings is risky for several reasons. First, foods for these events are typically cooked ahead of time and not consumed right away. Unfortunately, the foods end up remaining in the "danger zone" between 40°F and 140°F. Hot foods should be kept hot and cold foods cold, but proper refrigeration or heating equipment is not always available at these events. In addition, many people handle foods and serving utensils at large gatherings. It is important for everyone handling food or serving

utensils to thoroughly wash their hands before and after handling food and to avoid coughing or sneezing over foods. Also, when returning to the buffet table, always use a clean plate to avoid cross-contamination. Finally, the focus at these events is usually on having fun and socializing and not on food safety.

4. The main precaution that was ignored was keeping the food out of the "danger zone." Overall, it is risky to leave perishable items such as meat, fish, poultry, eggs, and dairy products at room temperature for more than 1 to 2 hours.

5. Ideally, the cooked food should have remained at room temperature for no longer than 1 hour. Thus, soon after Nicole and her husband took it out of the oven, it should have been separated into a few smaller pans to speed cooling and then refrigerated because they knew that it was not going to be served within 1 to 2 hours. Before leaving, they could have recombined the dish into one clean pan. Once they arrived at the party, the dish should have been refrigerated again and then thoroughly reheated when it was time to eat.

Further Readings

1. American Dietetic Association: Position of the American Dietetic Association: Food and water safety. *Journal of the American Dietetic Association* 109:1449, 2009.

2. Consumers Union: Dirty birds: Even "premium" chickens harbor dangerous bacteria. *Consumer Reports* p. 20, January 2007.

3. Dangour AD and others: Nutritional quality of organic foods: A systematic review. *American Journal of Clinical Nutrition* 90:680, 2009.

4. Geiger S: Eating seafood sustainably. *Today's Dietitian* 14(6):38, 2012.

5. Institute of Medicine Report: *Strategies to reduce sodium intake in the United States.* Consensus Report, Released: April 20, 2010 at **www.iom.edu/Reports/2010/Strategies-to-Reduce-Sodium-Intake-in-the-United-states.aspx.**

6. Kosa KM and others: Most Americans are not prepared to ensure food safety during power outages and other emergencies. *Food Protection Trends* 31:428, 2011.

7. National Center for Emerging and Zoonotic Infectious Diseases: *CDC estimates of foodborne illness in the United States: CDC 2011 estimates* February 2011, **www.cdc.gov/foodborneburden.**

8. Oken E and others: Which fish should I eat? Perspectives influencing fish consumption choices. *Environmental Health Perspectives* 120:790, 2012.

9. Scallan E and others: Foodborne illness acquired in the United States—Major pathogens. *Emerging Infectious Diseases* 17:7, 2011.

10. Schardt D: Safe at home: How to keep your kitchen from making you sick. *Nutrition Action Healthletter* p. 3, November 2011.

11. Schardt D: Going organic: What's the payoff? *Nutrition Action Healthletter* p. 1, October 2012.

12. Yeager D: Got organic? *Today's Dietitian* 10:60, October 2008.

 **To get the most out of your study of nutrition, visit McGraw-Hill Connect at www.mcgrawhillconnect.com where you will find NutritionCalc Plus, LearnSmart, and many other dynamic tools.**

Rate Your Plate

I. Take a Closer Look at Food Additives

Evaluate a food label of a convenience food item (e.g., frozen entree, ready-to-eat baked good) either in the supermarket or one you have available.

1. Write out the list of ingredients.

2. Identify the ingredients that you think may be food additives.

3. Based on the information available in this chapter, what are the functions of these food additives?

4. How might this food product differ without these ingredients?

II. Take a Closer Look at Organic Foods

Visit one or more supermarkets to see what organic foods are available. Note your findings below.

	Available	Not Available
Meat		
Poultry		
Milk		
Eggs		
Cheese		
Lettuce		
Apples		
Bananas		
Broccoli		
Other produce		
Breakfast cereal		
Snack chips		
Crackers		
Bread		
Pasta		

Do you currently purchase organic foods? Why or why not?

Find more Rate Your Plate activities for this chapter in Connect at www.mcgrawhillconnect.com.

Student Learning Outcomes

14.1 Describe how nutrition affects fertility.

14.2 Summarize the physiological changes of pregnancy, how they affect the nutrient requirements of a woman, and exemplify nutrients that may need to be supplemented during pregnancy.

14.3 Define "success" in pregnancy and identify lifestyle factors that promote a successful pregnancy for both the mother and the infant.

14.4 Specify optimal ranges of weight gain during pregnancy for women with low, healthy, or high prepregnancy BMI.

14.5 Outline guidance for exercise during pregnancy.

14.6 Describe the discomforts and complications of pregnancy that can be managed by dietary changes.

14.7 Summarize the physiological processes involved in breastfeeding and how breastfeeding affects the nutritional requirements of a woman.

Chapter 14
Nutrition During Pregnancy and Breastfeeding

14.8 Design an adequate, balanced meal plan for a pregnant or breastfeeding woman based on the Dietary Guidelines and MyPlate.

14.9 Enumerate several advantages of breastfeeding for both the mother and the infant.

14.10 Relate nutritional status of the parents to the risk of birth defects in the child.

The responsibility of nourishing and protecting a child is at once exhilarating and intimidating. Parents' desire to produce a healthy baby can arouse interest in nutrition and health information. They usually want to do everything possible to maximize their chances of having a robust, lively newborn.

Despite these intentions, the infant mortality rate in North America is higher than that seen in many other industrialized nations. In Canada, about 5 of every 1000 infants per year die before their first birthday, whereas in the United States, it is almost 6. These are alarming statistics for two countries that have such a high per-capita expenditure for health care compared to many other countries in the world. Comparatively, the rate of infant mortality in Sweden is less than 3 of every 1000 infants. In addition, in the United States, about 11% of pregnant women receive inadequate prenatal care.

Some aspects of fetal and newborn health are beyond our control. Still, as the comic in this chapter suggests, conscious decisions about social, health, environmental, and nutritional factors during pregnancy significantly affect the baby's future. Choosing to breastfeed the infant adds further benefits. Let us examine how eating well during pregnancy and breastfeeding can help a baby to have a healthy start in life.

14.1 Nutrition and Fertility

embryo In humans, the developing offspring in utero from about the beginning of the third week to the end of the eighth week after conception.

infertility Inability of a couple to conceive after 1 year of unprotected intercourse.

"Surprise, we're pregnant!" Considering the fact that only about half of all pregnancies are planned, a positive pregnancy test can be shocking news. Even when planned, women often do not suspect they are pregnant during the first few weeks after conception. They may not seek medical attention until 2 to 3 months after conception. Without fanfare, though, the **embryo** grows and develops daily. As you will learn in this chapter, the mother's nutritional status will affect the health of her baby before and long after birth. For that reason, the health and nutrition habits of a woman who is trying to become pregnant—or has the potential to become pregnant—are vitally important.

For up to 15% of couples who are planning for pregnancy, however, that pregnancy test shows a negative result month after month. **Infertility** refers to the inability of a couple to conceive after 1 year of unprotected intercourse. There are numerous possible causes for male and female infertility, many of which are outside the couple's control. In some cases, however, nutrition and lifestyle changes can improve a couple's chances of conceiving a child.

The nutritional status of both the mother- and the father-to-be can affect the likelihood of conception (see Further Reading 19). Some of the nutritional factors discussed in the following will affect the levels of hormones involved in reproduction. Others directly affect the viability of the egg or the sperm. Thus far, research clearly supports a link between body fat and fertility, and some evidence points to roles of certain dietary fats, carbohydrates, antioxidant nutrients, B vitamins, zinc, and iron.

ENERGY BALANCE

Recall from Chapter 7 that energy balance is the relationship between energy in and energy out. *Positive energy balance* describes a situation in which the amount of calories consumed exceeds the amount of calories required to support basic body processes and physical activity. Sustained over time, positive energy balance leads to gains in both lean mass and adipose tissue. *Negative energy balance* occurs when calorie intake falls short of calorie needs. Prolonged negative energy balance leads to loss of both lean and adipose tissue. At either extreme, prolonged energy imbalance can impair fertility.

It is important to note that adipose tissue serves as more than just a storage depot for energy; it also produces estrogen and other hormones and cellular signaling molecules with widespread effects. For example, leptin, a hormone produced primarily by adipose tissue, affects appetite, metabolic rate, immune function, growth, and reproduction.

Reproductive function is costly in terms of energy. Synthesis of reproductive hormones, maintenance of normal menstrual cycles, pregnancy, and breastfeeding require calories. With negative energy balance, little energy is available to maintain normal reproductive function. Consequently, many underweight women

Tina's Groove ©Rina Piccolo. Used with the permission of Rina Piccolo, King Features Syndicate and the Cartoonist Group. All rights reserved.

▲ Which diet and lifestyle habits contribute to a successful pregnancy? Which are likely to be harmful? Why should a woman begin to prepare for pregnancy months before conception of her new baby? When pregnant, does the mother need to "eat for two"? This chapter provides some answers.

experience amenorrhea, which is a sign of impaired ovulation. Some causes of low energy availability are undernutrition that stems from poverty (see Chapter 13), an eating disorder (see Chapter 11), or high levels of athletic training (see Chapter 10). During World War II, for example, famine in Holland drastically cut the calorie intake of women to about 1000 kcal per day. Many women became amenorrheic, and birth rates declined by about 50% during that time period. Studies in female athletes indicate that energy intake of at least 30 kcal per kilogram of lean mass is needed for normal reproductive function among women. For men, low body fat can decrease sex drive and sperm count.

On the opposite end of the spectrum, prolonged positive energy balance also decreases fertility. The extra adipose tissue affects the availability of reproductive hormones and induces insulin resistance. For women, these endocrine changes impair the success of ovulation and implantation. In fact, excess body fat is thought to cause about 25% of problems with ovulation that lead to infertility. Among men, excess body fat increases estrogen levels and decreases testosterone. Also, extra fat tissue increases the temperature of the testicular area. The changes in hormones and temperature result in lower sperm production. Excess body fat also increases oxidative stress, which damages DNA in both the egg and the sperm. For overweight or obese adults, studies show that losing just 5% to 10% of body weight can increase chances of conception.

▲ Several studies show a link between a low glycemic-index diet and improved fertility for women with PCOS.

POLYCYSTIC OVARY SYNDROME

As we discuss the ways energy imbalances influence fertility, it is important to mention **polycystic ovary syndrome** (PCOS). PCOS, characterized by many tiny cysts that surround the ovaries like a strand of pearls, is the leading cause of female infertility. Two main hormonal alterations are typical of PCOS: high testosterone and high insulin. All women secrete some testosterone, but women with PCOS secrete more than normal. The high levels of male hormones lead to some of the signs and symptoms of PCOS: excess hair growth on the face, acne, and a tendency to deposit fat around the waist. Insulin resistance is a common feature of the syndrome. Thus, women with PCOS are at higher risk for diabetes (see Chapter 4), high blood pressure (see Chapter 9), and cardiovascular disease (see Chapter 5). Importantly for our discussion here, women with PCOS have irregular or absent periods, difficulty becoming pregnant, and higher-than-average rates of miscarriage.

polycystic ovary syndrome (PCOS) A condition of hormonal imbalance (e.g., elevated testosterone and insulin) in a woman that can lead to infertility, weight gain in the abdominal region, excessive growth of body hair, and acne.

Many dietary and lifestyle changes have been studied to see how they can alter the course of PCOS. To date, evidence most clearly supports the importance of weight loss to improve metabolic and fertility issues among women with PCOS. If overweight women with PCOS lose just 5% of their body weight, regardless of diet composition, their chances of conception improve. Daily physical activity, known to improve insulin sensitivity, is a key component of any weight management strategy.

In addition to managing body weight, the quality and quantity of carbohydrates may make a difference in controlling PCOS and improving fertility. Some experts recommend reducing carbohydrate intake to the lower end of the range recommended by the Food and Nutrition Board (about 45% of total kcal) and choosing low glycemic-index carbohydrates. For example, women with PCOS are urged to choose whole grains instead of refined grains and whole fruits and vegetables rather than juices, and to steer clear of sugar-sweetened beverages. There are claims that following a low-carbohydrate diet (less than 45% of total kcal) is useful for women with PCOS, but these claims are not well supported by research. Severely restricting carbohydrate choices could limit the intake of important nutrients, such as B vitamins.

FOLATE

Routine intake of a daily multivitamin and mineral supplement has been linked to improved fertility in many studies. There are a variety of micronutrients that may contribute, but folic acid tops the list for both men and women. As you learned in Chapter 8, folate is involved in DNA synthesis and the metabolism of homocysteine. For no other cells is proper DNA synthesis so important as for the egg and sperm, which transmit genetic information from one generation to the next! Foods such as leafy green vegetables, strawberries, and orange juice are sources of natural folate. The synthetic form, folic acid, can be found in dietary supplements and fortified foods, such as ready-to-eat breakfast cereals.

ANTIOXIDANTS

The chemical reactions of metabolism produce free radicals (molecules with unpaired electrons) that can damage cell membranes and DNA. The body has some antioxidant mechanisms that limit the activity of free radicals, but when the production of free radicals exceeds the antioxidant capacity of the body, oxidative damage to cells is likely. Free radicals can damage egg and (especially) sperm cells and can affect how well a fertilized egg implants and matures. Research studies show that *diets* rich in antioxidant nutrients—vitamin E, vitamin C, selenium, zinc, beta-carotene, and some other plant pigments—are linked to improved fertility for both men and women. Foods of plant origin, including brightly colored fruits and vegetables, whole grains, and plant oils, are rich sources of antioxidant nutrients.

MINERALS

Iron and zinc are two minerals that have been linked to fertility. Zinc appears to be especially important for male fertility. Not only is it involved as a cofactor in antioxidant reactions that could protect sperm from oxidative damage, but zinc is also required for normal sexual maturation (recall the photo of the zinc-deficient teenager on page 379 of Chapter 9) and production of sperm and reproductive hormones. Men with poor zinc status have poor sperm quality, and studies have shown that zinc supplements can improve sperm quality.

For women, iron and zinc are needed for normal ovulation. Data from a large observational study of nurses showed that use of iron supplements before conception was linked to improved ovulatory function and therefore better fertility. Interestingly, in this study, higher intakes of nonheme iron (from plant sources) were specifically related to improved fertility.

DIETARY FAT

When trying to conceive, should you switch to decaf? There has been some concern that caffeine could decrease fertility, but the evidence is mixed. In some studies, intake of more than 4 cups of coffee (about 500 mg caffeine) has been linked to decreased fertility, but most studies show no clear link between moderate caffeine intake and fertility.

As recommended for the general population of healthy adults, men and women who are trying to conceive should try to limit sources of saturated and *trans* fats. For women, a diet rich in saturated and *trans* fat promotes insulin resistance and impairs ovulation. For men, high intakes of saturated and *trans* fat are linked to poor sperm quality. Instead of consuming the typical American fare of pizza and fast foods, men and women should emphasize plant oils and fish oils, which provide more unsaturated fats. Among men experiencing infertility, boosting intakes of omega-3 fatty acids, the type of polyunsaturated fatty acids found in fish oils and walnuts, can improve sperm quality.

ALCOHOL

While trying to get pregnant, it is safest to avoid alcohol. Some, but not all, studies show that higher intakes of alcohol (i.e., more than one or two drinks per day) are related to lower rates of conception. Alcohol can decrease levels of estrogen and testosterone and can therefore disrupt the normal cycles of ovulation and sperm production. In the Nutrition and Your Health section at the end of this chapter, you will learn about the devastating effects of alcohol on the developing babies *in utero*. Given the fact that many women do not realize they are pregnant until several weeks after conception, it is best to avoid alcohol.

☑ CONCEPT CHECK 14.1

1. How is energy balance related to fertility?
2. What is polycystic ovary syndrome? What hormonal changes are involved in PCOS, and how do they affect fertility?
3. List three nutrients that have been linked to fertility.

14.2 Prenatal Growth and Development

The length of a normal pregnancy is 38 to 42 weeks, measured from the first day of the woman's last menstrual period. For purposes of discussion, the duration of pregnancy is commonly divided into three periods, called **trimesters.** For 8 weeks after conception, a human embryo develops from a fertilized **ovum** into a **fetus.**

Until birth, the mother nourishes the fetus via a **placenta,** an organ that forms in her uterus to accommodate the growth and development of the fetus (Fig. 14-1). The role of the placenta is to exchange nutrients, oxygen and other gases, and waste products between the mother and the fetus. This occurs through a network of capillaries that bring the fetal blood close to the maternal blood supply, but the two blood supplies do not mix.

EARLY GROWTH—THE FIRST TRIMESTER IS A VERY CRITICAL TIME

In the formation of the human organism, egg and sperm unite to produce the **zygote** (Fig. 14-2). From this point, the reproductive process occurs very rapidly:

- Within 30 hours: zygote divides in half to form two cells.
- Within 4 days: cell number climbs to 128 cells.
- At 14 days: the group of cells is called an embryo.
- Within 35 days: heart is beating, embryo is 1/30 of an inch (8 millimeters) long, eyes and limb buds are clearly visible.
- At 8 weeks: the embryo is known as a fetus.
- At 13 weeks (end of first trimester): most organs are formed, and the fetus can move.

trimesters Three 13- to 14-week periods into which the normal pregnancy (on average, 40 weeks) is divided somewhat arbitrarily for purposes of discussion and analysis. Development of the offspring, however, is continuous throughout pregnancy, with no specific physiological markers demarcating the transition from one trimester to the next.

ovum The egg cell from which a fetus eventually develops if the egg is fertilized by a sperm cell.

fetus The developing life form from about the beginning of the ninth week after conception until birth.

placenta An organ that forms in the uterus in pregnant women. Through this organ, oxygen and nutrients from the mother's blood are transferred to the fetus, and fetal wastes are removed. The placenta also releases hormones that maintain the state of pregnancy.

zygote The fertilized ovum; the cell resulting from the union of an egg cell (ovum) and sperm until it divides.

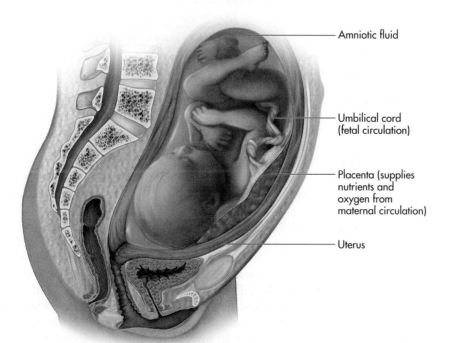

- Amniotic fluid
- Umbilical cord (fetal circulation)
- Placenta (supplies nutrients and oxygen from maternal circulation)
- Uterus

FIGURE 14-1 ◄ The fetus in relationship to the placenta. The placenta is the organ through which nourishment flows to the fetus.

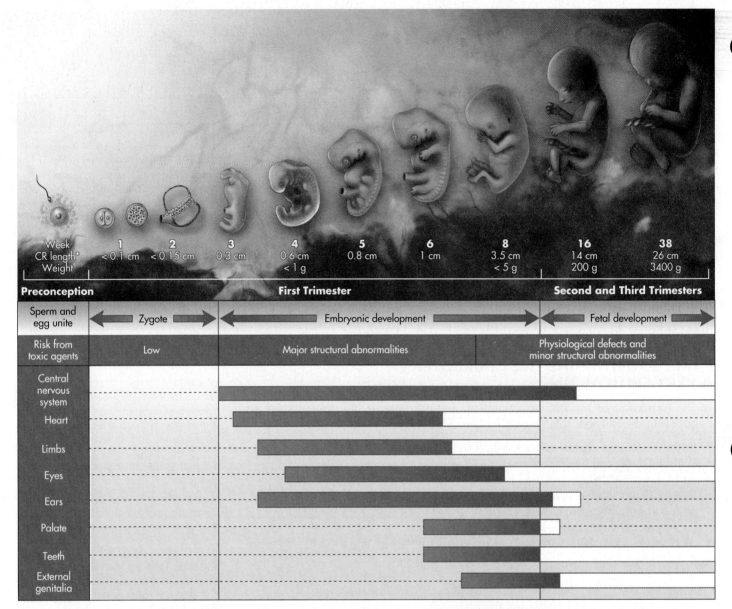

FIGURE 14-2 ▲ Harmful effects of toxic agents during pregnancy. Vulnerable periods of fetal development are indicated with purple bars. The purple shading indicates the time of greatest risk to the organ. The most serious damage to the fetus from exposure to toxins is likely to occur during the first 8 weeks after conception, two-thirds of the way through the first trimester. As the white bars in the chart show, however, damage to vital parts of the body—including the eyes, brain, and genitals—can also occur during the later months of pregnancy.

spontaneous abortion Cessation of pregnancy and expulsion of the embryo or nonviable fetus prior to 20 weeks' gestation. This is the result of natural causes, such as a genetic defect or developmental problem; also called *miscarriage*.

Growth begins in the first trimester with a rapid increase in cell number. This type of growth dominates embryonic and early fetal development. The newly formed cells then begin to grow larger. Further growth is a mix of increases in cell number and cell size. By the end of 13 weeks—the first trimester—most organs are formed and the fetus can move (see Fig. 14-2).

As the embryo or fetus develops, nutritional deficiencies, toxicities, and other harmful exposures have the potential to damage organ systems. For example, adverse reactions to medications, high intakes of vitamin A, exposure to radiation, or trauma can alter or arrest the current phase of fetal development, and the effects may last a lifetime (review Fig. 14-2). The most critical time for these potential problems is during the first trimester. Most **spontaneous abortions**—premature terminations of pregnancy that occur naturally—happen at this time. About one-half or more pregnancies end in this way, often so early that a woman does not even realize she was pregnant. (An additional 15% to 20% are lost before normal delivery.) Early spontaneous abortions usually result from a genetic defect or fatal error in fetal development. Smoking, alcohol abuse, use of aspirin and NSAIDs, and illicit drug use raise the risk for spontaneous abortion.

A woman should avoid substances that may harm the developing fetus, especially during the first trimester. This holds true, as well, for the time when a woman is trying to become pregnant. As previously mentioned, she is unlikely to be aware of her pregnancy for at least a few weeks. In addition, the fetus develops so rapidly during the first trimester that, if an essential nutrient is not available, the fetus may be affected even before evidence of the nutrient deficiency appears in the mother.

For this reason, the *quality*—rather than the *quantity*—of the woman's nutritional intake is most important during the first trimester. In other words, the mother should consume the same amount of calories as she did before she became pregnant, but she should focus on choosing more nutrient-dense foods. Although some women lose their appetite and feel nauseated during the first trimester, they should be careful to meet nutrient needs as much as possible.

SECOND TRIMESTER

By the beginning of the second trimester, a fetus weighs about 1 ounce. Arms, hands, fingers, legs, feet, and toes are fully formed. The fetus has ears and begins to form tooth sockets in its jawbone. Organs continue to grow and mature, and, with a stethoscope or Doppler instrument, physicians can detect the fetal heartbeat. Most bones are distinctly evident through the body. Eventually, the fetus begins to look more like an infant. It may suck its thumb and kick strongly enough to be felt by the mother. As was shown in Figure 14-2, the fetus can still be affected by exposure to toxins, but not to the degree seen in the first trimester.

During the second trimester, the mother's breast weight increases by approximately 30% due to the development of milk-producing cells and the deposition of 2 to 4 pounds of fat for **lactation.** This stored fat serves as a reservoir for the extra calories that will be needed to produce breast milk.

lactation The period of milk secretion following pregnancy; typically called *breastfeeding.*

THIRD TRIMESTER

By the beginning of the third trimester, a fetus weighs about 2 to 3 pounds. The third trimester is a crucial time for fetal growth. The fetus will double in length and increase its weight by three to four times. The fetus takes higher priority than the mother with regard to iron and will deplete the stores of the mother. If the mother is not meeting her iron needs, she can be severely depleted after delivery. An infant born after only about 26 weeks of **gestation** has a good chance of surviving if cared for in a nursery for high-risk newborns. However, the infant will not contain the stores of minerals (mainly iron and calcium) and fat normally accumulated during the last month of gestation. This and other medical problems, such as a poor ability to suck and swallow, complicate nutritional care for preterm infants.

By full term, the fetus usually weighs about 7 to 9 pounds (3 to 4 kilograms) and is about 20 inches (50 centimeters) long. Soft spots (fontanels) on top of the head indicate where the skull bones are growing together. The bones finally close by the time the baby is about 12 to 18 months of age.

gestation The period of intrauterine development of offspring, from conception to birth; in humans, normal gestation is 38 to 42 weeks.

✔ CONCEPT CHECK 14.2

1. What is the role of the placenta?
2. Describe how risks for fetal malformations vary throughout pregnancy.
3. During which trimester are most organs being formed?
4. During which trimester does fetal size increase the most?

fetal origins hypothesis A theory that links nutritional and other environmental insults that occur during gestation to the future health of the offspring.

low birth weight (LBW) Referring to any infant weighing less than 2.5 kilograms (5.5 pounds) at birth; most commonly results from preterm birth.

preterm An infant born before 37 weeks of gestation; also referred to as *premature*.

small for gestational age (SGA) Referring to infants who weigh less than the expected weight for their length of gestation. This corresponds to less than 2.5 kilograms (5.5 pounds) in a full-term newborn. A preterm infant who is also SGA will most likely develop some medical complications.

▼ A healthy newborn. At birth, a baby usually weighs about 7.5 pounds and is 20 inches long.

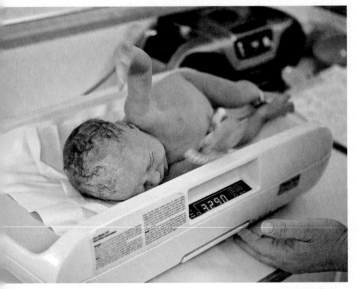

A goal of *Healthy People 2020* is to reduce low birth weight by 5% and preterm births by 10%. Currently, about 8% of live births are of low birth weight and about 12% are preterm.

14.3 Success in Pregnancy

The goal of pregnancy is to achieve optimal health for both the baby and the mother. For the mother, a successful pregnancy is one in which her physical and emotional health is protected so that she can return to her prepregnancy health status. For the infant, two widely accepted criteria are (1) a gestation period longer than 37 weeks and (2) a birth weight greater than 5.5 pounds (2.5 kilograms). Sufficient lung development, likely to have occurred by 37 weeks of gestation, is critical to the survival of a newborn. The longer the gestation (up to 42 weeks), the greater the ultimate birth weight and maturation state, leading to fewer medical problems and better quality of life for the infant.

As you read this chapter, you will notice frequent references to lifelong effects of maternal nutrition, physical activity, and other lifestyle practices on the child. This is called the *developmental origins of health and disease hypothesis,* or more simply, the **fetal origins hypothesis.** Emerging evidence links prenatal influences, such as famines, fasting, exposure to alcohol, and even environmental pollution, to the child's risks for disease later in life (see Newsworthy Nutrition on page 560). As we learn more about how environmental factors shape the ways our genes are expressed, it becomes increasingly evident that many aspects of our physical and mental health are programmed while we are yet *in utero.* What's more, these epigenetic changes may be passed down from generation to generation.

Overall, a successful pregnancy is the outcome of a complex interplay between genes, various lifestyle choices, and the environment. The decisions a mother and father make today can affect the health of their child for years to come. Although a mother's decisions, practices, and precautions during pregnancy contribute to the health of her fetus during all three trimesters, she cannot guarantee her fetus good health because some genetic and environmental factors are beyond her control. She and others involved in the pregnancy should not hold an unrealistic illusion of total control.

Low-birth-weight (LBW) infants are those weighing less than 5.5 pounds (2.5 kilograms) at birth. In the United States, 1 out of 12 newborns is LBW. Most commonly, LBW is associated with **preterm** birth. Medical costs during the first year of life for LBW infants are higher than those for normal-weight infants. In fact, hospital-related costs of caring for LBW newborns total more than $4 billion per year in the United States. Full-term and preterm infants who weigh less than the expected weight for their duration of gestation, the result of insufficient growth, are described as **small for gestational age (SGA).** Thus, a full-term infant weighing less than 5.5 pounds at birth is SGA but not preterm, whereas a preterm infant born at 30 weeks' gestation is probably LBW without being SGA. Infants who are SGA are more likely than normal-weight infants to have medical complications, including problems with blood glucose control, temperature regulation, growth, and development in the early weeks after birth.

PRENATAL CARE AND COUNSELING

Adequate prenatal care is a primary determinant of success in pregnancy. Ideally, women should receive examinations and counseling before becoming pregnant and continue regular prenatal care throughout pregnancy. If prenatal care is inadequate, delayed, or absent, untreated maternal nutritional deficiencies can deprive a developing fetus of needed nutrients. In addition, untreated health conditions, such as anemia, AIDS, hypertension, or diabetes, must be carefully addressed to minimize complications during pregnancy. Treating ongoing infections will also decrease risks of fetal damage. Without prenatal care, a woman is three times more likely to deliver an LBW baby—one who will be 40 times more likely to die during the first 4 weeks of life than a normal-birth-weight infant. According to the Physicians

Committee for Responsible Medicine, early and consistent prenatal care could reduce the number of LBW births by 12,600 per year in the United States. Although the ideal time to start prenatal care is before conception, about 20% of women in the United States receive *no* prenatal care throughout the first trimester—a critical time to positively influence the outcome of pregnancy.

Food habits cannot be predicted from income, education, or lifestyle. Although some women already have good dietary habits, most can benefit from nutritional advice. All should be reminded of habits that may harm the growing fetus, such as severe dieting or fasting. By focusing on appropriate prenatal care, nutrient intake, and health habits, parents give their fetus—and later, their infant—the best chance of thriving. Overall, the chances of producing a healthy baby are maximized with education, an adequate diet, and early and consistent prenatal medical care.

EFFECTS OF MATERNAL AGE

The age of the mother is another factor that determines pregnancy outcome. The ideal age for pregnancy is between 20 and 35 years of age. Outside that age range—at either extreme—complications are more likely to arise. The rates of teen pregnancy have declined since 1990; still, approximately 330,000 babies are born to teen mothers in the United State each year—the highest of any industrialized country. Teen pregnancy increases risk for negative outcomes for both mother and child (Table 14-1) and costs taxpayers an estimated $11 billion each year. A portion of these burdens stems from a disadvantaged background, but teenage pregnancy cuts across socioeconomic classes and the effects are evident even after controlling for background factors. Pregnant teens frequently exhibit a variety of risk factors that can complicate pregnancy and pose risk to the fetus. For instance, teenagers are more likely than adult women to be underweight at the start of pregnancy and to gain too little weight during pregnancy. In addition, their bodies lack the physical maturity needed to carry a fetus safely. Even with prenatal care—and 7% of teenage mothers receive none at all—10% of children born to teenage mothers are of low birth weight and 14.5% are preterm (see Further Reading 11).

Advanced maternal age also poses special risks for pregnancy. The likelihood of LBW and preterm delivery increase modestly, but progressively, with maternal age beyond 35 years. Given close monitoring, however, a woman older than 35 years has an excellent chance of producing a healthy infant.

CLOSELY SPACED AND MULTIPLE BIRTHS

Siblings born in succession to a mother with less than a year between birth and subsequent conception are more likely to be born with low birth weights than are those farther apart in age. The risks of low birth weight, preterm birth, or small size for gestational age are 30% to 40% higher for infants conceived less than 6 months following a birth compared to those conceived 18 to 23 months following a birth. These poor outcomes are probably related to a lack of enough time to rebuild nutrient stores depleted by the pregnancy. Similarly, multiple births (i.e., twins) increase the risk for preterm birth.

▲ To ensure optimal health and rapid treatment of medical conditions that develop during pregnancy, a pregnant woman should consult with her health care provider on a regular basis. Ideally, this consultation should begin before she becomes pregnant.

Women with acquired immune deficiency syndrome (AIDS) may pass the virus that causes this disease to the fetus during pregnancy or delivery. About one in three infected newborns will develop AIDS symptoms and die within just a few years. Studies show that these odds of mother-to-child transmission can be cut significantly if the woman begins taking highly active antiretroviral therapy (HAART) and receiving routine obstetrical care. Before HAART was available, up to 45% of babies born to mothers with the AIDS virus were infected. Now, transmission is less than 2%. Thus, screening pregnant women for AIDS and providing HAART to those with AIDS are advocated by many experts.

TABLE 14-1 ▶ The Burdens of Teenage Pregnancy

For Mother	For Child
↑ Depression and other mental health problems	↓ Birth weight
↑ Use of illicit drugs and alcohol	↑ Premature birth
↑ Poverty and reliance on public assistance	↑ Infant mortality
↓ Graduation rates from high school and college	↑ Hospital admissions during childhood
↑ Single parenthood	↓ Academic performance
	↓ Nutritional status
	↑ Rates of imprisonment during adulthood

EXPOSURE TO TOXIC CHEMICALS

Figure 14-2 depicts how and when toxic agents can harm the developing fetus. In preparation for and during pregnancy, the mother should undoubtedly eliminate alcohol, tobacco, and illicit drugs (e.g., marijuana and cocaine). During organ development, exposure to these toxic chemicals can cause malformations. The adverse effects of fetal exposure to alcohol during gestation are discussed in the Nutrition and Your Health section at the end of this chapter. Smoking is linked to preterm birth and appears to increase the risk of birth defects, sudden infant death, and childhood cancer.

Illicit drug use is particularly harmful during pregnancy. Many chemicals in recreational drugs cross the placenta and affect the fetus, whose detoxification systems are immature. Marijuana, the most common illegal drug used during the reproductive years, can result in reduced blood flow to the uterus and placenta, leading to poor fetal growth. Low birth weight and higher risk of premature delivery often are seen in infants whose mothers used marijuana during pregnancy. Use of psychoactive drugs, such as cocaine and methamphetamines, restricts fetal growth and brain development, the effects of which may plague the child for a lifetime.

Even prescribed and common over-the-counter medications could have harmful effects on the developing fetus. Problem drugs include aspirin (especially when used heavily), hormone ointments, nose drops and related "cold" medications, rectal suppositories, weight-control pills, antidepressants, and medications prescribed for preexisting illnesses. Some herbal therapies also have the potential to damage the fetus. Lower doses and/or safer alternatives should be substituted when a woman is planning to or has become pregnant.

In addition to these toxic chemicals, health hazards in the mother's environment, including job-related hazards and exposure to X-rays, should be minimized.

FOOD SAFETY

The USDA warns pregnant women to thoroughly cook (e.g., microwave) all ready-to-eat meats, including hot dogs and cold cuts, until they are steaming to reduce risk of *Listeria* infections.

Any foodborne illness during any stage of life is a concern. One type of foodborne illness that poses particular danger for pregnant women is caused by the bacterium *Listeria monocytogenes* (review Chapter 13). Infection with this microorganism typically causes mild flulike symptoms, such as fever, headache, and vomiting, about 7 to 30 days after exposure. However, pregnant women, newborn infants, and people with depressed immune function may suffer more severe symptoms, including spontaneous abortion and serious blood infections. In these high-risk groups, 25% of infections may be fatal. Unpasteurized milk, soft cheeses made from raw milk (e.g., brie, Camembert, feta, and blue cheeses), and some raw vegetables (e.g., cabbage and sprouts) can be sources of *Listeria* organisms, so it is especially important that pregnant women (and other people at high risk for infection) avoid these products. Experts advise consuming only pasteurized milk products and cooking meat, poultry, and seafood thoroughly to kill this and other foodborne organisms. It is unsafe in pregnancy to eat any raw meats or other raw animal products, uncooked hot dogs, or undercooked poultry. These food safety recommendations are included in the Dietary Guidelines for Americans discussed later in Section 14.4 (Table 14-3).

PREPREGNANCY BMI

Women should aim to achieve a healthy body weight prior to becoming pregnant. Infants born to women who begin pregnancy substantially above or below a healthy weight are more likely to experience problems than those born to women who begin pregnancy at a normal weight. For instance, babies born to obese women are at increased risk of having birth defects, death in the first few weeks after birth, and obesity in childhood. Many obese pregnant women experience high blood pressure, diabetes, and difficult deliveries.

At the other extreme, women who begin pregnancy underweight (BMI under 18.5) are more likely to have infants who are low birth weight and premature than women at a normal weight. These differences may be because underweight women tend to have lighter placentas and lower nutrient stores, especially iron, than heavier women, which can affect fetal growth negatively. An underweight woman can improve her nutrient stores and pregnancy outcome by gaining weight before pregnancy or gaining extra weight during pregnancy.

NUTRITIONAL STATUS

Is attention to good nutrition worth the effort? Yes; much research suggests that an adequate vitamin and mineral intake at least 8 weeks before conception and then during pregnancy can improve outcomes of pregnancy (see Further Readings 3 and 17). Extra nutrients and calories are used for fetal growth, as well as for the changes the mother's body undergoes to accommodate the fetus. Her uterus and breasts grow, the placenta develops, her total blood volume increases, the heart and kidneys work harder, and stores of body fat increase. In particular, meeting folate needs (400 micrograms of synthetic folic acid per day) helps to prevent birth defects such as neural tube defects (see the Nutrition and Your Health section at the end of this chapter) and decrease the risk of preterm delivery. Low intakes of calcium and iron or excessive intakes of vitamin A also are cause for concern during pregnancy.

Although it is difficult to predict to what degree poor nutrition will affect each pregnancy, a daily diet containing only 1000 kcal has been shown to greatly restrict fetal growth and development. Increased maternal and infant death rates seen in famine-stricken areas of Africa provide further evidence.

Genetic background can explain little of the observed differences in birth weight between developed and developing countries. Both environmental factors and nutritional factors are important. The worse the nutritional condition of the mother at the beginning of pregnancy, the more valuable a healthy prenatal diet and/or use of prenatal supplements are in improving the course and outcome of her pregnancy.

NUTRITION ASSISTANCE FOR LOW-INCOME FAMILIES

Poverty impacts pregnancy in many ways. Families of low socioeconomic status tend to receive inadequate health care. A lack of education and a lack of financial resources may contribute to poor health practices, such as diet patterns that fail to meet the mother's increased nutrient requirements.

Several U.S. government programs provide high-quality health care and foods to reduce infant mortality. These are designed to alleviate the negative impact of poverty, insufficient education, and inadequate nutrient intake on pregnancy outcome. An example of such a program is the Special Supplemental Nutrition Program for Women, Infants, and Children (WIC). This program offers health assessments and vouchers for foods that supply high-quality protein, calcium, iron, and vitamins A and C to pregnant women, infants, and children (up to age 5 years) from low-income populations. The WIC program is available in all areas of the United States and has a staff trained to help women have healthy babies. More than 9 million women, infants, and young children are benefiting from this program, but many eligible pregnant women are not participating.

▲ In the United States, low-income pregnant women and their infants (and children) benefit from the nutritional and medical attention provided by the WIC program.

✓ CONCEPT CHECK 14.3

1. In one or two sentences, how would you define success in pregnancy?
2. Define the terms *preterm, low birth weight,* and *small for gestational age.*
3. How is maternal age related to the outcome of pregnancy?
4. Is attention to good nutrition during pregnancy worth the effort? Why or why not?

14.4 Increased Nutrient Needs to Support Pregnancy

Pregnancy is a time of increased nutrient needs. It is important to recognize the need for individual assessment and counseling of mothers-to-be, as the nutritional and health status of each woman is different. Still, there are some general principles true for most women with regard to increased nutrient needs.

CALORIE NEEDS

To support the growth and development of the fetus, pregnant women need to increase their calorie intake. Calorie needs during the first trimester are essentially the same as for nonpregnant women. However, during the second and third trimesters, it is necessary for a pregnant woman to consume approximately 350 to 450 kcal more per day than her prepregnancy needs (the upper end of the range is needed in the third trimester).

Rather than seeing this as an opportunity to fill up on sugary desserts or fat-filled snacks, the woman should consume these extra calories in the form of nutrient-dense foods. For example, throughout the day, about six whole-wheat crackers, 1 ounce of cheese, and ½ cup of fat-free milk would supply the extra calories (and also some calcium). Although she "eats for two," the pregnant woman must not double her normal calorie intake. The "eating for two" concept refers more appropriately to increased needs for several vitamins and minerals. Micronutrient needs are increased by up to 50% during pregnancy, whereas calorie needs during the second and third trimesters represent only about a 20% increase.

If a woman is active during her pregnancy, she may need to increase her calorie intake by even more than the estimated 350 to 450 kcal per day. Her greater body weight requires more calories for activity. Many women find that they are inactive during the later months, partly because of their increased size, so an extra 350 to 450 kcal in their daily diets is usually enough.

Newsworthy Nutrition

Mothers' weight gain during pregnancy is linked to cognitive development of children

The Avon Longitudinal Study of Parents and Children (ALSPAC) includes 13,617 mother-offspring pairs in the United Kingdom who have been followed from pregnancy, through childbirth, and now into the offspring's young adulthood. In this analysis of gestational weight gain, children of mothers who gained less than recommended weight during pregnancy had lower School Entry Assessment scores at 4 years of age and lower performance on final examinations at 16 years of age. This study highlights the importance of adhering to the 2009 Institute of Medicine recommendations for gestational weight gain. Appropriate weight gain during pregnancy improves multiple aspects of health for both the mother and the child.

Source: Gage SH and others: Associations of maternal weight gain in pregnancy with offspring cognition in childhood and adolescence: Findings from the Avon Longitudinal Study of Parents and Children. *American Journal of Epidemiology* 177:402, 2013.

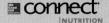 Check out the Connect site **www.mcgrawhillconnect.com** to further explore the fetal origins hypothesis.

STAYING ACTIVE DURING PREGNANCY

While pregnancy is not the time to begin an intense fitness regimen, women can generally take part in most low- or moderate-intensity activities during pregnancy. In fact, the American College of Obstetrics and Gynecology recommends that pregnant women should perform at least 30 minutes per day of moderate-intensity physical activity. Walking, cycling, swimming, or light aerobics for at least 150 minutes per week is generally advised. Such exercise may prevent pregnancy complications and promote an easier delivery (see Further Reading 16). Some research indicates that regular physical activity during pregnancy lowers a woman's risk of developing gestational diabetes by 50% and preeclampsia by 40%. These disorders of pregnancy are discussed further in Section 14.6. Figure 14-3 illustrates the broad range of benefits of exercise during pregnancy for both the mother and the baby. Indeed, preliminary research indicates that exercising during pregnancy could lower the risk for obesity in the child later in life.

Women who were highly active prior to pregnancy can maintain their activities as long as they remain healthy and review their plans with their health care providers. A few types of activities can potentially harm the fetus and should be avoided, especially those with inherent risk of falls and abdominal trauma. Examples of exercises to avoid, especially during the second and third trimesters, include downhill skiing, weightlifting, soccer, basketball, horseback riding, certain calisthenics (e.g., deep knee bends), any contact sports (e.g., hockey), and SCUBA diving.

Women with high-risk pregnancies, such as those experiencing premature labor contractions, may need to restrict their physical activity. To ensure optimal health for both herself and her infant, a pregnant woman should first consult her physician about physical activity and possible limitations.

OPTIMAL WEIGHT GAIN

Healthy prepregnancy weight and appropriate weight gain during pregnancy are excellent predictors of pregnancy outcome (see Further Readings 7, 9, and 11). The mother's diet should allow for approximately 2 to 4 pounds (0.9 to 1.8 kilograms)

FIGURE 14-3 ▲ Benefits of exercise during pregnancy for the mother and baby.

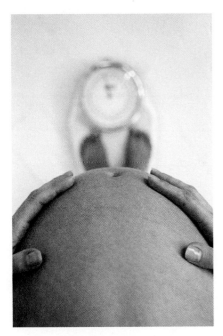

▲ Appropriate weight gain during pregnancy improves outcomes for both the mother and the baby.

of weight gain during the first trimester and then a subsequent weight gain of 0.8 to 1 pound (0.4 to 0.5 kilogram) weekly during the second and third trimesters (Fig. 14-4). A healthy goal for total weight gain for a woman of normal weight (based on BMI; Table 14-2) averages about 25 to 35 pounds (11.5 to 16 kilograms). Pregnant adolescents and women of various racial and ethnic groups are advised to meet these weight gain goals until further research determines any special recommendations. Women of normal prepregnancy BMI carrying twins should aim to gain within the range of 37 to 54 pounds, whereas overweight or obese women should gain less (31 to 50 pounds or 25 to 42 pounds, respectively).

For women who begin pregnancy with a low BMI, the goal increases to 28 to 40 pounds (12.5 to 18 kilograms). The goal decreases to 15 to 25 pounds (7 to 11.5 kilograms) for overweight women. Target weight gain for obese women is 11 to 20 pounds (5 to 9 kilograms). Figure 14-4 shows why the typical recommendation begins at 25 pounds.

A weight gain of between 25 and 35 pounds for a woman starting pregnancy at normal weight has repeatedly been shown to yield optimal health for both mother and fetus if gestation lasts at least 38 weeks. The weight gain should yield a birth weight of 7.5 pounds (3.5 kilograms). Although some extra weight gain during pregnancy is usually not harmful (about 5 to 10 pounds), it can set the stage for a pattern of weight gain during the child-bearing years if the mother does not return to her approximate prepregnancy weight after delivery. Key recommendations for weight management during pregnancy and lactation are included in the Dietary Guidelines for Americans (see Table 14-3).

Overweight and obesity do contribute to complications during pregnancy. Excess maternal body weight increases risk for diabetes, hypertension, blood clots, and spontaneous abortions during pregnancy. After childbirth, lasting effects of excess gestational weight gain include high BMI, central body fat distribution, and elevated blood pressure. For the baby, there is a greater chance of birth defects and macrosomia, in which the fetus grows larger than average in utero. Larger infants contribute

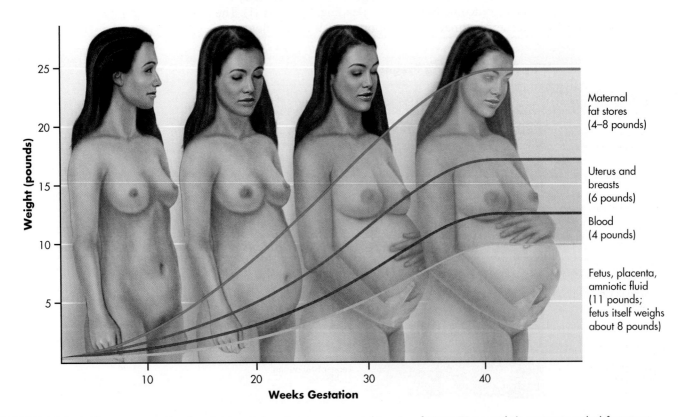

FIGURE 14-4 ▲ The components of weight gain in pregnancy. A weight gain of 25 to 35 pounds is recommended for most women. The various components total about 25 pounds.

TABLE 14-2 ▸ **Recommended Weight Gain in Pregnancy Based on Prepregnancy Body Mass Index (BMI)**

Prepregnancy BMI Category	Total Weight Gain*	
	(pounds)	(kilograms)
Low (BMI less than 18.5)	28 to 40	12.5 to 18
Normal (BMI 18.5 to 24.9)	25 to 35	11.5 to 16
High (BMI 25.0 to 29.9)	15 to 25	7 to 11.5
Obese (BMI greater than 30.0)	11 to 20	5 to 9

*The listed values are for pregnancies with one fetus. For women of normal BMI carrying twins, the range is 37 to 54 pounds (17 to 24.5 kilograms) or less for heavier women.

Reprinted in part with permission from *Weight Gain During Pregnancy: Reexamining the Guidelines,* Copyright 2009 by the Institute of Medicine and National Research Council of the National Academies. Courtesy of the National Academies Press, Washington, DC.

to a greater need for surgical delivery (i.e., Cesarean sections) among overweight and obese mothers. Over the long term, excessive maternal weight gain during pregnancy has been linked to increased risk for obesity and metabolic syndrome in the child.

Gestational weight gain is a key issue in prenatal care and a concern of many mothers-to-be. Even after the 2009 release of the weight gain guidelines summarized in Table 14-2, however, many women report receiving *no* guidance regarding weight gain from health care providers before or during pregnancy. Considering the extensive consequences of either inadequate or excessive weight gain during gestation, women's health professionals should take a more active role in educating pregnant mothers about what weight changes to expect, the consequences of too little or too much weight gain, and how to make corrections if the weight gain trajectory veers off course (see Further Reading 5). Weight gain during pregnancy should generally follow the pattern in Figure 14-4. Weekly monitoring of weight changes, especially on a chart that shows expected weight gains, can help a pregnant woman to assess how much to adjust her food intake and physical activity.

Realistic information about increased calorie requirements should be provided. Pregnant women need to understand that "eating for two" is more about increasing diet quality (i.e., choosing nutrient-dense foods) than diet quantity. Furthermore, as you learned in Chapter 7, successful weight management strategies involve a behavioral component; some women need to learn skills, such as self-monitoring of weight, dietary intake, and physical activity. The Internet and numerous smartphone apps offer specialized tools to make self-monitoring tasks simple and social.

If a woman deviates from the desirable pattern, she should make appropriate adjustments, but weight loss during pregnancy is never advised. For example, if a woman begins to gain too much weight during her pregnancy, she should not lose weight to get back on track. Even if a woman gains 35 pounds in the first 7 months of pregnancy, she must still gain more during the last 2 months. She should, however, slow the increase in weight to parallel the rise on the prenatal weight gain chart. In other words, the sources of the unnecessary calories should be found and minimized. Alternately, if a woman has not gained the desired weight by a given point in pregnancy, she should not gain the needed weight rapidly. Instead, she should slowly gain a little more weight than the typical pattern to meet the goal by the end of the pregnancy. A registered dietitian can help make any needed adjustments.

PROTEIN, CARBOHYDRATE, AND LIPID NEEDS

The RDA for protein increases by an additional 25 grams per day during pregnancy. (Fig. 14-5). All women should check to make sure they are eating enough protein as well as enough calories (so that protein can be spared for synthesis of new tissue). However, many women already consume protein in excess of their needs

During pregnancy, women in North America are more likely to gain excess weight and make poor food choices than to eat too little.

▲ During pregnancy, the Adequate Intake for total water increases 0.3 liters (1¼ cups) above prepregnancy needs to 3.0 liters (about 12½ cups) per day. For breastfeeding, consume 3.8 liters (16 cups) daily.

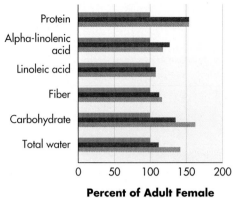

FIGURE 14-5 ▲ Relative macronutrient and water requirements for pregnancy and breastfeeding. Note that there is no RDA or AI for total fat; needs are based on 20% to 35% of overall energy intake.

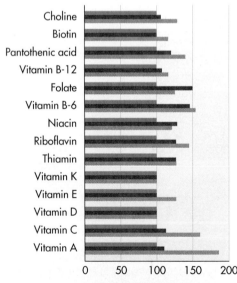

FIGURE 14-6 ▲ Relative vitamin requirements for pregnancy and breastfeeding.

and therefore do not need to focus specifically on increasing protein intake. Small changes are usually all that is necessary. For example, simply adding a cup of fat-free milk to the diet adds 90 nutrient-dense kcal and 8 grams of protein.

The RDA for carbohydrate increases to 175 grams daily, primarily to prevent ketosis. Ketone bodies, a by-product of metabolism of fat for energy, are thought to be poorly used by the fetal brain, implying possible slowing of fetal brain development. Carbohydrate intakes of most women, pregnant or not, already exceed the RDA (Fig. 14-5).

Fat intake should increase proportionally with calorie intake during pregnancy to maintain around 20% to 30% of total calories from fat. Pregnancy is not a time for a low-fat diet, as lipids are a source of extra calories and essential fatty acids needed during pregnancy. Recommendations for the types of lipids during pregnancy are generally the same as for nonpregnant adults. To reduce risk for cardiovascular disease, the American Heart Association recommends no more than 7% of total calories from saturated fat and no more than 1% from *trans* fat. Consumption of dietary cholesterol is not required, but keeping cholesterol intake at a maximum of 300 milligrams per day is also a good goal for maintenance of maternal cardiovascular health.

During pregnancy, it is particularly important to make sure to consume adequate essential fatty acids—linoleic acid (omega-6) and alpha-linolenic acid (omega-3). As you learned in Chapter 5, essential fatty acids cannot be synthesized in the body and must be consumed in the diet. For the developing fetus, essential fatty acids are required for growth, brain development, and eye development. Recommendations are slightly increased by pregnancy to 13 grams per day of omega-6 fatty acids and 1.4 grams per day of omega-3 fatty acids. These needs can be met by consuming 2 to 4 tablespoons per day of plant oils. Consumption within the range of 8 to 12 ounces per week of fish is recommended for meeting needs for essential fatty acids in the diet (see Further Reading 14). Some women may have difficulty incorporating the recommended amount of fish into their diets. Fish oil supplements are a good alternative, but consumers should choose brands that have been distilled to remove environmental contaminants. (See the Nutrition and Your Health section at the end of this chapter for a discussion of mercury in fish.)

VITAMIN NEEDS

Vitamin needs increase from prepregnancy RDAs/AIs by up to 30% for most of the B vitamins and even greater for vitamin B-6 (45%) and folate (50%) (Fig 14-6). Vitamin A needs only increase by 10%, so a specific focus on this vitamin is not needed. And remember, excess amounts of vitamin A are harmful to the developing fetus.

The extra amount of vitamin B-6 and other B vitamins (except folate) needed in the diet is easily met via wise food choices, such as a serving of a typical ready-to-eat breakfast cereal and some animal protein sources. Folate needs, however, often merit specific diet planning and possible vitamin supplementation. The synthesis of DNA, and therefore cell division, requires folate, so this nutrient is especially crucial during pregnancy. Ultimately, both fetal and maternal growth depend on an ample supply of folate. Red blood cell formation, which requires folate, increases during pregnancy. Serious folate-related anemia therefore can result if folate intake is inadequate. The RDA for folate increases during pregnancy to 600 micrograms DFE per day (review Chapter 8 for calculation of DFE). This is a critical goal in the nutritional care of a pregnant woman. Increasing folate intakes to meet 600 micrograms DFE per day for a pregnant woman can be achieved through dietary sources, a supplemental source of folic acid, or a combination of both. Choosing a diet rich in synthetic folic acid, such as from ready-to-eat breakfast cereals or meal replacement bars (look for approximately 50% to 100% of the Daily Value), is especially helpful in meeting folate needs. Recall from Chapter 8 that synthetic folic acid is much more easily absorbed than the various forms of folate found naturally in foods (see Further Reading 17).

Emerging evidence indicates that low maternal levels of vitamin D during pregnancy affect multiple health parameters in the offspring. About 54% of black women and 42% of white women have insufficient blood levels of the active form of vitamin D, even though many take a daily supplement containing 400 IU of vitamin D. Aside from its well-known roles in bone health and the incidence of rickets, poor vitamin D status is implicated in serious complications of pregnancy, including higher rates of diabetes and hypertension that occur during pregnancy, and a fourfold increased rate of Cesarean section. Vitamin D's role in immune regulation is underscored by higher rates of respiratory infections and mother-to-child transmission of HIV when its status is deficient. Because of the ability of vitamin D to modulate gene expression, vitamin D status is critically important during early fetal development through infancy. Diseases that develop later in childhood or adulthood, such as type 1 diabetes, multiple sclerosis, asthma, schizophrenia, and certain types of cancer, are associated with low vitamin D status during gestation. The RDA for pregnant women is 15 micrograms (600 IU) of vitamin D daily (the same as for nonpregnant women), but many experts advocate increasing this recommendation to 25 micrograms (1000 IU) or more (see Further Reading 6). Currently, there is no recommendation for universal vitamin D screening or supplementation, but the American College of Obstetricians and Gynecologists advises that for deficient individuals, 1000 to 2000 IU per day is a safe level of supplemental vitamin D during pregnancy.

MINERAL NEEDS

Mineral needs generally increase during pregnancy, especially the requirements for iodide, iron, and zinc (Fig. 14-7). (Calcium needs do not increase but still may deserve special attention because many women find it difficult to meet their calcium needs in general.)

Pregnant women need extra iodide (RDA of 220 micrograms per day) to support thyroid hormone synthesis (for the mother and the developing fetus) and fetal brain development. If a mother is deficient in iodide during pregnancy, she may develop a goiter (see Chapter 9), and her child may suffer from a devastating birth defect called **congenital hypothyroidism** (formerly called *cretinism*). The Nutrition and Your Health section at the end of this chapter provides further information on this and other nutrition-related birth defects. Typical iodide intakes are plentiful if the woman uses iodized salt.

The extra iron (RDA of 27 milligrams per day) is needed to synthesize a greater amount of hemoglobin during pregnancy and to provide iron stores for the fetus. About 2% of women begin pregnancy with clinically diagnosed iron-deficiency anemia, but an estimated one-third of pregnant women have poor iron stores, so heavy demands for iron during pregnancy cannot be met. The consequences of iron-deficiency anemia—especially during the first trimester—can be severe. Negative outcomes include preterm delivery, LBW infants, and increased risk for fetal death in the first weeks after birth.

Women often need a supplemental source of iron, especially if they do not consume iron-fortified foods, such as highly fortified breakfast cereals containing close to 100% of the Daily Value for iron (18 milligrams). The American College of Obstetricians and Gynecologists recommends screening for iron deficiency for all pregnant women and provision of iron supplements for women who are deficient. In addition, most prenatal supplements contain iron. A potential pitfall is that iron supplements can decrease appetite and can cause nausea and constipation. To alleviate these problems, it may be helpful to take these supplements between meals or just before going to bed. Milk, coffee, or tea should not be consumed with an iron supplement because these beverages have substances that interfere with iron absorption. Eating foods rich in vitamin C along with nonheme iron-containing foods and iron supplements helps increase iron absorption from those sources. Pregnant women who are not anemic may wait until the second trimester, when pregnancy-related nausea generally lessens, to start prenatal supplements if gastrointestinal side effects are a problem.

The RDA for zinc increases from 8 to 11 milligrams per day during pregnancy. Zinc is involved in many enzyme functions and protein synthesis. Zinc deficiency

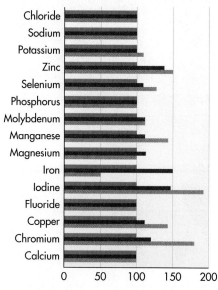

Percent of Adult Female RDA or AI

- ■ Adult female
- ■ Pregnancy
- ■ Breastfeeding

FIGURE 14-7 ▲ Relative mineral requirements for pregnancy and breastfeeding.

congenital hypothyroidism A birth defect that impairs thyroid hormone synthesis. If untreated, this can lead to mental retardation and stunting of growth.

▲ Green, leafy vegetables are rich food sources of folate, but folic acid—the synthetic form of the vitamin—is even more active in the body than the natural form. *Healthy People 2020* includes a goal of increasing by 10% the proportion of women of childbearing potential who consume at least 400 micrograms of folic acid per day. Currently, about 23% of women of childbearing potential consume adequate folic acid from fortified foods or dietary supplements.

during pregnancy has been linked to preterm and LBW births. The zinc intakes of pregnant women in the United States are generally adequate, but low-income women and those who follow vegan or vegetarian diets are more susceptible to poor zinc status. Also, because iron and zinc compete for absorption, high levels of iron supplementation during pregnancy may impair zinc absorption. Incorporating lean animal proteins or a fortified ready-to-eat breakfast cereal in the diet are good ways to obtain enough extra zinc during pregnancy.

USE OF PRENATAL VITAMIN AND MINERAL SUPPLEMENTS

With the exceptions of folate, iron, and vitamin D, the vitamin and mineral intakes of pregnant women in developed countries are generally adequate (see Further Reading 4). Research evidence supports routine supplementation with folic acid and iron during pregnancy. Beyond these two nutrients, some studies indicate that use of a multivitamin and mineral supplement is advantageous for reducing the number of LBW and SGA births. More research is needed to know if prenatal supplements are effective for reducing many other pregnancy complications. To date, there is not enough evidence to recommend prenatal multivitamin and mineral supplements for all pregnant women, but they are prescribed routinely by most physicians. Some supplements formulated for pregnancy are sold over the counter, while others are dispensed by prescription because of the high synthetic folic acid content (1000 micrograms), which could pose problems for others, such as older people (see Chapter 16). These supplements are high in iron (27 milligrams per pill). There is no evidence that use of such supplements causes significant health problems in pregnancy, aside perhaps from the combined amounts of supplementary and dietary vitamin A (see the Nutrition and Your Health section at the end of this chapter).

Instances when prenatal supplements may especially contribute to a successful pregnancy are those involving poor women, teenagers, women with a generally deficient diet, women carrying multiple fetuses, women who smoke or use

CASE STUDY Preparing for Pregnancy

Lily and her husband have decided that they are ready to prepare for Lily's first pregnancy. She is 25 years old, weighs 135 pounds, and is 67 inches tall. Lily has been reading everything she can find on pregnancy because she knows that her prepregnancy health is important to the success of her pregnancy.

She knows she should avoid alcohol, especially because alcohol is potentially toxic to the growing fetus in the first weeks of pregnancy, and she could become pregnant and not know about it right away. Lily is not a smoker, does not take any medications, and limits her coffee intake to 4 cups a day and soft drink intake to 3 colas per day. Based on her reading, she has decided to breastfeed her infant and has already inquired about childbirth classes. She has modified her diet to include some extra protein, along with more fruits and vegetables. She has also started taking an over-the-counter vitamin and mineral supplement. Lily has always kept in good shape, and she is admittedly worried about gaining too much weight during pregnancy. Recently, she started a running program 5 days a week, and she plans to continue running throughout her pregnancy.

Answer the following questions, and check your responses at the end of this chapter.

1. What recommendations do you have regarding Lily's use of dietary supplements?
2. What is Lily doing to prevent neural tube defects? What else could she do?
3. What recommendations would you make regarding Lily's caffeine consumption?
4. Lily is wise to pay attention to her protein intake to prepare for pregnancy and breastfeeding. Should she include fish as a source of protein in her diet? Why or why not?
5. Constipation is a common complaint during pregnancy. What suggestions do you have to help Lily avoid this health concern?
6. What information would you share with Lily about appropriate weight gain during pregnancy?

TABLE 14-3 ▶ **Targeted Recommendations for Pregnant and Breastfeeding Women from the Latest Dietary Guidelines for Americans**

Balancing Calories to Manage Weight

Women capable of becoming pregnant
- Achieve and maintain a healthy weight before becoming pregnant.

Women who are pregnant
- Gain weight within the 2009 Institute of Medicine gestational weight-gain guidelines (review Table 14-2).

Foods and Food Components to Reduce

Women capable of becoming pregnant and women who are pregnant
- During pregnancy or when conception is possible, avoid alcohol. No safe level of alcohol consumption during pregnancy has been established.

Women who are pregnant
- Only eat foods with seafood, meat, poultry, or eggs that have been cooked to recommended safe minimum internal temperatures.
- Do not consume unpasteurized (raw) juice or milk or foods made from unpasteurized milk, such as some soft cheeses (e.g., feta, queso blanco, queso fresco, Brie, Camembert, blue-veined cheeses, and Panela).
- Reheat deli and luncheon meats and hot dogs to steaming hot to kill *Listeria* and do not eat raw sprouts.

Women who are breastfeeding
- Wait at least 4 hours after drinking alcohol before breastfeeding.
- Alcohol should not be consumed at all until consistent latch-on and breastfeeding patterns are established.

Foods and Nutrients to Increase

Women capable of becoming pregnant
- Choose foods that supply heme iron, which is more readily absorbed by the body, additional iron sources, and enhancers of iron absorption such as vitamin C–rich foods.
- Consume 400 micrograms per day of synthetic folic acid (from fortified foods and/or supplements) in addition to food forms of folate from a varied diet.

Women who are pregnant
- Take an iron supplement, as recommended by an obstetrician or other health care providers.
- Consume 600 micrograms of dietary folate equivalents daily from all food sources.

Women who are pregnant or breastfeeding
- Consume 8 to 12 ounces of seafood per week from a variety of seafood types.
- Due to their methyl mercury content, limit white (albacore) tuna to 6 ounces per week and do not eat the following four types of fish: tilefish, shark, swordfish, and king mackerel.

alcohol or illegal drugs, and vegans. In other cases, healthy diets can provide the needed nutrients. When choosing a multivitamin, rely on brands that display the USP symbol on their label, signifying that the supplement meets the content, quality, purity, and safety standards of the United States Pharmacopeial Convention (see the Nutrition and Your Health section in Chapter 8). Of course, avoid megadoses of any nutrient. Skip supplements containing herbs, enzymes, and amino acids. Many of these ingredients have not been evaluated for safety during pregnancy or breastfeeding and may be toxic to the fetus. Furthermore, discard supplements that are past the expiration date, as some ingredients lose potency over time.

✓ CONCEPT CHECK 14.4

1. Patrice required 2200 kcal per day before she became pregnant. How many kilocalories will she need each day during her first, second, and third trimesters?
2. What is optimal weight gain during pregnancy for a woman who begins pregnancy at a healthy BMI? How does this differ for a woman who begins pregnancy underweight? Overweight? Obese?
3. List several nutrients that may need to be supplemented in the diet of a pregnant woman and the reason for each.

14.5 Food Plan for Pregnant Women

One dietary approach to support a successful pregnancy is based on MyPlate. For an active 24-year-old woman, about 2200 kcal is recommended during the first trimester (the same as recommended for such a woman when not pregnant). The plan should include:

- 3 cups of calcium-rich foods from the dairy group or use of calcium-fortified foods to make up for any gap between calcium intake and need
- 6 ounce-equivalents from the protein group
- 3 cups from the vegetables group
- 2 cups from the fruits group
- 7 ounce-equivalents from the grains group
- 6 teaspoons of vegetable oil

Specifically, choices from the dairy group should include low-fat or fat-free versions of milk, yogurt, and cheese. These foods supply extra protein, calcium, and carbohydrate, as well as other nutrients. Choices from the protein group should include both animal and vegetable sources. Besides protein, these foods help provide the extra iron and zinc needed. The vegetables and fruits group choices provide a variety of vitamins and minerals. One cup of fruit each day should be a good vitamin C source, and 1 cup of vegetables should be a green vegetable or other rich source of folate. Choices from the grains group should focus on whole-grain and enriched foods. One ounce of a whole-grain, ready-to-eat breakfast cereal significantly contributes to meeting many vitamin and mineral needs. Finally, inclusion of plant oils in the diet contributes essential fatty acids. Calories from solid fats and added sugars should be limited to 270 kcal per day.

In the second and third trimesters, about 2600 kcal is recommended for this woman. The plan should now include:

- 3 cups of calcium-rich foods from the dairy group or use of calcium-fortified foods
- 6 ½ ounce-equivalents from the protein group
- 3 ½ cups from the vegetables group
- 2 cups from the fruits group
- 9 ounce-equivalents from the grains group
- 8 teaspoons of vegetable oil

Solid fats and added sugars should contribute no more than 360 kcal per day. Table 14-4 illustrates one daily menu based on the 2600-kcal plan for pregnancy for women in the second or third trimesters. This menu meets the extra nutrient needs associated with pregnancy. Women who need to consume more than this—and some do for various reasons—should incorporate additional fruits, vegetables, and whole-grain breads and cereals, not poor nutrient sources such as desserts and sugared soft drinks.

It is a common myth that women instinctively know what to eat during pregnancy. Cravings during the last two trimesters are often related to hormonal changes in the mother or family traditions. Such "instinct" cannot be trusted, however, based on observations that some women crave nonfood items (called **pica**) such as laundry starch, chalk, cigarette ashes, and soil (clay). This practice can be extremely harmful to the mother and the fetus. Following the nutrition advice of respected health professionals, such as registered dietitians, is much more reliable than relying on cravings to meet nutrient needs.

PREGNANT VEGETARIANS

Women who are either lactoovovegetarians or lactovegetarians generally do not face special difficulties in meeting their nutritional needs during pregnancy. Like nonvegetarian women, they should be concerned primarily with meeting vitamin B-6, iron, folate, and zinc needs.

At **www.ChooseMyPlate.gov,** moms-to-be can find individualized dietary information by clicking on the link for Pregnant and Breastfeeding Women. Based on age, height, physical activity, and prepregnancy weight, a Daily Food Plan for Moms can be generated for each trimester.

pica The practice of eating nonfood items, such as dirt, laundry starch, or clay.

TABLE 14-4 ▶ Sample 2600 kcal Daily Menu That Meets the Nutritional Needs of Most Pregnant and Breastfeeding Women*

	Vitamin B-6	Folate	Iron	Zinc	Calcium
Breakfast					
1 cup Kellogg's Smart Start cereal	✓	✓	✓	✓	✓
1 cup orange juice		✓			
1 cup fat-free milk	✓				✓
Snack					
2 tbsp peanut butter	✓	✓	✓	✓	
2 stalks of celery		✓			
1 slice whole-wheat toast		✓	✓	✓	
1 cup plain low-fat yogurt	✓				✓
½ cup strawberries		✓			
Lunch					
2 cups spinach and fruit salad with 2 tbsp oil and vinegar dressing		✓			✓
2 slices whole-wheat toast		✓	✓	✓	
1 ½ ounces provolone cheese	✓				✓
Snack					
5 whole-wheat crackers		✓	✓	✓	
1 cup grape juice					
Dinner					
3 ounces lean hamburger, broiled (with condiments)	✓		✓	✓	
½ cup baked beans	✓	✓	✓	✓	
1 hamburger bun		✓	✓		
½ sliced tomato					
1 cup cooked broccoli		✓			✓
1 tsp soft margarine					
Iced tea					
Snack					
Granola bar (2 ounces)		✓	✓	✓	
½ banana	✓				
Foods or beverages containing solid fats or added sugars can be included (up to 360 kcal per day) to support adequate weight gain					

*Amount of solid fats and added sugars will vary based on the actual food choices made within each MyPlate group.

This diet meets nutrient needs for pregnancy and breastfeeding. Lack of a check (✓) indicates a poor source of the nutrient. The vitamin- and mineral-fortified breakfast cereal used in this example makes an important contribution to meeting nutrient needs. Fluids can be added as desired. Total intake of fluids, such as water, should be 10 cups per day for pregnant women or about 13 cups per day for breastfeeding women.

▲ How does this lunch of spinach and fruit salad and whole-wheat toast compare to MyPlate? Is iced tea a good beverage choice? Why or why not?

On the other hand, for a vegan, careful diet planning during preconception and pregnancy is crucial to ensure sufficient protein, vitamin D (particularly in the absence of sufficient sun exposure), vitamin B-6, iron, calcium, zinc, and especially a supplemental source of vitamin B-12. The basic vegan diet listed in Chapter 6 should be modified to include more grains, beans, nuts, and seeds to supply the necessary extra amounts of some of these nutrients. As mentioned, use of a prenatal multivitamin and mineral supplement also is generally advocated to help fill micronutrient gaps. However, although these are high in iron, this is not true for calcium (200 milligrams per pill). If iron and calcium supplements are used, they should not be taken together to avoid possible competition for absorption.

✓ CONCEPT CHECK 14.5

1. List three nutrients of special concern for a pregnant woman who practices a vegan lifestyle.
2. Modify the sample 2600-kcal daily menu in Table 14-4 so that it would be suitable for a pregnant vegan woman.
3. Michaela tells you about her pregnancy craving for ice cream. She says she has been eating one or two ice cream bars after lunch every day and usually stops for a milkshake at a fast-food restaurant on her way home from work. What nutrition information would you share with her?

14.6 Physiological Changes of Concern During Pregnancy

During pregnancy, the fetus's needs for oxygen and nutrients as well as excretion of waste products increase the burden on the mother's lungs, heart, and kidneys. Although a mother's digestive and metabolic systems work efficiently, some discomfort accompanies the changes her body undergoes to accommodate the fetus.

HEARTBURN, CONSTIPATION, AND HEMORRHOIDS

Hormones (such as progesterone) produced by the placenta relax muscles in the uterus and the gastrointestinal tract. This often causes heartburn as stomach acid refluxes into the esophagus (review Chapter 3). When this occurs, the woman should avoid lying down after eating, eat less fat so that foods pass more quickly from the stomach into the small intestine, and avoid spicy foods she cannot tolerate. She should also consume most liquids between meals to decrease the volume of food in the stomach after meals and thus relieve some of the pressure that encourages reflux. Women with more severe cases may need antacids or related medications.

Constipation often results as the intestinal muscles relax during pregnancy. It is especially likely to develop late in pregnancy, as the fetus competes with the GI tract for space in the abdominal cavity. To offset these discomforts, a woman should perform regular exercise and consume more fluid, fiber, and dried fruits, such as prunes (dried plums). The Adequate Intake for fiber in pregnancy is 28 grams, slightly more than for the nonpregnant woman. Fluid needs are 10 cups per day. These practices can help prevent constipation and a problem that frequently accompanies it, hemorrhoids. Straining during elimination can lead to hemorrhoids, which are more likely to occur during pregnancy because of other body changes. A re-evaluation of the need for and dose of iron supplementation should also be considered, as high iron intakes are linked to constipation.

EDEMA

Placental hormones cause various body tissues to retain fluid during pregnancy. Blood volume also greatly expands during pregnancy. The extra fluid normally causes some swelling (edema). There is no reason to restrict salt severely or use diuretics to limit mild edema. However, the edema may limit physical activity late in pregnancy and occasionally requires a woman to elevate her feet or wear compression stockings to control the symptoms. Overall, edema generally spells trouble only if hypertension and the appearance of extra protein in the urine accompany it (see "Hypertensive Disorders of Pregnancy").

NAUSEA AND VOMITING OF PREGNANCY

About 70% to 85% of pregnant women experience nausea during the early stages of pregnancy. This nausea may be related to the increased sense of smell induced by pregnancy-related hormones. Although commonly called "morning sickness," pregnancy-related nausea may occur at any time and persist all day. It is often the first signal to a woman that she is pregnant. To help control mild nausea, pregnant women can try the following: avoiding nauseating foods, such as fried or greasy foods; cooking with good ventilation to dissipate nauseating smells; eating saltine crackers or dry cereal before getting out of bed; avoiding large fluid intakes early in the morning; and eating smaller, more frequent meals. The iron in prenatal supplements triggers nausea in some women, so changing the type of supplement used or postponing use until the second trimester may provide relief in some cases. If a woman thinks her morning sickness is related to her prenatal supplement, she should discuss switching to another supplement with her physician.

Overall, if a food sounds good to a pregnant woman with nausea, whether it is broccoli, soda crackers, or lemonade, she should eat it and eat when she can, while also striving to follow her prenatal diet. The American College of Obstetricians and Gynecologists recommends the following for the prevention and treatment of nausea and vomiting of pregnancy:

- History of use of a balanced multivitamin and mineral supplement at the time of conception
- Use of megadoses of vitamin B-6 (10 to 25 milligrams taken three to four times a day), especially coupled with the antihistamine doxylamine (10 milligrams) with each dose
- Ginger may also be helpful (350 milligrams taken three times per day)

Usually, nausea stops after the first trimester; however, in about 10% to 20% of cases, it can continue throughout the entire pregnancy. In cases of serious nausea, the preceding practices offer little relief. Excessive vomiting can cause dangerous dehydration and must be avoided. When severe vomiting persists (about 0.5% to 2% of pregnancies), medical attention is needed.

▲ A few saltine crackers upon waking or between meals can help lessen pregnancy-related nausea.

ANEMIA

To supply fetal needs, the mother's blood volume expands to approximately 150% of normal. The number of red blood cells, however, increases by only 20% to 30%, and this occurs more gradually. As a result, a pregnant woman has a lower ratio of red blood cells to total blood volume in her system. This hemodilution is known as **physiological anemia.** It is a normal response to pregnancy, rather than the result of inadequate nutrient intake. If during pregnancy, however, iron stores and/or dietary iron intake are not sufficient to meet needs, any resulting iron-deficiency anemia requires medical attention. The Dietary Guidelines for Americans recommend that women who may become pregnant eat foods high in iron (Table 14-3).

physiological anemia The normal increase in blood volume in pregnancy that dilutes the concentration of red blood cells, resulting in anemia; also called *hemodilution.*

GESTATIONAL DIABETES

Hormones synthesized by the placenta decrease the efficiency of insulin. This leads to a mild increase in blood glucose, which helps supply calories to the fetus. If the rise in blood glucose becomes excessive, this leads to **gestational diabetes,** often beginning in weeks 20 to 28, particularly in women who have a family history of diabetes or who are obese. Other risk factors include maternal age over 35 and gestational diabetes in a prior pregnancy. In North America, gestational

gestational diabetes A high blood glucose concentration that develops during pregnancy and returns to normal after birth; one cause is the placental production of hormones that antagonize the regulation of blood glucose by insulin.

diabetes develops in about 4% of pregnancies (7% in the Caucasian population). Today, pregnant women with risk factors for type 2 diabetes (e.g., obesity, family history) should be screened for undiagnosed type 2 diabetes at the first prenatal visit. Pregnant women without type 2 diabetes should be screened for diabetes at 24 to 28 weeks by checking for elevated blood glucose concentration 1 to 2 hours after consuming 75 grams of glucose (see Further Reading 18). If gestational diabetes is detected, a special diet that distributes carbohydrates throughout the day needs to be implemented. Carbohydrate choices should be mostly whole, unprocessed grains, vegetables, fruits, and beans, which have a lower impact on blood glucose than refined grains or foods with lots of added sugars. Sometimes insulin injections or oral medications are also needed. Regular physical activity also helps control blood glucose.

The primary risk of uncontrolled diabetes during pregnancy is that the fetus can grow quite large. This is a result of the oversupply of glucose from maternal circulation coupled with an increased production of insulin by the fetus, which allows fetal tissues to take up an increased amount of building materials for growth. The mother may require a Cesarean section if the fetus is too large for a vaginal delivery. Another threat is that the infant may have low blood glucose at birth, because of the tendency to produce extra insulin that began during gestation. Other concerns are the potential for early delivery and increased risk of birth trauma and malformations. Beyond complications during pregnancy and childbirth, uncontrolled blood sugar during gestation may have some long-term repercussions for the child. High blood glucose induces changes in fetal metabolism and iron-binding proteins, such that iron stores in the fetus will be low.

Although gestational diabetes often disappears after the infant's birth, it increases the mother's risk of developing diabetes later in life, especially if she fails to maintain a healthy body weight. Studies show that infants of mothers with gestational diabetes may also have higher risks of developing obesity, metabolic syndrome, and type 2 diabetes as they grow to adulthood. For all these reasons, proper control of gestational diabetes (and any diabetes present in the mother before pregnancy) is extremely important.

HYPERTENSIVE DISORDERS OF PREGNANCY

Hypertension (blood pressure greater than 140/90 mmHg) occurs in about 6% to 8% of pregnancies in the United States. Sometimes, women enter pregnancy with chronic hypertension. However, when hypertension first appears after 20 weeks of gestation, it is termed **gestational hypertension** (formerly called *pregnancy-induced hypertension*).

About half of women with gestational hypertension eventually develop **preeclampsia** (mild form) or **eclampsia** (severe form). Early symptoms include a rise in blood pressure, excess protein in the urine, edema, changes in blood clotting, headache, and visual disturbances. Very severe effects, including convulsions, can occur in the second and third trimesters. If not controlled, eclampsia eventually damages the liver and kidneys, and mother and fetus may die.

The causes for hypertensive disorders of pregnancy are not well understood but likely involve interactions among genetics, certain environmental or lifestyle influences, and abnormal function of the placenta. The populations most at risk for these disorders are women under age 17 or over age 35, overweight or obese women, and those who have had multiple-birth pregnancies. A family history of gestational hypertension in the mother's or father's side of the family, diabetes, African-American race, and a woman's first pregnancy also raise risk.

Gestational hypertension resolves once the pregnancy ends, making delivery the most reliable treatment for the mother. However, if eclampsia develops before the fetus is ready to be born, physicians in many cases must use treatments to prevent the worsening of the disorder. Bed rest and administration of magnesium sulfate

gestational hypertension Blood pressure greater than 140/90 mmHg that is first diagnosed after 20 weeks of gestation. This may evolve into pre-eclampsia or eclampsia.

preeclampsia A form of gestational hypertension characterized by protein in the urine.

eclampsia A severe form of gestational hypertension characterized by protein in the urine and seizures (formerly called *toxemia*).

CRITICAL THINKING

Sandy, 4 months pregnant, has been having heartburn after meals, constipation, and difficult bowel movements. As a student of nutrition, you understand the digestive system and the role of nutrition in health. What remedies might you suggest to Sandy to relieve her problems?

are the most effective treatment methods. Magnesium likely acts to relax blood vessels and so leads to a fall in blood pressure. There is good evidence that adequate intakes of calcium and vitamin D are involved in reducing incidence of gestational hypertension. There is interest in the use of antioxidants to prevent or treat gestational hypertension, but results of a recent trial showed that vitamin C and E supplementation during pregnancy is not effective in preventing the disorder. Several other treatments, such as various antiseizure and antihypertensive medications, fish oils, and selenium, are under study (see Further Reading 20).

✔ CONCEPT CHECK 14.6

1. What diet and lifestyle strategies would you suggest to a pregnant woman who complains of frequent heartburn?
2. What diet and lifestyle strategies would you suggest to a pregnant woman who complains of constipation?
3. What diet and lifestyle strategies would you suggest to a pregnant woman who complains of morning sickness?
4. Define gestational diabetes. What are the potential consequences of this disorder for the mother and baby?
5. Differentiate between chronic hypertension and gestational hypertension. Define preeclampsia and eclampsia.

14.7 Breastfeeding

Breastfeeding the new infant further fosters his or her health and so complements the attention given to diet during pregnancy. The Academy of Nutrition and Dietetics (AND) and the American Academy of Pediatrics (AAP) recommend breastfeeding exclusively for the first 6 months, with the continued combination of breastfeeding and infant foods until 1 year (see Further Readings 2 and 12). The World Health Organization goes beyond that to recommend breastfeeding (with appropriate solid food introduction; see Chapter 15) for at least 2 years. Still, surveys show that only about 70% of North American mothers now begin to breastfeed their infants in the hospital, and at 4 and 6 months, only 33% and 20%, respectively, are still breastfeeding their infants. The number falls to 18% at 1 year of age. These statistics refer to Caucasian women; minority women are even less likely to be breastfeeding at these time intervals.

Women who choose to breastfeed usually find it an enjoyable, special time in their lives that strengthens the bond with their new infant. Although bottle feeding with an infant formula is also safe for infants, as discussed in Chapter 15, it does not equal the benefits derived from human milk in all aspects. If a woman does not breastfeed her child, breast weight returns to normal soon after birth.

PLANNING TO BREASTFEED

Almost all women are physically capable of breastfeeding their children (see later section "Medical Conditions Precluding Breastfeeding" for exceptions). In most cases, problems encountered in breastfeeding are due to a lack of appropriate information. Anatomical problems in breasts, such as inverted nipples, can be corrected during pregnancy. Breast size generally increases during pregnancy and is no indication of success in breastfeeding. Most women notice a dramatic increase in the size and weight of their breasts by the third or fourth day of breastfeeding. If these changes do not occur, a woman needs to speak with her physician or a lactation consultant.

▲ Breastfeeding is the preferred way to feed a young infant.

Many of the benefits of breastfeeding can be found in Table 14-5 and at **www.womenshealth.gov/Breastfeeding/index.cfm,** sponsored by the U.S. Surgeon General.

lobules Saclike structures in the breast that store milk.

prolactin A hormone secreted by the pituitary gland that stimulates the synthesis of milk in the breast.

let-down reflex A reflex stimulated by infant suckling that causes the release (ejection) of milk from milk ducts in the mother's breasts; also called *milk ejection reflex.*

oxytocin A hormone secreted by the pituitary gland. It causes contraction of the musclelike cells surrounding the ducts of the breasts and the smooth muscle of the uterus.

Breastfed infants must be followed closely over the first days of life to ensure that feeding and weight gain are proceeding normally. Monitoring is especially important with a mother's first child because the mother will be inexperienced with the technique of breastfeeding. Mothers and healthy infants are commonly discharged from the hospital 1 to 2 days after delivery, whereas 20 years ago, they stayed in the hospital for 3 or 4 days or longer. One result of such rapid discharge is a decreased period of infant monitoring by health care professionals. Incidents have been reported of infants developing dehydration and, in turn, blood clots soon after hospital discharge when breastfeeding did not proceed smoothly. Careful monitoring in this first week by a physician or lactation consultant is advised.

First-time mothers who plan to breastfeed should learn as much as they can about the process early in their pregnancy. Although it is the most natural way to feed a newborn child, the technique does not always come naturally. Gathering information on breastfeeding, what obstacles to anticipate, and how to respond to such obstacles will help new mothers and their babies to succeed with breastfeeding. It also helps to have an experienced friend or family member or a professional lactation consultant to call for advice when questions arise.

PRODUCTION OF HUMAN MILK

During pregnancy, cells in the breast form milk-producing cells called **lobules** (Fig. 14-8). Hormones from the placenta stimulate these changes in the breast. After birth, the mother produces more **prolactin** hormone to maintain the changes in the breast and therefore the ability to produce milk. During pregnancy, breast weight increases by about 1 to 2 pounds.

The hormone prolactin also stimulates the synthesis of milk. Infant suckling stimulates prolactin release from the pituitary gland. Milk synthesis then occurs as an infant nurses. The more the infant suckles, the more milk is produced. Because of this, even twins (and triplets) can be breastfed adequately.

Most protein found in human milk is synthesized by breast tissue. Some proteins also enter the milk directly from the mother's bloodstream. These proteins include immune factors (e.g., antibodies) and enzymes. Fats in human milk come from both the mother's diet and those synthesized by breast tissue. The sugar galactose is synthesized in the breast, whereas glucose enters from the mother's bloodstream. Together, these sugars form lactose, the main carbohydrate in human milk.

LET-DOWN REFLEX

An important brain-breast connection—commonly called the **let-down reflex**—is necessary for breastfeeding. The brain releases the hormone **oxytocin** to allow the breast tissues to let down (release) the milk from storage sites (Fig. 14-9). It then travels to the nipple area. A tingling sensation signals the let-down reflex shortly before milk flow begins. If the let-down reflex does not operate, little milk is available to the infant. The infant then gets frustrated, and this can frustrate the mother.

The let-down reflex is easily inhibited by nervous tension, a lack of confidence, and fatigue. Mothers should be especially aware of the link between tension and a weak let-down reflex. They need to find a relaxed environment where they can breastfeed.

- Milk-producing/storage cells (lobules)

- Ducts to carry milk to nipple

- Nipple

- Areolar margin

FIGURE 14-8 ▲ The anatomy of the breast. Many types of cells form a coordinated network to produce and secrete human milk.

After a few weeks, the let-down reflex becomes automatic. The mother's response can be triggered just by thinking about her infant or seeing or hearing another one cry. At first, however, the process can be a bit bewildering. A mother cannot measure the amount of milk the infant takes in, so she may fear that she is not adequately nourishing the infant.

As a general rule, a well-nourished breastfed infant should (1) have three to five wet diapers per day by 3 to 5 days of age and four to six wet diapers per day thereafter, (2) show a normal pattern of weight gain, and (3) pass at least one or two stools per day that look like lumpy mustard. In addition, softening of the breast during the feeding helps indicate that enough milk is being consumed. Parents who sense their infant is not consuming enough milk should consult a physician immediately because dehydration can develop rapidly. Losing more than 7% of birth weight is an indicator of a feeding problem that requires intervention.

It generally takes 2 to 3 weeks to fully establish the feeding routine: Infant and mother both feel comfortable, the milk supply meets infant demand, and initial nipple soreness disappears. Establishing the breastfeeding routine requires patience, but the rewards are great. The adjustments are easier if supplemental formula feedings are not introduced until breastfeeding is well established, after at least 3 to 4 weeks. Then it is fine if a supplemental bottle or two of infant formula per day is needed, but supplemental feedings will decrease milk production.

NUTRITIONAL QUALITIES OF HUMAN MILK

Human milk is different in composition from cow's milk. Unless altered, cow's milk should never be used in infant feeding until the infant is at least 12 months old. Cow's milk is too high in minerals and protein and does not contain enough carbohydrate to meet infant needs. In addition, the major protein in cow's milk is harder for an infant to digest than the major proteins in human milk. The proteins in cow's milk may also spur allergies in the infant. Finally, certain compounds in human milk presently under study show other possible benefits for the infant.

Colostrum. At the end of pregnancy, the first fluid made by the human breast is **colostrum.** This thick, yellowish fluid may leak from the breast during late pregnancy and is produced in earnest for a few days after birth. Colostrum contains

Healthy People 2020 has set a goal of 82% of women breastfeeding their infants at time of hospital discharge, 60% breastfeeding for 6 months, and 34% still breastfeeding at 1 year.

Disposable diapers can absorb so much urine that it is difficult to judge when they are wet. A strip of paper towel laid inside a disposable diaper makes a good wetness indicator. Alternatively, cloth diapers may be used for a day or two to assess whether nursing is supplying sufficient milk.

colostrum The first fluid secreted by the breast during late pregnancy and the first few days after birth. This thick fluid is rich in immune factors and protein.

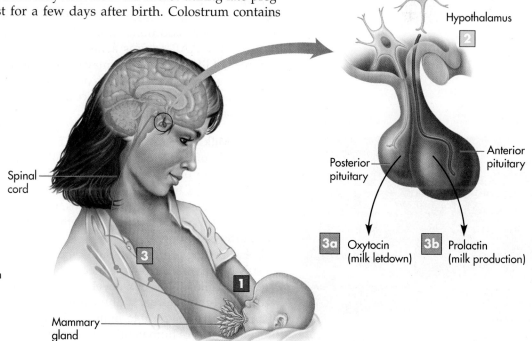

1 Suckling stimulates nerves in the nipple and areola that transmit impulses to the hypothalamus.

2 In response, the hypothalamus stimulates the posterior pituitary to release oxytocin and the anterior pituitary to release prolactin.

3 Oxytocin stimulates lobules in the breast to let down (release) milk from storage. Prolactin stimulates additional milk production.

Spinal cord

Mammary gland

Hypothalamus

Posterior pituitary

Anterior pituitary

3a Oxytocin (milk letdown)

3b Prolactin (milk production)

FIGURE 14-9 ▲ Let-down reflex. Suckling sets into motion the sequence of events that lead to milk let-down, the flow of milk into ducts of the breast.

antibodies, immune system cells, and growth factors, some of which pass unaltered through the infant's immature GI tract into the bloodstream. The first few months of life are the only time when we can readily absorb whole proteins across the GI tract. These immune factors and cells protect the infant from some GI tract diseases and other infectious disorders, compensating for the infant's immature immune system during the first few months of life.

One component of colostrum, the **Lactobacillus bifidus factor,** encourages the growth of *Lactobacillus bifidus* bacteria. These beneficial bacteria limit the growth of potentially toxic bacteria in the intestine. Overall, breastfeeding promotes the intestinal health of the breastfed infant in this way.

Lactobacillus bifidus **factor** A protective factor secreted in the colostrum that encourages growth of beneficial bacteria in the newborn's intestines.

Mature Milk. Human milk composition gradually changes until it achieves the normal composition of mature milk several days after delivery. Human milk looks very different from cow's milk. (Table 15-3 in Chapter 15 provides a direct comparison.) Human milk is thin and almost watery in appearance and often has a slightly bluish tinge. Its nutritional qualities, however, are impressive.

Human milk's proteins form a soft, light curd in the infant's stomach and are easy to digest. Some human milk proteins bind iron, reducing the growth of some bacteria that can cause diarrhea. Still other proteins offer the important immune protection already noted.

The lipids in human breast milk are high in linoleic acid and cholesterol, needed for brain development. Breast milk also contains long-chain omega-3 fatty acids, such as docosahexaenoic acid (DHA). This polyunsaturated fatty acid is used for the synthesis of tissues in the brain and the rest of the central nervous system and in the retina of the eye.

The fat composition of human milk changes during each feeding. The consistency of milk released initially (fore milk) resembles that of skim milk. It later has a greater fat proportion, similar to whole milk. Finally, the milk released after 10 to 20 minutes (hind milk) is essentially like cream. Babies need to nurse long enough (e.g., a total of 20 or more minutes) to get the calories in the rich hind milk to be satisfied between feedings and to grow well. The overall calorie content of human milk is about the same as that of infant formulas (67 kcal per 100 milliliters).

Human milk composition also allows for adequate fluid status of the infant, provided the baby is exclusively breastfed. A question commonly asked is whether the infant needs additional water if stressed by hot weather, diarrhea, vomiting, or fever. The AAP advises against supplemental water or juice during the first 6 months of life. The practice may unnecessarily introduce pathogens or allergens. Excessive water can lead to brain disorders, low blood sodium, and other problems. Thus, supplemental water and juice should be given only with a physician's guidance before 6 months of age.

Human Milk for Preterm Infants. There is no universal answer to whether a woman can breastfeed a preterm infant. In some cases, human milk is the most desirable form of nourishment, depending on infant weight and length of gestation. Feeding of human milk to preterm infants has been linked to lower infant mortality, decreased risk of infections, reduced stays in the neonatal intensive care unit, fewer hospital readmissions, better growth, and improved brain development (see Further Reading 2).

Breastfeeding a preterm infant demands great maternal dedication. Milk must usually be expressed from the breast and fed through a tube until the infant's sucking and swallowing reflex develops. Fortification of the milk with such nutrients as calcium, phosphorus, sodium, and protein is often necessary to meet the needs of a rapidly growing preterm infant. In some cases, special feeding problems may prevent the use of human milk or necessitate supplementing it with specialized formula. Sometimes total parenteral nutrition (intravenous feeding) is the only option. Working as a team, the pediatrician, neonatal nurses, and registered dietitian must guide the parents in this decision.

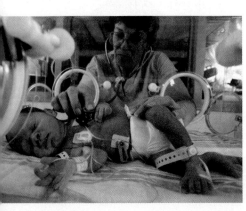

▲ If human milk is used to feed the preterm infant, fortification of the milk with certain nutrients is often needed.

FOOD PLAN FOR WOMEN WHO BREASTFEED

Nutrient needs for a breastfeeding mother change to some extent from those of the pregnant woman in the second and third trimester (see the inside cover of this book). There is a decrease in folate and iron needs and an increase in the needs for calories, vitamins A, E, and C, riboflavin, copper, chromium, iodide, manganese, selenium, and zinc. Still, these increased needs of the breastfeeding mother will be met by the general diet plan proposed for a woman in the latter stages of pregnancy. Recall that each day of this diet plan includes at least:

- 3 cups of calcium-rich foods such as from the dairy group or use of calcium-fortified foods to make up for any gap between calcium intake and need
- 6 ½ ounce-equivalents from the protein group
- 3 ½ cups from the vegetables group
- 2 cups from the fruits group
- 9 ounce-equivalents from the grains group
- 8 teaspoons of vegetable oil

Foods containing solid fats and added sugars (up to about 360 kcal) can then be added to allow for weight maintenance or gradual weight loss, whichever is needed.

Table 14-4 provided a menu for such a plan. Substituting a soyburger (veggie burger) for the hamburger in that menu would make this a practical guide for a lactovegetarian woman as well.

As in pregnancy, a serving of a highly fortified ready-to-eat breakfast cereal (or use of a balanced multivitamin and mineral supplement) is advised to help meet extra nutrient needs. And, as mentioned for pregnant women, breastfeeding mothers should consume 8 to 12 ounces of low-mercury fish per week (or 1 gram per day of omega-3 fatty acids from a fish oil supplement) because the omega-3 fatty acids present in fish are secreted into breast milk and are likely to be important for development of the infant's nervous system.

Milk production requires approximately 800 kcal every day. The estimated energy requirement during lactation is an extra 400 to 500 kcal daily above pre-pregnancy recommendations. The difference between that needed for milk production and the recommended intake—about 300 kcal—should allow a gradual loss of the extra body fat accumulated during pregnancy, especially if breastfeeding is continued for 6 months or more and the woman performs some physical activity. This shows just one of the natural benefits of following pregnancy with at least several months of breastfeeding.

After giving birth, women are often eager to shed the excess "baby fat." Breast-feeding, however, is no time for crash diets. A gradual weight loss of 1 to 4 pounds per month by the nursing mother is appropriate. At significantly greater rates of weight loss—when calories are restricted to less than about 1500 kcal per day—milk output decreases. A reasonable approach for a breastfeeding mother is to eat a balanced diet that supplies at least 1800 kcal per day; has moderate fat content; and includes a variety of dairy products, fruits, vegetables, and whole grains.

To promote the best possible feeding experience for the infant, there are several other dietary factors to consider. Hydration is especially important during breast-feeding; the woman should drink fluids every time her infant nurses. Drinking about 13 cups of fluids per day encourages ample milk production. Poor health habits, such as smoking cigarettes or drinking more than two alcoholic drinks a day, can decrease milk output. (Even less alcohol can have a deleterious effect on milk output in some women.) To avoid exposure to harmful levels of mercury, precautions concerning fish likely to contain mercury should extend past pregnancy for the breastfeeding mother. There is no evidence that maternal diet restrictions (e.g., peanuts, eggs, and fish) during pregnancy or breastfeeding prevent food allergies in infants. The AAP recommends exclusive breastfeeding and delaying introduction of solid foods until 4 to 6 months of age as the best approaches for preventing **atopic diseases** (see Further Reading 8).

▲ Eating 8 to 12 ounces (2 to 3 servings) of fish per week will help breastfeeding women ensure that their infants receive ample omega-3 fatty acids. It is important, however, to avoid those fish likely to be contaminated with mercury (swordfish, shark, king mackeral, and tile fish).

Most substances that the mother ingests are secreted into her milk. For this reason, she should limit intake of or avoid all alcohol and caffeine, and check all medications with a pediatrician. Some mothers believe that certain foods, such as garlic and chocolate, flavor the breast milk and upset the infant. If a woman notices a connection between a food she eats and the infant's later fussiness, she could consider avoiding that food. However, she might experiment again with it later, as infants become fussy for other reasons. Some researchers propose that the passage of flavors from the mother's diet into her milk affords an opportunity for the infant to learn about the flavor of the foods of its family long before solids are introduced. These researchers suspect that bottle-fed infants are missing significant sensory experiences that, until recent times in human history, were common to all infants.

atopic disease Condition resulting from an inappropriate immune response; examples include asthma, allergic rhinitis, food allergies, or eczema.

▲ Mothers who return to work outside the home can continue to breastfeed with the aid of a breastpump. In 2010, President Barack Obama signed an amendment to the Fair Labor Standards Act that requires most U.S. employers to allow break time and a private setting for breastfeeding mothers to express milk.

BREASTFEEDING TODAY

As noted, the vast majority of women are capable of breastfeeding and their infants benefit from it (see Table 14-5 and Further Reading 2). Nonetheless, some circumstances may make breastfeeding impractical or undesirable for a woman. Mothers who do not want to breastfeed their infants should not feel pressured to do so. Breastfeeding provides advantages, but none so great that a woman who decides to bottle feed should feel she is compromising her infant's well being.

Advantages of Breastfeeding. Human milk is tailored to meet infant nutrient needs for the first 4 to 6 months of life. However, there are some cases when infant dietary supplements, used under a pediatrician's guidance, are recommended:

- The AAP recommends *all* infants, including exclusively breastfed infants, be given 400 IU of vitamin D per day, beginning shortly after birth and continuing until the infant consumes that much from food (e.g., at least 2 cups (0.5 liters) of infant formula per day). Some sun exposure also helps in meeting vitamin D needs.
- Iron supplements are generally necessary for infants who were preterm, LBW, have blood disorders, or who were born with low iron stores.
- The AAP does not advise fluoride supplements before 6 months of age. After 6 months, the pediatrician or dentist may recommend supplemental fluoride if the infant's exposure to fluoride from drinking water, foods, and oral hygiene products is insufficient.
- Vitamin B-12 supplements are recommended for the breastfed infant whose mother is a complete vegetarian (vegan).

Fewer Infections. Due in part to the antibodies in human milk, breastfeeding reduces the infant's overall risk of developing infections. Breastfed infants also

TABLE 14-5 ▶ Advantages of Breastfeeding

For Infant
• Bacteriologically safe
• Always fresh and ready to go
• Provides antibodies and substances that contribute to maturation of the immune system
• Contributes to maturation of gastrointestinal tract via *Lactobacillus bifidus* factor
• Decreases risk of infections, such as diarrhea, respiratory disease, and ear infections
• Reduces risk of food allergies and intolerances, as well as some other allergies
• Reduces risk of celiac disease and inflammatory bowel diseases
• Establishes habit of eating in moderation, linked to 15% to 30% lower risk of obesity and 40% lower risk of type 2 diabetes later in life
• Contributes to proper development of jaws and teeth for better speech development
• May enhance nervous system development and eventual learning ability
• Decreases risk of childhood leukemia and lymphoma

For Mother
• Contributes to earlier recovery from pregnancy due to the action of hormones that promote a quicker return of the uterus to its prepregnancy state
• Decreases risk of several chronic diseases later in life, including hypertension, cardiovascular disease, and diabetes
• Decreases the risk of ovarian and premenopausal breast cancer
• Potential for quicker return to prepregnancy weight
• Potential for delayed ovulation, thus reducing chances of pregnancy in the short term

have fewer ear infections (otitis media) because they do not sleep with a bottle in their mouths. Experts strongly discourage allowing any infant to sleep with a bottle in his or her mouth; milk can pool in the mouth, throat, and inner ear, creating a growth medium for bacteria, which can lead to ear infections and dental caries. By reducing these common ailments, parents can decrease discomfort for the infant, avoid related trips to the doctor, and prevent possible hearing loss.

Lower Risk of Diseases.   Research now links breastfeeding with reduced risks for many diseases. Breastfed infants may learn to self-regulate food intake and avoid overeating, which may explain the connection to lowered risk for obesity and type 2 diabetes among adults who were breastfed as infants. The immunologic benefits of breastfeeding seem to be involved in lowered rates of type 1 diabetes, celiac disease, and inflammatory bowel diseases. Reductions in risks of childhood leukemia and lymphoma have also been observed.

Fewer Allergies and Intolerances.   Breastfeeding also reduces the chances of some allergies, especially in allergy-prone infants (see Chapter 15). The key time to attain this benefit from breastfeeding is during the first 4 to 6 months of an infant's life. A longer commitment is best, but breastfeeding for even a few weeks is beneficial. Infants are also better able to tolerate human milk than formulas. Formulas sometimes must be switched several times until caregivers find the best one for the infant.

Convenience and Cost.   Breastfeeding frees the mother from the time and expense involved in buying and preparing formula and washing bottles. Human milk is ready to go and sterile. This allows the mother to spend more time with her baby.

Possible Barriers to Breastfeeding. Widespread misinformation, the mother's need to return to a job, and social reticence serve as barriers to breastfeeding.

Misinformation.   The major barriers to breastfeeding are misinformation, such as the idea that one's breasts are too small, and the lack of role models. One positive note has been the widespread increase in the availability of lactation consultants over the past several years. First-time mothers who are interested in breastfeeding can find invaluable support from lactation consultants or by talking to women who have experienced it successfully. In almost every community, a group called La Leche League offers classes in breastfeeding and advises women who have problems with it (800-LALECHE or **www.lalecheleague.org**).

Return to an Outside Job.   Working outside the home can complicate plans to breastfeed. One possibility after a month or two of breastfeeding is for the mother to express and save her own milk. She can use a breast pump or manually express milk into a sterile plastic bottle or nursing bag (used in a disposable bottle system). Federal legislation passed in 2010 requires employers to provide reasonable breaks and a private space (other than a bathroom) for breastfeeding mothers to pump milk.

Saving human milk requires careful sanitation and rapid chilling. It can be stored in the refrigerator for 3 to 5 days or be frozen for 3 to 6 months. Thawed milk should be used within 24 hours.

There is a knack to learning how to express milk, but the freedom can be worth it, because it allows others to feed the infant the mother's milk. A schedule of expressing milk and using supplemental formula feedings is most successful if begun after 1 to 2 months of exclusive breastfeeding. After 1 month or so, the baby is well adapted to breastfeeding and probably feels enough emotional security and other benefits from nursing to feed both ways.

Some women can juggle both a job and breastfeeding, but others find it too cumbersome and decide to formula-feed. A compromise—balancing some breastfeedings, perhaps early morning and night, with infant formula feedings during the day—is possible. However, too many supplemental infant formula feedings decrease milk production.

▲ Breast milk can be expressed by the mother using a manual, battery-operated, or electric (shown) breast pump. The expressed milk can be stored for times when the mother is not available to breast-feed the infant.

Social Concerns. Another barrier for some women is embarrassment about nursing a child in public. Historically, our society has stressed modesty and has discouraged public displays of breasts—even for as good a cause as nourishing babies. In the United States, no state or territory has a law prohibiting breastfeeding. However, indecent exposure (including the exposure of women's breasts) has long been a common law or statutory offense. Now, 45 states, the District of Columbia, and the Virgin Islands have specific laws that protect a woman's right to breastfeed in any location. Women who feel reluctant should be reassured that they do have social support and that breastfeeding can be done discreetly.

Medical Conditions Precluding Breastfeeding. Breastfeeding may be ruled out by certain medical conditions in either the infant or the mother. For example, breastfeeding is contraindicated for infants with galactosemia, an inherited disorder in which the body cannot break down galactose. Recall from Chapter 4 that lactose, the main carbohydrate in human breast milk, is made of glucose and galactose. When galactose is not properly broken down, its by-products can damage body organs.

Certain medications, which pass into the milk and adversely affect the nursing infant, are best avoided while breastfeeding. In addition, a woman in North America or other developed region of the world who has a serious chronic disease (such as tuberculosis, AIDS, or HIV-positive status) or who is being treated with chemotherapy medications should not breastfeed.

> Phenylketonuria (PKU), a disorder of phenylalanine metabolism described in Chapter 6, was once thought to be a contraindication for breastfeeding. However, with complementary use of specialized phenylalanine-free formulas, infants with PKU can now enjoy the benefits of breastfeeding.

Cosmetic Alterations to the Breast. Nipple piercings should have no impact on a woman's ability to breastfeed, but the jewelry should be removed before each feeding. Repeated removal and reinsertion of the jewelry may be inconvenient and irritating, so it may be best to leave the jewelry out for the entire period of breastfeeding. Breast tattoos will not impair breastfeeding, either. However, getting a new nipple piercing or breast tattoo while breastfeeding is not advised due to the pain of healing and possibility of infection. Past breast augmentation or reduction surgeries may impair a woman's ability to breastfeed if milk-producing tissue was damaged during the surgery.

Environmental Contaminants in Human Milk. There is some legitimate concern over the levels of various environmental contaminants in human milk. However, the benefits of human milk are well established, and the risks from environmental contaminants are still largely theoretical. A few measures a woman could take to counteract some known contaminants are to (1) consume a variety of foods within each food group; (2) avoid freshwater fish from polluted waters; (3) carefully wash and peel fruits and vegetables (or choose organically raised produce, which has lower levels of pesticides than conventional produce); and (4) remove the fatty edges of meat, as pesticides concentrate in fat. In addition, a woman should not try to lose weight rapidly while nursing (more than ¾ to 1 pound per week) because contaminants stored in her fat tissue might then enter her bloodstream and affect her milk. If a woman questions whether her milk is safe, especially if she has lived in an area known to have a high concentration of toxic wastes or environmental pollutants, she should consult her local health department.

✓ CONCEPT CHECK 14.7

1. What do AND and AAP recommend for duration of breastfeeding?
2. Describe the physiological processes of milk production and letdown. Be sure to mention the hormones involved in these processes.
3. List three advantages of breastfeeding for the mother and three advantages for the infant.
4. Identify three micronutrients that may need to be supplemented in the diets of breastfed infants and give the rationale for each.
5. Describe three potential barriers to breastfeeding and suggest ways to overcome them.

Nutrition and Your Health
● Preventing Birth Defects

Eating well for a healthy pregnancy not only supplies materials for fetal growth and development but also helps to direct the amazing process of building a new life. Considering the complexity of the human body and its more than 20,000 genes, it is not surprising that abnormalities of structure, function, or metabolism are sometimes present at birth. Birth defects impact one of every 33 babies born in the United States. In some cases, they are so severe that a baby cannot survive or thrive. Defects in embryonic or fetal development are the presumed cause of many spontaneous abortions and are at the root of about 20% of infant deaths before 1 year of age. However, many babies with birth defects can go on to live healthy and productive lives.

A wide range of physical or mental disabilities result from birth defects. Heart defects are present in approximately 1 of every 100 to 200 newborn babies, accounting for a large proportion of infant deaths. Cleft lip and/or cleft palate are malformations of the lip or roof of the mouth and occur in approximately 1 in 700 to 1000 births. Neural tube defects are malformations of the brain or spinal cord that occur during embryonic development. Examples include spina bifida, in which all or part of the spinal cord is exposed, and anencephaly, in which some or all of the brain is missing (Fig. 14-10). Babies born with spina bifida can survive to adulthood but in many cases, have disabilities, such as paralysis, incontinence, and learning disabilities. Babies born with anencephaly die

soon after birth. Neural tube defects occur in 1 in 1000 births. Down syndrome, a condition in which an extra chromosome leads to mental retardation and other physical alterations, occurs in about 1 in 800 births. Other common birth defects include musculoskeletal defects, gastrointestinal defects, and metabolic disorders.

What causes a birth defect? About 15% to 25% of birth defects are known to be genetic (i.e., inherited or spontaneous mutations of the genetic code). Another 10% are due to environmental influences (e.g., exposure to **teratogens**). The specific cause of the remaining 65% to 75% of birth defects is unknown. Although the etiology of birth defects is multifactorial and many elements are beyond human control, good nutrition practices can positively influence the outcome of pregnancy.

Folic Acid

During the 1980s, researchers in the United Kingdom noticed a relationship between poor dietary habits and a high rate of neural tube defects among children of impoverished women. Subsequent intervention studies demonstrated that administration of a multivitamin supplement during the periconceptional period—the months before conception and during early pregnancy—reduced the recurrence of these birth defects. The specific link between dietary folic acid and neural tube defects was tested and confirmed in several follow-up studies. As you learned in Chapter 8, folate plays a leading role in the synthesis of DNA and the metabolism of amino acids. The rapid cell growth of pregnancy increases needs for folate during pregnancy to 600 micrograms DFE per day. Some women, for genetic reasons, may have an even higher requirement. Adequate folic acid in the periconceptional period decreases the risk of neural tube defects by about 70% and has also been associated with decreased risk of cleft lip/palate, heart defects, and Down syndrome.

In 1998, FDA mandated fortification of grain products to provide 140 micrograms of folic acid per 100 grams of grain consumed. In Canada, the level of fortification is 150 micrograms of folate per 100 grams of grain consumed. In general, this increases the average consumption of dietary folic acid by 100 micrograms per day.

In North America, maternal death as a result of childbirth is uncommon: only about 11 deaths in every 100,000 live births. The infant mortality rate, however, is much higher: for each 100,000 live births, about 600 to 700 infants die within the first year. The infant death rate among African-Americans is more than double the rates among whites and Hispanics in the United States.

teratogen A compound (natural or synthetic) that may cause or increase the risk of a birth defect. Exposure to a teratogen does not always lead to a birth defect; its effects on the fetus depend on the dose, timing, and duration of exposure.

Healthy Spine

Spine Affected by Spina Bifida

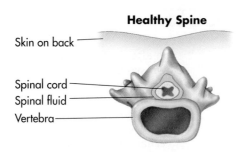
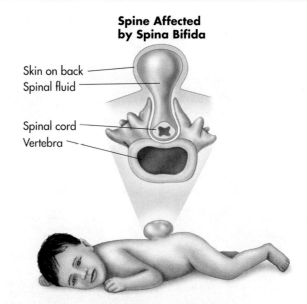

FIGURE 14-10 ▶ Spina bifida is one type of neural tube defect. Very early in fetal development, a ridge of neural-like tissue forms along the back of the embryo. As the fetus develops, this ridge develops into both the spinal cord and nerves at the lower end and the brain at the upper end. At the same time, the bones that make up the back gradually surround the spinal cord on all sides. In spina bifida, the backbones do not form a complete ring to protect the spinal cord. Deficient folate status in the mother during the beginning of pregnancy, especially in combination with a genetic abnormality in folate metabolism, greatly increases the risk of neural tube defects.

Adequate folate status is crucial for all women of childbearing age because the neural tube closes within the first 28 days of pregnancy, a time when many women are not even aware that they are pregnant. A well-planned diet can meet the RDA for folic acid, but the U.S. Public Health Service, March of Dimes, and Dietary Guidelines for Americans (Table 14-3) recommend that all women of childbearing age take a daily multivitamin and mineral supplement that contains 400 micrograms of folic acid. Women who have already had a child with a neural tube defect are advised to consume megadoses of folic acid—4000 micrograms per day. They are to begin supplementation at least 1 month before any future pregnancy. This must be done under strict physician supervision.

Iodide

Low iodide status during the first trimester of pregnancy—a critical period of brain development—may lead to congenital hypothyroidism (formerly called *cretinism:* see Further Reading 21). If left untreated, consequences (which may vary in severity) include mental retardation, stunting of growth, impaired hearing and speech, and infertility. Some other physical features are evident in Figure 14-11. When the defect is identified early (by newborn screening tests), these harmful effects can be prevented by treatment with thyroid hormones. With use of iodized salt, however, congenital hypothyroidism due to iodide deficiencies is rare.

Antioxidants

A case can be made for antioxidants in the prevention of birth defects as well. Free radicals are constantly generated within the body as a result of normal metabolic processes. An abundance of free radicals results in damage of cells and their DNA, which can lead to gene mutations or tissue malformations. Some research points to free radicals as a source of damage during embryo development and organogenesis. Antioxidant systems within the body act to minimize the damage caused by free radicals, and researchers hypothesize that dietary sources

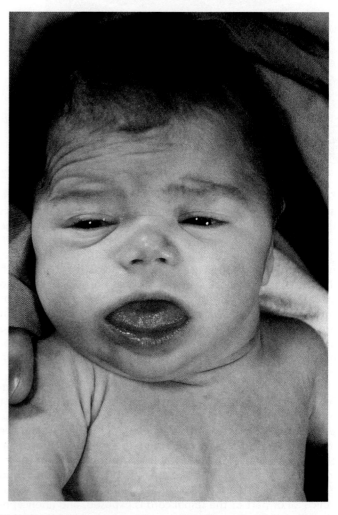

FIGURE 14-11 ▲ Child with congenital hypothyroidism due to maternal iodide deficiency. This birth defect leads to mental retardation, stunting of growth, and other physical defects, such as large tongue, enlarged head, and puffy eyes.

of antioxidants may aid in the prevention of birth defects. At this time, there is insufficient evidence to support supplementation of individual nutrients that participate in antioxidant systems—vitamin E, vitamin C, selenium, zinc, and copper—for the prevention of birth defects. However, use of a balanced multivitamin and mineral supplement while consuming a diet rich in whole grains, legumes, and a variety of fruits and vegetables will provide enough of these nutrients to meet current recommendations.

Vitamin A

While the needs for most vitamins and minerals increase by about 30% during pregnancy, the requirement for vitamin A increases by only 10%. Studies have shown the teratogenic potential of vitamin A in doses as low as approximately 3000 micrograms RAE per day. This is just over three times the RDA of 770 micrograms RAE per day for pregnant adult women.

Fetal abnormalities resulting from vitamin A toxicity primarily include facial and cardiac defects, but a wide range of defects have been reported. It is rare that food sources of vitamin A would lead to toxicity. Recall from Chapter 8 that preformed vitamin A is found in liver, fish, fish oils, fortified milk and yogurt, and eggs. Carotenoids, found in fruits and vegetables, are precursors of vitamin A that are converted into vitamin A in the small intestine. However, the efficiency of absorption of carotenoids decreases as intake increases. Vitamin A excesses typically arise from high-dose dietary supplements rather than food sources.

Typical North American diets supply adequate vitamin A from foods, so supplemental use is not generally necessary. During pregnancy, supplemental preformed vitamin A should not exceed 3000 micrograms RAE per day (15,000 IU per day). Most multivitamins and prenatal vitamins supply less than 1500 micrograms RAE per day. A balanced diet and prudent use of dietary supplements are actions that can sidestep potential problems with vitamin A toxicity.

Caffeine

Caffeine has been scrutinized for its safety during pregnancy, especially for any link with rate of birth defects. Caffeine decreases the mother's absorption of iron and may reduce blood flow through the placenta. In addition, the fetus is unable to detoxify caffeine. Research shows that as caffeine intake increases, so does the risk of miscarriage or delivering an LBW infant. Heavy caffeine use during pregnancy may also lead to caffeine withdrawal symptoms in the newborn. These risks are reported with caffeine intakes in excess of 500 milligrams, or the equivalent of about 5 cups of coffee per day. Moderate use of caffeine (up to 200 milligrams of caffeine, or the equivalent of 12 fluid ounces of regular coffee per day), however, is not associated with risk for birth defects. Accounting for caffeine intake from tea, over-the-counter medicines containing caffeine, and chocolate is also important.

Aspartame

Phenylalanine, a component of the artificial sweetener aspartame (NutraSweet® and Equal®), is a cause for concern for some pregnant women. High amounts of phenylalanine in maternal blood disrupt fetal brain development if the mother has a disease known as *phenylketonuria* (see next section). If the mother does not have this condition, however, it is unlikely that the baby will be affected by moderate aspartame use.

For most adults, diet soft drinks are the primary source of artificial sweeteners. Of greater concern than the safety of sweeteners during pregnancy is the quality of foods and beverages consumed. A high intake of diet soft drinks may crowd out healthier beverages, such as water and low-fat milk.

Obesity and Chronic Health Conditions

Even before becoming pregnant, women of childbearing age should have regular medical checkups to keep an eye on any health conditions that already exist or to identify any developing health problems. In some cases, the condition itself increases risk for birth defects. Obesity, high blood pressure, and uncontrolled diabetes are common health problems known to increase the risk for birth defects, including neural tube defects. In other cases, medications used to control illnesses may pose a risk to the developing fetus. Other health issues, such as seizure disorders and metabolic disorders, could also affect fetal development. A preconception visit with a health professional can help to sort out and make plans to minimize such risks. Once a woman has become pregnant, early and regular prenatal care can aid in the success of a pregnancy.

Women with diabetes are two to three times more likely to give birth to a baby with birth defects compared to women with normal glucose metabolism. Examples of birth defects common in this group include malformations of the spine, legs, and blood vessels of the heart. Some experts speculate that the mechanism by which diabetes increases birth defects is via excessive free radicals, which lead to oxidative damage of DNA during early gestation. Careful control of blood glucose drastically lowers risk for women with diabetes. Optimal blood glucose control can be achieved through a combination of dietary modifications and medications. Given that diabetes is on the rise among women of childbearing age, this elevated rate of birth defects has become an area of heightened concern.

Another health condition for which maternal nutritional control is of utmost importance is PKU. Recall from Chapter 6 that PKU is an error of metabolism in which the liver lacks the ability to process phenylalanine, leading to an accumulation of this amino acid and its metabolites in body tissues. Babies born to women who have phenylketonuria that is not controlled by diet are at heightened risk for brain defects, such as microcephaly and mental retardation (see Further Reading 1).

Alcohol

Conclusive evidence shows that repeated consumption of four or more alcoholic drinks at one sitting harms the fetus (see Further Reading 13). Such binge drinking is especially perilous during the first 12 weeks of pregnancy, as this is when critical early developmental events take place in utero. Recently, the media widely publicized the findings of a British research study showing that light drinking (up to one drink per day) during pregnancy was not harmful to neurological development of offspring, as measured later in childhood (see Further Reading 10). However, experts have not determined a safe level of alcohol intake during pregnancy. Until a safe level can be established, women are advised not to drink any alcohol—from beverages, foods, or medications (check the label)—during pregnancy or when there is a chance of conception. The embryo (and, at later stages, the fetus) has no means of detoxifying alcohol.

Women with chronic alcoholism produce children with a variety of physical and intellectual problems collectively called **fetal alcohol spectrum disorders (FASDs).** The most severe of these disorders is **fetal alcohol syndrome (FAS).** A diagnosis of FAS is based mainly on poor fetal and infant growth, physical deformities (especially of facial features), and mental retardation (Fig. 14-12). Irritability, hyperactivity, short attention span, and limited hand-eye coordination are other symptoms of FAS. Defects in vision, hearing, and mental processing may also develop over time. Other FASDs consist of some but not all of the defects of FAS. **Alcohol-related neurodevelopmental disorders (ARNDs)** include behavior and learning problems resulting from exposure to alcohol in utero. **Alcohol-related birth defects (ARBDs)** typically include malformations of the heart, kidneys, bones, and/or ears.

Exactly how alcohol causes these defects is not known. One line of research suggests that alcohol, or products of the metabolism of alcohol (e.g., acetaldehyde), cause faulty movement of cells in the brain during early stages of nerve cell

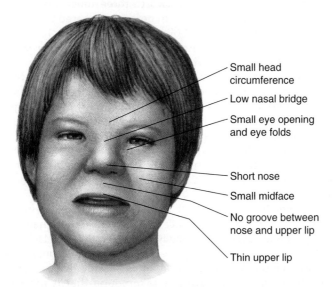

Small head circumference

Low nasal bridge

Small eye opening and eye folds

Short nose

Small midface

No groove between nose and upper lip

Thin upper lip

FIGURE 14-12 ▲ Fetal alcohol syndrome. The facial features shown are typical of affected children. Additional abnormalities in the brain and other internal organs accompany fetal alcohol syndrome but are not immediately apparent from simply looking at the child. Milder forms of alcohol-induced changes from a lower alcohol exposure to the fetus are known as alcohol-related neurodevelopmental disorders (ARNDs) and alcohol-related birth defects (ARBDs).

development or block the action of certain brain neurotransmitters. In addition, inadequate nutrient intake, reduced nutrient and oxygen transfer across the placenta, cigarette smoking commonly associated with alcohol intake, drug use, and possibly other factors contribute to the overall result.

The Dietary Guidelines for Americans recommend that alcoholic beverages not be consumed by pregnant women (Table 14-3). For more information about fetal alcohol syndrome, visit the website **www.cdc.gov/ncbddd/fasd/.**

Environmental Contaminants

There is little evidence to link birth defects with the amounts of pesticides, herbicides, or other contaminants in foods or public water supplies in North America. However, evaluating such a link can be methodologically difficult, and many would argue that regulations concerning contaminants in the food and water supply are too permissive. Thus, it seems prudent to take measures to decrease intake of pesticides and other contaminants wherever possible. For fruits and vegetables, peeling, removing outer leaves, and/or thoroughly rinsing and scrubbing with a brush under running water will remove the majority of contaminants. In animal products, toxins are most likely to accumulate in fatty tissues. Therefore, removing skin, discarding drippings, and trimming visible fat will decrease exposure from meat, poultry, and fish.

For fish, mercury is of particular concern because it can harm the nervous system of the fetus. Thus, FDA warns pregnant women to avoid swordfish, shark, king mackerel, and tile fish because of possible high mercury contamination. Largemouth bass are also implicated. In general, intake of other fish and shellfish should not exceed 12 ounces per week. Canned albacore tuna is a potential mercury source, so it should not be consumed in amounts exceeding 6 ounces per week. As a rule of thumb, consuming a *variety* of foods minimizes risk of exposure to any one contaminant from the food supply.

Summing Up

Although many risk factors for birth defects are beyond our control, a woman of childbearing potential can make some wise nutrition choices to improve her chances of having a healthy baby without birth defects. A varied and balanced diet, such as the food plan for pregnant women described in this chapter, along with a daily multivitamin and mineral

fetal alcohol spectrum disorders (FASDs) A group of irreversible physical and mental abnormalities in the infant that result from the mother's consuming alcohol during pregnancy.

fetal alcohol syndrome (FAS) Severe form of FASD that involves abnormal facial features and problems with development of the nervous system and overall growth as a result of maternal alcohol consumption during pregnancy.

A goal of *Healthy People 2020* is 98.3% abstinence from alcohol, cigarettes, and illicit drugs by pregnant women, a 10% improvement over current statistics.

supplement with 400 micrograms of folic acid will ensure adequate nutrient status. It is estimated that daily use of a multivitamin and mineral supplement containing folic acid will decrease the rate of all birth defects by 50%. Discuss use of any other dietary supplements with a physician to be sure that the fetus will not be exposed to toxic levels of vitamin A or other dangerous ingredients. Early and consistent prenatal care can help control obesity and any chronic health conditions that may complicate a pregnancy. Also, avoiding alcohol during pregnancy will eliminate any risk for fetal alcohol spectrum disorders.

Although it seems that advice for a healthy pregnancy is always directed at the mother, fathers-to-be are not off the hook! Health is a family affair, so encouraging healthy eating habits and avoiding smoking and alcohol are important for fathers, too. As you read in the beginning of this chapter, the genetics of the baby are certainly an outcome of both parents. Indeed, inadequate supplies of zinc, folate, antioxidants, and omega-3 fatty acids affect the quality of sperm. Overall, the periconceptional period is a time for good nutrition and careful lifestyle practices for mothers- and fathers-to-be.

Summary (Numbers refer to numbered sections in the chapter.)

14.1 Energy imbalances can adversely affect fertility by altering hormone levels and promoting oxidative damage. Polycystic ovary syndrome, which tends to co-occur with upper-body obesity, is a condition of hormonal imbalance that causes infertility. Other than managing body weight, nutritional factors that may improve male and/or female fertility include low glycemic index carbohydrates, unsaturated fats, antioxidants, folate, iron, and zinc.

14.2 Pregnancy is arbitrarily divided into three trimesters of 13 to 14 weeks each. The first trimester is characterized by a rapid increase in cell number as the zygote grows to be an embryo, then a fetus. During the first trimester, the growing organism is most susceptible to damage from exposure to toxic agents or nutrient deficiencies. By the start of the second trimester, the organs and limbs have formed and will continue to grow and develop. The third trimester is marked by rapid fetal growth and storage of nutrients in preparation for life outside the womb.

14.3 A successful pregnancy results in optimal health for both the infant and the mother. Pregnancy success is defined as (1) gestation longer than 37 weeks and (2) birth weight greater than 5.5 pounds (2.5 kilograms). Factors that predict pregnancy success include early and regular prenatal care, maternal age within the range of 20 to 35 years, and adequate nutrition. Factors that contribute to poor pregnancy outcome include inadequate prenatal care, obesity, underweight, teenage pregnancy, smoking, alcohol consumption, use of certain prescription medications and all illicit drugs, inadequate nutrition, heavy caffeine use, and various infections, such as listeriosis.

14.4 For women with a healthy prepregnancy BMI (18.5 to 24.9), total weight gain should be within the range of 25 to

35 pounds. Underweight women and those carrying multiple fetuses should gain more; overweight and obese women should gain less. During the first trimester, although she need not increase diet quantity, the woman should focus on diet quality to meet increased requirements for protein, carbohydrate, essential fatty acids, fiber, water, and many vitamins and minerals. A woman typically needs an additional 350 to 450 kcal per day during the second and third trimesters.

14.5 Following a plan based on the Dietary Guidelines for Americans as exemplified by MyPlate is recommended for pregnant and breastfeeding women. The mother-to-be should especially emphasize good sources of vitamin B-6, folate, vitamin D, iron, zinc, and calcium. Vegetarian diets are safe during pregnancy, but vegan mothers should specifically seek out good sources of vitamin B-12 and vitamin D. Prenatal multivitamin and mineral supplements are useful for meeting increased nutrient requirements during pregnancy.

14.6 Gestational hypertension, gestational diabetes, heartburn, constipation, nausea, vomiting, edema, and anemia are all possible discomforts and complications of pregnancy. Nutrition therapy can help minimize some of these problems.

14.7 Almost all women are able to breastfeed their infants. The nutritional composition of human milk is different from that of unaltered cow's milk and is much more desirable for the infant.

For the infant, the advantages of breastfeeding over formula feeding are numerous, including fewer intestinal, respiratory, and ear infections, and fewer allergies and food intolerances. Benefits for the mother include reduced risk of certain cancers, earlier recovery from pregnancy, and faster return to prepregnancy weight. For mothers who choose not to breastfeed or in medical situations that contraindicate breastfeeding (e.g., galactosemia), infants can be adequately nourished with formula.

NAYH There are several steps pregnant women can take to help prevent birth defects. Achieve a healthy body weight before pregnancy and strive to gain weight within the ranges

recommended by the Institute of Medicine. Adequate intakes of folic acid, iodide, and antioxidant nutrients are essential for prevention of many types of birth defects. Excesses of vitamin A and caffeine should be avoided. There is no safe level of alcohol intake known during pregnancy. Dietary control of diseases (e.g., diabetes and PKU) will also protect the fetus.

Check Your Knowledge (Answers to the following multiple choice questions are below.)

1. Which of the following nutrition interventions is most likely to improve fertility?
   a. taking a vitamin E supplement
   b. losing excess body fat
   c. consuming a low-carbohydrate diet
   d. taking an iron supplement

2. Increased carbohydrate needs during pregnancy are set to
   a. prevent ketosis.
   b. alleviate nausea.
   c. prevent gestational hypertension.
   d. supply adequate folate.

3. An infant born at 38 weeks' gestation weighing 5.0 pounds can be described as
   a. preterm.   c. SGA.
   b. LBW.   d. LBW and SGA.

4. If a woman is 5'2" and weighs 150 pounds before becoming pregnant, how much weight should she gain during pregnancy?
   a. 28 to 40 pounds (12.5 to 18 kilograms)
   b. 25 to 35 pounds (11.5 to 16 kilograms)
   c. 15 to 25 pounds (7 to 11.5 kilograms)
   d. As little as possible

5. Benefits of exercise during pregnancy include
   a. preventing excessive gestational weight gain.
   b. improved sleep.
   c. lower risk for gestational diabetes.
   d. All of the above.

6. Which of the following may help to alleviate nausea during pregnancy?
   a. postponing meals until the afternoon
   b. drinking large amounts of water
   c. postponing use of iron supplements until the second trimester
   d. All of the above.

7. Physiologically, milk production requires _____ kcals per day.
   a. 300   c. 800
   b. 500   d. 1000

8. A food plan for a woman in the third trimester of pregnancy differs from her prepregnancy diet in that
   a. fluid needs are higher.
   b. additional solid fats and added sugars are allowed.
   c. there are more servings from the grains group.
   d. All of the above.

9. Advantages of breastfeeding include
   a. decreased ear infections in the infant.
   b. decreased diarrheal diseases in the infant.
   c. decreased risk of breast cancer for the mother.
   d. All of the above.

10. Consuming one cup of coffee per day is associated with
    a. spontaneous abortions.   c. birth defects.
    b. LBW.   d. None of the above.

Answer Key: 1. b (LO 14.1), 2. a (LO 14.2), 3. d (LO 14.3), 4. c (LO 14.4), 5. d (LO 14.5), 6. c (LO 14.6), 7. c (LO 14.7), 8. d (LO 14.8), 9. d (LO 14.9), 10. d (LO 14.10)

Study Questions (Numbers refer to Learning Outcomes)

1. Provide three key pieces of nutrition advice for parents seeking to maximize their chances of conceiving. Why did you identify those specific factors? **(LO 14.1)**

2. Identify four key nutrients for which intake should be significantly increased during pregnancy. **(LO 14.2)**

3. Why does teenage pregnancy receive so much attention these days? At what age do you think pregnancy is ideal? Why? **(LO 14.3)**

4. Outline current weight-gain recommendations for pregnancy. What is the basis for these recommendations? **(LO 14.4)**

5. Suggest several safe exercises for a pregnant woman. **(LO 14.5)**

6. What nutrition advice would you give to a friend who suffers from morning sickness? **(LO 14.6)**

7. Describe the physiological mechanisms that stimulate milk production and release. How can knowing about these help mothers breastfeed successfully? **(LO 14.7)**

8. Describe a day's diet based on the Dietary Guidelines as exemplified by MyPlate that meets the increased nutrient needs of pregnancy. **(LO 14.8)**

9. Give three reasons a woman should give serious consideration to breastfeeding her infant. **(LO 14.9)**

10. Describe the importance of folic acid for conception and fetal development. **(LO 14.10)**

What the Dietitian Chose

The spinach salad meal with hard-boiled egg and whole-wheat roll provides only about 2 grams of iron. It is a healthy, nutrient-dense choice that provides some protein but does not make a significant contribution to meeting iron needs.

A hamburger is a source of iron—this meal would provide about 4.5 milligrams of iron—but it would not be a choice for a person who follows a vegetarian diet.

The bean and cheese taco option provides just as much iron (4.5 milligrams) as the hamburger meal and would be appropriate for a vegetarian. Even though the nonheme iron in beans is not as well absorbed as the heme iron in beef, the 20 milligrams of vitamin C in the stewed tomatoes will enhance iron absorption.

All of these meals, however, pale in comparison to the iron contribution of the fortified, ready-to-eat breakfast cereal. One cup of Kellogg's Smart Start cereal provides 18 milligrams of iron and the 72 milligrams of vitamin C in 1 cup of orange juice will enhance iron absorption. In one easy meal, any pregnant woman could obtain more than half of the 27-milligram RDA for iron during pregnancy.

CASE STUDY SOLUTION   Preparing for Pregnancy

1. Lily's use of an over-the-counter vitamin and mineral supplement provides ample synthetic folic acid and can help fill some gaps between dietary intakes and requirements for various nutrients. Still, she should discuss this supplement use with her physician and would eventually benefit more from a prenatal supplement, as this will have more iron than the over-the-counter supplement. She should be sure not to exceed 100% of the Daily Value for preformed vitamin A because toxicities of this vitamin could cause birth defects. Lily is wise to avoid use of herbal supplements during pregnancy unless her physician approves of them.

2. Folate deficiency is known to be linked to neural tube defects. The synthetic folic acid in vitamin and mineral supplements is well absorbed and will certainly help to meet her needs for this nutrient. Consuming more fruits and vegetables—especially green leafy vegetables—is also a wise choice. A fortified, ready-to-eat breakfast cereal also contributes to vitamin and mineral intake. Her physician may prescribe a prenatal supplement instead of the over-the-counter supplement.

3. Many experts would say that she is consuming too much caffeine and would be wise to cut down coffee and caffeine-containing soft drinks to a total of two servings or fewer per day.

4. If Lily likes fish, she could include two servings of fish per week in her diet plan for pregnancy and breastfeeding. Fish are a good source of protein and supply healthy fats, including omega-3 fatty acids, linked to brain and eye development in the fetus. However, Lily should avoid some types of fish that may have high mercury content, which can harm the nervous system of the developing fetus. Fish that typically have high levels of mercury include swordfish, shark, king mackerel, and tile fish.

5. The additional fruits and vegetables Lily is consuming will provide some fiber to help prevent constipation. Other ideas to increase her fiber intake include consuming a variety of whole grains and incorporating beans and legumes into her diet plan. Lily should also increase her fluid intake and be sure to maintain regular exercise, such as walking.

6. Lily expressed concern about her weight gain during pregnancy. Research clearly shows that adequate

weight gain during pregnancy is critical for growth and development of the fetus. Inadequate weight gain contributes to risk for delivering a low-birth-weight or small-for-gestational-age baby. On the other hand, excessive weight gain during pregnancy can present problems as well. Excessive weight gain may promote long-term health problems and complicate the pregnancy with gestational diabetes or large-for-gestational age babies. At her prepregnancy weight and height, Lily's BMI (21) is within the normal range, so she should aim to gain 25 to 35 pounds during the entire pregnancy. During the second and third trimesters, consuming an extra 350 to 450 kcal per day of nutrient-dense foods will help her to meet this goal. Regular, moderate exercise is also recommended throughout pregnancy. However, pregnancy is not a time to begin a vigorous exercise routine. Walking or riding a stationary bike are good activities to regulate weight gain. Lily should be reassured that her choice to breastfeed her baby will help her to lose those extra pounds after the baby is born.

Further Readings

1. AAP Committee on Genetics: Policy Statement: Maternal phenylketonuria. *Pediatrics* 122:445, 2008.

2. AAP Section on Breastfeeding: Policy Statement: Breastfeeding and the use of human milk. *Pediatrics* 129:e827, 2012.

3. Barger MK: Maternal nutrition and perinatal outcomes. *Journal of Midwifery and Women's Health* 55:502, 2010.

4. Blumfield ML and others: Micronutrient intakes during pregnancy in developed countries: Systematic review and meta-analysis. *Nutrition Reviews* 71: 118, 2013.

5. Deierlein AL and Siega-Riz AM: How can we assist women in managing gestational weight gain? *Women's Health* 8:603, 2012.

6. Dror DK and Allen LH: Vitamin D inadequacy in pregnancy: Biology, outcomes, and interventions. *Nutrition Reviews* 68:464, 2010.

7. Fraser A and others: Associations of gestational weight gain with maternal body mass index, waist circumference, and blood pressure measured 16 years after pregnancy: The Avon Longitudinal Study of Parents and Children (ALSPAC). *American Journal of Clinical Nutrition* 93:1285, 2011.

8. Greer FR and others: Effects of early nutritional interventions on the development of atopic disease in infants and children: The role of maternal dietary restriction, breastfeeding, timing of introduction of complementary foods, and hydrolyzed formulas. *Pediatrics* 121:183, 2008.

9. Han Z and others: Maternal underweight and the risk of preterm birth and low birth weight: A systematic review and meta-analysis. *International Journal of Epidemiology* 40:65, 2011.

10. Humphriss R and others: Prenatal alcohol exposure and childhood balance ability: Findings from a UK birth cohort study. *BMJ Open* 3:e002718, 2013.

11. Institute of Medicine and National Research Council: *Weight gain during pregnancy: Reexamining the guidelines.* Washington, DC: National Academies Press, 2009.

12. James DC and others: Position of the American Dietetic Association: Promoting and supporting breastfeeding. *Journal of the American Dietetic Association* 109:1926, 2009.

13. Jones KL: The effects of alcohol on fetal development. *Birth Defects Research (Part C)* 93:3, 2011.

14. Makrides M: Outcomes for mothers and their babies: Do n-3 long-chain polyunsaturated fatty acids and seafoods make a difference? *Journal of the American Dietetic Association* 108:1622, 2008.

15. March of Dimes Foundation: *Teenage pregnancy.* Updated 07/2012. Available at: **www.marchofdimes.com/materials/teenage-pregnancy.pdf.**

16. Prather H and others: Benefits of exercise during pregnancy. *Physical Medicine and Rehabilitation* 4: 845, 2012.

17. Proctor SB and Campbell CG: Position of the Academy of Nutrition and Dietetics: Nutrition and lifestyle for a healthy pregnancy outcome. *Journal of the Academy of Nutrition and Dietetics* 114:1099, 2014.

18. Rubin RC: Change is good—evidence to support lowering the diagnostic threshold for GDM. *Today's Dietitian* 13:10, 2011.

19. Swinney B: Eating to fuel fertility. *Today's Dietitian* 15:38, 2013.

20. Xu H and others: Role of nutrition in the risk of preeclampsia. *Nutrition Reviews* 67:639, 2009.

21. Zimmerman MB: Iodine deficiency in pregnancy and the effects of maternal iodine supplementation on the offspring: A review. *American Journal of Clinical Nutrition* 89(suppl):668S, 2009.

 **To get the most out of your study of nutrition, visit McGraw-Hill Connect at www.mcgrawhillconnect.com where you will find NutritionCalc Plus, LearnSmart, and many other dynamic tools.**

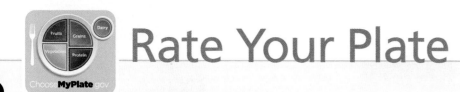

Rate Your Plate

I. Putting Your Knowledge About Nutrition and Pregnancy to Work

A college friend, Gayle, tells you that she is newly pregnant. You are aware that she usually likes to eat the following foods for her meals:

Breakfast

Skips this meal, or eats a granola bar
Coffee

Lunch

Sweetened yogurt, 1 cup
Small bagel with cream cheese
Occasional piece of fruit
Regular caffeinated soda, 12 ounces

Snack

Chocolate candy bar

Dinner

2 slices of pizza, macaroni and cheese, or 2 eggs with 2 slices
    of toast
Seldom eats a salad or vegetable
Regular caffeinated soda, 12 ounces

Snacks

Pretzels or chips, 1 ounce
Regular caffeinated soda, 12 ounces

1. Using NutritionCalc Plus software, evaluate Gayle's usual diet for protein, carbohydrate, folate, vitamin B-6, iron, and zinc. How does her intake compare with the recommended amounts for pregnancy?

    _____

    _____

    _____

2. Now redesign her diet and make sure that her intake meets pregnancy needs for protein, carbohydrate, folate, vitamin B-6, and zinc. (Hint: Fortified foods, such as breakfast cereal, are generally nutrient-rich foods, which can more easily help meet one's needs.) Increase the iron content as well, but it still may be below the RDA for pregnancy.

    _____

    _____

    _____

Student Learning Outcomes

Chapter 15 is designed to allow you to:

15.1 Describe the extent to which nutrition affects growth and physiological development from infancy through adolescence.

15.2 List specific nutrients often found to be lacking in the diets of infants, toddlers, preschoolers, and teenagers and make recommendations to remedy the problems.

15.3 Identify diet guidelines to meet the basic nutritional needs for normal growth and development for an infant and discuss some do's and don'ts associated with infant feeding.

15.4 Outline several challenges parents might face in dealing with eating habits during childhood and adolescence.

15.5 Describe the long-term effects of childhood obesity and suggest ways to prevent or treat the problem.

15.6 Identify common food allergens and suggest several practices that may reduce the risk of developing a food allergy.

Chapter 15

Nutrition from Infancy Through Adolescence

What Would You Choose?

Your brother and his wife have asked you to watch your 18-month-old niece, Lila, while they enjoy a weekend getaway at the lake to celebrate their anniversary. Before they drop her off, you make a trip to the grocery store to pick up some kid-friendly foods. Lila does not have any food allergies, but your brother warned you that she has gotten into some picky eating behaviors. Which of the following snack foods would you choose for your little houseguest?

a Reduced-fat popcorn

b Raw baby carrots with ranch dressing

c Fat-free light yogurt

d Whole-grain crackers with sliced cheddar cheese

connect Think about your choice as you read this chapter, |NUTRITION then see **What the Dietitian Chose** at the end of the chapter. To learn more about appropriate feeding practices for infants and preschoolers, check out the Connect site: **www.mcgrawhillconnect.com.**

Children born in North America have access to some of the best health care in the world. Even so, the numbers of children and teenagers with obesity and type 2 diabetes are on the rise. Kids spend more time in their computer chairs and less time on the playing field. Soft drinks and energy drinks have replaced much of the milk that children and teenagers previously consumed daily. Intakes of fruits, vegetables, and whole grains—although these are abundantly available—consistently fall short of recommendations. Evidently, there is room for improvement in the diets of most American children. How can we influence healthy eating behaviors?

During infancy and early childhood, the family mostly controls food intake, informs food preferences, and models eating behaviors. For school-age children and adolescents, peer and media influences take on more importance. Education designed to change nutrition habits should start early and involve the family. Family mealtimes help children to establish healthy eating habits, aid in prevention of childhood obesity, and, as children grow older, teach communication skills and improve self-esteem. This chapter examines the nutritional issues facing children and demonstrates how food choices should be tailored to adapt to the changing needs of growing children.

15.1 Assessing Growth

During infancy, attitudes toward foods and the eating process begin to take shape. If parents and other caregivers model good nutrition and are flexible, they can lead an infant into lifelong healthful food habits. A family environment that encourages healthy eating will provide the nutrients needed to optimize physical growth and development. However, these advantages do not guarantee that a child will thrive.

Children also need specific attention focused on them; they need to grow in a stimulating environment, and they need a sense of security. For example, children hospitalized for growth failure gain weight more quickly when more caring stimulation, such as holding and rocking the infant, accompanies needed nutrients.

THE GROWING INFANT

All babies seem to do is eat and sleep. There is a good reason for this. An infant's birth weight doubles in the first 4 to 6 months and triples within the first year. Never again is growth so rapid. Such rapid growth requires a lot of nourishment and sleep. After the first year, growth is slower; it takes 5 more years to double the weight seen at 1 year. An infant also increases in length in the first year by 50% and then continues to gain height through the teen years. These gains are not necessarily continuous; spurts of rapid growth alternate with periods of relatively little change. Height is essentially maximized by age 19, although increases of several inches may occur in the early twenties, especially for boys. Head size in proportion to total height shrinks from one-fourth to one-eighth during the climb from infancy to adulthood.

The human body needs a lot more food to support growth and development than it does to merely maintain its size once growth ceases. When nutrients are missing at critical phases of this process, growth and development may slow or even stop. To grow, children must consume adequate amounts of calories, protein, calcium, iron, zinc, and other nutrients. In countries of the developing world, about one-third of the children under 5 years of age are short and underweight for their ages. Poor nutrition—called **undernutrition**—is at the heart of the problem. Undernourished children are smaller versions of nutritionally fit children. In poorer countries, when breastfeeding ceases, children are often fed a high-carbohydrate, low-protein diet (see Chapter 12). This diet supports some growth but does not allow children to attain their full genetic potential. Undernutrition occurs in North America, as well, but **overnutrition** is much more prevalent.

EFFECT OF UNDERNUTRITION ON GROWTH

As with the fetus in utero, the long-term effects of nutritional problems in infancy and childhood depend on the severity, timing, and duration of the nutritional insult to cell processes.

The single best indicator of a child's nutritional status is growth: gains in weight are a good reflection of growth in the short term, whereas gains in height are an indicator of growth over the long term. Eating a poor diet during a critical stage of infancy or childhood hampers the cell division that occurs at that stage. Mild zinc deficiencies among North American children, for example, have been linked to poor growth. Consuming an adequate diet later usually will not compensate for lost growth, however, because the hormonal and other conditions needed for growth will not likely be present. In addition, growth ceases in girls and boys when the skeleton reaches its final size. This happens as growth plates at the ends of the bones fuse, which begins around 14 years of age in girls and 15 years of age in boys. The final stages of this process end at about 19 years of age in girls and 20 years of age in boys. Furthermore, muscles can increase in diameter later in life, but their linear growth is limited by the length of the bone.

For these reasons, a 15-year-old Central American girl who is 4 feet 8 inches tall cannot attain the adult height of a typical North American girl simply by eating

undernutrition Failing health that results from a long-standing dietary intake that is not enough to meet nutritional needs.

overnutrition A state in which nutritional intake greatly exceeds the body's needs.

better. Girls experience their peak rate of growth right before the onset of menses. Once the time for growth ceases (in women, this is about 5 years after they start menstruating), a sufficient nutrient intake helps maintain health and weight but cannot make up for lost growth in height.

USING GROWTH CHARTS

Health professionals assess a child's increases in height and weight by comparing them with typical growth patterns recorded on charts (Fig. 15-1). The charts contain **percentile** divisions, which represent the typical measurements for 90% to 96% of children. A percentile represents the rank of the person among 100 peers matched for age and gender. If a young boy, for example, is at the 90th percentile of height-for-age, he is shorter than 10% and taller than 89% of children his age. A child at the 50th percentile is considered average. Fifty of 100 children will be taller than this child; 49 will be shorter.

Individual growth charts are available for both males and females from the Centers for Disease Control and Prevention (CDC) (Fig. 15-1). For children from birth to age 2, CDC recommends using the growth charts developed by the

percentile Classification of a measurement of a unit into divisions of 100 units.

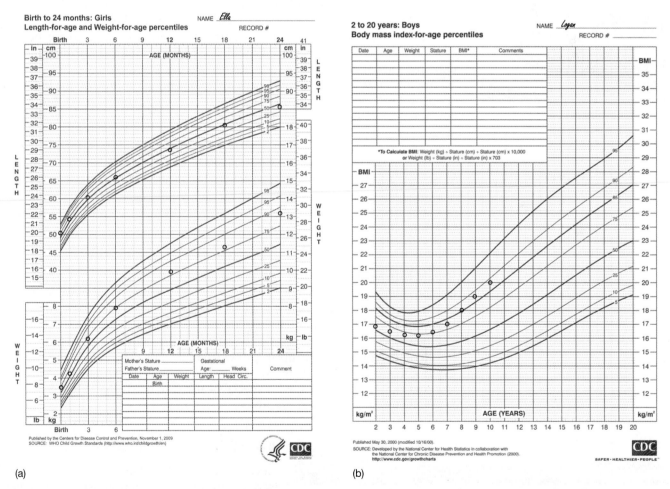

FIGURE 15-1 ▲ Growth charts for assessment of children in the growing years. The growth of a young girl and a boy are plotted to show how the charts are used in health care settings. For children from birth to age 2, CDC recommends using growth charts developed by WHO in 2006. For ages 2 to 20, use the CDC growth charts, developed by the National Center for Health Statistics and CDC in 2000. (a) Length-for-age and weight-for-age plotted for a young girl (Ella). A certain length (height) and weight correspond to a percentile value (i.e., a ranking of the individual among 100 peers). Ella has been tracking around the 50th percentile for length throughout the first 2 years of her life; her weight, which started out near the 50th percentile, is now between the 75th and 90th percentiles. (b) BMI-for-age plotted for a young boy (Logan) up to age 10. While Logan's BMI-for-age was around the 50th percentile when he was a preschooler, by age 10, he is above the 85th percentile. He would be classified as overweight according to his BMI-for-age.

Sources: (a) Centers for Disease Control and Prevention, based on WHO Child Growth Standards (2009); (b) Centers for Disease Control and Prevention (2000).

Children under 2 to 3 years of age are measured lying on their backs with knees unflexed, so the term *length* is used rather than *height* or *stature*.

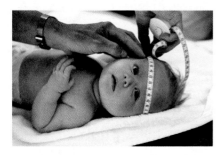

▲ Brain growth is faster in infancy than in any other stage of life. An infant's head needs to be large (about one-quarter of her body length) to allow for such growth. By the time she reaches adulthood, this girl's head will only be about one-eighth of her height.

World Health Organization (WHO) in 2006 to assess length-for-age, weight-for-age, weight-for-length, and head circumference-for-age. WHO growth charts are based on data collected from children from various regions of the world who were raised *under conditions for optimal growth and development.* This means that they were breast-fed as infants; had caregivers who followed recommended infant and child feeding practices set forth by WHO; had adequate health care; had mothers who did not use alcohol during pregnancy; and were not exposed to tobacco before or after birth. In contrast, older growth charts for young children were based on data from primarily white children who were mostly formula-fed during infancy. The WHO growth standards stress that breastfeeding is the biological norm for infant nutrition.

For children from ages 2 to 20, growth charts developed by the National Center for Health Statistics are available to assess weight-for-age and stature-for-age. However, the preferred growth chart for children and adolescents is body mass index (BMI)-for-age. For adults, BMI has fixed cutoff points (for example, a BMI of 25 for an adult is considered overweight). As Figure 15-1 shows, this is not true for children; BMI reference ranges are both gender and age specific.

In addition to taking measurements of weight and length, a health professional measures the infant's head circumference. Tracking head circumference-for-age on a growth chart (also available from CDC) is a means of assessing brain growth. The brain grows faster during the first year than at any other time of life. Unusual head circumference measurements can alert the health care team to a genetic disorder. A small head circumference could also be a result of malnutrition, infection, impaired mental development, or maternal substance abuse during pregnancy. An abnormally large measurement may be a sign of a tumor or fluid on the brain. Variations in head circumference may be due to harmless familial traits, but extreme deviations from the norm or rapid changes in percentiles over time warrant further investigation.

Infants and children should have their growth assessed during regular health checkups. It takes 1 to 3 years for the genetic potential (in terms of percentile ranking on growth charts) of infant growth to be established. By 3 years of age, a child's measurements, such as length (height) for age, should generally track along his or her established percentile. If the child's growth does not keep up with his or her length-for-age percentile, the physician needs to investigate whether a medical or nutritional problem is impeding the predicted growth. Likewise, when a child's BMI-for-age approaches the highest percentiles, caregivers should be concerned. A child between the 85th and 95th percentiles for BMI-for-age is considered overweight. At or above the 95th percentile, a child is considered obese (Table 15-1).

These growth charts are intended for healthy children, and they may not be appropriate for children born at very low birth weight or who have special health care needs. Preterm infants tend to be smaller than full-term infants in the early months but typically catch up in growth within 2 or 3 years. Catch-up growth requires that the child move up in the percentiles. Specialized growth charts have been developed for children with special health care needs that affect growth and development, such as Down syndrome.

TABLE 15-1 ▶ Weight Status Classifications for Children, Ages 2 to 20

Weight Status Classification	BMI-for-Age Percentile
Underweight	<5th percentile
Healthy weight	5th up to 85th percentile
Overweight	85th up to 95th percentile
Obese	≥95th percentile

Source: Centers for Disease Control and Prevention

ADIPOSE TISSUE GROWTH

Since 1970, researchers have speculated that over-feeding during infancy may increase the number of adipose tissue cells. Today, we know that the number of adipose cells can also increase as adulthood obesity develops. Still, if calorie intake is limited during infancy to keep down the number of adipose cells, the growth of other organ systems may also be severely restricted, especially brain and nervous system development. In addition, most overweight infants become normal-weight preschoolers without excessive diet restrictions. For these reasons, it is unwise to greatly restrict diet and especially fat intake in infants. After the first 12 months, fat intake can range from 30% to 40% of total kcal for ages 1 to 3 years and 25% to 35% of calorie intake for older children (and teenagers).

FAILURE TO THRIVE

About 5% to 10% of infants or children do not grow as expected. An infant may fail to reach important milestones, such as doubling birth weight by 6 months of age. On growth charts, weight-for-age may track below the fifth percentile. This condition of inadequate growth is termed **failure to thrive** (see Further Reading 6).

In some cases, failure to thrive has a specific medical cause. Physical problems that may limit calorie intake include poor oral cavity development or problems with breastfeeding (e.g., poor latch-on). If a child consumes enough food, growth may still falter if there are physical problems, such as celiac disease, which limit nutrient absorption. Lastly, some medical conditions, such as heart or lung disorders, lead to excessive energy expenditure.

However, about 80% of infants or children who fail to thrive have no apparent disease; environmental or social problems are at the root of their undernutrition. Poverty is the biggest environmental risk factor for undernutrition leading to failure to thrive. Sometimes the cause is poor infant-parent interaction, including situations of abuse or neglect. Either too little or too much concern over child feedings can lead to nutrition problems. Most poor interactions, however, arise from the parents' inexperience rather than intentional negligence. In the absence of a good parental role model or sound nutrition information, new parents may overdilute formula or provide foods (e.g., excessive fruit juice) that fill the child up without providing adequate calories.

Whatever the cause, the consequences of failure to thrive are serious and far-reaching. Possible outcomes include poor physical growth, impaired mental development, and behavioral problems. When health professionals encounter an infant failing to thrive, the true causes need to be identified and then treated. Counseling about proper nutrition and the importance of healthy parent-child interactions can help to get a child's growth back on track.

▲ Children older than 2 years are less likely to experience failure to thrive because they can often get food for themselves. Younger children, for the most part, are limited to what caregivers provide.

failure to thrive Condition of inadequate growth during infancy or early childhood caused by poor nutritional intake, inefficient nutrient absorption, or excessive energy expenditure; commonly defined as weight-for-age below the fifth percentile on multiple occasions or weight declining two or more major percentile lines on a standardized growth chart.

> ### ✓ CONCEPT CHECK 15.1
>
> 1. How do health care providers assess growth?
> 2. Define childhood overweight and obesity in terms of BMI-for-age.
> 3. If a female adolescent is short for her age due to a brief period of undernutrition between ages 10 to 12, can she catch up in growth after proper nutrition is restored? Why or why not?
> 4. What is failure to thrive? List some possible causes.

15.2 Infant Nutritional Needs

Infants' nutritional needs vary as they grow. Initially, human milk or infant formula supplies needed nutrients. Solid foods are not needed until around 6 months. Even after solid foods are added, the basis of an infant's diet for the first year is still human milk or infant formula. Because of the critical importance of adequate nutrition in infancy and the difficulties encountered in feeding some infants, there is more discussion in this chapter on this developmental period than on the later periods of childhood.

TABLE 15-2 ▶ Estimated Energy Requirements of Infants and Toddlers

Age	EER* Equation
0 to 3 months	(89 kcal × weight*) + 75
4 to 6 months	(89 kcal × weight) − 44
7 to 12 months	(89 kcal × weight) − 78
13 to 35 months	(89 kcal × weight) − 80

*Weight is in kilograms.

CALORIES

During infancy, energy needs per pound of body weight are the highest of any life stage (Table 15-2). For example, a 6-month-old infant requires two to four times more kcal per pound of body weight than an adult:

Healthy 6-month-old infant
700 kcal/15 pounds = 47 kcal/pound

Healthy 20-year-old woman
2200 kcal/135 pounds = 16 kcal/pound

Infants need a concentrated source of calories to meet these high demands. Exclusive feeding of either human milk or infant formula is ideal for the first 6 months of life; both are high in fat and supply about 640 kcal/quart (670 kcal/liter; Table 15-3). After 6 months of age, human milk or infant formula, supplemented by developmentally appropriate solid foods, can provide more calories, nutrients, and variety for the developing infant.

The infant's high calorie needs are primarily driven by rapid growth and high metabolic rate. The high metabolic rate is caused in part by the ratio of the infant's body surface area to its weight. More body surface area allows more heat loss from the skin; the body must use extra calories to replace that heat.

CARBOHYDRATE

Carbohydrate needs in infancy are 60 grams per day at 0 to 6 months and 95 grams per day at 7 to 12 months. These needs are based on the typical intakes of human milk by breastfed infants and their eventual use of solid foods. Both carbohydrate goals are satisfied by usual intakes of infants on a proper diet.

Do infants need fiber? There are no set AIs for fiber for infants and children younger than 2 years of age. For the first 6 months of life, breast milk or formula,

Looking at Table 15-3, you can see why unaltered cow's milk products are not recommended for infants. Cow's milk provides too much protein and minerals but not enough carbohydrates and fat. Fat-free cow's milk, especially, supplies inadequate calories to meet a growing infant's high nutritional demands.

TABLE 15-3 ▶ Composition of Human and Cow's Milk and Infant Formulas (per Liter)[a]

	Energy (kcal)	Protein (grams)	Fat (grams)	Carbohydrate (grams)	Minerals[b] (grams)
Milk					
Human milk	670[c]	11	45	70	2
Cow's milk, whole[d]	670	36	36	49	7
Cow's milk, fat-free[d]	360	36	1	51	7
Casein/Whey-Based Formulas					
Similac	680	14	36	71	3
Enfamil	670	15	37	69	3
Good Start	670	16	34	73	3
Soybean Protein-Based Formulas					
ProSobee	670	20	35	67	4
Isomil	680	16	36	68	4
Transition Formulas/Beverages[e]					
Similac Go and Grow Stage 3	630	17	34	68	3
EnfaGrow Toddler Next Step	492	33	10	66	5

[a]At 3 months of age, infants typically consume 0.75 to 1 liter of human milk or formula per day.
[b]Calcium, phosphorus, and other minerals.
[c]Rough estimate; ranges from 650 to 700 kcal per liter.
[d]Not appropriate for infant feeding, based primarily on high protein and mineral content.
[e]For use after 6 months of age or later (see label).

which contains no fiber, is adequate nutrition. As solid foods are introduced, include some fruits, vegetables, and whole grains. Some experts recommend working up to about 5 grams of fiber per day by 1 year of age. Keep in mind that too much fiber can limit nutrient absorption because it binds to some minerals and speeds the passage of food through the GI tract. Let the child's bowel habits be your guide. If the child is constipated, try increasing fiber and fluid intakes. On the other hand, if the child is uncomfortably gassy or is having many soft bowel movements per day, decrease the amount of fiber in the diet.

PROTEIN

Daily protein needs in infancy are about 9 grams per day for younger infants and about 11 grams per day for older infants. These needs also are based on the typical intakes of human milk by breastfed infants for 0 to 6 months and then on the needs for growth for older infants. About half of total protein intake should come from essential (indispensable) amino acids. As with carbohydrate, protein needs are easily satisfied by either human milk or infant formula. Protein intake should not greatly exceed this standard. Excess nitrogen and minerals supplied by high-protein diets would exceed the ability of an infant's kidneys to excrete the waste products of protein metabolism, thus putting much stress on overall kidney function.

In North America, infant protein deficiency is unlikely, except in cases of mistakes in formula preparation, such as when an infant's formula is excessively diluted with water. Protein deficiency may also be induced by elimination diets used to detect food **allergies** (hypersensitivities). As foods are eliminated from the diet, infants may not be offered enough protein to compensate for that supplied by the suspected food allergen (see the Nutrition and Your Health section at the end of this chapter).

allergy A hypersensitive immune response that occurs when immune bodies produced by us react with a protein we sense as foreign (an antigen).

FAT

Infants need about 30 grams of fat per day. Essential fatty acids should make up about 15% of total fat intake (about 5 grams per day). Both recommendations are again based on the typical intakes of human milk by breastfed infants and the eventual intake of solid foods. Fats are an important part of the infant's diet because they are vital to the development of the nervous system. As a concentrated source of calories, fat also helps to provide sufficient energy within the limits of the infant's small stomach capacity. Again, restriction of fat intake is not advised for infants or children under age 2 (Fig. 15-2).

Arachidonic acid (AA) and docosahexaenoic acid (DHA) are two long-chain fatty acids that have important roles in infant development. The nervous system, especially the brain and eyes, depends on these fatty acids for proper development. During the last trimester, DHA and AA provided by the mother accumulate in the brain and retinas of the eyes in the fetus. Breastfed infants are able to continue to acquire these fatty acids from human milk, especially if their mothers are regularly eating fish. Until recently, no infant formulas sold in the United States included AA or DHA, but many brands are now available with both AA and DHA. These are particularly useful for feeding preterm infants.

VITAMINS OF SPECIAL INTEREST

As noted in Chapter 8, vitamin K is routinely given by injection to all infants at birth. Formula-fed infants receive the other vitamins they need from formula. For bone health, immune function, and chronic disease prevention, the American Academy of Pediatrics recommends that all infants and children consume 400 IU vitamin D per day starting soon after birth. For all breastfed infants and formula-fed infants

Supplementing with more than 400 IU of vitamin D per day is not advised. No further benefits are seen beyond 400 IU per day and toxicity is possible. The UL for young infants is 1000 IU per day. Consequences of vitamin D toxicity among infants include vomiting, poor appetite, frequent urination, muscle pain and weakness, confusion, fatigue, and kidney damage.

FIGURE 15-2 ▶ The labels on infant foods, like those on adult foods, contain a Nutrition Facts panel. However, the information provided on infant food labels differs from that on adult food labels, especially with respect to total fat, saturated fat, and cholesterol content (see Fig. 2-13 for a comparison). Some cereal brands are fortified with various other micronutrients.

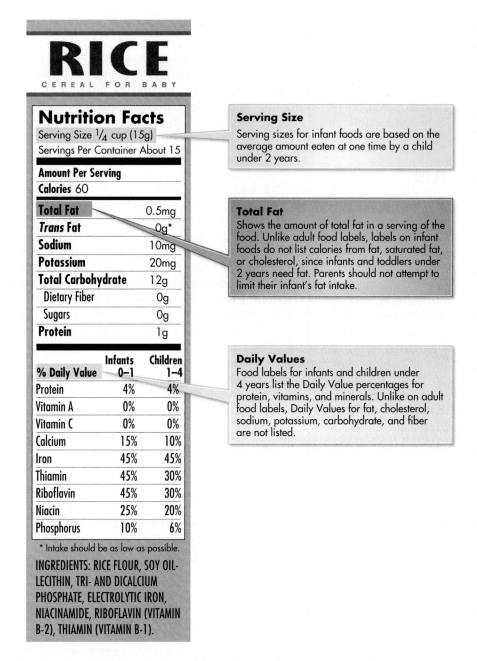

who consume less than 1 quart of formula per day, supplemental vitamin D is necessary until formula intake is at least 500 milliliters per day or vitamin D intake from food sources provides that much (see Further Reading 19). Breastfed infants whose mothers are total vegetarians (vegans) should also receive vitamin B-12 in supplement form.

MINERALS OF SPECIAL INTEREST

Infants are born with some internal stores of iron. However, if iron is not part of the diet, body iron stores will be depleted by about 6 months of age. If the mother was iron deficient during the pregnancy, these iron stores will be exhausted even sooner. As you will recall from Chapter 9, iron-deficiency anemia can lead to poor mental development in infants. Several studies indicate that iron deficiency anemia during infancy, even if corrected, has a lasting impact in terms of cognition, motor development, and behavior later in life. To maintain a desirable iron status, the American Academy of Pediatrics recommends that formula-fed infants should be given an iron-fortified formula from birth. Years ago, low-iron formulas were prescribed for infants with gastrointestinal distress. However, current evidence shows that they do not improve GI symptoms but rather place an infant at risk for iron deficiency. Low-iron infant formulas are still available, but their use is strongly discouraged. Breast milk is lower in iron than fortified infant formulas, but the form of iron in breast milk is much more bioavailable than the form in infant formula. Even so, by about 6 months of age, breastfed infants need solid foods to supply extra iron. This need for iron is a major consideration in the decision to introduce solid foods. Liquid iron supplements are rarely needed, except for preterm or LBW infants, those with blood disorders, or those born with inadequate stores of iron (i.e., due to maternal iron deficiency during pregnancy; see Further Reading 2).

Breast milk is a poor source of fluoride and formula manufacturers use fluoride-free water in formula preparation, so intake of this mineral during the first 6 months of life is low. However, fluoride supplementation is not advised before 6 months of age. After 6 months, the pediatrician or dentist may recommend fluoride supplements to aid in tooth development if fluoride supplied by tap water, foods, and toothpaste is inadequate.

Infants also need adequate amounts of zinc and iodide to support growth. However, when human milk and infant formula are provided in quantities to meet calorie needs, zinc and iodide requirements are generally met.

> The American Dental Association does not recommend fluoridated bottled water for use by infants because it heightens risk for enamel fluorosis during early tooth development.

WATER

An infant needs about 3 cups (700 to 800 milliliters) of water per day to regulate body temperature and transport oxygen, nutrients, and wastes throughout the body. For the vast majority of infants, human milk or formula supply enough water to keep the infant well hydrated.

Caregivers are especially concerned about dehydration in infants because it can occur very rapidly and have devastating consequences. In the first few days after birth, improper feeding techniques can leave an infant deprived of water and nutrients. Protracted episodes of vomiting or diarrhea can quickly deplete an infant of fluid and electrolytes.

To identify dehydration, look for these signs:

- More than 6 hours without a wet diaper
- Dark-yellow or strong-smelling urine
- Unusually tired and fussy
- Dry mouth and lips
- Absence of tears when crying
- Eyes and soft spot on the head appear sunken
- Cold and splotchy hands and feet

Severe dehydration can result in rapid loss of kidney function and warrants medical intervention. In some cases, hospitalization and intravenous rehydration may be necessary. Most of the time, dehydration can be corrected with special fluid-replacement formulas containing electrolytes, such as sodium and potassium. These oral rehydration solutions (e.g., Pedialyte®) are available in supermarkets and pharmacies to treat mild to moderate dehydration. A physician should guide any use of these products.

Caregivers often wonder if breast milk or formula is sufficient to keep an infant hydrated, especially in hot weather. They may be tempted to give an infant supplemental water or fruit juice. In some stores, bottled water products marketed specifically for infants may be placed alongside infant formulas and electrolyte-replacement solutions, giving the mistaken impression that bottled water products are an appropriate feeding supplement or substitute for fluid replacement for infants. Even in hot weather, the American Academy of Pediatrics does not recommend supplemental water or juice during the first 6 months. Excess water can cause hyponatremia in infants. It is important to remember that excessive fluid can be harmful, especially to the brain.

Overall, it is best to rely exclusively on breast milk or infant formula to meet infant fluid needs up to 6 months of age, unless a physician suggests otherwise. In sum, extremes in fluid intake—either too little or too much—can lead to health problems.

✓ CONCEPT CHECK 15.2

1. Using the EER equations listed in Table 15-2, calculate the calorie needs of a healthy 4-month-old infant who weighs 15 pounds (6.8 kilograms).

2. Why is unaltered cow's milk NOT recommended for infant feeding?

3. Do infants require any supplemental vitamins or minerals? If so, which ones? What is the rationale for their use?

4. You are planning a family trip to the zoo on a hot summer day. How can you keep your 3-month-old baby from becoming dehydrated?

15.3 Guidelines for Infant Feeding

Oral nutrition for infants comes in two forms: human breast milk or infant formula. As we introduced in Chapter 14, breastfeeding is the preferred method of infant feeding (review Table 14-5). Besides its benefits for immune development, improved mother-infant bonding, and lower long-term risk for chronic diseases, breast milk offers optimal nutrition that is uniquely suited for human infants. For mothers who do not breastfeed, whether due to necessity or preference, infant formula is a workable substitute. In fact, formula manufacturers model their products on human milk. In areas of the world with safe water supplies, formula feeding is a safe and nutritionally adequate alternative to breastfeeding.

BREAST MILK IS THE BEST MILK

Human milk is uniquely suited to meet the nutritional needs of human infants. Table 15-3 gives the composition of human milk, but recognize that these numbers are just estimates. Maternal diet and nutritional status may impact the composition of breast milk, particularly for fatty acids and some micronutrients. In addition, the true composition of human milk changes over time as the infant matures and even within a feeding.

Breast milk provides up to 55% of total calories as fat. Fat is a dense source of calories, so it can meet the high energy needs of the growing infant with a small volume of milk. Interestingly, the fat composition of breast milk changes within each feeding. When the baby first latches on and begins suckling, the consistency of breast milk is thin and watery; the infant takes in necessary carbohydrates, protein, vitamins, and minerals. As the feeding goes on, the fat content of the milk increases to fulfill calorie needs and satisfy the infant until the next feeding. The specific types of fat in human milk are ideal for infants, too. The short- and medium-chain fatty acids in breast milk are easily digested. Some of the fatty acids—AA and DHA—are essential for proper brain and eye development. If the mother's diet is rich in these fats, her milk will be a better source of them for her infant, as well.

Carbohydrates provide about 35% to 40% of the calories in human milk. The main carbohydrate in human milk is lactose, a disaccharide that tastes sweet and is easily digested by the human infant's digestive tract. Although lactase production tends to decline later in life, lactose intolerance is quite rare among infants (review Chapter 4). Human milk also contains some oligosaccharides that have a prebiotic effect on the community of beneficial microorganisms in the infant's gut. A healthy microbial population in the GI tract influences the development of the infant's immune system.

▲ Breastfeeding takes some skill and patience on the part of the mother, especially in the first few weeks, but the physical and emotional benefits are worth the effort.

Protein supplies less than 10% of the total calories in human milk. The kidneys of the newborn infant are still immature, so they can be stressed by high protein intakes. The proteins that are present are easily digested and unlikely to trigger food allergies. These proteins do more than supply calories and building blocks for tissue synthesis. They also promote the proper development of the immune system and enhance nutrient absorption.

For the most part, the micronutrient needs of the infant can be met by human milk. A notable exception is vitamin D. As you learned in Chapter 8, vitamin D needs can be met by sun exposure. However, many infants receive little or no sun exposure because of their geographic location or for fear of skin damage. Given that the vitamin D content of breast milk is quite low and sun exposure is limited, the American Academy of Pediatrics recommends that all infants (breastfed and formula-fed) should receive 400 IU per day of supplemental vitamin D until their dietary intake supplies this amount. For breastfed infants of mothers who follow a vegan diet, have had bariatric surgery, or have pernicious anemia, vitamin B-12 supplements are recommended. Even though the iron content of human milk is low, the form of iron in human milk is highly bioavailable, so healthy, full-term, breastfed infants rarely need iron supplementation. Any evidence of depleted iron stores, especially for preterm infants, warrants iron supplementation.

Breastfeeding during infancy impacts future feeding behaviors, as well. Interesting research shows that the flavors of the mother's diet are transferred into her milk. This can positively affect the infant's acceptance of a variety of solid foods later in life. Breastfeeding also reinforces the natural ability of the infant or child to self-regulate food intake. The infant has to do some work to get milk from the mother's breast, so it is unlikely that the infant will override satiety cues and over-eat from the breast. Much research supports the hypothesis that breastfed infants learn to eat based on internal cues of hunger and satiety, leading to better body weight management and lower risk for cardiovascular diseases and type 2 diabetes throughout life.

FORMULA FEEDING FOR INFANTS

Formula Composition. Infants cannot tolerate cow's milk as such because of its high protein and mineral content. Cow's milk is perfect for the growth needs of calves, but not for human infants. Thus, cow's milk must be altered by formula manufacturers to be safe for infant feeding. Altered forms of cow's milk, known as infant formulas, must conform to strict federal guidelines for nutrient composition and quality. Formulas generally contain lactose and/or sucrose for carbohydrate, heat-treated proteins from cow's milk, and vegetable oils for fat (review Table 15-3). Soy protein–based formulas are available for vegan infants or those who cannot tolerate lactose or the types of proteins found in cow's milk. Infants with milk protein allergies are often sensitive to soy as well, so the best choice for infants with allergies is a hydrolyzed protein formula. In this type of formula, the proteins have been broken down into small polypeptides and amino acids. A variety of other specialized formulas are also available for specific medical conditions. In any case, it is important to use an iron-fortified formula unless a physician recommends otherwise.

Some transition formulas/beverages have been introduced for older infants and toddlers (review Table 15-3). Some of these products are intended for use after 6 months of age if the infant is consuming solid foods, whereas others are intended for use only by toddlers. These transition products are lower in fat than human milk or standard infant formulas; their iron content is higher than that of cow's milk; and their overall mineral content is generally more like that of human milk than cow's milk. According to the manufacturers, the advantages of these transition formulas/beverages over standard formulas for older infants and toddlers include reduced cost and better flavor. Parents should consult their physician with regard to the use of these products.

▲ Bisphenol A (BPA) is a chemical used in the production of many plastics. Human exposure to BPA, mainly through leaching of the chemical from packaging into foods and beverages, is widespread. Concern about exposure stems from animal studies that link BPA with reproductive and developmental defects. However, the consensus among regulatory agencies in the United States and Canada is that current levels of BPA exposure are not harmful, even for infants. Nevertheless, in response to public concern, FDA banned use of BPA in the manufacture of baby bottles and sippy cups in 2012.

Formula Preparation. Some infant formulas come in ready-to-feed form. These are poured into a clean bottle and fed immediately. Room-temperature formula is acceptable for many infants. Otherwise, to warm a bottle of formula, a caregiver can run hot water over it or place it briefly in a pan of simmering water. Infant formulas should not be heated in a microwave oven because hot spots may develop, which can burn the infant's mouth and esophagus.

Powdered and concentrated fluid formula preparations are also commonly used. All utensils used to prepare formula should be washed and thoroughly rinsed. Powdered or concentrated formulas should be combined with clean, cold water, exactly following the directions on the formula label. The formula is then warmed, if desired, and fed immediately to the infant. Hot water from the faucet should not be used to make formula, since it poses a risk for high lead content (see Chapter 13). Cold water poses much less risk. For infants up to 6 months of age, pediatricians commonly recommend boiling (and then cooling) the water to be used in formula preparation and sterilizing bottles and utensils by immersion in boiling water.

Refrigerating prepared formula for 1 day is safe. However, formula left over from a feeding should be discarded because it will be contaminated by bacteria and enzymes from the infant's saliva. If well water is used, it should be boiled before making formula for at least the infant's first 3 months of life, and it should be analyzed for excessive concentration of naturally occurring nitrates, which can lead to a severe form of anemia. If nitrates are high in municipal water systems, consumers will be warned (such as in a local newspaper) not to use the water for making infant formula until the concentration falls to a safe amount. The American Dental Association does not recommend that formula be mixed with bottled nursery water, available alongside infant formula in most supermarkets, to limit risk of tooth discoloration from high fluoride levels.

▲ Careful attention during feeding allows the caregiver to notice the infant's signal as to when the feeding should cease.

FEEDING TECHNIQUE

Infants swallow a lot of air as they ingest either formula or human milk. To alleviate discomfort, it is important to burp an infant during feeding (every 1 to 2 ounces) and again at the end of the feeding. Spitting up a bit of milk is normal at this time.

When the infant begins acting full, bottle feeding should be stopped, even if some milk is left in the bottle. Common cues that signal that an infant has had enough include turning the head away, being inattentive, falling asleep, and becoming playful. Generally, the infant's appetite is a better guide than standardized recommendations concerning feeding amounts. Breastfeeding infants usually have had enough to eat after about 20 minutes. Although it is difficult to tell how much milk breastfed infants are getting, they also give signs when full. By carefully observing bottle-feeding or breastfeeding infants and responding to their cues appropriately, caregivers not only can be assured that the infants' calorie needs are being met but also can (1) foster a climate of trust and responsiveness and (2) help a child develop a habit of respecting internal cues of hunger and satiety.

EXPANDING THE INFANT'S MEALTIME CHOICES

By about 6 months of age, the infant is ready to start eating "table food." Initially, table foods add to—rather than replace—human milk or formula. In the first attempts to introduce solid foods, just getting the food into the infant's mouth may prove to be a challenge. By the end of the first year, though, the infant should be eating a variety of protein sources, vegetables, fruits, and grains so that the diet begins to reflect a balanced pattern (Table 15-4). Throughout the process of expanding the infant's mealtime choices, the caregiver must proceed slowly and respond to the infant's cues that he or she is hungry or has had enough to eat. Feeding habits developed through early exposures to food will set the stage for healthy eating to last a lifetime.

TABLE 15-4 ▶ Sample Daily Menu for a 1-Year-Old Child*

Breakfast	Snack
1 to 2 tbsp unsweetened applesauce	½ ounce cheddar cheese
¼ cup Cheerios	4 wheat crackers
½ cup whole milk	½ cup whole milk
Snack	**Dinner**
½ hard-cooked egg	1 ounce hamburger (crumbled)
½ slice wheat toast with ½ tsp margarine	1 to 2 tbsp mashed potatoes with ½ tsp margarine
½ cup mandarin orange segments	
½ cup water	1 to 2 tbsp cooked carrots (cut in strips, not coins)
	½ cup whole milk
Lunch	**Snack**
1 ounce roasted chicken, minced	½ banana
1 to 2 tbsp rice with ½ tsp margarine	2 oatmeal cookies (no raisins)
1 to 2 tbsp cooked peas	½ cup whole milk
½ cup whole milk	

Nutritional Analysis	
Total energy (kcal)	1100
% energy from	
Carbohydrate	40%
Protein	19%
Fat	41%

*This diet is just a start. A 1-year-old may need more or less food. In those cases, serving sizes should be adjusted. The milk can be fed by cup; some can be put into a bottle if the child has not been fully weaned from the bottle.

Recognizing the Infant's Readiness for Solid Foods. Parents may believe that the early addition of solid foods will help an infant sleep through the night. Actually, this achievement is a developmental milestone, and the amount of food consumed by the infant is of little relevance to the achievement of a good night's sleep. Before 4 to 6 months of age, infants are not physically mature enough to consume much solid food. Only occasionally does a rapidly growing infant need solid foods to meet calorie and nutrient needs before 6 months of age.

How does the caregiver know it is time to introduce solid foods? Infant size can serve as a rough indicator of readiness: reaching a weight of at least 13 pounds (6 kilograms) is a preliminary sign of readiness for solid foods. Another physiological cue is frequency of feeding, such as consuming more than 32 ounces (1 liter) of formula daily or breastfeeding more than 8 to 10 times within 24 hours. Underlying these noticeable signals are several important developmental factors:

1. *Nutritional need.* Before the infant is 6 months old, nutritional needs can generally be met with human milk and/or formula. After 6 months of age, however, many infants need the additional calories supplied by solid foods. In terms of individual nutrients, iron stores are exhausted by about 6 months of age. Either solid foods or iron supplements are then needed to supply iron if the child is breastfed or fed a low-iron or iron-free formula. (As previously mentioned, a vitamin D supplement should also be provided.)

2. *Physiological capabilities.* As the infant ages, the ability to digest and metabolize a wider range of food components improves. Before about 3 months of age, an infant's digestive tract cannot readily digest starch. Also, kidney function is limited until about 4 to 6 weeks of age. Until then, waste products from excessive amounts of dietary protein or minerals are difficult to excrete.

▲ Iron-fortified rice cereal is recommended as the first solid food to be fed to infants.

3. *Physical ability.* Three physical markers indicate that a child is ready for solid foods: (1) the disappearance of the extrusion reflex (thrusting the tongue forward and pushing food out of the mouth), (2) head and neck control, and (3) the ability to sit up with support. These usually occur around 4 to 6 months of age, but they vary with each infant.

4. *Allergy prevention.* An infant's intestinal tract is "leaky"; whole proteins can readily be absorbed from birth until 4 to 5 months of age. If the infant is exposed too early to some types of proteins—particularly those in cow's milk and egg whites—the infant may be predisposed to future allergies and other health problems, such as diabetes. For this reason, it is best to minimize the number of different types of proteins in an infant's diet, especially during the first 3 months (see the Nutrition and Your Health section on food allergies at the end of this chapter for details).

With these considerations in mind—nutritional need, physiological and physical readiness, and allergy prevention—the American Academy of Pediatrics recommends that solid foods not be introduced until about 6 months of age and that infants receive no unaltered cow's milk before 1 year.

Foods to Match Needs and Developmental Abilities During the First Year. If solid foods are introduced before 6 months of age, the primary goal of the food should be to meet iron needs. Therefore, the first solid foods should be iron-fortified cereals. Some pediatricians may recommend lean ground (strained) meats for more absorbable forms of iron. Rice is the best cereal to begin with because it is least likely to cause allergies.

When starting solid foods, it is important to start with a teaspoon serving size of a single-ingredient food item, such as rice cereal, and increase the serving size gradually. Once the new food has been fed for about a week without ill effects, another food can be added to the infant's diet. At first, this can be another type of cereal or perhaps a cooked and strained (or mashed) vegetable, meat, fruit, or egg yolk.

Waiting about 7 days between the introduction of each new food is important because it can take that long for evidence of an allergy or intolerance to materialize. Also, it is important to avoid introducing mixed foods until each component of the combination dish has been given separately without an adverse reaction. Signs of food allergies include diarrhea, vomiting, a rash, or wheezing. If one or more of these signs appears, the suspected problem food should be avoided for several weeks and then reintroduced in a small quantity. If the problem continues, a physician should be consulted. Fortunately, many babies outgrow food allergies later in childhood.

Until recently, parents and caregivers were advised to avoid feeding children a wide range of highly allergenic foods, including egg whites, chocolate, peanuts, tree nuts, fish, and other seafood. Now, the American Academy of Pediatrics acknowledges that there is no evidence that delaying introduction of solid foods—including these common food allergens—beyond 6 months of age is of any benefit for prevention of food allergies and other **atopic diseases.**

Many strained foods for infant feeding are available at the supermarket. Single-food items are more desirable than mixed dinners and desserts, which are less nutrient-dense. Most brands have no added salt, but some fruit desserts contain added sugar, which is not recommended for infant feeding.

As an alternative, plain, unseasoned cooked foods—vegetables, fruits, and meats—can be ground up in an inexpensive plastic baby food grinder. Another option is to puree a larger amount of food in a blender, freeze it in ice-cube portions, store in plastic bags, and defrost and warm as needed. Careful attention to cleanliness is necessary. Seasonings that may please the rest of the family should not be added to infant foods made at home. The infant does not notice the difference if

---

Typical Solid Food Progression, Starting at 6 Months*

Week 1	Rice cereal
Week 2	Add strained carrots
Week 3	Add applesauce
Week 4	Add oat cereal
Week 5	Add cooked egg yolk
Week 6	Add strained chicken
Week 7	Add strained peas
Week 8	Add plums

*Extending the rice cereal step for a month or so is advised if solid food introduction begins at 4 months of age. Also, if at any point signs of allergy or intolerance develop, substitute another similar food item.

atopic disease A condition involving an inappropriate immune response to environmental allergens; examples include asthma, eczema, and seasonal allergies.

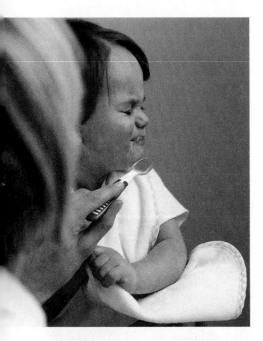

▲ Repeated exposure fosters acceptance of new tastes and textures.

salt, sugar, or spices are omitted. It is best to introduce infants to a variety of foods, so that by the end of the first year, the infant is consuming many foods—human milk or formula, meats, fruits, vegetables, and grains.

To ease early attempts at feeding solid foods, consider the following tips:

- Use a baby-sized spoon; a small spoon with a long handle is best.
- Hold the infant comfortably on the lap, as for breastfeeding or bottle feeding, but a little more upright to ease swallowing. When in this position, the infant expects food.
- Put a small dab of food on the spoon tip and gently place it on the infant's tongue.
- Convey a calm and casual approach to the infant, who needs time to get used to food.
- Expect the infant to take only two or three bites of the first meals.
- Present a new food on several consecutive days to aid an infant's acceptance of that food.

Self-feeding skills require coordination and can develop only if the infant is allowed to practice and experiment. By 6 to 7 months of age, the infant has learned to handle finger foods and transfer objects from one hand to the other with some dexterity. At about this time, teeth also begin to appear. By age 7 to 8 months, infants can push food around on a plate, play with a drinking cup, hold a bottle, and self-feed a cracker or a piece of toast. Through mastery of these manipulations, infants develop confidence and self-esteem. It is important that parents be patient and support these early self-feeding attempts, even though they appear inefficient.

At 9 to 10 months of age, the infant's desire to explore, experience, and play with foods may hinder feeding. Food is used as a means to explore the environment, and therefore, feeding time is often very messy—a bowl of macaroni may end up in the child's hair! Caregivers need to relax and take this phase of infant development in stride. By the end of the first year, finger-feeding becomes more efficient, and chewing is easier as more teeth erupt. Still, experimentation and unpredictability are to be expected.

WEANING FROM THE BREAST OR BOTTLE

Around the age of 6 months, expressed breast milk, formula, or water can be offered in a sippy cup with a wide, flat bottom. Drinking from a cup rather than from a bottle helps prevent **early childhood caries.** If an infant drinks continuously from a bottle, the carbohydrate-rich fluid bathes the teeth, providing an ideal growth medium for bacteria adhering to the teeth. These bacteria then make acids, which dissolve tooth enamel. To avoid dental caries, infants should not be put to bed with a bottle or placed in an infant seat with a bottle propped up.

By about 10 months of age, infants are learning to self-feed and likewise to drink independently from a cup. As children drink from a cup more frequently, fewer bottle feedings and/or breastfeedings are necessary. Infants should begin drinking from a cup by 1 year of age and should be completely weaned from a bottle by 18 months of age. The added mobility of crawling and walking should naturally lead to gradual weaning from the bottle or breast. Even so, getting a baby out of the bedtime-bottle habit can be difficult. Determined caregivers can either wince through a few nights of their baby's crying or slowly wean the baby away from the bottle with either a pacifier or water (for a week or so).

WHAT TO FEED AN INFANT

It can be difficult for new parents to make sense of nutrition goals for infants in the face of changing dietary recommendations from health authorities, cultural preferences, and outdated advice from friends and family. In response to various

early childhood caries Tooth decay that results from formula or juice (and even human milk) bathing the teeth as the child sleeps with a bottle in his or her mouth. The upper teeth are mostly affected as the lower teeth are protected by the tongue; formerly called *nursing bottle syndrome* and *baby bottle tooth decay.*

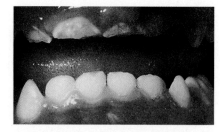

▲ Early childhood caries. An extreme example of tooth decay probably resulting from frequently putting the child to bed with a bottle. The upper teeth have decayed almost all the way to the gum line.

▲ Infants should begin drinking from a cup by 1 year of age. Cups with lids help to prevent spills, but allow a toddler to practice without a lid as his dexterity and coordination improve.

controversies surrounding infant feeding, the American Academy of Pediatrics has issued a number of statements concerning infant diets. The following guidelines are based on these statements:

- *Build to a variety of foods.* For the first 6 months of life, human milk (or infant formula) is all an infant needs. (Vitamin D supplementation is an exception.) When the infant is ready, start adding new foods, one at a time. During the first year, the goal is to teach an infant to enjoy a variety of nutritious foods. A lifetime of healthy eating habits begins with this important first step.
- *Pay attention to your infant's appetite to avoid overfeeding or underfeeding.* Feed infants when they are hungry. Never force an infant to finish an unwanted serving of food. Watch for signs that indicate hunger or fullness. This will reinforce the infant's natural capability to self-regulate food intake.
- *Infants need fat.* Although excessive intake of fat contributes to many adult health problems, it is an essential source of calories for growing infants. Fat also helps the nervous system to develop.
- *Choose fruits, vegetables, and grains, but do not overdo high-fiber foods.* During the second half of the first year, infants should be fed a variety of fruits and vegetables. However, studies show that by 1 year of age, vegetable choices are dominated by white potatoes. Continuing to offer choices of green and yellow vegetables during infancy and the toddler years will enhance intake of important vitamins, minerals, and phytochemicals. In terms of fiber, although many adults benefit from high-fiber diets, they are not good for infants. They are bulky, filling, and often low in calories. The natural amounts of fiber and nutrients in fruits, vegetables, and grains are appropriate as part of a healthy infant diet.
- *Infants need sugars in moderation.* Sugars are an additional source of calories for active, rapidly growing infants. Foods such as human milk, fruits, and small portions of 100% juices are natural sources of sugars and other nutrients as well. Foods that contain artificial sweeteners should be avoided; they do not provide the calories growing infants need. On the other hand, an excessive intake of sugars, particularly from sugar-sweetened beverages, contributes to the epidemic of childhood obesity.
- *Infants need sodium in moderation.* Sodium is a necessary mineral found naturally in almost all foods. As part of a healthy diet, infants need sodium for their bodies to work properly. However, average intakes of sodium among infants and toddlers are above the AI. Caregivers should delay introduction of cow's milk (a natural source of sodium) until 1 year of age and refrain from offering heavily seasoned and processed foods.
- *Choose foods containing iron, zinc, and calcium.* Infants need good sources of iron, zinc, and calcium for optimum growth in the first 2 years. These minerals are important for healthy blood, optimal growth, and strong bones. Many infant and toddler foods (e.g., cereal, crackers, and biter biscuits) are fortified with these minerals.

WHAT NOT TO FEED AN INFANT

Following are several foods and practices to avoid when feeding an infant:

- *Excessive infant formula or human milk.* After 6 to 8 months, solid foods should play a greater role in satisfying an infant's increasing appetite. The main reason to switch is that solid foods contain considerably more bioavailable iron than do human milk and low-iron formulas. About 24 to 32 ounces (¾ to 1 liter) of human milk or formula daily is ideal after 6 months, with food supplying the rest of the infant's calorie needs.

- *Foods that tend to cause choking.* Foods that are round or ovoid in shape, larger than a half-inch in diameter, or of a soft or sticky texture can easily get lodged in a child's throat. These foods include hot dogs, hard or gummy candies, whole nuts, grapes, coarsely cut meats, raw carrots, popcorn, and peanut butter. Caregivers should not allow younger children to gobble snack foods during playtime and should supervise all meals.
- *Potential food allergens before 6 months of age.* Cow's milk, egg whites, peanuts, tree nuts, soy, and wheat are responsible for 90% of food allergies in childhood. If any solid foods are introduced before 6 months of age, they should be iron-fortified rice or oat cereals, pureed meats, vegetables, or fruits.
- *Cow's milk, especially low-fat or fat-free cow's milk.* The American Academy of Pediatrics strongly urges parents not to give children under age 2 fat-reduced, 1%, or fat-free milk. Before age 2, the amount of this milk needed to meet calorie needs would supply too many minerals and in turn could overwhelm the kidneys' ability to excrete the excess. The lower fat intake might also harm nervous system development. Beyond 2 years, children can drink fat-reduced, 1%, or fat-free milk, because by this age they are consuming enough solid foods to supply calorie and fat needs.
- *Goat's milk.* Although perceived by some to pose lower risk for food allergies, goat's milk is low in folate, iron, vitamin C, and vitamin D, and should not be used as a source of nourishment for infants.
- *Feeding excessive fruit juice.* The fructose and sorbitol contained in some fruit juices, especially apple and pear juices, can lead to diarrhea because they are slowly absorbed. Also, if fruit juice or related drink products are replacing formula or milk in the diet, the infant may not be receiving adequate calories, calcium, or other nutrients essential for proper growth. Studies have shown a link between excessive amounts of fruit juice and failure to thrive, GI tract complications, obesity, short stature, and poor dental health. Thus, these substances should be used sparingly. Limit fruit juice to 4 to 6 ounces per day for infants from 6 months to 6 years of age.
- *Food safety hazards.* The immune system is still maturing during infancy and early childhood, so it is important to avoid potential sources of foodborne illness (review Chapter 13). For example, raw (unpasteurized) milk or soft cheeses (e.g., queso fresco) may be contaminated with bacteria or viruses. Meat, poultry, eggs, and seafood should be cooked to proper temperatures. In addition, honey may contain spores of *Clostridium botulinum,* which can lead to the potentially fatal foodborne illness known as *botulism.* Safe food handling starts with proper handwashing.
- *Excessive nutrient supplementation.* Intake of supplemental vitamins or minerals above 100% of the RDA or AI for age can increase risk for nutrient toxicities.

A Summary of Infant Feeding Recommendations

Breastfed Infants
- Breastfeed for 6 months or longer, if possible. Then introduce infant formula if and when breastfeeding declines or ceases. (Breast milk can also be pumped and placed in a bottle for later use.)
- Provide a vitamin D supplement (400 IU per day).
- Investigate the need for vitamin B-12, fluoride, and iron supplementation to prevent deficiencies.

Formula-Fed Infants
- Use infant formula for the first year of life, preferably an iron-fortified type.
- Provide a vitamin D supplement if formula intake provides less than 400 IU per day.
- Investigate the need for a fluoride supplement if the water supply is not fluoridated.

All Infants
- Add iron-fortified cereal at about 6 months of age.
- Provide a variety of basic, soft foods after 6 months of age, advancing to a varied diet.

✔ CONCEPT CHECK 15.3

1. List three similarities between human milk and infant formula. List three differences.
2. Describe four ways to assess an infant's readiness for solid foods.
3. Excessive intake of added sugars becomes common in late infancy. Describe several ways to limit intake of added sugars in an infant's diet.
4. List three foods to avoid giving to infants during the first year of life.

CASE STUDY Undernutrition During Infancy

Damon is a 7-month-old boy who has been taken into a clinic for a routine checkup. On examination, he was found to be moderately underweight relative to his age and body length. His physician scheduled a follow-up appointment in 3 months. At the 10-month visit, Damon appeared sluggish and was now even more underweight for his age and length.

A registered dietitian interviewed Damon's 16-year-old mother to collect information on Damon's dietary intake. His intake over the previous 24 hours consisted of two bottles of infant formula, three 8-ounce bottles of Kool-Aid, and a hot dog. Damon may have been fed some additional items on the nights that his mother left him with a neighbor, so that she could go out with friends for a few hours. Thus, she was not aware of all that he ate.

Answer the following questions, and check your responses at the end of this chapter.

1. Damon's mother did not specify what type of formula he takes. What questions would you ask about his formula?
2. What potential dangers await Damon if his growth continues to lag behind?
3. What foods should Damon's caregivers offer that are appropriate for his age and nutritional needs?
4. What problems might arise from consumption of sugary drinks from a bottle?
5. Does Damon need any vitamin or mineral supplements?

15.4 Toddlers and Preschool Children: Nutrition Concerns

The rapid growth rate that characterizes infancy tapers off during the toddler and preschool years. The average annual weight gain is only 4.5 to 6.6 pounds (2 to 3 kilograms) and the average annual height gain is only 3 to 4 inches (7.5 to 10 centimeters) between the ages of 2 and 5. As the growth rate tapers off, energy needs decrease and eating behavior changes. For example, among toddlers, the decreased growth rate leads to a decreased appetite, often called "picky eating."

Energy needs (relative to body weight) gradually decline from approximately 100 kcal per kilogram during infancy to about 90 kcal per kilogram for the preschooler. As you can see in Table 15-5, as the child gets older, physical activity level is a major determinant of energy requirements.

Except in cases of poverty or homelessness (review Chapter 12), the diets of toddlers and preschoolers in the United States and Canada are usually adequate. A few nutrients of particular concern among this age group are iron, calcium, and sodium (see Further Reading 14).

Iron. Childhood iron-deficiency anemia is most likely to appear in children between the ages of 6 and 24 months—a time when iron stores from gestation have run out, but intake of iron from food sources may be inadequate. It can lead to decreases in both stamina and learning ability because the oxygen supply to cells decreases. Another effect is lowered resistance to disease. The targeted efforts of the Special Supplemental Nutrition Program for Women, Infants, and Children (WIC) have helped to decrease the occurrence of iron deficiency among children, but it still remains a problem for almost 16% of toddlers and about 5% of preschoolers.

The RDA for iron is 7 milligrams per day for children ages 1 to 3 and 10 milligrams per day for children ages 4 to 8. The best way to prevent iron-deficiency anemia in children is to provide foods that are adequate sources of iron. Even though some animal products are high in saturated fat and cholesterol, the high proportion of heme iron in many animal foods allows the iron to be more readily absorbed than is iron from plant foods. Focus on lean cuts of meat, such as sirloin. Fortified

Quick Guide to Child Nutrition Needs

Carbohydrates
- 130 grams per day to supply energy for the central nervous system and prevent ketosis

Protein
- 13–19 grams per day (ages 1–3)
- 34–52 grams per day (older children)

Fat
- at least 5 grams per day of essential fatty acids
- 30%–40% of total kcal (ages 1–3)
- 25%–35% of total kcal (older children)

TABLE 15-5 ▶ Approximate Energy Needs of Toddlers and Preschoolers

	Boys				Girls		
	Physical Activity				Physical Activity		
Age	Less Than 30 Minutes per Day	30 to 60 Minutes per Day	More Than 60 Minutes per Day	Age	Less than 30 Minutes per Day	30 to 60 Minutes per Day	More Than 60 Minutes per Day
2	1000	1000	1000	2	1000	1000	1000
3	1200	1400	1400	3	1000	1200	1400
4	1200	1400	1600	4	1200	1400	1400
5	1200	1400	1600	5	1200	1400	1600

Source: Health and Nutrition Information for Preschoolers, **www.ChooseMyPlate.gov**.

breakfast cereals also contribute to meeting iron (and other nutrient) needs. Consuming a vitamin C source along with the less readily absorbed iron in plants and supplements aids absorption. While dietary changes can be effective for preventing iron-deficiency anemia, supplementation will be required to correct existing anemia (review Chapter 9).

Calcium. Childhood is a period of rapid bone growth and mineralization. As you learned in Chapter 9, bone mass can only accrue until the early twenties; after age 30, bone mass starts to decline. It is vital, therefore, to maximize bone mass during childhood and adolescence. The RDA for calcium for ages 1 to 3 is 700 milligrams per day. Between the ages of 4 and 8, calcium needs increase to 1000 milligrams per day. However, national surveys of food intake show that the diets of children fall short of the RDA for this important nutrient. Milk and other dairy products are the primary source of calcium in the diets of children, but unfortunately, milk consumption has declined as intake of sweetened beverages has increased. Two cups per day of milk will help toddlers and preschoolers meet their requirements for bone-building nutrients. Children up to 2 years of age should drink whole milk because they need the extra fat for energy, but after 2 years of age, reduced-fat or fat-free milk is preferred. For children who do not consume dairy products, whether due to choice or necessity, there are alternative sources of calcium and other bone-building nutrients. Fortified beverages, such as soy milk, almond milk, or orange juice, can supply as much calcium per serving as cow's milk. Some legumes and vegetables are sources of calcium, as well, but the mineral is not as bioavailable as it is from dairy foods.

Sodium. While iron and calcium intakes fall short of needs in preschool children, excessive sodium intake is a concern (see Further Reading 10). High intakes of fast foods and processed foods elevate sodium intakes to about 1000 milligrams per day more than preschoolers need. Caregivers can lower sodium intake by limiting salt added during cooking and at the table; cutting back on use of processed foods (e.g., luncheon meats and hot dogs); rinsing canned beans and vegetables before cooking; and encouraging consumption of fruits, vegetables, and whole grains in place of prepackaged snacks.

Feeding skills are an important part of physical and cognitive development. Young children explore their environment through the tastes and textures of foods, develop dexterity using utensils and drinking from a cup, and begin to express their autonomy by refusing certain foods. At this time in life, children are also testing boundaries to find out what is acceptable in their little corner of the world.

▲ Cow's milk is a bioavailable source of calcium and vitamin D for toddlers and preschoolers, but overreliance on milk can crowd out other nutrient-dense foods. Children who drink more than 3 cups of milk per day are likely to have poor intakes of iron and fiber.

MyPlate Kids' Place

ChooseMyPlate.gov

▲ USDA recently released a MyPlate website just for kids, featuring educational games, music, and other activities.

Renowned child nutrition expert Ellyn Satter, MS, RD, explains, "Parents are responsible for the *what, when,* and *where* of feeding; children are responsible for the *how much* and *whether* of eating."

Messy meal times, food refusals, and food jags can be sources of tension in families. Creating a more harmonious family atmosphere at mealtime is an important way to keep these behaviors from becoming serious feeding problems (see the next section). Caregivers must understand that these are normal phases of child development but should also be consistent about setting limits for behavior at the dinner table.

Because of the preschool child's reduced appetite, planning a diet that meets nutrient needs poses a special challenge to caregivers. Nutrient density is an important consideration for this age group. Overall, parents should focus on offering a variety of healthy choices, allowing the child to exert some autonomy over the specific type of food and the amount eaten.

MyPlate is a useful, easy-to-understand tool for children. The *proportions* apply to all ages, even though the *portions* will be smaller for children. Table 15-6 gives a general food plan that conforms to MyPlate proportions and is appropriate for preschool and school-age children. Until a child is about 5 years of age, portion sizes in the vegetables group, fruits group, and protein group should be about 1 tablespoon per year of life and can be increased as needed. The same advice does not apply to the grains or dairy groups, but consuming too much milk can leave the diet short on iron.

Luckily, normal-weight children have a built-in feeding mechanism that adjusts hunger to regulate food intake at each stage of growth. If a child is developing and growing normally and the caregiver is providing a variety of healthful foods, all can be confident of the child's well-being.

It is important to promote a healthy attitude about eating. While caregivers will want to focus on nutrient-dense foods, there is no reason to be overly restrictive about child food choices. In fact, when parents are extremely controlling about the family's food intake, children may be at risk for body dissatisfaction and disordered eating. There is room for occasional indulgences, a skipped meal or two, or once in a while "less than ideal" choices. It is eating and lifestyle habits over the course of a month (and lifetime) that matter. Children master their eating when adults set a good example, provide opportunities to learn, give support for exploration, and limit inappropriate behavior.

In the next few sections, we will consider some typical complaints and concerns of parents, explore the causes, and make suggestions for achieving optimal nutrition during the toddler and preschool years. Beyond these pages, USDA's MyPlate (**www.ChooseMyPlate.gov**) and Team Nutrition (**www.teamnutrition.usda.gov**) offer resources for planning nutritious, age-appropriate meals and snacks. Additional nutrition guidance, starting with pregnancy and extending through the preschool years, is available from the Start Healthy, Stay Healthy Resource Center at **www.gerber.com**.

UNDERSTANDING "PICKY EATING"

Many parents are baffled by their toddler's erratic eating behaviors. Toddlers and preschoolers tend not to eat as much or as regularly as infants. One day, young children may pick at their food and staunchly refuse to eat their green beans, but on the next day, they might ask for a second helping. Parents often need reminding that toddlers and preschoolers cannot be expected to eat as voraciously as infants or to eat adult-size portions. Because the growth rate slows after infancy, a toddler's drive to eat is not so intense. In addition, children are sometimes more interested in playing and exploring than eating!

Youngsters also tend to be wary of new foods. One reason is that they have more taste buds, and their taste buds are more sensitive than those of adults. A

TABLE 15-6 ▶ Food Plans for Children Based on MyPlate Daily Food Plans

Food Group	Serving Size	Approximate Number of Servings[1]				
		Age 2[2]	Age 5[3]	Age 8[3]	Age 12[3,4]	Age 16[3,4]
Grains	ounce	3	5	5	6–7	6–10
Vegetables	cup	1	1.5	2	2.5–3	2.5–3.5
Fruits	cup	1	1.5	1.5	2	2–2.5
Dairy	cup	2	2.5	3	3	3
Protein	ounce	2	4	5	5.5–6	5.5–7
Oils	teaspoon	3	4	5	6	6–8
Solid fats and added sugars	kcal	up to 140	up to 120	up to 120	up to 260–270	up to 260–400

[1]Log on to **www.chooseMyPlate.gov** for other ages and other physical activity levels.

[2]Based on less than 30 minutes of physical activity.

[3]Based on 30–60 minutes of physical activity.

[4]The lower amounts are appropriate for girls.

general distrust of unfamiliar things is common in this age group. Thus, familiarity plays an important role in food acceptance. Adults can encourage young children to broaden their food repertoire by repeating exposure to new food choices. It may take 10 or more exposures to a new food before a child finds it acceptable, but if adults can be patient and persevere, children will build good food habits.

Food preferences change rapidly in childhood and are influenced by food temperature, appearance, texture, and taste. The following are a few practical tips for improving acceptance of nutrient-dense foods.

- Build on what they know. Pairing a new food item with a familiar one can help to foster acceptance of the new food.
- Enlist the child's aid in food selection and preparation. For example, let the child pick out the tomatoes and squash from the local farmers market.
- Serve meals on a sectioned plate. Sometimes children object to having foods mixed, as in stews and casseroles, even if they normally like the ingredients separately.
- Keep it crunchy. Certain food characteristics, such as crisp textures and mild flavors, are appealing to children. Kids who reject mushy, cooked carrots may enjoy them raw or lightly steamed. (After about age 4, children can safely eat raw vegetables without fear of choking.)
- Finger foods are fun. Preschoolers eventually develop skill with spoons and forks and can even use dull knives, but it is still a good idea to serve some finger foods, especially with healthy dips such as yogurt sauce or hummus.
- Save the best for last. If a child is prone to leave his or her chicken on the plate untouched, serve the chicken first. Hunger is the best means of getting a child to eat!

The dinner table should not become a battleground. Caregivers should avoid nagging, forcing, and bribing children to encourage eating. Using dessert as leverage to get a child to eat vegetables tends to elevate the status of dessert, while making the vegetables seem less palatable. Indirectly, such tactics reinforce picky-eating behaviors due to the added attention given to them. In addition, pressuring children to clean their plates may teach them to override their internal cues for

▲ Letting a child help select or prepare her food can increase the likelihood she will eat her food.

Food jags, common among preschoolers, are no cause for alarm. A child may switch from one specific food focus to another with equal intensity (older infants may also act this way). If the caregiver continues to offer choices, the child will soon begin to eat a wider variety of foods again, and the specific food focus will disappear as suddenly as it appeared.

satiety. One reasonable policy is the one-bite rule: children should take at least one bite or taste of the foods presented to them. Also, caregivers should realize that what they *do* speaks louder than what they *say*; if caregivers eat a variety of nutrient-dense foods, children will imitate their behavior. When a child refuses to eat, it is best not to overreact. Doing so may give the child the idea that not eating is a means of getting attention or manipulating a scene. Most children do not starve themselves to any point approaching physical harm. When children refuse to eat, have them sit at the table for a while; if they still are not interested in eating, remove the food and wait until the next scheduled meal or snack. Finally, instead of giving attention to negative behaviors, focus on positive behaviors when they occur. Praise a child for trying something new. Overall, mealtime should be a relaxed, pleasant opportunity to enjoy healthful foods and good company (see Further Reading 1).

Although picky eating is usually just a manifestation of slower growth and a desire for autonomy, a child's sudden loss of appetite may be reason for concern. A poor appetite may be a sign of underlying illness, such as an infection or gastrointestinal problem. Be alert for signs of eating disorders, as well. Extreme, self-imposed dietary restrictions could be an early sign of anorexia nervosa (review Chapter 11). Alternatively, **avoidant/restrictive food intake disorder** (ARFID) is an eating disorder primarily diagnosed among children. With this disorder, a child lacks interest in eating specific foods or all foods in general, which leads to weight loss or failure to grow as expected, as well as many nutrient deficiencies. The restricted food intake is much more severe than the picky eating or intermittent food refusal typical of childhood; children with this disorder become malnourished and may need tube feeding or intravenous feeding. It is most likely related to stress, anxiety, or depression. In some cases, the disorder may be a response to an adverse experience, such as choking or vomiting after eating a certain food.

REDEFINE SNACKING

Parents may be concerned that frequent snacking will prevent children from eating well at mealtimes. However, children have small stomachs and need to eat every 3 to 4 hours. Sticking to three meals a day offers no special nutritional advantages; it is just a social custom. Instead, offering five or six small meals can help children meet their nutritional requirements more successfully than limiting them to three meals each day. If the stretch between lunch and dinner is 6 hours, an afternoon snack about 2 hours before dinner could provide some needed nutrients and may preempt a cranky attitude at the evening meal.

When we eat is not nearly as important as *what* we eat. Perhaps families simply need to redefine snacking altogether. A snack should not be synonymous with an indulgent dessert. Rather, a snack should be a small meal of nutrient-dense foods. Let hunger—not the clock—guide the timing of meals and snacks. It is important that these snack choices be planned ahead in order to have healthy choices available (Table 15-7). Fruits and vegetables (fresh, frozen, or canned) and whole-grain breads and crackers are good snack choices. Working parents should make sure their children are provided with nutritious snacks to tide them over until dinnertime.

The location of snacking is important. Sitting calmly at the table instead of running around the house will decrease the risk for choking. Limiting distractions by turning off the television will help to prevent mindless eating. The caregiver could offer two or three nutrient-dense options and allow the child to choose one; responsibility for food choices by the child should start at an early age. Lastly, caregivers should promote handwashing and good oral hygiene just as for a meal.

avoidant/restrictive food intake disorder Eating disorder characterized by failure to meet energy or nutrient needs, resulting in significant weight loss, nutritional deficiencies, or dependence on tube or intravenous feeding; the eating disturbance is not better explained by lack of available food, a medical problem, or another eating disorder.

Choking is a very preventable hazard for young children. Some suggestions for caregivers include:

- Set a good example at the table by taking small bites and chewing foods thoroughly.
- Have children sit at the table, take their time, and focus on their food during meals and snacks.
- Avoid giving children any foods that are round, firm, sticky, or cut into large chunks, especially before molars emerge (around age 4). Some examples of foods to avoid are nuts, grapes, raisins, popcorn, peanut butter, and hard pieces of raw fruits or vegetables.

TABLE 15-7 ▶ Twenty Healthy Snack Ideas for Children

	Iron	Zinc	Calcium	Vitamin C	Fiber
Almonds (1 oz)*			✓		✓
Unsweetened applesauce (1/2 cup)				✓	✓
Bean and cheese burrito (1)	✓	✓	✓		✓
Cheese (1 oz) and whole-wheat crackers (6)	✓	✓	✓		✓
Dried cranberries (1/4 cup)				✓	✓
Frozen fruit pieces (1 cup)				✓	✓
Fruit salad (1 cup)				✓	✓
Fruit smoothie with bananas and strawberries (1 cup)				✓	✓
Hard-boiled egg	✓	✓			
Hummus (2 tbsp) with bell pepper rings (1 cup)				✓	✓
Low-fat microwave popcorn (3 tbsp unpopped)*					✓
Mini-pizzas on whole-grain English muffins (2)	✓	✓	✓	✓	✓
Peanut butter (2 tbsp) and apple slices (1 cup)*		✓			✓
Quick breads, such as banana bread, 1 slice	✓				✓
String cheese (1 stick)		✓	✓		
Trail mix (1/4 cup)*	✓	✓			✓
Tuna salad (1/2 cup) in whole-wheat pita pocket	✓	✓			✓
Whole-grain cereal (1 cup)	✓	✓		✓	✓
Whole-wheat pasta salad with veggies (1 cup)	✓	✓			✓
Yogurt (8 oz) with granola (2 tbsp)			✓		✓

*Snack items that are best suited for children older than age 4 due to potential for choking.

CHOOSE DIETARY SUPPLEMENTS CAREFULLY

Major scientific groups, such as the Academy of Nutrition and Dietetics and the American Society for Nutrition, state that multivitamin and mineral supplements are generally unnecessary for healthy children; it is better to emphasize good foods. In fact, consuming fortified foods and supplements may lead to intakes above the UL for some nutrients, such as vitamin A and zinc. Children's supplements that are made to look like candy may result in accidental overdose, particularly of iron. Fortified ready-to-eat breakfast cereals with milk are especially helpful in closing any gap between current micronutrient intake and needs, such as for folate, vitamin D, vitamin E, iron, or zinc.

For a child who is ill, has a very erratic food preference pattern or appetite, or is on a weight-loss diet, the American Academy of Pediatrics states that the child may benefit from a children's multivitamin and mineral supplement not exceeding 100% of Daily Values on the label. Still, as mentioned many times in this textbook, such a practice does not substitute for an otherwise healthy diet—children included. If current childhood feeding practices are to become more healthful, the focus should be on whole-grain breads and cereals, fruits, vegetables, and low-fat milk and milk products.

REDUCE LEAD POISONING

Humans may be exposed to lead from drinking contaminated water, consuming or inhaling lead dust (e.g., from cracked and peeling lead paint), contaminated dietary supplements (e.g., calcium supplements derived from bone meal), or foods stored or prepared in lead-containing vessels. In the United States, nearly half a million children between the ages of 1 and 5 have unacceptably high blood lead levels. Young children are particularly susceptible to lead poisoning because they are small, absorb lead quickly, spend a lot of time on the floor, and are apt to put objects in their mouths. In the short-term, symptoms of lead poisoning include gastrointestinal distress, lack of appetite, irritability, fatigue, and anemia. Over the long-term, devastating effects include intellectual and behavioral impairments and increased risk for several chronic diseases in adulthood.

Although it does not address the source of exposure, proper nutrition can reduce the risks of lead poisoning for children. Consuming regular meals, moderating fat intake, and ensuring adequate iron and calcium status are dietary practices known to reduce lead absorption. Adequate zinc, thiamin, and vitamin E intakes also reduce the harmful effects of absorbed lead. For lowest lead levels, only cold water should be used for drinking and preparation of formula or food. Letting cold water run from the tap for 2 to 3 minutes after a long period of inactivity (e.g., overnight) will limit the amount of lead that has accumulated in tap water. If the public water supply contains a high concentration of lead, bottled water is a safer alternative, particularly for formula preparation. Overall, a balanced meal plan that offers a variety of whole grains, lean meats, and low-fat dairy products is especially useful for protecting children from lead poisoning.

OVERCOME CONSTIPATION WITH LIFESTYLE CHANGES

Constipation, a common problem among children, can be defined as hard, dry stools that are difficult to pass. Typically, a 4-year-old child has one bowel movement per day, but normal bowel habits vary widely. Therefore, the frequency of bowel movements is not as important as the consistency of stools. Pediatricians diagnose constipation after two or more weeks of delayed or difficult bowel movements. In rare situations, constipation can be a sign of a serious problem. If a child has a fever or vomiting along with constipation, if there is blood in the stool, or if the abdomen becomes swollen, caregivers should seek immediate medical attention.

What causes constipation? Although there could be a serious medical problem, most cases are related to lifestyle. Lack of physical activity contributes to constipation. In addition, on average, children (and adults) in the United States barely obtain half of the AI for fiber. Altered bowel habits also may be a sign of a food allergy or intolerance to a food component such as cow's milk. The majority of the time, however, constipation results from the child withholding bowel movements. For children, a painful bowel movement can be so traumatic that they try to resist subsequent bowel movements. The longer they hold their stools, the harder and drier they get, leading to another painful movement. This cycle disrupts regular bowel habits, leading to distress and, if not treated, **fecal impaction.**

When presented with a constipated child, a physician first has to rule out a medical cause, such as an intestinal blockage. Treatment of fecal impaction may require evacuation of the bowels (e.g., with an enema). Once bowels have been evacuated, lifestyle changes are necessary to prevent future problems. Although various types of laxatives may be prescribed by the physician in the short term, diet and lifestyle are the safest strategies over the long term. First, regular bowel habits must be established. For example, parents should set aside time for the child to use the toilet, without rushing, after each meal. Rewards, such as stickers on a chart, may be used to reinforce good habits. Increasing physical activity while cutting back on sedentary activities (e.g., watching television or playing video games) can help to promote regular bowel movements. The primary dietary interventions to alleviate constipation include eating more fiber and drinking more fluids. In the initial stages of treatment, providing certain fruit juices (e.g., prune, grape, and apple) and substituting soy milk for cow's milk may relieve constipation.

Objectives of *Healthy People 2020* include reducing mean blood lead level by 10% and eliminating elevated blood lead levels in children.

fecal impaction The presence of a mass of hard, dry feces that remains in the rectum as a result of chronic constipation.

Ultimately, whole fruits (e.g., plums, peaches, and apricots) are better choices than juices because whole fruits are less concentrated sources of calories. Other foods to emphasize for fiber include vegetables, whole-grain breads and cereals, and beans. The daily fiber goals for children set by the Food and Nutrition Board vary by age (see margin). Few children meet these goals. It is important to increase fluid consumption along with fiber to avoid another fecal impaction. Accompanying fluid recommendations are 4 cups (900 milliliters) per day for toddlers and about 5 cups (1200 milliliters) per day for older children.

PLAN VEGETARIAN DIETS APPROPRIATELY

Vegetarian diets can pose several risks for young children. These include the possibility of developing iron-deficiency anemia, a deficiency of vitamin B-12, and rickets from a vitamin D deficiency. During the first few years of life, children also may not consume enough calories when following a bulky vegetarian diet. These known pitfalls are easily avoided by informed diet planning (see Nutrition and Your Health: Vegetarian and Plant-Based Diets in Chapter 6). Diets for children who eat totally vegetarian fare should focus on protein, vitamin B-12, iron, and zinc content, with additional emphasis on vitamin D (or regular sun exposure) and calcium. Some of these dietary inadequacies can be compensated for by increasing oils, nuts, seeds, ready-to-eat breakfast cereals, and fortified soy milk in the diet.

PROMOTE GOOD ORAL HEALTH

A proper diet goes a long way in reducing the risk for dental caries in young children. In addition to beginning oral hygiene when teeth start to appear and seeking early pediatric dental care, the following diet-related tips can help reduce dental problems in children:

- Drink fluoridated water (or any water) as opposed to carbohydrate-rich or acidic beverages (e.g., fruit juice, soft drinks, sports drinks, and energy drinks). If sugary or acidic beverages are consumed, it is better to drink them *with* meals rather than *between* meals. Sipping juice continuously between meals (e.g., from a sippy cup) exposes teeth to caries-promoting sugars and acids.
- Use small amounts of fluoridated toothpaste twice daily.
- Snack in moderation. Constant exposure of teeth to sugars and acids throughout the day (i.e., grazing) tends to promote caries.
- Make wise snack choices. We automatically think of sticky, sugary snacks as promoters of dental caries, but foods such as pretzels and popcorn provide a source of carbohydrates for oral bacteria, as well. In contrast, crunchy fruits and vegetables, such as apples or celery, can help to brush away sticky food particles. Snacking on dairy products, such as cheese, can actually buffer the acids that lead to tooth decay.
- If toddlers or preschoolers are chewing gum, sugarless gum is the best choice, as this has been shown to reduce the incidence of dental caries.

LINKS BETWEEN AUTISM AND NUTRITION

Autism spectrum disorder (ASD) is characterized by a range of problems with social interaction, verbal and nonverbal communication, and/or unusual, repetitive, or limited activities and interests. These disorders usually are diagnosed in early childhood and affect an estimated 1 in every 68 children, with higher prevalence in boys than girls. The causes for ASD are not well understood, but there is a definite genetic component.

Food and Nutrition Board Fiber Recommendations for Children

Young Children
1–3 years	19 grams/day
4–8 years	25 grams/day

Boys
9–13 years	31 grams/day
14–18 years	38 grams/day

Girls
9–13 years	26 grams/day
14–18 years	26 grams/day

Chapter 4 noted that it is unlikely that the use of sugar is the cause of hyperactivity or antisocial behavior in most children.

ASD can both affect and be affected by nutritional status (see Further Reading 5). In addition to developmental and behavioral abnormalities, many children with ASD also experience GI disorders, such as constipation, diarrhea, or reflux disease. Such disorders may impair nutrient intake or absorption. Medications used to treat behavioral problems may alter appetite. Some autistic children may have feeding problems related to developmental impairments. Also, picky eating behaviors may affect nutrient intake. Children with ASD can be very rigid with their food selections, rejecting foods or entire food groups based on sensory qualities such texture, color, and temperature. Thus, careful attention to nutrient-dense food choices is of prime importance.

There are many nutrient-based theories concerning the causes and treatment of ASD. Nutritional interventions, such as dietary restrictions or nutrient supplements, are commonly employed by families affected by ASD. A widely used nutritional intervention is the gluten-free, casein-free (GFCF) diet, which eliminates all wheat, barley, rye, and milk products. Proponents of this treatment propose that sensitivities to certain food proteins may have effects on neurotransmitter synthesis, thereby altering nervous system function. Clinical evidence supporting the effectiveness of the GFCF diet is limited, but research continues and anecdotal reports abound. At this time, the American Academy of Pediatrics does not endorse the GFCF diet as a treatment for autism. Imposing additional dietary restrictions on a child who is already picky about food choices may set the child up for nutrient deficiencies. If the GFCF diet is used, families should enlist the aid of a registered dietitian to ensure dietary adequacy, particularly for protein, calcium, vitamin D, folic acid, and some B vitamins.

Other popular therapies for ASD include supplementation with probiotics, vitamins B-6 and B-12, folic acid, magnesium trimethylglycine and dimethylglycine, melatonin, and omega-3 fatty acids. Although sparse, research on these therapies is encouraging. There is evidence of altered absorption or metabolism of nutrients among children with autism, so even with adequate nutrient intake, availability of some nutrients for metabolic processes may be low. Even though these supplements carry a low risk of adverse effects, caution is warranted to avoid overdoses. Because of the rising incidence of ASD and the lack of curative treatments, nutritional interventions for ASD will continue to be an active area of research.

Despite a popularly held belief, scientific evidence does not support a causal relationship between mercury in vaccines and autism. Although small risks are inherent with routine vaccinations, the risks of infectious diseases are far greater.

✔ CONCEPT CHECK 15.4

1. Picky eating is a common complaint among parents of preschoolers. Why is picky eating common at this age? Provide three or more suggestions to help a preschooler choose nutritious foods.

2. How often do preschoolers need to eat throughout the day? List three nutrient-dense snack ideas that would be appropriate for a 3-year-old child.

3. Should toddlers and preschoolers take a multivitamin and mineral supplement? Why or why not?

4. Explain the connections between nutrition and oral health. List three ways to reduce risk for dental caries with healthy eating habits.

5. What is autism spectrum disorder? What are the nutrition concerns of children with ASD?

▲ How does this boy's dinner of broiled fish, broccoli, green beans, carrots, and squash compare to MyPlate? Which groups are missing?

15.5 School-Age Children: Nutrition Concerns

The diets of many school-age students can stand general improvement, particularly with regard to fruit, vegetable, whole-grain, and dairy choices. Drinking minimal amounts of sugared soft drinks is also advised. One survey of U.S. schoolchildren revealed that on the day of the survey, 40% of the children ate no vegetables, except for potatoes or tomato sauce, and 20% ate no fruits. Less than 20% of school-age girls consume adequate calcium. In general, the nutritional concerns and goals applicable to school-age children are the same as those discussed in relation to preschoolers. However, with the added pressures of peers, health messages from the media, and an increasing desire for independence, these goals may be harder to achieve as children grow older. MyPlate's daily food plans, which are tailored to age, gender, height, weight, and activity level, continue to be a good basis for diet planning, with an emphasis on moderating fat and sugar intake and ensuring adequate iron, zinc, and calcium intake (Fig. 15-3). Now let us look at several nutritional issues of particular concern during the school-age years.

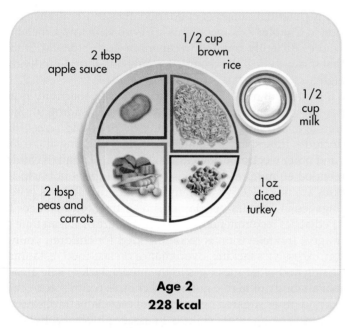

2 tbsp apple sauce
1/2 cup brown rice
1/2 cup milk
2 tbsp peas and carrots
1 oz diced turkey

Age 2
228 kcal

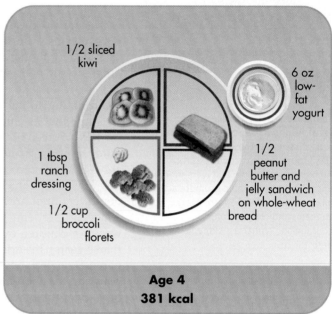

1/2 sliced kiwi
6 oz low-fat yogurt
1 tbsp ranch dressing
1/2 peanut butter and jelly sandwich on whole-wheat bread
1/2 cup broccoli florets

Age 4
381 kcal

12 grapes
1 cup skim milk
3/4 cup corn
beef taco with lettuce, tomato, cheese, and sour cream

Age 8
520 kcal

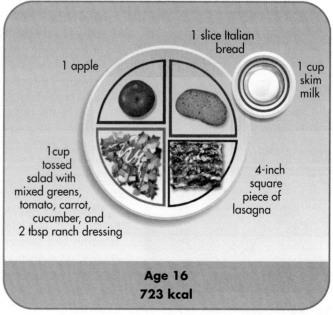

1 apple
1 slice Italian bread
1 cup skim milk
1 cup tossed salad with mixed greens, tomato, carrot, cucumber, and 2 tbsp ranch dressing
4-inch square piece of lasagna

Age 16
723 kcal

FIGURE 15-3 ▲ Using MyPlate to build a healthy meal for children. MyPlate is useful tool for all Americans, ages 2 and older. MyPlate proportions apply to children as well as adults, but portion sizes and food choices may vary by age.

▲ To get kids involved in exercise, new physical education classes have been introduced into schools. These classes provide fitness lessons in such activities as rock climbing, in-line skating, and recreational jogging. These classes help promote activity because they take the focus away from teams and competition, which often discourage and embarrass kids who lack athletic talent.

▲ Supersized portions of foods such as hamburgers and sugared soft drinks are fueling a nation of supersized kids.

REVERSING TRENDS FOR OVERWEIGHT AND OBESITY

By far, the most troublesome nutritional problem facing children today is the rise in childhood obesity. Since the 1970s, the incidence of childhood overweight and obesity has more than tripled such that about one-third of U.S. school-age children are now classified as overweight or obese. The number of cases is increasing, especially in minority populations. In the short run, ridicule, embarrassment, possibly depression, and short stature linked to early puberty are the main consequences of childhood obesity. In the long run, significant health problems associated with obesity, such as cardiovascular disease, type 2 diabetes, and hypertension, usually will appear in adulthood or earlier. Childhood obesity is a serious health threat because about 40% of obese children (and about 80% of obese adolescents) become obese adults. Significant weight gain generally begins between ages 5 and 7, during puberty, or during the teenage years.

Research points to many potential causes of childhood obesity. Recall the nature versus nurture discussion in Chapter 7. Some infants are born with lower metabolic rates; they use calories more efficiently and in turn can more easily store fat. Studies also suggest, though, that this genetic link accounts for only one-third of individual differences in body weight.

Researchers believe that, although diet is an important factor, inactivity is also a contributor to the increase in childhood obesity (see Further Reading 9). Studies show that as children age, physical activity steadily declines and screen time increases. Screen time includes time spent watching television, working at the computer, or using phones and other electronic devices. Only about one-half of children are getting the recommended 60 minutes of exercise per day. It does not help that physical education classes are now elective in many high schools. Today's generation of children now engages in 7 hours per day of screen time, on average. The American Academy of Pediatrics recommends a limit of 14 hours of screen time per week for children over age 2 (no television is recommended for children younger than 2 years). In addition, excessive snacking, overreliance on fast-food restaurants, parental neglect, advertising and other messages in the media, lack of safe areas to play, and the abundant availability of high-calorie food choices contribute to childhood obesity. Soft drinks and other sugared beverages are especially implicated.

The initial approach in treating an obese child is to assess physical activity. If a child spends much free time in sedentary activities (such as watching television or playing video games), more physical activities should be encouraged. The Physical Activity Guidelines for Americans recommend 60 minutes or more of moderate to vigorous physical activity per day for children and adolescents. Learning to engage in and enjoy regular physical activity will help children not only to attain a healthy body weight but also to keep a similar body weight later in life. An increase in physical activity will not just happen; parents and other caregivers need to plan for it. Getting the family together for a brisk walk after dinner encourages healthy habits for all involved. Age-appropriate activities for elementary school-age children include walking, dancing, jumping rope, and participation in organized sports that focus on fun rather than intense competition. For middle school-age children, more complex organized sports (e.g., football and basketball) are of interest, and some weight training with small weights can also be beneficial.

Moderation in calorie intake is important, especially the limitation of high-calorie foods, such as sugared soft drinks and whole milk. The focus should be on more vitamin- and mineral-dense foods and healthy snacks. An emphasis on appropriate portion sizes may help youth learn to curb excessive food intake. Making small changes such as substituting low-fat for whole milk or canned fruit in its own juice instead of heavy syrup can moderately cut calories without sacrificing taste or disrupting normal eating patterns. To specifically address the increased burden of overweight and obesity among minority populations, health professionals must become versed in varied cultural food preferences.

Resorting to a weight-loss diet is usually not necessary; it is best to emphasize changing habits that allow for weight maintenance. Children have an advantage over adults in dealing with obesity: their bodies can use stored energy for growth. An overweight child who maintains his body weight through a growth spurt will end up with a more favorable body composition. This is one reason it is desirable to treat obesity in childhood. If weight loss is necessary in younger children, it should be gradual, about ½ to 1 pound per week. The child should be watched closely to ensure that the rate of growth continues to be normal. The child's calorie intake should not be so low that gains in height diminish. In some cases, medications may be prescribed under a physician's care (e.g., orlistat [Xenical]). For the 1% to 2% of American children who are morbidly obese, bariatric surgery is an option for weight management (see Further Readings 17 and 18).

EARLY SIGNS OF CARDIOVASCULAR DISEASE

Also paralleling the increase in childhood obesity, early signs of cardiovascular disease have become increasingly prevalent among children and adolescents. One out of five youths between the ages of 12 and 19 years has abnormal blood lipids. Therefore, lifestyle modifications to delay the progression of the disease are important throughout the lifespan. The American Academy of Pediatrics now recommends universal blood lipid screening for all children around the ages of 9 to 11 and even earlier screening for "at-risk" children, who are overweight, have high blood pressure, smoke, or have diabetes; have a family history of cardiovascular disease; or whose family history is unknown. For children whose cholesterol is elevated, lifestyle approaches, such as weight management through dietary modification and increased physical activity, are the first line of therapy (see Further Reading 12). Following a meal plan that is based on the Dietary Guidelines and conforms to MyPlate's Daily Food Plan would be appropriate for prevention of cardiovascular disease. Some high-risk children may be candidates for cholesterol-lowering medications.

TYPE 2 DIABETES AMONG YOUTH

Type 2 diabetes was once regarded as an adult condition. As reviewed in the Nutrition and Your Health section, Diabetes—When Blood Glucose Regulation Fails, in Chapter 4, it frequently occurs in overweight people older than 40. However, physicians have noted an alarming increase in the frequency of the disease among children (and teenagers). This is primarily due to the rise in obesity in this age group, coupled with minimal physical activity. Up to 85% of children with the disease are overweight at diagnosis.

Starting at age 10, children who are overweight or obese and who have risk factors for type 2 diabetes should be screened for type 2 diabetes every 2 years. Besides obesity and a sedentary lifestyle, examples of risk factors include having a close relative with the disease or belonging to a nonwhite population. In 2013, the American Academy of Pediatrics released the first-ever guidelines for management of type 2 diabetes in children (see Further Reading 8). These guidelines provide recommendations for monitoring of blood glucose, use of medications, weight management, and physical activity. Dietary management strategies include a regular schedule of meals and snacks; education on appropriate portion sizes; limiting sweetened beverages, high-fat foods, snacks, and fast foods; and focusing on incorporating more fruits, vegetables, and low-fat or fat-free dairy products. For physical activity, experts advise children to engage in moderate- or vigorous-intensity physical activity for at least 60 minutes each day.

▲ For kids who complain about waking up early to make time for breakfast, consider preparing a portable breakfast (e.g., a sandwich bag full of dry cereal, nuts, and dried cranberries) the night before, for grab-and-go convenience the next morning.

START THE DAY WITH BREAKFAST

You have heard it before: *Breakfast is the most important meal of the day.* Yet, as many as one-third of school-age children do not eat breakfast, and the problem gets worse as children reach the teenage years. Children who skip breakfast are missing out on

Newsworthy Nutrition

Skip Soft Drinks, Not Breakfast, to Curb Childhood Obesity

In a 2-year observational study, data were collected from 271 school-aged children to examine links between diet and physical activity habits and childhood obesity. Four main factors emerged as predictors of healthy weight among children: eating breakfast with the family, reducing intake of sweetened beverages, engaging in regular physical activity, and avoiding screen-based sedentary behaviors. The results of this study can guide public policy efforts aimed at preventing childhood obesity.

Source: Carlson JA and others: Dietary-related and physical activity-related predictors of obesity in children: A 2-year prospective study. *Childhood Obesity* 8: 110, 2012.

connect
|NUTRITION

Check out the Connect site **www.mcgrawhillconnect.com** to further explore healthy eating behaviors.

▲ Excessive consumption of fruit juice is linked to obesity, unhealthy blood lipids, fat accumulation in the liver, and dental caries among children. The American Academy of Pediatrics recommends no more than 4 to 6 ounces per day for children 1 to 6 years (and 8 to 12 ounces per day for ages 7 to 18 years).

important nutrients that fuel the brain and the body. A fortified ready-to-eat breakfast cereal is typically the greatest source of iron, vitamin A, and folic acid for children ages 2 to 18. Although there is controversy over the true benefit of breakfast for cognitive ability, children who eat breakfast are more likely to meet their daily needs for vitamins and minerals compared to children not eating breakfast. Also, a growing body of research shows that starting the day with breakfast reduces risk for obesity (see Newsworthy Nutrition).

Fortified, ready-to-eat breakfast cereals offer lots of nutrition in a tasty and convenient package. Consumption of breakfast cereal is responsible for improved intakes of vitamin A and iron. Eating breakfast in general is linked to better intakes of vitamin A, vitamin C, calcium, iron, and fiber. Note that breakfast menus need not be limited to traditional fare. A little imagination can spark the interest of even the most reluctant child. Instead of conventional breakfast foods, parents can offer leftovers from dinner, such as pizza, spaghetti, soups, yogurt topped with trail mix, chili with beans, or sandwiches. For lasting energy and satiety, combine traditional carbohydrate-rich breakfast foods with a source of protein, such as low-fat cheese, nuts, or eggs.

CHOOSE HEALTHY FATS

Diets of school-age children should include a variety of foods from each major group, not necessarily excluding any specific food because of its fat content. Overemphasis on fat-reduced diets during childhood has been linked to an increase in eating disorders and encourages an inappropriate "good food, bad food" attitude.

However, surveys of dietary intake among children show that they are consuming too much saturated fat, most of which comes from whole milk, full-fat dairy products, and fatty meats. Furthermore, few children (or adults) meet recommendations to include two servings of fish per week to ensure adequate intake of omega-3 fats. Emphasizing low-fat dairy products (after age 2), offering broiled or baked fish, choosing leaner cuts of meat, trimming visible fat from meats, and removing the skin from poultry before serving foods will establish heart-healthy eating habits to last a lifetime. Snacks for children should include only moderate fat and sugar and emphasize fruit, vegetables, whole grain, and dairy choices. Ideas for healthy snacks are found in Table 15-7.

SELECT APPROPRIATE BEVERAGES

Maintaining proper hydration is important for children. The fluid needs of school-age children range from 1.7 to 2.4 liters per day, depending on age and gender. However, over the past 30 years, beverage choices have shifted from calorie-free water and nutrient-dense milk to the empty calories provided by sweetened beverages. In fact, sugar-sweetened beverages (e.g, soft drinks, flavored fruit drinks, and sports drinks) and the sweeteners added to flavored milk account for about 200 empty kcal per day for school-age children (see Further Reading 3). The 135% increase in sweetened beverage consumption has paralleled the three-fold rise in childhood obesity since the 1970s. Such high intakes of sugar-laden beverages are not only contributing excess calories, but they are also linked to increased levels of inflammation and worsened blood lipid profiles among children (see Further Reading 13). Even 100% fruit juices, which are perceived by many to be an important source of vitamin C and potassium in children's diets, have been linked to obesity, fatty liver, and metabolic syndrome (see Further Reading 20).

Replacing sugar-sweetened beverages with water and choosing unflavored, low-fat, or fat-free milk instead of flavored milk would reduce sugar intake by about 10.5 teaspoons per day and shrink overall calorie intake by about 10%. Furthermore, replacing 100% fruit juices with whole fruits would supply important nutrients in a lower-calorie package for children. Overall, children should be given water and low-fat or fat-free milk as primary beverage choices. Fruit juice should be limited to 4 to

6 fluid ounces per day for young children or 8 to 12 fluid ounces per day for older children.

PROMOTE SOUND NUTRITION IN SCHOOLS

Children spend the majority of their waking hours in school, so it is a great place to learn about and practice positive, healthy eating habits (see Further Reading 4). A strong emphasis on nutrition education in schools can help children understand why healthy diet habits will make them feel more energetic, look better, and work more efficiently. USDA's Team Nutrition initiative supports child nutrition programs with education materials that promote healthy food choices and physical activity (see **www.teamnutrition.usda.gov**). Most schools have included nutrition education in their health or science curricula, but until recently, these healthy nutrition messages were not consistently backed up by the food offerings in school cafeterias.

In 2010, President Barack Obama signed into law the Healthy Hunger-Free Kids Act, which extended funding for the National School Lunch Program, School Breakfast Program, and several other federal nutrition programs. The law also authorized USDA to make some big changes to the nutritional quality of foods provided in schools. By the beginning of the 2012–2013 school year, public school food service programs had to meet a new set of nutrition standards that stipulate how many servings of fruits, vegetables, and whole grains are offered and replace whole milk with skim or 1% milk. These standards also will gradually reduce the sodium content of meals over the next decade.

Breakfasts and lunches prepared by school cafeterias are not the only targets of school nutrition reforms. In 2014, new standards for the quality of competitive foods sold on school campuses (e.g., from snack bars and vending machines) went into effect. These guidelines set calorie limits on snacks and restrict the levels of saturated fat and sodium in foods that can be sold to students.

These new school nutrition regulations are based on research studies that show how changing the quality of foods offered to students at school can stem the rise in children's BMI. Students' food choices, however, depend a lot on how the food choices taste. School food service programs, although they receive some government reimbursement, often rely on cafeteria, snack bar, or vending machine sales to break even. It remains to be seen how these school nutrition reforms will influence children's eating behaviors.

If we are to trim children's waistlines, improving the nutrition know-how of children is just part of the solution. Only about one-third of total daily calories from Monday through Friday are consumed at school, so positive nutrition influences must extend beyond the classroom and into the home to change persistent trends toward obesity and chronic diseases. Caregivers and other adult role models need to create safe opportunities for children to be active and must practice what they preach when it comes to healthy habits at home.

School breakfasts and lunches provided a nutritional safety net for more than 31 million students during the 2012–2013 school year. More than two-thirds of those meals were provided free or at reduced prices for children from low-income families.

✔ CONCEPT CHECK 15.5

1. Provide an example of a meal that resembles MyPlate and is appropriate for a 7-year-old child.
2. List three lifestyle changes to reduce childhood obesity. In what big way do weight management strategies for children differ from those for adults?
3. Describe appropriate beverage choices for school-age children. What are the implications of excessively consuming sugar-sweetened beverages?
4. Tim refuses to eat breakfast before school. He doesn't like cereal, toast, or any of the other usual breakfast foods. What can Tim's parents do to ensure that he eats nutritious foods before leaving for school?

The physical changes of puberty cause body dissatisfaction for some adolescents. Late-blooming boys may be frustrated with slow gains in height and muscularity, whereas girls can be dissatisfied with gains in fat mass, which are a normal part of development. Be alert for signs of eating disorders (see Chapter 11).

15.6 Teenage Years: Nutrition Concerns

Teenagers are on the cusp of adulthood; parents and schools may still be providing healthful food choices for them, yet they are capable of acquiring and preparing food for themselves. They pursue their independence, experience identity crises, seek peer acceptance, and worry about physical appearance. Advertisers push a vast array of products—candy, fast foods, soft drinks, and energy drinks—at the teenage market. Frequently, these foods crowd out nutrient-dense foods, thus limiting intake of calcium, iron, zinc, fat-soluble vitamins, and folate.

Teens often do not think about the long-term benefits of good health. Developmentally, they have a hard time relating today's actions to tomorrow's health outcomes. Still, healthful teen food habits do not require giving up favorite foods. On occasion, small portions of fast foods and sweet treats can fit in with a diet based on abundant fruits and vegetables, lean sources of protein, fat-free and reduced-fat dairy products, and whole-grain products.

One of the most salient nutritional changes among adolescents is an increase in calorie intake. Most girls begin a rapid growth spurt between the ages of 10 and 13, and most boys experience rapid growth between the ages of 12 and 15. Early-maturing girls may begin their growth spurt as early as age 7 to 8, whereas early-maturing boys may begin growing by age 9 to 10. Nearly every organ and bone in the body grows during this adolescent growth spurt. Girls gain about 10 inches (25 centimeters) in height, and boys gain about 12 inches (30 centimeters). Girls tend to accumulate both lean and fat tissue, whereas boys tend to gain mostly lean tissue. This growth spurt provides about 50% of ultimate adult weight and about 15% of ultimate adult height (review Fig. 15-1).

As the growth spurt begins, teenagers begin to eat more. Physically active teenage boys, especially, seem driven to consume everything in sight! Teenage girls need 1800 to 2400 kcal per day, whereas teenage boys require 2200 to 3200 kcal per day. If teens choose nutrient-dense foods, they can take advantage of their ravenous appetites to easily satisfy their increased requirements for calcium, iron, and zinc. As discussed for younger age groups, MyPlate can serve as a guide (Table 15-6 and Fig. 15-3). Unfortunately, teens very often meet (or exceed) their increased energy needs with empty calories rather than nutrient-dense food options. About one-third of their calories come from solid fats and added sugars.

Fruit and vegetable intake among teenagers is generally dismal; only about one-fourth of high school students regularly consume a minimum of five servings per day of fruits and vegetables. Sadly, potato chips and French fries make up more than one-third of the vegetable servings consumed by teens. Low consumption of fruits and vegetables correlates with inadequate intakes of vitamin A, vitamin C, vitamin E, folate, magnesium, and fiber. Also, as teens (especially girls) trade their glasses of milk for bottles of soft drinks and other sweetened beverages, their intakes of calcium, phosphorus, and vitamin D fall short of recommendations. On the other hand, intakes of saturated fat, cholesterol, sodium, and sugars exceed the recommendations set by the American Heart Association, placing teens at risk for obesity and cardiovascular disease.

Childhood obesity, introduced in Section 15.5, continues to be a major nutritional problem into adolescence. Among children between the ages of 12 and 19, 18.4% are obese. There are gender and ethnic disparities for childhood obesity: black females and Hispanic males are at highest risk. Overweight and obese teens are very likely to become obese adults and to develop comorbid conditions, such as type 2 diabetes, hypertension, cardiovascular diseases, sleep apnea, and joint problems. If a teen is still growing, he or she has an advantage in terms of weight management; by holding body weight steady while height gains are achieved, body mass index will decrease over time. However, if a teen attains ultimate adult height and is still obese, a weight-loss regimen may be necessary. Weight loss should be gradual, perhaps 1 pound per week, and generally follow the advice in Chapter 7.

▲ An active lifestyle coupled with a healthy diet should be part of the teen years. Both habits contribute to bone development and bone strength.

Calcium and Vitamin D. Over the last 20 years, soft drinks have been replacing milk as the preferred beverage among children. Figure 2-1 in Chapter 2 shows the stark contrast between milk and typical soft drinks with respect to calcium and other nutrients. Intake of milk is especially poor among adolescent females, who may view dairy products as a source of unwanted calories. This trend in milk consumption begins early in childhood, but we discuss it here because the gap between needs and actual intake of bone-building nutrients is greatest during the adolescent growth spurt. Less than 10% of girls and less than 25% of boys meet recommendations for calcium intake. To make matters worse, one out of five children is deficient in vitamin D. Meanwhile, the adolescent growth spurt marks a critical time for bone development. Calcium requirements for 14- to 18-year-old girls and boys are 1300 milligrams per day—higher than during any other time of life. As you learned in Chapter 9, failure to maximize bone mineralization during childhood sets the stage for development of osteoporosis later in life.

Three servings per day from the dairy group are recommended for all teenagers and young adults to meet calcium needs. If dairy products are not consumed, alternative calcium sources need to be included. Nondairy sources of calcium include almonds, legumes, some green vegetables, and fortified foods (e.g., fruit juices, cereal, and granola bars). However, it is important to note that these alternative sources of calcium may not provide other important nutrients supplied by dairy products, such as protein, vitamin A, vitamin D, and vitamin B-12.

Iron. About 10% of teenagers have low iron stores or iron-deficiency anemia. Iron-deficiency anemia is a highly undesirable condition for a teen. It can lead to fatigue and a decreased ability to concentrate and learn, such that academic and physical performance suffers. Iron-deficiency anemia sometimes appears in boys during their growth spurt, but adolescent females are at greatest risk of deficiency due to heavy menstrual flow and poor dietary intake. It is important that teenagers choose good food sources of iron, such as lean meats and enriched cereals. Teenage girls, in particular, need to eat good sources of iron (or regularly consume a balanced multivitamin and mineral supplement with iron).

Many of the nutritional issues of adolescents—obesity, snacking, beverage choices, and skipping meals—have been adequately described with reference to younger children. Here, we present a few nutrition dilemmas that pertain especially to teenagers.

▲ The teenage years are noted for snacking. With reasonable food choices, teenagers can have healthful diets.

BREAK THE FAST-FOOD HABIT

It is convenient, casual, inexpensive, and all their friends work there. These are reasons why, on any given day, about 40% of the nation's youth eat food from a fast-food restaurant. Unfortunately, the average trip to a fast-food establishment yields about 300 extra calories, 14 additional grams of fat, and 400 milligrams of sodium *in excess* of typical home-prepared meals for teenagers (see Further Reading 16).

By making some small changes in their food selections, teens can still enjoy the company of friends without detriment to their health. When building a sandwich, opt for one layer of meat instead of double or triple patties and select grilled instead of fried meat. For deli sandwiches, choose moderate portions of lean meats, such as roasted turkey or beef, rather than fatty slices of bologna and salami. Skip the condiments or request them on the side; the mayonnaise on a typical fast-food sandwich supplies about 100 fat-laden kcal. Each slice of cheese supplies another 80 to 100 kcal. When it comes to choosing a side dish, a small baked potato or a garden salad with reduced-fat dressing will provide fewer calories and more nutrients than the typical 500-kcal large serving of fries. Calories from regular soft drinks—especially when free refills are available—can quickly add up. Teens should choose reduced-fat or fat-free milk as a nutrient-dense alternative or opt for calorie-free beverages, such as water or diet soft drinks. For pizza, choose veggie toppings instead of sausage and pepperoni. If possible, order a pizza with low-fat cheese and whole-grain crust.

When burgers are measured in pounds instead of ounces, portion control is an issue. While already large, portion sizes at fast-food establishments continue to grow. Choosing items from the kids' menu can lessen the impact of dining out on adolescent wallets and waistlines. Supersized meals, while they may seem economical, should be avoided unless they are to be divided and shared among friends. By choosing wisely and eating moderate portions, teens can still enjoy the social aspects of dining at fast-food restaurants with friends.

CURB CAFFEINE INTAKE

The combined demands of school, work, extracurricular activities, social commitments, and late-night screen time leave many adolescents looking for a quick pick-me-up. Commonly, they are turning to caffeine, the most widely used stimulant on the planet. Soft drinks, a common choice among youth, provide about 25 milligrams of caffeine per serving. On average, 30% of adolescents report consuming energy beverages, which typically contain between 100 and 200 milligrams of caffeine per serving. Consumption of coffee and tea, which yield about 100 milligrams of caffeine per cup, is on the rise among teens. Various foods, including chocolate and some types of candies or sports nutrition products, contain caffeine as well. Average caffeine intake from all sources is just over 100 milligrams per day among teens. Many consumers are unaware of how much caffeine they are consuming; the exact amount of caffeine is not always listed on energy drink labels because (1) it is not currently required by food labeling laws and (2) some manufacturers (especially of energy drinks) consider it to be part of a "proprietary blend."

For children, the American Academy of Pediatrics advises limiting caffeine intake to 100 milligrams per day, if it is used at all. Some of the negative effects of caffeine at any age are gastrointestinal distress, sleep disturbances, anxiety, increased blood pressure, and irregular heartbeat. For children, in particular, there is concern that excessive caffeine intake could affect normal neurological and cardiovascular development. Furthermore, disturbances in normal sleep patterns could affect growth and learning ability. Alarmingly, there have been thousands of reports of caffeine poisoning—and even some deaths—as a result of excessive intake of energy drinks. Clearly, excessive caffeine intake has no place in the diets of children (see Further Reading 7).

CHOOSING VEGETARIAN DIETS

Teenagers, who strive to forge an identity by adopting dietary patterns different from those of their families, may choose to follow vegetarian diets. As discussed in Chapter 6, vegetarians enjoy many health benefits, including lower body weight and better control of blood glucose and cholesterol. Indeed, an increased focus on plant foods is needed in this population who often misses out on their recommended daily servings of fruits and vegetables. However, teens may not know enough about the alternative diet pattern to keep from developing health problems, such as iron-deficiency anemia. The bulk of a plant-based diet is not as much of a concern for teens as it is for younger children with smaller stomach capacity, but a strictly vegetarian diet must be monitored for adequate energy, protein, iron, vitamin B-12, calcium, and vitamin D (the latter if sun exposure is not sufficient) at any age. These nutrients are particularly important in teenagers, as their diets are often already nutrient-poor.

Teens often cite respect for humane treatment of animals as their main reason for choosing vegetarian diets, but be observant of teens who choose vegetarianism as a method of weight loss. Vegetarianism is sometimes used as a socially acceptable way to restrict the diet and, for some, can be an early sign of disordered eating.

ALCOHOL ABUSE AMONG TEENS

In Section 15.5, we discussed how the beverage choices of school-age children are in need of improvement because they provide too much sugar and not enough micronutrients. The nutrient density of beverages continues to be a problem among teenagers, but a new problem arises: alcohol abuse. Developmentally, adolescents are prone to experimentation, rebellion, and risk-taking, so use of this illegal and dangerous substance is common among teenagers. Results of the national Youth Risk Behavior Survey demonstrate that approximately 20% of teenagers have tried

The proposed Food Labeling Modernization Act of 2013 would require food manufacturers to clearly label caffeine content of foods and beverages.

alcohol by the age of 13. At some point throughout the teenage years, about 70% of teens report drinking alcohol at least once, and 22% report binge drinking.

It is just harmless fun, right? Wrong! The statistics show that alcohol use beginning in adolescence has severe consequences (see Further Reading 15). The adolescent's body and brain are still developing. Exposure to alcohol can decrease brain mass in the area of the brain involved in decision-making, memory, and learning. This is evidenced by academic problems and poor decision-making, which can lead to legal troubles, physical assault, and risky sexual behaviors. The most dangerous consequence of poor judgment is drinking and driving. About 1 in 10 teenagers admits to drinking and driving, a risky behavior that is implicated in about one-third of fatal motor vehicle accidents involving teens. Alcohol also contributes to other causes of accidental injuries and deaths, such as drowning, falls, and burns.

Adolescent alcohol abuse exacts a toll on long-term physical health, as well. Studies show that alcohol abuse beginning during adolescence is a strong predictor of alcohol abuse during adulthood. In Chapter 16, you will learn more about the effects of alcohol consumption on adults. Nutritional status can be affected because alcohol abuse is often accompanied by nutrient-poor diets. Also, weight gain from empty calories increases the risk for obesity-related diseases, such as hypertension and cardiovascular disease. These physical consequences may not surface until later in life, but it is certain that the effects of alcohol on the liver, brain, and cardiovascular system can start early.

We should not turn a blind eye on this problem; alcohol use by teenagers should not be viewed as a normal part of growing up. On the contrary, the physical, emotional, and intellectual consequences of underage drinking can be durable and devastating. Parents and other caregivers should talk to their children about the consequences of alcohol abuse, set clear rules, and monitor their children's behavior.

☑ CONCEPT CHECK 15.6

1. Which two minerals are most likely to be deficient in teen diets? Exemplify two rich food sources of each of these minerals that could help teenagers meet their nutrient needs.

2. Design a meal for a teen that resembles MyPlate and can be purchased from a fast-food restaurant.

3. Are energy drinks safe for consumption by children of any age? Why or why not?

4. List three consequences of alcohol abuse that are specific to adolescents.

Newsworthy Nutrition

Glycemic Index of Food Choices May Influence Acne

About 80% to 90% of teens experience acne to some degree. Although it is popularly believed that nuts, chocolate, and pizza contribute to acne, scientific studies have failed to show a strong role for any of these dietary factors. Based on the results of observational research, two dietary factors stand out as possible contributors to acne: foods with high glycemic index and high consumption of dairy products. Both of these dietary patterns are hypothesized to increase insulin levels, which through various mechanisms lead to increased sebum production and altered skin cell growth. To date, the highest-quality evidence supports the link between glycemic index of food choices and acne among males, but research on dairy products as well as on the fatty acid content of the diet is still underway. Whether or not it improves acne, changing the diet to include more whole grains, fruits, vegetables, and legumes would have multiple health benefits.

Source: Burris J and others: Acne: The role of medical nutrition therapy. *Journal of the Academy of Nutrition and Dietetics* 113: 416, 2013.

 Check out Connect Checkout Connect **www.mcgrawhillconnect.com**
to further explore diet and acne.

Nutrition and Your Health
Food Allergies and Intolerances

Food allergies are on the rise. Between 1997 and 2007, food allergies among children increased by 18%. What used to be a rare medical incident is now the cause for 30,000 emergency room visits and 150 deaths per year. Accounting for direct medical costs, special foods, and time lost from work, food allergies cost Americans $25 billion per year. Today, food allergies affect about 8% (5.9 million) of children in the United States, one-third of whom suffer from multiple food allergies (see Further Reading 11).

Adverse reactions to foods—indicated by sneezing, coughing, nausea, vomiting, diarrhea, hives, and other rashes—are broadly classed as **food allergies** (also called hypersensitivities) or **food intolerances.** The term *food sensitivity* is ill-defined but generally refers to any symptom that is perceived to be food-related. In our discussion, we group adverse food reactions into just two categories. Those caused by an immune response are termed food allergies, and those not caused by an immune response are described as food intolerances. Let us examine each process, first allergies and then intolerances.

Food Allergies: Symptoms and Mechanism

Allergic reactions to foods are common (Fig. 15-4) and occur more frequently in females than males. Food allergies occur most often during infancy and young adulthood.

Symptoms of food allergies may affect the:

- Skin: itching, tingling*, redness, hives, and swelling
- GI tract: nausea, vomiting, diarrhea, intestinal gas, bloating, pain, constipation, and indigestion
- Respiratory tract: runny nose, wheezing, congestion, and difficulty breathing*
- Cardiovascular system: low blood pressure* and rapid heart rate*

These symptoms usually set in shortly after consuming the offending food protein and may last for a few seconds or a few days. The symptoms marked with an asterisk (*) are signs of a very rapid and potentially fatal type of allergic response called **anaphylaxis.** This severe allergic response results in low blood pressure and respiratory distress. A person who is extremely sensitive to a food may not be able to touch the food or even be in the same room where it is being cooked without reacting to it. Although any food can trigger anaphylaxis, the most common culprits are peanuts (a legume, not a nut), tree nuts (e.g., walnuts, pecans, etc.), shellfish, milk (look for an ingredient called casein on the label), eggs (look for the ingredient albumin on the label), soybeans, wheat, and fish. Other foods frequently identified with adverse reactions include meat and meat products, fruits, and cheese. For a small number of people, avoiding foods such as peanuts or shellfish is a matter of life and death.

Basically, allergies are an inappropriate response of the immune system. When immune cells identify a harmful foreign protein (**antigen**), they destroy it and produce antibodies to it, so that the next response to the harmful substance will be swift and effective. Almost all food allergies are caused by proteins in foods that act as antigens (also called **allergens**). In these cases, the immune system mistakes the food protein for a harmful substance and mounts an immune response, leading to symptoms such as hives, runny nose, and GI disturbances.

No one is sure why the immune system sometimes overreacts to harmless proteins. The early introduction (e.g., before 4 to 6 months of age) of solid foods to infants may trigger food allergies. The reasoning is that the infant's GI tract is immature and "leaky," allowing some undigested proteins to be absorbed into the bloodstream. This is beneficial for breastfed infants, who can absorb immune proteins from breast milk. However, if some food proteins are introduced before the GI

food allergy An adverse reaction to food that involves an immune response; also called *food hypersensitivity.*

food intolerance An adverse reaction to food that does not involve an allergic reaction.

anaphylaxis A severe allergic response that results in lowered blood pressure and respiratory distress. This can be fatal.

antigen Any substance that induces a state of sensitivity and/or resistance to microorganisms or toxic substances after a lag period; substance that stimulates a specific aspect of the immune system.

allergen A foreign protein, or antigen, that induces excess production of certain immune system antibodies; subsequent exposure to the same protein leads to allergic symptoms. While all allergens are antigens, not all antigens are allergens.

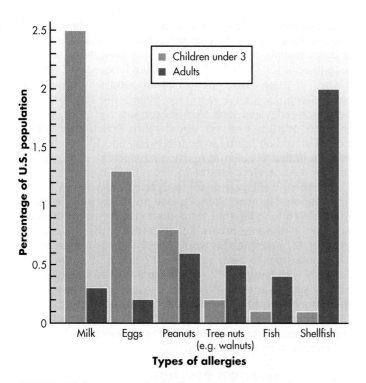

FIGURE 15-4 ▲ Types of allergies and percentage of U.S. population with those allergies.

Source: Journal of Allergy and Clinical Immunology (study done by Mount Sinai School of Medicine).

tract has matured, these antigens may enter the bloodstream and stimulate an immune response.

The *hygiene hypothesis* offers another interesting explanation: in our "germophobic" society, with the protection of antibiotics, hand sanitizers, and antimicrobial soaps and cleaners, our immune systems are not vigorously challenged by antigens. As a result, the immune system may become sensitized to innocuous substances, such as food proteins. Recent research supports the hygiene hypothesis. Children who grow up on farms or who have pets and are thereby exposed to many antigens have fewer allergies and asthma than children who grow up in more sterile environments.

Recently, researchers have proposed a link between low levels of vitamin D and food allergies. The rise in incidence of food allergies has paralleled an increase in vitamin D deficiency. The relationship between vitamin D and food allergies may be mediated by the vitamin's role in immune function.

Whatever the explanation, one thing is clear: food allergies are an important concern for many parents and others who work with children.

TESTING FOR A FOOD ALLERGY

The diagnosis of a food allergy can be a difficult task (Table 15-8). It requires the participation of a skilled physician. The first step in determining whether a food allergy is present is to record a detailed history of symptoms, including the time from ingestion to onset of symptoms, duration of symptoms, most recent reaction, food suspected of causing a reaction, and quantity and nature of food needed to produce a reaction. A family history of allergic diseases can also help, as allergies tend to run in families. A physical examination may reveal evidence of an allergy, such as skin diseases and asthma. Various diagnostic tests can rule out other conditions.

The first step in diagnosing a food allergy is to eliminate from the diet (for 1 to 2 weeks) all food components that appear to cause allergic symptoms. The person generally starts out eating foods to which almost no one reacts, such as rice, vegetables, noncitrus fruits, and fresh meats and poultry. If symptoms are still present, the person can more severely restrict the diet or even use special formula diets that are hypoallergenic.

TABLE 15-8 ▶ Assessment Strategies for Food Allergies

History	Include description of symptoms, time between food ingestion and onset of symptoms, duration of symptoms, most recent allergic episode, quantity of food required to produce reaction, suspected foods, and allergic diseases in other family members.
Physical examination	Look for signs of an allergic reaction (rash, itching, intestinal bloating, etc.).
Elimination diet	Establish a diet lacking the suspected offending foods and stay on it for 1 to 2 weeks or until symptoms clear.
Food challenge	Add back small amounts of excluded foods, one at a time, as long as anaphylaxis is not a possible consequence.
Blood test	Determine the presence of antibodies in blood that bind to food antigens tested.
Skin test	Place a sample of the suspected allergen under the skin and watch for an inflammatory reaction.

Once a diet is found that causes no symptoms, called an **elimination diet,** foods can be added back one at a time. This type of food challenge is an option only when the culprit foods are known to pose no risk of anaphylaxis in the person. Doses of ½ to 1 teaspoon (2.5 to 5 milliliters) are given at first. The amount is increased until the dose approximates usual intake. Any reintroduced food that causes significant symptoms to appear is identified as an allergen for the person.

Laboratory tests can also aid in diagnosis of food allergies. Skin testing involves pricking the skin with a small amount of purified food extract and observing any allergic response (e.g., a red eruption at the prick site). These types of tests are easy and safe, even for infants, but they may not clearly diagnose a food allergy. A positive skin-prick test merely indicates that a person has been sensitized to a food; it cannot definitively identify if that food is the cause for the symptoms in question. Newer types of blood testing, however, have more diagnostic value. Blood tests estimate the blood concentration of antibodies that bind certain foodborne antigens.

LIVING WITH FOOD ALLERGIES

Once potential allergens are identified, dietary modifications must be made. In some cases, small amounts of the offending food can be consumed without an observable reaction. Also, some food allergens are destroyed by heating, so cooking may eliminate the allergic response. This is effective primarily for allergies to fruits or vegetables, not for the more common allergies to milk, peanuts, or seafood. For most cases, though, complete avoidance of allergy-causing food ingredients is the safest course of action. This makes careful reading of food labels essential. The Food Allergen Labeling and Consumer Protection Act of 2006 requires manufacturers to clearly identify the presence of major food allergens (milk, eggs, fish, shellfish, peanuts, tree nuts, wheat, and soy) on food product labels.

A major challenge when treating a person with a food allergy is to make sure that what remains in the diet can still provide essential nutrients. The small food intake of children permits less leeway in removing offending foods that may contain numerous nutrients. A registered dietitian can help guide the diet-planning process to ensure that the remaining food choices still meet nutrient needs or to guide supplement use, if that is necessary.

elimination diet A restrictive diet that systematically tests foods that may cause an allergic response by first eliminating them for 1 to 2 weeks and then adding them back, one at a time.

People with a history of serious allergic reactions and those who have asthma should carry a self-administered form of epinephrine, such as EpiPen, to subside an episode of anaphylaxis, should it occur.

Free information on food allergies is available by contacting Food Allergy Research and Education (FARE). The telephone number is (800) 929-4040; the website is **www.foodallergy.org**.

About 80% of young children with food allergies outgrow them before 3 years. Parents should be made aware of this and not assume the allergy will be long-lived. Food allergies diagnosed after 3 years of age are often longer-lived, but not always. In these cases, about 33% of people outgrow their food allergies within 3 years. For others, the condition may be prolonged; some food allergies can last a lifetime, such as those for peanuts, tree nuts, and shellfish. Periodic reintroduction of offending foods can be tried every 6 to 12 months or so to see whether the allergic reaction has decreased. If no symptoms appear, tolerance to the food has developed.

Several strategies are under study to ease the dietary restrictions imposed by food allergies. One possibility includes treatment with antibodies that will increase the threshold at which an allergic response occurs. For a person with an allergy to peanuts, for example, this would alleviate some anxiety about severe reactions to trace amounts of peanuts found in foods. Similarly, immunotherapy, which exposes allergic individuals to very small but progressively larger amounts of food allergens, may help some people build up a tolerance to certain food components. Vaccines are another area of research. Also, scientists are working on genetically engineered foods that do not contain common allergens.

PREVENTING FOOD ALLERGIES

With the rising number of cases of food allergies, many new parents wonder when and how to introduce new foods during infancy and early childhood. It is evident that introducing foods other than human milk or infant formula before 4 months of age is associated with higher risk of allergic diseases. Most experts, including the American Academy of Pediatrics and the American Academy of Allergy, Asthma, and Immunology, advise waiting to introduce solid foods until 4 to 6 months of age for lowest risk of food allergies. Delaying introduction of solid foods beyond 6 months of age is not advised.

Any food can contain a potential allergen, but certain foods have been found to have high allergenic potential. The eight leading food allergens in the United States are milk, eggs, peanuts, tree nuts, fish, shellfish, wheat, and soy. The American Academy of Pediatrics and the National Institute of Allergy and Infectious Diseases no longer advise delaying the introduction of highly allergenic foods beyond 6 months of age, even for infants with a family history of food allergies.

Until recently, allergy-prone women were advised to avoid highly allergenic foods during pregnancy and breastfeeding. Allergens can cross the placenta during pregnancy and are secreted in breast milk. However, research does not demonstrate a benefit of maternal dietary restrictions in preventing food allergies in infants.

The best course of action is to breastfeed the infant exclusively for 6 months and continue to breastfeed through 12 months with appropriate introduction of solid foods to meet the infant's nutritional needs. Human milk contains factors that play a role in the maturation of the small intestine. Formula-fed infants, especially those on cow's milk–based formulas, have a greater risk for developing food allergies. There is evidence that hydrolyzed infant formulas (in which the large proteins have been broken down into smaller peptides) may be useful in allergy prevention for infants at risk of developing food allergies. However, hydrolyzed formulas are about three times as costly as iron-fortified cow's milk–based formulas.

FOOD INTOLERANCES

Food intolerances are adverse reactions to foods that do not involve immunologic mechanisms. Generally, larger amounts of an offending food are required to produce the symptoms of an intolerance than to trigger allergic symptoms. Common causes of food intolerances include:

- Constituents of certain foods (e.g., red wine, tomatoes, and pineapples) that have a druglike activity, causing physiological effects such as changes in blood pressure
- Certain synthetic compounds added to foods, such as sulfites, food-coloring agents, and monosodium glutamate (MSG)

INGREDIENTS: ENRICHED FLOUR (WHEAT FLOUR, MALTED BARLEY FLOUR, NIACIN, REDUCED IRON, THIAMIN MONONITRATE, RIBOFLAVIN, FOLIC ACID), CHEDDAR AND ASIAGO CHEESES (MILK, SALT, CHEESE CULTURES, ENZYMES), UNSALTED BUTTER, CHOPPED SCALLIONS, SUGAR, YEAST, SALT, MIXED TOCOPHEROLS (PRESERVATIVE).

CONTAINS WHEAT AND MILK

▲ In 2006, the Food Allergen Labeling and Consumer Protection Act mandated that food manufacturers make consumers aware of the presence of highly allergenic ingredients in their products.

- Food contaminants, including antibiotics and other chemicals used in the production of livestock and crops, as well as insect parts not removed during processing
- Toxic contaminants, which may be ingested with improperly handled and prepared foods containing *Clostridium botulinum, Salmonella* bacteria, or other foodborne microorganisms (see Chapter 13)
- Deficiencies in digestive enzymes, such as lactase (review Chapter 4)

Almost everyone is sensitive to one or more of these causes of food intolerance, many of which produce GI tract symptoms.

Sulfites, added to foods and beverages as antioxidants, cause flushing, spasms of the airway, and a loss of blood pressure in susceptible people. Wine, dehydrated potatoes, dried fruits, gravy, soup mixes, and restaurant salad greens commonly contain sulfites. A reaction to MSG may include an increase in blood pressure, numbness, sweating, vomiting, headache, and facial pressure. MSG is commonly found in restaurant food and many processed foods (e.g., soups). A reaction to tartrazine, a food-coloring additive, includes spasm of the airway, itching, and reddening skin. Tyramine, a derivative of the amino acid tyrosine, is commonly found in "aged" foods, such as cheeses and red wines. This natural food constituent can cause high blood pressure in people taking monoamine oxidase (MAO) inhibitor medications, which may be prescribed for clinical depression.

The basic treatment for food intolerances is to avoid specific offending components. However, total elimination often is not required because people generally are not as sensitive to compounds causing food intolerances as they are to allergens.

Summary (Numbers refer to numbered sections in this chapter.)

15.1 Growth is rapid during infancy; birth weight doubles in 4 to 6 months, and length increases by 50% in the first year. An adequate diet, especially in terms of calories, protein, and zinc, is essential to support normal growth. Growth charts aid nutritional assessment of infants and children by tracking changes in body weight, height (or length), head circumference, and body mass index over time.

15.2 The energy needs of infants are highest per kilogram of body weight compared to any other life stage, around 100 kcal/kg. Fat should make up about 50% of total energy intake. DHA and AA are important fatty acids for nervous system development. Carbohydrate needs range from 60 grams per day for younger infants to 95 grams per day for older infants. Protein needs are 9 grams per day for younger infants and 11 grams per day for older infants. Supplementation with vitamin D, iron, and fluoride may be appropriate for some infants. Adequate hydration can be maintained using only breast milk or formula; supplemental water is not recommended during the first 6 months of life.

15.3 Infant nutrient needs can be met by human milk or iron-fortified infant formula for the first 6 months of life. Most infants do not need solid foods before 6 months of age. Introduction of solid foods should be based on an infant's nutritional needs, physical abilities, and developmental readiness. Solid foods should be introduced one at a time, starting with iron-fortified infant cereals or ground meats (sources of iron). Some foods to avoid giving infants in the first year include honey, cow's milk (especially fat-reduced varieties), foods with added salt or sugar, and foods that may cause choking.

15.4 A slower growth rate results in decreased appetite among preschool children. Other common nutrition-related concerns include iron-deficiency anemia, constipation, and dental caries. With smaller portion sizes and picky eating behaviors, it is crucial to offer several small meals and snacks with a variety of nutrient-dense foods. Follow the example set forth by MyPlate, but use smaller portions (e.g., 1 tablespoon of food per year of life). For autism spectrum disorders, several nutritional interventions, including the gluten-free, casein-free diet, and some dietary supplements, are under study.

15.5 Among school-age children, excessive energy and fat intakes coupled with low levels of physical activity have led to an alarming increase in overweight, obesity, type 2 diabetes, and cardiovascular disease. Parents can provide healthful food choices and encourage at least 60 minutes of physical activity per day. When controlled early through diet and exercise interventions, the problem of obesity may correct itself as the child continues to grow in height. Other important nutrition strategies for school-age children include starting the day with breakfast and selecting low-fat or fat-free milk or water instead of sugary beverages. Recent changes in meal offerings through schools are aimed at curtailing the rise in childhood obesity.

15.6 During the adolescent growth spurt, both boys and girls have increased needs for iron, calcium, and overall calories. Inadequate calcium intake by teenage girls is a major concern because it can set the stage for the development of osteoporosis later in life. Adolescents need to limit their intakes of high-fat and high-sugar fast foods and snacks and use caffeine in moderation (if at all). Alcohol abuse during adolescence has many severe consequences, including impaired brain development and increased risk for liver and cardiovascular diseases in adulthood.

NAYH The most common food allergies are associated with peanuts, tree nuts, shellfish, milk, eggs, soybeans, wheat, and fish. Food allergies occur most often during infancy and young adulthood.

Check Your Knowledge (Answers to the following questions are below.)

1. Inadequate intake of which of the following results in poor growth?
   a. calories
   b. iron
   c. zinc
   d. All of the above.

2. Milk is a nutrient-dense source of all of the following except
   a. protein.
   b. iron.
   c. calcium.
   d. zinc.

3. To ensure adequate vitamin and mineral intake for a picky eater,
   a. provide a fortified breakfast cereal.
   b. promise dessert as a reward for eating meats and vegetables.
   c. use a multivitamin and mineral supplement.
   d. None of the above.

4. An 11-month-old girl who weighs 19 pounds needs approximately _____ kcal per day.
   a. 690
   b. 810
   c. 845
   d. 930

5. Introduction of cow's milk should be delayed until 12 months of age because it
   a. contains too much fat.
   b. supplies too much lactose.
   c. contains too much protein.
   d. All of the above.

6. Your niece breaks out in hives and feels nauseous after eating a salad containing mango. She probably has a food
   a. sensitivity.
   b. allergy.
   c. intolerance.
   d. All of the above.

7. Which of the following are benefits of consuming a fortified, ready-to-eat breakfast cereal instead of skipping breakfast?
   a. Improved academic performance
   b. Meeting RDAs for iron and calcium
   c. Lower risk for childhood obesity
   d. All of the above.

8. A gluten-free, casein-free diet is under study for treatment of
   a. rickets.
   b. anemia.
   c. lead poisoning.
   d. autism.

9. For treatment of overweight, school-age children should
   a. eat fewer meals.
   b. follow a low-carbohydrate eating plan.
   c. exercise for 60 minutes per day or more.
   d. avoid dairy products.

10. You are trying to introduce an apple and blueberry puree to a 7-month-old infant, but she rejects it. You should
    a. assume she doesn't like apples and blueberries.
    b. offer the food again on another day.
    c. force a spoonful into her mouth.
    d. None of the above.

Answer key: 1. d (LO 15.1), 2. b (LO 15.2), 3. a (LO 15.2), 4. a (LO 15.3), 5. c (LO 15.3), 6. b (LO 15.6), 7. d (LO 15.4), 8. d (LO 15.4), 9. c (LO 15.5), 10. b (LO 15.3)

Study Questions (Numbers refer to Learning Outcomes)

1. List two factors that limit "catch-up" growth when a nutrient-deficient diet has been consumed throughout childhood. **(LO 15.1)**

2. A 3-month-old infant is taken to a clinic with failure to thrive. What are two possible explanations? **(LO 15.1)**

3. Which two nutrients are of particular concern in planning diets for teenagers? Why does each deserve to be singled out? **(LO 15.2)**

4. List three nutrients of concern for a child who chooses a vegetarian lifestyle. **(LO 15.2)**

5. Outline three key factors that help determine when to introduce solid foods into an infant's diet. **(LO 15.3)**

6. Compare the guidelines for infant feeding summarized in Chapter 15 with the 2010 Dietary Guidelines for Americans for children over age 2 and adults discussed in Chapter 2. Which guidelines are similar? Do any contradict each other? If so, why? **(LO 15.3)**

7. Describe the pros and cons of snacking. What is the basic advice for healthful snacking from childhood through the teenage years? **(LO 15.4)**

8. List three reasons why preschoolers are noted for "picky" eating. For each, describe an appropriate parent response. **(LO 15.4)**

9. What three factors are likely to contribute to obesity in a typical 10-year-old child? **(LO 15.5)**

10. List evidence-based strategies to reduce the risk of food allergies in infants. **(LO 15.6)**

What the Dietitian Chose

At 18 months of age, Lila's diet should include small portions of a variety of foods from each MyPlate food group. She should have several teeth by now, but her chewing ability is still limited. As a young child, Lila has high nutrient needs and yet has a small stomach capacity, so snacking makes an important contribution to meeting overall nutrient needs.

Reduced-fat popcorn is a healthy, low-fat snack for older children and adults. Feeding children bulky foods with low energy density, however, can lead to fullness without meeting nutritional needs. Furthermore, popcorn is a choking hazard.

Although baby carrots with ranch dressing would be a good source of beta-carotene and healthy vegetable oils, toddlers are not expert chewers just yet, and large pieces of hard, crunchy foods present a choking hazard. Slicing the carrots into long, thin strips would be a better idea.

Fat-free, light yogurt is a good source of calcium and vitamin D for building healthy bones and certainly does not present choking risk. However, infants and toddlers need the fat and sugar in dairy products to supply calories for their growing bodies.

Whole-milk yogurt, such as YoBaby®, would be a better choice.

Whole-grain crackers with sliced cheddar cheese would be an excellent snacking option for Lila. This snack provides a healthy balance of carbohydrates, protein, and fat. Using whole-grain crackers, rather than those made with refined flour, will boost the vitamin and mineral content (review Chapter 4).

Toddlers are known for picky eating behaviors. Lila's slower pace of growth may reduce her appetite and she may be more interested in exploring the world around her than sitting down for a meal. Make sure Lila sits down to eat and supervise her while she is eating. Finally, be a good role model. If she sees you eat the healthy snacks you have offered, she will be more likely to accept them.

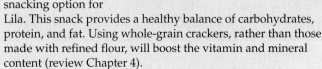

CASE STUDY SOLUTION ## Undernutrition During Infancy

1. Ask if the formula is iron-fortified. Also, verify that Damon's caregivers are appropriately preparing his formula by adding the correct amount of powder or concentrated liquid to water. Be sure that Damon's caregivers plan to continue giving him formula until 1 year of age.

2. The scenario described puts Damon at risk for failure to thrive, likely due to his caregivers' lack of knowledge about appropriate infant feeding practices. His diet is probably lacking in total calories as well as calcium, iron, and zinc. Inadequacies may lead to stunted growth, cognitive problems, behavioral problems, and weak bones.

3. In addition to an iron-fortified infant formula, Damon should be eating iron-fortified infant cereals, pureed fruits, vegetables, and meats, and even some table foods that are soft and cut into small pieces. After a child reaches 6 months of age, solid foods help to meet increased nutrient demands. Processed foods, such as the hot dog described in the scenario, are usually high in sodium, fat, or sugars. Hot dogs and other foods that are round, compressible, and small in diameter are choking hazards for infants and children under age 4. These are not ideal for meeting Damon's nutritional needs. Allowing Damon to self-feed to some extent will help him to develop motor skills and build confidence.

4. Infants should not be drinking sugary beverages such as colas or fruit-flavored drinks, as these items lack the nutrient density essential to meeting Damon's needs. If his caregivers provide juice at all, the total amount should be limited to 6 fluid ounces of 100% fruit juice. To prevent early childhood caries and develop self-feeding skills, Damon should be learning to drink from a cup rather than a bottle.

5. The combination of an appropriately prepared, iron-fortified infant formula and a variety of solid foods should provide Damon all the nutrients he needs. However, if he is anemic, he may need an iron supplement.

Further Readings

1. Academy of Nutrition and Dietetics: Position of the American Dietetic Association: Nutrition guidance for healthy children ages 2 to 11 years. *Journal of the Academy of Nutrition and Dietetics* 114:1257, 2014.

2. Baker RD and others: Clinical report—Diagnosis and prevention of iron deficiency and iron-deficiency anemia in infants and young children (0–3 years of age). *Pediatrics* 126:1040, 2010.

3. Briefel RR and others: Reducing calories and added sugars by improving children's beverage choices. *Journal of the Academy of Nutrition and Dietetics* 113:269, 2013.

4. Briggs M and others: Position of the American Dietetic Association, School Nutrition Association, and Society for Nutrition Education: Comprehensive school nutrition services. *Journal of the American Dietetic Association* 110:1738, 2010.

5. Cermak SA and others: Food selectivity and sensory sensitivity in children with autism spectrum disorders. *Journal of the American Dietetic Association* 110:238, 2010.

6. Cole SZ and Lanham JS: Failure to thrive: An update. *American Family Physician* 83:829, 2011.

7. Committee on Nutrition and the Council on Sports Medicine and Fitness: Sports drinks and energy drinks for children and adolescents: Are they appropriate? *Pediatrics* 127, 1182, 2011.

8. Copeland KC and others: Management of newly diagnosed type 2 diabetes mellitus (T2DM) in children and adolescents. *Pediatrics* 131:364, 2013.

9. Council on Communications and Media: Children, adolescents, obesity, and the media. *Pediatrics* 128:201, 2011.

10. Ford CN and others: Trends in dietary intake among US 2–6-year-old children, 1989–2008. *Journal of the Academy of Nutrition and Dietetics* 113:35, 2013.

11. Gupta RS and others: Childhood food allergies: Current diagnosis, treatment, and management strategies. *Mayo Clinic Proceedings* 88:512, 2013.

12. Kavey RW and others: Expert panel on integrated guidelines for cardiovascular health and risk reduction in children and adolescents: Summary report. *Pediatrics* 128:S213, 2011.

13. Kosova EC and others: The relationship between sugar-sweetened beverage intake and cardiometabolic markers in young children. *Journal of the Academy of Nutrition and Dietetics* 113:219, 2013.

14. Moag-Stahlberg A: The state of family nutrition and physical activity: Are we making progress? Report of the American Dietetic Association and American Dietetic Association Foundation, 2011. Available at **http://www.eatright.org/foundation/fnpa.**

15. Office of Juvenile Justice and Delinquency Prevention: Effects and consequences of underage drinking. *Juvenile Justice Bulletin,* 2012. Available at **http://www.ojjdp.gov/pubs/237145.pdf.**

16. Powell LM and Nguyen BT: Fast-food and full-service restaurant consumption among children and adolescents: Effect on energy, beverage, and nutrient intake. *JAMA Pediatrics* 167:14, 2013.

17. Rao G: Childhood obesity: Highlights of AMA expert committee recommendations. *American Family Physician* 78:56, 2008.

18. Spruiit-Metz D: Etiology, treatment, and prevention of obesity in childhood and adolescence: A decade in review. *Journal of Research on Adolescence* 21:129, 2011.

19. Wagner CL and others: Prevention of rickets and vitamin D deficiency in infants, children, and adolescents. *Pediatrics* 122(5):1142, 2008.

20. Wojcicki JM and Heyman MB: Reducing childhood obesity by eliminating 100% fruit juice. *American Journal of Public Health* 102:1630, 2012.

 **To get the most out of your study of nutrition, visit McGraw-Hill Connect at www.mcgrawhillconnect.com where you will find NutritionCalc Plus, LearnSmart, and many other dynamic tools.**

Rate Your Plate

Getting Young Bill to Eat

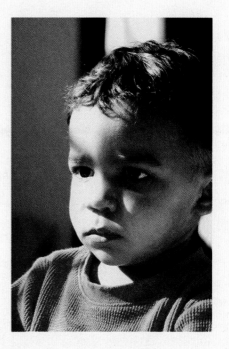

Bill is 3 years old, and his mother is worried about his eating habits. He refuses to eat vegetables, meat, and dinner in general. Some days he eats very little food. He wants to eat snacks most of the time. Mealtime is a battle because Bill says he is not hungry, and his mother wants him to eat a sit-down lunch and dinner to make sure he gets all the nutrients he needs and to eat everything served on his plate. He drinks five or six glasses of whole milk per day because that is the one food he likes.

When his mother prepares dinner, she makes plenty of vegetables, boiling them until they are soft, hoping this will appeal to Bill. Bill's dad waits to eat his vegetables last, regularly telling the family that he eats them only because he has to. He also regularly complains about how dinner has been prepared. Bill saves his vegetables until last and usually gags when his mother orders him to eat them. Bill has been known to sit at the dinner table for an hour until the war of wills ends. Bill's mother serves casseroles and stews regularly because they are convenient. Bill likes to eat breakfast cereal, fruit, and cheese, and regularly requests these foods for snacks. However, his mother tries to deny his requests, so that he will have an appetite for dinner. Bill's mother comes to you and asks you what she should do to get Bill to eat.

Analysis

1. List four mistakes Bill's parents are making that contribute to Bill's poor eating habits.

   _____

   _____

   _____

   _____

2. List four strategies they might try to promote good eating habits in Bill.

   _____

   _____

   _____

   _____

 **Find more Rate Your Plate activities for this chapter in Connect at www.mcgrawhillconnect.com.**

Student Learning Outcomes

Chapter 16 is designed to allow you to:

16.1 Discuss demographic trends among adults in North America and how they impact health care.

16.2 List several hypotheses about the causes of aging.

16.3 Describe how physiological changes of aging affect nutritional status of adults.

16.4 Describe how psychosocial changes of aging affect nutritional status of adults.

16.5 Compare the dietary intake of adults with current recommendations.

16.6 Identify nutrition-related health conditions of the adult years and describe the prevention and treatment options.

16.7 List several nutritional programs available to help meet nutritional needs of older adults.

16.8 Compare benefits of moderate alcohol use to the risks of alcohol abuse.

Chapter 16
Nutrition During Adulthood

Eating is one of our great pleasures. Guided by common sense and moderation, eating well is also a means to good health. Most of us want a long, productive life, free of illness. Unfortunately, many people from early middle age onward suffer from obesity, cardiovascular disease, hypertension and strokes, type 2 diabetes, osteoporosis, and other chronic diseases. We can slow the development of—and in some cases, even prevent—these diseases by following a dietary pattern such as that exemplified by MyPlate or the Mediterranean diet. The effect of such a diet is most profitable if we begin early and continue throughout adulthood. We serve ourselves best—as individuals and as a nation—by striving to maintain vitality even in the later decades of life. This concept was first explored in Chapter 1 and is discussed again in this chapter, in light of the nutrition needs of adults.

Keep in mind that today's behaviors can significantly influence your health years from now. Although genes do play a role, as discussed in Chapter 3, many of the health problems that occur with age are not inevitable; they result from diet-related disease processes that influence physical health. Much can be learned from healthy older people whose attention to a healthy diet and physical activity—along with a little help from heredity—keeps them active and vibrant well beyond typical retirement years. Successful aging is the goal. Age quickly or slowly: it is partly your choice.

16.1 The Graying of North America

Due to advances in health care and sanitation, the demographics of developed countries are shifting so that, as a population, we are getting older. In North America, the group constituting those aged 85+ years is the fastest growing segment. Between 1997 and 2050, the population aged 85+ years in the United States is expected to increase from 3.4 to 19 million (Fig. 16-1). Even more amazing, 1 million or more people in the United States could be over 100 years old in 2050.

This "graying" of North America poses some problems. Although people older than age 65 account for just 13% of the U.S. population, they account for more than 25% of all prescription medications used, 40% of acute hospital stays, and 50% of the federal health budget. Hip fractures alone cost the nation about $12 billion per year. Of older persons, 80% or more have chronic conditions, such as cardiovascular disease, type 2 diabetes, hypertension, and osteoporosis.

Postponing these chronic diseases for as long as possible will help control health care costs. Health and independence contribute quality—not just quantity—to life and lessen the load on an already overburdened health care system. Keep in mind that aging is not a disease. Furthermore, diseases that commonly accompany old age—osteoporosis and atherosclerosis, for example—are not an inevitable part of aging. Many can be prevented or managed. Some people *do* die of old age, not as a direct result of disease.

THE CAUSES OF AGING

Adulthood, the longest stage of the life cycle, begins when an adolescent completes his or her physical growth. Unlike earlier stages of the life cycle, nutrients are used primarily to maintain the body rather than support physical growth. (Pregnancy is the only time during adulthood when substantial amounts of nutrients are used for growth.) As adults get older, nutrient needs change. For example, vitamin D needs increase for older adults.

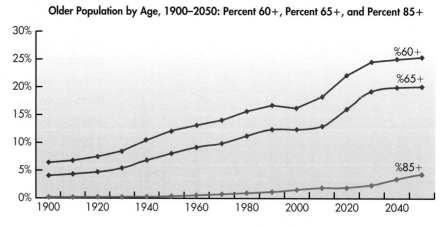

FIGURE 16-1 ▲ Growth of the U.S. Population of Older Adults. This chart shows that the proportion of the total U.S. population composed of older adults has been steadily increasing over the past century and how these trends are expected to continue. The 85+ demographic group, although still the smallest group in total numbers, is experiencing the most rapid rate of growth. Enhancements in health care and sanitation, as well as the baby boom that occurred after World War II, are contributing to the graying of North America. Conversely, most demographic groups under the age of 45 are shrinking as a percentage of overall population (not shown here). This means that fewer young people will be available to care for a growing population of older adults in years to come.

Source: U.S. Administration on Aging.

Aging can be defined as the time-dependent physical and physiological changes in body structure and function that occur normally and progressively throughout adulthood as humans mature and become older. One view of aging describes it as a process of slow cell death, beginning soon after fertilization. When we are young, aging is not apparent because the major metabolic activities are geared toward growth and maturation. We produce plenty of active cells to meet physiological needs. During late adolescence and adulthood, the body's major task is to maintain cells. From the beginning of adulthood until age 30 or so, the body operates at peak performance: stature, stamina, strength, endurance, efficiency, and health are at their lifetime highs. Rates of cell synthesis and breakdown are balanced in most tissues. Inevitably, though, cells age and die. After about age 30, the rate of cell breakdown slowly begins to exceed the rate of cell renewal, leading to a gradual decline in organ size and efficiency. Eventually, the body cannot adjust to meet all physiological demands, and body functioning begins to decrease (Fig. 16-2). Still, body systems and organs usually retain enough **reserve capacity** to handle normal, everyday demands throughout one's entire lifetime. Problems caused by diminished capacity typically do not arise unless severe demands are placed on the aging body. For example, alcohol intake can overtax an aging liver. The stress of shoveling a snow-covered sidewalk can exceed the capacity of the heart and lungs. Coping with an illness also can push an older body beyond its capacity.

The causes of aging remain a mystery. Most likely, the physiological changes of aging are the sum of automatic cellular changes, lifestyle practices, and environmental influences, as listed in Table 16-1. Even with the most supportive environment and healthy lifestyle, cell structure and function still decline over time. The eventual death of deteriorating cells is actually beneficial because it likely prevents diseases such as cancer. Unfortunately, there are negative consequences to this natural cell progression because, as more and more cells in an organ die, organ function decreases. For example, **kidney nephrons** are continually lost as we age. In some people, this loss exhausts the kidneys' reserve capacity and ultimately leads to kidney failure. Most people, however, maintain sufficient kidney function throughout life.

The diseases and degenerative processes commonly observed in older people have long been assumed to be unavoidable consequences of aging. Certainly, some of the declines we blame on aging may be inevitable, such as gradual reductions in tissue and organ cell numbers, graying hair, and reduced lung capacity. However, many of

▲ The baby boom turned 65 in 2011, resulting in an increase in demand for attention to the health concerns of this aging population.

aging Time-dependent physical and physiological changes in body structure and function that occur normally and progressively throughout adulthood as humans mature and become older.

reserve capacity The extent to which an organ can preserve essentially normal function despite decreasing cell number or cell activity.

kidney nephrons The units of kidney cells that filter wastes from the bloodstream and deposit them into the urine.

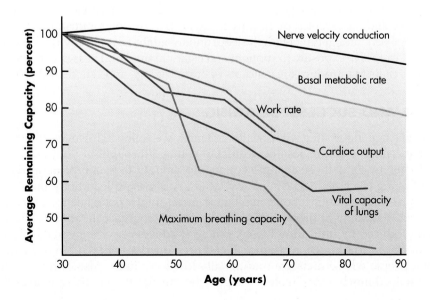

FIGURE 16-2 ◄ Declines in physiological function seen with aging. The decline in many body functions is especially evident in sedentary people.

▲ Adopting diet and lifestyle practices that minimize a decline in body function is an investment in your future health. Visit **www.nia.nih.gov** for many free resources on healthy aging.

glycosylation The process by which glucose attaches to (glycates) other compounds, such as proteins.

TABLE 16-1 ▶ Current Hypotheses About the Causes of Aging

Errors occur in copying the genetic blueprint (DNA).
Once sufficient errors in DNA copying accumulate, a cell can no longer synthesize the major proteins needed to function, and it therefore dies.

Connective tissue stiffens.
Parallel protein strands, found mostly in connective tissue, cross-link to each other. This decreases flexibility in key body components.

Electron-seeking compounds damage cell parts.
Electron-seeking free radicals can break down cell membranes and proteins. One way to limit damage from these compounds is to consume adequate amounts of vitamins E and C, selenium, and carotenoids.

Hormone function changes.
The blood concentration of many hormones, such as testosterone in men, falls during the aging process. Replacement of these and other hormones is possible, but the resulting risks and benefits are largely unknown.

Glycosylation of proteins.
Blood glucose, when chronically elevated, attaches to (glycates) various blood and body proteins. This decreases protein function and can encourage immune system attack on such altered proteins.

The immune system loses some efficiency.
The immune system is most efficient during childhood and young adulthood, but with advancing age, it is less able to recognize and counteract foreign substances, such as viruses, that enter the body. Nutrient deficiencies, particularly of protein, vitamin E, vitamin B-6, and zinc, also hamper immune function.

Autoimmunity develops.
Autoimmune reactions occur when white blood cells and other immune system components begin to attack body tissues in addition to foreign proteins. Many diseases, including some forms of arthritis, involve this autoimmune response.

Death is programmed into the cell.
Each human cell can divide only about 50 times. Once this number of divisions occurs, the cell automatically succumbs.

Excess calorie intake speeds body breakdown.
Experimentally, underfed animals, such as spiders, mice, and rats, live longer than their well fed counterparts. Usual calorie intake must be reduced by about 30% to see this effect. This approach is the only proven way to substantially slow the aging process (see Further Readings 6 and 12).

the so-called usual or degenerative age-related changes can, in fact, be minimized, prevented, and/or reversed by healthy lifestyles (e.g., eating a nutritious diet, exercising regularly, and getting enough sleep) and avoiding adverse environmental factors (e.g., excessive exposure to sunlight and cigarette smoke). These discoveries have led researchers to introduce the concepts of "usual aging" and "successful aging."

USUAL AND SUCCESSFUL AGING

Body cells age, no matter what health practices we follow. However, to a considerable extent, you can choose how quickly you age throughout your adult years. *Usual aging* refers to those changes commonly thought to be a typical or expected part of aging, such as increasing body fatness, decreasing lean body mass, rising blood pressure, declining bone mass, and increasingly poor health. Researchers point out that many of these changes really represent an acceleration of the aging process induced by unhealthy lifestyle choices, adverse environmental exposures, and/or chronic disease. For instance, blood pressure does not tend to rise with age among people whose diets are traditionally low in sodium. Also, lean body mass is maintained much better in older people who exercise than in those who do not.

Keep in mind that extending life without delaying onset of chronic disease prolongs suffering in many cases. In addition, the greater number of disabled years is costly to all North Americans. For these reasons, prolonging life without compressing the number of disabled years is called the "failure of success."

Successful aging, on the other hand, describes physical and physiological function declines that occur only because one grows older, not because lifestyle choices, environmental exposures, and chronic disease have assaulted body tissues. Those who are successful agers experience age-related declines at a slower rate and the onset of chronic disease symptoms at a later age than usual agers. Striving to have the greatest number of healthy years and the fewest years of illness is often referred to as **compression of morbidity.** In other words, a person tries to delay the onset of disabilities caused by chronic disease and to compress significant sickness related to aging into the last few years—or months—of life.

FACTORS AFFECTING THE RATE OF AGING

Life span refers to the maximum number of years a human can live. As far as we know, this has not changed in recorded time. Aside from some unverifiable reports of a 160-year-old man in Ethiopia, the longest human life documented to date is 122 years for a woman and 116 years for a man. **Life expectancy,** alternatively, is the time an average person born in a specific year can expect to live. Life expectancy in North America is about 76 years for men and about 81 years for women, with a span of "healthy years" of about 64. Furthermore, most people who survive to age 80 can expect to live an additional 7 to 10 years.

The rate at which one ages is individual; it is determined by heredity, lifestyle, and environment. With the exception of heredity, most of the factors that influence the rate of aging are directly linked to choices that are under our control.

Heredity. Living to an old age tends to run in some families. If your parents and grandparents lived a long time, you are likely to have the potential to live to an old age, too. Studies of twins indicate that about 20% to 30% of longevity can be attributed to genetics.

One of the most obvious genetic characteristics influencing longevity is gender. In the case of humans, as well as most other species, females tend to live longer than males. Another genetic characteristic that can influence longevity is metabolic efficiency. Individuals with a **thrifty metabolism** require fewer calories for metabolic processes and are able to store body fat more easily than those with faster metabolic rates. Throughout history, it was the individuals with thrifty metabolism who tended to live the longest because they efficiently stored fat during times of plenty and thus had the energy stores needed to survive frequent periods of food scarcity. In today's environment of labor-saving devices and abundant, energy-dense foods, however, a thrifty metabolism may actually reduce longevity. Accumulation of excessive body fat increases the risk of developing health problems that reduce life expectancy (e.g., heart disease, hypertension, and certain cancers).

Heredity is largely unchangeable, but let us take a look at how you can exert control over your lifestyle and environment.

Lifestyle. Lifestyle is one's pattern of living; it includes food choices, exercise patterns, and substance use (e.g., alcohol, drugs, and tobacco). Lifestyle choices can have a major impact on health and longevity, as well as on the expression of genetic potential. If individuals have a family history of premature heart disease, they can adjust their diet, exercise, and tobacco use patterns to slow the progression of the disease, get needed medical care, and possibly extend their lifetime. The converse is true, too. That is, lifestyle choices (e.g., high-fat diet and couch potato mentality) can increase susceptibility to diseases that hasten the rate of aging, ultimately shortening life expectancy, even if a person's genetic potential is for a very long life.

In an effort to unlock the secrets to living a long and healthy life, researchers have been very interested in studying lifestyle patterns of communities in which life expectancy is higher than average. Several communities in which people quite

---

compression of morbidity Delay of the onset of disabilities caused by chronic disease.

life span The potential oldest age a person can reach.

life expectancy The average length of life for a given group of people born in a specific year.

thrifty metabolism A genetic tendency toward efficient use of energy that results in below-average energy requirements and increased storage of calories as fat.

Besides having other long-lived family members, people who live to 100 years generally:

- Do not smoke and do not drink heavily
- Gain little weight in adulthood
- Eat many fruits and vegetables
- Perform daily physical activity
- Challenge their minds
- Have a positive outlook
- Maintain close friendships
- Are (or were) married (especially true for men)
- Have a healthy rate of HDL-cholesterol production

▲ A plant-based diet, exemplified by this dinner of couscous, chickpeas, and vegetables, is a common feature among populations who enjoy long, healthy lives.

often live to see their 90th and 100th birthdays include Okinawa, some parts of the Mediterranean region, and areas of California that are home to members of the Seventh-day Adventist religious denomination.

Worldwide, the highest average life expectancy is in Okinawa, a stretch of islands off the coast of Japan: 86 years for women and 78 years for men. The Okinawan diet is based on rice, fish, vegetable protein sources, fruits, vegetables, tea, herbs for seasonings, and small amounts of meat. Alcohol and salt intake is minimal. The low energy density translates into a generally low calorie intake, and the average BMI is 21. A desire to mimic their weight management success, compression of morbidity, and longevity has given rise to several popular books and websites that promote Okinawan dietary practices.

Followers of the traditional Mediterranean diet enjoy some of the lowest recorded rates of chronic disease in the world. Described in detail in Chapter 5, the Mediterranean diet features abundant daily intake of fruits, vegetables, whole grains, beans, nuts, and seeds. Olive oil, a source of heart-healthy monounsaturated fat, is the main dietary fat. Beans and fish are emphasized as sources of protein, whereas dairy products, eggs, poultry, and meats are used less frequently. Daily exercise is a way of life. In addition, many Mediterraneans partake of wine in moderation at mealtimes.

Loma Linda, California is a town with a large population of Seventh-day Adventists, a Christian religious denomination that emphasizes stewardship of the mind and body as a way to honor God. Men in this group live about 7 years longer and women live about 4 years longer than average Americans. About 30% of Seventh-day Adventists follow a vegetarian diet, but even for those who occasionally

Newsworthy Nutrition

Calorie restriction: Is your diet a fountain of youth?

Animal studies show that long-term calorie restriction increases longevity. Researchers for the Comprehensive Assessment of the Long-Term Effects of Reducing Intake of Energy study explored the possibility that 25% calorie restriction for 6 months would alter biomarkers of longevity in humans. Forty-six overweight adults completed the study. Subjects were randomized to one of four groups: control (no caloric restriction), CR (25% calorie restriction), CREX (12.5% calorie restriction plus 12.5% increased energy expenditure by structured exercise), and LCD (low-calorie diet followed by weight-maintenance diet after 15% weight loss was achieved). Subjects followed controlled diets that were deficient in calories but otherwise nutritionally adequate and underwent numerous measurements, including body composition, calorimetry (review Chapter 7), body temperature, and various blood tests. Fasting levels of insulin, energy expenditure (i.e., metabolic rate), and a marker of DNA damage decreased in all three intervention groups. Among the CR and CREX groups, core body temperature also decreased. These results indicate that prolonged calorie restriction in humans reduces metabolic rate and decreases fasting insulin and body temperature, two biomarkers of longevity. Further research is needed to distinguish effects of calorie restriction on other biomarkers and to measure life span. For this study, diet was carefully controlled and monitored frequently, and all foods were provided. If calorie restriction does prove to increase longevity, could you stick with it?

Source: Heilbronn LK and others: Effect of 6-month calorie restriction on biomarkers of longevity, metabolic adaptation, and oxidative stress in overweight subjects. *Journal of the American Medical Association* 295:1539, 2006.

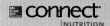 **connect** | NUTRITION   Check out the Connect site **www.mcgrawhillconnect.com** to learn more about theories of aging.

include animal products in their diets, the focus is on unprocessed plant foods. In particular, regular consumption of nuts and whole grains have emerged as dietary factors that predict the lower rates of chronic diseases observed in this population. Unlike the Mediterraneans, most Seventh-day Adventists avoid alcohol. Other salient aspects of the Seventh-day Adventist lifestyle that are associated with longevity include daily physical activity, avoidance of tobacco, strong faith, and strong social support within a tight-knit community.

Did you notice some common behaviors among these groups of people who enjoy extended quality and quantity of life? They all tend to focus on unprocessed, fiber-rich foods, healthy sources of fat (e.g., vegetable oils and fish), and lean sources of protein. Besides diet, physical activity is a major part of their daily routines. Compare these lifestyle choices with those of typical North Americans.

Environment. Some aspects of the environment that exert a powerful influence on the rate of aging are income, education level, health care, shelter, and psychosocial factors. For instance, being able to purchase nutritious foods, quality health care, and safe housing helps decrease the rate of aging. Having the education needed to earn a sufficient income, as well as the knowledge required to select a nutritious diet and make wise lifestyle choices, also can slow the aging process. In addition, the willingness to seek health care promptly when it is needed, the capacity to follow the instructions of a health care provider, and the desire to accept the responsibility for one's own health can slow the rate of aging. Likewise, shelter that protects individuals from physical danger, climatic extremes, and solar radiation helps slow the aging process. Allowing people to make at least some decisions for themselves and control their own activities (autonomy), and providing psychosocial support (informational and emotional resources) promote successful aging and psychological well-being. In contrast, aging is likely to accelerate if any or all of the converse (i.e., insufficient income, low education level, lack of health care, inadequate shelter, and/or lack of autonomy and psychosocial support) are true.

✔ CONCEPT CHECK 16.1

1. Describe three current hypotheses about the causes of aging.
2. What is the difference between *usual* and *successful* aging?
3. Discuss how heredity, lifestyle, and environment influence aging.

16.2 Nutrient Needs During Adulthood

The challenge of the adult years is to maintain the body, preserve its function, and avoid chronic disease; that is, to age successfully. A healthy diet can help achieve this goal. One blueprint for a healthy diet comes from the 2010 Dietary Guidelines for Americans, discussed in Chapter 2. The advice from those guidelines can be summarized into three main goals:

1. Balance calories with physical activity to manage weight.
2. Consume more of certain foods and nutrients, such as fruits, vegetables, whole grains, fat-free and low-fat dairy products, and seafood.
3. Consume fewer foods with sodium, saturated fats, *trans* fats, cholesterol, added sugars, and refined grains.

Overall, good nutrition benefits adults in many ways. Meeting nutrient needs delays the onset of certain diseases; improves the management of some existing diseases; speeds recovery from many illnesses; increases mental, physical, and social well-being; and often decreases the need for and length of hospitalization (see Further Readings 2 and 8). As you know, American adults are fairly well nourished, although some dietary excesses and inadequacies do exist. For instance, common dietary excesses are calories, fat, sodium, and, for some, alcohol. The diets of adult women tend to fall short of the recommended amounts of vitamins D and E, folate, magnesium, calcium, zinc, and fiber. The diets of adult men tend to be low in the same nutrients, except vitamin D, which does not become problematic until age 50. The iron intake of most women during their childbearing years (19 to 50) is insufficient to meet their needs; however, due to a reduced iron need after **menopause,** older women do get enough iron (see Further Reading 3).

People age 65 and up, particularly those in long-term care facilities and hospitals, are the single largest group at risk of malnutrition. They may become underweight and show signs of numerous micronutrient deficiencies (e.g., vitamins B-6 and B-12 and folate). To pinpoint those over age 65 at risk of nutrient deficiencies, the American Academy of Family Physicians, the Academy of Nutrition and Dietetics (formerly the American Dietetic Association), and the National Council on Aging developed the Nutrition Screening Initiative checklist (Fig. 16-3). Older Americans, family members, and health care providers can use the checklist to identify those at nutritional risk *before* health deteriorates significantly. If problems arise in consuming a healthful diet, registered dietitians can offer professional and personalized advice.

menopause Cessation of menses in women, usually beginning at about age 50.

FIGURE 16-3 ► A nutrition checklist for older adults.

From the Nutrition Screening Initiative, a project of the American Academy of Family Physicians, the Academy of Nutrition and Dietetics (formerly the American Dietetic Association), and the National Council on Aging, Inc., and funded in part by a grant from Abbott Nutrition (formerly Ross Products Division, Abbott Laboratories).

A Nutrition Test for Older Adults

Here's a nutrition check for anyone over age 65. Circle the number of points for each statement that applies. Then compute the total and check it against the nutritional score.

Points	
2	1. The person has a chronic illness or current condition that has changed the kind or amount of food eaten.
3	2. The person eats fewer than two full meals per day.
2	3. The person eats few fruits, vegetables, or milk products.
2	4. The person drinks three or more servings of beer, liquor, or wine almost every day.
2	5. The person has tooth or mouth problems that make eating difficult.
4	6. The person does not have enough money for food.
1	7. The person eats alone most of the time.
1	8. The person takes three or more different prescription or over-the-counter drugs each day.
2	9. The person has unintentionally lost or gained 10 pounds within the last 6 months.
2	10. The person cannot always shop, cook, or feed himself or herself.
Total	

Nutritional score:

0–2: Good. Recheck in 6 months.

3–5: Marginal. A local agency on aging has information about nutrition programs for the elderly. The National Association of Area Agencies on Aging can assist in finding help; call (800) 677-1116. Recheck in 6 months.

6 or more: High risk. A doctor should review this test and suggest how to improve nutritional health.

The DRIs for adults are divided by gender and by age: ages 19 to 30, 31 to 50, 51 to 70, and beyond 70 years of age. The intervals encompassing ages 19 through 50 are often referred to as *young adulthood*, 51 to 70 as *middle adulthood*, and beyond 70 years of age as *older adulthood*. These changes in nutrient needs take into consideration aging-related physiological alterations in body composition, metabolism, and organ function.

CALORIES

After age 30 or so, total calorie needs of physically inactive adults fall steadily throughout adulthood. There are a variety of explanations for the lower calorie requirements of older adults. Basal metabolic rate declines by about 2% per decade after age 30, such that overall energy needs of a 70-year-old man are reduced by 100 to 150 kcal per day compared to those of a 30-year-old man. Losses of lean mass and decreases in physical activity also tend to accompany aging. To a considerable extent, adults can exert control over this reduction in calorie need by exercising. Exercise can halt, slow, and even reverse reductions in lean body mass and subsequent declines in calorie need. Also, keeping calorie needs high makes it easier to meet one's nutrient needs and avoid becoming overweight.

PROTEIN

The protein intake of adults of all ages in North America typically exceeds the current RDA (0.8 grams per kilogram of body weight) and falls within the recommended range of 10% to 35% of total kilocalories. However, several studies indicate that consuming protein in amounts slightly higher than the RDA (in the range of 1.0 to 1.3 grams per kilogram of body weight) may help preserve muscle and bone mass among older adults (see Further Readings 1 and 5). Adults who have limited food budgets, have difficulty chewing meat, or are lactose intolerant may not get enough protein. Recall from Chapter 6 that any protein consumed in excess of that needed for the maintenance of body tissue will be broken down and used as energy or stored as fat. The waste products of metabolism of protein must be removed by the kidneys; excessive protein intake may accelerate kidney function decline.

FAT

The fat intake of adults of all ages is often at or above the 20% to 35% of total kilocalories recommended by the Food and Nutrition Board. It is a good idea for almost all adults to reduce their fat intake because of the strong link between high-fat diets and obesity, heart disease, and certain cancers. In addition, reducing fat intake "frees up" some calories that can be better "spent" on complex carbohydrates (see Further Readings 11 and 13).

CARBOHYDRATES

The total carbohydrate intake of adults of all ages in North America is often lower than recommended. In addition, many adults need to shift the carbohydrate composition of their diets to emphasize complex carbohydrates while minimizing intake of sugary, simple carbohydrates. While intakes of grains are generally adequate, only about 35% of Americans follow the advice to "make half your grains whole grains." A diet rich in complex carbohydrates helps us achieve nutrient needs and stay within calorie bounds because many highly sweetened foods are low in nutrients and high in calories. Replacing sweets and refined carbohydrates with foods rich in complex carbohydrates also improves blood glucose management. This is particularly helpful because inactivity and increasing body fatness are connected to insulin resistance. Dysfunctions of carbohydrate metabolism are so common that

ostomy Surgically created short circuit in intestinal flow where the end point usually opens from the abdominal cavity rather than the anus; for example, a colostomy. Short circuiting the intestinal flow means more water is lost in fecal matter than would be if the intestinal tract were intact.

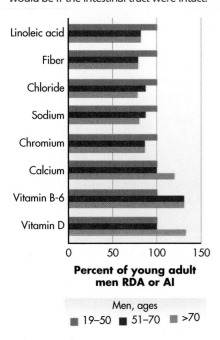

Percent of young adult men RDA or AI

Men, ages
■ 19–50 ■ 51–70 ■ >70

FIGURE 16-4 ▲ Relative nutrient requirements for aging adult men. Only nutrients that vary by age are shown.

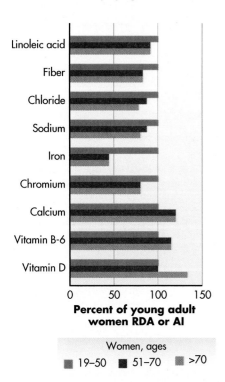

Percent of young adult women RDA or AI

Women, ages
■ 19–50 ■ 51–70 ■ >70

FIGURE 16-5 ▲ Relative nutrient requirements for aging adult women. Only nutrients that vary by age are shown.

more than 25% of those age 65 years or older have diabetes. A diet rich in fiber helps adults reduce their risk of colon cancer and heart disease, lower their blood cholesterol levels, and avoid constipation. The typical American adult gets slightly more than half the recommended amount of dietary fiber.

WATER

Many adults, especially those in the later years, fail to consume adequate quantities of water. In fact, many are in a constant state of mild dehydration and at risk of electrolyte imbalances. Low fluid intakes in older adults may be caused by a fading sensitivity to thirst sensations, chronic diseases, and/or conscious reductions in fluid intake in order to reduce the frequency of urination. Some also may have increased fluid output because they are taking certain medications (i.e., diuretics and laxatives), have an **ostomy,** and/or experience an age-related decline in the kidneys' ability to concentrate urine. Dehydration is very dangerous and, among other symptoms, can cause disorientation and mental confusion, constipation, fecal impaction, and death.

MINERALS AND VITAMINS

Dietary requirements for many nutrients change throughout the adult years (Figs. 16-4 and 16-5). The micronutrients that need special attention because they tend to be present in less than optimal amounts in the diets of many adults are calcium, vitamin D, iron, zinc, magnesium, folate, and vitamins B-6, B-12, and E. Adults with impaired absorption or who are unable to consume a nutritious diet may benefit from mineral or vitamin supplements matched with their needs. In fact, many nutrition experts recommend a daily balanced multivitamin and mineral supplement for older adults, especially for those 70 years of age and older. Supplements or fortified foods can be especially helpful when it comes to meeting vitamin D and vitamin B-12 needs.

Calcium and Vitamin D. These bone-building nutrients tend to be low in the diets of all adults. They become particularly problematic after age 50. Inadequate intake of these nutrients, combined with their reduced absorption, the reduced synthesis of vitamin D in the skin, and the kidneys' decreased ability to put vitamin D in its active form, greatly contributes to the development of osteoporosis. Getting enough of these nutrients is a challenge for many older adults because food sources of vitamin D are limited in the North American diet, and the major sources—fatty fish and fortified milk—are not widely consumed by older adults. Plus, with increasing age, lactase production frequently decreases. As you know, one of the richest and most absorbable sources of these nutrients, milk, contains lactose. To get the vitamin D and calcium needed, many with lactose intolerance can consume small amounts of milk at mealtime with no ill effects. Calcium-fortified foods, cheese, yogurt, fish eaten with bones (e.g., canned sardines or salmon), and dark green leafy vegetables can help those with lactose intolerance meet calcium needs—but these foods often do not provide vitamin D. Just 10 to 15 minutes per day of sunlight can make a large difference in vitamin D status.

Iron. Iron deficiency anemia, the most common type of malnutrition during the adult years, is found most frequently in women in their reproductive years because their diets do not provide enough iron to compensate for the iron lost monthly during menstruation. Other common causes of iron deficiency in adults of all ages include digestive tract injuries that cause bleeding (i.e., bleeding ulcers or hemorrhoids) and the use of medicines, such as aspirin, that cause blood loss. Impaired iron absorption due to age-related declines in stomach acid production may contribute to iron deficiency in older adults.

Zinc. In addition to less than optimal dietary zinc intake during adulthood, zinc absorption declines as stomach acid production diminishes with age. Poor zinc status may contribute to the taste sensation losses, mental lethargy, declines in immune function, and delayed wound healing many elderly adults experience.

Magnesium. This mineral tends to be low in adults' diets. Inadequate magnesium intakes may contribute to the bone loss, muscular weakness, and mental confusion seen in some elderly adults. It also can lead to sudden death from poor heart rhythm and is linked to cardiovascular disease, osteoporosis, and diabetes. The best source of magnesium is the diet; supplements can cause diarrhea.

Sodium. Average sodium intakes of American adults are about 3400 milligrams per day—more than two times the AI (see Further Reading 9)! A heavy reliance on processed foods and meals prepared in restaurants is mostly to blame for the high sodium intakes of Americans. The declining sense of taste that typically accompanies aging could also contribute to a preference for highly salted foods. The Dietary Guidelines advise adults to consume no more than 2300 milligrams of sodium per day. For adults over age 50; those with existing hypertension, pre-hypertension, diabetes, or kidney diseases; and all African-Americans, the upper limit is even lower: 1500 milligrams per day. These groups of people are especially sensitive to the effects of sodium on blood pressure. Less than 10% of adults, however, keep their sodium intakes below the UL. The most widely recognized consequence of high sodium intake is hypertension (see Chapter 9), but high sodium intake has also been linked to osteoporosis (increased calcium excretion in the urine) and may overtax the poorly functioning kidneys of older adults.

Even though excessive sodium receives most of the attention, low blood sodium (hyponatremia) is also a concern for older adults. Adults older than age 70, especially those who take diuretic medications or who have poor kidney function, are at increased risk for hyponatremia. The consequences of mild hyponatremia include lightheadedness, confusion, and unsteady gait, which can certainly increase the risk for falls among older adults. Other problems include fatigue, muscle cramps, and lack of appetite. There is no reason to severely restrict sodium among older adults who do not have kidney disease, but lowering sodium intakes closer to the AI would improve health for most adults.

▲ Sodium needs for older adults are 1200 to 1300 milligrams per day. Older adults routinely consume at least this much sodium. Potassium needs are 4700 milligrams per day. Many older adults do not meet this goal.

Folate and Vitamins B-6 and B-12. Sufficient folate, because of its role in prevention of neural tube defects, is very important to women during the childbearing years. In later years, folate and vitamins B-6 and B-12 are especially important because they are required to clear homocysteine from the bloodstream. Introduced in Chapter 5, elevated blood concentrations of homocysteine are associated with the increased risk of cardiovascular disease, stroke, bone fracture, and neurological decline seen in some elderly people. Vitamin B-12 is a particular problem for the older population because a deficiency may exist even when intake appears to be adequate. As people age, the stomach slows its production of acid and intrinsic factor, which leads to poor absorption of vitamin B-12. If vitamin B-12 is depleted, anemia and nerve damage could result. Adults age 51 years and older need to meet vitamin B-12 needs with supplements or foods fortified with synthetic vitamin B-12 because it is more readily absorbed than natural forms of B-12.

Vitamin E. The dietary intake of most of the population falls short of recommendations for vitamin E. Low vitamin E intake means that the body has a reduced supply of antioxidants, which may increase the degree of cell damage caused by free radicals, promote the progression of chronic diseases and cataracts, and speed aging. In addition, low vitamin E levels can lead to declines in physical abilities.

Carotenoids. Dietary intakes of certain carotenoids have been shown to have a variety of important anti-aging and health protective effects. Specifically, lutein and zeaxanthin have been linked with the prevention of cataracts and age-related

▲ How does this breakfast of whole-grain cereal, fruit, and milk compare to MyPlate?

macular degeneration. Diets high in fruit and vegetables, the major sources of carotenoids and other beneficial phytochemicals, are consistently shown to be protective against a wide variety of age-related conditions.

ARE ADULTS FOLLOWING CURRENT DIETARY RECOMMENDATIONS?

In general, adults in North America are trying to follow many of the diet recommendations described in this chapter. Since the mid-1950s, they have consumed less saturated fat as more people substitute fat-free and low-fat milk for cream and whole milk. However, they eat more cheese, usually a concentrated form of saturated fat. Since 1963, they have eaten less butter, fewer eggs, less animal fat, and more vegetable oils and fish. These changes generally comply with recommendations to reduce the intake of saturated fat and cholesterol in favor of unsaturated fat choices. Animal breeders are raising much leaner cattle and hogs than those produced in 1950, which also helps reduce saturated fat intake.

Other aspects of the average adult diet are still in need of improvement. The latest nutrition survey of eating habits in the United States shows that the major contributors of calories to the adult diet are white bread, beef, doughnuts, cakes and cookies, soft drinks, milk, chicken, cheese, alcoholic beverages, salad dressing, mayonnaise, potatoes, and sugars/syrups/jams. If Americans were truly lowering their intakes of sugar, saturated fat, and sodium while increasing their intakes of fiber, many of these foods would not appear at the top of the list.

✔ CONCEPT CHECK 16.2

1. Is the current adult RDA for protein appropriate for older adults? Why or why not?
2. Which nutrients should be limited in the diets of most American adults? Suggest three specific dietary changes that would help to limit these food components.
3. Name three nutrients that are commonly lacking in the diets of adults. Suggest one rich food source of each of these nutrients.

Researchers believe that maintaining lean muscle mass may be the most important strategy for successful aging. This is because maintaining lean muscle mass:

- Maintains basal metabolic rate, which helps decrease the risk of obesity.
- Keeps body fat low, which helps control blood cholesterol levels and avoid the onset of type 2 diabetes.
- Maintains body water, which decreases the risk of dehydration and improves body temperature regulation.

16.3 Physiological Factors Related to Nutritional Status of Adults

As for people in any age group, the food choices and nutritional adequacy of adults' diets depend on the interplay of physiological, psychosocial, and economic factors. Alterations in any one of these factors can result in deteriorations in the quality of dietary intake, nutritional status, and health. The implications of many of the physiological changes that occur during adulthood on dietary intake and nutrient needs are summarized in Table 16-2. Some of the changes listed (e.g., tooth loss and loss in taste and smell perceptions) can influence dietary intake. Other changes (e.g., loss of lean body tissue) can alter nutrient and/or calorie needs. Still other changes (e.g., reduced stomach acidity, diminished kidney function) can cause changes in nutrient utilization. Chronic diseases and the need for medications are additional physiological changes many adults experience that can influence food intake and nutrient needs. The following section details the influences of several of these factors on nutritional status during adulthood.

TABLE 16-2 ▶ **Typical Physiological Changes of Aging and Recommended Diet and Lifestyle Responses**

Physiological Changes	Recommended Responses
Appetite ⬇	• Monitor weight and strive to eat enough to maintain healthy weight. • Use meal replacement products, such as Boost® and Ensure Plus®. • Choose energy-dense foods, such as plant sources of fat.
Sense of taste and smell ⬇	• Vary the diet. • Experiment with salt-free herbs and spices.
Chewing or swallowing ability ⬇	• Work with a dentist to maximize chewing ability. • Modify food consistency as necessary.
Sense of thirst ⬇	• Monitor fluid intake. • Stay alert for evidence of dehydration (e.g., minimal output or dark-colored urine).
Stomach acidity ⬇	• Include some lean meat and iron-fortified foods in the diet. • Ask physician to monitor blood iron status. • Consume iron-rich foods with a source of vitamin C. • Choose foods fortified with vitamin B-12 or use a supplemental source of vitamin B-12.
Bowel function ⬇	• Consume enough fiber daily, choosing primarily fruits, vegetables, and whole-grain breads and cereals. • Meet fluid needs.
Lactase production ⬇	• Limit milk serving size at each use. • Substitute yogurt or cheese for milk. • Use reduced-lactose or lactose-free products. • Seek nondairy calcium sources.
Liver function ⬇	• Consume alcohol in moderation, if at all. • Avoid consuming dietary supplements that contain more than 100% of the Daily Value of nutrients, especially vitamin A.
Insulin function ⬇	• Maintain healthy body weight. • Choose carbohydrates with low glycemic index (see Chapter 4). • Perform regular physical activity.
Kidney function ⬇	• If necessary, work with physician and registered dietitian to modify protein and other nutrients in diet.
Immune function ⬇	• Meet nutrient needs, especially protein, vitamin E, vitamin B-6, and zinc. • Perform regular physical activity.
Lung function ⬇	• Avoid tobacco products. • Perform regular physical activity.
Vision ⬇	• Regularly consume sources of carotenoids, vitamin C, vitamin E, and zinc (e.g., fruits, vegetables, and whole-grain breads and cereals). • Moderate total fat intake. • Wear sunglasses in sunny conditions. • Avoid tobacco products. • Perform regular physical activity (to lessen insulin resistance). • In the case of diagnosed moderate macular degeneration, talk with a physician about following a protocol of zinc, copper, vitamin E, vitamin C, and beta-carotene supplementation.
Lean tissue ⬇	• Meet nutrient needs, especially protein and vitamin D. • Perform regular physical activity, including strength training.
Cardiovascular function ⬇	• Use diet modifications or physician-prescribed medications to keep blood lipids and blood pressure within desirable ranges. • Stay physically active. • Achieve and maintain a healthy body weight.

(continued)

TABLE 16-2 ► Typical Physiological Changes of Aging and Recommended Diet and Lifestyle Responses *(continued)*

Physiological Changes	Recommended Responses
Bone mass ⊥	• Meet nutrient needs, especially protein, calcium, and vitamin D (regular sun exposure helps meet needs for vitamin D). • Perform regular physical activity, especially weight-bearing exercise. • Women should consider use of approved osteoporosis medications at menopause. • Remain at a healthy weight (especially avoid unneeded weight loss).
Mental function ⊥	• Meet nutrient needs (e.g., vitamin E, vitamin C, vitamin B-6, folate, and vitamin B-12) and consume seafood twice a week. • Strive for lifelong learning. • Perform regular physical activity. • Obtain adequate sleep.
Fat stores ↑	• Avoid overeating. • Perform regular physical activity.

BODY COMPOSITION

The primary changes in body composition that occur with aging are diminished lean body mass, increased fat stores, and decreased body water. A focus on exercise, as detailed in a later section, can attenuate many of these unwanted changes.

The loss of lean body mass is termed **sarcopenia.** Some muscle cells shrink and others are lost as muscles age; some muscles lose their elasticity as they accumulate fat and collagen. Loss of muscle mass leads to a decrease in basal metabolism, muscle strength, and energy needs. Less muscle mass also leads to lower physical activity, which makes the prognosis for maintaining muscle even worse. Clearly, it is best to avoid this downward spiral.

As lean tissue declines with age, body fat often increases, a condition called **sarcopenic obesity** (see Further Reading 14). Much of this increase in body fat results from overeating and limited physical activity, although even athletic men and lean women typically gain some degree of midsection fat after age 50. A small fat gain in adulthood may not compromise health, but large gains are problematic. Recall that obesity can raise blood pressure and blood glucose, and diminish a person's ability to perform daily tasks, such as carrying groceries from the car.

Decreases in body weight can be a problem for adults age 70 and older, of whom about 2% are underweight (BMI less than 18.5). Unintended weight loss is troublesome for older people in particular because it increases the risk of undernutrition, which alters an individual's ability to cope with illnesses and injuries and could ultimately lead to death. Potential causes for unintended weight loss among older adults (many of which can be detected using the nutrition screening tool in Figure 16-3) include:

- Illness
- Depression/social isolation
- Side effects of medications
- Changes in taste or smell
- Reduced chewing ability
- Limited financial resources
- Decreased dexterity or strength

BONES AND JOINTS

Recall from Chapter 9 that bone loss is an expected consequence of aging. In women, bone loss rapidly occurs after menopause. For men, bone loss is slow and steady from middle age throughout later life. Many older people may suffer from undiagnosed osteomalacia, a condition mainly caused by insufficient vitamin D. Osteoporosis can limit the ability of older people to shop, prepare food, and exercise. Consuming adequate vitamin D, calcium, and protein; not smoking; drinking alcohol moderately or not at all; and engaging in weight-bearing exercises can help preserve bone mass. Medications also can help lessen bone loss.

sarcopenia In general, loss of muscle tissue. Among older adults, this loss of lean mass greatly increases their risk of illness and death.

sarcopenic obesity Loss of muscle mass accompanied by gains in fat mass.

There are over 100 forms of arthritis, a disease that causes the degeneration and roughening of the cartilage that covers and cushions the joints. Such changes in the joints cause them to ache and become inflamed and painful to move. Osteoarthritis, which increases in prevalence with advancing age, is the leading cause of disability among older persons. Rheumatoid arthritis, which is not as common, is more prevalent in younger adults.

Although precise causes or cures are unknown, many unproven arthritis "remedies" have been publicized. Unusual diets, food restrictions, and nutrient supplements are some of the more popular treatments. However, no special diet, food, or nutrient has been proven to reliably prevent, relieve, or cure arthritis in humans. As far as supplements go, glucosamine and/or chondroitin have been most extensively studied in relation to arthritis. Many (but not all) studies demonstrate that these supplements can relieve pain, slow the progression of joint degeneration, or rebuild cartilage. Maintaining a healthy weight, which reduces stress on painful arthritic joints, is the only diet-related treatment known to offer some relief. MyPlate can be used as a guide in making healthy food choices for weight management.

EXERCISE

Many physical changes of aging can be traced back to a sedentary lifestyle. As you might predict, an active lifestyle helps preserve muscle mass and decrease body fat. Physical activity increases muscle strength and mobility, improves balance and decreases the risk of falling, eases daily tasks that require some strength, improves sleep, slows bone loss, and increases joint movement, thus reducing injuries. It also has a positive impact on a person's mental outlook. Ideally, an active lifestyle should be maintained throughout life and include activities to build endurance, strength, balance, and flexibility. The 2008 Physical Activity Guidelines for Americans provide the following guidelines specifically for older adults.

Aerobic Exercise. All adults should engage in moderate-intensity aerobic exercise for at least 150 minutes per week, vigorous-intensity aerobic exercise for 75 minutes per week, or an equivalent combination of the two. This amount of aerobic activity improves endurance and aids in prevention of chronic diseases. Longer duration of daily exercise may be required for weight loss or weight maintenance. Weight-bearing exercises are particularly helpful for preservation of bone mass. For older adults who have not been physically active, it is important to increase the pace gradually and encourage even small bouts (e.g., 10 minutes) of physical activity.

▲ Endurance, strength, balance, and flexibility are the four elements of the Go4Life campaign from the National Institute on Aging. Download a free copy of *Exercise & Physical Activity: Getting Fit for Life* at **www.nia.nih.gov**.

Strength Training. To maintain lean tissue and basal metabolic rate, strength training should include 8 to 10 different exercises (each with two sets of 8 to 15 repetitions), performed 2 to 3 times per week. Exercises that involve the large muscle groups (e.g., arms, back, and legs) and exercises that enhance grip strength should be emphasized. Start slowly, concentrate on breathing, rest between sets, avoid locking joints in the arms and legs, and stop an exercise if it becomes painful.

Balance Exercises. For those over age 65 who are at risk for falling, exercises that improve balance are recommended. Tai chi and yoga are a few types of exercise that can improve balance, but even standing on one foot or getting up from a chair without using hands can be a good start.

Flexibility Exercises. Stretching exercises for each major muscle group should accompany aerobic or strength exercises at least 2 days per week. Improving flexibility can make it easier to perform many simple tasks, such as tying one's shoes.

All older adults should avoid inactivity. Having an exercise plan, developed with a health professional to accommodate individual health risks and needs, will enhance success for older adults. Men older than age 40; women older than age 50;

those with heart conditions, diabetes, or joint problems; and anyone who has been sedentary should consult a physician before beginning an exercise program. The Go4Life campaign from the National Institute of Aging encourages older adults to become more active and provides many practical tips for exercising safely. Learn more at **go4life.nia.nih.gov** and see Further Reading 4.

DIGESTIVE SYSTEM

As you recall from Chapter 3, digestion begins in the mouth. About 25% of older adults have no natural teeth, and many more are missing some teeth. The problem of tooth loss is worse among low-income populations. Even with properly fitting dentures, chewing ability may be limited. Older adults with poor dentition may avoid meats or crunchy fruits and vegetables, thereby missing out on important nutrients such as protein, iron, and zinc (from meat) as well as potassium and fiber (from fruits and vegetables). Pureed meats and cooked vegetables are easier options for older adults with chewing problems (see Further Reading 15).

Further along the GI tract, the production of HCl, intrinsic factor, and some digestive enzymes (e.g., lactase) declines with advancing age. In addition, some medications affect acid production. As a result of low acid production, absorption of some minerals, such as iron, is impaired. Low levels of acid and intrinsic factor reduce the digestion and absorption of vitamin B-12. Thus, even with adequate intakes of iron and vitamin B-12, older adults may become anemic. Symptoms of lactose intolerance can lead to avoidance of dairy products, which can limit the availability of bone-building nutrients. Fortified foods or supplements can help older adults overcome these problems with digestion and absorption of nutrients.

Constipation is the main intestinal problem for older people. To prevent constipation, older people should meet fiber needs, drink enough fluids, and exercise. Fiber medications are generally unnecessary but may be useful when total energy consumption does not allow for enough fiber intake. Because some medications can cause constipation, a physician should be consulted to determine if a laxative or stool softener is needed.

In addition to changes in the GI tract, the functions of the accessory organs decline as we age. For instance, the liver functions less efficiently. A history of significant alcohol consumption or liver disease will intensify any existing problems with liver function. As liver efficiency declines, its ability to detoxify many substances, including medications, alcohol, and vitamin and mineral supplements, drops. The possibility for vitamin toxicity increases.

The gallbladder also functions less efficiently in later years. Gallstones can block the flow of bile out of the gallbladder into the small intestine, thereby interfering with fat digestion. Obesity is a major risk factor for gallbladder disease, especially in older women. A low-fat diet or surgery to remove the gallbladder may be necessary.

Although pancreatic function may decline with age, this organ has a large reserve capacity. One sign of a failing pancreas is high blood glucose, although this can occur as the result of several conditions. The pancreas may be secreting less insulin, or cells may be resisting insulin action (as is commonly seen in obese people with upper-body fat storage). Where appropriate, improved nutrient intake, regular physical activity, and loss of excess body weight can improve insulin action and blood glucose regulation.

NERVOUS SYSTEM

A gradual loss of nerve cells that transmit signals may decrease taste and smell perceptions and impair neuromuscular coordination, reasoning, and memory. Both hearing and vision decline with age. Hearing impairment is the greatest in those who have been exposed constantly to loud noises, such as urban traffic, aircraft noise, and music. Because they cannot hear well, older people may avoid social contacts, which increases their risk of poor dietary intake.

Declining eyesight, frequently caused by retina degeneration and cataracts, can affect a person's abilities to grocery shop, locate desired foods, read labels for

▲ Poor dental health contributes to decreased food intake and digestive problems. Serving softer, easier-to-chew foods and allowing extra time for chewing and swallowing improve food intake.

nutritional content, and prepare foods at home. Vision losses also may cause people to curtail social contacts, reduce physical activity, and not practice daily personal health and grooming routines. Macular degeneration, one form of failing eyesight in old age, is quite common, affecting about 9.1 million adults in the United States. A major risk factor is cigarette smoking. Diets rich in carotenoids help reduce the risk of macular degeneration. The risk of developing cataracts is decreased by consuming a diet rich in fruits and vegetables.

Neuromuscular coordination losses may make it difficult to shop for and prepare food. Physical tasks as simple as opening food packages can become so difficult that individuals restrict dietary intake to foods that require little preparation and depend on others to provide food that is ready to eat. Eating may become difficult, too. Loss of coordination makes it a challenge to grasp cup handles and manipulate eating utensils. As a result, older adults may avoid foods that can be easily spilled (e.g., soups and juices) or that need to be cut (e.g., meats, large vegetable pieces) and restrict food intake to easy-to-eat finger foods. Some may even withdraw from social interaction and eat alone, which can lead to inadequate nutrient intake.

▲ Immune function declines with age, so food safety becomes increasingly important for older adults. Chapter 13 provides advice on this topic, such as washing one's hands and work surfaces before preparing food.

IMMUNE SYSTEM

With age, the immune system often operates less efficiently. Consuming adequate protein, vitamins (especially folate and vitamins A, D, and E), iron, and zinc helps maximize immune system function (see Chapters 8 and 9). Recurrent sicknesses and poor wound healing are warning signs that a deficient diet (especially of protein and zinc) may be hindering the function of the immune system. On the other hand, overnutrition appears to be equally harmful to the immune system. For example, obesity and excessive fat, iron, and zinc intakes can suppress immune function.

ENDOCRINE SYSTEM

As adulthood progresses, the rate of hormone synthesis and release can slow. A decrease in insulin release or sensitivity to insulin, for instance, means that it takes longer for blood glucose levels to return to normal after a meal. Maintaining a healthy weight, exercising regularly, eating a diet low in fat and high in fiber, and avoiding foods with a high glycemic index can enhance the body's ability to use insulin and restore elevated blood glucose levels to normal after a meal.

CHRONIC DISEASE

The prevalence of obesity, heart disease, osteoporosis, cancer, hypertension, and diabetes rises with age. More than 8 out of every 10 elderly people have one of these chronic and potentially debilitating diseases. Half of all elderly people have at least two chronic conditions. Chronic diseases may have a strong impact on dietary intake. For instance, obesity, heart disease, and osteoporosis may impair physical mobility to the extent that victims are unable to shop for and prepare food. Chronic disease also can influence nutrient and calorie needs. Cancer, for example, boosts both nutrient and calorie needs. Hypertension may indicate a need to lower sodium intake. Nutrient utilization can be affected by chronic disease, too. For instance, diabetes alters the body's ability to utilize glucose. In addition, the effects of heart disease on the kidneys may impair their ability to reabsorb glucose, amino acids, and vitamin C.

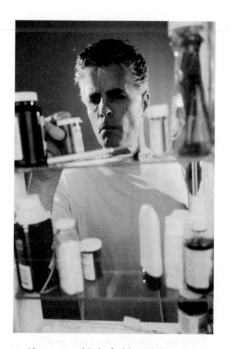

▲ About one-third of older adults take eight or more medications each day. In some cases, drugs can affect nutrient status. For example, certain diuretics may increase excretion of minerals in the urine. Other times, nutrients and other food components can impact the effectiveness of drugs. Vitamin K, for example, can affect the action of anti-clotting medications.

MEDICATIONS

Older adults are major consumers of medications (prescription and over-the-counter) and nutrient supplements. About 90% of older adults take at least one prescription medication daily, and half of all people over age 65 take several medicines each day. The rate of supplement use increases throughout adulthood such that by age 50, approximately half of all adults are using supplements daily. Physiological declines that occur during aging (e.g., reduced body water and reduced liver and

Numerous reports have documented significant health risks associated with the use of some herbal and alternative remedies, sometimes resulting in death. Studies especially implicate germander, pokeroot, sassafras, mandrake, penny-royal, comfrey, chaparral, yohimbe, lobelia, jin bu huan, kava kava, products containing stephanie and magnolia, senna, hai gen fen, paraguay tea, kombucha tea, tung shueh (Chinese black balls), and willow bark.

kidney function) exaggerate and prolong the effects of medications and nutrient supplements in older adults.

Medications can eradicate infections and control chronic diseases, but some also adversely affect nutritional status, particularly of those who are older and/or take many different medications. For instance, some medications depress taste and smell acuity or cause anorexia or nausea that can blunt interest in eating and lead to reduced dietary intake. Some medications alter nutrient needs. Aspirin, for example, increases the likelihood of stomach bleeding, so long-term use may elevate the need for iron, as well as other nutrients. Antibiotics kill beneficial bacteria along with pathogens, so they can limit the amount of vitamin K that is synthesized by bacteria in the large intestine (see Chapter 8). Some medications may impair nutrient utilization; diuretics and laxatives may cause excessive excretion of water and minerals. Even vitamin and mineral supplements may have unanticipated effects on nutritional status. Iron supplements taken in large doses can interfere with the functioning of zinc and copper. Folate supplements can mask vitamin B-12 deficiencies.

People who must take medications should eat nutrient-dense foods and avoid any specific food or supplement that interferes with the function of their medications. For example, vitamin K can reduce the action of oral anticoagulants, aged cheese can interfere with certain drugs used to treat hypertension and depression, and grapefruit can interfere with medications such as tranquilizers and those that lower cholesterol levels. A physician or pharmacist should be consulted about any restrictions on food and/or supplements.

COMPLEMENTARY AND ALTERNATIVE MEDICINE

At least half of adults report using some kind of dietary or herbal supplement. Recall from Chapter 2 that the safety, purity, and effectiveness of dietary supplements are not tightly monitored by FDA. Table 16-3 reviews some popular herbal remedies used by older adults. Note from the table that these products can pose health risks in certain people. In addition, they may be expensive ($100 per month or more in some cases) and are not covered by health insurance plans. Use of many herbal products has declined because of expense and questionable benefits. Instead, consumers are opting to spend their money on tried-and-true products, such as multivitamins, vitamin D, omega-3 fatty acids, fiber, and probiotics.

A rational approach to the use of herbal products is to use only one product at a time, keep a diary of symptoms, and check with one's physician first before discontinuing a prescribed medication. In addition, FDA advises anyone who experiences adverse side effects from an herbal remedy to contact a physician. Physicians are then encouraged to report such adverse events to FDA, state and local health departments, and consumer protection agencies.

For additional information about herbal remedies, access the following websites. These are regularly updated and cover the herbals listed in Table 16-3, as well as many others.

- National Institutes of Health National Center for Complementary and Alternative Medicine (NCCAM)
  www.nccam.nih.gov
- American Botanical Council
  abc.herbalgram.org
- Complementary and Alternative Medicine Program at Stanford (CAMPS)
  camps.stanford.edu
- National Institutes of Health Office of Dietary Supplements
  ods.od.nih.gov
- Natural Medicines Comprehensive Database
  naturaldatabase.therapeuticresearch.com

CRITICAL THINKING

Jamila went to her local pharmacy yesterday to look for a product to help her stay awake while studying. On the shelves, she found a dietary supplement claiming to be a Chinese herbal remedy for sleepiness and fatigue. She thought that because a pharmacy carried the product, it should be safe and work as indicated on the label.

Is she correct in these assumptions? Are there specific risks associated with taking such herbal remedies?

TABLE 16-3 ▶ A Closer Look at Some Popular Herbal Remedies

Product	Purported Effects*	Side Effects	Who Should Especially Seek Physician Guidance Before Use
Black cohosh	• Mild reduction of postmeno-pausal symptoms (shown to be effective in some but not all women; use should not go beyond 6 months in general)	• Nausea • Liver damage	• Women who have had breast cancer • Pregnant women • Anyone taking estrogen, hypertension medications, or blood-thinning medications** • Anyone with abnormal liver function
Chondroitin sulfate	• Pain relief from osteoarthritis (possibly effective)	• GI tract upset • Swelling • Hair loss • Irregular heartbeat	• People with asthma • Men with prostate cancer • Anyone who takes blood-thinning medications
Cranberry	• Prevention or treatment of urinary tract infections (possibly effective) • Prevention of *Helicobacter pylori* infections of the stomach, with the aim of reducing the risk for ulcers (some evidence of efficacy)	• Gastrointestinal upset and diarrhea • Use of concentrated tablets may increase the risk of kidney stones	• People susceptible to kidney stones • Anyone taking antidepressants or prescription painkillers • Anyone taking blood thinners
Echinacea	• Prevention or treatment of colds or other infections (possibly effective)	• Nausea • Skin irritation • Allergic reactions • Minor GI tract upset • Increased urination	• Anyone with an autoimmune disease • Pre- or postsurgical patients • Anyone with allergies to daisies
Garlic	• Antifungal properties (possibly effective) • Reduction of blood cholesterol or blood pressure (possibly effective)	• GI tract upset (e.g., heartburn, flatulence) • Unpleasant odor • Allergic reactions	• Pre- or postsurgical patients • Perinatal women • Anyone taking blood-thinning medications or AIDS medications
Ginger	• Relief of nausea and vomiting (possibly effective)	• Heartburn • Diarrhea • Increased menstrual bleeding	• Although some studies show ginger can help to relieve nausea and vomiting associated with pregnancy, it should be used cautiously due to possible risks for bleeding. • People with bleeding disorders or who take blood-thinning medications • Anyone with a heart condition • People taking blood glucose-lowering medications
Ginkgo biloba	• Increased circulation (possibly effective) • Improvement of memory (especially for people with Alzheimer's disease; low evidence of efficacy)	• Mild headache • GI tract upset • Allergic reactions • Irritability • Reduced blood clotting • Seizures (if contaminated with toxic ginkgo seeds)	• People with bleeding disorders • Pre- or postsurgical patients • Anyone with allergies to the plant • Concurrent use of feverfew, garlic, ginseng, dong quai, or red clover • Anyone taking diabetes medications, blood-thinning medications, vitamin E supplements, antidepressants, or diuretics
Ginseng	• Lowering blood sugar (possibly effective) • Increased energy (possibly ineffective)	• Hypertension • Asthma attacks • Irregular heartbeat • Hypoglycemia • Insomnia • Headache • Nervousness • GI tract upset • Reduced blood clotting • Menstrual irregularities and breast tenderness	• Anyone who takes a prescription drug • Women who have had breast cancer • Anyone with chronic GI tract disease • Anyone with uncontrolled hypertension

(continued)

TABLE 16-3 ▶ **A Closer Look at Some Popular Herbal Remedies** *(continued)*

Product	Purported Effects*	Side Effects	Who Should Especially Seek Physician Guidance Before Use
Glucos-amine sulfate	• Prevention and treatment of osteoarthritis (likely effective)	• GI tract upset	• People with asthma or shellfish allergies
St. John's wort	• Alleviation of mild to moderate depression (likely effective)	• Mild GI tract upset • Rash • Tiredness • Restlessness • Increased sensitivity to sunlight	• Anyone who takes a prescription drug • People with UV sensitivity, including that induced by medications or other treatments*** • People with bipolar disorder, major depression, schizophrenia, and Alzheimer's disease • Anyone recovering from a graft or organ transplant
Turmeric	• Pain relief from osteoarthritis (possibly effective)	• GI tract upset • Dizziness	• People with gallbladder problems • People with GERD • Pre- or postsurgical patients
Valerian	• Alleviation of insomnia (possibly effective) • Reduction of anxiety (insufficient evidence to rate effectiveness)	• Impaired attention • Headache • Morning grogginess • Irregular heartbeat • GI tract upset • Disagreeable odor • Withdrawal delerium	• Anyone taking anesthetics or central nervous system depressants**** • Anyone who drinks alcohol • People about to operate heavy machinery or drive

Pregnant or breastfeeding women, children under 2 years of age, anyone over the age of 65 years, and anyone with a chronic disease should never take supplements unless under the guidance of a physician. A concern has been raised with regard to patients who abruptly end alternative medicines at the start of hospital treatments or deny that they are using alternative therapy. Interactions between alternative therapies and pharmaceutical drugs can be drastic and include complications such as delirium, clotting abnormalities, and rapid heartbeat, resulting in the need for intensive care. Full disclosure of all prescription and nonprescription treatments aids in prevention of such complications. Experts recommend that, if time permits, patients stop taking herbal products for about a week before a scheduled surgery or otherwise take all original supplement containers to the hospital, so that the anesthesiologist can evaluate what was taken.

*Ratings of effectiveness from MedlinePlus and/or NCCAM.

**Coumadin, aspirin, Heparin, Lovenox, or Fragmin.

***Sulfa medications, anti-inflammatory medications, or acid-reflux medications.

****Valium, halcion, and seconal.

☑ CONCEPT CHECK 16.3

1. What is sarcopenia? What diet and lifestyle changes would you suggest to avoid this condition?

2. Describe three ways aging affects the processes of digestion, absorption, and utilization of nutrients.

3. List two types of complementary and alternative medicine adults may use to improve immune function. List two that adults may use to cope with arthritis.

16.4 Psychosocial Factors Related to Nutritional Status of Adults

The potential for possible embarrassment caused by deteriorating physical capabilities may cause older adults to withdraw from social interaction and eat alone rather than with others. Those who eat alone, regardless of the reason, seldom eat as much or as nutritiously as they should. Both young and old people who eat without

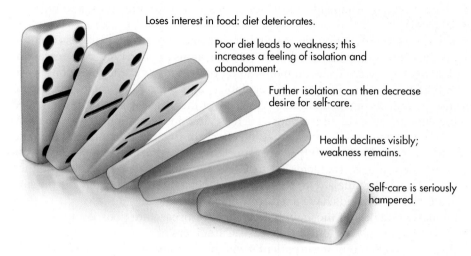

Social isolation; perhaps spouse has died.

Loses interest in food: diet deteriorates.

Poor diet leads to weakness; this increases a feeling of isolation and abandonment.

Further isolation can then decrease desire for self-care.

Health declines visibly; weakness remains.

Self-care is seriously hampered.

FIGURE 16-6 ◄ The decline of health often seen in older adults. A small change leads to a chain of events or "domino effect" that results in poor health. Early intervention can prevent declines in physical health related to psychosocial factors.

companionship tend to feel unmotivated to shop for or prepare foods. Many become apathetic toward life, which over time causes health and nutritional status to decline. As you will see in Section 16.5, several nutrition assistance programs can help older people obtain the food and social support needed for good health.

DEPRESSION

A positive outlook on life and intact support networks help make food and eating interesting and satisfying. In contrast, social isolation, grief, chronic pain and infirmity, or a change in lifestyle can lead to depression, loss of appetite, lack of interest in food, and disability. Major depression affects 5% to 8% of adults in the United States. This statistic increases to about 11% of men and 16% of women over the age of 65. Left untreated, depression can lead to a continual decline in appetite, which results in weakness, poor nutrition, mental confusion, and increased feelings of isolation and loneliness (Fig. 16-6). In contrast, some people cope by overeating, which can lead to obesity and its associated problems. Depression may signal an underlying illness and can also impair recovery from other illnesses or injuries. As many as 15% of cases end in suicide. For these reasons, early detection of depression is important in older adults. Depression is often treatable, but medication alone will not help those experiencing major life changes, such as the death of a spouse. Adequate social support and/or psychological intervention are also essential.

Just as depression affects diet, dietary intake may influence depression. As yet, there is insufficient evidence to make nutrition recommendations to alleviate symptoms of depression, but it has been linked to poor intakes of omega-3 fatty acids, vitamin D, and some B vitamins.

ALZHEIMER'S DISEASE

Alzheimer's disease is an irreversible, abnormal, progressive deterioration of the brain that causes victims to steadily lose the ability to remember, reason, and comprehend. Alzheimer's disease often takes a terrible toll on the mental and eventual physical health of older people. About 5.1 million adults in the United States have the disease.

The 10 warning signs of Alzheimer's disease are listed in the margin. No one is sure exactly what causes this disease, but scientists have proposed various causes, including alterations in cell development or protein production in

Ten Warning Signs of Alzheimer's Disease

1. Recent memory loss that affects job performance
2. Difficulty performing familiar tasks
3. Problems with language
4. Disorientation to time and place
5. Faulty or decreased judgment
6. Problems with abstract thinking
7. Tendency to misplace things
8. Changes in mood or behavior
9. Changes in personality
10. Loss of initiative

the brain, strokes, altered blood lipoprotein composition, obesity, poor blood glucose regulation (e.g., diabetes), high blood pressure, viral infections, and high free radical levels.

Early efforts at prevention are of prime importance, because the process of cognitive decline begins 10 to 20 years before warning signs appear. Preventive measures for Alzheimer's disease focus on maintaining brain activity through lifelong learning, eating a diet rich in fruits and vegetables, and taking ibuprofen. The role of nutrition in preventing or minimizing the risk of this disease is being investigated. Getting enough antioxidant nutrients, such as vitamin C, vitamin E, and selenium, helps protect the body from the damaging effects of free radicals. Adequate intakes of folate and vitamins B-6 and B-12 are especially important because elevated blood homocysteine is also a risk factor. Dietary fats, too, may play a role in keeping this disease at bay. Individuals with diets rich in omega-3 fatty acids and low in saturated and *trans* fatty acids have a reduced risk of Alzheimer's disease.

The dietary intakes of those with Alzheimer's disease are poor, compared with those of a similar age without this disease. Caregivers of those who have Alzheimer's disease need to monitor the patient's weight to ensure maintenance of a healthy weight and nutritional state. Other tips are serving omega-3-rich fish in meals twice per week and making sure eating habits do not pose a health risk (e.g., holding food in one's mouth or forgetting to swallow). Regular physical activity has also been shown to improve mental status in people afflicted with this disease (see Further Reading 7).

ECONOMIC FACTORS

The amount of money available for purchasing food can have a great impact on the types and amounts of food one eats. Unemployment, underemployment, retirement, death of a wage earner, or anything else that limits income makes it difficult to get sufficient quantities of healthful foods and can diminish nutritional status and health. Insufficient income is a particular problem among those ages 65 and up, and as a result, they frequently have trouble making sure they remain well nourished. The Commodity Supplemental Food Program and Supplemental Nutrition Assistance Program are two federal U.S. programs that can help low-income individuals of all ages procure the foods they need.

Warning signs of undernutrition in older people that form the acronym DETERMINE:
- **D**isease
- **E**ating poorly
- **T**ooth loss or mouth pain
- **E**conomic hardship
- **R**educed social contact and interaction
- **M**ultiple medications
- **I**nvoluntary weight loss or gain
- **N**eed for assistance with self-care
- **E**lder at an advanced age

✔ CONCEPT CHECK 16.4

1. How does depression influence nutritional status?
2. List at least two chronic illnesses that are linked to poor intake of omega-3 fatty acids.
3. Poor economic status can limit the ability to purchase foods. Beyond this, how can limited financial resources affect nutritional status?

16.5 Ensuring a Healthful Diet for the Adult Years

Recommended dietary practices for later years would be to increase the diet's nutrient density and to make sure fiber and fluid intakes are adequate. In addition, some protein should come from lean meats to help meet protein, vitamin B-6, vitamin B-12, iron, and zinc needs.

Singles of all ages face logistical problems with food: purchasing, preparing, storing, and using food with minimal waste are challenging. Value-priced packages of meats and vegetables are normally too large to be useful for a single person. Many singles live in small dwellings, some without kitchens and freezers. Creating a diet to accommodate a limited budget and facilities and a single appetite requires special considerations. The following are some practical suggestions for diet planning for singles:

- If one owns a freezer, cook large amounts, divide into portions, and freeze.
- Buy only what one will use; small containers may be expensive, but letting food spoil is also costly.
- Ask the grocer to break open a family-sized package of wrapped meat or fresh vegetables and separate it into smaller units.
- Buy only several pieces of fruit—perhaps a ripe one, a medium-ripe one, and an unripe one—so that the fruit can be eaten over a period of several days.
- Keep a box of dry milk handy to add nutrients to recipes for baked foods and other foods for which this addition is acceptable.

Table 16-4 provides more ideas for healthful eating in later years.

Nutritional deficiencies and protein-calorie malnutrition have been identified among some aging populations, particularly those in hospitals, nursing homes, or long-term care facilities. These nutritional problems increase the risk for many

TABLE 16-4 ▶ Guidelines for Healthful Eating in Later Years

- Eat regularly; small, frequent meals may be best. Use nutrient-dense foods as a basis for menus.
- Use labor-saving devices and some convenience foods, but try to incorporate some fresh foods into daily menus.
- Try new foods, new seasonings, and new ways of preparing foods. Use canned goods in moderation or choose those low in sodium.
- Keep easy-to-prepare foods on hand for times when you feel tired.
- Have a treat occasionally, perhaps an expensive cut of meat or a favorite fresh fruit.
- Eat in a well-lit or sunny area; serve meals attractively; and use foods with different flavors, colors, shapes, textures, and smells.
- Arrange kitchen and eating area so that food preparation and cleanup are easier.
- Eat with friends, relatives, or at a senior center when possible.
- Share cooking responsibilities with a neighbor.
- Use community resources for help with shopping and other daily care needs.
- Stay physically active.
- If possible, take a walk before eating to stimulate the appetite.
- When necessary, chop, grind, or blend hard-to-chew foods. Softer, protein-rich foods (e.g., shredded meats, eggs) can be substituted for whole pieces of meat when poor dental function limits normal food intake. Prepare soups, stews, cooked whole-grain cereals, and casseroles.
- If your dexterity is limited, cut the food ahead of time (perhaps with the help of a friend or family member), use utensils with deep sides or handles, and obtain more specialized utensils if needed.

diseases, including bed sores (pressure ulcers), and compromise recovery from illness and surgery. Friends, relatives, and health care professionals should monitor nutrient intake in all older people, including those who live in long-term care facilities. Family members can play a valuable role in making sure nutrient needs are met by looking for weight maintenance based on regular, healthful meal patterns. If problems arise in consuming a healthful diet, registered dietitians can offer professional and personalized advice.

Overall, older adults benefit from good nutrition in many ways. Meeting nutrient needs delays the onset of some diseases; improves the management of some existing diseases; hastens recovery from many illnesses; increases mental, physical, and social well-being; and often decreases the need for and length of hospitalization. Older adults should use MyPlate as a guide to healthy meals but emphasize nutrients of special concern (e.g., calcium, vitamin D, and vitamin B-12). Daily Food Plans are available from **www.ChooseMyPlate.gov** for older adults, based on age, weight, height, gender, and activity level. A focus on nutrient-dense food choices, plenty of fluids, modified physical activity goals, and use of certain dietary supplements address the unique needs of older adults.

Obtaining enough food may be difficult for some older persons, especially if they are unable to drive and relatives do not live close enough to help with cooking or shopping. For an older person, a request for help may be equated to a loss of independence. Pride or fear of being victimized by those they hire may stand in the way of much-needed aid. In these cases, friends can be a big help. Special transportation arrangements may also be available through a local transit company or taxi service.

Many eligible older people are missing meals and are poorly nourished because they do not realize that programs are available to help them. Irregular meal patterns and weight loss, often caused by difficulties in preparing food, are warning signs that undernutrition may be developing. An effort should be made to identify poorly nourished people and inform them of community services.

COMMUNITY NUTRITION SERVICES FOR OLDER PEOPLE

Health care advice and services for older people can come from clinics, private practitioners, hospitals, and health maintenance organizations (see Further Reading 10). Home health care agencies, adult day-care programs, adult overnight-care programs, and **hospice care** (for the terminally ill) can provide daily care.

The Older Americans Act Nutrition Program serves about 242 million meals each year to adults over the age of 60 in the United States. Federal standards mandate that these meals supply at least one-third of adult energy and nutrient requirements.

Some meals (e.g., Meals on Wheels) are delivered directly to older adults in their homes. Although home-delivered meals can make a valuable contribution to the nutritional status of home-bound older adults, services are usually limited to one or two meals per day. If a recipient has a poor appetite, the food may end up stored for later or simply thrown away. If foods are not eaten on delivery and not stored properly, risk for foodborne illness could be a concern.

Other meals are provided by congregate meal programs, which usually serve lunch at a central location. With congregate meals, the social aspect of eating tends to improve nutritional intake. However, programs generally provide just one meal per day on just 5 days peer week. Where can older adults with limited resources find additional nutrition assistance?

In addition to congregate and home-delivered meals, federal commodity distribution is available in some areas of the United States to low-income older people. Older people whose incomes are below the poverty level can benefit from the SNAP program (food stamps) (see Chapter 12 for details on these programs). Food cooperatives and a variety of clubs and religious and social organizations provide additional aid.

hospice care A program offering care that emphasizes comfort and dignity at the end of life.

To learn more about resources available for older adults in your area, check out the following websites:
Elder Care Locator
www.eldercare.gov
National Institute on Aging
www.nia.nih.gov
American Geriatrics Society
www.americangeriatrics.org
Administration on Aging
www.aoa.gov

✔ CONCEPT CHECK 16.5

1. Gerald is a 76-year-old man whose wife recently passed away. He now lives alone for the first time in his life and is not accustomed to preparing meals for himself. What three pieces of advice would you give Gerald about eating well for an older, single man?

2. List three possible nutrition resources for an older adult with limited financial means.

CASE STUDY Dietary Assistance for an Older Adult

Frances is a 78-year-old woman who suffers from macular degeneration, osteoporosis, and arthritis. Since her husband died a year ago, she has moved from their family house to a small one-bedroom apartment. Her eyesight is progressively getting worse, making it hard to go to the grocery store or even to cook (for fear of burning herself). She is often lonely; her only son lives 1 hour away and works two jobs, but he visits her as often as he can. Frances has lost her appetite and, as a result, often skips meals during the week. She has resorted to eating mostly cold foods. These are simple to prepare but seriously limit the variety and palatability of her diet. Also, she wears dentures and has trouble chewing tough meats and foods with crisp textures. She is slowly losing weight as a result of her dietary changes and loss of appetite.

Her typical diet usually consists of a breakfast that may include 1 slice of wheat toast with margarine, honey, and cinnamon, and 1 cup of hot tea. If she has lunch, she normally has ½ can of peaches, half of a turkey and cheese sandwich, and ½ glass of water. For dinner, she might have half of a tuna fish sandwich made with mayonnaise and 1 cup of iced tea. She usually includes one or two cookies at bedtime.

Answer the following questions and check your responses at the end of this chapter.

1. What nutrients are likely to be inadequate in her current diet?
2. What potential effects will Frances' poor dietary pattern have on her health status?
3. Which physiological changes of aging will add to the effects of her poor diet (review Table 16-2)?
4. What other questions would you ask Frances to get a better picture of her nutrition status?
5. What services are available in the community that could help Frances improve her diet?
6. What other convenience foods could be included in her diet to make it more healthful and more varied?

Given the wide spectrum of alcohol use and abuse, knowledge of alcohol consumption and its relationship to overall health is essential to the study of nutrition. Alcoholic beverages contain the chemical form of alcohol known as **ethanol.** Although not a nutrient *per se,* alcohol is a source of calories (about 7 kcal per gram) for approximately half of adults, constituting about 3% of total calories in the average North American diet (Table 16-5).

A "standard drink" contains about 14 grams of alcohol. For beer or wine coolers, this equates to a 12-fluid ounce serving. Most cans of beer are 12 fluid ounces, but some cans or bottles may contain as much as 40 fluid ounces. Malt liquor has a slightly higher alcohol content than beer, so the standard drink size is 8 to 9 fluid ounces. For wine, a 5-fluid ounce glass is a standard drink. A standard drink of hard liquor, such as whiskey or rum, is the size of a shot glass—1.5 fluid ounces. These standard drink sizes are depicted in Figure 16-7.

Now that you know what a standard drink is, let us define some commonly referenced patterns of drinking. **Moderate drinking** is defined by single-day and weekly amounts; to be classified as a moderate drinker, an individual must meet both criteria. For men, moderate drinking is no more than 4 drinks in 1 day AND no more than 14 drinks per week. For women, moderate drinking is no more than 3 drinks in 1 day AND no more than 7 drinks per week. **Heavy drinking** encompasses

all patterns of alcohol consumption in excess of moderate drinking. **Binge drinking,** introduced in the Chapter 1 Nutrition and Your Health section, is characterized by drinking so much within a short period of time (usually within 2 hours) that blood alcohol concentration rises above the legal limit of 0.08 grams per deciliter. This usually correlates to five or more drinks in a row for men or four or more drinks in a row for women. Binge drinking is certainly linked to negative effects on physical and emotional health, but it is not necessarily an alcohol use disorder, which will be discussed in a later section.

Moderate consumption of alcohol by a person of legal age is an acceptable practice and even has some health benefits (see Further Reading 3). However, only about half of alcohol consumed is done so in moderation. About one in five people over the age of 12 report binge drinking in the last 30 days. Furthermore, about 18 million people in the United States suffer from **alcohol use disorders.** By far, alcohol is the most commonly abused drug.

How Alcoholic Beverages Are Produced

The basis of alcohol production is fermentation, a process by which microorganisms break down simple sugars (e.g., glucose or maltose) to alcohol, carbon dioxide, and water in the absence of oxygen. High-carbohydrate foods especially encourage the growth of yeast, the microorganism responsible for alcohol production. Wine is formed by the fermentation of grape or other fruit juices. Beer is made from malted cereal grain. Distilled spirits (e.g., vodka, gin, and whiskey) are made from any number of fruits, vegetables, and grains. Production temperatures, the composition of the food used for fermentation, and aging techniques determine the characteristics of the product.

Alcohol proof represents twice the volume of alcohol in percentage terms. Thus, 80 proof vodka is 40% alcohol.

ethanol Chemical term for the form of alcohol found in alcoholic beverages.

moderate drinking For men, consuming no more than 4 drinks in 1 day AND no more than 14 drinks per week; for women, consuming no more than 3 drinks in 1 day AND no more than 7 drinks per week.

heavy drinking Any pattern of alcohol consumption in excess of 4 drinks in 1 day or 14 drinks per week for men or 3 drinks in 1 day or 7 drinks per week for women.

binge drinking Drinking sufficient alcohol within a 2-hour period to increase blood alcohol content to 0.08 grams per deciliter or higher; for men, consuming five or more drinks in a row; for women, consuming four or more drinks in a row.

alcohol use disorder A psychiatric disorder characterized by a problematic pattern of alcohol use that leads to significant impairment or distress.

TABLE 16-5 ► Alcohol, Carbohydrate, and Calorie Content of Alcoholic Beverages*

Beverage	Amount (fluid ounces)	Alcohol (grams)	Carbohydrates (grams)	Calories (kcal)
Beer				
Regular	12.0	13	13	146
Light	12.0	11	5	99
Distilled Spirits				
Gin, rum, vodka, bourbon, whiskey (80 proof), brandy, and cognac	1.5	14	—	96
Wine				
Red	5	14	2	102
White	5	14	1	100
Dessert and sweet	5	23	17	225
Rosé	5	14	2	100
Mixed Drinks				
Manhattan	3.0	26	3	191
Martini	3.0	27	—	189
Bourbon and soda	3.0	11	—	78
Whiskey sour	3.0	13	14	147

*There is little to no fat or protein contribution to calorie content.
Source: USDA.

FIGURE 16-7 ► What is a "standard drink"?

The standard drink sizes shown here each provide about 14 grams of alcohol. Keep in mind that alcoholic beverages served in bars and restaurants can be 20% to 45% larger than a standard drink.

Source: NIAAA.

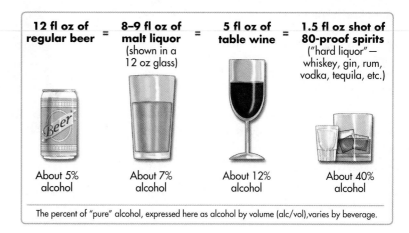

| 12 fl oz of regular beer | = | 8–9 fl oz of malt liquor (shown in a 12 oz glass) | = | 5 fl oz of table wine | = | 1.5 fl oz shot of 80-proof spirits ("hard liquor"— whiskey, gin, rum, vodka, tequila, etc.) |

About 5% alcohol   About 7% alcohol   About 12% alcohol   About 40% alcohol

The percent of "pure" alcohol, expressed here as alcohol by volume (alc/vol), varies by beverage.

Absorption and Metabolism of Alcohol

Alcohol requires no digestion. It is absorbed rapidly from the GI tract by diffusion, making it the most efficiently absorbed of all calorie sources. Once absorbed, alcohol is freely distributed into all the fluid compartments within the body. About 1% to 3% of alcohol is excreted via urine and about 1% to 5% evaporates via the breath, the basis for the breathalyzer test. Most alcohol (90% to 98%), however, is metabolized. The liver is the primary site for alcohol metabolism, and some may also be metabolized by the cells lining the stomach. The main pathway of alcohol metabolism involves the enzymes **alcohol dehydrogenase** and **acetaldehyde dehydrogenase**. Alcohol cannot be stored in the body, so it takes absolute priority over other energy sources for metabolism.

alcohol dehydrogenase An enzyme used in alcohol (ethanol) metabolism that converts alcohol into acetaldehyde.

acetaldehyde dehydrogenase An enzyme used in ethanol metabolism that eventually converts acetaldehyde into carbon dioxide and water.

TABLE 16-6 ▶ **Blood Alcohol Concentration and Symptoms**

Concentration*	Sporadic Drinker	Chronic Drinker	Hours for Alcohol to Be Metabolized**
50 (party high) (0.05%)	Congenial euphoria; decreased tension; and noticeable impairment in driving and coordination	No observable effect	2–3
75 (0.075%)	Gregarious	Often no effect	3–4
80 to 100 (0.08%–0.1%)	Uncoordinated; 0.08% is legally drunk (as in drunk driving) in the United States and Canada.	Minimal signs	4–6
125–150 (0.125%–0.15%)	Unrestrained behavior; episodic uncontrolled behavior	Pleasurable euphoria or beginning of uncoordination	6–10
200–250 (0.2%–0.25%)	Alertness lost; lethargic	Effort is required to maintain emotional and motor control.	10–24
300–350 (0.3%–0.35%)	Stupor to coma	Drowsy and slow	10–24
>500 (>0.5%)	Some will die	Coma	>24

*Milligrams of alcohol per 100 milliliters of blood.

**For a social drinker; alcohol metabolism is somewhat faster in chronic alcohol abusers.

Modified from Goldman L and Schafer AI: *Goldman's Cecil Medicine*, 24th edition, Philadelphia, 2012, Elsevier Health Sciences. Used with permission.

As a person's alcohol consumption exceeds the body's capacity to metabolize it, blood alcohol concentration rises, the brain is exposed to alcohol, and symptoms of intoxication appear (see Table 16-6). Absorption and metabolism of alcohol depend on numerous factors: genetics, gender, body size, physical condition, meal composition, rate of gastric emptying, the alcohol content of the beverage, use of certain drugs, chronic alcohol use, and even how much sleep one has had. Women absorb and metabolize alcohol differently than men. The amount of alcohol metabolized by the cells lining the stomach is greater in men than in women. Women also have less body water in which to dilute the alcohol than do men. Overall, women develop alcohol-related ailments, such as cirrhosis of the liver, more rapidly than men do with the same alcohol-consumption habits.

Benefits of Moderate Alcohol Use

When used in moderation, alcohol is linked to several health benefits. Benefits of alcohol use are associated with specific intakes of about one drink per day for men and slightly less than one for women. Socialization and relaxation are among the intangible benefits of moderate alcohol use by people of legal drinking age. In terms of physiological benefits, moderate drinkers experience lower risk of developing cardiovascular diseases and type 2 diabetes. Previous consumers of alcohol no longer experience the benefits of alcohol when consumption ceases. See Table 16-7 for additional benefits of moderate alcohol consumption.

▲ Of all the alcohol sources, red wine in moderation is often singled out as the best choice because of the added bonus of the many phytochemicals present (e.g., resveratrol). These are leached out from the grape skins as the red wine is fermented. Dark beer is also a source of phytochemicals.

Risks of Heavy Drinking

An alcohol use disorder is a formal psychiatric diagnosis defined in the latest edition of the *Diagnostic and Statistical Manual of Mental Disorders (DSM-5)* as a problematic pattern of alcohol use leading to significant impairment or distress. According to *DSM-5*, diagnosis depends on meeting two or more of the following criteria within the past year:

CRITICAL THINKING

For many people, drinking and smoking go hand-in-hand. What health problems arise from the combination of these behaviors?

- Use of alcohol in larger amounts or over a longer period than intended
- Persistent desire or unsuccessful efforts at cutting down or controlling alcohol use
- Spending a great deal of time obtaining, using, and recovering from the effects of alcohol
- Experiencing cravings for alcohol
- Repeated use of alcohol that results in failure to fulfill major obligations at school, work, or home
- Continued use of alcohol despite interpersonal problems created by the effects of heavy drinking
- Giving up important social, occupational, or recreational activities due to use of alcohol
- Recurrent alcohol use in situations in which it is physically hazardous
- Continuing to use alcohol even after realizing one has a problem caused by heavy drinking
- Developing a **tolerance** to the effects of alcohol
- Experiencing symptoms of **withdrawal** in the absence of alcohol use

Alcohol abuse disorders affect about 17% of adult men and about 8% of adult women at some point in their lives. Studies suggest that about 40% of a person's risk for developing these disorders is genetic. Certainly, genetic variations coding for the enzymes of alcohol metabolism are involved, but many other genes are also under study. Therefore, people with a family history of heavy drinking, particularly children of alcoholics, should be especially aware of their alcohol consumption.

Early diagnosis of alcohol use disorders can prevent multiple health problems and save millions in health care costs. Asking a person about the quantity and frequency of alcohol consumption is an important means of detecting problematic behaviors (see the margin note on the CAGE questionnaire on this page). Observable warning signs of an alcohol use disorder may include an alcohol odor on the breath; flushed face and reddened skin; nervous system disorders, such as tremors; unexplained work absences; frequent accidents; and falls or injuries of vague origin. Laboratory evidence (e.g., impaired liver function, enlarged red blood cells, and elevated triglycerides) is also helpful for diagnosis of alcohol use disorders.

tolerance Needing more of a substance to achieve the desired effect (e.g., intoxication) or experiencing diminished effects of a given amount of a substance after repeated use.

withdrawal Physical symptoms related to cessation of substance use, such as sweating, rapid pulse, shakiness, insomnia, nausea and vomiting, anxiety, and even seizures.

Despite the few benefits of regular, moderate use, the risks of heavy drinking are more numerous and harmful. Although it is one of the most preventable health problems, excessive consumption of alcohol contributes significantly to 5 of the 10 leading causes of death in North America: heart failure, certain forms of cancer, **cirrhosis** of the liver, motor vehicle and other accidents, and suicides (review Table 16-7 for additional health risks). In the United States, about $225 billion is spent annually in terms of lost productivity, medical care, and property damage associated with alcohol use disorders. Overall, alcohol use disorders typically reduce a person's life expectancy by 15 years.

Alcohol is most damaging to the liver. Cirrhosis develops in up to 20% of cases of alcoholism and is the second leading reason for liver transplants, affecting about 2 million people in the United States. This chronic and usually relentlessly progressive disease is characterized by fatty infiltration of the liver. Fatty liver occurs in response to increased synthesis of fat and decreased use of it for energy by the liver. Eventually, the enlarged fat deposits choke off the blood supply, depriving the liver cells of oxygen and nutrients. Liver cells can accumulate so much fat that they burst, die, and are replaced by connective (scar) tissue. At this stage, the liver is deemed cirrhotic (Fig. 16-8). Early stages of alcoholic liver injury are reversible,

FIGURE 16-8 ▲ Effects of alcohol on the liver. Alcohol is particularly damaging to the liver. Pictured are (a) healthy liver and (b) liver with cirrhosis. There is no cure for this disease except a liver transplant.

cirrhosis A loss of functioning liver cells, which are replaced by nonfunctioning connective tissue. Any substance that poisons liver cells can lead to cirrhosis. The most common cause is a chronic, excessive alcohol intake. Exposure to certain industrial chemicals also can lead to cirrhosis.

The CAGE questionnaire is used to identify alcohol use disorders. More than one positive response suggests an alcohol problem.

C: Have you ever felt you ought to *cut* down on drinking?

A: Have people *annoyed* you by criticizing your drinking?

G: Have you ever felt bad or *guilty* about your drinking?

E: Have you ever had a drink first thing in the morning to steady your nerves or get rid of a hangover (*eye-opener*)?

TABLE 16-7 ▶ A Summary of Effects of Moderate Versus Heavy Drinking

	Moderate Drinking	Heavy Drinking
Coronary heart disease	Decreased risk of death in those at high risk for coronary heart disease–related death, primarily by increasing HDL-cholesterol in some people, decreasing blood clotting, and relaxing blood vessels	Heart rhythm disturbances, heart muscle damage, increased blood triglycerides, and increased blood clotting
Hypertension and stroke	Mild decrease in blood pressure; less ischemic stroke in people with normal blood pressure	Increased blood pressure (hypertension); more ischemic and hemorrhagic stroke
Peripheral vascular disease	Decreased risk due to reduced blood clotting	No benefit
Blood glucose regulation and type 2 diabetes	Decreased risk of developing type 2 diabetes; decreased risk of death from cardiovascular disease among diabetics	Hypoglycemia; reduced insulin sensitivity; and damage to pancreas (site of insulin production)
Bone and joint health	Some increase in bone mineral content in women, linked to estrogen output	Loss of active bone-forming cells and eventual osteoporosis (many nutrient deficiencies also contribute to this problem); increased risk of gout
Brain function	Enhanced brain function and decreased risk of dementia by increasing blood circulation in the brain	Brain tissue damage and decreased memory
Skeletal muscle health	No benefit	Skeletal muscle damage
Cancer	No benefit	Increased risk of oral, esophageal, stomach, liver, lung, colorectal, and breast cancer, to name a few (especially if the diet is deficient in the vitamin folate)
Liver function	No benefit	Fatty infiltration and eventual liver cirrhosis, especially if a person is also infected with hepatitis C; iron toxicity
GI tract disease	Decreased risk of certain bacterial infections in the stomach	Inflammation of the stomach (and pancreas); absorptive cell damage leading to malabsorption of nutrients
Immune system function	No benefit	Reduced function and increased infections
Nervous system function	No benefit	Loss of nerve sensation and nervous system control of muscles
Sleep disturbances	Some relaxation	Fragmented sleep patterns; worsens sleep apnea
Impotence and decreased libido	No benefit	Contributes to the problem in both men and women
Drug overdose	No benefit	Contributes to the problem, especially in combination with sedatives
Obesity	No benefit	Increased abdominal fat deposition; contributes to weight gain as calories from alcoholic beverages quickly add up
Nutrient intake	May supply some B vitamins and iron	Leads to numerous nutrient deficiencies: protein, vitamins, and minerals
Fetal health	No benefit	Variety of toxic effects on the fetus when alcohol is consumed by pregnant women (see Chapter 14)
Socialization and relaxation	Provides some benefit to socialization and leads to relaxation by increasing brain neurotransmitter activity	Contributes to violent behavior and agitation
Traffic deaths and other violent deaths	No benefit	Contributes to both traffic death and violent death

Ethnicity plays an important role in both the probability of and health risks associated with heavy drinking. Native Americans suffer the highest rates of unintentional injuries, suicide, homicide, and domestic abuse related to alcohol use. African-American alcoholics are at greater risk than other racial groups for tuberculosis, hepatitis C, HIV/AIDS, and other infectious diseases. Hispanic Americans are at particular risk for cirrhosis-related death.

but advanced stages are not. Once a person has cirrhosis, there is a 50% chance of death within 4 years, a far worse prognosis than many forms of cancer. While no specific level of alcohol consumption guarantees cirrhosis, some evidence suggests that damage is caused by a dose as low as 40 grams per day for men (3 beers) and 20 grams per day for women (1½ beers).

Alcoholic beverages have little nutritional value and, thus, nutrient deficiencies are a common result of alcohol use disorders. The protein and vitamin content is extremely low, except in beer, where it is marginal. Iron content varies from drink to drink, with red wine ranking especially high in iron. Deficiencies arise mostly from poor nutrient intakes, but increased urinary losses and fat malabsorption (linked to poor pancreatic function) are also to blame. Vitamins most susceptible to depletion from heavy drinking include vitamins A, D, E, and K; thiamin; niacin; folate; vitamins B-6 and B-12; and vitamin C. Mineral deficiencies of calcium, phosphorus, potassium, magnesium, zinc, and iron are possible. On the other hand, vitamin and mineral toxicity is also of concern. Damage to the GI tract and liver, as well as high levels of some minerals in alcoholic beverages, may lead to toxicity of vitamin A, iron, lead, or cobalt. In nutritional treatment of alcohol use disorders, the immediate aim is eliminating alcohol intake, followed by replenishment of nutrient stores.

Older adults are uniquely vulnerable to alcohol use disorders, perhaps due to an abundance of free time, social events involving drinking, loneliness, or depression. Common symptoms of alcohol use disorders—trembling hands, slurred speech, sleep problems, memory loss, and unsteady gait—can be easily overlooked as signs of old age. Slower alcohol metabolism and decreased body water allow older adults to become intoxicated from a smaller amount of alcohol than their younger counterparts. Even moderate alcohol consumption can exacerbate some chronic health conditions, such as diabetes and osteoporosis. As well, even small amounts of alcohol can react negatively with various medications used by older persons. The adverse health effects of drinking may be amplified in older adults, so people over the age of 65 should limit alcohol consumption to no more than one drink per day.

Once a diagnosis of an alcohol use disorder is established, a physician can arrange appropriate treatment and counseling for the person and his or her family. Treatment often includes the use of certain medications, counseling, and social support. Total abstinence must be the ultimate objective. Alcoholics Anonymous (AA) or other reputable therapy programs can support alcoholics and their families as they recover from this devastating disease.

Guidance Regarding Alcohol Use

No government agencies recommend drinking alcohol. The 2010 Dietary Guidelines for Americans provide the following advice regarding use of alcoholic beverages:

- Those who choose to drink alcoholic beverages should do so sensibly and in moderation—defined as the consumption of up to one drink per day for women and up to two drinks per day for men.
- Alcoholic beverages should not be consumed by some individuals, including those who cannot restrict their alcohol intake, women who are pregnant or may become pregnant, children and adolescents, individuals taking medications that can interact with alcohol, and those with specific medical conditions.
- Alcoholic beverages should be avoided by individuals engaging in activities that require attention, skill, or coordination, such as driving or operating machinery.
- For weight management, monitor calorie intake from alcoholic beverages.

As the understanding of the relationship between drinking alcohol and health grows, registered dietitians and other health professionals can promote healthy lifestyles—not by encouraging indiscriminate drinking, but rather by reassuring adults that moderate alcohol consumption may have beneficial health outcomes.

To learn more about alcohol use disorders, visit these websites:

- National Institute on Alcohol Abuse and Alcoholism; **www.niaaa.nih.gov**
- American Society of Addiction Medicine; **www.asam.org**
- American Self-Help Clearinghouse; **www.mentalhelp.net/ selfhelp**

▲ The limit for alcohol intake for older adults is one drink per day.

Summary (Numbers refer to numbered sections in the chapter.)

16.1 Although maximum life span has not changed, life expectancy has increased dramatically over the past century. For many societies, this means that an increasing proportion of the population is over 65 years of age. As health care costs rise, the goal of delaying disease becomes ever important.

Adulthood is characterized by body maintenance and gradual physiological transitions, often referred to as "aging." The physiologic changes of aging are the sum of cellular changes, lifestyle practices, and environmental influences. Many of these changes can be minimized, prevented, and/or reversed by healthy lifestyles. Usual aging refers to the age-related physical and physiological changes that are commonly thought to be typical of aging. Successful aging describes the declines in physical and physiological function that occur because one grows older. Striving to have the greatest number of healthy years and the fewest of illness is referred to as compression of morbidity.

16.2 A healthy diet based on MyPlate and the 2010 Dietary Guidelines for Americans can help one to preserve body function, avoid chronic disease, and age successfully. These guidelines recommend that individuals eat a variety of foods; balance the food eaten with physical activity to maintain or improve weight; choose a diet with plenty of whole grains, vegetables, and fruits; choose a diet low in saturated fat, *trans* fat, and cholesterol; limit sugars and salt; and moderate alcoholic beverage intake. Regular physical activity is also important, as is safe food preparation. American adults are fairly well nourished, although common dietary excesses are calories, fat, sodium, and, for some, alcohol. Common dietary inadequacies include vitamins D and E, folate, magnesium, calcium, zinc, and fiber. People ages 65 and older, particularly those in long-term care facilities and hospitals, are at risk for malnutrition. The Nutrition Screening Initiative checklist can help identify older adults at risk of nutrient deficiencies. The DRIs for adults are divided by gender and age to reflect how nutrient needs change as adults grow older. These changes in nutrient needs take into consideration the aging-related physiological alterations in body composition, metabolism, and organ function.

16.3 The food choices and nutritional adequacy of adults' diets depend on physiological, psychosocial, and economic factors. Alterations in any of these factors can result in deteriorations in the quality of dietary intake, nutritional status, and health. Physiological factors that influence dietary intake and alter nutrient needs and/or nutrient utilization include changes in body composition, body systems, and chronic diseases. Of particular concern is sarcopenia, or loss of muscle mass that frequently accompanies aging. The use of medications and supplements can improve health and quality of life, but some also adversely affect nutritional status. Herbal products should be used with caution.

16.4 Psychosocial influences on nutritional status include changes in lifestyle and social interaction, and the presence of mental health issues, such as depression and Alzheimer's disease. Economic factors may also have a great impact on the types and amounts of food one eats.

16.5 Diet plans for adults should be based on nutrient-dense foods and need to be individualized for existing health problems, physical abilities, the presence of drug-nutrient interactions, possible depression, and economic constraints. A balanced multivitamin and mineral supplement can be used to help meet needs, especially for adults age 70 years and older. Most communities have congregate or home-delivered meal systems, food stamps, and other provisions to provide nutrition aid for those who qualify.

NAYH Alcohol requires no digestion and is metabolized in the liver and other tissues. The benefits of alcohol use are associated with low to moderate alcohol consumption. These benefits include the pleasurable and social aspects of alcohol use, a reduction in various forms of cardiovascular disease, increase in insulin sensitivity, and protection against some harmful stomach bacteria. Heavy drinking, however, contributes significantly to five of the 10 leading causes of death in North America. Alcohol increases the risk of developing certain forms of heart damage, inflammation of the pancreas, GI tract damage, vitamin and mineral deficiencies, cirrhosis of the liver, certain forms of cancer, hypertension, and hemorrhagic stroke—to name a few. If alcohol is consumed, it should be consumed in moderation with meals. Women (and older adults, in general) are advised to drink no more than 3 drinks in 1 day and no more than 7 drinks per week; men should drink no more than 4 drinks in 1 day and no more than 14 drinks per week.

Check Your Knowledge (Answers to the following questions are on the next page.)

1. The reason the incidence of obesity increases with age is that
   a. the basal metabolic rate decreases with age.
   b. physical activity often decreases with age.
   c. energy intake exceeds energy expenditure.
   d. All of the above.

2. Nutrition programs such as congregate meals or home-delivered meals provide which of the following?
   a. An improved nutritional status
   b. A social atmosphere
   c. An economical meal for low-income elderly
   d. All of the above.

3. Among the older population of the United States, the age of the fastest growing segment is _____ years.
   a. 65
   b. 74
   c. 79
   d. 85+

4. Which of the following accurately portrays a theory about the causes of aging?
   a. Increases in testosterone and estrogen affect cell processes.
   b. Blood sugar decreases, failing to supply adequate energy to brain cells.
   c. Inadequate calorie intake speeds body breakdown.
   d. Excess free radicals damage cell components.

5. The immune system becomes less efficient with age, so it is especially important to consume adequate _____ and _____, nutrients that contribute to immune function.
   a. vitamin A, potassium
   b. protein, zinc
   c. zinc, iodide
   d. vitamin A, vitamin K

6. Which of the following is a useful strategy to prevent or delay the onset of Alzheimer's disease?
   a. Avoid stressing the brain with challenging mental tasks.
   b. Limit intake of dairy products.
   c. Consume adequate B vitamins, such as folate, vitamin B-6, and vitamin B-12.
   d. Increase the ratio of omega-6 to omega-3 fatty acids in the diet.

7. To maintain optimal nutritional status and healthy weight, the diet of an older person should have a _____ nutrient density and be _____ in energy content.
   a. low, high
   b. low, low
   c. high, moderate
   d. high, high

8. Donald has been diagnosed with cirrhosis. To cope with his impaired liver function, he will need to
   a. take high doses of vitamin and mineral supplements.
   b. minimize alcohol intake.
   c. limit his intake of dietary fiber.
   d. All of the above.

9. Alcohol is digested in the
   a. stomach.
   b. small intestine.
   c. liver.
   d. None of the above; alcohol requires no digestion.

10. Alcohol is most damaging to the
   a. brain cells because alcohol can be used as an energy source even before glucose.
   b. kidney cells because this is where alcohol is excreted.
   c. cells of the gastrointestinal tract because they are in direct contact with ingested alcohol.
   d. liver cells because this is where alcohol is metabolized.

Answer Key: 1. d (LO 16.3), 2. d (LO 16.7), 3. d (LO 16.1), 4. d (LO 16.2), 5. b (LO 16.3), 6. c (LO 16.4), 7. c (LO 16.5), 8. b (LO 16.6), 9. d (LO 16.8), 10. d (LO 16.8)

Study Questions (Numbers refer to Learning Outcomes)

1. What is the difference between life span and life expectancy? **(LO 16.1)**

2. Describe two hypotheses proposed to explain the causes of aging, and note evidence for each in your daily life experiences. **(LO 16.2)**

3. List four organ systems that can decline in function in later years, along with a diet/lifestyle response to help cope with the decline. **(LO 16.3)**

4. Defend the recommendation for regular physical activity during older adulthood, including some resistance activity (weight training). **(LO 16.3)**

5. List four warning signs of undernutrition in older people that are part of the acronym DETERMINE. Briefly justify the inclusion of each. **(LO 16.4)**

6. List three important points made by the Dietary Guidelines for Americans for the general population and give an example of why each one may be difficult for older adults to implement. What are some suggestions for overcoming these barriers? **(LO 16.5)**

7. How might the nutritional needs of older people differ from those of younger people? How are their needs similar? Be specific. **(LO 16.5)**

8. List three common herbal remedies. What are the possible benefits and risks of each one? If your grandmother were considering using any of these herbal remedies, what advice would you give her? **(LO 16.6)**

9. What three resources in a community are widely available to aid older adults in maintaining nutritional health? **(LO 16.7)**

10. List two benefits of moderate alcohol intake. List two risks of heavy drinking. Should a nondrinker take up drinking for the health benefits? **(LO 16.8)**

What the Dietitian Chose

As far as nutrient deficiencies go, vitamin B-12 deficiency is relatively common. About 6% of older adults are deficient in vitamin B-12, and many more likely have marginal vitamin status. Older adults are at particular risk of vitamin B-12 deficiency because stomach production of acid and intrinsic factor, which are required for vitamin B-12 absorption, tends to decline with age. In addition, certain medications can reduce acid production and thereby affect B-12 absorption.

As you learned in Chapter 8, foods of animal origin are the only natural sources of vitamin B-12. Although the leafy, green vegetables would be an excellent source of folate—another B vitamin that is required for red blood cell health—they are not a good food source of vitamin B-12.

A 3-ounce serving of pot roast provides about 100% of the 2.4 micrograms of vitamin B-12 needed per day. However, recall that stomach acid is needed to cleave vitamin B-12 from protein in food sources. Even though beef is a rich source of vitamin B-12, with reduced stomach acid secretion, not much of the vitamin B-12 will be absorbed.

Synthetic vitamin B-12—the form found in dietary supplements and fortified foods—is actually better absorbed than the natural form of the vitamin. This is because synthetic vitamin B-12 does not have to be cleaved from food proteins before it can bind with intrinsic factor and be absorbed in the ileum of the small intestine. A fortified breakfast cereal, such as Cheerios®, provides about 80% of the RDA for vitamin B-12, and because this vitamin B-12 is not protein-bound, it is efficiently absorbed. Thus, the Cheerios® would be a good way to boost vitamin B-12 status.

A dietary supplement formulated for older adults, such as Centrum® Silver® Adults 50+, provides 25 micrograms of efficiently-absorbed, synthetic vitamin B-12. That's 10 times the RDA! Is it a problem to consume this megadose of vitamin B-12? Actually, no Upper Level has been set for vitamin B-12 and no toxicity has ever been reported. To correct your existing vitamin B-12 deficiency, the high dose of vitamin B-12 in this supplement would be most effective. After your nutrition status is back to normal, you could switch to a fortified food, such as breakfast cereal, to keep your red blood cells healthy with adequate B-12 and other micronutrients.

CASE STUDY SOLUTION   Dietary Assistance for an Older Adult

1. Frances' typical diet is low in many nutrients, including protein, calcium, iron, zinc, and vitamins B-12 and D, as well as fiber.
2. Her poor diet will cause a decline in her lean tissue and bone mass. With low intakes of iron and vitamin B-12, she may experience anemia. Her low intake of fiber may lead to problems with constipation.
3. Physiological changes that will tend to worsen her appetite include decreases in senses of taste and smell, diminished sense of thirst, and poor chewing ability (related to tooth loss). Bowel function tends to decline with age, which would only be further complicated by her low fiber intake. Low production of stomach acid and intrinsic factor might increase her risk of anemia, which is already a concern due to poor intakes of iron and vitamin B-12.
4. It would be helpful to have additional information about her medical history, including any current diagnoses and medications. Medical conditions, such as high blood cholesterol, would likely affect diet recommendations. Many prescription and over-the-counter medications can affect appetite and nutrient availability. Also, knowing her height and weight could assist in setting diet goals. Information regarding her economic resources could be useful when making dietary recommendations that she can actually follow.
5. Frances could contact a local government agency that offers congregate meal programs at a central location. She could inquire about location and availability of transportation to the site. This would give her social contact with other older persons, probably an important element missing in her life. This could help alleviate her loneliness. She could also request Meals on Wheels (if available) to provide one hot meal per day. One hot meal per day prepared for her may be what she needs to help stimulate her appetite. If her funds are limited, she

may qualify for nutrition assistance from SNAP or local food pantries.

6. Nutrient-dense yet convenient foods that could improve her nutrient intake include milk; peanut butter; fortified breakfast cereals; canned chicken or deli meats; yogurt; sliced cheese; cottage cheese; calcium-fortified orange juice; canned or frozen fruits and vegetables; and some fresh fruits and vegetables that do not require preparation, such as prewashed lettuce and bananas. A further possibility is a nutrition bar or a liquid nutritional supplement, such as a can of Ensure ® Plus. The resulting increase in her nutrient intake would help prevent disease in the future and increase her sense of well-being.

Further Readings

1. Berner LA and others: Characterization of dietary protein among older adults in the United States: Amount, animal sources, and meal patterns. *Journal of the Academy of Nutrition and Dietetics* 113: 809, 2013.

2. Bernstein M and others: Position of the Academy of Nutrition and Dietetics: Food and nutrition for older adults: Promoting health and wellness. *Journal of the Academy of Nutrition and Dietetics* 112: 1255, 2012.

3. Brannon CA: Alcohol: Functional food or addictive drug? *Today's Dietitian* 10(12):8, 2008.

4. Chodzko-Zajko WJ and others: American College of Sports Medicine Position Stand: Exercise and physical activity for older adults. *Medicine and Science in Sports and Exercise* 41:1510, 2009.

5. De Souza Genaro P and Martini LA: Effect of protein intake on bone and muscle mass in the elderly. *Nutrition Reviews* 68:616, 2010.

6. Esposito K and others: Long-term effect of Mediterranean-style diet and calorie restriction on biomarkers of longevity and oxidant stress in overweight men. *Cardiology Research and Practice* 2011:1, 2010.

7. Getz L: The Mediterranean diet and cognition. *Today's Dietitian* 16:26, 2014.

8. Grieger L: Dietary tips for baby boomers: Ageless advice for an aging generation. *Today's Dietitian* 10(3):38, 2008.

9. Institute of Medicine: Sodium intake in populations: Assessment of evidence. National Academy of Sciences, 2013. Available at www.iom.edu/Reports/2013/Sodium-Intake-in-Populations-Assessment-of-Evidence.aspx Accessed 07/09/1013

10. Kamp BJ and others: Position of the American Dietetic Association, American Society for Nutrition, and Society for Nutrition Education: Food and nutrition programs for community-residing older adults. *Journal of the American Dietetic Association* 110: 463, 2010.

11. Kushi LH and others: American Cancer Society guidelines on nutrition and physical activity for cancer prevention: Reducing the risk of cancer with healthy food choices and physical activity. *CA: A Cancer Journal for Clinicians* 62:30, 2012.

12. Redman LM and others: Effect of calorie restriction in non-obese humans on physiological, psychological, and behavioral outcomes. *Physiology and Behavior* 94:643, 2008.

13. Riediger ND and others: A systemic review of the roles of n-3 fatty acids in health and disease. *Journal of the American Dietetic Association* 109:668, 2009.

14. Stenholm S and others: Sarcopenic obesity— definition, etiology, and consequences. *Current Opinions in Clinical Nutrition and Metabolic Care* 11:693, 2008.

15. Touger-Decker R and Mobley C: Position of the American Academy of Nutrition and Dietetics: Oral health and nutrition. *Journal of the Academy of Nutrition and Dietetics* 113:693, 2013.

 **To get the most out of your study of nutrition, visit McGraw-Hill Connect at www.mcgrawhillconnect.com where you will find NutritionCalc Plus, LearnSmart, and many other dynamic tools.**

Rate Your Plate

I. Am I Aging Healthfully?

Take Control of Your Aging by Dr. William B. Malarkey (Wooster Book Co., Wooster, OH, 1999) includes a plan that incorporates various diet and lifestyle factors associated with successful aging. Indicate the degree to which you are following such a plan (or alternatively, fill this out with a parent or another older relative in mind).

Physical: Do you eat a well-balanced diet, exercise on a regular basis, remain free of illness, abstain from smoking, refrain from drinking alcohol excessively, and experience refreshing sleep?

_____

_____

Intellectual: Are you analytical, do you read regularly, do you learn new things each day, do you engage your mental ability at work (or at school), and do you often reflect on your life?

_____

_____

_____

Emotional: Are you at peace, do you like who you are, are you optimistic, and do you laugh and relax regularly?

_____

_____

_____

Relational: Are you a good listener, do you feel supported by friends, do you attend social functions, do you talk with family members often, and do you feel close to coworkers (or fellow students)?

_____

_____

_____

Spiritual: Do you appreciate nature, give to or serve others, meditate or seek religious worship, and feel life has meaning?

_____

_____

_____

The more of these factors that you include in your life, the more well-rounded your plan is for maintaining overall health. Any one of the five areas in which you are not achieving success should show you characteristics to work on in the future.

II. Helping Older Adults Eat Better

During their lifetimes, most people usually eat meals with families or loved ones. As people reach older adulthood, many of them are faced with living and eating alone. In a study of the diets of 4400 older adults in the United States, one man in every five living alone and over age 55 ate poorly. One of four women between the ages of 55 and 64 years followed a low-quality diet. These poor diets can contribute to deteriorating mental and physical health. Consider the following example of the living situation of an older adult.

Neal, a 70-year-old man, lives alone in a home in a local suburb. His wife died 1 year ago. He does not have many friends; his wife was his primary confidante. His neighbors across the street and next door are friendly, and Neal used to help them with yard projects in his spare time. Neal's health has been good, but he has had trouble with his teeth recently. His diet has been poor, and in the past 3 months, his physical and mental vigor has deteriorated. He has been slowly lapsing into a depression, so he keeps the shades drawn and rarely leaves his house. Neal keeps very little food in the house because his wife did most of the cooking and shopping, and he just is not that interested in food.

If you were one of Neal's relatives and learned of Neal's situation, what six things could you do or suggest to help improve his nutritional status and mental outlook? Look back into this chapter to get some ideas.

1. _____

2. _____

3. _____

4. _____

5. _____

6. _____

Appendix A

Daily Values Used on Food Labels

Daily Values Used on Food Labels in the United States, with a Comparison to the Latest RDAs and Other Nutrient Standards*

Dietary Constituent	Unit of Measure	Current Daily Values for People Over 4 Years of Age	RDA or Other Current Dietary Standard	
			Males 19–30 Years Old	Females 19–30 Years Old
Total fat[†]	g	<65	—	—
Saturated fatty acids[†]	g	<20	—	—
Protein[†]	g	50	56	46
Cholesterol[§]	mg	<300	—	—
Carbohydrate[†]	g	300	130	130
Dietary fiber	g	25	38	25
Vitamin A	µg Retinol activity equivalents	1000	900	700
Vitamin D	International units	400	600	600
Vitamin E	International units	30	22–33	22–33
Vitamin K	µg	80	120	90
Vitamin C	mg	60	90	75
Folate	µg	400	400	400
Thiamin	mg	1.5	1.2	1.1
Riboflavin	mg	1.7	1.3	1.1
Niacin	mg	20	16	14
Vitamin B-6	mg	2	1.3	1.3
Vitamin B-12	µg	6	2.4	2.4
Biotin	µg	300	30	30
Pantothenic acid	mg	10	5	5
Calcium	mg	1000	1000	1000
Phosphorus	mg	1000	700	700
Iodide	µg	150	150	150
Iron	mg	18	8	18
Magnesium	mg	400	400	310
Copper	mg	2	0.9	0.9
Zinc	mg	15	11	8
Sodium[‡]	mg	<2400	1500	1500
Potassium[‡]	mg	3500	4700	4700
Chloride[‡]	mg	3400	2300	2300
Manganese	mg	2	2.3	1.8
Selenium	µg	70	55	55
Chromium	µg	120	35	25
Molybdenum	µg	75	45	45

Abbreviations: g = gram, mg = milligram, and µg = microgram

*Daily Values are generally set at the highest nutrient recommendation in a specific age and gender category. Many Daily Values exceed current nutrient standards. This is in part because aspects of the Daily Values were originally developed in the early 1970s using estimates of nutrient needs published in 1968. The Daily Values have yet to be updated to reflect the current state of knowledge.

[†]These Daily Values are based on a 2000-kcal diet, instead of RDAs, with a caloric distribution of 30% from fat (and one-third of this total from saturated fat), 60% from carbohydrate, and 10% from protein.

[‡]The considerably higher Daily Values for sodium and chloride are there to allow for more diet flexibility, but the extra amounts are not needed to maintain health.

[§]Based on recommendations of U.S. federal agencies.

Appendix B

Diabetes Menu-Planning Tools

Lists, Choices, and Exchanges: Making Sense of Meal Planning for Diabetes

Registered dietitians and other diabetes educators work closely with their patients to help them understand how the foods they eat directly impact their day-to-day quality of life. **Food lists** (also known as **exchange lists**) are one-way diabetes educators can help people to plan their diets to better manage their blood sugar. The first exchange lists for people with diabetes were developed by the American Dietetic Association (now the Academy of Nutrition and Dietetics), American Diabetes Association, and U.S. Public Health Service more than 50 years ago. They have been revised over the years to reflect advances in nutrition recommendations and the ever-expanding variety of foods in the marketplace. The most recent version, *Choose Your Foods: Food Lists for Diabetes*, was published in 2014.

Food lists organize the many details of the nutrient composition of foods into a manageable framework based on calorie and macronutrient content. In the *Food Lists for Diabetes*, individual foods are placed into three broad groups: carbohydrates, proteins, and fats. Within these groups are lists that contain foods of similar macronutrient composition: various types of milk and milk substitutes, fruits, vegetables, starches, other carbohydrates, proteins, and fats. There are even lists that show how to account for alcohol, combination foods (e.g., casseroles), and a wide variety of fast foods. These lists are designed so that, when the given serving size is observed, each food on a list provides roughly the same amount of carbohydrate, protein, fat, and calories. The patient and a registered dietitian first tailor a healthy eating plan to meet the patient's energy and specific macronutrient needs. Then, the patient can select **choices** (i.e., **exchanges**) from each of the various lists that fit into the plan without having to look up or memorize the nutrient values of numerous foods.

Because the *Food Lists for Diabetes* offer a quick way to estimate the calorie, carbohydrate, protein, and fat content in any food or meal, they are a valuable menu-planning tool for people without diabetes, as well. In fact, the Academy of Nutrition and Dietetics and the American Diabetes Association have published a related guide, *Choose Your Foods: Food Lists for Weight Management*.

Table B-1 summarizes the basic nutrient composition of foods in each food list. The serving sizes of individual foods in a list may vary, but general estimates are given. The protein and milk and milk substitutes lists are divided into subclasses, which vary in fat content and, thus, in the amount of calories they provide. You can see that each food list is unique in the calories and macronutrients it supplies. A healthy meal plan should include foods from each of the lists to ensure nutrient adequacy. Study Table B-1 to become familiar with the food groupings, the approximate sizes of choices on each food list, and the amounts of carbohydrate, protein, fat, and calories per choice.

Recognize that the *Food Lists for Diabetes* group foods somewhat differently than MyPlate. For the *Food Lists*, we care more about nutrient composition and the food's eventual effect on blood sugar than its botanical origin. For example, the starch list includes not only bread, dry cereal, cooked cereal, rice, and pasta but also baked beans, corn on the cob, and potatoes. Although potatoes and corn are vegetables, their macronutrient composition resembles that of bread more than that of broccoli.

Food Lists A system for classifying foods into numerous lists based on the foods' macronutrient composition, and establishing serving sizes, so that one serving of each food on a list contains the same amount of carbohydrate, protein, fat, and calorie content; also called exchange lists.

choice The serving size of a food on a specific exchange list; formerly called exchange.

TABLE B-1 ▸ **Approximate Nutrient Composition of Food Choices from** *Choose Your Foods: Food Lists for Diabetes, 2014*

Groups/Lists	Household Measures*	Carbohydrate (g)	Protein (g)	Fat (g)	Energy (kcal)
Carbohydrates					
Starch (e.g., bread, cereal, pasta, rice, crackers, and beans)	1 slice, ¾ cup raw, or ½ cup cooked	15	3	1 or less[†]	80
Fruit	1 small/medium piece	15	—	—	60
Milk and milk substitutes	1 cup				
Fat-free, low-fat		12	8	0–3[†]	90
Reduced-fat		12	8	5	120
Whole		12	8	8	160
Nonstarchy vegetables	1 cup raw or ½ cup cooked	5	2	—	25
Sweets, desserts, and other carbohydrates	Varies	15	Varies	Varies	Varies
Proteins	1 ounce				
Lean		—	7	2	45
Medium-fat		—	7	5	75
High-fat		—	7	8	100
Plant-based		Varies	7	Varies	Varies
Fats	1 teaspoon	—	—	5	45
Alcohol	Varies	Varies	—	—	100

*An estimate; see food lists for actual amounts.

[†]Calculated as 1 gram for purposes of calorie contribution.

Source of data: Choose Your Foods: Food Lists for Diabetes, 2014 which is the basis of a meal planning system designed by a committee of the American Diabetes Association and the Academy of Nutrition and Dietetics.

In addition, many foods that would traditionally be categorized as dairy products do not appear with the milk and milk substitutes list. Instead, cheeses are grouped as proteins, whereas cream and cream cheese show up on the fats lists.

In some instances, a food counts for more than one choice at a time. In the category of sweets, desserts, and other carbohydrates, you will find a variety of snack items and condiments that count as carbohydrates and fats. The food lists also provide the user with some guidance on accounting for a wide variety of combination foods, such as pizza, casseroles, and soups. A list of free foods includes choices such as reduced-fat or fat-free foods, condiments, seasonings, and sugar-free drinks that, when consumed in moderation, have little or no impact on energy intake and blood sugar.

USING THE FOOD LISTS TO DEVELOP DAILY MENUS

Now let us use the *Food Lists* to plan a 1-day menu. Our kilocalorie target will be 2000 kcal, with 55% derived from carbohydrates (1100 kcal), 15% from protein (300 kcal), and 30% from fat (600 kcal). This can be achieved with 2 reduced-fat milk choices, 3 nonstarchy vegetable choices, 5 fruit choices, 11 starch choices, 4 lean protein choices, and 6 fat choices (Table B-2). This is just one of many possible combinations; the *Food Lists* offer great flexibility.

Table B-3 arbitrarily distributes these choices into breakfast, lunch, dinner, and a snack. Breakfast includes one reduced-fat milk choice, two fruit choices, two starch choices, and one fat choice. Drawing from the food lists, this plan could be achieved with ¾ cup of a ready-to-eat breakfast cereal, 1 cup of reduced-fat milk, 1 slice of bread with 1 teaspoon margarine, and 1 cup of orange juice.

Lunch consists of two fat choices, four starch choices, one nonstarchy vegetable choice, one reduced-fat milk choice, and two fruit choices. One possible combination would be a bacon, lettuce, and tomato sandwich consisting of two slices of bread, one slice of bacon, 1 teaspoon mayonnaise, a slice of tomato, and some lettuce (a free food). Add to this meal one large banana (two fruit choices), 1 cup of reduced-fat milk, and six graham crackers (2½ inches by 2½ inches). Later add a snack of ¾ ounce of pretzels for another starch choice.

TABLE B-2 ▶ Possible Food Choice Patterns That Yield 55% of Calories as Carbohydrate, 30% as Fat, and 15% as Protein

Food List	kcal/day						
	1200*	1600*	2000	2400	2800	3200	3600
Milk (reduced-fat)	2	2	2	2	2	2	2
Nonstarchy Vegetable	3	3	3	4	4	4	4
Fruit	3	4	5	6	8	9	9
Starch	5	8	11	13	15	18	21
Protein (lean)	4	4	4	5	6	7	8
Fat	2	4	6	8	10	11	13

This is just one set of options. More protein choices could be included if less milk were used, for example.

*Calorie intakes of 1200 and 1600 kcal contain 20% of calories as protein and 50% of calories as carbohydrate to allow for greater flexibility in diet planning.

Dinner consists of four lean meat choices, one fruit choice, two vegetable choices, one fat choice, and two starch choices. One possible combination consists of a 4-ounce broiled steak (meat only, no bone), one medium baked potato (1 choice = 1 small baked potato) with 1 teaspoon of margarine, 1 cup of broccoli, and one kiwi fruit. Coffee (if desired) is not counted because it contains no appreciable calories. Finally, we have a bedtime snack containing 2 starch choices and 2 fat choices. This could be achieved with 1 bagel and 2 tablespoons of regular cream cheese.

TABLE B-3 ▶ Sample 1-Day 2000-kcal Menu Based on the Food Lists for Diabetes Plan*

Breakfast

1 reduced-fat milk choice	1 cup reduced-fat milk (some on cereal)
2 fruit choices	1 cup orange juice
2 starch choices	¾ cup ready-to-eat breakfast cereal and 1 piece whole-wheat toast
1 fat choice	1 teaspoon soft margarine on toast

Lunch

4 starch choices	2 slices whole-wheat bread and 6 graham crackers (2½ inches by 2½ inches)
2 fat choices	1 slice bacon, 1 teaspoon mayonnaise
1 nonstarchy vegetable choice	1 sliced tomato
2 fruit choices	1 banana (9 inches)
1 reduced-fat milk choice	1 cup reduced-fat milk

Snack

1 starch choice	¾ ounce pretzels

Dinner

4 lean protein	4 ounces lean steak (well trimmed)
2 starch choices	1 medium baked potato
1 fat choice	1 teaspoon soft margarine
2 nonstarchy vegetable choices	1 cup cooked broccoli
1 fruit choice	1 kiwi fruit
	Coffee (if desired)

Snack

2 starch choices	1 bagel
2 fat choices	2 tablespoons regular cream cheese

*The target plan was a 2000-kcal intake, with 55% of calories from carbohydrate, 15% from protein, and 30% from fat. Computer analysis indicates that this menu yielded 2040 kcal, with 53% of calories from carbohydrate, 16% from protein, and 31% from fat—in close agreement with the targeted goals.

FIGURE B-1 ▶ Record the Food Lists for Diabetes pattern you have chosen in the left-hand column. Then distribute the food choices throughout the day, noting the food to be used and the serving size.

Food List	Total Food Choices to be Consumed Daily	Food Choices Consumed at Each Meal		
		Breakfast	Lunch	Dinner
Milk and milk substitutes				
Nonstarchy vegetables				
Fruits				
Starch				
Proteins				
Fats				

This 1-day menu is only one of endless possibilities with the *Food Lists for Diabetes*. Apple juice could replace the orange juice; two apples could be exchanged for the banana. For simplicity, we have used a variety of individual foods to achieve the total number of choices for this healthy eating plan. However, the *Food Lists* also include some commonly-used combination foods. For instance, a 1-cup serving of lasagna typically provides two medium-fat meat choices plus two carbohydrate choices. With practice, you will be able to estimate the choices from complex foods on your own (Fig. B-1). For now, using individual foods makes learning the *Food Lists* much easier. Finally, you might want to prove to yourself that the food choices listed in Table B-3 really fulfill the plan set forth in Table B-2.

Examples of Food Choices from Food Lists for Diabetes

In this section, you will find just a few examples of the many food choices that are included in the most recent edition of *Choose Your Foods: Food Lists for Diabetes*. For complete information, you can order your own copy of the booklet from the Academy of Nutrition and Dietetics for under $5.00.

Starches

Starches provide 15 grams of carbohydrate, 0 to 3 grams of protein, 0 to 1 gram of fat, and about 80 kcal per serving. Keep in mind that the serving sizes for starch choices on these food lists are usually smaller than those recommended by MyPlate. Also, some foods with high fat content may be counted as 1 starch plus 1 or 2 fats. Beans, peas, and lentils count as 1 starch plus 1 lean protein choice.

BREAD

Serving Size	Food
¼	Bagel, large (about 4 oz)
1 slice	Bread
½	English muffin
1	Pancake (4-inches diameter)

Serving Size	Food
1	Tortilla, flour (6-inches diameter)
1/3	Tortilla, flour (10-inches diameter)
3¼-inch square	Naan
½	Hamburger bun

CEREALS

Serving Size	Food
½ cup	Cooked cereal (e.g., oatmeal)
¼ cup	Granola cereal
1½ cups	Puffed cereal (e.g., puffed rice)

Serving Size	Food
½ cup	Sweetened cereal (e.g., Frosted Flakes®)
¾ cup	Unsweetened ready-to-eat cereal (e.g., Cheerios®)

GRAINS

Serving Size	Food
1/3 cup	Rice, cooked (e.g., white and brown)
1/3 cup	Pasta, cooked

Serving Size	Food
½ cup	Wild rice, cooked
1/3 cup	Quinoa, cooked

STARCHY VEGETABLES

Serving Size	Food
½ cup	Corn
1 cup	Mixed vegetables (e.g., corn, peas, and carrots)
½ cup	Spaghetti sauce
¼ large	Potato, baked

Serving Size	Food
½ cup	Potatoes, mashed
1 cup	Winter squash (e.g., acorn and butternut)
½ cup	Sweet potato

CRACKERS AND SNACKS

Serving Size	Food
8	Animal crackers
3 2½-inch squares	Graham crackers
6	Saltines
6	Butter crackers (e.g., Ritz®; count as 1 starch + 1 fat)

Serving Size	Food
3 cups	Popcorn
¾ ounce	Pretzels
8	Baked snack chips
13	Tortilla or potato chips (count as 1 starch + 2 fats)

BEANS, PEAS, AND LENTILS (count as 1 starch + 1 lean protein)

Serving Size	Food	Serving Size	Food
⅓ cup	Baked beans	½ cup	Lentils, cooked
½ cup	Beans, cooked or canned (e.g., black, garbanzo, and kidney)	½ cup	Peas, cooked (e.g., black-eyed and split)

Fruits

Once choice from the Fruits list provides 15 grams of carbohydrate, 0 grams of protein, 0 grams of fat, and 60 kcal. Typically, 1 fruit choice is equal to ½ cup of canned or frozen fruit, 1 small fresh fruit, ½ cup of unsweetened fruit juice, or 2 tablespoons of dried fruit. Recognize that the fruit you buy at the grocery store may amount to more than one fruit choice; a large banana, for example, counts for two fruit choices. The serving sizes of fruit juices and dried fruit are small because these are more concentrated sources of carbohydrates and energy.

FRUITS

Serving Size	Food	Serving Size	Food
1 small	Apple (about 4 oz)	½ cup	Kiwi, sliced
½ cup	Applesauce, unsweetened	1 medium	Orange
1 extra small	Banana (about 4 inches)	½ cup	Pineapple, canned
1 cup	Blackberries	3	Prunes
¾ cup	Blueberries	½ cup	Pomegranate seeds (arils)
12	Cherries	1¼ cup	Watermelon, diced
17	Grapes	1¼ cup	Strawberries, whole
1 cup	Honeydew melon, diced		

FRUIT JUICE

Serving Size	Food	Serving Size	Food
½ cup	Apple juice or apple cider	½ cup	Orange juice
⅓ cup	Grape juice	⅓ cup	Prune juice

Milk and Milk Substitutes

Milk and milk substitutes are divided into subcategories based on their fat content. All milk and yogurt products provide 12 grams of carbohydrate and 8 grams of protein but may vary in fat content from 0 to 8 grams per choice. The subcategory of other milk foods and milk substitutes includes some products that may be used in place of milk in the diet (e.g., soy milk), but have a slightly different nutrient profile than traditional milk and yogurt products. Those foods, as indicated below, are counted as a combination of carbohydrate (15 grams of carbohydrate, 70 kcal) and fat (5 grams of fat, 45 kcal) choices. Please note that other products used as dairy alternatives (e.g., almond milk) are listed with fat choices.

FAT-FREE (SKIM) AND LOW-FAT MILK AND YOGURT (12 grams of carbohydrate, 8 grams of protein, 0–3 grams of fat, and 100 kcal)

Serving Size	Food
1 cup	Fat-free (skim) milk, 1% milk, or buttermilk
½ cup	Canned, evaporated, fat-free milk
¾ cup	Yogurt (fat-free plain or fat-free Greek, unsweetened or artificially sweetened)

REDUCED-FAT (2%) MILK AND YOGURT
(12 grams of carbohydrate, 8 grams of protein, 5 grams of fat, and 120 kcal)

Serving Size	Food
1 cup	2% milk, acidophilus milk, or kefir
⅔ cup	Yogurt (reduced-fat, plain)

WHOLE MILK AND YOGURT
(12 grams of carbohydrate, 8 grams of protein, 8 grams of fat, and 160 kcal)

Serving Size	Food
1 cup	Whole milk, buttermilk, or goat's milk
½ cup	Evaporated whole milk
1 cup	Yogurt (whole milk, plain)

OTHER MILK FOODS AND MILK SUBSTITUTES

Serving Size	Food	
⅓ cup	Eggnog (made from whole milk)	1 carbohydrate + 1 fat
1 cup	Rice drink, plain, fat-free	1 carbohydrate
1 cup	Rice drink, flavored, low-fat	2 carbohydrates
1 cup	Soy milk, plain, low-fat	½ carbohydrate + ½ fat
1 cup	Soy milk, regular, plain	½ carbohydrate + 1 fat
⅔ cup	Yogurt with fruit, low-fat	1 fat-free milk + 1 carbohydrate

Nonstarchy Vegetables

Nonstarchy vegetables still provide carbohydrates, but not as much as their starchy counterparts. One nonstarchy vegetable provides 5 grams of carbohydrate, 2 grams of protein, 0 grams of fat, and 25 kcal. Typically, a choice is equal to ½ cup of cooked vegetables or 1 cup of raw vegetables. Large servings of nonstarchy vegetables (i.e., three choices) should be counted as one carbohydrate choice (15 grams of carbohydrate, 70 kcal) rather than multiple nonstarchy vegetables. Because of their low carbohydrate content, salad greens (e.g., iceberg, romaine, and endive) actually count as free foods. To comply with advice from the Dietary Guidelines for Americans, it is important to select a variety of starchy and nonstarchy vegetables each day because each has a distinct micronutrient and phytonutrient profile. Take extra care to select vegetables with deep colors, such as spinach, carrots, and beets.

Serving Size	Food
½ cup	Asparagus, cooked
1 cup	Baby carrots, raw
½ cup	Beets, cooked
½ cup	Broccoli, cooked
½ cup	Collard greens, cooked

Serving Size	Food
1 cup	Cucumber, raw slices
½ cup	Green beans, cooked
½ cup	Summer squash, cooked
½ cup	Tomatoes, stewed

Sweets, Desserts, and Other Carbohydrates

Foods on this list may not match the nutrient profiles of other starches, but they are commonly consumed and must be accounted for in diet planning. Sweetened beverages, desserts, and sweeteners and condiments that we add to foods can be counted as a combination of carbohydrate (15 grams of carbohydrates, 70 kcal) and fat (5 grams of fat, 45 kcal) choices.

BEVERAGES, SODA, AND SPORTS DRINKS

Serving Size	Food	
½ cup	Cranberry juice cocktail	1 carbohydrate
1 cup	Fruit drink or lemonade	2 carbohydrates
1 can (12 oz)	Soft drink, regular	2½ carbohydrates
1 cup	Sports drink (e.g., Gatorade®)	1 carbohydrate

BROWNIES, CAKE, COOKIES, GELATIN, PIE, AND PUDDING

Serving Size	Food	
1¼-inch square	Brownie, unfrosted	1 carbohydrate + 1 fat
¹⁄₁₂ cake	Angel food cake, unfrosted	2 carbohydrates
2-inch square	Cake, frosted	2 carbohydrates + 1 fat
2	Chocolate chip cookies	1 carbohydrate + 2 fats
5	Vanilla wafers	1 carbohydrate + 1 fat
½ cup	Gelatin, regular	1 carbohydrate
⅛	Pumpkin pie	1½ carbohydrates + 1½ fats
½ cup	Pudding, regular, made with 2% milk	2 carbohydrates

CANDY, SPREADS, SWEETS, SWEETENERS, SYRUPS, AND TOPPINGS

Serving Size	Food	
5	Chocolate kisses	1 carbohydrate + 1 fat
2 tbsp	Liquid nondairy coffee creamer	1 carbohydrate
1 tbsp	Honey	1 carbohydrate
1 tbsp	Jam or jelly, regular	1 carbohydrate
1 tbsp	Pancake syrup, regular	1 carbohydrate

CONDIMENTS AND SAUCES

Serving Size	Food	
2 tbsp	Barbecue sauce	1 carbohydrate
½ cup	Gravy	½ carbohydrate + ½ fat
3 tbsp	Salad dressing, fat-free, cream-based	1 carbohydrate

DOUGHNUTS, MUFFINS, PASTRIES, AND SWEET BREADS

Serving Size	Food	
1	Glazed doughnut	2 carbohydrates + 2 fats
1 (4 oz)	Muffin, regular	4 carbohydrates + 2½ fats
1 (2½ oz)	Danish	2½ carbohydrates + 2 fats

FROZEN BARS, FROZEN DESSERTS, FROZEN YOGURT, AND ICE CREAM

Serving Size	Food	
1 (3 oz)	Frozen 100% fruit juice bar	1 carbohydrate
½ cup	Ice cream, no sugar added	1 carbohydrate + 1 fat
½ cup	Ice cream, regular	1 carbohydrate + 2 fats
½ cup	Sherbet	2 carbohydrates
½ cup	Greek frozen yogurt, low-fat	1½ carbohydrates

Protein

Similar to the choices on the Milk and Milk Substitutes list, protein choices vary in fat and calorie content. Lean protein choices, such as egg whites and skinless poultry, provide 0 gram of carbohydrate, 7 grams of protein, 2 grams of fat, and 45 kcal. Medium-fat protein choices, such as whole eggs and poultry with skin, provide 0 gram of carbohydrate, 7 grams of protein, 5 grams of fat, and 75 kcal. High-fat protein choices, including many types of sausage and bacon, provide 0 gram of carbohydrate, 7 grams of protein, 8 grams of fat, and 100 kcal. Plant-based protein choices usually contain some carbohydrates, so they count as a combination of carbohydrate or starch and protein choices. Note that choices are very small; a typical hamburger would count as three or four protein choices.

LEAN PROTEIN (0 gram of carbohydrate, 7 grams of protein, 2 grams of fat, and 45 kcal)

Serving Size	Food	Serving Size	Food
1 oz	Beef with 10% or lower fat (e.g., round and sirloin)	1 oz	Lean pork (e.g., ham and tenderloin)
1 oz	Cheese with 3 grams of fat or less (e.g., fat-free mozzarella)	1 oz	Poultry, without skin
1 oz	Fish, not fried (e.g., catfish, cod, and tuna canned in water)	1 oz	Deli meats with 3 grams of fat or less per serving (e.g., turkey and ham)
2	Egg whites	1 oz	Shellfish (e.g., shrimp and crab)
1 oz	Wild game (e.g., buffalo and venison)		

MEDIUM-FAT PROTEIN (0 gram of carbohydrate, 7 grams of protein, 5 grams of fat, and 75 kcal)

Serving Size	Food	Serving Size	Food
1 oz	Beef with 15% or lower fat (e.g., rib roast and ground beef)	1 oz	Fish, fried
1 oz	Cheese with 4 to 7 grams of fat per ounce (e.g., feta and mozzarella)	1 oz	Pork (e.g., cutlet and shoulder roast)
1	Egg	1 oz	Poultry, with skin

HIGH-FAT PROTEIN (0 gram of carbohydrate, 7 grams of protein, 8 grams of fat, and 100 kcal)

Serving Size	Food	Serving Size	Food
2 slices	Bacon, pork	1 oz	Deli meats with 8 grams of fat or more per serving (e.g., bologna and salami)
1 oz	Cheese (e.g., American, cheddar, Parmesan, and Swiss)	1 oz	Sausage (e.g., bratwurst and summer sausage)
1	Hot dog		

PLANT-BASED PROTEIN

Serving Size	Food	
⅓ cup	Baked beans	1 starch + 1 lean protein
½ cup	Beans, cooked or canned (e.g., black, kidney, and pinto)	1 starch + 1 lean protein
½ cup	Edamame, shelled	½ carbohydrate + 1 lean protein
⅓ cup	Hummus	1 carbohydrate + 1 medium-fat protein
3 oz	Meatless burger, soy-based	½ carbohydrate + 2 lean proteins
½ cup	Tofu	1 medium-fat protein

Fats

One fat choices is 5 grams of fat and 45 kcal. Fats are subdivided into unsaturated fats, which come mainly from plant sources, and saturated fats, which come mainly from animal sources. In line with recommendations from other major health authorities, the Food Lists for Diabetes advises people to choose unsaturated fats in place of saturated fats and to avoid *trans* fats.

UNSATURATED FATS – MONOUNSATURATED FATS (5 grams of fat and 45 kcal)

Serving Size	Food
1 cup	Almond milk, unsweetened
2 tbsp	Avocado
1½ tsp	Nut butter (e.g., almond and peanut)
6	Almonds

Serving Size	Food
10	Peanuts
16	Pistachios
1 tsp	Oil (e.g., canola and olive)

UNSATURATED FATS – POLYUNSATURATED FATS (5 grams of fat and 45 kcal)

Serving Size	Food
1 tbsp	Low-fat vegetable oil spread
1 tsp	Margarine
1 tbsp	Low-fat mayonnaise
1 tsp	Mayonnaise

Serving Size	Food
1 tsp	Oil (e.g., corn, safflower, and sunflower)
2 tbsp	Salad dressing, reduced fat (may contain carbohydrate)
1 tbsp	Salad dressing, regular
1 tbsp	Flaxseed, ground

SATURATED FATS (5 grams of fat and 45 kcal)

Serving Size	Food
1 slice	Bacon
1 tbsp	Butter, reduced-fat

Serving Size	Food
1 tsp	Butter, regula
2 tbsp	Coconut, shredded

Free Foods

A *free food* is any food or drink that contains less than 20 kcal or less than 5 grams of carbohydrate per serving. When eaten in small amounts throughout the day, these foods have little impact on blood sugar control. Foods with a serving size listed should be limited to three servings per day. Foods listed without a serving size can be eaten as often as you like. However, many free foods are high in sodium, so moderation is important.

LOW-CARBOHYDRATE FOODS

Serving Size	Food	Serving Size	Food
1 piece	Candy, hard or sugar-free		Sugar substitutes
2 tsp	Jam or jelly, light or no-sugar-added	½ cup	Raw nonstarchy vegetables (e.g., broccoli, carrots, cucumber, and tomato)
	Gelatin, sugar-free	¼ cup	Cooked nonstarchy vegetables (e.g., carrots, cauliflower, and green beans)
	Salad greens		

REDUCED-FAT OR FAT-FREE FOODS

Serving Size	Food	Serving Size	Food
1 tbsp	Cream cheese, fat-free	1 tbsp	Mayonnaise, fat-free
4 tsp	Coffee creamer, liquid, sugar-free, flavored	1 tbsp	Salad dressing, fat-free
1 tsp	Margarine spread, reduced-fat	2 tbsp	Whipped topping, light or fat-free

CONDIMENTS

Serving Size	Food	Serving Size	Food
2 tsp	Barbecue sauce	1 tbsp	Parmesan cheese, grated
1 tbsp	Ketchup	1½	Dill pickles (medium)
	Hot pepper sauce	1 tbsp	Soy sauce
	Mustard (e.g., brown, Dijon, or yellow)		

DRINKS/MIXES

Serving Size	Food	Serving Size	Food
	Bouillon or broth		Coffee, unsweetened or artificially sweetened
	Club soda		Water
	Diet soft drinks, sugar-free		Water, flavored, sugar-free

SEASONINGS

Serving Size	Food	Serving Size	Food
	Garlic, fresh or powder		Spices
	Herbs, fresh or dried		

Combination Foods

These foods contain a mixture of ingredients and cannot be grouped into one food list. Many of these foods are high in sodium.

ENTREES

Serving Size	Food	
1 cup (8 oz)	Casserole-type entrees (e.g., tuna noodle, lasagna, and spaghetti with meatballs)	2 carbohydrates + 2 medium-fat proteins
1 cup (8 oz)	Stews (meat and vegetables)	1 carbohydrate + 1 medium-fat protein + 0 to 3 fats

FROZEN MEALS/ENTREES

Serving Size	Food	
1 (5 oz)	Burrito (beef and bean)	3 carbohydrates + 1 lean protein + 2 fats
9 to 12 oz	Dinner-type healthy meal (less than 400 kcal)	2 to 3 carbohydrates + 1 to 2 lean proteins + 1 fat
¼ of a 12-inch	Pizza with thin crust and meat toppings	2 carbohydrates + 2 medium-fat proteins + 1 ½ fats
1 (4½ oz)	Pocket sandwich	3 carbohydrates + 1 lean protein + 1 to 2 fats

SALADS (DELI-STYLE)

Serving Size	Food	
½ cup	Coleslaw	1 carbohydrate + 1½ fats
½ cup	Macaroni salad	2 carbohydrates + 3 fats
½ cup (3½ oz)	Tuna salad or chicken salad	½ carbohydrate + 2 lean proteins + 1 fat

SOUPS

Serving Size	Food	
1 cup (8 oz)	Bean, lentil, or split pea soup	1½ carbohydrates + 1 lean protein
1 cup (8 oz)	Chowder (made with milk)	1 carbohydrate + 1 lean protein + 1½ fats
1 cup (8 oz)	Ramen noodle soup	2 carbohydrates + 2 fats
1 cup (8 oz)	Tomato soup (made with water)	1 carbohydrate
1 cup (8 oz)	Broth-based soups with vegetables and meat	1 carbohydrate + 1 lean protein

Fast Foods

Fast foods are high in sodium and fat. These should be consumed in moderation, if at all. It is much easier to control carbohydrate, fat, sodium, and calorie intake when you prepare your own foods at home rather than relying on restaurants.

MAIN DISHES/ENTREES

Serving Size	Food	
1 (7 oz)	Chicken breast, breaded and fried	1 carbohydrate + 6 medium-fat proteins
6	Chicken nuggets or tenders	1 carbohydrate + 2 medium-fat proteins + 1 fat
1 (2 oz)	Chicken wing, breaded and fried	½ carbohydrate + 2 medium-fat proteins
⅛ of 14-inch	Pizza, thick crust with or without meat toppings	2½ carbohydrates + 1 high-fat protein + 1 fat

ASIAN

Serving Size	Food	
1 cup (6 oz)	Meat with vegetables in sauce	1 carbohydrate + 2 lean proteins + 1 fat
1 (3 oz)	Egg roll with meat filling	1½ carbohydrates + 1 lean protein + 1½ fats
1 cup	Fried rice, meatless	2½ carbohydrates + 2 fats
1 cup	Lo mein or chow mein	2 carbohydrates + 2 fats

MEXICAN

Serving Size	Food	
1 (6 oz)	Burrito with beans and cheese	3½ carbohydrates + 1 medium-fat protein + 1 fat
8	Nachos with cheese	2½ carbohydrates + 1 high-fat protein + 2 fats
1 (3 oz)	Crisp taco with meat and cheese	1 carbohydrate + 1 medium-fat protein + ½ fat
1 lb	Taco salad with chicken and tortilla bowl	3½ carbohydrates + 4 medium-fat proteins + 3 fats

SANDWICHES

Serving Size	Food	
1 (4 oz)	Breakfast burrito with sausage, egg, and cheese	1½ carbohydrates + 2 high-fat proteins
1 (7 oz)	Grilled chicken sandwich	3 carbohydrates + 4 lean proteins
1 (5 oz)	Fried fish fillet sandwich with cheese and tartar sauce	2½ carbohydrates + 2 medium-fat proteins + 1 ½ fats
1 (8 oz)	Cheeseburger (4 oz) with condiments	3 carbohydrates + 4 medium-fat proteins + 2½ fats
1 6-inch	Submarine sandwich (no cheese or sauce)	3 carbohydrates + 2 lean proteins + 1 fat

SIDES/APPETIZERS

Serving Size	Food	
1 medium (5 oz)	French fries	3½ carbohydrates + 3 fats
8 (5 oz)	Onion rings	3½ carbohydrates + 4 fats
1 small	Side salad (no cheese, croutons, or dressing)	1 nonstarchy vegetable

BEVERAGES AND DESSERTS

Serving Size	Food	
12 fl oz	Coffee, latte, with fat-free milk	1 fat-free milk
16 fl oz	Milkshake	7 carbohydrates + 4 fats
1 small	Ice cream cone	2 carbohydrates + ½ fat

Alcohol

For people with diabetes, up to 1 or 2 drinks per day for women and men, respectively, can safely fit into a healthy eating plan. Alcohol itself does not raise blood glucose, but alcoholic drinks often contain carbohydrates that must be counted. One alcohol equivalent provides 100 kcal. One carbohydrate choice provides 15 grams of carbohydrate and 70 kcal. Alcohol should be consumed with a meal to lower the risk of hypoglycemia.

12 fl oz	Beer, regular	1 alcohol equivalent + 1 carbohydate
1½ fl oz	Distilled spirits (e.g., rum and vodka)	1 alcohol equivalent
5 fl oz	Champagne	1 alcohol equivalent
3½ fl oz	Dessert wine	1 alcohol equivalent + 1 carbohydrate

Appendix C
Dietary Intake and Energy Expenditure Assessment

Although it may seem overwhelming at first, it is easy to track the foods you eat. One tip is to record foods and beverages consumed as soon as possible after consumption.

I. **Fill in the food record form that follows.** Appendix C contains a blank copy (see the completed example in Table C-1). Then, to estimate the nutrient values of the foods you are eating, consult food labels and use the NutritionCalc Plus software (CD or online) available with this book. If these resources do not have the serving size you need, adjust the value. If you drink ½ cup of orange juice, for example, but a table has values only for 1 cup, halve all values before you record them. Then, consider pooling all the same food to save time; if you drink a cup of 1% milk three times throughout the day, enter your milk consumption only once as 3 cups. As you record your intake for use on the nutrient analysis form that follows, consider the following tips:

- Measure and record the amounts of foods eaten in portion sizes of cups, teaspoons, tablespoons, ounces, slices, or inches (or convert metric units to these units).
- Record brand names of all food products, such as "Quick Quaker Oats."
- Measure and record all those little extras, such as gravies, salad dressings, taco sauces, pickles, jelly, sugar, ketchup, and margarine.
- For beverages
  - List the type of milk, such as whole, fat-free, 1%, evaporated, chocolate, or reconstituted dry.
  - Indicate whether fruit juice is fresh, frozen, or canned.
  - Indicate type for other beverages, such as fruit drink, fruit-flavored drink, Kool-Aid, and hot chocolate made with water or milk.
- For fruits
  - Indicate whether fresh, frozen, dried, or canned.
  - If whole, record number eaten and size with approximate measurements (such as 1 apple—3 inches in diameter).
  - Indicate whether processed in water, light syrup, or heavy syrup.
- For vegetables
  - Indicate whether fresh, frozen, dried, or canned.
  - Record as portion of cup, teaspoon, or tablespoon, or as pieces (such as carrot sticks—4 inches long, ½ inch thick).
  - Record preparation method.
- For cereals
  - Record cooked cereals in portions of tablespoon or cup (a level measurement after cooking).
  - Record dry cereal in level portions of tablespoon or cup.
  - If margarine, milk, sugar, fruit, or something else is added, measure and record amount and type.

- For breads
  - Indicate whether whole wheat, rye, white, and so on.
  - Measure and record number and size of portion (biscuit—2 inches across, 1 inch thick; slice of homemade rye bread—3 inches by 4 in., ¼ inch thick).
  - Sandwiches: list all ingredients (lettuce, mayonnaise, tomato, and so on).
- For meat, fish, poultry, and cheese
  - Give size (length, width, and thickness) in inches or weight in ounces after cooking for meat, fish, and poultry (such as cooked hamburger patty—3 inches across, ½ inch thick).
  - Give size (length, width, and thickness) in inches or weight in ounces for cheese.
  - Record measurements only for the cooked, edible part—without bone or fat left on the plate.
  - Describe how meat, poultry, or fish was prepared.
- For eggs
  - Record as soft or hard cooked, fried, scrambled, poached, or omelet.
  - If milk, butter, or drippings are used, specify types and amount.
- For desserts
  - List commercial brand or "homemade" or "bakery" under brand.
  - Purchased candies, cookies, and cakes: specify kind and size.
  - Measure and record portion size of cakes, pies, and cookies by specifying thickness, diameter, and width or length, depending on the item.

TABLE C-1 ▶ Example of a One-Day Food Record

Time	Minutes Spent Eating	M or S*	H† (0–3)	Activity While Eating	Place of Eating	Food and Quantity	Others Present	Reason for Choice
7:10 a.m.	15	M	2	Standing, fixing lunch	Kitchen	Orange juice, 1 cup	—	Health
						Crispix, 1 cup		Habit
						Nonfat milk, ½ cup		Health
						Sugar, 2 tsp		Taste
						Black coffee, 1 cup		Habit
10:00 a.m.	4	S	1	Sitting, taking notes	Classroom	Diet cola, 12 oz	Class	Weight control
12:15 p.m.	40	M	2	Sitting, talking	Student union	Chicken sandwich with lettuce and mayonnaise (3 oz chicken, 2 slices of white bread, and 2 tsp mayonnaise)	Friends	Taste
						Pear, 1 medium		Health
						Nonfat milk, 1 cup		Health
2:30 p.m.	10	S	1	Sitting, studying	Library	Regular cola, 12 oz	Friend	Hunger
6:30 p.m.	35	M	3	Sitting, talking	Kitchen	Pork chop, 1	Boyfriend	Convenience
						Baked potato, 1		Health
						Margarine, 2 tbsp		Taste
						Lettuce and tomato salad, 1 cup		Health
						Ranch dressing, 2 tbsp		Taste
						Peas, ½ cup		Health
						Whole milk, 1 cup		Habit
						Cherry pie, 1 piece		Taste
						Iced tea, 12 oz		Health
9:10 p.m.	10	S	2	Sitting, studying	Living room	Apple, 1 medium	—	Weight control
						Water, 1 cup		Weight control

*M or S: Meal or snack.
†H: Degree of hunger (0 = none; 3 = maximum).

Time	Minutes Spent Eating	M or S*	H† (0–3)	Activity While Eating	Place of Eating	Food and Quantity	Others Present	Reason for Choice

*M or S: Meal or snack.

†H: Degree of hunger (0 = none; 3 = maximum).

II. **Now complete the nutrient analysis form as shown, using your food record.**
A blank copy of this form for your use is on the following pages.
NutritionCalc Plus will create such a table for you if you enter all food eaten.

Nutrient Analysis Form (Sample)

Name	Quantity	kcal	Protein (g)	Carbohydrates (g)	Fiber (g)	Total fat (g)	Monounsaturated fat (g)	Polyunsaturated fat (g)	Saturated fat (g)	Cholesterol (g)	Calcium (mg)	Iron (mg)
Egg bagel, 3.5-inches diameter	1 ea.	180	7.45	34.7	0.748	1.00	0.286	0.400	0.171	44.0	20.0	2.10
Jelly	1 tbsp	49.0	0.018	12.7	—	0.018	0.005	0.005	0.005	—	2.00	0.120
Orange juice, prepared fresh or frozen	1½ cup	165	2.52	40.2	1.49	0.210	0.037	0.045	0.025	—	33.0	0.411
Cheeseburger, McDonald's	2 ea.	636	30.2	57.0	0.460	32.0	12.2	2.18	13.3	80.0	338	5.68
French fries, McDonald's	1 order	220	3.00	26.1	4.19	11.5	4.37	0.570	4.61	8.57	9.10	0.605
Cola beverage, regular	1½ cup	151	—	38.5	—	—	—	—	—	—	9.00	0.120
Pork loin chop, broiled, lean	4 oz	261	36.2	—	—	11.9	5.35	1.43	4.09	112	5.67	1.04
Baked potato with skin	1 ea.	220	4.65	51.0	3.90	0.200	0.004	0.087	0.052	—	20.0	2.75
Peas, frozen, cooked	½ cup	63.0	4.12	11.4	3.61	0.220	0.019	0.103	0.039	—	19.0	1.25
Margarine, regular or soft, 80% fat	20 g	143	0.160	0.100	—	16.1	5.70	6.92	2.76	—	5.29	—
Iceberg lettuce, chopped	2 cup	14.6	1.13	2.34	1.68	0.212	0.008	0.112	0.028	—	21.2	0.560
French dressing	2 oz	300	0.318	3.63	0.431	32.0	14.2	12.4	4.94	—	7.10	0.227
Reduced-fat (i.e., 2%) milk	1 cup	121	8.12	11.7	—	4.78	1.35	0.170	2.92	22.0	297	0.120
Graham crackers	2 ea.	60.0	1.04	10.8	1.40	1.46	0.600	0.400	0.400	—	6.00	0.367
Totals		2584	99.0	300	17.9	112	44.1	24.8	33.4	266	792	15.4
RDA or related nutrient standard*		2900	58	130	38						1000	8
% of nutrient needs		89	170	230	47						79	193

Abbreviations: g = grams, mg = milligrams, μg = micrograms.

*The values listed are for a male age 19 years. The number of kcal is a rough estimate. It is better to base energy needs on actual energy output.

‡Amounts refer to actual folate content, rather than dietary folate equivalents (DFEs). This difference is important to consider if the food contains added synthetic folic acid as part of enrichment or fortification. Any such folic acid is absorbed about twice as much as the folate present naturally in foods. So the total contribution of folate in the food in comparison to human needs will be greater than if all the folate was naturally in the food product. Nutrient analysis tables have yet to be updated to reflect the dietary folate equivalents of products.

Nutrient Analysis Form (Sample) cont'd

Magnesium (mg)	Phosphorus (mg)	Potassium (mg)	Sodium (mg)	Zinc (mg)	Vitamin A (RAE)	Vitamin C (mg)	Vitamin E (mg)	Thiamin (mg)	Riboflavin (mg)	Niacin (mg)	Vitamin B-6 (mg)	Folate (µg)	Vitamin B-12 (µg)
18.0	61.0	65.0	300	0.612	7.00	—	1.80	2.58	0.197	2.40	0.030	16.3	0.065
0.720	1.00	16.0	4.00	—	0.200	0.710	0.016	0.002	0.005	0.036	0.005	2.00	—
36.0	60.0	711	3.00	0.192	28.5	145	0.714	0.300	0.060	0.750	0.165	163	
45.8	410	314	1460	5.20	134	4.10	0.560	0.600	0.480	8.66	0.230	42.0	1.82
26.7	101	564	109	0.320	5.00	12.5	0.203	0.122	0.020	2.26	0.218	19.0	0.027
3.00	46.0	4.00	15.0	0.049	—	—	—	—	—	—	—	—	—
34.0	277	476	88.2	2.54	3.15	0.454	0.405	1.30	0.350	6.28	0.535	6.77	0.839
55.0	115	844	16.0	0.650	—	26.1	0.100	0.216	0.067	3.32	0.701	22.2	—
23.0	72.0	134	70.0	0.750	53.4	7.90	0.400	0.226	0.140	1.18	0.090	46.9	—
0.467	4.06	7.54	216	0.041	199	0.028	2.19	0.002	0.006	0.004	0.002	0.211	0.017
10.1	22.4	177	10.1	0.246	37.0	4.36	0.120	0.052	0.034	0.210	0.044	62.8	—
5.81	3.63	7.03	666	0.045	0.023	—	15.9	—	—	—	0.006	—	—
33.0	232	377	122	0.963	140	2.32	0.080	0.095	0.403	0.210	0.105	12.0	0.888
6.00	20.0	36.0	86.0	0.113	—	—	—	0.020	0.030	0.600	0.011	1.80	—
298	1425	3732	3165	11.7	607	204	22.5	5.52	1.79	25.9	2.14	395	3.65
400	700	4700	1500	11	900	90	15	1.2	1.3	16	1.3	400‡	2.4
75	204	80	210	106	67	226	150	450	138	162	160	99	152

Nutrient Analysis Form

Name	Quantity	kcal	Protein (g)	Carbohydrates (g)	Fiber (g)	Total fat (g)	Monounsaturated fat (g)	Polyunsaturated fat (g)	Saturated fat (g)	Cholesterol (g)	Calcium (mg)	Iron (mg)
Totals												
RDA or related nutrient standard*												
% of nutrient needs												

*The number of kcals is a rough estimate. It is better to base energy needs on actual energy output.

†Use RAE values.

‡Use DFE values.

Nutrient Analysis Form cont'd

Magnesium (mg)	Phosphorous (mg)	Potassium (mg)	Sodium (mg)	Zinc (mg)	Vitamin A (RAE)	Vitamin C (mg)	Vitamin E (mg)	Thiamin (mg)	Riboflavin (mg)	Niacin (mg)	Vitamin B-6 (mg)	Folate (µg)	Vitamin B-12 (µg)

III. **Complete the following table as you summarize your dietary intake.**

Percentage of kcal from Protein, Fat, Carbohydrate, and Alcohol

Intake

Protein (P): _____g/day × 4 kcal per gram = (P)_____kcal per day
Fat (F): _____g/day × 9 kcal per gram = (F)_____kcal per day
Carbohydrate (C): _____g/day × 4 kcal per gram = (C)_____kcal per day
Alcohol (A): = (A)_____kcal per day*

Total kcal (T)/day = (T)_____kcal per day

Percentage of kcal from protein:

$\frac{(P)}{(T)}$ × 100 =_____%

Percentage of kcal from fat:

$\frac{(F)}{(T)}$ × 100 =_____%

Percentage of kcal from carbohydrate:

$\frac{(C)}{(T)}$ × 100 =_____%

Percentage of kcal from alcohol:

$\frac{(A)}{(T)}$ × 100 =_____%

Note: The four percentages can total 99, 100, or 101, depending on the way in which figures were rounded off earlier.

*To calculate how many kcal in a beverage are from alcohol, first look up the beverage using the Nutrition Calc software. Then determine how many kcal are from carbohydrate (multiply carbohydrate grams times 4), fat (fat grams times 9), and protein (protein grams times 4). The remaining kcal are from alcohol.

IV. **Use the table on the following page to again record your food intake for one day, placing each food item in the correct category of MyPlate, with the correct number of servings (see Chapter 2).** A food such as a turkey and cheese sandwich contributes to three categories: grains, protein foods, and dairy. You can expect that many food choices will contribute to more than one group. Indicate the number of servings from MyPlate that each food yields.

Indicate the Number of Servings from MyPlate That Each Food Yields

Food or Beverage	Amount Eaten	Dairy	Protein	Fruits	Vegetables	Grains	SoFAS*
Group totals							
Recommended servings from www.Choose MyPlate.gov's Daily Food Plan							
Overages/shortages in numbers of servings							

*SoFAS: solid fats and added sugars.

V. **Evaluation.** Are there weaknesses suggested in your nutrient intake that correspond to missing servings in MyPlate? Consider adjusting your food choices to comply with MyPlate standards to improve your nutrient intake.

VI. **For the same day you keep your food record, also keep a 24-hour record of your activities.** Include sleeping, sitting, and walking, as well as the obvious forms of exercise. Calculate your energy expenditure for these activities using Table 7-5 in Chapter 7 or the diet analysis software available with this book. Try to substitute a similar activity if your particular activity is not listed. Calculate the total kcal you used for the day (total for column 3). Following is an example of an activity record. A blank form follows for your use. Ask your professor whether you are to turn in the form or the activity printout from NutritionCalc Plus.

Weight (kg)*: 70 kg

Activity	Time (Minutes): Convert to Hours	Energy Cost		
		Column 1 kcal/kg/hr (from Table 7-5)	Column 2 (Column 1 × Time)	Column 3 (Column 2 × Weight in kg)
Brisk walking	(60 min) 1 hr	4.4	(× 1) = 4.4	(× 70) = 308

*lb/2.2.

Weight (kg)*:

Activity	Time (Minutes): Convert to Hours	Energy Cost		
		Column 1 kcal/kg/hr (from Table 7-5)	Column 2 (Column 1 × Time)	Column 3 (Column 2 × Weight in kg)
Total kcal used (from adding all of column 3).				

*lb/2.2.

Appendix D

Chemical Structures Important in Nutrition

Amino Acids

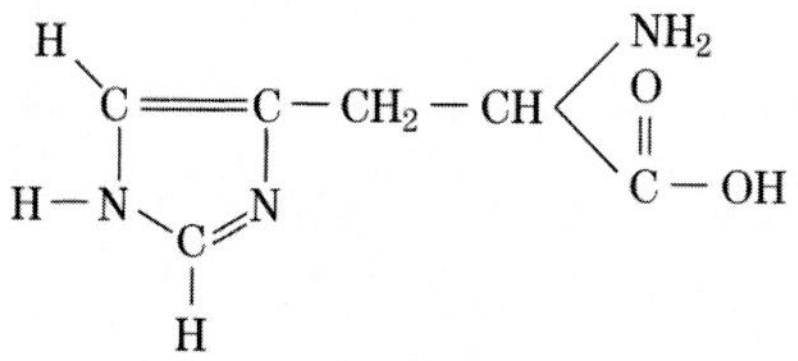

Histidine (His)
(essential)

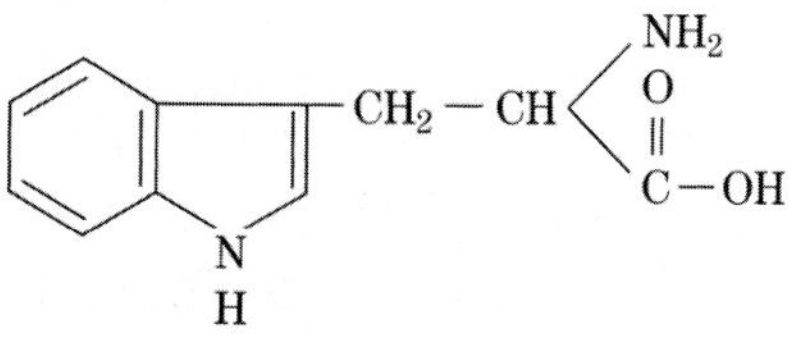

Tryptophan (Trp)
(essential)

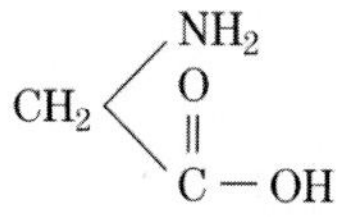

Glycine (Gly)

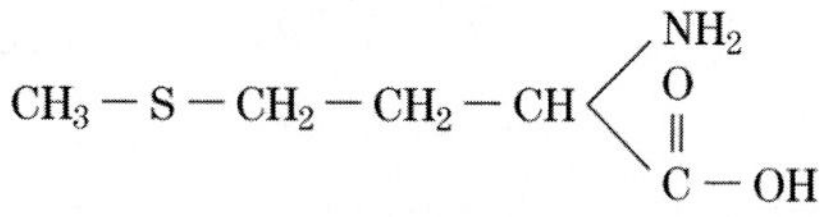

Methionine (Met)
(essential)

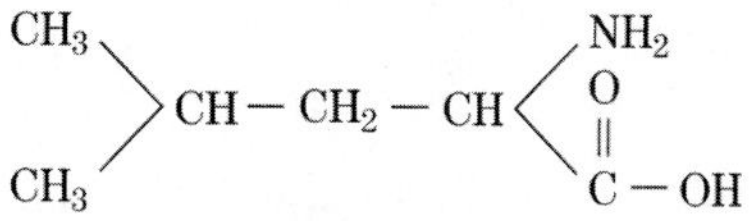

Leucine (Leu)
(essential)

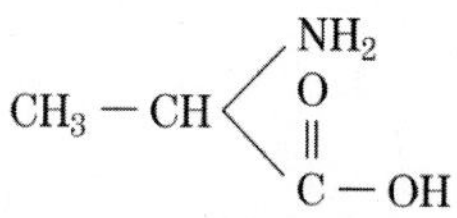

Alanine (Ala)

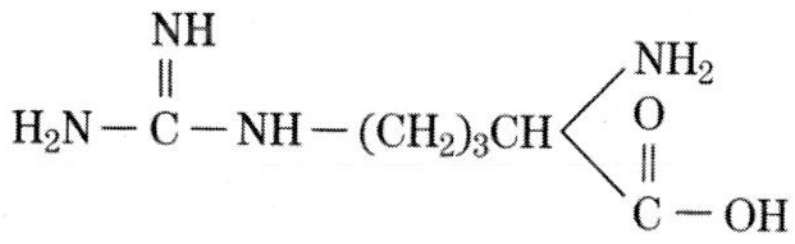

Arginine (Arg)
(essential in infancy)

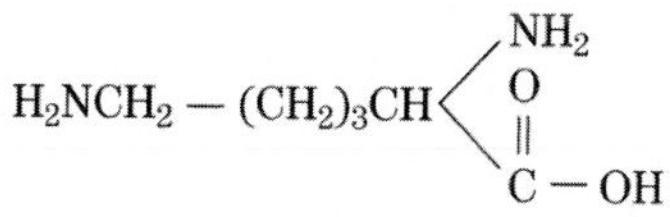

Lysine (Lys)
(essential)

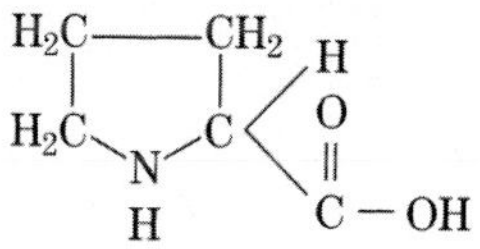

Proline (Pro)

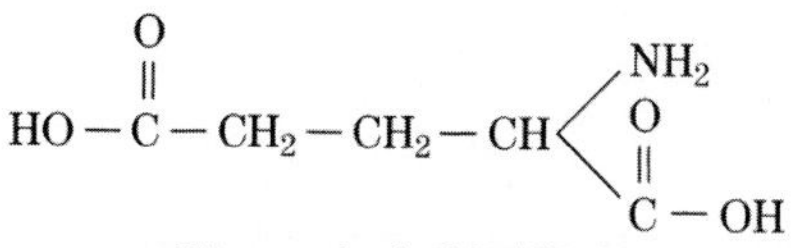

Glutamic Acid (Glu)

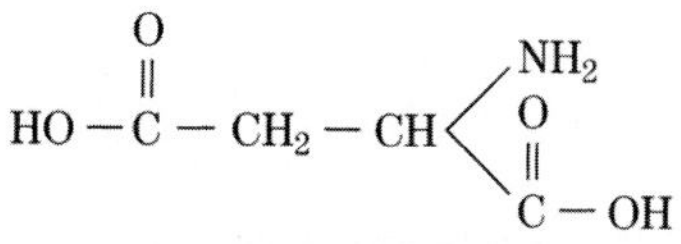

Aspartic Acid (Asp)

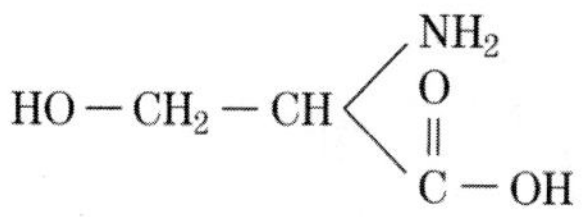

Serine (Ser)

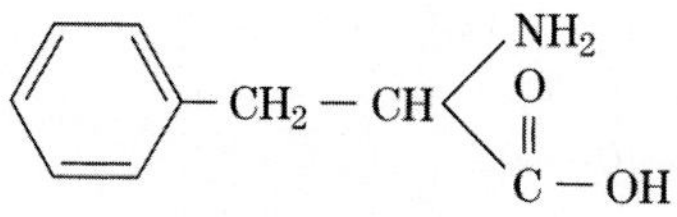

Phenylalanine (Phe)
(essential)

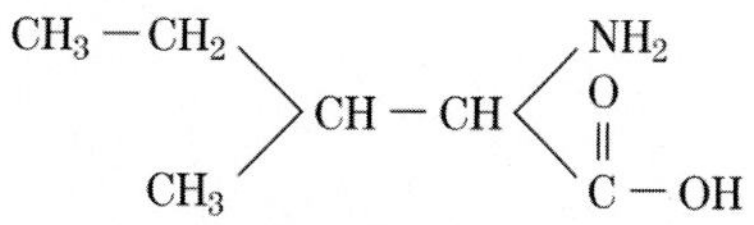

Isoleucine (Ile)
(essential)

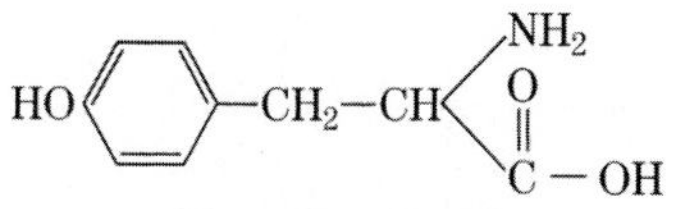

Tyrosine (Tyr)

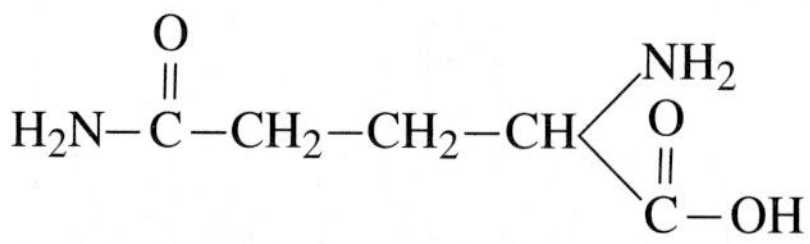

Glutamine (Gln)

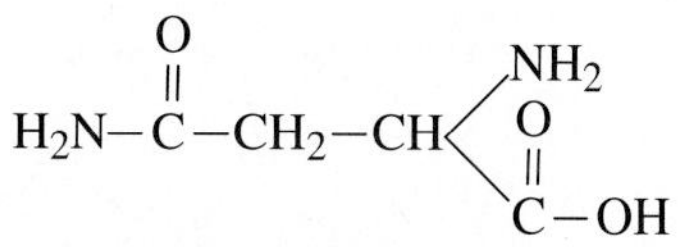

Asparagine (Asn)

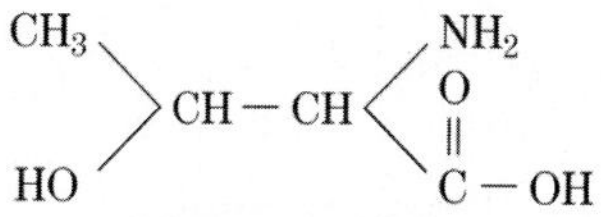

Threonine (Thr)
(essential)

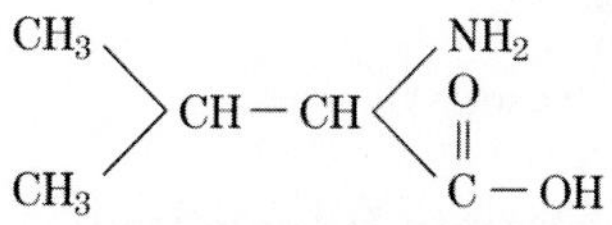

Valine (Val)
(essential)

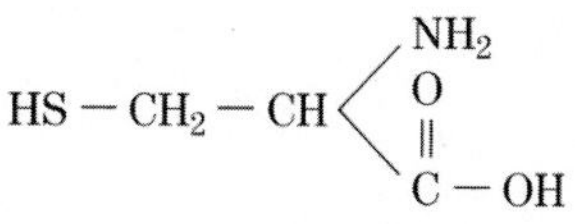

Cysteine (Cys)

Vitamins

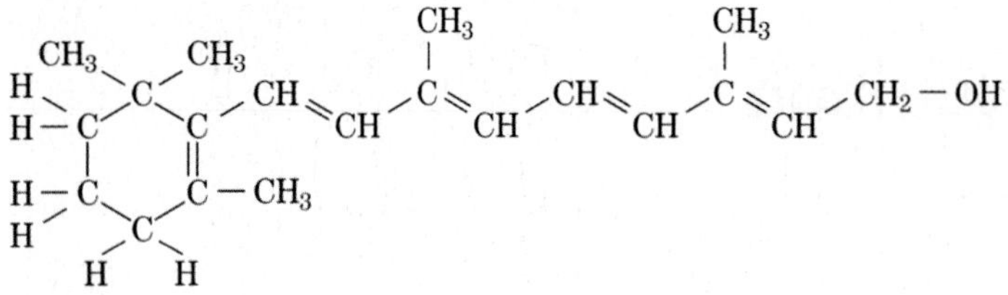

Vitamin A: retinol

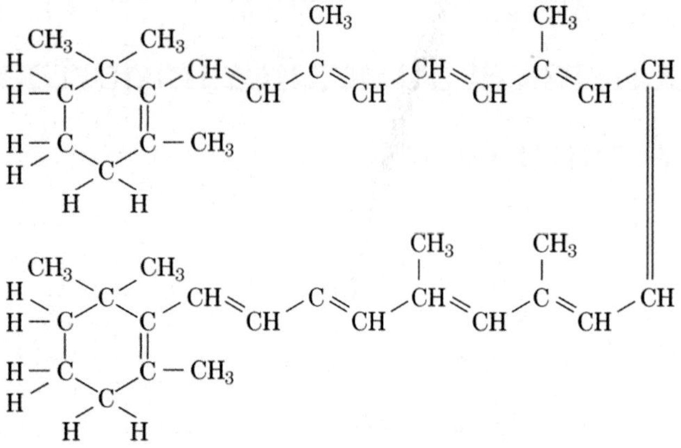

Beta-carotene

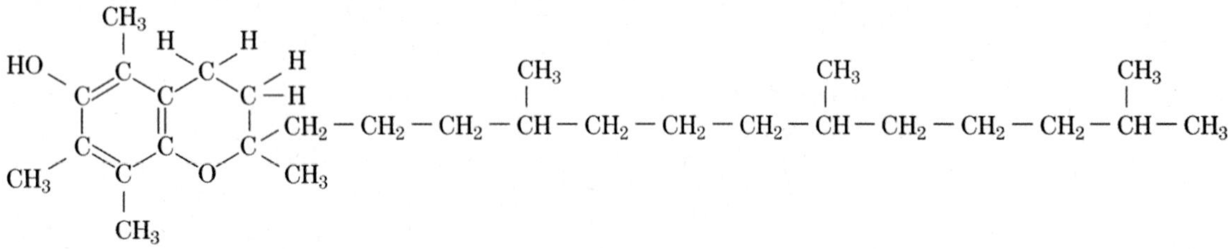

Vitamin E

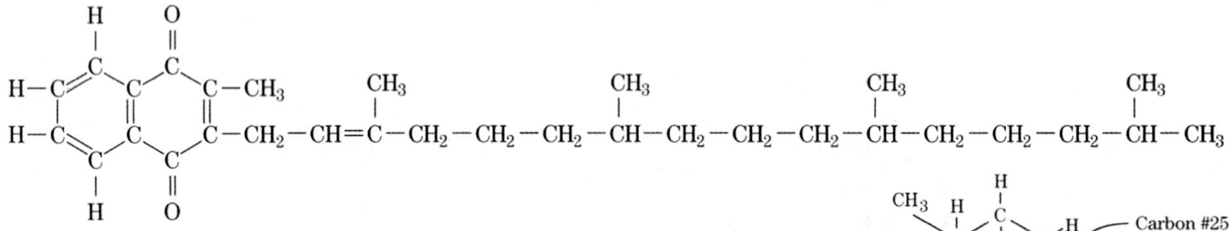

Vitamin K

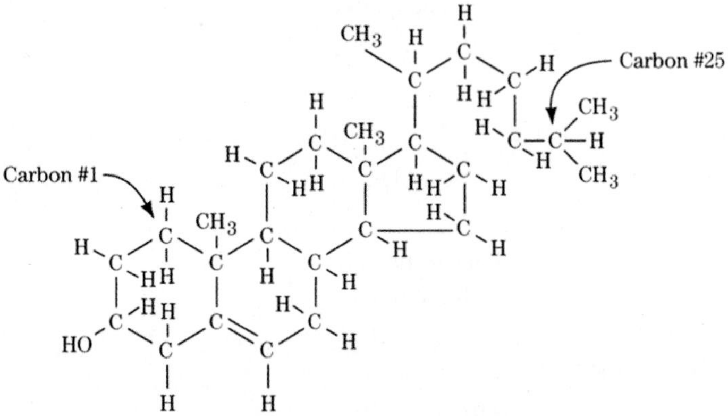

7-Dehydrocholesterol

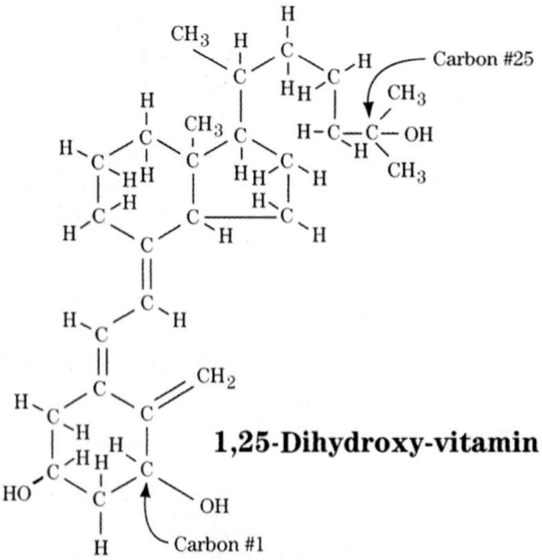

1,25-Dihydroxy-vitamin D$_3$ (calcitriol)

Active vitamin D (calcitriol) and its precursor 7-dehydrocholesterol

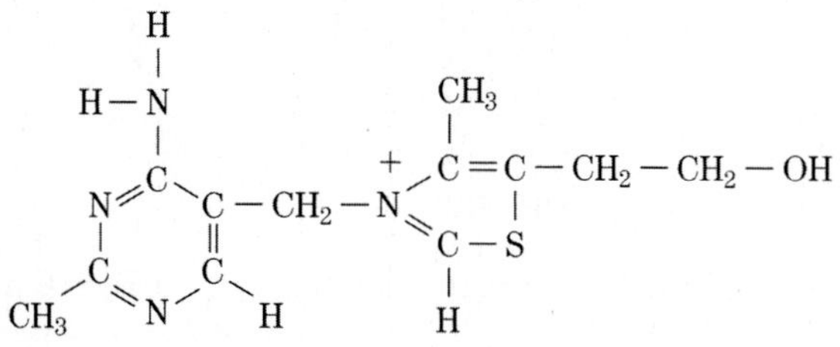

Thiamin

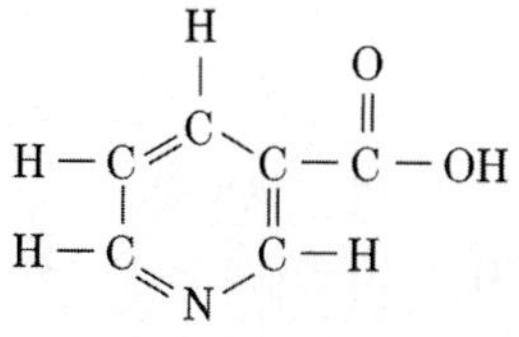

Niacin (nicotinic acid and nicotinamide)

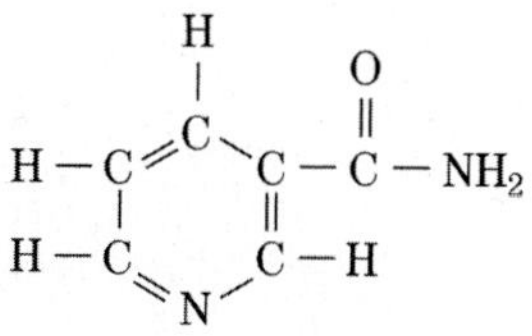

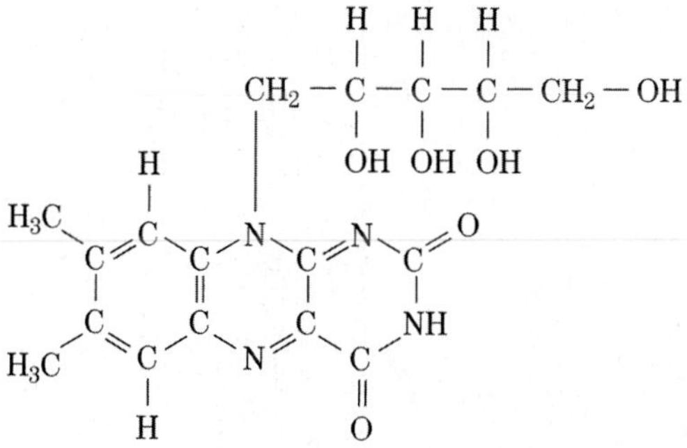

Nicotinic acid

Nicotinamide

Riboflavin

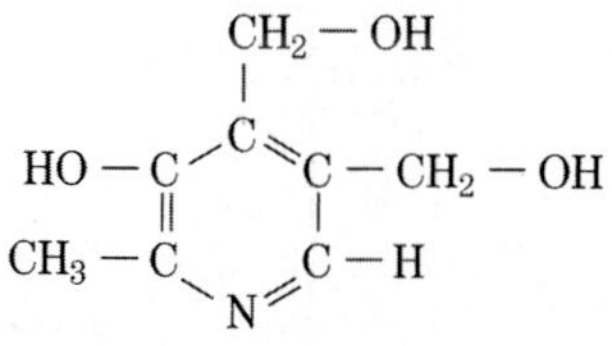

Pyridoxine

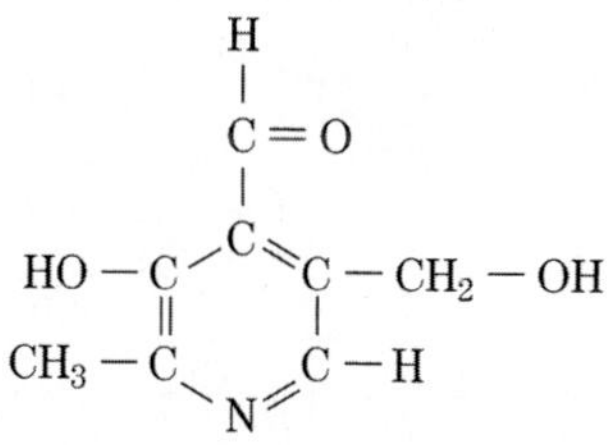

Pyridoxal

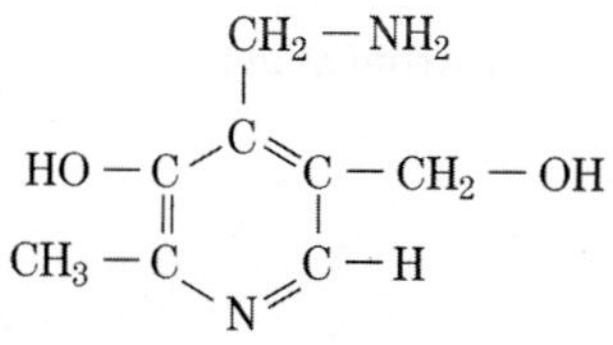

Pyridoxamine

Vitamin B-6 (a general name for three compounds—pyridoxine, pyridoxal, and pyridoxamine)

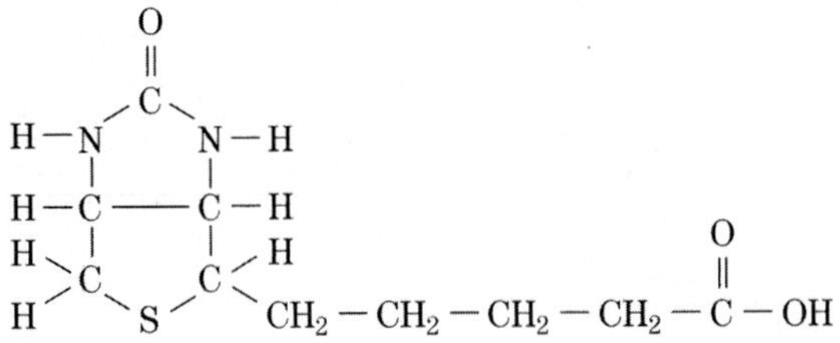

Biotin

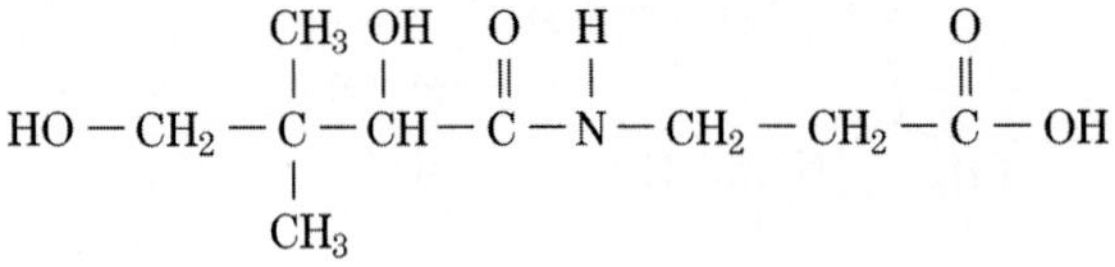

Pantothenic acid

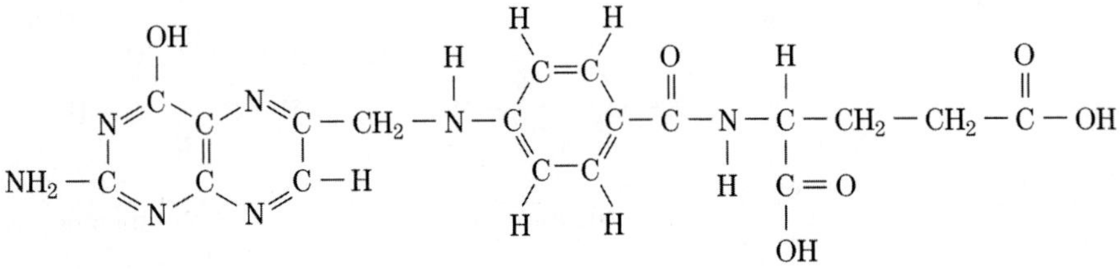

Folate (folic acid form)

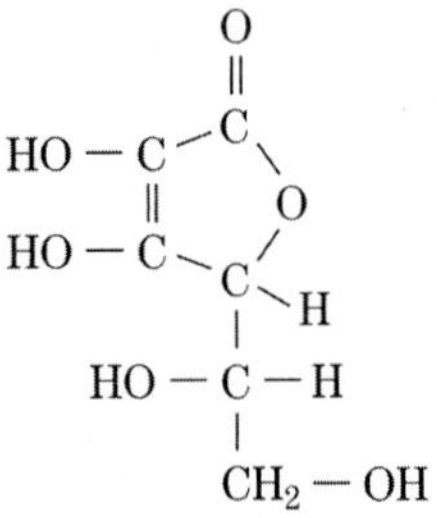

Vitamin C (ascorbic acid)

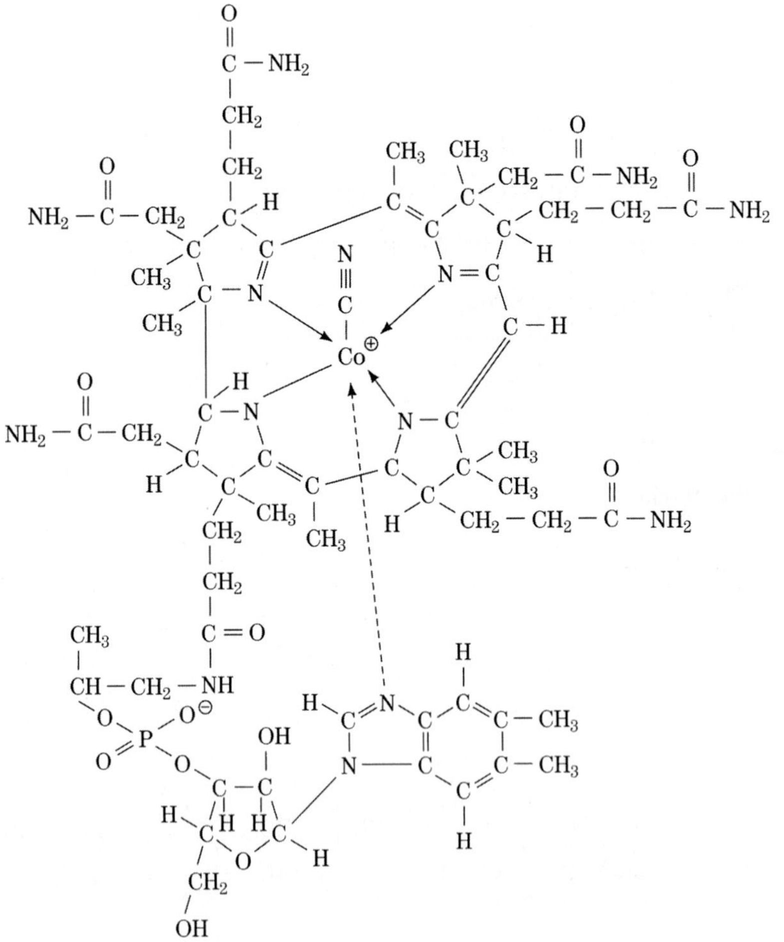

Vitamin B-12 (cyanocobalamin) The arrows in this diagram indicate that the spare electrons on the nitrogens are attracted to the cobalt atom.

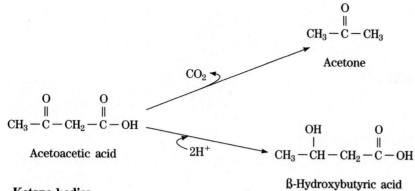

Ketone bodies

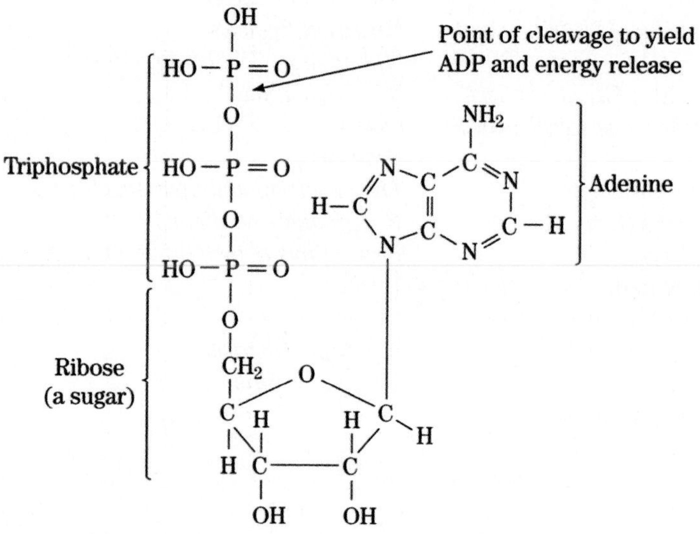

Acetone

Acetoacetic acid

CO_2

$2H^+$

ß-Hydroxybutyric acid

Point of cleavage to yield
ADP and energy release

Triphosphate

Adenine

Ribose
(a sugar)

**Adenosine triphosphate
(ATP)**

Appendix E
Sources of Nutrition Information

Consider the following reliable sources of food and nutrition information:

Journals That Regularly Cover Nutrition Topics

*American Family Physician**
American Journal of Clinical Nutrition
American Journal of Epidemiology
American Journal of Medicine
American Journal of Nursing
American Journal of Obstetrics and Gynecology
American Journal of Public Health
American Scientist
Annals of Internal Medicine
Annual Review of Medicine
Annual Review of Nutrition
Archives of Disease in Childhood
British Journal of Nutrition
BMJ (British Medical Journal)
Canadian Journal of Dietetic Practice and Research
Cancer
Cancer Research
Circulation
Critical Reviews in Food Science and Nutrition
Diabetes
Diabetes Care

Disease-a-Month
FASEB Journal
Food and Chemical Toxicology
Food Engineering
Food Technology
Gastroenterology
Gut
International Journal of Obesity
*Journal of the American College of Nutrition**
*Journal of the Academy of Nutrition and Dietetics**
Journal of the American Geriatrics Society
JAMA (Journal of the American Medical Association)
Journal of Applied Physiology
Journal of Clinical Investigation
Journal of Food Science
Journal of Human Nutrition and Dietetics
JNCI (Journal of the National Cancer Institute)
Journal of Nutrition
*Journal of Nutrition Education and Behavior**

Journal of Nutrition in Gerontology and Geriatrics
Journal of Pediatrics
The Lancet
Mayo Clinic Proceedings
Medicine & Science in Sports & Exercise
Nature
The New England Journal of Medicine
Nutrition
Nutrition & Dietetics
Nutrition Reviews
*Nutrition in Clinical Practice**
*Nutrition Today**
Obesity
Pediatrics
The Physician and Sportsmedicine
*Postgraduate Medicine**
Proceedings of the Nutrition Society
Science
*Science News**
Scientific American
Today's Dietitian

The majority of these journals are available in college and university libraries or in a specialty library on campus, such as one designated for health sciences. Most of them are now available online. As indicated, a few journals will be filed under their abbreviations, rather than the first word in their full name. A reference librarian can help you locate any of these sources. The journals with an asterisk (*) are ones you may find especially interesting and useful because of the number of nutrition articles presented each month or the less technical nature of the presentation.

Magazines for the Consumer That Cover Nutrition Topics

Better Homes and Gardens
Good Housekeeping

Health
Men's Health

Parents
Self

Textbooks and Other Sources for Advanced Study of Nutrition Topics

Erdman JW, MacDonald IA, Zeisel SH: *Present knowledge in nutrition.* Washington, DC: International Life Sciences and Wiley-Blackwell, 2012.

Gropper SS, Smith JL: *Advanced nutrition and human metabolism.* 6th ed. Belmont, CA: Wadsworth, Cengage, 2013.

Mahan LK, Escott-Stump S, Raymond JL: *Krause's food and the nutrition care process,* 13th ed. St. Louis: Elsevier Saunders, 2012.

Murray RK and others: *Harper's illustrated biochemistry.* 29th ed. New York: McGraw-Hill, 2012.

Ross AC and others: *Modern nutrition in health and disease.* 11th ed. Philadelphia: Lippincott, Williams & Wilkins, 2014.

Stipanuk MH, Caudill MA: *Biochemical, physiological, and molecular aspects of human nutrition.* 3rd ed. St. Louis: Philadelphia: Elsevier Saunders, 2013.

Newsletters That Cover Nutrition Issues on a Regular Basis

Beef Insights
Cattlemen's Beef Board
www.beefnutrition.org

Berkeley Wellness
University of California at Berkeley
www.berkeleywellness.com

Consumer Health Digest
www.consumerhealthdigest.com

The Dairy Download
National Dairy Council
www.nationaldairycouncil.org

Environmental Nutrition
www.environmentalnutrition.com

Harvard Health Letter (and others)
Harvard Medical School
www.health.harvard.edu/newsletters

Health and Nutrition Letter
Tufts University
www.nutritionletter.tufts.edu

Mayo Clinic Health Letter
Mayo Clinic
healthletter.mayoclinic.com

Nutrition Action Healthletter
Center for Science in the Public Interest
www.cspinet.org

Nutrition Unscrambled Blog
Egg Nutrition Center
www.enc-online.org

Soy Connection
United Soybean Board
www.soyconnection.com

Women's Nutrition Connection Newsletter
Weill Cornell Medical College
www.womensnutritionconnection.com

Professional Organizations

Academy of Nutrition and Dietetics
www.eatright.org

American Academy of Pediatrics
www.aap.org

American Cancer Society
www.cancer.org

American College of Sports Medicine
www.acsm.org

American Dental Association
www.ada.org

American Diabetes Association
www.diabetes.org

American Geriatrics Society
www.americangeriatrics.org

American Heart Association
www.americanheart.org

American Institute for Cancer Research
www.aicr.org

American Medical Association
www.ama-assn.org

American Public Health Association
www.apha.org

American Society for Nutrition
www.nutrition.org

Canadian Diabetes Association
www.diabetes.ca

Canadian Nutrition Society
www.cns-scn.ca

Dietitians of Canada
www.dietitians.ca

Environmental Working Group
www.ewg.org

Food and Nutrition Board of the Institute of Medicine
www.iom.edu/About-IOM/Leadership-Staff/Boards/Food-and-Nutrition-Board.aspx

Institute of Food Technologists
www.ift.org

National Council on Aging
www.ncoa.org

National Osteoporosis Foundation
www.nof.org

Society for Nutrition Education and Behavior
www.sneb.org

Professional Organizations with a Commitment to Nutrition Issues

Bread for the World Institute
www.bread.org

Food Research and Action Center
frac.org

Institute for Food and Development Policy
www.foodfirst.org

La Leche League International
www.llli.org

March of Dimes
www.marchofdimes.org

National Council Against Health Fraud
www.ncahf.org

National WIC Association
www.nwica.org

Overeaters Anonymous
www.oa.org

Oxfam America
www.oxfamamerica.org

Local Resources for Advice on Nutrition Issues

Registered dietitians (RDs) or Registered Dietitian Nutritionists (RDNs) in health care, city, county, or state agencies, as well as in private practice

Cooperative extension agents in county extension offices

Nutrition faculty affiliated with departments of food and nutrition, and dietetics

Government Agencies Concerned with Nutrition Issues or That Distribute Nutrition Information

United States

Agricultural Research Service
United States Department of Agriculture
www.ars.usda.gov

Food and Drug Administration
www.fda.gov

Food Safety and Inspection Service
United States Department of
    Agriculture
www.fsis.usda.gov

MyPlate
www.choosemyplate.gov

National Agricultural Library
www.nal.usda.gov

National Cancer Institute
www.cancer.gov

National Center for Health Statistics
www.cdc.gov/nchs

National Heart, Lung, and Blood
    Institute
www.nhlbi.nih.gov

National Institute on Aging
www.nia.nih.gov

Publication.USA.gov
http://publications.usa.gov

U.S. Government Printing Office
www.gpo.gov

Canada

Canadian Food Inspection Agency
www.inspection.gc.ca

Health Canada
www.hc-sc.gc.ca

United Nations

Food and Agriculture Organization
www.fao.org

World Health Organization
www.who.int

Trade Organizations and Companies That Distribute Nutrition Information

Abbott Nutrition
www.abbottnutrition.com

American Institute of Baking
www.aibonline.org

American Meat Institute
www.meatami.com

Beech-Nut Nutrition
www.beechnut.com

Campbell Soup Company
www.campbellsoup.com

Dannon Company
www.dannon.com

Del Monte Foods
www.delmonte.com

DSM Nutritional Products
www.dsm.com

General Mills/Pillsbury
www.generalmills.com

Gerber Products Company
www.gerber.com

H.J. Heinz
www.heinzbaby.com

Idaho Potato Commission
www.idahopotatoes.com

Kellogg Company
www.kelloggs.com/us/

Kraft Foods Group, Inc.
www.kraftrecipes.com

Mead Johnson Nutrition
www.meadjohnson.com

National Dairy Council
www.nationaldairycouncil.org

Sunkist Growers
www.sunkist.com

Appendix F
English-Metric Conversions and Metric Units

Metric-English Conversions

LENGTH

English (USA)	Metric
inch (in)	= 2.54 cm, 25.4 mm
foot (ft)	= 0.30 m, 30.48 cm
yard (yd)	= 0.91 m, 91.4 cm
mile (statute) (5280 ft)	= 1.61 km, 1609 m
mile (nautical) (6077 ft, 1.15 statute mi)	= 1.85 km, 1850 m

Metric	English (USA)
millimeter (mm)	= 0.039 in (thickness of a dime)
centimeter (cm)	= 0.39 in
meter (m)	= 3.28 ft, 39.37 in
kilometer (km)	= 0.62 mi, 1091 yd, 3273 ft

WEIGHT

English (USA)	Metric
grain	= 64.80 mg
ounce (oz)	= 28.35 g
pound (lb)	= 453.60 g, 0.45 kg
ton (short—2000 lb)	= 0.91 metric ton (907 kg)

Metric	English (USA)
milligram (mg)	= 0.002 grain (0.000035 oz)
gram (g)	= 0.04 oz ($\frac{1}{28}$ of an oz)
kilogram (kg)	= 35.27 oz, 2.20 lb
metric ton (1000 kg)	= 1.10 tons

VOLUME

English (USA)	Metric
cubic inch	= 16.39 cc
cubic foot	= 0.03 m^3
cubic yard	= 0.765 m^3
teaspoon (tsp)	= 5 ml
tablespoon (tbsp)	= 15 ml
fluid ounce	= 0.03 liter (30 ml)*
cup (c)	= 237 ml
pint (pt)	= 0.47 liter
quart (qt)	= 0.95 liter
gallon (gal)	= 3.79 liters

Metric	English (USA)
milliliter (ml)	= 0.03 oz
liter (L)	= 2.12 pt
liter	= 1.06 qt
liter	= 0.27 gal

1 liter ÷ 1000 = 1 milliliter or 1 cubic centimeter (10^{-3} liter)
1 liter ÷ 1,000,000 = 1 microliter (10^{-6} liter)

*Note: 1 ml = 1 cc.

Metric and Other Common Units

Unit/Abbreviation	Other Equivalent Measure
milligram/mg	$\frac{1}{1000}$ of a gram
microgram/μg	$\frac{1}{1,000,000}$ of a gram
deciliter/dl	$\frac{1}{10}$ of a liter (about ½ cup)
milliliter/ml	$\frac{1}{1000}$ of a liter (5 ml is about 1 tsp)
International Unit/IU	Crude measure of vitamin activity generally based on growth rate seen in animals

Fahrenheit-Celsius Conversion Scale

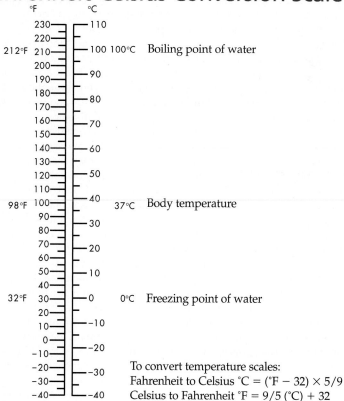

212°F 210 — 100 100°C Boiling point of water
98°F 100 — 40 37°C Body temperature
32°F 30 — 0 0°C Freezing point of water

To convert temperature scales:
Fahrenheit to Celsius °C = (°F − 32) × 5/9
Celsius to Fahrenheit °F = 9/5 (°C) + 32

Household Units

3 teaspoons	= 1 tablespoon	= 15 grams
4 tablespoons	= ¼ cup	= 60 grams
5⅓ tablespoons	= ⅓ cup	= 80 grams
8 tablespoons	= ½ cup	= 120 grams
10⅔ tablespoons	= ⅔ cup	= 160 grams
16 tablespoons	= 1 cup	= 240 grams
1 tablespoon	= ½ fluid ounce	= 15 milliliters
1 cup	= 8 fluid ounces	= 15 milliliters
1 cup	= ½ pint	= 240 grams
2 cups	= 1 pint	= 480 grams
4 cups	= 1 quart	= 960 grams = 1 liter
2 pints	= 1 quart	= 960 grams = 1 liter
4 quarts	= 1 gallon	= 3840 grams = 4 liters

GLOSSARY Medical Terminology to Aid in the Study of Nutrition

1,25-dihydroxyvitamin D₃ (calcitriol) Biologically active form of vitamin D.

7-dehydrocholesterol Precursor of vitamin D found in the skin.

25-hydroxyvitamin D₃ (calcidiol or calcifediol) Form found in blood. Sometimes shortened to 25(OH)D₃.

absorption The process by which substances are taken up from the GI tract and enter the bloodstream or the lymph.

absorptive cells Also known as *enterocytes;* intestinal cells that line the villi and participate in nutrient absorption.

acceptable daily intake (ADI) Estimate of the amount of a sweetener that an individual can safely consume daily over a lifetime. ADIs are given as milligrams per kilograms of body weight per day.

acesulfame K Alternative sweetener that yields no energy to the body; 200 times sweeter than sucrose.

acetaldehyde dehydrogenase An enzyme used in ethanol metabolism that eventually converts acetaldehyde into carbon dioxide and water.

acquired immune deficiency syndrome (AIDS) A disorder in which a virus (human immunodeficiency virus [HIV]) infects specific types of immune system cells. This leaves the person with reduced immune function and, in turn, defenseless against numerous infectious agents; typically contributes to the person's death.

adaptive thermogenesis This term encompasses the ability of humans to regulate body temperature within narrow limits (thermoregulation). Two visible examples of thermogenesis are fidgeting and shivering when cold.

added sugars Sugars or syrups that are added to foods during processing or preparation.

additives Substances added to foods, either intentionally or incidentally.

adenosine diphosphate (ADP) A breakdown product of ATP. ADP is synthesized into ATP using energy from foodstuffs and a phosphate group (abbreviated Pᵢ).

adenosine triphosphate (ATP) The main energy currency for cells. ATP energy is used to promote ion pumping, enzyme activity, and muscular contraction.

Adequate Intake (AI) Nutrient intake amount set for any nutrient for which insufficient research is available to establish an RDA. AIs are based on estimates of intakes that appear to maintain a defined nutritional state in a specific life stage.

adjustable gastric banding A restrictive procedure in which the opening from the esophagus to the stomach is reduced by a hollow gastric band.

aerobic Requiring oxygen.

aging Time-dependent physical and physiological changes in body structure and function that occur normally and progressively throughout adulthood as humans mature and become older.

air displacement A method for estimating body composition that makes use of the volume of space taken up by a body inside a small chamber.

alcohol dehydrogenase An enzyme used in alcohol (ethanol) metabolism that converts alcohol into acetaldehyde.

alcohol Ethyl alcohol or ethanol (CH_3CH_2OH) is the compound in alcoholic beverages.

alcohol use disorder A psychiatric disorder characterized by a problematic pattern of alcohol use that leads to significant impairment or distress.

aldosterone A hormone produced by the adrenal glands when blood volume is low. It acts on the kidneys to conserve sodium (and therefore water) to increase blood volume.

allergen A foreign protein, or antigen, that induces excess production of certain immune system antibodies; subsequent exposure to the same protein leads to allergic symptoms. While all allergens are antigens, not all antigens are allergens.

allergy A hypersensitive immune response that occurs when immune bodies produced by us react with a protein we sense as foreign (an antigen).

alpha-linolenic acid An essential omega-3 fatty acid with 18 carbons and 3 double bonds.

amino acid The building block for proteins containing a central carbon atom with nitrogen and other atoms attached.

amphetamine A group of medications that induce stimulation of the central nervous system and have other effects in the body. Abuse is linked to physical and psychological dependence.

amylase A starch-digesting enzyme produced by salivary glands and the pancreas.

amylopectin A digestible branched-chain type of starch composed of glucose units.

amylose A digestible straight-chain type of starch composed of glucose units.

anaerobic Not requiring oxygen.

anal sphincters A group of two sphincters (inner and outer) that help control expulsion of feces from the body.

anaphylaxis A severe allergic response that results in lowered blood pressure and respiratory distress. This can be fatal.

anemia A decreased oxygen-carrying capacity of the blood. This can be caused by many factors, such as iron deficiency or blood loss.

anencephaly Birth defect characterized by the absence of some or all of the brain and skull.

angiotensin A hormone produced by the liver and activated by enzymes from the kidneys. It signals the adrenal glands to produce aldosterone and also directs the kidneys to conserve sodium (and therefore water). Both of these actions have the effect of increasing blood volume.

animal model Use of animals to study disease to understand more about human disease.

anorexia nervosa An eating disorder characterized by extreme restriction of energy intake relative to requirements leading to significantly low body weight.

anthropometric assessment Measurement of body weight and the lengths, circumferences, and thicknesses of parts of the body.

antibody Blood protein that binds foreign proteins found in the body; also called *immunoglobulin.* This helps to prevent and control infections.

antidiuretic hormone A hormone secreted by the pituitary gland when blood concentration of solutes is high. It causes the kidneys to decrease water excretion, which increases blood volume.

antigen Any substance that induces a state of sensitivity and/or resistance to micro-organisms or toxic substances after a lag period; foreign substance that stimulates a specific aspect of the immune system.

antioxidant Generally a compound that stops the damaging effects of reactive substances seeking an electron (i.e., oxidizing agents). This prevents breakdown (oxidizing) of substances in foods or the body, particularly lipids.

anus Last portion of the GI tract; serves as an outlet for the digestive system.

appetite The primarily psychological (external) influences that encourage us to find and eat food, often in the absence of obvious hunger.

arachidonic acid An omega-6 fatty acid made from linoleic acid with 20 carbon atoms and 4 carbon–carbon double bonds.

artery A blood vessel that carries blood away from the heart.

aseptic processing A method by which food and container are separately and simultaneously sterilized; it allows manufacturers to produce boxes of milk that can be stored at room temperature.

aspartame Alternative sweetener made of two amino acids and methanol; about 200 times sweeter than sucrose.

atherosclerosis A buildup of fatty material (plaque) in the arteries, including those surrounding the heart.

atopic disease A condition involving an inappropriate immune response to environmental allergens; examples include asthma, eczema, and seasonal allergies.

atypical anorexia nervosa A subthreshold eating disorder in which a person meets most of the criteria for diagnosis of anorexia nervosa, except weight is within a normal range.

avoidant/restrictive food intake disorder Eating disorder characterized by failure to meet energy or nutrient needs, resulting in significant weight loss, nutritional deficiencies, or dependence on tube or intravenous feeding; the eating disturbance is not better explained by lack of available food, a medical problem, or another eating disorder.

bacteria Single-cell microorganisms; some produce poisonous substances, which cause illness in humans. Bacteria can be carried by water, animals, and people. They survive on skin, clothes, and hair, and thrive in foods at room temperature. Some can live without oxygen and survive by means of **spore** formation.

bariatrics The medical specialty focusing on the treatment of obesity.

basal metabolism The minimal amount of calories the body uses to support itself in a fasting state when resting and awake in a warm, quiet environment. It amounts to roughly 1 kcal per kilogram per hour for men and 0.9 kcal per kilogram per hour for women; these values are often referred to as *basal metabolic rate (BMR)*.

benign Noncancerous; tumors that do not spread.

beriberi The thiamin-deficiency disorder characterized by muscle weakness, loss of appetite, nerve degeneration, and sometimes edema.

BHA, BHT Butylated hydroxyanisole and butylated hydroxytoluene—two common synthetic antioxidants added to foods.

bile A liver secretion stored in the gall-bladder and released through the common bile duct into the first segment of the small intestine. It is essential for the digestion and absorption of fat.

binge drinking Drinking sufficient alcohol within a 2-hour period to increase blood alcohol content to 0.08 grams per deciliter or higher; for men, consuming 5 or more drinks in a row; for women, consuming 4 or more drinks in a row.

binge eating Consuming an abnormally large amount of food within a short time period (e.g., 2 hours).

binge-eating disorder An eating disorder characterized by recurrent episodes of binge eating that are associated with marked distress and lack of control over behavior, but not followed by inappropriate compensatory behaviors to prevent weight gain.

bioavailability The degree to which a consumed nutrient is absorbed and used by the body.

biochemical assessment Measurement of biochemical functions (e.g., concentrations of nutrient by-products or enzyme activities in the blood or urine) related to a nutrient's function.

bioelectrical impedance The method to estimate total body fat that uses a low-energy electrical current. The more fat storage a person has, the more impedance (resistance) to electrical flow will be exhibited.

biological pest management Control of agricultural pests by using natural predators, parasites, or pathogens. For example, ladybugs can be used to control an aphid infestation.

biotechnology A collection of processes that involve the use of biological systems for altering and, ideally, improving the characteristics of plants, animals, and other forms of life.

bisphosphonates Drugs that bind minerals and prevent osteoclast breakdown of bone. Examples are alendronate (Fosamax) and risedronate (Actonel).

body mass index (BMI) Weight (in kilograms) divided by height (in meters) squared; a value of 25 and above indicates overweight, and a value of 30 and above indicates obesity.

bolus A moistened mass of food swallowed from the oral cavity into the pharynx.

bomb calorimeter An instrument used to determine the calorie content of a food.

bond A linkage between two atoms, such as that formed by the sharing of electrons.

branched-chain amino acids Amino acids with a branching carbon backbone; these are leucine, isoleucine, and valine. All are essential amino acids.

brown adipose tissue A specialized form of adipose tissue that produces large amounts of heat by metabolizing energy-yielding nutrients without synthesizing much useful energy for the body. The unused energy is released as heat.

buffers Compounds that cause a solution to resist changes in acid–base conditions.

bulimia nervosa An eating disorder characterized by recurrent episodes of binge eating followed by inappropriate compensatory behaviors to prevent weight gain.

capillary A microscopic blood vessel that connects the smallest arteries and veins; site of nutrient, oxygen, and waste exchange between body cells and the blood.

capillary bed Network of one-cell-thick vessels that create a junction between arterial and venous circulation. It is here that gas and nutrient exchange occurs between body cells and the blood.

carbohydrate A compound containing carbon, hydrogen, and oxygen atoms. *Sugars, starches,* and *fibers* are forms of *carbohydrates.*

carbohydrate loading A process in which a high-carbohydrate diet is consumed for several days before an athletic event while tapering exercise duration in an attempt to increase muscle glycogen stores.

carbon skeleton Amino-acid structure that remains after the amino group ($—NH_2$) has been removed.

cardiovascular system The body system consisting of the heart, blood vessels, and blood. This system transports nutrients, waste products, gases, and hormones throughout the body and plays an important role in immune responses and regulation of body temperature.

carotenoids Precursors of vitamin A found in plant foods.

case-control study A study in which individuals who have a disease or condition, such as lung cancer, are compared with individuals who do not have the condition.

celiac disease Chronic, immune-mediated disease precipitated by exposure to dietary gluten in genetically predisposed people.

cell nucleus An organelle bound by its own double membrane and containing chromosomes, the genetic information for cell protein synthesis and cell replication.

cell The structural basis of plant and animal organization. Cells have the ability to take up compounds from and excrete compounds into their surroundings.

cellular differentiation The process of a less specialized cell becoming a more specialized type. Think of stem cells in the bone marrow becoming red and white blood cells.

cellulose An undigestible nonfermentable straight-chain polysaccharide made of glucose molecules.

cerebrovascular accident (CVA) Death of part of the brain tissue due typically to a blood clot. Also termed a *stroke.*

ceruloplasmin Copper-containing protein in the blood; functions in the transport of iron.

chain-breaking Breaking the link between two or more behaviors that encourage overeating, such as snacking while watching television.

chemical reaction An interaction between two chemicals that changes both chemicals.

cholesterol A waxy lipid found in all body cells. It has a structure containing multiple chemical rings that is found only in foods that contain animal products.

chromosome A single, large DNA molecule and its associated proteins; contains many genes to store and transmit genetic information.

chylomicron Lipoprotein made of dietary fats surrounded by a shell of cholesterol, phospholipids, and protein. Chylomicrons are formed in the absorptive cells of the small intestine after fat absorption and travel through the lymphatic system to the bloodstream.

chyme A mixture of stomach secretions and partially digested food.

cirrhosis A loss of functioning liver cells, which are replaced by nonfunctioning connective tissue. Any substance that poisons liver cells can lead to cirrhosis. The most common cause is a chronic, excessive alcohol intake. Exposure to certain industrial chemicals also can lead to cirrhosis.

cis **fatty acid** A form of an unsaturated fatty acid that has the hydrogens lying on the same side of the carbon–carbon double bond.

clinical assessment Examination of general appearance of skin, eyes, and tongue; evidence of rapid hair loss; sense of touch; and ability to cough and walk.

coenzyme A compound (e.g., water-soluble vitamin) that combines with an inactive enzyme to form a catalytically active form. In this manner, coenzymes aid in enzyme function.

cognitive behavioral therapy Psychological therapy in which the person's assumptions about dieting, body weight, and related issues are confronted. New ways of thinking are explored and then practiced by the person. In this way, an individual can learn new ways to control disordered eating behaviors and related life stress.

cognitive restructuring Changing one's frame of mind regarding eating; for example, instead of using a difficult day as an excuse to overeat, substituting other pleasures for rewards, such as a relaxing walk with a friend.

colostrum The first fluid secreted by the breast during late pregnancy and the first few days after birth. This thick fluid is rich in immune factors and protein.

community-supported agriculture (CSA) Farms that are supported by a community of growers and consumers who provide mutual support and share the risks and benefits of food production, usually including a system of weekly delivery or pick up of vegetables and fruit, and sometimes dairy products and meat.

compensatory behaviors Actions taken to rid the body of excess calories and/or to alleviate guilt or anxiety associated with a binge; examples include vomiting, misuse of laxatives, or excessive exercise.

complementary proteins Two food protein sources that make up for each other's inadequate supply of specific essential amino acids; together, they yield a sufficient amount of all nine and so provide high-quality (complete) protein for the diet.

complex carbohydrate Carbohydrate composed of many sugar units (polysaccharide). Examples include glycogen, starch, and fiber.

compression of morbidity Delay of the onset of disabilities caused by chronic disease.

conditionally essential amino acids Amino acids that must be made from essential amino acids if insufficient amounts are eaten.

congenital hypothyroidism The stunting of body growth and poor development in the offspring that result from inadequate maternal intake of iodide during pregnancy (formerly called cretinism).

congenital lactase deficiency Birth defect resulting in the inability to produce lactase, such that a lactose-free diet is required from birth.

connective tissue Protein tissue that holds different structures in the body together. Some body structures are made up of connective tissue—notably, tendons and cartilage. Connective tissue also forms part of bone and the nonmuscular structures of arteries and veins.

contingency management Forming a plan of action to respond to a situation in which overeating is likely, such as when snacks are within arm's reach at a party.

control group Participants in an experiment who are not given the treatment being tested.

cortical bone The compact or dense bone found on the outer surfaces of bone.

creatine An organic (i.e., carbon-containing) molecule in muscle cells that serves as a part of a high-energy compound (termed creatine phosphate or phosphocreatine) capable of synthesizing ATP from ADP.

cytoplasm The fluid and organelles (except the nucleus) in a cell.

dehydration Inadequate intake of water to replace losses.

Delaney Clause A clause to the 1958 Food Additives Amendment of the Pure Food and Drug Act in the United States that prevents the intentional (direct) addition to foods of a compound shown to cause cancer in laboratory animals or humans.

dementia A general loss or decrease in mental function.

denaturation Alteration of a protein's three-dimensional structure, usually because of treatment by heat, enzymes, acid or alkaline solutions, or agitation.

dental caries Erosions in the surface of a tooth caused by acids made by bacteria as they metabolize sugars.

deoxyribonucleic acid (DNA) The site of hereditary information in cells; DNA directs the synthesis of cell proteins.

diastolic blood pressure The pressure in the arterial blood vessels when the heart is between beats.

dietary assessment Estimation of typical food choices relying mostly on the recounting of one's usual intake or a record of one's previous days' intake.

dietary fiber Fiber found in food.

Dietary Guidelines for Americans General goals for nutrient intakes and diet composition set by the USDA and the U.S. Department of Health and Human Services.

Dietary Reference Intakes (DRIs) Term used to encompass nutrient recommendations made by the Food and Nutrition Board of the Institute of Medicine. These include RDAs, AIs, EERs, and ULs.

digestion Process by which large ingested molecules are mechanically and chemically broken down to produce basic nutrients that can be absorbed across the wall of the GI tract.

digestive system System consisting of the GI tract and accessory structures (liver, gallbladder, and pancreas). This system performs the mechanical and chemical processes of digestion, absorption of nutrients, and elimination of wastes.

diglyceride A breakdown product of a triglyceride consisting of two fatty acids bonded to a glycerol backbone.

direct calorimetry A method of determining a body's energy use by measuring heat released from the body. An insulated chamber is usually used.

disaccharide Class of sugars formed by the chemical bonding of two monosaccharides.

disordered eating Mild and short-term changes in eating patterns that occur in relation to a stressful event, an illness, or a desire to modify one's diet for a variety of health and personal appearance reasons.

diuretic A substance that increases urinary fluid excretion.

diverticula Pouches that protrude through the exterior wall of the large intestine.

diverticulitis An inflammation of the diverticula caused by acids produced by bacterial metabolism inside the diverticula.

diverticulosis The condition of having many diverticula in the large intestine.

docosahexaenoic acid (DHA) An omega-3 fatty acid with 22 carbons and 6 carbon–carbon double bonds. It is present in large amounts in fatty fish and is slowly synthesized in the body from alpha-linolenic acid. DHA is especially present in the retina and brain.

double-blind study An experimental design in which neither the participants nor the researchers are aware of each participant's assignment (test or placebo) or the outcome of the study until it is completed. An independent third party holds the code and the data until the study has been completed.

dual energy X-ray absorptiometry (DEXA) A highly accurate method of measuring body composition and bone mass and density using multiple low-energy X rays.

duodenum First segment of the small intestine that receives chyme from the stomach and digestive juices from the pancreas and gallbladder. This is the site of most chemical digestion of nutrients; approximately 10 inches in length.

early childhood caries Tooth decay that results from formula or juice (and even human milk) bathing the teeth as the child sleeps with a bottle in his or her mouth. The upper teeth are mostly affected as the lower teeth are protected by the tongue; formerly called *nursing bottle syndrome* and *baby bottle tooth decay.*

eating disorder Severe alterations in eating patterns linked to physiological changes. The alterations are associated with food restriction, binge eating, inappropriate compensatory behaviors, and fluctuations in weight. They also involve a number of emotional and cognitive changes that affect the way a person perceives and experiences his or her body.

eating pattern A combination of foods and beverages that constitutes an individual's complete dietary intake over time.

eclampsia A severe form of gestational hypertension characterized by protein in the urine and seizures (formerly called *toxemia*).

edema The buildup of excess fluid in extracellular spaces.

eicosanoids A class of hormone compounds, including the prostaglandins, derived from polyunsaturated fatty acids such as arachidonic acid and involved in the cellular activity affecting practically all important functions in the body.

eicosapentaenoic acid (EPA) An omega-3 fatty acid with 20 carbons and 5 carbon–carbon double bonds. It is present in large amounts in fatty fish and is slowly synthesized in the body from alpha-linolenic acid.

electrolytes Substances that separate into ions in water and, in turn, are able to conduct an electrical current. These include sodium, chloride, and potassium.

elimination diet A restrictive diet that systematically tests foods that may cause an allergic response by first eliminating them for 1 to 2 weeks and then adding them back, one at a time.

embryo In humans, the developing offspring in utero from about the beginning of the third week to the end of the eighth week after conception.

empty calories Calories from solid fats and/or added sugars. Foods with empty calories supply energy but few or no other nutrients.

emulsifier A compound that can suspend fat in water by isolating individual fat droplets, using a shell of water molecules or other substances to prevent the fat from coalescing.

endocrine gland A hormone-producing gland.

endocrine system The body system consisting of the various glands and the hormones these glands secrete. This system has major regulatory functions in the body, such as reproduction and cell metabolism.

endometrium The membrane that lines the inside of the uterus. It increases in thickness during the menstrual cycle until ovulation occurs. The surface layers are shed during menstruation if conception does not take place.

endoplasmic reticulum (ER) An organelle composed of a network of canals running through the cytoplasm. Part of the ER contains ribosomes.

endorphins Natural body tranquilizers that may be involved in the feeding response and function in pain reduction.

energy balance The state in which energy intake, in the form of food and beverages, matches the energy expended, primarily through basal metabolism and physical activity.

energy density A comparison of the calorie (kcal) content of a food with the weight of the food. An energy-dense food is high in calories but weighs very little (e.g., potato chips), whereas a food low in energy density has few calories but weighs a lot, such as an orange.

enterohepatic circulation A continual recycling of compounds such as bile acids between the small intestine and the liver.

environmental assessment Includes details about living conditions, education level, and the ability of the person to purchase, transport, and cook food. The person's weekly budget for food purchases is also a key factor to consider.

enzyme A compound that speeds the rate of a chemical reaction but is not altered by the reaction. Almost all enzymes are proteins (some are made of genetic material).

epidemiology The study of how disease rates vary among different population groups.

epigenetics Changes in gene function that are independent of DNA sequence. For example, malnutrition during pregnancy may modify gene expression in the fetus and affect long-term body weight regulation in the offspring.

epigenome The way that the genome is marked and packaged inside the cell nucleus.

epiglottis The flap that folds down over the trachea during swallowing.

epinephrine A hormone also known as *adrenaline*; it is released by the adrenal glands (located on each kidney) and various nerve endings in the body. It acts to increase glycogen breakdown in the liver, among other functions.

epithelial tissue The surface cells that line the outside of the body and all external passages within it.

ergogenic Work-producing. An ergogenic aid is a mechanical, nutritional, psychological, pharmacological, or physiological substance or treatment intended to directly improve exercise performance.

erythropoietin A hormone secreted mostly by the kidneys that enhances red blood cell synthesis and stimulates red blood cell release from bone marrow.

esophagus A tube in the GI tract that connects the pharynx with the stomach.

essential amino acids The amino acids that cannot be synthesized by humans in sufficient amounts or at all and therefore must be included in the diet; there are nine essential amino acids. These are also called *indispensable amino acids.*

essential fatty acids Fatty acids that must be supplied by the diet to maintain health. Currently, only linoleic acid and alpha-linolenic acid are classified as essential.

essential nutrient In nutritional terms, a substance that, when left out of a diet, leads to signs of poor health. The body either cannot produce this nutrient or cannot produce enough of it to meet its needs. If added back to a diet before permanent damage occurs, the affected aspects of health are restored.

Estimated Energy Requirement (EER) Estimate of the energy (kcal) intake needed to match the energy use of an average person in a specific life stage.

ethanol Chemical term for the form of alcohol found in alcoholic beverages.

exercise Physical activities that are planned, repetitive, and intended to improve physical fitness.

extracellular fluid Fluid present outside the cells; represents about one-third of body fluid.

extracellular space The space outside cells; represents one-third of body fluid.

failure to thrive Condition of inadequate growth during infancy or early childhood caused by poor nutritional intake, inefficient nutrient absorption, or excessive energy expenditure; commonly defined as weight-for-age below the fifth percentile on multiple occasions or weight declining two or more major percentile lines on a standardized growth chart.

famine An extreme shortage of food, which leads to massive starvation in a population; often associated with crop failures, war, and political unrest.

fat adaptation Manipulating the diet and physical training regimen so that muscles become more efficient at metabolizing fat as fuel during aerobic activity.

fat-soluble vitamins Vitamins that dissolve in fat and such substances as ether and benzene but not readily in water. These vitamins are A, D, E, and K.

fecal impaction The presence of a mass of hard, dry feces that remains in the rectum as a result of chronic constipation.

feces Mass of water, fiber, tough connective tissues, bacterial cells, and sloughed intestinal cells that passes through the large intestine and is excreted through the anus; also called stool.

female athlete triad A condition characterized by disordered eating, lack of menstrual periods (amenorrhea), and osteoporosis.

fermentation The conversion of carbohydrates into alcohols, acids, and carbon dioxide without the use of oxygen.

fetal alcohol spectrum disorders (FASDs) A group of irreversible physical and mental abnormalities in the infant that result from the mother's consuming alcohol during pregnancy.

fetal alcohol syndrome (FAS) Severe form of FASD that involves abnormal facial features and problems with development of the nervous system and overall growth as a result of maternal alcohol consumption during pregnancy.

fetal origins hypothesis A theory that links nutritional and other environmental insults that occur during gestation to the future health of the offspring.

fetus The developing life form from about the beginning of the ninth week after conception until birth.

fiber Substances in plant foods not digested by the processes that take place in the human stomach or small intestine. These add bulk to feces. Fiber naturally found in foods is also called *dietary fiber.*

fluorosis Discoloration of tooth enamel sometimes accompanied with pitting due to consuming a large amount of fluoride for an excessive period.

foam cells Lipid-loaded white blood cells that have surrounded large amounts of a fatty substance, usually cholesterol, on the blood vessel walls.

food allergy An adverse reaction to food that involves an immune response; also called *food hypersensitivity.*

foodborne illness Sickness caused by the ingestion of food containing harmful substances.

food desert An area where 33% or 500 people, whichever is less, live more than a mile from a grocery store in an urban area or more than 10 miles away in a rural area.

food insecure Condition in which the quality, variety, and/or desirability of the diet is reduced and there is difficulty at times providing enough food for everyone in the household.

food insecurity A condition of anxiety regarding running out of either food or money to buy more food.

food intolerance An adverse reaction to food that does not involve an allergic reaction.

fructose A six-carbon monosaccharide that usually exists in a ring form; found in fruits and honey; also known as *fruit sugar.*

fruitarian A person who primarily eats fruits, nuts, honey, and vegetable oils.

functional fiber Fiber added to foods that has been shown to provide health benefits.

functional foods Foods that provide health benefits beyond those supplied by the traditional nutrients they contain.

fungi Simple parasitic life forms, including molds, mildews, yeasts, and mushrooms. They live on dead or decaying organic matter. Fungi can grow as single cells, like yeast, or as a multicellular colony, as seen with molds.

galactose A six-carbon monosaccharide that usually exists in a ring form; closely related to glucose.

gallbladder An organ attached to the underside of the liver; site of bile storage, concentration, and eventual secretion.

gastroesophageal reflux disease (GERD) Disease that results from stomach acid backing up into the esophagus. The acid irritates the lining of the esophagus, causing pain.

gastrointestinal (GI) tract The main sites in the body used for digestion and absorption of nutrients. It consists of the mouth, esophagus, stomach, small intestine, large intestine, rectum, and anus. Also called the *digestive tract.*

gastroplasty Gastric bypass surgery performed on the stomach to limit its volume to approximately 30 milliliters. Also referred to as stomach stapling.

gender and development (GAD) approach Understanding the roles and responsibilities of both men and women in the process of sustainable development.

gene A specific segment on a chromosome. Genes provide the blueprint for the production of cell proteins.

gene expression Use of DNA information on a gene to produce a protein. Thought to be a major determination of cell development.

generally recognized as safe (GRAS) A list of food additives that in 1958 were considered safe for consumption.

Manufacturers were allowed to continue to use these additives, without special clearance, when needed for food products. FDA bears responsibility for proving they are not safe but can remove unsafe products from the list.

genes A specific segment on a chromosome. Genes provide the blueprints for the production of all body proteins.

genetically modified organism (GMO) Any organism created by genetic engineering.

genetic engineering Manipulation of the genetic makeup of any organism with recombinant DNA technology.

gestational diabetes A high blood glucose concentration that develops during pregnancy and returns to normal after birth; one cause is the placental production of hormones that antagonize the regulation of blood glucose by insulin.

gestational hypertension Blood pressure greater than 140/90 mmHg that is first diagnosed after 20 weeks of gestation. This may evolve into preeclampsia or eclampsia.

gestation The period of intrauterine development of offspring, from conception to birth; in humans, normal gestation is 38 to 42 weeks.

glucagon A hormone made by the pancreas that stimulates the breakdown of glycogen in the liver into glucose; this ends up increasing blood glucose. Glucagon also performs other functions.

glucose A six-carbon monosaccharide that usually exists in a ring form; found as such in blood and in table sugar bonded to fructose; also known as *dextrose*.

glycemic index (GI) The blood glucose response of a given food, compared to a standard (typically, glucose or white bread). Glycemic index is influenced by starch structure; fiber content; food processing; physical structure; and macronutrients in the meal, such as fat.

glycerol A three-carbon alcohol used to form triglycerides.

glycogen A carbohydrate made of multiple units of glucose with a highly branched structure. It is the storage form of glucose in humans and is synthesized (and stored) in the liver and muscles.

glycosylation The process by which glucose attaches to (glycates) other compounds, such as proteins.

goiter An enlargement of the thyroid gland; this is often caused by insufficient iodide in the diet.

Golgi complex The cell organelle near the nucleus that processes newly synthesized protein for secretion or distribution to other organelles.

green revolution This refers to increases in crop yields that accompanied the introduction of new agricultural technologies in less-developed countries, beginning in the 1960s. The key technologies were high-yielding, disease-resistant strains of rice, wheat, and corn; greater use of fertilizer and water; and improved cultivation practices.

gruels A thin mixture of grains or legumes in milk or water.

hard water Water that contains high levels of calcium, magnesium, and sometimes iron.

heart attack Rapid fall in heart function caused by reduced blood flow through the heart's blood vessels. Often part of the heart dies in the process. Technically called a myocardial infarction.

heavy drinking Any pattern of alcohol consumption in excess of 4 drinks in 1 day or 14 drinks per week for men or 3 drinks in 1 day or 7 drinks per week for women.

helminth Parasitic worm that can contaminate food, water, feces, animals, and other substances.

hematocrit The percentage of blood made up of red blood cells.

heme iron Iron provided from animal tissues in the form of hemoglobin and myoglobin. Approximately 40% of the iron in meat, fish, and poultry is heme iron; it is readily absorbed.

hemicellulose A nonfermentable fiber containing xylose, galactose, glucose, and other monosaccharides bonded together.

hemochromatosis A disorder of iron metabolism characterized by increased iron absorption and deposition in the liver and heart. This eventually poisons the cells in those organs.

hemoglobin The iron-containing part of the red blood cell that carries oxygen to the cells and carbon dioxide away from the cells. The heme iron portion is also responsible for the red color of blood.

hemorrhagic stroke Damage to part of the brain resulting from rupture of a blood vessel and subsequent bleeding within or over the internal surface of the brain.

hemorrhoid A pronounced swelling of a large vein, particularly veins found in the anal region.

hepatic portal circulation The portion of the circulatory system that uses a large vein (portal vein) to carry nutrient-rich blood from capillaries in the intestines and portions of the stomach to the liver.

hepatic portal vein Large vein leaving the intestine and stomach and connecting to the liver.

high-density lipoprotein (HDL) The lipoprotein in the blood that picks up cholesterol from dying cells and other sources and transfers it to the other lipoproteins in the bloodstream, as well as directly to the liver; low HDL increases the risk for cardiovascular disease.

high-fructose corn syrup Corn syrup that has been manufactured to contain between 42% and 90% fructose.

high-quality (complete) proteins Dietary proteins that contain ample amounts of all nine essential amino acids.

histamine A breakdown product of the amino acid histidine that stimulates acid secretion by the stomach and has other effects on the body, such as contraction of smooth muscles, increased nasal secretions, relaxation of blood vessels, and changes in relaxation of airways.

homocysteine An amino acid that arises from the metabolism of methionine. Vitamin B-6, folate, vitamin B-12, and choline are required for its metabolism. Elevated levels are associated with an increased risk of cardiovascular disease.

hospice care A program offering care that emphasizes comfort and dignity at the end of life.

human immunodeficiency virus (HIV) The virus that leads to **acquired immune deficiency syndrome (AIDS)**.

hunger The primarily physiological (internal) drive to find and eat food, mostly regulated by internal cues to eating.

hydrogenation The addition of hydrogen to a carbon–carbon double bond, producing a single carbon–carbon bond with two hydrogens attached to each carbon.

hyperglycemia High blood glucose, above 125 milligrams per 100 milliliters of blood.

hypertension High blood pressure. The pressure inside the blood vessels exceeds 140/90 mm Hg.

hypoglycemia Low blood glucose, below 40 to 50 milligrams per 100 milliliters of blood for nondiabetics.

hyponatremia Dangerously low blood sodium level.

hypothalamus A region at the base of the brain that contains cells that play a role in the regulation of energy intake, respiration, body temperature, and other body functions.

hypotheses Tentative explanations by a scientist to explain a phenomenon.

identical twins Two offspring that develop from a single ovum and sperm and, consequently, have the same genetic makeup.

ileocecal sphincter The ring of smooth muscle between the end of the small intestine and the beginning of the large intestine.

ileum Last segment of the small intestine; approximately 5 feet in length.

incidental food additives Additives that appear in food products indirectly, from environmental contamination of food ingredients or during the manufacturing process.

indirect calorimetry A method to measure energy use by the body by measuring oxygen uptake and carbon dioxide output. Formulas are then used to convert this gas exchange value into energy use, estimating the proportion of energy nutrients that are being oxidized for energy in the fuel mix.

infertility Inability of a couple to conceive after 1 year of unprotected intercourse.

infrastructure The basic framework of a system of organization. For a society, this includes roads, bridges, telephones, and other basic technologies.

inorganic Any substance lacking carbon atoms bonded to hydrogen atoms in the chemical structure.

insulin A hormone produced by the pancreas. Among other processes, insulin increases the synthesis of glycogen in the liver and the movement of glucose from the bloodstream into body cells.

intentional food additives Additives knowingly (directly) incorporated into food products by manufacturers.

international unit (IU) A crude measure of vitamin activity, often based on the growth rate of animals in response to the vitamin. Today IUs have largely been replaced by more precise milligram or microgram measures.

intracellular fluid Fluid contained within a cell; it represents about two-thirds of body fluid.

intrinsic factor A proteinlike compound produced by the stomach that enhances vitamin B-12 absorption in the ileum.

irradiation A process in which radiation energy is applied to foods, creating compounds (free radicals) within the food that destroy cell membranes, break down DNA, link proteins together, limit enzyme activity, and alter a variety of other proteins and cell functions of microorganisms that can lead to food spoilage. This process does not make the food radioactive.

jejunum Middle segment of the small intestine; approximately 4 feet in length.

ketone bodies Partial breakdown products of fat that contain three or four carbons.

ketosis The condition of having a high concentration of ketone bodies and related breakdown products in the bloodstream and tissues.

kidney nephrons The units of kidney cells that filter wastes from the bloodstream and deposit them into the urine.

kilocalorie (kcal) Kilocalorie (kcal) is a unit that describes the energy content of food. Specifically, a kcal is the heat energy needed to raise the temperature of 1000 grams (1 liter) of water 1° Celsius. Remember that although kcal refers to a 1000 calorie unit of measurement, it is commonly referred to as calories. "Calories" is a familiar term for the energy content of a food, so we use it in this book.

kwashiorkor A disease occurring primarily in young children who have an existing disease and consume a marginal amount of calories and insufficient protein in relation to needs. The child generally suffers from infections and exhibits edema, poor growth, weakness, and an increased susceptibility to further illness.

kyphosis Abnormally increased bending of the spine.

lactase An enzyme made by absorptive cells of the small intestine; this enzyme digests lactose to glucose and galactose.

lactation The period of milk secretion following pregnancy; typically called *breastfeeding.*

lactic acid A three-carbon acid formed during anaerobic cell metabolism; a partial breakdown product of glucose; also called *lactate.*

Lactobacillus bifidus **factor** A protective factor secreted in the colostrum that encourages growth of beneficial bacteria in the newborn's intestines.

lactoovovegetarian A person who consumes plant products, dairy products, and eggs.

lactose Glucose bonded to galactose; also known as *milk sugar.*

lactose intolerance A condition in which symptoms such as abdominal gas, bloating, and diarrhea appear as a result of severe lactose maldigestion.

lactose maldigestion (primary and secondary) Primary lactose maldigestion occurs when production of the enzyme lactase declines for no apparent reason. Secondary lactose maldigestion occurs when a specific cause, such as long-standing diarrhea, results in a decline in lactase production. When significant symptoms develop after lactose intake, it is then called lactose intolerance.

lactovegetarian A person who consumes plant products and dairy products.

lanugo Downlike hair that appears after a person has lost much body fat through semi-starvation. The hair stands erect and traps air, acting as insulation for the body to compensate for the relative lack of body fat, which usually functions as insulation.

laxative A medication or other substance that stimulates evacuation of the intestinal tract.

lean body mass Body weight minus fat storage weight equals lean body mass. This includes organs such as the brain, muscles, and the liver, as well as bone and blood and other body fluids.

lecithin A group of compounds that are major components of cell membranes.

leptin A hormone made by adipose tissue in proportion to total fat stores in the body that influences long-term regulation of fat mass. Leptin also influences release of the hormone insulin.

let-down reflex A reflex stimulated by infant suckling that causes the release (ejection) of milk from milk ducts in the mother's breasts; also called *milk ejection reflex.*

life expectancy The average length of life for a given group of people born in a specific year.

life span The potential oldest age a person can reach.

lignins A nonfermentable fiber made up of a multiringed alcohol (noncarbohydrate) structure.

limiting amino acid The essential amino acid in lowest concentration in a food or diet relative to body needs.

linoleic acid An essential omega-6 fatty acid with 18 carbons and 2 double bonds.

lipase Fat-digesting enzyme produced by the salivary glands, stomach, and pancreas.

lipid A compound containing much carbon and hydrogen, little oxygen, and sometimes other atoms. Lipids do not dissolve in water and include fats, oils, and cholesterol.

lipoprotein A compound found in the bloodstream containing a core of lipids with a shell composed of protein, phospholipid, and cholesterol.

lipoprotein lipase An enzyme attached to the cells that form the inner lining of blood vessels; it breaks down triglycerides into free fatty acids and glycerol.

lobules Saclike structures in the breast that store milk.

locavore Someone who eats food grown or produced locally or within a certain radius such as 50, 100, or 500 miles.

long-chain fatty acid A fatty acid that contains 12 or more carbons.

low birth weight (LBW) Referring to any infant weighing less than 2.5 kilograms (5.5 pounds) at birth; most commonly results from preterm birth.

low-density lipoprotein (LDL) The lipoprotein in the blood containing primarily cholesterol; elevated LDL is strongly linked to cardiovascular disease risk.

lower-body obesity The type of obesity in which fat storage is primarily located in the buttocks and thigh area. Also known as gynoid or gynecoid obesity.

lower esophageal sphincter A circular muscle that constricts the opening of the esophagus to the stomach. Also called the *gastroesophageal sphincter* or the *cardiac sphincter.*

lower-quality (incomplete) proteins Dietary proteins that are low in or lack one or more essential amino acids.

lumen The hollow opening inside a tube, such as the GI tract.

lymph A clear fluid that flows through lymph vessels; carries most forms of fat after their absorption by the small intestine.

lymphatic system A system of vessels and lymph that accepts fluid surrounding cells and large particles, such as products of fat absorption. Lymph eventually passes into the bloodstream from the lymphatic system.

lysosome A cellular organelle that contains digestive enzymes for use inside the cell for turnover of cell parts.

macronutrient A nutrient needed in gram quantities in a diet.

macular degeneration A painless condition leading to disruption of the central part of the retina (in the eye) and, in turn, blurred vision.

major mineral Vital to health, a mineral required in the diet in amounts greater than 100 milligrams per day.

malignant Malicious; in reference to a tumor, the property of spreading locally and to distant sites.

malnutrition Failing health that results from long-standing dietary practices that do not coincide with nutritional needs.

maltase An enzyme made by absorptive cells of the small intestine; this enzyme digests maltose to two glucoses.

maltose Glucose bonded to glucose.

marasmus A disease resulting from consuming a grossly insufficient amount of protein and calories; one of the diseases classed as protein–calorie malnutrition. Victims have little or no fat stores, little muscle mass, and poor strength. Death from infections is common.

megadose Intake of a nutrient beyond estimates of needs to prevent a deficiency or what would be found in a balanced diet; 2 to 10 times human needs is a starting point for such a dosage.

megaloblast A large, immature red blood cell that results from the inability of the cell to divide normally (*mega-lo* = large; *blast* = primitive or immature).

megaloblastic (macrocytic) anemia Anemia characterized by the presence of abnormally large red blood cells.

menopause The cessation of the menstrual cycle in women, usually beginning at about 50 years of age.

metabolic syndrome A condition in which a person has poor blood glucose regulation, hypertension, increased blood triglycerides, and other health problems. This condition is usually accompanied by obesity, lack of physical activity, and a diet high in refined carbohydrates. Also called Syndrome X.

metabolic water Water formed as a by-product of carbohydrate, lipid, and protein metabolism.

metabolism Chemical processes in the body by which energy is provided in useful forms and vital activities are sustained.

metastasize The spreading of disease from one part of the body to another, even to parts of the body that are remote from the site of the original tumor. Cancer cells can spread via blood vessels, the lymphatic system, or direct growth of the tumor.

micronutrient A nutrient needed in milligram or microgram quantities in a diet.

microvilli Extensive folds on the muscosal surface of the absorptive cells.

mineral Element used in the body to promote chemical reactions and to form body structures.

mitochondria Organelles that are the main sites of energy production in a cell. They contain the pathway for oxidizing fat for fuel, among other metabolic pathways.

moderate drinking For men, consuming no more than 4 drinks in 1 day AND no more than 14 drinks per week; for women, consuming no more than 3 drinks in 1 day AND no more than 7 drinks per week.

moderate-intensity aerobic physical activity Aerobic activity that increases a person's heart rate and breathing to some extent (4–6 on RPE scale). Examples include brisk walking, dancing, swimming, or bicycling on level terrain.

monoglyceride A breakdown product of a triglyceride consisting of one fatty acid attached to a glycerol backbone.

monosaccharide Simple sugar, such as glucose, that is not broken down further during digestion.

monounsaturated fatty acid A fatty acid containing one carbon–carbon double bond.

motility Generally, the ability to move spontaneously. It also refers to movement of food through the GI tract.

mucilages A viscous fiber consisting of chains of galactose, mannose, and other monosaccharides; characteristically found in seaweed.

mucus A thick fluid secreted by many cells throughout the body. It contains a compound that has both carbohydrate and protein parts. It acts as a lubricant and means of protection for cells.

muscle-strengthening activity Physical activity that increases skeletal muscle strength, power, endurance, and mass. Examples include lifting weights, using weight machines, and calisthenics (e.g., push-ups).

muscle tissue A type of tissue adapted to contract to cause movement.

myelin A lipid and protein combination (lipoprotein) that covers nerve fibers.

myocardial infarction Death of part of the heart muscle. Also termed a *heart attack.*

myoglobin Iron-containing protein that binds oxygen in muscle tissue.

negative energy balance The state in which energy intake is less than energy expended, resulting in weight loss.

negative protein balance A state in which protein intake is less than related protein losses, as is often seen during acute illness.

neotame General-purpose, nonnutritive sweetener that is approximately 7000 to 13,000 times sweeter than table sugar. It has a chemical structure similar to aspartame's.

nervous system The body system consisting of the brain, spinal cord, nerves, and sensory receptors. This system detects sensations, directs movements, and controls physiological and intellectual functions.

nervous tissue Tissue composed of highly branched, elongated cells that transport nerve impulses from one part of the body to another.

neural tube defect A defect in the formation of the neural tube occurring during early fetal development. This type of defect results in various nervous system disorders, such as spina bifida. Folate deficiency in the pregnant woman increases the risk that the fetus will develop this disorder.

neuron The structural and functional unit of the nervous system. Consists of a cell body, dendrites, and an axon.

neurotransmitter A compound made by a nerve cell that allows for communication between it and other cells.

night blindness Vitamin A–deficiency disorder that results in loss of the ability to see under low-light conditions.

night eating syndrome Eating a lot of food in the late evening and nocturnal awakenings with ingestion of food.

nitrosamine A carcinogen formed from nitrates and breakdown products of amino acids; can lead to stomach cancer.

nonceliac gluten sensitivity One or more of a variety of immune-related conditions with symptoms similar to celiac disease that are precipitated by the ingestion of gluten in people who do not have celiac disease.

nonessential amino acids Amino acids that can be synthesized by a healthy body in sufficient amounts; there are 11 nonessential amino acids. These are also called *dispensable amino acids*.

nonfermentable fiber A fiber that is not easily metabolized by intestinal bacteria; also called *insoluble fiber*.

nonheme iron Iron provided from plant sources, supplements, and animal tissues other than in the forms of hemoglobin and myoglobin. Nonheme iron is less efficiently absorbed than heme iron; absorption is closely dependent on body needs.

nonspecific immunity Defenses that stop the invasion of pathogens; requires no previous encounter with a pathogen; also called *innate immunity*.

norepinephrine A neurotransmitter from nerve endings and a hormone from the adrenal gland. It is released in times of stress and is involved in hunger regulation, blood glucose regulation, and other body processes.

NSAIDs Nonsteroidal anti-inflammatory drugs; includes aspirin, ibuprofen (Advil®), and naproxen (Aleve®).

nutrient density The ratio derived by dividing a food's nutrient content by its calorie content. When the food's contribution to our nutrient need for that nutrient exceeds its contribution to our calorie need, the food is considered to have a favorable nutrient density.

nutrients Chemical substances in food that contribute to health, many of which are essential parts of a diet. Nutrients nourish us by providing calories to fulfill energy needs, materials for building body parts, and factors to regulate necessary chemical processes in the body.

nutrigenetics Study of the effects of genes on nutritional health, such as variations in nutrient requirements and responsiveness to dietary modifications.

nutrigenomics Study of how food impacts health through its interaction with our genes and its subsequent effect on gene expression.

nutritional genomics Study of interactions between nutrition and genetics; includes nutrigenetics and nutrigenomics.

nutritional state The nutritional health of a person as determined by anthropometric measurements (height, weight, circumferences, and so on), biochemical measurements of nutrients or their by-products in blood and urine, a clinical (physical) examination, a dietary analysis, and economic evaluation; also called nutritional status.

nutrition security Secure access to a nutritious diet coupled with a sanitary environment and adequate health services and care.

oleic acid An omega-9 fatty acid with 18 carbons and 1 double bond.

omega-3 (ω-3) fatty acid An unsaturated fatty acid with the first double bond on the third carbon from the methyl end ($-CH_3$).

omega-6 (ω-6) fatty acid An unsaturated fatty acid with the first double bond on the sixth carbon from the methyl end ($-CH_3$).

organ A group of tissues designed to perform a specific function; for example, the heart, which contains muscle tissue, nerve tissue, and so on.

organelles Compartments, particles, or filaments that perform specialized functions within a cell.

organ system A collection of organs that work together to perform an overall function.

osmosis The passage of water through a semipermeable membrane from a less concentrated compartment to a more concentrated compartment.

osteomalacia Adult form of rickets. The bones have low mineral density and subsequently are at risk for fracture.

osteopenia A bone disease defined by low mineral density.

osteoporosis The presence of a stress-induced fracture or a T-score of −2.5 or lower. The bones are porous and fragile owing to low mineral density.

ostomy Surgically created short circuit in intestinal flow where the end point usually opens from the abdominal cavity rather than the anus; for example, a colostomy. Short circuiting the intestinal flow means more water is lost in fecal matter than would be if the intestinal tract were intact.

overnutrition A state in which nutritional intake greatly exceeds the body's needs.

ovum The egg cell from which a fetus eventually develops if the egg is fertilized by a sperm cell.

oxalic acid (oxalate) An organic acid found in spinach, rhubarb, and sweet potatoes that can depress the absorption of certain minerals present in the food, such as calcium.

oxidize In the most basic sense, an electron has been lost or an oxygen has been gained by a chemical substance. This change typically alters the shape and/or function of the substance.

oxytocin A hormone secreted by the pituitary gland. It causes contraction of the musclelike cells surrounding the ducts of the breasts and the smooth muscle of the uterus.

parasite An organism that lives in or on another organism and derives nourishment from it.

parathyroid hormone (PTH) A hormone made by the parathyroid gland that increases synthesis of the active form of vitamin D. This hormone works with vitamin D to increase blood levels of calcium.

pasteurizing The process of heating food products to kill pathogenic microorganisms and reduce the total number of bacteria.

pectin A viscous fiber containing chains of galacturonic acid and other monosaccharides; characteristically found between plant cell walls.

pellagra Niacin-deficiency disease characterized by dementia, diarrhea, and dermatitis, and possibly leading to death.

pepsin A protein-digesting enzyme produced by the stomach.

peptide bond A chemical bond formed between amino acids in a protein.

percentile Classification of a measurement of a unit into divisions of 100 units.

periodization Cycling the volume, intensity, and activities of workouts throughout the training season.

peristalsis A coordinated muscular contraction used to propel food down the GI tract.

pernicious anemia The anemia that results from a lack of vitamin B-12 absorption; it is *pernicious* because of associated nerve degeneration that can result in eventual paralysis and death.

peroxisome A cell organelle that destroys toxic products within the cell.

phagocytosis Process in which a cell forms an indentation, and particles or fluids enter the indentation and are engulfed by the cell.

pH A measure of relative acidity or alkalinity of a solution. The pH scale is 0 to 14. A pH of 7 is neutral; a pH below 7 is acidic; and a pH above 7 is alkaline.

pharynx The organ of the digestive tract and respiratory tract located at the back of the oral and nasal cavities, commonly known as the throat.

phenylketonuria (PKU) Disease caused by a defect in the liver's ability to metabolize the amino acid phenylalanine into the amino acid tyrosine; untreated, toxic by-products of phenylalanine build up in the body and lead to mental retardation.

phosphocreatine (PCr) A high-energy compound that can be used to re-form ATP. It is used primarily during bursts of activity, such as lifting and jumping.

phospholipid Any of a class of fat-related substances that contain phosphorus, fatty acids, and a nitrogen-containing base. The phospholipids are an essential part of every cell.

photosynthesis Process by which plants use energy from the Sun to synthesize energy-yielding compounds, such as glucose.

physical activity Any movement of skeletal muscles that requires energy.

physical fitness The ability to perform moderate to vigorous physical activity without undue fatigue.

physiological anemia The normal increase in blood volume in pregnancy that dilutes the concentration of red blood cells, resulting in anemia; also called *hemodilution.*

phytic acid (phytate) A constituent of plant fibers that binds positive ions to its multiple phosphate groups.

phytochemical A chemical found in plants. Some phytochemicals may contribute to a reduced risk of cancer or cardiovascular disease in people who consume them regularly.

pica The practice of eating nonfood items, such as dirt, laundry starch, or clay.

placebo Generally a fake medicine or treatment used to disguise the treatments given to the participants in an experiment.

placenta An organ that forms in the uterus in pregnant women. Through this organ, oxygen and nutrients from the mother's blood are transferred to the fetus, and fetal wastes are removed. The placenta also releases hormones that maintain the state of pregnancy.

plaque A cholesterol-rich substance deposited in the blood vessels; it contains various white blood cells, smooth muscle cells, various proteins, cholesterol and other lipids, and eventually calcium.

plasma The fluid, extracellular portion of the circulating blood. This includes the blood serum plus all blood-clotting factors. In contrast, serum is the fluid that remains after clotting factors have been removed from plasma.

polycystic ovary syndrome (PCOS) A condition of hormonal imbalance (e.g., elevated testosterone and insulin) in a woman that can lead to infertility, weight gain in the abdominal region, excessive growth of body hair, and acne.

polypeptide A group of amino acids bonded together, from 50 to 2000 or more.

polysaccharides Carbohydrates containing many glucose units, from 10 to 1000 or more.

polyunsaturated fatty acid A fatty acid containing two or more carbon–carbon double bonds.

pool The amount of a nutrient stored within the body that can be mobilized when needed.

positive energy balance The state in which energy intake is greater than energy expended, generally resulting in weight gain.

positive protein balance A state in which protein intake exceeds related protein losses, as is needed during the times of growth.

prebiotic Substance that stimulates bacterial growth in the large intestines.

preeclampsia A form of gestational hypertension characterized by protein in the urine.

preservatives Compounds that extend the shelf life of foods by inhibiting microbial growth or minimizing the destructive effect of oxygen and metals.

preterm An infant born before 37 weeks of gestation; also referred to as *premature.*

primary hypertension Blood pressure of 140/90 mm Hg or higher with no identified cause; also called *essential hypertension.*

probiotic Product that contains specific types of bacteria. Use is intended to colonize the large intestine with the specific bacteria in the product. An example is yogurt.

progression Incremental increase in frequency, intensity, and time spent in each type of physical activity over several weeks or months.

prolactin A hormone secreted by the pituitary gland that stimulates the synthesis of milk in the breast.

prostate gland A solid, chestnut-shaped organ surrounding the first part of the urinary tract in the male. The prostate gland secretes substances into the semen.

protease Protein-digesting enzyme produced by the stomach, small intestine, and pancreas.

protein-calorie malnutrition (PCM) A condition resulting from regularly consuming insufficient amounts of calories and protein. The deficiency eventually results in body wasting, primarily of lean tissue, and an increased susceptibility to infections. Also known as *protein–energy malnutrition.*

protein equilibrium A state in which protein intake is equal to related protein losses; the person is said to be in protein balance.

protein Food and body compounds made of amino acids; proteins contain carbon, hydrogen, oxygen, nitrogen, and sometimes other atoms in a specific configuration. Proteins contain the form of nitrogen most easily used by the human body.

protein turnover The process by which cells break down old proteins and resynthesize new proteins. In this way, the cell will

have the proteins it needs to function at that time.

protozoa One-celled animals that are more complex than bacteria. Disease-causing protozoa can be spread through food and water.

provitamin A A substance that can be converted into vitamin A.

purging disorder An eating disorder characterized by repeated purging (e.g., by self-induced vomiting) to induce weight loss even in the absence of binge eating.

pyloric sphincter Ring of smooth muscle between stomach and small intestine.

pyruvic acid A three-carbon compound formed during glucose metabolism; also called *pyruvate.*

radiation Literally, energy that is emitted from a center in all directions. Various forms of radiation energy include X rays and ultra-violet rays from the Sun.

rancid Containing products of decomposed fatty acids that have an unpleasant flavor and odor.

receptor A site in a cell at which compounds (such as hormones) bind. Cells that contain receptors for a specific compound are partially controlled by that compound.

recombinant DNA technology A test tube technology that rearranges DNA sequences in an organism by cutting the DNA, adding or deleting a DNA sequence, and rejoining DNA molecules with a series of enzymes.

Recommended Dietary Allowance (RDA) Nutrient intake amount sufficient to meet the needs of 97% to 98% of the individuals in a specific life stage.

rectum Terminal portion of the large intestine.

registered dietitian (RD) or registered dietitian nutritionist (RDN) A person who has completed a baccalaureate degree program approved by the Accreditation Council for Education in Nutrition and Dietetics (ACEND), performed at least 1200 hours of supervised professional practice, passed a registration examination, and complied with continuing education requirements.

relapse prevention A series of strategies used to help prevent and cope with weight-control lapses, such as recognizing high-risk situations and deciding beforehand on appropriate responses.

reserve capacity The extent to which an organ can preserve essentially normal function despite decreasing cell number or cell activity.

resting metabolism The amount of calories the body uses when the person has not eaten in 4 hours and is resting (e.g., 15 to 30 minutes) and awake in a warm, quiet environment. It is usually slightly higher than basal metabolism owing to the less strict criteria for the test; often referred to as *resting metabolic rate (RMR)*.

retina A light-sensitive lining in the back of the eye. It contains retinal.

retinal Aldehyde form of vitamin A.

retinoic acid Acid form of vitamin A.

retinoids Chemical forms of preformed vitamin A; one source is animal foods.

retinol Alcohol form of vitamin A.

retinyl Storage form of vitamin A.

ribonucleic acid (RNA) The single-stranded nucleic acid involved in the transcription of genetic information and translation of that information into protein structure.

ribosomes Cytoplasmic particles that mediate the linking together of amino acids to form proteins; may exist freely in the cytoplasm or attached to endoplasmic reticulum.

rickets A disease characterized by poor mineralization of newly synthesized bones because of low calcium content. Arising in infants and children, this deficiency is caused by insufficient amounts of the vitamin D hormone in the body.

R-proteins Proteins produced by the salivary glands that bind to free vitamin B-12 in the stomach and protect it from stomach acid.

saccharin Alternative sweetener that yields no energy to the body; 300 times sweeter than sucrose.

saliva Watery fluid, produced by the salivary glands in the mouth, that contains lubricants, enzymes, and other substances.

salt Compound of sodium and chloride in a 40:60 ratio.

sarcopenia In general, loss of muscle tissue. Among older adults, this loss of lean mass greatly increases their risk of illness and death.

sarcopenic obesity Loss of muscle mass accompanied by gains in fat mass.

satiety A state in which there is no longer a desire to eat; a feeling of satisfaction.

saturated fatty acid A fatty acid containing no carbon–carbon double bonds.

scavenger cells Specific form of white blood cells that can bury themselves in the artery wall and accumulate LDL. As these cells take up LDL, they contribute to the development of atherosclerosis.

scurvy The vitamin C–deficiency disease characterized by weakness, fatigue, slow wound healing, opening of previously healed wounds, bone pain, fractures, sore and bleeding gums, diarrhea, and pinpoint hemorrhages on the skin.

secondary hypertension Blood pressure of 140/90 mm Hg or higher as a result of disease (e.g., kidney dysfunction or sleep apnea) or drug use.

secretory vesicles Membrane-bound vesicles produced by the Golgi complex; contain protein and other compounds to be secreted by the cell.

self-monitoring Tracking foods eaten and conditions affecting eating; actions are usually recorded in a diary, along with location, time, and state of mind. This is a tool to help people understand more about their eating habits.

sequestrants Compounds that bind free metal ions. By so doing, they reduce the ability of ions to cause rancidity in foods containing fat.

set point Often refers to the close regulation of body weight. It is not known what cells control this set point or how it functions in weight regulation. There is evidence, however, that mechanisms exist that help regulate weight.

sickle cell disease (sickle cell anemia) An illness that results from a malformation of the red blood cell because of an incorrect structure in part of its hemoglobin protein chains.

simple sugar Carbohydrate composed of only one (monosaccharide) or two (disaccharide) sugar units.

Sleeve gastrectomy Surgical reduction of stomach by about 75% (to about the size of a banana). Also known as the gastric sleeve.

small for gestational age (SGA) Referring to infants who weigh less than the expected weight for their length of gestation. This corresponds to less than 2.5 kilograms (5.5 pounds) in a full-term newborn. A preterm infant who is also SGA will most likely develop some medical complications.

soft water Water that contains little or no calcium or magnesium. Soft water may contain sodium and other minerals.

solid fats Fats that are solid at room temperature, such as butter and margarine. Foods containing solid fats tend to be high in saturated fatty acids or *trans* fatty acids.

solvent A liquid substance in which other substances dissolve.

sorbitol Alcohol derivative of glucose that yields about 3 kcal per gram but is slowly absorbed from the small intestine; used in some sugarless gums and dietetic foods.

specific immunity Function of white blood cells directed at specific antigens; also called *adaptive immunity.*

spina bifida Birth defect resulting from improper closure of the neural tube

during embryonic development. The spinal cord or fluid may bulge outside the spinal column.

spontaneous abortion Cessation of pregnancy and expulsion of the embryo or nonviable fetus prior to 20 weeks' gestation. This is the result of natural causes, such as a genetic defect or developmental problem; also called *miscarriage.*

spores Dormant reproductive cells capable of turning into adult organisms without the help of another cell. Various bacteria and fungi form spores.

starch A carbohydrate made of multiple units of glucose attached together in a form the body can digest; also known as *complex carbohydrate.*

sterol A compound containing a multi-ring (steroid) structure and a hydroxyl group (–OH). Cholesterol is a typical example.

stevia Alternative sweetener derived from South American shrub; 100 to 300 times sweeter than sucrose.

stimulus control Altering the environment to minimize the stimuli for eating; for example, removing foods from site and storing them in kitchen cabinets.

stress fracture A fracture that occurs from repeated jarring of a bone. Common sites include bones of the foot.

stroke A decrease or loss in blood flow to the brain that results from a blood clot or other change in arteries in the brain. This in turn causes the death of brain tissue. Also called a *cerebrovascular accident.*

subclinical Stage of a disease or disorder not severe enough to produce symptoms that can be detected or diagnosed.

sucralose Alternative sweetener that has chlorines in place of three hydroxyl (—OH) groups on sucrose; 600 times sweeter than sucrose.

sucrase An enzyme made by absorptive cells of the small intestine; this enzyme digests sucrose to glucose and fructose.

sucrose Fructose bonded to glucose; table sugar.

sugar A simple carbohydrate with the chemical composition $(CH_2O)_n$. The basic unit of all sugars is glucose, a six-carbon ring structure. The primary sugar in the diet is sucrose, which is made up of glucose and fructose.

sustainable agriculture Agricultural system that provides a secure living for farm families; maintains the natural environment and resources; supports the rural community; and offers respect and fair treatment to all involved, from farm workers to consumers to the animals raised for food.

sustainable development Economic growth that will simultaneously reduce poverty, protect the environment, and preserve natural capital.

symptom A change in health status noted by the person with the problem, such as stomach pain.

synapse The space between one neuron and another neuron (or cell).

systolic blood pressure The pressure in the arterial blood vessels associated with the pumping of blood from the heart.

teratogen A compound (natural or synthetic) that may cause or increase the risk of a birth defect. Exposure to a teratogen does not always lead to a birth defect; its effects on the fetus depend on the dose, timing, and duration of exposure.

tetany A body condition marked by sharp contraction of muscles and failure to relax afterward; usually caused by abnormal calcium metabolism.

theory An explanation for a phenomenon that has numerous lines of evidence to support it.

therapeutic phlebotomy Periodic blood removal, as a blood donation, for the purpose of ridding the body of excess iron.

thermic effect of food (TEF) The increase in metabolism that occurs during the digestion, absorption, and metabolism of energy-yielding nutrients. This represents 5% to 10% of calories consumed.

thrifty metabolism A genetic tendency toward efficient use of energy that results in below-average energy requirements and increased storage of calories as fat.

thyroid hormones Hormones produced by the thyroid gland that regulate growth and metabolic rate.

tissue saturation The limited storage capacity of water-soluble vitamins in the tissues.

tissues Collections of cells adapted to perform a specific function.

Tolerable Upper Intake Level (UL) Maximum chronic daily intake level of a nutrient that is unlikely to cause adverse health effects in almost all people in a specific life stage.

tolerance Needing more of a substance to achieve the desired effect (e.g., intoxication) or experiencing diminished effects of a given amount of a substance after repeated use.

total parenteral nutrition The intravenous feeding of all necessary nutrients, including the most basic forms of protein, carbohydrates, lipids, vitamins, minerals, and electrolytes.

toxins Poisonous compounds produced by an organism that can cause disease.

trabecular bone The less dense, more open structure bone found in the inner layer of bones.

trace mineral Vital to health, a mineral required in the diet in amounts less than 100 milligrams per day.

transcription Process by which genetic information stored as DNA within the nucleus is copied to RNA during protein synthesis.

trans fatty acid A form of an unsaturated fatty acid, usually a monounsaturated one when found in food, in which the hydrogens on both carbons forming the double bond lie on opposite sides of that bond.

transgenic organism Organism that contains genes originally present in another organism.

translation Process by which genetic information copied onto RNA dictates the amino acid sequence to form a protein.

triglyceride The major form of lipid in the body and in food. It is composed of three fatty acids bonded to glycerol, an alcohol.

trimesters Three 13- to 14-week periods into which the normal pregnancy (on average, 40 weeks) is divided somewhat arbitrarily for purposes of discussion and analysis. Development of the offspring, however, is continuous throughout pregnancy, with no specific physiological markers demarcating the transition from one trimester to the next.

trypsin A protein-digesting enzyme secreted by the pancreas to act in the small intestine.

tumor Mass of cells; may be cancerous (malignant) or noncancerous (benign).

type 1 diabetes A form of diabetes characterized by total insulin deficiency due to destruction of insulin-producing cells of the pancreas. Insulin therapy is required.

type 1 osteoporosis Porous trabecular bone characterized by rapid bone demineralization following menopause.

type 2 diabetes A form of diabetes characterized by insulin resistance and often associated with obesity. Insulin therapy can be used but is often not required.

type 2 osteoporosis Porous trabecular and cortical bone observed in men and women after the age of 70.

ulcer Erosion of the tissue lining, usually in the stomach or the upper small intestine. As a group, these are generally referred to as *peptic ulcers.*

ultratrace mineral A mineral present in the human diet in trace amounts but that has not been shown to be essential to human health.

umami A brothy, meaty, savory flavor in some foods. Monosodium glutamate enhances this flavor when added to foods.

undernutrition Failing health that results from a long-standing dietary intake that is not enough to meet nutritional needs.

underwater weighing A method of estimating total body fat by weighing the individual on a standard scale and then weighing him or her again submerged in water. The difference between the two weights is used to estimate total body volume.

underweight A body mass index below 18.5. The cutoff is less precise than for obesity because this condition has been less studied.

upper-body obesity The type of obesity in which fat is stored primarily in the abdominal area; defined as a waist circumference more than 40 inches (102 centimeters) in men and more than 35 inches (89 centimeters) in women. Also known as *android obesity.*

urea Nitrogenous waste product of protein metabolism; major source of nitrogen in the urine, chemically.

ureter Tube that transports urine from the kidney to the urinary bladder.

urethra Tube that transports urine from the urinary bladder to the outside of the body.

urinary system The body system consisting of the kidneys, urinary bladder, and the ducts that carry urine. This system removes waste products from the circulatory system and regulates blood acid–base balance, overall chemical balance, and water balance in the body.

vegan A person who eats only plant foods.

vein A blood vessel that carries blood to the heart.

very-low-calorie diet (VLCD) Known also as *protein-sparing modified fast (PSMF),* this diet allows a person 400 to 800 kcal per day, often in liquid form. Of this, 120 to 480 kcal is carbohydrate, and the rest is mostly high-quality protein.

very-low-density lipoprotein (VLDL) The lipoprotein created in the liver that carries cholesterol and lipids that have been taken up or newly synthesized by the liver.

vigorous-intensity aerobic physical activity Aerobic activity that greatly increases a person's heart rate and breathing (7–8 on RPE scale). Examples include jogging, singles tennis, swimming continuous laps, or bicycling uphill.

villi (singular, villus) The fingerlike protrusions into the small intestine that participate in digestion and absorption of food.

virus The smallest known type of infectious agent, many of which cause disease in humans. A virus is essentially a piece of genetic material surrounded by a coat of protein. They do not metabolize, grow, or move by themselves. They reproduce only with the aid of a living cellular host.

viscous fiber A fiber that is readily fermented by bacteria in the large intestine; also called *soluble fiber*.

vitamin Compound needed in very small amounts in the diet to help regulate and support chemical reactions in the body.

vitamin D$_2$ (ergocalciferol) Form found in nonanimal sources, such as in some mushrooms.

vitamin D$_3$ (cholecalciferol) Previtamin form found naturally in some animal sources, including fish and egg yolks.

water intoxication Potentially fatal condition that occurs with a high intake of water, which results in severe dilution of the blood and other fluid compartments.

water-soluble vitamins Vitamins that dissolve in water. These vitamins are the B vitamins and vitamin C.

water The universal solvent; chemically, H$_2$O. The body is composed of about 60% water.

white blood cells One of the formed elements of the circulating blood system; also called *leukocytes*. White blood cells are able to squeeze through intracellular spaces and migrate. They phagocytize bacteria, fungi, and viruses, as well as detoxify proteins that may result from allergic reactions, cellular injury, and other immune system cells.

whole grains Grains containing the entire seed of the plant, including the bran, germ, and endosperm (starchy interior). Examples are whole wheat and brown rice.

Wilson's disease A genetic disorder that results in accumulation of copper in the tissues; characterized by damage to the liver, nervous system, and other organs.

withdrawal Physical symptoms related to cessation of substance use, such as sweating, rapid pulse, shakiness, insomnia, nausea and vomiting, anxiety, and even seizures.

xerophthalmia Hardening of the cornea and drying of the surface of the eye, which can result in blindness.

xylitol Alcohol derivative of the five-carbon monosaccharide xylose.

zygote The fertilized ovum; the cell resulting from the union of an egg cell (ovum) and sperm until it divides.

Photo Credits

Chapter 14

Opener: ©Getty Images RF; p. 551: ©McGraw-Hill Education/Jill Braaten, photographer; p. 556: ©Don Bayley/Getty Images RF; p. 557: ©Blend Images RF; p. 559: ©UpperCut Images/Getty Images RF; 14.3: ©Tanya Constantine/Blend Images LLC RF; p. 562: ©John Slater/Getty Images RF; p. 563: ©Getty Images RF; p. 566(top): ©Brand X Pictures/JupiterImages RF; (bottom): ©Hill Street Studios/Blend Images LLC RF; p. 569: ©Getty Images/Jonelle Weaver RF; p. 571: ©Royalty-Free/Corbis; p. 573: ©Diane McDonald/Getty Images RF; p. 576 & 577: ©Royalty-Free/Corbis; p. 578: ©Photodisc Collection/Getty Images RF; p. 579: ©2008 Medela AG.; p. 581: ©Floortje/Getty Images RF; 14.11: ©Medical-on-Line/Alamy RF; p. 587: ©Digital Vision/PunchStock RF; p. 589: ©Andersen Ross/Getty Images RF.

Chapter 15

Opener: ©BananaStock/PunchStock RF; 15.1(both): CDC; p. 594: ©Pixtal/age fotostock RF; p. 595: ©PhotoAlto/PictureQuest RF; p. 600: ©Getty Images RF; p. 601: ©McGraw-HIll Education/Mark Dierker, photographer; p. 602: ©Steve Mason/Getty Images RF; p. 603: ©E. Dygas/Getty Images RF; p. 604: ©Corbis/PictureQuest RF; p. 605(top): ©Paul Casamassimo, DDS, MS; (bottom): ©Digital Vision/Getty Images RF; p. 608: ©Jules Frazier/Getty Images RF; p. 609: ©Andrew Olney/age fotostock RF; p. 610: U.S. Department of Agriculture, Center for Nutrition Policy and Promotion; p. 611: ©Indeed/Aflo/Getty Images RF; p. 613: ©Ingram Publishing/SuperStock RF; p. 616: ©BananaStock/PunchStock RF; p. 618(top): ©Stockbyte/PunchStock RF; (bottom): ©Valerie Loiseleux/Getty Images RF; p. 619: ©Ingram Publishing/SuperStock RF; p. 620: ©Creatas/PunchStock RF; p. 622: ©SW Productions/Getty Images RF; p. 623: ©SW Productions/Getty Images RF; p. 626: ©Corbis Super/Alamy RF; p. 629: ©FoodIngredients/Alamy RF; p. 631: ©Dynamic Graphics/JupiterImages RF; p. 633: ©McGraw-Hill Education/Jill Braaten, photographer.

Chapter 16

Opener: ©Digital Vision/PunchStock RF; p. 637: ©Royalty-Free/Corbis; p. 638: ©Stockbyte/Getty Images RF; p. 640: ©Getty Images/Jonelle Weaver RF; 16.4, 16.5: Administration for Community Living U.S. Department of Health and Human Services; p. 645: ©Blend Images/Getty Images RF; p. 646: ©Image Source/Corbis RF; p. 649: ©Anne Smith; p. 650: ©Stockbyte/PunchStock RF; p. 651(top): ©Photodisc/Getty Images RF; (bottom): ©Ingram Publishing RF; p. 659: ©Getty Images RF; p. 660: ©Getty Images/Digital Vision RF; 16.7: ©National Institute on Alcohol Abuse and Alcoholism; p. 662: ©Ingram Publishing/Alamy RF; 16.8a,b: ©Science Source; p. 665: ©Steve Mason/Getty Images RF; p. 668: ©LWA/Dann Tardif/age fotostock RF; p. 670: ©Royalty-Free/Corbis; p. 671: ©Mel Curtis/Getty Images RF.

Index

A

A

Dietary Reference Intakes (DRIs): Recommended Intakes for Individuals, Vitamins
Food and Nutrition Board, Institute of Medicine, National Academies

Life Stage Group	Vitamin A (μg/d)[a]	Vitamin C (mg/d)	Vitamin D (μg/d)[b,c]	Vitamin E (mg/d)[d]	Vitamin K (μg/d)	Thiamin (mg/d)	Riboflavin (mg/d)	Niacin (mg/d)[e]	Vitamin B-6 (mg/d)	Folate (μg/d)[f]	Vitamin B-12 (μg/d)	Pantothenic Acid (mg/d)	Biotin (μg/d)	Choline (mg/d)[g]
Infants														
0–6 mo	400*	40*	10	4*	2.0*	0.2*	0.3*	2*	0.1*	65*	0.4*	1.7*	5*	125*
7–12 mo	500*	50*	10	5*	2.5*	0.3*	0.4*	4*	0.3*	80*	0.5*	1.8*	6*	150*
Children														
1–3 y	300	15	15	6	30*	0.5	0.5	6	0.5	150	0.9	2*	8*	200*
4–8 y	400	25	15	7	55*	0.6	0.6	8	0.6	200	1.2	3*	12*	250*
Males														
9–13 y	600	45	15	11	60*	0.9	0.9	12	1.0	300	1.8	4*	20*	375*
14–18 y	900	75	15	15	75*	1.2	1.3	16	1.3	400	2.4	5*	25*	550*
19–30 y	900	90	15	15	120*	1.2	1.3	16	1.3	400	2.4	5*	30*	550*
31–50 y	900	90	15	15	120*	1.2	1.3	16	1.3	400	2.4	5*	30*	550*
51–70 y	900	90	15	15	120*	1.2	1.3	16	1.7	400	2.4[h]	5*	30*	550*
>70 y	900	90	20	15	120*	1.2	1.3	16	1.7	400	2.4[h]	5*	30*	550*
Females														
9–13 y	600	45	15	11	60*	0.9	0.9	12	1.0	300	1.8	4*	20*	375*
14–18 y	700	65	15	15	75*	1.0	1.0	14	1.2	400[i]	2.4	5*	25*	400*
19–30 y	700	75	15	15	90*	1.1	1.1	14	1.3	400[i]	2.4	5*	30*	425*
31–50 y	700	75	15	15	90*	1.1	1.1	14	1.3	400[i]	2.4	5*	30*	425*
51–70 y	700	75	15	15	90*	1.1	1.1	14	1.5	400	2.4[h]	5*	30*	425*
>70 y	700	75	20	15	90*	1.1	1.1	14	1.5	400	2.4[h]	5*	30*	425*
Pregnancy														
≤18 y	750	80	15	15	75*	1.4	1.4	18	1.9	600[i]	2.6	6*	30*	450*
19–30 y	770	85	15	15	90*	1.4	1.4	18	1.9	600[i]	2.6	6*	30*	450*
31–50 y	770	85	15	15	90*	1.4	1.4	18	1.9	600[i]	2.6	6*	30*	450*
Lactation														
≤18 y	1200	115	15	19	75*	1.4	1.6	17	2.0	500	2.8	7*	35*	550*
19–30 y	1300	120	15	19	90*	1.4	1.6	17	2.0	500	2.8	7*	35*	550*
31–50 y	1300	120	15	19	90*	1.4	1.6	17	2.0	500	2.8	7*	35*	550*

mg = milligram, mg = microgram

NOTE: This table (taken from the DRI reports; see www.nap.edu) presents Recommended Dietary Allowances (RDAs) in **bold type** and Adequate Intakes (AIs) in ordinary type followed by an asterisk (*). RDAs and AIs may both be used as goals for individual intake. RDAs are set to meet the needs of almost all (97 to 98%) individuals in a group. For healthy breastfed infants, the AI is the mean intake. The AI for other life stage and gender groups is believed to cover needs of all individuals in the group, but lack of data or uncertainty in the data prevents being able to specify with confidence the percentage of individuals covered by this intake.

[a] As retinol activity equivalents (RAEs). 1 RAE = 1 μg retinol, 12 μg β-carotene, 24 μg α-carotene, or 24 μg β-cryptoxanthin. To calculate RAEs from REs of provitamin A carotenoids in foods, divide the REs by 2. For preformed vitamin A in foods or supplements and for provitamin A carotenoids in supplements, 1 RE = 1 RAE.

[b] cholecalciferol. 1 μg cholecalciferol = 40 IU vitamin D.

[c] In the absence of adequate exposure to sunlight.

[d] As α-tocopherol. α-Tocopherol includes RRR-α-tocopherol, the only form of α-tocopherol that occurs naturally in foods, and the 2R-stereoisomeric forms of α-tocopherol (RRR-, RSR-, RRS-, and RSS-α-tocopherol) that occur in fortified foods and supplements. It does not include the 2S-stereoisomeric forms of α-tocopherol (SRR-, SSR-, SRS-, and SSS-α-tocopherol), also found in fortified foods and supplements.

[e] As niacin equivalents (NE). 1 mg of niacin = 60 mg of tryptophan; 0–6 months = preformed niacin (not NE).

[f] As dietary folate equivalents (DFE). 1 DFE = 1 μg food folate = 0.6 μg of folic acid from fortified food or as a supplement consumed with food = 0.5 μg of a supplement taken on an empty stomach.

[g] Although AIs have been set for choline, there are few data to assess whether a dietary supply of choline is needed at all stages of the life cycle, and it may be that the choline requirement can be met by endogenous synthesis at some of these stages.

[h] Because 10 to 30% of older people may malabsorb food-bound B-12, it is advisable for those older than 50 years to meet their RDA mainly by consuming foods fortified with B-12 or a supplement containing B-12.

[i] In view of evidence linking folate intake with neural tube defects in the fetus, it is recommended that all women capable of becoming pregnant consume 400 μg from supplements or fortified foods in addition to intake of food folate from a varied diet.

It is assumed that women will continue consuming 400 μg from supplements or fortified food until their pregnancy is confirmed and they enter prenatal care, which ordinarily occurs after the end of the periconceptional period—the critical time for formation of the neural tube.

Adapted from the Dietary Reference Intakes series, National Academies Press. Copyright 1997, 1998, 2000, 2001, 2011, by the National Academy of Sciences. The full reports are available from the National Academies Press at www.nap.edu.

Dietary Reference Intakes (DRIs): Recommended Intakes for Individuals, Elements
Food and Nutrition Board, Institute of Medicine, National Academies

Life Stage Group	Calcium (mg/d)	Chromium (µg/d)	Copper (µg/d)	Fluoride (mg/d)	Iodine (µg/d)	Iron (mg/d)	Magnesium (mg/d)	Manganese (mg/d)	Molybdenum (µg/d)	Phosphorus (mg/d)	Selenium (µg/d)	Zinc (mg/d)
Infants												
0–6 mo	200*	0.2*	200*	0.01*	110*	0.27*	30*	0.003*	2*	100*	15*	2*
7–12 mo	260*	5.5*	220*	0.5*	130*	11	75*	0.6*	3*	275*	20*	3
Children												
1–3 y	700	11*	340	0.7*	90	7	80	1.2*	17	460	20	3
4–8 y	1000	15*	440	1*	90	10	130	1.5*	22	500	30	5
Males												
9–13 y	1300	25*	700	2*	120	8	240	1.9*	34	1250	40	8
14–18 y	1300	35*	890	3*	150	11	410	2.2*	43	1250	55	11
19–30 y	1000	35*	900	4*	150	8	400	2.3*	45	700	55	11
31–50 y	1000	35*	900	4*	150	8	420	2.3*	45	700	55	11
51–70 y	1000	30*	900	4*	150	8	420	2.3*	45	700	55	11
>70 y	1200	30*	900	4*	150	8	420	2.3*	45	700	55	11
Females												
9–13 y	1300	21*	700	2*	120	8	240	1.6*	34	1250	40	8
14–18 y	1300	24*	890	3*	150	15	360	1.6*	43	1250	55	9
19–30 y	1000	25*	900	3*	150	18	310	1.8*	45	700	55	8
31–50 y	1000	25*	900	3*	150	18	320	1.8*	45	700	55	8
51–70 y	1200	20*	900	3*	150	8	320	1.8*	45	700	55	8
>70 y	1200	20*	900	3*	150	8	320	1.8*	45	700	55	8
Pregnancy												
≤18 y	1300	29*	1000	3*	220	27	400	2.0*	50	1250	60	12
19–30 y	1000	30*	1000	3*	220	27	350	2.0*	50	700	60	11
31–50 y	1000	30*	1000	3*	220	27	360	2.0*	50	700	60	11
Lactation												
≤18 y	1300	44*	1300	3*	290	10	360	2.6*	50	1250	70	13
19–30 y	1000	45*	1300	3*	290	9	310	2.6*	50	700	70	12
31–50 y	1000	45*	1300	3*	290	9	320	2.6*	50	700	70	12

NOTE: This table presents Recommended Dietary Allowances (RDAs) in **bold type** and Adequate Intakes (AIs) in ordinary type followed by an asterisk (*). RDAs and AIs may both be used as goals for individual intake. RDAs are set to meet the needs of almost all (97 to 98%) individuals in a group. For healthy breastfed infants, the AI is the mean intake. The AI for other life stage and gender groups is believed to cover needs of all individuals in the group, but lack of data or uncertainty in the data prevents being able to specify with confidence the percentage of individuals covered by this intake.

Sources: Dietary Reference Intakes for Calcium, Phosphorus, Magnesium, Vitamin D, and Fluoride (1997); Dietary Reference Intakes for Thiamin, Riboflavin, Niacin, Vitamin B-6, Folate, Vitamin B-12, Pantothenic Acid, Biotin, and Choline (1998); Dietary Reference Intakes for Vitamin C, Vitamin E, Selenium, and Carotenoids (2000); Dietary Reference Intakes for Vitamin A, Vitamin K, Arsenic, Boron, Chromium, Copper, Iodine, Iron, Manganese, Molybdenum, Nickel, Silicon, Vanadium, and Zinc (2001); and Dietary Reference Intakes for Calcium and Vitamin D (2011). These reports may be accessed via www.nap.edu.

Adapted from the Dietary Reference Intake series, National Academies Press. Copyright 1997, 1998, 2000, 2001, and 2011 by the National Academy of Sciences. The full reports are available from the National Academies Press at www.nap.edu.

B

Dietary Reference Intakes (DRIs): Recommended Intakes for Individuals, Macronutrients
Food and Nutrition Board, Institute of Medicine, National Academies

Life Stage Group	Carbohydrate (g/d)	Total Fiber (g/d)	Fat (g/d)	Linoleic Acid (g/d)	α-Linolenic Acid (g/d)	Protein[a] (g/d)
Infants						
0–6 mo	60*	ND	31*	4.4*	0.5*	9.1*
7–12 mo	95*	ND	30*	4.6*	0.5*	11.0
Children						
1–3 y	130	19*	ND[b]	7*	0.7*	13
4–8 y	130	25*	ND	10*	0.9*	19
Males						
9–13 y	130	31*	ND	12*	1.2*	34
14–18 y	130	38*	ND	16*	1.6*	52
19–30 y	130	38*	ND	17*	1.6*	56
31–50 y	130	38*	ND	17*	1.6*	56
51–70 y	130	30*	ND	14*	1.6*	56
>70 y	130	30*	ND	14*	1.6*	56
Females						
9–13 y	130	26*	ND	10*	1.0*	34
14–18 y	130	26*	ND	11*	1.1*	46
19–30 y	130	25*	ND	12*	1.1*	46
31–50 y	130	25*	ND	12*	1.1*	46
51–70 y	130	21*	ND	11*	1.1*	46
>70 y	130	21*	ND	11*	1.1*	46
Pregnancy						
14–18 y	175	28*	ND	13*	1.4*	71
19–30 y	175	28*	ND	13*	1.4*	71
31–50 y	175	28*	ND	13*	1.4*	71
Lactation						
14–18 y	210	29*	ND	13*	1.3*	71
19–30 y	210	29*	ND	13*	1.3*	71
31–50 y	210	29*	ND	13*	1.3*	71

NOTE: This table presents Recommended Dietary Allowances (RDAs) in **bold type** and Adequate Intakes (AIs) in ordinary type followed by an asterisk (*). RDAs and AIs may both be used as goals for individual intake. RDAs are set to meet the needs of almost all (97 to 98%) individuals in a group. For healthy breastfed infants, the AI is the mean intake. The AI for other life stage and gender groups is believed to cover needs of all individuals in the group, but lack of data or uncertainty in the data prevents being able to specify with confidence the percentage of individuals covered by this intake.

[a]Based on 0.8g protein/kg body weight for reference body weight.
[b]ND = not determinable at this time.

Sources: Dietary Reference Intakes for Energy, Carbohydrate, Fiber, Fat, Fatty Acids, Cholesterol, Protein, and Amino Acids (2002). This report may be accessed via www.nap.edu.
Adapted from the Dietary Reference Intake series, National Academies Press. Copyright 1997, 1998, 2000, 2001, by the National Academy of Sciences. The full reports are available from the National Academies Press at www.nap.edu.

D

Dietary Reference Intakes (DRIs): Recommended Intakes for Individuals, Electrolytes and Water
Food and Nutrition Board, Institute of Medicine, National Academies

Life Stage Group	Sodium (mg/d)	Potassium (mg/d)	Chloride (mg/d)	Water (L/d)
Infants				
0–6 mo	120*	400*	180*	0.7*
7–12 mo	370*	700*	570*	0.8*
Children				
1–3 y	1000*	3000*	1500*	1.3*
4–8 y	1200*	3800*	1900*	1.7*
Males				
9–13 y	1500*	4500*	2300*	2.4*
14–18 y	1500*	4700*	2300*	3.3*
19–30 y	1500*	4700*	2300*	3.7*
31–50 y	1500*	4700*	2300*	3.7*
51–70 y	1300*	4700*	2000*	3.7*
> 70 y	1200*	4700*	1800*	3.7*
Females				
9–13 y	1500*	4500*	2300*	2.1*
14–18 y	1500*	4700*	2300*	2.3*
19–30 y	1500*	4700*	2300*	2.7*
31–50 y	1500*	4700*	2300*	2.7*
51–70 y	1300*	4700*	2000*	2.7*
> 70 y	1200*	4700*	1800*	2.7*
Pregnancy				
14–18 y	1500*	4700*	2300*	3.0*
19–50 y	1500*	4700*	2300*	3.0*
Lactation				
14–18 y	1500*	5100*	2300*	3.8*
19–50 y	1500*	5100*	2300*	3.8*

NOTE: The table is adapted from the DRI reports. See www.nap.edu. Adequate Intakes (AIs) are followed by an asterisk (*). These may be used as a goal for individual intake. For healthy breastfed infants, the AI is the average intake. The AI for other life stage and gender groups is believed to cover the needs of all individuals in the group, but lack of data prevent being able to specify with confidence the percentage of individuals covered by this intake; therefore, no Recommended Dietary Allowance (RDA) was set.

Source: *Dietary Reference Intakes for Water, Potassium, Sodium, Chloride, and Sulfate* (2005). This report may be accessed via www.nap.edu.

Acceptable Macronutrient Distribution Ranges

	Range (percent of energy)		
Macronutrient	Children, 1–3 y	Children, 4–18 y	Adults
Fat	30–40	25–35	20–35
omega-6 polyunsaturated fats (linoleic acid)	5–10	5–10	5–10
omega-3 polyunsaturated fats[a] (α-linolenic acid)	0.6–1.2	0.6–1.2	0.6–1.2
Carbohydrate	45–65	45–65	45–65
Protein	5–20	10–30	10–35

[a]Approximately 10% of the total can come from longer-chain n-3 fatty acids.

SOURCE: *Dietary Reference Intakes for Energy, Carbohydrate, Fiber, Fat, Fatty Acids, Cholesterol, Protein, and Amino Acids* (2002). The report may be accessed via www.nap.edu.

Adapted from the Dietary Reference Intakes series, National Academies Press. Copyright 1997, 1998, 2000, 2001, 2011, by the National Academy of Sciences. The full reports are available from the National Academies Press at www.nap.edu.

E

Dietary Reference Intakes (DRIs): Tolerable Upper Intake Levels (ULᵃ), Vitamins
Food and Nutrition Board, Institute of Medicine, National Academies

Life Stage Group	Vitamin A (μg/d)[b]	Vitamin C (mg/d)	Vitamin D (μg/d)	Vitamin E (mg/d)[c],[d]	Vitamin K	Thiamin	Riboflavin	Niacin (mg/d)[d]	Vitamin B-6 (mg/d)	Folate (μg/d)[d]	Vitamin B-12	Pantothenic Acid	Biotin	Choline (g/d)	Carotenoids[e]
Infants															
0–6 mo	600	ND	25	ND	ND	ND	ND	ND	ND	ND	ND	ND	ND	ND	ND
7–12 mo	600	ND	38	ND	ND	ND	ND	ND	ND	ND	ND	ND	ND	ND	ND
Children															
1–3 y	600	400	63	200	ND	ND	ND	10	30	300	ND	ND	ND	1.0	ND
4–8 y	900	650	75	300	ND	ND	ND	15	40	400	ND	ND	ND	1.0	ND
Males, Females															
9–13 y	1700	1200	100	600	ND	ND	ND	20	60	600	ND	ND	ND	2.0	ND
14–18 y	2800	1800	100	800	ND	ND	ND	30	80	800	ND	ND	ND	3.0	ND
19–70 y	3000	2000	100	1000	ND	ND	ND	35	100	1000	ND	ND	ND	3.5	ND
>70 y	3000	2000	100	1000	ND	ND	ND	35	100	1000	ND	ND	ND	3.5	ND
Pregnancy															
≤18 y	2800	1800	100	800	ND	ND	ND	30	80	800	ND	ND	ND	3.0	ND
19–50 y	3000	2000	100	1000	ND	ND	ND	35	100	1000	ND	ND	ND	3.5	ND
Lactation															
≤18 y	2800	1800	100	800	ND	ND	ND	30	80	800	ND	ND	ND	3.0	ND
19–50 y	3000	2000	100	1000	ND	ND	ND	35	100	1000	ND	ND	ND	3.5	ND

[a] UL = The maximum level of daily nutrient intake likely to pose no risk of adverse effects. Unless otherwise specified, the UL represents total intake from food, water, and supplements. Due to lack of suitable data, ULs could not be established for vitamin K, thiamin, riboflavin, vitamin B-12, pantothenic acid, biotin, or carotenoids. In the absence of ULs, extra caution may be warranted in consuming levels above recommended intakes.

[b] As preformed vitamin A only.

[c] As α-tocopherol; applies to any form of supplemental α-tocopherol.

[d] The ULs for vitamin E, niacin, and folate apply to synthetic forms obtained from supplements, fortified foods, or a combination of the two.

[e] β-Carotene supplements are advised only to serve as a provitamin A source for individuals at risk of vitamin A deficiency.

ND = Not determinable due to lack of data of adverse effects in this age group and concern with regard to lack of ability to handle excess amounts. Source of intake should be from food only to prevent high levels of intake.

SOURCES: Dietary Reference Intakes for Calcium, Phosphorous, Magnesium, Vitamin D, and Fluoride (1997); Dietary Reference Intakes for Thiamin, Riboflavin, Niacin. Vitamin B-6, Folate, Vitamin B-12, Pantothenic Acid, Biotin, and Chlorine (1998); Dietary Reference Intakes for Vitamin C, Vitamin E, Selenium, and Carotenoids (2000); and Dietary Reference Intakes for Vitamin A, Vitamin K, Arsenic, Boron, Chromium, Copper, Iodine, Iron, Manganese, Molybdenum, Nickel, Silicon, Vanadium, and Zinc (2001). These reports may be accessed via www.nap.edu.

Adapted from the Dietary Reference Intakes series, National Academies Press. Copyright 1997, 1998, 2000, 2001, 2011, by the National Academy of Sciences. The full reports are available from the National Academies Press at www.nap.edu.

Dietary Reference Intakes (DRIs): Tolerable Upper Intake Levels (UL[a]), Elements and Electrolytes[b,c]
Food and Nutrition Board, Institute of Medicine, National Academies

Life Stage Group	Arsenic[b]	Boron (mg/d)	Calcium (g/d)	Copper (μg/d)	Fluoride (mg/d)	Iodine (μg/d)	Iron (mg/d)	Magnesium (mg/d)[d]	Manganese (mg/d)	Molybdenum (μg/d)	Nickel (mg/d)	Phosphorus (g/d)	Selenium (μg/d)	Vanadium (mg/d)[e]	Zinc (mg/d)	Sodium (mg/d)	Chloride (mg/d)
Infants																	
0–6 mo	ND[f]	ND	1	ND	0.7	ND	40	ND	ND	ND	ND	ND	45	ND	4	ND	ND
7–12 mo	ND	ND	1.5	ND	0.9	ND	40	ND	ND	ND	ND	ND	60	ND	5	ND	ND
Children																	
1–3 y	ND	3	2.5	1000	1.3	200	40	65	2	300	0.2	3	90	ND	7	1500	2300
4–8 y	ND	6	2.5	3000	2.2	300	40	110	3	600	0.3	3	150	ND	12	1900	2900
Males, Females																	
9–13 y	ND	11	3	5000	10	600	40	350	6	1100	0.6	4	280	ND	23	2200	3400
14–18 y	ND	17	3	8000	10	900	45	350	9	1700	1.0	4	400	ND	34	2300	3600
19–70 y	ND	20	2.5[g]	10000	10	1100	45	350	11	2000	1.0	4	400	1.8	40	2300	3600
>70 y	ND	20	2	10000	10	1100	45	350	11	2000	1.0	3	400	1.8	40	2300	3600
Pregnancy																	
≤18 y	ND	17	3	8000	10	900	45	350	9	1700	1.0	3.5	400	ND	34	2300	3600
19–50 y	ND	20	2.5	10000	10	1100	45	350	11	2000	1.0	3.5	400	ND	40	2300	3600
Lactation																	
≤18 y	ND	17	3	8000	10	900	45	350	9	1700	1.0	4	400	ND	34	2300	3600
19–50 y	ND	20	2.5	10000	10	1100	45	350	11	2000	1.0	4	400	ND	40	2300	3600

[a]UL = The maximum level of daily nutrient intake that is likely to pose no risk of adverse effects. Unless otherwise specified, the UL represents total intake from food, water, and supplements. Due to lack of suitable data, ULs could not be established for arsenic, chromium, and silicon. In the absence of ULs, extra caution may be warranted in consuming levels above recommended intakes.

[b]Although a UL was not determined for arsenic, there is no justification for adding arsenic to food or supplements.

[c]Although silicon has not been shown to cause adverse effects in humans, there is no justification for adding silicon to supplements.

[d]The ULs for magnesium represent intake from a pharmacological agent only and do not include intake from food and water.

[e]Although vanadium in food has not been shown to cause adverse effects in humans, there is no justification for adding vanadium to food and vanadium supplements should be used with caution. The UL is based on adverse effects in laboratory animals and this data could be used to set a UL for adults but not children and adolescents.

[f]ND = Not determinable due to lack of data of adverse effects in this age group and concern with regard to lack of ability to handle excess amounts. Source of intake should be from food only to prevent high levels of intake.

[g]Upper Limit declines to 2 after age 50.

SOURCES: Dietary Reference Intakes for Calcium and Vitamin D (2011); Dietary Reference Intakes for Calcium, Phosphorus, Magnesium, Vitamin D, and Fluoride (1997); Dietary Reference Intakes for Thiamin, Riboflavin, Niacin, Vitamin B-6, Folate, Vitamin B-12, Pantothenic Acid, Biotin, and Choline (1998); Dietary Reference Intakes for Vitamin C, Vitamin E, Selenium, and Carotenoids (2000); Dietary Reference Intakes for Vitamin A, Vitamin K, Arsenic, Boron, Chromium, Copper, Iodine, Iron, Manganese, Molybdenum, Nickel, Silicon, Vanadium, and Zinc (2001); and *Dietary Reference Intakes for Water, Potassium, Sodium, Chloride, and Sulfate* (2004). These reports may be accessed via www.nap.edu.

Adapted from the Dietary Reference Intakes series, National Academies Press. Copyright 1997, 1998, 2000, 2001, 2011, by the National Academy of Sciences. The full reports are available from the National Academies Press at www.nap.edu.

F

ISBN 978-1-259-91820-9
MHID 1-259-91820-3

EAN

90000

9 781259 918209

mheducation.com/highered